DIGITAL **UPDATE**

QUANTITATIVE LITERACY

3rd Edition

Thinking Between the Lines

Bruce C. Crauder
Oklahoma State University

Benny Evans
Oklahoma State University

Jerry A. Johnson
University of Nevada

Alan V. Noell
Oklahoma State University

Austin • Boston • New York • Plymouth

For Doug, Robbie, and especially Laurie —**Bruce**
For Simon, Lucas, Rosemary, Isaac, Bear, Grace, and Astrid —**Benny**
For Jeanne —**Jerry**
For Evelyn, Scott, Philip, Alyson, Laura, Stephen, Tamara, and especially Liz —**Alan**

Senior Vice President, STEM: Daryl Fox
Program Director: Andrew Dunaway
Program Manager: Nikki Miller Dworsky
Marketing Manager: Leah Christians
Director of Development, STEM: Debbie Hardin
Executive Project Manager, Content, STEM: Katrina Mangold
Editorial Project Manager: Andy Newton
Marketing Assistant: Elizabet Cabrera
Directors of Content: Catriona Kaplan, Daniel Lauve
Executive Media Editor: Alexandra Gordon
Senior Media Editor: Holly Floyd
Media Editor: Doug Newman
Senior Director of Content Management Enhancement: Tracey Kuehn
Senior Managing Editor: Lisa Kinne
Senior Content Project Managers: Edward Dionne, Edgar Doolan
Senior Workflow Project Manager: Paul Rohloff
Director of Design, Content Management: Diana Blume
Senior Design Services Manager: Natasha Wolfe
Senior Cover Design Manager: John Callahan
Cover Illustrator: Eli Ensor
Text Designer: Patrice Sheridan
Art Manager: Matt McAdams
Illustrations: Emily Cooper, Eli Ensor
Director of Digital Production: Keri deManigold
Advanced Media Project Manager: Hanna Squire
Permissions Editor: Michael McCarty
Executive Permissions Editor: Robin Fadool
Composition: Lumina Datamatics, Inc.
Printing and Binding: King Printing Co., Inc.
iPad Photo: Radu Bercan/Shutterstock

Library of Congress Control Number: 2021947300

Student Edition Paperback:
ISBN-13: 978-1-319-24446-0
ISBN-10: 1-319-24446-7

Student Edition Loose-leaf:
ISBN-13: 978-1-319-41603-4
ISBN-10: 1-319-41603-9

Printed in the United States of America

1 2 3 4 5 6 26 25 24 23 22 21

Macmillan Learning
120 Broadway
New York, New York 10271
www.macmillanlearning.com

In 1946, William Freeman founded W. H. Freeman and Company and published Linus Pauling's *General Chemistry*, which revolutionized the chemistry curriculum and established the prototype for a Freeman text. W. H. Freeman quickly became a publishing house where leading researchers can make significant contributions to mathematics and science. In 1996, W. H. Freeman joined Macmillan and we have since proudly continued the legacy of providing revolutionary, quality educational tools for teaching and learning in STEM.

About the Authors

Jason Wallace

BRUCE CRAUDER graduated from Haverford College in 1976, receiving a B.A. with honors in mathematics, and received his Ph.D. from Columbia in 1981, writing a dissertation on algebraic geometry. After post-doctoral positions at the Institute for Advanced Study, the University of Utah, and the University of Pennsylvania, Bruce went to Oklahoma State University in 1986 where he is now a professor of mathematics and also associate dean.

Bruce's research in algebraic geometry has resulted in 10 refereed articles in as many years in his specialty: three-dimensional birational geometry. He has since worked on the challenge of the beginning college math curriculum, resulting in the creation of two new courses with texts to support the courses. He is especially pleased with these texts, as they combine scholarship with his passion for teaching.

Bruce has two sons, one a process engineer and volunteer firefighter and the other a video producer in sports media. He and his wife, Laurie, enjoy birdwatching, traveling, teaching, and Oklahoma sunsets.

Benny Evans

BENNY EVANS received his Ph.D. in mathematics from the University of Michigan in 1971. After a year at the Institute for Advanced Study in Princeton, New Jersey, he went to Oklahoma State University where he is currently professor emeritus of mathematics. In his tenure at OSU, he has served as undergraduate director, associate head, and department head. His career has included visiting appointments at the Institute for Advanced Study, Rice University, Texas A&M, and University of Nevada at Reno.

Benny's research interests are in topology and mathematics education. His record includes 28 papers in refereed journals, numerous books and articles (some in collaboration with the other three authors), and 25 grants from the National Science Foundation, Oklahoma State Board of Regents, and private foundations.

Benny's hobbies include fishing and spending time with his grandchildren. Fortunately, it is possible to do both at the same time.

Jeanne Johnson

JERRY JOHNSON received his Ph.D. in mathematics in 1969 from the University of Illinois, Urbana. From 1969 until 1993 he taught at Oklahoma State University, with a year off in 1976 for a sabbatical at Pennsylvania State University. In 1993, he moved to the University of Nevada, Reno, to become the founding director of their Mathematics Center as well as director of a project called Mathematics across the Curriculum. From 1995 to 2001, he served as chairman of the Mathematics Department. He is now professor of mathematics emeritus.

Over the years, Jerry has published 17 research articles in prominent mathematics journals and more than 35 articles in various publications related to mathematics education. He wrote and published a commercially successful animated 3-D graphing program called GyroGraphics and was a founding member of the National Numeracy Network, whose goal is to promote quantitative literacy.

Jerry's wife, Jeanne, is a folklorist with a Ph.D. from Indiana University, Bloomington. They enjoy traveling (Jerry has been to all 50 states), gardening (especially tomatoes), cooking (you fry it, I'll try it), and cats (the kind that meow).

Elizabeth Noell

ALAN NOELL received his Ph.D. in mathematics from Princeton University in 1983. After a postdoctoral position at Caltech, he joined the faculty of Oklahoma State University in 1985. He is currently a professor of mathematics and an associate head. His scholarly activities include research in complex analysis and curriculum development.

Alan and his wife, Liz, have four children, two daughters-in-law, and one son-in-law. They are very involved with their church and also enjoy reading and cats (the same kind as Jerry).

Brief Contents

Contents

Preface

Will Rogers, a popular humorist and actor in the 1920s and 1930s, famously quipped, "All I know is just what I read in the papers, and that's an alibi for my ignorance." Today we get our information from a wider variety of sources, including television broadcasts, magazines, and Internet sites, as well as newspapers. But the sentiment remains the same—we rely on public media for the information we require to understand the world around us and to make important decisions in our daily lives.

In today's world, the vast amount of information that is available in the form of print and images is overwhelming. Interpreting and sorting this information, trying to decide what to believe and what to reject, is ever more daunting in the face of rapidly expanding sources. It is such an issue that the term "fake news" has become common recently. In addition, the important news of the day often involves complicated topics such as economics, finance, political polls, and an array of statistical data—all of which are quantitative in nature.

Our Goals and Philosophy

The overall goal of this book is to present and explain the quantitative tools necessary to understand issues arising in the popular media and in our daily lives. Through contemporary real-world applications, our aim is to teach students the practical skills they will need throughout their lives to be critical thinkers, informed decision makers, and intelligent consumers of the quantitative information that they see every day. This goal motivates our choice of topics and our use of numerous articles from the popular media as illustrations.

Being an intelligent reader and consumer requires critical thinking. This fact is conveyed by the book's subtitle, *Thinking Between the Lines.* For example, visual displays of data such as graphs can be misleading. How can you spot this when it happens? A poll often presents, along with its results, a "margin of error." What does this mean? Banks often use the terms APR and APY. What do these mean? In these and many other practical situations, math matters. A nice example of the role in the text of critical thinking can be found in a discussion of how to graphically display federal defense spending.

Similarly, instead of simply defining logarithms by formulas, we discuss how they relate to the decibel and Richter scales. Instead of just presenting graph theory, we discuss how it relates to spell checkers and the latest technology in air traffic control. This "in context" philosophy guides the presentation throughout the text; therefore, it should always be easy to answer a student if he or she asks the famous question, "What is this stuff good for?"

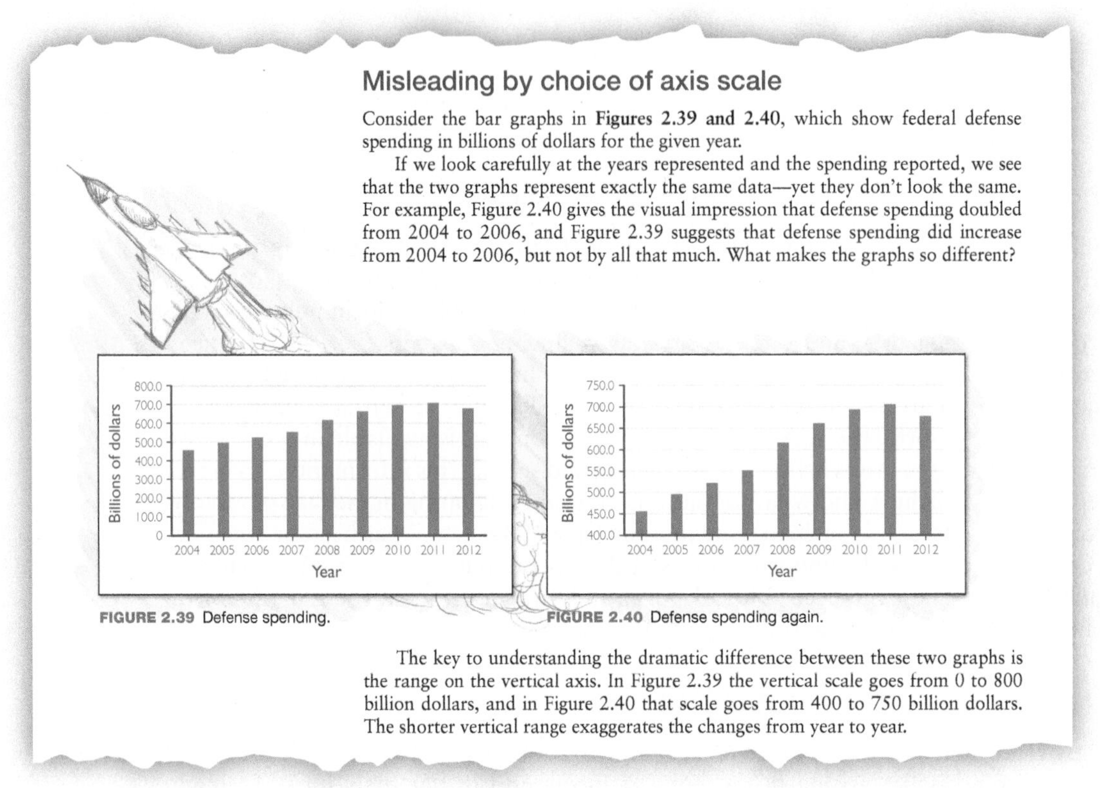

Misleading by choice of axis scale

Consider the bar graphs in **Figures 2.39 and 2.40**, which show federal defense spending in billions of dollars for the given year.

If we look carefully at the years represented and the spending reported, we see that the two graphs represent exactly the same data—yet they don't look the same. For example, Figure 2.40 gives the visual impression that defense spending doubled from 2004 to 2006, and Figure 2.39 suggests that defense spending did increase from 2004 to 2006, but not by all that much. What makes the graphs so different?

FIGURE 2.39 Defense spending.

FIGURE 2.40 Defense spending again.

The key to understanding the dramatic difference between these two graphs is the range on the vertical axis. In Figure 2.39 the vertical scale goes from 0 to 800 billion dollars, and in Figure 2.40 that scale goes from 400 to 750 billion dollars. The shorter vertical range exaggerates the changes from year to year.

We want to do more than just help students succeed in a liberal arts math course. We want them to acquire skills and experience material that they will be able to use throughout their lives as intelligent consumers of information.

Theoretical probability

To discuss theoretical probability, we need to make clear what we mean by an event. An *event* is a collection of specified outcomes of an experiment.[2] For example, suppose we roll a standard die (with faces numbered 1 through 6). One event is that the number showing is even. This event consists of three outcomes: a 2 shows, a 4 shows, or a 6 shows.

To define the probability of an event, it is common to describe the outcomes corresponding to that event as *favorable* and all other outcomes as *unfavorable*. If the event is that an even number shows when we toss a die, there are three favorable outcomes (2, 4, or 6) and three unfavorable outcomes (1, 3, or 5).

KEY CONCEPT

If each outcome of an experiment is *equally likely*, the probability of an event is the fraction of favorable outcomes. We can express this definition using a formula:

$$\text{Probability of an event} = \frac{\text{Number of favorable outcomes}}{\text{Total number of possible equally likely outcomes}}$$

We will often shorten this to

$$\text{Probability} = \frac{\text{Favorable outcomes}}{\text{Total outcomes}}$$

Balanced Coverag

Our audience for this book is stude who have not taken math courses p elementary algebra. Unlike some li eral arts math books that may foc too much on either the quantitati or qualitative aspects of key mat ematical concepts, we offer studer and teachers a book that strikes t right balance between reading text a doing calculations. For example, Chapter 5, we discuss the probab ity of an event not occurring. T usual formula is presented, but it also clearly described in words and familiar contexts. This balance of cc cepts and calculation engages studer in thinking about interesting ideas a also gives them the opportunity apply and practice the math under ing those ideas.

Abundant Use of Applications: Real, Contemporary, and Meaningful

Wherever possible, we use real applications, real scenarios, and real data. For example, in Chapter 1, we use the Berkeley gender discrimination case to introduce the concept of paradox.

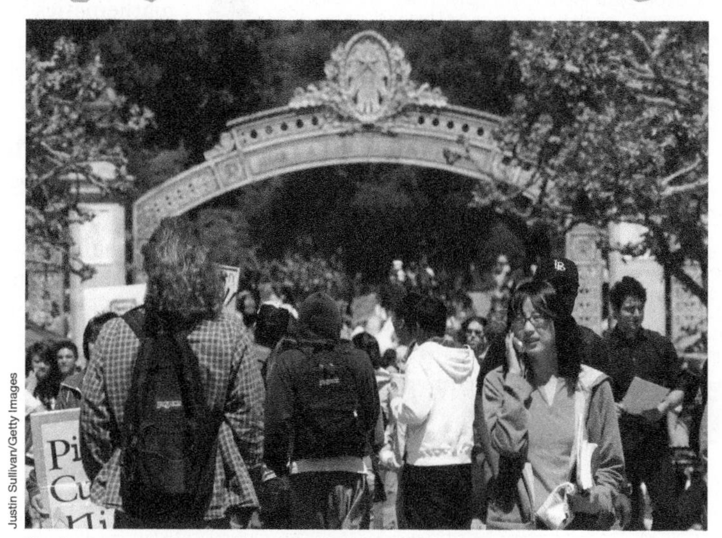

Justin Sullivan/Getty Images

Students on the University of California at Berkeley campus.

These numbers may seem to be conclusive evidence of discrimination. When we look more closely, however, we find that graduate school admissions are based on departmental selections. There are many departments at Berkeley, but to keep things simple, we will assume that the university consists of only two departments: Math and English. Here is how the male and female applicants might have been divided between these two departments.

In Chapter 5, we use the prevalence of the human papillomavirus, HPV, to introduce the concept of positive predictive value.

Positive predictive value

A common misconception is that if one tests positive, and the test has a high sensitivity and specificity, then one should conclude that the disease is almost certainly present. But a more careful analysis shows that the situation is not so simple, and that *prevalence* plays a key role here.

KEY CONCEPT

The prevalence of a disease in a given population is the percentage of the population that has the disease.

For example, 5.1% of U.S. females aged 14 to 19 years have HPV, the most common sexually transmitted infection. That means the prevalence of HPV among women in the United States aged 14 to 19 is 5.1%. (The prevalence was much higher before the HPV vaccination was added to the routine immunization schedule.) For comparison, the prevalence of epilepsy in the United States is less than 1%.

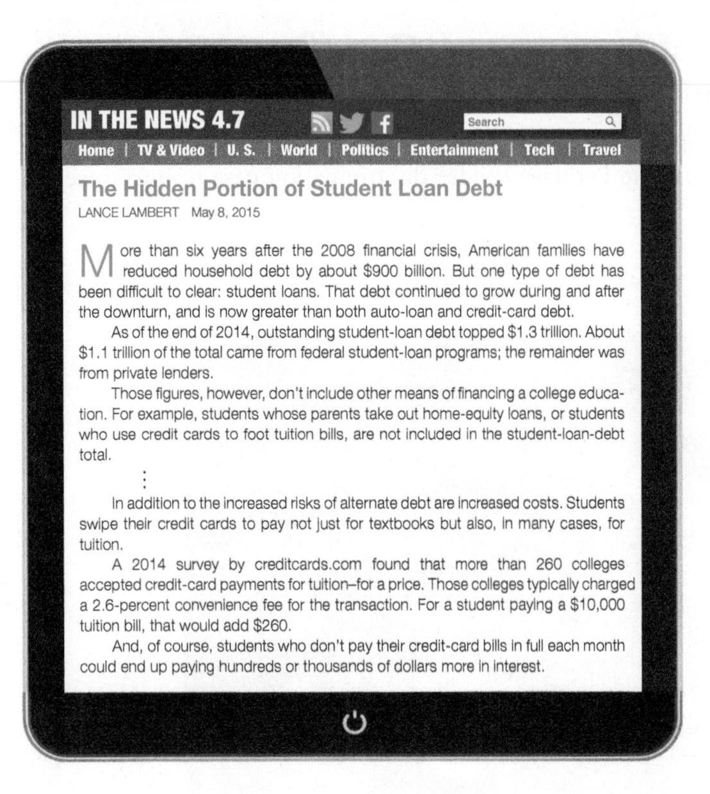

IN THE NEWS 4.7

Home | TV & Video | U. S. | World | Politics | Entertainment | Tech | Travel

The Hidden Portion of Student Loan Debt

LANCE LAMBERT May 8, 2015

More than six years after the 2008 financial crisis, American families have reduced household debt by about $900 billion. But one type of debt has been difficult to clear: student loans. That debt continued to grow during and after the downturn, and is now greater than both auto-loan and credit-card debt.

As of the end of 2014, outstanding student-loan debt topped $1.3 trillion. About $1.1 trillion of the total came from federal student-loan programs; the remainder was from private lenders.

Those figures, however, don't include other means of financing a college education. For example, students whose parents take out home-equity loans, or students who use credit cards to foot tuition bills, are not included in the student-loan-debt total.

⋮

In addition to the increased risks of alternate debt are increased costs. Students swipe their credit cards to pay not just for textbooks but also, in many cases, for tuition.

A 2014 survey by creditcards.com found that more than 260 colleges accepted credit-card payments for tuition–for a price. Those colleges typically charged a 2.6-percent convenience fee for the transaction. For a student paying a $10,000 tuition bill, that would add $260.

And, of course, students who don't pay their credit-card bills in full each month could end up paying hundreds or thousands of dollars more in interest.

In the News

Throughout the text, explanations, examples, and exercises are often motivated by everyday events reported in the popular media. Each chapter and each section opens with real articles gleaned from newspapers, magazines, and Internet sites. They are presented under the feature titled In the News. These lead naturally to discussions of mathematics as part of everyday life. For example, In the News 4.7 launches a discussion of credit cards and paying off consumer debt.

An Emphasis on Solving Problems

Concepts are reinforced with worked Examples that model problem solving and are often followed by Try it Yourself exercises for students to practice on their own.

Examples

Examples in the text are selected as much as possible for their relevance to student experience. For instance, Example 4.26 shows students how to calculate the finance charge on clothes purchased on credit.

The solution for this example includes a clear and complete explanation that students can follow and reference when solving similar types of problems.

EXAMPLE 4.26 Calculating finance charges: Buying clothes

Suppose your Visa card calculates finance charges using an APR of 22.8%. Your previous statement showed a balance of $500, in response to which you made a payment of $200. You then bought $400 worth of clothes, which you charged to your Visa card. Complete the following table:

	Previous balance	Payments	Purchases	Finance charge	New balance
Month 1					

SOLUTION

We start by entering the information given.

	Previous balance	Payments	Purchases	Finance charge	New balance
Month 1	$500.00	$200.00	$400.00		

The APR is 22.8%, and we divide by 12 to get a monthly rate of 1.9%. In decimal form, this calculation is 0.228/12 = 0.019. Therefore,

$$\text{Finance charge} = 0.019 \times (\text{Previous balance} - \text{Payments} + \text{Purchases})$$
$$= 0.019 \times (\$500 - \$200 + \$400)$$
$$= 0.019 \times \$700 = \$13.30$$

That gives a new balance of

$$\text{New balance} = \text{Previous balance} - \text{Payments} + \text{Purchases} + \text{Finance charge}$$
$$= \$500 - \$200 + \$400 + \$13.30 = \$713.30$$

Your new balance is $713.30. Here is the completed table.

	Previous balance	Payments	Purchases	Finance charge	New balance
Month 1	$500.00	$200.00	$400.00	1.9% of $700 = $13.30	$713.30

Try It Yourself

Many worked examples are followed by a very similar problem that gives students the opportunity to test their understanding. This feature serves to actively engage students in the learning process rather than just reading passively. The answers to the Try It Yourself problems are provided at the end of the section so that students can check their solutions.

TRY IT YOURSELF 4.26

This is a continuation of Example 4.26. You make a payment of \$300 to reduce the \$713.30 balance and then charge a TV costing \$700. Complete the following table:

	Previous balance	Payments	Purchases	Finance charge	New balance
Month 1	\$500.00	\$200.00	\$400.00	\$13.30	\$713.30
Month 2					

The answer is provided at the end of this section.

Exercise Sets

The exercise sets are designed to help students appreciate the relevance of section topics to everyday life, and they often include direct references to popular media.

Each exercise set begins with **Test Your Understanding** exercises, which are multiple choice, fill in the blank, or simply quick checks on basic understanding.

Exercise Set 3.1

Test Your Understanding

1. If a quantity is growing as a linear function with a constant rate of change of m per year, how do you use this year's amount to find next year's amount?

2. A function is linear if: **a.** it has a constant rate of change **b.** its graph is a straight line **c.** its formula is $y =$ Growth rate $\times$ x + Initial value **d.** all of the above.

3. The slope or growth rate of a linear function is calculated as ________ divided by ________.

4. True or false: Data may be appropriately modeled by a linear function if the points in a data plot lie on a straight line.

5. True or false: A proper use of units can help us to determine the meaning of the slope of a linear function.

6. True or false: The most important information from the trend line is the initial value.

Problems

Note: In some exercises, you are asked to find a formula for a linear function. If no letters are assigned to quantities in an exercise, you are expected to choose appropriate letters and also give appropriate units.

7. **Interpreting formulas.** The height of a young flower is increasing in a linear fashion. Its height t weeks after the first of this month is given by $H = 2.1t + 5$ millimeters. Identify the initial value and growth rate, and explain in practical terms their meaning.

8. **Interpreting formulas.** I am driving west from Nashville at a constant speed. My distance west from Nashville t hours after noon is given by $D = 65t + 30$ miles. Identify the initial value and growth rate, and explain in practical terms their meaning.

9. **Finding linear formulas.** A child is 40 inches tall at age 6. For the next few years, the child grows by 2 inches per year. Explain why the child's height after age 6 is a linear function of his age. Identify the growth rate and initial value. Using t for time in years since age 6 and H for height in inches, find a formula for H as a linear function of t.

18. Make another table. Suppose you put $4000 in a savings account at an APR of 6% compounded monthly. Fill in the table below. (Calculate the interest and compound it each month rather than using the compound interest formula.)

Month	Interest earned	Balance
		$4000.00
1	$	$
2	$	$
3	$	$

19. Compound interest calculated. Assume that we invest $2000 for one year in a savings account that pays an APR of 10% compounded quarterly.

a. Make a table to show how much is in the account at the end of each quarter. (Calculate the interest and compound it each quarter rather than using the compound interest formula.)

b. Use your answer to part a to determine how much total interest the account has earned after one year.

c. Compare the earnings to what simple interest or semi-annual compounding would yield.

The exercise sets continue with a group of questions called **Problems**, which are very similar to worked examples and easily accessible to all students. These Problems form the heart of the exercise set.

These are followed by **Writing About Mathematics** exercises, which challenge students to seek answers beyond the textbook. These exercises are suitable for reports, group work, or other activities that engage students in deeper thinking.

to find the APR. *Note*: This can also be solved without technology, using algebra.

40. Find the compounding period. Suppose a CD advertises an APR of 5.10% and an APY of 5.20%. Solve the equation

$$0.052 = \left(1 + \frac{0.051}{n}\right)^n - 1$$

for n to determine how frequently interest is compounded.

Writing About Mathematics

41. Current rates. Look up some current interest rates being paid on CDs by different financial institutions and write a report on your findings.

42. Bond rates. Interest rates on bonds can differ significantly among nations. Look up some current interest rates being paid on bonds by different countries and write a report on your findings.

43. Hamilton. Alexander Hamilton was the first U.S. Treasury secretary. Write a report on Hamilton and his accomplishments.

44. Usury. Look up the meaning of the word *usury* and write a report on how it is viewed in different cultures and religions.

What Do You Think?

56. Malware. The mathematical notion of exponential growth refers to a precise type of growth that happens to be (eventually) very rapid. But the phrase is often applied to any fast-growing phenomenon. The headline at tech.fortune.cnn.com/2011/11/17/androids-growing-malware-problem says this: "Android's malware problem is growing exponentially." Do you think this headline is really describing exponential growth, or is it just using a figure of speech? Explain your thinking.

57. Paypal. The mathematical notion of exponential growth refers to a precise type of growth that happens to be (eventually) very rapid. But the phrase is often applied to any fast-growing phenomenon. The headline at venturebeat.com/2012/01/10/paypals-mobile-payments-4b-2011/ says this: "Paypal's mobile payments growing exponentially, reached $4B in 2011." Do you think this headline is really describing exponential growth, or is it just using a figure of speech? Explain your thinking.

58. Is it exponential? You see a news report about taxes, inflation, or other economic issues and wonder whether you are seeing exponential growth. How would you make a judgment about this question? What factors might lead you to believe that you are seeing exponential growth, and what

The final part of each exercise set are the **What Do You Think?** exercises, which are open-ended questions that encourage students to apply their critical thinking skills as well as their verbal and written skills.

A Wealth of Features to Help Students Engage, Study, and Learn

The book includes many pedagogical features that will help students learn and review key concepts. These include a cutting-edge illustration program, Quick Review, Key Concepts, Chapter Summaries, and Chapter Quizzes. Each section concludes with abundant exercises that are graded from simple to more complex. We review each of the features in turn.

Artwork

Many students are visual learners, and we have used graphs, photos, cartoons, and sketches liberally throughout the text. Many illustrations are paired so that students can easily compare and contrast similar concepts.

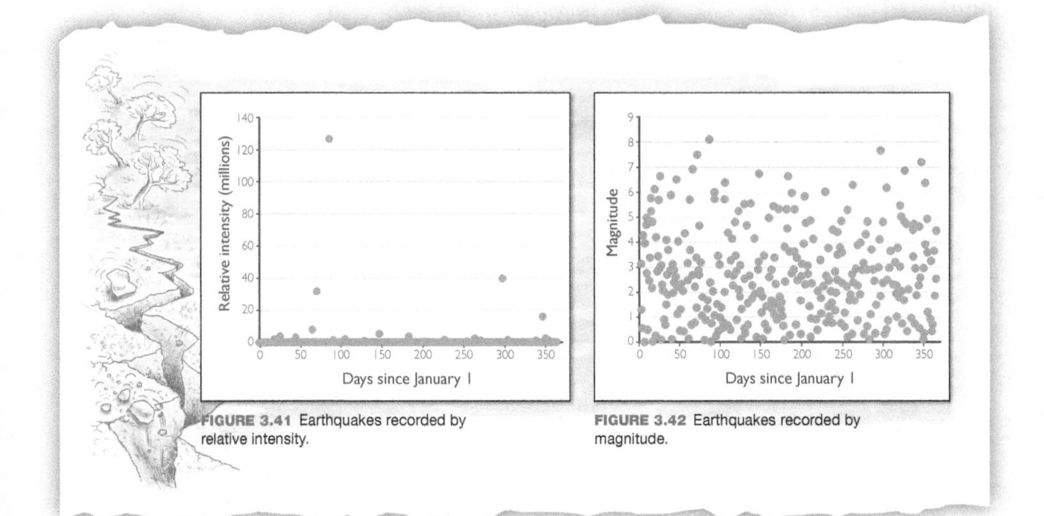

FIGURE 3.41 Earthquakes recorded by relative intensity.

FIGURE 3.42 Earthquakes recorded by magnitude.

Cartoons are used to lighten things up!

"We were going over some of your returns from a past life and..."

The disparity in part c of Example 4.35 is referred to as the "marriage penalty." Many people believe the marriage penalty is unfair, but there are those who argue that it makes sense to tax a married couple more than two single people. Can you think of some arguments to support the two sides of this question?

Claiming deductions lowers your tax by reducing your taxable income. Another way to lower your tax is to take a *tax credit*. Here is how to apply a tax credit: Calculate the tax owed using the tax tables (making sure first to subtract from the total income any deductions) and then subtract the tax credit from the tax determined by the tables. Because a tax credit is subtracted directly from the tax you owe, a tax credit of $1000 has a much bigger impact on lowering your taxes than a deduction of $1000. That is the point of the following example.

Key Concepts

Most mathematics books state definitions of terms formally. We have chosen to define terms in a concise and informal way and to label each as a Key Concept. We do so to avoid intimidating students and to focus on the concepts rather than dry terminology. Examples of this will be found on page 400 in Chapter 6 on statistics, where "outliers" and "quartiles" are explained.

> **KEY CONCEPT**
>
> An outlier is a data point that is significantly different from most of the data. Typically, outliers have a greater effect on the mean, but a lesser effect on the median.

> **KEY CONCEPT**
>
> - The first quartile of a list of numbers is the median of the lower half of the numbers in the list.
> - The second quartile is the same as the median of the list.
> - The third quartile is the median of the upper half of the numbers in the list.
>
> If the list has an even number of entries, it is clear what we mean by the "lower half" and "upper half" of the list. If the list has an odd number of entries, eliminate the median from the list. This new list now has an even number of entries. The "lower half" and "upper half" refer to this new list.

Summaries

Students may occasionally fee overwhelmed by new concepts so throughout each chapter w provide brief boxed Summarie of core concepts. This is a opportunity for students to pause collect their thoughts, and rein force what they have learned Summaries also provide referenc points when students are solv ing exercises. For instance, i Section 1.2, a discussion of som common types of logical falla cies is offered. These are groupe together with brief descriptions o each in Summary 1.2.

> **SUMMARY 1.2**
> Some Common Informal Fallacies
>
> **Appeal to ignorance** A statement is either accepted or rejected because of a lack of proof.
>
> **Dismissal based on personal attack** An argument is dismissed based on an attack on the proponent rather than on its merits.
>
> **False authority** The validity of a claim is accepted based on an authority whose expertise is irrelevant.
>
> **Straw man** A position is dismissed based on the rejection of a distorted or different position (the "straw man").
>
> **Appeal to common practice** An argument for a practice is based on the popularity of that practice.
>
> **False dilemma** A conclusion is based on an inaccurate or incomplete list of alternatives.
>
> **False cause** A causal relationship is concluded based on the fact that two events occur together or that one follows the other.
>
> **Circular reasoning** This fallacy simply draws a conclusion that is really a restatement of the premise.
>
> **Hasty generalization** This fallacy occurs when a conclusion is drawn based on a few examples that may be atypical.

Calculation Tips and Technology Support

Tips are provided to help students avoid common errors in doing calculations. One kind of tip involves advice in preparing to make a calculation. For example, Calculation Tip 5.1 helps students deal with probability calculations. Others may give helpful hints for carrying out a complex calculation, such as Calculation Tip 6.1, which summarizes how to calculate a standard deviation. Some may also include tips for performing calculations on a calculator or with statistical software, such as Calculation Tip 5.3.

Where appropriate, a screen shot of a calculator is used to reinforce correct data input when solving problems.

CALCULATION TIP 5.3 Factorials with a Calculator or Computer

- On the TI-83 and TI-84 graphing calculators, there's a factorial command ! located in the math probability menu. To find 10!, enter 10 MATH PRB 4.
- In the spreadsheet program Excel©, the command for n! is FACT(n). For example, to find 10!, enter FACT(10).
- You can also find factorials on the calculator that comes with Microsoft Windows©. Choose the View menu in the calculator and click on Scientific to get the full size. You can see the n! button in the fourth column, just above the 1/x button.

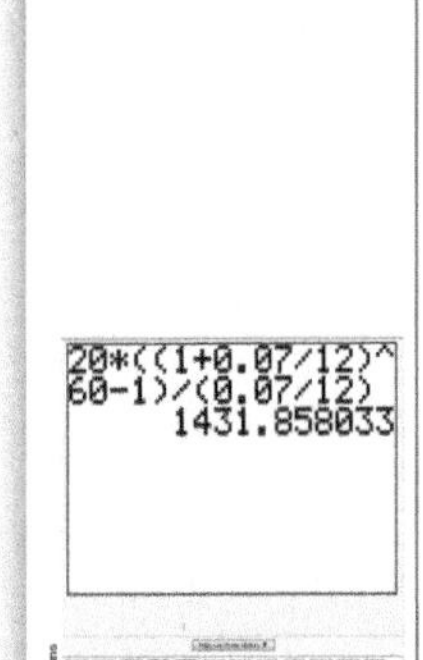

EXAMPLE 4.21 Using the balance formula: Saving money regularly

Suppose we have a savings account earning 7% APR. We deposit \$20 to the account at the end of each month for five years. What is the future value for this savings arrangement? That is, what is the account balance after five years?

SOLUTION

We use the regular deposits balance formula (Formula 4.8). The monthly deposit is \$20, and the monthly interest rate as a decimal is

$$r = \frac{\text{APR}}{12} = \frac{0.07}{12}$$

The number of deposits is $t = 5 \times 12 = 60$. The regular deposits balance formula gives

$$\text{Balance after 60 deposits} = \frac{\text{Deposit} \times ((1+r)^t - 1)}{r}$$
$$= \frac{\$20 \times ((1 + 0.07/12)^{60} - 1)}{(0.07/12)}$$
$$= \$1431.86$$

The future value is \$1431.86.

TRY IT YOURSELF 4.21

Suppose we have a savings account earning 4% APR. We deposit \$40 to the account at the end of each month for 10 years. What is the future value of this savings arrangement?

The answer is provided at the end of this section.

Rule of Thumb Statements

One facet of quantitative literacy is the ability to make estimates. The Rule of Thumb feature provides hints on making such "ballpark" estimates without appealing to complicated formulas for exact answers. For example, Rule of Thumb 4.2 shows how to estimate certain monthly payments on loans. These estimates provide a simple and effective check to catch possible errors in exact calculations.

RULE OF THUMB 4.2 Monthly Payments for Short-Term Loans

For all loans, the monthly payment is *at least* the amount we would pay each month if no interest were charged, which is the amount of the loan divided by the term (in months) of the loan. This would be the payment if the APR were 0%. It's a rough estimate of the monthly payment for a short-term loan if the APR is not large.

Algebraic Spotlights

An effort is made throughout the text to motivate the use of complicated formulas, but some derivations require algebra skills that may be beyond the scope of a course. Such derivations are presented in a special discussion called Algebraic Spotlight. An example is the derivation of the monthly payment formula in Chapter 4. While some instructors like to present derivations of formulas to their students, others would prefer to avoid them. Therefore, these spotlights are intended to be used at the instructor's discretion.

ALGEBRAIC SPOTLIGHT 4.3 Derivation of the Monthly Payment Formula, Part II

Now we use the account balance formula to derive the monthly payment formula. Suppose we borrow B_0 dollars at a monthly interest rate of r as a decimal and we want to pay off the loan in t monthly payments. That is, we want the balance to be 0 after t payments. Using the account balance formula we derived in Algebraic Spotlight 4.2, we want to find the monthly payment M that makes $B_t = 0$. That is, we need to solve the equation

$$0 = B_t = B_0(1+r)^t - M\frac{(1+r)^t - 1}{r}$$

for M. Now

$$0 = B_0(1+r)^t - M\frac{(1+r)^t - 1}{r}$$

$$M\frac{(1+r)^t - 1}{r} = B_0(1+r)^t$$

$$M = \frac{B_0 r(1+r)^t}{((1+r)^t - 1)}$$

Because B_0 is the amount borrowed and t is the term, this is the monthly payment formula we stated earlier.

Quick Review Making Graphs

Typically, we locate points on a graph using a table such as the one below, which shows the number of automobiles sold by a dealership on a given day.

Day (independent variable)	1	2	3	4
Cars sold (dependent variable)	3	6	5	2

When we create a graph of a function, the numbers on the horizontal axis correspond to the independent variable, and the numbers on the vertical axis correspond to the dependent variable. For this example, we put the day (the independent variable) on the horizontal axis, and we put the number of cars sold (the dependent variable) on the vertical axis.

The point corresponding to day 3 is denoted by (3, 5) and is represented by a single point on the graph. To locate the point (3, 5), we begin at the origin and move 3 units to the right and 5 units upward. This point is displayed on the graph as one of the four points of that graph. The aggregate of the plotted points is the graph of the function.

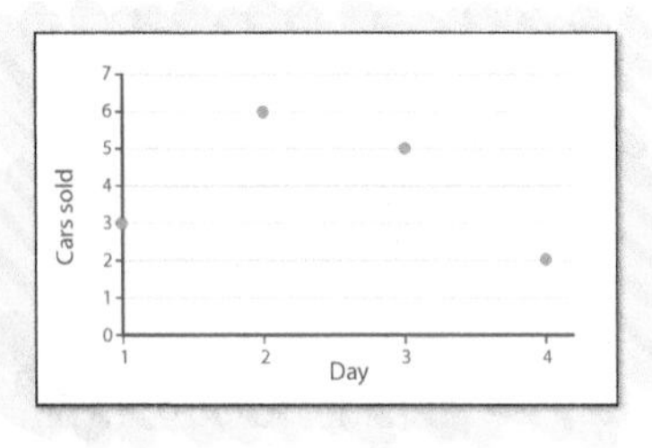

Quick Review

Some students may need occasional reminders about basic facts. These refreshers are provided at key points throughout the text in Quick Reviews. For example, in Chapter 2 there is a Quick Review on making graphs.

Chapter Summaries

Chapter summaries are included to wrap up the discussion and give an overview of the major points in the chapter. For example, at the end of Chapter 2, the summary brings together the major points about visualization of data.

CHAPTER SUMMARY

In our age, we are confronted with an overwhelming amount ofinformation. Analyzing quantitative data is an important aspect of our lives. Often this analysis requires the ability to understand visual displays, such as graphs, and to make sense of tables of data. A key point to keep in mind is that growth rates are reflected in graphs and tables.

Measurements of growth: How fast is it changing?

Data sets can be presented in many ways. Often the information is presented using a table. To analyze tabular information, we choose an independent variable. From this point of view, the table presents a function, which tells how one quantity (the dependent variable) depends on another (the independent variable).

A simple tool for visualizing data sets is the *bar graph*. See **Figure 2.67** for an example of a bar graph. A bar graph is often appropriate for small data sets when there is nothing to show between data points.

Both absolute and percentage changes are easily calculated from data sets, and such calculations often show behavior not readily apparent from the original data. We calculate the percentage change using

$$\text{Percentage change} = \frac{\text{Change in function value}}{\text{Previous function value}} \times 100\%$$

One example of the value of calculating such changes is the study of world population in Example 2.3.

To find the average growth rate of a function over an interval, we calculate the change in the function divided by the change in the independent variable:

$$\text{Average growth rate} = \frac{\text{Change in function value}}{\text{Change in independent variable}}$$

Keeping track of the units helps us to interpret the average growth rate in a given context.

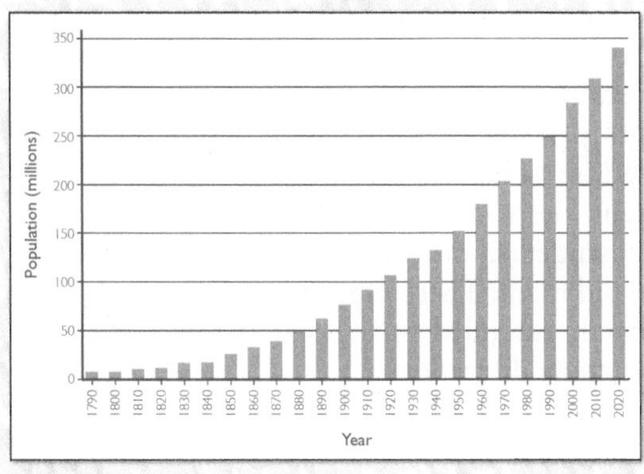

FIGURE 2.67 A bar graph.

KEY TERMS

logic, p. 17
premises, p. 17
hypotheses, p. 17
conclusion, p. 17
valid, p. 17
informal fallacy, p. 18
formal fallacy, p. 18
deductive, p. 24
inductive, p. 24
or, p. 33
logically equivalent, p. 40
set, p. 46
elements, p. 46
subset, p. 46
disjoint, p. 46
Venn diagram, p. 47

Key Terms

Prior to each Chapter Quiz is a list of Key Terms from the chapter along with the page number on which each term is introduced. This allows students to review vocabulary from the chapter to be sure they have not missed any important terms.

CHAPTER QUIZ

1. If I draw a card at random from a standard deck of 52 cards, what is the probability that I draw a red queen? Write your answer as a fraction.

Answer 2/52 = 1/26

If you had difficulty with this problem, see Example 5.1.

2. Suppose I flip two identical coins. What is the probability that I get a head and a tail? Write your answer as a fraction.

Answer 1/2

If you had difficulty with this problem, see Example 5.2.

3. Suppose a softball team has 25 players. Assume that 13 bat over .300, 15 are infielders, and 5 are both. What is the probability that a player selected at random is either batting over .300 or an infielder? Write your answer as a fraction.

Answer 23/25

If you had difficulty with this problem, see Example 5.6.

4. A new medical test is given to a population whose disease status is known. The results are in the table below.

	Has disease	Does not have disease
Test positive	300	100
Test negative	10	200

Use these data to estimate the sensitivity and specificity of this test. Write your answers in percentage form rounded to one decimal place.

Answer Sensitivity 96.8%; specificity 66.7%
If you had difficulty with this problem, see Example 5.10.

5. The table below shows test results for a certain population.

	Has disease	Does not have disease
Test positive	10	30
Test negative	5	40

Chapter Quiz

Each chapter ends with a Chapter Quiz, which consists of a representative sample of exercises. These are intended to give students a review of the major points in the chapter and a chance to test themselves on retention of the concepts they have learned. These exercises are accompanied by answers and references to worked examples in the text. For instance, the quiz at the end of Chapter 5 contains 12 exercises.

Topics and Organization

Chapter 1 covers logic. The message is straightforward: We need to critically analyze the input we get from the popular media rather than accept it at face value. The basic tools needed to distinguish valid arguments from invalid ones are provided. The closing section of this chapter focuses on number sense, bringing home the meaning of very large numbers like the national debt and very small numbers like those used to measure the size of a bacterium. Number sense is also useful in making ballpark estimates that help us judge whether figures we encounter in daily life are accurate.

Familiar topics such as interest rates, inflation, and population growth rates are prime examples of issues that are understood in terms of rates of change. Chapter 2 discusses their general significance and shows how they are reflected in tabular and graphical presentations. The presentation and treatment of data in Chapter 2 are cast in terms of rates of change, along with the tabular and visual consequences of presenting such data. The treatment of data in terms of statistics, in contrast, is presented in Chapter 6.

In Chapter 3, we focus on constant growth rates (linear functions) and constant percentage growth rates (exponential functions) because these are the ones most commonly encountered in the popular media. We believe that laying this foundation in Chapters 2 and 3, prior to discussing financial issues in Chapter 4, is an efficient and natural way to approach things. Furthermore, it provides students with a foundation for understanding other kinds of growth rates (both positive and negative) such as inflation (or deflation) and population (growth or decline).

The basic concept that exponential functions change at a constant percentage rate unifies the discussion of applications. Rather than treat compound interest as a topic separate from exponential population growth and other exponential models, our unified treatment emphasizes that there is one basic idea with several important applications. It also highlights the contrast between linear and exponential growth, surely one of the most important distinctions for quantitative literacy. The final section of the chapter presents the basics of quadratic functions and features their use in modeling real-word phenomena.

Chapter 4, Personal Finance, covers the basics of borrowing, saving, credit, and installment loans. Chapters 2 and 3 are recommended as a foundation for this material, but Chapter 4 is structured so that it can stand on its own. Many college students are already facing some of the complexities of financial life through their use of credit cards and loans to pay for their education. Soon enough they will be looking at large purchases such as cars or homes and facing other significant financial issues that are covered in this chapter. A key feature is an emphasis on how compound interest affects both savings and debt. Illustrations are offered to show the importance of an early start to long-term savings.

Chapter 5 focuses on probability and Chapter 6 covers statistics. From weather reports to medical tests to risk management and gambling, probability plays an important role in our lives. And it lays a foundation for statistics, which is probably the single most used and misused mathematical tool employed by the popular media. We see statistics in such diverse guises as SAT scores, political polls, median home prices, and batting averages. But any attempt to distill the meaning of large amounts of data into a few numbers or phrases is a risky business, and statistics, if improperly used or understood, can mislead rather than inform. Our treatment gives students the tools they need to view the statistics dispensed by the popular media with a critical eye.

Chapter 7 provides an introduction to graph theory, where we explore connections to topics such as efficient routes and spell checkers. The most salient connection is to social networks such as Facebook, which are familiar to virtually every student these days.

Chapter 8 examines voting and social choice. Students may be surprised to learn that elections and voting are more complex than might be suspected. Multiple candidates and voting blocs lead to unexpected complications. Inheritance or divorce can lead to difficult issues involving fair division of money, property, and items of sentimental value. We look at several schemes for accomplishing this division. Finally, the issue of how members of the U.S. House of Representatives are apportioned among the states is explained.

The book closes with Chapter 9, a look at geometry. We discuss measurements, symmetry, and tilings. Proportionality arising from symmetry is key to understanding many things, from the relative cost of pizzas to why King Kong is (unfortunately?) an impossibility. Geometry is important in art and architecture, but also has its share of controversy, which we explore.

For good or ill, mathematics is part of our daily lives. One need not be a professional mathematician to employ it effectively. A few basic mathematical ideas and the confidence to employ them can be important tools that help us understand and engage in the world we inhabit.

New to the Third Edition

Over 20% of examples and exercises have been updated and new ones added. In addition, by popular demand, explanations and presentation of certain topics have been improved, including a new subsection treating linear versus exponential functions (see Section 3.2); a new, more careful treatment of finance charges for credit cards (see Section 4.4); a new subsection on frequency tables and histograms (see Section 6.1); more extensive treatment of the interpretation, calculation, and use of z-scores (see Section 6.2); and a revised, clearer way of calculating fair division using the Knaster procedure (see Section 8.3).

The exercise sets have been extensively reworked. A new addition to the exercise sets, Test Your Understanding, present multiple choice and fill-in-the-blank exercises for a quick check on understanding. The main portion of the exercise sets, Problems, has been reworked so that the multiple parts of multi-part exercises now fall as lettered items under a single problem number, simplifying the experience for both instructors creating assignments and students completing them. Another addition to the exercises set is Writing About Mathematics, with open-ended research questions great for extra assignments or group work. The What Do You Think? conceptual exercises have been brought into the exercise sets, to serve as a capstone, rather than sitting outside as an afterthought. And, of course, many exercises have had data updated.

Go Digital with Achieve

Macmillan's groundbreaking online platform Achieve contains assessment, a fully accessible e-book, and dynamic resources designed to engage students before, during, and after class. Achieve was co-designed with students, educators, and Macmillan's Learning Science team to provide instructors with flexible tools to engage their students within and outside of the classroom.

Resources in Achieve include:

- **Dynamic Activities**, powered by Desmos, provide interactive examples that let the student manipulate figures based on the text in order to better understand the visual aspects and dimensions of the concepts being presented.
- **MathClips** are animated whiteboard videos that work out individual problems in detail.
- **LearningCurve** adaptive quizzing helps students deepen their conceptual understanding.
- The **e-book** is fully accessible and boasts handy features like note taking, highlighting, offline reading, and screen reader functionality.
- **Homework exercises** feature built-in coaching tools—detailed and error-specific feedback, hints, and fully worked solutions—meant to guide students toward the correct answers.
- Easy-to-use **Instructor Activity Guides** and iClicker integration help engage students in active learning.

Achieve can be integrated with popular LMS systems for ease of use. Read more about integration at www.macmillanlearning.com/achievelamath.

For Instructors

The following resources are available within Achieve to provide instructor support:

- **Image Slides** containing all textbook figures and tables
- **Lecture Slides**
- **Instructor's Guide with Solutions**
- **Test Bank** containing hundreds of multiple-choice, fill-in-the-blank, and short answer questions.

iClicker is a two-way radio-frequency classroom response solution developed by educators for educators. Each step of iClicker's development has been informed by teaching and learning. iClicker can be used via handheld remote or mobile app; mobile app access is included with student access to Achieve. To learn more about iClicker, visit www.iclicker.com.

Acknowledgments

We are grateful for the thoughtful comments and insights from the reviewers who helped us shape the third edition. Their reviews were invaluable, helping us see our materials through the eyes of others.

We remain grateful to the many instructors who made important contributions to the creation of the first and second editions.

Third edition reviewers

Darry Andrews, *The Ohio State University*
Mary Ann Barber, *University of North Texas*
Timothy Beaver, *Isothermal Community College*
Jordan Bertke, *Central Piedmont Community College*
Chris Black, *Central Washington University*
Mark Branson, *Stevenson University*
Jonathan Brucks, *University of Texas at San Antonio*
Shawn Clift, *Eastern Kentucky University*
Cherlyn Converse, *California State University, Fullerton*
Shari Davis, *Old Dominion University*
Mindy Diesslin, *Wright State University*
Kathleen Donoghue, *Robert Morris University*
Nataliya Doroshenko, *University of Memphis*
Jonathan David Farley, *Morgan State University*
Lauren Fern, *University of Montana*
David French, *Tidewater Community College*
Deborah Fries, *Wor–Wic Community College*
Sydia Gayle-Fenner, *Central Piedmont Community College*
Joseph Huber, *University of Kansas*
Brian Karasek, *Glendale Community College*
Susan Keith, *Perimeter College at Georgia State University*
Jane Kirchner West, *Trident Technical College*
Winifred A. Mallam, *Texas Woman's University*
Stacy Martig, *St. Cloud State University*
Maggie May, *Moraine Park Technical College*
Cynthia McGinnis, *Northwest Florida State College*
David E. Meel, *Bowling Green State University*
Junalyn Navarra-Madsen, *Texas Woman's University*
LeAnn Neel-Romine, *Ball State University*
Cynthia Piez, *University of Idaho*
Charlotte Pisors, *Baylor University*
Robin Rufatto, *Ball State University*
Haazim Sabree, *Perimeter College at Georgia State University*
Jason C. Stone, *Cleveland State University*
Vicki Todd, *Southwestern Community College*
Michael Traylor, *Wake Technical Community College*
Cynthia Vanderlaan, *Indiana University–Purdue University Fort Wayne*
Janet Yi, *Ball State University*

Second edition reviewers

Kimberly Alacan, *Daytona State College*
Susan P. Bradley, *Angelina College*
Marcela Chiorescu, *Georgia College & State University*
Richard DeCesare, *Southern Connecticut State University*
Lauren Fern, *University of Montana*
Wilson B. Grab, *Forsyth Technical Community College*
Pat Humphrey, *Georgia Southern University*
Andy D. Jones, *Prince George's Community College*
William Thomas Kiley, *Georgia Mason University*
Michael Leen, *New England College*
Andrew Long, *Northern Kentucky University*
Crystal Lorch, *Ball State University*
Jonathan Loss, *Catawba Valley Community College*
Laura Lynch, *College of Coastal Georgia*
Hee Seok Nam, *Washburn University*
Cornelius Nelan, *Quinnipiac University*
Kathy Rodgers, *University of Southern Indiana*
Hugo Rossi, *University of Utah*
Hilary Seagle, *Southwestern Community College*
Nic Sedlock, *Framingham State University*
Christopher Shaw, *Columbia College Chicago*
John T. Zerger, *Catawba College*

First edition reviewers

Reza Abbasian, *Texas Lutheran University*
Lowell Abrams, *George Washington University*
Marwan Abu-Sawwa, *University of North Florida*
Tom Adamson, *Phoenix College*
Margo Alexander, *Georgia State University*
K. T. Arasu, *Wright State University*
Kambiz Askarpour, *Essex Community College*
Wahab Baouchi, *Regis University*
Erol Barbut, *University of Idaho*
Ronald Barnes, *University of Houston—Downtown*
Wes Barton, *Southwestern Community College*
David Baughman, *University of Illinois at Chicago*
Jonathan Bayless, *Husson University*
Mary Beard, *Kapiolani Community College*
Jaromir Becan, *University of Texas at San Antonio*
Rudy Beharrysingh, *Haywood Community College*
Curtis Bennett, *Loyola Marymount University*
David Berry, *Xavier University*
Dip Bhattacharya, *Clarion University of Pennsylvania*
Phil Blau, *Shawnee State University*
Tammy Blevins, *Hillsborough Community College*
Karen Blount, *Frederick Community College*
Russell Blyth, *Saint Louis University*
Mark Brenneman, *Mesa Community College*

Pat Brislin, *Millersville University of Pennsylvania*
Albert Bronstein, *University of Illinois at Chicago*
Joanne Brunner, *Joliet Junior College*
Corey Bruns, *University of Colorado at Boulder*
Stanislaw Buchcic, *Wright College*
Gerald Burton, *Virginia State University*
Dale Buske, *St. Cloud State University*
Jason Callahan, *St. Edward's University*
Jeff Cathrall, *University of Colorado at Boulder*
Christine Cedzo, *Gannon University*
Jiang-Ping Chen, *St. Cloud State University*
Lee Chiang, *Trinity Washington University*
Madeline Chowdhury, *Mesa Community College*
Lars Christensen, *Texas Tech University*
Boyd Coan, *Norfolk State University*
Shelley Cook, *Stephen F. Austin State University*
Grace Coulombe, *Bates College*
Richard Coyne, *Arizona State University*
Tony Craig, *Paradise Valley Community College*
Kenneth Cramm, *Riverside City College*
Karen Crossin, *George Mason University*
Cheryl Cunnington, *North Idaho University*
Shari Davis, *Old Dominion University*
Megan Deeney, *Western Washington University*
Dennis DeJong, *Dordt College*
Sloan Despeaux, *Western Carolina University*
Mindy Diesslin, *Wright State University*
Qiang Dotzel, *University of Missouri—St. Louis*
William Drury, *University of Massachusetts Lowell*
Mark Ellis, *Central Piedmont Community College*
Solomon Emeghara, *Bloomfield College*
Scott Erway, *Itasca Community College*
Brooke Evans, *Metropolitan State College of Denver*
Cathy Evins, *Roosevelt University*
Scott Fallstrom, *University of Oregon*
Vanessa Farren, *Seneca College of Applied Arts and Technology*
Lauren Fern, *University of Montana*
Marc Formichella, *University of Colorado at Boulder*
Monique Fuguet, *University of Massachusetts—Boston*
Joe Gallegos, *Salt Lake City Community College*
Kevin Gammon, *Cumberland University*
Stephen Gendler, *Clarion University*
Antanas Gilvydis, *City Colleges of Chicago—Richard J. Daley College*
Paul Glenn, *Catholic University of America*
Cliona Golden, *Bard College*
Marilyn Grapin, *Metropolitan Center/Empire State College*
Charlotte Gregory, *Trinity College*
Susan Hagen, *Virginia Polytechnic and State University*
C. Hail, *Union University*
Susan Haller, *St. Cloud State University*
Richard Hammond, *St. Joseph's College*
David Hartz, *College of St. Benedict*
William Haver, *Virginia Commonwealth University*
Dan Henschell, *Douglas College*
Ann Herbst, *Santa Rosa Junior College*
Carla Hill, *Marist College*
Frederick Hoffman, *Florida Atlantic University*
Fran Hopf, *University of South Florida*
Mark Hunacek, *Iowa State University*
Jerry Ianni, *CUNY—La Guardia Community College*
Kelly Jackson, *Camden County College*
Nancy Jacqmin, *Carlow University*
Benny John, *University of Houston—Downtown*
Clarence Johnson, *Cleveland State University*
Terrence Jones, *Norfolk State University*
Richard Kampf, *Great Basin College*
Robert Keller, *Loras College*
Deborah Kent, *Hillsdale College*
Jane Kessler, *Quinnipiac University*
Margaret Kiehl, *Rensselaer Polytechnic Institute*
Sung Eun Kim, *Judson College*
Jared Knittel. *Texas State University*
Marshall Kotzen, *Worcester State University*
Kathryn Kozak, *Coconino County Community College*
Karole Kurnow, *Dominican College*
Latrice Laughlin, *University of Alaska—Fairbanks*
Namyong Lee, *Minnesota State University*
Raymond Lee, *University of North Carolina at Pembroke*
Richard Leedy, *Polk Community College*
Warren Lemerich, *Laramie City Community College*
Kathy Lewis, *California State University—Fullerton*
Michael Little Crow, *Scottsdale Community College*
Antonio Magliaro, *Quinnipiac University*
Iryna Mahlay, *Cleveland State University*
Antoinette Marquard, *Cleveland State University*
Jeannette Martin, *Washington State University*
John Martin, *Santa Rosa Junior College*
Christopher Mason, *Community College of Vermont*
Kathleen McDaniel, *Buena Vista University*
Karol McIntosh, *University of South Florida*
Farzana McRae, *Catholic University of America*
Ryan Melendez, *Arizona State University*
Phyllis Mellinger, *Hollins University*
Tammy Muhs, *University of Central Florida*
Ellen Mulqueeny, *Baldwin-Wallace College*
Judy Munshower, *Clarke University*
Bishnu Naraine, *St. Cloud State University*
Jennifer Natoli, *Manchester Community College*
Lars Neises, *Spokane Falls Community College*
Donald Neth, *Kent State University*
Cao Nguyen, *Central Piedmont Community College*
Stephen Nicoloff, *Paradise Valley Community College*
John Noonan, *Mount Vernon Nazarene University*
Douglas Norton, *Villanova University*
Jackie Nygaard, *Brigham Young University*
Jon Oaks, *Henry Ford Community College*
Michael Oppedisano, *Onondaga Community College*
Anne O'Shea, *North Shore Community College*
Andrew Pak, *Kapiolani Community College*
Kenny Palmer, *Tennessee Technological University*

Donna Passman, *Bristol Community College*
Laramie Paxton, *Coconino County Community College*
Shawn Peterson, *Texas State University*
Michael Petrie, *University of Northern Colorado*
Yevgeniy Ptukhin, *Texas Tech University*
Evelyn Pupplo-Cody, *Marshall University*
Parthasarathy Rajagopal, *Kent State University*
Crystal Ravenwood, *Whatcom Community College*
Carolynn Reed, *Austin Community College*
Rochelle Ring, *City College of New York*
Nancy Rivers, *Wake Technical Community College*
Leanne Robertson, *Seattle University*
Joe Rody, *Arizona State University*
Cosmin Roman, *The Ohio State University—Lima*
Tracy Romesser, *Erie Community College*
Martha Rooney, *University of Colorado at Boulder*
M. E. Rosar, *William Paterson University*
Anita Ross, *North Park University*
Katherine Safford-Ramus, *Saint Peter's College*
Matt Salomone, *Bridgewater State University*
Paula Savich, *Mayland Community College*
Susan Schmoyer, *Worcester State University*
Alison Schubert, *Wake Technical Community College*
Kathy Schultz, *Pensacola Junior College—Pensacola*
Sal Sciandra, *Niagara County Community College*
Brian Scott, *Cleveland State University*
Stacy Scudder, *Marshall University*
Tsvetanka Sendova, *Bennett College*
Sandra Sikorski, *Baldwin-Wallace College*
Laura Smithies, *Kent State University*
Mark Snavely, *Carthage College*
Lawrence Somer, *Catholic University of America*
Sandy Spears, *Jefferson Community and Technical College*
Dori Stanfield, *Davidson Community College*
Eryn Stehr, *Minnesota State University Mankato*
Kim Steinke, *College of the North Atlantic*
Donna Stevenson, *The State University of New York—Jefferson Community College*
Robert Storfer, *Florida International University*
Sarah Stovall, *Stephen F. Austin State University*
Joseph Tahsoh, *South Carolina State University*
Susan Talarico, *University of Wisconsin—Stevens Point*
Louis Talman, *Metropolitan State College of Denver*
Julie Theoret, *Johnson State College*
Jamie Thomas, *University of Wisconsin, Manitowoc*
Mike Tindall, *Seattle Pacific University*
Karl Ting, *Mission College*
Hansun To, *Worchester State*
Emilio Toro, *University of Tampa*
Preety Tripathi, *The State University of New York at Oswego*
David Tucker, *Midwestern State University*
Ulrike Vorhauer, *Kent State University*
Walter Wallis, *Southern Illinois University*
Tom Walsh, *Kean University*
Richard Watkins, *Tidewater Community College*
Leigh Ann Wells, *Western Kentucky University*
Ed Wesly, *Harrington College of Design*
Cheryl Whitelaw, *Southern Utah University*
Mary Wiest, *Minnesota State University*
David Wilson, *The State University of New York—College at Buffalo*
Antoine Wladina, *Farleigh Dickinson University*
Alma Wlazlinski, *McLennan Community College*
Jim Wolper, *Idaho State University*
Debra Wood, *University of Arizona*
Judith Wood, *Central Florida Community College*
Bruce Woodcock, *Central Washington University*
Jennifer Yuen, *Illinois Institute of Art*
Abbas Zadegan, *Florida International University*
Paulette Zizzo, *Wright State University—Main Campus*
Cathleen Zucco-Teveloff, *Rowan University*
Marc Zucker, *Marymount Manhattan College*

We wish to thank the dedicated and professional staff at Macmillan Learning who have taken such excellent care of our project. We owe a debt of gratitude to Ruth Baruth, who believed in us and this book from the very beginning; to Terri Ward, who guided this project through development; to Nikki Miller Dworsky and Katrina Mangold, who were invaluable in helping us with this third edition; and to the many skilled individuals involved in the entire production process: Edgar Doolan, Robin Fadool, Michael McCarty, Vicki Tomaselli, and Susan Wein. We are entirely in debt to Randi Blatt Rossignol, our development editor for the first and second editions, whose fine editorial eye, intelligence, humor, and wisdom utterly transformed our text, refining it over and over until all involved were finally satisfied. Our text is the beneficiary of her gentle but insistent ministrations.

Bruce Crauder

Benny Evans

Jerry Johnson

Alan Noell

Index of Applications

Education

Finance

Geography and Population Dynamics

Life Sciences

Medicine

Physical Sciences

Political Science

Social Networking

Social Sciences

Sports

Technology

Transportation

QUANTITATIVE LITERACY

1 Critical Thinking

There is no need to make a case for the importance of literacy in contemporary society. The ability to read with comprehension and the ability to communicate clearly are universally recognized as necessary skills for the average citizen. But there is another kind of literacy, one that may be just as important in a world filled with numbers, charts, diagrams, and even formulas: *quantitative literacy*.

In contemporary society we are confronted with quantitative issues at every turn. We are required to make choices and decisions about purchases, credit cards, savings plans, investments, and borrowing money. In addition, politicians and advertisers throw figures and graphs at us in an attempt to convince us to vote for them or buy their product. If we want to be informed citizens and savvy consumers, we must be quantitatively literate.

At the heart of quantitative literacy is critical thinking, which is necessary to deal intelligently with the information and misinformation that come to us via television, computers, newspapers, and magazines.

The following excerpt is from an article that appeared in *The Chronicle of Higher Education*. It shows the importance of critical thinking in education.

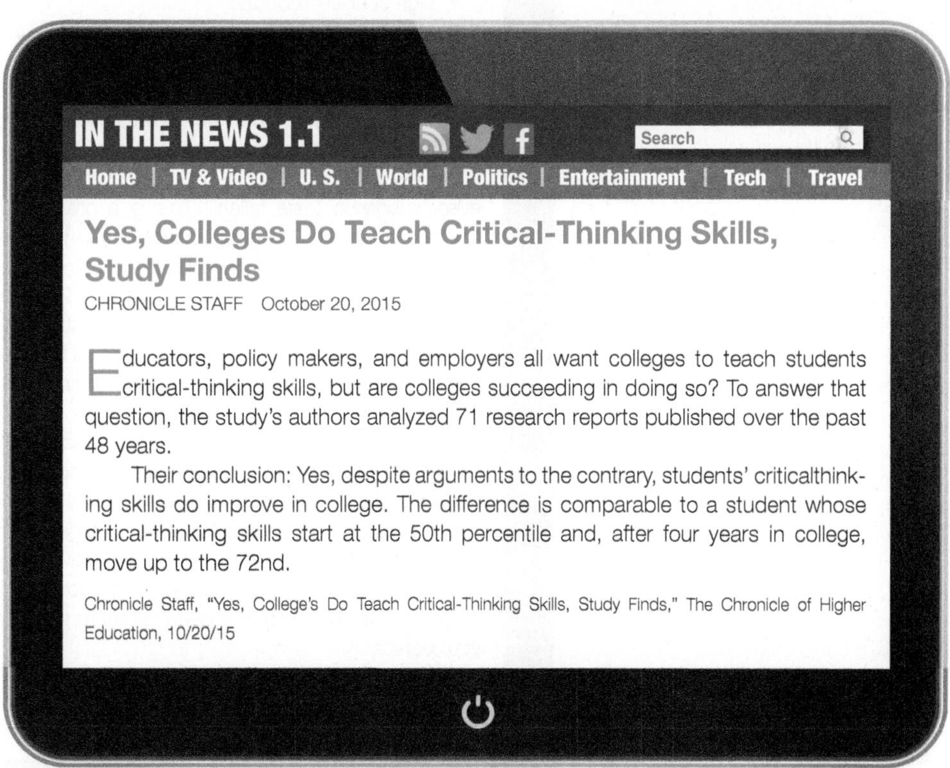

Yes, Colleges Do Teach Critical-Thinking Skills, Study Finds

CHRONICLE STAFF October 20, 2015

Educators, policy makers, and employers all want colleges to teach students critical-thinking skills, but are colleges succeeding in doing so? To answer that question, the study's authors analyzed 71 research reports published over the past 48 years.

Their conclusion: Yes, despite arguments to the contrary, students' criticalthinking skills do improve in college. The difference is comparable to a student whose critical-thinking skills start at the 50th percentile and, after four years in college, move up to the 72nd.

Chronicle Staff, "Yes, College's Do Teach Critical-Thinking Skills, Study Finds," The Chronicle of Higher Education, 10/20/15

This chapter opens with a phenomenon called *Simpson's paradox*, which illustrates in a very interesting way the need for critical thinking. We next discuss basic logic, the use of Venn diagrams, and number sense.

1.1 Public policy and Simpson's paradox: Is "average" always average?

TAKE AWAY FROM THIS SECTION
View with a critical eye conclusions based on averages.

News headlines often summarize complicated economic data using a simple statement. For example, public policy analysts are concerned about wages within economic groups and how wages change over periods of economic instability. It would be encouraging to read that overall wages (adjusted for inflation) in the United States increased from 2000 to 2013. That statement is true, but it does not fully describe a complicated situation. The article below from the blog *Revolutions* explains why. It refers to the *median*, which is the middle value in a list of data.[1]

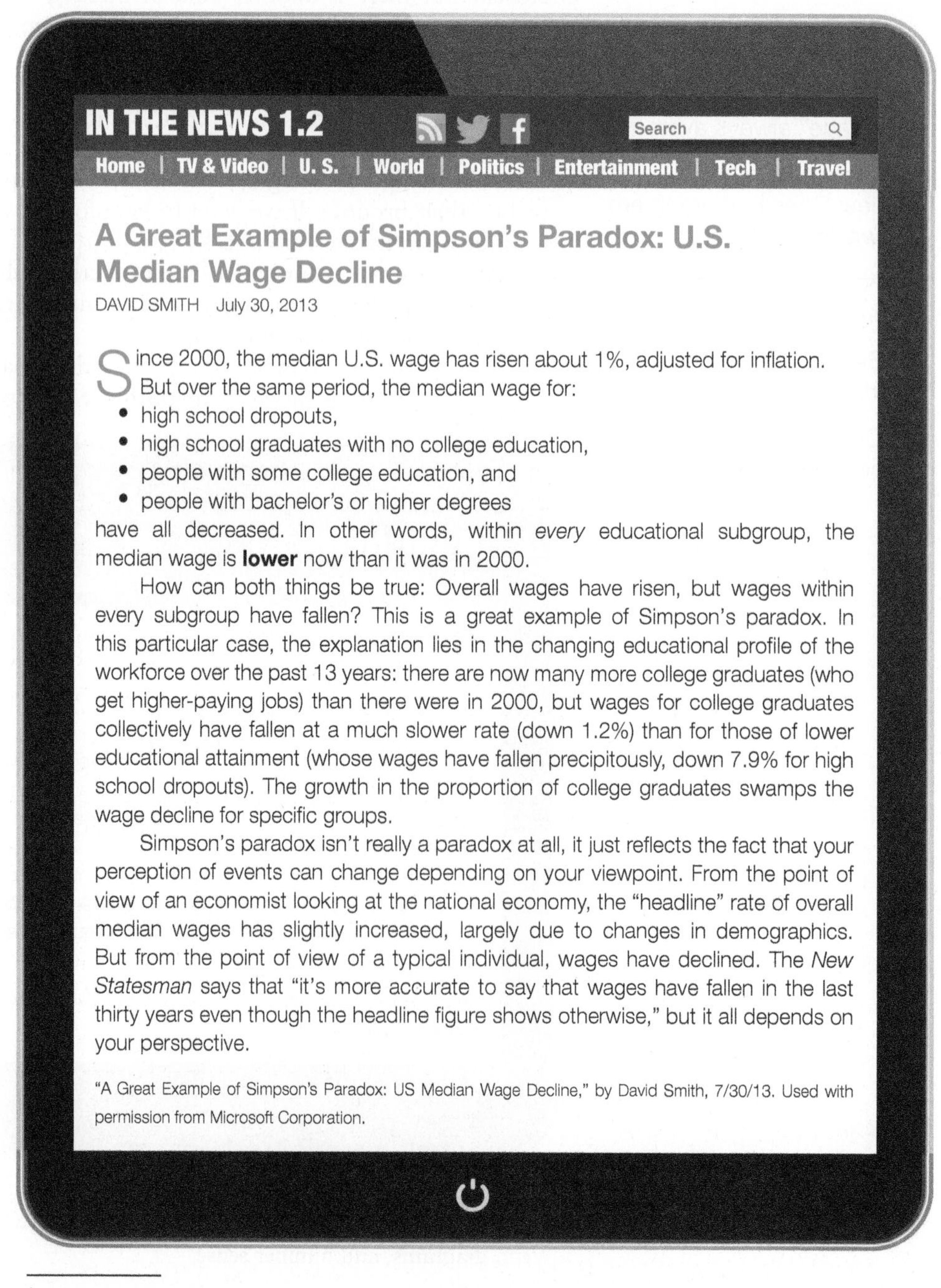

IN THE NEWS 1.2

Home | TV & Video | U. S. | World | Politics | Entertainment | Tech | Travel

A Great Example of Simpson's Paradox: U.S. Median Wage Decline

DAVID SMITH July 30, 2013

Since 2000, the median U.S. wage has risen about 1%, adjusted for inflation. But over the same period, the median wage for:

- high school dropouts,
- high school graduates with no college education,
- people with some college education, and
- people with bachelor's or higher degrees

have all decreased. In other words, within *every* educational subgroup, the median wage is **lower** now than it was in 2000.

How can both things be true: Overall wages have risen, but wages within every subgroup have fallen? This is a great example of Simpson's paradox. In this particular case, the explanation lies in the changing educational profile of the workforce over the past 13 years: there are now many more college graduates (who get higher-paying jobs) than there were in 2000, but wages for college graduates collectively have fallen at a much slower rate (down 1.2%) than for those of lower educational attainment (whose wages have fallen precipitously, down 7.9% for high school dropouts). The growth in the proportion of college graduates swamps the wage decline for specific groups.

Simpson's paradox isn't really a paradox at all, it just reflects the fact that your perception of events can change depending on your viewpoint. From the point of view of an economist looking at the national economy, the "headline" rate of overall median wages has slightly increased, largely due to changes in demographics. But from the point of view of a typical individual, wages have declined. The *New Statesman* says that "it's more accurate to say that wages have fallen in the last thirty years even though the headline figure shows otherwise," but it all depends on your perspective.

"A Great Example of Simpson's Paradox: US Median Wage Decline," by David Smith, 7/30/13. Used with permission from Microsoft Corporation.

[1] *Simpson's paradox applies to the median as well as the average.*

According to the article, it is possible to have the overall wages in the United States increase over a period of time but have the wages within every economic subgroup decrease over the same period. This phenomenon seems impossible, but the following example shows that it can happen.

Simpson's paradox and wages

The following example shows the data behind the article at the beginning of this section.

EXAMPLE 1.1 Calculating averages: Wages

The following tables show the number of workers (in thousands) and the total weekly earnings (in thousands of dollars) in the first quarter of the given year. These figures are for workers 25 years and older in the United States in two educational subgroups: those without a bachelor's degree and those with a bachelor's degree. The data are adapted from tables compiled by the U.S. Bureau of Labor Statistics and are adjusted for inflation. (The original data show the median for each educational subgroup, but for simplicity we use the average. The difference is discussed in Section 6.1.)

	Year 2000	
	Number of workers (thousands)	Total weekly earnings (thousands)
Less than bachelor's	61,166	$39,500,093
Bachelor's or higher	27,564	$33,169,777
Total	88,730	$72,669,870

	Year 2013	
	Number of workers (thousands)	Total weekly earnings (thousands)
Less than bachelor's	57,154	$33,867,652
Bachelor's or higher	37,013	$44,008,457
Total	94,167	$77,876,109

Show that from 2000 to 2013 the average weekly earnings per worker decreased in each educational subgroup but increased overall. Round each average to the nearest dollar.

SOLUTION

First we find the average weekly earnings per worker in 2000 for those with less than a bachelor's degree:

$$\text{Average weekly earnings} = \frac{\text{Total weekly earnings}}{\text{Number of workers}}$$

$$= \frac{\$39{,}500{,}093 \text{ thousand}}{61{,}166 \text{ thousand}}$$

This average is about $646 per worker.

We calculate the remaining averages in a similar fashion. The results are shown in the following table.

	Year 2000 average earnings	Year 2013 average earnings
Less than bachelor's	$39,500,093/61,166 or $646	$33,867,652/57,154 or $593
Bachelor's or higher	$33,169,777/27,564 or $1203	$44,008,457/37,013 or $1189
Total	$72,669,870/88,730 or $819	$77,876,109/94,167 or $827

Thus, for both educational subgroups the average weekly earnings decreased from 2000 to 2013. But overall the average weekly earnings *increased*.

TRY IT YOURSELF 1.1

Here are the data from a hypothetical town over the same period.

	Year 2000	
	Number of workers	Total weekly earnings
Less than bachelor's	400	$260,000
Bachelor's or higher	700	$532,000
Total	1100	$792,000
	Year 2013	
	Number of workers	Total weekly earnings
Less than bachelor's	200	$128,000
Bachelor's or higher	1100	$825,000
Total	1300	$953,000

Make a table showing the average weekly earnings per worker for each educational subgroup and overall in both 2000 and 2013. (Round each average to the nearest dollar.) Interpret your results.

The answer is provided at the end of the section.

It is easy to see why Simpson's paradox is called a paradox. Such a result seems impossible. But the example above shows that combining, or aggregating, data can mask underlying patterns.

Even though this phenomenon is called a paradox, these results are not so surprising if we look more carefully. The article notes that the relative number of college graduates increased significantly from 2000 to 2013. In 2000, the ratio of college graduates to total workers was 27,564/88,730, or about 31%, whereas in 2013 that ratio was 37,013/94,167, or about 39%. This shift in the level of educational attainment, plus the fact that better-educated workers are paid higher wages, helps to explain this phenomenon.

The Berkeley gender discrimination case

A famous but relatively simple case provides a classic illustration of Simpson's paradox. Data from a 1973 study seemed to give persuasive evidence that the University of California at Berkeley was practicing gender discrimination in graduate school admissions. The following table shows that Berkeley accepted 44.0% of male applicants but only 35.0% of female applicants. (All of the percentages in this table and the next are rounded to one decimal place. Consult the end of this section for a quick review of percentages.)

	Applicants	Accepted	% accepted
Male	8442	3714	44.0%
Female	4321	1512	35.0%

For example, here is the way to calculate the percentage of male applicants who were accepted:

$$\text{Percentage of male applicants accepted} = \frac{\text{Males accepted}}{\text{Male applicants}} \times 100\%$$
$$= \frac{3714}{8442} \times 100\%$$

This is about 44.0%.

Justin Sullivan/Getty Images

Students on the University of California at Berkeley campus.

These numbers may seem to be conclusive evidence of discrimination. When we look more closely, however, we find that graduate school admissions are based on departmental selections. There are many departments at Berkeley, but to keep things simple, we will assume that the university consists of only two departments: Math and English. Here is how the male and female applicants might have been divided between these two departments.

	Males			Females		
	Applicants	Accepted	% Accepted	Applicants	Accepted	% Accepted
Math	2000	500	25.0%	3000	780	26.0%
English	6442	3214	49.9%	1321	732	55.4%
Total	8442	3714	44.0%	4321	1512	35.0%

Note that each of the two departments actually accepted a larger percentage of female applicants than male applicants: The Math Department accepted 25.0% of males and 26.0% of females, and the English Department accepted 49.9% of males and 55.4% of females. The explanation of why a larger percentage of men overall were accepted lies in the fact that the Math Department accepts significantly fewer applicants than does English, both for men (25.0% compared to 49.9%) and for women (26.0% compared to 55.4%). Most women applied to the Math Department, where it is more difficult to be accepted, but most men applied to English, where it is less difficult.

A similar type of mind-bending phenomenon occurs in sports, as the following example shows. These instances of Simpson's paradox are taken from a post on the blog *Natural Blogarithms*.

EXAMPLE 1.2 Comparing averages: Batting records

The following table shows the batting records in 2010 of two Major League Baseball teams, the Chicago Cubs and the Oakland Athletics, both for games played indoors and those played outdoors.

	Indoors		Outdoors	
	At-bats	Hits	At-bats	Hits
Chicago Cubs	462	134	5050	1280
Oakland Athletics	113	33	5335	1363

To calculate the batting average we divide the number of hits by the number of at-bats. Which team had the higher average playing indoors? Which had the higher average playing outdoors? Which had the higher average overall? Round all figures to three decimal places.

SOLUTION

We calculate the average for the Chicago Cubs playing indoors as follows:

$$\begin{aligned}\text{Average indoors} &= \frac{\text{Hits}}{\text{At-bats}}\\ &= \frac{134}{462}\end{aligned}$$

This is about 0.290. We calculate averages for the Chicago Cubs and the Oakland Athletics both indoors and outdoors in a similar fashion. The results are shown in the following table.

	Indoors	Outdoors
Average for Chicago Cubs	134/462 or 0.290	1280/5050 or 0.253
Average for Oakland Athletics	33/113 or 0.292	1363/5335 or 0.255

Now we find the overall averages. Altogether the Chicago Cubs got a total of $134 + 1280 = 1414$ hits out of $462 + 5050 = 5512$ trips to the plate. Therefore,

$$\text{Overall average for Chicago Cubs} = \frac{\text{Hits}}{\text{At-bats}} = \frac{1414}{5512}$$

This is about 0.257.

Jonathan Daniel/Getty Images

Anthony Rizzo (44) of the Chicago Cubs bats against the Philadelphia Phillies at Wrigley Field on July 24, 2015, in Chicago, Illinois.

A similar calculation for the Oakland Athletics gives the average overall of 1396/5448, or about 0.256. Hence, the Oakland Athletics had the higher batting average both indoors and outdoors, but the Chicago Cubs had the higher batting average overall.

TRY IT YOURSELF 1.2

The following table shows the batting records in 2010 of the Colorado Rockies and the Milwaukee Brewers, both for games played indoors and those played outdoors.

	Indoors		Outdoors	
	At-bats	Hits	At-bats	Hits
Colorado Rockies	272	66	5258	1386
Milwaukee Brewers	1075	271	4531	1200

Which team had the higher average playing indoors? Which had the higher average playing outdoors? Which had the higher average overall? Round all figures to three decimal places.

The answer is provided at the end of the section.

As we have seen in the other examples in this section, the explanation for this seemingly paradoxical situation with batting averages lies in the absolute numbers involved in calculating the averages. Each team had a higher batting average indoors than outdoors. But the Chicago Cubs had a higher percentage of their at-bats indoors, $462/(462+5050)$, or about 8%, than did the Oakland Athletics, $113/(113+5335)$, or about 2%. Simpson's paradox makes the use of overall batting averages alone somewhat suspect in the debate about which team is better.

Simpson's paradox occurs only when the categories involved are of different sizes. That is what happened in Example 1.2, in which one team had more at-bats indoors (where batting averages were higher) than the other. The same thing happened in the Berkeley case, where males applied more to the English Department (which accepted a higher percentage of applicants) than did females.

This paradox cannot occur when category sizes are the same. For example, suppose that Candidate A and Candidate B are running for the same office in the same state. If Candidate A has a percentage lead over Candidate B in all possible categories (say Republican, Democrat, and Independent voters), then Candidate B cannot possibly lead overall. The reason is that the three categories, Republican, Democrat, and Independent, are the same size for each of the two candidates. It may help in this context to think in terms of total numbers of supporters rather than percentages.

The examples of Simpson's paradox presented here show the importance of critical thinking. The use of a single number such as an average to describe complex data can be perilous. Further, combining data can mask underlying patterns when some factor distorts the overall picture but not the underlying patterns. To understand information presented to us, it is important for us to know at least the basics of how the information is produced and what it means.

SUMMARY 1.1
Simpson's Paradox

Simpson's paradox occurs when data are aggregated. Percentages from larger data sets are compared with percentages from smaller data sets. For example, one basketball team may have a better shooting percentage each year than a second team even though the second team has a better overall shooting percentage. Without the additional information of the number of shots attempted, the percentages mean little.

The following Quick Review is provided as a refresher for doing percentage calculations.

Quick Review Calculating Percentages

To calculate $p\%$ of a quantity, we multiply the quantity by $\frac{p}{100}$:

$$p\% \text{ of Quantity} = \frac{p}{100} \times \text{Quantity}$$

For example, to find 45% of 500, we multiply 500 by $\frac{45}{100}$:

$$45\% \text{ of } 500 = \frac{45}{100} \times 500 = 225$$

Therefore, 45% of 500 is 225.

To find what percentage of a whole a part is, we use

$$\text{Percentage} = \frac{\text{Part}}{\text{Whole}} \times 100\%$$

For example, to find what percentage of 140 is 35, we proceed as follows:

$$\begin{aligned}\text{Percentage} &= \frac{\text{Part}}{\text{Whole}} \times 100\% \\ &= \frac{35}{140} \times 100\% \\ &= 0.25 \times 100\% \\ &= 25\%\end{aligned}$$

Thus, 35 is 25% of 140.
For more review on calculating percentages, see Appendix 4.

WHAT DO YOU THINK?

Batting average: You read on the sports page that your favorite softball player had a lower batting percentage each of the last five years than a star player on another team. But your friend told you that over the five-year period your favorite softball player led the league in batting percentage. Does this mean that either the newspaper article or your friend is wrong? If not, explain how both could be right.

Questions of this sort may invite many correct responses. We present one possibility: This situation is similar to that involving batting averages in Example 1.2. Such a result for batting averages is entirely possible. Aggregate averages can be influenced by unusual subgroups of data, and five-year batting averages can be biased by an odd year's average. As an example, suppose player 1 has one at-bat and one hit per year for the first four years. In the fifth year, she has 100 at-bats and only one hit. Her overall average is very low because of the poor performance in the fifth year. Compare this with player 2, who has 100 at-bats and 99 hits for each of the first four years and then has no hit in her one at-bat in the fifth year. Her overall average is very good, but the average each year is not as good as player 1's average. The apparent contradiction comes from the fact that almost all of player 1's at-bats came in the fifth year but most of player 2's at-bats were in the first four years.

Further questions of this type may be found in the *What Do You Think?* section of exercises.

Try It Yourself answers

Try It Yourself 1.1: Calculating averages: Wages

	Year 2000 average earnings	Year 2013 average earnings
Less than bachelor's	$650	$640
Bachelor's or higher	$760	$750
Total	$720	$733

For both educational subgroups, the average weekly earnings decreased from 2000 to 2013. But overall the average weekly earnings increased.

Try It Yourself 1.2: Calculating averages: Batting averages The Milwaukee Brewers had a higher average than the Colorado Rockies indoors, 0.252 to 0.243, and also outdoors, 0.265 to 0.264. But the Colorado Rockies had the higher average overall, 0.263 to 0.262.

Exercise Set 1.1

In these exercises, round all answers in percentage form to one decimal place unless you are instructed otherwise.

Test Your Understanding

1. Explain in your own words what Simpson's paradox means.

2. True or false: It is possible for one basketball player's shooting percentage to be higher than a second player's each season, yet for the second player to have a higher overall shooting percentage.

3. True or false: Percentages in each possible category determine the overall percentage for the totality of all categories.

4. True or false: Simpson's paradox is a counterintuitive phenomenon that can mislead one who is not thinking critically.

Problems

5. Assume that 30 of 40 students passed a certain exam. What percentage passed?

6. Assume that 50 of 200 students caught the flu. What percentage caught the flu?

7. Assume that 32% of the people in my town of 5000 voted for my brother in the mayoral election. How many people voted for my brother?

8. Assume that 6% of a workforce of 4500 are currently unemployed. How many are unemployed?

9–15. Simpson's paradox or not? *In Exercises 9 through 15, determine whether the scenario described is possible. If it is possible, determine whether the scenario is an example of Simpson's paradox.*

9. In a gubernatorial election in a certain state, the percentage of supporters of Candidate A is larger than the corresponding percentage of supporters of Candidate B in each category of Democrat, Republican, and Independent. But Candidate B has a larger overall percentage of supporters.

10. On each math test, my score is higher than my friend's score. We are in the same class, and her overall test average is higher than mine.

11. The average age of men in Town A is higher than the average age of men in Town B, and the average age of women in Town A is higher than the average age of women in Town B. But the average age of people in Town A is lower than the average age of people in Town B.

12. Consider again the scenario in Exercise 11, but assume now that Towns A and B have equal numbers of men and equal numbers of women.

13. The batting averages from both Little League and Peewee Leagues in my town are higher than for the corresponding leagues in your town, and the overall batting average in my town is higher than the overall batting average in your town.

14. Consider again the scenario in Exercise 13, but assume now that the Little Leagues in both towns have the same numbers of at-bats, and the Peewee Leagues in both towns have the same numbers of at-bats.

15. In Hospitals A and B, all patients are classified as ill or very ill. A larger percentage of ill patients in Hospital A recover than in Hospital B, and a larger percentage of very ill patients in Hospital B recover than in Hospital A. Overall, a larger percentage of patients in Hospital A recover than in Hospital B.

16. Test scores. The following table shows test scores on a statewide exam.

	Local school		Statewide	
	Students tested	**Passed**	**Students tested**	**Passed**
Low income	500	300	100,000	61,000
High income	7000	4725	400,000	272,000
Total	7500	5025	500,000	333,000

a. Make a table that shows percentages of students in each category and total number that passed the exam. Round the percentages to one decimal place.

b. Did each category of student in the local school do better or worse than statewide students?

c. Did the aggregate of all students in the local school perform better or worse than all statewide students?

17. Batting averages. The following table shows at-bats and hits for two batters in two years.

	2019		2020	
	At-bats	**Hits**	**At-bats**	**Hits**
Batter 1	500	150	12	3
Batter 2	20	8	300	78

a. Find the batting average for each batter in each year and the average over the two-year period.

b. Which batter had the higher average in 2019?

c. Which batter had the higher average in 2020?

d. Which batter had the higher average over the two-year period?

18. Which hospital would you choose? The local newspaper examined a town's two hospitals and found that, over the last six years, Mercy Hospital had a patient survival rate of 79.0% while County Hospital's survival rate was 90.0%. The following table summarizes the findings.

	Lived	Died	Total	% who lived
Mercy Hospital	790	210	1000	79%
County Hospital	900	100	1000	90%

Patients were categorized upon admission as being in fair condition (or better) versus being in worse than fair condition.

a. The following table shows the survival numbers for patients admitted in fair condition:

Patients admitted in fair condition (or better)				
	Lived	Died	Total	% who lived
Mercy Hospital	580	10	590	
County Hospital	860	30	890	

Fill in the last column for this table.

b. The following table shows the survival numbers for patients admitted in worse than fair condition:

Patients admitted in worse than fair condition				
	Lived	Died	Total	% who lived
Mercy Hospital	210	200	410	
County Hospital	40	70	110	

Complete the last column for this table.

c. The head of County Hospital argued that, based on the overall figures, her hospital came out well ahead of Mercy by 90.0% to 79.0% and, therefore, is doing a far better job. If you were head of Mercy Hospital, how would you respond?

d. You are a parent trying to decide to which hospital to send your sick child. On the basis of the preceding three tables, which hospital would you choose? Explain.

19. Hiring. In a recent hiring period, a hypothetical department store, U-Mart, hired 62.0% of the males who applied and 14.1% of the females. A lawsuit was contemplated because these numbers seemed to indicate that there was gender discrimination. On closer examination, it was found that U-Mart's hiring was only for two of the store's departments: hardware and ladies' apparel. The hardware department hired 60 out of 80 male applicants and 15 out of 20 female applicants. The ladies' apparel department hired 2 out of 20 male applicants and 30 out of 300 female applicants. This information is summarized in the accompanying table.

	Males applying	Males hired	%	Females applying	Females hired	%
Hardware	80	60		20	15	
Ladies' apparel	20	2		300	30	
Total	100	62	62.0%	320	45	14.1%

a. Complete the preceding table to show the percentages of males and females hired in each department.

b. You are an attorney for a female plaintiff. How would you argue that there is gender discrimination?

c. You are an attorney for U-Mart. How would you argue that there is no gender discrimination based on the results of parts a and b?

d. How would you vote if you were on the jury based on the results of parts a through c?

20. Success and poverty. The success rates of two schools were compared in two categories: students from families above the poverty line and students from families below the poverty line. The results are in the following tables.

Students from families above the poverty line				
	Passed	Failed	Total	% passing
School A	400	77	477	
School B	811	190	1001	

Students from families below the poverty line				
	Passed	Failed	Total	% passing
School A	199	180	379	
School B	38	67	105	

a. Complete the table for students above the poverty line.

b. Complete the table for students below the poverty line.

c. What percentage of students from School A are from families above the poverty line?

d. What percentage of students from School B are from families above the poverty line?

e. Complete the following table for all students:

All students in both groups				
	Passed	Failed	Total	% passing
School A				
School B				

f. The principal of School B argued that, based on the overall figures in the table from part e, her school came out ahead and therefore is doing a better job than School A. How would you respond?

g. You are a parent trying to decide to which school to send your child. On the basis of the tables in parts a, b, and e, which school would you choose? Explain.

21. Children in South Africa. In 1990, scientists initiated a study of the development of children in the greater Johannesburg/Soweto metropolitan area of South Africa.[2] They collected information on certain children at birth, and five years later invited these children and their caregivers to be interviewed about health-related issues. Relatively few participated in the follow-up interviews, and the researchers wanted to know whether those who participated had similar characteristics to those who did not participate. In the following tables, these are referred to as the *five-year group* and the *children not traced*, respectively. One factor used to compare these groups was whether the mother had health insurance at the time the child was born. Here are the results for this factor, including only participants who were either white or black.

All participants				
	Had insurance	No insurance	Total	% with insurance
Children not traced	195	979		
Five-year group	46	370		

a. Fill in the blanks in the preceding table. Round all percentages to whole numbers.

b. In the summary of the cited study, the author noted, "If the five-year sample is to be used to draw conclusions about the entire birth cohort, the five-year group should have characteristics similar to those who were not traced from the initial group." Are the percentages of those with insurance similar between the two groups in the table? If not, what conclusion would you draw on the basis of this quotation?

c. Here is the table showing the results for whites. Complete the table. Round all percentages to whole numbers.

White participants				
	Had insurance	No insurance	Total	% with insurance
Children not traced	104	22		
Five-year group	10	2		

d. Here is the table showing the results for blacks. Complete the table. Round all percentages to whole numbers.

Black participants				
	Had insurance	No insurance	Total	% with insurance
Children not traced	91	957		
Five-year group	36	368		

e. For whites, compare the percentages having insurance between the two groups: children not traced and the five-year group. For blacks, compare the percentages having insurance between the two groups: children not traced and the five-year group.

f. On the basis of your answer to part e, revisit the question at the end of part b. Do the tables for whites and blacks provide any evidence that the five-year sample cannot be used to draw conclusions about the entire group? *Note*: The author of the summary observed that in South Africa whites generally were much more likely to have health insurance than were blacks. Further, relatively few whites were included in the original study. As a result, there were relatively few in the first of the preceding tables who had health insurance. Another factor in the paradoxical outcome is that, in contrast to blacks, relatively few of the whites who were initially involved participated in the interviews and were included in the five-year group.

22. Editorial. An editorial in the *Reno Gazette-Journal* on February 1, 2015, addressed a major argument that critics used against Nevada Governor Brian Sandoval's plan to raise taxes to improve Nevada's public education system. The argument was that school funding has increased for years but performance hasn't improved, so more money is not the answer. In particular, critics said that Nevada education funding has doubled since the 1960s but average student performance has remained flat. The editorial asserted that this reasoning is not correct. It went on to point out explicitly how Simpson's paradox comes into play because the balance between white and Hispanic students has changed.

a. The editorial used the following numbers for simplicity: Assume the number of students in Nevada to be only 100, both in 1960 and today. Suppose that in 1960, 90 students were white and 10 were Hispanic. In addition, suppose that 80% of white students graduated and 60% of Hispanic students graduated. What was the overall graduation rate in 1960?

b. Suppose that the demographic has changed to be 50% white and 50% Hispanic today. Also assume that both groups improved: The graduation rate for white students increased to 88%, and the graduation rate for Hispanic students increased to 66%. What is the overall graduation rate now?

c. Explain how your answers to parts a and b illustrate Simpson's paradox.

Writing About Mathematics

23. History of Simpson's paradox. Research the history of Simpson's paradox and describe some of the early examples given.

[2] *Christopher H. Morrell, "Simpson's Paradox: An Example from a Longitudinal Study in South Africa,"* Journal of Statistics Education *7 (1999).*

24. Edward H. Simpson. The phenomenon of Simpson's paradox was first described by Edward H. Simpson.[3] Write about the life of Simpson.

25. Your own experience. Write about a case of Simpson's paradox, either one that you are familiar with or one that you have researched.

What Do You Think?

26. PTA. You are considering attending the local PTA meeting tonight. But you have heard that the students in your school perform better overall on standardized exams than do students from the school across town, so perhaps you won't bother. Do you need more information to conclude that the students in your school are doing just fine, so your participation is not needed? What additional information might you need? **Exercise 20 is relevant to this topic.**

27. You choose. As governor, you are considering two plans for distributing money to cities in your state. Your hope is to reward cities that have demonstrated success in reducing crime. Plan A is simple. It looks at the percentage decrease in crime rates over the past few years. Plan B is more complicated. It includes the features of Plan A, but it also takes into account the demographics of each city, including income levels. Which plan might have the best chance of achieving your goal? Explain your reasoning.

28. Percents. The sales tax in your friend's state just went from 5% to 6%. Your friend says that's not bad—it's only a 1 percent increase. In what way is your friend correct? In what way is your friend incorrect?

29. Class average. There are 52 students in your class. The top 26 students averaged 90% on the last test, and the bottom 26 averaged 70%. Is the class average 80%? Explain.

30. Class average again. The men in your class had an average score of 60% on the last test, and the women had an average of 80%. Is the class average necessarily 70%? Explain.

31. Simpson's paradox. Use your own words to state Simpson's paradox. Explain why it is called a paradox, and give an example.

32. Percentage of what? One question to ask when you need to calculate a percentage is, "Percentage of what?" Give two reasonable answers to the following problem: The difference between 100 and 110 is 10, but what is the percentage difference between 100 and 110?

33. Combining percentages. You are a purchasing agent. You always buy 1000 items from one source, namely 10 widgets and 990 doodads. In the past, widgets and doodads cost $1 each, so you paid $1000. Now the price of widgets and the price of doodads have both increased by 10% to $1.10. By how much has the total you pay increased? What percentage of $1000 is that increase? (This number is the percentage increase in the total paid.) Is the answer found simply by adding 10% to 10% to get 20%?

34. Combining percentages again. You are a purchasing agent. You always buy 1000 items from one source, namely 10 widgets and 990 doodads. In the past, widgets and doodads cost $1 each, so you paid $1000. Now the price of widgets has increased by 10% to $1.10, and the price of doodads has increased by 5% to $1.05. By how much has the total you pay increased? What percentage of $1000 is that increase? (This number is the percentage increase in the total paid.) Is it closer to 5% or to 10%? Explain how your answer to this third question is related to Simpson's paradox.

1.2 Logic and informal fallacies: Does that argument hold water?

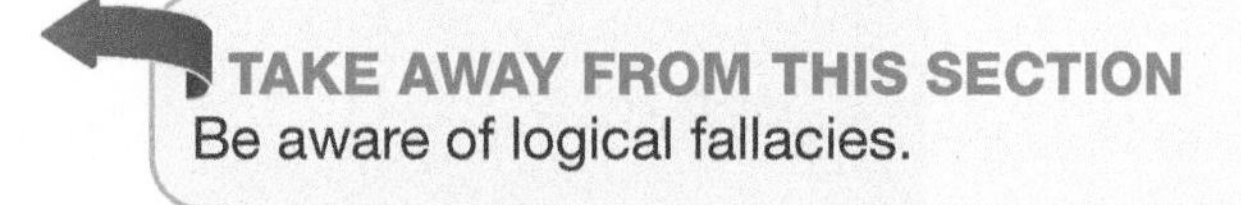

Voters must judge the quality of arguments in political debates. The following excerpt is from an article that appeared in *The Philadelphia Inquirer*.

[3] *E. H. Simpson, "The Interpretation of Interaction in Contingency Tables,"* Journal of the Royal Statistical Society. Series B *13 (1951), 238–241.*

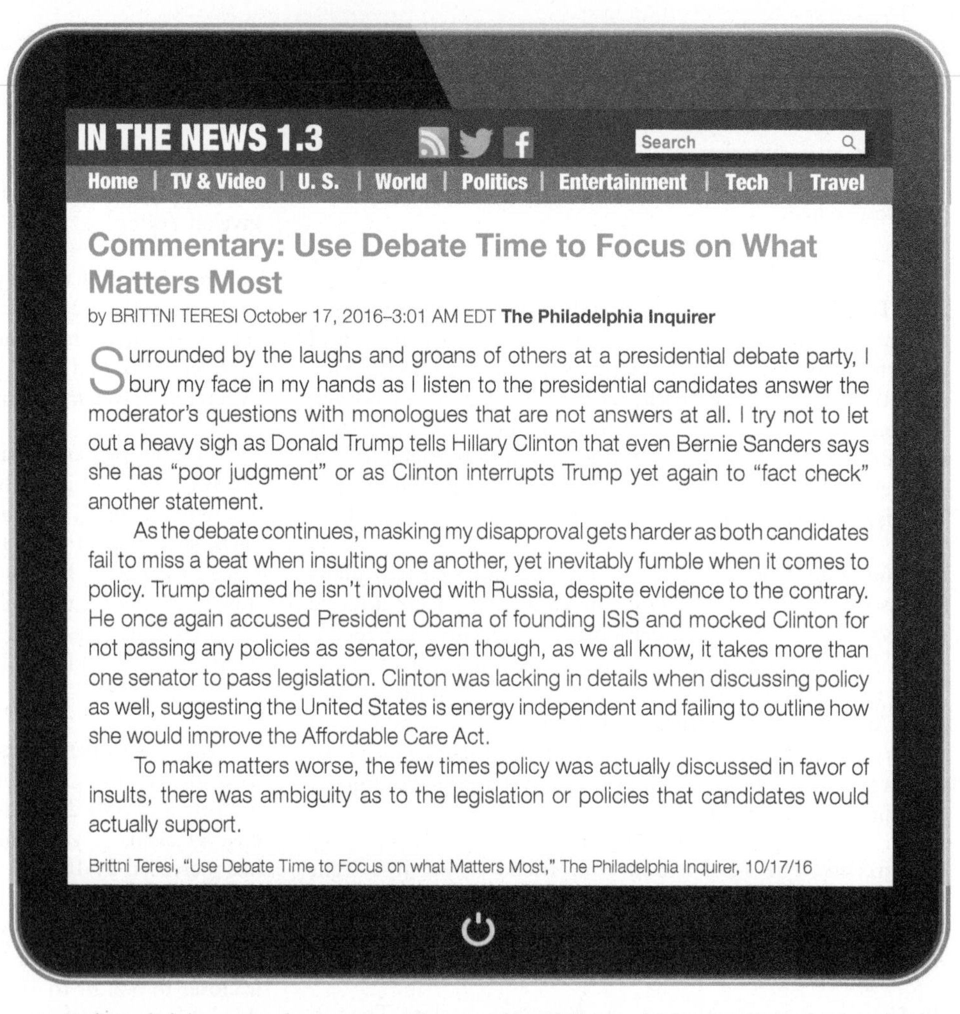

IN THE NEWS 1.3

Search

Home | TV & Video | U. S. | World | Politics | Entertainment | Tech | Travel

Commentary: Use Debate Time to Focus on What Matters Most

by BRITTNI TERESI October 17, 2016–3:01 AM EDT **The Philadelphia Inquirer**

Surrounded by the laughs and groans of others at a presidential debate party, I bury my face in my hands as I listen to the presidential candidates answer the moderator's questions with monologues that are not answers at all. I try not to let out a heavy sigh as Donald Trump tells Hillary Clinton that even Bernie Sanders says she has "poor judgment" or as Clinton interrupts Trump yet again to "fact check" another statement.

As the debate continues, masking my disapproval gets harder as both candidates fail to miss a beat when insulting one another, yet inevitably fumble when it comes to policy. Trump claimed he isn't involved with Russia, despite evidence to the contrary. He once again accused President Obama of founding ISIS and mocked Clinton for not passing any policies as senator, even though, as we all know, it takes more than one senator to pass legislation. Clinton was lacking in details when discussing policy as well, suggesting the United States is energy independent and failing to outline how she would improve the Affordable Care Act.

To make matters worse, the few times policy was actually discussed in favor of insults, there was ambiguity as to the legislation or policies that candidates would actually support.

Brittni Teresi, "Use Debate Time to Focus on what Matters Most," The Philadelphia Inquirer, 10/17/16

Political debates and commercials are often delivered in sound bites to evoke an emotional reaction. Critical examination of what these "pitches" actually say can reveal exaggerations, half-truths, or outright fallacies.

In the face of such heated rhetoric and tenuous logic, we appeal to an older source, one that remains relevant today.

Quote from Thomas Jefferson

In a republican nation, whose citizens are to be led by reason and persuasion and not by force, the art of reasoning becomes of first importance.

In this section, we begin a study of logic and examine some informal logical fallacies.

JEWEL SAMAD/Getty Images

The 2016 Republican presidential candidate Donald Trump, left, and Democratic presidential candidate Hillary Clinton participate in their first televised debate on September 26, 2016, in Hempstead, NY.

Logical arguments and fallacies

KEY CONCEPT

Logic is the study of methods and principles used to distinguish good (correct) from bad (incorrect) reasoning.

A logical argument involves one or more *premises* (or *hypotheses*) and a *conclusion*. The premises are used to justify the conclusion. The argument is *valid* if the premises properly support the conclusion. Here is an example.

Premises: (1) *Smoking increases the risk of heart disease*, and (2) *My grandfather is a smoker*.

Conclusion: *My grandfather is at increased risk of heart disease.*

This is a valid argument because the premises justify the conclusion. In determining whether an argument is valid, we do not question the premises. An argument is valid if the conclusion follows from the assumption that the premises are true. The truth or falsity of the premises is a separate issue. Nor is it simply the fact that the conclusion is true that makes an argument valid—the important thing is that the premises support the conclusion.

Consider, for example, the following invalid argument:

Premises: (1) *My grandfather smoked*, and (2) *he suffered from heart disease.*

Conclusion: *Therefore, smoking is a contributing factor to heart disease.*

Here, the conclusion may indeed be true, but the case of a single smoker is not sufficient evidence from which to draw the conclusion. Even if it were, this example does not show that smoking contributed to the heart disease—it is established only that the two events of smoking and having heart disease both occurred.

KEY CONCEPT

A logical argument consists of premises (also called hypotheses) and a conclusion. The premises are assumptions that we accept as a starting point. The argument is valid if the premises justify the conclusion.

EXAMPLE 1.3 Identifying parts of an argument: Wizards and beards

Identify the premises and conclusion of the following argument. Is this argument valid?

All wizards have white beards. Gandalf is a wizard. Therefore, Gandalf has a white beard.

SOLUTION

The premises are (1) *All wizards have white beards* and (2) *Gandalf is a wizard*. The conclusion is *Gandalf has a white beard*. This is a valid argument because the premises support the conclusion: If in fact all wizards have white beards and Gandalf is a wizard, it must be the case that Gandalf has a white beard.

TRY IT YOURSELF 1.3

Identify the premises and conclusion of the following argument. Is this argument valid?

All wizards have white beards. Gandalf has a white beard. Therefore, Gandalf is a wizard.

The answer is provided at the end of the section.

The term *fallacy*, as used by logicians, refers to an argument that may on the surface seem to be correct but is in fact incorrect. There are many types of fallacies. Some are easy to spot, and others are more subtle. Although there are many ways in which incorrect inferences can be constructed, fallacies fall into two general categories: informal and formal.

KEY CONCEPT

An informal fallacy is a fallacy that arises from the content of an argument, not its form or structure. The argument is incorrect because of *what* is said, not *how* it is said. A formal fallacy arises from the form or structure of an argument. The formal fallacy is independent of the content of the argument.

In this section, we focus on informal fallacies. We reserve the treatment of formal fallacies for the next section. Our discussion generally follows the development of Copi's classic work[4] but includes other sources as well. We divide informal fallacies into two categories: those mistakes in reasoning that arise from appeals to irrelevant factors and those that arise from unsupported assumptions.

Fallacies of relevance

In fallacies of relevance, the premises are logically irrelevant to, and hence incapable of, establishing the truth of their conclusions. We offer several examples, but this collection is by no means exhaustive.

Appeal to ignorance This type of fallacy occurs whenever it is argued that a statement is true or not true simply on the basis of a lack of proof. This is a fallacy of relevance because the lack of proof of a statement is not by itself relevant to the truth or falsity of that statement. The structure of this fallacy is as follows:

- A certain statement is unproven.
- Therefore, the statement must be false.

Example: *For over 75 years, people have tried and failed to show that aliens have not visited Earth. So we must finally accept the fact that at least some of the UFO reports are based on actual alien visits.*

In this example, the inability to disprove alien visits is not relevant to the existence of such visits. Perhaps it would help to consider the opposite argument: *For over 75 years, people have not been able to show that aliens have visited Earth. So we must finally accept the fact that no such visits have ever occurred.* The same set of facts is used to establish both that aliens have visited Earth and that they have not. The point is that an absence of proof cannot establish either that aliens have visited or that they have not.

Dismissal based on personal attack This type of fallacy occurs when we simply attack the person making an assertion instead of trying to prove or disprove the assertion itself. This is a fallacy of relevance because the character of the person making a claim is not necessarily relevant to the claim being made. This fallacy has the following structure:

---

[4] *Irving M. Copi and Carl Cohen,* Introduction to Logic, *13th ed. (Upper Saddle River, NJ: Prentice-Hall, 2008).*

- A person presents an argument or point of view.
- The character of that person is brought into question.
- Based on the character attack, it is concluded that the argument or point of view is incorrect.

Example: *City council member Smith advocates allocating funds to repair the Puddle Creek Bridge. As we know, the restaurant she owns has been accused of several health department violations, so she cannot be trusted. The city council should not spend the money required to repair the bridge.*

In this argument, the conclusion is that a proposal to repair a bridge should be opposed, but no information is provided regarding the merits of the proposal itself. Rather, the argument is that the proposal should be opposed because of the claim that one of its supporters is suspect. Even if that accusation is true, it does not necessarily follow that the proposal is bad. The proposal cannot be evaluated without relevant information, such as the current condition of the bridge, the cost to repair it, etc.

False authority This fallacy arises whenever the validity of an argument is accepted based solely on the fact that it is supported by someone, but that person's credentials or expertise are not relevant to the argument. This is clearly a fallacy of relevance. This fallacy has the following structure:

- A person makes a claim based on his or her authority.
- The claim is outside the scope of that person's authority.
- The truth of the argument is concluded based on the authority of the claimant.

Example: *Over the past few years, I have starred in a number of the most popular movies in America, so I can assure you that Johnson and Johnson's new anti-nausea drug is medically safe and effective.*

The fallacy here is obvious—starring in movies in no way qualifies anyone to judge the efficacy of any medical product. The prevalence of commercials where sports figures or movie stars tell us which clothes to buy, which foods to eat, and which candidates to vote for speaks volumes about the effectiveness of such fallacious arguments.

Of course, *appropriate* authority does indeed carry logical weight. For example, the authority of Neil deGrasse Tyson lends credence to a claim regarding astrophysics because he is a world-renowned astrophysicist. On the other hand, his authority would probably add nothing to a toothpaste commercial.

Straw man This fallacy occurs whenever an argument or position is dismissed based on the refutation of a distorted or completely different argument or position (the straw man)—presumably one that is easier to refute. This is a fallacy of relevance because the refutation does not even address the original position. The structure of this fallacy is as follows:

- A position or point of view is presented.
- The case for dismissing a distorted or *different* position or point of view (the straw man) is offered.
- The original position is dismissed on the basis of the refutation of the straw-man position.

Example: *A group of my fellow senators is proposing a cut in military expenditures. I cannot support such a cut because leaving our country defenseless in these troubled times is just not acceptable to me.*

Tsgt. Effrain Lopez/ZUMA Press/Newscom

A U.S. Air Force MQ-1 Predator unmanned aerial vehicle assigned to the California Air National Guard's 163rd Reconnaissance Wing flies near the Southern California Logistics Airport in Victorville, California.

The "straw man" here is the proposition that our military defense program should be dismantled. It is easy to reach an agreement that this is a bad idea. The fallacy is that no such proposal was made. The proposal was to cut military expenditures, not to dismantle the military.

Political speech is often full of this type of blatant fallacy. Rather than engaging in honest, reasoned debate of the issues, we often see distortions of an opponent's position (the straw man) refuted. The politicians who do this are trying to appeal to emotion rather than reason. Critical thinking is our best defense against such tactics.

Appeal to common practice This fallacy occurs when we justify a position based on its popularity. It is a fallacy of relevance because common practice does not imply correctness. This fallacy has the following structure:

- The claim is offered that a position is popular.
- The validity of the claim is based on its popularity.

Example: *It is OK to cheat on your income tax because everybody does it.*

Of course, it isn't true that everybody cheats on their income taxes, but even if it were, that is not justification for cheating. (That argument certainly will not carry much weight with the IRS if you're audited.) This type of argument is often used in advertising.

EXAMPLE 1.4 Identifying types of fallacies of relevance: Two fallacies

Identify the type of each of the following fallacies of relevance.

1. *I have scoured the library for information on the Abominable Snowman. I cannot find a single source that proves that anyone has actually seen it. This shows that there is no Abominable Snowman.*

2. *The popular comedian Ricky Gervais made a commercial in which he asserts that Verizon is the top network in the country. He's a famous star who would never say that unless it were true.*

SOLUTION

The first argument relies on a lack of proof to draw the conclusion. This is an appeal to ignorance.

The second argument involves false authority. The fact that Ricky Gervais is a popular comedian does not lend authority to his opinion about a telephone network.

TRY IT YOURSELF 1.4

Identify the fallacy of relevance in the following argument:

I don't see why I should vote in the election. Quite often, less than half of the adult population casts a vote.

The answer is provided at the end of the section.

Fallacies of presumption

In fallacies of presumption, false or misleading assumptions are either tacitly or explicitly made, and these assumptions are the basis of the conclusion.

False dilemma This fallacy occurs when a conclusion is based on an inaccurate or incomplete list of choices. This is a fallacy of presumption because the (incorrect) assumption is being made that the list offered is complete. This fallacy has the following structure:

- An incomplete or inaccurate list of consequences of not accepting an argument is presented.
- A conclusion is drawn based on the best (or least bad) of these consequences.

Example: *You'd better buy this car or your wife will have to walk to work and your kids will have to walk to school. I know you don't want to inconvenience your family, so let's start the paperwork on the automobile purchase.*

The salesperson is telling us we have two choices, "Buy this car or inconvenience your family." In fact, we probably have more than these two choices. There may be a less expensive car that will serve my family's needs, or perhaps public transportation is an option in my neighborhood. The dilemma presumed by the salesperson is not necessarily true. The conclusion that we should purchase the car is based on choosing the best of the options offered by the salesperson, but in fact there are additional choices available.

An argument of this sort is valid only if the list of choices is indeed complete. This type of argument is commonly presented by public officials: *We must either increase your taxes or cut spending on education. So you must support an increase in taxes.*

False cause The fallacy of false cause occurs whenever we conclude a causal relationship based solely on common occurrence. It is an error of presumption because we are tacitly assuming that a causal relationship must exist. The structure of this fallacy is as follows:

- Two events occur together, or one follows the other.
- The fact that the events are related is used to conclude that one *causes* the other.

Example: *Studies have shown that many people on a high-carbohydrate diet lose weight. Therefore, a high-carbohydrate diet leads to weight loss.*

The premise of this argument is that the two events of losing weight and eating a high-carbohydrate diet sometimes occur together. The fallacy is that if two events occur together, then one must cause the other. It is just as reasonable to conclude that the weight loss causes the high-carbohydrate diet!

Consider this example: *Each morning my dog barks, and there are never any elephants in my yard. Therefore, my dog is keeping the elephants away from my home.* It is probably true that the dog barks and that there are no elephants in the yard, but these facts have nothing to do with each other.

When two events occur together, there are three distinct possibilities:

1. *There is no causal relationship.* Example: My dog barks and there are no elephants in my yard.

2. *One event causes the other.* Example: Ice on the sidewalk causes the sidewalk to be slippery.

3. *Both events are caused by something else.* Example: A mother observes that her child often has a fever and a runny nose at the same time. She doesn't conclude that the runny nose causes the fever. Instead, she concludes that the child has a cold, which causes both symptoms.

In general, causality is a difficult thing to establish, and we should view such claims skeptically. We will return to the topic of causality when we discuss *correlation* in Section 4 of Chapter 6.

Circular reasoning or begging the question This fallacy occurs when the repetition of a position is offered as its justification. This is a fallacy of presumption because it is assumed that a statement can prove itself. The structure of this fallacy is as follows:

- A position or argument is offered.
- The position or argument is concluded to be true based on a restatement of the position.

Example: *Giving undocumented immigrants a pathway to citizenship would be a mistake, because it is just flat wrong to do it.*

Larry Downing/Reuters/Newscom

U.S. President Barack Obama speaks in support of the U.S. Senate's bipartisan immigration reform bill while in the East Room of the White House in Washington, June 11, 2013.

This argument says essentially that this kind of immigration reform is wrong because it is wrong. Sometimes circular arguments may be longer and more convoluted than the simple example presented here and therefore more difficult to spot.

Another example is this: *I have listed Mary as a reference—she can vouch for me. Tom will verify her trustworthiness, and I can vouch for Tom.*

Hasty generalization The fallacy of hasty generalization occurs when a conclusion is drawn based on a few examples that may be atypical. This is a fallacy of presumption because we are assuming that the few cases considered are typical of all cases. This fallacy has the following form:

- A statement is true in several cases that may be atypical.
- The conclusion that it is generally true or always true is drawn based on the few examples.

Example: *I know the quarterback, the tight end, and the center on our football team, and all three are excellent students. The athletes at our university do not shirk their scholarly duties.*

Here we are making a judgment about athletes at the university based on the good qualities of three football players. The exemplary behavior of these three athletes tells us nothing about how the remaining athletes are behaving.

Finding three athletes who are scholars does not mean that all athletes are, but these examples might prompt us to explore further. Examples cannot prove the general case, but they can help us formulate ideas that may merit further investigation. For example, I might be very pleased to learn that three of my daughter's friends are honor students. Perhaps she is wise in her choice of friends. It is certainly worth further attention.

SUMMARY 1.2
Some Common Informal Fallacies

Appeal to ignorance A statement is either accepted or rejected because of a lack of proof.

Dismissal based on personal attack An argument is dismissed based on an attack on the proponent rather than on its merits.

False authority The validity of a claim is accepted based on an authority whose expertise is irrelevant.

Straw man A position is dismissed based on the rejection of a distorted or different position (the "straw man").

Appeal to common practice An argument for a practice is based on the popularity of that practice.

False dilemma A conclusion is based on an inaccurate or incomplete list of alternatives.

False cause A causal relationship is concluded based on the fact that two events occur together or that one follows the other.

Circular reasoning This fallacy simply draws a conclusion that is really a restatement of the premise.

Hasty generalization This fallacy occurs when a conclusion is drawn based on a few examples that may be atypical.

There are other types of fallacies. Some of these will be examined in the exercises.

EXAMPLE 1.5 Identifying types of informal fallacies: Two fallacies

Classify each of the following fallacies:

1. *He says we should vote in favor of lowering the sales tax, but he has a criminal record. So I think decreasing the sales tax is a bad idea.*

2. *My dad is a professor of physics, and he says Dobermans make better watchdogs than collies.*

SOLUTION

In the first argument, we dismiss a position based on a personal attack. In the second, knowledge of physics does not offer qualification for judging dogs. This is a use of false authority.

TRY IT YOURSELF 1.5

Identify the type of fallacy illustrated in the following argument. *Do you want to buy this MP3 player or do you want to do without music?*

The answer is provided at the end of the section.

Inductive reasoning and pattern recognition

The fallacies we have looked at are usually flawed arguments in which the premises are *supposed* to provide conclusive grounds for accepting the conclusion, but in fact do not. An argument with premises and a conclusion is called a *deductive* argument because it *deduces* a specific conclusion from the given premises. The classic example is:

Premises: (1) *All men are mortal* and (2) *Socrates is a man.*

Conclusion: *Therefore, Socrates is mortal.*

In a valid deductive argument, the premises provide *conclusive* grounds for accepting the conclusion.

An *inductive* argument, unlike a deductive argument, draws a general conclusion from specific examples. It does not claim that the premises give irrefutable proof of the conclusion (that would constitute the fallacy of hasty generalization), only that the premises provide plausible support for it.

An inductive argument related to the deductive argument above is as follows:

Premise: Socrates is mortal. Plato is mortal. Aristotle is mortal. Archimedes is mortal.

Inductive conclusion: Therefore, all men are mortal.

In this argument, the conclusion that all men are mortal is based on knowledge of a few individuals. A proper deductive argument can establish a conclusion with certainty. But an inductive argument draws a conclusion based on only partial evidence in support of it.

KEY CONCEPT

A deductive argument draws a conclusion from premises based on logic. An inductive argument draws a conclusion from specific examples. The premises provide only partial evidence for the conclusion.

In some cases inductive arguments are extremely convincing. Take this one for example:

Premise: Every human being who has ever lived has eventually died.

Inductive conclusion: All human beings are mortal.

Because the evidence is so overwhelming, some people might say that the conclusion is deduced from the premise. Technically speaking, however, this is still an inductive argument. The conclusion is based on evidence, not on deductive logic.

Here's another inductive argument with lots of evidence:

Premise: Every NFL referee I have ever seen is male.

Inductive conclusion: All NFL referees are male.

Sarah Thomas officiating a game.

This argument may sound convincing, but in 2015 Sarah Thomas became the first full-time female official in NFL history.

The weight of an inductive argument depends to a great extent on just how much evidence is in the premises. Such argumentation also has value in that it may suggest further investigation. In mathematics, this is often how we discover things—we observe patterns displayed by specific examples. This may lead us to propose a more general result, but in mathematics no such result is accepted as true until a proper deductive argument is found.

To show what we mean, let's look at the following table, which shows the squares of the first few odd integers:

Odd integer	1	3	5	7	9
Odd integer squared	1	9	25	49	81

Note that each of the squares turns out to be odd. Reasoning inductively, we propose that the square of *every* odd integer is odd. Note that the pattern in the table supports this claim, but the table does not prove the claim. The table does not, for example, guarantee that the square of 163 is an odd integer. Mathematicians are indeed able to give a (relatively simple) deductive proof of this proposal. Thus, the pattern that we saw in the table is, in fact, true for all odd integers.

Inductive reasoning of this type can occasionally lead us astray. Recall that a prime number is a number (other than 1) that has no divisors other than itself and 1. So 7 is prime because its only divisors are 1 and 7. On the other hand, 15 is not prime because both 3 and 5 are divisors of 15. It has divisors other than 1 and 15. Let's consider the expression n^2+n+41. The table below shows some values for this expression.

n	1	2	3	4	5	6	7
n^2+n+41	43	47	53	61	71	83	97

After we observe that 41, 43, 47, 53, 61, 71, 83, and 97 are all prime, we might be led to propose that the expression n^2+n+41 always gives a prime number. If you explore further, you will find that the result is prime for $n=1$ up through $n=39$. But if we put in $n=40$, the result is not prime:

$$40^2+40+41=1681=41\times 41$$

In this case, the pattern we observed in the table is not true for all numbers. Inductive reasoning can lead us to important discoveries. But as we have seen, it can sometimes lead us to wrong conclusions.

The idea of an inductive argument is closely tied to pattern recognition. We explore this connection further in the following example.

EXAMPLE 1.6 Pattern recognition: Inductive logic

A certain organism reproduces by cell division. The following table shows the number of cells observed to be present over the first few hours:

Hours	0	1	2	3	4	5	6
Number of cells	1	2	4	8	16	32	64

Describe the pattern shown by the table and suggest a general rule for finding the number of cells in terms of the number of hours elapsed.

SOLUTION

Looking closely at the table, we see a pattern. The number of cells present is always a power of 2.

Hours	0	1	2	3	4	5	6
Number of cells	$1 = 2^0$	$2 = 2^1$	$4 = 2^2$	$8 = 2^3$	$16 = 2^4$	$32 = 2^5$	$64 = 2^6$

The pattern suggests that, in general, we find the number of cells present by raising 2 to the power of the number of hours elapsed.

TRY IT YOURSELF 1.6

An Internet guide gives a table showing how much fertilizer I need to cover a yard with a given square footage. Measurements are in square feet.

Yard size	100 sq ft	400 sq ft	900 sq ft	1600 sq ft	2500 sq ft
Pounds needed	1	4	9	16	25

Use this table to determine a pattern that leads to a general relationship between yard size and fertilizer needed.

The answer is provided at the end of the section.

WHAT DO YOU THINK?

A political debate: In the debate last night, Candidate A said that Candidate B's parents were both drug users. Candidate B said that Candidate A was soft on crime because his neighbor had a lot of parking tickets. Which candidate, if either, won your vote? Explain your reasoning.

Questions of this sort may invite many correct responses. We present one possibility: This is an exaggerated version of a familiar scene. Both candidates are engaging in personal attack rather than discussing important issues. There are many reasons why we might vote for one candidate or the other, but a neighbor's parking tickets are not among them. The arguments they present are fallacious, and it is all too common to see political campaigns that focus on irrelevant or invalid assertions. A debate of this character might convince some that neither candidate is worthy of support. It is perhaps at least part of the reason that relatively few Americans vote in national elections.

Further questions of this type may be found in the *What Do You Think?* section of exercises.

Try It Yourself answers

Try It Yourself 1.3: Identifying parts of an argument: Wizards and beards The premises are (1) *All wizards have white beards* and (2) *Gandalf has a white beard.* The conclusion is *Gandalf is a wizard.* This argument is not valid.

Try It Yourself 1.4: Identifying types of fallacies of relevance: Types of fallacies This is an appeal to common practice.

Try It Yourself 1.5: Identifying types of informal fallacies: Types of fallacies This is a false dilemma.

Try It Yourself 1.6: Pattern recognition: Fertilizer To find the number of pounds needed, we divide the yard size in square feet by 100.

Exercise Set 1.2

Test Your Understanding

1. State in your own words the meaning of the fallacy of appeal to ignorance.

2. In the straw man fallacy, a position is rejected based on: **a.** a distorted or unrelated position **b.** false authority **c.** circular reasoning **d.** none of the above.

3. In the fallacy of false dilemma, on what is the conclusion based?

4. What is the difference between a fallacy of relevance and a fallacy of presumption?

5. Is the following argument a case of inductive reasoning or of deductive reasoning? *Every college freshman takes math. Johnny is a college freshman. Therefore, Johnny takes math.*

6. Is the following argument a case of inductive reasoning or of deductive reasoning? *Yesterday there were twice as many tribbles as the day before. Today there are twice as many tribbles as yesterday. Tomorrow there will be twice as many tribbles as today.*

Problems

7. Classifying fallacies. Identify the type of the following fallacy.

My mom is an accountant. She says life cannot exist where there is no sunlight.

8. Inductive reasoning. A chart at the nursery shows the coverage I can get from bags of grass seed.

Bags of seed	1	2	3	4
Coverage	200 sq ft	400 sq ft	600 sq ft	800 sq ft

I bought 5 bags of seed. How much coverage can I expect from my purchase?

9–15. Premises and conclusion. *In Exercises 9 through 15, identify the premises and conclusion of the given argument.*

9. *All dogs go to heaven, and my terrier is a dog. So my terrier will go to heaven.*

10. *If you take this new drug, you will lose weight.*

11. *Nobody believes the mayor anymore because he was caught in a lie.*

12. *Today's children do not respect authority. A disdain for authority leads to social unrest. As a result, the social fabric is at risk.*

13. *A people free to choose will always choose peace.* (Ronald Reagan)

14. *He who would live in peace and at ease must not speak all that he knows or all he sees.* (Benjamin Franklin)

15. *Bad times have a scientific value. These are occasions a good learner would not miss.* (Ralph Waldo Emerson)

16–21. Valid arguments. *In Exercises 16 through 21, decide whether or not the given argument is valid.*

16. *All dogs go to heaven. My pet is a dog. Therefore, my pet will go to heaven.*

17. *The senate has a number of dishonest members. We just elected a new senator. Therefore, that senator is dishonest.*

18. *If I have diabetes, my blood-glucose level will be abnormal. My blood-glucose level is normal, so I do not have diabetes.*

19. *People who drive too fast get speeding tickets. Because you got a speeding ticket, you were driving too fast.*

20. *If you follow the yellow brick road, it will lead you to the Wizard of Oz. I have met the Wizard of Oz, so I must have followed the yellow brick road.*

21. *The job of every policeman is to protect and serve the public. Because I am dedicated to protecting and serving the public, I am a policeman.*

22–39. Types of fallacies. *Classify the fallacies in Exercises 22 through 39 as one of the types of fallacies discussed in this section.*

22. *Every time a Democrat gets elected, our taxes go up. Raising taxes is a Democratic policy.*

23. *I don't think we should believe him because I heard he dropped out of school.*

24. *My math teacher says Cheerios is the healthiest breakfast cereal.*

25. *The Republican Party has been in power for the past eight years, and during that period the number of jobs has declined. A decrease in jobs is a Republican policy.*

26. *We can't afford the more expensive car because it will cost more.*

27. *In this war you are either for us or against us. Because you do not support the war effort, you are our enemy.*

28. *My congressman accepted large amounts of money from a political action group. All congressmen do it, so he didn't do anything wrong.*

29. *More people use Crest than any other toothpaste. So it must be the best toothpaste on the market.*

30. *I visited three state parks in California and saw redwood trees in all three. There must be redwood trees in all California state parks.*

31. *No one in my family has tested positive for HIV. So my family is free of AIDS.*

32. *My history teacher claims that accepted statistics for the population of South America before Columbus are, in fact, far too small. I have heard that he is something of a racist with regard to Native Americans. His conclusions about the Indian population cannot be valid.*

33. *My math teacher says studying can improve my grade. I don't believe I will make an "A" no matter how hard I study, so I'm skipping the review session tonight.*

34. *Everybody wants to enroll in Professor Smith's section. He must be the best teacher on campus.*

35. *If you don't go with me, you are going to miss the biggest party of the year. So hop in, and let's go.*

36. *I bought that product, and I lost 25 pounds. It is the best weight-loss product on the market.*

37. *I know that my income tax is too high because I have to pay too much.*

38. *The three smartest students in my math class are of Asian descent. Asians are better at mathematics than Americans.*

39. *The president says we need to help people in Third World countries who don't have sufficient food or medical care. Because all of our money is going overseas, I don't support foreign aid.*

40–43. Additional informal fallacies. *Exercises 40 through 43 introduce additional informal fallacies. In each case, explain why the example given is a fallacy.*

40. Appeal to fear. *You should support my point of view. I am having dinner with your boss tonight, and I might have to tell her what you said last week.*

41. Fallacy of accident. *All men are created equal. So you can throw the football just as well as the quarterback for the Dallas Cowboys.*

42. Slippery slope. *They keep cutting state taxes. Eventually, the state won't have enough revenue to repair the roads.*

43. Gambler's fallacy. *I have lost on 20 straight lottery tickets. I'm due for a win, so I'm buying extra tickets today.* Note: This fallacy is explored further in Section 5 of Chapter 5.

44. Currency conversion. The euro is the currency used by most member states of the European Union. Here is a table of values from the Web for converting euros to American dollars.

Euro	1	2	3	4	5	6	7	8	9
Dollar	1.13	2.26	3.30	4.52	5.65	6.78	7.81	9.04	10.17

Based on this table, what would be the cost in American dollars of a bottle of wine marked 16 euros?

45. Children's blocks. The following chart shows the number of children's blocks that will fit in cubes of various sizes. Size measurements are in inches.

Size of cube	1 by 1 by 1	2 by 2 by 2	3 by 3 by 3	4 by 4 by 4	5 by 5 by 5
Number of blocks	1	8	27	64	125

What pattern is shown in this chart? In general, how can you find the number of children's blocks that will fit in a cube of a given size? How many blocks will fit in a cube that is 50 inches on a side?

46–48. Finding patterns. *In Exercises 46 through 48, find a pattern suggested by the given sequence.*

46. 2, 4, 6, 8, 10, . . .

47. 1, 3, 5, 7, 9, . . .

48. $1, \frac{1}{2}, \frac{1}{4}, \frac{1}{8}, \frac{1}{16}, \frac{1}{32}, \ldots$

Writing About Mathematics

49. Research. Find and report on an argument presented in the popular media that is an informal fallacy.

50. Irving Copi. We noted at the beginning of this section that our discussion follows a classic text on logic by Irving Copi. Report on his life and accomplishments.

51. Aristotle. The ancient Greek philosopher Aristotle is noted for his influence on logic. Report on his life and accomplishments.

52. More fallacies. We noted in the text that our list of fallacies is by no means exhaustive. Report on a type of fallacy recognized by logicians that is not included in this section.

What Do You Think?

53. Arguments. A classmate says that every argument having premises and a conclusion is a sound argument. How would you respond? **Exercise 16 is relevant to this topic.**

54. Appeal to common practice. Explain, using your own words, why an argument depending on an appeal to common practice is not compelling. Give your own example of an argument depending on an appeal to common practice.

55. False dilemma. Explain, using your own words, why an argument presenting a false dilemma is not compelling. Give your own example of an argument presenting a false dilemma.

56. City councilman. You heard the following on television from your city councilman: "Last week I saw three people run the stop sign at 4th and Duck. Nobody stops there, so I am going to have the sign taken down." Does this argument convince you that the stop sign should be removed? Why or why not? **Exercise 30 is relevant to this topic.**

57. Authority. Professor X at Harvard is a nationally known scholar with two PhD degrees. He says that the methods used to estimate the age of dinosaur fossils are all wrong. Do you accept his assertion? (By the way, his degrees are in Music and English.)

58. Astrophysics. Astrophysical theory has been successful at explaining most observed events in the universe, but there are things it has yet to explain (like "dark matter"). Should the theory be rejected because of this? Explain.

59. Warming. I read about a witch doctor in South America who believes that global warming is real and is caused by evil spirits. That shows me that global warming is obviously false. Do you agree? Explain.

60. Terrorists. Every terrorist attempt I've read about since 9/11 involved a Muslim. That proves to me that all Muslims are terrorists. Do you agree? Explain.

61. Hasty induction? Explain the difference between the fallacy of hasty generalization and the sort of reasoning used in pattern recognition. Give an example (not necessarily from mathematics) of pattern recognition that is useful.

1.3 Formal logic and truth tables: Do computers think?

TAKE AWAY FROM THIS SECTION
Understand the formal logic used by computers.

The following excerpt is from an article that appeared in *Chemistry World*. It refers to the use of logic systems for information processing in the context of medicine. Formal logic focuses on the structure or form of a statement. We as humans make judgments based on information we gather. But computers have a strict set of rules that they follow, and there is no room for judgment. The strict rules by which computers operate constitute the formal version of logic that we investigate here.

Larry Pribyl/Oregon State University

This new chip developed by researchers at Oregon State University may be a relatively easy and cost-effective way to monitor vital signs.

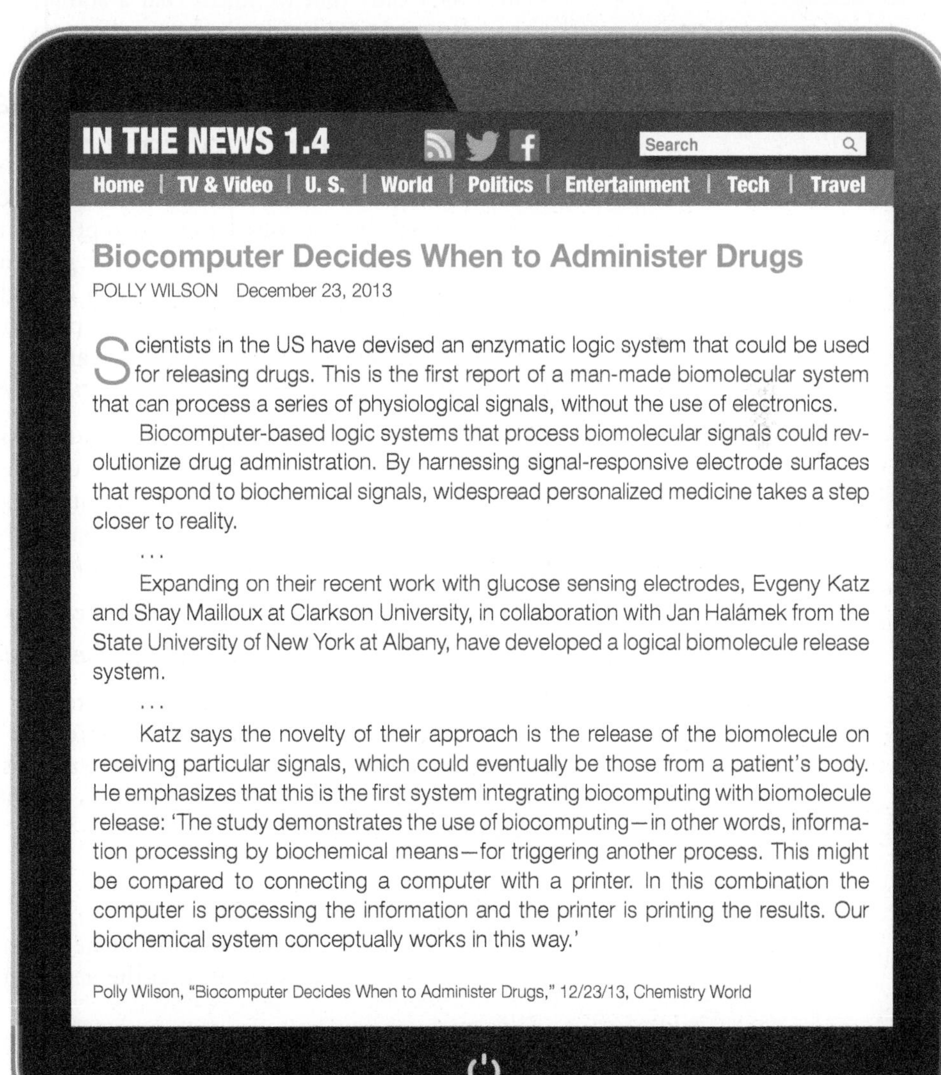

Biocomputer Decides When to Administer Drugs

POLLY WILSON December 23, 2013

Scientists in the US have devised an enzymatic logic system that could be used for releasing drugs. This is the first report of a man-made biomolecular system that can process a series of physiological signals, without the use of electronics.

Biocomputer-based logic systems that process biomolecular signals could revolutionize drug administration. By harnessing signal-responsive electrode surfaces that respond to biochemical signals, widespread personalized medicine takes a step closer to reality.

. . .

Expanding on their recent work with glucose sensing electrodes, Evgeny Katz and Shay Mailloux at Clarkson University, in collaboration with Jan Halámek from the State University of New York at Albany, have developed a logical biomolecule release system.

. . .

Katz says the novelty of their approach is the release of the biomolecule on receiving particular signals, which could eventually be those from a patient's body. He emphasizes that this is the first system integrating biocomputing with biomolecule release: 'The study demonstrates the use of biocomputing—in other words, information processing by biochemical means—for triggering another process. This might be compared to connecting a computer with a printer. In this combination the computer is processing the information and the printer is printing the results. Our biochemical system conceptually works in this way.'

Polly Wilson, "Biocomputer Decides When to Administer Drugs," 12/23/13, Chemistry World

Symbols and rules can be used to analyze logical relationships in much the same way they are used to analyze relationships in algebra. This formalism has the advantage of allowing us to focus on the structure of logical arguments without being distracted by the actual statements themselves.

Formal logic deals only with statements that can be classified as either true or false (but not both). To clarify what we mean, consider the following sentences:

This sentence is false. Is this sentence true or is it false? If the sentence is true, then it's telling us the truth when it says that it's false—so the sentence must be false. However, if the sentence is false, then it's not telling us the truth when it says that it's false—so the sentence must be true. This is rather paradoxical: It appears that there is no reasonable way to classify this sentence as either true or false. Therefore, the sentence does not qualify as a statement to which the rules of logic can be applied.

Please turn down the music. This sentence is a request for action. It cannot be classified as either true or false.

I am 6 feet tall. Although you may not know my height, it is the case that either I am 6 feet tall or I am not. The statement is either true or false (and not both), so it is a statement in the formal logical sense. We emphasize that it is necessary only that we know that a statement can be classified as either true or false—it is not necessary that we know which of the two applies.

Operations on statements: Negation, conjunction, disjunction, and implication

In order to focus on the formal structure of logic, we use letters to represent statements just as algebra uses letters to represent numbers. For example, we might represent the statement *He likes dogs* by p. In algebra we combine variables that represent numbers using operations such as addition $(+)$, multiplication $(\times)$, and so forth. In formal logic we use logical operations to combine symbols that stand for statements. But the logical operations are not multiplication or division. Instead, they are *negation*, *conjunction*, *disjunction*, and *implication*.

Negation The *negation* of a statement is the assertion that means the opposite of the original statement: One statement is true when the other is false. As in common English usage, we will indicate the negation using NOT. For example, if p represents the statement *He likes dogs*, then the statement NOT p means *It is not the case that he likes dogs* or more simply *He does not like dogs.*

$$p = \textit{He likes dogs}$$
$$\text{NOT } p = \textit{He does not like dogs}$$

By its definition, the negation of a true statement is false, and the negation of a false statement is true. For example, if the statement p that *He likes dogs* is true, then the statement NOT p that *He does not like dogs* is false. We show this below using a *truth table*. This table lists the possibilities for the truth of p and the corresponding value for the truth of NOT p.

p	NOT p
T	F
F	T

This means that the negation of p is false when p is true.
This means that the negation of p is true when p is false.

In common usage, we sometimes encounter a *double negative*. Consider the following statement:

I enjoy dogs, but it's just not true that I don't like other animals as well.

In this statement, the speaker is telling us that, in fact, he does like other animals. If q is the statement that *I like animals other than dogs*, then the preceding quote is represented by NOT NOT q, which has the same meaning as q.

The Internet search engine Google will perform the NOT operation, but it uses the minus sign (preceded by a space) to indicate negation. For example, if you search on **Syria**, you will get a number of hits, including at least one from the Central Intelligence Agency. If you want to refine your search to exclude references to the CIA, you use **Syria -CIA.** (The space before the minus sign is necessary.) The result is a list of pages about Syria that don't mention the CIA.

Advanced Search

Find pages with...		To do this in the search box
all these words:		Type the important words: tricolor rat terrier
this exact word or phrase:		Put exact words in quotes: "rat terrier"
any of these words:		Type OR between all the words you want: miniature OR standard
none of these words:		Put a minus sign just before words you don't want: -rodent, -"Jack Russell"
numbers ranging from:	to	Put 2 periods between the numbers and add a unit of measure: 10..35 lb, $300..$500, 2010..2011

Then narrow your results by...		
language:	any language	Find pages in the language you select.
region:	any region	Find pages published in a particular region.
last update:	anytime	Find pages updated within the time you specify.
site or domain:		Search one site (like wikipedia.org) or limit your results to a domain like .edu, .org or .gov
terms appearing:	anywhere in the page	Search for terms in the whole page, page title, or web address, or links to the page you're looking for.
SafeSearch:	Show most relevant results	Tell SafeSearch whether to filter sexually explicit content.
file type:	any format	Find pages in the format you prefer.
usage rights:	not filtered by license	Find pages you are free to use yourself.

Advanced Search

You can also...

Find pages that are similar to, or link to, a URL
Search pages you've visited
Use operators in the search box
Customize your search settings

Google advanced search page.

Conjunction The *conjunction* of two statements is the assertion that *both* are true. In common English usage, *and* signifies conjunction. We will use the symbol AND to denote conjunction. The following example shows how:

Assume that

p = *Baseball is 90% mental*

q = *The other half of baseball is physical*

The conjunction of these two statements is one of the many humorous sayings of Yogi Berra. In symbols, p AND q represents *Baseball is 90% mental* **and** *the other part is physical.*

Because the conjunction of two statements is the assertion that both are true, the conjunction is true when, and only when, both statements are true. For example, Yogi Berra's preceding statement cannot be true because the truth of either of the two statements implies that the other is false. Hence, one or the other is false—so the conjunction is false.

The truth table for the conjunction shows the truth value of p AND q for each possible combination of the truth values of p and q.

p	q	p AND q	
T	T	T	This means that p AND q is true when p is true and q is true.
T	F	F	This means that p AND q is false when p is true and q is false.
F	T	F	This means that p AND q is false when p is false and q is true.
F	F	F	This means that p AND q is false when p is false and q is false.

The following four statements illustrate the four different cases in the conjunction truth table:

True conjunction: *A fever may accompany a cold* **and** *a headache may accompany a cold.*

False conjunction: *You are required to pay income tax* **and** *pigs often fly.*

False conjunction: *Aspirin cures cancer* **and** *water is wet.*

False conjunction: *The president's policies are always best for America* **and** *Congress never passes an unwise bill.*

Internet search engines understand the use of the AND operator. If you search for **Syria and Turkey,** you will see sites that mention both Syria and Turkey—for example, refugees from Syria who have fled to Turkey.

EXAMPLE 1.7 Determining truth in a conjunction: Health insurance policies

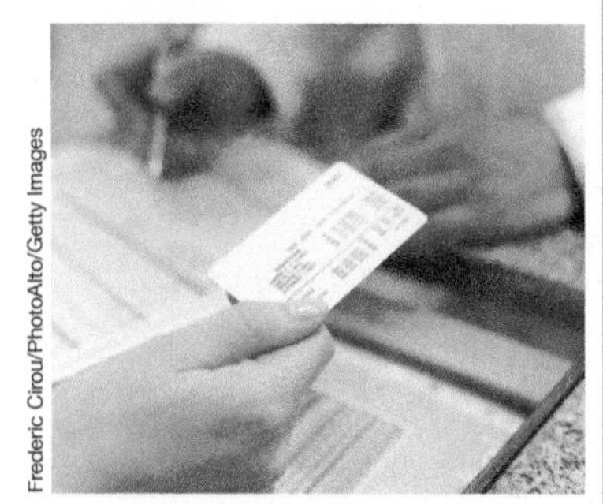

Frederic Cirou/PhotoAlto/Getty Images

Health insurance card.

In the past, many health insurance policies did not cover preexisting conditions. That is, they did not cover illness that existed prior to the purchase of the policy. A salesman for such a policy stated, *If you buy this policy, it will cover cases of flu in your family next winter, and it will cover treatment for your wife's chronic arthritis.* Was the salesman telling the truth?

SOLUTION

The policy did not cover the preexisting arthritic condition, so that part of the conjunction was not true. Conjunctions are true only when both parts are true. The salesman did not speak truthfully.

TRY IT YOURSELF 1.7

I scored 62% on a difficult exam where 60% was a passing grade. I told my friend *The test was hard, and I passed it.* Did I speak truthfully?

The answer is provided at the end of this section.

Disjunction The *disjunction* of two statements is the assertion that either one or the other is true (or possibly both). Compare this with the conjunction, where both statements *must* be true. Disjunction in common English usage is indicated with *or*. For symbolic representations, we use OR. Here is an example.

Assume that

p = *This medication may cause dizziness*

q = *This medication may cause fatigue*

We use the disjunction to state the warning on my medication: p OR q stands for the statement *This medication may cause dizziness or fatigue.*

In common English usage, there is no doubt as to what is meant when we combine statements using **and**. There can be a question, however, when we consider the usage of the word **or**. In the sample warning, it is clearly intended that the medicine may cause either or *both* conditions. This is the *inclusive* use of **or**, which means either or both of the combined statements may be true. By way of contrast, in April of 2020, Democrats in Wisconsin had the opportunity to vote for Joe Biden or Bernie Sanders (the other contenders had withdrawn). That definitely did not mean they could vote for both. This is an illustration of what is called the *exclusive or*, meaning that one choice or the other is allowed, but not both.

KEY CONCEPT

In mathematics and logic, the word or is always used in the inclusive sense: one or the other, or possibly both.

In logic, a disjunction is true when either one part is true or both parts are true. It is false only when *both* parts are false. Here is the truth table for disjunction.

p	q	p OR q	
T	T	T	This means that p OR q is true when p is true and q is true.
T	F	T	This means that p OR q is true when p is true and q is false.
F	T	T	This means that p OR q is true when p is false and q is true.
F	F	F	This means that p OR q is false when p is false and q is false.

The following four statements illustrate the four different cases in the disjunction truth table:

True disjunction: *A fever may accompany a cold* **or** *a headache may accompany a cold.*

True disjunction: *You are required to pay income tax* **or** *pigs often fly.*

True disjunction: *Aspirin cures cancer* **or** *water is wet.*

False disjunction: *The president's policies are always best for America* **or** *Congress never passes an unwise bill.*

Internet searches use OR in the inclusive sense. For example, **red or blue Toyota** will bring up information about Toyotas that have something to do with red, or blue, or *both*.

EXAMPLE 1.8 Determining the meaning of a disjunction: Menu options

In my favorite restaurant, the waiter asks if I want butter or sour cream on my baked potato. Is this the inclusive or exclusive **or**?

SOLUTION

His statement surely means you can have butter, sour cream, or both. Therefore, it is the inclusive **or**.

TRY IT YOURSELF 1.8

A waiter says, *Your meal comes with fries or a baked potato.* Is this the inclusive or exclusive **or**?

The answer is provided at the end of this section.

Implication A statement of the form "If p then q" is called an *implication* or a *conditional* statement. We use the same terminology as in the preceding section: p is called the *premise* or the *hypothesis*, and q is called the *conclusion*. We represent the conditional symbolically as $p \to q$. Consider the following example:

Assume that

$p =$ *Your average is 90% or more*

$q =$ *You get an A for the course*

The string of symbols $p \to q$ represents the conditional **If** *your average is 90% or more*, **then** *you get an A for the course*. In this conditional, the statement *Your average is 90% or more* is the premise or hypothesis, and the statement *You will get an A* is the conclusion.

You can think of a conditional as a promise that the conclusion will happen upon the condition that the premise happens. (This is why such statements are called conditional.) It is important to note that when the teacher promised an A for an average of 90% or better, nothing was said about what might happen if your average is less than 90%. So, for example, if your average is 88%, you may still get an A, or you may get a lower grade. Neither outcome would indicate that the grading promise had been broken. In fact, the only situation where the grading promise is broken is when a student gets an average of 90% or more but does not get an A for the course.

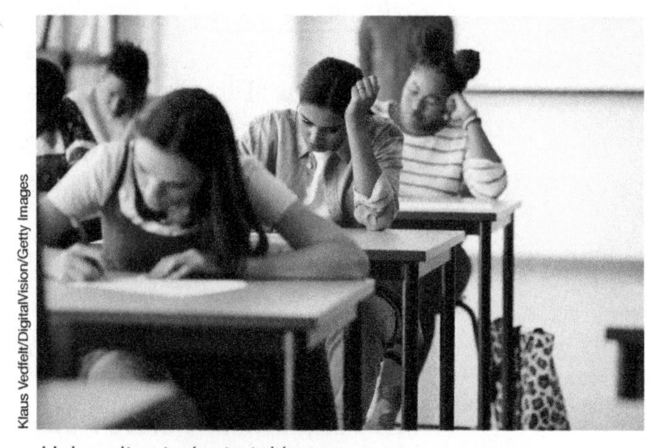
Klaus Vedfelt/DigitalVision/Getty Images

University students taking an exam.

In general, we regard a conditional statement to be true as long as the promise it involves is not broken. The promise is not broken if the premise is not true, so a conditional statement is regarded as true any time the premise is false.

The fact that a conditional statement is true whenever the hypothesis is false may seem counterintuitive. Think about it this way: If the premise is not true, then the promise has not been broken for the simple reason that it has never been tested.

Therefore, we cannot say the conditional statement is false. But if the conditional statement is not false, then it must be true, because a statement in logic must be either true or false.

Here's an old joke that uses the mathematical interpretation of the conditional statement:

Sue: "If you had two billion dollars, would you give me one billion?"

Bill: "Sure."

Sue: "If you had two million dollars, would you give me one million?"

Bill: "Of course."

Sue: "If you had twenty dollars, would you give me ten?"

Bill: "No!"

Sue: "Why not?"

Bill: "Because I *have* twenty dollars!"

The point of the joke is that one can promise anything based on an unfulfilled condition and not be accused of lying.

The truth table for the conditional statement is as follows:

p	q	$p \to q$	
T	T	T	This means that $p \to q$ is true when p is true and q is true.
T	F	F	This means that $p \to q$ is false when p is true and q is false.
F	T	T	This means that $p \to q$ is true when p is false and q is true.
F	F	T	This means that $p \to q$ is true when p is false and q is false.

Here are examples of the truth table entries.

True conditional: **If** *Earth is spherical,* **then** *you can go from Spain to Japan by traveling west.*

False conditional: **If** *the Moon orbits Earth,* **then** *the Moon is a planet.*

True conditional: **If** *Earth is the planet nearest to the Sun,* **then** *Earth orbits the Sun once each year.*

True conditional: **If** *Earth is the planet farthest from the Sun,* **then** *the Moon is made of green cheese.*

EXAMPLE 1.9 Determining the meaning of the conditional: Presidential promise

Between 2009 and 2012, President Obama repeatedly made the following promise in connection with the Affordable Care Act: *If you like your health care plan, then you can keep it.*

Under which of the following scenarios would you think the president's promise was kept?

Scenario 1: You like your health care plan, but you can't keep it.

Scenario 2: You don't like your health care plan, but you can't keep it.

Scenario 3: You don't like your health care plan, and you can keep it.

SOLUTION

In Scenario 1, the premise is true but the conclusion is false, so the implication is false. The president's promise was not kept. In Scenarios 2 and 3, the premise of the conditional is false. Thus the conditional is true, and the promise should be considered to have been kept.

Conditional statements can be expressed in a variety of ways. Sometimes the phrase *only if* is used. For example, consider the statement *You get a ticket only if you are speeding*. This means *If you get a ticket, then you are speeding*. In general, "p only if q" means "If p then q" and is represented by $p \rightarrow q$.

EXAMPLE 1.10 Representing statements and finding the truth: Majors

Let m represent the statement *Bill is a math major*, and let c represent the statement *Bill is a chemistry major*. First, represent each of these combinations of statements in symbolic form. Second, assume that Bill is indeed a math major but not a chemistry major and determine which of the statements are true and which are false.

a. Bill is a math major or a chemistry major.

b. Bill is not a math major and is a chemistry major.

c. If Bill is not a math major, then he's a chemistry major.

SOLUTION

a. This is a disjunction, and we represent it using m OR c. This statement is true (even though c is false) because m is true.

b. Note that NOT m means *Bill is not a math major.* So we use (NOT m) AND c to represent the conjunction. Observe that we are careful to include the parentheses here. It is always advisable to use parentheses to make sure that the use of NOT is properly understood. The conjunction is false because one of the statements is not true. (In fact, neither is true.)

c. This is a conditional. Note that NOT m means *Bill is not a math major*. So (NOT m) $\rightarrow c$ is the symbolic statement we seek. The premise *Bill is not a math major* is false, so the implication is true (even though the conclusion is false).

TRY IT YOURSELF 1.10

Represent these combinations of statements in symbolic form. Then determine which are true under the assumption that Bill is a math major but not a chemistry major.

a. Bill is neither a math major nor a chemistry major.

b. If Bill is a math major, then he's a chemistry major.

The answer is provided at the end of this section.

Using parentheses in logic

Let's look more closely at the use of parentheses in part b of the preceding example. Without explanation, the expression NOT m AND c is ambiguous and could be interpreted using parentheses in at least two ways: (NOT m) AND c; NOT (m AND c).

Interpretation 1: (NOT m) AND c. This expression means, "It is not true that he is a math major. In addition, he is a chemistry major." This is the same as the statement *He is not a math major and is a chemistry major.*

Interpretation 2: NOT (m AND c). This expression means, "It is not the case that he has the two majors math and chemistry."

These two interpretations say entirely different things. Thus, the use of parentheses is very important. If we leave them out and just write NOT m AND c, the meaning is ambiguous. It could mean either of the two preceding interpretations.

This illustrates why parentheses are often necessary to avoid ambiguity. You should use parentheses when you want to view a whole collection of symbols as one single entity. When a complex statement contains one or more parts that are themselves combinations of statements, the parentheses are nearly always vital to a proper understanding of the overall meaning. *When in doubt, use parentheses!*

Truth tables for complex statements

We can use the basic truth tables presented earlier to analyze more complex statements. Consider, for example, the following quote from the philosopher George Santayana: "*Those who do not remember the past are condemned to repeat it.*" We can rephrase this statement as **If** *you do not remember the past*, **then** *you are condemned to repeat it.*

Assume that

p = *You remember the past*

q = *You are condemned to repeat the past*

We can express the given statement symbolically as (NOT p) $\rightarrow q$.

To make a truth table, we begin with a table that lists the possible truth values for p and q. We include columns for NOT p, and for the implication $\rightarrow$.

p	q	NOT p	(NOT p) $\rightarrow q$	
T	T			p is true, and q is true.
T	F			p is true, and q is false.
F	T			p is false, and q is true.
F	F			p is false, and q is false.

When parentheses are present, we always work from "the inside out," so the next step is to complete the column corresponding to the "inside," that is, NOT p. The truth values of this statement are the opposite of those for p.

p	q	NOT p	(NOT p) $\rightarrow q$	
T	T	F		NOT p is false.
T	F	F		NOT p is false.
F	T	T		NOT p is true.
F	F	T		NOT p is true.

We complete the table by applying the truth table for $\rightarrow$ to the last column. Remember that the conditional is true except when the premise is true but the conclusion is false.

p	q	NOT p	(NOT p) $\rightarrow q$	
T	T	F	T	NOT p is false, and q is true, so the conditional is true.
T	F	F	T	NOT p is false, and q is false, so the conditional is true.
F	T	T	T	NOT p is true, and q is true, so the conditional is true.
F	F	T	F	NOT p is true, and q is false, so the conditional is false.

The truth value of the compound statement (NOT p) $\rightarrow q$ is shown in the last column. Note, for example, that the table tells us that (NOT p) $\rightarrow q$ is true when p is true and q is true. This fact is relevant to those people who remember the past and are condemned to repeat it. In these cases, Santayana's statement is true.

EXAMPLE 1.11 Using truth tables for complex statements: Proposition 187

In 1994 California voters approved Proposition 187, which included the statement *A person shall not receive any public social services until he or she has been verified as a United States citizen or as a lawfully admitted alien.* The law was later judged unconstitutional by a federal court.

Let

C = *Citizenship has been verified*

A = *Lawfully admitted alien status has been verified*

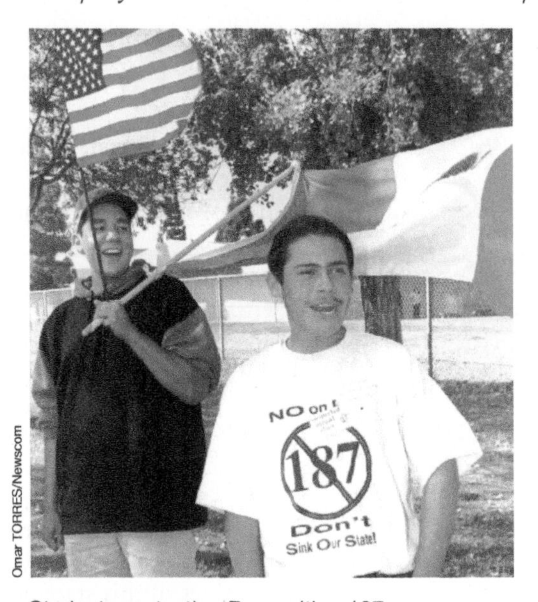

Students protesting Proposition 187.

Then under Proposition 187, the condition that would deny services is NOT (C OR A). Use a truth table to analyze this statement.

SOLUTION

To begin making the truth table, we list the four possibilities for C and A.

C	A	C OR A	NOT (C OR A)	
T	T			C is true, and A is true.
T	F			C is true, and A is false.
F	T			C is false, and A is true.
F	F			C is false, and A is false.

Working from the inside out, we next fill in the column for C OR A.

C	A	C OR A	NOT (C OR A)	
T	T	T		C is true, and A is true, so C OR A is true.
T	F	T		C is true, and A is false, so C OR A is true.
F	T	T		C is false, and A is true, so C OR A is true.
F	F	F		C is false, and A is false, so C OR A is false.

We take the negation of this column to obtain the final answer.

C	A	C OR A	NOT (C OR A)	
T	T	T	F	C OR A is true, so NOT (C OR A) is false.
T	F	T	F	C OR A is true, so NOT (C OR A) is false.
F	T	T	F	C OR A is true, so NOT (C OR A) is false.
F	F	F	T	C OR A is false, so NOT (C OR A) is true.

The final truth value of the statement NOT (C OR A) is the last column in the truth table.

TRY IT YOURSELF 1.11
Make the truth table for NOT (p AND q).

The answer is provided at the end of this section.

More conditional statements: Converse, inverse, contrapositive

The conditional is the combination of statements that is most often misinterpreted, so it deserves further attention.

Assume that

p = *You support my bill*

q = *You are a patriotic American*

© Hipix/Alamy

This conditional is a statement that a politician might use to garner support for a bill he or she wants to see passed:

$p \to q$: **If** *you support my bill*, **then** *you are a patriotic American.*

There are three common variations of the conditional that may lead to confusion.

The converse: $q \to p$. **If** *you are a patriotic American*, **then** *you support my bill.* The converse reverses the roles of the premise and conclusion.

The inverse: (NOT p) $\to$ (NOT q). **If** *you do not support my bill*, **then** *you are not a patriotic American.* The inverse replaces both the premise and conclusion by their negations.

The contrapositive: (NOT q) $\to$ (NOT p). **If** *you are not a patriotic American*, **then** *you do not support my bill.* The contrapositive both reverses the roles of and negates the premise and conclusion.

The original conditional $p \to q$ says that all supporters of the bill are patriotic Americans, but it allows for the possibility that some nonsupporters of the bill are also patriotic Americans. The contrapositive, (NOT q) $\to$ (NOT q), says that if you are not a patriotic citizen, then you do not support the bill. This is just another way of saying the same thing as $p \to q$, that is, that supporters of the bill are patriotic but nonsupporters may be patriotic as well. The conditional and its contrapositive are really just different ways of saying the same thing.

Similarly, the converse $q \to p$ and the inverse (NOT p) $\to$ (NOT q) of a conditional are also different ways of saying the same thing, but what they say is *not* the same as the original conditional. They say that nonsupporters are unpatriotic.

One way to compare the various forms is to look at a truth table showing all four.

p	q	NOT p	NOT q	Conditional $p \rightarrow q$	Converse $q \rightarrow p$	Inverse (NOT p) $\rightarrow$ (NOT q)	Contrapositive (NOT q) $\rightarrow$ (NOT p)
T	T	F	F	T	T	T	T
T	F	F	T	F	T	T	F
F	T	T	F	T	F	F	T
F	F	T	T	T	T	T	T

This table shows that the conditional and its contrapositive have the same truth value. Thus, they are termed *logically equivalent*. Similarly, the converse and inverse are logically equivalent.

KEY CONCEPT

Two statements are logically equivalent if they have the same truth tables. Logically equivalent statements are just different ways of saying the same thing.

The fact that a conditional is logically equivalent to its contrapositive is useful whenever it may be easier to observe consequences of a phenomenon than the phenomenon itself. For example, it is not easy to determine whether heavier objects fall faster than lighter objects when air resistance is ignored. Aristotle believed they did. But if this were true, then heavier and lighter objects dropped from a high place would strike the ground at different times. Famously, Galileo observed that this was not true, and an important fact about gravity was verified.

The contrapositive arises in thinking about the results of medical tests. For example, people suffering from diabetes have abnormal blood sugar (*If you have diabetes, then you have abnormal blood sugar*). When a doctor finds your blood sugar to be normal, the contrapositive is used to conclude that you do not have diabetes (*If you do not have abnormal blood sugar, then you do not have diabetes*).

Whether intentional or not, the error of confusing a conditional statement with its converse or inverse is commonplace in politics and the popular media. The statement *The perpetrator has blood type A* can be restated as *If you are the perpetrator, then you have blood type A*. If the prosecutor then tells us that the defendant has blood type A, he or she may be hoping the jury will think the converse, namely *If you have blood type A, then you are the perpetrator*. This is not a valid conclusion because many innocent people also have blood type A.

SUMMARY 1.3 Conditional Statements

Consider the conditional statement **If** *statement A*, **then** *statement B*.

- The contrapositive statement is

  If *statement B is false*, **then** *statement A is false.*

  A conditional is logically equivalent to its contrapositive.

- The converse statement is

  If *statement B*, **then** *statement A*.

- The inverse statement is

  If *statement A is false*, **then** *statement B is false.*

  The inverse is logically equivalent to the converse.

- Replacing a conditional statement with its inverse or converse is *not* logically valid.

EXAMPLE 1.12 Formulating variants of the conditional: Politics

Formulate the converse, inverse, and contrapositive of each of the following conditional statements:

a. *If you vote for me, your taxes will be cut.*

b. *All Democrats are liberals.*

SOLUTION

a. The converse of the statement is *If your taxes are cut, you voted for me.* The inverse is *If you do not vote for me, your taxes will not be cut.* The contrapositive is *If your taxes are not cut, you did not vote for me.*

b. The statement *All Democrats are liberals* is another way of saying *If you are a Democrat, you are a liberal.* So the converse is *If you are a liberal, you are a Democrat* or *All liberals are Democrats.* The inverse is *If you are not a Democrat, you are not a liberal.* The contrapositive is *If you are not a liberal, you are not a Democrat.*

TRY IT YOURSELF 1.12

Formulate the converse, inverse, and contrapositive of the statement *Every Republican is a conservative.*

The answer is provided at the end of this section.

Logic and computers

Computers use *logic gates.* These gates perform functions corresponding to the logical operations such as negation and conjunction that we have studied in this section. Here is the connection: In classical computing, information is stored using the two numbers 0 and 1. The connection with logic is made by thinking of 0 as "false" and 1 as "true."

There are several types of logic gates. The logic gate NOT converts 0 to 1 and 1 to 0, just as in formal logic the negation of a false statement is true and the negation of a true statement is false. The logic gate OR takes two inputs, each of them 0 or 1. It returns 1 if at least one of the inputs is 1. This corresponds to the fact that in logic the disjunction of two statements is true (corresponding to 1) if at least one of the statements is true.

EXAMPLE 1.13 Using logic gates and logical operations: Input and output

The logic gate AND takes two inputs, each of them 0 or 1. It corresponds to the logical operation of conjunction. If the inputs to the AND gate are 1 and 1, what is the output?

SOLUTION

This situation corresponds to the conjunction p AND q when both p and q are true. In that case, p AND q is true, so the output of the logic gate is 1.

TRY IT YOURSELF 1.13

The logic gate XOR takes two inputs, each of them 0 or 1. It corresponds to the *exclusive* use of **or**, in which we ask whether one or the other, but not both, is true. If the inputs to the XOR gate are 1 and 1, what is the output?

The answer is provided at the end of this section.

WHAT DO YOU THINK?

Do computers think? Computers can perform logical operations. Does this fact mean that computers can think in the way humans can? Take the first three sections of this textbook into account as you formulate your answer.

Questions of this sort may invite many correct responses. We present one possibility: Are logical operations alone the same as thought? It is not a simple question. It is undeniably the case that computers can sort through mountains of data at lightning speed, and they can make decisions based on available information. They can even learn. Once they know a song or two that appeals to us, they are very good at selecting more music that we like. All of these and other operations performed by computers show aspects of thought. However, many associate thought with awareness, and this leads to science fiction computers like HAL from the movie *2001* or the *Terminator*. Awareness is not a feature of today's computers, but what of the future?

Further questions of this type may be found in the *What Do You Think?* section of exercises.

Try It Yourself answers

Try It Yourself 1.7: Determining truth in a conjunction: Grade on exam The statement is true.

Try It Yourself 1.8: Determining the meaning of a disjunction: Menu options This is an example of the exclusive **or**.

Try It Yourself 1.10: Representing statements and finding the truth: Majors

a. (NOT m) AND (NOT c). The conjunction is false.

b. $m \rightarrow c$. The conditional is false.

Try It Yourself 1.11: Using truth tables for complex statements: Proposition 187

p	q	p AND q	NOT (p AND q)
T	T	T	F
T	F	F	T
F	T	F	T
F	F	F	T

Try It Yourself 1.12: Formulating variants of the conditional: Politics The converse is *If you are a conservative, then you are a Republican.* The inverse is *If you are not a Republican, then you are not a conservative.* The contrapositive is *If you are not a conservative, then you are not a Republican.*

Try It Yourself 1.13: Using logic gates and logical operations: Input and output The output is 0.

Exercise Set 1.3

Test Your Understanding

1. Match the following logical operations with the appropriate symbol.

1. Negation	a. AND
2. Conjunction	b. NOT
3. Implication	c. OR
4. Disjunction	d. $\rightarrow$

2. If the statement p OR q is true, then: **a.** both p and q must be true **b.** either p or q must be true, but not both **c.** either p or q must be true, and possibly both **d.** p must be true but not q.

3. True or false: Computers use the logical rules similar to those presented in this section.

4. True or false: Computer search engines, such as Google, use the logical rules presented in this section.

5. The only time the conditional statement $p \rightarrow q$ is false is when ______.

6. The converse of the implication $p \rightarrow q$ is ______.

Problems

7. **Representing statements.** Let p represent the statement *You drive too fast*, and let q represent the statement *You get a traffic ticket.* Using symbols, express the statement *If you don't drive too fast, you will not get a traffic ticket.*

8. **Conditional.** Write the statement *Stop talking or I'll leave* as a conditional statement.

9. **A research project using the Internet.** You need to write a report on France or England in the eighteenth OR nineteenth century. Propose an Internet search phrase that will help you.

10. **Converse.** Consider the statement *If it rains, then I carry an umbrella.* Formulate the converse. The original statement is quite reasonable. Is the converse reasonable?

11. **Inverse.** Consider the statement *If I buy this MP3 player, then I can listen to my favorite songs.* Formulate the inverse. The original statement seems reasonable. Is the inverse reasonable?

12. **Politicians.** Congresswoman Jones says, "If we pass my bill, the economy will recover." A year later, Congressman Smith says to Congresswoman Jones, "Well, you were wrong, we didn't pass your bill and the economy has recovered just fine." Is there anything wrong here? Explain.

13. **Taxing implications.** Consider the statement *If you don't pay your taxes by April 15, then you owe a penalty.* A taxpayer is confused because she paid her taxes by April 15 but she still owes a penalty. Explain her confusion.

14–18. More on representing statements. *Let p represent the statement You live in Utah, and let q represent the statement You work in Nevada. Use this information in Exercises 14 through 18 to express the given statement in symbolic form.*

14. You do not live in Utah and you work in Nevada.

15. If you work in Nevada, then you cannot live in Utah.

16. You do not live in Utah or you do not work in Nevada.

17. Nobody who works in Nevada can live in Utah.

18. You do not live in Utah only if you work in Nevada.

19–25. Translating into symbolic form. *In Exercises 19 through 25, translate the given sentence into symbolic form by assigning letters to statements that do not contain a negative. State clearly what your letters represent.*

19. If we pass my bill, the economy will recover.

20. We didn't pass your bill, and the economy recovered.

21. For breakfast I want cereal and eggs.

22. I want cereal or eggs for breakfast.

23. If you don't clean your room, I'll tell your father.

24. I'll give you a dollar if you clean your room and take out the trash.

25. I'll go downtown with you if we can go to a movie or a restaurant.

26–28. Translating into English. *Let p represent the statement He's American, and let q represent the statement He's Canadian. Each of Exercises 26 through 28 has two logically equivalent statements. That is, they are just different ways of saying the same thing. Write each of these statements in plain English.*

26. (NOT p) $\rightarrow q$ and p OR q

27. NOT (p OR q) and (NOT p) AND (NOT q)

28. NOT (p AND q) and (NOT p) OR (NOT q)

29–43. True or false. *In Exercises 29 through 43, determine whether the given statement is true or false. Be sure to explain your answer.*

29. $3+3=7$ or $2+2=5$.

30. $3+3=6$ or $2+2=4$.

31. If pigs fly, then Mickey Mouse is president.

32. If George Washington was the first U.S. president, then Mickey Mouse is the current U.S. president.

33. I will be president if pigs fly.

34. If pigs can fly, then Earth is flat.

35. If Earth is flat, then pigs can fly.

36. If pigs can fly, then oceans are water.

37. If oceans are water, then pigs can fly.

38. Mickey Mouse is president of the United States and $5+5=10$.

39. Mickey Mouse is president of the United States or $5+5=10$.

40. If Mickey Mouse is president of the United States, then $5+5=10$.

41. Mickey Mouse is a cartoon character and $5+5=10$.

42. Mickey Mouse is president of the United States or $5+5=11$.

43. If $5+5=12$, then Mickey Mouse is president of the United States.

44–51. The inverse. *In Exercises 44 through 51, formulate the inverse of the given statement.*

44. *All dogs chase cats.*

45. *If you don't clean your room, I'll tell your father.*

46. *I'll give you a dollar if you clean your room and take out the trash.*

47. *I'll go downtown with you if you agree to go to a movie or a restaurant.*

48. *If I am elected, then taxes will be cut.*

49. *All math courses are important.*

50. *If you don't exercise regularly, your health will decline.*

51. *If you take your medicine, you will get well.*

52–59. The converse. *In Exercises 52 through 59, formulate the converse of the given statement.*

52. *All dogs chase cats.*

53. *If you don't clean your room, I'll tell your father.*

54. *I'll give you a dollar if you clean your room and take out the trash.*

55. *I'll go downtown with you if you agree to go to a movie or a restaurant.*

56. *If I am elected, then taxes will be cut.*

57. *All math courses are important.*

58. *If you don't exercise regularly, your health will decline.*

59. *If you take your medicine, you will get well.*

60–71. Contrapositives. *In Exercises 60 through 71, formulate the contrapositive of the following statements:*

60. *If there is life on Mars, then there must be traces of water there.*

61. *If you are neutral in situations of injustice, you have chosen the side of the oppressor.* (Desmond Tutu)

62. *If you have ten thousand regulations, you destroy all respect for the law.* (Winston Churchill)

63. *If you drink and drive, you will get arrested.*

64. *If aliens had visited Earth, we would have found concrete evidence by now.*

65. *If you don't clean your room, I'll tell your father.*

66. *I'll give you a dollar if you clean your room and take out the trash.*

67. *I'll go downtown with you if you agree to go to a movie or a restaurant.*

68. *If we continue to produce greenhouse gases, then global warming will occur.*

69. *My friends all drive Porsches.*

70. *Pneumonia leads to severe respiratory distress.*

71. *If thunderstorms are approaching, they will be seen on radar.*

72–75. Truth tables. *In Exercises 72 through 75, make the truth table for the given compound statement.*

72. p AND (NOT q)

73. $p \rightarrow$ (p OR q)

74. $p \rightarrow$ (p AND q)

75. p AND (p OR q)

76–78. Equivalence. *In each of Exercises 76 through 78, use truth tables to show that the given two statements are logically equivalent. That is, show that the two statements have the same truth values.*

76. (NOT p) $\rightarrow q$ and p OR q

77. NOT (p AND q) and (NOT p) OR (NOT q)

78. (NOT p) AND q and NOT (p OR (NOT q))

79. Logic gate AND. If the inputs to the AND gate are 0 and 1, what is the output?

80. Logic gate NAND. The logical operation NAND corresponds to negating the conjunction of two statements: p NAND q means NOT (p AND q). The corresponding logic gate NAND is one of two basic gates, in the sense that every other logic gate can be constructed using it. (The other basic gate NOR is defined in Exercise 81.) The gate NAND takes two inputs, each of them 0 or 1. If the inputs to the NAND gate are 0 and 1, what is the output?

81. Logic gate NOR. The logical operation NOR corresponds to negating the disjunction of two statements: p NOR q means NOT (p OR q). The corresponding logic gate NOR is one of two basic gates, in the sense that every other logic gate can be constructed using it. (The other basic gate NAND is defined in Exercise 80.) The gate NOR takes two inputs, each of them 0 or 1. If the inputs to the NOR gate are 0 and 0, what is the output?

Writing About Mathematics

82. History of truth tables. Research the development and use of truth tables in the late nineteenth and early twentieth centuries. Be sure to include the contributions of Charles Lutwidge Dodgson (better known as Lewis Carroll).

83. Converse. Find and report on an argument presented in the popular media that makes a fallacious argument using the converse.

84. Artificial intelligence. Report on what computer scientists mean when they use the term *artificial intelligence.*

What Do You Think?

85. The jury is out. The district attorney was able to prove that the robber had red hair. The defendant has red hair. Did the D.A. prove his case? Explain your reasoning. **Exercise 10 is relevant to this topic.**

86. It never happens. *It never happens that we don't beat State U. at football.* Can this statement be expressed without all the negatives? If so, how would you do it?

87. A complicated situation. A complicated paragraph has lots of if's, then's, not's, and's, and or's. It is difficult to sort out the meaning. A friend tells me we can use a truth table to help understand what is meant. Do you think a truth table would help? Explain your reasoning.

88. Martian. Someone said, *The Libertarian Party would win the presidential election if they would nominate a Martian.* Is this statement true?

89. Iowa. An old joke goes like this: *What did you do during the war? I was in the Iowa National Guard defending the state from enemy invasion. What? But there wasn't any enemy in Iowa! Right—I did a great job, didn't I?* Discuss the National Guard member's assertion from a logical viewpoint.

90. NOR. Exercise 81 in Section 1.3 defines the logical operation NOR. Does this usage really have the same meaning as the usual word "nor"? Explain. **Exercise 81 is relevant to this topic.**

91. Conditional statements. A friend of yours thinks that the following statement should be regarded as false because the conclusion is false: *If Mickey Mouse was the first U.S. president, then Donald Duck was the second U.S. president.* How would you answer your friend? **Exercise 34 is relevant to this topic.**

92. Contrapositive. The contrapositive of a conditional statement is logically equivalent to the original statement. Give an example in which the contrapositive is more useful than the original statement for a specific purpose. **Exercise 68 is relevant to this topic.**

93. Converse confusion. Why are fallacious arguments involving the converse made so often?

1.4 Sets and Venn diagrams: Pictorial logic

TAKE AWAY FROM THIS SECTION
Produce visual displays that show categorical relationships as well as logical statements.

In this section, we study graphical representations of sets known as *Venn diagrams*. The following excerpt from Education Week illustrates the usefulness of Venn diagrams.

IN THE NEWS 1.5

Search

Home | TV & Video | U. S. | World | Politics | Entertainment | Tech | Travel

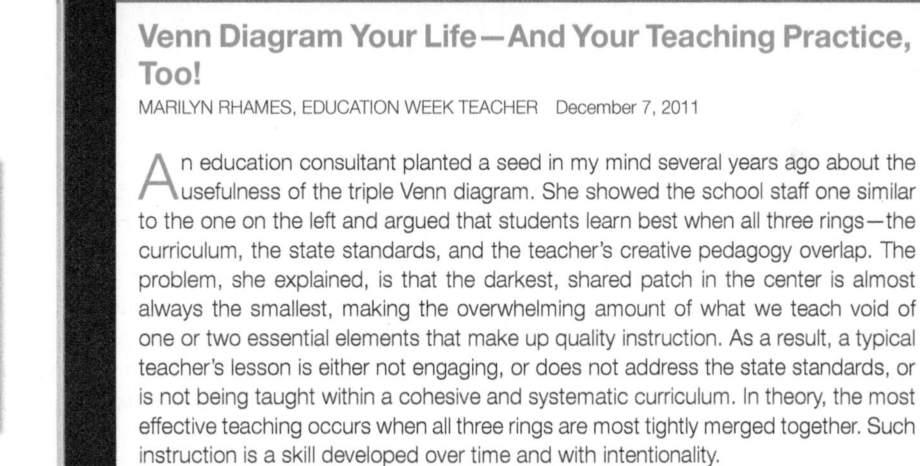

Venn Diagram Your Life—And Your Teaching Practice, Too!

MARILYN RHAMES, EDUCATION WEEK TEACHER December 7, 2011

An education consultant planted a seed in my mind several years ago about the usefulness of the triple Venn diagram. She showed the school staff one similar to the one on the left and argued that students learn best when all three rings—the curriculum, the state standards, and the teacher's creative pedagogy overlap. The problem, she explained, is that the darkest, shared patch in the center is almost always the smallest, making the overwhelming amount of what we teach void of one or two essential elements that make up quality instruction. As a result, a typical teacher's lesson is either not engaging, or does not address the state standards, or is not being taught within a cohesive and systematic curriculum. In theory, the most effective teaching occurs when all three rings are most tightly merged together. Such instruction is a skill developed over time and with intentionality.

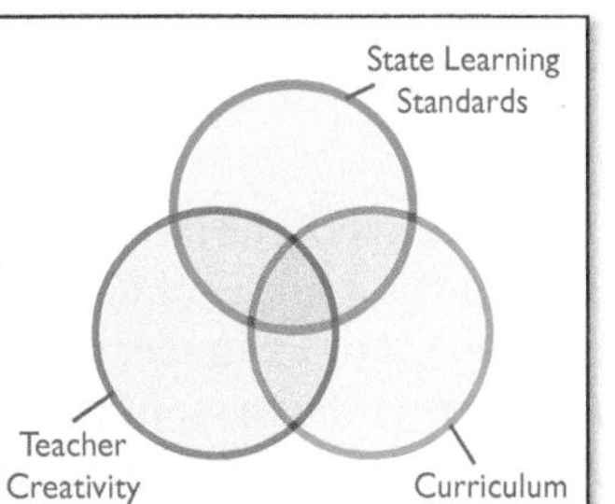

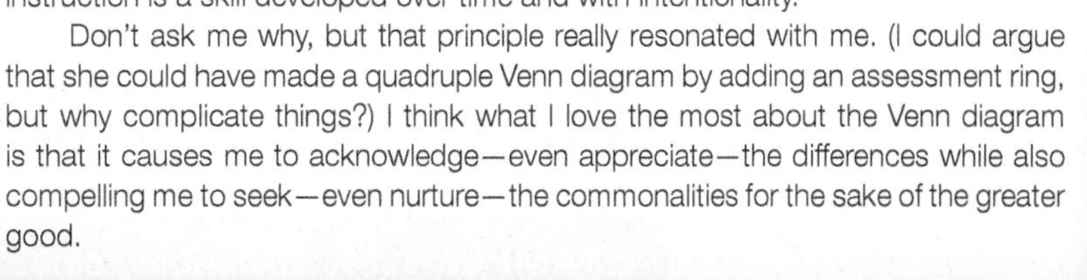

Don't ask me why, but that principle really resonated with me. (I could argue that she could have made a quadruple Venn diagram by adding an assessment ring, but why complicate things?) I think what I love the most about the Venn diagram is that it causes me to acknowledge—even appreciate—the differences while also compelling me to seek—even nurture—the commonalities for the sake of the greater good.

As the article indicates, the Venn diagram is an important tool for critical thinking.

Sets

Underlying the Venn diagram discussed in the preceding article is the idea of a *set*, which is nothing more than a collection of objects. This is where we begin.

KEY CONCEPT

A set is a collection of objects, which are called elements of the set.

A set is specified by describing its elements. This is sometimes done in words and sometimes simply by listing the elements and enclosing them in braces. For example, we can specify the set of all U.S. states that touch the Pacific Ocean by listing the elements of this set as

{Alaska, California, Hawaii, Oregon, Washington}

California is an element of this set, but Nebraska is not. If we let P stand for the set of Pacific Ocean states, we would write

$P =$ {Alaska, California, Hawaii, Oregon, Washington}

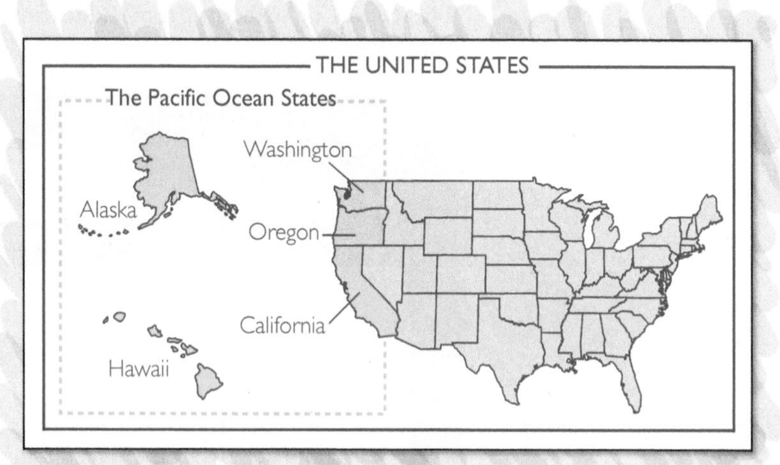

The Pacific Ocean states.

The states in the set P of Pacific Ocean states are also elements of the larger set U of all 50 U.S. states. This is an example of the idea of a *subset*: P is a subset of U.

At the other extreme, some sets have nothing in common. For example, the set of U.S. states that touch the Atlantic Ocean shares no members with the set of Pacific states. Whenever two sets have no elements in common, they are said to be *disjoint*.

KEY CONCEPT

The set B is a subset of the set A if every element of B is also an element of A. Two sets are called disjoint if they have no elements in common.

EXAMPLE 1.14 Describing sets: The U.S. Congress

In parts a, b, and c, we let S denote the set of U.S. senators, R the set of members of the U.S. House of Representatives, and G the set of all elected U.S. government officials.

a. How many elements are in the set S?

b. Of the three sets, which are subsets of others?

c. Which pairs of the three sets S, R, and G are disjoint?

d. If V is the collection of all vowels in the alphabet, specify the set as a list of elements.

SOLUTION

a. There are 2 senators from each of the 50 states, making a total of 100 U.S. senators. Therefore, there are 100 elements in S.

b. Both senators and representatives are elected officials. So S is a subset of G and R is a subset of G.

c. The Senate and House have no members in common, so S and R are disjoint.

d. $V = \{a, e, i, o, u\}$. (Sometimes y is a vowel, depending on its usage.)

TRY IT YOURSELF 1.14

a. Let S be the set of all U.S. states. Determine which of the following are elements of S: Hawaii, Guam, Alaska, Puerto Rico.

b. Let O be the set of odd whole numbers from 1 through 9. Write O as a list enclosed in braces.

c. Let N denote the set of North American countries, let S denote the set of South American countries, and let A be the set {United States, Canada, Mexico}. Which of these are subsets of others? Which pairs are disjoint?

The answer is provided at the end of this section.

Introduction to Venn diagrams

In many cases, pictures improve our understanding. A Venn diagram is a way of picturing sets.

KEY CONCEPT

A Venn diagram is a visual device for representing sets. It usually consists of circular regions inside a rectangle.

A Venn diagram typically consists of either two or three circles inside a rectangle. The area inside the rectangle represents all the objects that are under consideration at the time, and the areas inside the circles represent certain sets of these objects.

These diagrams can be a useful tool for answering questions about the number of elements in a set and for understanding relationships between the elements. For example, **Figure 1.1** shows a Venn diagram[5] in which the rectangle represents all elected government officials. One circle represents senators, and the other circle represents members of the House of Representatives. As we noted in the solution to part c of Example 1.14, these two sets are disjoint, and that relationship is indicated by the fact that the circles do not overlap. The region outside the two circles represents elected officials who are neither senators nor representatives. The president and vice president would go in this region.

In **Figure 1.2**, we have added a circle that represents Republican senators. This new circle is inside the "Senators" circle, indicating that the set of Republican senators is a subset of the set of all senators.

---

[5] *Technically, Figure 1.1 and Figure 1.2 are known as* Euler *diagrams, but we will not bother with that distinction here.*

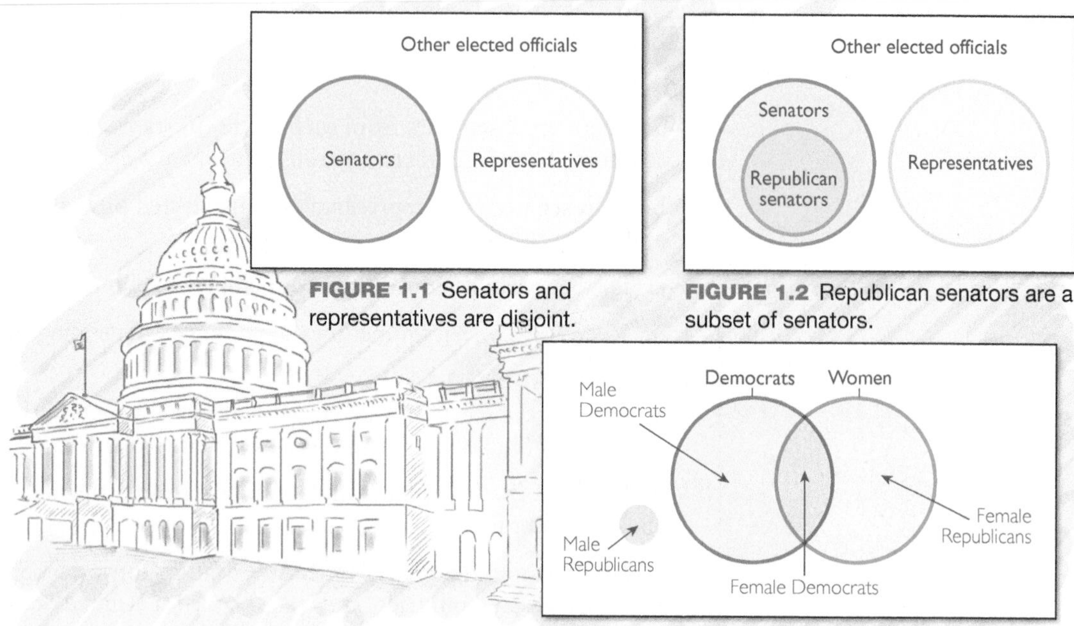

FIGURE 1.1 Senators and representatives are disjoint.

FIGURE 1.2 Republican senators are a subset of senators.

FIGURE 1.3 House of Representatives.

A more common example of a Venn diagram is shown in **Figure 1.3**, where the two circles overlap. In this diagram, the category under discussion is all U.S. representatives. One circle represents the set of Democratic members, and the other circle represents female members. There are four distinct regions in the diagram. (At the time of this writing, all members of the House of Representatives were either Democrats or Republicans.)

- The region common to both circles represents female Democrats.
- The region inside the "Democrat" circle but outside the "Women" circle represents male Democrats.
- The region inside the "Women" circle but outside the "Democrat" circle represents female Republicans.
- The region outside both circles represents members who are neither Democrats nor female. These are male Republicans.

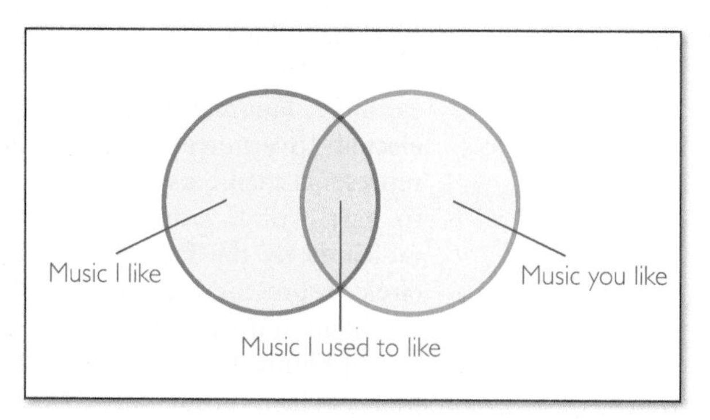

Some Venn diagrams are just for fun.

To further clarify what the various regions in a Venn diagram represent, let the category be the whole numbers from 1 to 10. Let $A = \{2, 3, 4, 5, 6\}$ and

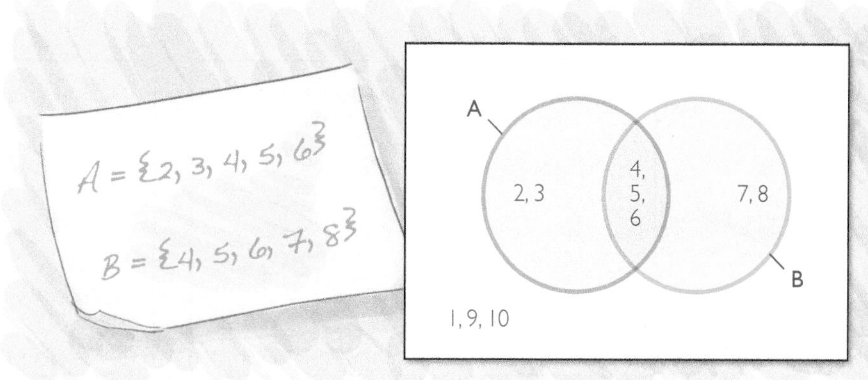

FIGURE 1.4 Parts of a Venn diagram.

$B = \{4, 5, 6, 7, 8\}$. The corresponding Venn diagram is shown in **Figure 1.4**. It shows which elements go in each part of the diagram.

EXAMPLE 1.15 Making Venn diagrams: Medical tests

Medical tests do not always produce accurate results. For this example, we use the category of all people who have undergone a medical test for hepatitis C. Let H denote the set of all patients who have hepatitis C, and let P be the set of all people who test positive:

$$H = \{\text{Patients with hepatitis C}\}$$

$$P = \{\text{Those who test positive}\}$$

Make a Venn diagram using these sets, and locate in it the following regions:

- **True positives:** These are people who test positive and have hepatitis C.
- **False positives:** These are people who test positive but do not have hepatitis C.
- **True negatives:** These are people who test negative and do not have hepatitis C.
- **False negatives:** These are people who test negative but do have hepatitis C.

SOLUTION

Because the two sets C and P overlap, we make the diagram shown in **Figure 1.5**. We locate the four regions as follows:

- True positives both test positive and have hepatitis C. They are common to both sets, so they lie in the overlap between the two circles.
- False positives are inside the "Positive" circle because they test positive but outside the "Hepatitis" circle because they do not have hepatitis C.
- True negatives lie outside both circles.
- False negatives are inside the "Hepatitis" circle but outside the "Positive" circle.

The completed diagram is shown in **Figure 1.6**.

TRY IT YOURSELF 1.15

If the category is states of the United States, make a Venn diagram showing Gulf States (states that touch the Gulf of Mexico) and states west of the Mississippi River. Locate the following in your diagram: Florida, New York, Oregon, and Texas.

The answer is provided at the end of this section.

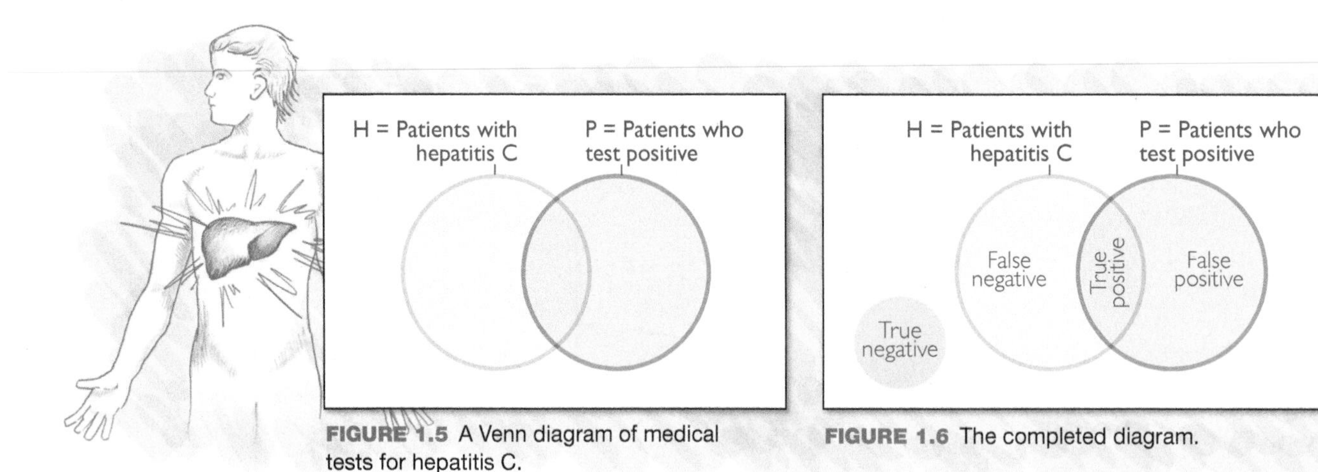

FIGURE 1.5 A Venn diagram of medical tests for hepatitis C.

FIGURE 1.6 The completed diagram.

Venn diagrams can be useful in helping to count the number of items in a given category.

EXAMPLE 1.16 Making Venn diagrams with numbers: Drug trial

A double-blind drug trial separated a group of 100 patients into a test group of 50 and a control group of 50. The test group got the experimental drug, and the control group got a placebo.[6] The result was that 40% of the test group improved and only 20% of the control group improved. Let T be the test group and I the set of patients who improved:

$$T = \{\text{Test group}\}$$
$$I = \{\text{Patients who improved}\}$$

Make a Venn diagram that shows the number of people in each of the following categories:

- Test group who improved
- Test group who did not improve
- Control group who improved
- Control group who did not improve

SOLUTION

The basic Venn diagram is shown in **Figure 1.7**. The rectangle represents all who were in the trial. The numbers and location of each group are as follows.

- **Test group who improved:** This is the overlap of the two circles. There are 40% of 50 or 20 people in this region.
- **Test group who did not improve:** This is inside the T circle but outside the I circle. There are $50 - 20 = 30$ people in this group.
- **Control group who improved:** This is in the I circle but outside the T circle. There are 20% of 50 or 10 patients in this category.
- **Control group who did not improve:** This is outside both circles. There are $50 - 10 = 40$ patients in this category.

We have entered these numbers in the completed Venn diagram shown in **Figure 1.8**.

[6] *A* placebo *is a harmless substance that contains no medication and that is outwardly indistinguishable from the real drug. We discuss clinical trials in Section 4 of Chapter 6.*

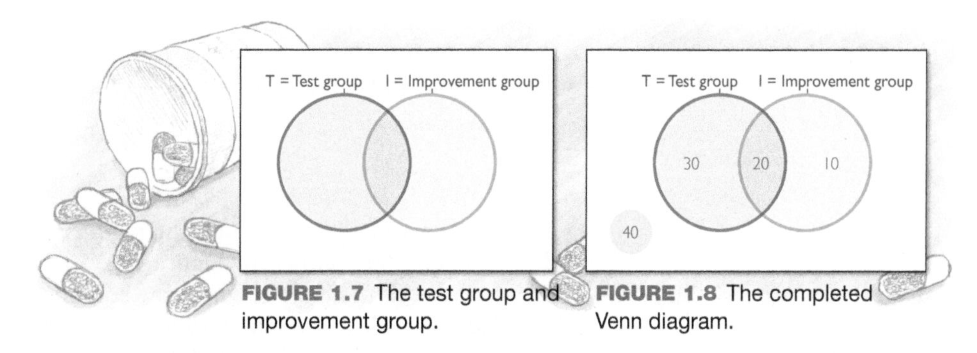

FIGURE 1.7 The test group and improvement group.

FIGURE 1.8 The completed Venn diagram.

TRY IT YOURSELF 1.16

A group of 50 children from an inner-city school and 50 children from a suburban school were tested for reading proficiency. The result was that 60% of children from the inner-city school passed and 80% of children from the suburban school passed. Make a Venn diagram showing the number of children in each category.

The answer is provided at the end of this section.

Logic illustrated by Venn diagrams

We noted at the beginning of this section that Venn diagrams can be used to analyze logical statements. Consider, for example, the conditional statement *If you are stopped for speeding, then you will get a traffic ticket.* Let S be the set of all people who are stopped for speeding and T the set of people who get traffic tickets:

$$S = \{\text{People stopped for speeding}\} \qquad T = \{\text{People who get traffic tickets}\}$$

Our conditional can be recast as *everyone who is stopped for speeding will get a traffic ticket.* This means that S is a subset of T. Thus, to represent this statement using a Venn diagram, we put the circle for S inside the circle for T. The result is shown in **Figure 1.9**. This figure makes it clear why we cannot necessarily conclude the converse: *If you get a traffic ticket, then you were stopped for speeding.* Any driver in between the two circles is a driver who got a ticket but was not stopped for speeding (**Figure 1.10**). The Venn diagram for the converse is shown in **Figure 1.11**.

FIGURE 1.9 Venn diagram for a conditional: If you are stopped for speeding, you get a ticket.

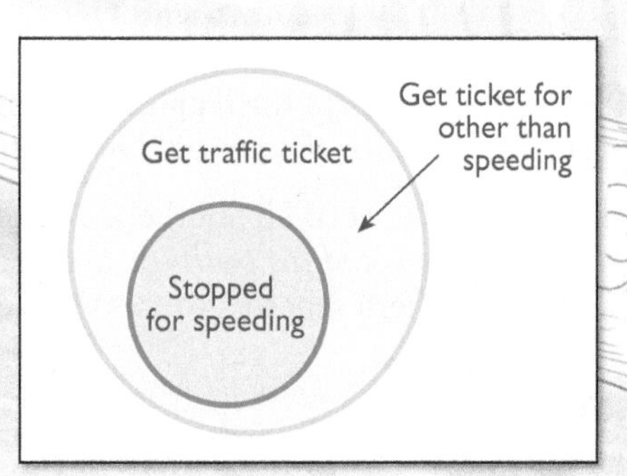

FIGURE 1.10 Why the converse may not be true.

FIGURE 1.11 Diagram for the converse: If you get a traffic ticket, then you were stopped for speeding.

Note that the roles of the circles are reversed. The only time a statement and its converse are both true is when the two circles in the Venn diagram are the same.

We can also represent the conjunction and disjunction using Venn diagrams. For the conjunction *He is a math major and a music major*, let A denote the set of all math majors and B the set of all music majors:

$$A = \{\text{Math majors}\}$$
$$B = \{\text{Music majors}\}$$

To say he is a math *and* a music major means that he belongs to *both* sets. These double majors are shown in **Figure 1.12**. The shaded region in that figure is the region of the Venn diagram where the conjunction is true.

The disjunction: *Joe is a math major or a music major* means that Joe belongs to one set or the other. **Figure 1.13** illustrates the corresponding region of the Venn diagram.

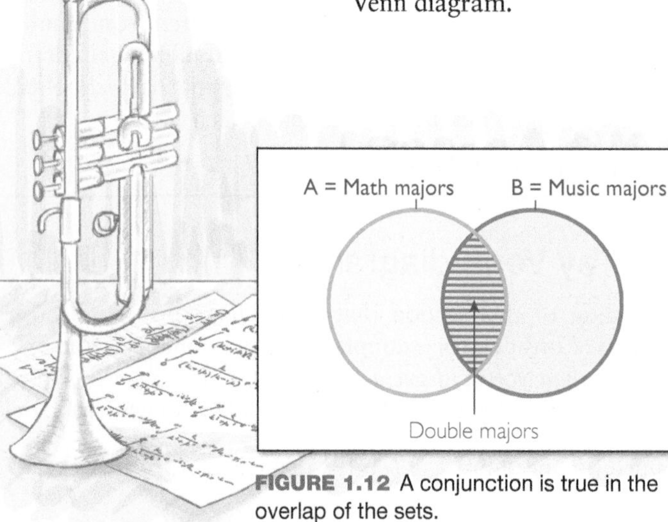

FIGURE 1.12 A conjunction is true in the overlap of the sets.

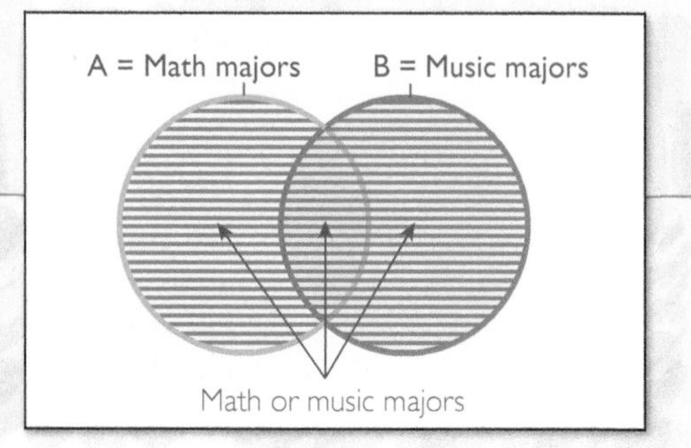

FIGURE 1.13 A disjunction is true in either one or both sets.

EXAMPLE 1.17 Matching logical statements and Venn diagrams: Farm bill

Let F denote the set of senators who voted for the farm bill and T the set of senators who voted for the tax bill:

$$F = \{\text{Senators who voted for the farm bill}\}$$
$$T = \{\text{Senators who voted for the tax bill}\}$$

a. Make a Venn diagram for F and T, and shade in the region for which the statement *He voted for the tax bill but not the farm bill* is true.

b. Let S denote the set of all Southern senators. Suppose it is true that *all Southern senators voted for both the tax and farm bills.* Add the circle representing Southern senators to the Venn diagram you made in part a.

SOLUTION

a. The statement is true for all senators who are in the set T but not in the set F. That group of senators is inside the "Tax" circle but outside the "Farm" circle. We have shaded this region in the Venn diagram shown in **Figure 1.14**.

b. Because all Southern senators voted for both bills, they all belong to both F and T. That makes S a subset of the overlap of the two circles. Thus, the circle representing S is inside the region common to both F and T, as shown in **Figure 1.15**.

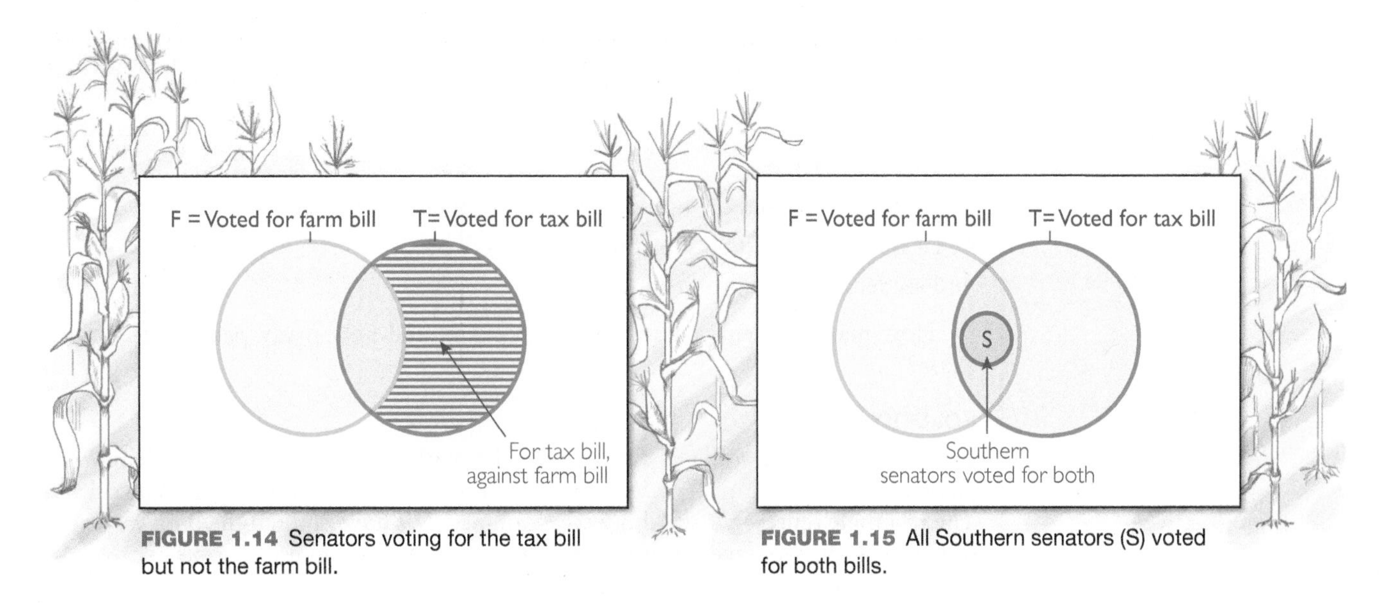

FIGURE 1.14 Senators voting for the tax bill but not the farm bill.

FIGURE 1.15 All Southern senators (S) voted for both bills.

Counting using Venn diagrams

Venn diagrams are useful in counting. To illustrate this, suppose that 17 tourists travel to Mexico or Costa Rica. Some visit Mexico, some visit Costa Rica, and some visit both. If 10 visit Mexico but not Costa Rica and 5 visit both, how many people visit Costa Rica? Let

$$M = \{\text{Tourists visiting Mexico}\}$$
$$C = \{\text{Tourists visiting Costa Rica}\}$$

Using these sets, we make the Venn diagram shown in **Figure 1.16**. The rectangle represents all the tourists.

The next step is to fill in the numbers we know. Now 10 tourists visit Mexico and not Costa Rica, so this number goes in the "Mexico" circle but outside the "Costa Rica" circle. Similarly, 5 people visit both countries, and this is the number that goes in the overlap of the two circles. In **Figure 1.17**, we have entered these numbers and used n to denote the number who visited Costa Rica but not Mexico. Because the sum of all the numbers in the two circles gives the total number of tourists, we find

$$10 + 5 + n = 17$$

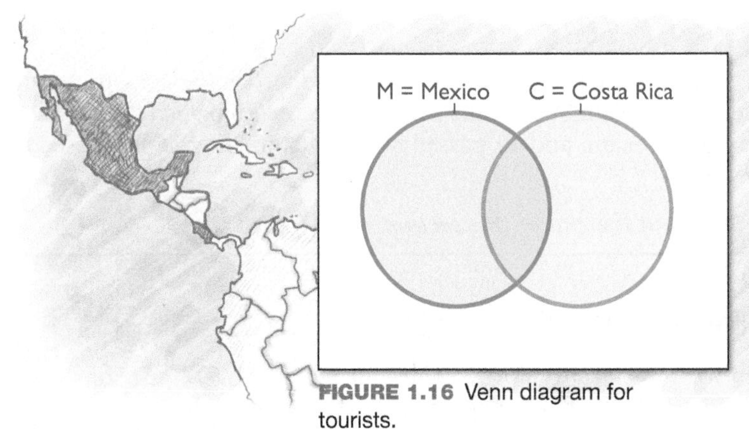

FIGURE 1.16 Venn diagram for tourists.

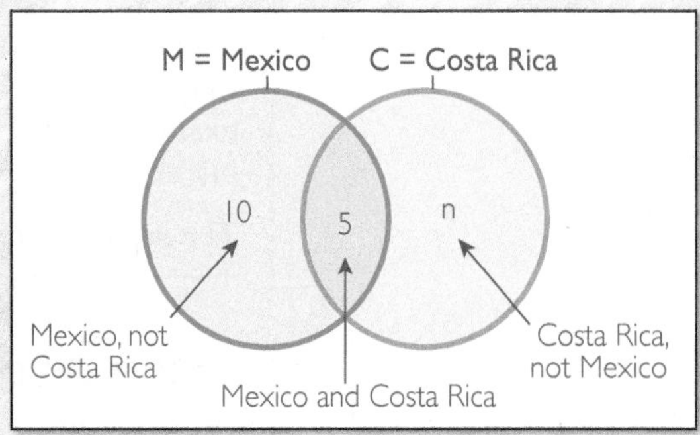

FIGURE 1.17 Entering known values.

which gives $n = 2$. So 2 tourists visited Costa Rica but not Mexico. Adding this number to the 5 tourists who visited both countries gives 7 visitors to Costa Rica.

EXAMPLE 1.18 Counting using Venn diagrams: Exam results

Suppose that 100 high school students are examined in science and mathematics. Here are the results.

- 41 students passed the math exam but not the science exam.
- 19 students passed both exams.
- 8 students failed both exams.

How many students passed the math exam, and how many passed the science exam?

SOLUTION

Let M denote those students who passed the math exam and S those students who passed the science exam. The Venn diagram for these exams is shown in **Figure 1.18**. The rectangle represents all who were examined. We have entered the data we know in **Figure 1.19**. We have put n in the "Passed science, not math" region because we don't know that number yet.

If we add the numbers in all the regions, we get the total number of students, which is 100. Thus,

$$8 + 19 + 41 + n = 100$$

This gives $n = 32$. We know that 41 passed the math exam but not the science exam and that 19 passed both. Hence, $41 + 19 = 60$ passed the math exam. Also, 32 students passed the science exam but not the math exam, and 19 passed both. That gives $32 + 19 = 51$ who passed the science exam.

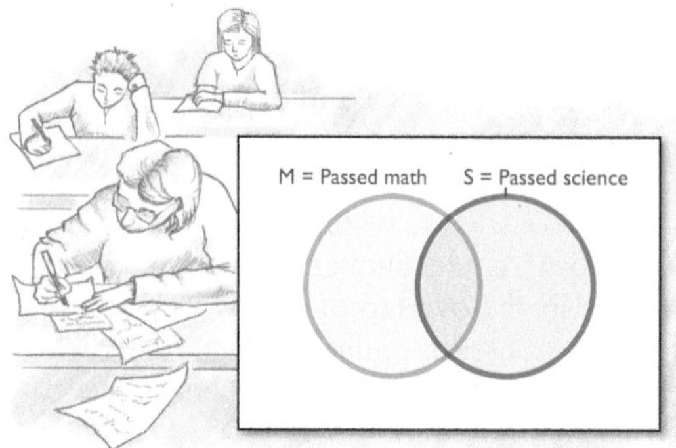

FIGURE 1.18 Venn diagram for science and math exams.

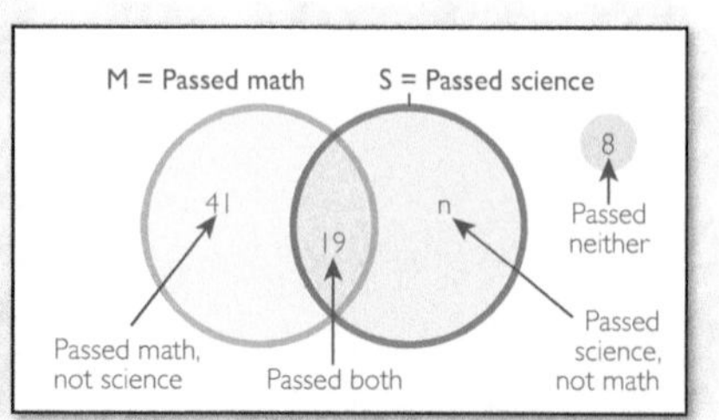

FIGURE 1.19 Entering given data.

TRY IT YOURSELF 1.18

Suppose that 100 students are given a history exam and a sociology exam. Assume that 32 passed the history exam but not the sociology exam, 23 passed the sociology exam but not the history exam, and 12 passed neither exam. How many passed the history exam?

The answer is provided at the end of this section.

Venn diagrams with three sets

A Venn diagram for three sets typically shows all possible overlaps. Consider the Venn diagram in **Figure 1.20**, where the rectangle represents all highway collisions. The three circles show the percentage of collisions resulting from one or a combination of three factors: the vehicle, human error, and the road. For example, a collision may result from faulty brakes, an impaired driver, or a curve that is too tight for the prevailing speed. In the nation, 12% of all collisions involve some factor of the vehicle traveling the roadway; 34% involve some characteristic of the roadway; and 93% are due to human factors.[7]

The various regions of the Venn diagram are labeled in Figure 1.20. For example, the number in the region labeled "Road and human, not vehicle" means that 27% of accidents are due to a combination of road conditions and human error but do not involve vehicular problems.

Having three sets makes our calculations a bit more complicated, but the underlying concept is the same as for two sets. The following example illustrates the procedure.

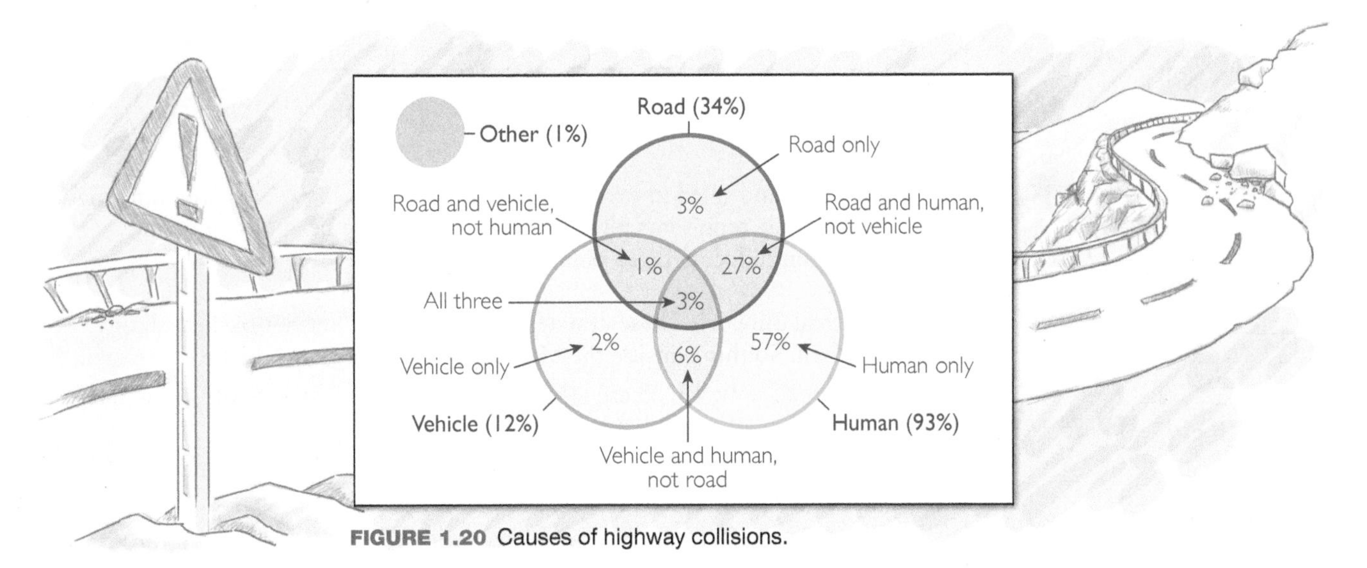

FIGURE 1.20 Causes of highway collisions.

EXAMPLE 1.19 Using three sets: Sports

A survey asked 110 high school athletes which sports they played among football, basketball, and baseball. It showed the following results:

Football only	24
Basketball only	13
Baseball only	12
Football and baseball, but not basketball	9
Football and basketball, but not baseball	5
Basketball and baseball, but not football	10
All three sports	6

a. How many played none of these three sports?

b. How many played exactly two of the three sports?

c. How many played football?

SOLUTION

a. In this example, we denote by F the set of those students who played football, by B the set of those who played basketball, and by S the set of those who played baseball. Using these sets, we have made the Venn diagram in

[7] *This information is taken from the* California 2006 Five Percent Report. *It is available at the Federal Highway Administration Web site.*

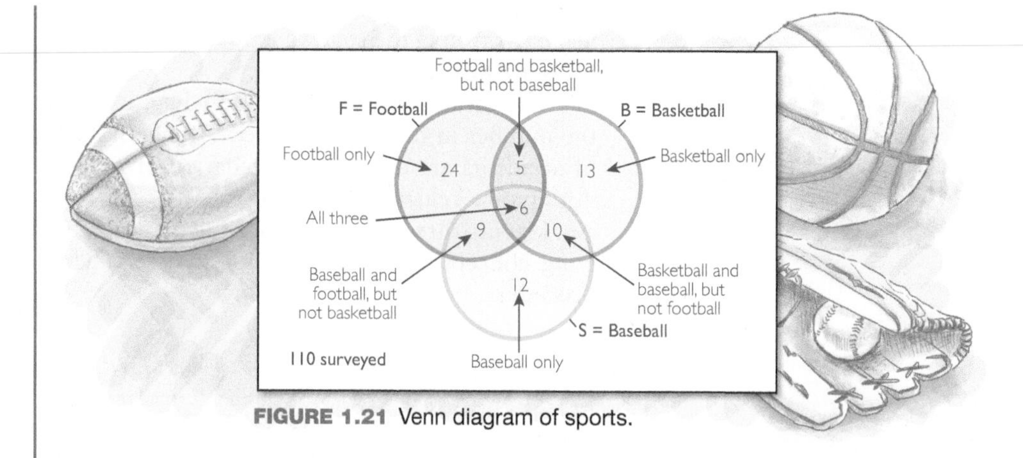

FIGURE 1.21 Venn diagram of sports.

Figure 1.21 and filled in each of the regions for which we are given information. (The rectangle represents all students surveyed.) We used the following information to locate these numbers:

- Football only, 24. These athletes played football but not basketball or baseball. So this is inside the "Football" circle but outside the other two.
- Basketball only, 13. These athletes are inside the "Basketball" circle but outside the other two.
- Baseball only, 12. These athletes are inside the "Baseball" circle but outside the other two.
- Football and baseball, but not basketball, 9. These athletes played football and baseball and so are in the overlap of the "Football" and "Baseball" circles. Because they did not play all three sports, this number goes in the overlap of the "Football" and "Baseball" circles but outside the region common to all three circles.
- Football and basketball, but not baseball, 5. This number goes in the overlap of the "Football" and "Basketball" circles but outside the region common to all three circles.
- Basketball and baseball, but not football, 10. This number goes in the overlap of the "Basketball" and "Baseball" circles but outside the region common to all three circles.
- All three sports, 6. These athletes participated in all three sports, so this number goes in the region common to all three circles.

Now we can answer the question in part a. There is one region that has no number in it. That is the one that is outside all three circles—the number who played none of these three sports. To find this number, we note that if we add up all the numbers, including those who played none of these sports, we get the total number surveyed: 110. So the number of students who played none of these sports is

$$110 - 24 - 13 - 12 - 9 - 5 - 10 - 6 = 31$$

We have added this number to the diagram in **Figure 1.22**.

b. From the completed diagram in Figure 1.22, we see that the numbers 9, 5, and 10 lie in exactly two of the three sets F, B, and S. (So 9 played football and baseball, but not basketball; 5 played football and basketball, but not baseball;

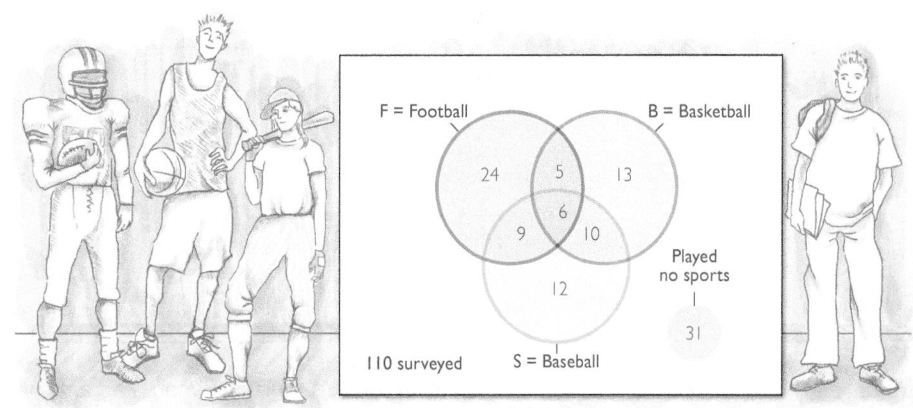

FIGURE 1.22 Adding the number who play none of these sports.

and 10 played basketball and baseball, but not football.) The sum of these is $9+5+10=24$, which is the number of students who play exactly two of the three sports.

c. From Figure 1.22, we see that the set F is divided into regions labeled by four numbers: 24, 9, 6, and 5. When we add these numbers, we obtain the number of athletes who played football: $24+9+6+5=44$.

WHAT DO YOU THINK?

False positives: You have just gotten a positive result from a medical test for a disease, but you are not sure that you really are sick. That is, you think the result may be a false positive. Part of the literature you got from the doctor includes a Venn diagram. It has a circle for "tests positive" and a circle for "has the disease." Each area of the diagram has a percentage associated with it. You notice that there is a large percentage inside the "tests positive" circle but outside the "has the disease" circle. Does this fact reinforce your concern about the test results? Explain your reasoning.

Questions of this sort may invite many correct responses. We present one possibility: Medical tests are not 100% accurate. Sometimes they indicate illness when none is present, and sometimes they fail to detect real illness. Also, some tests are more reliable than others. The percentage in the diagram inside the "tests positive" circle but outside the "has the disease" circle represents the percentage of positive results for healthy people. A large percentage in this category is cause for concern. Positive results for healthy people are called *false positives*. Tests with a large percentage of false positives are to some degree unreliable.

Further questions of this type may be found in the *What Do You Think?* section of exercises.

Try It Yourself answers

Try It Yourself 1.14: Describing sets: States, numbers, and countries

a. Hawaii and Alaska are U.S. states and so are elements of S, but Guam and Puerto Rico are not.

b. $O=\{1, 3, 5, 7, 9\}$.

c. Here A is a subset of N, A and S are disjoint, and N and S are disjoint.

Try It Yourself 1.15: Making Venn diagrams: U.S. states

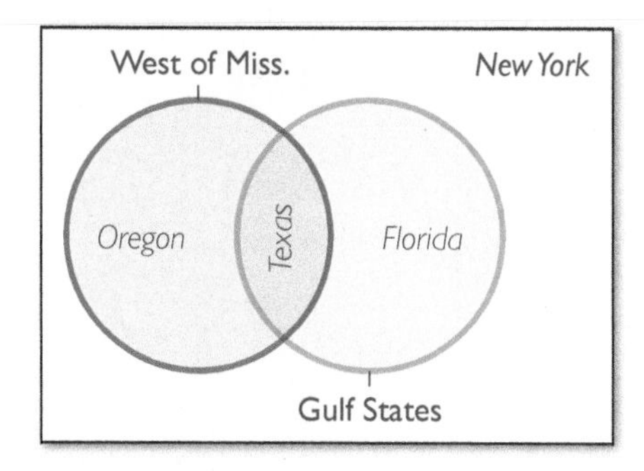

Try It Yourself 1.16: Making Venn diagrams with numbers: Reading proficiency The rectangle represents all who were tested.

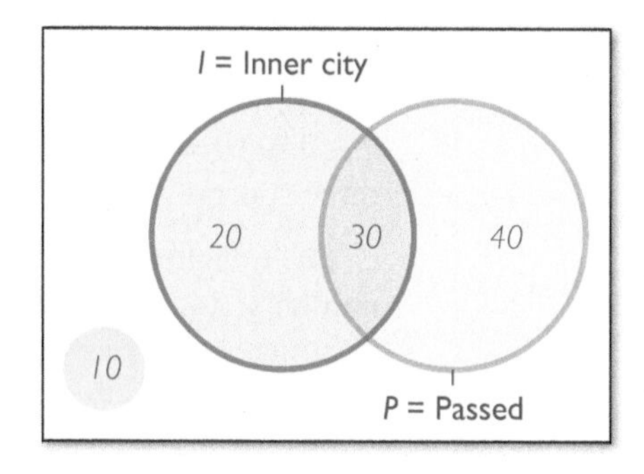

Try It Yourself 1.18: Counting using Venn diagrams: Exam results Sixty-five students passed the history exam.

Exercise Set 1.4

Test Your Understanding

1. Two sets are disjoint provided ____________.

2. A set A is a subset of a set B provided: **a.** every element of B is also an element of A **b.** every element of A is also an element of B **c.** A and B have no elements in common **d.** A has fewer elements than B does.

3. In a Venn diagram, the overlapping region between two circles represents ____________.

4. Consider a Venn diagram where the circle representing the set A is inside the circle representing the set B. How does one describe the relationship between the sets A and B?

Problems

5–8. Listing set elements. *In Exercises 5 through 8, write the given set as a list of elements.*

5. The set of whole numbers between 5 and 9 inclusive

6. The set of even whole numbers between 5 and 9 inclusive

7. The set of consonants in the alphabet

8. The set of states that border the one in which you live

9–11. Relationships among sets. *Exercises 9 through 11 deal with relationships among pairs of sets. Determine which pairs are disjoint and for which pairs one of the sets is a subset of the other.*

9. $A = \{a, b, c, d\}$, $B = \{b, c\}$, and $C = \{d\}$.

10. A is the set of oceans, B is the set of continents, and C is the set {Asia, Pacific}.

11. A is the set of Republican governors, B is the set of governors, and C is the set of elected state officials.

12. **Miscounting.** A store sells only bottled water and sodas (some of which come in bottles). A clerk wants to know how many items a customer wants to buy, so he asks how many sodas and how many bottles there are. The customer answers truthfully that there are 10 sodas and 10 bottles. Should the clerk conclude that there are 20 items altogether? Can the clerk determine the number of items using the given information? Use a Venn diagram to explain your answers.

13. **Countries.** Make a Venn diagram where the category is countries of the world. Use A to denote countries that touch the Atlantic Ocean and N to denote countries of the Northern Hemisphere. On your diagram, locate these countries: United States, Brazil, Japan, and Australia.

14. **Plants.** Make a Venn diagram where the category is animals. Use M for mammals and P for predators. Locate the following on your diagram: tiger, elephant, turtle, and shark.

15. Fruit. For this exercise, the category is fruit. Make a Venn diagram using the set Y of yellow fruit and the set C of citrus fruit. Locate the following fruits in your diagram: banana, cherry, lemon, and orange.

16–23. Using a Venn diagram. *Figure 1.23 shows a Venn diagram for students at a certain college. Let H denote the set of history majors, M the set of male students, and A the set of students receiving financial aid. Exercises 16 through 23 refer to this diagram.*

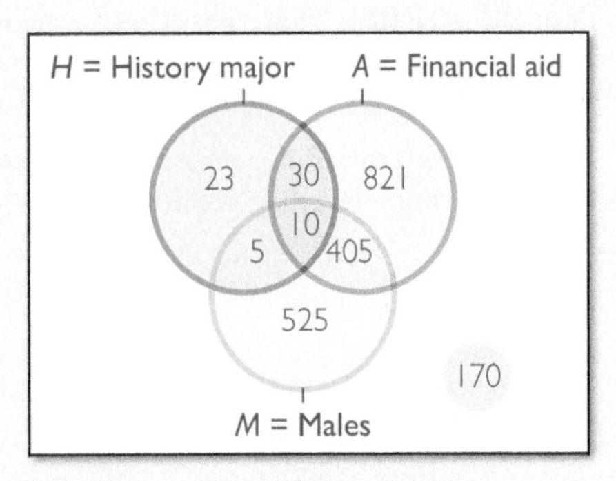

FIGURE 1.23 Students.

16. How many male history majors receive financial aid?

17. How many history majors are there?

18. How many males receive financial aid?

19. How many females receive financial aid?

20. How many history majors do not receive financial aid?

21. How many females are not history majors?

22. How many male students are there?

23. How many students are at this college?

24–28. Logic and Venn diagrams. *Exercises 24 through 28 refer to the Venn diagram in* **Figure 1.24**, *which shows the set H of history majors and the set P of political science majors at a certain college.*

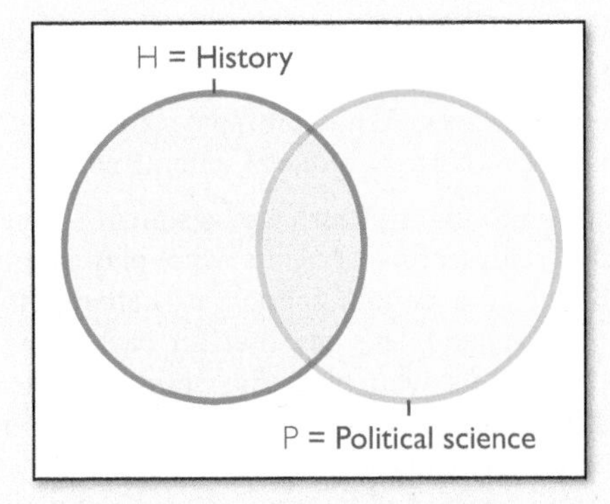

FIGURE 1.24 History and political science majors.

24. Shade in the region of Figure 1.24 where the statement *She is a political science major or a history major* is true.

25. Shade in the region of Figure 1.24 where the statement *She is a political science major but not a history major* is true.

26. Shade in the region of Figure 1.24 where the statement *She is a political science major and a history major* is false.

27. Let G be the set of all students from Greece. Add a circle for G to the Venn diagram in Figure 1.24 so that the statement *All Greek students are history majors but none are political science majors* is true.

28. Let G be the set of all students from Greece. Add a "circle" for G to the Venn diagram in Figure 1.24 so that the statement *Some Greek students are political science majors and some are history majors but none are double majors* is true. (This "circle" will not be round!)

29. PSA test. The following table presents partial data concerning the accuracy of the PSA (Prostate Specific Antigen) test for prostate cancer.[8]

	Has prostate cancer	Does not have prostate cancer
Test positive		2790
Test negative	220	6210

a. Make a Venn diagram in which you locate (without numbers) the following regions. (Various choices for the sets are possible.)

- **True positives:** These are people who test positive and have prostate cancer.
- **False positives:** These are people who test positive but do not have prostate cancer.
- **True negatives:** These are people who test negative and do not have prostate cancer.
- **False negatives:** These are people who test negative but do have prostate cancer.

b. Use the table to fill in the number of people in as many of the regions of your Venn diagram as you can. For which region can you not determine a number?

c. Assume that the total number of people tested was 10,000. Fill in the number you were missing in part b.

d. The *sensitivity* of a medical test is defined to be the number of true positives divided by the number who have the disease. Find the sensitivity of the PSA test. Express your answer as a percentage. *Note*: The sensitivity is one measure of the accuracy of a medical test—accurate tests have a high sensitivity. These topics are discussed in Section 2 of Chapter 5.

30. A drug trial. In a drug trial, a test group of 50 patients is given an experimental drug, and a control group of 40 patients is given a placebo. Assume that 15 patients from the control group improved and 30 from the test group improved. Let C denote the control group and I the group of patients who improved. Make a Venn diagram using C and I. Show the number of patients in each region of the diagram.

[8] *The data are adapted from* www.devicelink.com/ivdt/archive/96/01/005.html.

31. Weather. Over a period of 100 days, conditions were recorded as sunny or cloudy and warm or cold. Let S denote sunny days and W warm days. Of 75 sunny days, 55 were warm. There were only 10 cold and cloudy days. Make a Venn diagram showing the numbers in each region of the diagram. How many days were warm?

32. Cars. Suppose that 178 cars were sorted by model and color. Assume that 10 were Fords but not blue, 72 were blue but not Fords, and 84 were neither blue nor Fords. Make a Venn diagram using Fords and blue cars, and include the numbers in each region of the diagram. How many blue cars were observed?

33. Employed freshmen. A certain college has 11,989 students. Assume that 6120 are unemployed sophomores, juniors, or seniors and that 4104 are employed sophomores, juniors, or seniors. There are 745 employed freshmen. Use F for freshmen and E for employed students, and make a Venn diagram marking the number of students in each region of the diagram. How many freshmen are there?

34. Beverage survey. A survey of 150 students asked which of the following three beverages they liked: sodas, coffee, or bottled water. The results are given in the following table.

Sodas only	28
Coffee only	17
Water	51
Sodas and water	26
Coffee and water	27
Sodas and coffee but not water	9
Coffee and water, not sodas	6
All three	21

a. Use S for soda, W for water, and C for coffee, and make a Venn diagram for drinks. The completed diagram should show the number in each region.

b. How many liked coffee only?

c. How many liked water but not coffee?

d. How many liked none of the three?

e. How many liked coffee and water?

f. How many liked exactly two of the three?

g. How many liked exactly one of the three?

35. News. A survey asked a group of people where they got their news: a newspaper, TV, or the Internet. The responses are recorded in the following table.

Newspaper and TV	46
TV only	20
Internet only	12
Newspaper and Internet	57
TV and Internet	50
Newspapers	81
TV and Internet, but not newspapers	19
All three	31
None of the three	8

a. Using N for a newspaper, T for TV, and I for the Internet, make a Venn diagram for news sources. The completed diagram should give the number in each region of the diagram.

b. How many were surveyed?

c. How many got their news from only one source (among those sources given)?

d. How many got some of their news from the Internet?

e. How many got none of their news from TV?

Writing About Mathematics

36. John Venn. Venn diagrams were introduced in about 1880 by John Venn. Write a report on him.

37. More than three circles. There have been various efforts to make a diagram similar to a Venn diagram for more than three sets. Do some research and report on these efforts.

38. Euler. Look up the terms *Euler diagram* and *Euler circles*. Write a report describing what they are and how they relate to Venn diagrams.

What Do You Think?

39. What does it mean? You see a Venn diagram in an article that shows the demographics of some of your state representatives. You notice that the circle representing African Americans and the circle representing state senators do not meet. What do you conclude from this? **Exercise 26 is relevant to this topic.**

40. Make an example. "All men are mortal. Socrates is a man. Therefore, Socrates is mortal." Make a Venn diagram that shows the logic of this argument. Locate Socrates in your diagram. **Exercise 27 is relevant to this topic.**

41. Four circles? All the Venn diagrams we encountered had two or three circles. Can you explain why you don't see one with four circles?

42. Four categories. The Venn diagram in Figure 1.23 in this section involves the categories of history majors, males, and financial aid. There are eight subsets in this Venn diagram. What if there were a fourth category—varsity athletes? Even though you may not be able to draw a Venn diagram with four circles, how many subsets would there be if you could?

43. Squares. Would squares work as well as circles to make Venn diagrams? Explain your answer.

44. Advantages. What are some advantages of using Venn diagrams to represent sets?

45. More advantages. What are some advantages of using Venn diagrams to represent logical statements?

46. Counting too many. Your class is shown a Venn diagram with circles representing students who play the only three sports offered at a certain school: basketball, hockey, and soccer. For simplicity, let's say that for each sport there are exactly 40 students who play that sport. A fellow student concludes that there are 120 students who play a sport at the school. Explain why that is not necessarily a valid conclusion, both in terms of logic and in terms of the Venn diagram.

47. Counting. Your class is shown a Venn diagram with circles representing students who play the only three sports offered at a certain school: basketball, hockey, and soccer. For simplicity, let's say that for each sport there are exactly 40 students who play that sport. What is the smallest possible number of students who play a sport at the school? What is the largest possible number of students who play a sport at the school? Draw Venn diagrams to represent each extreme.

1.5 Critical thinking and number sense: What do these figures mean?

TAKE AWAY FROM THIS SECTION
Cope with the myriad measurements the average consumer encounters every day.

The following is an excerpt from an article prepared by the Congressional Budget Office. It refers to the deficit in the federal budget.

IN THE NEWS 1.6

Home | TV & Video | U. S. | World | Politics | Entertainment | Tech | Travel

The Budget and Economic Outlook: 2016 to 2026

January 25, 2016

The Budget Deficit for 2016 Will Increase After Six Years of Decline

The 2016 deficit will be $544 billion, CBO estimates, $105 billion more than the deficit recorded last year. At 2.9 percent of gross domestic product (GDP), the expected shortfall for 2016 will mark the first time that the deficit has risen in relation to the size of the economy since peaking at 9.8 percent in 2009. About $43 billion of this year's increase in the deficit results from a shift in the timing of some payments that the government would ordinarily have made in fiscal year 2017, but that will instead be made in fiscal year 2016, because October 1, 2016—the first day of fiscal year 2017—falls on a weekend. If not for that shift, the projected deficit in 2016 would be $500 billion, or 2.7 percent of GDP.

The 2016 deficit that CBO currently projects is $130 billion higher than the one that the agency projected in August 2015. That increase is largely attributable to legislation enacted since August—in particular, the retroactive extension of a number of provisions that reduce corporate and individual income taxes.

An artist's depiction of 1 billion dollars. The pallets contain 10 million hundred-dollar bills.

Amounts of money like the $544 billion deficit are so large that they may become incomprehensible.[9] Perhaps it helps to note that if every man, woman, and

[9] *Everett Dirksen, a senator from Illinois several decades ago, is often quoted as saying, "A billion here, a billion there, and pretty soon you're talking real money."*

child in the state of Oklahoma bought a \$25,000 automobile, they wouldn't come close to spending \$544 billion. In fact, there would be enough money left over to buy \$25,000 automobiles for each living soul in the states of Iowa, Kansas, Louisiana, Minnesota, and Nebraska.

Making informed decisions often depends on understanding what very large or very small numbers really mean. Putting such numbers into a personal context helps us get a sense of their true scale. We can also use estimation to sort our way through otherwise intimidating calculations. For a discussion about problem solving, see Appendix 5.

Magnitudes, very large numbers, and very small numbers

Often we convey information about numbers by using relative sizes or *magnitudes* rather than specific amounts. For example, the number of dollars in your wallet may be counted in tens, the number of students on your campus may be counted in thousands, Earth's population is measured in billions, and the national debt is expressed in trillions of dollars. These examples of magnitudes are integral powers of 10. Some examples will show what powers of 10 mean.

Examples of powers of 10

Positive powers of 10

- $10^3 = 1000$ (three zeros) is a thousand.
- $10^6 = 1{,}000{,}000$ (six zeros) is a million, which is a thousand thousands.
- $10^9 = 1{,}000{,}000{,}000$ (nine zeros) is a billion, which is a thousand millions.
- $10^{12} = 1{,}000{,}000{,}000{,}000$ (twelve zeros) is a trillion, which is a thousand billions.

Negative powers of 10

- $10^{-2} = 0.01$ (two digits to the right of the decimal) is a hundredth.
- $10^{-3} = 0.001$ (three digits to the right of the decimal) is a thousandth.
- $10^{-6} = 0.000\ 001$ (six digits to the right of the decimal) is a millionth.
- $10^{-9} = 0.000\ 000\ 001$ (nine digits to the right of the decimal) is a billionth.

The gigantic number 10^{100} is a "one" followed by 100 zeros and actually has the funny name *googol.* (It's an interesting fact that the name of the company Google is a misspelling of googol.) Physicists believe there are only about 10^{80} atoms in the universe—far fewer than a googol.

To get a better grasp of these magnitudes, we refer to the figures on the facing page. The first figure is a view of space through a frame that is 10^{21} meters on each side.[10] That frame is large enough to show the entire Milky Way galaxy, which has a diameter of about 100,000 lightyears. (A lightyear is the distance light travels in a year, about 10^{16} meters.) The pictures decrease in scale by powers of 10 until we finally view the structure of a DNA molecule through a frame that is 10^{-9} meter on a side.

Dealing with powers of 10 sometimes requires a bit of algebra with exponents. The following quick review is included for those who need a refresher. Additional information about exponents, including scientific notation, is provided in Appendix 2.

[10] *One meter is about 3.28 feet.*

Quick Review Exponents

Negative exponents: a^{-n} is the reciprocal of a^n.

Definition	Example
$a^{-n} = \dfrac{1}{a^n}$	$10^{-3} = \dfrac{1}{10^3} = \dfrac{1}{1000} = 0.001$

Zero exponent: If $a \neq 0$, then $a^0 = 1$.

Definition	Example
$a^0 = 1$	$10^0 = 1$

Basic properties of exponents:

Property	Example
$a^p a^q = a^{p+q}$	$10^2 \times 10^3 = 10^{2+3} = 10^5 = 100{,}000$
$\dfrac{a^p}{a^q} = a^{p-q}$	$\dfrac{10^6}{10^4} = 10^{6-4} = 10^2 = 100$
$(a^p)^q = a^{pq}$	$(10^3)^2 = 10^{3\times 2} = 10^6 = 1{,}000{,}000$

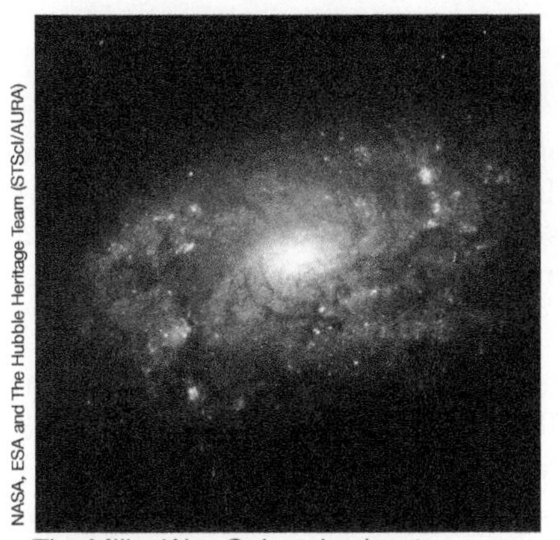
NASA, ESA and The Hubble Heritage Team (STScI/AURA)

The Milky Way Galaxy is about 10^{21} meters across.

NASA Goddard Space Flight Center

The diameter of Earth is just over 10^7 meters.

NASA

San Francisco Bay is about 10^5 meters, or 100 kilometers, long.

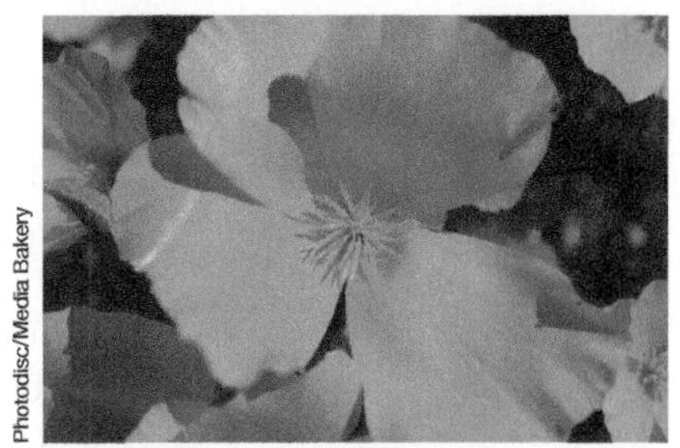
Photodisc/Media Bakery

Some lily flowers are about 0.1, or 10^{-1}, meter across.

Darlyne A. Murawski/National Geographic Creative

The eye of a bee is about 10^{-3} meter across.

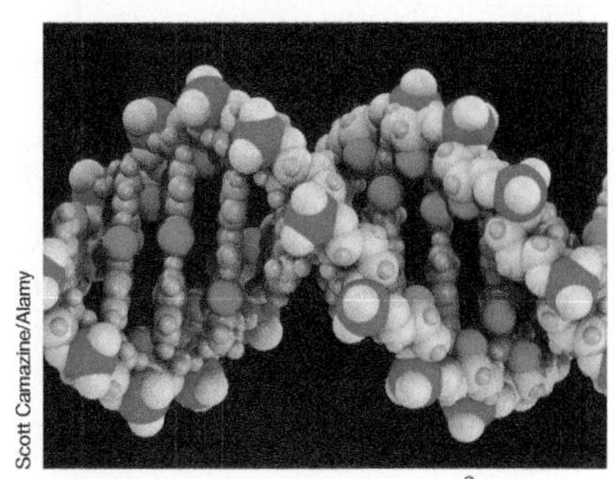
Scott Camazine/Alamy

A DNA molecule is about 10^{-9} meter across.

A familiar example of magnitudes is provided by computer memory, which is measured in terms of *bytes*. A single byte holds one character (such as a letter of the alphabet). Modern computing devices have such large memories that we measure them using the following terms:

A *kilobyte* is a thousand $= 10^3$ bytes, usually abbreviated *KB*.
A *megabyte* is a million $= 10^6$ bytes, usually abbreviated *MB* or *meg*.
A *gigabyte* is a billion $= 10^9$ bytes, usually abbreviated *GB* or *gig*.
A *terabyte* is a trillion $= 10^{12}$ bytes, usually abbreviated *TB*.

To find out how many megabytes are in a gigabyte, we divide:

$$\begin{aligned}\text{Numbers of megs in a gig} &= \frac{10^9 \text{ bytes}}{10^6 \text{ bytes}} \\ &= 10^{9-6} \\ &= 10^3\end{aligned}$$

Thus, there are 1000 megabytes in a gigabyte. Similarly, we calculate that there are a million megabytes in a terabyte.

EXAMPLE 1.20 Comparing sizes using powers of 10: Computer memory

In the 1980s, one of this book's authors owned a microcomputer that had a memory of 64 kilobytes. The computer on his desk today has 8 gigabytes of RAM. Is the memory of today's computer tens, hundreds, or thousands of times as large as the old computer's memory?

SOLUTION

We express the size of the memory using powers of 10. The old computer had a memory of 64×10^3 bytes, and today's computer has 8×10^9 bytes of memory. To compare the sizes, we divide:

$$\begin{aligned}\frac{\text{New memory size}}{\text{Old memory size}} &= \frac{8 \times 10^9 \text{ bytes}}{64 \times 10^3 \text{ bytes}} \\ &= \frac{8}{64} \times 10^{9-3} \\ &= 0.125 \times 10^6 \\ &= 125{,}000\end{aligned}$$

The new computer has 125 thousand times as large a memory as the old computer.

TRY IT YOURSELF 1.20

One kilobyte of computer memory will hold about 2/3 of a page of typical text (without formatting). A typical book has about 400 pages. Can the 9-gig flash drive I carry in my pocket hold tens, hundreds, or thousands of such books?

The answer is provided at the end of this section.

Taming those very large and very small numbers

Getting a handle on large numbers (or very small numbers) often means expressing them in familiar terms. For example, to get a feeling for how much a billion is, let's change the distance of a billion inches into more meaningful terms.[11] There are

[11] *Unit conversion is reviewed in Appendix 1.*

12 inches in a foot and 5280 feet in a mile. So, we change inches to feet by dividing by 12 and feet to miles by dividing by 5280. A billion inches is 10^9 inches. Therefore, a billion inches equals

$$\begin{aligned} 10^9 \text{ inches} &= 10^9 \times \frac{1}{12} \text{ feet} \\ &= 10^9 \times \frac{1}{12} \times \frac{1}{5280} \text{ miles} \end{aligned}$$

This is about 16,000 miles. The length of the equator of Earth is about 25,000 miles, so 1 billion inches is about 2/3 of the length of the equator of Earth.

As another illustration, suppose you have a lot of dollar bills—1 billion of them. If you put your dollars in a stack on the kitchen table, how tall would the stack be? A dollar bill is about 0.0043 or 4.3×10^{-3} inch thick. A billion is 10^9, so the height of the stack would be

$$4.3 \times 10^{-3} \times 10^9 = 4.3 \times 10^{-3+9} = 4.3 \times 10^6 = 4.3 \text{ million inches}$$

As before, we divide by 12 to convert to feet and then by 5280 to convert to miles:

$$\begin{aligned} 4.3 \times 10^6 \text{ inches} &= 4.3 \times 10^6 \times \frac{1}{12} \text{ feet} \\ &= 4.3 \times 10^6 \times \frac{1}{12} \times \frac{1}{5280} \text{ miles} \end{aligned}$$

This is about 67.9 miles, so 1 billion dollar bills make a stack almost 68 miles high.

EXAMPLE 1.21 Understanding large numbers: The national debt

As of 2020, the national debt was about 27.8 trillion dollars, and there were about 330 million people in the United States. The national debt is not just an abstract figure. The American people actually owe it. Determine how much each person in the United States owes. Round your answer to the nearest $100.

SOLUTION

We use powers of 10 to express both the national debt ($27.8 \text{ trillion} = 27.8 \times 10^{12}$) and the population of the United States ($330 \text{ million} = 330 \times 10^6$):

$$\text{Debt per person} = \frac{27.8 \times 10^{12} \text{ dollars}}{330 \times 10^6}$$
$$= \frac{27.8}{330} \times 10^{12-6} \text{ dollars}$$
$$= \frac{27.8}{330} \times 10^6 \text{ dollars}$$

This is about 0.0842×10^6 dollars, so each person owes $84,200.

TRY IT YOURSELF 1.21

The U.S. Census Bureau estimated that there were approximately 129,000,000 households in the United States in 2020. How much does each household owe on the national debt? Round your answer to the nearest $100.

The answer is provided at the end of this section.

At the other end of the scale from numbers describing the national debt or computer memory are extremely small numbers. Microchips that can be implanted in pets are about 0.0015 meter in diameter. That is 1.5 millimeters—about the size of an uncooked grain of rice. The eye of a bee, which is about 1 millimeter across (see the figure on p. 63), may provide additional context for this magnitude.

A smaller magnitude is a *micron*, which is one-millionth of a meter, or 10^{-6} meter. In this scale, you could get a close-up view of a bacterium.

Even smaller are the particles used in the developing field of *nanotechnology*, which is the control of matter on an atomic or molecular scale. It may be possible to develop *nanoparticles* in the 10- to 100-*nanometer* range that can seek out and selectively destroy cancer cells. One nanometer is 10^{-9} meter, a billionth of a meter. On this scale, we can view the structure of DNA, as we saw in the illustration on p. 63. We further develop our sense of these very small sizes in the next example.

EXAMPLE 1.22 Understanding very small numbers: Nanoparticles

Human hair can vary in diameter, but one estimate of the average diameter is 50 microns, or 50×10^{-6} meter. How many 10-nanometer particles could be stacked across the diameter of a human hair?

SOLUTION

Now, 10 nanometers is 10×10^{-9} meter. We divide to find the number of 10-nanometer particles that can be stacked across the diameter of a human hair:

$$\frac{\text{Diameter of hair}}{\text{Diameter of nanoparticle}} = \frac{50 \times 10^{-6} \text{ meter}}{10 \times 10^{-9} \text{ meter}}$$
$$= 5 \times 10^{-6+9}$$
$$= 5 \times 10^3$$

Thus, 5000 of the nanoparticles would be needed to span the diameter of a human hair.

TRY IT YOURSELF 1.22

Suppose the universe expanded so that the 10-nanometer particle grew to the size of a 15-millimeter (15×10^{-3} meter) diameter marble. If a 15-millimeter marble grew in the same proportion, how many meters long would its new diameter be? How many miles long would the new diameter be? (Use the fact that 1 kilometer is about 0.62 mile, and round your answer to the nearest mile.)

The answer is provided at the end of this section.

Goddard/cartoonstock.com

Decisions based on estimates

There are many occasions when a quick, ballpark figure will suffice—an exact calculation is not necessary. In fact, many of the examples already discussed in this section involve estimates. What we will pursue here are examples that a person would tend to encounter in everyday life.

For instance, if a salesperson is showing us a car, we need to know whether it really is in our price range. Suppose we are looking at a car with a price tag of \$23,344 and we plan to pay off the car in 48 monthly payments at an annual interest rate of 6%. We may need help figuring out the exact monthly payment, especially considering the interest that will be charged for the loan. But we can get a rough estimate by approximating the numbers by ones that make the calculation easier.

Suppose we estimate the cost of the car to be \$25,000 and round off the 48 monthly payments to 50 monthly payments. These numbers are much easier to handle, and we should be able to figure in our heads what the monthly payments using them would be. If we ignore interest, we find

$$\text{Estimated monthly payment} = \frac{\text{Total cost}}{\text{Number of payments}} = \frac{\$25{,}000}{50} = \$500$$

In Chapter 4, we will learn a formula for the monthly payment in such a situation. According to that formula, borrowing \$23,344 at an annual interest rate of 6% and repaying in 48 monthly payments result in a monthly payment of \$548.23. So our estimate was a bit low. But if we can afford (say) only \$450 per month, our estimate warns us to look for a less expensive car.

EXAMPLE 1.23 Estimating costs: Buying gold

You are interested in investing in gold. Vendor 1 offers 1 ounce of gold for \$1400. Vendor 2 offers 30 grams of gold for \$1410. You look on the Internet and find that 1 ounce is 28.35 grams.

a. How would you make a quick estimate to compare the two offers?

b. Find the price per gram for each offer to get an accurate comparison of the two offers.

SOLUTION

a. Because 30 grams is more than 28.35 grams, the second vendor is selling almost 2 additional grams of gold for an additional price of only $10. Vendor 2 appears to be offering the lower price.

b. Vendor 1 sells 28.35 grams for $1400. To get the cost of 1 gram of gold, we divide by 28.35:

$$\text{Price per gram: Vendor 1} = \frac{1400}{28.35} = 49.38 \text{ dollars}$$

Vendor 2 sells 30 grams of gold for $1410. To get the cost of 1 gram of gold, we divide by 30:

$$\text{Price per gram: Vendor 2} = \frac{1410}{30} = 47.00 \text{ dollars}$$

This shows that our estimate from part a would have led us to the correct conclusion. Vendor 2 has the lower price.

TRY IT YOURSELF 1.23

You are shopping for a washing machine, and you can afford about $500. A machine that lists for $550 is on sale for 11% off. How would you decide quickly whether this washer is in your price range?

The answer is provided at the end of this section.

EXAMPLE 1.24 Estimating costs: Buying gas in the United States and Canada

You are traveling to Canada, and you wonder whether to fill up before you cross the border. You see a sign at a gas station on the U.S. side of the border showing $2.30, and on the Canadian side of the border, you see a sign showing $1.19. One might think that the Canadian gas is much cheaper; however, let's use critical thinking and look a little closer.

In Canada (and most of the rest of the world), gasoline is measured in liters rather than gallons. Also, the dollar sign refers not to U.S. dollars but to *Canadian* dollars. So the sign on the Canadian side of the border means that gasoline costs 1.19 Canadian dollars per liter.

In parts a and b, use an exchange rate of 0.78 U.S. dollar per Canadian dollar.

a. Use the fact that a quart and a liter are nearly the same and that the Canadian dollar is worth a little less than the U.S. dollar to get a quick estimate of the cost of gasoline in Canada measured in U.S. dollars per gallon.

b. There are 3.79 liters in a gallon. Use this fact to find the actual cost in U.S. dollars per gallon of gasoline at the Canadian station. Was your estimate in part a good enough to tell you where you should buy your gas?

SOLUTION

a. Because 1 Canadian dollar is worth about 1 U.S. dollar and 1 liter costs about 1 Canadian dollar, gas costs about 1 U.S. dollar per liter. Because a quart is about the same as a liter, there are about 4 liters in a gallon. This means that gas in Canada costs approximately 4 U.S. dollars per gallon. This estimate suggests that gas is cheaper at the station in the United States.

b. Because there are 3.79 liters in a gallon, gasoline in Canada costs

$$3.79\frac{\cancel{\text{liters}}}{\text{gallon}} \times 1.19\frac{\text{Canadian dollars}}{\cancel{\text{liter}}} = 4.5101\frac{\text{Canadian dollars}}{\text{gallon}}$$

One way to check that the units are converted correctly is to include them in the computation, as we did here, and combine or cancel them using the usual rules of algebra. Note that the units involving liters cancel. Further information on unit conversion is in Appendix 1.

Because each Canadian dollar is worth 0.78 U.S. dollar, we multiply by 0.78 to get the cost in U.S. dollars of a gallon of gas in Canada:

$$\text{Cost in U.S. dollars} = 4.5101 \times 0.78$$

This is about 3.52, so the actual cost of gas on the Canadian side of the border is 3.52 U.S. dollars per gallon. This calculation confirms that gas is cheaper at the station in the United States.

Our estimate in part a was good enough to tell us that gas on the American side was the better buy.

Areas and volumes are important measurements for consumers. For example, carpet is priced in both square feet and square yards. There are 3 feet in a yard. How many square feet are in a square yard? In **Figure 1.25**, we have made a picture of a square yard. It has a length and width of 3 feet. Recall that the area of a rectangle is the length times the width. So the area in square feet of a square yard is

$$\begin{aligned}\text{Area of 1 square yard} &= \text{Length} \times \text{Width}\\ &= 3\text{ feet} \times 3\text{ feet}\\ &= 9\text{ square feet}\end{aligned}$$

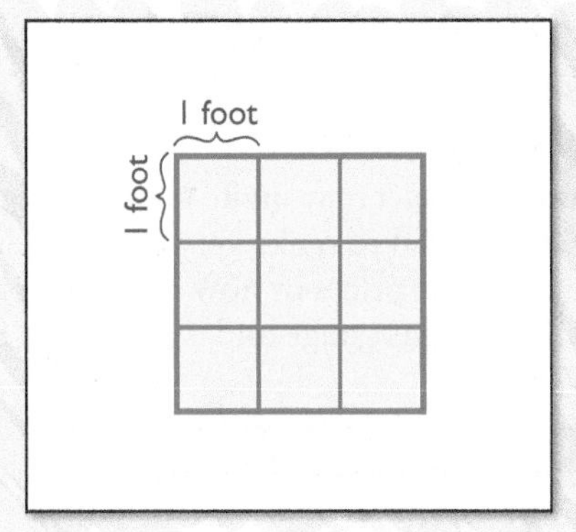

FIGURE 1.25 1 square yard is 9 square feet.

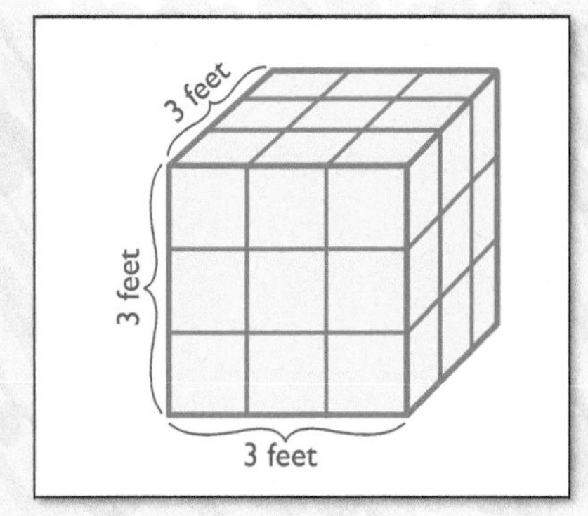

FIGURE 1.26 1 cubic yard is 27 cubic feet.

Concrete is usually priced by the cubic yard.[12] **Figure 1.26** shows that 1 cubic yard is a cube that is 3 feet on each side. Recall that the volume of a box is the length times the width times the height. So the volume in cubic feet of a cubic yard is

$$\begin{aligned}\text{Volume of 1 cubic yard} &= \text{Length} \times \text{Width} \times \text{Height}\\ &= 3 \text{ feet} \times 3 \text{ feet} \times 3 \text{ feet}\\ &= 27 \text{ cubic feet}\end{aligned}$$

So 1 square yard is 9 square feet, and 1 cubic yard is 27 cubic feet.

The next example illustrates that savvy consumers need both the skill of estimation and the ability to measure areas and volumes.

EXAMPLE 1.25 Estimating with areas and volumes: Flooring

You want to redo your living room floor. Hardwood costs $10.40 per square foot, and the carpet you like costs $28.00 per (square) yard.[13]

a. You need to decide right away whether you should be looking at hardwood or carpet for your floor. Use the fact that the cost of hardwood is about $10 per square foot to estimate the cost of a square yard of hardwood. Use your estimate to decide how the cost of hardwood compares with the cost of carpet.

b. Find the actual cost of a square yard of hardwood.

SOLUTION

a. A common (and ultimately expensive) consumer error is to think that, because 1 yard is 3 feet, 1 square yard is 3 square feet. But as we noted above, 1 square yard is, in fact, 9 square feet. Because hardwood costs about $10 per square foot, 9 square feet costs about $90. Because the carpet is $28 per yard, we estimate that hardwood is more than three times as expensive as carpet.

b. To find the actual cost of a square yard of hardwood, we multiply the exact cost of $10.40 per square foot by 9 square feet:

$$\$10.40 \times 9 = \$93.60$$

The result is a bit higher than the estimate of $90 we found in part a, and it confirms our conclusion that hardwood is much more expensive than carpet.

TRY IT YOURSELF 1.25

A German sports car advertises an engine displacement of 3500 cubic centimeters. An American sports car boasts a 302-cubic-inch engine. Use the fact that 1 centimeter is 0.39 inch to determine which displacement is larger.

The answer is provided at the end of this section.

WHAT DO YOU THINK?

Estimation: Calculators are now readily available. Is there still any value in being able to make estimations?

Questions of this sort may invite many correct responses. We present one possibility: Many times we need to make very quick decisions. Exit here or take the next exit? It may depend on gas prices or how much gas is left in our tank. We may need to take a quick look at the gas gauge and think about our mileage to decide if we need to stop

[12] *Often sellers of concrete quote the price "per yard." What they really mean is the price per* cubic *yard.*

[13] *Often sellers of carpet quote the price "per yard." What they really mean is the price per* square *yard.*

or proceed to our destination. Calculators or iPhones may not be available and use of them is certainly not advisable while driving. Moreover, although they can give accurate answers, we may need to make a decision before we can access technology.

Further questions of this type may be found in the *What Do You Think?* section of exercises.

Try It Yourself answers

Try It Yourself 1.20: Comparing sizes using powers of 10: Computer memory Thousands: The flash drive will hold about 15,000 such books.

Try It Yourself 1.21: Understanding large numbers: The national debt $215,500.

Try It Yourself 1.22: Understanding very small numbers: Nanoparticles 22,500 meters or 14 miles.

Try It Yourself 1.23: Estimating costs: Washing machine 11% off is a bit more than 10% or $55 off. Thus, the washer is in your price range.

Try It Yourself 1.25: Estimating with areas and volumes: Engine displacement The American engine is larger because the German engine is about 208 cubic inches.

Exercise Set 1.5

Test Your Understanding

1. Explain in your own words why scientific notation is used to express very large and very small numbers.

2. Explain how you could use powers of 10 to obtain a rough estimate of

$$\frac{1{,}002{,}001{,}236}{10{,}101{,}344}$$

3. At European service stations, you will see gasoline priced by the *liter*. You can estimate the price per gallon by using which of the following? **a.** A gallon is about the same as a liter. **b.** A liter is about the same as a quart. **c.** There are two liters in a gallon.

4. Raising the number 10 to a negative power gives: **a.** a negative number. **b.** a very large number. **c.** a very small number.

5. You found hardwood flooring priced by the square foot. To find the price per square yard, you should: **a.** multiply the price per square foot by 9. **b.** multiply the price per square foot by 3. **c.** divide the price per square foot by 9. **d.** divide the price per square foot by 3.

6. If you know the price of a car, explain how you can find a quick estimate of your monthly payment.

Problems

7. **Rhinoviruses.** Rhinoviruses, which are the leading cause of the common cold, have a diameter of 20 nanometers (20×10^{-9} meter). Pneumococcus, a bacterium causing pneumonia, measures about 1 micron (10^{-6} meter) in diameter. How many times as large in diameter as the virus is the bacterium?

8. **Trillion.** How many millions are in a trillion?

9. **Pennies.** Some high school students want to collect a million pennies for charity. How many dollars is that?

10. **Oklahoma.** In the spring of 2016 the state of Oklahoma was facing a budget deficit of $1.3 billion. The population of the state was approximately 3.9 million. How much would each person in the state have to contribute to pay a $1.3 billion deficit? Round your answer to the nearest dollar.

11–16. **A billion dollars.** *A 1-dollar bill is 6.14 inches long, 2.61 inches wide, and 0.0043 inch thick. This information is used in Exercises 11 through 16.*

11. If you had a billion dollars in 100-dollar bills, how many bills would you have?

12. If you laid a billion 1-dollar bills side by side, how many miles long would the trail be? Round your answer to the nearest mile.

13. If you laid a billion 1-dollar bills end to end, how many miles long would the trail be? (Round your answer to the nearest mile.) About how many times around Earth would this string of bills reach? (The circumference of Earth is approximately 25,000 miles.)

14. If you spread a billion 1-dollar bills over the ground, how many square miles would they cover? Round your answer to the nearest square mile.

15. How many cubic feet would a billion dollars occupy? Round your answer to the nearest cubic foot.

16. Measure your classroom. How many such rooms would a billion 1-dollar bills fill?

17. Counting to a billion. Let's say it takes about 3 seconds to count to 10 out loud. How many years will it take to count out loud to a billion? Ignore leap years and round your answer to one decimal place.

18. Queue. Suppose we lined up all the people in the world (about 7.8 billion people). Assume that each person occupied 12 inches of the line. How many times around Earth would they reach? (Earth has a circumference of about 25,000 miles.) Round your answer to one decimal place.

19. Use of social media. The use of social media has grown rapidly. According to one report in August 2015, about 1 million active users of social media via mobile devices were added each day.[14] How many users were added, on average, each second? Round your answer to the nearest whole number.

20. Lightyear. A lightyear is the distance light travels in a year. Light travels 186,000 miles in a second. How many trillion miles long is a lightyear? Round your answer to one decimal place.

21. Mars rovers. The Mars rovers Curiosity and Perseverance are remotely controlled from Earth, and the signal is delayed due to the great distance. The signal travels at the speed of light, which is 186,000 miles in a second. Assume that Mars is about 140 million miles from Earth when a controller sends a command to a rover. How long will it take for the controller to know what actually happened? In other words, how long will it take for the signal to make the 140-million-mile trip to Mars and back? Give your answer to the nearest minute.

22. Light. Light travels 186,000 miles in a second. How many feet does it travel in a nanosecond? (A nanosecond is a billionth of a second.) Round your answer to the nearest foot.

23. You try. Several ways to convey the size of a billion were explored in this section. Try to find other effective ways to convey the size of a billion.

24. Loan I. You are borrowing $7250 to buy a car. If you pay the loan off in 3 years (36 months), estimate your monthly payment by ignoring interest and rounding down the amount owed.

25. Loan II. You have a $3100 student loan. Suppose you pay the loan off in two and a half years (30 months). Estimate your monthly payment by ignoring interest and rounding down the amount owed.

26. Gas. You're going on a 1211-mile trip, and your car gets about 29 miles per gallon. Gas prices along your route average $3.10. Estimate the cost of gasoline for this trip by rounding the distance to 1200 miles, the mileage to 30 miles per gallon, and the cost of gas to $3.00 per gallon. Calculate the answer without using estimation and compare the two results.

27. State population. In 2020, the population of Iowa was about 3.2 million, and the population of the United States was about 330 million. We want to determine what percentage of the U.S. population was the population of Iowa. First estimate this percentage by rounding the population of Iowa to 3.3 million. Then calculate the percentage without using estimation, and compare the results. Round your answers in percentage form to two decimal places.

28. Oil. In 2020, the United States consumed 18,100,000 barrels of oil per day. A barrel of oil is 42 U.S. gallons. In 2020, the population of the United States was about 330,000,000. Estimate how many gallons each person in the United States consumed each day in 2020 by rounding down all three numbers, using 300,000,000 for the population. Calculate the answer without using estimation and compare the two results. Round your answers to two decimal places.

29. Carpet. You need to order carpet. One store advertises carpet at a price of $1.50 per square foot. Another store has the same carpet advertised at $12.00 "per yard." (What the store really means is the price per *square* yard.) Which is the better buy?

30–32. World population. *As of 2020, the world's population was about 7.8 billion people. The population of the United States was 330 million. This information is used in Exercises 30 through 32.*

30. In 2020, what percent of the world's population was the U.S. population? Round your answer as a percentage to one decimal place.

[14] *See* www.wearesocial.com.

31. The land area of the state of Texas is about 261,797 square miles. If in 2020 all the world's people were put into Texas, how many people would there be in each square mile? Round your answer to the nearest whole number.

32. The area of New York City is 305 square miles. If in 2020 every person in the world moved into New York City, how many square feet of space would each person have? Round your answer to one decimal place. *Note*: There are 5280 feet in a mile. First determine how many square feet are in a square mile.

33. A zettabyte. A recent article refers to a statement by H. Sebastian Seung that, using a three-dimensional representation, the amount of information at the cellular level of the human brain and all its connections would be a *zettabyte*.[15] This newly coined word means a trillion gigabytes, or about 75 billion 16-gigabyte iPads. How many bytes are in a zettabyte? Express your answer as a power of 10.

Writing About Mathematics

34. Nanotechnology. Write about modern developments in nanotechnology.

35. Disk drives. Write about the development of disk drives in computers. Pay particular attention to the evolution in drive capacity.

36. Moore's law. Gordon Moore was one of the founders of Intel Corporation. He proposed something that is often called *Moore's law*. Look up Moore's law and write a report on it as well as on the man.

37. National debt. Do some research to find historical levels of the national debt as well as projections for the future. Report on your findings.

What Do You Think?

38. That's a lot. Physicists estimate that there are between 10^{78} and 10^{82} atoms in the universe. How do these two numbers compare with one another? If the larger number is right, how many universes with the smaller number of atoms would fit inside the larger universe? **Exercise 8 is relevant to this topic.**

39. The iPad. As of 2020, the iPad comes in six available memory sizes: 32 GB, 64 GB, 128 GB, 256 GB, 512 GB, and 1 TB. Do some research to figure out how many movies you could put on each model.

40. A grain of rice. How long, in meters, is a grain of rice?

41. Deficit. The cartoon just above Example 1.21 on page 65 jokes that the government will reduce the deficit by moving the decimal two points to the left. If the deficit were originally 1.5 trillion dollars, what would the deficit be if the decimal were moved two points to the left?

42. Days? A friend asks you how many days there are in a lightyear. What do you say? **Exercise 20 is relevant to this topic.**

43. Millionaires. According to one Web site, in 2015 there were about 10.4 million households in the United States with a net worth of a million dollars or more. How much total wealth would they have if they each had exactly a million dollars?

44. Powers of 10. In this section we compare magnitudes using powers of 10. Would it matter if we used powers of some other number, such as 2? Can you think of any subject, or type of object, that does make comparisons based on powers of 2?

45. Flooring. You ask a sales representative about the cost of flooring, and the answer is \$10.50. What crucial piece of information is missing?

46. Debt. How would you use what you learned in this section to help someone understand the size of the U.S. federal debt?

[15] www.nytimes.com/2014/05/27/science/all-circuits-are-busy.html?

CHAPTER SUMMARY

Today's media sources bombard us with information and misinformation, often with the aim of selling us products, ideas, or political positions. Sorting the good from the bad is a constant challenge that requires a critical eye. Solid reasoning is a crucial tool for success.

Public policy and Simpson's paradox: Is "average" always average?

Simpson's paradox is a striking example of the need for critical thinking skills. An average based on combining categories can be skewed by some factor that distorts the overall picture but not the underlying patterns. Considering only the overall average may lead to invalid conclusions. A simple, but hypothetical, example is given by the following table, which shows applicants for medical school and law school.

	Males			Females		
	Applicants	Accepted	% accepted	Applicants	Accepted	% accepted
Medical school	10	2	20.0%	20	4	20.0%
Law school	20	5	25.0%	12	3	25.0%
Total	30	7	23.3%	32	7	21.9%

Note that the medical school accepts 20% of male applicants and 20% of female applicants. The law school accepts 25% of male applicants and 25% of female applicants. But overall 23.3% of male applicants are accepted and only 21.9% of female applicants are accepted. The discrepancy is explained by the fact that females tended to apply to the medical school and males tended to apply to the law school—and the medical school accepts fewer applicants (on a percentage basis) than does the law school. The overall average might lead one to suspect gender bias where none exists.

Logic and informal fallacies: Does that argument hold water?

Logic is the study of methods and principles used to distinguish good reasoning from bad. A logical argument involves one or more *premises* (or *hypotheses*) and a *conclusion*. The premises are used to justify the conclusion. The argument is *valid* if the premises properly support the conclusion.

People who try to sway our opinions regarding products, politics, or public policy sometimes use flawed logic to bolster their case. The well-informed citizen can recognize phony arguments. We consider two main types of fallacies: *fallacies of relevance* and *fallacies of presumption*. A classic fallacy of presumption is *false cause*. For example, a commercial may tell us that people who drive our sports cars have lots of friends. The implication is that you will have more friends if you buy their car. But, in fact, no evidence is offered to show that either event *causes* the other. The fact that two events occur together does not automatically mean that one causes the other. There are several other informal fallacies.

An argument of the form *if premises, then conclusion* is a *deductive* argument. An *inductive* argument draws a general conclusion from specific examples. It does not claim that the premises give irrefutable proof of the conclusion, only that the premises provide plausible support for it. The idea of an inductive argument is closely tied to pattern recognition.

Formal logic and truth tables: Do computers think?

Logic has a formal structure that may remind us of algebra. This structure focuses on the way statements are combined, and it is the basis on which computers operate. One feature of formal logic is the *truth table*, which gives the truth or falsity of a compound statement based on the truth value of its components.

Operations on statements include *negation*, *conjunction*, and *disjunction*. Of particular interest is the *conditional statement* or *implication*. Such statements have the form "if *hypothesis*, then *conclusion*." One example is, "If I am elected, then your taxes will decrease." Analyzing conditional statements in terms of the converse, the inverse, and the contrapositive can be useful. The politician who made the preceding statement may wish you to believe that if she is not elected, then your taxes will not decrease. No such conclusion is warranted: Replacing a statement by its inverse is not valid.

It is important to remember that a conditional statement is considered to be true if the hypothesis is false. For example, consider again the following statement: *If I am elected, then your taxes will decrease.* This statement is true if the politician is not elected—regardless of what happens to your taxes.

Computers use *logic gates* that perform functions corresponding to logical operations. In classical computing, information is stored using the two numbers 0 and 1. The connection with logic is made by thinking of 0 as "false" and 1 as "true."

Sets and Venn diagrams: Pictorial logic

Sets can be used, among other things, to analyze logical statements. The basic tool for this purpose is the Venn diagram. Consider this statement: *If you speed, then you will get a ticket.* We can represent it using the Venn diagram shown in **Figure 1.27**. One circle represents the collection of all speeders, and the other represents those who get a ticket. The fact that the "Speeders" circle is inside the "Get tickets" circle means that all speeders get tickets.

We also use Venn diagrams to show how categories relate to each other. Suppose, for example, that our state legislature consists only of Democrats and Republicans. **Figure 1.28** shows how the House divides by party and gender: One circle represents females; the outside of that circle represents males. The other circle represents Democrats; the outside of that circle represents Republicans. So the intersection of the two circles shows female Democrats. Venn diagrams can also be used to count complicated sets.

Critical thinking and number sense: What do these figures mean?

Relative sizes of numbers are often indicated using *magnitudes* or powers of 10. For example, a city may have a population in hundreds of thousands (10^5), and world population is measured in billions (10^9).

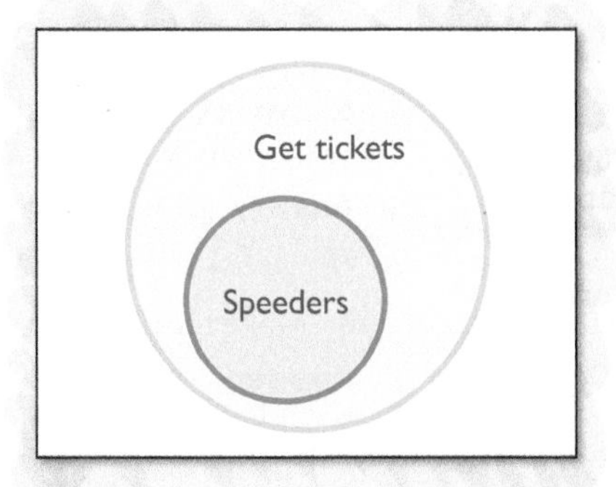

FIGURE 1.27 Venn diagram illustrating a statement.

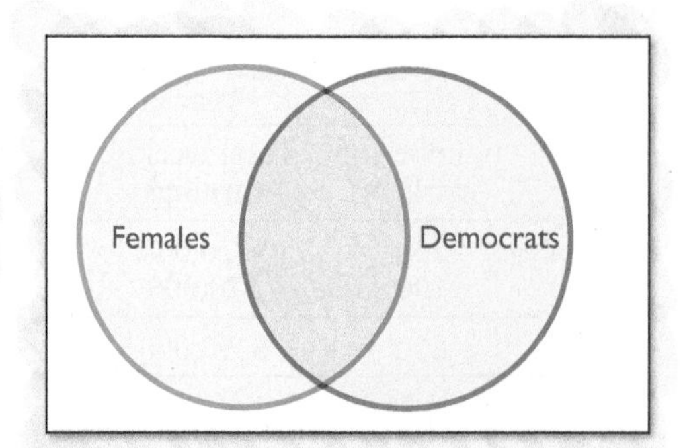

FIGURE 1.28 House members divided by sex and party.

We can get a feeling for very large and very small numbers by putting them into a more familiar context. For example, the national debt of 27.8 trillion dollars means that every man, woman, and child in the United States owes about $84,200.

One way to avoid complicated computations is to estimate by rounding the given numbers to make the arithmetic easier. Another method of estimation is to ignore features that complicate the calculation.

KEY TERMS

logic, p. 17
premises, p. 17
hypotheses, p. 17
conclusion, p. 17
valid, p. 17
informal fallacy, p. 18
formal fallacy, p. 18
deductive, p. 24
inductive, p. 24
or, p. 33
logically equivalent, p. 40
set, p. 46
elements, p. 46
subset, p. 46
disjoint, p. 46
Venn diagram, p. 47

CHAPTER QUIZ

1. The following tables show, for two declining towns, the number of workers and the total weekly earnings at the beginning of the given year.

	Year 2005		
	Number of workers	Total weekly earnings	Average weekly earnings
Town A	500	$320,000	
Town B	100	$70,000	
Total			

	Year 2015		
	Number of workers	Total weekly earnings	Average weekly earnings
Town A	50	$31,000	
Town B	60	$42,000	
Total			

Complete the table by filling in the blank spaces. Round each average to the nearest dollar.

Answer

	Year 2005		
	Number of workers	Total weekly earnings	Average weekly earnings
Town A	500	$320,000	$640
Town B	100	$70,000	$700
Total	600	$390,000	$650

	Year 2015		
	Number of workers	Total weekly earnings	Average weekly earnings
Town A	50	$31,000	$620
Town B	60	$42,000	$700
Total	110	$73,000	$664

If you had difficulty with this problem, see Example 1.1.

2. In Exercise 1, did the average weekly earnings for Town A increase or decrease over this period? What happened for Town B? What happened overall?

Answer The average weekly earnings for Town A decreased. The average stayed the same for Town B. Overall the average increased.

If you had difficulty with this problem, see Example 1.1.

3. Identify the premises and conclusion of the following argument. Is this argument valid?
Most basketball players are tall. Tom is tall. Therefore, Tom is a basketball player.

Answer The premises are (1) *Most basketball players are tall.* (2) *Tom is tall.* The conclusion is *Tom is a basketball player.* The argument is not valid because the premises do not support the conclusion.

If you had difficulty with this problem, see Example 1.3.

4. Identify the type of the fallacy in the following argument. *John is a smoker, so you shouldn't trust his advice about a healthy lifestyle.*

Answer This is dismissal based on personal attack. John's habits are attacked rather than his argument.

If you had difficulty with this problem, see Example 1.5.

5. Let p represent the statement *You study for your math exam* and q the statement *You pass your math exam.* Use these letters to express symbolically the statement *If you don't study for your math exam, you fail the math exam.*

Answer (NOT p) $\rightarrow$ (NOT q).

If you had difficulty with this problem, see Example 1.10.

6. Make the truth table for $p \rightarrow$ (NOT q).

Answer

p	q	NOT q	$p \rightarrow$ (NOT q)
T	T	F	F
T	F	T	T
F	T	F	T
F	F	T	T

If you had difficulty with this problem, see Example 1.11.

7. Tests for drug use are not always accurate. Make a Venn diagram using the categories of drug users and those who test

positive. Shade the area of the diagram that represents *false negatives*, that is, drug users who test negative.

Answer The region for false negatives is shaded.

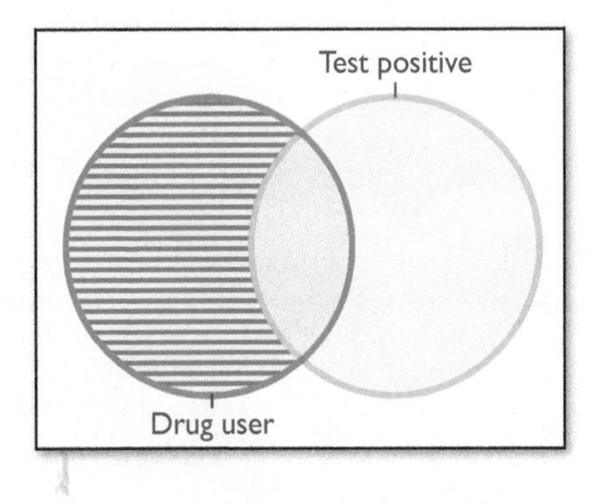

If you had difficulty with this problem, see Example 1.15.

8. The national debt in a country with a population of 1.1 million is 4.2 billion dollars. Calculate the amount each person owes. Round your answer to the nearest $100.

Answer Each person owes

$$\frac{4.2 \text{ billion dollars}}{1.1 \text{ million}} = \frac{4.2 \times 10^9 \text{ dollars}}{1.1 \times 10^6} = \frac{4.2}{1.1} \times 10^{9-6} \text{ dollars}$$
$$= \frac{4.2}{1.1} \times 10^3 \text{ dollars}$$

This is about 3.8×10^3 dollars, so each person owes about $3800.

If you had difficulty with this problem, see Example 1.21.

9. A group of 19 people buys $207 worth of pizza. Estimate the amount each owes for lunch by rounding the number of people to 20 and the price to $200. What is the exact answer?

Answer The estimate is $200/20 = $10. The exact amount is $207/19 or about $10.89.

If you had difficulty with this problem, see Example 1.25.

Exit
& 42 Street
Open M-F 6:30am-7:30pm

2 Analysis of Growth

The following article appears at the Gigaom Web site.

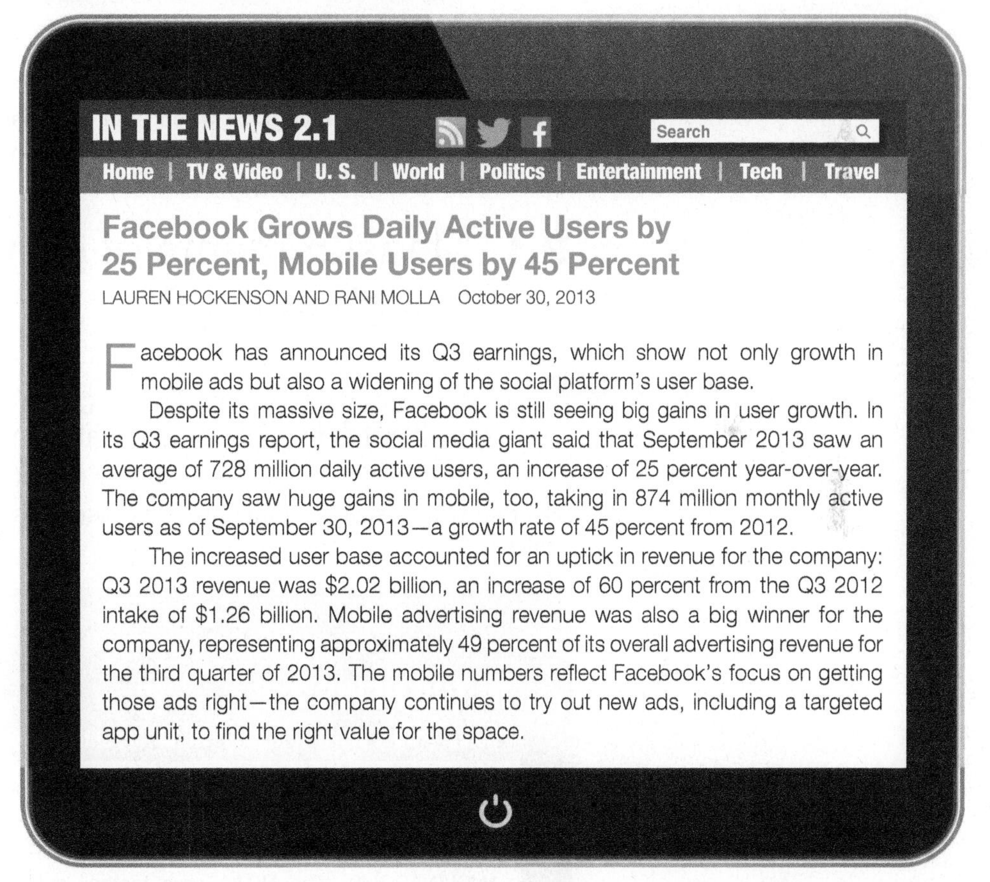

IN THE NEWS 2.1

Home | TV & Video | U. S. | World | Politics | Entertainment | Tech | Travel

Facebook Grows Daily Active Users by 25 Percent, Mobile Users by 45 Percent

LAUREN HOCKENSON AND RANI MOLLA October 30, 2013

Facebook has announced its Q3 earnings, which show not only growth in mobile ads but also a widening of the social platform's user base.

Despite its massive size, Facebook is still seeing big gains in user growth. In its Q3 earnings report, the social media giant said that September 2013 saw an average of 728 million daily active users, an increase of 25 percent year-over-year. The company saw huge gains in mobile, too, taking in 874 million monthly active users as of September 30, 2013—a growth rate of 45 percent from 2012.

The increased user base accounted for an uptick in revenue for the company: Q3 2013 revenue was $2.02 billion, an increase of 60 percent from the Q3 2012 intake of $1.26 billion. Mobile advertising revenue was also a big winner for the company, representing approximately 49 percent of its overall advertising revenue for the third quarter of 2013. The mobile numbers reflect Facebook's focus on getting those ads right—the company continues to try out new ads, including a targeted app unit, to find the right value for the space.

This article, along with **Figure 2.1**, makes clear that the growth rate of Facebook is dramatic and has resulted in a tremendous increase in its value—but without the figure, the growth might not be as vivid.

In this technological age, it is easy to be overwhelmed with information, which frequently comes in the form of numerical data: the number of cases of flu each year, the variation in the inflation rate over the past decade, the growth rate of Facebook, and so forth. Presenting data graphically, as in Figure 2.1, makes them easier to digest.

In this chapter, we will explore further how information can be presented using different kinds of graphs and tables, and we will examine their advantages and disadvantages. We will learn to analyze data, especially growth rates, and to make

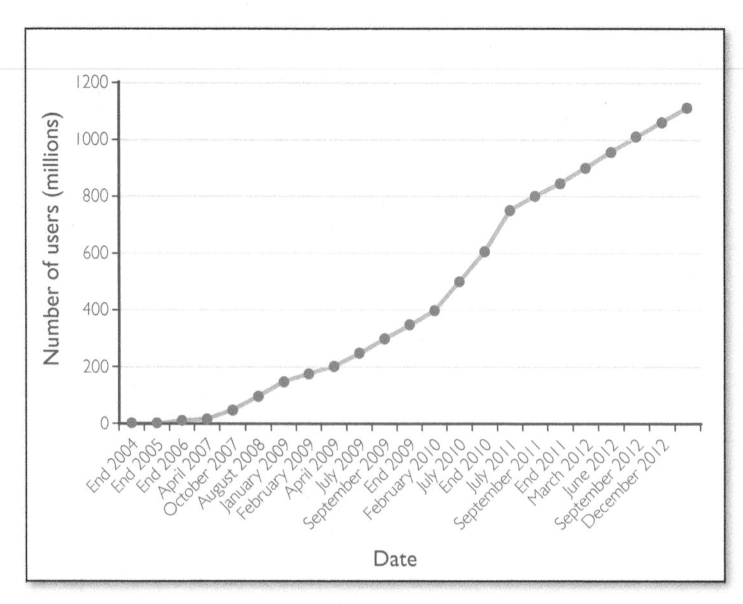

FIGURE 2.1 Growth of Facebook.

predictions. We will also see that one should learn to be wary of graphical presentations and to recognize when data are misrepresented and when inaccurate conclusions are drawn. Empowering you to analyze tables and graphs critically is the ultimate goal of this chapter.

2.1 Measurements of growth: How fast is it changing?

TAKE AWAY FROM THIS SECTION
Use growth rates to analyze quantitative information.

The following article from the ESPN website deals with college graduation rates for athletes.

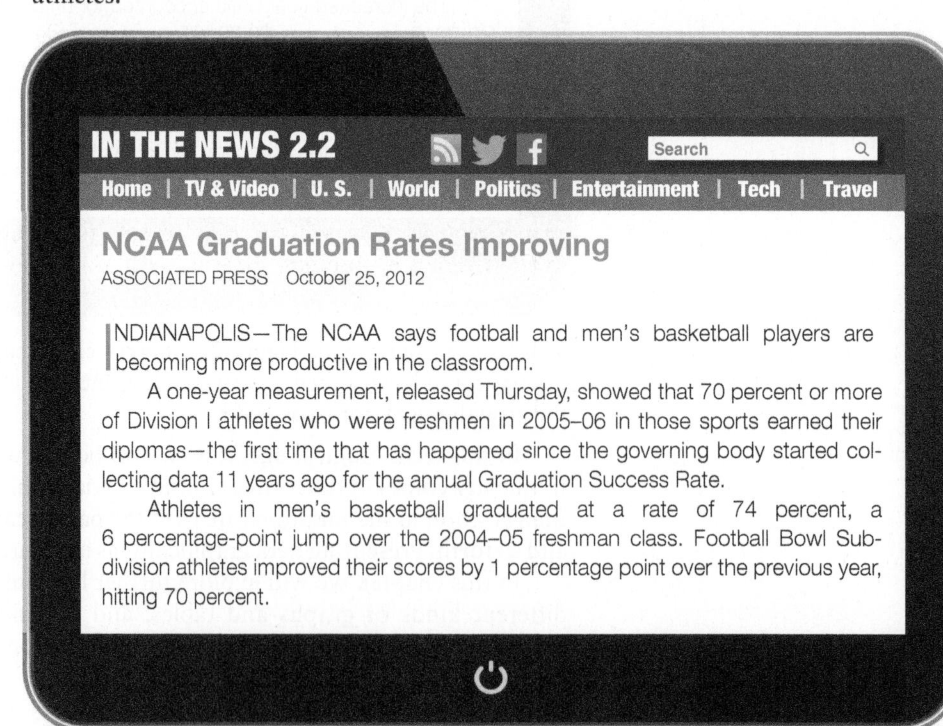

IN THE NEWS 2.2

Search

Home | TV & Video | U. S. | World | Politics | Entertainment | Tech | Travel

NCAA Graduation Rates Improving

ASSOCIATED PRESS October 25, 2012

INDIANAPOLIS—The NCAA says football and men's basketball players are becoming more productive in the classroom.

A one-year measurement, released Thursday, showed that 70 percent or more of Division I athletes who were freshmen in 2005–06 in those sports earned their diplomas—the first time that has happened since the governing body started collecting data 11 years ago for the annual Graduation Success Rate.

Athletes in men's basketball graduated at a rate of 74 percent, a 6 percentage-point jump over the 2004–05 freshman class. Football Bowl Subdivision athletes improved their scores by 1 percentage point over the previous year, hitting 70 percent.

The focus of this article is the trend in graduation rates—how they change over time, which is crucial for understanding the current situation.

Although raw information can seem overwhelming, certain basic principles are helpful for mining data. In this section, we examine and analyze quantitative information presented in tables and bar graphs. The basic measures of growth are important keys for analysis.

Intuitive notion of functions

If we work for an hourly wage, our pay for the week depends on the number of hours we work. We can think of the number of hours we work as one variable and our weekly pay as another variable, one that *depends* on the number of hours we work. In mathematics it is helpful to formalize relationships of this sort. Because the weekly pay depends on the number of hours worked, we refer to the weekly pay as the *dependent variable*. The number of hours worked is the *independent variable*.[1] The *process* used to get the salary from hours worked is a *function*. We say that *the weekly pay is a function of the number of hours worked*. The function tells how one quantity depends on another. More advanced mathematics texts give a precise definition of this important idea, but for our purposes an informal notion of a function is sufficient.

KEY CONCEPT

When one quantity, or variable, depends on another, the latter is referred to as the independent variable and the former is referred to as the dependent variable. A function describes *how* the dependent variable depends on the independent variable.

Examples of functions abound in our everyday experience. Here are a few:

- Your height is a function of your age. Age is the independent variable, and height is the dependent variable because height depends on age.
- The amount of income tax you owe is a function of your taxable income. Taxable income is the independent variable, and tax owed is the dependent variable because it depends on your income.
- The number of euros you can get for a dollar is a function of the date. The date is the independent variable, and the number of euros you can get for a dollar is the dependent variable.

The independent variable is sometimes called the *input* value and the dependent variable the *output* value of a function. With this terminology, we can think of a function as a "machine" that accepts the input variable and produces the output variable. A common illustration is shown in **Figure 2.2.** We should note that

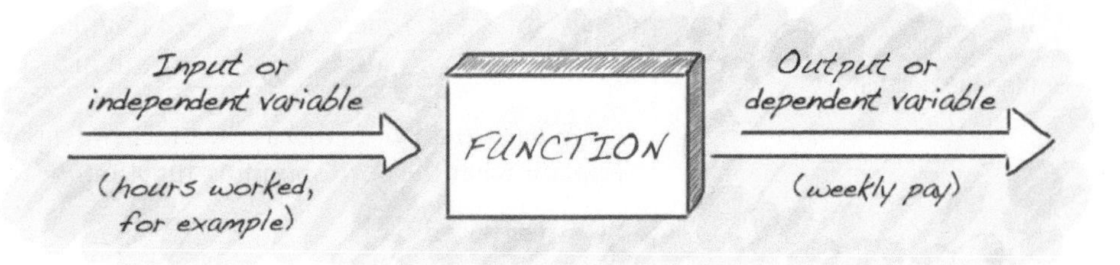

FIGURE 2.2 Visual representation of a function.

[1] *The terminology of independent and dependent variables should not be interpreted to imply that one variable somehow causes the other.*

a function is required to have *exactly one* output for every input. It would not do, for example, to have two different weekly earnings for a person who worked 40 hours this week. Only one paycheck will be produced.

Tables and bar graphs

Tables are common ways of presenting functions. Consider **Table 2.1**, which shows population data from the U.S. Census Bureau.

TABLE 2.1 Population of the United States

Year	Population (millions)	Year	Population (millions)
1790	3.93	1910	92.23
1800	5.31	1920	106.02
1810	7.24	1930	123.20
1820	9.64	1940	132.16
1830	12.87	1950	151.33
1840	17.07	1960	179.32
1850	23.19	1970	203.30
1860	31.44	1980	226.54
1870	38.56	1990	248.71
1880	50.19	2000	281.42
1890	62.98	2010	308.75
1900	76.21	2020	330 (est.)*

*Estimated as census figures for 2020 were not yet complete at time of publication.

When we look at the census table, it is natural to pick the date and then look up the population for that date. This makes sense because the population depends on the date. In the case of the census, the date is the independent variable and the population is a function of this variable because it depends on the date. In this table, the first column lists the values of the independent variable and the second column lists the values of the dependent variable. When data tables list the variables vertically in columns, it is customary for the first column to represent the independent variable and the second column to represent the dependent variable. When a data table lists the values of the variables in rows, it is typical for the first row to represent the independent variable and the second row to represent the dependent variable.

EXAMPLE 2.1 Choosing variables: Swimming speed and length

Matt9122/Shutterstock
Bottlenose dolphin

The following table shows how swimming speed is related to length for some sea creatures.

Length	Top swimming speed	Animal name
2 meters	70 km/hr	Bluefin tuna
3 meters	35 km/hr	Bottlenose dolphin
3.5 meters	40 km/hr	Great white shark
8.5 meters	49 km/hr	Killer whale

What would you label as the independent and dependent variables for this table? Explain what the corresponding function means.

SOLUTION

We could choose the length to be the independent variable and the top swimming speed to be the dependent variable. The function would then give the top swimming speed for an animal of a given length (among those listed).

However, in this example, it is also reasonable to let the top swimming speed be the independent variable and the length be the dependent variable. That is because in this table, each swimming speed corresponds to exactly one length. With this choice, the function shows the length of an animal with a given top swimming speed (among those listed).

TRY IT YOURSELF 2.1

The following table shows the most medals won by any country in the given year of the Olympic Winter Games:

Year	Most medals won by a country	Country
1992	26	Germany
1994	26	Norway
1998	29	Germany
2002	36	Germany
2006	29	Germany
2010	37	United States
2014	33	Russia
2018	29	Norway

What would you label as the independent and dependent variables for this table? Explain what the corresponding function means.

The answer is provided at the end of this section.

Data tables are meant to convey information. But the tables themselves are sometimes dry and difficult to decipher. There are important tools that help us understand information presented through a visual display. One of these is the *bar graph*. Bar graphs are useful in displaying data from tables of relatively modest size. For large data tables, other types of visual representation may be more appropriate. To make a bar graph of the census data in Table 2.1, we let the horizontal axis represent the independent variable (the date in this case) and the heights of the vertical bars show the value of the function (the U.S. population, in millions) for the corresponding date. The completed bar graph is in **Figure 2.3**. It shows how the population increased with time.

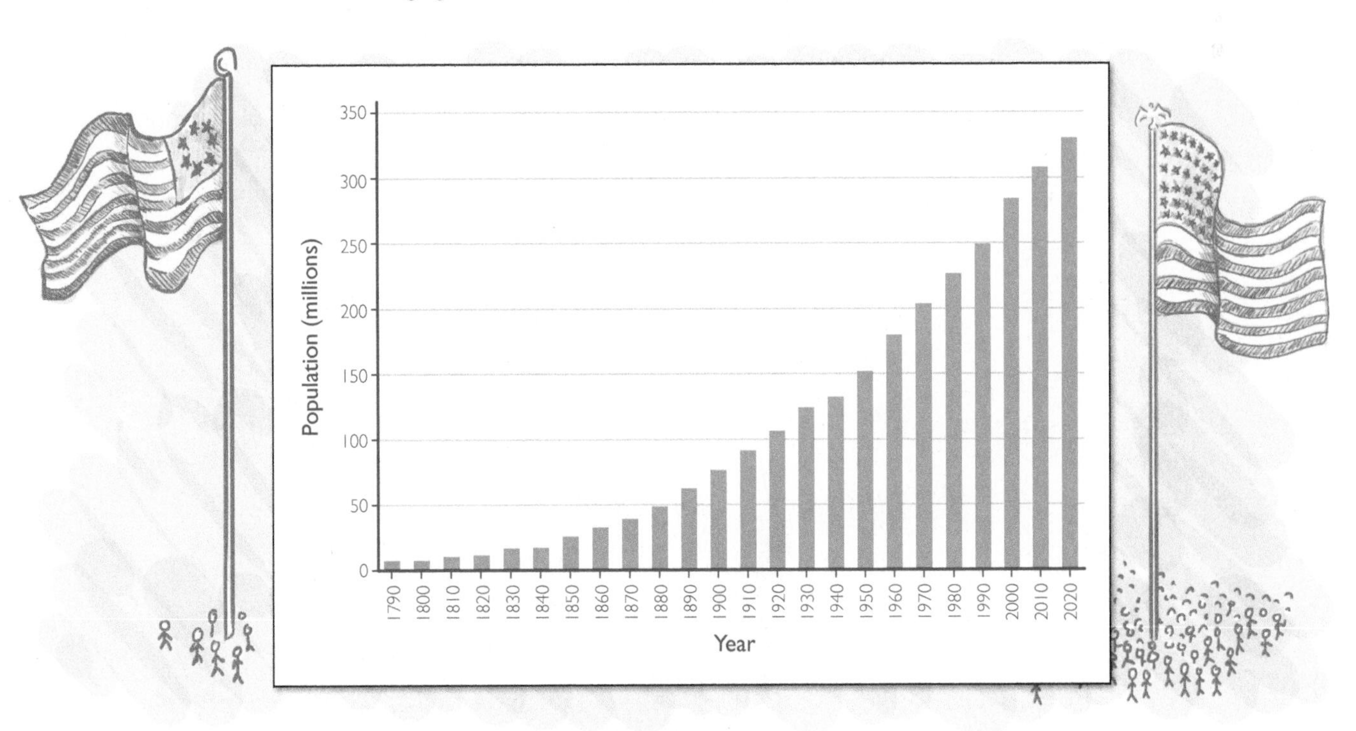

FIGURE 2.3 Bar graph of U.S. population.

Percentage change

The apparent regularity of the graph in Figure 2.3 hides some important information. We can find out more about U.S. population growth by calculating the *percentage change*, that is, the *percentage increase* or *percentage decrease*. Population growth is often measured in this way; in the article opening this section, the growth rate of Facebook was measured in terms of percentages.

KEY CONCEPT

The percentage change or relative change in a function is the percentage increase in the function from one value of the independent variable to another.

The formula for percentage change from one function value to another is

$$\text{Percentage change} = \frac{\text{Change in function value}}{\text{Previous function value}} \times 100\%$$

In the case of the census data, this change is the *percentage* increase in population from decade to decade. The following Quick Review is provided as a reminder of how to make percentage calculations.

Quick Review Percentage Change

Suppose the function value changes from 4 to 5. The *change*, or *absolute change*, from 4 to 5 is simply $5 - 4 = 1$. The *percentage change*, or *relative change*, from 4 to 5 is the change in the function value divided by the previous function value, written as a percentage:

$$\text{Percentage change} = \frac{\text{Change in function value}}{\text{Previous function value}} \times 100\% = \frac{1}{4} \times 100\% = 25\%$$

This percentage change of 25% says that 5 is 25% more than 4.

If, however, the function value decreases from 5 to 4, then the change from 5 to 4 is $4 - 5 = -1$. The percentage change from 5 to 4 is the change in the function value divided by the previous function value, written as a percentage:

$$\text{Percentage change} = \frac{\text{Change in function value}}{\text{Previous function value}} \times 100\% = \frac{-1}{5} \times 100\% = -20\%$$

This percentage change of −20% says that 4 is 20% less than 5.

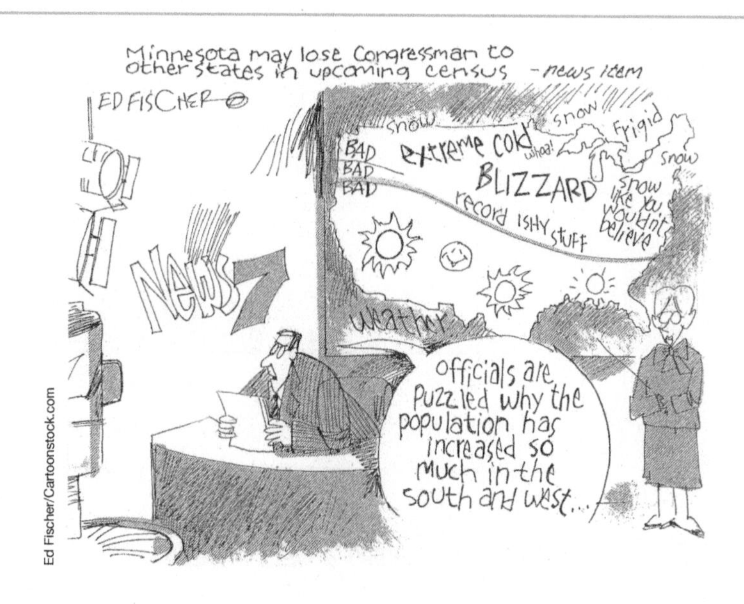

Ed Fischer/Cartoonstock.com

EXAMPLE 2.2 Calculating percentage change: U.S. population

Calculate the percentage change in the U.S. population from 1790 to 1800. Round your answer in percentage form to the nearest whole number. The data you need are found in Table 2.1.

SOLUTION

To find the percentage change, we divide the change in the population from 1790 to 1800 by the population in 1790 and then multiply by 100%:

$$\begin{aligned}\text{Percentage change from 1790 to 1800} &= \frac{\text{Change in function value}}{\text{Previous function value}} \times 100\% \\ &= \frac{\text{Change from 1790 to 1800}}{\text{Population in 1790}} \times 100\% \\ &= \frac{5.31 \text{ million} - 3.93 \text{ million}}{3.93 \text{ million}} \times 100\% \\ &= \frac{1.38}{3.93} \times 100\%\end{aligned}$$

This is about $0.35 \times 100\%$, or 35%. Between 1790 and 1800, the population increased by 35%.

TRY IT YOURSELF 2.2

Calculate the percentage change in the U.S. population from 1800 to 1810. Round your answer in percentage form to the nearest whole number.

The answer is provided at the end of this section.

The percentage change during each decade is shown in the accompanying table and as a bar graph in **Figure 2.4**.

Decade	Percentage change	Decade	Percentage change
1790–1800	35%	1910–1920	15%
1800–1810	36%	1920–1930	16%
1810–1820	33%	1930–1940	7%
1820–1830	34%	1940–1950	15%
1830–1840	33%	1950–1960	18%
1840–1850	36%	1960–1970	13%
1850–1860	36%	1970–1980	11%
1860–1870	23%	1980–1990	10%
1870–1880	30%	1990–2000	13%
1880–1890	25%	2000–2010	10%
1890–1900	21%	2010–2020	7%
1900–1910	21%		

As we look at each percentage change, some things stand out. Note that between 1790 and 1860, the percentage changes in each decade are approximately the same—between 33% and 36%. That is, the percentage increase in population was nearly constant over this period.

The decade of the Civil War (1860–1870) jumps out. Even if we didn't know the historical significance of this decade, the bar graph of percentage change would alert us that something important may have occurred in that decade. Observe that the significance of this decade is not apparent from the graph of population shown in Figure 2.3. It is interesting to note that percentage population growth never returned

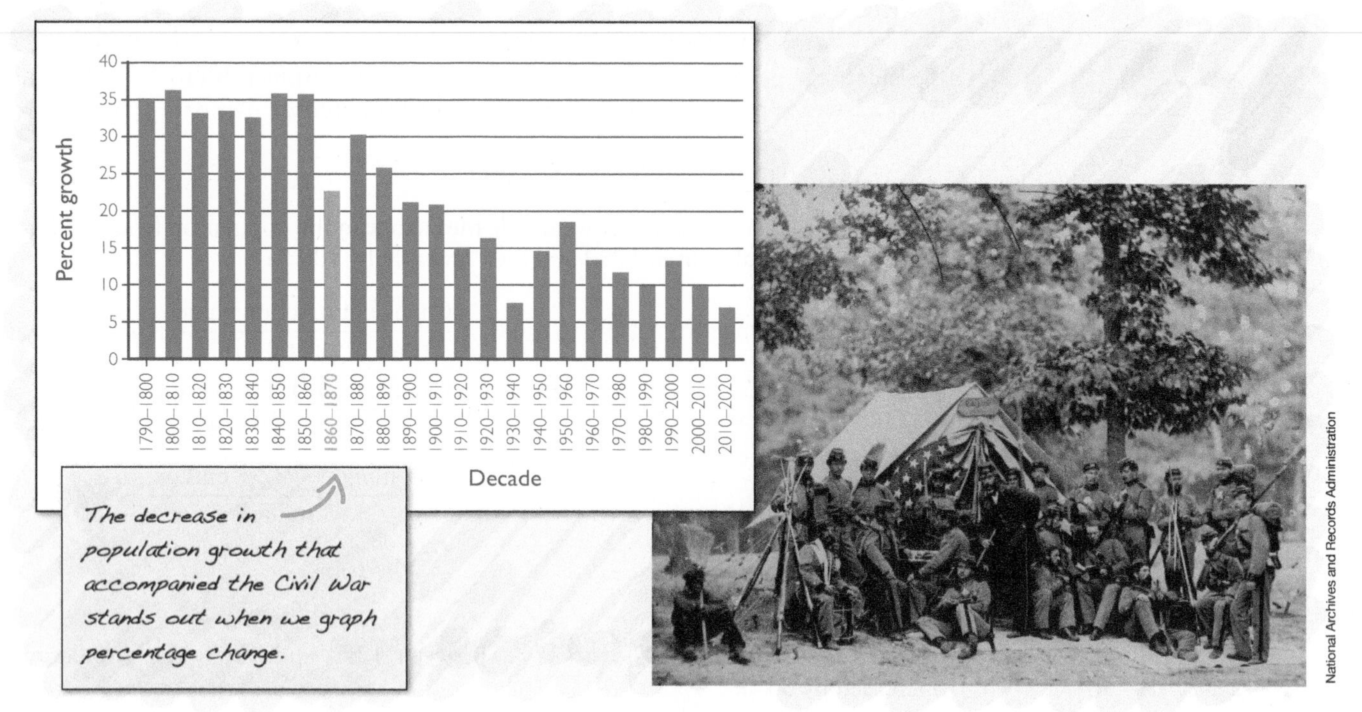

FIGURE 2.4 Left: Percentage change in U.S. population. Right: Engineers of the 8th N.Y. State Militia, 1861.

to its pre–Civil War levels. The percentage change generally declined to its current level of 7% to 13% per decade.

SUMMARY 2.1 Tables and Percentage Change

1. A data table shows a relationship between two quantities. We designate one quantity as the independent variable and think of the other quantity, the dependent variable, as a function of the independent variable.

2. A bar graph provides a visual display of a table that can help us understand what the information shows.

3. Percentage change can provide information that may not be readily apparent from the raw data. We calculate percentage change from one function value to another using

$$\text{Percentage change} = \frac{\text{Change in function value}}{\text{Previous function value}} \times 100\%$$

EXAMPLE 2.3 Creating and analyzing data: World population

The following table shows the world population (in billions) on the given date:

Date	1950	1960	1970	1980	1990	2000	2010	2020
Population (billions)	2.56	3.04	3.71	4.45	5.29	6.09	6.85	7.80

a. Identify the independent variable and the function, and make a bar graph that displays the data.

b. Make a table and bar graph showing percentage changes between decades.

c. Comment on what your graphs tell you. Compare the second graph with the corresponding graph for U.S. population growth.

SOLUTION

a. The independent variable is the date, and the function gives the world population (in billions) on that date. To create a bar graph, we list dates across the horizontal axis and make bars representing population for each date. The finished product is in **Figure 2.5**.

b. We calculate the percentage change from one date to the next using

$$\text{Percentage change} = \frac{\text{Change in population}}{\text{Previous population}} \times 100\%$$

So the percentage increase from 1950 to 1960 is

$$\text{Percentage change} = \frac{3.04 - 2.56}{2.56} \times 100\%$$

or about 19%.

The percentage changes for other decades are calculated in a similar fashion. The table is given below, and the bar graph is in **Figure 2.6**.

Decade	1950–60	1960–70	1970–80	1980–90	1990–2000	2000–10	2010–20
% change	19%	22%	20%	19%	15%	12%	14%

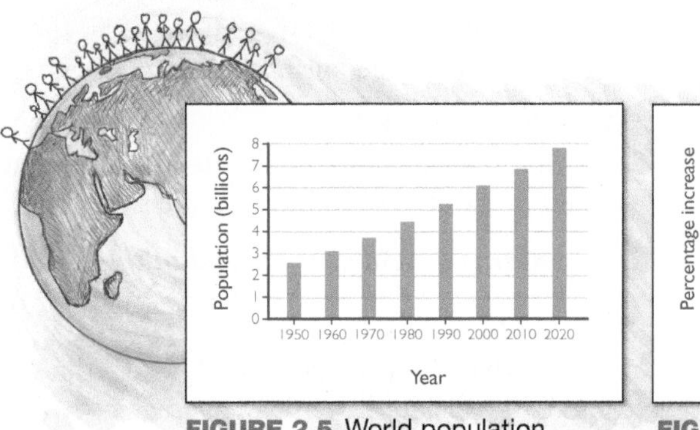

FIGURE 2.5 World population.

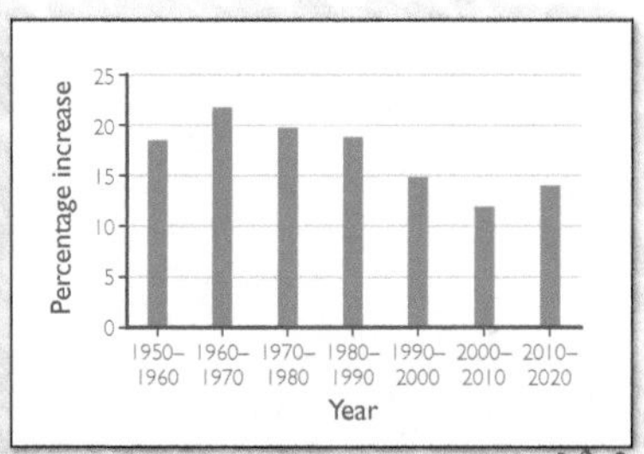

FIGURE 2.6 Percentage growth of world population.

c. The world population is steadily increasing, but from 1960 on, the percentage change is not increasing. Starting from 1960, the percentage change goes from 22% to a low of 12% per decade. The percentage increase for the world population is still larger than that for the United States.

Average rate of change

Percentage change focuses on the dependent variable. Note that when we calculated the percentage change in U.S. population from 1790 to 1800, only population figures were used. The actual calculation did not involve the dates at all. This way of measuring change makes it difficult to compare growth over time periods of different lengths. To tell at what rate the function changes with respect to the independent variable, we calculate the *average rate of change*. It is the most useful measure of change that we will encounter.

The average rate of change is always calculated over a given interval, often a time period (such as 1790 to 1800). To calculate the average rate of change, we divide the change in the function value over that interval by the change in the independent variable.

KEY CONCEPT

The average rate of change of a function over an interval is the change in the function value divided by the change in the independent variable.

We calculate the average rate of change using the formula

$$\text{Average rate of change} = \frac{\text{Change in function value}}{\text{Change in independent variable}}$$

Intuitively, we can think of the average rate of change of a function as the *average growth rate*, and we will often employ this descriptive term.

EXAMPLE 2.4 Calculating average growth rate: U.S. population

Calculate the average growth rate in the population of the United States from 1940 to 1950. Explain its meaning.

SOLUTION

The change in time (the independent variable) is 10 years. According to Table 2.1 on page 82, the change in population (the function) over that interval of time is $151.33 - 132.16 = 19.17$ million people. So the average growth rate in population over this decade is

$$\frac{\text{Change in function value}}{\text{Change in independent variable}} = \frac{19.17}{10} = 1.917 \text{ million people per year}$$

This result tells us that from 1940 to 1950 the U.S. population grew, *on average*, by about 1.92 million people each year.

TRY IT YOURSELF 2.4

Show that the average growth rate in the population of the United States from 1950 to 1960 is about 2.80 million people per year. Explain the meaning of this rate.

The answer is provided at the end of this section.

Note that the average growth rate from 1940 to 1950 is not the same as that from 1950 to 1960. In general, average growth rates vary from decade to decade. In some cases, the average growth rate may be negative. This indicates a decline rather than an increase. The following example illustrates this possibility.

EXAMPLE 2.5 Calculating a negative growth rate: Population of Russia

The population of Russia declined from about 146 million in 2000 to about 143 million in 2007. Calculate the average growth rate over this period and explain its meaning.

SOLUTION

The change in population is negative, $143 - 146 = -3$ million. The change in time is 7 years. Thus,

$$\text{Average growth rate} = \frac{\text{Change in function value}}{\text{Change in independent variable}} = \frac{-3}{7}$$

or about -0.429 million per year. This means that over this interval the population declined, on average, by about 429,000 people per year.

TRY IT YOURSELF 2.5

It is projected that the population of Russia will decline from 143 million in 2007 to about 111 million in 2050. Use this projection to predict the average growth rate from 2007 to 2050. Explain the meaning of this rate.

The answer is provided at the end of this section.

It is always important to pay attention to the units of measurement. This is particularly true for average growth rates. The units of the average growth rate are the units of the function divided by the units of the independent variable. See the preceding example, where the average growth rate is measured in millions of people per year.

In practical settings, the average growth rate almost always has a familiar meaning, and proper use of units can help determine that meaning. For example, if the function gives the distance driven in an automobile and the independent variable is time, the average growth rate is the average velocity, which typically is measured in miles per hour. Thus, if you make a 100-mile trip and the trip takes 2 hours, the average growth rate is 50 miles per hour. In other settings, the average growth rate may represent the average number of traffic accidents per week, the average number of births per thousand women, or the average number of miles per gallon.

EXAMPLE 2.6 Interpreting the average growth rate: Tuition cost

Assume that the independent variable is the year and the function gives the tuition cost, in dollars, at your university. Give the units of the average growth rate and explain in practical terms what that rate means.

SOLUTION

The change in the independent variable is the elapsed time measured in years, and the change in the function value is the tuition increase measured in dollars. So the units of the average growth rate are dollars per year. The average growth rate tells how much we expect the tuition to increase each year. (A tuition decrease would result in a negative growth rate but, unfortunately, tuition usually goes up these days, not down.)

TRY IT YOURSELF 2.6

Assume that the independent variable is the day and the function is the cumulative number of flu cases reported to date. Give the units of the average growth rate and explain in practical terms what that rate means.

The answer is provided at the end of this section.

SUMMARY 2.2
Three Types of Change

There are three types of change associated with any data table defining a function.

1. **Absolute change,** which is the change from one entry to the next. It is calculated using

$$\text{Absolute change} = \text{Current function value} - \text{Previous function value}$$

2. **Percentage change,** which is the percentage change from one entry to the next. It is calculated using

$$\text{Percentage change} = \frac{\text{Change in function value}}{\text{Previous function value}}$$

3. **Average rate of change, or average growth rate,** which is the average change from one entry to the next. It is calculated using

$$\text{Average growth rate} = \frac{\text{Change in function value}}{\text{Change in independent variable}}$$

Interpolation

Once we know the average rate of change, we can obtain more information from the data. Suppose we want to estimate the U.S. population in the year 1945. According to Example 2.4, between 1940 and 1950, the U.S. population grew at an average rate of 1.917 million people per year. So it is reasonable to assume that over the 5-year period from 1940 to 1945, the population increased by approximately $5 \times 1.917 = 9.585$ million. We know that the population in 1940 was 132.16 million. We can then estimate that the population in 1945 was

$$\text{Population in } 1940 + 5 \text{ years of increase} = 132.16 + 9.585 = 141.745 \text{ million}$$

or about 141.75 million.

Using average growth rates to estimate values between data points is known as *interpolation*.

KEY CONCEPT

Interpolation is the process of estimating unknown values between known data points using the average growth rate.

EXAMPLE 2.7 Interpolating data: Opinions on legalizing marijuana

In 2009, 45.6% of college freshmen in the United States believed that marijuana should be legalized.[2] In 2015, that figure was 56.4%. Use these figures

a. to estimate the percentage in the fall of 2010, and

b. to estimate the percentage in the fall of 2011. The actual figure for 2011 was 49.1%. What does this say about how the percentage growth rate varied over time?

Krista Kennell/Sipa/AP Photo

Medical marijuana dispensary in Los Angeles.

SOLUTION

a. We want to interpolate, so first we find the average growth rate. The change in the independent variable from 2009 to 2015 is 6 years. The change in the dependent variable over that period is $56.4 - 45.6 = 10.8$ percentage points. Hence, the average growth rate from 2009 to 2015 was:

$$\frac{\text{Change in function value}}{\text{Change in independent variable}} = \frac{10.8}{6} = 1.8 \text{ percentage points per year}$$

[2] *Data from the annual CIRP Freshman Surveys by the Higher Education Research Institute at UCLA.*

This means that, on average, the increase was 1.8 percentage points each year over this period. The year 2010 represents 1 year of increase from 2009, so the estimated percentage in 2010 is about

$$\begin{aligned}\text{Percentage in 2009} + \text{1 year of increase} &= 45.6 + 1 \times 1.8 \\ &= 47.4\%\end{aligned}$$

b. The year 2011 represents 2 years of increase from 2009, so the estimated percentage in 2011 is about

$$\begin{aligned}\text{Percentage in 2009} + \text{2 years of increase} &= 45.6 + 2 \times 1.8 \\ &= 49.2\%\end{aligned}$$

Our estimate of 49.2% for 2011 is a little more than the actual figure of 49.1%. This means that the growth rate of the percentage was a little lower from 2009 to 2011 than it was from 2011 to 2015.

TRY IT YOURSELF 2.7

In 2018, the number of broadband Internet subscribers for one provider was 15.2 million. In 2021, the number of subscribers was 14.6 million. Estimate the number of subscribers in 2020.

The answer is provided at the end of this section.

EXAMPLE 2.8 Interpolating data: Ebola

Ebola is a deadly virus. The outbreak of Ebola in West Africa from 2014 to 2016 resulted in over 11,300 deaths. The following table from the World Health Organization shows the cumulative number of Ebola cases reported on certain dates in October 2015:

Date	October 1	October 16	October 22	October 26
Number of cases	28,408	28,468	28,504	28,528

Abbas Dulleh/AP/REX/Shutterstock

Health workers wash their hands after taking a blood specimen from a child to test for the Ebola virus in Liberia.

a. Calculate the average growth rate of cases from October 1 to October 16. Be sure to express your answer using proper units.

b. Use your answer from part a to estimate the cumulative number of Ebola cases by October 9. The actual cumulative number on October 9 was 28,429. What does this say about how the growth rate varied over time?

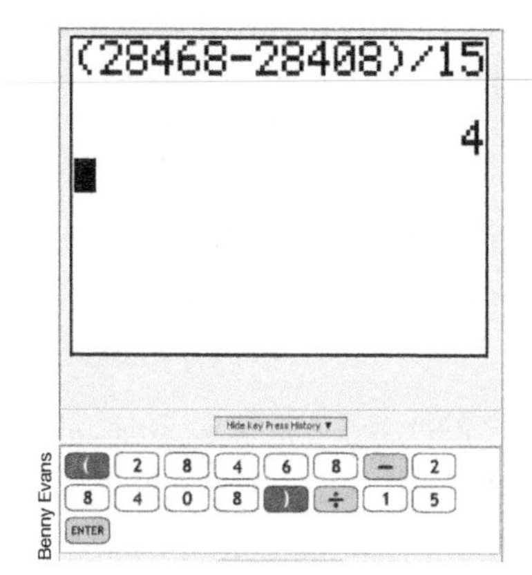

Benny Evans

SOLUTION

a. The independent variable is the date, and the function is the cumulative number of Ebola cases reported. The average growth rate from October 1 to October 16 is

$$\frac{\text{Change in reported cases}}{\text{Elapsed time}} = \frac{28{,}468 - 28{,}408}{15} = 4 \text{ new cases per day}$$

b. There were 28,408 cases on October 1, and we expect to see about 4 new cases on each subsequent day until October 16. Thus, we make our estimate as follows:

$$\begin{aligned}&\text{Estimated cases on October 9}\\ &\quad = \text{Cases on October } 1 + 8 \times \text{Average new cases per day}\\ &\quad = 28{,}408 + 8 \times 4\\ &\quad = 28{,}440\end{aligned}$$

We estimate there were 28,440 cases by October 9. Our estimate using interpolation is relatively good but somewhat higher than the actual value of 28,429. The fact that interpolation gives an overestimate indicates that the growth rate between October 1 and October 9 was lower than the growth rate between October 1 and October 16—the number of new cases per day was lower over the first part of the 15-day period.

We should be aware that sometimes interpolation works well and sometimes it does not. We will return to this question in Chapter 3, but for now it suffices to say that interpolation works better when the values in the table are closer together.

Extrapolation

It is also possible to use average growth rates to make estimates beyond the limits of available data and thus to forecast trends. This process is called *extrapolation*. As with interpolation, extrapolation requires a little judgment and common sense.

KEY CONCEPT

Extrapolation is the process of estimating unknown values beyond known data points using the average growth rate.

As an example, let's extrapolate beyond the limits of the table of Ebola cases to estimate the number of cases on October 27. We begin by calculating the average number of new cases per day from October 22 to October 26:

$$\text{Average number of new cases per day} = \frac{28{,}528 - 28{,}504}{4} = 6 \text{ cases per day}$$

So, on October 27th we expect 6 more cases than on October 26, for a total of $28{,}528 + 6 = 28{,}534$. We estimate 28,534 cases on October 27.

EXAMPLE 2.9 Extrapolating data: An epidemic

In a certain epidemic, there is an average of 12 new cases per day. If a total of 500 cases have occurred as of today, what cumulative number of cases do we expect 3 days from now?

SOLUTION

We expect 12 new cases for each of 3 days, for a total of 36 new cases. So the cumulative number of cases expected 3 days from now is $500 + 36 = 536$.

TRY IT YOURSELF 2.9

In a certain epidemic, there is an average of 13 new cases per day. If a total of 200 cases have occurred as of today, what cumulative number of cases do we expect 4 days from now?

The answer is provided at the end of this section.

Now let's see what happens if we try to extrapolate farther beyond the limits of the Ebola data by predicting the number of cases on November 25, 30 days after October 26. We start with 28,528 cases on October 26 and use the rate of change of 6 new cases per day for 30 days to obtain

$$\text{Expected cases on November 25} = 28{,}528 + 30 \times 6 = 28{,}708 \text{ cases}$$

The actual number of Ebola cases on November 25, as reported by the World Health Organization, was 28,601. That's 107 fewer cases than our estimate.

This example illustrates why one should be cautious in using extrapolation to make estimates well beyond the limits of the data.

SUMMARY 2.3
Interpolation and Extrapolation

1. The average growth rate of a function over an interval is calculated using

$$\text{Average growth rate} = \frac{\text{Change in function value}}{\text{Change in independent variable}}$$

2. To make an estimate by interpolation or extrapolation from a function value, multiply the average growth rate (e.g., cases per day, accidents per week, miles per hour) by the number of increments (e.g., days, weeks, hours) in the independent variable and add the result to the function value.

3. At best, interpolation and extrapolation give estimates. They are most reliable over short intervals.

EXAMPLE 2.10 Interpolating and extrapolating data: Age of first-time mothers

The following table shows the average age, in years, of first-time mothers in the given year:

Year	1980	1990	2000	2010	2020
Average age	22.7	24.2	24.9	25.4	26.0

a. Estimate the average age of first-time mothers in 2007.

b. Predict the average age of first-time mothers in 2025.

c. Predict the average age of first-time mothers in the year 3000. Explain why the resulting figure is not to be trusted.

SOLUTION

a. We estimate the average age in 2007 by interpolating. For this we need the average growth rate between 2000 and 2010. The average age changed from 24.9 years to 25.4 years over a 10-year period, so

$$\text{Average growth rate} = \frac{25.4 - 24.9}{10} = 0.05 \text{ year per year}$$

Thus, the average age of first-time mothers increased at a rate of 0.05 year per year over this decade. There are 7 years from 2000 to 2007, and the average age increased, on average, by 0.05 year each year. We therefore expect the average age in 2007 to be $24.9 + (7 \times 0.05) = 25.25$ years, or about 25.3 years.

b. In this case, we estimate the average age by extrapolating. For this we again need the average growth rate between 2010 and 2020, which is $(26.0 - 25.4)/10$ or 0.06 year per year. We therefore estimate an increase of 0.06 year over each of the 5 years from 2020 to 2025:

$$\text{Estimated average age in 2025} = 26.0 + (5 \times 0.06) = 26.3 \text{ years}$$

c. This is exactly like part b. The average growth rate from 2010 to 2020 is 0.06 year per year. Because the year 3000 is 980 years after the year 2020, this growth rate gives a prediction for the year 3000 of $26.0 + (980 \times 0.06) = 84.8$ years. Our projection for the average age in the year 3000 of first-time mothers is 85 years. This number is absurd and clearly illustrates the danger of extrapolating too far beyond the limits of the given data.

WHAT DO YOU THINK?

Interpolation: Stock market prices are an example of data that can be very volatile. Explain why care must be used with interpolation in such a case.

Questions of this sort may invite many correct responses. We present one possibility: When we use interpolation, we use two data points to make an estimate of what happened in between. For example, if a stock price is \$7.00 a share now and \$7.05 later, interpolation would suggest a price of between \$7.00 and \$7.05 in the period between. But in a rapidly changing market, the actual price per share might be far different. Interpolation gives accurate answers when there is very little change over the interval in question. But when dramatic changes occur, information provided by interpolation may be no better than a guess.

Further questions of this type may be found in the *What Do You Think?* section of exercises.

Try It Yourself answers

Try It Yourself 2.1: Choosing variables: Medals at Olympic Winter Games We choose the year to be the independent variable and the most medals won by any country to be the dependent variable. Then the most medals won is a function of the year: The function gives the most medals won by any country in a given year.

Try It Yourself 2.2: Calculating percentage change: U.S. population 36%.

Try It Yourself 2.4: Calculating average growth rate: U.S. population The average growth rate over this decade is

$$\frac{179.32 - 151.33}{10} = 2.799 \text{ million people per year}$$

or about 2.80 million people per year. This says that from 1950 to 1960, the U.S. population grew, on average, by about 2.80 million people each year.

Try It Yourself 2.5: Calculating a negative growth rate: Population of Russia -0.744 million per year. If the estimate is correct, the population will decline, on average, by 744,000 per year.

Try It Yourself 2.6: Interpreting the average growth rate: Flu cases The units of the average growth rate are flu cases per day. The average growth rate measures the number of new flu cases reported each day.

Try It Yourself 2.7: Interpolating data: National debt
14.8 million subscribers.

Try It Yourself 2.9: Extrapolating data: An epidemic 252 total cases.

Exercise Set 2.1

Test Your Understanding

1. True or false: A function shows how one quantity depends on another.

2. Tools that aid in understanding raw data include: **a.** percentage change **b.** average growth rate **c.** a bar graph **d.** all of the above.

3. The average growth rate equals: **a.** the change in the function value divided by the change in the independent variable **b.** the change in the function value divided by the change in the dependent variable **c.** the change in the independent variable divided by the change in the dependent variable.

4. The method that uses the average growth rate to estimate unknown values between data points is called ______.

5. The method that uses the average growth rate to estimate unknown values beyond known data points is called ______.

6. True or false: The units associated with the average growth rate are not particularly important in practical settings.

Problems

7. Percentage change. The following shows the value of an investment on January 1 of the given year.

Date	2017	2018	2019	2020	2021
Value	\$100.00	\$110.00	\$121.00	\$145.00	\$159.50

a. Make a table that shows percentage increase from each year to the next.

b. Make a bar graph that shows percentage increase from each year to the next.

c. Over what period was percentage growth the greatest?

8–10. Units for average growth rate.

8. The independent variable is the day of the year. The function is the cumulative number of accidents since the first of the year at a certain dangerous intersection up to that day. State the units of the average growth rate and give its practical meaning.

9. The independent variable is the day of the year. The function is the cumulative number of cases of a disease since the first of the year. State the units of the average growth rate and give its practical meaning.

10. The independent variable is the hour of the day. The function is the distance west of New York City of an automobile since it left the city on the freeway. State the units of the average growth rate and give its practical meaning.

11. Home runs. The following table shows the total number of home runs scored by major league baseball players in the given year:

Year	Home runs
2007	4957
2008	4878
2009	5042
2010	4613
2011	4552
2012	4934
2013	4661
2014	4186
2015	4909
2016	5610
2017	6105
2018	5585
2019	6776

Choose independent and dependent variables for this table and explain what the corresponding function means.

12. Basketball. The following table shows the number of bids to the NCAA Men's Basketball Tournament for the Big 12 Conference in the given year:

Year	Number of bids by Big 12
2014	7
2015	7
2016	7
2017	6
2018	7
2019	6
2020	canceled
2021	7

Choose independent and dependent variables for this table and explain what the corresponding function means.

13. Children in preprimary programs. The following table shows the percentage of children in the United States between the ages of 3 and 5 who are enrolled in preprimary programs.

Date	2008	2010	2012	2014
Percentage	63.0	63.7	64.3	64.7

a. Use interpolation to estimate the percentage of children enrolled in 2013.

b. Use extrapolation to estimate the percentage of children enrolled in 2016.

14. Customer complaints. The following table shows the number of customer complaints against airlines operating in the United States during the given year.

Year	2002	2004	2006	2008	2010
Number of complaints	9466	7452	8325	10,648	16,508

a. Use interpolation to estimate the number of complaints in 2007.

b. Complete the table below with the absolute numbers and the percentages by which the complaints changed over each two-year period. Round your final answers to the nearest whole number.

Range	2002–2004	2004–2006	2006–2008	2008–2010
Change in complaints				
Percentage change				

c. Use the second row of the table to estimate the number of complaints in 2001. The actual number was 16,508. What events in 2001 might cause the actual number to be higher than your estimate? Explain.

15. Grades. The bar graph in **Figure 2.7** shows the percentage of college freshmen in the United States, as of the fall of the given year, whose average grade in high school was an A-, an A, or an A+.

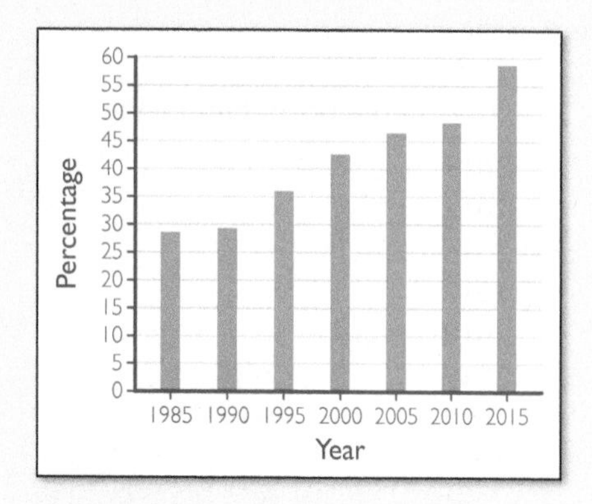

FIGURE 2.7 Freshmen with high school grades of "A."

a. Describe in general how the percentage changed over time.

b. Find the average yearly growth rate in the percentage from 2010 to 2015. Round your answer to the nearest whole number.

c. Estimate the percentage in the year 2012. Round your answer to the nearest whole number.

d. The actual percentage in 2012 was 49.5%. Compare this with your answer to part c. What does this say about how the rate of growth in the percentage varied from 2010 to 2015?

16. College expenditures. The table below shows total expenditures for colleges and universities in millions of dollars. All data are in current dollars.

Year	1970	1980	1990	2000	2010
Expenditures	21,043	56,914	134,656	236,784	461,000

a. What, in general, does the table tell you about spending on colleges and universities from 1970 through 2010?

b. Use interpolation to complete the following table:

Year	1970	1975	1980	1990	1991	2000	2010
Expenditures	21,043		56,914	134,656		236,784	461,000

c. Make a table that shows the absolute change in expenditures over each 10-year period.

d. Estimate expenditures in 2015. Explain your reasoning.

17. Inflation. Inflation is a measure of the buying power of your dollar. For example, if inflation this year is 5%, then a given item would be expected to cost 5% more at the end of the year than at the beginning of the year. **Figure 2.8** shows the rate of inflation in the United States for the given years.

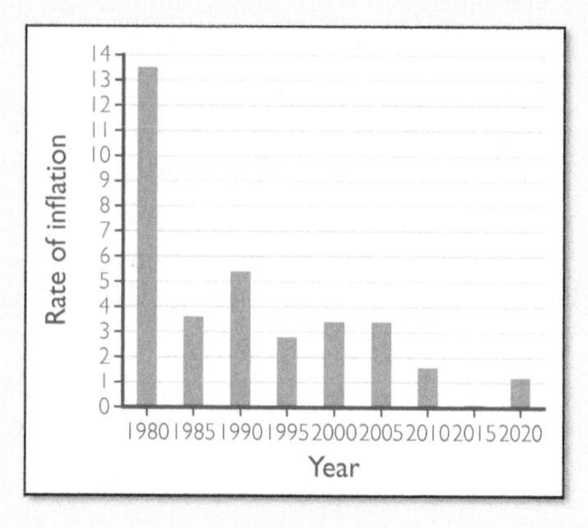

FIGURE 2.8 Rate of inflation.

a. Use interpolation to estimate the rate of inflation in the United States for each of the years 1996 through 1999. Round your answer to one decimal point as a percentage.

b. Assume that men's clothing followed the rate of inflation. If a man's suit cost \$300 at the beginning of 1995, what did it cost at the end of 1995? Round your answer to the nearest dollar.

c. It is a fact that in 1949 the rate of inflation in the United States was negative, and the value was −2%. If a man's suit cost \$100 at the beginning of 1949, what did it cost at the end of 1949?

d. Extreme inflation can have devastating effects. In 1992, inflation was 100% in Lebanon and Bulgaria.[3] If a man's suit cost \$100 at the beginning of a year with that rate of inflation, what would it cost at the end of the year?

Exercise 18 is suitable for group work.

18. School costs. The following table shows the average total cost, measured as dollars expended per pupil, of operating a high school with the given enrollment. This table is based on a study of economies of scale in high school operations from the 1960s.

Enrollment	170	250	450	650
Average total cost	532	481	427	413

a. Estimate the average cost per student for a high school with an enrollment of 350 students.

b. The actual cost per student with an enrollment of 350 is \$446 per pupil. Use this information to compare how the average cost changes from an enrollment of 250 to 350 with how the average cost changes from an enrollment of 350 to 450.

[3] *See* www.theodora.com/wfb/1992/rankings/inflation_rate_pct_1.html

c. What is the average growth rate of the average cost per additional student enrolled, when moving from an enrollment of 450 students to an enrollment of 650 students?

d. What would you estimate for the average cost per student at an enrollment of 750 students?

e. Some policy analysts use tables such as this to argue that small schools should be consolidated to make the operations more cost-effective. On the basis of your answers to previous parts, determine which is likely to save more money (in terms of average total cost): consolidation that results in increasing the enrollment by 100 starting at the low end of the scale (say, from 250 to 350) or such consolidation starting at a higher level (say, from 350 to 450). *Note*: These computations, of course, represent only one side of a complex problem. School consolidation can be devastating to small rural communities.

The following exercises are designed to be solved using technology such as calculators or computer spreadsheets.

19. Enrollment in French. The table below shows course enrollments (in thousands) in French in U.S. institutions of higher education for the given year.

Year	1995	1998	2002	2006	2009
Enrollment	205.4	199.1	202.0	206.4	216.4

a. Explain in general terms what the data on enrollments in French show.

b. For each period shown in the table (1995–1998, 1998–2002, etc.), calculate the average growth rate per year in enrollments in French. Note that the time periods are not of the same length.

c. Make a bar graph displaying the average growth rate per year in enrollments in French.

20. AP examinations. The following table shows the percentage of Advanced Placement examinations in which students achieved the scores of 1 through 5 for the years 2010, 2015, and 2020.

Score	5	4	3	2	1
2010	15.1	19.5	23.4	20.5	21.6
2015	13.3	19.5	25.1	22.1	19.9
2020	15.9	21.9	26.5	20.7	15.0

a. Make a bar graph showing the data. *Suggestion:* Use the different scores (1 through 5) for categories and put a bar for each year.

b. For which year were the scores above 3 the highest?

21. Buying power. The following table shows how the buying power of a dollar is reduced with inflation. The second line in the table gives the percent reduction of a dollar's buying power in one year due to the corresponding inflation rate.

Annual inflation rate	10%	20%	30%	40%	50%
Buying power reduction	9.1%	16.7%	23.1%	28.6%	33.3%

Annual inflation rate	60%	70%	80%	90%	100%
Buying power reduction	37.5%	41.2%	44.4%	47.4%	50.0%

a. Make a bar graph for this table with the inflation rate on the horizontal axis.

b. If the buying power of a dollar declined by 25% during a year, what was the inflation rate during that year? *Suggestion*: First note that 25% is between 23.1% and 28.6%, so the answer must be an inflation rate between 30% and 40%. Interpolate to find the answer more accurately.

Writing About Mathematics

22. History and population I. The table on page 85 shows a percentage increase from 1930 to 1940 that is significantly less than other decades. Investigate what may have caused this, and write a report on your findings. In particular, you might include a discussion of an event called "Black Tuesday" that may have been a factor.

23. History and population II. The table on page 85 shows that the population increased by a smaller percentage from 1910 to 1920 than the preceding or following decades. There were two historical events that may have accounted for this. One was a great pandemic of influenza. Investigate the history of this pandemic, and write a brief report on it. What other great historical event may have influenced the population growth in this decade?

24. History and population III. The table on page 85 shows that the population increased by a larger percentage from 1950 to 1960 than the preceding or following decades. This growth is part of what is often referred to as the "baby boom." Write a report on the baby boom and what caused it.

25. Your own interests I. Write a report on a topic of interest to you that uses the percentage change as a tool for analysis.

26. Your own interests II. Write a report on a topic of interest to you that uses the average rate of change as a tool for analysis.

What Do You Think?

27. Analyzing data. You are given a table of data, and it is important for you to understand the situation represented by the data. What kinds of things might you do that would help you analyze the data? Your answer might include calculations that you would make as well as visual displays.

28. Average growth rate. A newspaper article tells you that flu cases in your city are increasing by 36 new cases. Is information missing from this report? What additional information would you need in order to make sense of the report?

29. Long-term extrapolation. You read two articles about unemployment in the United States. One article uses recent data to estimate unemployment rates one year from now. The other article uses recent data to estimate unemployment rates 10 years in the future. Do you have reason to trust one article over the other? Explain your reasoning.

30. Sales tax. Harry notes that the state sales tax went from 5% to 5.5%, which he says is not too bad because it's just a one-half percent increase. But Linda says that it really is bad because it's a 10% increase. Who's right, and why?

31. Bar graphs. You see a bar graph where the horizontal axis represents time in months and the vertical axis represents the growth rate of a stock account during that month. You notice that one of the bars is below the horizontal axis. Explain what this means.

32. Bar graphs and functions. Consider the table in Exercise 12, which shows bids to the NCAA Men's Basketball Tournament for the Big 12. Try to make a bar graph by letting the horizontal axis represent the number of bids and letting the heights of the vertical bars represent the year. Explain any difficulties you encounter.

33. Percentage change. When we studied population growth in the United States and the world in this section, we calculated the percentage change. If you were given a table showing how food prices in the United States changed with time, would it be helpful to calculate the percentage change? Can you think of other situations in which the percentage change gives important information?

34. No change? The population of a town is 500. Then the population decreases by 100 over the year, and the population grows by 100 over the next year. What is the average yearly growth rate in the population over the two-year period? Does this growth rate fully represent what happened over this period? If not, what calculations would you make to give a better representation?

35. Which would you choose? One pair of jeans is advertised at 10% off the regular price. At another store, the same pair of jeans is advertised at \$5.00 off the regular price. Which is the better buy? Do you need additional information to make your decision? If so, what additional information?

2.2 Graphs: Picturing growth

TAKE AWAY FROM THIS SECTION
Interpret growth rates displayed by graphs.

The following *In the News 2.3* discusses climate change and a graph sometimes used to model it.

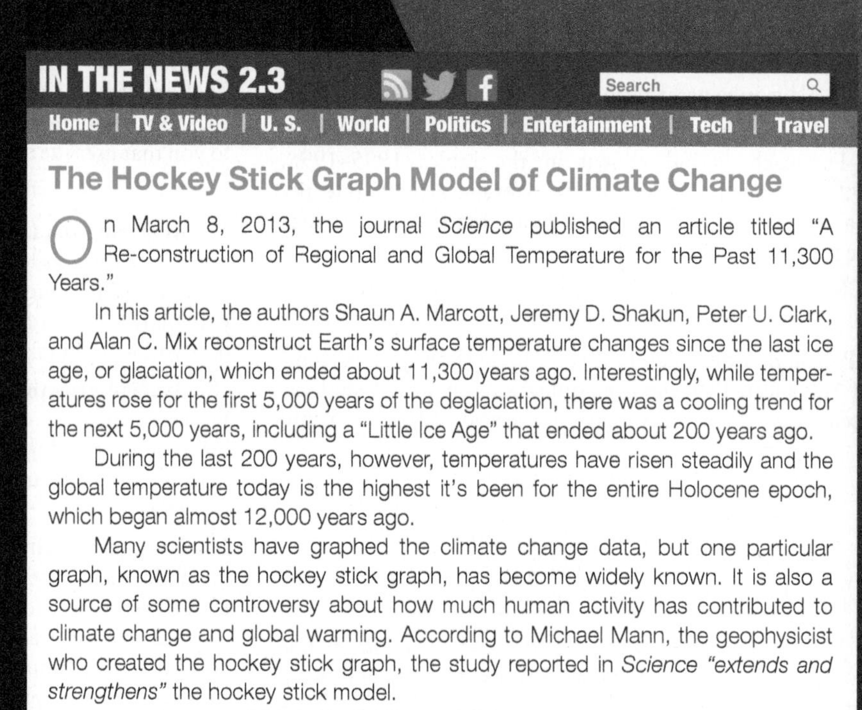

IN THE NEWS 2.3

Search

Home | TV & Video | U. S. | World | Politics | Entertainment | Tech | Travel

The Hockey Stick Graph Model of Climate Change

On March 8, 2013, the journal *Science* published an article titled "A Re-construction of Regional and Global Temperature for the Past 11,300 Years."

In this article, the authors Shaun A. Marcott, Jeremy D. Shakun, Peter U. Clark, and Alan C. Mix reconstruct Earth's surface temperature changes since the last ice age, or glaciation, which ended about 11,300 years ago. Interestingly, while temperatures rose for the first 5,000 years of the deglaciation, there was a cooling trend for the next 5,000 years, including a "Little Ice Age" that ended about 200 years ago.

During the last 200 years, however, temperatures have risen steadily and the global temperature today is the highest it's been for the entire Holocene epoch, which began almost 12,000 years ago.

Many scientists have graphed the climate change data, but one particular graph, known as the hockey stick graph, has become widely known. It is also a source of some controversy about how much human activity has contributed to climate change and global warming. According to Michael Mann, the geophysicist who created the hockey stick graph, the study reported in *Science "extends and strengthens"* the hockey stick model.

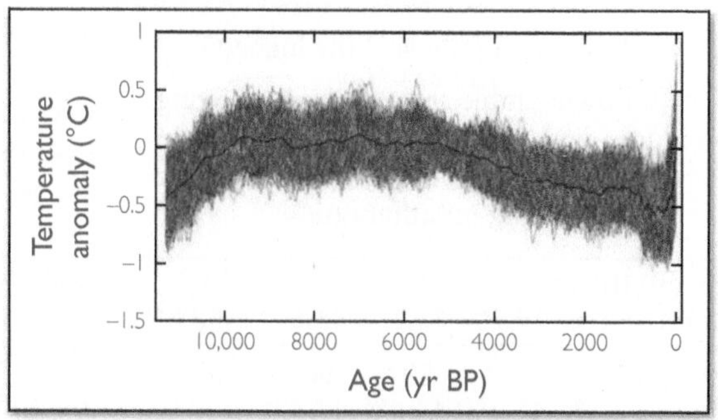

FIGURE 2.9 Global warming. **Source:** Marcott et al. (2013). Reprinted by permission of the American Association for the Advancement of Science.

At the center of the public controversy over global warming is the graph shown in **Figure 2.9**. In this graph the black line represents the temperature relative to a long-term average. This black line is produced by a complicated method that involves averaging certain measurements, which are shown by colored lines.

There is controversy regarding both the accuracy of the graph and its interpretation. But at the heart of the controversy is how the graph changes and what that change means. From 1900 on (the "blade" portion of the "hockey stick," at the extreme right-hand side of the graph), the character of the graph is different from the earlier part—the "hockey stick handle." It is this change in character that concerns scientists, politicians, and the general public.

Why are graphs so often used by both the scientific community and the popular media? A graph is a good example of the old saying that a picture is worth a thousand words. Graphs allow us to visualize information, which helps us see patterns that may not be readily apparent from a table. For example, a graph reveals visually where the function is increasing or decreasing and how rapid the change is. In this section, we study various types of graphs and learn how to interpret these patterns.

In the preceding section, we used bar graphs to present data from tables. In this section, we consider two other ways to present data: scatterplots and line graphs. These ways of presenting data are often seen in the media, and it is important to be familiar with them, to interpret them, to analyze them critically, and to be aware of their advantages and disadvantages.

Quick Review Making Graphs

Typically, we locate points on a graph using a table such as the one below, which shows the number of automobiles sold by a dealership on a given day.

Day (independent variable)	1	2	3	4
Cars sold (dependent variable)	3	6	5	2

When we create a graph of a function, the numbers on the horizontal axis correspond to the independent variable, and the numbers on the vertical axis correspond to the dependent variable. For this example, we put the day (the independent variable) on the horizontal axis, and we put the number of cars sold (the dependent variable) on the vertical axis.

The point corresponding to day 3 is denoted by (3, 5) and is represented by a single point on the graph. To locate the point (3, 5), we begin at the origin and move 3 units to the right and 5 units upward. This point is displayed on the graph as one of the four points of that graph. The aggregate of the plotted points is the graph of the function.

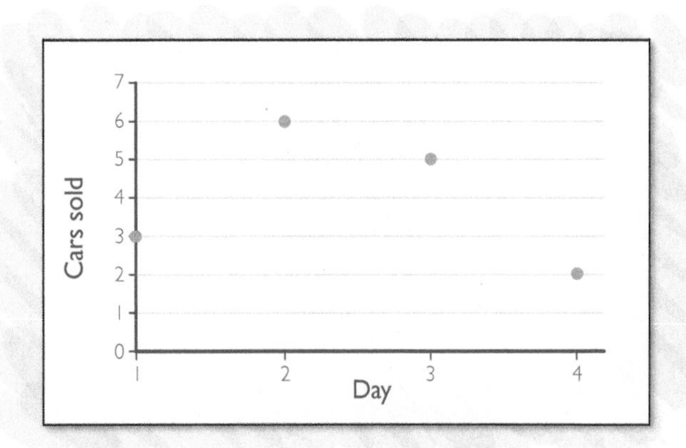

Scatterplots

Often we have a collection of isolated data points and want to make a visual display in order to search for patterns. When the values of the independent variable are evenly spaced, a bar graph may be used. When the values of the independent variable are not evenly spaced, a scatterplot is often used to represent the data. Scatterplots are also used to represent large numbers of data points that cannot be reasonably displayed using a bar graph.

KEY CONCEPT

A scatterplot is a graph consisting of isolated points, with each dot corresponding to a data point.

An interesting example comes from recording information on more than 100 eruptions of the Old Faithful geyser at Yellowstone National Park. These data show the waiting time between eruptions in terms of the duration of the eruption. The graph is shown in **Figure 2.10**. Because each eruption is a separate data point, the scatterplot is simply a plot of the 100 or more separate points.

Certainly, it would be impractical and messy to try to represent these data using a bar graph. Moreover, the scatterplot neatly indicates that there are two types of eruptions: those with short duration and short waiting time, and those with long duration and long waiting time.

Scatterplots are often used to discover patterns in the data. For example, the line segments that have been added in Figure 2.10 suggest such patterns.

Line graphs and smoothed line graphs

Many times connecting the data points on a scatterplot makes patterns easier to spot. The result of connecting the dots is called a *line graph*.

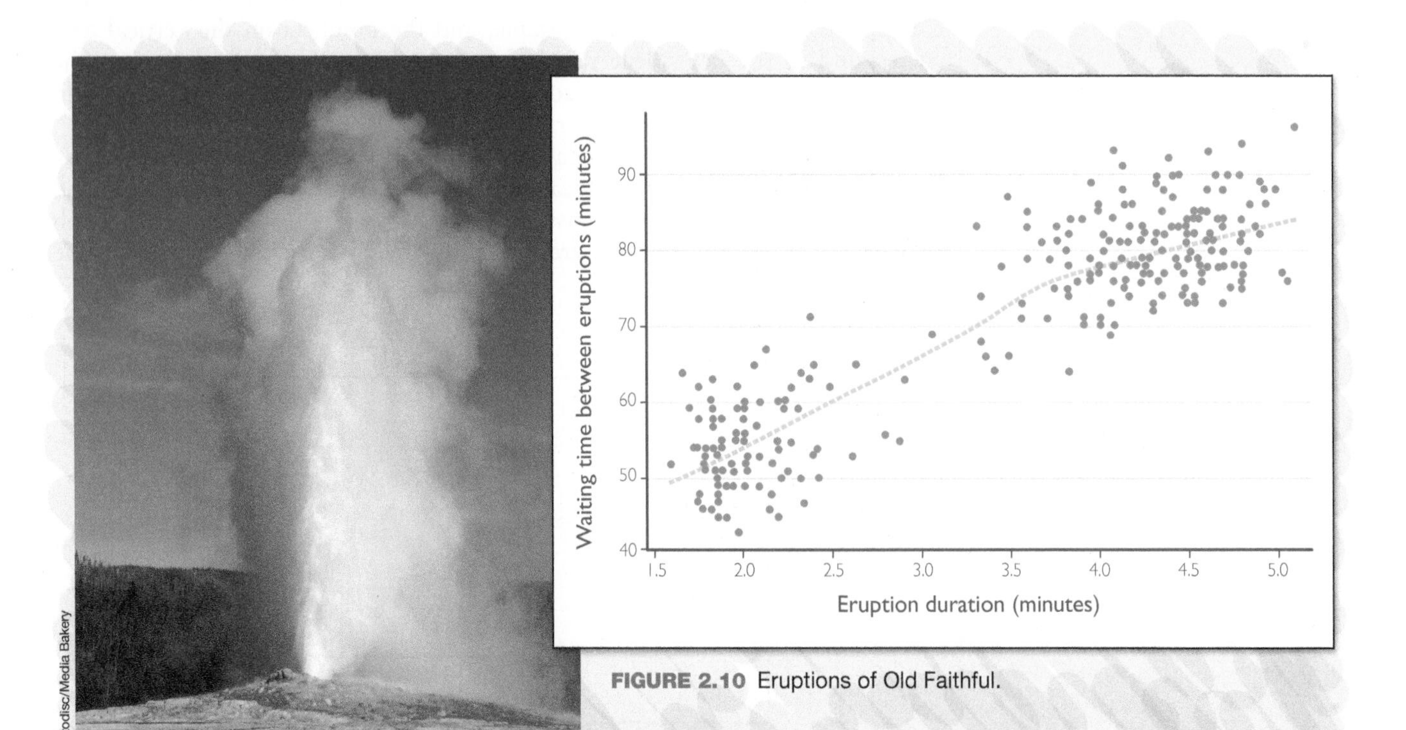

FIGURE 2.10 Eruptions of Old Faithful.

KEY CONCEPT

To make a line graph, we begin with a scatterplot and join the adjacent points with straight line segments.

An interesting example comes from the following table, which shows running speed of various animals as a function of length:[4]

Animal	Length (inches)	Speed (feet per second)
Deermouse	3.5	8.2
Chipmunk	6.3	15.7
Desert crested lizard	9.4	24.0
Grey squirrel	9.8	24.9
Red fox	24.0	65.6
Cheetah	47.0	95.1

We make the line graph in two steps. First we make the scatterplot shown in **Figure 2.11.** Then we join adjacent points with line segments to get the line graph shown in **Figure 2.12.**

One advantage of a line graph over a scatterplot is that it allows us to estimate the values between the points in the scatterplot. Suppose, for example, we want to know how fast a 15-inch-long animal can run. There is no data point in the table corresponding to 15 inches, but we can use the graph to estimate it. In **Figure 2.13,** we find the 15-inch point on the horizontal axis (halfway between 10 and 20) and then go up to the line graph. We cross the graph at a level corresponding to about 40 feet per second on the vertical axis. So, these segments allow us to fill gaps in data by estimation.

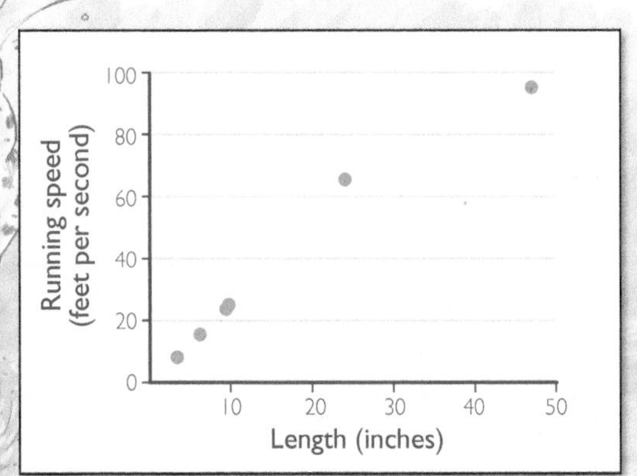

FIGURE 2.11 Scatterplot of running speed versus length.

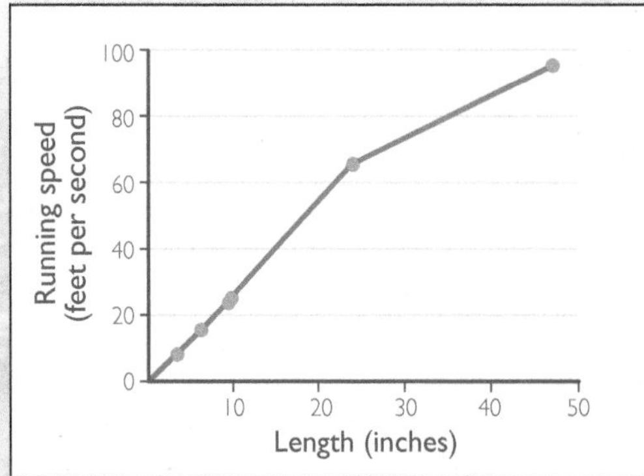

FIGURE 2.12 Line graph produced by adding segments.

[4] *This table is adapted from J. T. Bonner,* Size and Cycle *(Princeton, NJ: Princeton University Press, 1965). We will encounter it again in Chapter 3 when we study trend lines.*

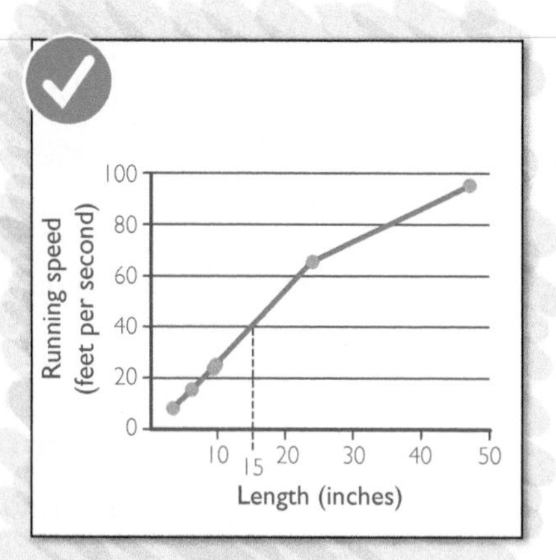

FIGURE 2.13 Information added by the line graph.

An interesting fact about filling in gaps using line segments is that the value of the function (running speed in this case) determined in this way is exactly the same as that obtained by interpolation, which we discussed in Section 2.1. Line graphs provide a graphical representation of the average growth rate from one point to the next.

One must take care not to use a line graph when there is nothing between categories. For example, **Figure 2.14** shows two line graphs. These graphs show the number of people playing various card games at the Yahoo! Web site on a certain Wednesday and a certain Sunday.[5] The line segment joining blackjack to bridge would indicate that there are card games in between them. A bar graph would be appropriate here. We will encounter graphs where points are connected to make line graphs, even though mathematically a scatterplot or bar graph would be more appropriate. One such example is Figure 2.17–see the discussion there (footnote 6, on page 104).

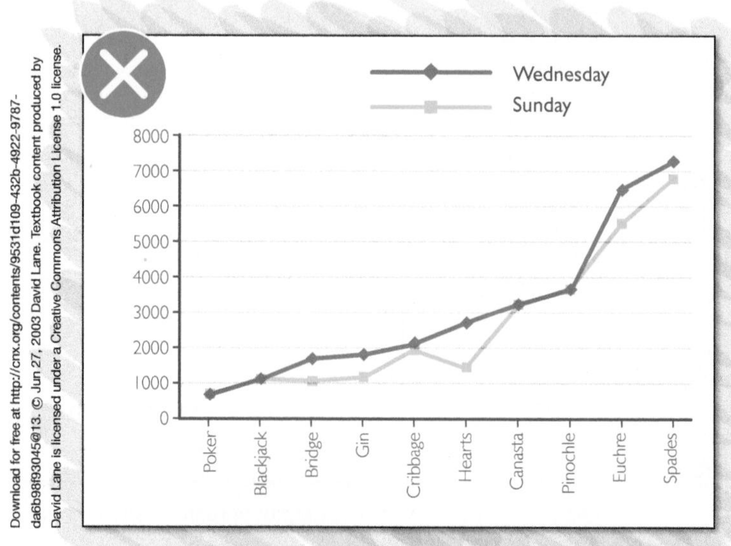

FIGURE 2.14 When a line graph is not appropriate.

[5] *This figure is taken from* `http://cnx.org/content/m10927/latest/`, *where it is used to illustrate the misuse of line graphs.*

EXAMPLE 2.11 Making a scatterplot and a line graph: Running speed and temperature

off the mark.com by Mark Parisi

THE BOSS WOULDN'T LET ANYONE LEAVE THE MEETING UNTIL SOMEONE EXPLAINED HOW THE WINDOW GOT BROKEN

The running speed of ants varies with the ambient temperature. Here are data collected at various temperatures:

Temperature (degrees Celsius)	Speed (centimeters per second)
25.6	2.62
27.5	3.03
30.4	3.56
33.0	4.17

First make a scatterplot of the data showing the speed as the function and the temperature as the independent variable, then make a line graph using these data.

SOLUTION

To make a scatterplot, we plot the data points, each of which corresponds to a point on the graph: (25.6, 2.62), . . . , (33.0, 4.17). The scatterplot is the graph of these points, as shown in **Figure 2.15**. We join the points with line segments to get the line graph in **Figure 2.16**.

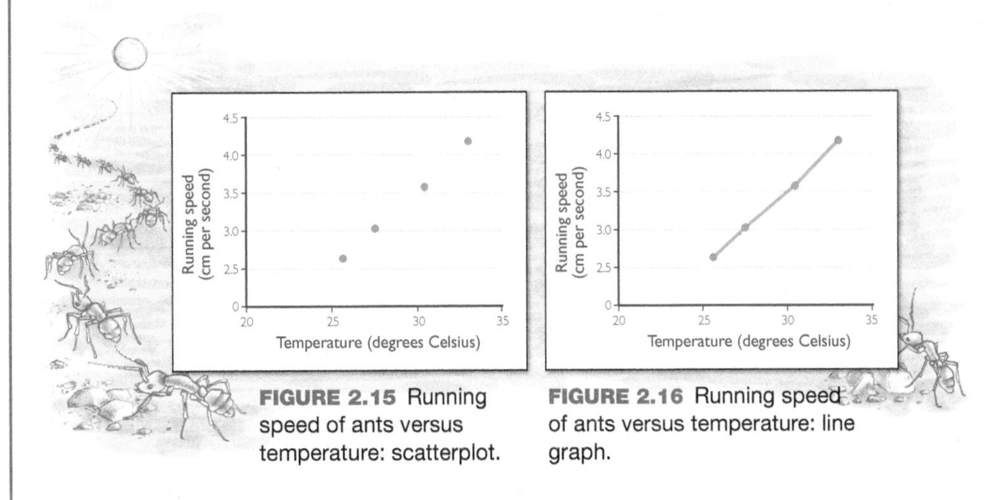

FIGURE 2.15 Running speed of ants versus temperature: scatterplot.

FIGURE 2.16 Running speed of ants versus temperature: line graph.

Age (days)	Height (centimeters)
5	17.43
35	100.88
41	128.68
61	209.84
70	231.23

TRY IT YOURSELF 2.11

Use the table (in the margin) to make a scatterplot of the height of a sunflower as a function of its age. Then make a line graph.

The answer is provided at the end of this section.

Interpreting line graphs: Growth rates

A key feature of line graphs is that they display maximums and minimums as well as visual evidence of growth rates. Consider the following example, which is taken from the Web site of the National Cancer Institute. The table below shows the five-year cancer mortality rates per 100,000 person-years in the United States. The data are sorted by race and sex.

Period	Five-year cancer mortality rates			
	Black male	White male	Black female	White female
1950–54		216.7		180.2
1955–59		225.0		172.5
1960–64		230.0		165.6
1965–69		239.6		162.2
1970–74	304.0	250.4	177.7	160.6
1975–79	334.7	259.1	179.9	161.3
1980–84	365.1	265.8	191.2	166.1
1985–89	383.9	269.1	199.3	170.6
1990–94	392.6	268.7	205.7	172.8
1995–99	363.2	252.7	200.7	167.9
2000–04	324.2	235.9	189.7	162.0
2005–09	283.8	216.7	174.1	151.8
2010–14	247.3	199.8	161.8	141.9

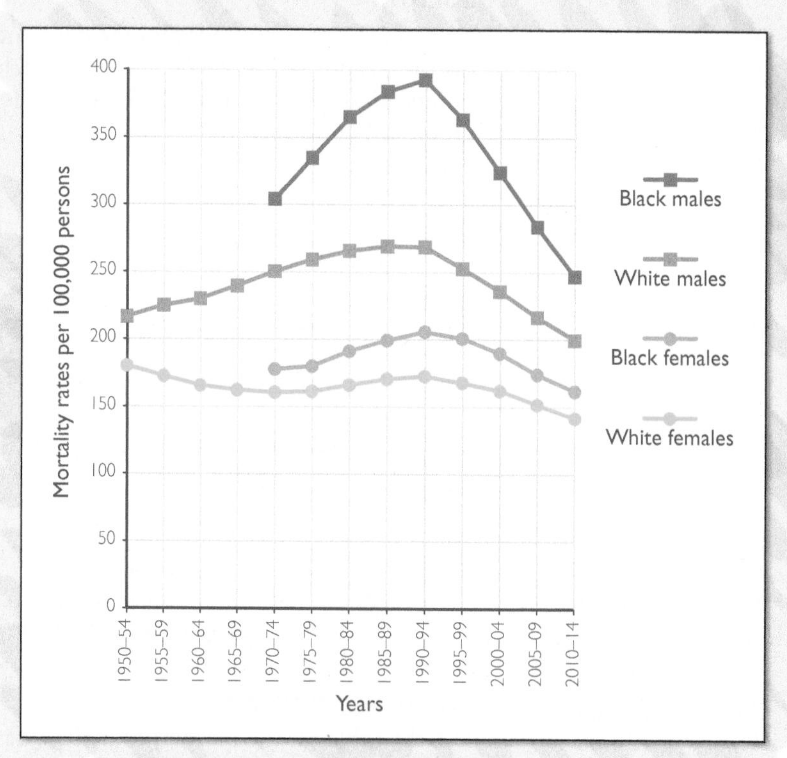

FIGURE 2.17 Five-year cancer mortality rates.

Figure 2.17 shows line graphs of these data.[6] This figure reveals several trends that are not readily apparent from the table. For one, it is immediately clear that cancer mortality rates among black males are much higher than for any of the other categories presented. These rates were on the rise from 1970–74 to 1990–94 and then fell sharply through 2010–14, ending at a level below that of 1970–74.

The graph shows that cancer mortality rates among white women reached a minimum of about 160 per 100,000 at the point representing the period 1970–74.

[6]*The lines in this line graph have no meaning as such, and a scatter plot or a bar graph would be better mathematically. This graph is based on a similar line graph from the National Cancer Institute; certainly the lines help us visualize the data.*

Those rates began increasing slightly after that time and then decreased to the lowest level in 2010–14. Also, we see that cancer mortality rates for white males reached their peak at the point representing the period 1985–89. It would take considerable effort to see these trends using only the table.

It is important to note not only how high a graph is but also how steep it is: A steeper graph indicates a growth rate of greater magnitude. For example, we observe that the graph of mortality rates for black males begins to level off from 1980–84 to 1990–94. That is, the *rate* of increase appears to be slowing. Rates for black women appear to have increased at a fairly steady rate from 1975–79 to 1990–94, and then decreased at a steady rate from 1990–94 to 2010–14. The line graph makes apparent these changes in growth rates, far more so than the numerical data themselves do.

The sign of the growth rate for a graph can be seen by examining where the graph increases or decreases. A positive growth rate is indicated when the graph increases as we move from left to right along the horizontal axis, and a negative growth rate is indicated by a decreasing graph. For example, the mortality rate for white women is decreasing from 1950–54 to 1970–74 and also from 1990–94 to 2010–14, indicating negative growth rates during those periods. The increasing graph from 1970–74 to 1990–94 indicates a positive growth rate.

SUMMARY 2.4
Growth Rates and Graphs

1. The growth rate of data is reflected in the steepness of the graph. Steeper graphs indicate a growth rate of greater magnitude.
2. An increasing graph indicates a positive growth rate, and a decreasing graph indicates a negative growth rate.

EXAMPLE 2.12 Interpreting line graphs: Income

The line graph in **Figure 2.18** shows the yearly gross income in thousands of dollars for a small business from 2012 through 2021. Explain what this graph says about the yearly income. Pay particular attention to the rate of growth of income.

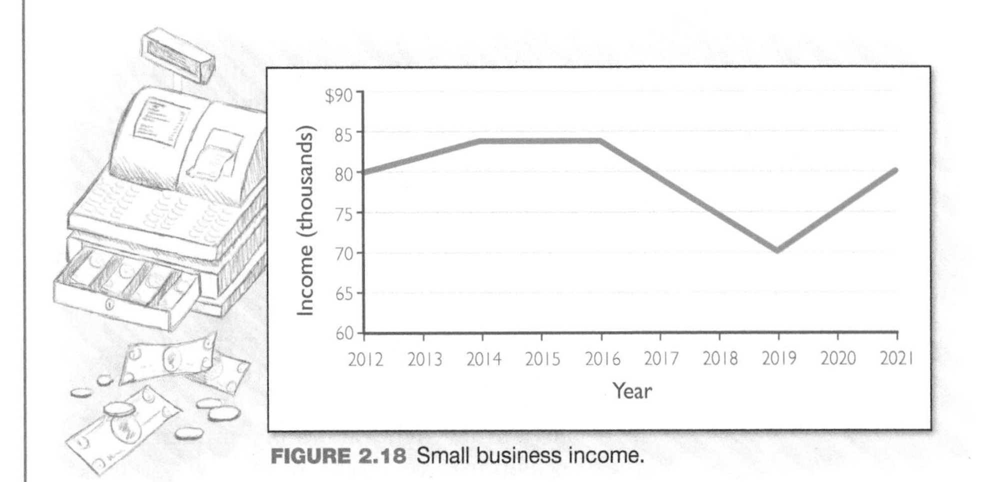

FIGURE 2.18 Small business income.

SOLUTION

Income increased between the years 2012 and 2014. After that it remained constant until 2016, when it began to decline. Income reached a minimum in 2019 and increased after that.

The growth rate of income was positive from 2012 to 2014 and from 2019 to 2021. The fact that the graph is steeper from 2019 to 2021 than from 2012 to 2014 means that income was growing at a faster rate from 2019 to 2021.

Because income declined from 2016 to 2019, the growth rate was negative over that period. From 2014 to 2016, income did not change, meaning there was no growth. So the growth rate over that period was zero.

TRY IT YOURSELF 2.12

The graph in **Figure 2.19** shows sales in thousands of dollars from 2012 to 2021. Describe the rate of growth in sales over this period.

The answer is provided at the end of this section.

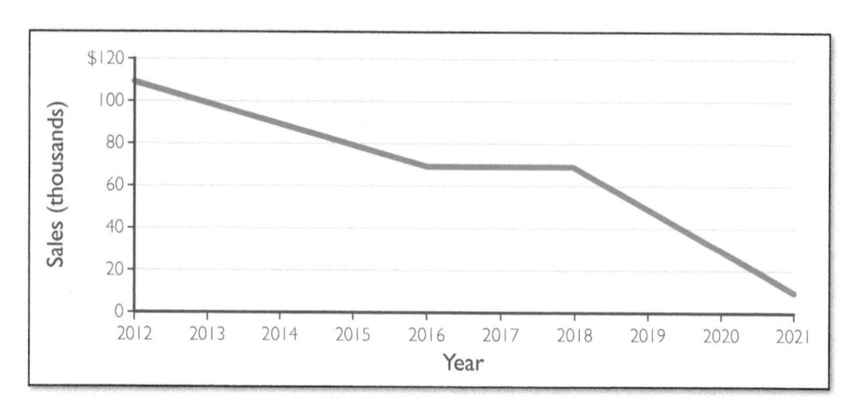

FIGURE 2.19 Sales.

Growth rates for smoothed line graphs

Sometimes it is more appropriate to join points with a curve rather than a straight line. The result is a *smoothed line graph*.

KEY CONCEPT

A smoothed line graph is made from a scatterplot by joining data points smoothly with curves instead of line segments.

A smoothed line graph may be appropriate when the growth rate is continuous. For example, **Figure 2.20** shows the weight of a baby over the first 12 months of life. Weight changes continuously, not in jerks. Hence, a smooth graph is appropriate.

Smoothed line graphs that are accurate contain more information about growth rates than do straight line graphs. For example, **Figure 2.21** shows the growth of money in a savings account, and **Figure 2.22** shows a man's height as a function of his age. Both graphs are increasing and so display a positive growth rate. But the *shapes* of the two curves are different. The graph in Figure 2.21 gets steeper as we move to the right, indicating an increasing growth rate. That is, the value of the account is increasing at an increasing rate. This is typical of how savings accounts grow, as we shall see in Chapter 4.

In contrast, the graph in Figure 2.22 gets less steep as we move to the right, indicating a decreasing growth rate. That is, the graph is increasing at a decreasing rate. Practically speaking, this means that our growth in height slows as we age.

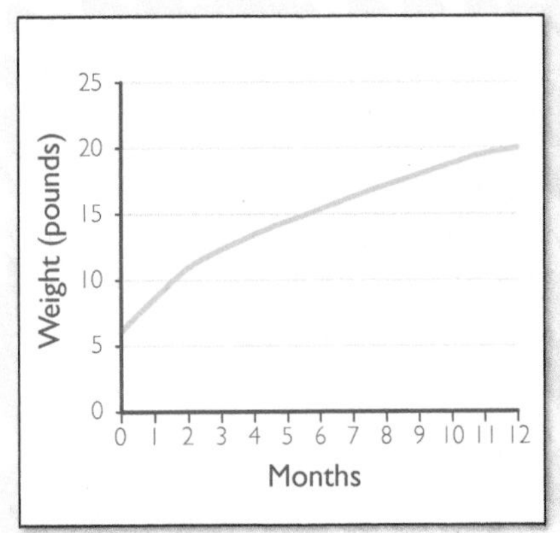

FIGURE 2.20 Smoothed graph showing a child's weight.

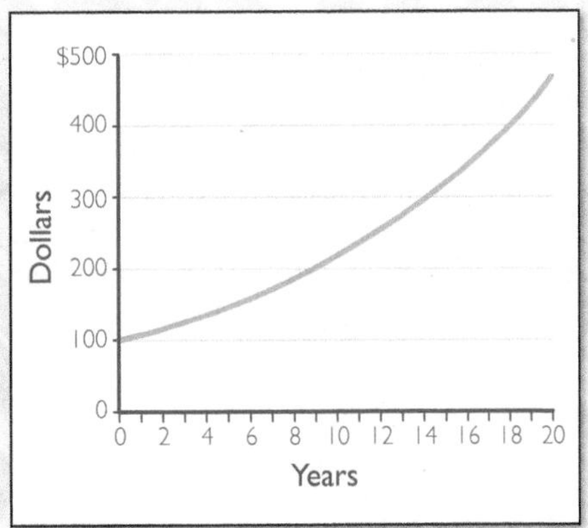

FIGURE 2.21 Money in a savings account.

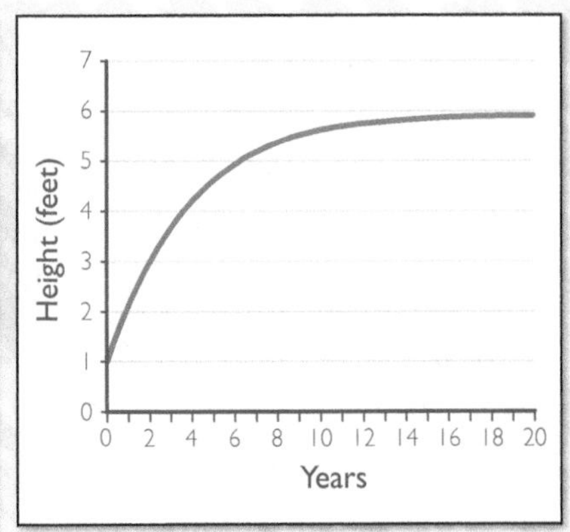

FIGURE 2.22 Height versus age.

EXAMPLE 2.13 Making graphs with varying growth rates: Cleaning a waste site

In cleaning toxic waste sites, typically the amount of waste eliminated decreases over time. That is, the graph of the amount of toxic waste remaining as a function of time decreases at a decreasing rate. Sketch an appropriate graph for the amount of toxic waste remaining as a function of time.

SOLUTION

The graph should be decreasing, but it should become less steep as we move to the right. One possibility is shown in **Figure 2.23**.

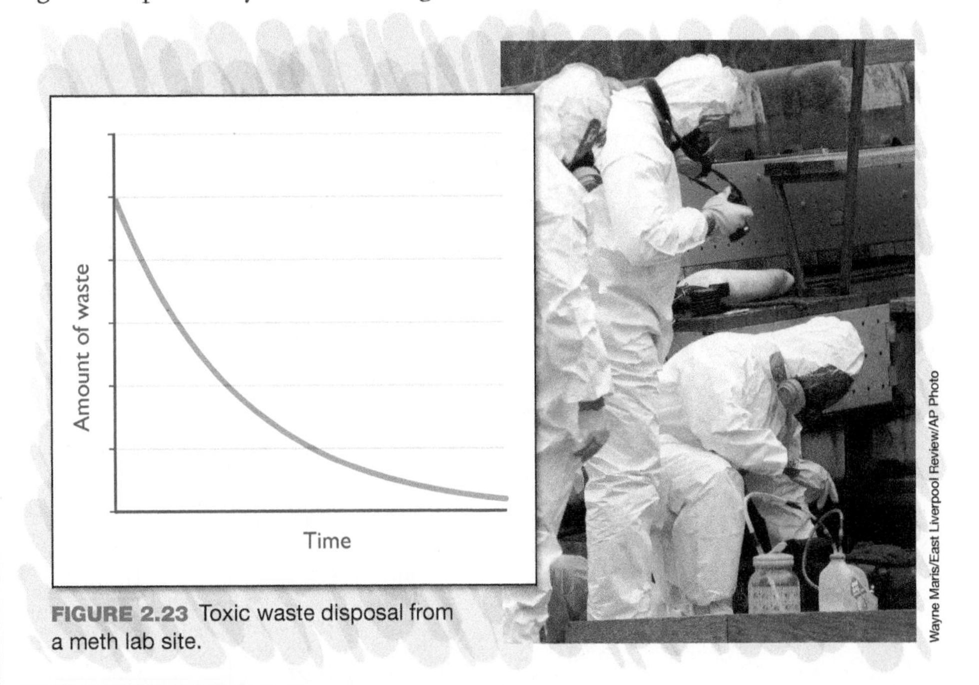

FIGURE 2.23 Toxic waste disposal from a meth lab site.

TRY IT YOURSELF 2.13

As an automobile ages, gas mileage typically decreases, and it decreases at an increasing rate. Sketch a possible graph showing gas mileage as a function of time.

The answer is provided at the end of this section.

Growth rates in practical settings

In practical settings, the growth rate has a familiar meaning. For example, the graph in **Figure 2.24** shows a population that increases for a time and then begins to decrease. The growth rate in this context is, as the name suggests, the rate of population growth. In this graph, the growth rate is positive in Year 0. As we move toward Year 3, it remains positive but becomes less and less so until we reach Year 3, where the growth rate is 0. Beyond Year 3 the growth rate gets progressively more and more negative through Year 5.

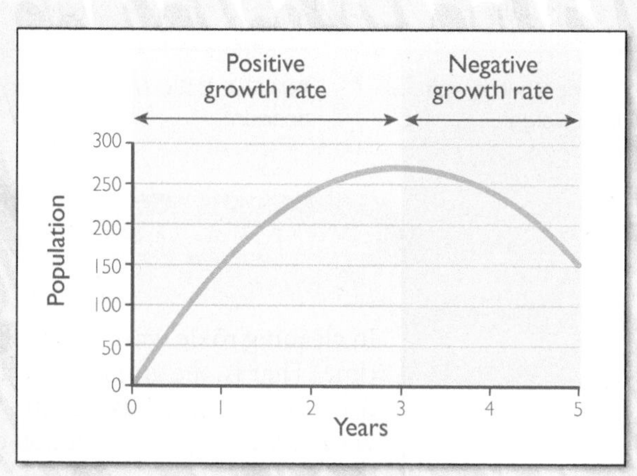

FIGURE 2.24 A population graph.

EXAMPLE 2.14 Describing growth rates using graphs: Deer population

Ecologists have closely studied the deer population on the George Reserve in Michigan.[7] A graph of the population over time is shown in **Figure 2.25**. Here, the horizontal axis represents the number of years since observations began, and the vertical axis represents the number of deer present.

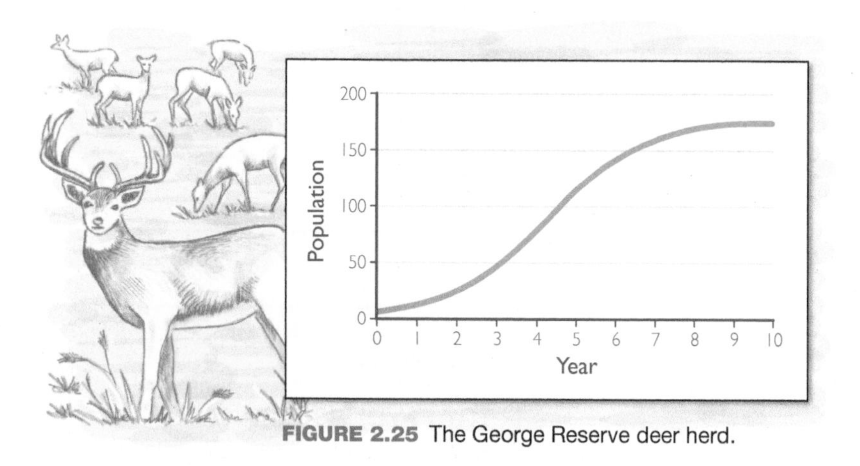

FIGURE 2.25 The George Reserve deer herd.

[7] *One interesting source is Dale R. McCullough,* The George Reserve Deer Herd *(Ann Arbor: University of Michigan Press, 1979).*

a. Give a brief general description of how the deer population changes over the 10-year period shown. Your description should include the growth rate of the population and how that rate changes. Explain what this means for the population in the long run.

b. Estimate the time at which the population is growing at the fastest rate.

SOLUTION

a. The population grew throughout the 10-year observation. Over about the first four years of the observation, the graph is increasing and getting steeper, so the population shows increasingly rapid growth over this period. This means that not only is the population increasing over this period, but also the *rate* of population growth is increasing. From about four years on, we see that the graph continues to increase but becomes less steep. Thus, the population continues to grow, but at a decreasing rate. The graph levels off as we approach Year 10. This flattening of the graph indicates that the growth rate is declining to near 0, and the population is stabilizing at about 175 deer.

b. The population is growing at the fastest rate where the graph is the steepest. We estimate that this occurs at about Year 4.

WHAT DO YOU THINK?

Which stock? You are trying to decide which of two stocks to purchase. Both are increasing in value, but Stock 1 is increasing at an increasing rate while Stock 2 is increasing at a decreasing rate. Based solely on this information, which stock would you purchase? Explain why you chose the stock you did.

Questions of this sort may invite many correct responses. We present one possibility: Stock 1 is probably the better bet. Because the stock price is increasing at an increasing rate, it is growing ever faster in value. If that trend continues, an investor can expect a tidy profit. Stock 2 is growing in value, but the growth rate is slowing. This would seem to indicate a leveling-off in value. Thus, the current trend suggests only a minimal profit. (As reasonable as this may seem, stock market prices remain unpredictable).

Further questions of this type may be found in the *What Do You Think?* section of exercises.

Try It Yourself answers

Try It Yourself 2.11: Making a scatterplot and a line graph: Sunflower growth

Scatterplot:

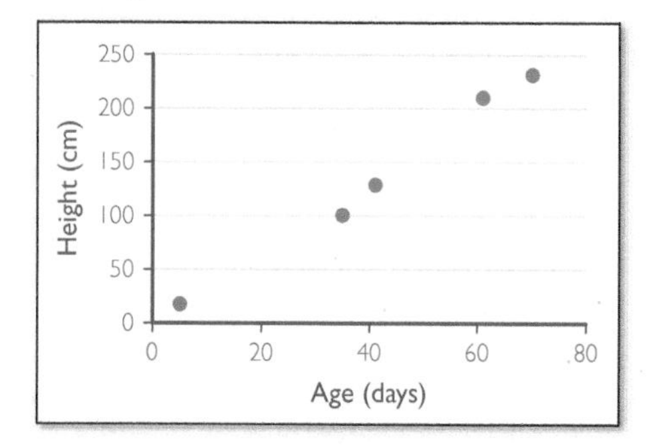

Line graph:

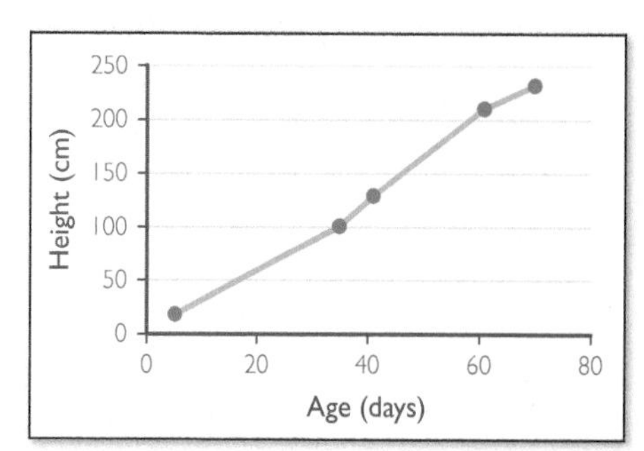

Try It Yourself 2.12: Interpreting line graphs: Sales growth The rate of growth in sales is negative from 2012 to 2016 and from 2018 to 2021. The growth rate is more negative over the second period. From 2016 to 2018, the growth rate is 0.

Try It Yourself 2.13: Making graphs with varying growth rates: Gas mileage The graph should be decreasing and become steeper as we move to the right. The accompanying figure gives one example.

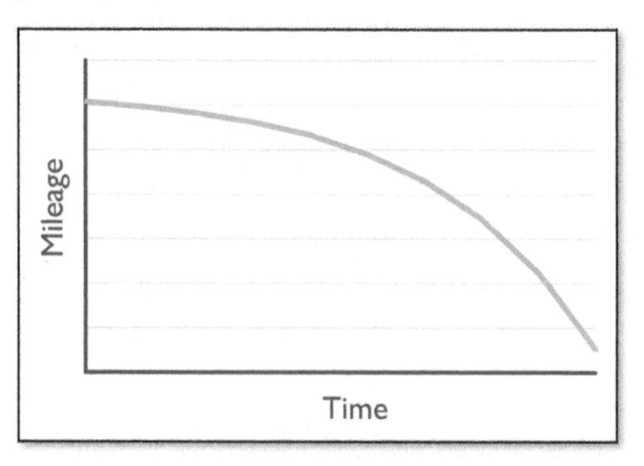

Exercise Set 2.2

Test Your Understanding

1. Matching. Match the type of graph with an important feature.

A. Scatterplot	I. May show continuous growth rate
B. Line graph	II. Shows many isolated data points
C. Smoothed line graph	III. Connects data points with straight lines

2. More matching. Match information about the rate of change with the corresponding feature of the graph.

A. Positive growth rate	I. Decreasing graph
B. Negative growth rate	II. Increasing graph

3. Using a line graph to estimate unknown function values is the same as: **a.** making a smoothed line graph **b.** using the percentage change **c.** interpolating using the average growth rate.

4. True or false: A graph can show features that may not be apparent from the underlying data.

Problems

5. Expenses. The line graph in **Figure 2.26** shows expenses in thousands of dollars for a small business from 2010 through 2020. Explain what the graph says about yearly expenses. Pay particular attention to the growth rate of expenses.

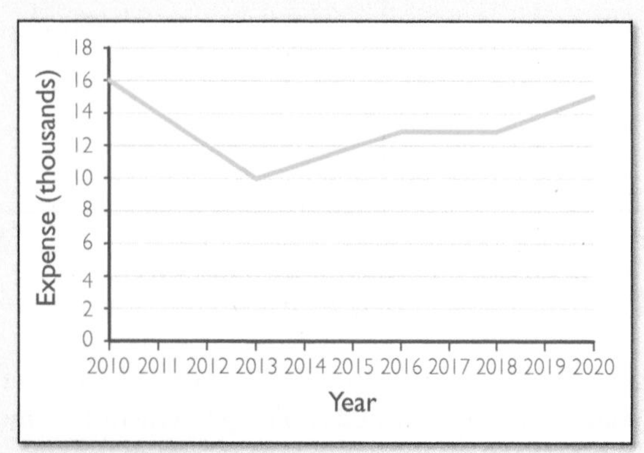

FIGURE 2.26 Figure for Exercise 5.

6. Varying growth rates. A savings account increases in value at an ever increasing rate over the period from 2010 through 2020. Sketch a graph of the value of the account versus time that shows growth of this type.

7. An endangered population. The graph in **Figure 2.27** shows the population of a certain endangered species. Give a general description of how the population changes over the 10-year period shown. Your description should include the growth rate of the population and how the rate changes.

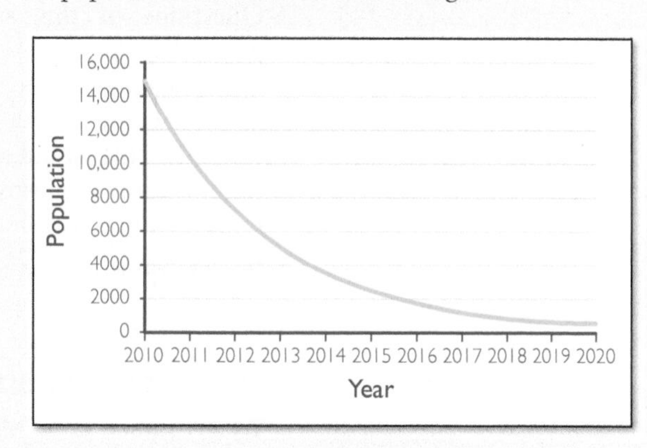

FIGURE 2.27 Figure for Exercise 7.

8. A graph with given growth rates. A certain population has a negative growth rate for a time, but the growth rate is eventually positive. Sketch a possible graph that shows population levels with this property.

9. Prescription drugs. Here is a table showing retail prescription drug sales in billions of dollars for different U.S. sales outlets.

Sales outlet	Traditional chain	Independent	Mass merchant	Super-markets	Mail order
Sales (billions of dollars)	106.6	44.7	26.6	25.9	62.6

Represent these data with a bar graph, a scatterplot, and a line graph. Which type of graph do you think best represents the data?

10. Graduation rates. According to the NCAA,[8] 86.1% of female athletes at Division I schools who entered college in 1998 graduated within six years. This increased to 87.9% in 2002 and by 2006 it had reached 88.4%. Of their male counterparts, only 69.8% of students who entered in 1998 graduated within six years. This number increased to 72.1% in 2002 and by 2006 it had reached 75.7%. On one set of axes, sketch two line graphs displaying this information, one for females and another for males.

11. An investment. You invest $1000 in a risky fund. Over the first six months, your investment loses half its value. Over the next six months, the value increases until it reaches $1000. Make a smoothed line graph that shows your investment over the 12-month period.

12. Paper airplanes. The following table shows the gliding distance in meters of a certain type of paper airplane and its wingspan in centimeters:

Wingspan (centimeters)	5	10	15	20
Glide distance (meters)	5	8	9	13

Make a scatterplot of these data with wingspan on the horizontal axis.

13. A man's height. A child is 38 inches tall at age 3. He grows to 72 inches at age 17, at which time he has achieved his adult height. Make a line graph showing his height as a function of his age over the first 20 years.

14. A population of foxes. Sketch a graph of a population of foxes that reflects the following information: The population grew rapidly from 1990 to 2000. In 2000, a disease caused the rate of growth to slow until the population reached a maximum in 2010. From 2010 onward, the fox population declined very slowly.

15. A flu epidemic. During a certain outbreak of the flu, 20 new cases were initially reported. The number of new cases increased each day for five days until a maximum of 35 new cases was reported. From Day 5 on, the number of new cases declined. Sketch a line graph of new cases of flu as a function of time in days.

16. Making a graph. Sketch a graph that starts from the vertical axis at 10 and has the following properties: The graph is initially decreasing at an increasing rate, but after a bit the graph decreases at a decreasing rate.

17. Making a graph. Sketch a graph that is increasing at a decreasing rate from 0 to 3, decreasing at an increasing rate from 3 to 6, and decreasing at a decreasing rate from 6 on.

18. A newspaper. Suppose a certain newspaper has recorded daily sales beginning in January 2021. On the first day of that month, 200 copies were sold. From January through May, sales increased each day until they hit a high of 800 copies on June 1. From there sales decreased each day to 400 copies on December 1. Sales declined most rapidly at some point in August. Sketch a graph reflecting this information. *Suggestion*: Locate the points you know (January 1, June 1, and December 1) first.

19. Bears. A population of bears is introduced into a game preserve. Over the first five years, the bear population shows a negative growth rate. The growth rate is positive over the next five years. Sketch a possible graph of the bear population as a function of time.

20. Wolves. A population of wolves is introduced into a game preserve. Over the first five years, the wolf population shows a positive growth rate. The growth rate is negative over the next five years. Sketch a possible graph of the wolf population as a function of time.

21–23. Language enrollments. *The line graph in* **Figure 2.28** *shows total course enrollments in languages other than English in U.S. institutions of higher education from 1960 to 2009. (Enrollments in ancient Greek and Latin are not included.) Exercises 21 through 23 refer to this figure.*

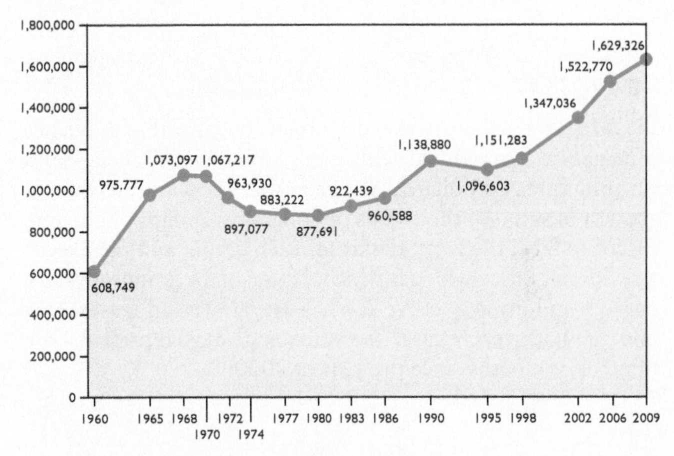

FIGURE 2.28 Enrollments in languages other than English in U.S. institutions of higher education (2009).

21. During which time periods did the enrollments decrease?

22. Calculate the average growth rate per year in enrollments over the two periods 1960–1965 and 2006–2009. Note that the time periods are not of the same length.

23. *This exercise uses the results of Exercise 22.* Use your results from Exercise 22 to determine during which period, 1960–1965 or 2006–2009, the enrollments grew more quickly on average. Relate your answer to the steepness, or slope, of the corresponding line segments.

24. Population growth. The graphs in **Figures 2.29 and 2.30** show the population (in thousands) of a certain animal in two different reserves, reserve A and reserve B. In each case, the horizontal axis represents the number of years since observation began.

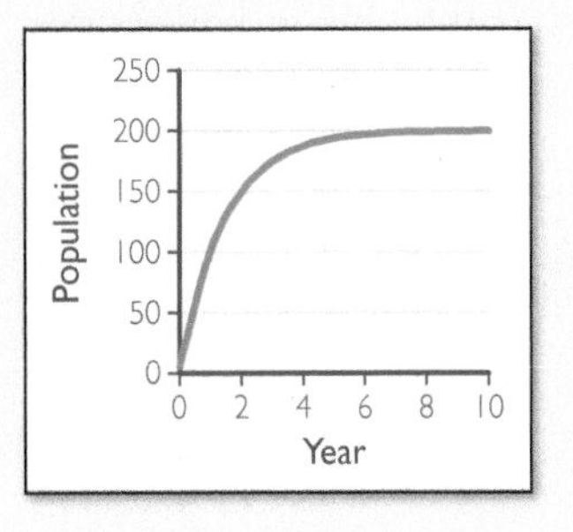

FIGURE 2.29 Population in reserve A.

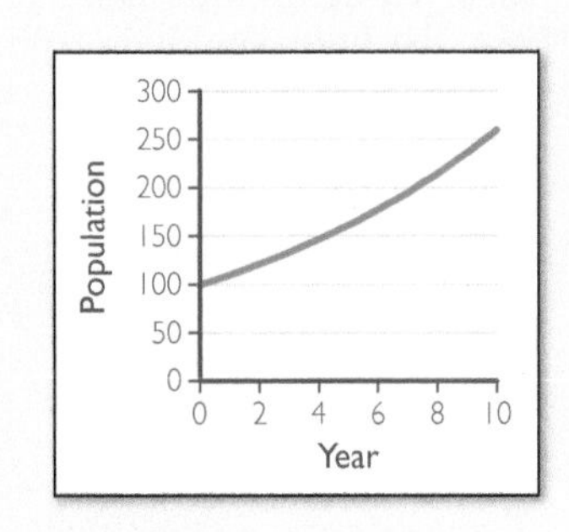

FIGURE 2.30 Population in reserve B.

[8] www.ncaa.org

Figures 2.31 and 2.32 show two graphs of population growth rates. Let's call them growth rate I and growth rate II. One of the graphs describes the population in reserve A, and the other describes the population in reserve B. Which is which? For each reserve, give a careful explanation of how the graph of population is related to the appropriate graph of growth rate.

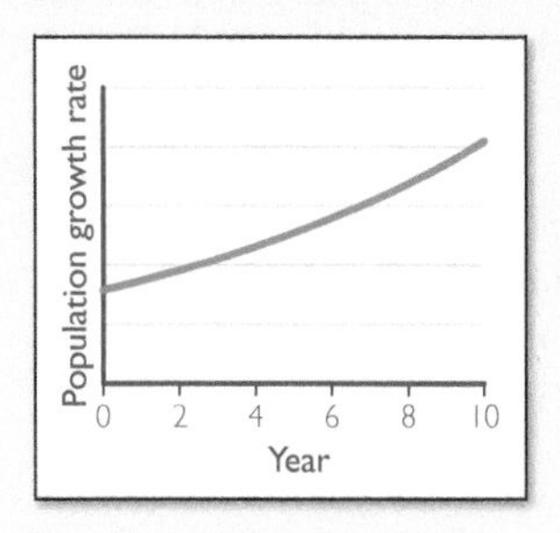

FIGURE 2.31 Population growth rate I.

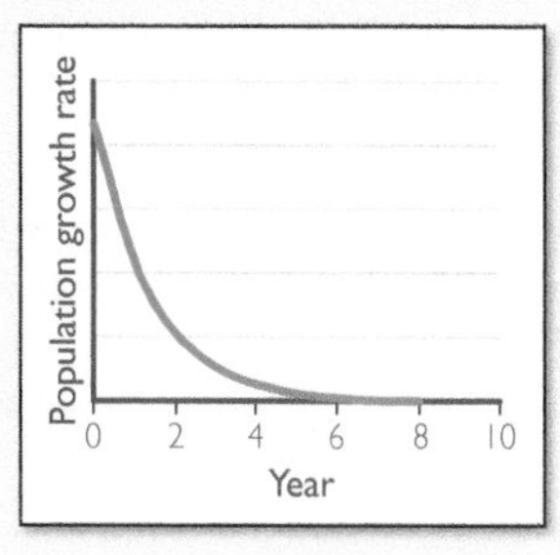

FIGURE 2.32 Population growth rate II.

25. Magazine sales. An executive for a company that publishes a magazine is presented with two graphs, one showing sales (in thousands of dollars), and the other showing the rate of growth in sales (in thousands of dollars per month). Someone forgot to label the vertical axis for each graph, and the executive doesn't know which is which. Let's call the graphs mystery curve 1 and mystery curve 2. They are shown in **Figures 2.33 and 2.34.** (In each case, the horizontal axis represents the number of months since the start of 2020.)

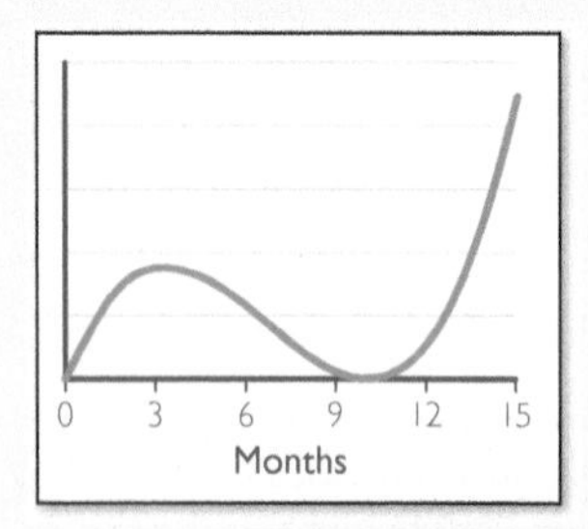

FIGURE 2.33 Mystery curve 1.

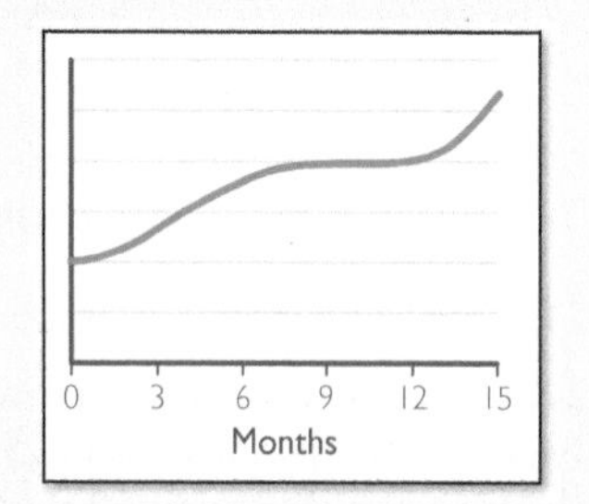

FIGURE 2.34 Mystery curve 2.

Help the executive out by identifying which curve shows sales and which shows the rate of growth in sales. Give a careful explanation of how the graph of sales is related to the graph of growth in sales.

26. Location and speed. The graph in **Figure 2.35** shows the location of a vehicle moving east on a straight road. The location is measured as the distance east of a fixed observation point. The graph in **Figure 2.36** shows the speed of the same vehicle, as measured by speedometer readings. In each case, the horizontal axis represents the time in hours since measurements began.

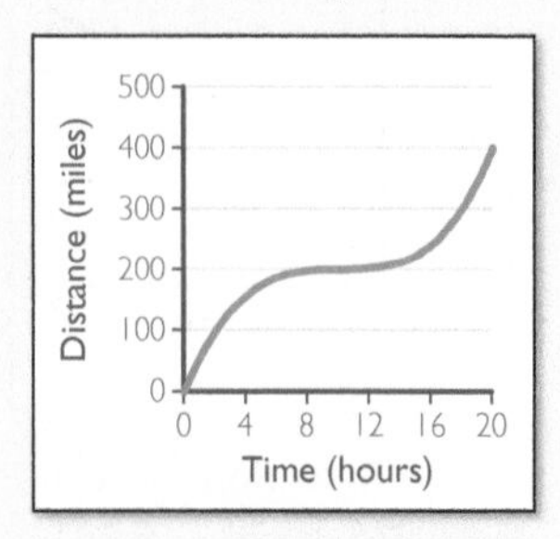

FIGURE 2.35 Location of vehicle.

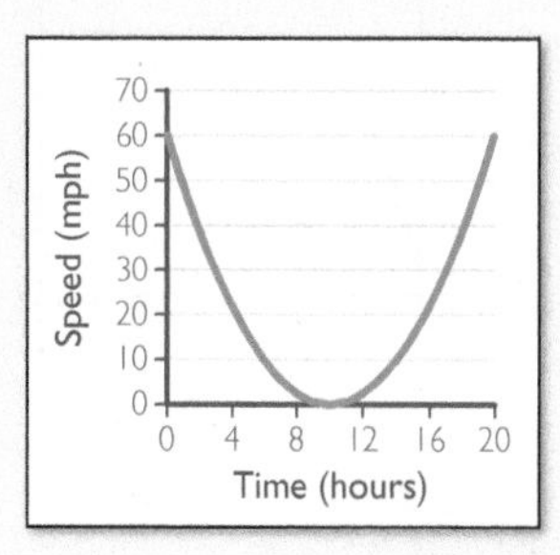

FIGURE 2.36 Speed of vehicle.

In this situation, the speedometer reading measures the growth rate of distance from your starting point. In Example 2.14, we saw how the graph of population growth rate is related to the graph of population. Explain how the graph of speed in Figure 2.36 is related to the graph of location in Figure 2.35, and write a travel story that matches these graphs.

Writing About Mathematics

27. Historical figures: Descartes and Fermat. Graphs and charts have been used throughout recorded history. Their incorporation into mathematics should be credited to any number of early mathematicians. Certainly, two on any such list are René Descartes and Pierre de Fermat. Write a brief report on their development of graphs in mathematics.

28. Historical figures: Oresme. Descartes and Fermat are famous early mathematicians, but lesser known is Nicole Oresme, who preceded both Fermat and Descartes by 300 years. He employed graphs in mathematics long before either. Write a brief report on his use of graphs in mathematics.

What Do You Think?

29. What kind of graph? As club treasurer, you want to make a chart that will give a clear picture of club income over the past year. You are considering a scatterplot, a bar graph, a line graph, and a smoothed line graph. Explain the advantages and disadvantages of each. Which one would you choose?

30. Wise graph choice? Your favorite charity uses a line graph to show its 10 largest donors. Would there have been a better choice for the display? Explain your reasoning.

31. Population. Refer to **Figure 2.37**, which shows the population growth rate. Is the population itself increasing or decreasing? Explain your answer. **Exercise 24 is relevant to this topic.**

32. A trip. Refer to **Figure 2.38**. Explain what happened at the five-hour mark. Where did the trip end?

33. Scatterplots and line graphs. A fellow student has found a scatterplot of data in a newspaper. He wants to create a line graph by connecting the dots. Describe ways in which the resulting line graph could be more useful than the original scatterplot.

34. Scatterplots and line graphs again. A fellow student has found a scatterplot of data in a newspaper. He wants to create a line graph by connecting the dots. What cautions would you give about doing that?

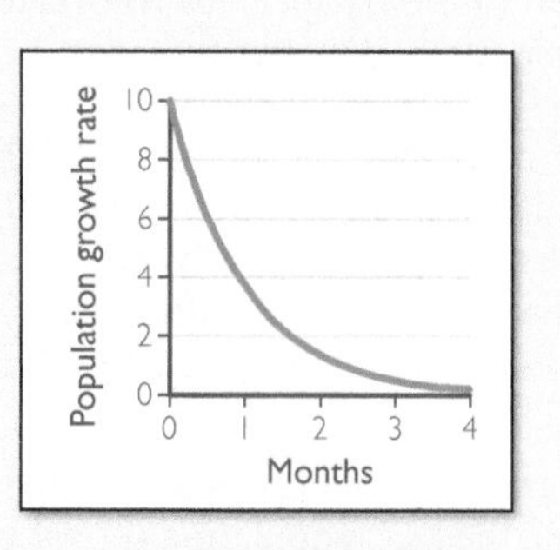

FIGURE 2.37 Population.

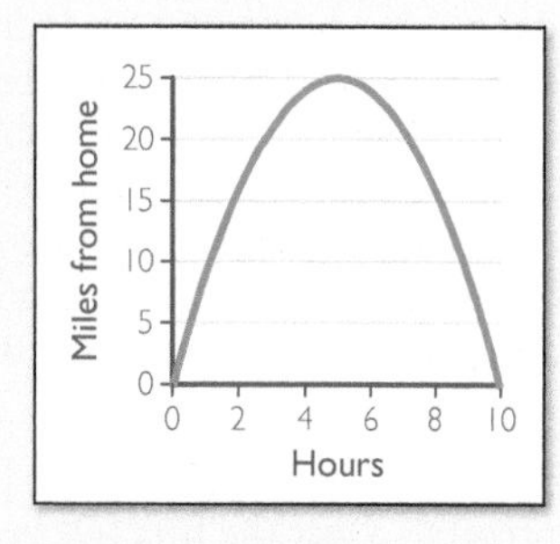

FIGURE 2.38 Trip.

35. Extrapolation. In this section we noted that filling in gaps in a scatterplot using line segments gives estimates for function values that agree with those obtained by interpolation. What graphical procedure corresponds to extrapolation? Draw a graph to illustrate your answer.

36. Death rate. You and a friend are discussing the good news that the annual number of deaths due to a certain disease has been decreasing over the last decade. Your friend says that it is even more encouraging that the rate of decrease of this annual number is decreasing. How would you reply?

37. Endangered species. A line graph shows the population of an endangered species over the past few years. Explain what features you would look for in the graph that might offer encouraging news about the species.

2.3 Misleading graphs: Should I believe my eyes?

TAKE AWAY FROM THIS SECTION
View graphical presentations with a critical eye.

The following article appears at the Web site *Quartz*.

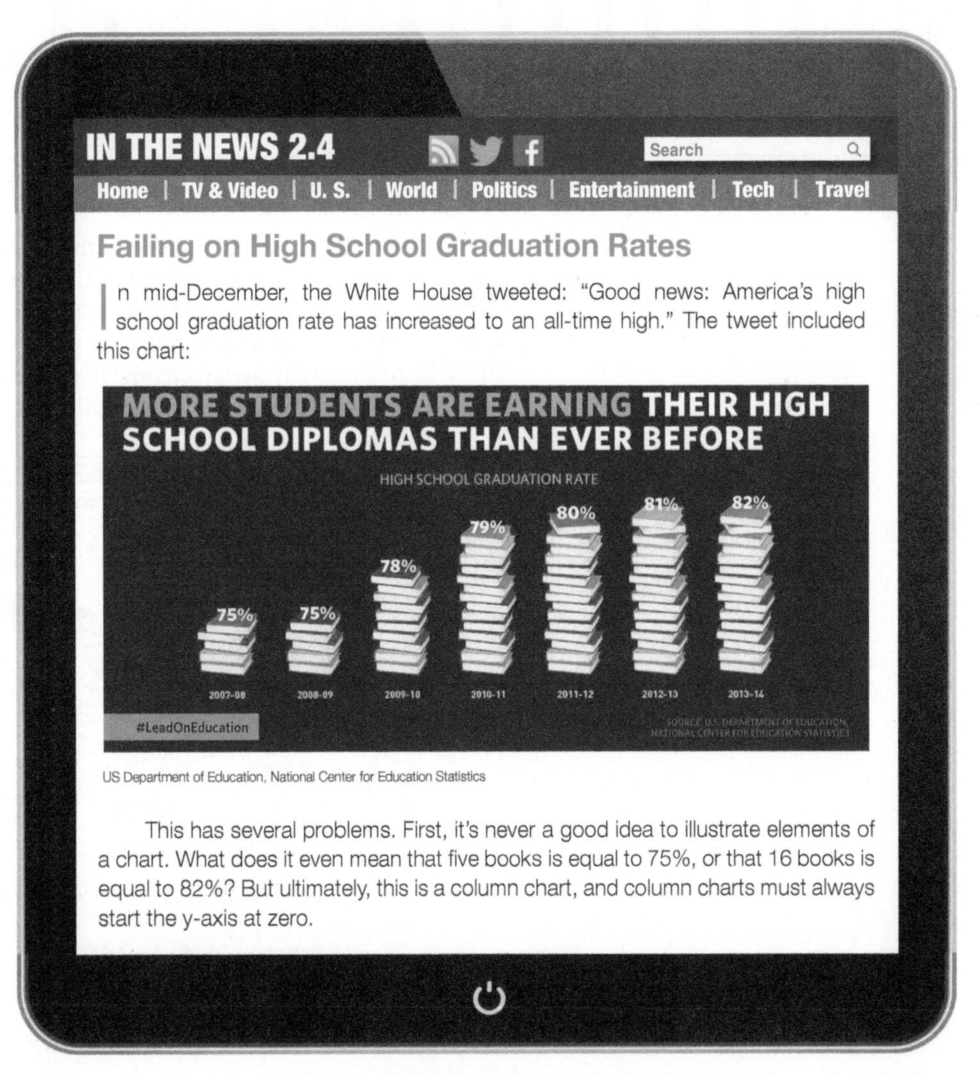

IN THE NEWS 2.4

Search

Home | TV & Video | U. S. | World | Politics | Entertainment | Tech | Travel

Failing on High School Graduation Rates

In mid-December, the White House tweeted: "Good news: America's high school graduation rate has increased to an all-time high." The tweet included this chart:

US Department of Education, National Center for Education Statistics

This has several problems. First, it's never a good idea to illustrate elements of a chart. What does it even mean that five books is equal to 75%, or that 16 books is equal to 82%? But ultimately, this is a column chart, and column charts must always start the y-axis at zero.

This article refers to a highly misleading graph. Graphs can be helpful visual tools, but sometimes they can be misleading (intentionally or otherwise).

There are many ways that graphs can mislead us into drawing incorrect, inaccurate, or inappropriate conclusions about data.[9] We illustrate some of the most common types of misleading graphs in this section.

Misleading by choice of axis scale

Consider the bar graphs in **Figures 2.39 and 2.40**, which show federal defense spending in billions of dollars for the given year.

If we look carefully at the years represented and the spending reported, we see that the two graphs represent exactly the same data—yet they don't look the same. For example, Figure 2.40 gives the visual impression that defense spending doubled from 2004 to 2006, and Figure 2.39 suggests that defense spending did increase from 2004 to 2006, but not by all that much. What makes the graphs so different?

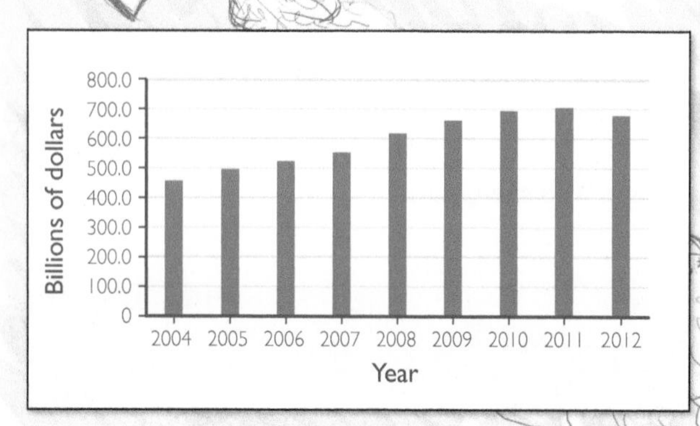

FIGURE 2.39 Defense spending.

FIGURE 2.40 Defense spending again.

The key to understanding the dramatic difference between these two graphs is the range on the vertical axis. In Figure 2.39 the vertical scale goes from 0 to 800 billion dollars, and in Figure 2.40 that scale goes from 400 to 750 billion dollars. The shorter vertical range exaggerates the changes from year to year.

One way to assess the actual change in defense spending is to find the percentage change. Let's consider the change from 2004 to 2006. In both graphs, we see that

"How close to the truth do you want to come, sir?"

the amount spent in 2004 was about 456 billion dollars, and the amount spent in 2006 was about 522 billion dollars. First we find the absolute change:

$$\text{Change} = 522 - 456 = 66 \text{ billion dollars}$$

[9] *A classic reference here is Darrell Huff's* How to Lie with Statistics *(New York: W.W. Norton, 1954).*

Now, to find the percentage change, we divide by the spending in 2004:

$$\text{Percentage change} = \frac{66}{456} \times 100\%$$

This is about 14.5%. The percentage increase is fairly significant, but it is not nearly as large as that suggested by Figure 2.40, which gives the impression that spending has almost doubled—that is, increased by 100%.

It is striking to see how easily someone with a particular point of view can make visual displays to suit his or her own position. If we wanted others to believe that defense spending remained nearly stable over this time period, which graph would we use? On the other hand, which graph would we use if we wanted others to believe that defense spending showed a dramatic increase over this time period?

It is important to note that, as with many of the comparisons we make in this section, both graphs are accurate representations of the underlying data. But the graph in Figure 2.40 may be considered visually misleading.

EXAMPLE 2.15 Analyzing a choice of scale: A misleading graph

An article from the *Media Matters*[10] Web site asserts that Fox Business's *Cavuto* displayed a graphic showing what the wealthiest Americans in the top marginal income tax bracket would pay if the tax cuts enacted during the George W. Bush administration were allowed to expire. The chart exaggerated the increase, which would go from 35%, the current top rate, to 39.6%—less than five percentage points.[11]

The graphic referred to in the article is shown in **Figure 2.41**. The same Web site presents the graph in **Figure 2.42** as a more accurate way to represent the same information.

a. What was the top tax bracket at the time indicated by "Now" in the graphs? What would the top tax bracket have been if the Bush tax cuts had been allowed to expire?

b. Fill in the blank in this sentence: The top tax bracket if the tax cuts expired was ___% more than the top tax bracket labeled "Now." (Round your answer to the nearest whole number.)

c. What impression does Fox's original graph give about how the top tax bracket would change?

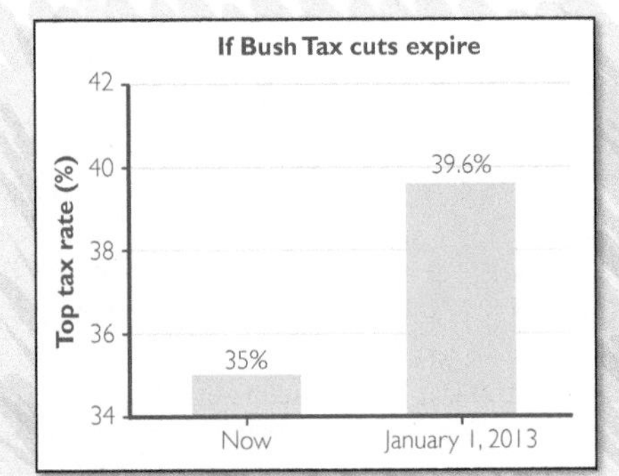

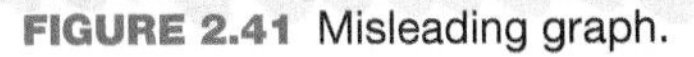

FIGURE 2.41 Misleading graph.

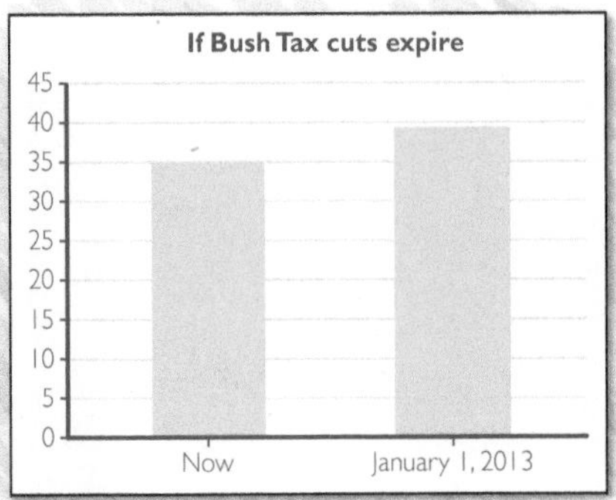

FIGURE 2.42 Replacement graph.

---

[10] *This article and the graphics shown here appear at the Web site* http://mediamatters.org/research/2012/10/01/a-history-of-dishonest-fox-charts/190225

[11] *We will discuss marginal tax rates and tax brackets in Section 4.5.*

SOLUTION

a. The graph shows that the top tax bracket at the time labeled "Now" in the graphs is 35% and the top tax bracket if the Bush tax cuts had been allowed to expire is 39.6%.

b. To find the answer, we divide the difference of $39.6 - 35 = 4.6$ percentage points by the bracket "Now":

$$\text{Percentage change} = \frac{\text{Difference in percentages}}{\text{Percentage "Now"}} \times 100\%$$
$$= \frac{4.6}{35} \times 100\%$$

or about 13%. We conclude that the bracket if the rates expired is about 13% larger than the bracket "Now."

c. In the original graph, the bar representing expired tax cuts is about 6 units high, and the bar representing "Now" is 1 unit high. This gives the impression that the rate for the higher tax bracket is 6 times as large as that "Now," which is 500% more instead of 13% more.

TRY IT YOURSELF 2.15

The graphs in Figure 2.41 and Figure 2.42 represent exactly the same data. Explain what accounts for the distortion in the first graph, as compared with the second one.

The answer is provided at the end of this section.

Default ranges on graphs generated by calculators and computers

Graphing software for calculators and computers have default methods for deciding how to scale the axes for graphs. These defaults often give good results, but sometimes the graphs will be misleading unless they are adjusted. Here is an excerpt from an article that appeared in the *Daily Oklahoman* concerning scores on the ACT college entrance exam in Oklahoma and the nation. We note that ACT scores range from 1 to 36.

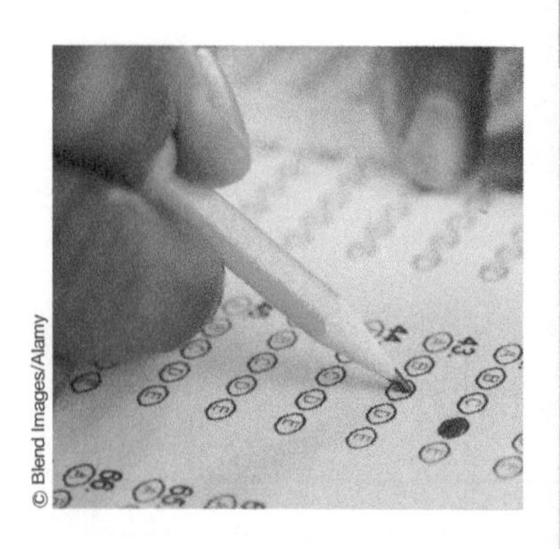

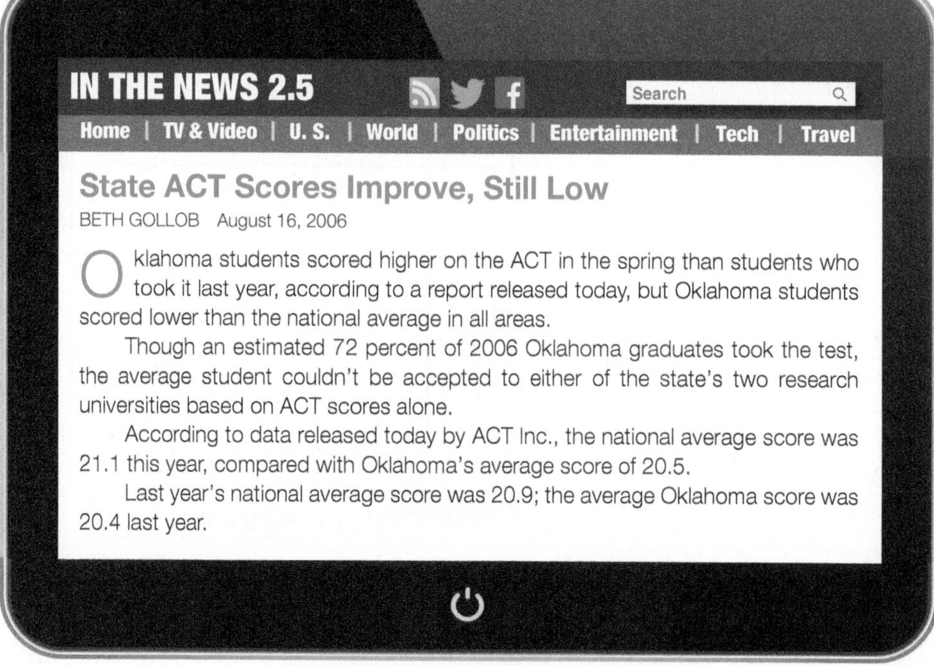

IN THE NEWS 2.5

Search

Home | TV & Video | U. S. | World | Politics | Entertainment | Tech | Travel

State ACT Scores Improve, Still Low

BETH GOLLOB August 16, 2006

Oklahoma students scored higher on the ACT in the spring than students who took it last year, according to a report released today, but Oklahoma students scored lower than the national average in all areas.

Though an estimated 72 percent of 2006 Oklahoma graduates took the test, the average student couldn't be accepted to either of the state's two research universities based on ACT scores alone.

According to data released today by ACT Inc., the national average score was 21.1 this year, compared with Oklahoma's average score of 20.5.

Last year's national average score was 20.9; the average Oklahoma score was 20.4 last year.

The article is accurate. It states, among other things, that the national average was 21.1 and the Oklahoma average was 20.5. Now consider **Figure 2.43**, which shows a graph of the ACT scores for these Oklahoma students. This bar graph is based on data that compare the subject area test scores in Oklahoma to those in the nation as a whole. It certainly gives the impression that Oklahoma scores are far below national scores—especially in math.

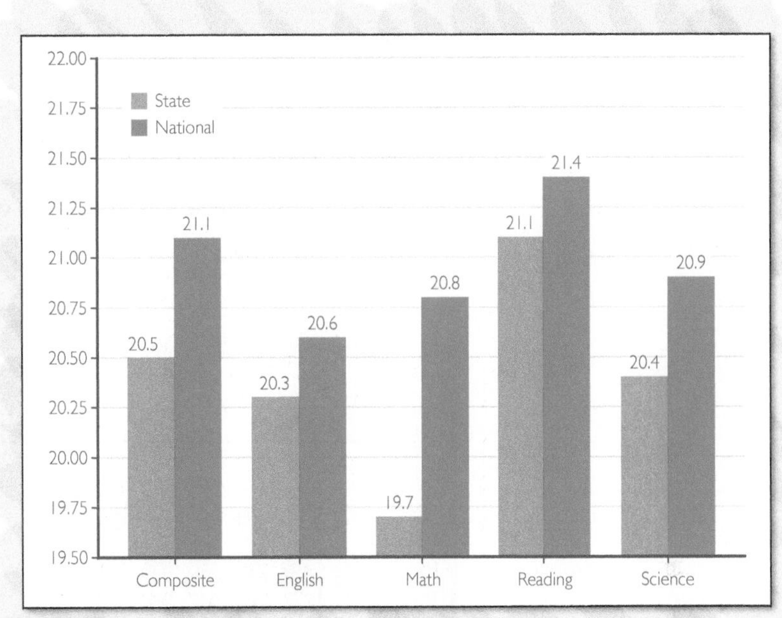

FIGURE 2.43 Bar graph of ACT scores.

Is this impression accurate? Are the Oklahoma scores really as far below the national scores as the bar graph makes it appear? We'll explore that question further in Exercise 16, where the scale on the vertical axis is examined more closely.

Even though the graph may be a misleading representation of the data, the reporter might be trying to give a fair and accurate presentation. It is worth noting that the graph shown in Figure 2.43 is exactly the graph obtained by accepting the default ranges offered by one of the most popular graphing packages.

We emphasize that, although a graph may or may not be *intended* to deceive, the effect is the same. It is up to us to think critically so as not to allow ourselves to be misled.

CALCULATION TIP 2.1 Graphing on Calculators

A graphing calculator or computer software usually has default settings for the scales it uses on the two axes. When we plot a graph with a calculator or computer, we need to be alert to those scales. We might need to adjust them manually to obtain a graph that conveys an appropriate picture.

EXAMPLE 2.16 Choosing scales for graphs: Presenting a point of view on gasoline prices

The following table shows the average price of a gallon of regular gasoline in January of the given year:

Year	2010	2011	2012	2013	2014	2015	2016
Average price	2.77	3.15	3.44	3.39	3.39	2.21	2.06

Your debate team must be prepared to present a case for the following proposition and against it.

Proposition: The average price of a gallon of regular gasoline showed a significant decrease from 2010 to 2016.

Make a bar graph of the data that you would use for an argument in support of the proposition. Make a second bar graph that you would use for an argument against the proposition.

SOLUTION

To support the proposition, you want a graph that emphasizes the decrease in gas prices. A narrow range on the vertical scale emphasizes the differences in data. In **Figure 2.44**, we have made a graph to support the proposition using a vertical range from \$1.75 to \$3.50. To argue against the proposition, we want a graph that deemphasizes the differences in data. A wider vertical range will accomplish this. In **Figure 2.45**, we have used a vertical range from \$0 to \$15.00. This graph would be used in an argument against the proposition.

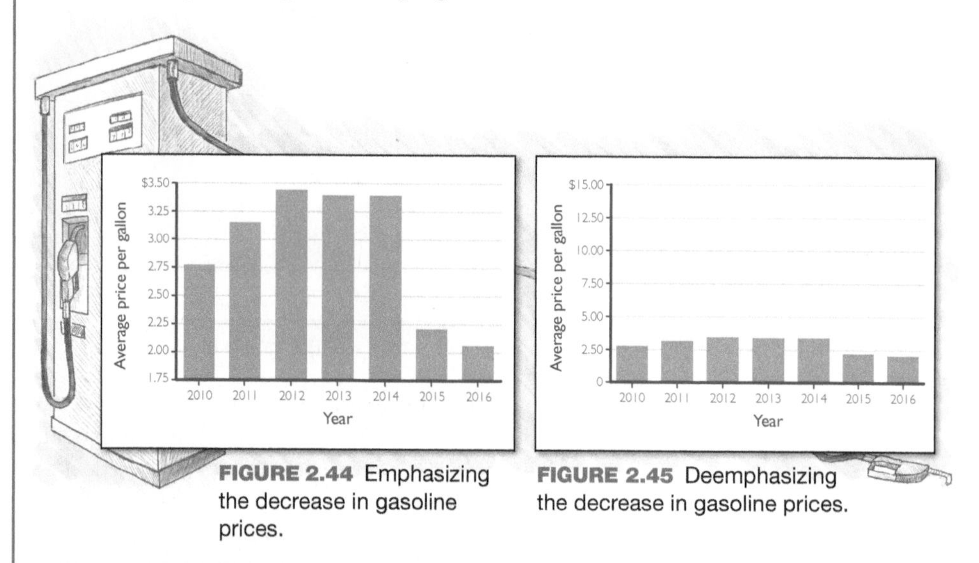

FIGURE 2.44 Emphasizing the decrease in gasoline prices.

FIGURE 2.45 Deemphasizing the decrease in gasoline prices.

TRY IT YOURSELF 2.16

The table below shows average ACT scores for 1990 through 2000, which, according to the ACT, "is the first decade ever in which the national average increased substantially."

Year	1990	1991	1992	1993	1994	1995	1996	1997	1998	1999	2000
ACT scores	20.6	20.6	20.6	20.7	20.8	20.8	20.9	21.0	21.0	21.0	21.0

Make a bar graph that you would use to argue that average ACT scores increased significantly during this period. Make a second bar graph that you would use to argue that average ACT scores did not increase significantly over this period.

The answer is provided at the end of this section.

Misleading by misrepresentation of data: Inflation

The scatterplot in **Figure 2.46** appeared in an article by Frederick Klein in the July 5, 1979, edition of the *Wall Street Journal*. The graph shows the total amount of currency (in billions of dollars) in circulation. The plot appeared with the caption, "Over a quarter-century the currency pile-up steepens." The article uses this graph to bolster the case that "the amount of currency in individual hands is soaring." On the surface, the graph makes a strong case for the author, and his conclusion is technically accurate—that is, the amount of currency in circulation was increasing. But this conclusion ignores the *value* of the currency.

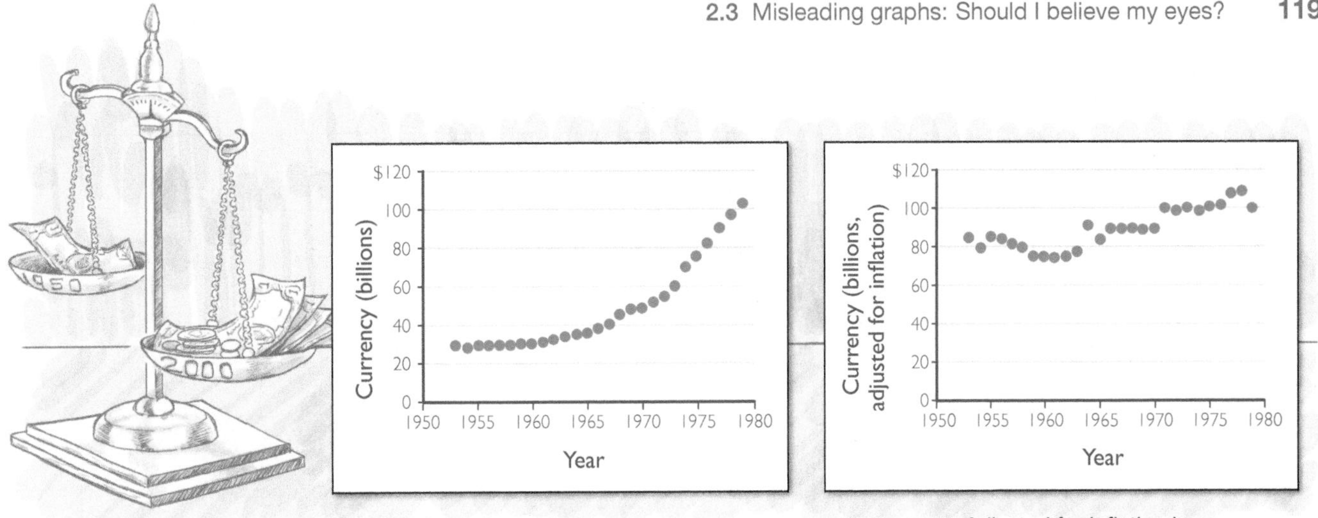

FIGURE 2.46 Amount of currency in circulation.

FIGURE 2.47 Adjusted for inflation in constant 1979 dollars.

The price of many commodities increases over time. For instance, the median price of an existing one-family house in 1968 was \$20,100, the median price of a house in 1982 was \$67,800, and the median price of a house in 2020 was \$334,000.[12] This increase in home values is for the most part a reflection of increases in prices overall. The overall increase in prices is measured by the *inflation rate*.[13]

Another way of looking at this question is in terms of the value of a dollar. For example, from 1950 to 2000, prices, on average, increased by 596%. This means that a dollar would buy the same amount of goods in 1950 as the dollar plus \$5.96, or \$6.96, would buy in 2000. So a "1950 dollar" has the same buying power as 6.96 "2000 dollars." When a graph is *adjusted for inflation*, all currency amounts are converted to their dollar values for the same year, which are then called *constant dollars*. For example, \$1.00 in year 1950 would be \$6.96 in constant year-2000 dollars.

Let's recap how this adjustment works. If inflation from Year 1 to Year 2 is r as a decimal, then 1 dollar in Year 1 had the same purchasing power as $1 + r$ dollars in Year 2. In the above example, Year 1 was 1950, Year 2 was 2000, and $r = 5.96$.

In general, to convert Year-1 dollars to Year-2 dollars, just multiply Year-1 dollars by $1 + r$. For example, if inflation is 5% from Year 1 to Year 2, then to convert 100 Year-1 dollars to Year-2 dollars, multiply \$100.00 by $1 + 0.05 = 1.05$ to get \$105.00 Year-2 dollars. This means that an item costing \$100.00 in Year 1 would cost \$105.00 in Year 2.

In **Figure 2.47**, we have reproduced the scatterplot after adjusting for inflation by using constant 1979 dollars. This means that we adjusted the currency values for each of the years to their value for 1979. Using constant dollars means we are using the same measure of value in 1950 and 1980. Before adjusting for inflation, the vertical axis represents the amount of currency. After the adjustment, the vertical axis represents the buying power of the currency, which is a better measure of wealth. We see that, in terms of buying power, the amount of currency increased much less dramatically than the original graph suggested. It could be argued that the graph in Figure 2.46 presents a somewhat distorted view of the amount of currency because of the changing value of the dollar over time.

Another example is shown in **Figure 2.48**. This figure shows the federal minimum wage by year. The blue (lower) graph shows the minimum wage (referred to as *nominal*), which increased from 1938 through 2015. The red graph shows the minimum wage in constant 2015 dollars. It is the blue graph adjusted to 2015 dollars. Because the red graph shows the minimum wage in terms of a fixed value

[12] *Prices don't always go up. Housing prices fell quite sharply from 2007 to 2009.*

[13] *Inflation is calculated using the* consumer price index. *We will look more closely at inflation in Chapter 4.*

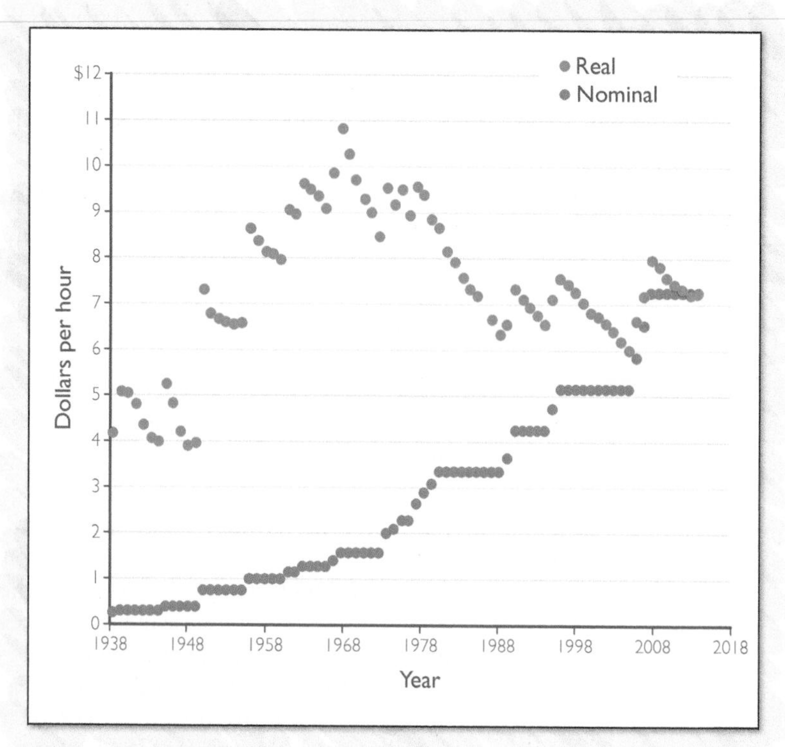

FIGURE 2.48 Minimum wage.

of the dollar, it provides a better comparison of minimum wages earned from year to year than does the blue graph. It is interesting that the value of the minimum wage did not always increase (as one might be led to believe by the blue graph) but actually peaked in 1968.

Let's see how we adjust the graph in terms of constant 2015 dollars for the year 1968. The nominal minimum wage in 1968 was \$1.60. According to the U.S. Bureau of Labor, the inflation rate from 1968 to 2015 was 566%. Therefore, to convert \$1.60 in 1968 to year-2015 dollars, we multiply \$1.60 by $1 + r = 1 + 5.66 = 6.66$:

$$\$1.60 \text{ in year-1968 dollars is } \$1.60 \times 6.66 = \$10.66 \text{ in year-2015 dollars.}$$

Note that this result agrees with the location of the peak of the red graph in Figure 2.48.

SUMMARY 2.5
Adjusting for Inflation

To make valid comparisons of currency values over a period of time, it is important to know whether the values are adjusted for inflation. Here is how this adjustment is done. If inflation from Year 1 to Year 2 is r as a decimal, then to convert Year-1 dollars to Year-2 dollars we simply multiply Year-1 dollars by $1 + r$. For example, if inflation is 8% from Year 1 to Year 2, then to convert 10 Year-1 dollars to Year-2 dollars multiply \$10.00 by $1 + 0.08 = 1.08$ to get \$10.80.

EXAMPLE 2.17 Adjusting for inflation: Gasoline prices

The following table shows the average cost of a gallon of regular gasoline in the given year. These data are plotted in **Figure 2.49.**

Year	1970	1980	1990	2000	2010
Price per gallon	\$0.36	\$1.25	\$1.16	\$1.51	\$2.78

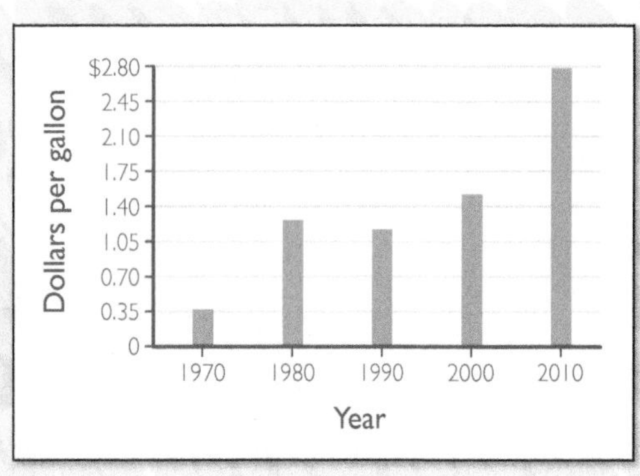

FIGURE 2.49 Gasoline prices.

Inflation rates are shown in the following table:

Time span	1970–2010	1980–2010	1990–2010	2000–2010
Inflation	451%	154%	64%	26%

a. Use the graph in Figure 2.49 to determine, without adjusting for inflation, in what year gasoline was most expensive.

b. Complete the following table showing gasoline prices in constant 2010 dollars:

Year	1970	1980	1990	2000	2010
Price per gallon	$0.36	$1.25	$1.16	$1.51	$2.78
Price in year-2010 dollars					

c. Make a bar graph showing gasoline prices in constant 2010 dollars.

d. In what year were gasoline prices, adjusted for inflation, the highest?

SOLUTION

a. Gasoline was most expensive in 2010.

b. To find the price per gallon of 1970 gas in terms of "2010 dollars," we use the inflation rate of 451%. Expressed as a decimal, this number is $r = 4.51$. Using Summary 2.5, we convert $0.36 in 1970 dollars to 2010 dollars by multiplying $0.36 by $1 + r = 5.51$. Thus,

$$\$0.36 \text{ in 1970 dollars is } \$0.36 \times 5.51 = \$1.98 \text{ in 2010 dollars}$$

So the price of gas in 1970 was $1.98 in 2010 dollars. This gives the first entry in the table. We find the remaining entries in a similar fashion. The table below shows the result.

Year	1970	1980	1990	2000	2010
Price per gallon	$0.36	$1.25	$1.16	$1.51	$2.78
Price in year-2010 dollars	0.36 × 5.51 = $1.98	1.25 × 2.54 = $3.18	1.16 × 1.64 = $1.90	$1.51 × 1.26 = $1.90	$2.78

c. The bar graph giving prices in 2010 dollars is shown below.

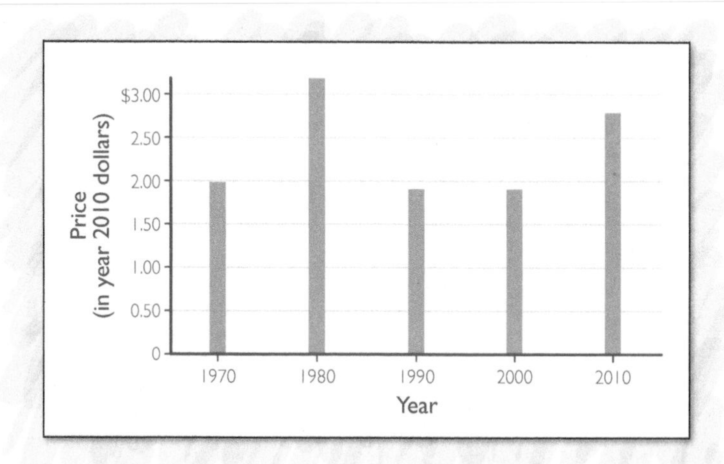

d. The price of gasoline, adjusted for inflation, was highest in 1980. Comparing this answer with the answer to part a shows the value of adjusting for inflation.

These examples lead us to a question about the example on defense spending we considered at the beginning of this section. Were the graphs in Figures 2.39 and 2.40 adjusted for inflation? The answer is that they were not. You will have the opportunity to look further at this question in Exercise 23.

Misleading by using insufficient data

A graph can give an accurate picture—provided sufficient data are used to produce it. If the data are insufficient, graphs can mislead us into drawing inaccurate conclusions. In **Figure 2.50**, we show a bar graph of U.S. passenger car production by year. The graph seems to show that automobile production declined steadily from 1960 to 2010. If we look closely at the labels on the horizontal axis, we see that we have recorded data points only once per decade. In **Figure 2.51**, we have added data points for each five-year interval. This graph does show an overall pattern of decline in car production, but it also shows that there were fluctuations in production. It is clear what makes the graphs different: We used more data points to make the graph in Figure 2.51.

In the case of bar graphs or scatterplots, we can tell how many data points have been plotted, but in the case of smoothed line graphs, there may be no indication of how many data points were used. To illustrate this point, we consider again the data on the production of passenger cars, but now we use smoothed line graphs.

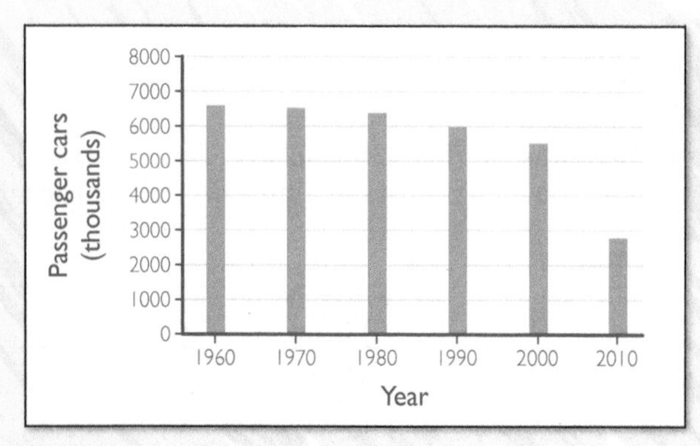

FIGURE 2.50 Car production.

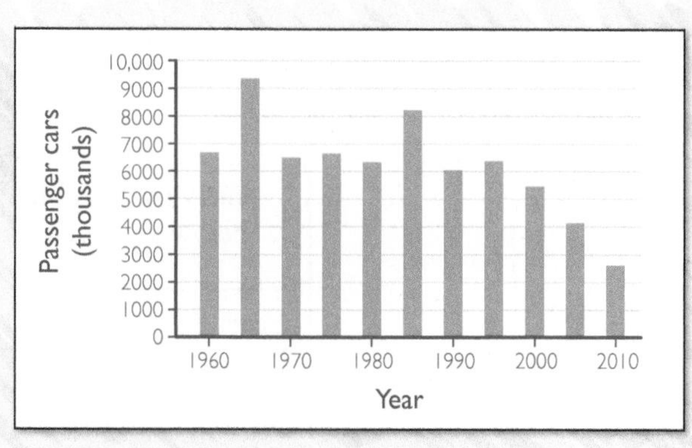

FIGURE 2.51 Car production revisited.

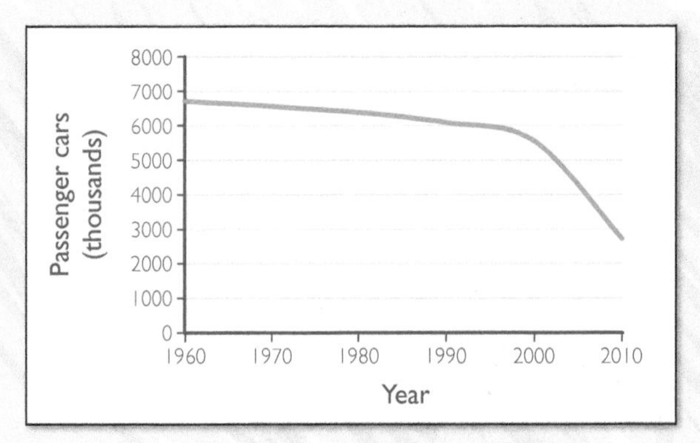

FIGURE 2.52 Smoothed line graph over 10-year periods.

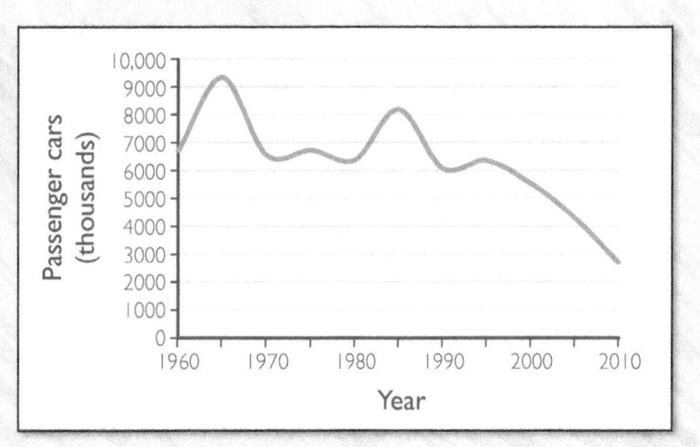

FIGURE 2.53 Smoothed line graph over 5-year periods.

A smoothed line graph using data points once per decade is shown in **Figure 2.52**, and a graph using data from five-year intervals is shown in **Figure 2.53**. Unlike the preceding bar graphs, there is no way to tell how many data points were used to make the new graphs, and hence there is no way to determine which gives a more accurate picture of automobile production.

Workers transport a Porsche Macan car body during the official production start of the new Porsche Macan, in Leipzig, Germany, 2014.

If we see a smoothed line graph in the media, we may not know how many data points were used in producing it. If there's a reason to question the accuracy, the wisest course of action is to look for the original data source.

Pictorial representations

Often data are presented using images designed to grab our attention. These images can be helpful if they are properly constructed. But they can be misleading if they are not properly presented. One common graphical device is the *pie chart*. Pie charts are used to display how constituent parts make up a whole. They are circular regions divided into slices like pieces of a pie. The pie represents the whole, and the slices represent individual parts of the whole. For example, if one item is 30% of the whole, then its corresponding slice should make up about 30% of the area of the pie.

EXAMPLE 2.18 Making a pie chart: Undergraduate enrollment at UC Davis

Make a pie chart showing the following enrollment data from the University of California, Davis, for fall 2015:

UC Davis Undergraduates	28,257
Agricultural and Environmental Sciences	6,984
Engineering	4,445
Letters and Science	11,186
Biological Sciences	5,642

SOLUTION

First we calculate the percentage of the whole represented by each category of student. For example, the 6984 students in Agricultural and Environmental Sciences represent

$$\frac{6{,}984}{28{,}257} \times 100\%$$

or about 24.7% of the whole undergraduate population of 28,257. The resulting percentages are shown in the table below.

UC Davis Undergraduates	28,257	100%
Agricultural and Environmental Sciences	6,984	24.7%
Engineering	4,445	15.7%
Letters and Science	11,186	39.6%
Biological Sciences	5,642	20.0%

Our pie must be divided in these proportions. The result is shown in **Figure 2.54**.

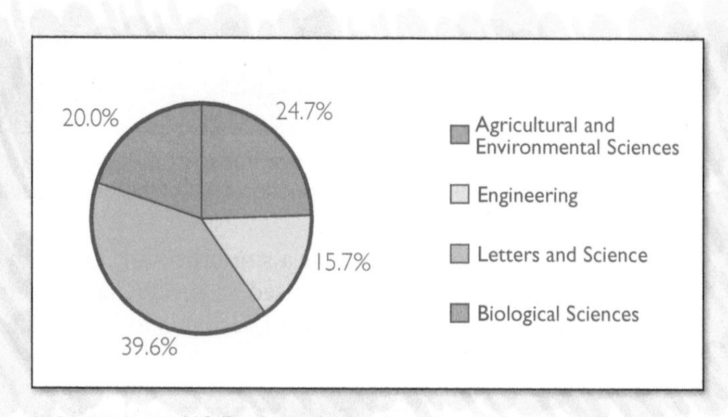

FIGURE 2.54 UC Davis undergraduate enrollment.

1958—Eisenhower

1963—Kennedy

1968—Johnson

1973—Nixon

1978—Carter

1984—Reagan

1990—Bush

1994—Clinton

2006—Bush

Pause/Shutterstock

A visual currency representation claiming to show the value of a dollar.

Pie charts such as the one in Figure 2.54 can be helpful in understanding data. As a contrast, consider the pie chart on the next page. It presumably tells us something about repeat offenders. Recall that the sections of a pie chart represent percentages of a whole. It is unclear what the sections in this chart are intended to represent.

Other types of graphical descriptions can also be misleading. For example, the currency representation on the left is supposed to show the value of the dollar under various presidents. But notice that the "Carter dollar" is labeled as being worth 44 cents, compared to $1.00 for the "Eisenhower dollar." Hence, the "Carter dollar" should be almost half the size of the "Eisenhower dollar." In fact, it is less than an eighth of the size in area of the "Eisenhower dollar." The picture itself gives a false impression of the relative values of currency during these two presidential administrations.

There is a similar difficulty in a fuel economy chart. Note that the length of the line segment corresponding to 18 miles per gallon is less than a quarter of the length of the line segment for 27.5 miles per gallon. The figure gives an inflated view of the progress in automobile efficiency.

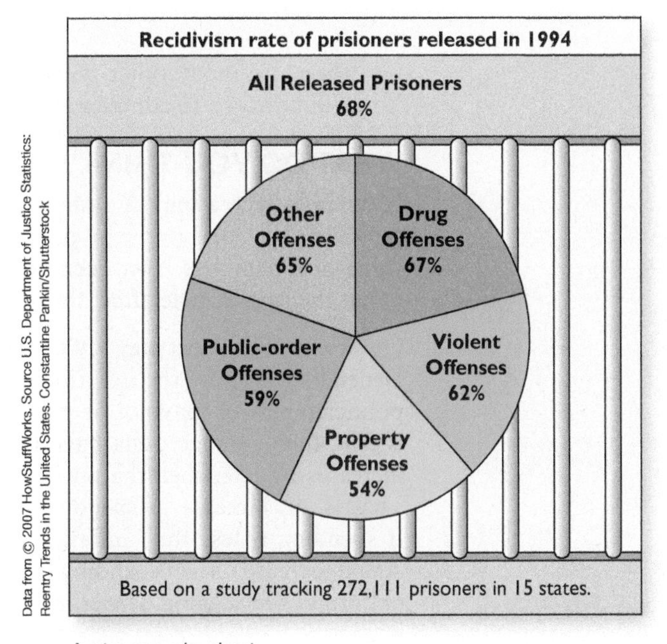

A strange pie chart.

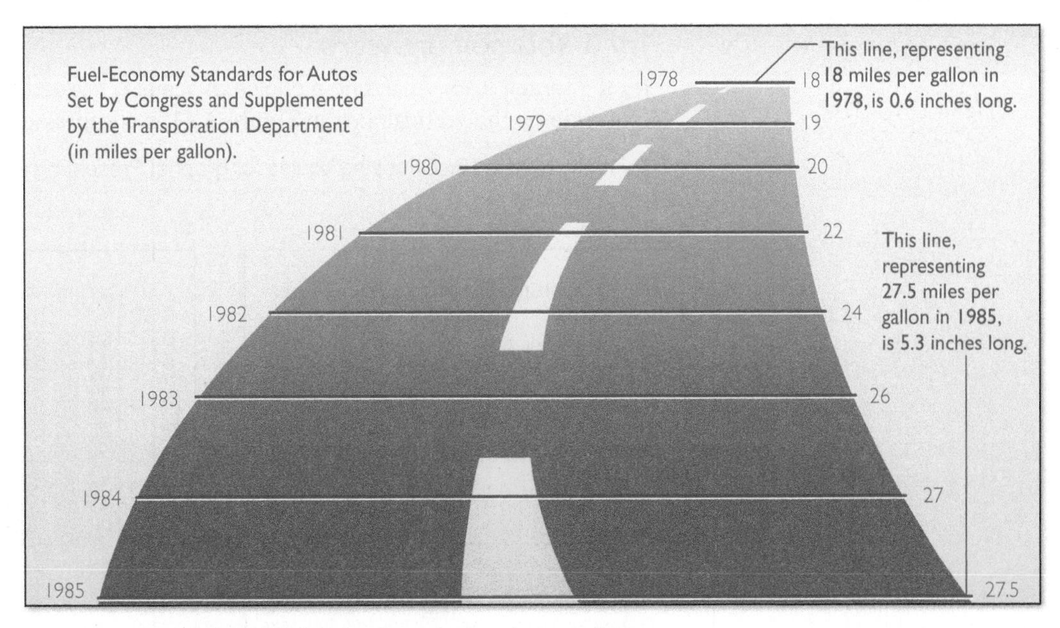

A visual representation claiming to show changes in gas mileage.

SUMMARY 2.6
Cautionary Notes

1. The range of the axes can distort graphs.
2. Be careful that data are accurately represented. For example, data that depend on the value of currency may have to be adjusted for inflation.
3. Any graph can be misleading if insufficient data are plotted. It may be particularly difficult to determine how many data points were used to produce a smoothed line graph.
4. Visual representations can be misleading if the relative sizes of the items pictured do not correspond to the relative sizes of the data.
5. If in doubt, we should always go to the original data source to determine the accuracy of a pictorial representation.

There are many other ways of presenting graphs that distort our perception. You will be asked to comment on some of these in the exercises.

WHAT DO YOU THINK?

Is it really increasing? A politician tells you that educational spending has been growing each year over the past few years. But a news article tells you that the value of money has been declining due to inflation over the past few years. How might the news article affect the accuracy of the politician's statement?

Questions of this sort may invite many correct responses. We present one possibility: Depending on how the inflation and percentage growth in spending compare, the politician may or may not be painting an accurate picture. For example, if inflation is 3%, then average costs have increased by 3%. So if educational spending also increases by 3%, then the amount of goods and services purchased are the same. That is, an increase in spending that matches inflation is necessary to stay even. If spending is less than inflation, then there is an effective decrease. There is an actual increase only if spending is greater than inflation. This example illustrates the importance of digging deeper into news reports and doing your own research.

Further questions of this type may be found in the *What Do You Think?* section of exercises.

Try It Yourself answers

Try It Yourself 2.15: Analyzing a choice of scale: A misleading graph from Fox Business
The scale on the vertical axis in Figure 2.41 is compressed.

Try It Yourself 2.16: Choosing scales for graphs: Presenting a point of view on ACT scores

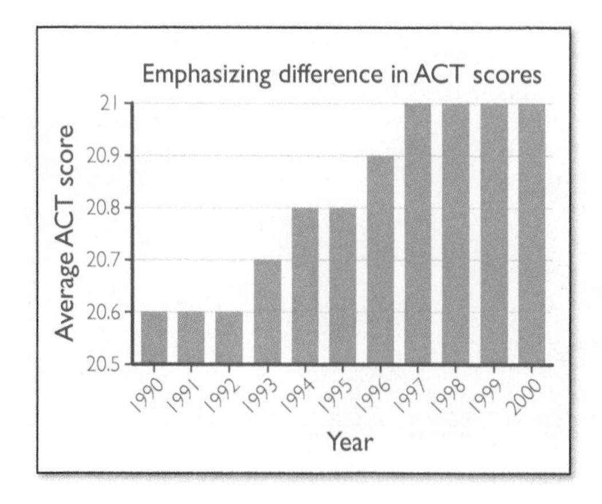

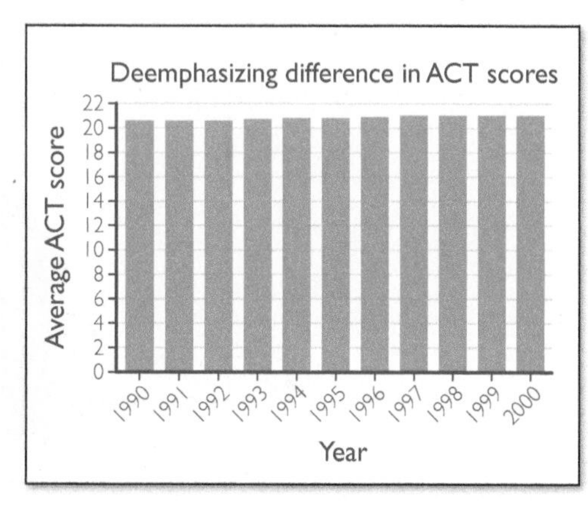

Exercise Set 2.3

Test Your Understanding

1. Graphs may appear to misrepresent data in case: **a.** scales on the axes are not properly chosen **b.** an insufficient number of data points are used to make the graph **c.** data are not accurately represented **d.** all of the above.

2. Does a smaller vertical scale emphasize change from one point to the next or deemphasize the change?

3. True or false: A line graph is better than a bar graph even if some points on the graph don't correspond to any real item.

4. True or false: In order to understand a graph, it is necessary to have a clear understanding of what the underlying data actually represent.

5. Pie charts: **a.** show how constituent parts make up a whole **b.** may show only a small percentage of the whole **c.** always show an excess.

6. True or false: For visual representations of data, it is important to check whether the relative sizes of the items pictured correspond to the relative sizes of the data.

Problems

7. **Female graduation rates.** According to the NCAA,[14] 86.1% of female athletes at Division I schools who entered college in 1998 graduated within six years. This increased to 87.9% in 2002 and by 2006 it had reached 88.4%. Make a line graph that emphasizes the increase in graduation rates.

8. **Male graduation rates.** According to the NCAA, 69.8% of male athletes who entered in 1998 graduated within six years. This number increased to 72.1% in 2002 and by 2006 it had reached 75.7%. Make a line graph that deemphasizes the variation in graduation rates.

9. **Cancer mortality rates.** Consider the graph of cancer mortality rates for white males shown in Figure 2.17 in Section 2.2. The original table of data appears on page 104. Make a bar graph of the data that emphasizes the variation in mortality rates from 1950 through 2014.

10. **Female coaches.** The following table is taken from data provided in the report *Women in Intercollegiate Sport: A Longitudinal, National Study; Twenty-Nine Year Update*, by L. J. Carpenter and R. V. Acosta. It shows the percentage of female head coaches for women's teams (all divisions, all sports) in the given year. Make a line graph of the data that deemphasizes the decline in the percentage of female head coaches from 1980 through 2010.

Year	1986	1992	1998	2004	2010
Percent	50.6	48.3	47.4	44.1	42.6

11. **Informative graph?** Figure 2.55 shows the distribution of education level among the staff positions at a certain university. Discuss the usefulness of this graph. *Suggestion*: Look at the first bar.

[14]www.ncaa.org

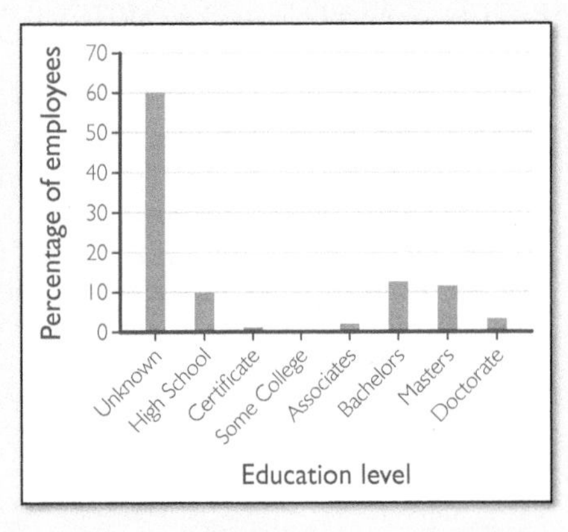

FIGURE 2.55 Total university education distribution.

12. **Donations.** An article in the Oklahoma State University newspaper *The Daily O'Collegian* looked at the seven largest gifts to the university. The graph in **Figure 2.56** accompanied the article. In a few sentences, explain what you think the graph is trying to convey. In particular, what information do the lines connecting the dots convey? Can you suggest a better way to present these data?

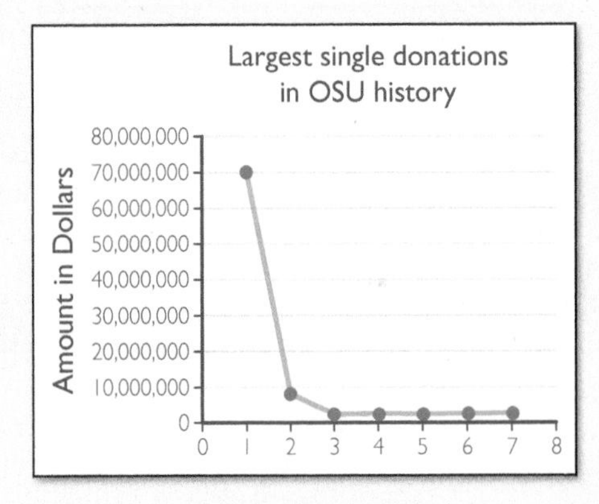

FIGURE 2.56 Donations to OSU.

13. **Pets.** The graph in **Figure 2.57** shows the number of pets of various types on my street. What information do the lines connecting the dots convey? Can you suggest a better way to present these data?

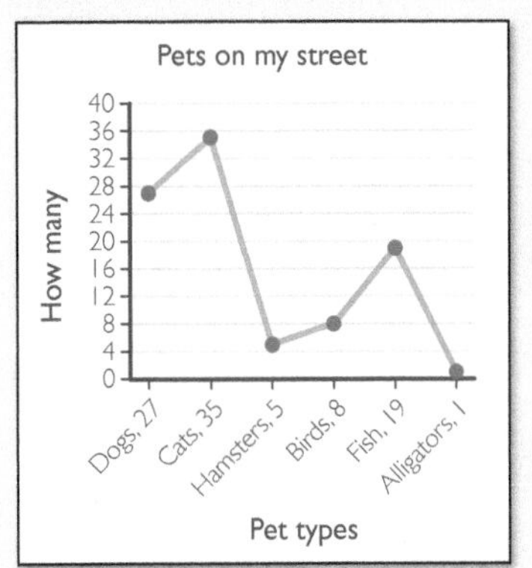

FIGURE 2.57 Pets.

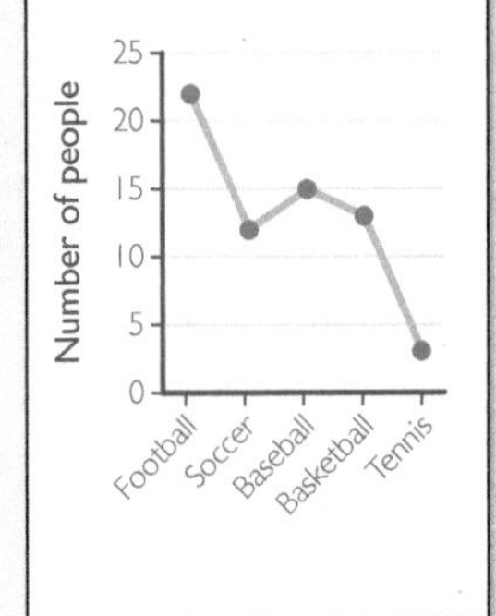

FIGURE 2.58 Sports.

14. Sports. The graph in **Figure 2.58** shows the favorite sports of people in my class. What information do the lines connecting the dots convey? Can you suggest a better way to present these data?

15. Drivers. Figure 2.59 shows the number (in tens) of fatally injured drivers involved in automobile accidents in the state of New York. Comment on the grouping of ages. The graph shows many more fatal accidents for drivers 30 to 39 than for drivers 65 and older. Do you suspect that there are more drivers in one age group than in another? What additional information would you need to make a judgment as to which age group represented the safest drivers?

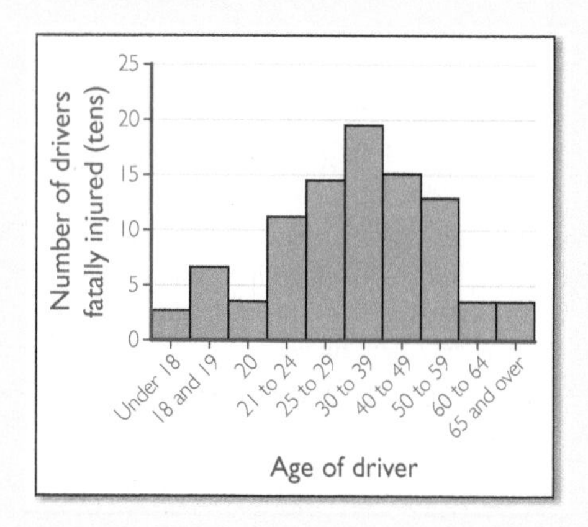

FIGURE 2.59 Fatal automobile accidents.

16. Oklahoma ACT scores. This exercise refers to the In the News feature 2.5 on page 116 and the graph in Figure 2.43, which accompanied the original article.

a. Calculate, for the composite and each of the subject areas, the difference between the national average and the Oklahoma scores, and then express each as a percentage of the national score.

b. Sketch a bar graph of the data in Figure 2.43 using a scale of 0 to 22 on the vertical axis.

c. Explain how someone might prefer the original bar graph or the bar graph you made in part b, depending on the impression to be conveyed. Which graph do you think more accurately reflects the percentages you found in part a?

17. Wedding costs. The data in the graph in **Figure 2.60** were taken from the Web site of *The Wedding Report*. The graph shows average wedding costs in thousands of dollars for the given year.

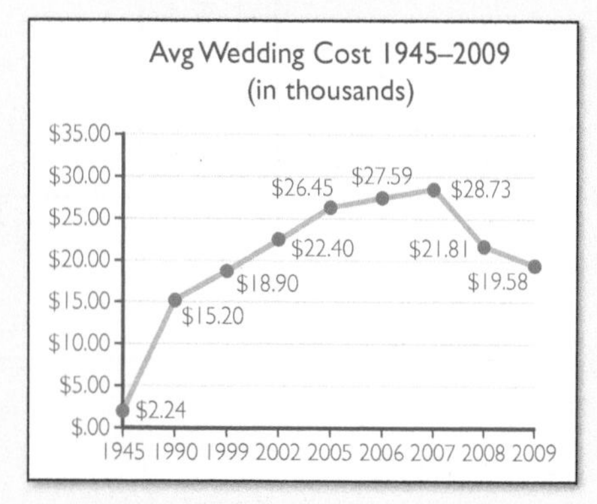

FIGURE 2.60 Average wedding costs.

a. Over which of the time periods indicated by the horizontal axis do wedding costs appear to be increasing most rapidly?

b. What is the average growth rate per year in wedding costs from 1945 to 1990? (First express your answer in thousands of dollars per year rounded to two decimal places, and then express it in dollars per year.)

c. What is the average growth rate per year in wedding costs from 2005 to 2006?

d. In view of your answers to parts b and c, do you think the graph in Figure 2.60 is misleading?

18. Web conferencing. The data in the graph in **Figure 2.61** show the worldwide revenue in millions of dollars for the Web conferencing market from 2001 through 2007. The data appeared in the December 2003/January 2004 edition of *Technology Review*.

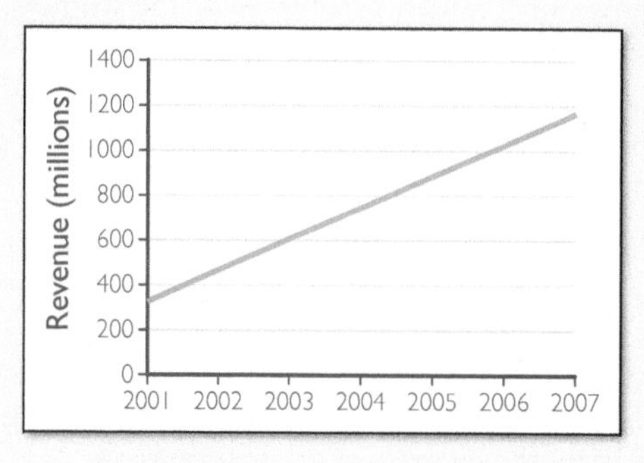

FIGURE 2.61 Revenue for Web conferencing.

a. In light of the date of publication of the article, comment on your estimation of the reliability of the graph in Figure 2.61.

b. Can you think of a reason why the person who produced the graph in Figure 2.61 made it a straight line?

19. Stock charts. Consider the data in the graph in **Figure 2.62**, which tracks the Dow Jones Industrial Average over a specific week. What do you notice about the scale on the vertical axis? It is common for graphs that track the Dow (and other stock market indices) over relatively short periods of time to use a similar vertical scale. What is the advantage of such a scale in this situation?

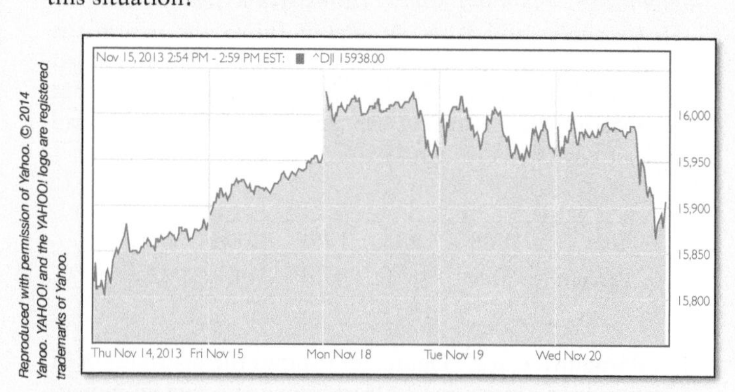

FIGURE 2.62 Dow Jones Industrial Average.

20. Tulsa State Fair. The data in the bar graph in **Figure 2.63** appeared in a report of the Tulsa County Public Facilities Authority. The graph shows Tulsa State Fair attendance in millions from 1998 through 2002.

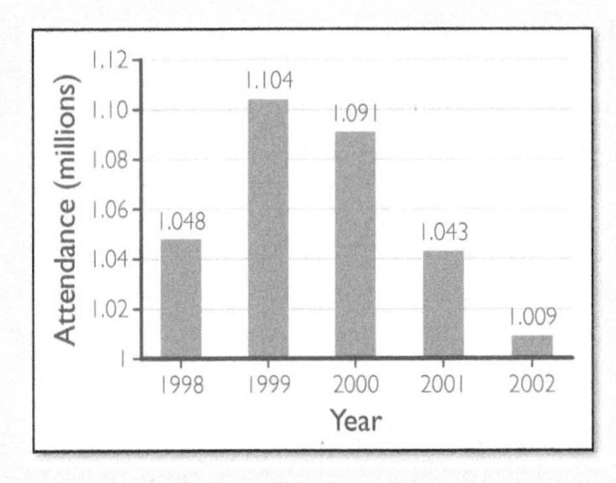

FIGURE 2.63 Tulsa State Fair attendance.

a. In a few sentences, explain what you think the graph in Figure 2.63 appears to show regarding State Fair attendance.

b. Make a table of percentage change in attendance for each year.

c. Sketch a bar graph using 0 to 1.2 million on the vertical axis. Does this give a more accurate display of the data? Why or why not?

21. **Minorities employed.** The bar graph in **Figure 2.64** shows how the percentage of federal employees who are minorities has changed with time.[15] By adjusting the vertical range, make a bar graph that might accompany an article with the title "Federal employment of minorities shows little change."

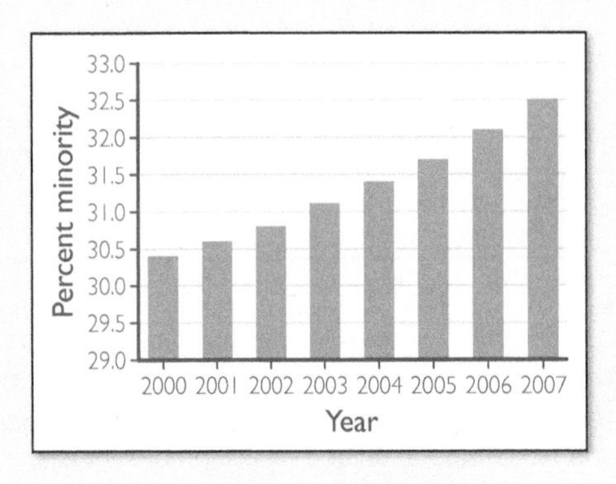

FIGURE 2.64 Minorities employed at federal level.

22. **Aging employees.** The following table shows the average age in years of federal employees at various times.

Year	2000	2001	2002	2003	2004	2005	2006	2007
Average age	46.3	46.5	46.5	46.7	46.8	46.9	46.9	47.0

a. Make a bar graph that emphasizes the increase in average age.

b. Make a bar graph that gives an impression of relative stability in average age.

c. By what percentage did the average age increase from 2000 to 2007?

d. In light of the result of part c, which of the graphs that you made in parts a and b is a better representation of the change in average age?

23. **Adjusting defense spending for inflation.** The amount in billions of constant 2004 dollars spent on defense in the given year is shown in the table below. Make a bar graph of the data and compare the information provided by this graph with the graphs in Figures 2.39 and 2.40.

Year	2004	2005	2006	2007	2008	2009	2010	2011	2012
2004 dollars	455.8	480.4	490.5	501.7	542.2	581.7	603.3	592.7	555.9

24–27. **Adjusting for inflation.** *Exercises 24 through 27 use the following tables, which show inflation over the given period:*

Time span	1980–2020	1990–2020	2000–2020	2010–2020
Inflation	214%	98%	50%	19%

Time span	2005–11	2007–11	2009–11
Inflation	15%	8%	5%

24. The following table shows the average starting salary for geologists for the given year:

Year	2005	2007	2009	2011
Salary	$67,800	$82,200	$83,600	$93,000

a. Make a bar graph of starting salaries for geologists.

b. Complete the following table showing starting salaries in constant 2011 dollars:

Year	2005	2007	2009	2011
Salary	$67,800	$82,200	$83,600	$93,000
2011 dollars				

c. Make a bar graph of salaries in constant 2011 dollars.

d. Compare the graphs from parts a and c.

25. The following table shows the average price per ounce of gold in the given year:

Year	1980	1990	2000	2010	2020
Price per ounce	$612.56	$383.51	$279.11	$1224.53	$1769.64

a. Make a bar graph of gold prices.

b. Complete the following table showing gold prices in constant 2020 dollars:

Year	1980	1990	2000	2010	2020
Price per ounce	$612.56	$383.51	$279.11	$1224.53	$1769.64
2020 dollars					

c. Make a bar graph of gold prices in constant 2020 dollars.

[15] *Data, but not graph, from* Statistical Abstract of the United States. *Only federal civilian nonpostal employees included.*

26. The following table shows the average price per barrel of crude oil in the given year:

Year	1980	1990	2000	2010	2020
Price per barrel	$37.42	$23.19	$27.39	$71.21	$41.59

a. Complete the following table showing oil prices in constant 2020 dollars:

Year	**1980**	**1990**	**2000**	**2010**	**2020**
Price per barrel	$37.42	$23.19	$27.39	$71.21	$41.59
2020 dollars					

b. Make a bar graph of oil prices in constant 2020 dollars.

c. Was crude oil more expensive in 1990 or 2000? Explain your answer.

27. The following table shows the average price of a ticket to the Super Bowl in the given year:

Year	1980	1990	2000	2010	2020
Ticket price	$30	$125	$325	$1000	$3488

a. Complete the following table showing ticket prices in constant 2020 dollars:

Year	**1980**	**1990**	**2000**	**2010**	**2020**
Ticket price	$30	$125	$325	$1000	$3488
2020 dollars					

b. Make a bar graph of ticket prices in constant 2020 dollars.

c. Compare ticket prices in constant dollars for 1980 and 2020.

28. Federal receipts. The pie chart in **Figure 2.65** shows government revenue in the United States in 2015. The total amount of federal revenue was about 3.2 trillion dollars. Complete the table below.

Source	Income tax	Payroll tax	Corporate income tax	Other taxes
Percent				
Amount				

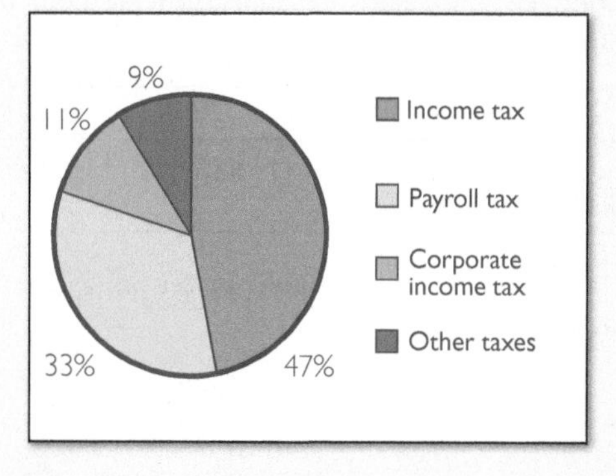

FIGURE 2.65 Government revenue in 2015.

29. A cake. The following table shows ingredients in a hypothetical recipe for a cake as a percent of the total by weight. (The authors cannot recommend actual use of this recipe.)

Ingredients	Flour	Milk	Sugar	Eggs
Percent	50	25	15	10

Represent the data in this table in two ways: by sketching a bar graph and by sketching a pie chart.

Exercise 30 is suitable for group work.

30. Health care and longevity. The graphical presentation in **Figure 2.66** compares health-care spending with life expectancy in various countries. Where on this graph is the best place to be (lower left, lower right, upper left, upper right)? Where on this graph is the worst place to be (lower left, lower right, upper left, upper right)? Make a plot with the axes reversed and answer the previous questions. See if you can find another way to present the data, either of your own devising or from the Web.

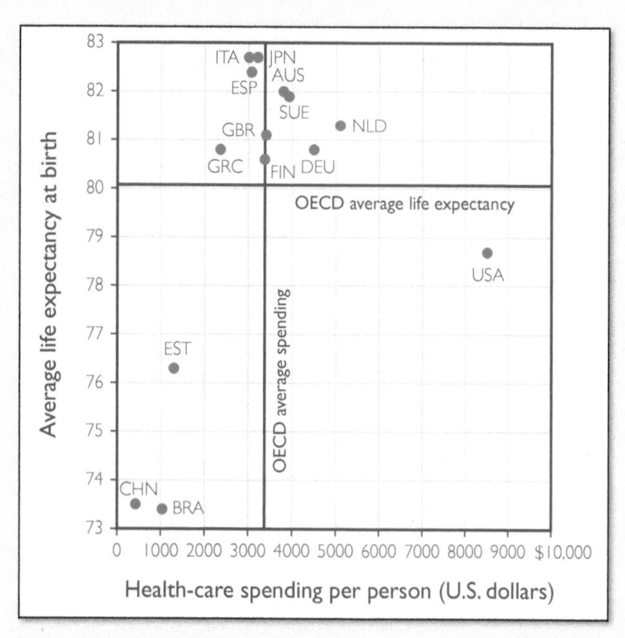

FIGURE 2.66 Health-care spending. **Data from:** Organisation for Economic Cooperation and Development (OECD) Health Data 2013.

Writing About Mathematics

31. Your own experience. Find a graph in the popular media that you consider to be misleading. Explain why it is misleading and construct an alternative graph that is a more accurate representation.

32. How to lie with statistics. The title of this exercise is also the title of a 1954 book by Darrell Huff. Investigate this book and its author and report on your findings.

What Do You Think?

33. Do I trust this report? A blogger bolsters his position on government spending with a graph. But when you look closely, you find that no scale is shown on the vertical axis. Does that give you cause to mistrust the graph? Explain your reasoning.

34. People who vote. A report on television shows a graph to accompany a story that the percentage of people in your city who vote in state elections has declined substantially over the past few years. As a civic-minded citizen, you are disappointed by the report. Then you notice that the scale on the vertical axis ranges from 36% to 38%. Might this affect your reaction to the story? Explain your reasoning. **Exercise 21 is relevant to this topic.**

35. Misleading percentages. A report states that the number of inmates in the Mayberry County jail has increased by 50% over the past year, and presents this as an alarming fact. Is it necessarily so? What if there were only two inmates in the county jail last year? How many would there be now?

36. Danger signs. You see a graph where the vertical scale begins with a large number rather than zero. Explain why this should always give you reason to be suspicious.

37. Pie charts. When you see a pie chart that you might be suspicious about, you could try adding up the numerical values assigned to the pieces of the pie. Explain why this might help detect an error or something misleading.

38. Better graph? Elvis says he's going to use a pie chart to help illustrate the growth of social media over the last four years. How would you explain to Elvis that this is probably not the right kind of chart to use? What would you recommend as a better choice?

39. Poverty. A county reports that the number of people in the county living below the poverty threshold has increased from 100 to 200 over a decade. This looks like bad news. The report also notes, however, that the population of the county has increased from 1000 to 2000. Use percentages to describe in another way how the poverty rate has changed. How is this procedure similar to adjusting monetary quantities for inflation?

40. Pictorial representations. In this section we have noted several ways in which pictorial representations can be misleading. What good reasons are there for using pictorial representations?

41. How to do it. If you were going to make a graph that emphasizes a certain point of view, perhaps by distorting real data, what techniques might you employ?

CHAPTER SUMMARY

In our age, we are confronted with an overwhelming amount of information. Analyzing quantitative data is an important aspect of our lives. Often this analysis requires the ability to understand visual displays, such as graphs, and to make sense of tables of data. A key point to keep in mind is that growth rates are reflected in graphs and tables.

Measurements of growth: How fast is it changing?

Data sets can be presented in many ways. Often the information is presented using a table. To analyze tabular information, we choose an independent variable. From this point of view, the table presents a function, which tells how one quantity (the dependent variable) depends on another (the independent variable).

A simple tool for visualizing data sets is the *bar graph*. See **Figure 2.67** for an example of a bar graph. A bar graph is often appropriate for small data sets when there is nothing to show between data points.

Both absolute and percentage changes are easily calculated from data sets, and such calculations often show behavior not readily apparent from the original data. We calculate the percentage change using

$$\text{Percentage change} = \frac{\text{Change in function value}}{\text{Previous function value}} \times 100\%$$

One example of the value of calculating such changes is the study of world population in Example 2.3.

To find the average growth rate of a function over an interval, we calculate the change in the function divided by the change in the independent variable:

$$\text{Average growth rate} = \frac{\text{Change in function value}}{\text{Change in independent variable}}$$

Keeping track of the units helps us to interpret the average growth rate in a given context.

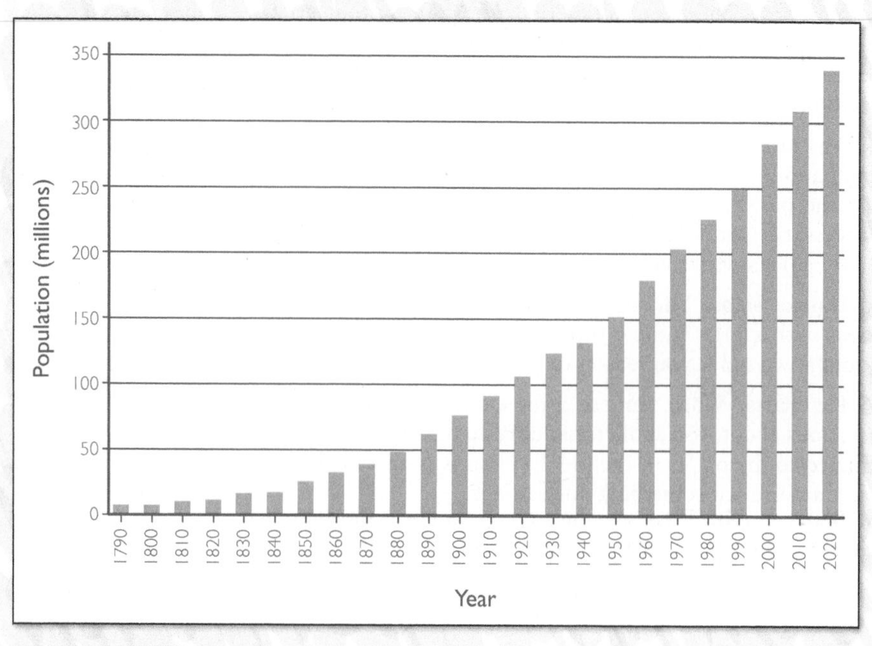

FIGURE 2.67 A bar graph.

Data sets often have gaps. With no further information, there is no way to accurately fill in the gaps. But *interpolation* using the average growth rate is available to provide estimates. Such estimates should always be viewed critically.

Extending data sets by *extrapolation* can be helpful in predicting function values, but doing so is a risky business, especially if we go far beyond the limits of the data set. This is illustrated in Example 2.10, a study of the age of first-time mothers.

Graphs: Picturing growth

Graphical presentations provide a visual display that can significantly aid understanding, usually by conveying information about growth rates. There are various types of graphical displays, and each has its advantages as well as its limitations. The three most common types of graphical representations are *bar graphs* (mentioned above), *scatterplots*, and *line graphs*. A commonly used variation on the third of these is the *smoothed line graph*. Examples of the second and third type are shown in **Figures 2.68 and 2.69.** A scatterplot can be used to display and study the relationship between any two sets of quantitative data. A line graph, if properly used, enhances the picture given by a scatterplot by filling gaps in the data with estimates.

In general, these four types of graphical displays allow us to locate not only intervals of increase and decrease but also rates of increase and decrease.

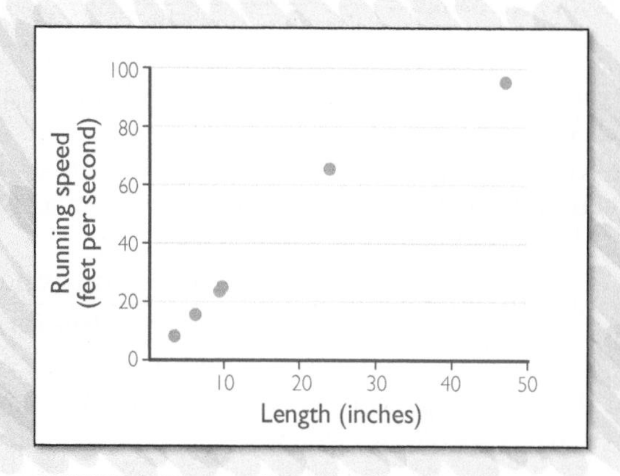

FIGURE 2.68 A scatterplot.

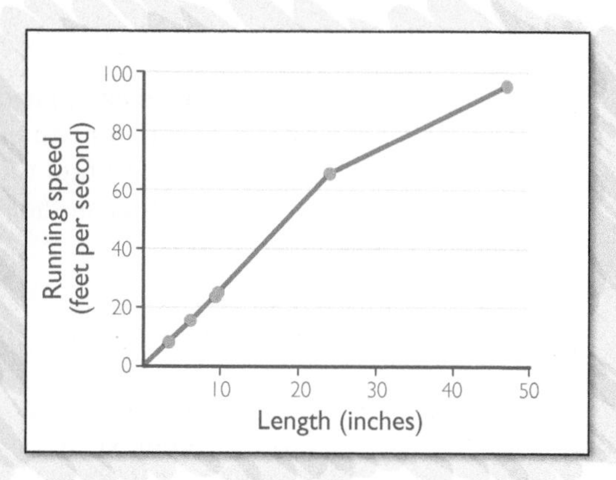

FIGURE 2.69 A line graph.

Misleading graphs: Should I believe my eyes?

Graphical representations are sometimes misleading, whether or not this is intentional. A typical case is shown in the graphs of defense spending in **Figures 2.70 and 2.71**. The graphs represent identical data. The difference in appearance is due to the scale on the vertical axes.

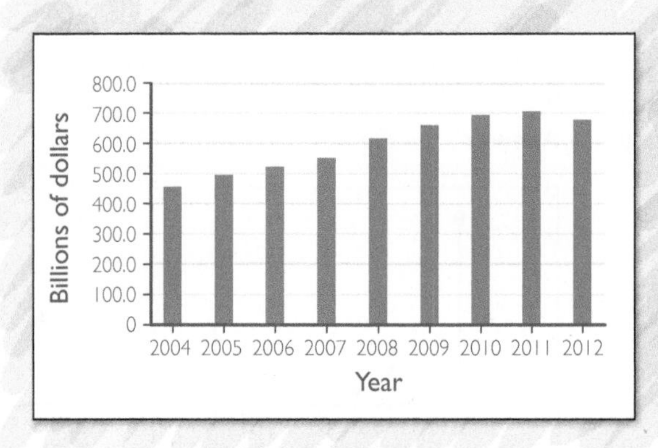

FIGURE 2.70 Defense spending.

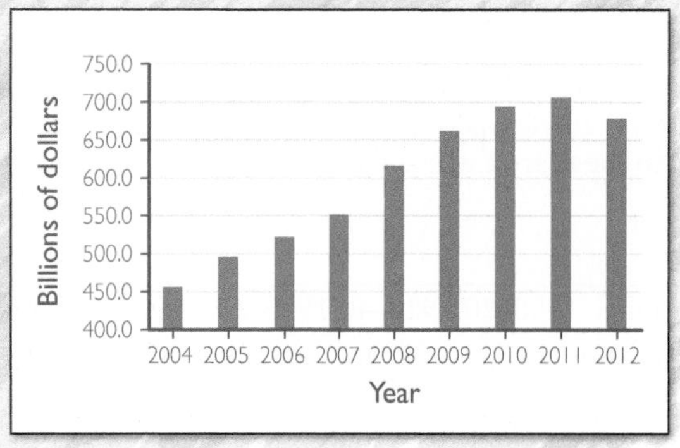

FIGURE 2.71 The same defense spending data.

When the data in a graph depend on the value of currency, it is important to know whether the currency values have been adjusted for inflation. For example, the graph in **Figure 2.72** shows the average price in dollars per barrel of crude oil in the given year. **Figure 2.73** shows that price in constant 2020 dollars. Using constant dollars gives a more accurate way of comparing the true cost of crude oil at different times.

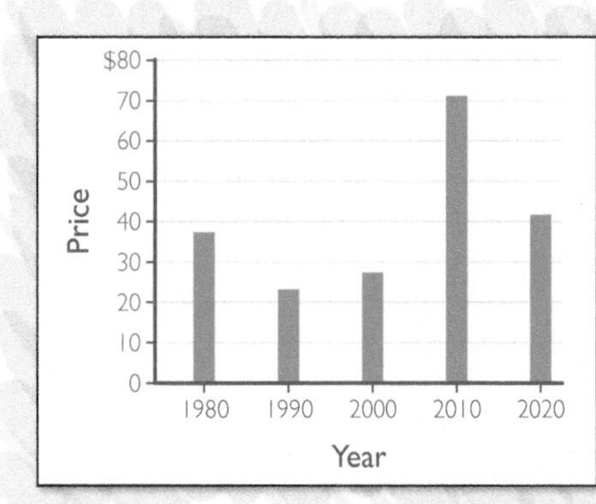

FIGURE 2.72 Price per barrel of oil.

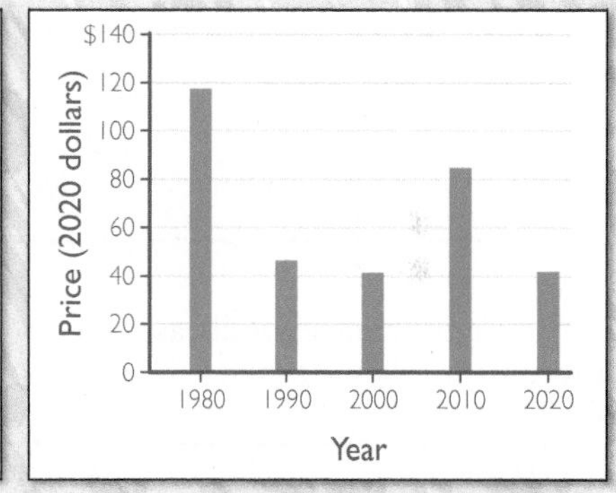

FIGURE 2.73 Crude oil price adjusted for inflation.

Any graph can be misleading if insufficient data are plotted. It may be particularly difficult to determine how many data points were used to produce a smoothed line graph.

Often data are presented using images designed to grab our attention. One common graphical representation is the *pie chart*, which is used to display how constituent parts make up a whole. In general, visual representations can be misleading if the relative sizes of the items pictured do not correspond to the relative sizes of the data.

KEY TERMS

independent variable, p. 81
dependent variable, p. 81
function, p. 81
percentage change, p. 84
relative change, p. 84
average rate of change, p. 87
interpolation, p. 90
extrapolation, p. 92
scatterplot, p. 100
line graph, p. 101
smoothed line graph, p. 106

CHAPTER QUIZ

1. The following table shows the population (in millions) of greater New York City in the given year:

Date	1900	1920	1940	1960	1980	2000
Population (millions)	3.44	5.62	7.46	7.78	7.07	8.01

Make a table and a bar graph showing the percentage change over each 20-year period. Round your answers in percentage form to the nearest whole number.

Answer

Date range	1900–20	1920–40	1940–60	1960–80	1980–2000
Percentage change	63%	33%	4%	−9%	13%

If you had difficulty with this problem, see Example 2.3.

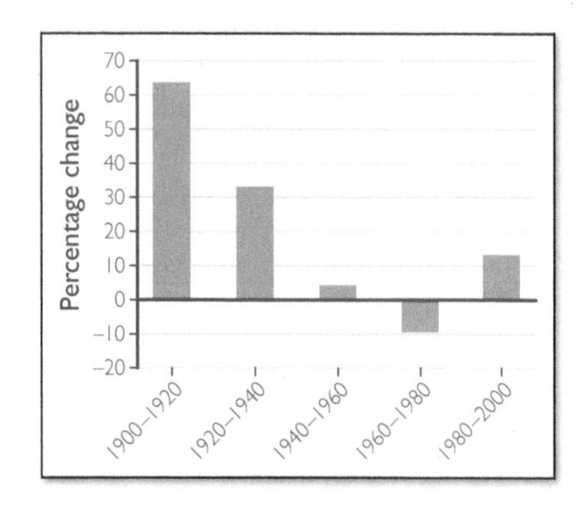

2. The cumulative number of disease cases from an epidemic are shown in the following table:

Day	1	5	8	12	15
Number of cases	22	35	81	91	96

a. Make a new table showing the average number of new cases per day over each period. Round your answers to two decimal places.

b. Use your answer from part a to estimate the cumulative number of cases by Day 11.

Answer

a.

Period	1–5	5–8	8–12	12–15
Average new cases per day	3.25	15.33	2.50	1.67

b. Interpolation gives 88.5, so the cumulative number of cases is about 89.

If you had difficulty with this problem, see Example 2.8.

3. The following table shows the barometric pressure in (millimeters of mercury) at various altitudes above sea level:

Altitude (thousand feet)	0	20	40	60	80	100
Pressure (mm mercury)	760	350	141	54	21	8

Make a scatterplot and a line graph of the data. Use the altitude for the independent variable and the pressure for the dependent variable.

Answer See **Figures 2.74 and 2.75**.

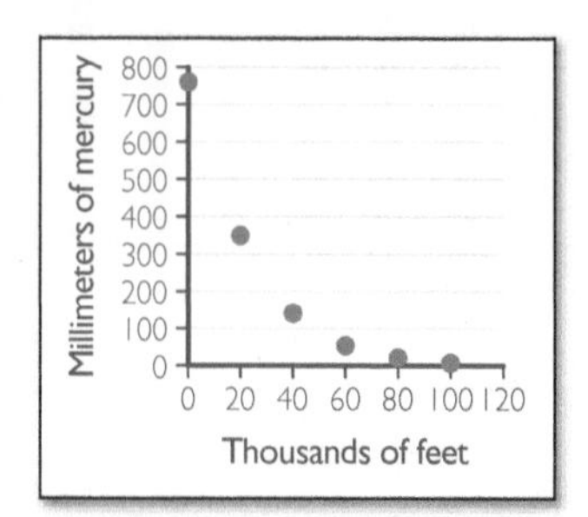

FIGURE 2.74 Pressure versus altitude: scatterplot.

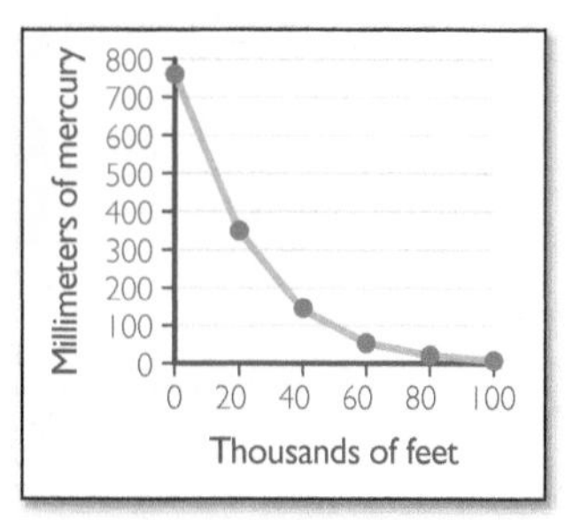

FIGURE 2.75 Pressure versus altitude: line graph.

If you had difficulty with this problem, see Example 2.11.

4. **Figure 2.76** shows a population of elk. Describe the growth rate in the elk population over the 10-year period. Approximately when was the population declining fastest?

Answer The population is declining throughout the 10-year period, so the growth rate is negative. The rate of decline increases until about Year 4, when the rate starts to level off. The time of fastest decline is around Year 4.

If you had difficulty with this problem, see Example 2.14.

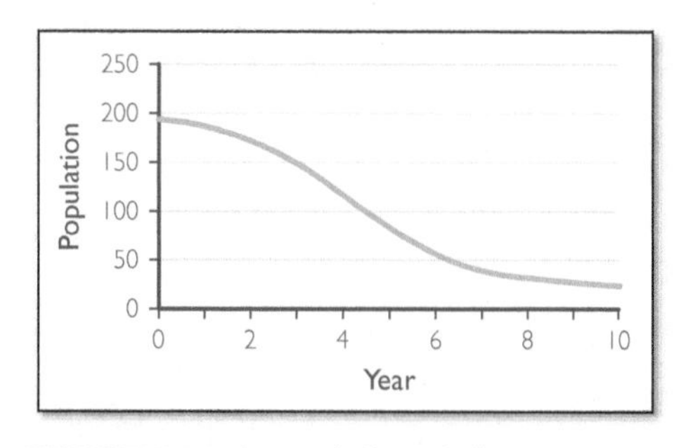

FIGURE 2.76 A population of elk.

5. The following table shows total sales in the given month:

Month	Jan.	Feb.	March	April	May
Sales	$5322	$5483	$6198	$6263	$6516

Make a bar graph that emphasizes the increase in sales over the five-month period, and make a second bar graph that de-emphasizes the increase.

Answer To emphasize the increase, we use a narrow range on the vertical axis. One possibility is shown in **Figure 2.77**. To deemphasize the increase, we use a wide range on the vertical axis. One possibility is shown in **Figure 2.78**.

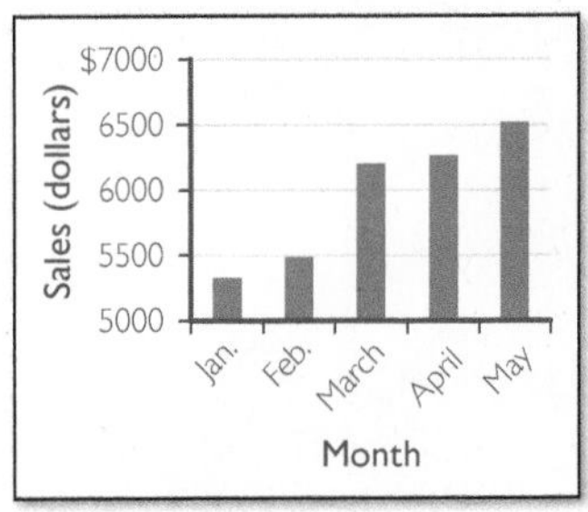

FIGURE 2.77 Emphasizing growth in sales.

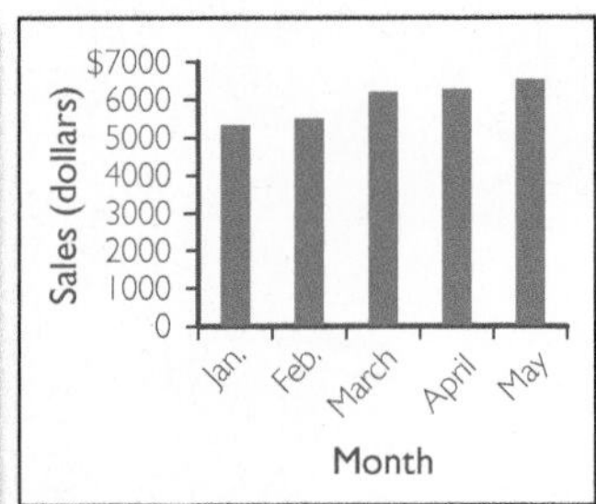

FIGURE 2.78 De-emphasizing growth in sales.

If you had difficulty with this problem, see Example 2.16.

6. The following table shows the cost of a certain item in the given year:

Year	1980	1990	2000	2010	2020
Cost	$23.00	$35.00	$44.00	$53.00	$65.00

Time span	1980–2020	1990–2020	2000–2020	2010–2020
Inflation	214%	98%	50%	19%

a. Complete the following table showing the cost in constant 2020 dollars:

Year	1980	1990	2000	2010	2020
Cost	$23.00	$35.00	$44.00	$53.00	$65.00
2020 dollars					

b. Make a bar graph showing the cost in constant 2020 dollars.

c. When was the item most expensive, taking inflation into account?

Answer

a.

Year	1980	1990	2000	2010	2020
Cost	$23.00	$35.00	$44.00	$53.00	$65.00
2020 dollars	$72.22	$69.30	$66.00	$63.07	$65.00

b.

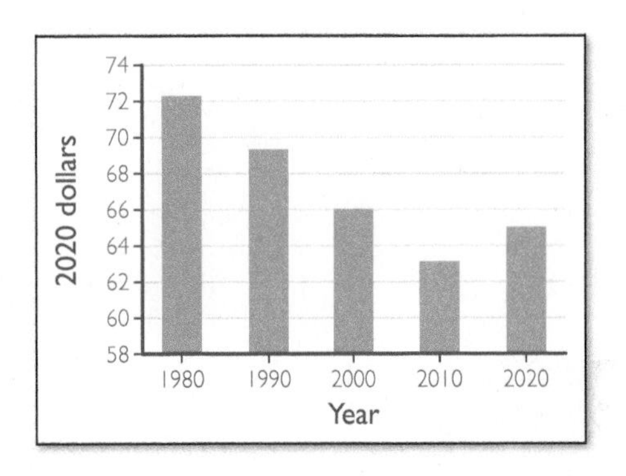

c. The maximum cost, adjusted for inflation, occurred in 1980.

If you had difficulty with this problem, see Example 2.17.

John Moore/Getty Images

3 Linear and Exponential Change:

Comparing Growth Rates

The following article is about hunger in the United States.

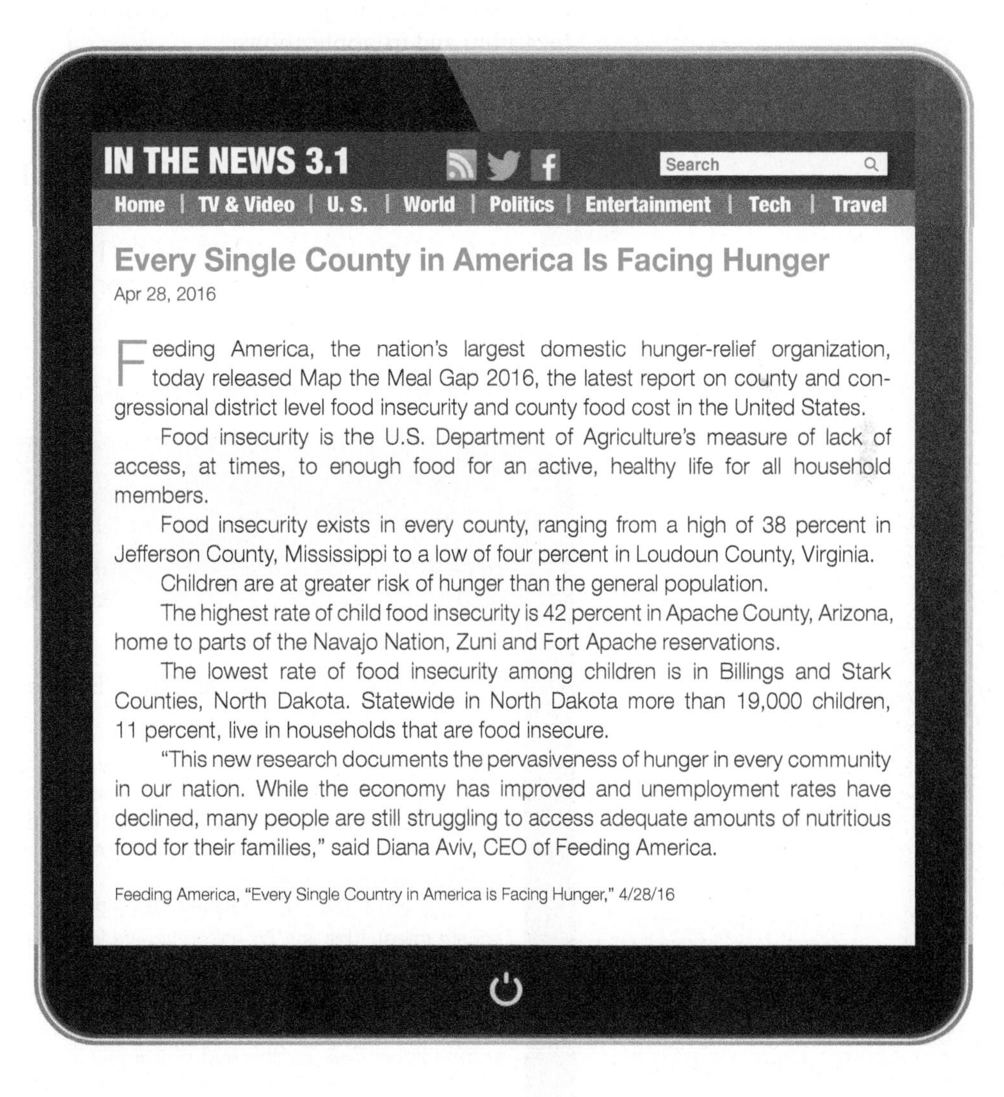

Every Single County in America Is Facing Hunger

Apr 28, 2016

Feeding America, the nation's largest domestic hunger-relief organization, today released Map the Meal Gap 2016, the latest report on county and congressional district level food insecurity and county food cost in the United States.

Food insecurity is the U.S. Department of Agriculture's measure of lack of access, at times, to enough food for an active, healthy life for all household members.

Food insecurity exists in every county, ranging from a high of 38 percent in Jefferson County, Mississippi to a low of four percent in Loudoun County, Virginia.

Children are at greater risk of hunger than the general population.

The highest rate of child food insecurity is 42 percent in Apache County, Arizona, home to parts of the Navajo Nation, Zuni and Fort Apache reservations.

The lowest rate of food insecurity among children is in Billings and Stark Counties, North Dakota. Statewide in North Dakota more than 19,000 children, 11 percent, live in households that are food insecure.

"This new research documents the pervasiveness of hunger in every community in our nation. While the economy has improved and unemployment rates have declined, many people are still struggling to access adequate amounts of nutritious food for their families," said Diana Aviv, CEO of Feeding America.

Feeding America, "Every Single Country in America is Facing Hunger," 4/28/16

Thomas Robert Malthus, 1766–1834.

Concern about hunger in the United States, and indeed throughout the world, is nothing new. The English demographer and political economist Thomas Robert Malthus (1766–1834) is best known for his influential views on population growth. In 1798 he wrote the following in *An Essay on the Principle of Population*:

> I say that the power of population is indefinitely greater than the power in the earth to produce subsistence for man. Population, when unchecked, increases in a geometrical ratio whereas the food-supply grows at an arithmetic rate.
>
> The power of population is so superior to the power of the earth to produce subsistence for man, that premature death must in some shape or other visit the human race.

Note the distinction Malthus makes between a "geometrical ratio" and an "arithmetic rate." In contemporary terms, we would say that population growth is *exponential* and food production growth is *linear*. Inherent properties of these two types of growth led Malthus to his dire predictions. His analysis is regarded today as an oversimplification of complex issues, but these specific types of growth remain crucial tools in analyzing demand versus availability of Earth's resources.

Change can follow various patterns, but linear and exponential patterns are of particular importance because they occur so often and in such significant applications. We will examine what they mean in this chapter. We will also consider the logarithm and its applications.

3.1 Lines and linear growth: What does a constant rate mean?

TAKE AWAY FROM THIS SECTION
Understand linear functions and consequences of a constant growth rate.

The following article is from the National Snow and Ice Data Center.

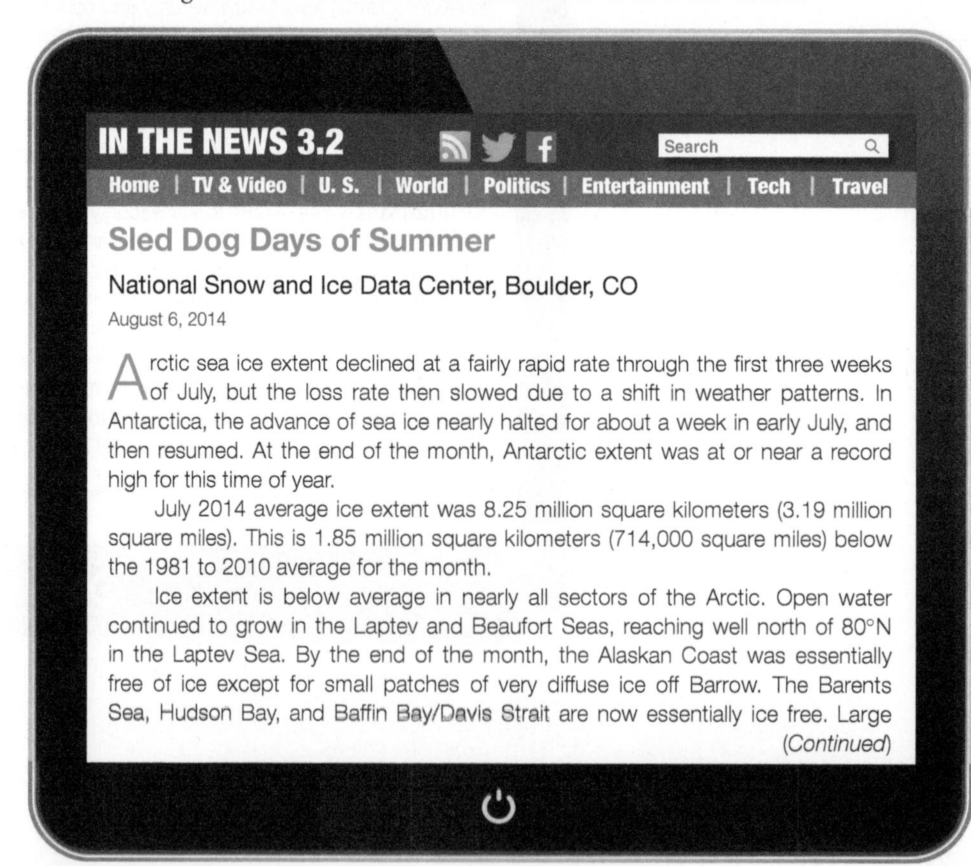

Sled Dog Days of Summer

National Snow and Ice Data Center, Boulder, CO

August 6, 2014

Arctic sea ice extent declined at a fairly rapid rate through the first three weeks of July, but the loss rate then slowed due to a shift in weather patterns. In Antarctica, the advance of sea ice nearly halted for about a week in early July, and then resumed. At the end of the month, Antarctic extent was at or near a record high for this time of year.

July 2014 average ice extent was 8.25 million square kilometers (3.19 million square miles). This is 1.85 million square kilometers (714,000 square miles) below the 1981 to 2010 average for the month.

Ice extent is below average in nearly all sectors of the Arctic. Open water continued to grow in the Laptev and Beaufort Seas, reaching well north of 80°N in the Laptev Sea. By the end of the month, the Alaskan Coast was essentially free of ice except for small patches of very diffuse ice off Barrow. The Barents Sea, Hudson Bay, and Baffin Bay/Davis Strait are now essentially ice free. Large

(Continued)

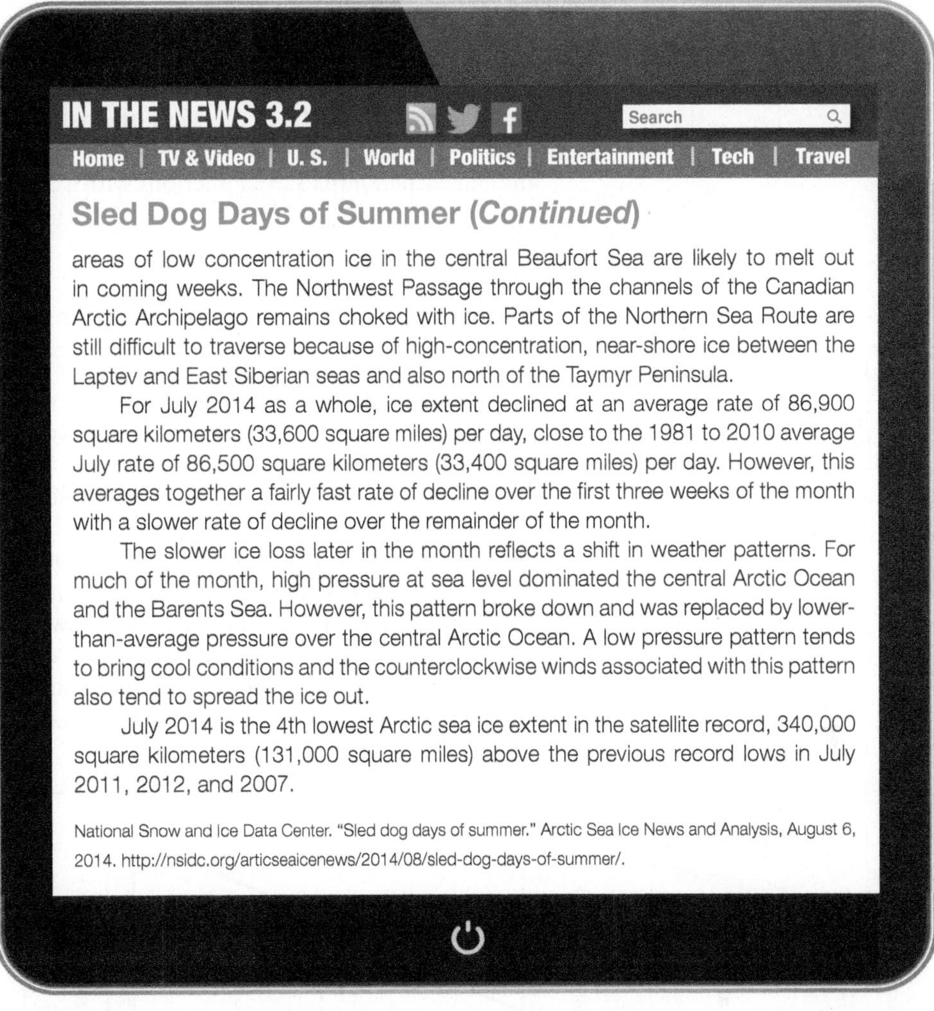

IN THE NEWS 3.2

Search

Home | TV & Video | U. S. | World | Politics | Entertainment | Tech | Travel

Sled Dog Days of Summer (*Continued*)

areas of low concentration ice in the central Beaufort Sea are likely to melt out in coming weeks. The Northwest Passage through the channels of the Canadian Arctic Archipelago remains choked with ice. Parts of the Northern Sea Route are still difficult to traverse because of high-concentration, near-shore ice between the Laptev and East Siberian seas and also north of the Taymyr Peninsula.

For July 2014 as a whole, ice extent declined at an average rate of 86,900 square kilometers (33,600 square miles) per day, close to the 1981 to 2010 average July rate of 86,500 square kilometers (33,400 square miles) per day. However, this averages together a fairly fast rate of decline over the first three weeks of the month with a slower rate of decline over the remainder of the month.

The slower ice loss later in the month reflects a shift in weather patterns. For much of the month, high pressure at sea level dominated the central Arctic Ocean and the Barents Sea. However, this pattern broke down and was replaced by lower-than-average pressure over the central Arctic Ocean. A low pressure pattern tends to bring cool conditions and the counterclockwise winds associated with this pattern also tend to spread the ice out.

July 2014 is the 4th lowest Arctic sea ice extent in the satellite record, 340,000 square kilometers (131,000 square miles) above the previous record lows in July 2011, 2012, and 2007.

National Snow and Ice Data Center. "Sled dog days of summer." Arctic Sea Ice News and Analysis, August 6, 2014. http://nsidc.org/articseaicenews/2014/08/sled-dog-days-of-summer/.

The analysis in this article is accompanied by **Figure 3.1**, which shows how the monthly July ice extent declined from 1979 to 2014. The figure uses a straight line to help us understand the decline. Straight lines and the relationships they represent frequently occur in the media, and every critical reader needs a basic understanding of them.

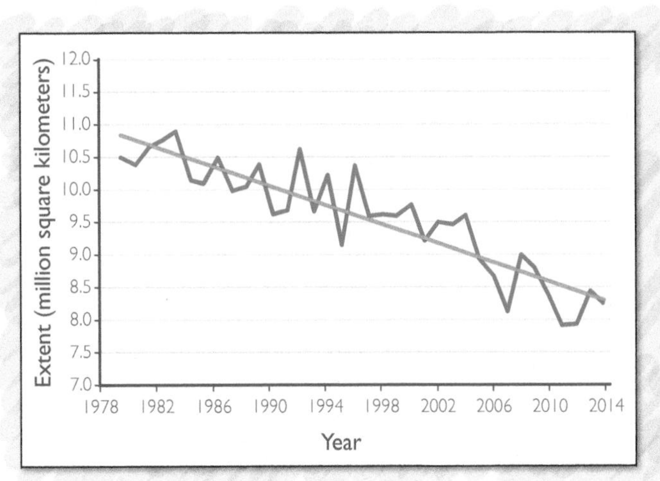

FIGURE 3.1 Arctic sea ice extent in July, 1979 to 2014.

Recognizing linear functions

We have noted that in the media we often see graphics showing straight lines, such as the line in Figure 3.1. A straight line represents a special type of function, called a *linear* function.

A simple example will help us understand linear functions. Suppose we start with \$100 in a cookie jar and each month we add \$25. We can think of the number of months as the independent variable and the amount of money in the cookie jar as the dependent variable. In this view, the balance in the cookie jar is a function of the number of months since we began saving. In this case, the function grows by the same amount each month (\$25). Functions with a constant growth rate are *linear functions*.

KEY CONCEPT

A function is called linear if it has a constant growth rate.

We note that the growth rate may be a negative number, which indicates a decreasing function. For example, suppose we start with money in a cookie jar and withdraw \$5 each month. Then the balance is changing by the same amount each month, so it is a linear function of the number of months since we started withdrawing money. In this case, the growth rate is −\$5 (negative 5 dollars) per month.

Not all functions are linear. A person's height is not a linear function of her age. If it were, she would grow by the same amount each year of her life.

An important feature of linear functions is that their graphs are straight lines. This is illustrated in **Figures 3.2 and 3.3**, where we have graphed the functions in our preceding examples. For comparison, a graph of typical height versus age for females is shown in **Figure 3.4**. It is not a straight line, and this is further evidence that height as a function of age is not a linear function.

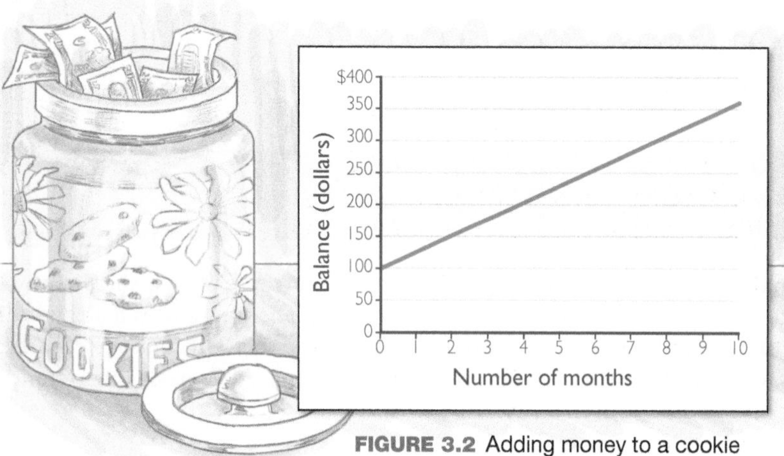

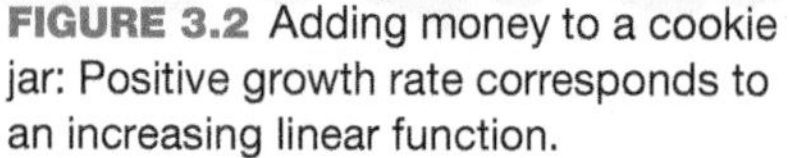

FIGURE 3.2 Adding money to a cookie jar: Positive growth rate corresponds to an increasing linear function.

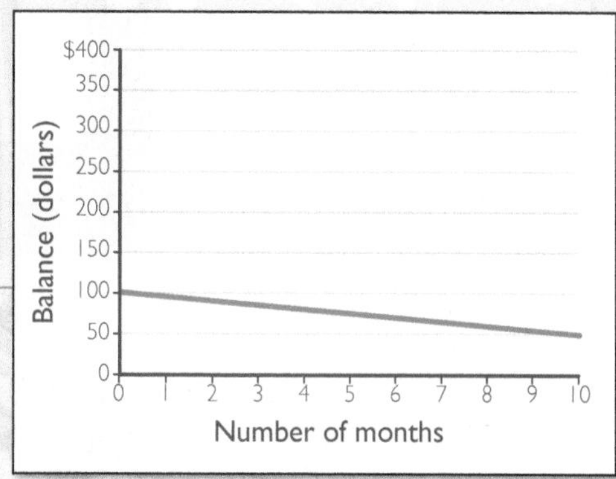

FIGURE 3.3 Taking money from a cookie jar: Negative growth rate corresponds to a decreasing linear function.

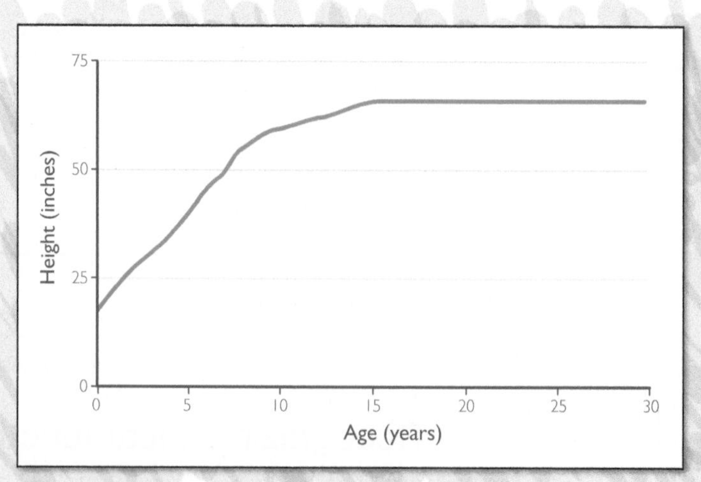

FIGURE 3.4 Height is not a linear function of age.

Trust me, it's just an expression: I'm a crow and I can tell you I don't always fly in a straight line...

EXAMPLE 3.1 Determining what is linear and what is not: Total cost and salary

In each part below, a function is described. Find the growth rate of the function and give its practical meaning. Make a graph of the function. Is the function linear?

a. For my daughter's wedding reception, I pay \$500 rent for the building plus \$15 for each guest. This describes the total cost of the reception as a function of the number of guests.

b. My salary is initially \$30,000, and I get a 10% salary raise each year for several years. This describes my salary as a function of time.

SOLUTION

a. The growth rate in this case is the extra cost incurred for each additional guest, which is \$15. The additional cost is the same for each additional guest, so the growth rate is constant. Because the growth rate is constant, the total cost of the reception is a linear function of the number of guests. The graph of cost versus number of guests is shown in **Figure 3.5**. It is a straight line, as expected.

b. The growth rate in this case is the increase per year in my salary. The first year my salary increased by 10% of \$30,000, or \$3000, so my new annual salary is \$33,000. The second year my salary increased by 10% of \$33,000, or \$3300. The growth rate is not the same each year, so my salary is not a linear function of time in years. A graph showing salary over several years with a 10% increase each year is illustrated in **Figure 3.6**. The graph is not a straight line, which is further verification that salary is not a linear function of time in years.

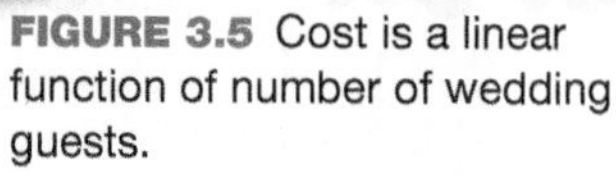

FIGURE 3.5 Cost is a linear function of number of wedding guests.

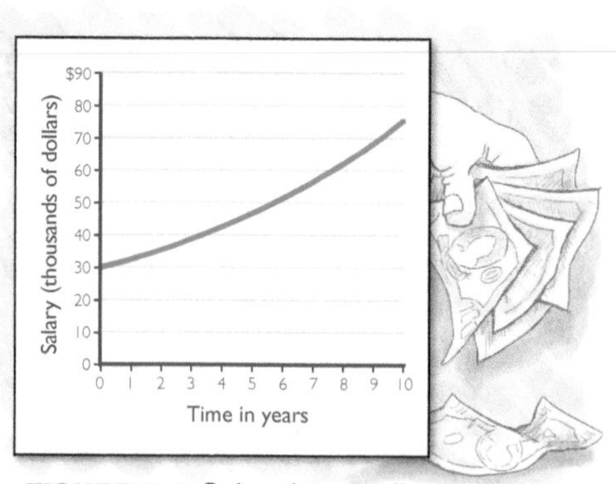

FIGURE 3.6 Salary is not a linear function of time.

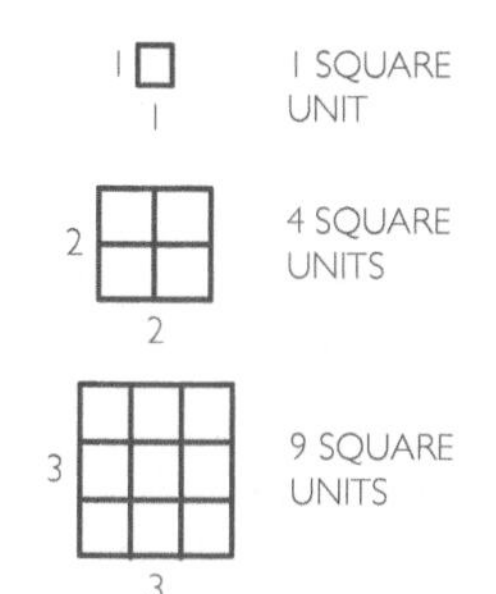

FIGURE 3.7 Areas of squares of different sizes.

TRY IT YOURSELF 3.1

One of the following is a linear function, and one is not. In each case, determine the practical meaning of the growth rate, then determine whether the given function is linear.

a. One inch is the same as 2.54 centimeters. Consider distance in centimeters to be a function of distance in inches.

b. Consider the area of a square to be a function of the length of a side. **Figure 3.7** may be helpful here.

The answer is provided at the end of this section.

Formulas for linear functions

Let's look again at the cookie jar example, in which we start with \$100 and add \$25 each month. We use x to denote the number of months since we started saving and y to denote the balance in dollars after x months.

After $x = 1$ month, there is $y = 25 \times 1 + 100 = \$125$ in the cookie jar
After $x = 2$ months, there is $y = 25 \times 2 + 100 = \$150$ in the cookie jar
After $x = 3$ months, there is $y = 25 \times 3 + 100 = \$175$ in the cookie jar

Following this pattern, we see that after x months, there will be

$$y = 25x + 100$$

dollars in the cookie jar.

In the preceding formula, the number \$25 per month is the growth rate. The number \$100 is the *initial value* because it is the amount with which we start. In other words, \$100 is the function value (the value of y) when the independent variable x is 0. This way of combining the initial value with the growth rate is typical of how we obtain formulas for linear functions. The general formula for a linear function is

$$y = \text{Growth rate} \times x + \text{Initial value}$$

It is customary to use m to denote the growth rate and b to denote the initial value. With these choices, the formula for a linear function is

$$y = mx + b$$

Figure 3.8 shows how the growth rate m determines the steepness of a line. Moving one unit to the right corresponds to a rise of m units of the graph. That is why the growth rate m is often called the *slope*. For linear functions, we will use the terms *slope* and *growth rate* interchangeably.

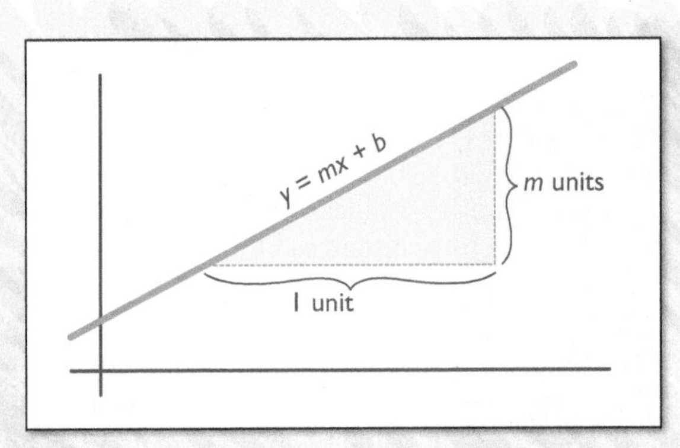

FIGURE 3.8 The growth rate m is the slope of the line.

SUMMARY 3.1
Linear Functions

- A linear function is a function with a constant growth rate. The growth rate is often called the slope.
- The graph of a linear function is a straight line.
- The formula for a linear function takes the form

$$y = \text{Growth rate} \times x + \text{Initial value}$$

If m is the growth rate or slope and b is the initial value, the formula becomes

$$y = mx + b$$

EXAMPLE 3.2 Interpreting formulas: Athletic records

Let L denote the length in meters of the winning long jump in the early years of the modern Olympic Games. We think of L as a function of the number n of Olympic Games since 1900. (Thus, because the Olympic Games were held every four years, $n = 0$ corresponds to the 1900 Olympics, $n = 1$ to the 1904 Olympics, and so on.) An approximate linear formula is $L = 0.14n + 7.20$. Identify the initial value and growth rate, and explain in practical terms their meaning.

The Olympic symbol.

SOLUTION

The initial value of the linear function $y = mx + b$ is b, and the growth rate or slope is m. So, for $L = 0.14n + 7.20$, the initial value is 7.20 meters. It is the (approximate) length of the winning long jump in the 1900 Olympic Games. The growth rate is 0.14 meter per Olympic Game. This means that the length of the winning long jump increased by approximately 0.14 meter from one game to the next.

TRY IT YOURSELF 3.2

Let H denote the height in meters of the winning pole vault in the early years of the modern Olympic Games. We think of H as a function of the number n of Olympic Games since 1900. An approximate linear formula is $H = 0.20n + 3.3$. Identify the initial value and growth rate, and explain in practical terms the meaning of each.

The answer is provided at the end of this section.

Note that in Example 3.2, we used the variables n and L rather than x and y. There is nothing special about the names x and y for the variables, and it is common to use variable names that help us remember their practical meanings.

If we know the initial value and growth rate for a linear function, we can immediately write its formula. The method is shown in the following example.

NASA

The Saturn V carried the first men to the moon in 1969.

EXAMPLE 3.3 Finding linear formulas: A rocket

A rocket starting from an orbit 30,000 kilometers above the surface of Earth blasts off and flies at a constant speed of 1000 kilometers per hour away from Earth. Explain why the function giving the rocket's distance from Earth in terms of time is linear. Identify the initial value and growth rate, and find a linear formula for the distance.

SOLUTION

First we choose letters to represent the function and variable. Let d denote the distance in kilometers from Earth after t hours. The growth rate is the velocity, 1000 kilometers per hour. (The velocity is positive because the distance is increasing.) Because the growth rate is constant, d is a linear function of t. The initial value is 30,000 kilometers, which is the height above Earth at blastoff (that is, at time $t = 0$). We find the formula for d using

$$d = \text{Growth rate} \times t + \text{Initial value}$$
$$d = 1000t + 30{,}000$$

TRY IT YOURSELF 3.3

Suppose that a stellar object is first detected at 1,000,000 kilometers from Earth and that it is traveling toward Earth at a speed of 2000 kilometers per hour. Explain why the function giving the object's distance from Earth in terms of time is linear. Identify the initial value and growth rate, and find a linear formula for the distance.

The answer is provided at the end of this section.

Finding and interpreting the slope

We have seen that it is easy to find the formula for a linear function if we know the initial value and the growth rate, or slope. But sometimes in linear relationships the growth rate m is not given. In such a situation, we calculate the average growth rate in the same way as we did in Chapter 2. We use the formula

$$\text{Slope} = \text{Growth rate} = \frac{\text{Change in function value}}{\text{Change in independent variable}}$$

When we write the equation of a linear function as $y = mx + b$, we can write the formula for the slope as

$$m = \text{Slope} = \text{Growth rate} = \frac{\text{Change in } y}{\text{Change in } x}$$

A geometric interpretation of this formula is shown in **Figure 3.9**.

Recall from Section 2.1 that the units of the average growth rate are the units of the function divided by the units of the independent variable. In practical settings, the slope has a familiar meaning, and proper use of units can help us to determine that meaning.

Suppose, for example, that it is snowing at a steady rate, which means that the depth of snow on the ground is a linear function of the time since it started snowing. At some point during the snowfall, we find that the snow is 8 inches deep. Four hours later we find that the snow is 20 inches deep. Then the depth has increased by 12 inches over a 4-hour period. (See **Figure 3.10**.) That is a slope of

$$m = \frac{\text{Change in depth}}{\text{Change in time}} = \frac{12 \text{ inches}}{4 \text{ hours}} = 3 \text{ inches per hour}$$

The slope is measured in inches per hour, which highlights the fact that the slope represents the rate of snowfall.

$$m = \text{slope} = \frac{\text{change in } y}{\text{change in } x}$$

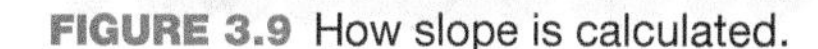

FIGURE 3.9 How slope is calculated.

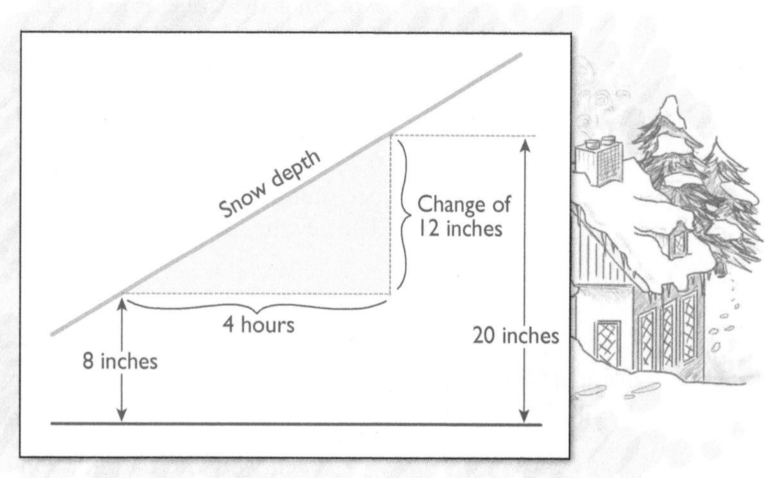

FIGURE 3.10 Growth rate of snow accumulation.

EXAMPLE 3.4 Calculating the slope: A gas tank

Suppose your car's 20-gallon gas tank is full when you begin a road trip. Assume that you are using gas at a constant rate, so the amount of gas in your tank (in gallons) is a linear function of the time in hours you have been driving. After traveling for two hours, your fuel gauge reads three-quarters full. Find the slope of the linear function and explain in practical terms what it means.

SOLUTION

We need to know the change in the amount of gas in the tank. When the tank is three-quarters full, there are 15 gallons of gas left in the tank. The amount of gas in the tank has decreased by 5 gallons. This means that the change in gas is -5 gallons. This occurred over two hours of driving, so the slope is

$$\begin{aligned} m &= \frac{\text{Change in gas}}{\text{Change in time}} \\ &= \frac{-5 \text{ gallons}}{2 \text{ hours}} = -2.5 \text{ gallons per hour} \end{aligned}$$

In practical terms, this means we are using 2.5 gallons of gas each hour.

Pallava Bagla/Corbis News//Getty Images

An ice core from Antarctica.

TRY IT YOURSELF 3.4

Scientists believe that thousands of years ago the depth of ice in a glacier was increasing at a constant rate, so the depth in feet was a linear function of time in years. Using a core sample, they measured a depth of 25 feet at one time and a depth of 28 feet five years later. What is the slope of the linear function? Explain in practical terms the meaning of the slope.

The answer is provided at the end of this section.

If an airplane is flying away from us along a straight line at a constant speed of 300 miles per hour, its distance from us increases by 300 miles for each hour that passes. This is a typical description of a linear function: If the function has the formula $y = mx + b$, a 1-unit increase in the independent variable x corresponds to a change in the function value y by an amount equal to the slope m. This interpretation is key to the way we use the slope in practical settings. For example, the frequency of chirping by crickets is related to the temperature. The relationship that has been suggested between the temperature T in degrees Fahrenheit and the number n of cricket chirps per minute is

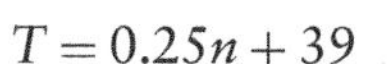

$$T = 0.25n + 39$$

Note that the slope of this linear function is 0.25. The slope is measured in degrees Fahrenheit per chirp per minute.

Suppose initially the temperature is 50 degrees Fahrenheit. An hour later we find that the number of chirps per minute has increased by 20. What is the new temperature? An increase of one chirp per minute corresponds to an increase of 0.25 degree Fahrenheit because 0.25 is the slope. Therefore, an increase by 20 chirps per minute corresponds to an increase in temperature of $20 \times 0.25 = 5$ degrees. The new temperature is $50 + 5 = 55$ degrees Fahrenheit.

SUMMARY 3.2
Interpreting and Using the Slope

- We calculate the slope of a linear function using

$$\text{Slope} = \text{Growth rate} = \frac{\text{Change in function value}}{\text{Change in independent variable}}$$

- When we write the equation of a linear function as $y = mx + b$, the formula for the slope becomes

$$m = \text{Slope} = \text{Growth rate} = \frac{\text{Change in } y}{\text{Change in } x}$$

- For a linear function, a 1-unit increase in the independent variable corresponds to a change in the function value by an amount equal to the slope. For the linear function $y = mx + b$, each 1-unit increase in x corresponds to a change of m units in y.

EXAMPLE 3.5 Using the slope: Temperature and speed of sound

The speed of sound in air depends on the temperature. If T denotes the temperature in degrees Fahrenheit and S is the speed of sound in feet per second, the relation is given by the linear formula

$$S = 1.1T + 1052.3$$

What is the speed of sound when the temperature is 0 degrees Fahrenheit? If the temperature increases by 1 degree Fahrenheit, what is the corresponding increase in the speed of sound?

SOLUTION

Note that the initial value of the linear function is 1052.3. Therefore, when the temperature is 0 degrees, the speed of sound is 1052.3 feet per second.

The slope of the linear function is 1.1 feet per second per degree Fahrenheit. So a 1-degree increase corresponds to a 1.1-foot-per-second increase in the speed of sound.

TRY IT YOURSELF 3.5

Consider again the linear formula in the example. Suppose a bright sun comes out one day and raises the temperature by 3 degrees. What is the corresponding increase in the speed of sound?

The answer is provided at the end of this section.

The next example illustrates the importance of correctly interpreting the slope of a linear function.

EXAMPLE 3.6 Interpreting the slope: Temperature conversion

Measuring temperature using the Fahrenheit scale is common in the United States, but use of the Celsius (sometimes referred to as *centigrade*) scale is more common in most other countries. The temperature in degrees Fahrenheit is a linear function of the temperature in degrees Celsius. On the Celsius scale, 0 degrees is the freezing temperature of water. This occurs at 32 degrees on the Fahrenheit scale. Also, 100 degrees Celsius is the boiling point (at sea level) of water. This occurs at 212 degrees Fahrenheit.

a. What is the slope of the linear function giving the temperature in degrees Fahrenheit in terms of the temperature in degrees Celsius? Use your answer to determine what increase in degrees Fahrenheit corresponds to a 1-degree increase on the Celsius scale.

b. Choose variable and function names, and find a linear formula that converts degrees Celsius to degrees Fahrenheit. Make a graph of the linear function.

c. A news story released by Reuters on March 19, 2002, said that the Antarctic peninsula had warmed by 36 degrees Fahrenheit over the past half-century. Such temperature increases would result in catastrophic climate changes worldwide, and it is surprising that an error of this magnitude could have slipped by the editorial staff of Reuters. We can't say for certain how the error occurred, but it is likely that the British writer saw a report that the temperature had increased by 2.2 degrees Celsius. Verify that a temperature of 2.2 degrees Celsius is about 36 degrees Fahrenheit.

d. The report described in part c did not say that the temperature was 2.2 degrees Celsius but that the temperature *increased* by 2.2 degrees Celsius. What increase in Fahrenheit temperature should the writer have reported?

SOLUTION

a. To find the slope, we need the change in degrees Fahrenheit for a given change in degrees Celsius. An increase on the Celsius scale from 0 to 100 degrees corresponds to an increase on the Fahrenheit scale from 32 to 212, a 180-degree increase. Therefore,

$$\begin{aligned}\text{Slope} &= \frac{\text{Change in degrees Fahrenheit}}{\text{Change in degrees Celsius}}\\ &= \frac{180}{100} = 1.8 \text{ degrees Fahrenheit per degree Celsius}\end{aligned}$$

Therefore, a 1-degree increase on the Celsius scale corresponds to a 1.8-degree increase on the Fahrenheit scale.

b. We use F for the temperature in degrees Fahrenheit and C for the temperature in degrees Celsius. The linear relationship we seek is expressed by the formula $F = \text{Slope} \times C + \text{Initial value}$. We found in part a that the slope is 1.8. Now we need the initial value, which is the Fahrenheit temperature when the Celsius temperature is 0 degrees. That is 32 degrees Fahrenheit. Therefore, the formula is $F = 1.8C + 32$. We have graphed this function in **Figure 3.11**.

c. We put 2.2 degrees Celsius into the formula for converting Celsius to Fahrenheit:

$$\begin{aligned}F &= 1.8C + 32\\ &= 1.8 \times 2.2 + 32 = 35.96 \text{ degrees Fahrenheit}\end{aligned}$$

Rounding gives 36 degrees Fahrenheit.

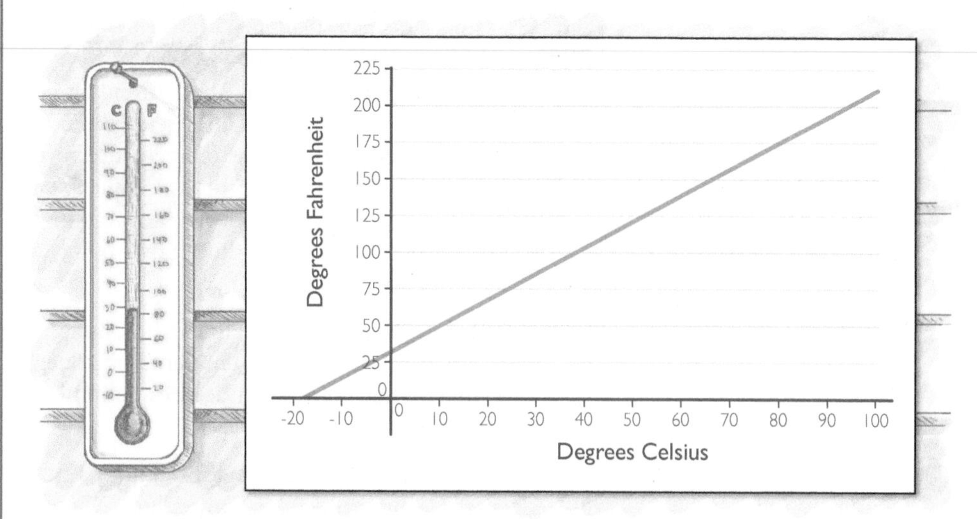

FIGURE 3.11 Fahrenheit temperature is a linear function of Celsius temperature.

d. The slope of the linear function that converts from Celsius to Fahrenheit is 1.8 degrees Fahrenheit per degree Celsius. As we noted in the solution to part a, this tells us that each 1-degree increase on the Celsius scale corresponds to a 1.8-degree increase on the Fahrenheit scale. So an increase of 2.2 degrees Celsius corresponds to an increase of $2.2 \times 1.8 = 3.96$ degrees Fahrenheit. The writer should have reported a warming of about 4 degrees Fahrenheit.

Linear data and trend lines

The following table is taken from the 2020 federal income tax table provided by the Internal Revenue Service.[1] It applies to a married couple filing jointly.

Taxable income	$41,000	$41,100	$41,200	$41,300	$41,400
Tax owed	$4528	$4540	$4552	$4564	$4576

In **Figure 3.12**, we have plotted these data points. The data points appear to fall on a straight line, which indicates that we can use a linear function to describe taxable income, at least over the range given in the table. To verify this, we calculate the average growth rate over each of the income ranges. Recall from Chapter 2 that we find the average growth rate using

$$\text{Average growth rate} = \frac{\text{Change in function value}}{\text{Change in independent variable}} = \frac{\text{Change in tax owed}}{\text{Change in taxable income}}$$

These calculations are shown in the following table:

Interval	**41,000 to 41,100**	**41,100 to 41,200**	**41,200 to 41,300**	**41,300 to 41,400**
Change in income	$41{,}100 - 41{,}000 = 100$	$41{,}200 - 41{,}100 = 100$	$41{,}300 - 41{,}200 = 100$	$41{,}400 - 41{,}300 = 100$
Change in tax owed	$4540 - 4528 = 12$	$4552 - 4540 = 12$	$4564 - 4552 = 12$	$4576 - 4564 = 12$
Average growth rate	$\frac{12}{100} = 0.12$	$\frac{12}{100} = 0.12$	$\frac{12}{100} = 0.12$	$\frac{12}{100} = 0.12$

[1] *The actual table shows taxable income in increments of $50 rather than $100.*

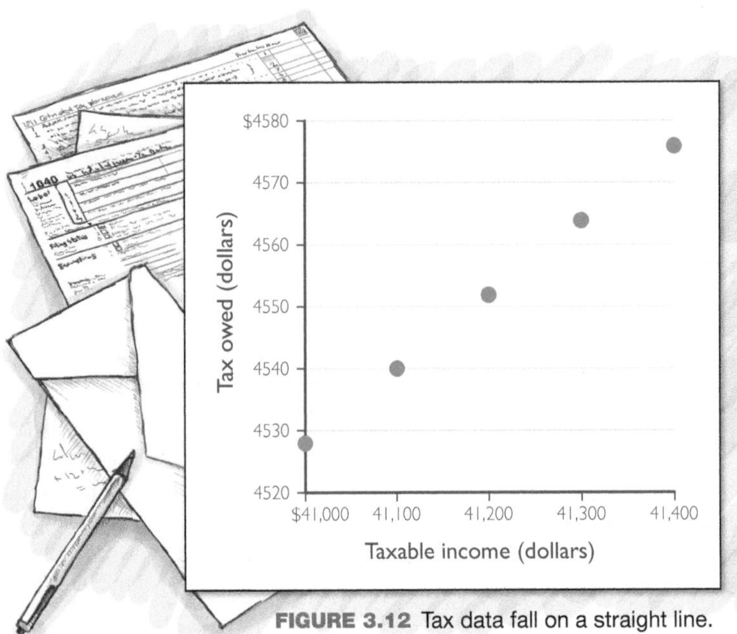

FIGURE 3.12 Tax data fall on a straight line.

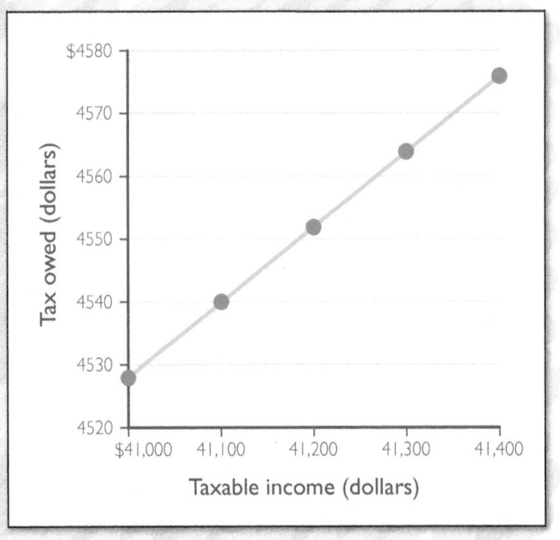

FIGURE 3.13 Line added to plot of tax data.

The fact that the average growth rate over each interval is the constant 0.12 tells us that the tax owed is a linear function of income with slope \$0.12 per dollar. (If we had obtained different values for the average growth rate over different intervals, we would have concluded that the data are not related by a linear formula.)

This slope is known as the *marginal tax rate*. It tells us the additional tax owed on each additional dollar of income. For example, if you have \$100 more in taxable income than someone else, your tax burden will be greater by $0.12 \times 100 = 12$ dollars. Let T be the tax owed in dollars, and let I be the taxable income (also in dollars) in excess of \$41,000. The tax owed when $I = 0$ is \$4528, and this is the initial value. Therefore,

$$T = \text{Slope} \times I + \text{Initial value}$$
$$T = 0.12I + 4528$$

In **Figure 3.13**, we have added the graph of the line to the plot of the tax data. The graph passes through each of the data points, which verifies the validity of our formula.

The linear function fits the tax data exactly. Many data sets are not perfectly linear like this, but they may be fairly well approximated by a linear function. And we may be able to get useful information from such an approximation. This is especially true when we have good reason to suspect that the relationship is approximately linear.

As a case in point, let's return to the example of running speed versus length discussed in Section 2.2. We recall the following table, which shows the running speed of various animals versus their length:[2]

Animal	Length (inches)	Speed (feet per second)
Deer mouse	3.5	8.2
Chipmunk	6.3	15.7
Desert crested lizard	9.4	24.0
Grey squirrel	9.8	24.9
Red fox	24.0	65.6
Cheetah	47.0	95.1

Figure 3.14 shows the scatterplot. We note that the points do not fall on a straight line, so the data in the table are not exactly linear. Nevertheless, the points seem to fall *nearly* on a straight line, so it may make sense to *approximate* the data using a linear function. Statisticians, scientists, and engineers frequently use a line called a *regression line* or *trend line* to make such approximations.

KEY CONCEPT

Given a set of data points, the regression line or trend line is a line that comes as close as possible (in a certain sense) to fitting those data.

The procedure for calculating a trend line for given data is called *linear regression.* An explanation of this procedure is outside the scope of this book, but linear regression is programmed into most calculators and commercial spreadsheets. All one has to do is enter the data.

In **Figure 3.15**, we have added the trend line produced by the spreadsheet program Excel©. We see that the points lie close to the line, so it appears to be a reasonable approximation to the data. Note that the spreadsheet gives both the graph and the linear formula.

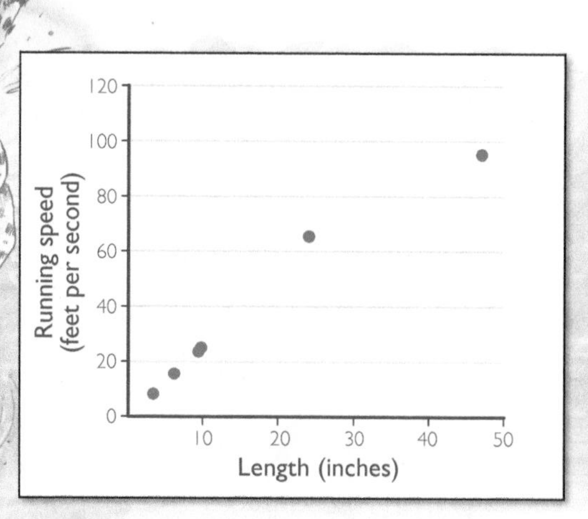

FIGURE 3.14 Scatterplot of running speed versus length.

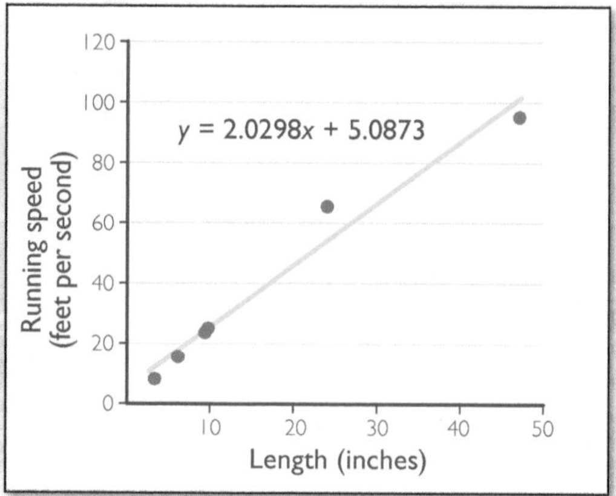

FIGURE 3.15 Trend line added.

The equation of the trend line is $y = 2.03x + 5.09$. (We have rounded the initial value and the slope to two decimal places.) This means that running speed S in feet per second can be closely estimated by

$$S = 2.03L + 5.09$$

---

[2] *This table is adapted from J. T. Bonner,* Size and Cycle *(Princeton, NJ: Princeton University Press, 1965).*

where L is the length measured in inches. We can use this formula to make estimates about animals not shown in the table. For example, we would expect a 20-inch-long animal to run

$$S = 2.03 \times 20 + 5.09 = 45.69 \text{ feet per second}$$

or about 45.7 feet per second. The most important bit of information we obtain from the trend line is its slope. In this case, that is 2.03 feet per second per inch. This value for the slope means that, in general, an animal that is 1 inch longer than another would be expected to run about 2.03 feet per second faster.

We can also gain information by studying the position of the data points relative to the trend line. For example, in Figure 3.15, the point for the red fox is above the trend line, so the red fox is faster than one might expect for an animal of its size.

EXAMPLE 3.7 Meaning of the equation for the trend line: Home runs

Figure 3.16 shows a scatterplot of the average number of home runs per game in Major League Baseball from 2000 through 2020. In **Figure 3.17**, we have added the trend line along with its equation. Identify the slope of the trend line, and explain in practical terms its meaning.

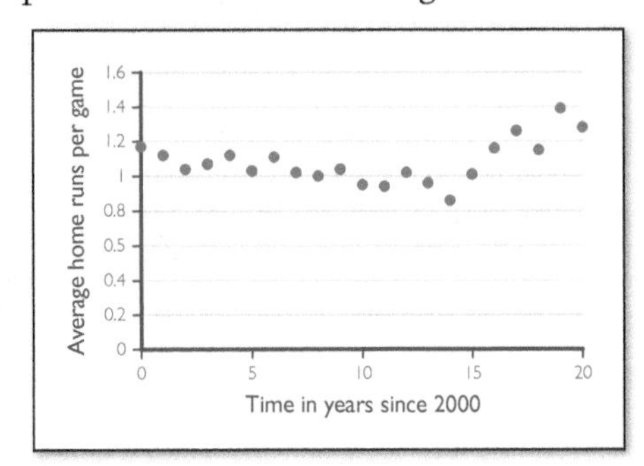

FIGURE 3.16 Average number of home runs per game.

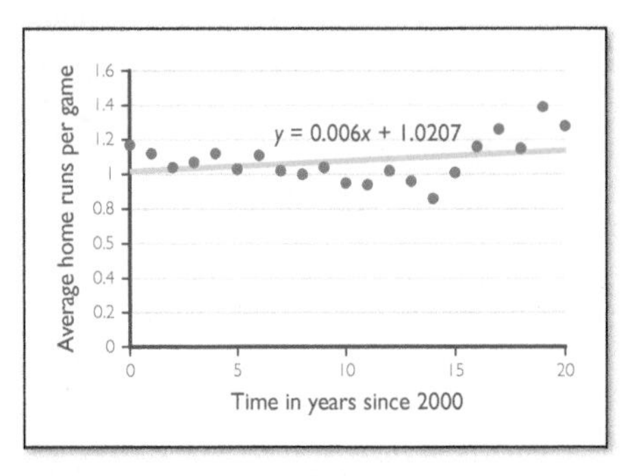

FIGURE 3.17 Trend line and equation added.

SOLUTION

The slope of the trend line is 0.006 home runs per game per year. The fact that the slope is positive means that the average number of home runs per game increased from 2000 through 2020. A good way to express the slope is to say that the average number of home runs increased by 6 per thousand games per year.

TRY IT YOURSELF 3.7

Suppose the trend line for walks per 500 plate appearances is $W = 0.2t + 10.3$, where W is the number of walks per 500 plate appearances and t is the time in years since 1919. Explain in practical terms the meaning of the slope.

The answer is provided at the end of this section.

It is important to note that the trend line in Figure 3.17 adds clarity to the scatterplot in Figure 3.16.

WHAT DO YOU THINK?

Is it linear? You see a news report about taxes, inflation, or other economic issues and wonder if you are seeing linear growth. How would you make a judgment about this question? What factors might lead you to believe you are seeing linear growth, and what might lead you to believe you are not seeing linear growth?

Questions of this sort may invite many correct responses. We present one possibility: The key characteristic of linear growth is constant rate of change. If the increase is the same from year to year, then the growth is linear, and so short-term predictions are practical. In contrast, if there was a large increase last year but a small increase this year, then growth is not linear, and accurate predictions may be more difficult to make. In practical settings growth is rarely exactly linear. But as long as the increases from year to year are nearly the same, a linear model may be appropriate.

Further questions of this type may be found in the *What Do You Think?* section of exercises.

Try It Yourself answers

Try It Yourself 3.1: Determining what is linear and what is not: Distance and area

a. The growth rate for distance is the increase in distance measured in centimeters for each 1-inch increase in distance. That increase is 2.54 centimeters. The distance in centimeters is a linear function of the distance in inches.

b. The growth rate for the area is the increase in area for each unit increase in the length of a side. The area of a square is not a linear function of the length of a side.

Try It Yourself 3.2: Interpreting formulas: Athletic records The initial value of 3.3 meters is the (approximate) height of the winning pole vault in the 1900 Olympic Games. The growth rate of 0.20 meter per Olympic game means that the height of the winning pole vault increased by approximately 0.20 meter from one game to the next.

Try It Yourself 3.3: Finding linear formulas: Distance from Earth Let d denote the distance from Earth in kilometers after t hours. Then d is a linear function of t because the growth rate (the velocity) is constant. The initial value is 1,000,000 kilometers, and the growth rate is −2000 kilometers per hour (negative because the object is traveling toward Earth, so the distance is decreasing). The formula is $d = -2000t + 1{,}000{,}000$.

Try It Yourself 3.4: Calculating the slope: Ice depth 0.6 foot per year. The depth increased by 0.6 foot each year.

Try It Yourself 3.5: Using the slope: Temperature and speed of sound 3.3 feet per second.

Try It Yourself 3.7: Meaning of the equation for the trend line: Walks Each year the number of walks per 500 plate appearances increases, on average, by 0.2.

Exercise Set 3.1

Test Your Understanding

1. If a quantity is growing as a linear function with a constant rate of change of m per year, how do you use this year's amount to find next year's amount?

2. A function is linear if: **a.** it has a constant rate of change **b.** its graph is a straight line **c.** its formula is $y = \text{Growth rate} \times x + \text{Initial value}$ **d.** all of the above.

3. The slope or growth rate of a linear function is calculated as ________ divided by ________.

4. True or false: Data may be appropriately modeled by a linear function if the points in a data plot lie on a straight line.

5. True or false: A proper use of units can help us to determine the meaning of the slope of a linear function.

6. True or false: The most important information from the trend line is the initial value.

Problems

Note: In some exercises, you are asked to find a formula for a linear function. If no letters are assigned to quantities in an exercise, you are expected to choose appropriate letters and also give appropriate units.

7. Interpreting formulas. The height of a young flower is increasing in a linear fashion. Its height t weeks after the first of this month is given by $H = 2.1t + 5$ millimeters. Identify the initial value and growth rate, and explain in practical terms their meaning.

8. Interpreting formulas. I am driving west from Nashville at a constant speed. My distance west from Nashville t hours after noon is given by $D = 65t + 30$ miles. Identify the initial value and growth rate, and explain in practical terms their meaning.

9. Finding linear formulas. A child is 40 inches tall at age 6. For the next few years, the child grows by 2 inches per year. Explain why the child's height after age 6 is a linear function of his age. Identify the growth rate and initial value. Using t for time in years since age 6 and H for height in inches, find a formula for H as a linear function of t.

10. Finding linear formulas. The temperature at 6:00 A.M. on a certain summer day is 75 degrees. Over the morning the temperature increases by 1 degree per hour. Explain why temperature over the morning is a linear function of time since 6:00 A.M. Identify the slope and initial value. Using t for time in hours since 6:00 A.M. and T for temperature, find a formula for T as a linear function of t.

11. Calculating the slope. Your tire air pressure is decreasing at a constant rate so that the air pressure is a linear function of time. Initially there is 34 pounds per square inch of air pressure. Two hours later there is 28 pounds per square inch of air pressure. What is the slope of the linear function that is air pressure?

12. Using the slope. The depth D, in inches, of snow in my yard t hours after it started snowing this morning is given by $D = 1.5t + 4$. If the depth of the snow is 7 inches now, what will be the depth one hour from now?

13–24. Linear or not? *In Exercises 13 through 24, you are asked to determine whether the relationship described represents a function that is linear. In each case, explain your reasoning. If the relationship can be represented by a linear function, find an appropriate linear formula, and explain in practical terms the meaning of the growth rate.*

13. Is the relationship between the height of a staircase and the number of steps linear?

14. A cable TV company charges a flat rate of \$10 per month plus \$4.99 for each on-demand movie ordered. Explain why the amount of money you pay in a month is a linear function of the number of on-demand movies you order.

15. A company called Blurb (www.blurb.com/volume-printing) prints books. According to its advertisement, you get a 15% discount on all orders of 51 or more copies of the same book (up to 100 copies). Is the cost per order a linear function of the number of copies you order (up to 100)?

16. Suppose that I wrote 150 words of my English paper yesterday and today I begin typing at a rate of 35 words per minute. Is the total number of words typed a linear function of the number of minutes since I began typing today?

17. My savings account pays 2% per year, and the interest is compounded yearly. That is, each year accrued interest is added to my account balance. Is the amount of money in my savings account a linear function of the number of years since it was opened? *Suggestion*: Compare the interest earned the first year with the interest earned the second year.

18. If we drop a rock near the surface of Earth, its velocity increases by 32 feet per second for each second the rock falls. Is the velocity of the rock a linear function of the time since it was dropped?

19. If we drop a rock near the surface of Earth, physics tells us that the distance in feet that it falls is given by $16t^2$, where t is the number of seconds since the rock was dropped. Is the distance a rock falls a linear function of the time since it was dropped? *Suggestion*: Compare the distance fallen over the first second to that fallen over the next second.

20. In order to ship items I bought on eBay, I pay a delivery company a flat fee of \$10, plus \$1.50 per pound. Is the shipping cost a linear function of the number of pounds shipped?

21. My car gets 25 miles per gallon at 60 miles per hour. Is the number of miles I can drive at 60 miles per hour a linear function of the number of gallons of gas in the tank?

22. A certain type of bacterium reproduces by dividing each hour. If we start with just one bacterium, is the number of bacteria present a linear function of time?

23. On a typical day, the temperature increases as the sun rises. It is hottest around midday, and then things cool down toward evening. Is the relationship between temperature and time of day linear?

24. The method of calculating the compensation paid to retired teachers in the state of Oklahoma is as follows.[3] Yearly retirement income is 0.02 times the number of years of service times the average of the last three years of salary. Is retirement income for Oklahoma teachers a linear function of the number of years of service?

25. Income tax. One federal tax schedule states that if you are single and your taxable income is between \$37,450 and \$90,750, then you owe \$5156.25 plus 25% of your taxable income over \$37,450. Is the tax you owe a linear function of your taxable income if the taxable income is between \$37,450 and \$90,750? If so, find a formula giving tax owed T as a linear function of taxable income I over \$37,450. Both T and I are measured in dollars.

26. Working on commission. A certain man works in sales and earns a base salary of \$1000 per month, plus 5% of his total sales for the month. Is his monthly income a linear function of his monthly sales? If so, find a linear formula that gives monthly income in terms of sales.

27. Driving up a hill. Suppose that you are at the base of a hill and see a sign that reads "Elevation 3500 Feet." The road you are on goes straight up the hill to the top, which is 3 horizontal miles from the base. At the top, you see a sign that reads "Elevation 4100 Feet." What is the growth rate in your elevation with respect to horizontal distance as you drive up the road? Use E for elevation in feet and h for horizontal distance in miles, and find a formula that gives your elevation as a linear function of your horizontal distance from the base of the hill.

28. Price of Amazon's Kindle. Figure 3.18 shows how the price of Amazon's Kindle has decreased over time. The green line is a trend line. (You can check that the hash marks on the horizontal axis are one-month intervals.)

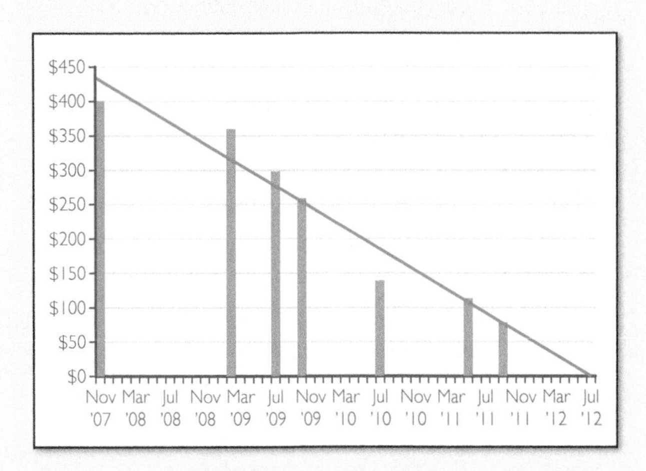

FIGURE 3.18 Price of the Kindle. **Source:** Company Filings (Sept 2011). Reprinted by permission of Business Insiders, Inc.

[3] *The actual calculation is somewhat more complicated.*

a. The trend line estimates that the price was \$432 in November 2007 and \$80 in September 2011. (That is an interval of 46 months.) Use this information to determine the slope of the trend line. Round your answer in dollars per month to two decimal places. Be careful about the sign.

b. Use your answer to part a to determine by how much the price of a Kindle should decrease every six months.

c. Use the information in part a to find a formula for the price, in dollars, as a linear function of the time in months since November 2007.

d. Use your answer to part c to determine the prediction of the trend line for the price in July 2012 (56 months after November 2007). What do you conclude from this unreasonable result?

29. Speed of sound in the ocean. Measuring the speed of sound in the ocean is an important part of marine research.[4] One application is the study of climate change. The speed of sound depends on the temperature, salinity, and depth below the surface. For a fixed temperature of 25 degrees Celsius and salinity of 35 parts per thousand, the speed of sound is a function of the depth. At the surface, the speed of sound is 1534 meters per second. For each increase in depth by 1 kilometer, the speed of sound increases by 17 meters per second.

a. Explain why the function expressing the speed of sound in terms of depth is linear.

b. Identify the slope and initial value of the linear function that gives the speed of sound in terms of depth. Explain in practical terms what each means.

c. What increase in the speed of sound is caused by a 2-kilometer increase in depth?

d. Use your answer to part c to determine the speed of sound when the depth is 2 kilometers.

e. Use D for depth (in kilometers) and S for the speed of sound (in meters per second), and find a linear formula for S as a function of D.

f. Use your formula from part e to calculate the speed of sound when the depth is 2 kilometers.

30. Speed on a curve. On rural highways, the average speed S (in miles per hour) is related to the amount of curvature C (in degrees) of the road.[5] On a straight road ($C = 0$), the average speed is 46.26 miles per hour. This decreases by 0.746 mile per hour for each additional degree of curvature.

a. Explain why the function expressing S in terms of C is linear.

b. Find the slope of the linear function expressing S in terms of C. Explain in practical terms its meaning. *Suggestion*: Be careful about the sign of the slope.

c. Find a formula expressing S as a linear function of C.

d. What is the average speed if the degree of curvature is 15 degrees?

31. The IRS. In 2003, the Internal Revenue Service collected about 1.952 trillion dollars. Based on the trend line for the data from 2003 through 2008, collections increased by (approximately) 178 billion (or 0.178 trillion) dollars per year. In this exercise, we assume that collections increased by *exactly* 178 billion dollars per year from 2003 through 2008.

a. Explain how the preceding assumption would show that the function expressing tax collections in terms of time from 2003 through 2008 is linear. Why do we have to limit the time interval to 2003 through 2008?

b. Find a formula for tax collections as a linear function of time from 2003 through 2008.

c. What would you estimate for the collections in 2006?

32. A real estate agency. Suppose that a real estate agency makes its money by taking a percentage commission of total sales. It also has fixed costs of \$15,257 per month associated with rent, staff salaries, utilities, and supplies. In the month of August, total sales were \$832,000 and net income (after paying fixed costs) was \$9703. Because each dollar increase in sales results in the same increase (a certain percentage of that dollar) in net income, net income is a linear function of total sales. Find a linear formula that gives net income in terms of total sales. What percentage commission on sales is this company charging?

33. Meaning of marginal tax rate. Recall from the discussion on page 149 that the marginal tax rate is the additional tax you expect to pay for each additional dollar of taxable income. The marginal tax rate is usually expressed as a percentage. In the example, the marginal tax rate was 12%. Does that mean that your total tax is 12% of your taxable income? Specifically, if your taxable income is \$41,300, is your total tax 12% of \$41,300? Explain why or why not.

34. More on tax tables. The following table shows the federal income tax owed by a single taxpayer for the given level of taxable income for 2020. Both are measured in dollars.

Taxable income	97,000	97,050	97,100	97,150	97,200	97,250	97,300
Tax owed	17,366	17,378	17,390	17,402	17,414	17,426	17,438

a. Show that, over the range of taxable income shown in the table, the tax owed is a linear function of taxable income.

b. What tax do you owe if you have a taxable income of \$97,000?

c. How much additional tax is due on each dollar of taxable income over \$97,000?

d. Let A denote the amount (in dollars) of your taxable income in excess of \$97,000 and T the tax (in dollars) you owe. Find a linear formula[6] that gives T in terms of A.

35. The flu. The following table shows the total number of patients diagnosed with the flu in terms of the time in days since the outbreak started:

Time in days	0	5	10	15	20	25
Number of flu patients	35	41	47	53	59	65

[4] *For further background, see* www.dosits.org.

[5] *A. Taragin, "Driver Performance on Horizontal Curves,"* Proceedings of the Highway Research Board 33 *(Washington, DC: Highway Research Board, 1954), 446–466.*

[6] *The formula does not apply for taxable incomes above \$100,000.*

a. Show that the function giving the number of diagnosed flu cases in terms of time is linear.

b. Find the slope of the linear function from part a. Explain in practical terms the meaning of the slope.

c. Find a formula for the linear function in part a.

d. What would you expect to be the total number of diagnosed flu cases after 17 days?

36. High school graduates. The following table shows the number (in thousands) graduating from public high schools in the United States in the given year.

1980	1981	1982	1983	1984	1985
2747.7	2725.3	2704.8	2597.6	2494.8	2414.0
1986	**1987**	**1988**	**1989**	**1990**	**1991**
2382.6	2428.8	2500.2	2458.8	2320.3	2234.9
1992	**1993**	**1994**	**1995**	**1996**	**1997**
2226.0	2233.2	2220.8	2273.5	2273.1	2358.4
1998	**1999**	**2000**	**2001**	**2002**	**2003**
2439.1	2485.6	2553.8	2569.0	2621.5	2719.9
2004	**2005**	**2006**	**2007**	**2008**	
2753.4	2799.3	2815.5	2892.4	2999.5	

The scatterplot of the data is shown in **Figure 3.19.** In **Figure 3.20,** the trend line has been added. Does the trend line appear to offer an appropriate way to analyze the data? Explain your reasoning.

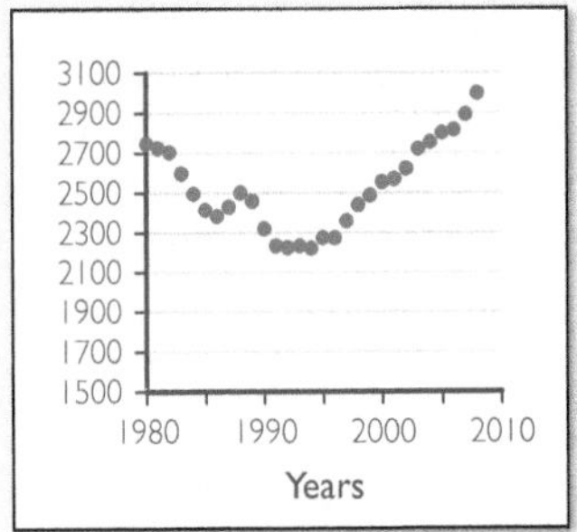

FIGURE 3.19 Scatterplot of public high school graduates.

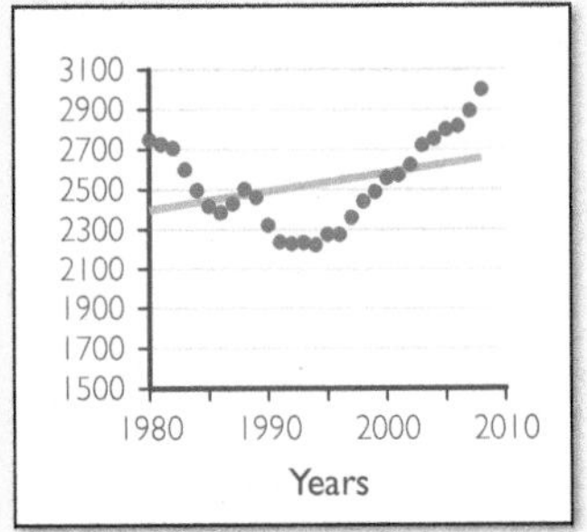

FIGURE 3.20 Trend line added.

37–39. Consumer Price Index. *The Consumer Price Index (CPI) is calculated by the U.S. Department of Labor. It is a measure of the relative price of a typical market basket of goods in the given year. The percentage increase in the CPI from one year to the next gives one measure of the annual rate of inflation. The following table shows the CPI in December of the given year. Here, C represents the Consumer Price Index, and t is the time in years since December 2013.*

t	0	1	2	3	4	5	6	7
C	233.0	234.8	236.5	241.4	246.5	251.2	257.0	260.5

*In **Figure 3.21**, we have plotted the data, and in **Figure 3.22**, we have added the trend line, which is given by* $C = 4.20t + 230.42$. *This information is used in Exercises 37 through 39.*

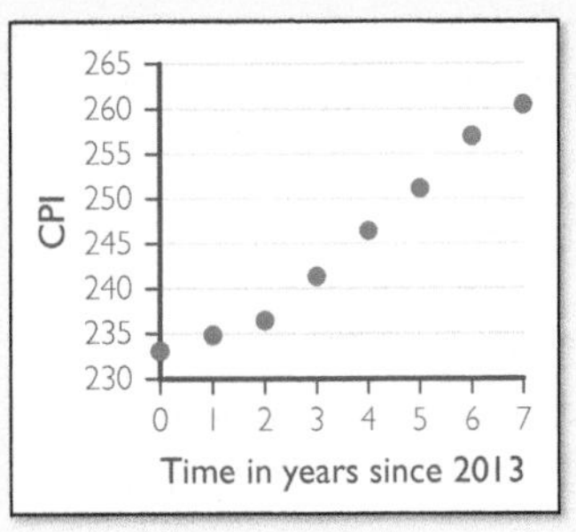

FIGURE 3.21 Plot of consumer price index.

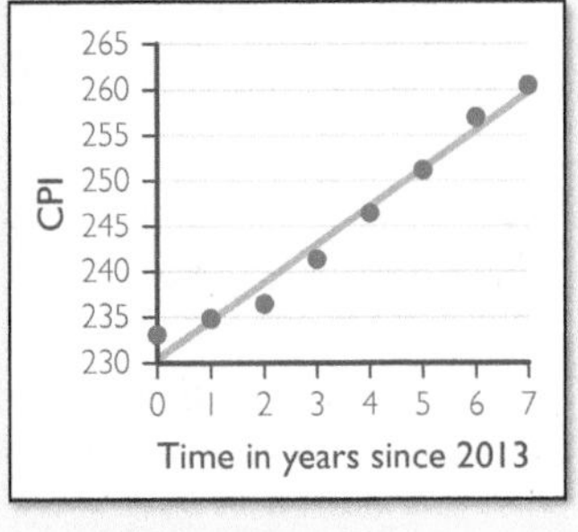

FIGURE 3.22 Plot of consumer price index with trend line added.

37. Determine the slope of the trend line and explain in practical terms the meaning of the slope.

38. Which year after 2013 shows a significantly larger CPI than would be expected from the trend line? Interpret your answer.

39. What CPI does the trend line predict for December 2021? Round your answer to one decimal place.

Exercises 40 and 41 are suitable for group work.

40. A closer look at tax tables. The table below gives selected entries from the 2014 federal income tax table. It applies to a married couple filing jointly. Taxable income and tax owed are both in dollars.

Taxable income	73,300	73,500	73,700	73,900	74,100	74,300
Tax owed	10,091	10,121	10,151	10,194	10,244	10,294
Taxable income	74,500	74,700	74,900	75,100	75,300	75,500
Tax owed	10,344	10,394	10,444	10,494	10,544	10,594

a. Make a table that shows the additional tax owed over each income span.

b. Your table in part a should show that the data are not linear over the entire income range. But it should also show that the data are linear over each of two smaller ranges. Identify these ranges. *Note*: This is an example of a *piecewise-linear function*. The term just means that two linear functions are pasted together.

c. For each of the two ranges you identified in part b, calculate the marginal tax rate.

41. The table below gives selected entries from the 2015 federal income tax table. It applies to a married couple filing jointly. Taxable income and tax owed are both in dollars.

Taxable income	73,300	73,500	73,700	73,900	74,100	74,300
Tax owed	10,076	10,106	10,136	10,166	10,196	10,226
Taxable income	74,500	74,700	74,900	75,100	75,300	75,500
Tax owed	10,256	10,286	10,319	10,369	10,419	10,469

Repeat Exercise 40 for this table. (See also Exercise 46.)

42–45. Life expectancy. *The following table shows the average life expectancy, in years, of a child born in the given year:*

Year	1995	2000	2005	2010	2015	2020
Life expectancy	75.8	76.8	77.4	78.7	78.7	77.8

If t denotes the time in years since 1995 and E the life expectancy in years, then it turns out that the trend line for these data is given by

$$E = 0.10t + 76.32$$

The data and the trend line are shown in **Figure 3.23**. *This information is used in Exercises 42 through 45.*

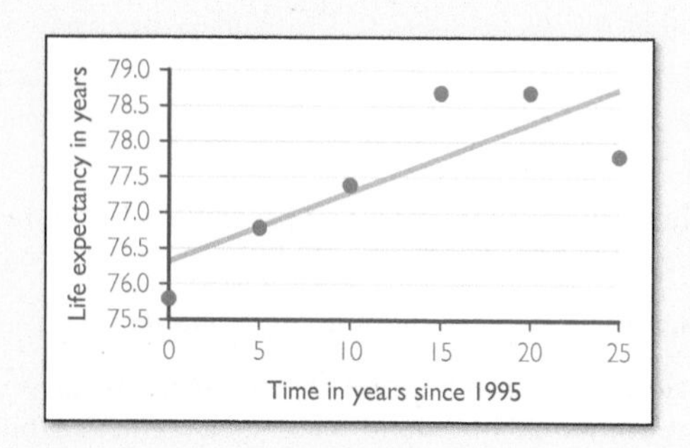

FIGURE 3.23 Life expectancy by year of birth.

42. What is the slope of the trend line, and what is its meaning in practical terms?

43. During which year after 1995 was life expectancy lower than would have been expected from the linear trend? Explain.

44. If the linear trend established by these data persisted through 2025, what would be the life expectancy of a child born in 2025? (Round your answer to one decimal place.)

45. If the linear trend established by these data persisted through 2300, what would be the life expectancy of a child born in 2300? (Round your answer to one decimal place.) Does this age seem reasonable to you?

Writing About Mathematics

46. Bracket creep. First complete Exercises 40 and 41. In those exercises, consider the income level where the transition from one marginal tax rate to the next takes place. Examine tax tables from other years to find how this level changed from year to year, and report your findings. How does the term *bracket creep* come up in your investigation?

47. In the news. Find a news article that incorporates trend lines into its presentation. Write a report summarizing the article and explaining how the trend line is used.

48. Malthus. Write a report on the life of Thomas Robert Malthus and the influence of his ideas.

What Do You Think?

49. Doubling. For some linear relationships, doubling the independent variable results in doubling the function value. For example, doubling the number of steps in a staircase doubles the height of the staircase. Does this doubling property hold for every linear relationship? If so, explain why. If not, give an example of a linear function for which this property does not hold. Try to determine all linear functions for which this property holds. **Exercise 13 is relevant to this topic.**

50. Is it linear? You see an article about an interesting biological phenomenon. The article is accompanied by a graph. What properties of the graph might lead you to believe that the phenomenon is expressed using a linear relationship? What properties of the graph might lead you to believe that the phenomenon is not expressed using a linear relationship?

51. When does it apply? There are many settings in which linear functions model relationships over limited ranges only. For example, a factory owner says he has determined that the number of units U produced is related to the number W of workers by the linear function $U = 3W + 2$. Is this relationship reasonable if there are no workers? Use the Internet to investigate other settings where linear functions apply over limited ranges only. *Suggestion*: There are examples of this sort in this section. **Exercise 28 is relevant to this topic.**

52. Marginal tax rate. Recall that the *marginal tax rate* is the slope of the linear function giving your tax as a function of your taxable income. Often the marginal tax rate is expressed as a percentage. For example, a slope of \$0.15 per dollar corresponds to a marginal tax rate of 15%. Does this fact mean that if your taxable income is \$100, then your tax will be 15% of \$100, or \$15? If not, explain carefully how this percentage can be used to describe how your tax depends on your taxable income. **Exercise 34 is relevant to this topic.**

53. Appropriate use of trend lines. A trend line can be found for *any* set of numerical data points. What are some ways to determine whether the trend line offers an appropriate way to analyze the data? **Exercise 36 is relevant to this topic.**

54. Trend lines. Explain how an appropriately used trend line can help us understand data. **Exercise 38 is relevant to this topic.**

55. Piecewise linear functions. Sometimes a linear growth rate changes abruptly to a larger or smaller linear growth rate. When this happens, the graph takes the form of a series of straight lines pasted together, as shown in **Figure 3.24**. Find

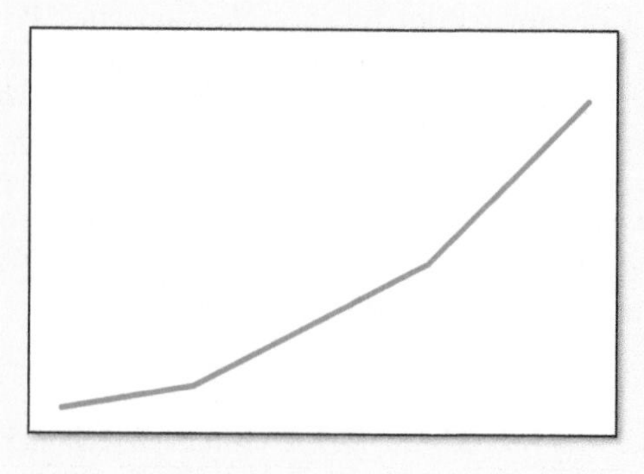

FIGURE 3.24 The graph of a piecewise-linear function.

examples of common relationships that are piecewise-linear. *Suggestion*: You might start by looking at federal income tax rates or postal rates.

56. Is it linear? Look at historical records of tuition rates at your university. Determine whether it is reasonable to model these rates by a linear function. If so, propose one.

57. Positive, negative, or zero. Suppose you are looking at three separate relationships that are modeled by linear functions. One linear model has a positive slope, one has a negative slope, and one has a slope of zero. What do these slopes tell you about the relationships they model? Find practical examples of each of the three.

3.2 Exponential growth and decay: Constant percentage rates

TAKE AWAY FROM THIS SECTION
Understand exponential functions and consequences of constant percentage change.

The following article is from the *Seattle Times*.

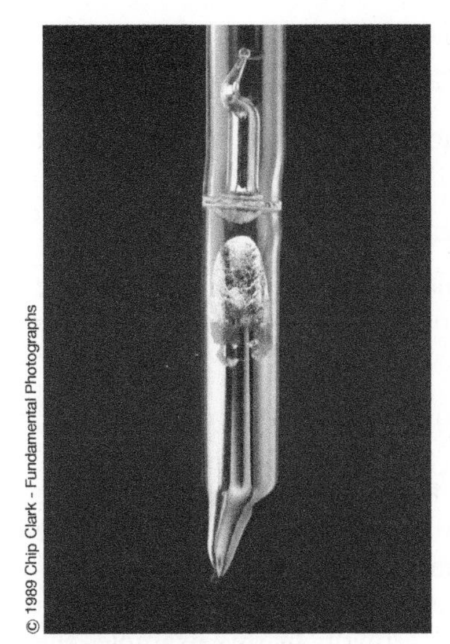

Cesium is liquid at room temperature and highly reactive, rapidly oxidizing in air.

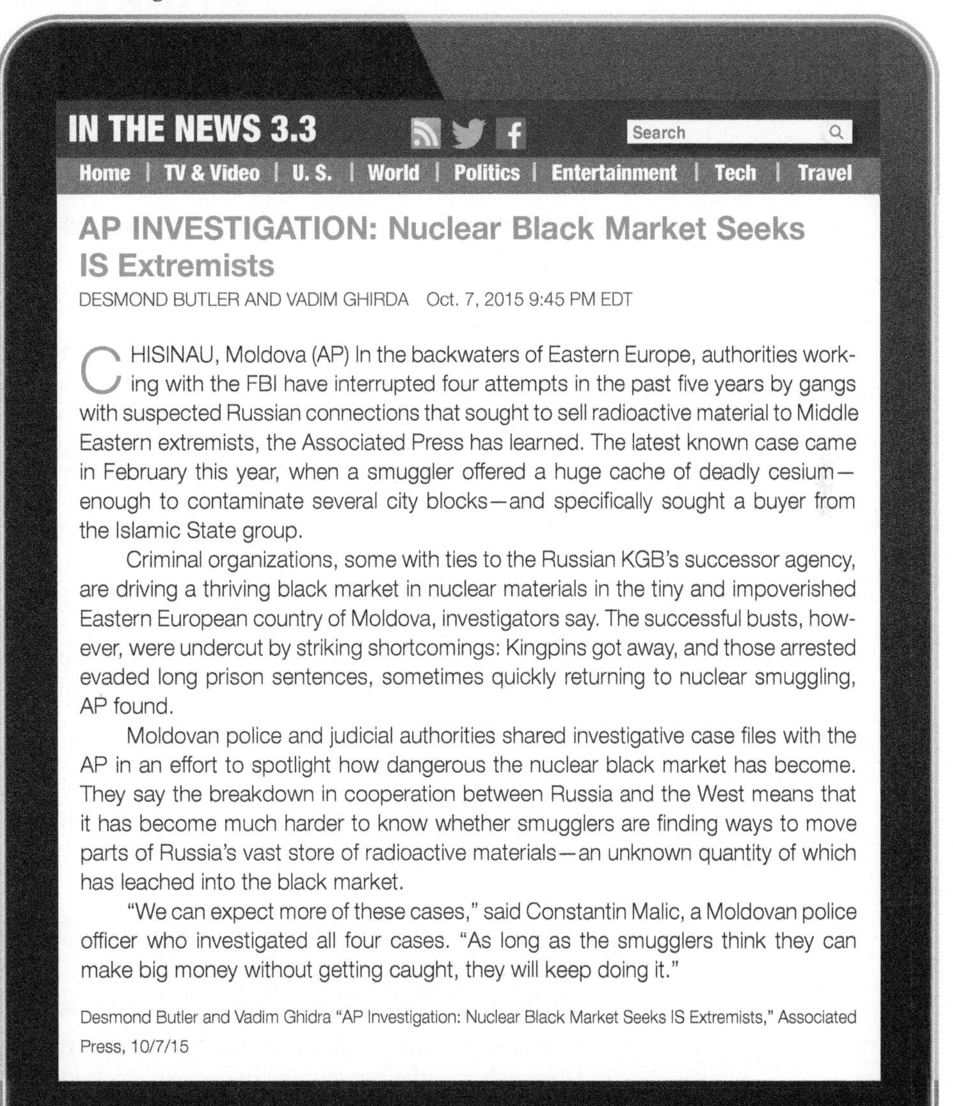

IN THE NEWS 3.3

Search

Home | TV & Video | U. S. | World | Politics | Entertainment | Tech | Travel

AP INVESTIGATION: Nuclear Black Market Seeks IS Extremists

DESMOND BUTLER AND VADIM GHIRDA Oct. 7, 2015 9:45 PM EDT

CHISINAU, Moldova (AP) In the backwaters of Eastern Europe, authorities working with the FBI have interrupted four attempts in the past five years by gangs with suspected Russian connections that sought to sell radioactive material to Middle Eastern extremists, the Associated Press has learned. The latest known case came in February this year, when a smuggler offered a huge cache of deadly cesium—enough to contaminate several city blocks—and specifically sought a buyer from the Islamic State group.

Criminal organizations, some with ties to the Russian KGB's successor agency, are driving a thriving black market in nuclear materials in the tiny and impoverished Eastern European country of Moldova, investigators say. The successful busts, however, were undercut by striking shortcomings: Kingpins got away, and those arrested evaded long prison sentences, sometimes quickly returning to nuclear smuggling, AP found.

Moldovan police and judicial authorities shared investigative case files with the AP in an effort to spotlight how dangerous the nuclear black market has become. They say the breakdown in cooperation between Russia and the West means that it has become much harder to know whether smugglers are finding ways to move parts of Russia's vast store of radioactive materials—an unknown quantity of which has leached into the black market.

"We can expect more of these cases," said Constantin Malic, a Moldovan police officer who investigated all four cases. "As long as the smugglers think they can make big money without getting caught, they will keep doing it."

Desmond Butler and Vadim Ghidra "AP Investigation: Nuclear Black Market Seeks IS Extremists," Associated Press, 10/7/15

Dangerous radioactive materials, such as those referred to in the article, will eventually decay into relatively harmless matter. The problem is that this process takes an extremely long time because the way in which these materials decay is exponential, not linear. We will return to this issue later in this section.

Recall that linear change occurs at a constant rate. Exponential change, in contrast, is characterized by a constant *percentage* rate. The rates may be positive or negative. In certain idealized situations, a population of living organisms, such as bacteria, grows at a constant percentage rate. Radioactive substances provide an example of a negative rate because the amount decays rather than grows.

The nature of exponential growth

To illustrate the nature of exponential growth, we consider the simple example of bacteria that reproduce by division. Suppose there are initially 2000 bacteria in a petri dish and the population doubles every hour. We refer to 2000 as the *initial value* because it is the amount with which we start. After one hour the population doubles, so

$$\begin{aligned}\text{Population after 1 hour} &= 2 \times \text{Initial population}\\ &= 2 \times 2000 = 4000 \text{ bacteria}\end{aligned}$$

After one more hour the population doubles again, so

$$\begin{aligned}\text{Population after 2 hours} &= 2 \times \text{Population after 1 hour}\\ &= 2 \times 4000 = 8000 \text{ bacteria}\end{aligned}$$

In general, we find the population for the next hour by multiplying the current population by 2:

$$\text{Population next hour} = 2 \times \text{Current population}$$

Let's find the percentage change in the population. Over the first hour the population grows from 2000 to 4000, an increase of 2000. The percentage change is then

$$\text{Percentage change} = \frac{\text{Change in population}}{\text{Previous population}} \times 100\% = \frac{2000}{2000} \times 100\% = 100\%$$

Over the second hour the population grows from 4000 to 8000, another increase of 100%. In fact, doubling always corresponds to a 100% increase. Thus, the population shows a 100% increase each hour. This means that the population grows not by a constant number but by a constant *percentage*, namely 100%. Growth by a constant percentage characterizes exponential growth. In this example, the population size is an exponential function of time.

KEY CONCEPT

An exponential function is a function that changes at a constant percentage rate.

EXAMPLE 3.8 Determining constant percentage growth: Tripling and population size

If a population triples each hour, does this represent constant percentage growth? If so, what is the percentage increase each hour? Is the population size an exponential function of time?

SOLUTION

The population changes each hour according to the formula

$$\text{Population next hour} = 3 \times \text{Current population}$$

To get a clear picture of what is happening, suppose we start with 100 individuals:

$$\text{Initial population} = 100$$
$$\text{Population after 1 hour} = 3 \times 100 = 300$$
$$\text{Population after 2 hours} = 3 \times 300 = 900$$

Let's look at this in terms of growth:

$$\text{Growth over first hour} = 300 - 100 = 200 = 200\% \text{ increase over } 100$$
$$\text{Growth over second hour} = 900 - 300 = 600 = 200\% \text{ increase over } 300$$

Multiplying by 3 results in a 200% increase. So the population is growing at a constant percentage rate, namely 200% each hour. Therefore, the population is an exponential function of time.

TRY IT YOURSELF 3.8

If a population quadruples each hour, does this represent constant percentage growth? If so, what is the percentage? Is the population size an exponential function of time?

The answer is provided at the end of this section.

Formula for exponential functions

Let's return to the example of an initial population of 2000 bacteria that doubles every hour. Some further calculations will show a simple but important pattern. If we let N denote the population t hours after we begin the experiment, then

$$\text{When } t = 0, N = 2000 = 2000 \times 2^0$$
$$\text{When } t = 1, N = 2000 \times 2 = 2000 \times 2^1$$
$$\text{When } t = 2, N = 2000 \times 2 \times 2 = 2000 \times 2^2$$
$$\text{When } t = 3, N = 2000 \times 2 \times 2 \times 2 = 2000 \times 2^3$$
$$\text{When } t = 4, N = 2000 \times 2 \times 2 \times 2 \times 2 = 2000 \times 2^4$$

The pattern is evident. To calculate the population after t hours, we multiply the initial value 2000 by 2 raised to the power t:

$$N = 2000 \times 2^t$$

In **Figure 3.25**, we have plotted the data points we calculated for the population. In **Figure 3.26**, we have added the graph of $N = 2000 \times 2^t$. This curve passes through each of the data points. This gives further evidence that the formula we proposed is correct.

This formula allows us to make calculations that would be difficult or impossible if we didn't have it. For example, we can find the population four and a half hours after the experiment begins. We just put 4.5 in place of t:

$$N = 2000 \times 2^{4.5}$$

The result is 45,255 bacteria. (We rounded to the nearest whole number because we are counting bacteria, which don't occur as fractions.)

It is important to note the following two facts about this example:

- We find next hour's population by multiplying the current population by 2. The number 2 is called the *base* of this exponential function.
- The formula for the population is

$$N = \text{Initial value} \times \text{Base}^t$$

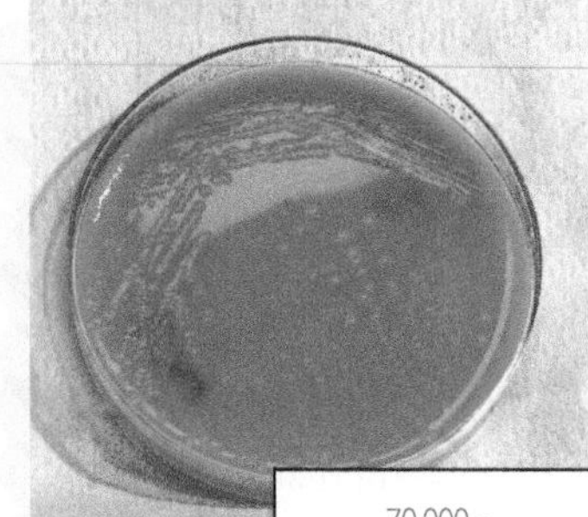

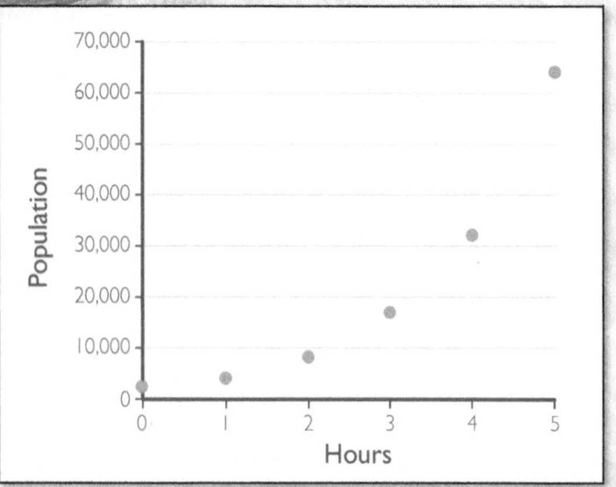

FIGURE 3.25 A population doubles each hour.

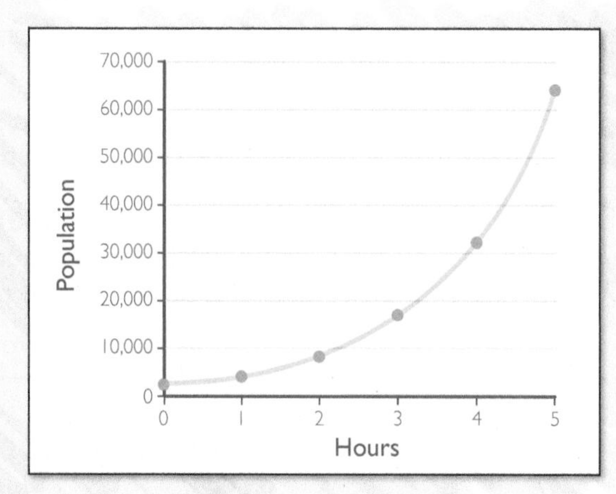

FIGURE 3.26 Adding the graph of the exponential function.

These observations are typical of exponential functions, as we summarize below.

SUMMARY 3.3
Exponential Formulas

The formula for an exponential function y of t is

$$y = \text{Initial value} \times \text{Base}^t$$

An exponential function y of t is characterized by the following property: When t increases by 1, to find the new value of y, we multiply the current value by the base. In symbols,

$$y\text{-value for } t+1 = \text{Base} \times y\text{-value for } t$$

EXAMPLE 3.9 Finding an exponential formula: An investment

The value of a certain investment grows according to the rule

$$\text{Next year's balance} = 1.07 \times \text{Current balance}$$

Find the percentage increase each year, and explain why the balance is an exponential function of time. Assume that the original investment is \$800. Find an exponential formula that gives the balance in terms of time. What is the balance after 10 years?

SOLUTION

We multiply this year's balance by 1.07 to get next year's balance. So next year's balance is 107% of this year's balance. That is an increase of 7% per year. Because the balance grows by the same percentage each year, it is an exponential function of time.

Let B denote the balance in dollars after t years. We use the formula

$$B = \text{Initial value} \times \text{Base}^t$$

The initial value is \$800 because that is the original investment. The base is 1.07 because that is the multiplier we use to find next year's balance. This gives the formula

$$B = 800 \times 1.07^t$$

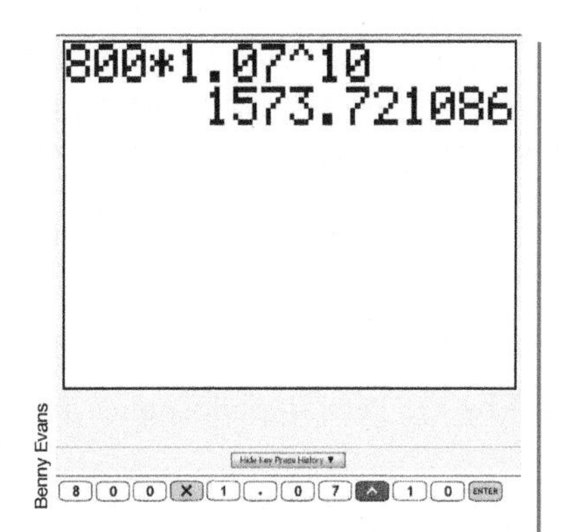

To find the balance after 10 years, we substitute $t = 10$ into this formula:

$$\text{Balance after } t \text{ years} = 800 \times 1.07^t$$

$$\text{Balance after 10 years} = 800 \times 1.07^{10} = \$1573.72$$

Figure 3.27 shows the graph of the balance in Example 3.9.

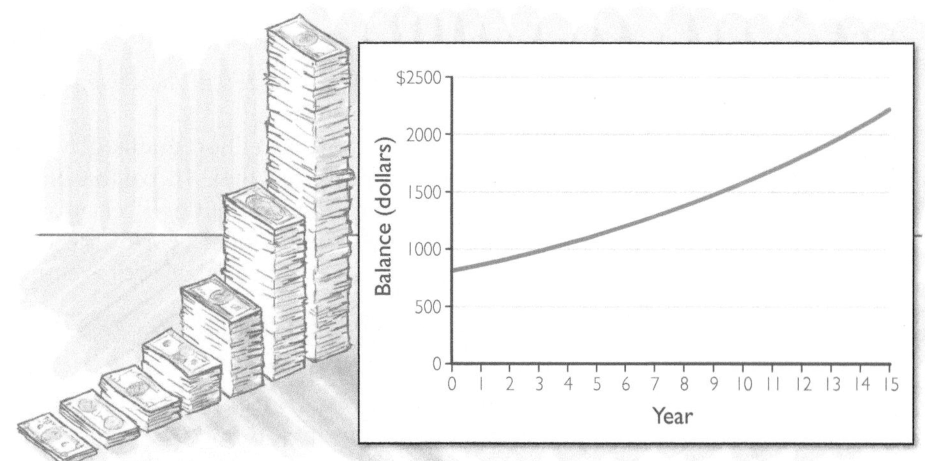

FIGURE 3.27 The balance of an investment.

TRY IT YOURSELF 3.9

An investment grows according to the rule

$$\text{Next month's balance} = 1.03 \times \text{Current balance}$$

Find the percentage increase each month. If the initial investment is \$450, find an exponential formula for the balance as a function of time. What is the balance after two years?

The answer is provided at the end of this section.

The rapidity of exponential growth

Let's look again at the graph of the bacteria population we examined earlier. The graph of $N = 2000 \times 2^t$ is in **Figure 3.28**. The shape of this graph is characteristic of graphs of exponential growth: It is increasing at an *increasing* rate—unlike linear graphs, which have a *constant* growth rate. The rate of growth is rather slow to begin with, but as we move to the right, the curve becomes very steep, showing more rapid growth. Referring to **Figure 3.29**, we see that the growth from the fifth to the sixth hour is almost the same as the growth over the first five hours.

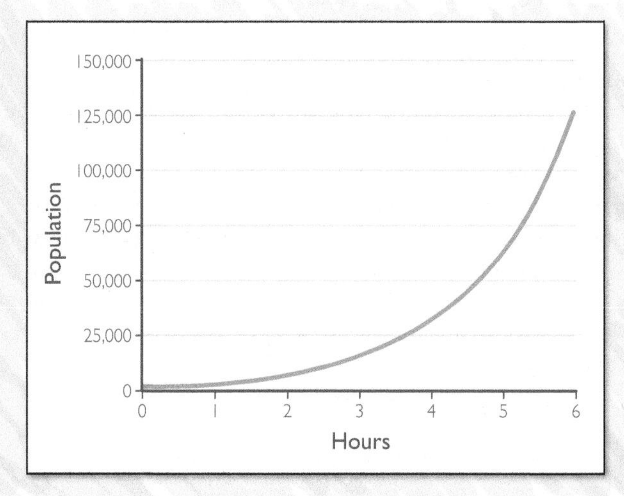

FIGURE 3.28 Graph of population growth.

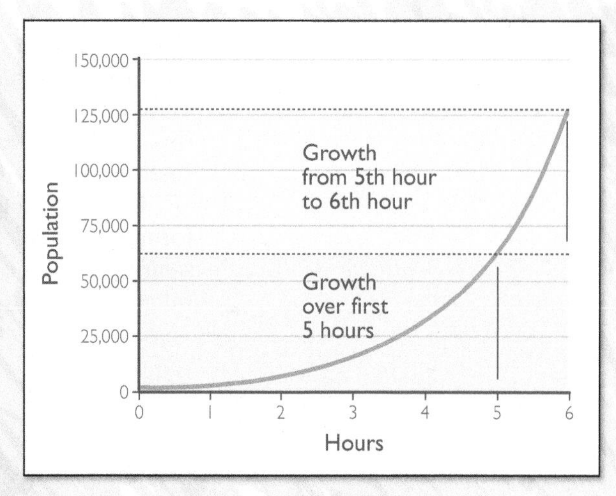

FIGURE 3.29 Comparing growth rates.

A useful illustration of the eventual high rate of exponential growth is provided by an ancient tale. A grateful (but naive) king grants a reward to a loyal subject by promising to give him two grains of wheat on the first square of a chess board, four grains on the second, eight on the third, and so on. In general, there will be 2^n grains of wheat on the nth square. On the last square of the chess board, the 64th, the king has promised to place

$$2^{64} = 18{,}446{,}744{,}073{,}709{,}551{,}616 \text{ grains of wheat}$$

This number was calculated in the eleventh century by the Persian philosopher and mathematician al-Biruni.

There are about a million grains of wheat in a bushel, so the king's promise forces him to put more than 18 trillion bushels of wheat on the 64th square. At \$3.50 per bushel, that amount of wheat would be worth about 65 trillion dollars—about five times the national debt of the United States on January 1, 2010. This illustrates how all exponential growth eventually reaches gargantuan proportions.

EXAMPLE 3.10 Calculating growth: An investment

Consider the investment from Example 3.9, in which we found that the balance B after t years is given by $B = 800 \times 1.07^t$ dollars. What is the growth of the balance over the first 10 years? Compare this with the growth from year 40 to year 50.

SOLUTION

The initial value of the investment is \$800. In Example 3.9, we found that the balance after 10 years was \$1573.72. This gives the growth over the first 10 years as

$$\text{Growth over first 10 years} = \$1573.72 - \$800 = \$773.72$$

To calculate the growth from year 40 to year 50, we put $t = 40$ and $t = 50$ into the formula $B = 800 \times 1.07^t$:

$$\text{Balance after 40 years} = 800 \times 1.07^{40} = \$11{,}979.57$$
$$\text{Balance after 50 years} = 800 \times 1.07^{50} = \$23{,}565.62$$

That is an increase of \$23,565.62 − \$11,979.57 = \$11,586.05. This is almost 15 times the growth over the first 10 years.

TRY IT YOURSELF 3.10

After t months an investment has a balance of $B = 400 \times 1.05^t$ dollars. Compare the growth over the first 20 months with the growth from month 50 to month 70.

The answer is provided at the end of this section.

Relating percentage growth and base

Suppose that a certain job opportunity offers a starting salary of \$50,000 and includes a 5% salary raise each year. In evaluating this opportunity, you may wish to know what your salary will be after 15 years. A common error would be to assume that a 5% raise each year for 15 years is the same as a 15 × 5% = 75% raise. As we shall see, the actual increase is much more.

Our knowledge of exponential functions will allow us to find an appropriate formula and calculate the salary after 15 years. We reason as follows. Because the salary increases by the same percentage (namely 5%) every year, it is an exponential function of time in years. The initial value is \$50,000. Now we want to find the base.

"No, we're not eliminating your position, Fischer. We're just eliminating your salary."

A 5% annual raise means that next year's salary is 105% of the current salary. To calculate 105% of a quantity, we multiply that quantity by 1.05. Therefore,

$$\text{Next year's salary} = 1.05 \times \text{Current salary}$$

This says that the base is 1.05.

We now know that your salary is given by an exponential formula with initial value \$50,000 and base 1.05. Then your salary S in dollars after t years is given by

$$S = \text{Initial value} \times \text{Base}^t$$
$$S = 50{,}000 \times 1.05^t$$

To find the salary after 15 years, we put $t = 15$ into this formula:

$$\text{Salary after 15 years} = 50{,}000 \times 1.05^{15} = \$103{,}946.41$$

We note that this salary represents more than a 100% increase over the initial salary of \$50,000—much more than the 75% increase that some might expect.

The key observations in this example are that the salary is given by an exponential function and that the base corresponding to 5% is 1.05. If we write 5% as a decimal, we get 0.05. We can think of the base 1.05 as $1 + 0.05$. In general, suppose a quantity is growing at a constant percentage rate r in decimal form. Then we find the new value by multiplying the old value by $1 + r$. Thus, the amount after t periods (years in this case) is an exponential function with base $1 + r$. For exponential growth, the base is always greater than 1. The exponential formula for the amount is

$$\text{Amount} = \text{Initial value} \times (1 + r)^t$$

SUMMARY 3.4
Exponential Growth

1. A quantity grows exponentially when it increases by a constant percentage over a given period (day, month, year, etc.).

2. If r is the percentage growth per period, expressed as a decimal, then the base of the exponential function is $1 + r$. For exponential growth, the base is always greater than 1. The formula for exponential growth is

$$\text{Amount} = \text{Initial value} \times (1 + r)^t$$

Here, t is the number of periods.

3. Typically, exponential growth starts slowly and then increases rapidly.

EXAMPLE 3.11 Finding a formula from the growth rate: Health care

U.S. health-care expenditures in 2010 reached 2.47 trillion dollars. It is difficult to predict future health-care costs. One prediction is that, in the near term, they will grow by 6.5% each year. Assuming that this growth rate continues, find a formula that gives health-care expenditures as a function of time. If this trend continues, what will health-care expenditures be in 2030?

SOLUTION

Because health-care expenditures are increasing at a constant percentage rate, they represent an exponential function of time. Let H denote the expenditures, in trillions of dollars, t years after 2010. We find the formula using

$$H = \text{Initial value} \times (1 + r)^t$$

The initial value is the expenditures in 2010, namely 2.47 trillion dollars. Also, 6.5% as a decimal is 0.065, and this is the value we use for r:

$$1 + r = 1 + 0.065 = 1.065$$

So the formula is

$$H = 2.47 \times 1.065^t \text{ trillion dollars}$$

Now, 2030 is 20 years after 2010. Thus, to predict health-care expenditures in 2030, we use $t = 20$ in the formula for H:

$$\text{Expenditures in 2030} = 2.47 \times 1.065^{20} \text{ trillion dollars}$$

The result is about 8.7 trillion dollars.

In **Figure 3.30**, we have plotted the graph of health-care expenditures from Example 3.11.

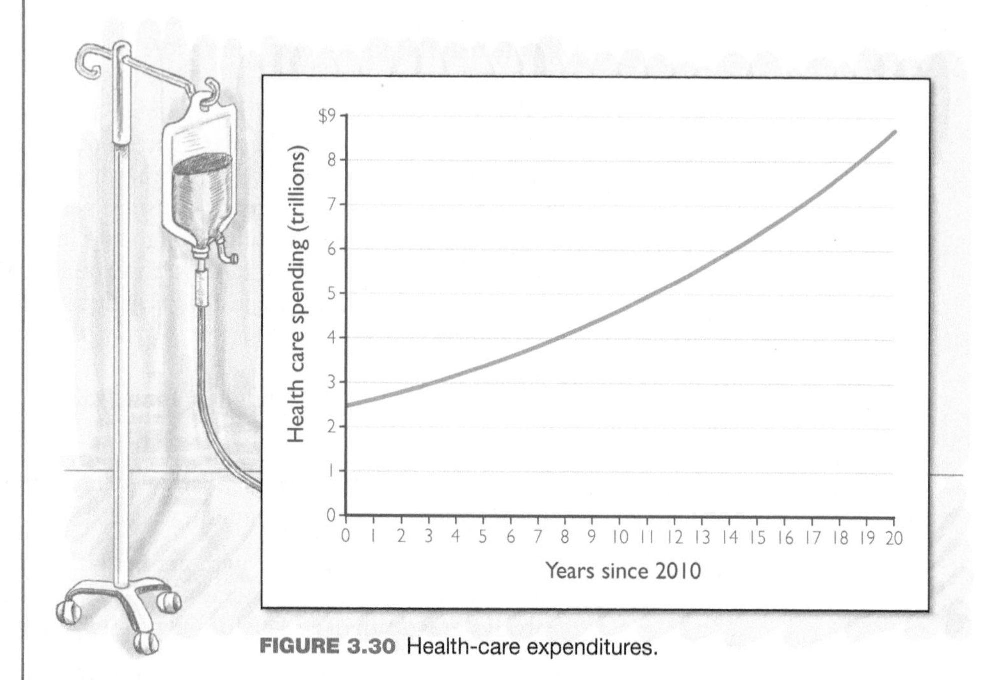

FIGURE 3.30 Health-care expenditures.

TRY IT YOURSELF 3.11

From *globalsecurity.org:* "China plans to raise its defense budget by 10.7 percent to 720.2 billion yuan ($114.3 billion) in 2013, according to a budget report to be reviewed by the national legislature." Assume that this percentage growth rate continues, and find a formula that gives China's defense spending (in billion yuan) as a function of time from 2013 onward. What does this formula predict for China's defense spending in 2016?

The answer is provided at the end of this section.

Exponential decay

The storage of nuclear waste is a perennial concern. The decay of nuclear material is an example of exponential change that is decreasing instead of increasing.

Let's examine how this works out in a simple example. Suppose we have a bacteria population that is dying off due to contaminants. Assume that initially the population size is 2000 and that the population is declining by 25% each hour. This means that 75% of the present population will still be there one hour from now. Because the change is by a constant percentage, this is an example of an exponential function. But because the population is declining, we call this process *exponential decay*.

Because the population after one hour is 75% of the current population, we find the population for the next hour by multiplying the current population by 0.75:

$$\text{Population next hour} = 0.75 \times \text{Current population}$$

This means that the population as an exponential function of time in hours has a base of 0.75. We obtain the formula for exponential decay just as we did for exponential growth. If N denotes the population size after t hours,

$$N = \text{Initial value} \times \text{Base}^t$$
$$N = 2000 \times 0.75^t$$

We used the formula for this population to obtain the graph in **Figure 3.31**. Its shape is characteristic of graphs of exponential decay. Note that the graph is decreasing. It declines rapidly at first but then the graph levels off, showing a much smaller rate of decrease.

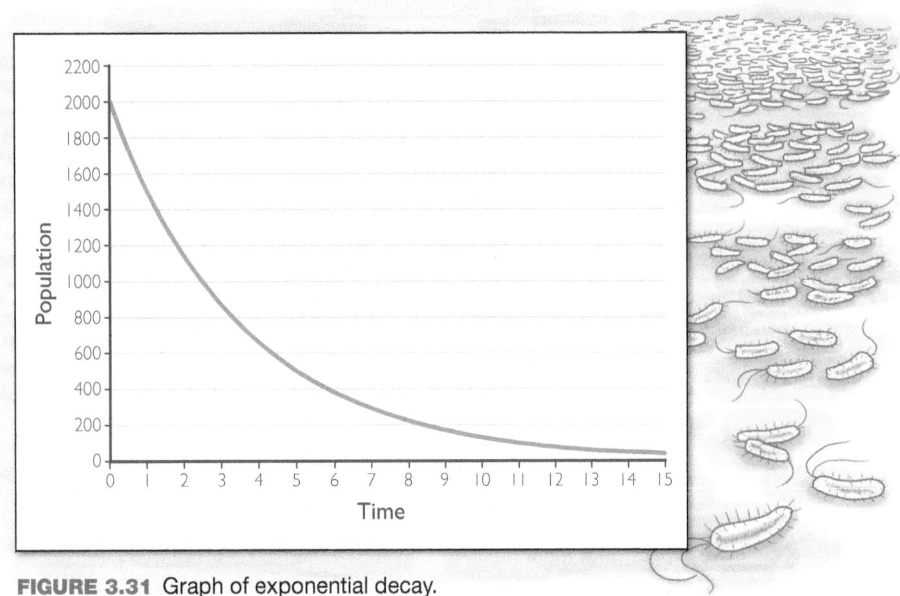

FIGURE 3.31 Graph of exponential decay.

A key observation from this example of population decline is that if the rate of decline is 25%, we find the base of 0.75 for the resulting exponential function using $1 - 0.25 = 0.75$. This procedure is typical of exponential decay.

Recall Summary 3.4, where the percentage increase per period was r as a decimal. In that case, the base was $1 + r$. If a quantity is declining at a constant percentage rate of r as a decimal, we use the same idea. But now the rate r is replaced by $-r$, which means that the base is $1 - r$. For exponential decay, the base is always less than 1.

In summary, we see that the formula for the amount left after t periods is

$$\text{Amount} = \text{Initial value} \times (1 - r)^t$$

SUMMARY 3.5
Exponential Decay

- A quantity decays exponentially when it decreases by a constant percentage over a given period (day, month, year, etc.).
- Assume that a quantity decays exponentially, and let r denote the percentage decay rate per period, expressed as a decimal. Then the base of the exponential function is $1 - r$, and the formula for the function is

$$\text{Amount} = \text{Initial value} \times (1 - r)^t$$

Here, t is the number of periods. For exponential decay, the base is always less than 1.

- Typically, exponential decay is rapid at first but eventually slows.

EXAMPLE 3.12 Finding a formula for exponential decay: Antibiotics in the bloodstream

After antibiotics are administered, the concentration in the bloodstream declines over time. Amoxicillin is a common antibiotic used to treat certain types of infections. Suppose that 70 milligrams of amoxicillin are injected and that the amount of the drug in the bloodstream declines by 49% each hour. Find an exponential formula that gives the amount of amoxicillin in the bloodstream as a function of time since the injection. Another injection will be required when the level declines to 10 milligrams. Will another injection be required before five hours?

SOLUTION

Let A denote the amount (in milligrams) of amoxicillin in the bloodstream after t hours. When expressed as a decimal, 49% is 0.49. Hence, the base of the exponential function is

$$1 - r = 1 - 0.49 = 0.51$$

Because the initial value is 70 milligrams, the formula is

$$A = \text{Initial value} \times (1 - r)^t$$
$$A = 70 \times 0.51^t$$

To find the amount of amoxicillin in the blood after five hours, we put $t = 5$ into this formula:

$$\text{Amount remaining after 5 hours} = 70 \times 0.51^5 \text{milligrams}$$

The result is about 2.4 milligrams, which is less than the minimum of 10 milligrams. Therefore, another injection will be needed before five hours.

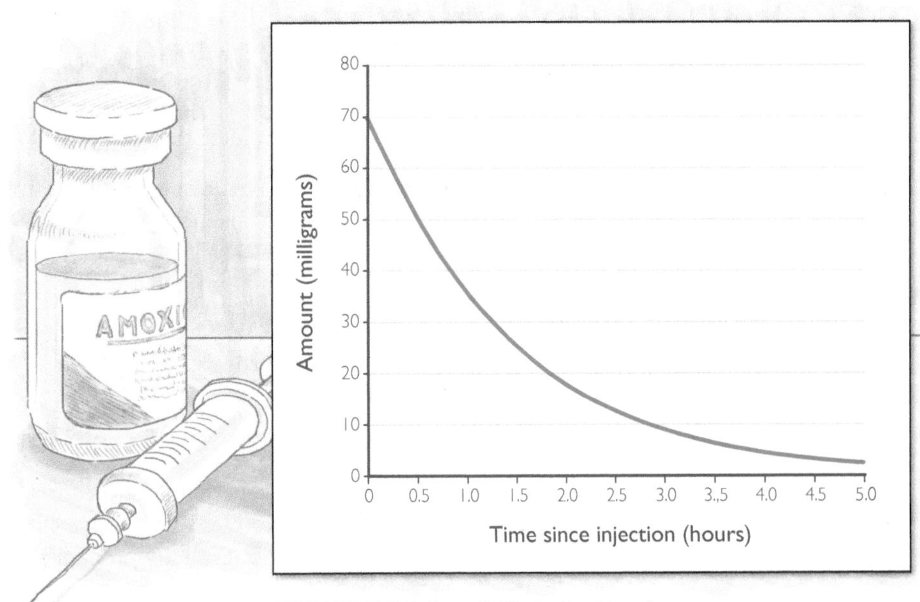

FIGURE 3.32 Amoxicillin in the blood.

TRY IT YOURSELF 3.12

A certain population is initially 4000 and declines by 10% per year. Find an exponential formula that gives the population as a function of time. What will be the population after 10 years?

The answer is provided at the end of this section.

Figure 3.32 shows the amount of amoxicillin in the blood as a function of time for Example 3.12. The graph shows that the level declines to 10 milligrams in about 3 hours.

EXAMPLE 3.13 Using the formula for exponential decay: The Beer-Lambert-Bouguer law

Light intensity decreases with depth.

When light strikes the surface of water, its intensity decreases with depth. The *Beer-Lambert-Bouguer law* states that the percentage of decrease in intensity is the same for each additional unit of depth, so intensity is an exponential function of depth. In the waters near Cape Cod, Massachusetts, light intensity decreases by 25% for each additional meter of depth.

a. Use I_0 for light intensity at the surface, and find an exponential formula for light intensity I at a depth of d meters in the waters off Cape Cod.

b. What is the percentage decrease in light intensity from the surface to a depth of 10 meters?

c. There is sufficient light for photosynthesis to occur in marine phytoplankton at ocean depths where 1% or more of surface light is available. Can photosynthesis by marine phytoplankton occur at a depth of 10 meters off Cape Cod? *Note*: During photosynthesis, marine phytoplankton take in carbon dioxide and release oxygen. This process is crucial for maintaining life on Earth.

SOLUTION

a. Expressed as a decimal, 25% is $r = 0.25$. So the base of the exponential function is $1 - r = 1 - 0.25 = 0.75$. Using I_0 as initial intensity in the formula for percentage decrease gives

$$I = \text{Initial value} \times (1 - r)^d$$
$$I = I_0 \times 0.75^d$$

b. We find the intensity at 10 meters using $d = 10$ in the formula from part a:

$$\text{Intensity at 10 meters} = I_0 \times 0.75^{10}$$

This is about $0.06I_0$. This result means that to find the intensity at a depth of 10 meters, we multiply surface intensity by 0.06. Therefore, 6% of surface intensity is left at a depth of 10 meters, so the intensity has decreased by 94%.

c. From part b we know that at a depth of 10 meters, the light intensity is 6% of light intensity at the surface. That is more than 1%, so the intensity is sufficient for photosynthesis.

Radioactive decay and half-life

The main reason for the perennial concern about the storage of nuclear waste is that such waste is radioactive. Radioactive substances decay over time by giving off radiation or nuclear particles. This process is called *radioactive decay*. The rate of decay of a radioactive substance is normally measured in terms of its *half-life*.

KEY CONCEPT

The half-life of a radioactive substance is the time it takes for half of the substance to decay.

The long half-life of radioactive waste materials is one of the major concerns associated with the use of nuclear fuels for energy generation. Elements have different forms, called *isotopes*. The isotope Pu-239 of plutonium has a half-life of 24,000 years. Suppose that we start with 100 grams of plutonium-239. Then after 24,000 years there will be half of the original amount remaining, or 50 grams. After another 24,000 years half that amount will again decay, leaving 25 grams. This is summarized by the following formula:

$$\text{Amount left at end of a half-life} = \frac{1}{2} \times \text{Current amount}$$

This formula means that the amount of plutonium-239 remaining after h half-lives is an exponential function with base $\frac{1}{2}$. We started with 100 grams, so after h half-lives the amount remaining is

$$\text{Amount remaining} = 100 \times \left(\frac{1}{2}\right)^h \text{ grams}$$

We show the graph of remaining plutonium-239 in **Figure 3.33**.

Note that the horizontal axis represents time in half-lives, not years. The shape of the graph in Figure 3.33 epitomizes the concerns about nuclear waste. The leveling-off of the curve means that dangerous amounts of the radioactive substance will be present for years to come. See also Exercise 38.

The formula

$$\text{Amount remaining} = 100 \times \left(\frac{1}{2}\right)^h \text{ grams}$$

describes exponential decay with a percentage decrease (as a decimal) of $r = 1 - 1/2 = 0.5$ when time is measured in half-lives. In practice, we may want to know the amount remaining when time is measured in years, not in half-lives. For example, suppose we ask how much Pu–239 will remain after 36,000 years. The key here is to determine how many half-lives 36,000 years represents. We know that one half-life is 24,000 years and $36{,}000/24{,}000 = 1.5$, so 36,000 years is 1.5 half-lives. Thus, to find the amount remaining after 36,000 years, we use $h = 1.5$ in the preceding formula:

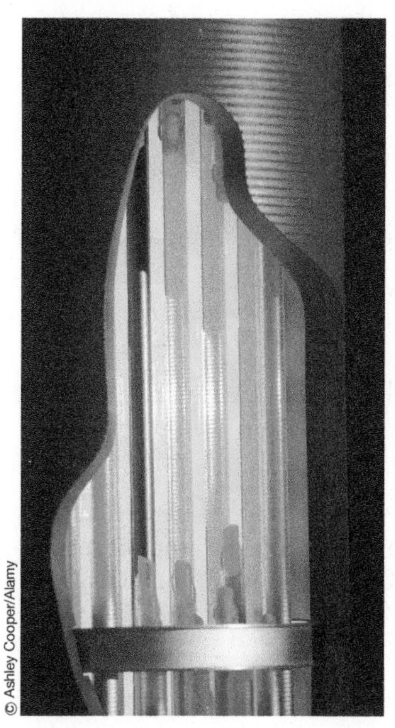

A cross-section display of nuclear fuel rods.

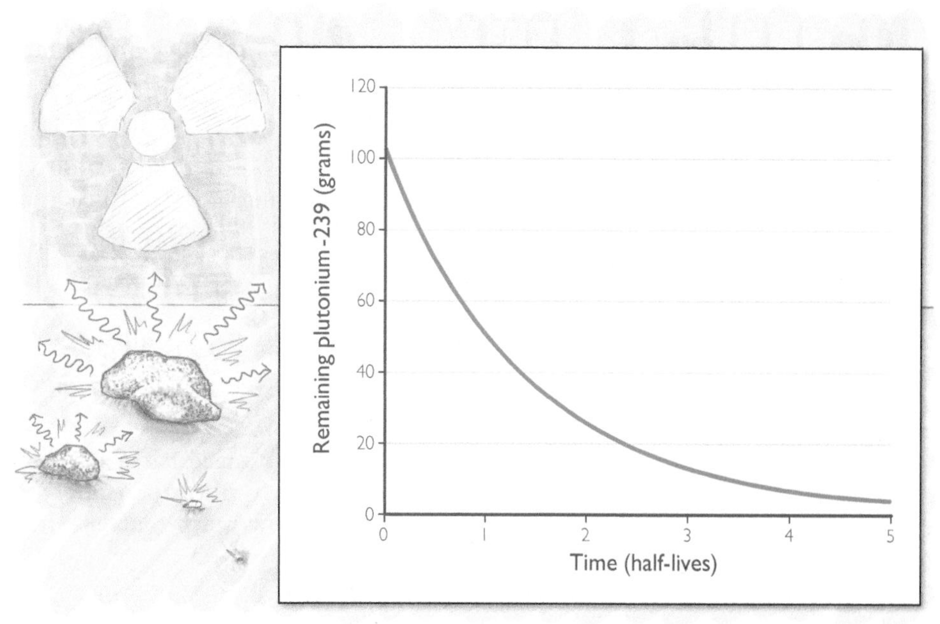

FIGURE 3.33 Plutonium-239 decay.

$$\text{Amount remaining after 36,000 years} = 100 \times \left(\frac{1}{2}\right)^{1.5} \text{ grams}$$

This is about 35.4 grams.

Plutonium decays in a fashion typical of all radioactive substances. The half-life is different for each radioactive substance.

SUMMARY 3.6
Half-life

1. After h half-lives, the amount of a radioactive substance remaining is given by the exponential formula

$$\text{Amount remaining} = \text{Initial amount} \times \left(\frac{1}{2}\right)^{h}$$

2. We can find the amount remaining after t years by first expressing t in terms of half-lives and then using the formula above.

Now we look at one important application of our study of radioactive decay. Measurements of a radioactive isotope of carbon can be used to estimate the age of organic remains. The procedure is called *radiocarbon dating.*

EXAMPLE 3.14 Applying radioactive decay: Radiocarbon dating

The isotope known as carbon-14 is radioactive and will decay into the stable form nitrogen-14. Assume that the percentage of carbon-14 in the air over the past 50,000 years has been about constant. As long as an organism is alive, it ingests air, and the level of carbon-14 in the organism remains the same. When it dies, it no longer absorbs carbon-14 from the air, and the carbon-14 in the organism decays, with a half-life of 5770 years.

Now, when a bit of charcoal or bone from a prehistoric site is found, the percent of carbon-14 remaining is measured. By comparing this with the percentage of carbon-14 in a living tree today, scientists can estimate the time of death of the tree from which the charcoal came.

Suppose a tree contained C_0 grams of carbon-14 when it was cut down. What percentage of the original amount of carbon-14 would we find if it was cut down 30,000 years ago?

SOLUTION

The amount C (in grams) remaining after h half-lives is given by

$$C = C_0 \times \left(\frac{1}{2}\right)^h$$

Now 5770 years is one half-life, so 30,000 years is 30,000/5770 or about 5.20 half-lives. We use this value for h to calculate the amount remaining after 30,000 years:

$$\text{Amount after 30,000 years} = C_0 \times \left(\frac{1}{2}\right)^{5.2} \text{ grams}$$

This is about $0.027C_0$ grams. Thus, about 2.7% of the original amount of carbon-14 remains after 30,000 years.

TRY IT YOURSELF 3.14

A certain radioactive substance has a half-life of 450 years. If there are initially 20 grams, find a formula for the exponential function that gives the amount remaining after h half-lives. How much remains after 720 years? Round your answer to one decimal place.

The answer is provided at the end of this section.

In Example 3.14, we determined the percentage of carbon-14 from the age of the sample. In the next section, we return to the topic of radiocarbon dating and see how to estimate the age of the sample based on the percentage of carbon-14.

"Don't worry if you can't remember your age grandad, we'll get you carbon dated!"

Linear or exponential?

In Section 3.1 we discussed linear functions, which are characterized by a constant rate of change. In this section we have studied exponential functions, which are characterized by a constant *percentage* rate of change. Being able to recognize this difference as encountered in everyday life is an important quantitative literacy skill.

EXAMPLE 3.15 Examining phenomena: Linear or exponential?

Determine which of the following situations are best described by a linear function and which are best described by an exponential function.

a. An investment that is growing by 3.25% each month

b. The depth of water in a pool that is being filled so that the depth increases by the same amount each hour

c. The amount of a drug remaining in the bloodstream if that amount decreases by the same percentage each hour

d. The amount of money in your piggy bank if you add \$1 per week

e. The amount of money in your piggy bank if you take out \$1 per week

f. The value of a bad investment that has been decreasing in value by half every year

g. The value of Facebook stock if it triples every year

SOLUTION

The key is to remember that linear functions change by the same amount over each period, whereas exponential functions change by the same percentage over each period.

a. The investment is growing by the same percentage, 3.25%, over each period. This indicates an exponential function.

b. The water depth is growing by the same amount each period. This indicates a linear function.

c. The amount of the drug is decreasing by the same percentage over each period. This fact indicates exponential decay, which corresponds to an exponential function.

d. Adding the constant amount \$1 per week indicates a linear function.

e. Subtracting the constant amount \$1 per week indicates a decreasing linear function.

f. A decrease by half every year is a percentage decline of 50% per year. Decreasing by the same percentage over each period indicates exponential decay, which corresponds to an exponential function. (An alternative explanation is that this situation describes an exponential function with base 1/2.)

g. If the value triples every year, the growth rate is 200%. Growth by the same percentage over each period indicates an exponential function. (An alternative explanation is that this situation describes an exponential function with base 3.)

TRY IT YOURSELF 3.15

The amount of a radioactive substance decreases by the same percentage each year. Is a linear or exponential function indicated?

EXAMPLE 3.16 Identifying graphs: Linear or exponential?

Match the graphs in **Figure 3.34** through **Figure 3.37** with the appropriate function in parts a through d.

a. $y = 2^x$

b. $y = 2x$

c. $y = -2x$

d. $y = (1/2)^x$

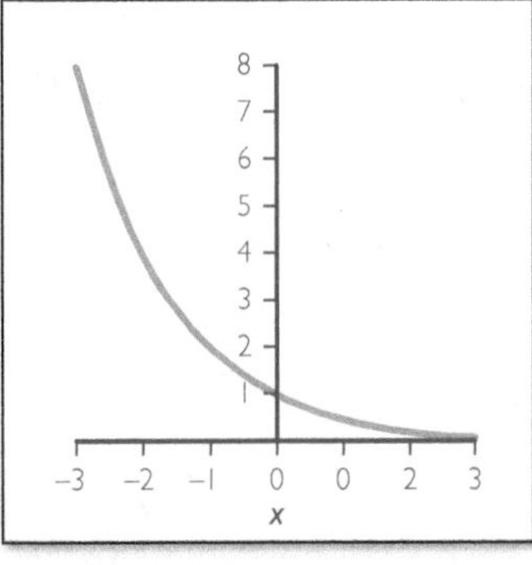

FIGURE 3.34 I.

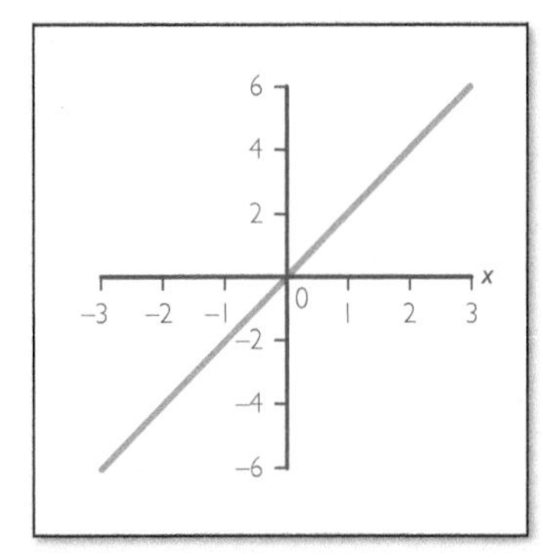

FIGURE 3.35 II.

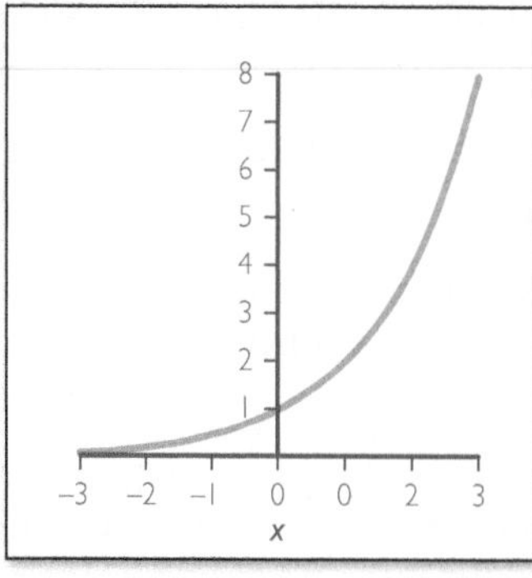

FIGURE 3.36 III.

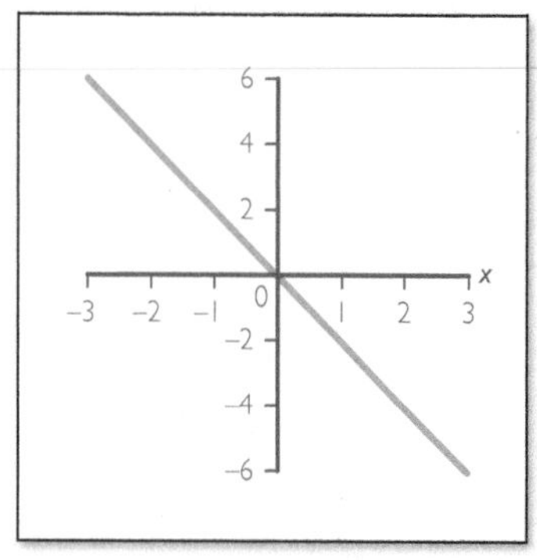

FIGURE 3.37 IV.

SOLUTION

a. This is an exponential function with base greater than 1. The graph is increasing, but it is not a straight line. Thus, III is the proper graph.

b. This is a linear function with positive slope. The graph is an increasing straight line. Thus, II is the proper graph.

c. This is a linear function with negative slope. The graph is a decreasing straight line. Thus, IV is the correct graph.

d. This is an exponential function with base less than 1. The graph is decreasing, but it is not a straight line. Thus, graph number I is the correct graph.

TRY IT YOURSELF 3.16

Which of the graphs above shows exponential decay?

WHAT DO YOU THINK?

Nuclear waste: In this section we saw that the amount of a radioactive substance remaining decreases as an exponential function of time. Explain why this fact makes handling of radioactive waste a long-term problem. In particular, what is it about the shape of the graph in Figure 3.33 that makes this handling a long-term problem? Would the problem be the same if the amount remaining decreased as a linear function of time?

Questions of this sort may invite many correct responses. We present one possibility: Constant percentage change may result in significant absolute decrease initially, but in the long term the absolute decrease is negligible. If there are 100 items present subject to a constant 50% decrease, then over the first term the number is decreased by 50. But over the second term the number is decreased by only 25. Because the decrease is exponential, low levels of nuclear waste persist almost indefinitely. Depending on the waste product in question, they may remain dangerous for hundreds or even thousands of years. The graph referred to here is decreasing at a decreasing rate, and it gives a visual display of this type of decay. In the case of linear decay, we can point to a time when all of the substance will be gone. That never occurs with exponential decay.

Further questions of this type may be found in the *What Do You Think?* section of exercises.

Try It Yourself answers

Try It Yourself 3.8: Determining constant percentage growth: Quadrupling and population size Yes: Quadrupling represents 300% growth, so this is exponential.

Try It Yourself 3.9: Finding an exponential formula: An investment 3% per month. If B denotes the balance in dollars after t months, $B = 450 \times 1.03^t$. The balance after two years (24 months): \$914.76.

Try It Yourself 3.10: Calculating growth: An investment Growth over the first 20 months: \$661.32. Growth from month 50 to month 70: \$7583.61, about 11.5 times the earlier growth.

Try It Yourself 3.11: Finding a formula from the growth rate: Defense spending If D denotes defense spending (in billions of yuan) t years after 2013, $D = 720.2 \times 1.107^t$. In 2016: 977.0 billion yuan.

Try It Yourself 3.12: Finding a formula for exponential decay: Population decline If N denotes the population after t years, $N = 4000 \times 0.90^t$. After 10 years, the population will be about 1395.

Try It Yourself 3.14: Applying radioactive decay: Half-life If A is the amount (in grams) remaining, $A = 20 \times (1/2)^h$. After 720 years, there will be about 6.6 grams remaining.

Try It Yourself 3.15: Examining phenomena: Linear or exponential? A constant percentage decrease indicates an exponential function.

Try It Yourself 3.16: Identifying graphs: Linear or exponential? Graph number I shows exponential decay.

Exercise Set 3.2

Test Your Understanding

1. If a quantity is growing exponentially (as a function of time in years) with base b, how does one use this year's amount to find next year's amount?

2. If a quantity doubles each week, then the base of the exponential function describing the quantity in terms of weeks is ________.

3. True or false: An exponential function with base greater than 1 eventually shows rapid growth.

4. If a quantity has a percentage growth rate of r as a decimal, then the base of the corresponding exponential function is: **a.** r **b.** $1 + r$ **c.** $1 - r$ **d.** none of the above.

5. If a quantity has a percentage decay rate of r as a decimal, then the base of the corresponding exponential function is: **a.** r **b.** $1 + r$ **c.** $1 - r$ **d.** none of the above.

6. The half-life of a radioactive substance is the time required for ________.

Problems

Note: In some exercises you are asked to find a formula for an exponential function. If no letters are assigned to quantities in an exercise, you are of course expected to choose them and give the appropriate units.

7. **Determining constant percentage change.** A Web site goes viral. The number of users quadruples each day. What is the constant percentage increase each day?

8. **Determining constant percentage change.** A population is declining. Each year, only a quarter of last year's population remains. What is the constant percentage decrease per year?

9. **Finding an exponential formula.** The balance of an investment is increasing according to the rule

$$\text{Next year's balance} = 1.15 \times \text{Current balance}$$

If the original value of the investment is \$500, find a formula that gives the balance B after t years.

10. **Finding an exponential formula.** The balance of an investment is decreasing according to the rule

$$\text{Next year's balance} = 0.85 \times \text{Current balance}$$

If the original value of the investment is \$1000, find an exponential formula that gives the balance B after t years.

11. **Finding formula from growth rate.** You owe \$300 on your credit card. For every month that you fail to make a payment, your balance increases by 2%. Find a formula for the balance B owed after t months with no payment.

12. **Finding formula from decay rate.** You pay off a debt of \$500 by paying 20% of the remaining balance each month. Find a formula that gives the remaining debt D after t months.

13. **A drug.** After the initial injection, the amount of a drug in the bloodstream decreases by 5% each hour. Using D_0 for the initial amount, find a formula that gives the amount D of the drug in the bloodstream t hours after injection.

14. **Half-life.** The half-life of sodium-22 is 2.6 years. If there are initially 20 grams of sodium-22, find a formula that gives the amount S, in grams, remaining after t years.

15. **A poor investment.** You invested \$1000 in an account that loses money at a rate of 3% per year.

a. Find a formula that gives the amount B, in dollars, left after t years.

b. How much will remain after five years?

16–23. Linear or exponential? *In Exercises 16 through 23, determine whether the situation described is best modeled by a linear function or by an exponential function.*

16. Bob's salary. Bob's salary grows by \$3000 each year.

17. Mary's salary. Mary's salary grows by 3% each year.

18. Water in a tank. Water is pumped into a tank at a rate of 10 gallons per minute. (We are considering the volume of water in the tank.)

19. Ants. An ant colony is growing at a rate of 2% each month.

20. Radioactivity. A certain radioactive substance has a half-life of three years.

21. A piggy bank. You put \$20 into your nephew's piggy bank each week.

22. Moore's law. Moore's law says that the number of transistors in an integrated circuit will double approximately every two years. (Gordon Moore was a co-founder of Intel.)

23. A plant. A plant grows in height by one inch per week.

24. Graphs. Determine which of the graphs in **Figure 3.38** can represent (1) an exponential function, (2) a linear function, or (3) neither.

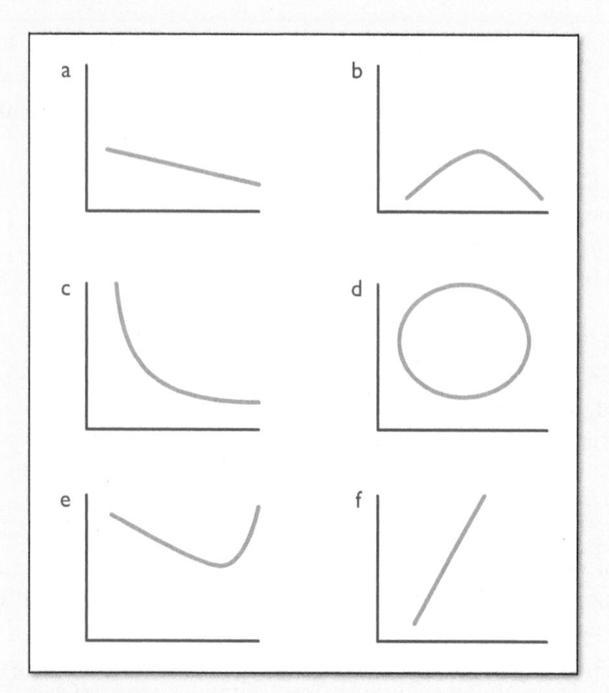

FIGURE 3.38

25. Heart health. One study measured the benefits of more sleep for healthier hearts.[7] Heart health was measured in terms of calcification of the coronary arteries. The main conclusion of the study was, "One hour more of sleep decreased the estimated odds of calcification by 33%." Rephrase this conclusion using the terminology of this section.

26. Large percent change. Chinese textile imports went up by 1000%. Were the imports

a. 10 times the old figure?
b. 11 times?
c. 100 times?
d. 110 times?
e. none of these?

27. Balance. An investment grows according to the rule

$$\text{Next month's balance} = 1.002 \times \text{Current balance}$$

Find the percentage increase each month, and explain why the balance is an exponential function of time.

28. Internet domain hosts. With new technologies and new types of successful business ventures, it is common for growth to be approximately exponential. An example is the growth in the number of Internet domain hosts. One model is that the number of domain hosts grew according to the rule

$$\text{Next year's number} = 1.43 \times \text{Current number}$$

a. Find the percentage increase each year, and explain why the number of hosts is an exponential function of time.

b. The number of domain hosts in 1995 was 8.2 million. Find an exponential formula that gives the number of hosts in terms of the time, in years since 1995.

c. What number of Internet domains does your formula in part b give for 2005? Round your answer to the nearest million.

29. German gold bonds. A certain gentleman acquired 54 gold bonds issued by the Weimar Republic in 1934. The value of each was 2500 troy ounces of gold. The bonds matured in 1954, at which time the value of each was 2500 troy ounces of gold at 1954 prices. The price of gold at that time was around \$350 per troy ounce. The bonds could not be redeemed because in 1954 there was no single German country—it had been divided into two nations. But provisions of the bond stated that from 1954 onward the unredeemed bonds grew in value according to the exponential formula

$$\text{Value} = \text{Initial value} \times 1.00019^{365t}$$

where t is the number of years since 1954. What was the total value of this man's 54 bonds in 2017? If the bonds remain unredeemed in 2024, what will be their total value? Give your answers in billions of dollars rounded to two decimal places.

30. Headway. For traffic that flows on a highway, the *headway* is the average time between vehicles. On four-lane highways, the probability P (as a decimal) that the headway is at least t seconds when there are 500 vehicles per hour traveling one way is given by[8]

$$P = 0.87^t$$

Under these circumstances, what is the probability that the headway is at least 20 seconds? Round your answer to two decimal places.

[7] *C. R. King et al., "Short Sleep Duration and Incident Coronary Artery Calcification,"* Journal of the American Medical Association *300 (2008), 2859–2866.*

[8] *See Institute of Traffic Engineers,* Transportation and Traffic Engineering Handbook, *ed. John E. Baerwald (Englewood Cliffs, NJ: Prentice-Hall 1976), p. 102.*

31. Newton's law of heating. If we put a cool object into a preheated oven, Newton's law of heating tells us that the difference between the temperature of the oven and the temperature of the object decreases exponentially with time. The percentage rate of decrease depends on the material that is being heated. Suppose a potato initially has a temperature of 75 degrees and the oven is preheated to 375 degrees. Use the formula

$$D = 300 \times 0.98^t$$

where D is the temperature difference between the oven and the potato, t is the time in minutes the potato has been in the oven, and all temperatures are measured in degrees Fahrenheit. Round your answers to the nearest degree.

a. What is the temperature difference after 30 minutes?

b. What is the temperature of the potato after 30 minutes?

32. Tsunami waves. Crescent City, California, is in an area that is subject to tsunamis. Its harbor was severely damaged by the tsunami caused by the earthquake off the coast of Japan on March 11, 2011. The probability P (as a decimal) that no tsunami with waves of 15 feet or higher will strike Crescent City over a period of t years is given by the formula[9]

$$P = 0.98^t$$

a. If you move to Crescent City and stay there for 20 years, what is the probability that you will witness no tsunami waves of 15 feet or higher? Round your answer to two decimal places.

b. What is the percentage decrease of the probability for each one-year increase in the time interval?

33. Population growth. Initially, a population is 500, and it grows by 2% each year. Explain why the population is an exponential function of time. Find a formula for the population at any time, and determine the population size after five years.

34. World population. The world population in 2020 was estimated to be 7.8 billion people, and it is increasing by about 1.1% per year. Assume that this percentage growth rate remains constant through 2025.[10] Explain why the population is an exponential function of time. What would you expect the world population to be in 2025?

35. United States population. According to the U.S. Census Bureau, the population of the United States in 2020 was 330 million people. The rate of growth in population was 0.4% per year. Assume that this rate of growth remains the same through 2025. Explain why the population is an exponential function of time. What would you predict the U.S. population to be in 2025? Round your answer to the nearest million.

36. Epidemics. In many epidemics, the cumulative number of cases grows exponentially, at least over a limited time. A recent example is the coronavirus (COVID-19) pandemic. On March 6, 2020, there were 237 cumulative cases of coronavirus in the United States. Over the next week, the cumulative number of cases in the United States increased by about 34% each day. If this growth rate had continued, what would have been the cumulative number of cases in the United States reported 30 days after March 6?

37. Library costs. The library at a certain university reported that journal prices had increased by 150% over a period of 10 years. The report concluded that this represented a price increase of 15% each year. If journal prices had indeed increased by 15% each year, what percentage increase would that give over 10 years? Round your answer as a percentage to the nearest whole number.

38. A nuclear waste site. Cesium-137 is a particularly dangerous by-product of nuclear reactors. It has a half-life of 30 years. It can be readily absorbed into the food chain. Suppose we place 3000 grams of cesium-137 in a nuclear waste site.

a. How much cesium-137 will be present after 30 years, or one half-life? After 60 years, or two half-lives?

b. Find an exponential formula that gives the amount of cesium-137 remaining in the site after h half-lives.

c. How many half-lives of cesium-137 is 100 years? Round your answer to two decimal places.

d. Use your answer to part c to determine how much cesium-137 will be present after 100 years. Round your answer to the nearest whole number.

39. Folding paper. Each time you fold an ordinary sheet of paper in half, you double the thickness.

a. A single sheet of 20-pound bond paper is about a tenth of a millimeter thick. How would you come up with this estimate? (*Hint:* Paper often comes in reams of 500 pages.)

b. Find an exponential formula that gives the thickness of the sheet of paper in part a after f folds in millimeters and in kilometers.

c. What would be the thickness in kilometers if you were physically able to fold it 50 times? *Note*: For comparison, the distance from Earth to the Sun is about 149,600,000 kilometers.

40–41. Inflation. *The rate of inflation measures the percentage increase in the price of consumer goods. The rate of inflation in the year 2020 was 1.2% per year. To get a sense of what this rate would mean in the long run, let's suppose that it persists through 2040. Exercises 40 and 41 refer to this inflation.*

40. What would be the cost in 2021 of an item that costs $100 in 2020?

41. What would be the cost in 2040 of an item that costs $100 in 2020?

---

[9] *Taken from data provided on p. 294 of Robert L. Wiegel, ed.,* Earthquake Engineering *(Englewood Cliffs, NJ: Prentice-Hall, 1970).*

[10] *In fact, for almost all of human history, the human population growth rate has been slowly increasing. However, in 1962 the rate reached a peak of 2.2% and declined steadily to its current level of 1.1%.*

42–43. Deflation. *From 1929 through the early 1930s, the prices of consumer goods actually decreased. Economists call this phenomenon* deflation. *The rate of deflation during this period was about 7% per year, meaning that prices decreased by 7% per year. To get a sense of what this rate would mean in the long run, let's suppose that this rate of deflation persisted over a period of 20 years. This is the situation for Exercises 42 and 43.*

42. What would be the cost after one year of an item that costs $100 initially?

43. What would be the cost after 20 years of an item that costs $100 initially?

44. Blogosphere growth. The number of blogs (or weblogs) grew rapidly for several years. According to one report, there were about 2 million blogs in March 2004, after which the number of blogs doubled approximately every six months. *Note:* Growth tapered off after 2007, and at the end of 2011 there were about 175 million blogs.

a. Let N denote the number (in millions) of blogs and d the number of doubling periods after March 2004. Use a formula to express N as an exponential function of d.

b. How many blogs does the formula derived in part a project for September 2006?

c. The Nielsen Company[11] says that the actual number of blogs in October 2006 was about 35,800,000. What conclusions do you draw from this about the reliability of projections over time?

45. Retirement options. This exercise illustrates just how fast exponential functions grow in the long term. Suppose you start work for a company at age 25. You are offered two rather unlikely retirement options:

Retirement option 1: When you retire, you will receive $20,000 for each year of service.

Retirement option 2: When you start work, the company deposits $2500 into a savings account that pays a monthly rate of 1.5%. When you retire, the account will be closed and the balance given to you.

How much will you have under the second plan at age 55? At age 65? Which retirement option is better if you plan to retire at age 55? Which if you plan to retire at age 65?

46. Cleaning dirty water. Many physical processes are exponential in nature. A typical way of cleaning a tank of contaminated water is to run in clean water at a constant rate, stir, and let the mixture run out at the same rate. Suppose there are initially 100 pounds of a contaminant in a large tank of water. Assume that the cleaning method described here removes 10% of the remaining contaminant each hour. Round your answers to one decimal place.

a. Find an exponential formula that gives the number of pounds of contaminant left in the tank after t hours.

b. How much contaminant is *removed* during the first three hours?

c. How much contaminant is *removed* from the tenth ($t = 10$) to the thirteenth ($t = 13$) hour of the cleaning process?

47. Cleaning waste sites. In many cases, removing dangerous chemicals from waste sites can be modeled using exponential decay. This is a key reason why such cleanups can be dramatically expensive. Suppose for a certain site there are initially 20 parts per million of a dangerous contaminant and that our cleaning process removes 5% of the remaining contaminant each day. Round your answers to two decimal places.

a. How much contaminant (in parts per million) is removed during the first two days?

b. How much contaminant is removed on days 10 and 11?

48. Ponzi schemes. Charles Ponzi became infamous in the early 1920s for running a classic "pyramid" investment scheme. Since that time such schemes have borne his name and are illegal. Even so, in 2009 a man named Bernard Madoff pleaded guilty to bilking investors out of billions of dollars using just such a scheme. Pyramid schemes work roughly like this. I get "rounds" of investors by promising a fat return on their money. I actually have no product or service to offer. I simply take money from second-round investors to pay off first-round investors while keeping a tidy sum for myself. Third-round investors pay off second-round investors, and so on. Suppose that there are 10 first-round investors, and for each investor in a round, I need 10 investors in the next round to pay them off. How many fifth-round investors do I need to pay off the fourth-round investors? After paying off the ninth-round investors, I take the money and run. How many investors get cheated out of their money?

49. Radiocarbon dating. Radiocarbon dating was discussed in Example 3.14. Recall from that example that the half-life of carbon-14 is 5770 years. This exercise refers to an organic sample that is 12,000 years old.

a. How many half-lives is 12,000 years? Round your answer to two decimal places.

b. Use your answer to part a to determine what percentage of the original amount of carbon-14 remains after 12,000 years. Round your answer as a percentage to the nearest whole number.

50. Growth of social networks. A 2012 article in the *Los Angeles Times* reported that the number of weekly visits to Pinterest's Web site by North American users was 1.27 million in 2011 and that this number increased by 2183% over the ensuing year.

a. Assuming that growth continued at this percentage rate, make a model that shows the number P, in millions, of North American Pinterest users t years after 2011.

b. How many Pinterest users does your model give for 2013?

c. Look up the population of North America, and compare the result with your answer to part b.

The following exercises are designed to be solved using technology such as calculators or computer spreadsheets.

51. Economic growth. The CIA *World Factbook*[12] estimates the economic performance of countries. One such figure is the *gross domestic product* (GDP), which is the annual market value of the goods and services produced in that country.

[11] www.nielsen.com/us/en/newswire/2012/buzz-in-the-blogosphere-millions-more-bloggers-and-blog-readers.html

[12] www.cia.gov/library/publications/the-world-factbook/

a. In 2015 the GDP of India was about 2.1 trillion dollars. It is predicted to grow by about 7.5% each year. Use this prediction and a spreadsheet to determine how long it will take for the economy of India to double in size from its level in 2015. (Round your answer to the nearest year.)

b. In 2015 the GDP of China was about 11.0 trillion dollars. It is predicted to grow by about 7% each year. Use this prediction and a spreadsheet to determine how long it will take for the economy of China to double in size from its level in 2015. (Round your answer to the nearest year.)

Writing About Mathematics

52. Research project. The method of radiocarbon dating described in Example 3.14 is an example of a general technique known as *radiometric dating*. Investigate this general technique. What radioactive substances are used, and what are their half-lives? What are the limitations of this technique?

53. In the news I. Find a news article that includes mention of exponential growth or decay. Write a report summarizing the article and explaining how exponential growth or decay comes up.

54. In the news II. Look up an article concerning the rapid growth of a company in the news and write a report about it. In particular, try to ascertain whether exponential growth is involved. Examples of such a company might be Facebook, Instagram, or Uber.

55. Radioactive threats. The article at the beginning of this section described the danger of radioactive substances getting into the wrong hands. Do some further investigation of this issue, and include some information on which substances are most dangerous and why.

What Do You Think?

56. Malware. The mathematical notion of exponential growth refers to a precise type of growth that happens to be (eventually) very rapid. But the phrase is often applied to any fast-growing phenomenon. The headline at tech.fortune.cnn.com/2011/11/17/androids-growing-malware-problem says this: "Android's malware problem is growing exponentially." Do you think this headline is really describing exponential growth, or is it just using a figure of speech? Explain your thinking.

57. Paypal. The mathematical notion of exponential growth refers to a precise type of growth that happens to be (eventually) very rapid. But the phrase is often applied to any fast-growing phenomenon. The headline at venturebeat.com/2012/01/10/paypals-mobile-payments-4b-2011/ says this: "Paypal's mobile payments growing exponentially, reached $4B in 2011." Do you think this headline is really describing exponential growth, or is it just using a figure of speech? Explain your thinking.

58. Is it exponential? You see a news report about taxes, inflation, or other economic issues and wonder whether you are seeing exponential growth. How would you make a judgment about this question? What factors might lead you to believe that you are seeing exponential growth, and what might lead you to believe that you are not seeing exponential growth?

59. Is it exponential? You see an article about an interesting biological phenomenon. The article is accompanied by a graph. What properties of the graph might lead you to believe that the phenomenon is expressed using an exponential relationship? What properties of the graph might lead you to believe that the phenomenon is not expressed using an exponential relationship?

60. Shape of exponential growth. The graph of exponential growth gets steeper as we move to the right. What is it about the definition of exponential growth that guarantees this shape?

61. Base and percentage growth. If a population grows by 5% each year, then 1.05 is the base for the exponential formula expressing the population as a function of time in years. Often beginning students ask, "Where does the '1' come from in the expression 1.05?" How would you answer that question in terms understandable to beginning students? **Exercise 33 is relevant to this topic.**

62. Exponential decay. Many phenomena, radioactivity for example, display exponential decay with time. For such a phenomenon, would you expect the change to be greater from the fifth to the sixth year or from the fifty-ninth to the sixtieth year? Explain your answer.

63. Examples of exponential decay. The decay of nuclear material is a prime example of exponential decay. Use the Internet to find other examples and report on them. *Suggestion*: You might look at cleanup of sites contaminated by dangerous chemicals. **Exercise 38 is relevant to this topic.**

64. A long-term investment. You have the option of making a long-term investment in either of two companies. Company A shows rapid linear growth. Company B shows moderate exponential growth. Which do you think would be the better investment? What additional information might you need in order to make your decision? **Exercise 45 is relevant to this topic.**

3.3 Logarithmic phenomena: Compressed scales

TAKE AWAY FROM THIS SECTION

Understand the use of logarithms in compressed scales and in solving exponential equations.

The following appeared in a report issued by the *United States Geological Survey*.

IN THE NEWS 3.4

Search

Home | TV & Video | U. S. | World | Politics | Entertainment | Tech | Travel

Record Number of Oklahoma Tremors Raises Possibility of Damaging Earthquakes

Updated USGS-Oklahoma Geological Survey Joint Statement on Oklahoma Earthquakes

May 2, 2014

Oklahoma Earthquakes Magnitude 3.0 and greater

Number of Earthquakes per year

160
140
120
100
80
60
40
20
0

~1.6/year

Earthquakes in all of 2013

As of May 2, 2014

1978 to 1999 2001 2003 2005 2007 2009 2011 2013

Year

The rate of earthquakes in Oklahoma has increased remarkably since October 2013—by about 50 percent—significantly increasing the chance for a damaging magnitude 5.5 or greater quake in central Oklahoma.

A new U.S. Geological Survey and Oklahoma Geological Survey analysis found that 145 earthquakes of magnitude 3.0 or greater occurred in Oklahoma from January 2014 (through May 2; see accompanying graphic). The previous annual record, set in 2013, was 109 earthquakes, while the long-term average earthquake rate, from 1978 to 2008, was just two magnitude 3.0 or larger earthquakes per year. Important to people living in central and north-central Oklahoma is that the likelihood of future, damaging earthquakes has increased as a result of the increased number of small and moderate shocks.

The analysis suggests that a likely contributing factor to the increase in earthquakes is triggering by wastewater injected into deep geologic formations. This phenomenon is known as injection-induced seismicity, which has been documented for nearly half a century, with new cases identified recently in Arkansas, Ohio, Texas, and Colorado. A recent publication by the USGS suggests that a magnitude 5.0 foreshock to the 2011 Prague, Okla., earthquake was human-induced by fluid injection; that earthquake may have then triggered the mainshock and its aftershocks. OGS studies also indicate that some of the earthquakes in Oklahoma are due to fluid injection. The OGS and USGS continue to study the Prague earthquake sequence in relation to nearby injection activities.

In this section we study the *logarithm*. The logarithm is a mathematical function that is used to define compressed scales for certain measurements, including the magnitude of earthquakes referred to in the preceding article. The effect of compressing the scale is to expand or decompress the data so that important features become visible. Logarithms are also the basis for measuring the loudness of sound, acidity of substances, and brightness of stars. Logarithms are closely related to exponential functions, as we will see shortly.

In addition, the logarithm is the mathematical tool needed to solve certain exponential equations. Solving such equations is important for the applications that were discussed in the preceding section, such as the growth of investments and populations and finding the age of archaeological artifacts.

What is a logarithm?

We have noted that population growth is often modeled using an exponential function of time. But we could turn the model around and think of the time as a function of the population. This function would tell us, for example, when the population reaches 1000 and when it reaches 1,000,000. This kind of function, which reverses the effect of an exponential function, is called a *logarithmic function.* The corresponding graphs for a population of bacteria that doubles each hour are shown in **Figures 3.39 and 3.40**. As expected, the exponential graph shown in Figure 3.39 gets steeper as we move to the right. For example, over the first hour the population grows by 2 (from 2 to 4), but over the last hour shown the population grows by 8 (from 8 to 16). If we think of time as a function of population, this says that growth in population from 2 to 4 takes one hour, and growth in population from 8 to 16 also takes one hour. This observation explains the growth rate of the logarithmic function shown in Figure 3.40: Growth is rapid at first but slows as we move to the right.

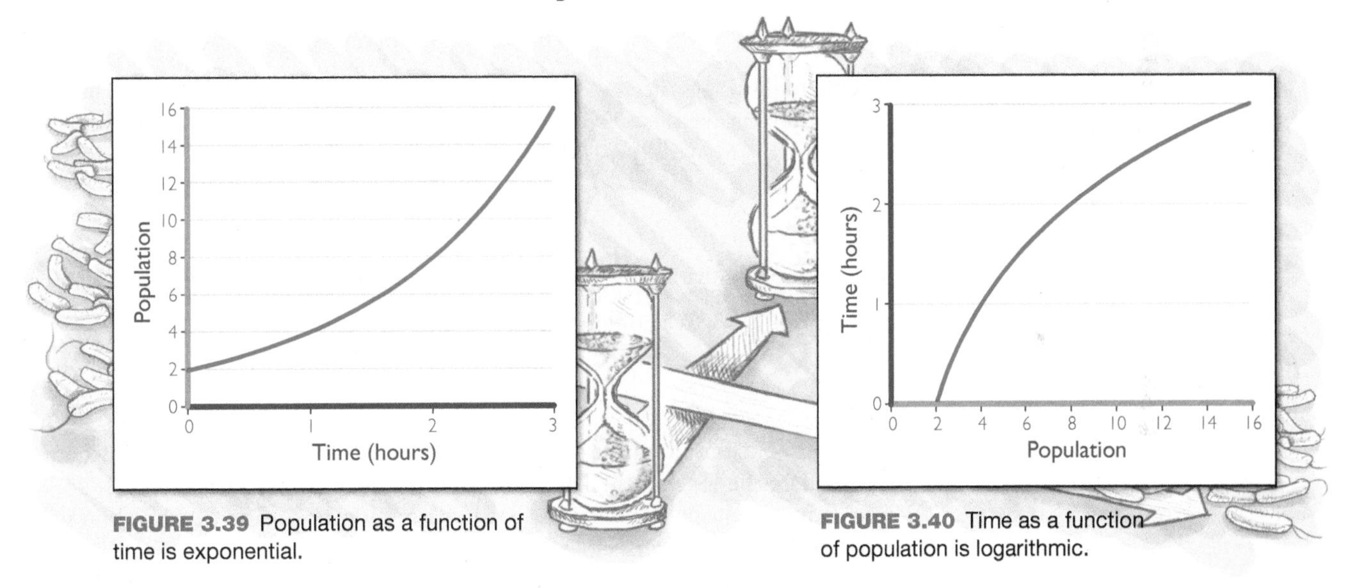

FIGURE 3.39 Population as a function of time is exponential.

FIGURE 3.40 Time as a function of population is logarithmic.

We have said that the logarithm reverses the action of an exponential function. Here is a more precise definition: The *common logarithm*[13] of a number, often shortened to just "the log," is the power of 10 that equals that number. For example, the log of 100 is 2 because $100 = 10^2$. There are other kinds of logarithms besides the common logarithm, but in this section, we will use the term *log* or *logarithm* to mean the common logarithm.

KEY CONCEPT

The common logarithm or log of a positive number x, written $\log x$, is the power to which 10 must be raised in order to equal x. Formally,

$$\log x = t \text{ if and only if } 10^t = x$$

Intuitively, we can think of $\log x$ as the *power of 10 that equals x.*

[13] *The common logarithm is also referred to as the "base 10 logarithm."*

Because the logarithmic equation $\log x = t$ means exactly the same thing as the exponential equation $x = 10^t$, we will often write both forms side by side to remind us of this fact. Thus, for example,

$$\log 10 = 1 \text{ because } 10^1 = 10\text{; that is, to get 10, we raise 10 to the power 1}$$
$$\log 100 = 2 \text{ because } 10^2 = 100\text{; that is, to get 100, we raise 10 to the power 2}$$
$$\log 1000 = 3 \text{ because } 10^3 = 1000\text{; that is, to get 1000, we raise 10 to the power 3}$$
$$\log \frac{1}{10} = -1 \text{ because } 10^{-1} = \frac{1}{10}\text{; that is, to get } \frac{1}{10}\text{, we raise 10 to the power } -1$$

For quantities that are not whole-number powers of 10, we use a scientific calculator or computer to find the logarithm.

EXAMPLE 3.17 Calculating logarithms: By hand and by calculator

What is the logarithm of

a. 1 million?

b. one thousandth?

c. 5?

SOLUTION

a. One million is written as a 1 followed by 6 zeros, so it equals 10^6. That is, to get one million we raise 10 to the power 6. This means that the logarithm of one million is 6. We express this fact with a formula as

$$\log 1{,}000{,}000 = \log 10^6 = 6$$

b. One thousandth is $\frac{1}{1000} = 0.001 = 10^{-3}$. That is, to get $\frac{1}{1000}$ we raise 10 to the power -3. We conclude that

$$\log \frac{1}{1000} = -3$$

c. Because 5 is not a whole-number power of 10, we use a scientific calculator to calculate the logarithm. The result to three decimal places is 0.699. It's interesting to check this answer by using a calculator to find $10^{0.699}$. You will see that the result is very close to 5.

TRY IT YOURSELF 3.17

What is the logarithm of 1 billion?

The answer is provided at the end of this section.

The Richter scale

Now that we know what the logarithm is, let's see how it is used in real-life examples of compressed scales. A familiar example is the Richter scale used in measuring the *magnitude* of earthquakes. Charles Richter and Beno Gutenburg pioneered this method in the 1930s.[14]

Direct measurement of earthquakes is done by *seismometers*, which record the *relative intensity* of ground movement. **Figure 3.41** shows a typical year of earthquakes measured by relative intensity. The graph gives the impression that there were only a few earthquakes over the period of a year, but in fact there were

[14] *There are other methods for measuring the magnitude of earthquakes. See Exercise 46 for one alternative method.*

many thousands. The difficulty in graphing relative intensity is that the scale is very broad, stretching from 0 to 140,000,000, so only the very largest earthquakes appear above the horizontal axis. This phenomenon is why Richter and Gutenburg used the logarithm to change the scale.

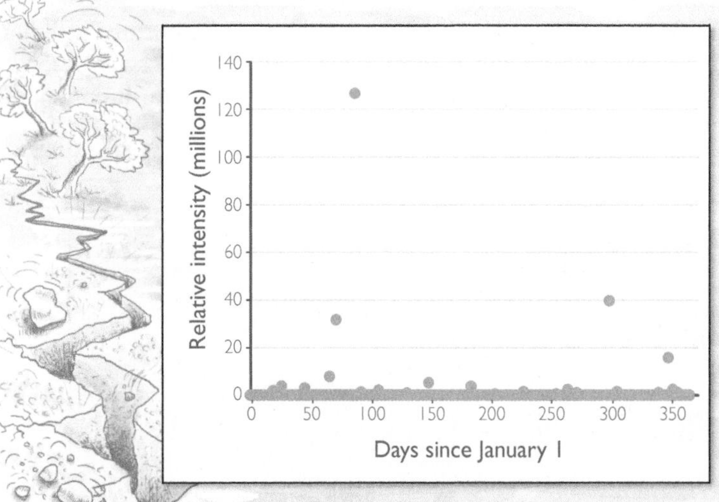

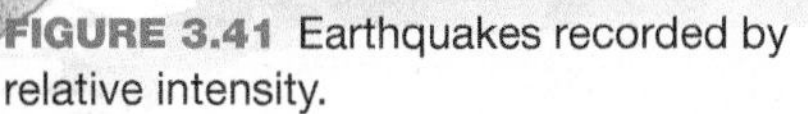

FIGURE 3.41 Earthquakes recorded by relative intensity.

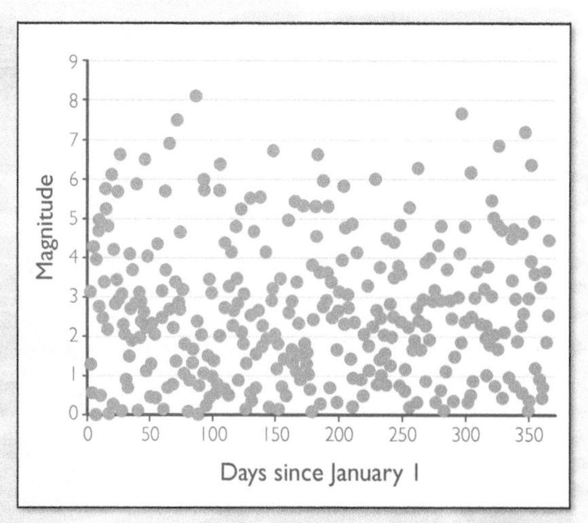

FIGURE 3.42 Earthquakes recorded by magnitude.

If we measure earthquakes by magnitude rather than relative intensity, we obtain the graph in **Figure 3.42**, which shows that many earthquakes occurred. The graphs in Figures 3.41 and 3.42 clearly show how compressing the scale results in a decompression of the data so that invisible features become visible.

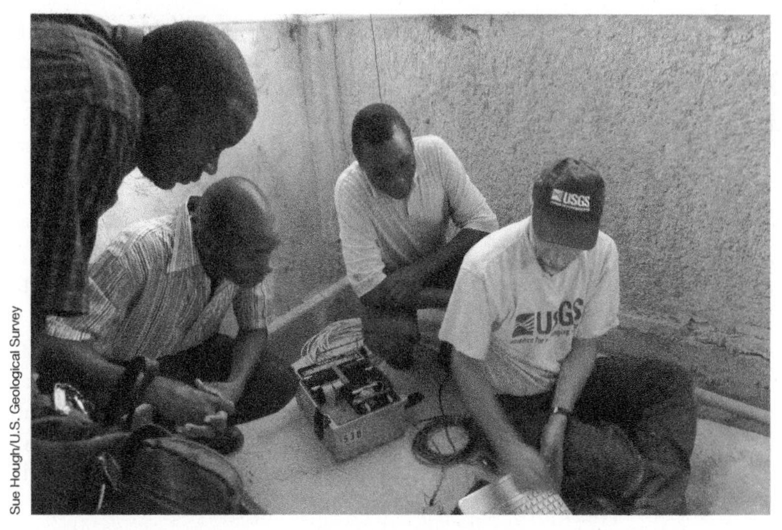

Sue Hough/U.S. Geological Survey

A seismometer being used in Haiti after the 2010 earthquake.

KEY CONCEPT

The relative intensity of an earthquake is a measurement of ground movement. The magnitude of an earthquake is the logarithm of relative intensity:

$$\text{Magnitude} = \log(\text{Relative intensity}), \quad \text{Relative intensity} = 10^{\text{Magnitude}}$$

An earthquake of magnitude 5.6 on the Richter scale would generally be considered moderate, and a 6.3-magnitude earthquake would be considered fairly strong. An earthquake with magnitude of 2.5 or less probably would go unnoticed by people in the area.

On March 11, 2011, an earthquake of magnitude 9.0 struck the Pacific Ocean off the coast of Japan. It caused a destructive tsunami; more than 10,000 people were killed. The most powerful shock ever recorded was the Great Chilean Earthquake of 1960, which registered 9.5 on the Richter scale. On December 26, 2004, an earthquake of magnitude 9.1 struck the Indian Ocean near Sumatra, Indonesia; the resulting tsunami killed about 230,000 people. The earthquake that struck Haiti on January 12, 2010, was as devastating in terms of loss of life but much smaller in size (see Exercise 45).

Walter Mooney/U.S. Geological Survey

Aftermath of the 2010 Haiti earthquake.

The following table provides more information on interpreting the Richter scale in practical terms. Note that a small increase in the magnitude yields dramatic effects. That is one result of using a compressed scale.

Richter magnitude	**Effects**	**Estimated number of earthquakes per year**
Less than 2.0	Not felt	Continual
2.0 to 2.9	Generally not felt, but recorded	1.3 million
3.0 to 3.9	Often felt, but rarely cause damage	130,000
4.0 to 4.9	Noticeable shaking of indoor items. Significant damage unlikely	13,000
5.0 to 6.0	May cause major damage to poorly constructed buildings	1300
6.1 to 6.9	May cause lots of damage in populated areas	100
7.0 to 7.9	Major quake, serious damage	20
8.0 to 8.9	Can totally destroy communities	1
9.0 or higher	Rare, great quake. May cause major damage to areas as much as 1000 miles away	1 every 20 years

EXAMPLE 3.18 Calculating magnitude and relative intensity: Earthquakes

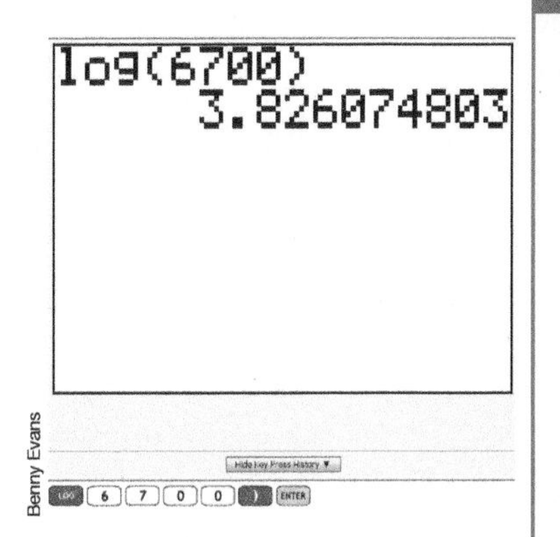

Benny Evans

If an earthquake has a relative intensity of 6700, what is its magnitude?

SOLUTION

The magnitude is the logarithm of relative intensity:

$$\text{Magnitude} = \log(\text{Relative intensity})$$
$$= \log 6700$$

Using a calculator, we find the magnitude to be 3.8. (The magnitude is usually rounded to one decimal place as we have done here.)

TRY IT YOURSELF 3.18

What is the relative intensity of an earthquake of magnitude 3.4? Round your answer to the nearest whole number.

The answer is provided at the end of this section.

Interpreting change on the Richter scale

To interpret the effect of a change in magnitude, we apply what we have learned about the growth of exponential functions. Recall Summary 3.3 from the preceding section: If y is an exponential function of t, increasing t by 1 unit causes y to be multiplied by the base. If we apply this fact to the exponential relationship

$$\text{Relative intensity} = 10^{\text{Magnitude}}$$

which has a base of 10, we see that a 1-point increase in magnitude corresponds to multiplying the relative intensity by 10. More generally, increasing the magnitude by t points multiplies relative intensity by 10^t. For example, a magnitude 6.3 quake is 10 times as intense as a magnitude 5.3 quake, and a magnitude 7.3 quake is $10^2 = 100$ times as intense as a magnitude 5.3 quake. This shows why, as noted in the preceding table, a relatively small increase on the Richter scale corresponds to a large increase in the relative intensity of a quake.

SUMMARY 3.7
Meaning of Magnitude Changes

- An increase of 1 unit on the Richter scale corresponds to increasing the relative intensity by a factor of 10.
- An increase of t units in magnitude corresponds to increasing the relative intensity by a factor of 10^t.

This relationship lets us compare some historical earthquakes.

EXAMPLE 3.19 Interpreting magnitude changes: Comparing some important earthquakes

a. In 2011 an earthquake measuring 5.6 on the Richter scale occurred near Prague, Oklahoma. The quake was thought to be related to wastewater injection wells nearby.[15] In 2012 a magnitude 8.6 earthquake occurred off the coast of Indonesia. How did the intensity of the two quakes compare?

b. In 1964 a magnitude 9.2 earthquake was recorded in Anchorage, Alaska. Compare the relative intensity of this quake with the magnitude 9.5 earthquake in 1960 near Valdivia, Chile.

[15] *Wastewater is a result of the process of hydraulic fracturing.*

SOLUTION

a. The Indonesian quake was $8.6 - 5.6 = 3.0$ points higher on the Richter scale. Increasing the magnitude by 3 points multiplies the relative intensity by 10^3. Thus, the Indonesian quake was 1000 times as intense as the quake in Oklahoma.

b. The Chilean quake was $9.5 - 9.2 = 0.3$ point higher on the Richter scale. Increasing the magnitude by 0.3 point multiplies the relative intensity by $10^{0.3}$. Using a calculator, and rounding to two decimal places, we find that $10^{0.3}$ is about 2.0. Thus, the Chilean quake was about twice as intense as the quake in Alaska. This example shows how a small change in magnitude indicates a much more powerful earthquake.

TRY IT YOURSELF 3.19

In May 2008, an earthquake struck Sichuan, China. It had a magnitude of 8.0 and reportedly killed 70,000 people. In 1994 an earthquake measuring 6.7 occurred in Northridge, California. How did the intensities of the two quakes compare?

The answer is provided at the end of this section.

The decibel as a measure of sound

The human ear can hear sounds over a huge range of relative intensities—from rustling leaves to howling jet engines. Furthermore, the brain perceives large changes in sound intensity as smaller changes in loudness. Hence, it is useful to have a scale that measures sound like the human brain does. This is the *decibel* scale, abbreviated dB.

KEY CONCEPT

The decibel rating of a sound is 10 times the logarithm of its relative intensity:[16]

$$\text{Decibels} = 10 \log(\text{Relative intensity})$$

The corresponding exponential equation is

$$\text{Relative intensity} = 10^{0.1 \times \text{Decibels}}$$

This is used in the approximate form

$$\text{Relative intensity} = 1.26^{\text{Decibels}}$$

Decibel readings are given as whole numbers.

If a sound doubles in intensity, we do not hear it as being twice as loud. As shown in the following table, a sound 100 times as intense as another sound is heard as 20 decibels louder. Record decibel readings at rock concerts and football stadiums exceed 135. Physicians warn that sustained exposure to noise levels over 85 decibels can lead to hearing impairment.

[16] *The relative intensity of a sound is a measure of the sound power per unit area as compared to the power per unit area of a barely audible sound. Physicists use 10^{-12} watts per square meter as this base intensity.*

The following table lists the decibel reading and relative intensity of some familiar sounds.[17] It illustrates the compressed nature of the decibel scale.

Sound	Decibels	Relative intensity
Threshold of audibility	0	1
Rustling leaves	10	10
Whisper	20	100
Normal conversation	60	1,000,000
Busy street traffic	70	10,000,000
Vacuum cleaner	80	10^8
Large orchestra	100	10^{10}
Front row at rock concert	110	10^{11}
Pain threshold	130	10^{13}
Jet takeoff	140	10^{14}
Perforation of eardrum	160	10^{16}

The graph in **Figure 3.43** shows the decibel reading in terms of the relative intensity. This graph shows how our brains actually perceive loudness. This means that a unit increase in the relative intensity has more effect on our perception of soft sounds than it does on our perception of louder sounds. Our brains can distinguish small differences in soft sounds but find the same difference almost indistinguishable for louder sounds. That explains why doubling the relative intensity does not double our perception of a sound.

Because relative intensity is an exponential function of decibels, we can determine what happens to relative intensity when decibels increase. The base of the exponential equation

$$\text{Relative intensity} = 1.26^{\text{Decibel}}$$

is 1.26. Hence, an increase of one decibel multiplies relative intensity by 1.26, and an increase of t decibels multiplies relative intensity by 1.26^t.

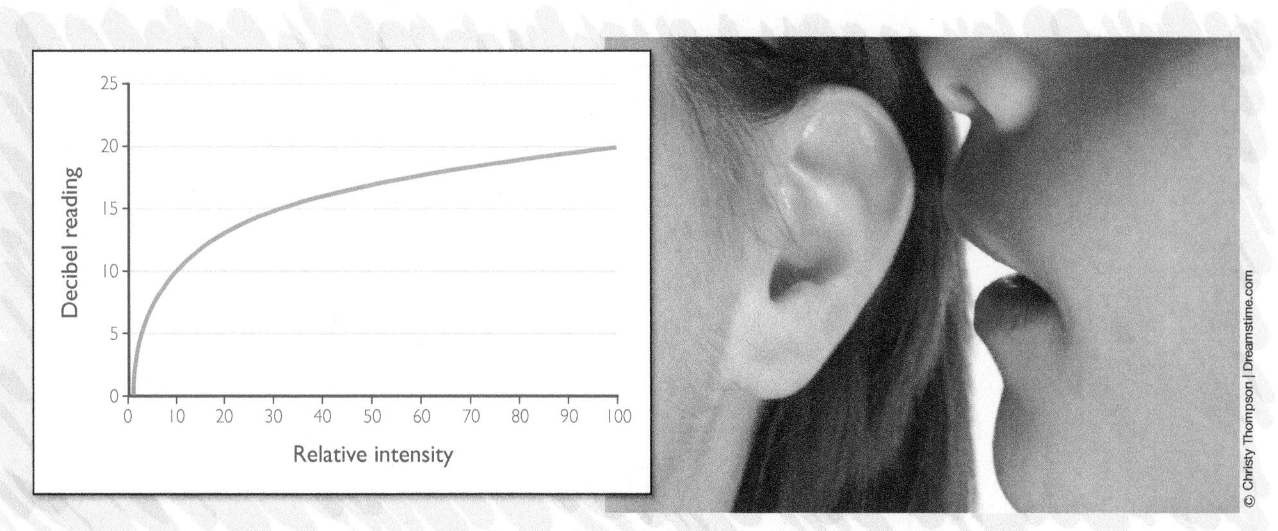

FIGURE 3.43 Left: Decibel reading in terms of relative intensity. Right: A whisper is about 20 decibels.

[17] *The loudness of a sound decreases as we move away from the source. See Exercise 28 for a discussion.*

SUMMARY 3.8
Increasing Decibels

- An increase of one decibel multiplies relative intensity by 1.26.
- An increase of t decibels multiplies relative intensity by 1.26^t.

EXAMPLE 3.20 Interpreting decibel changes: Vacuum cleaner to bulldozer

According to the preceding table, the sound from a vacuum cleaner is about 80 decibels. An idling bulldozer produces a sound that is about 85 decibels. How does the relative intensity of the idling bulldozer compare with that of the vacuum cleaner?

SOLUTION

According to Summary 3.8, an increase of t decibels multiplies relative intensity by 1.26^t. Therefore, increasing the number of decibels by five multiplies the intensity by 1.26^5 or about 3.2. Thus, the sound from the bulldozer is about 3.2 times as intense as that of the vacuum cleaner.

TRY IT YOURSELF 3.20

When a bulldozer is working, the noise can reach 93 decibels. How does the intensity of the sound of the working bulldozer compare with that of the vacuum cleaner? Round your answer to the nearest whole number.

The answer is provided at the end of this section.

Now we see in a practical example how doubling the intensity affects the loudness.

EXAMPLE 3.21 Calculating doubled intensity: Adding a speaker

Suppose we have a stereo speaker playing music at 60 decibels. What decibel reading would be expected if we added a second speaker?

SOLUTION

One might expect the music to be twice as loud, but we have doubled the *intensity* of the sound, not the loudness. The relative intensity of a 60-decibel speaker is

$$\begin{aligned}\text{Relative intensity} &= 1.26^{\text{Decibels}} \\ &= 1.26^{60}\end{aligned}$$

With the second speaker added, the new relative intensity is doubled:

$$\text{New relative intensity} = 2 \times 1.26^{60}$$

Therefore, the decibel reading of the pair of speakers is given by

$$\begin{aligned}\text{Decibels} &= 10 \log(\text{Relative intensity}) \\ &= 10 \log(2 \times 1.26^{60})\end{aligned}$$

This is about 63 decibels.

Does adding a second speaker double the sound?

The conclusion of this example may seem counter intuitive: Adding a second 60-decibel speaker does not double the decibel level, which is the perceived loudness. Far from doubling loudness, adding another speaker only increases the decibel reading from 60 to 63. This interesting phenomenon is roughly true of all five human senses, including sight, and is known as *Fechner's law*.[18] This law states that psychological sensation is a logarithmic function of physical stimulus.

Other common logarithmic phenomena, such as apparent brightness of stars and acidity of solutions, will be introduced in the exercises.

Solving exponential equations

When we introduced the logarithm, we said that it reversed the effect of an exponential function. This means that logarithms allow us to find solutions to exponential equations—solutions that we could not otherwise find. To proceed, we first need to be aware of some basic rules of logarithms. We state them here without proof. Derivations can be found in most books on college algebra.

Properties of Logarithms

Logarithm rule 1: $\log(A^t) = t\log(A)$

Logarithm rule 2: $\log(AB) = \log(A) + \log(B)$

Logarithm rule 3: $\log\left(\frac{A}{B}\right) = \log(A) - \log(B)$

Now let's return to the problem of population growth to illustrate how logarithms are used to solve an equation.

Suppose we have a population that is initially 500 and grows at a rate of 0.5% per month. How long will it take for the population to reach 800?

Because the population is growing at a constant percentage rate, it is an exponential function. The monthly percentage growth rate as a decimal is $r = 0.005$. Hence, the base of the exponential function is $1 + r = 1.005$. Therefore, the population size N after t months is given by

$$\begin{aligned} N &= \text{Initial value} \times (1 + r)^t \\ &= 500 \times 1.005^t \end{aligned}$$

[18] *Gustav T. Fechner lived from 1801 to 1887.*

To find out when the population N reaches 800, we need to solve the equation

$$800 = 500 \times 1.005^t$$

We divide both sides by 500 to simplify:

$$800 = 500 \times 1.005^t$$
$$\frac{800}{500} = \frac{500}{500} \times 1.005^t$$
$$1.6 = 1.005^t$$

To complete the solution, we need to find the unknown exponent t. This is where logarithms play a crucial role. We apply the logarithm function to both sides:

$$\log 1.6 = \log(1.005^t)$$

Now, according to logarithm rule 1,

$$\log(1.005^t) = t \log 1.005$$

Therefore,

$$\log 1.6 = t \log 1.005$$

Finally, dividing by log 1.005 gives

$$t = \frac{\log 1.6}{\log 1.005}$$

To complete the calculation requires a scientific calculator, which gives about 94.2 months for t. Thus, the population reaches 800 in about seven years and 10 months.

This method of solution applies to any equation of the form $A = B^t$.

SUMMARY 3.9 Solving Exponential Equations

The solution for t of the exponential equation $A = B^t$ is

$$t = \frac{\log A}{\log B}$$

EXAMPLE 3.22 Solving exponential equations: Growth of an investment

An investment is initially \$5000 and grows by 10% each year. How long will it take the account balance to reach \$20,000? Round your answer in years to one decimal place.

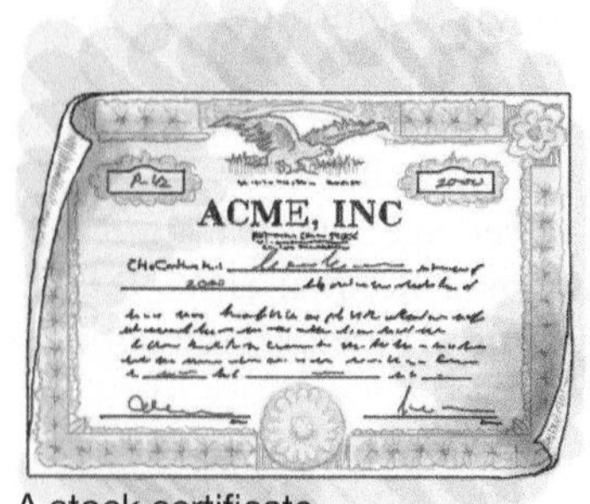

A stock certificate.

SOLUTION

The balance is growing at a constant percentage rate, so it is an exponential function. Because 10% as a decimal is $r = 0.1$, the base is $1 + r = 1.1$. The initial investment is \$5000, so we find the balance B (in dollars) after t years using

$$B = \text{Initial value} \times (1 + r)^t$$
$$B = 5000 \times 1.1^t$$

To find when the balance is \$20,000, we need to solve the equation

$$20{,}000 = 5000 \times 1.1^t$$

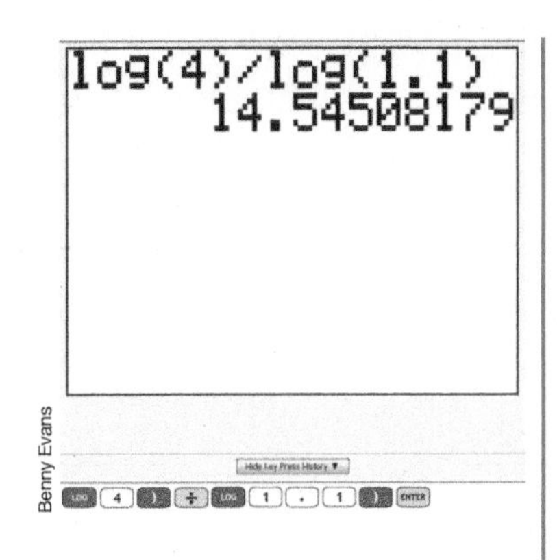

We first divide both sides by 5000 to get

$$\frac{20{,}000}{5000} = \frac{5000}{5000} \times 1.1^t$$
$$4 = 1.1^t$$

This is an exponential equation of the form $A = B^t$ with $A = 4$ and $B = 1.1$. We use Summary 3.9 to find the solution:

$$t = \frac{\log A}{\log B} = \frac{\log 4}{\log 1.1}$$

This is about 14.5 years.

TRY IT YOURSELF 3.22

An investment is initially \$8000 and grows by 3% each month. How long does it take for the value of the investment to reach \$10,000? Round your answer in months to one decimal place.

The answer is provided at the end of this section.

Doubling time and more

It is a fact that when a quantity grows exponentially, it will eventually double in size. (See Exercise 16 for more information.) Further, the doubling time depends only on the base—it does not depend on the initial amount. Doubling occurs over and over, and always over the same time span. Suppose we want to find how long it takes an exponential function to double. Some may find the algebra confusing, so we show how to do it with the specific example

$$y = 100 \times 3^t$$

side by side with the general case

$$y = \text{Initial value} \times \text{Base}^t$$

For the specific example, we want to find t when $y = 200$. For the general case, we want to find t when $y = 2\times$ Initial value. The calculation is the same for both:

Special case	Action	General case
$200 = 100 \times 3^t$	Solve this equation for t.	$2 \times \text{Initial value} = \text{Initial value} \times \text{Base}$
$2 = 3^t$	Divide by the initial value.	$2 = \text{Base}^t$
$\text{Doubling time} = t = \dfrac{\log 2}{\log 3}$	Use Summary 3.9 to solve.	$\text{Doubling time} = t = \dfrac{\log 2}{\log(\text{Base})}$

A similar formula applies for any multiple of the initial value.

SUMMARY 3.10
Doubling Time and More

Suppose a quantity grows as an exponential function with a given base. The time t required to multiply the initial value by K is

$$t = \frac{\log K}{\log(\text{Base})}$$

The special case $K = 2$ gives the doubling time:

$$\text{Doubling time} = \frac{\log 2}{\log(\text{Base})}$$

EXAMPLE 3.23 Finding doubling time: An investment

Suppose an investment is growing by 7% each year. How long does it take the investment to double in value? Round your answer in years to one decimal place.

SOLUTION

Because there is constant percentage growth, the balance is an exponential function. The base is $1 + r = 1.07$. To find the doubling time, we use the formula in Summary 3.10 with Base = 1.07:

$$\text{Doubling time} = \frac{\log 2}{\log(\text{Base})} = \frac{\log 2}{\log 1.07}$$

This is about 10.2 years.

TRY IT YOURSELF 3.23

An investment is growing by 2% per month. How long does it take for the investment to double in value? Round your answer in months to one decimal place.

The answer is provided at the end of this section.

Investments may grow exponentially, but radioactive substances decay exponentially, as we saw in Section 3.2. There, we discussed at some length the half-life of radioactive substances. The first formula in Summary 3.10 can be used to answer questions about radioactive decay as well as growth of investments. The next example does this.

EXAMPLE 3.24 Radiocarbon dating

In Example 3.14, we discussed how carbon-14, with a half-life of 5770 years, is used to date objects. Recall that the amount of a radioactive material decays exponentially with time, and the base of that exponential function is 1/2 if time is measured in half-lives.

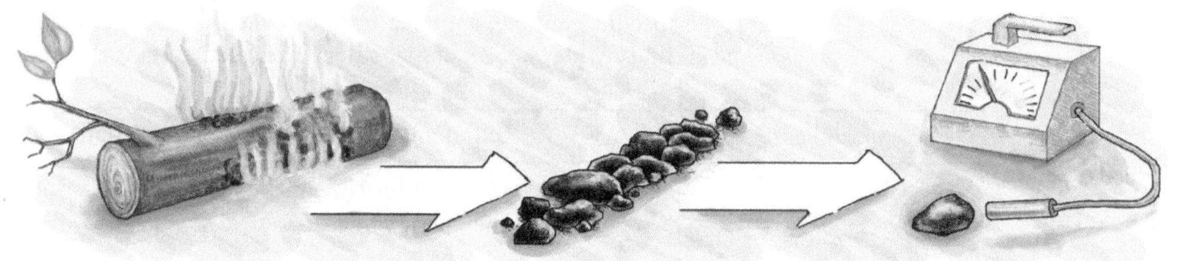

Living organisms absorb carbon-14 (radiocarbon) during their lifetimes. *Radiocarbon decays at a known rate. Paleontologists are able to determine the age of charcoal by measuring the amount of carbon-14 it contains.* *Carbon-14 decays into nitrogen-14, emitting an electron. A radiation counter records the number of electrons emitted when a small piece of the charcoal is burned.*

Suppose the charcoal from an ancient campfire is found to contain only one-third of the carbon-14 of a living tree. How long ago did the tree that was the source of the charcoal die? Give the answer first in half-lives and then in years rounded to the nearest hundred.

SOLUTION

The amount of carbon-14 present is an exponential function of time in half-lives, and the base is 1/2. Because we want to know when only one-third of the original amount is left, we use $K = 1/3$ in the formula from Summary 3.10:

$$\text{Time to multiply by } 1/3 \text{ is } t = \frac{\log K}{\log(\text{Base})} = \frac{\log(1/3)}{\log(1/2)}$$

This is about 1.58, so the time required is 1.58 half-lives. Because each half-life is 5770 years, 1.58 half-lives is

$$1.58 \times 5770 = 9116.6 \text{ years}$$

Thus, the tree died about 9100 years ago.

WHAT DO YOU THINK?

Why use logarithmic scales? We use logarithms to measure earthquakes, the loudness of sound, acidity, and apparent brightness of stars. Explain why logarithmic scales are used to measure these and other quantities.

Questions of this sort may invite many correct responses. We present one possibility: We use the logarithm to measure quantities that have a very broad range. Earthquake intensity is a prime example. Relative intensity varies from 0 up to a billion or more with almost all in the range of 1000 or less. A graphical display of relative intensity would hide almost all earthquakes. The logarithm compresses this scale so that differences in smaller intensities become apparent, but we can still grasp the intensity of larger quakes. The same is true of other phenomena such as intensities of light or sound. It seems to be the case that the human brain performs a logarithmic adjustment for light and sound. Thus, doubled light intensity does not appear to us to be doubly bright. Scales such as the logarithmic decibel scale accurately reflect how we perceive loudness.

Further questions of this type may be found in the *What Do You Think?* section of exercises.

Try It Yourself answers

Try It Yourself 3.17: Calculating logarithms: By hand and by calculator The logarithm of 1 billion is 9.

Try It Yourself 3.18: Calculating magnitude and relative intensity: Earthquakes $10^{3.4}$ is about 2512.

Try It Yourself 3.19: Interpreting magnitude changes: Comparing some important earthquakes The Sichuan quake was about 20 times as intense as the Northridge quake.

Try It Yourself 3.20: Interpreting decibel changes: Vacuum cleaner to bulldozer The sound from the working bulldozer is about 20 times as intense as the sound of the vacuum cleaner.

Try It Yourself 3.22: Solving exponential equations: Growth of an investment 7.5 months.

Try It Yourself 3.23: Finding doubling time: An investment 35.0 months.

Exercise Set 3.3

Test Your Understanding

1. In words, the logarithm of x is ________.

2. The logarithmic equation $\log x = t$ is equivalent to the exponential equation ________.

3. True or false: The logarithm is used to define compressed scales, resulting in a decompression of the data.

4. The magnitude of an earthquake is related to the relative intensity by the formula **a.** Magnitude = log(Relative intensity) **b.** Magnitude = 10log(Relative intensity) **c.** Magnitude = Relative intensity **d.** Magnitude = $10^{\text{Relative intensity}}$.

5. True or false: A relatively large increase on the Richter scale corresponds to a small increase in the relative intensity of an earthquake.

6. Decibels are related to the relative intensity of a sound by the formula **a.** Decibels = log(Relative intensity) **b.** Decibels = 10log(Relative intensity) **c.** Decibels = Relative intensity **d.** Decibels = $10^{\text{Relative intensity}}$.

Problems

7. Calculating logarithms by hand. Without the aid of a calculator, find the logarithm of:

a. 10,000

b. 1 billion

c. one-tenth

8. Using a calculator to find logarithms. Use a calculator or computer to find the values of the following. Round your answer to two decimal places.

a. log 7

b. log 70

c. log 700

9. Finding the magnitude. If the relative intensity of an earthquake is 7450, find the magnitude (rounded to one decimal place).

10. Finding the relative intensity. If the magnitude of an earthquake is 4.1, find the relative intensity (rounded to the nearest whole number).

11. Decibel changes. If an adjustment raises the the sound on your television by 6 decibels, how is the relative intensity of the sound affected? Round your answer to one decimal place.

12. Solving exponential equations. The value of a certain account after t years is given by $B = 300 \times 1.3^t$ dollars. How long will it take for the value of the account to increase to \$500? Round your answer to one decimal place.

13. Tripling time. A certain population is growing according to the formula $N = 1350 \times 1.05^t$. What is the tripling time for this population? Round your answer to one decimal place.

14. Radiocarbon dating. A support pole for an ancient structure is found to have one-fifth of the carbon-14 of a modern, living tree. How long ago did the tree used to make the support pole die? Give your answer to the nearest whole number.

15. Log of a negative number? Use your calculator to attempt to find the log of -1. What does your calculator say? Explain why, according to our definition, the logarithm of a negative number does not exist.

16. Not doubling. In connection with the discussion of doubling times, we stated that when a quantity grows exponentially, it will eventually double in size. This property does not hold for all increasing functions. Draw a graph that starts at the point $x = 1, y = 1$ and grows forever but for which y never doubles, that is, y never reaches 2.

17. Kansas and California. In 1812, a magnitude 7.1 earthquake struck Ventura, California. In 1867, a magnitude 5.1 quake struck Manhattan, Kansas. How many times as intense as the Kansas quake was the California quake?

18. Ohio earthquake. In 1986, an earthquake measuring 5.0 on the Richter scale hit northeast Ohio. How many times as intense as a magnitude 2.0 quake was it?

19. Alaska and California. In 1987, a magnitude 7.9 quake hit the Gulf of Alaska. In 1992, a magnitude 7.6 quake hit Landers, California. How do the relative intensities of the two quakes compare?

20. New Madrid quake. On December 16, 1811, an earthquake occurred near New Madrid, Missouri, that temporarily reversed the course of the Mississippi River. The quake had a magnitude of 8.8. On October 17, 1989, a calamitous quake measuring 7.1 on the Richter scale occurred in San Francisco during a World Series baseball game on live TV. How many times as intense as the San Francisco quake was the New Madrid quake?

21–23. Energy of earthquakes. *The energy released by an earthquake is related to the magnitude by an exponential function. The formula is*

$$\text{Energy} = 25{,}000 \times 31.6^{\text{Magnitude}}$$

The unit of energy in this equation is a joule. *One joule is approximately the energy expended in lifting 3/4 of a pound 1 foot. Exercises 21 through 23 refer to this formula.*

21. On April 19, 1906, San Francisco was devastated by a 7.9-magnitude earthquake. How much energy was released by the San Francisco earthquake?

22. Use the fact that energy is an exponential function of magnitude to determine how a 1-unit increase in magnitude affects the energy released by an earthquake.

23. We noted earlier that on December 26, 2004, an earthquake of magnitude 9.1 struck the Indian Ocean near Sumatra, Indonesia. How did the energy released by the Indonesian quake compare with that of the San Francisco quake of Exercise 21?

24. Jet engines. A jet engine close up produces sound at 155 decibels. What is the decibel reading of a pair of nearby jet engines? In light of the table on page 185, what does your answer suggest about those who work on the tarmac to support jet aircraft?

25. Stereo speakers. A speaker is playing music at 80 decibels. A second speaker playing the same music at the same decibel reading is placed beside the first. What is the decibel reading of the pair of speakers?

26. More on stereo speakers. Consider the lone stereo speaker in Exercise 25. Someone has the none-too-clever idea of doubling the decibel level. Would 100 of these speakers do the job? How about 1000? You should see a pattern developing. If you don't see it yet, try 10,000 speakers. How many speakers are required to double the decibel level?

27. Adding two speakers. Suppose a speaker is playing music at 70 decibels. If two more identical speakers (playing the same music) are placed beside the first, what is the resulting decibel level of the sound?

28. Decibels and distance. The loudness of a sound decreases as the distance from the source increases. Doubling the distance from a sound source such as a vacuum cleaner multiplies the relative intensity of the sound by 1/4. If the sound from a vacuum cleaner has a reading of 80 decibels at a distance of 3 feet, what is the decibel reading at a distance of 6 feet?

29. Apparent brightness. The apparent brightness on Earth of a star is measured on the *magnitude* scale.[19] The apparent magnitude m of a star is defined by

$$m = 2.5 \log I$$

where I is the relative intensity. The accompanying exponential formula is

$$I = 2.51^m$$

To find the relative intensity, we divide the intensity of light from the star Vega by the intensity of light from the star we are studying, with both intensities measured on Earth. Thus, apparent magnitude is a logarithmic function of relative intensity. The star Fomalhaut has a relative intensity of 2.9, in the sense that on Earth light from Vega is 2.9 times as intense as light from Fomalhaut. What is the apparent magnitude of Fomalhaut? Round your answer to two decimal places. *Note*: The magnitude scale is perhaps the reverse of what you might think. The higher the magnitude, the dimmer the star. Some very bright stars have a negative apparent magnitude.

30. Finding relative brightness. *This is a continuation of Exercise 29.* A star has an apparent magnitude of 4. Use the exponential formula in Exercise 29 to find its relative intensity. Round your answer to one decimal place.

31. Acid rain. Acid rain can do serious damage to the environment as well as human health. Acidity is measured according to the *pH scale*. (For more information on the pH scale, see Exercise 32.) Pure water has a pH of 7, which is considered neutral. A pH less than 7 indicates an *acidic* solution, and a pH higher than 7 indicates a *basic* solution. (Lye and baking soda solutions are very basic; vinegar and lemon juice are acidic.) An increase of 1 on the pH scale causes acidity to be multiplied by a factor of 1/10. Equivalently, a decrease of 1 on the pH scale causes acidity to be multiplied by a factor of 10. If rainfall has a pH of 4, how many times as acidic as pure water is it? (It is worth noting that most acid rain in the United States has a pH of about 4.3.)

32. More on pH. Acidity is measured on the pH scale, as mentioned in Exercise 31. Here, pH stands for "potential of hydrogen." Here is how the scale works.

The acidity of a solution is determined by the concentration H of hydrogen ions.[20] The formula is

$$\text{pH} = -\log H$$

The accompanying exponential formula is

$$H = 0.1^{\text{pH}}$$

Lower pH values indicate a more acidic solution. Normal rainfall has a pH of 5.6. Rain in the eastern United States often has a pH level of 3.8. How many times as acidic as normal rain is this?

33. Saving for a computer. You have \$200 and wish to buy a computer. You find an investment that increases by 0.6% each month, and you put your \$200 into the account. When will the account enable you to purchase a computer costing \$500?

34–35. Spent fuel rods. *The half-life of cesium-137 is 30 years. Suppose that we start with 50 grams of cesium-137 in a storage pool. Exercises 34 and 35 refer to this cesium.*

34. Find a formula that gives the amount C of cesium-137 remaining after h half-lives.

35. How many half-lives will it take for there to be 10 grams of cesium-137 in the storage pool? (Round your answer to two decimal places.) How many years is that?

36. Cobalt-60. Cobalt-60 is subject to radioactive decay, and each year the amount present is reduced by 12.3%.

a. The amount of cobalt-60 present is an exponential function of time in years. What is the base of this exponential function?

b. What is the half-life of cobalt-60? Round your answer to one decimal place.

37. Inflation. Suppose that inflation is 2.5% per year. This means that the cost of an item increases by 2.5% each year. Suppose that a jacket cost \$100 in the year 2022.

a. Find a formula that gives the cost C in dollars of the jacket after t years.

b. How long will it take for the jacket to cost \$200? Round your answer to the nearest year.

c. How long will it take for the jacket to cost \$400? *Suggestion*: Once you have completed part b, you do not need a calculator for this exercise because you have calculated the doubling time for this exponential function.

38. World population. If the per capita growth rate of the world population continues to be what it was in the year 2020, the world population t years after July 1, 2020, will be

$$7.8 \times 1.011^t \text{ billion}$$

According to this formula, when will the world population reach 9 billion?

39. A poor investment. Suppose you make an investment of \$1000 that you are not allowed to cash in for 10 years. Unfortunately, the value of the investment decreases by 10% per year.

a. How much money will be left after the end of the 10-year term?

b. How long will it be before your investment decreases to half its original value? Round your answer to one decimal place.

40. Economic growth. The CIA *World Factbook*[21] estimates the economic performance of countries. One such figure is the *gross domestic product* (GDP), which is the annual market value of the goods and services produced in that country.

a. In 2015, the GDP of the United States was about 18.0 trillion dollars. It is predicted to grow by about 2% each year. What is the projected value of the GDP in 2030 (to the nearest trillion dollars)?

b. In 2015, the GDP of China was about 11.0 trillion dollars. It is predicted to grow by about 7% each year.

[19] *The notion of magnitude for stars goes back to the ancient Greeks, who grouped the stars into six magnitude classes, with the brightest stars being of the first magnitude. The use of the term "magnitude" for stars motivated Richter to use the same term in his scale for earthquakes.*

[20] *The concentration is measured in moles per liter of the solution.*

[21] www.cia.gov/the-world-factbook/

What is the projected value of the GDP in 2030 (to the nearest trillion dollars)? Use this prediction and your answer to part a to determine whether China will have a larger economy than the United States by the year 2030. If so, determine in what year the economy of China will reach the level projected in part a. (Round up to the next year.)

41. Negative decibels? Our discussion of decibels did not mention that they can be negative. On the Redmond, Washington campus of Microsoft is an *anechoic chamber* that, according to Guinness World Records, is the quietest place in the world. It measures −20.3 decibels.

a. What is the relative intensity of a sound measuring −1 decibels? Round your answer to one decimal place.

b. What is the relative intensity of a sound measuring −10 decibels? Round your answer to one decimal place.

c. What is the relative intensity of Microsoft's anechoic chamber? Round your answer to three decimal places.

d. Is it possible for the relative intensity to actually be zero?

42. Stadium noise. According to Guinness World Records, on September 29, 2014, the fans of the Kansas City Chiefs football team set a record for a crowd roar of 142.2 decibels at Arrowhead Stadium in Kansas City, Missouri. The Seattle Seahawks had previously held the record after their fans produced a roar of 137.6 decibels at CenturyLink Field in Seattle on December 2, 2013.

a. To what sound does the table on page 185 compare the record crowd roar?

b. How many times as intense as the noise in Seattle was the noise in Kansas City? Round your answer to the nearest whole number.

Exercises 43 and 44 are suitable for group work.

43. Yahoo stock graphs. If you go to Yahoo's finance site at http://finance.yahoo.com and look up a stock, you will see a chart, or graph, showing the price plotted over time. Among the chart settings is the option to set the chart scale to either logarithmic or linear. (Google has the same feature.) Selecting the logarithmic option yields a plot of the logarithm of the stock price (note the scale on the vertical axis). Try this with Starbucks' stock (symbol SBUX). Print a copy of the graphs of SBUX over a five-year period using the Linear option and using the Log option. Compare the graphs and discuss the differences. Pick out a couple of other stocks and do the same thing with them.

44. The logarithmic Dow. For graphs that have sharp jumps, applying the logarithm can dampen the jumps and give a clearer visual presentation of the data. As mentioned in Exercise 43, Yahoo and Google finance sites both allow this. **Figure 3.44** shows the logarithmic Dow from 1897 through 2019.[22]

a. In the late 1920s through the mid-1930s, there is a sharp drop. What historical event corresponds to this drop?

b. Lay a straightedge on the graph and comment on the hypothesis that the logarithmic Dow is almost linear.

c. If we agree that the logarithmic Dow is almost linear, then it can be shown that the Dow itself is growing exponentially. Explain what this would mean in the long term for a diversified stock portfolio.

[22] © G. *William Schwert, 2000–2019.* www.billschwert.com/dj.pdf

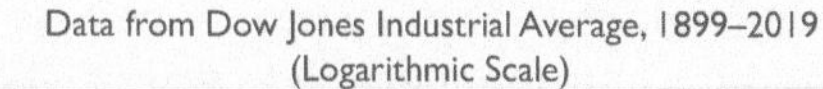

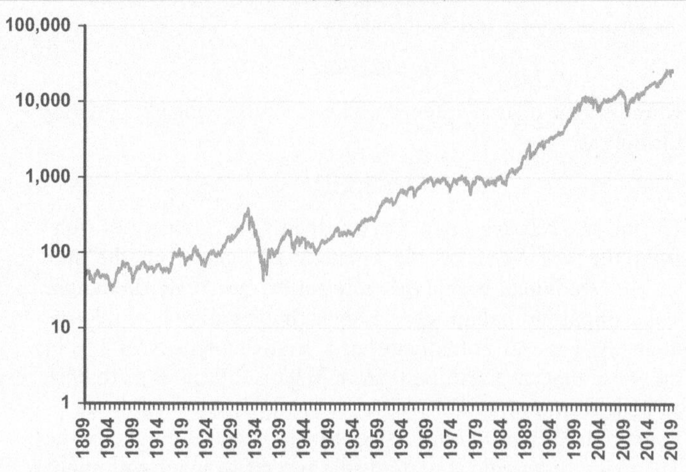

FIGURE 3.44 The logarithmic Dow.

Writing About Mathematics

45. Earthquake in Haiti. The earthquake that devastated Haiti on January 12, 2010, had a magnitude of 7.0 and killed tens of thousands of people. Six weeks later a quake of magnitude 8.8 struck southern Chile and killed about 500 people. Investigate why the much smaller earthquake in Haiti was so much more devastating.

46. Beyond Richter. There are other methods than the Richter scale for measuring the magnitude of earthquakes. One alternative is the *moment magnitude scale*. Investigate how this scale is defined. Does the definition involve a logarithm?

47. Historical figure: John Napier. John Napier is credited with inventing the idea of logarithms, but of a different kind than the common logs discussed here. Write a brief report on Napier's life and accomplishments.

48. Historical figure: Henry Briggs. Henry Briggs was an English mathematician notable for changing Napier's logarithm into what is called the common logarithm. Write a brief report on Briggs's life and accomplishments.

49. Historical figure: Gustav T. Fechner. Gustav T. Fechner was mentioned in connection with Fechner's law. Write a brief report on Fechner's life and accomplishments.

50. Historical figure: Alexander J. Ellis. The name of Alexander J. Ellis is associated with the *cent*, a unit of measure for musical intervals that involves the logarithm. Write a brief report on Ellis and his work.

What Do You Think?

51. Magnitude. You and a friend hear a report comparing two earthquakes, one with magnitude 6 and the other a much more intense quake with magnitude 7. Your friend says that scientists should stop using the magnitude scale to compare earthquakes because that scale makes significant differences seem insignificant. How would you explain to your friend why it is important to use the magnitude scale?

52. Compressed scales. In this section we have said that the Richter scale and other logarithmic scales are compressed.

Explain in a way that a fellow student could understand what is meant by a compressed scale, and use a couple of examples to show how the compression works in practice. **Exercise 22 is relevant to this topic.**

53. **Fechner's law.** In this section we stated Fechner's law, which says that psychological sensation is a logarithmic function of physical stimulus. For example, the eye perceives brightness as a logarithmic function of the intensity of light. Draw a graph of perceived brightness as a function of the intensity of light. Explain how the shape of the graph reflects the underlying logarithm. **Exercise 29 is relevant to this topic.**

54. **Logarithmic and exponential functions.** Explain the relationship between logarithmic and exponential functions. **Exercise 29 is relevant to this topic.**

55. **Spent fuel rods.** There is currently a great deal of controversy regarding the disposal of spent nuclear fuel rods. Why does this issue cause such controversy? Would you be willing to have a nuclear disposal site in your state? **Exercise 35 is relevant to this topic.**

56. **Damage from earthquakes.** In 2010, a 7.0-magnitude earthquake struck Haiti. While far from the most powerful of earthquakes, the Haiti quake devastated the island, taking a huge toll in terms of both damage to physical structures and loss of life. Do some research to determine the factors, in addition to the magnitude, that contribute to damage from earthquakes. Report on some moderate quakes, like the Haiti quake, that resulted in catastrophic damage. **Exercise 45 is relevant to this topic.**

57. **Measuring earthquakes.** Over the years scientists have used a number of scales, not just the Richter scale, to measure earthquakes. Investigate some of the other scales that have been used. Which are in current use? **Exercise 46 is relevant to this topic.**

58. **The logarithmic Dow.** The Dow Jones Industrial Average from 1897 through 2019 is graphed on a logarithmic scale in Figure 3.44. Use the Internet to find a graph of the Dow over the same period on an uncompressed (regular) scale. Compare the two graphs. For example, which graph shows the trends more clearly, and which shows significant economic events more clearly? **Exercise 44 is relevant to this topic.**

59. **A mathematical question.** In this section we have restricted our attention to the *common logarithm*, but there are in fact many other logarithms (just as there are many exponential functions). Mathematicians prefer the *natural logarithm*. On your calculator it is likely to be displayed as ln. Investigate the natural logarithm and why mathematicians prefer it.

3.4 Quadratics and parabolas

TAKE AWAY FROM THIS SECTION
Understand quadratic functions and the shape and uses of parabolas.

The following excerpt is taken from an article published by the American Society of Mechanical Engineers (ASME).

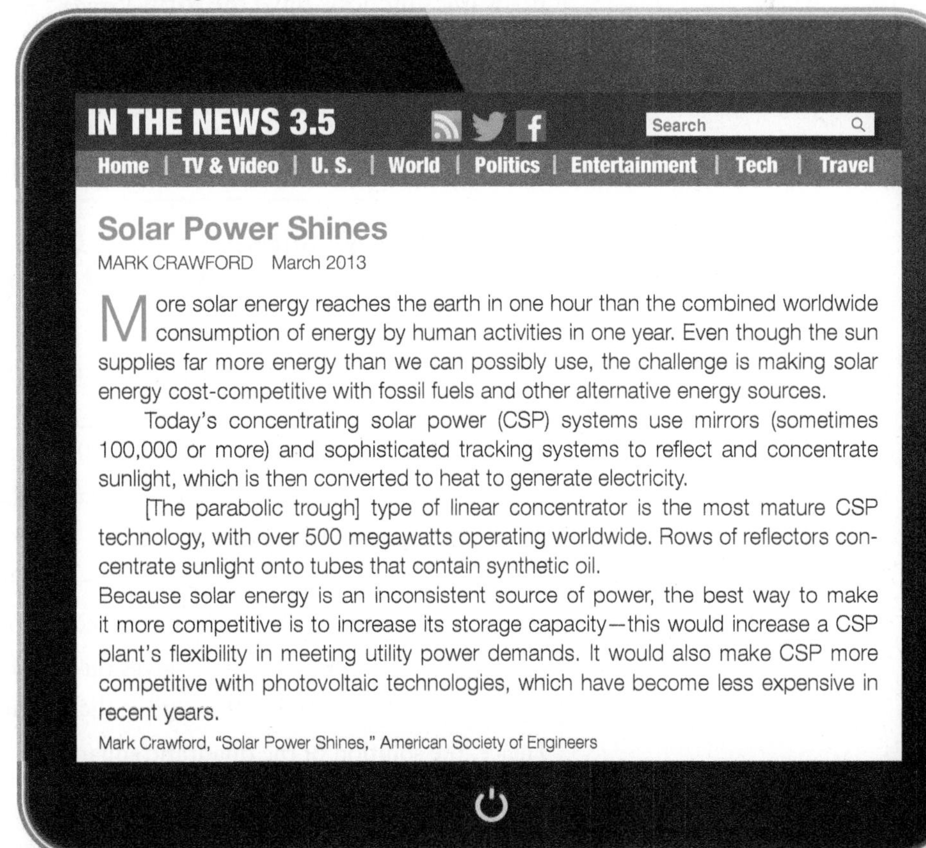

Solar Power Shines

MARK CRAWFORD March 2013

More solar energy reaches the earth in one hour than the combined worldwide consumption of energy by human activities in one year. Even though the sun supplies far more energy than we can possibly use, the challenge is making solar energy cost-competitive with fossil fuels and other alternative energy sources.

Today's concentrating solar power (CSP) systems use mirrors (sometimes 100,000 or more) and sophisticated tracking systems to reflect and concentrate sunlight, which is then converted to heat to generate electricity.

[The parabolic trough] type of linear concentrator is the most mature CSP technology, with over 500 megawatts operating worldwide. Rows of reflectors concentrate sunlight onto tubes that contain synthetic oil.

Because solar energy is an inconsistent source of power, the best way to make it more competitive is to increase its storage capacity—this would increase a CSP plant's flexibility in meeting utility power demands. It would also make CSP more competitive with photovoltaic technologies, which have become less expensive in recent years.

Mark Crawford, "Solar Power Shines," American Society of Engineers

Victor de Schwanberg/Science Source

Line focus water heater. This system uses curved mirrors to form a parabolic trough that focuses sunlight onto a tube (which contains water) running above the mirrors.

This article refers to parabolic trough collectors, which use curved mirrors to collect the energy of the sun in order to produce electricity (see the photo of the line focus water heater). These troughs have a *parabolic* shape because mirrors with that shape focus incoming sunlight at a single point. Parabolas are graphs of *quadratic functions*, and we start by looking at those functions.

Quadratic functions and their zeros

Quadratic functions, which may be familiar from high school mathematics, are functions that have the form $y = ax^2 + bx + c$ with $a \neq 0$. We are interested in solving *quadratic equations*, which have the form $ax^2 + bx + c = 0$.

KEY CONCEPT

Quadratic functions have the form $y = ax^2 + bx + c$.
Quadratic equations have the form $ax^2 + bx + c = 0$.

You might recall the process of solving quadratic equations by *factoring*. The method is based on the observation that if the product of two numbers is 0, then at least one of the factors must be 0. The following reminder is offered.

Quick Review Solving Quadratic Equations by Factoring

To solve the quadratic equation

$$ax^2 + bx + c = 0$$

we factor the left-hand side and set each factor equal to zero.

For example, to solve $x^2 - 5x + 6 = 0$, we factor to obtain

$$x^2 - 5x + 6 = (x - 2)(x - 3)$$

Thus, our equation becomes

$$(x - 2)(x - 3) = 0$$

We finish by setting each factor, in turn, equal to 0:

$x - 2 = 0$ gives the solution $x = 2$

$x - 3 = 0$ gives the solution $x = 3$

Hence, the solution of the quadratic equation is $x = 2$ or $x = 3$.

Some quadratic expressions are difficult to factor. For these, we need the *quadratic formula*.

KEY CONCEPT

To solve the quadratic equation $ax^2 + bx + c = 0$, we use the quadratic formula

$$x = \frac{-b \pm \sqrt{b^2 - 4ac}}{2a}$$

If $b^2 - 4ac < 0$, then there is no (real) solution.

A derivation of this formula is given at the end of this section in Algebraic Spotlight 3.1.

Suppose, for example, that we wish to solve the quadratic equation $x^2 - 3x - 2 = 0$. The factors are not apparent, so we use the quadratic formula with a, b, and c as indicated:

$$\overset{a}{1}\,x^2 \overset{b}{-3}\,x \overset{c}{-2} = 0$$

$$x = \frac{-\overset{b}{(-3)} \pm \sqrt{\overset{b^2}{(-3)^2} - \overset{4\times a\times c}{4 \times 1 \times (-2)}}}{\underset{2\times 1}{2\times a}}$$

$$= \frac{3 \pm \sqrt{9+8}}{2}$$

$$= \frac{3 \pm \sqrt{17}}{2}$$

It is worth emphasizing that the $\pm$ symbol means that there are two solutions:

$$x = \frac{3 + \sqrt{17}}{2} \quad \text{or} \quad x = \frac{3 - \sqrt{17}}{2}$$

These are the exact solutions. Using a calculator, we can see that they are about $x = 3.56$ or $x = -0.56$.

EXAMPLE 3.25 Solving quadratic equations: Factoring and using the quadratic formula

a. Solve $x^2 + 4x - 5 = 0$ by factoring.

b. Use the quadratic formula to solve $x^2 + 2x - 2 = 0$.

SOLUTION

a. The first step is to factor:

$$x^2 + 4x - 5 = (x+5)(x-1)$$

Rewriting the equation gives

$$(x+5)(x-1) = 0$$

Now we set each factor equal to 0:

$x + 5 = 0$ gives the solution $x = -5$

$x - 1 = 0$ gives the solution $x = 1$

Hence, the solution is $x = 1$ or $x = -5$.

b. The quadratic formula gives:

$$
\begin{aligned}
x^2 + 2x - 2 &= 0 \\
x &= \frac{-2 \pm \sqrt{2^2 - 4 \times 1 \times (-2)}}{2 \times 1} \\
&= \frac{-2 \pm \sqrt{12}}{2} \\
&= \frac{-2 \pm \sqrt{4 \times 3}}{2} \\
&= \frac{-2 \pm 2\sqrt{3}}{2} \\
&= -1 \pm \sqrt{3}
\end{aligned}
$$

Hence, the solution is $x = -1 + \sqrt{3}$ or $x = -1 - \sqrt{3}$.

TRY IT YOURSELF 3.25

a. Solve $x^2 + 6x - 27 = 0$ by factoring.

b. Use the quadratic formula to solve $x^2 + x - 1 = 0$.

The answer is provided at the end of this section.

There are many practical phenomena that are appropriately modeled by quadratic functions, and solving quadratic equations is an important skill for analysis of such models. This fact is illustrated in the following two examples.

EXAMPLE 3.26 Factoring: Throwing a rock

If a rock is thrown downward with an initial speed of 32 feet per second, the distance D, in feet, that the rock travels in t seconds is given by

$$D = 16t^2 + 32t$$

If the rock is thrown from the top of a tower that is 128 feet tall, how long does it take for the rock to strike the ground?

SOLUTION

We need to know how long it takes for the rock to travel 128 feet. That is, we need to solve the equation $16t^2 + 32t = 128$ or $16t^2 + 32t - 128 = 0$. This equation is easier to solve if we first divide both sides by 16 to get $t^2 + 2t - 8 = 0$. Now we factor the left-hand side to obtain

$$t^2 + 2t - 8 = (t + 4)(t - 2)$$

The equation to be solved becomes

$$(t + 4)(t - 2) = 0$$

We set each factor, in turn, equal to 0:

$$
\begin{array}{lll}
t + 4 = 0 & \text{gives the solution} & t = -4 \\
t - 2 = 0 & \text{gives the solution} & t = 2
\end{array}
$$

We discard the negative solution because it makes no sense in this context. We conclude that the rock hits the ground after 2 seconds.

TRY IT YOURSELF 3.26

Solve this problem if the tower is 48 feet tall.

The answer is provided at the end of this section.

Often in real settings, we encounter quadratic equations that cannot be solved easily by factoring. For these, we need the quadratic formula.

EXAMPLE 3.27 Using the quadratic formula: Stable population levels

For a certain population, the growth rate G, in thousands of individuals per year, depends on the size N, in thousands, of the population. The relation is

$$G = 1 + N - 0.2N^2$$

The population level is *stable* when it is neither increasing nor decreasing—that is, when the growth rate is 0. At what level is the population stable?

SOLUTION

We need to find when $G = 0$, which says $1 + N - 0.2N^2 = 0$. This is a quadratic equation, but it is not written in the standard form. We rearrange it to get $-0.2N^2 + N + 1 = 0$. Now we use the quadratic formula to solve the equation:

$$-0.2N^2 + N + 1 = 0$$

$$N = \frac{-1 \pm \sqrt{1^2 - 4 \times (-0.2) \times 1}}{2 \times (-0.2)}$$

$$N = \frac{-1 \pm \sqrt{1.8}}{-0.4}$$

$$N = -0.85 \quad \text{or} \quad 5.85$$

Here, we have rounded to two decimal places. The negative solution makes no sense for a population size, so we discard it. We conclude that the population level is stable when there are 5.85 thousand individuals present.

TRY IT YOURSELF 3.27

At what level is the population stable if the growth rate is given by

$$G = 2 + 2N - 0.3N^2$$

Once again, assume that G is measured in thousands of individuals per year and N is measured in thousands.

The answer is provided at the end of this section.

Parabolas

The graph of a quadratic function has a special shape known as a *parabola*. Typical parabolas are shown in **Figures 3.45 and 3.46**. Note that for $y = ax^2 + bx + c$, the parabola opens upward when a is positive and downward when a is negative.

A key bit of information about a parabola is the location of its tip, called the *vertex*.

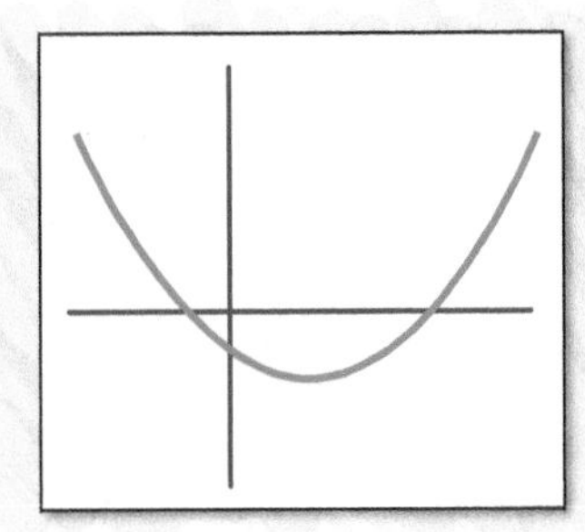

FIGURE 3.45 The graph $y = ax^2 + bx + c$ is an upward-opening parabola when a is positive.

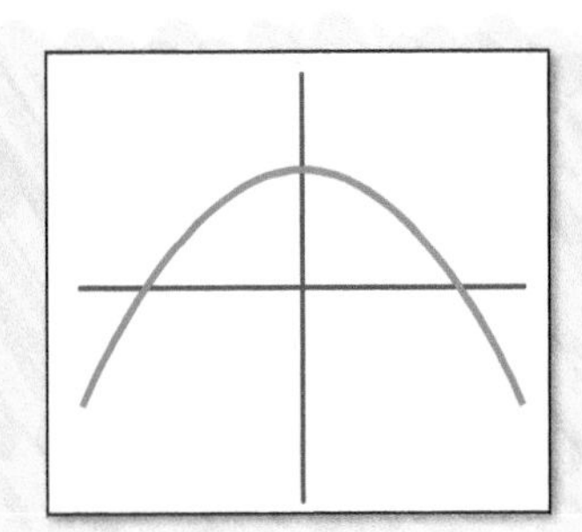

FIGURE 3.46 The graph $y = ax^2 + bx + c$ is a downward-opening parabola when a is negative.

KEY CONCEPT

The graph of a quadratic function is a parabola. The vertex of a parabola is located at

- the minimum point if the parabola opens upward
- the maximum point if the parabola opens downward

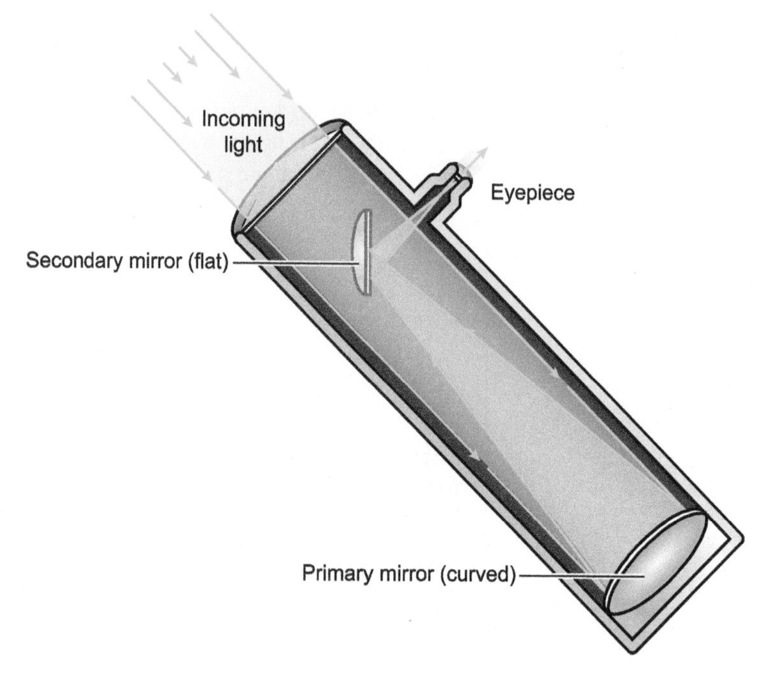

A reflector telescope employs a parabolic mirror.

We can use the symmetry of a parabola to locate the x-value of the vertex. We start with a typical upward-opening parabola, the top graph in **Figure 3.47**. It is the graph of $y = ax^2 + bx + c$ with a positive. If we remove the constant term c, the graph is shifted up or down (depending on the sign of c), but the x-value of the vertex does not change. This fact is illustrated in Figure 3.47.

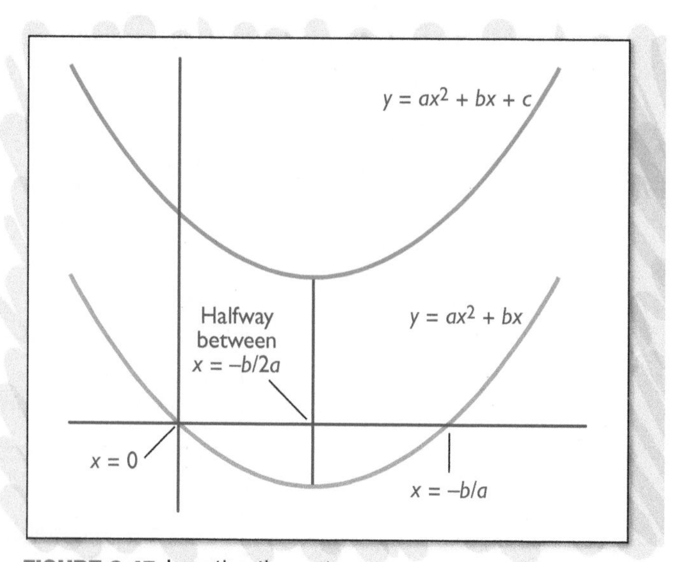

FIGURE 3.47 Locating the vertex.

Therefore, we may as well consider the graph of $y = ax^2 + bx$. Next we find where this graph crosses the horizontal axis. The solutions of $ax^2 + bx = 0$ are easy to find by factoring:

$$
\begin{aligned}
ax^2 + bx &= 0 \\
x(ax + b) &= 0 \\
x = 0 \quad &\text{or} \quad ax + b = 0 \\
x = 0 \quad &\text{or} \quad x = -\frac{b}{a}
\end{aligned}
$$

Finally, the symmetry of the parabola means that the x-value of the vertex is halfway between the two solutions, $x = 0$ and $x = -b/a$. Therefore, the x-value of the vertex is $x = \frac{1}{2} \times \left(-\frac{b}{a}\right) = -\frac{b}{2a}$. The same formula holds if the parabola opens downward.

SUMMARY 3.11
Vertex of a Parabola

The x-value of the vertex of the parabola $y = ax^2 + bx + c$ is

$$x = -\frac{b}{2a}$$

EXAMPLE 3.28 Finding the vertex: x- and y-values

Find both the x- and y-values for the vertex of $y = 3x^2 - 24x + 10$.

SOLUTION

We find the x-value by using the formula $x = -\frac{b}{2a}$. We put in 3 for a and -24 for b:

$$
\begin{aligned}
x\text{-value} &= -\frac{b}{2a} \\
&= -\frac{-24}{2 \times 3} \\
&= 4
\end{aligned}
$$

To find the y-value, we put the x-value of 4 in the function $y = 3x^2 - 24x + 10$:

$$y\text{-value} = 3 \times 4^2 - 24 \times 4 + 10 = -38$$

TRY IT YOURSELF 3.28

Find the x- and y-values of the vertex of $y = 2x^2 - 28x + 12$.

The answer is provided at the end of this section.

We can use our knowledge about the vertex of a parabola to solve some classic max-min problems.

EXAMPLE 3.29 A classic max-min problem: Constructing a pen

A rectangular pen is to be constructed using the side of a barn as one boundary and 100 feet of fence to make the other three sides. See **Figure 3.48.** Our goal is to find the length and width of the rectangle that encloses the largest area.

Here is our approach. In Figure 3.48, we have used x to denote the width of the pen and y to denote the length, both in feet. Because we are using 100 feet of fence, $y + 2x = 100$. Therefore, $y = 100 - 2x$. Now recall that the area of a rectangle is the width times the length:

$$\begin{aligned}\text{Area} &= \text{Width} \times \text{Length} \\ &= xy \\ &= x(100 - 2x) \text{ square feet}\end{aligned}$$

Use this information to find the dimensions of the rectangle with maximum area.

SOLUTION

We need to find the maximum value for Area $= x(100 - 2x) = -2x^2 + 100x$. The parabola $y = -2x^2 + 100x$ opens downward, so the maximum value for the area occurs at the vertex of the parabola. The x-value of the vertex is

$$\begin{aligned}x\text{-value} &= -\frac{b}{2a} \\ &= -\frac{100}{2 \times (-2)} \\ &= 25\end{aligned}$$

Therefore, to enclose the maximum area, we use $x = 25$ feet for the width. That leaves $100 - 2x = 100 - 2 \times 25 = 50$ feet for the length.

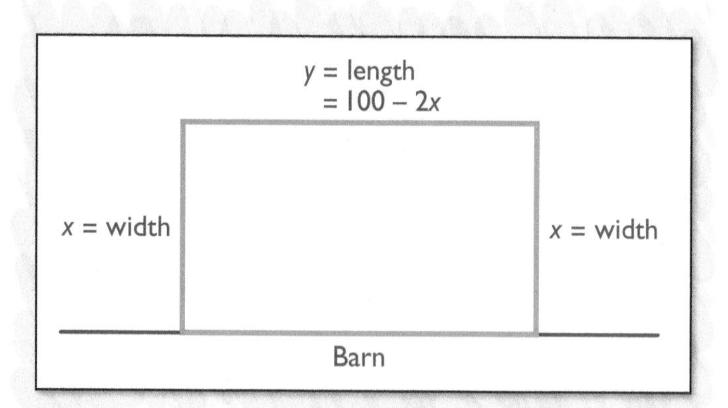

FIGURE 3.48 Building a pen next to a barn.

TRY IT YOURSELF 3.29

Solve this problem if 200 feet of fence are used.

The answer is provided at the end of this section.

ALGEBRAIC SPOTLIGHT 3.1 The Quadratic Formula

The derivation of the quadratic formula uses an algebraic method known as *completing the square*. See Exercises 34 and 35 for illustrations of key steps in the derivation.

We want to solve the quadratic equation $ax^2 + bx + c = 0$:

$$ax^2 + bx = -c$$ Move c to the right side.

$$x^2 + \frac{b}{a}x = -\frac{c}{a}$$ Divide by a.

$$x^2 + \frac{b}{a}x + \frac{b^2}{4a^2} = \frac{b^2}{4a^2} - \frac{c}{a}$$ Add $\frac{b^2}{4a^2}$ to each side in order to make the left side a perfect square.

$$\left(x + \frac{b}{2a}\right)^2 = \frac{b^2 - 4ac}{4a^2}$$ Write the left side as a square and combine the right into a single fraction.

$$x + \frac{b}{2a} = \frac{\pm\sqrt{b^2 - 4ac}}{2a}$$ Take the square root of both sides.

$$x = \frac{-b \pm \sqrt{b^2 - 4ac}}{2a}$$ Move $\frac{b}{2a}$ to the right side and combine fractions.

Thus, the solution of $ax^2 + bx + c = 0$ is $x = \dfrac{-b \pm \sqrt{b^2 - 4ac}}{2a}$.

WHAT DO YOU THINK?

Discriminant. The **discriminant** of the quadratic expression $ax^2 + bx + c$ is the part of the quadratic formula that is "under the square root" symbol:

$$b^2 - 4ac$$

It may be positive, negative, or zero. What can you say about the graph of $y = ax^2 + bx + c$ in each of these three cases? *Suggestion*: Think about the relationship between solutions given by the quadratic formula and places where the graph crosses the horizontal axis.

Questions of this sort may invite many correct responses. We present one possibility: The real solutions of the quadratic equation $a^2 + bx + c = 0$ occur at the points where the graph of $y = ax^2 + bx + c$ touches or crosses the horizontal axis. If the discriminant is negative, then the quadratic formula asks us to take the square root of a negative number. There is no real number that works for that. Hence, there are no real solutions of the quadratic equation. That means the graph does not cross the horizontal axis. Hence, it lies entirely above or entirely below the horizontal axis.

If the discriminant is 0, then the quadratic equation gives exactly one solution

$$x = \frac{-b}{2a}$$

In this instance the graph touches the horizontal exactly once, at its vertex.

When the discriminant is positive, the quadratic formula gives two solutions, so that the graph crosses the horizontal axis twice.

Further questions of this type may be found in the *What Do You Think?* section of exercises.

Try It Yourself answers

Try It Yourself 3.25: Solving quadratic equations: Factoring and using the quadratic formula

a. $x = 3$ or $x = -9$

b. $x = \dfrac{-1+\sqrt{5}}{2}$ or $x = \dfrac{-1-\sqrt{5}}{2}$, or about $x = 0.62$ or $x = -1.62$

Try It Yourself 3.26: Factoring: Throwing a rock The rock hits the ground after 1 second.

Try It Yourself 3.27: Using the quadratic formula: Stable population levels At a population level of 7.55 thousand individuals.

Try It Yourself 3.28: Finding the vertex: x- and y-values x-value $= 7$, y-value $= -86$

Try It Yourself 3.29: A classic max-min problem: Constructing a pen The width is 50 feet, and the length is 100 feet.

Exercise Set 3.4

Test Your Understanding

1. The graph of a quadratic function is a ______.

2. The formula that solves an equation of the form $ax^2 + bx + c = 0$ is called ______.

3. An equation of the form $ax^2 + bx + c = 0$ has: **a.** at least one solution **b.** at most one solution **c.** exactly two solutions **d.** at most two solutions.

4. The vertex of the parabola $y = ax^2 + bx + c$ has x-value ______.

5. When a is positive, the vertex of the parabola $y = ax^2 + bx + c$ occurs at: **a.** the maximum point **b.** the minimum point **c.** the origin **d.** a positive value.

Problems

6. **Life expectancy versus health-care spending.** One article shows a relationship between per capita health-care spending h (in thousands of dollars per person per year) and life expectancy E (in years).[23] The following formula is adapted from that article:

$$E = 72.23 + 3.85h - 0.37h^2$$

a. According to this model, what health-care expenditure corresponds to maximum longevity? Round your answer in thousands of dollars per person per year to two decimal places.

b. Use your answer to part a to determine the optimum life expectancy given by this model. (Round your answer to two decimal places.) We note that in fact per capita health-care spending is $8745 per year, and life expectancy is 78.7 years.

7. **Practice with solving quadratic equations.** Solve the following equations.

a. $2 - x^2 = x$

b. $y(y-3) = 18$

c. $(r-1)(r-2) = 12$

d. $2x = 1 + \dfrac{10}{x}$ *Suggestion*: First multiply both sides of the equation by x.

e. $x^2 + 4x = 1$

8. **Reciprocal I.** Find a number that is exactly 1 more than 6 times its reciprocal. That is, solve the equation $x = 6 \times \dfrac{1}{x} + 1$. Is there more than one answer? If so, find all answers.

9. **Reciprocal II.** Find a *positive* number that is exactly 1 more than its reciprocal. That is, solve the equation $x = \dfrac{1}{x} + 1$. Is there more than one answer? If so, find all answers. Give exact answers, not numerical approximations.

10. **Finding vertices.**

a. Find both the x- and y-values of the vertex of the parabola whose equation is $y = 2x^2 - 12x + 5$.

b. What value of x makes $3x^2 - 18x + 5$ a minimum? What is this minimum value?

c. What value of x makes $7 + 8x - x^2$ a maximum? What is this maximum value?

[23] Cody Steele, How to Use Brushing to Investigate Outliers on a Graph. *The Minitab Blog (July 10, 2014).*

11. Thrown stone. If a stone is thrown downward from the top of a building with an initial speed of 2 feet per second, the distance d, in feet, that it falls in t seconds is

$$d = 16t^2 + 2t$$

How many seconds will it take for the rock to go a distance of 27.5 feet?

12–15. **A ball.** *If a ball is thrown upward from the ground with an initial speed of 96 feet per second, its height h, in feet, after t seconds is*

$$h = -16t^2 + 96t$$

Exercises 12 through 15 refer to this situation.

12. How many seconds will it take for the ball to fall back to the ground?

13. How many seconds will it take for the ball to reach its maximum height?

14. What is the maximum height the ball will reach?

15. How many seconds will it take for the ball to reach a height of 80 feet? There are two answers. Explain why.

16–19. **A cannonball.** *If air resistance is ignored, a projectile such as a rifle bullet or a cannonball follows a parabolic path. Suppose a cannonball fired from the origin follows the graph*

$$y = x - 0.0005x^2$$

where both x and y are measured in feet. This information is used in Exercises 16 through 19.

16. The 20-foot-high wall of the enemy fort is located 1970 feet downrange. Will the cannonball clear the wall?

17. How far downrange will the cannonball hit the ground?

18. How far downrange will the cannonball reach its maximum height?

19. What is the maximum height the cannonball reaches?

20. Women employed. The number of women employed outside the home reached a maximum during World War II. Let W denote the number, in millions, of women employed outside the home t years after 1940. A quadratic model is $W = -0.74t^2 + 3.1t + 16$. Find the value of t for which the number of women employed is the greatest. Round your answer to the nearest year. As a date, what year is this?

21. Speed and safety. For commercial vehicles driving at night on city streets, the accident rate A (per 100,000,000 vehicle-miles) depends on the speed s in miles per hour. A quadratic model is $A = 7.8s^2 - 514s + 8734$. Find the speed at which the accident rate is at a minimum. Round your answer in miles per hour to the nearest whole number.

22. Parking. The number N, in thousands, of vehicles parked downtown in a city depends on the time h in hours since 9:00 A.M. A quadratic model is $N = -0.16h^2 + h + 6$. Find the value of h for which the number of vehicles parked is the greatest. Round your answer to the nearest hour. What time of day is this?

23. A box. You want to make an open box (it has no lid) with a square bottom, a height of 4 inches, and a volume of 324 cubic inches. To do this, you take a square piece of cardboard, cut 4-inch-long squares from each corner and fold up the edges. What should the dimensions of the cardboard be before you cut it? *Suggestion*: Refer to **Figures 3.49 and 3.50**, where x is in inches. Remember that for a box, Volume = Length × Width × Height. Use the dimensions shown in Figure 3.50 to calculate the volume of the box in terms of x, and set that equal to 324 cubic inches.

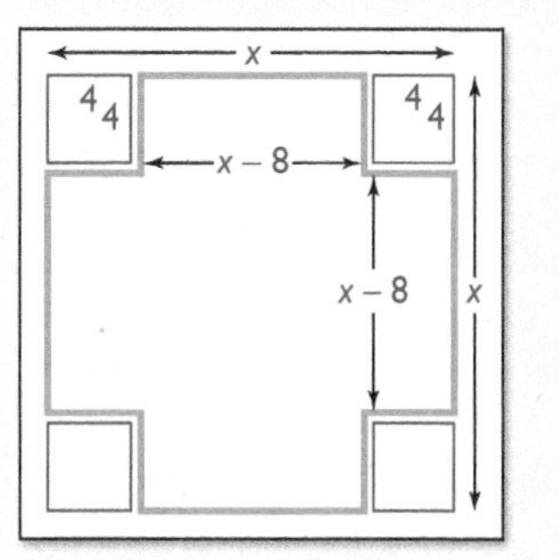

FIGURE 3.49 Cutting squares from a piece of cardboard.

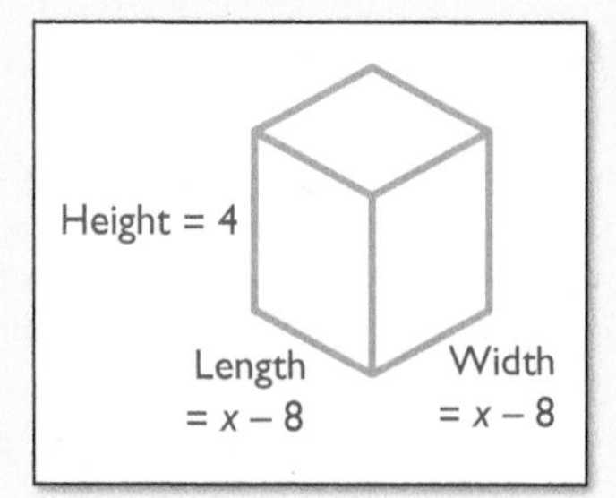

FIGURE 3.50 Cardboard folded to make a box.

24. A garden. A farmer wants to enclose a rectangular garden with an area of 1000 square feet. She wants the plot to be divided into three sections, as shown in **Figure 3.51**. If she uses 200 feet of fencing, what should the dimensions of the garden be? Is there more than one answer? Round your results in feet to one decimal place. *Suggestion*: Refer to Figure 3.51, where x is in feet and the formula for the length uses the fact that the area is 1000 square feet. Use the dimensions shown in the figure to calculate the total amount of fence used, and set that equal to 200 feet.

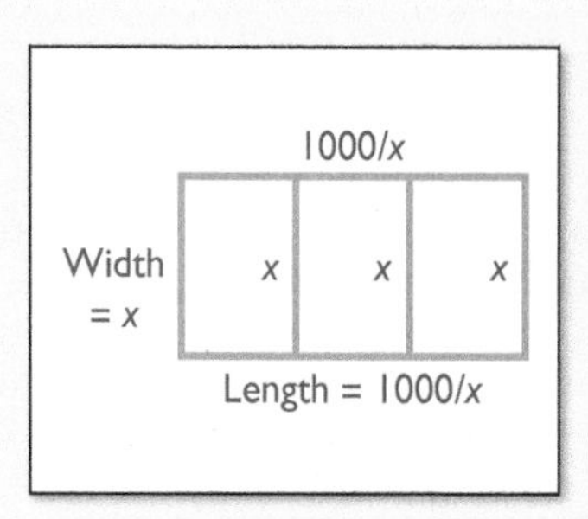

FIGURE 3.51

25. Three sides. A landscaper wants to enclose just three sides of a rectangular garden with an area of 1000 square feet. He wants the plot to be divided into four sections, as shown in **Figure 3.52**. If he uses 200 feet of fencing, what should the dimensions of the garden be? Is there more than one answer? Round your results in feet to one decimal place. *Suggestion*: Refer to Figure 3.52, where x is in feet and the formula for the length uses the fact that the area is 1000 square feet. Use the dimensions shown in the figure to calculate the total amount of fence used, and set that equal to 200 feet. Keep in mind that only three sides are enclosed.

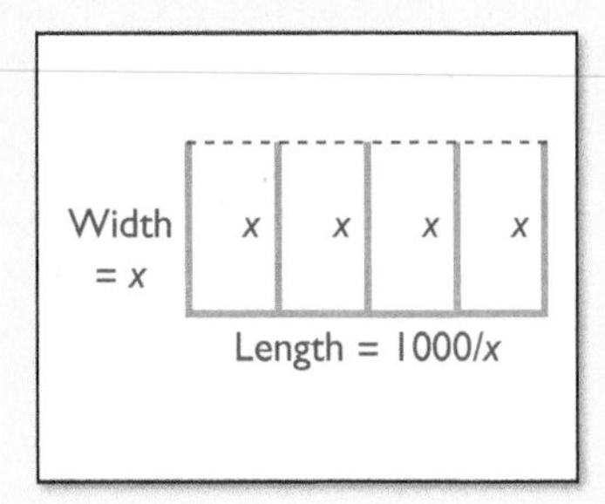

FIGURE 3.52

26. Minimum area of the pen? Refer to Example 3.29, where we considered a rectangular pen of maximum area to be constructed using the side of a barn as one boundary and 100 feet of fence to make the other three sides. In this setting, is there a pen of minimum area? Answer this question both in terms of the parabola found in the solution to Example 3.29 and in practical terms.

27. Carpet. A carpet is 4 yards longer than it is wide. If it has an area of 25 square yards, what are its dimensions? Round your results in yards to one decimal place. *Suggestion*: Refer to **Figure 3.53**, where x is in yards. Use the dimensions shown in the figure to calculate the area, and set that equal to 25 square yards.

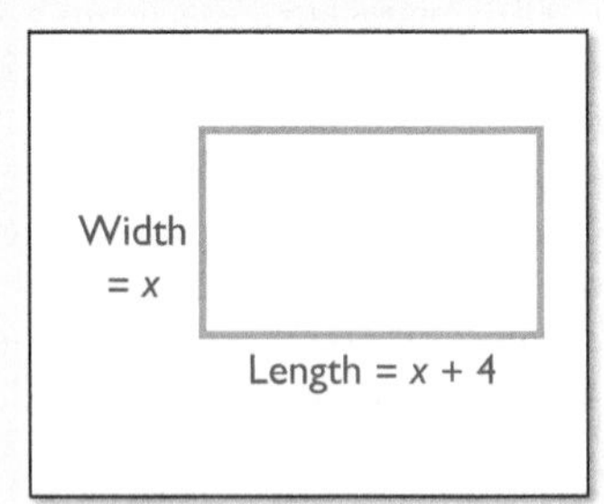

FIGURE 3.53

28. A picture. A picture is 3 feet wide by 4 feet tall. You want to put it on a rectangular matting for framing so that it has equal margins all the way around. If the total area of the matting is 20 square feet, how wide are the margins? *Suggestion*: Refer to **Figure 3.54**, where x is in feet. Use the dimensions shown in the figure to calculate the area of the matting, and set that equal to 20 square feet.

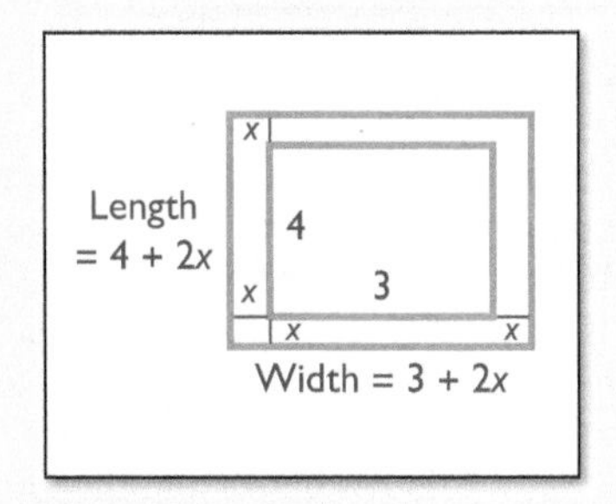

FIGURE 3.54

29. A path. A garden measuring 10 feet by 14 feet is to have a path built around it. The total area of the path and garden together will then be 221 square feet. What will the width of the path be? *Suggestion*: Refer to **Figure 3.55**, where x is in feet. Use the dimensions shown in the figure to calculate the total area, and set that equal to 221 square feet.

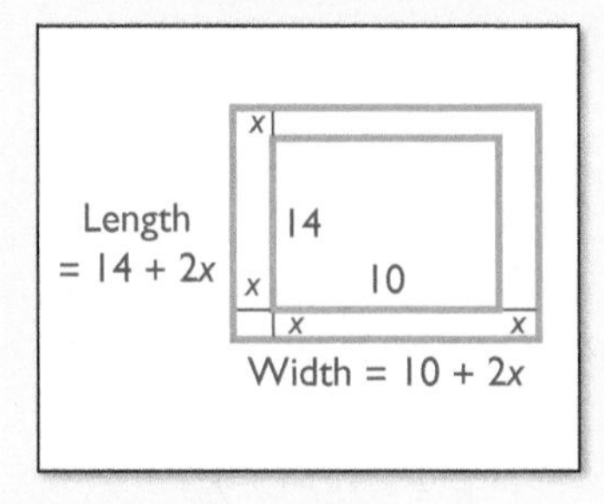

FIGURE 3.55

30. Circular path. You measured the area of a circular garden (in square meters) and then measured its circumference (in meters). You discovered that the numerical value of the area was exactly 5 more than the numerical value of the circumference. What is the area of the garden? Round your answer to the nearest square meter. *Suggestion*: The area of the garden is πr^2 if the radius is r. First find r by solving $\pi r^2 = 2\pi r + 5$.

31. Ladder I. One end of a ladder is resting on the top of a vertical wall that is 7 feet high. The distance from the foot of the ladder to the base of the wall is 2 feet less than the length of the ladder. How long is the ladder? *Suggestion*: Recall the Pythagorean theorem, which says for legs a and b and hypotenuse c of a right triangle that $a^2 + b^2 = c^2$. Apply this theorem to **Figure 3.56**, where x is measured in feet.

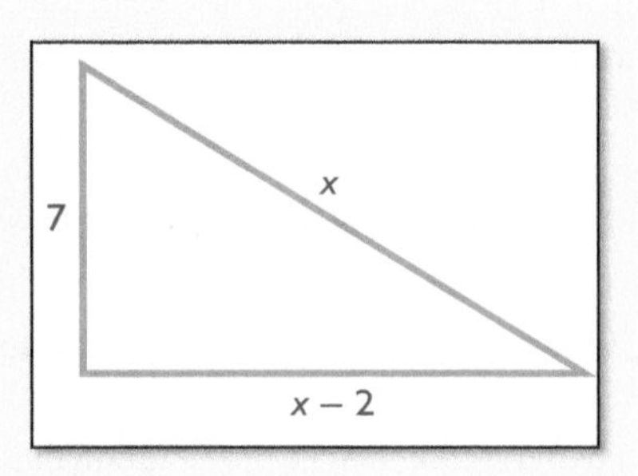

FIGURE 3.56

32. Ladder II. One end of a 13-foot-long ladder is resting on the top of a vertical wall. The distance from the foot of the ladder to the base of the wall is 7 feet less than the height of the wall. How high is the wall? *Suggestion*: Recall the Pythagorean theorem, which says for legs a and b and hypotenuse c of a right triangle that $a^2 + b^2 = c^2$. Apply this theorem to **Figure 3.57**, where x is measured in feet.

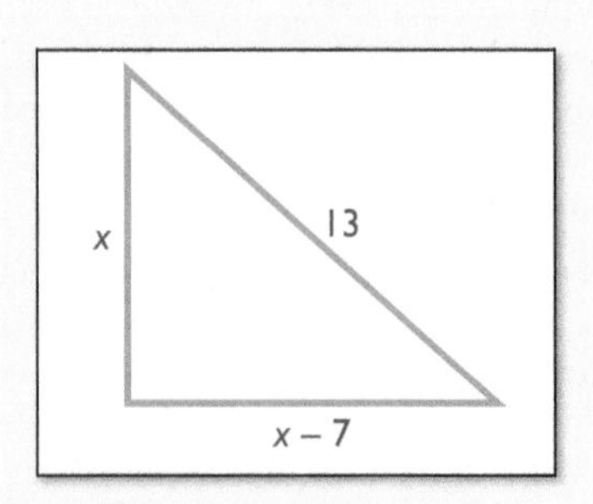

FIGURE 3.57

33. Correct method? A friend tells you that you can solve the equation

$$(x-1)(x-2)=12$$

by saying $x-1=12$, so $x=13$, or $x-2=12$, so $x=14$. Is your friend correct?

34. Taking square roots. An important step in deriving the quadratic formula is to take square roots when the left-hand side of the equation is a perfect square. Consider the equation

$$x^2-4x+4=7$$

Write the left-hand side as a square, then take square roots of both sides to solve the equation. Give exact answers, not numerical approximations.

35. Completing the square. A key step in deriving the quadratic formula is to *complete the square*. If we have x^2+bx, what do we need to add to make a perfect square? The answer turns out to be $\frac{b^2}{4}$. For example, if we add $\frac{4^2}{4}=4$ to x^2+4x, we get x^2+4x+4, which is the perfect square $(x+2)^2$. What do you need to add to x^2+8x in order to make it a perfect square? Add the number you find, and write the result as a perfect square.

Writing About Mathematics

36. History. Write a report on the history of the quadratic formula.

37. Cubic equations. An equation of the form $ax^3+bx^2+cx+d=0$ is called a *cubic equation*. There is a formula for solving cubic equations, but it is much more complex than the quadratic formula. Write a report on the history of attempts to solve cubic equations.

38. Parabolas. As we have seen, parabolas occur in interesting places like reflecting telescopes and thrown objects. Find occurrences of parabolas in settings not already mentioned in this section and write a report on your findings.

What Do You Think?

39. The Hubble Space Telescope. The mirror for the Hubble Space Telescope was improperly made. It should have had a perfect parabolic shape but did not. Do some research and report on this incident. Your report should include the nature of the flaw in the mirror and steps taken to correct the flaw.

40. Complex numbers. Sometimes when we solve quadratic equations we get *complex numbers* as solutions. Do some research to find out what complex numbers are and why they are used in advanced mathematics.

41. Air resistance. As mentioned in Exercises 16–19, projectiles will trace out a parabolic path if there is no air resistance. Do some research on the Internet to see what happens if air resistance is taken into account.

42. Orbits. Many objects in our solar system orbit the sun in *elliptical* orbits, but some have parabolic orbits. Research objects with parabolic orbits. What is the eventual fate of these objects?

43. Two solutions. Figures 3.45 and 3.46 show parabolas. Explain in terms of these figures why the quadratic formula gives at most two solutions for a quadratic equation.

44. Max-min for linear or exponential functions? In this section we solve max-min problems for phenomena modeled by quadratic functions. Does it make sense to solve max-min problems for phenomena modeled by linear or exponential functions?

45. Dish TV antennas. Below is a photo of Dish TV antennas. Discuss what you think they have in common with reflecting telescopes and solar collectors.

Edgars Dubrovskis/Shutterstock

46. Vomit Comet. Part of the training of astronauts is to take them on an airplane that flies a parabolic path. The result is that the passengers feel weightless. (The plane is affectionately called the Vomit Comet because the passengers usually get sick.) Read this article: www.avweb.com/aviation-news/nasas-vomit-comet-hitchin-a-ride-on-a-buckin-kc-135/ and write a brief report on what you learned.

47. Braking distance. The term *braking distance* refers to the distance your car travels after you have hit your brakes. Obviously your braking distance increases as your speed increases. Suppose you double your speed, say from 30 to 60 mph. What effect do you think it will have on your braking distance? Will it double? It may surprise you to learn that your braking distance increases by a factor of four! In general, your braking distance is proportional to the *square* of your speed (a quadratic relationship). That means, for example, that if you triple your speed, your braking distance increases by a factor of nine! The *Nevada State Driver's Handbook* (www.dmvnv.com/pdfforms/dlbook.pdf) has a discussion of this. Examine what it has to say and check your own state's manual to see what it says about this issue.

CHAPTER SUMMARY

This chapter discusses four basic types of functions: *linear*, *exponential*, *logarithmic*, and *quadratic*. All four are important and occur naturally.

Lines and linear growth: What does a constant rate mean?

A linear function is one with a constant growth rate. The formula for a linear function is $y = mx + b$, where m is the growth rate or *slope* and b is the initial value. A 1-unit increase in x corresponds to a change of m units in y.

The growth rate, or slope, of a linear function can be calculated using

$$\text{Slope} = \text{Growth rate} = \frac{\text{Change in function value}}{\text{Change in independent variable}}$$

In practical settings, the slope always has an important meaning, and understanding that meaning is often the key to analyzing linear functions. When a linear function is given in a real-world context, proper use of the units of the slope (such as inches per hour) can help us determine the meaning of the slope.

The graph of a linear function is always a straight line. When data points almost fall on a straight line, it may be appropriate to approximate the data with a linear function. We use the terms *regression line* or *trend line* for graphs of these linear approximations.

Exponential growth and decay: Constant percentage rates

Exponential functions are characterized by a constant *percentage* growth (or decay). For example, if a population grows by 5% each year, the population size is an exponential function of time: The percentage growth rate of this function is constant, 5% per year. The graph of an increasing exponential function has the characteristic shape shown in **Figure 3.58**. If a population is decreasing by 5% each year, the population shows exponential decay, which has the characteristic shape shown in **Figure 3.59**.

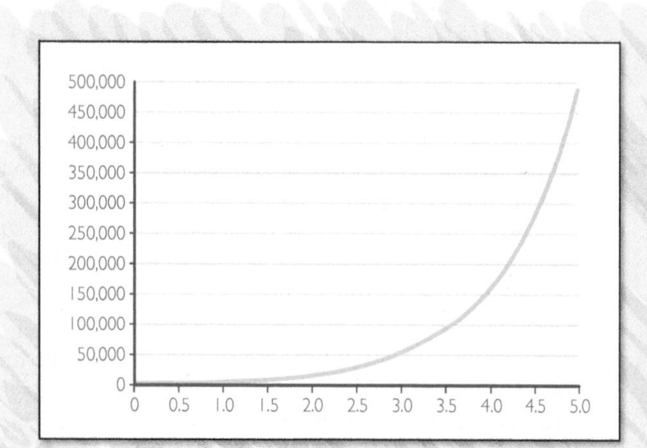

FIGURE 3.58 Increasing exponential function: exponential growth.

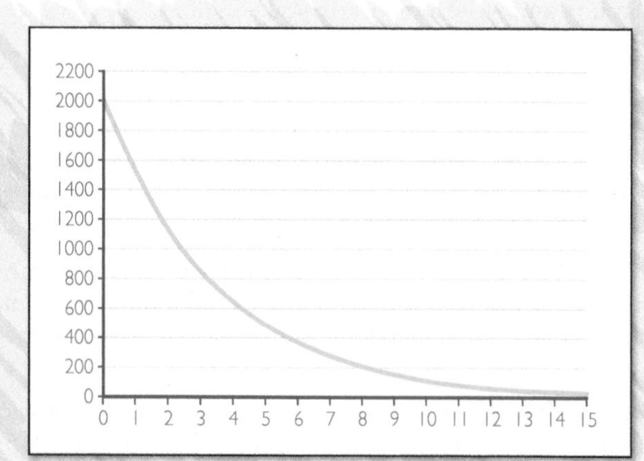

FIGURE 3.59 Decreasing exponential function: exponential decay.

The formula for an exponential function is

$$\text{Amount} = \text{Initial value} \times \text{Base}^t$$

An exponential function y of t is characterized by the following property: When t increases by 1, to find the new value of y, we multiply the current value by the base.

For constant percentage increase of r as a decimal, the base is $1 + r$. This gives the formula

$$\text{Amount} = \text{Initial value} \times (1 + r)^t$$

For constant percentage decrease of r as a decimal, the base is $1 - r$. This gives the formula

$$\text{Amount} = \text{Initial value} \times (1 - r)^t$$

The amount of a radioactive material decays over time as an exponential function. The base is $1/2$ if time is measured in half-lives.

Logarithmic phenomena: Compressed scales

The *common logarithm* of a positive number x is the power to which 10 must be raised to equal x. The common logarithm of x is written $\log x$. Formally,

$$\log x = t \text{ if and only if } 10^t = x$$

One of the most familiar uses of logarithmic functions is the Richter scale, which is applied in the measurement of earthquakes. An increase of one point on the Richter scale corresponds to multiplying the relative intensity of the earthquake by 10. The logarithm provides the compression needed for the Richter scale.

The decibel scale is another logarithmic scale. It is used in the measurement of sound. In fact, according to *Fechner's law*, for all five of the human senses the perceived magnitude of a sensation can be described using the logarithm.

The graph in **Figure 3.60** shows the decibel reading in terms of the relative intensity. This figure shows the characteristic shape of the graph of a logarithmic function. It illustrates the decreasing growth rate that is expected from the applications to the Richter scale and human sensation.

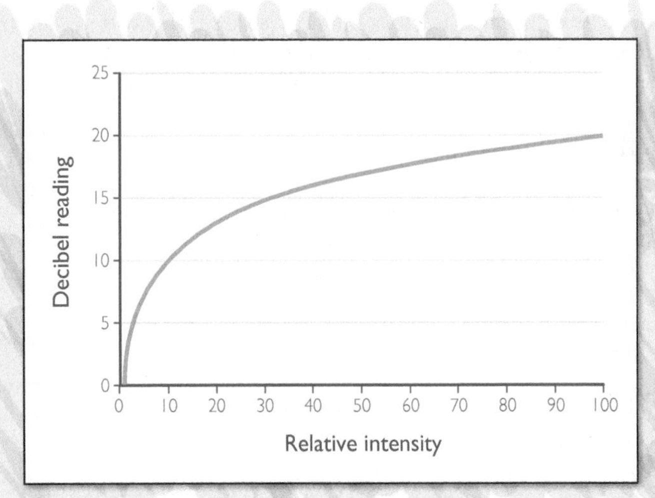

FIGURE 3.60 Decibel reading in terms of relative intensity.

Logarithms can be used to solve exponential equations. This solution yields exact formulas for the doubling time, tripling time, etc., of exponential growth.

Quadratics and parabolas

Quadratic functions have the form $y = ax^2 + bx + c$. Solutions to *quadratic equations* $ax^2 + bx + c = 0$ are given by the famous *quadratic formula*

$$x = \frac{-b \pm \sqrt{b^2 - 4ac}}{2a}$$

Graphs of quadratic functions are *parabolas*, which have shapes as shown in **Figure 3.45** or **Figure 3.46**. The *vertex* of a parabola is the minimum point if the parabola opens upward (and the maximum point if it opens downward). The x-value of the vertex occurs at $x = \frac{-b}{2a}$. Quadratic functions arise in some max-min problems, in which the vertex corresponds to a solution.

KEY TERMS

linear, p. 140
regression line, p. 150
trend line, p. 150
exponential function, p. 158
half-life, p. 168
common logarithm, p. 179
relative intensity, p. 181
magnitude, p. 181
decibel, p. 184
quadratic function, p. 196
quadratic equation, p. 196
quadratic formula, p. 197
parabola, p. 200
vertex, p. 200

CHAPTER QUIZ

1. For the following scenarios, explain the practical meaning of the growth rate, and determine which functions are linear.

   a. A gym charges a \$100 initiation fee and \$25 per month. Is the total amount spent on the gym membership a linear function of the number of months you go?

   b. Another gym charges a \$50 initiation fee plus \$30 per month for the first six months and \$25 per month for each month over six. Is the total amount spent on the gym membership a linear function of the number of months you go?

Answer In both cases, the growth rate is the extra cost for another month. The first scenario determines a linear function because that monthly cost is constant. The second scenario does not determine a linear function because the monthly cost is not constant: it varies depending on how many months you go.

If you had difficulty with this problem, see Example 3.1.

2. It begins raining at 10:00 A.M. and continues raining at a constant rate all day. The rain gauge already contains 2 inches of water when the rain begins. Between 12:00 P.M. and 2:00 P.M., the level of the rain gauge increases by 1 inch. Find a linear formula that gives the height of water in the gauge as a function of the time since 10:00 A.M.

Answer If H denotes the height (in inches) of water in the gauge t hours after 10:00 A.M., $H = 2 + 0.5t$.

If you had difficulty with this problem, see Example 3.6.

3. Data for the yearly best 100-meter dash time for a small college from 2010 through 2015 were collected and fit with a trend line. The equation for the trend line is $D = 12 - 0.03t$, where t is the time in years since 2010 and D is the best time (in seconds) for that year. Explain in practical terms the meaning of the slope of the trend line.

Answer The slope is -0.03 second per year. It means that the best time was decreasing by about 0.03 second each year.

If you had difficulty with this problem, see Example 3.7.

4. My salary in 2013 was \$48,000. I got a 3% raise each year for eight years. Find a formula that gives my salary as a function of the time since 2013. What was my salary in 2021?

Answer If we use S for my salary (in dollars) and t for the time in years since 2013, $S = 48{,}000 \times 1.03^t$. In 2021 my salary was \$60,804.96.

If you had difficulty with this problem, see Example 3.11.

5. Uranium-235, which can be used to make atomic bombs, has a half-life of 713 million years. Suppose 2 grams of uranium-235 were placed in a "safe storage location" 150 million years ago. Find an exponential formula that gives the amount of uranium-235 remaining after h half-lives. How many half-lives is 150 million years? (Round your answer to two decimal places.) How much uranium-235 still remains in "safe storage" today?

Answer If A is the amount (in grams) remaining, $A = 2 \times (1/2)^h$. 150 million years is 0.21 half-life. The amount remaining is about 1.7 grams.

If you had difficulty with this problem, see Example 3.14.

6. One earthquake has a magnitude of 3.5, and another has a magnitude of 6.5. How do their relative intensities compare?

Answer The larger earthquake is 1000 times as intense as the smaller quake.

If you had difficulty with this problem, see Example 3.19.

7. An idling bulldozer makes a sound of about 85 decibels. What would be the decibel reading of two idling bulldozers side by side?

Answer 88 decibels

If you had difficulty with this problem, see Example 3.21.

8. My salary is increasing by 4% each year. How long will it take for my salary to double?

Answer About 17.7 years.

If you had difficulty with this problem, see Example 3.23.

9. Use the quadratic formula to solve $3x^2 - 3x - 1 = 0$.

Answer $x = \dfrac{3 + \sqrt{21}}{6}$ or $x = \dfrac{3 - \sqrt{21}}{6}$

If you had difficulty with this problem, see Example 3.25.

10. For a certain population, the growth rate G, in millions of individuals per year, depends on the size of the population N, in millions. The relation is

$$G = 7 + 2N - 0.7N^2$$

For what population level is the population stable?

Answer $N = 4.90$ million

If you had difficulty with this problem, see Example 3.27.

11. Find both the x- and y-values for the vertex of $y = -2x^2 + 12x - 3$.

Answer The x-value is 3, and the y-value is 15.

If you had difficulty with this problem, see Example 3.28.

Financial Aid
Entrance

4 Personal Finance

4.1 Saving money: The power of compounding

4.2 Borrowing: How much car can you afford?

4.3 Saving for the long term: Build that nest egg

4.4 Credit cards: Paying off consumer debt

4.5 Inflation, taxes, and stocks: Managing your money

The following article appeared at CBSNews.com.

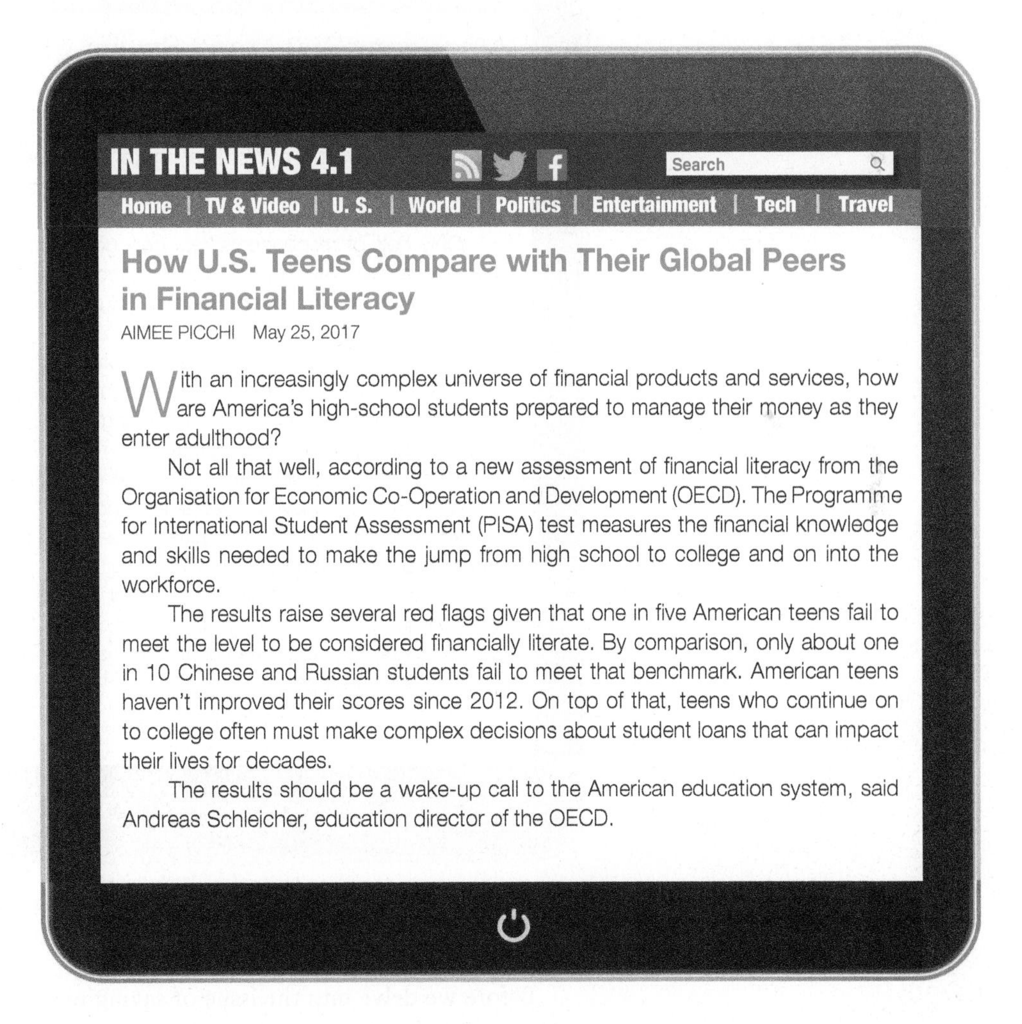

How U.S. Teens Compare with Their Global Peers in Financial Literacy

AIMEE PICCHI May 25, 2017

With an increasingly complex universe of financial products and services, how are America's high-school students prepared to manage their money as they enter adulthood?

Not all that well, according to a new assessment of financial literacy from the Organisation for Economic Co-Operation and Development (OECD). The Programme for International Student Assessment (PISA) test measures the financial knowledge and skills needed to make the jump from high school to college and on into the workforce.

The results raise several red flags given that one in five American teens fail to meet the level to be considered financially literate. By comparison, only about one in 10 Chinese and Russian students fail to meet that benchmark. American teens haven't improved their scores since 2012. On top of that, teens who continue on to college often must make complex decisions about student loans that can impact their lives for decades.

The results should be a wake-up call to the American education system, said Andreas Schleicher, education director of the OECD.

In this chapter, we explore basic financial terminology and mechanisms. In Section 4.1, we examine the basics of compound interest and savings, and in Section 4.2, we look at borrowing. In Section 4.3, we consider long-term savings plans such as retirement funds. In Section 4.4, we focus on credit cards, and in Section 4.5, we discuss financial terms heard in the daily news.

4.1 Saving money: The power of compounding

TAKE AWAY FROM THIS SECTION
Understand compound interest and the difference between APR (annual percentage rate) and APY (annual percentage yield).

The following article from CNNMoney says that many Americans are not saving enough money.

Americans Still Aren't Saving Enough

NEW YORK (CNNMoney)

MELANIE HICKEN @MELHICKEN February 26, 2013

Many Americans are still failing to sock away enough money to pay for retirement—or emergencies, for that matter—according to a survey of more than 1,000 people released Monday.

Only half of respondents reported good savings habits, including having a spending and/or saving plan in place, according to the Consumer Federation of America and the American Savings Education Council, which conducted the survey.

While more than half of all Baby Boomers and Gen-Xers will be able to retire with enough money to cover basic retirement needs, including health care costs, a significant number are at risk of running short, according to projections by the Employee Benefit Research Institute.

The survey also found that only 49% of non-retired respondents feel they are saving enough to achieve a "desirable standard of living" in retirement.

When it came to overall household savings, the survey found there was little improvement compared to last year.

"Clearly the great recession has had a lingering effect on many Americans," said Stephen Brobeck, executive director of the Consumer Federation of America. "Millions of families have been unable to make progress in rebuilding their finances, especially their savings."

Courtesy CNN.

Money management begins with saving, and (as the preceding article notes) Americans generally don't save enough. In this section, we see how to measure the growth of savings accounts.

Before we delve into the issue of saving money, we should say a little bit about interest rates. As you can see from the chart in **Figure 4.1**, interest rates on certificates of deposit (CDs) have fluctuated over the years, and in recent years they have been at historic lows. As of April 2021, a one-year CD was paying about 0.5% and a five-year CD 1.0%. While we want to acknowledge realistic rates, many of the examples

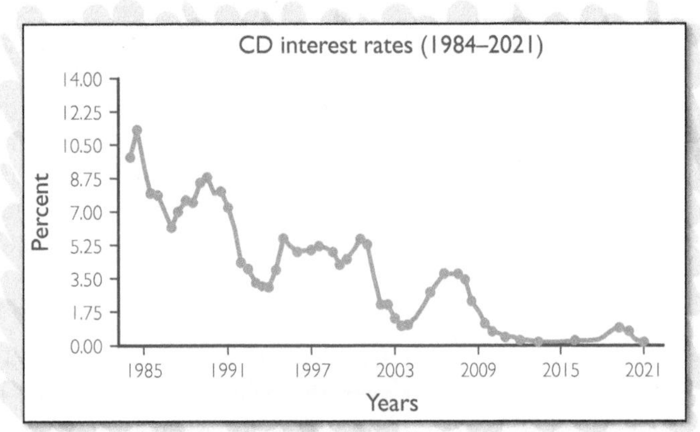

FIGURE 4.1 Certificate of deposit interest rates from 1984 to 2021.

and exercises in this chapter sometimes use rates like 6%, 10%, or 12% to help make concepts easy to understand by avoiding complex and distracting calculations.

Some readers may wish to take advantage of the following Quick Review of linear and exponential functions before proceeding.

Quick Review Linear Functions

Linear functions play an important role in the mathematics of finance.

Linear functions: A linear function has a constant growth rate, and its graph is a straight line. The growth rate of the function is also referred to as the *slope*.

We find a formula for a linear function of t using

$$\text{Linear function} = \text{Growth rate} \times t + \text{Initial value}$$

Example: If we initially have \$1000 in an account and add \$100 each year, the balance is a linear function because it is growing by a constant amount each year. After t years, the balance is

$$\text{Balance after } t \text{ years} = \$100t + \$1000$$

The graph of this function is shown in **Figure 4.2**. Note that the graph is a straight line.

For additional information, see Section 3.1 of Chapter 3.

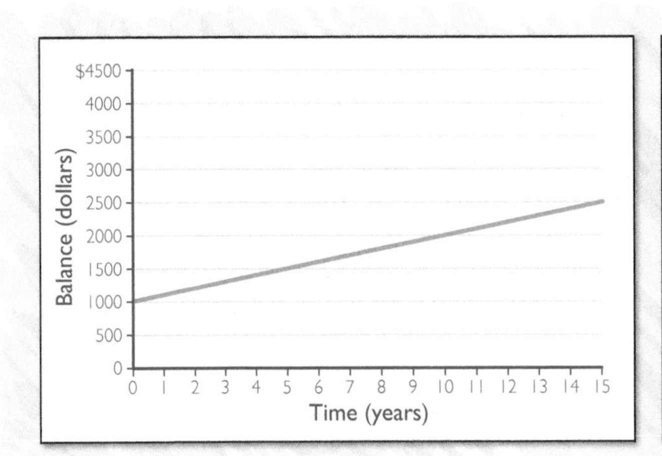

FIGURE 4.2 The graph of a linear function is a straight line.

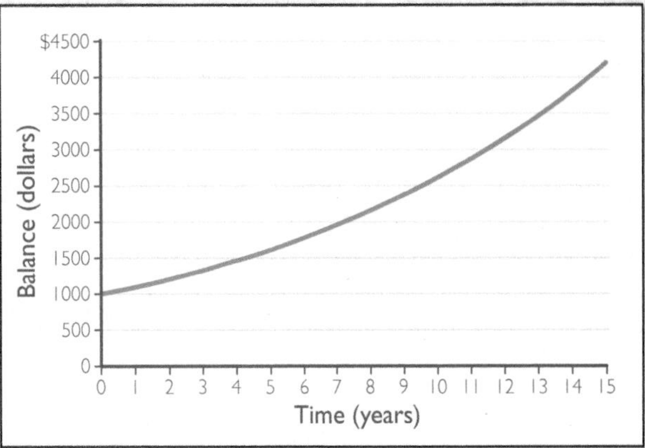

FIGURE 4.3 The graph of an exponential function (with a base greater than 1) gets steeper as we move to the right.

Quick Review Exponential Functions

Exponential functions play an important role in the mathematics of finance.

Exponential functions: An increasing exponential function exhibits a constant percentage growth rate. If r is this percentage growth rate per period expressed as a decimal, the base of the exponential function is $1+r$. We find a formula for an exponential function of the number of periods t using

$$\text{Exponential function} = \text{Initial value} \times (1+r)^t$$

Example: If we initially have \$1000 in an account that grows by 10% per year, the balance is exponential because it is growing at a constant percentage rate. Now 10% as a decimal is $r = 0.10$, so $1 + r = 1.10$. After t years, the balance is

$$\text{Balance after } t \text{ years} = \$1000 \times 1.10^t$$

The graph of this function is shown in **Figure 4.3**. The increasing growth rate is typical of increasing exponential functions.

For additional information, see Section 3.2 of Chapter 3.

Investments such as savings accounts earn interest over time. Interest can be earned in different ways, and that affects the growth in value of an investment.

Simple interest

The easiest type of interest to understand and calculate is *simple interest*.

KEY CONCEPT

The initial balance of an account is the principal. Simple interest is calculated by applying the interest rate to the principal only, not to interest earned.

cartoonstock.com

"This mattress fits up to \$120,000 dollars."

Suppose, for example, that we invest \$1000 in an account that earns simple interest at a rate of 10% per year. Then we earn 10% of \$1000 or \$100 in interest each year. If we hold the money for six years, we get \$100 in interest each year for six years. That comes to \$600 interest. If we hold it for only six months, we get a half-year's interest or \$50.

FORMULA 4.1 Simple Interest Formula

$$\text{Simple interest earned} = \text{Principal} \times \text{Yearly interest rate (as a decimal)} \times \text{Time in years}$$

EXAMPLE 4.1 Calculating simple interest: An account

We invest \$2000 in an account that pays simple interest of 4% each year. Find the interest earned after five years.

SOLUTION

The interest rate of 4% written as a decimal is 0.04. The principal is \$2000, and the time is 5 years. We find the interest by using these values in the simple interest formula (Formula 4.1):

$$\text{Simple interest earned} = \text{Principal} \times \text{Yearly interest rate} \times \text{Time in years}$$
$$= \$2000 \times 0.04/\text{year} \times 5 \text{ years} = \$400$$

TRY IT YOURSELF 4.1

We invest \$3000 in an account that pays simple interest of 3% each year. Find the interest earned after six years.

The answer is provided at the end of this section.

Compound interest

Situations involving simple interest are fairly rare and usually occur when money is loaned or borrowed for a short period of time. More often, interest payments are made in periodic installments during the life of the investment. The interest payments are credited to the account periodically, and future interest is earned not only on the original principal but also on the interest earned to date. This type of interest calculation is referred to as *compounding*.

KEY CONCEPT

Compound interest means that accrued interest is periodically added to the account balance. This interest is paid on the principal and on the interest that the account has already earned. In short, compound interest includes *interest on the interest*.

To see how compound interest works, let's return to the \$1000 investment earning 10% per year we looked at earlier, but this time let's assume that the interest is compounded annually (at the end of each year). At the end of the first year, we earn 10% of \$1000 or \$100—the same as with simple interest. When interest is compounded, we add this amount to the balance, giving a new balance of \$1100. At the end of the second year, we earn 10% interest on the \$1100 balance:

$$\text{Second year's interest} = 0.10 \times \$1100 = \$110$$

This amount is added to the balance, so after two years the balance is

$$\text{Balance after 2 years} = \$1100 + \$110 = \$1210$$

An illustration of growth due to compound interest.

For comparison, we can use the simple interest formula to find out the simple interest earned after two years:

$$\text{Simple interest after two years} = \text{Principal} \times \text{Yearly interest rate} \times \text{Time in years}$$
$$= \$1000 \times 0.10/\text{year} \times 2 \text{ years} = \$200$$

The interest earned is \$200, so the balance of the account is \$1200.

After two years, the balance of the account earning simple interest is only $1200, but the balance of the account earning compound interest is $1210. Compound interest is always more than simple interest, and this observation suggests a rule of thumb for estimating the interest earned.

RULE OF THUMB 4.1 Estimating Interest

Interest earned on an account with compounding is always at least as much as that earned from simple interest. If the money is invested for a short time, simple interest can be used as a rough estimate.

The following table compares simple interest and annual compounding over various periods. It uses $1000 for the principal and 10% for the annual rate. This table shows why compounding is so important for long-term savings.

	Simple interest			Yearly compounding		
End of year	Interest	Balance	Growth	Interest	Balance	Growth
1	10% of $1000 = $100	$1100	$100	10% of $1000 = $100	$1100	$100
2	10% of $1000 = $100	$1200	$100	10% of $1100 = $110	$1210	$110
3	10% of $1000 = $100	$1300	$100	10% of $1210 = $121	$1331	$121
10	$100	$2000		$235.79	$2593.74	
50	$100	$6000		$10,671.90	$117,390.85	

To better understand the comparison between simple and compound interest, observe that for simple interest the balance is growing by the same *amount*, $100, each year. This means that the balance for simple interest is showing linear growth. For compound interest, the balance is growing by the same *percent*, 10%, each year. This means that the balance for compound interest is growing exponentially. The graphs of the account balances are shown in **Figure 4.4**. The widening gap between the two graphs shows the power of compounding.

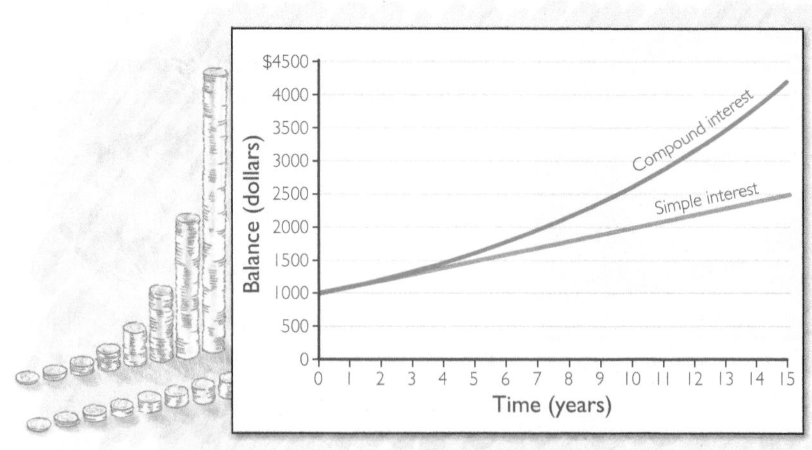

FIGURE 4.4 Balance for simple interest is linear, and balance for compound interest is exponential.

EXAMPLE 4.2 Calculating compound interest: Annual compounding

You invest \$500 in an account that pays 6% compounded annually. What is the account balance after two years?

SOLUTION

Now 6% expressed as a decimal is 0.06. The first year's interest is 6% of \$500:

$$\text{First year's interest} = 0.06 \times \$500 = \$30.00$$

This interest is added to the principal to give an account balance at the end of the first year of \$530.00. We use this figure to calculate the second year's interest:

$$\text{Second year's interest} = 0.06 \times \$530.00 = \$31.80$$

We add this to the balance to find the balance at the end of two years:

$$\text{Balance after 2 years} = \$530.00 + \$31.80 = \$561.80$$

TRY IT YOURSELF 4.2

Find the balance of this account after four years.

The answer is provided at the end of this section.

Other compounding periods and the APR

Interest may be compounded more frequently than once a year. For example, compounding may occur semi-annually, in which case the *compounding period* is half a year. Compounding may also be done quarterly, monthly, or even daily (**Figure 4.5**). To calculate the interest earned, we need to know the *period interest rate*.

KEY CONCEPT

The period interest rate is the interest rate for a given compounding period (for example, a month). Financial institutions report the annual percentage rate or APR. To calculate this, they multiply the period interest rate by the number of periods in a year.

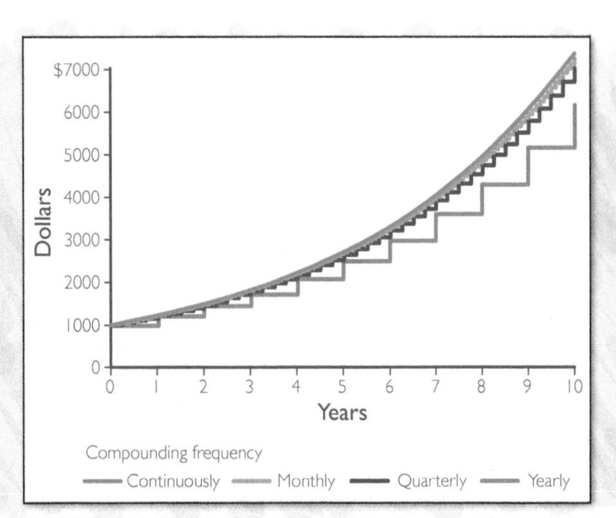

FIGURE 4.5 Effect of compounding at various frequencies at 20% APR and an initial investment of \$1000.

The following formula shows how the APR is used in calculations to determine interest rates.

FORMULA 4.2 Period Interest Rate Formula

$$\text{Period interest rate} = \frac{\text{APR}}{\text{Number of periods in a year}}$$

Suppose, for example, that we invest \$500 in a savings account that has an APR of 6% and compounds interest monthly. Then there are 12 compounding periods each year. We find the monthly interest rate using

$$\text{Monthly interest rate} = \frac{\text{APR}}{12} = \frac{6\%}{12} = 0.5\%$$

"We can give you a 12% rate if you never withdraw it."

Each month we add 0.5% interest to the current balance. The following table shows how the account balance grows over the first few months:

End of month	Interest earned	New balance	Percent increase
1	0.5% of \$500.00 = \$2.50	\$502.50	0.5%
2	0.5% of \$502.50 = \$2.51	\$505.01	0.5%
3	0.5% of \$505.01 = \$2.53	\$507.54	0.5%
4	0.5% of \$507.54 = \$2.54	\$510.08	0.5%

Compound interest formula

So far we have calculated the interest step-by-step to see how the balance grows due to compounding. Now we simplify the process by giving a formula for the balance.

If r is the period interest rate expressed as a decimal, we find the balance after t periods using the following:

FORMULA 4.3 Compound Interest Formula

$$\textbf{Balance after } t \textbf{ periods} = \textbf{Principal} \times (1 + r)^t$$

Alternatively, we can write the formula as

$$\text{Balance after } y \text{ years} = \text{Principal} \times \left(1 + \frac{\text{APR}}{n}\right)^{(n \times y)}$$

where interest is compounded n times per year and y is the number of years. This form is equivalent because $r = \frac{\text{APR}}{n}$ and the number of periods is $t = n \times y$.

Here is an explanation for the formula: Over each compounding period, the balance grows by the same percentage, so the balance is an exponential function of t. That percentage growth is r as a decimal, and the initial value is the principal. Using these values in the standard exponential formula

$$\text{Exponential function} = \text{Initial value} \times (1 + r)^t$$

gives the compound interest formula.

Let's find the formula for the balance if \$500 is invested in a savings account that pays an APR of 6% compounded monthly. The APR as a decimal is 0.06, so in decimal form the monthly rate is $r = 0.06/12 = 0.005$. Thus, $1 + r = 1.005$. By Formula 4.3,

$$\begin{aligned}\text{Balance after } t \text{ months} &= \text{Principal} \times (1 + r)^t \\ &= \$500 \times 1.005^t\end{aligned}$$

We can use this formula to find the balance of the account after five years. Five years is 60 months, so we use $t = 60$ in the formula:

$$\text{Balance after 60 months} = \$500 \times 1.005^{60} = \$674.43$$

The APR by itself does not determine how much interest an account earns. The number of compounding periods also plays a role, as the next example illustrates.

EXAMPLE 4.3 Calculating values with varying compounding periods: Value of a CD

Suppose we invest \$10,000 in a five-year certificate of deposit (CD) that pays an APR of 6%.

a. What is the value of the mature CD if interest is compounded annually? (*Maturity* refers to the end of the life of a CD. In this case, maturity occurs at five years.)

b. What is the value of the mature CD if interest is compounded quarterly?

c. What is the value of the mature CD if interest is compounded monthly?

d. What is the value of the mature CD if interest is compounded daily?

e. Compare your answers from parts a–d.

SOLUTION

a. The annual compounding rate is the same as the APR. Now 6% as a decimal is $r = 0.06$. We use $1 + r = 1.06$ and $t = 5$ years in the compound interest formula (Formula 4.3):

$$\begin{aligned}\text{Balance after 5 years} &= \text{Principal} \times (1 + r)^t \\ &= \$10{,}000 \times 1.06^5 \\ &= \$13{,}382.26\end{aligned}$$

b. Again, we use the compound interest formula. To find the quarterly rate, we divide the APR by 4. The APR as a decimal is 0.06, so as a decimal, the quarterly rate is

$$r = \text{Quarterly rate} = \frac{\text{APR}}{4} = \frac{0.06}{4} = 0.015$$

Thus, $1 + r = 1.015$. Also five years is 20 quarters, so we use $t = 20$ in the compound interest formula:

$$\begin{aligned}\text{Balance after 20 quarters} &= \text{Principal} \times (1+r)^t \\ &= \$10{,}000 \times 1.015^{20} \\ &= \$13{,}468.55\end{aligned}$$

c. This time we want the monthly rate, so we divide the APR by 12:

$$r = \text{Monthly rate} = \frac{\text{APR}}{12} = \frac{0.06}{12} = 0.005$$

Also, five years is 60 months, so

$$\begin{aligned}\text{Balance after 60 months} &= \text{Principal} \times (1+r)^t \\ &= \$10{,}000 \times 1.005^{60} \\ &= \$13{,}488.50\end{aligned}$$

d. We assume that there are 365 days in each year, so as a decimal, the daily rate is

$$r = \text{Daily rate} = \frac{\text{APR}}{365} = \frac{0.06}{365}$$

This is $r = 0.00016$ to five decimal places, but for better accuracy, we won't round this number. (See Calculation Tip 1.) Five years is $5 \times 365 = 1825$ days, so

$$\begin{aligned}\text{Balance after 1825 days} &= \text{Principal} \times (1+r)^t \\ &= \$10{,}000 \times \left(1 + \frac{0.06}{365}\right)^{1825} \\ &= \$13{,}498.26\end{aligned}$$

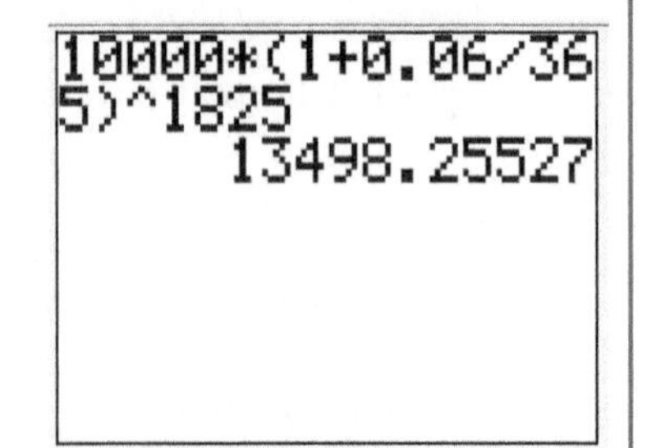

Benny Evans

e. We summarize these results in the following table:

Compounding period	Balance at maturity
Yearly	\$13,382.26
Quarterly	\$13,468.55
Monthly	\$13,488.50
Daily	\$13,498.26

This table shows that increasing the number of compounding periods increases the interest earned even though the APR and the number of years stay the same.

CALCULATION TIP 4.1 Rounding

Some financial calculations are very sensitive to rounding. To obtain accurate answers, when you use a calculator, it is better to keep all the decimal places rather than to enter parts of the formula that you have rounded. You can do this by either entering the complete formula or using the memory key on your calculator to store numbers with lots of decimal places.

For instance, in part d of the preceding example, we found the balance after 1825 days to be \$13,498.26. But if we round the daily rate to 0.00016, we get $\$10{,}000 \times 1.00016^{1825} = \$13{,}390.72$. As this shows, rounding significantly affects the accuracy of the answer.

More information on rounding is given in Appendix 3.

SUMMARY 4.1
Compound Interest

1. With compounding, accrued interest is periodically added to the balance. As a result interest is earned each period on both the principal and whatever interest has already accrued.

2. Financial institutions advertise the annual percentage rate (APR).

3. If interest is compounded n times per year, to find the period interest rate, we divide the APR by n:

$$\text{Period rate} = \frac{\text{APR}}{n}$$

4. We can calculate the account balance after t periods using the compound interest formula (Formula 4.3):

$$\text{Balance after } t \text{ periods} = \text{Principal} \times (1 + r)^t$$

Here, r is the period interest rate expressed as a decimal.
Alternatively, we can write the formula as

$$\text{Balance after } y \text{ years} = \text{Principal} \times \left(1 + \frac{\text{APR}}{n}\right)^{(n \times y)}$$

where interest is compounded n times per year and y is the number of years.
Many financial formulas, including this one, are sensitive to round-off error, so it is best to do all the calculations and then round.

APR versus APY

Suppose we see one savings account offering an APR of 4.32% compounded quarterly and another one offering 4.27% compounded monthly. What we really want to know is which one pays us more money at the end of the year. This is what the *annual percentage yield* or APY tells us.

KEY CONCEPT

The annual percentage yield or APY is the actual percentage return earned in a year. Unlike the APR, the APY tells us the actual percentage growth per year, including returns on investment due to compounding.

A federal law passed in 1991 requires banks to disclose the APY.

To understand what the APY means, let's look at a simple example. Suppose we invest \$100 in an account that pays 10% APR compounded semi-annually. We want to see how much interest is earned in a year. The period interest rate is

$$\frac{\text{APR}}{2} = \frac{10\%}{2} = 5\%$$

so we take $r = 0.05$. The number of periods in a year is $t = 2$. By the compound interest formula (Formula 4.3), the balance at the end of one year is

$$\text{Principal} \times (1 + r)^t = \$100 \times 1.05^2 = \$110.25$$

We have earned a total of \$10.25 in interest. As a percentage of \$100, that is 10.25%. This number is the APY. It is the actual percent interest earned over the period of one year. The APY is always at least as large as the APR.

"A FIVE YEAR CD? WHO'S GOT TIME TO LISTEN TO THAT?!"

Here is a formula for the APY.

FORMULA 4.4 APY Formula

$$\text{APY} = \left(1 + \frac{\text{APR}}{n}\right)^n - 1$$

Here both APY and APR are in decimal form, and n is the number of compounding periods per year.

A derivation of this formula is shown in Algebraic Spotlight 4.1. Let's apply this formula in the preceding example with 10% APR compounded semi-annually. Compounding is semi-annual, so the number of periods is $n = 2$, and the APR as a decimal is 0.10:

$$\begin{aligned}\text{APY} &= \left(1 + \frac{\text{APR}}{n}\right)^n - 1\\ &= \left(1 + \frac{0.10}{2}\right)^2 - 1\\ &= 0.1025\end{aligned}$$

As a percent, this is 10.25%—the same as we found above.

ALGEBRAIC SPOTLIGHT 4.1 Calculating APY

Suppose we invest money in an account that is compounded n times per year. Then the period interest rate is

$$\text{Period rate} = \frac{\text{APR}}{n}$$

The first step is to find the balance after one year. Using the compound interest formula (Formula 4.3), we find the balance to be

$$\text{Balance after 1 year} = \text{Principal} \times \left(1 + \frac{\text{APR}}{n}\right)^n$$

How much money did we earn? We earned

$$\text{Balance minus Principal} = \text{Principal} \times \left(1 + \frac{\text{APR}}{n}\right)^n - \text{Principal}$$

If we divide these earnings by the amount we started with, namely, the principal, we get the percentage increase:

$$\text{APY as a decimal} = \left(1 + \frac{\text{APR}}{n}\right)^n - 1$$

EXAMPLE 4.4 Calculating APY: An account with monthly compounding

We have an account that pays an APR of 10%. If interest is compounded monthly, find the APY. Round your answer as a percentage to two decimal places.

SOLUTION

We use the APY formula (Formula 4.4). As a decimal, 10% is 0.10, and there are $n = 12$ compounding periods in a year. Therefore,

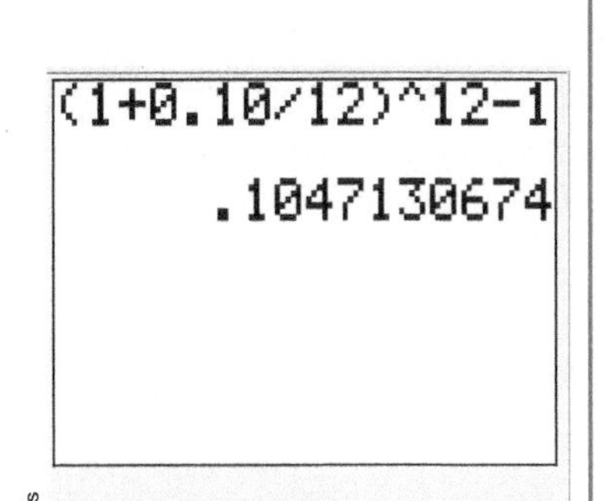

Benny Evans

$$\text{APY} = \left(1 + \frac{\text{APR}}{n}\right)^n - 1$$
$$= \left(1 + \frac{0.10}{12}\right)^{12} - 1$$

To four decimal places, this is 0.1047. Thus, the APY is about 10.47%.

TRY IT YOURSELF 4.4

We have an account that pays an APR of 10%. If interest is compounded daily, find the APY. Round your answer as a percentage to two decimal places.

The answer is provided at the end of this section.

Using the APY

We can use the APY as an alternative to the APR for calculating compound interest. **Table 4.1** gives both the APR and the APY for CDs from First Command

TABLE 4.1 Rates from First Command Bank (2008–2009)

CD rates with quarterly compounding[1]		
30 Day	APR	APY
$1000 – $99,999.99	3.21%	3.25%
$100,000+	3.26%	3.30%
90 Day	APR	APY
$1000 – $9999.99	3.22%	3.25%
$10,000 – $99,999.99	3.26%	3.30%
$100,000+	3.35%	3.40%
1 Year	APR	APY
$1000 – $9999.99	3.31%	3.35%
$10,000 – $99,999.99	3.35%	3.40%
$100,000+	3.55%	3.60%
18 Month	APR	APY
$1000 – $9999.99	3.56%	3.60%
$10,000 – $99,999.99	3.59%	3.65%
$100,000+	3.74%	3.80%
2 Year	APR	APY
$1000 – $9999.99	3.80%	3.85%
$10,000 – $99,999.99	3.84%	3.90%
$100,000+	3.98%	4.05%

[1] *According to federal regulations, the tolerance of 1/8 of 1 percentage point above or below the annual percentage rate applies to any required disclosure of the annual percentage rate. See* `https://www.fdic.gov/regulations/laws/rules/6500-2280.html.`

Bank during late 2008 to early 2009. We use these old rates because they demonstrate the point we are trying to make better than the current lower rates do.

The APY can be used directly to calculate a year's worth of interest. For example, suppose we purchase a one-year \$100,000 CD from First Command Bank. Table 4.1 gives the APY for this CD as 3.60%. So after one year the interest we will earn is

$$\text{One year's interest} = 3.60\% \text{ of Principal} = 0.036 \times \$100{,}000 = \$3600$$

EXAMPLE 4.5 Using APY to find the value of a CD

a. Suppose we purchase a one-year CD from First Command Bank for \$25,000. According to Table 4.1, what is the value of the CD at the end of the year?

b. In March 2014, GE Capital Retail Bank advertised a one-year \$25,000 CD at 1.04% compounded daily. They said this was an APY of 1.05%. If we purchased this one-year CD for \$25,000, what would its value be at the end of the year?

SOLUTION

a. According to Table 4.1, the APY for this CD is 3.40%. This means that the CD earns 3.40% interest over the period of one year. Therefore,

$$\text{One year's interest} = 3.4\% \text{ of Principal} = 0.034 \times \$25{,}000 = \$850$$

We add this to the principal to find the balance:

$$\text{Value after 1 year} = \$25{,}000 + \$850 = \$25{,}850$$

b. The APY for this CD is 1.05%, meaning the CD earns 1.05% interest over the period of one year. Therefore,

$$\text{One year's interest} = 1.05\% \text{ of Principal} = 0.0105 \times \$25{,}000 = \$262.50$$

We add this to the principal to find the balance:

$$\text{Value after 1 year} = \$25{,}000 + \$262.50 = \$25{,}262.50$$

TRY IT YOURSELF 4.5

Suppose we purchase a one-year CD from First Command Bank for \$125,000. According to Table 4.1, what is the value of the CD at the end of the year?

The answer is provided at the end of this section.

We can also use the APY rather than the APR to calculate the balance over several years if we wish. The APY tells us the actual percentage growth per year, including interest we earned during the year due to periodic compounding.[2] Thus, we can think of compounding annually (regardless of the actual compounding period) using the APY as the annual interest rate. Once again, the balance is an exponential function. In this formula, the APY is in decimal form.

FORMULA 4.5 APY Balance Formula

$$\text{Balance after } y \text{ years} = \text{Principal} \times (1 + \text{APY})^y$$

[2] *The idea of an APY normally applies only to compound interest. It is not used for simple interest.*

EXAMPLE 4.6 Using APY balance formula: CD balance at maturity

In April 2021, a bank offered a five-year CD at 1.09% APY. If you buy a \$100,000 CD, calculate the balance at maturity.

SOLUTION

The APY in decimal form is 0.0109 and the CD matures after five years, so we use $y = 5$ in the APY balance formula (Formula 4.5):

$$\begin{aligned}\text{Balance after 10 years} &= \text{Principal} \times (1 + \text{APY})^y \\ &= \$100{,}000 \times 1.0109^5 \\ &= \$105{,}570.11\end{aligned}$$

TRY IT YOURSELF 4.6

Suppose we earn 1.07% APY on a six-year \$50,000 CD. Calculate the balance at maturity.

The answer is provided at the end of this section.

SUMMARY 4.2
APY

1. The APY gives the true (effective) annual interest rate. It takes into account money earned due to compounding.
2. If n is the number of compounding periods per year,

$$\text{APY} = \left(1 + \frac{\text{APR}}{n}\right)^n - 1$$

Here, the APR and APY are both in decimal form.

3. The APY is always at least as large as the APR. When interest is compounded annually, the APR and APY are equal. When compounding is more frequent, the APY is larger than the APR. The more frequent the compounding, the greater the difference.
4. The APY can be used to calculate the account balance after y years:

$$\text{Balance after } t \text{ years} = \text{Principal} \times (1 + \text{APY})^y$$

Here, the APY is in decimal form.

Future and present value

Often we invest with a goal in mind, for example, to make a down payment on the purchase of a car. The amount we invest is called the *present value*. The amount the account is worth after a certain period of time is called the *future value* of the original investment. Sometimes we know one of these two and would like to calculate the other.

KEY CONCEPT

The present value of an investment is the amount we initially invest. The future value is the value of that investment at some specified time in the future.

If the account grows only by compounding each period at a constant interest rate after we make an initial investment, then the present value is the principal, and the future value is the balance given by the compound interest formula (Formula 4.3):

$$\text{Balance after } t \text{ periods} = \text{Principal} \times (1 + r)^t$$

so

$$\text{Future value} = \text{Present value} \times (1 + r)^t$$

We can rearrange this formula to obtain

$$\text{Present value} = \frac{\text{Future value}}{(1 + r)^t}$$

In these formulas, t is the total number of compounding periods, and r is the period interest rate expressed as a decimal.

Which of these two formulas we should use depends on what question we are trying to answer. By the way, if we make regular deposits into the account, then the formulas are much more complicated. We will examine that situation in Section 4.3.

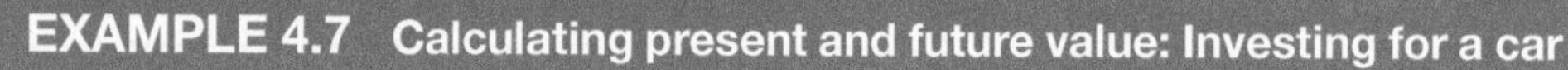

EXAMPLE 4.7 Calculating present and future value: Investing for a car

You would like to have \$10,000 to buy a car three years from now. How much would you need to invest now in a savings account that pays an APR of 9% compounded monthly?

SOLUTION

In this problem, we know the future value (\$10,000) and would like to know the present value. The monthly rate is $r = 0.09/12 = 0.0075$ as a decimal, and the number of compounding periods is $t = 36$ months. Thus,

$$\begin{aligned}\text{Present value} &= \frac{\text{Future value}}{(1 + r)^t} \\ &= \frac{\$10{,}000}{1.0075^{36}} \\ &= \$7641.49\end{aligned}$$

Therefore, you should invest \$7641.49 now.

TRY IT YOURSELF 4.7

Find the future value of an account after four years if the present value is \$900, the APR is 8%, and interest is compounded quarterly.

The answer is provided at the end of this section.

Doubling time for investments

Exponential functions eventually get very large. This means that even a modest investment today in an account that pays compound interest will grow to be very large in the future. In fact, your money will eventually double and then double again.

In Section 3.3 of Chapter 3, we used logarithms to find the doubling time. Now we introduce a quick way of estimating the doubling time called the *Rule of 72*.

KEY CONCEPT

The Rule of 72 says that the doubling time in years is about 72/APR. Here, the APR is expressed as a percentage, not as a decimal. The estimate is fairly accurate if the APR is 15% or less.

SUMMARY 4.3
Doubling Time Revisited

- The exact doubling time is given by the formula

$$\text{Number of periods to double} = \frac{\log 2}{\log(\text{Base})} = \frac{\log 2}{\log(1+r)}$$

Here, r is the period interest rate as a decimal.

- We can approximate the doubling time using the *Rule of 72*:

$$\text{Estimate for doubling time} = \frac{72}{\text{APR}}$$

Here, the APR is expressed as a percentage, not as a decimal, and time is measured in years. The estimate is fairly accurate if the APR is 15% or less, but we emphasize that this is only an approximation.

EXAMPLE 4.8 Computing doubling time: An account with quarterly compounding

Suppose an account earns an APR of 8% compounded quarterly. First estimate the doubling time using the Rule of 72. Then calculate the exact doubling time and compare the result with your estimate.

SOLUTION

The Rule of 72 gives the estimate

$$\text{Estimate for doubling time} = \frac{72}{\text{APR}} = \frac{72}{8} = 9 \text{ years}$$

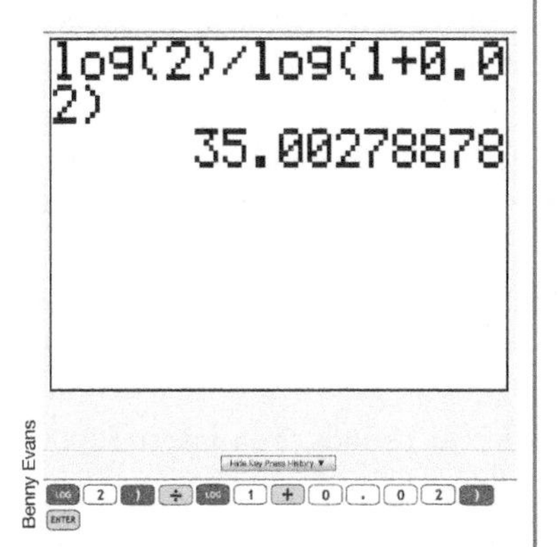

Benny Evans

To find the exact doubling time, we need the period interest rate r. The period is a quarter, so $r = 0.08/4 = 0.02$. Putting this result into the doubling time formula, we find

$$\begin{aligned}\text{Number of periods to double} &= \frac{\log 2}{\log(1+r)} \\ &= \frac{\log 2}{\log(1+0.02)}\end{aligned}$$

The result is about 35.0. Therefore, the actual doubling time is 35.0 quarters, or eight years and nine months. Our estimate of nine years was three months too high.

TRY IT YOURSELF 4.8

Suppose an account has an APR of 12% compounded monthly. First estimate the doubling time using the Rule of 72, and then calculate the exact doubling time in years and months.

The answer is provided at the end of this section.

WHAT DO YOU THINK?

APR and APY: In this section we discussed the APR and the APY. A fellow student says that the APY gives a better indication of the return you can expect on an investment. Do you agree? Explain.

Questions of this sort may invite many correct responses. We present one possibility: The APY is indeed a better indication of the return you can expect on an investment than is the APR, just as it is a better indication of the interest you will pay on a consumer loan. The APR does not take into account the number of compounding periods, so this figure alone cannot tell you the interest that will be earned. The APY uses both the APR and the number of compounding periods to give the actual percentage earnings for an investment. The APY is always at least as large as the APR.

Further questions of this type may be found in the *What Do You Think?* section of exercises.

Try It Yourself answers

Try It Yourself 4.1: Calculating simple interest: An account $540.

Try It Yourself 4.2: Calculating compound interest: Annual compounding $631.24.

Try It Yourself 4.4: Calculating APY: An account with daily compounding 10.52%.

Try It Yourself 4.5: Using APY to find the value of a CD $129,500.

Try It Yourself 4.6: Using APY balance formula: CD balance at maturity $53,297.10.

Try It Yourself 4.7: Calculating present and future value: An account with quarterly compounding $1235.51.

Try It Yourself 4.8: Computing doubling time: An account with monthly compounding The Rule of 72 gives an estimate of six years. The exact formula gives 69.7 months, or about five years and 10 months.

Exercise Set 4.1

Test Your Understanding

1. True or false: Simple interest means your money earns the same amount no matter how long it stays in an account.

2. True or false: Simple interest never earns more than compound interest.

3. To say that interest is compounded means ________.

4. To find the APR, we multiply the period interest rate by ________.

5. True or false: Under some conditions, the APR can be more than the APY.

6. True or false: To say the interest earned by a savings account is compounded semi-annually means that equal amounts of money are deposited twice during the year.

7. The Rule of 72 tells you approximately: **a.** how much money you will have after 72 months **b.** how long it takes money earning compound interest to double **c.** what interest rate you need to double your money **d.** how much money you will have in a year if you invest 72 dollars.

8. The principal for a savings account is: **a.** the amount that you will eventually have in your account **b.** the interest rate **c.** the person who sets up the account **d.** the initial amount that is deposited.

Problems

In exercises for which you are asked to calculate the APR or APY as the final answer, round your answer as a percentage to two decimal places.

9. If the interest earned by a savings account paying an APR of 8% is compounded quarterly, what percent of the current balance is deposited every three months during the year? **a.** 2% **b.** 4% **c.** 6% **d.** 8%

10. If an investment pays an APR of 12% compounded monthly, what percentage of the current balance is added to the investment each month?

11. Simple interest. We invest $4000 in an account that pays simple interest of 3% each year. How much interest is earned after 10 years?

12. Compound interest. You invest $1000 in an account that pays 5% compounded annually. What is the balance after two years?

13. Compounding using different periods. You invest $2000 in an account that pays an APR of 6%.

a. What is the value of the investment after three years if interest is compounded yearly? Round your answer to the nearest cent.

b. What is the value of the investment after three years if interest is compounded monthly? Round your answer to the nearest cent.

14. Calculating APY. Find the APY for an account that pays an APR of 12% if interest is compounded monthly.

15. Simple interest. Assume that a three-month CD purchased for $2000 pays simple interest at an annual rate of 10%. How much total interest does it earn? What is the balance at maturity?

16. More simple interest. Assume that a 30-month CD purchased for $3000 pays simple interest at an annual rate of 5.5%. How much total interest does it earn? What is the balance at maturity?

17. Make a table. Suppose you put $3000 in a savings account at an APR of 8% compounded quarterly. Fill in the table below. (Calculate the interest and compound it each quarter rather than using the compound interest formula.)

Quarter	Interest earned	Balance
		$3000.00
1	$	$
2	$	$
3	$	$
4	$	$

18. Make another table. Suppose you put $4000 in a savings account at an APR of 6% compounded monthly. Fill in the table below. (Calculate the interest and compound it each month rather than using the compound interest formula.)

Month	Interest earned	Balance
		$4000.00
1	$	$
2	$	$
3	$	$

19. Compound interest calculated. Assume that we invest $2000 for one year in a savings account that pays an APR of 10% compounded quarterly.

a. Make a table to show how much is in the account at the end of each quarter. (Calculate the interest and compound it each quarter rather than using the compound interest formula.)

b. Use your answer to part a to determine how much total interest the account has earned after one year.

c. Compare the earnings to what simple interest or semi-annual compounding would yield.

20. Using the compound interest formula. *This is a continuation of Exercise 19.* In Exercise 19, we invested $2000 for one year in a savings account with an APR of 10% compounded quarterly. Apply the compound interest formula (Formula 4.3) to see whether it gives the answer for the final balance obtained in Exercise 19.

21. The difference between simple and compound interest. Suppose you invest $1000 in a savings account that pays an APR of 0.6%. If the account pays simple interest, what is the balance in the account after 20 years? If interest is compounded monthly, what is the balance in the account after 20 years?

22. Calculating interest. Assume that an investment of $7000 earns an APR of 6% compounded monthly for 18 months.

a. How much money is in your account after 18 months?

b. How much interest has been earned?

23. Retirement options. At age 25, you start work for a company and are offered two retirement options.

Retirement option 1: When you retire, you will receive a lump sum of $30,000 for each year of service.

Retirement option 2: When you start to work, the company deposits $15,000 into an account that pays a monthly interest rate of 1%, and interest is compounded monthly. When you retire, you get the balance of the account.

Which option is better if you retire at age 65? Which is better if you retire at age 55?

24. Compound interest. Assume that an 18-month CD purchased for $7000 pays an APR of 6% compounded monthly. What is the APY? Would the APY change if the investment were $11,000 for 30 months with the same APR and with monthly compounding?

25. More compound interest. Assume that a 24-month CD purchased for $7000 pays an APY of 4.25% (and of course interest is compounded). How much do you have at maturity?

26. Interest and APY. Assume that a one-year CD purchased for $2000 pays an APR of 8% that is compounded semi-annually. How much is in the account at the end of each compounding period? (Calculate the interest and compound it each period rather than using the compound interest formula.) How much total interest does it earn? What is the APY?

27. More interest and APY. Assume that a one-year CD purchased for $2000 pays an APR of 8% that is compounded quarterly. How much is in the account at the end of each compounding period? (Calculate the interest and compound it each period rather than using the compound interest formula.) How much total interest does it earn? What is the APY?

28. A CD. Suppose you bought a two-year CD for $10,000 at an APR of 3%.

a. What is the APY, assuming monthly compounding?

b. What is the total interest paid on the CD over the two-year period?

29. Some interest and APY calculations. Parts b and c refer to the 2008–2009 rates at First Command Bank shown in Table 4.1.

a. Assume that a one-year CD for $5000 pays an APR of 8% that is compounded quarterly. How much total interest does it earn? What is the APY?

b. If you purchased a one-year CD for $150,000 from First Command Bank, how much interest would you have received at maturity? Was compounding taking place? Explain.

c. If you purchased a two-year CD for $150,000 from First Command Bank, the APY (4.05%) was greater than the APR (3.98%) because compounding was taking place. Use the APR to calculate what the APY would be with daily compounding. How does your answer compare to the APY in the table?

30. Interest and APR. Assume that a two-year CD for $4000 pays an APY of 8%. How much interest will it earn? Can you determine the APR?

31. Find the APR. Sue bought a six-month CD for $3000. She said that at maturity it paid $112.50 in interest. Assume that this was simple interest, and determine the APR.

32. Getting rich. The *Dilbert* cartoon shown above jokes about making a million dollars by investing $100 at an APR of 5% and waiting for 190 years. Assuming that interest is compounded annually, how much does this investment really produce?

33. Future value. What is the future value of a 10-year investment of $1000 at an APR of 9% compounded monthly? Explain what your answer means.

34. Present value. What is the present value of an investment that will be worth $2000 at the end of five years? Assume an APR of 6% compounded monthly. Explain what your answer means.

35. Doubling again. You have invested $2500 at an APR of 9%. Use the Rule of 72 to estimate how long it will be until your investment reaches $5000, and how long it will be until your investment reaches $10,000.

36. Getting APR from doubling rate. A friend tells you that her savings account doubled in 12 years. Use the Rule of 72 to estimate what the APR of her account was.

37. Find the doubling time. Consider an investment of $3000 at an APR of 6% compounded monthly. Use the formula that gives the exact doubling time to determine exactly how long it will take for the investment to double. (See the first part of Summary 4.3. Be sure to use the monthly rate for r.) Express your answer in years and months. Compare this result with the estimate obtained from the Rule of 72.

Exercises 38 through 40 are designed to be solved using technology such as calculators or computer spreadsheets.

38. Find the rate. A news article published in 2004 said that the U.S. House of Representatives passed a bill to distribute funds to members of the Western Shoshone tribe. Here is an excerpt from that article:

> The Indian Claims Commission decided the Western Shoshone lost much of their land to gradual encroachment. The tribe was awarded $26 million in 1977. That has grown to about $145 million through compound interest, but the tribe never took the money.

Assume monthly compounding and determine the APR that would give this growth in the award over the 27 years from 1977 to 2004. *Note*: This can also be solved without technology, using algebra.

39. Solve for the APR. Suppose a CD advertises an APY of 8.5%. Assuming that the APY was a result of monthly compounding, solve the equation

$$0.085 = \left(1 + \frac{\text{APR}}{12}\right)^{12} - 1$$

to find the APR. *Note*: This can also be solved without technology, using algebra.

40. Find the compounding period. Suppose a CD advertises an APR of 5.10% and an APY of 5.20%. Solve the equation

$$0.052 = \left(1 + \frac{0.051}{n}\right)^{n} - 1$$

for n to determine how frequently interest is compounded.

Writing About Mathematics

41. Current rates. Look up some current interest rates being paid on CDs by different financial institutions and write a report on your findings.

42. Bond rates. Interest rates on bonds can differ significantly among nations. Look up some current interest rates being paid on bonds by different countries and write a report on your findings.

43. Hamilton. Alexander Hamilton was the first U.S. Treasury secretary. Write a report on Hamilton and his accomplishments.

44. Usury. Look up the meaning of the word *usury* and write a report on how it is viewed in different cultures and religions.

What Do You Think?

45. Present and future value. Explain the meaning of the terms *future value* and *present value*.

46. Simple formula? You and a friend are discussing the return on an account with an initial balance of $100 and an APY of 5%. You want to use the APY balance formula (Formula 4.5). Your friend says that he has a simpler approach: Each year add 5% of $100 or $5 to the balance. Is your friend's approach valid? What term is used to describe the interest calculation employed by your friend? Under what conditions will your friend's approach give a reasonably good solution?

47. Simple versus compound interest. What features would you expect to see if you graph the balance of an account earning simple interest? What features would you expect to see if the account earns compound interest? **Exercise 21 is relevant to this topic.**

48. Which would you choose? You have the option of borrowing money from one source that charges simple interest or from another source that charges the same APR but compounds the interest monthly. Which would you choose, and why? **Exercise 21 is relevant to this topic.**

49. Which would you choose? You have the option of loaning money to one friend who promises to pay simple interest or to another friend who promises to pay the same APR but compound the interest. Which would you choose, and why? **Exercise 21 is relevant to this topic.**

50. Given APR. You are considering two banks. Both banks offer savings accounts with an APR of 2%. What other information would you want to know so that you can choose between the two banks?

51. Doubling the number of compounding periods. If a bank doubles the number of times per year that it compounds interest, does that double the amount you earn in a year on your savings account? Explain.

52. Doubling principal invested. If you double the principal you invest in your savings account, does that double the interest you earn in a year? Explain.

53. APR = APY. A bank tells you that its APR and APY are the same. What does that tell you about compounding?

4.2 Borrowing: How much car can you afford?

TAKE AWAY FROM THIS SECTION
Be able to calculate the monthly payment on a loan.

The following article appeared at the site COLLEGEdata.

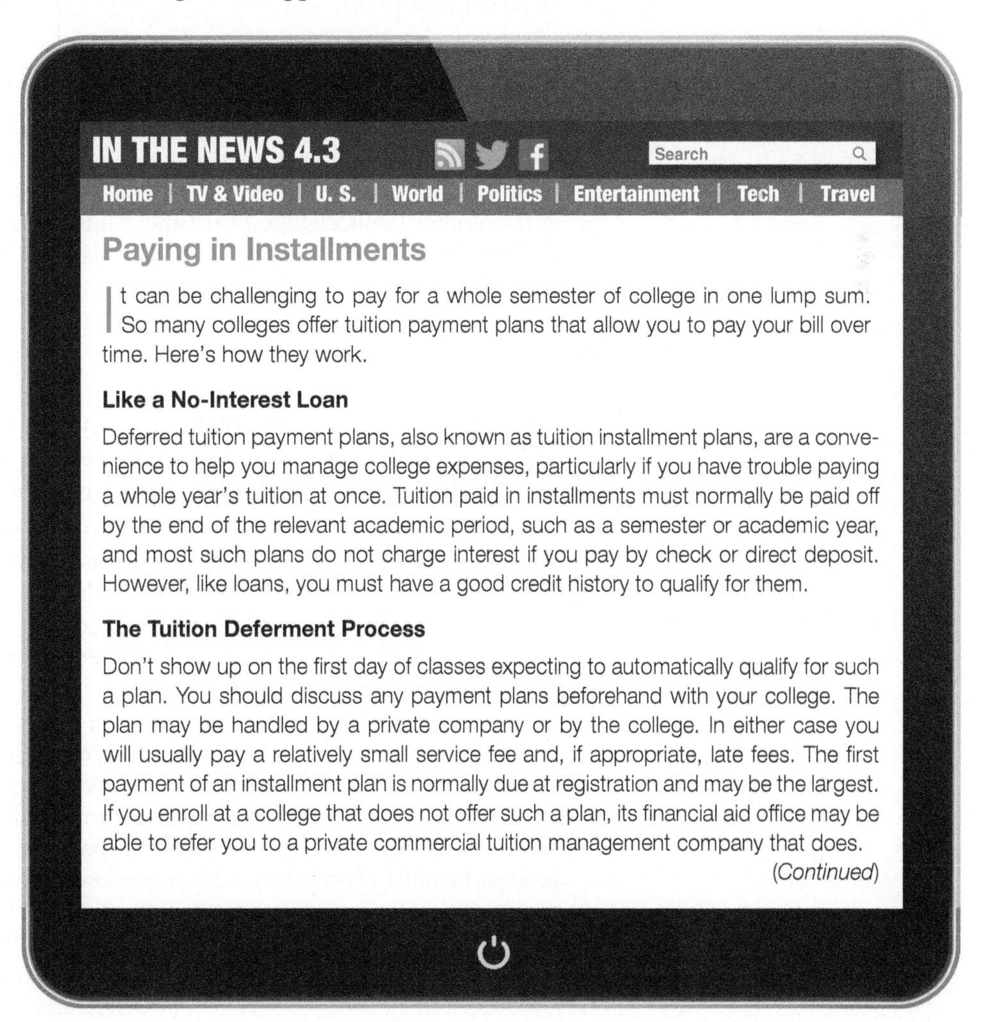
IN THE NEWS 4.3

Search

Home | TV & Video | U. S. | World | Politics | Entertainment | Tech | Travel

Paying in Installments

It can be challenging to pay for a whole semester of college in one lump sum. So many colleges offer tuition payment plans that allow you to pay your bill over time. Here's how they work.

Like a No-Interest Loan

Deferred tuition payment plans, also known as tuition installment plans, are a convenience to help you manage college expenses, particularly if you have trouble paying a whole year's tuition at once. Tuition paid in installments must normally be paid off by the end of the relevant academic period, such as a semester or academic year, and most such plans do not charge interest if you pay by check or direct deposit. However, like loans, you must have a good credit history to qualify for them.

The Tuition Deferment Process

Don't show up on the first day of classes expecting to automatically qualify for such a plan. You should discuss any payment plans beforehand with your college. The plan may be handled by a private company or by the college. In either case you will usually pay a relatively small service fee and, if appropriate, late fees. The first payment of an installment plan is normally due at registration and may be the largest. If you enroll at a college that does not offer such a plan, its financial aid office may be able to refer you to a private commercial tuition management company that does.

(Continued)

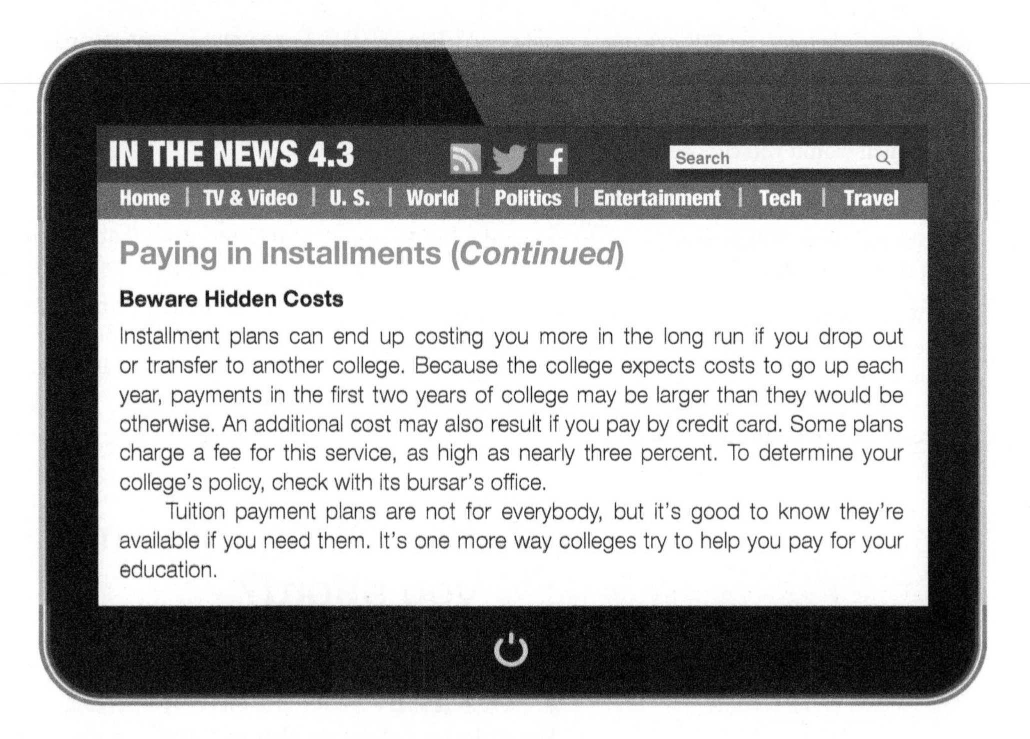
IN THE NEWS 4.3

Home | TV & Video | U. S. | World | Politics | Entertainment | Tech | Travel

Paying in Installments (*Continued*)

Beware Hidden Costs

Installment plans can end up costing you more in the long run if you drop out or transfer to another college. Because the college expects costs to go up each year, payments in the first two years of college may be larger than they would be otherwise. An additional cost may also result if you pay by credit card. Some plans charge a fee for this service, as high as nearly three percent. To determine your college's policy, check with its bursar's office.

Tuition payment plans are not for everybody, but it's good to know they're available if you need them. It's one more way colleges try to help you pay for your education.

Installment payments are a part of virtually everyone's life. The goal of this section is to explain how such payments are calculated and to see the implications of installment plans.

Installment loans and the monthly payment

When we borrow money to buy a car or a house, the lending institution typically requires that we pay back the loan plus interest by making the same payment each month for a certain number of months. Note that interest rates for mortgages and various other loans change over time. Thus, in many of the examples and exercises in this section we use interest rates that may not be current, just as we did with CD interest rates in Section 4.1.

KEY CONCEPT

With an installment loan you borrow money for a fixed period of time, called the term of the loan, and you make regular payments (usually monthly) to pay off the loan plus interest accumulated during that time.

To see how the monthly payment is calculated, we consider a simple example. Suppose you need \$100 to buy a calculator but don't have the cash available. Your sister is willing, however, to lend you the money at a rate of 5% per month, provided you repay it with equal payments over the next two months. (By the way, this is an astronomical APR of 60%.)

How much must you pay each month? Because the loan is for only two months, you need to pay at least half of the \$100, or \$50, each of the two months. But that doesn't account for the interest. If we were to pay off the loan in a lump sum at the end of the two-month term, the interest on the account would be calculated in the same way as for a savings account that pays 5% per month. Using the compound interest formula (Formula 4.3 from Section 4.1), we find

$$\begin{aligned}\text{Balance owed after 2 months} &= \text{Principal} \times (1+r)^t \\ &= \$100 \times 1.05^2 = \$110.25\end{aligned}$$

This would be two monthly payments of approximately \$55.13. This amount overestimates your monthly payment, however: in the second payment, you shouldn't have to pay interest on the amount you have already repaid.

We will see shortly that the correct payment is \$53.78 for each of the two months. This is not an obvious amount, but it is reasonable—it lies between the two extremes of \$50 and \$55.13.

When we take out an installment loan, the amount of the payment depends on three things: the amount of money we borrow (sometimes called the *principal*), the interest rate (or APR), and the term of the loan.

Each payment reduces the balance owed, but at the same time interest is accruing on the outstanding balance. This makes the calculation of the monthly payment fairly complicated, as you might surmise from the following formula.[3]

FORMULA 4.6 Monthly Payment Formula

$$\text{Monthly payment} = \frac{\text{Amount borrowed} \times r(1+r)^t}{((1+r)^t - 1)}$$

Here, t is the term in months and $r = \text{APR}/12$ is the monthly interest rate as a decimal.

Alternatively, we can write the formula as

$$\text{Monthly payment} = \frac{\text{Amount borrowed} \times \frac{\text{APR}}{12}\left(1+\frac{\text{APR}}{12}\right)^{(12y)}}{\left(\left(1+\frac{\text{APR}}{12}\right)^{(12y)} - 1\right)}$$

where y is the term in years.

A derivation of this formula is presented at the end of this section in Algebraic Spotlight 4.2 and Algebraic Spotlight 4.3.

To give a sense of these monthly payments, here are two examples that might arise when getting a mortgage:

- If you borrowed \$100,000 at 6% APR for 30 years, your monthly payment would be approximately \$600.
- If you borrowed \$100,000 at 3% APR for 15 years, your monthly payment would be approximately \$690.

If you want to borrow money, the monthly payment formula allows you to determine in advance whether you can afford that car or home you want to buy. The formula also lets you check the accuracy of any figure that a potential lender quotes.

Let's return to the earlier example of a \$100 loan to buy a calculator. We have a monthly rate of 5%. Expressed as a decimal, 5% is 0.05, so $r = 0.05$ and $1 + r = 1.05$. Because we pay off the loan in two months, we use $t = 2$ in the monthly payment formula (Formula 4.6):

$$\begin{aligned}\text{Monthly payment} &= \frac{\text{Amount borrowed} \times r(1+r)^t}{((1+r)^t - 1)} \\ &= \frac{\$100 \times 0.05 \times 1.05^2}{(1.05^2 - 1)} \\ &= \$53.78\end{aligned}$$

[3] *This formula is built in as a feature on some hand-held calculators and most computer spreadsheets. Loan calculators are also available on the Web.*

EXAMPLE 4.9 Using the monthly payment formula: College loan

You need to borrow \$5000 so you can attend college next fall. You get the loan at an APR of 6% to be paid off in monthly installments over three years.[4]

a. Calculate your monthly payment.

b. What is the total of all payments?

c. How much interest was paid in all?

SOLUTION

a. The monthly rate as a decimal is

$$r = \text{Monthly rate} = \frac{\text{APR}}{12} = \frac{0.06}{12} = 0.005$$

This gives $1 + r = 1.005$. We want to pay off the loan in three years or 36 months, so we use a term of $t = 36$ in the monthly payment formula:

$$\begin{aligned}\text{Monthly payment} &= \frac{\text{Amount borrowed} \times r(1+r)^t}{((1+r)^t - 1)} \\ &= \frac{\$5000 \times 0.005 \times 1.005^{36}}{(1.005^{36} - 1)} \\ &= \$152.11\end{aligned}$$

```
5000*0.005*1.005
^36/(1.005^36-1)
        152.1096873
```

Benny Evans

b. There are 36 payments of \$152.11 each, so the total of all payments is

$$36 \times \$152.11 = \$5475.96$$

c. Because the total payments are \$5475.96, of which \$5000 was the amount borrowed, the difference,

$$\$5475.96 - 5000 = \$475.96,$$

is the amount of interest paid over the life of the loan.

TRY IT YOURSELF 4.9

You borrow \$8000 at an APR of 9% to be paid off in monthly installments over four years. Calculate your monthly payment.

The answer is provided at the end of this section.

Suppose you can afford a certain monthly payment and you'd like to know how much you can borrow to stay within that budget. The monthly payment formula can be rearranged to answer that question.

[4] *In April 2021, federally subsidized student loans were around 2.75% APR, and some private loans were advertised at about 4.25% APR.*

FORMULA 4.7 Companion to the Monthly Payment Formula (Formula for Amount Borrowed)

$$\text{Amount borrowed} = \frac{\text{Monthly payment} \times ((1+r)^t - 1)}{(r \times (1+r)^t)}$$

where t is the term of the loan in months.

Alternatively, we can write the formula as

$$\text{Amount borrowed} = \frac{\text{Monthly payment} \times \left(\left(1 + \frac{\text{APR}}{12}\right)^{(12y)} - 1\right)}{\left(\frac{\text{APR}}{12} \times \left(1 + \frac{\text{APR}}{12}\right)^{(12y)}\right)}$$

where y is the term of the loan in years.

EXAMPLE 4.10 Computing how much I can borrow: Buying a car

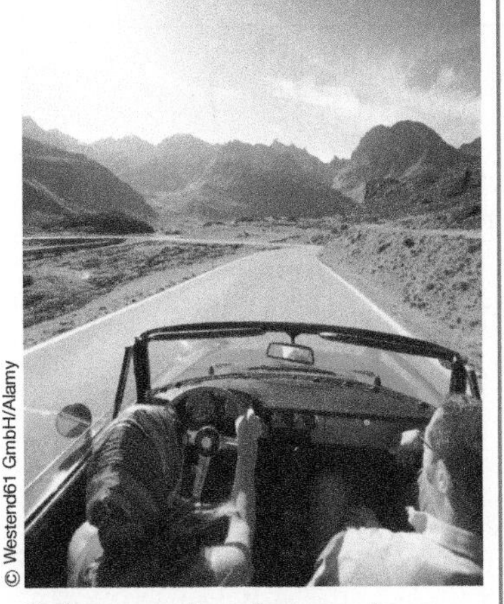

Before you shop for a car, know how much car you can afford.

```
250*((1+0.05/12)
^36-1)/((0.05/12
)*(1+0.05/12)^36
)
        8341.425321
```

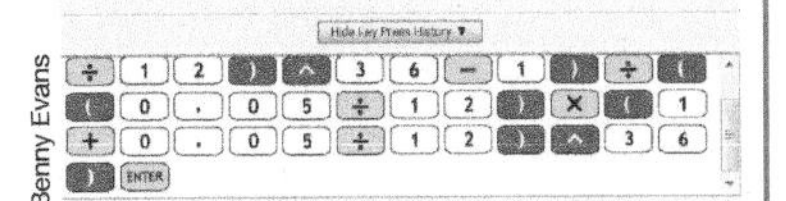

We can afford to make payments of \$250 per month for three years. Our car dealer is offering us a loan at an APR of 5%. What price automobile should we be shopping for?

SOLUTION

The monthly rate as a decimal is

$$r = \text{Monthly rate} = \frac{0.05}{12}$$

To four decimal places, this is 0.0042, but for better accuracy, we won't round this number. Now, three years is 36 months, so we use $t = 36$ in the companion payment formula (Formula 4.7):

$$\begin{aligned}\text{Amount borrowed} &= \frac{\text{Monthly payment} \times ((1+r)^t - 1)}{(r \times (1+r)^t)} \\ &= \frac{\$250 \times ((1+0.05/12)^{36} - 1)}{((0.05/12) \times (1+0.05/12)^{36})} \\ &= \$8341.43\end{aligned}$$

We should shop for cars that cost \$8341.43 or less.

TRY IT YOURSELF 4.10

We can afford to make payments of \$300 per month for four years. We can get a loan at an APR of 4%. How much money can we afford to borrow?

The answer is provided at the end of this section.

SUMMARY 4.4
Monthly Payments

In parts 1 and 2, the monthly rate r is the APR in decimal form divided by 12, and t is the term in months.

1. The monthly payment is

$$\text{Monthly payment} = \frac{\text{Amount borrowed} \times r(1+r)^t}{((1+r)^t - 1)}$$

2. A companion formula gives the amount borrowed in terms of the monthly payment:

$$\text{Amount borrowed} = \frac{\text{Monthly payment} \times ((1+r)^t - 1)}{(r \times (1+r)^t)}$$

3. Alternatively, we can write these formulas in term of y, the term of the loan in years:

$$\text{Monthly payment} = \frac{\text{Amount borrowed} \times \frac{\text{APR}}{12}\left(1+\frac{\text{APR}}{12}\right)^{(12y)}}{\left(\left(1+\frac{\text{APR}}{12}\right)^{(12y)} - 1\right)}$$

and

$$\text{Amount borrowed} = \frac{\text{Monthly payment} \times \left(\left(1+\frac{\text{APR}}{12}\right)^{(12y)} - 1\right)}{\left(\frac{\text{APR}}{12} \times \left(1+\frac{\text{APR}}{12}\right)^{(12y)}\right)}$$

4. These formulas are sensitive to round-off error, so it is best to do calculations all at once, keeping all the decimal places rather than doing parts of a computation and entering the rounded numbers.

EXAMPLE 4.11 Calculating monthly payment and amount borrowed: A new car

Suppose we need to borrow \$15,000 at an APR of 9% to buy a new car.

a. What will the monthly payment be if we borrow the money for $3\frac{1}{2}$ years? How much interest will we have paid by the end of the loan?

b. We find that we cannot afford the \$15,000 car because we can only afford a monthly payment of \$300. What price car can we shop for if the dealer offers a loan at a 9% APR for a term of $3\frac{1}{2}$ years?

SOLUTION

a. The monthly rate as a decimal is

$$r = \text{Monthly rate} = \frac{\text{APR}}{12} = \frac{0.09}{12} = 0.0075$$

We are paying off the loan in $3\frac{1}{2}$ years, so $t = 3.5 \times 12 = 42$ months. Therefore, by the monthly payment formula (Formula 4.6),

$$\text{Monthly payment} = \frac{\text{Amount borrowed} \times r(1+r)^t}{((1+r)^t - 1)}$$

$$= \frac{\$15{,}000 \times 0.0075 \times 1.0075^{42}}{(1.0075^{42} - 1)}$$

$$= \$417.67$$

Now let's find the amount of interest paid. We will make 42 payments of \$417.67 for a total of $42 \times \$417.67 = \$17{,}542.14$. Because the amount we borrowed is \$15,000, this means that the total amount of interest paid is $\$17{,}542.14 - \$15{,}000 = \$2542.14$.

b. The monthly interest rate as a decimal is $r = 0.0075$, and there are still 42 payments. We know that the monthly payment we can afford is \$300. We use the companion formula (Formula 4.7) to find the amount we can borrow on this budget:

$$\text{Amount borrowed} = \frac{\text{Monthly payment} \times ((1+r)^t - 1)}{(r \times (1+r)^t)}$$

$$= \frac{\$300 \times (1.0075^{42} - 1)}{(0.0075 \times 1.0075^{42})}$$

$$= \$10{,}774.11$$

This means that we can afford to shop for a car that costs no more than \$10,774.11.

The next example shows how saving compares with borrowing.

EXAMPLE 4.12 Comparing saving versus borrowing: A loan and a CD

a. Suppose we borrow \$5000 for one year at an APR of 7.5%. What will the monthly payment be? How much interest will we have paid by the end of the year?

b. Suppose we were able to buy a one-year \$5000 CD at an APR of 7.5% compounded monthly. How much interest will be paid at the end of the year?

c. In part a, the financial institution loaned us \$5000 for one year, but in part b we loaned the financial institution \$5000 for one year. What is the difference in the amount of interest paid? Explain why the amounts are different.

SOLUTION

a. In this case, the principal is \$5000, the monthly interest rate r as a decimal is $0.075/12 = 0.00625$, and the number t of payments is 12. We use the monthly payment formula (Formula 4.6):

$$\text{Monthly payment} = \frac{\text{Amount borrowed} \times r(1+r)^t}{((1+r)^t - 1)}$$

$$= \frac{\$5000 \times 0.00625 \times 1.00625^{12}}{(1.00625^{12} - 1)}$$

$$= \$433.79$$

We will make 12 payments of \$433.79 for a total of $12 \times \$433.79 = \$5205.48$. Because the amount we borrowed is \$5000, the total amount of interest paid is \$205.48.

b. To calculate the interest earned on the CD, we use the compound interest formula (Formula 4.3 from the preceding section). The monthly rate is the same as in part a. Therefore,

$$\begin{aligned} \text{Balance} &= \text{Principal} \times (1 + r)^t \\ &= \$5000 \times 1.00625^{12} \\ &= \$5388.16 \end{aligned}$$

This means that the total amount of interest we earned is \$388.16.

c. The interest earned on the \$5000 CD is \$182.68 more than the interest paid on the \$5000 loan in part a.

Here is the explanation for this difference: When we save money, the financial institution credits our account with an interest payment—in this case, every month. We continue to earn interest on the full \$5000 and on those interest payments for the entire year. When we borrow money, however, we repay the loan monthly, thus decreasing the balance owed. We are being charged interest only on the balance owed, not on the full \$5000 we borrowed.

One common piece of financial advice is to reduce the amount you borrow by making a *down payment*. Exercise 22 explores the effect of making a down payment on the monthly payment and the total interest paid.

Estimating payments for short-term loans

The formula for the monthly payment is complicated, and it's easy to make a mistake in the calculation. Is there a simple way to give an estimate for the monthly payment? There are, in fact, a couple of ways to do this. We give one of them here and another when we look at home mortgages.

One obvious estimate for a monthly payment is to divide the loan amount by the term (in months) of the loan. *This would be our monthly payment if no interest were charged.* For a loan with a relatively short term and not too large an interest rate, this gives a rough lower estimate for the monthly payment.[5]

RULE OF THUMB 4.2 Monthly Payments for Short-Term Loans

For all loans, the monthly payment is *at least* the amount we would pay each month if no interest were charged, which is the amount of the loan divided by the term (in months) of the loan. This would be the payment if the APR were 0%. It's a rough estimate of the monthly payment for a short-term loan if the APR is not large.

EXAMPLE 4.13 Estimating monthly payment: Can we afford it?

The largest monthly payment we can afford is \$800. Can we afford to borrow a principal of \$20,000 with a term of 24 months?

SOLUTION

The rule of thumb says that the monthly payment is at least $\$20{,}000/24 = \$833.33$. This is more than \$800, so we can't afford this loan.

[5] *If the term is at most five years and the APR is less than 7.5%, the actual payment is within 20% of the ratio. If the term is at most two years and the APR is less than 9%, the actual payment is within 10% of the ratio.*

TRY IT YOURSELF 4.13

The largest monthly payment we can afford is \$450. Can we afford to borrow a principal of \$18,000 with a term of 36 months?

The answer is provided at the end of this section.

For the loan in the preceding example, our rule of thumb says that the monthly payment is *at least* \$833.33. Remember that this amount does not include the interest payments. If the loan has an APR of 6.6%, for example, the actual payment is \$891.83.

Once again, a rule of thumb gives an estimate—not the exact answer. It can at least tell us quickly whether we should be shopping on the BMW car lot.

Amortization tables and equity

When you make payments on an installment loan, part of each payment goes toward interest, and part goes toward reducing the balance owed. An *amortization table* is a running tally of payments made and the outstanding balance owed.

KEY CONCEPT

An amortization table or amortization schedule for an installment loan shows for each payment made the amount applied to interest, the amount applied to the balance owed, and the outstanding balance.

EXAMPLE 4.14 Making an amortization table: Buying a computer

Jim Stern/Bloomberg/Getty Images

Suppose we borrow \$1000 at 12% APR to buy a computer. We pay off the loan in 12 monthly payments. Make an amortization table showing payments over the first six months.

SOLUTION

The monthly rate is $12\%/12 = 1\%$. As a decimal, this is $r = 0.01$. The monthly payment formula with $t = 12$ gives

$$\text{Monthly payment} = \frac{\text{Amount borrowed} \times r(1+r)^t}{((1+r)^t - 1)}$$

$$= \frac{\$1000 \times 0.01 \times 1.01^{12}}{(1.01^{12} - 1)}$$

$$= \$88.85$$

Because the monthly rate is 1%, each month we pay 1% of the outstanding balance in interest. When we make our first payment, the outstanding balance is \$1000, so we pay 1% of \$1000, or \$10.00, in interest. Thus, \$10 of our \$88.85 goes toward interest, and the remainder, \$78.85, goes toward the outstanding balance. So after the first payment, we owe:

$$\text{Balance owed after 1 payment} = \$1000.00 - \$78.85 = \$921.15$$

When we make a second payment, the outstanding balance is \$921.15. We pay 1% of \$921.15 or \$9.21 in interest, so $\$88.85 - \$9.21 = \$79.64$ goes toward the balance due. This gives the balance owed after the second payment:

$$\text{Balance owed after 2 payments} = \$921.15 - \$79.64 = \$841.51$$

If we continue in this way, we get the following table:

Payment number	Payment	Applied to interest	Applied to balance owed	Outstanding balance
				\$1000.00
1	\$88.85	1% of \$1000.00 = \$10.00	\$78.85	\$921.15
2	\$88.85	1% of \$921.15 = \$9.21	\$79.64	\$841.51
3	\$88.85	1% of \$841.51 = \$8.42	\$80.43	\$761.08
4	\$88.85	1% of \$761.08 = \$7.61	\$81.24	\$679.84
5	\$88.85	1% of \$679.84 = \$6.80	\$82.05	\$597.79
6	\$88.85	1% of \$597.79 = \$5.98	\$82.87	\$514.92

TRY IT YOURSELF 4.14

Suppose we borrow \$1000 at 30% APR and pay it off in 24 monthly payments. Make an amortization table showing payments over the first three months.

The answer is provided at the end of this section.

The loan in Example 4.14 is used to buy a computer. The amount you have paid toward the actual cost of the computer (the principal) at a given time is referred to as your *equity* in the computer. For example, the preceding table tells us that after four payments we still owe \$679.84. This means we have paid a total of \$1000.00 − \$679.84 = \$320.16 toward the principal. That is our equity in the computer.

KEY CONCEPT

If you borrow money to pay for an item, your equity in that item at a given time is the part of the principal you have paid.

EXAMPLE 4.15 Calculating equity: Buying land

You borrow \$150,000 at an APR of 6% to purchase a plot of land. You pay off the loan in monthly payments over 10 years.

a. Find the monthly payment.

b. Complete the four-month amortization table below.

Payment number	Payment	Applied to interest	Applied to balance owed	Outstanding balance
				\$150,000.00
1				
2				
3				
4				

c. What is your equity in the land after four payments?

SOLUTION

a. The monthly rate is APR/12 = 6%/12 = 0.5%. As a decimal, this is $r =$ 0.005. We use the monthly payment formula with $t = 10 \times 12 = 120$ months:

$$\text{Monthly payment} = \frac{\text{Amount borrowed} \times r(1+r)^t}{((1+r)^t - 1)}$$

$$= \frac{\$150{,}000 \times 0.005 \times 1.005^{120}}{(1.005^{120} - 1)}$$

$$= \$1665.31$$

b. For the first month, the interest we pay is 0.5% of the outstanding balance of \$150,000:

$$\text{First month interest} = \$150{,}000 \times 0.005 = \$750$$

Now \$750 of the \$1665.31 payment goes to interest and the remainder, \$1665.31 − \$750 = \$915.31, goes toward reducing the principal. The balance owed after one month is

$$\text{Balance owed after 1 month} = \$150{,}000 - \$915.31 = \$149{,}084.69$$

This gives the first row of the table. The completed table is shown below.

Payment number	Payment	Applied to interest	Applied to balance owed	Outstanding balance
				\$150,000.00
1	\$1665.31	0.5% of \$150,000.00 = \$750.00	\$915.31	\$149,084.69
2	\$1665.31	0.5% of \$149,084.69 = \$745.42	\$919.89	\$148,164.80
3	\$1665.31	0.5% of \$148,164.80 = \$740.82	\$924.49	\$147,240.31
4	\$1665.31	0.5% of \$147,240.31 = \$736.20	\$929.11	\$146,311.20

c. The table from part b tells us that after four payments we still owe \$146,311.20. So our equity is

$$\text{Equity after 4 months} = \$150,000 - \$146,311.20 = \$3688.80$$

The graph in **Figure 4.6** shows the percentage of each payment from Example 4.15 that goes toward interest, and **Figure 4.7** shows how equity is built. Note that in the early months, a large percentage of the payment goes toward interest. For long-term loans, an even larger percentage of the payment goes toward interest early on. This means that equity is built slowly at first. The rate of growth of equity increases over the life of the loan. Note in Figure 4.7 that when you have made half of the payments (60 payments), you have built an equity of just over \$60,000—much less than half of the purchase price.

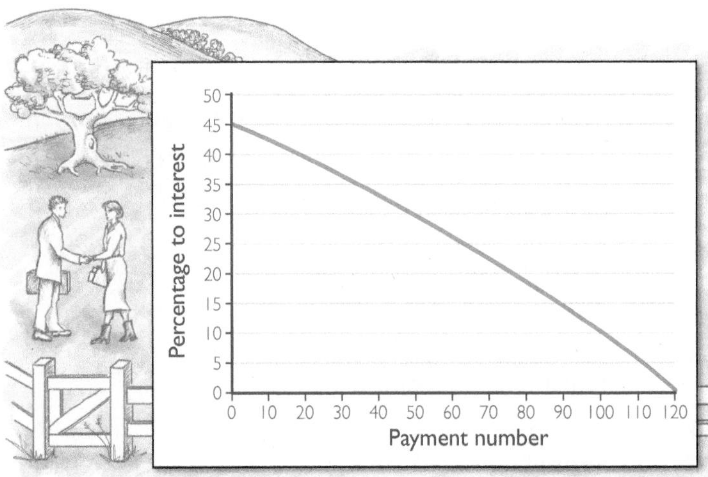

FIGURE 4.6 Percentage of payment that goes to interest: 10-year loan.

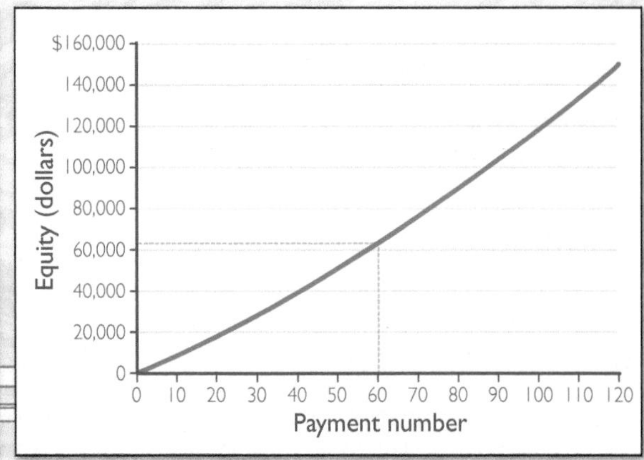

FIGURE 4.7 Equity built: 10-year loan.

Home mortgages

Figure 4.8 shows that, like other interest rates, mortgage rates have fluctuated over the years, and have recently been at historic lows. As of April 2021, the average rate for a 30-year mortgage was 3.25%.

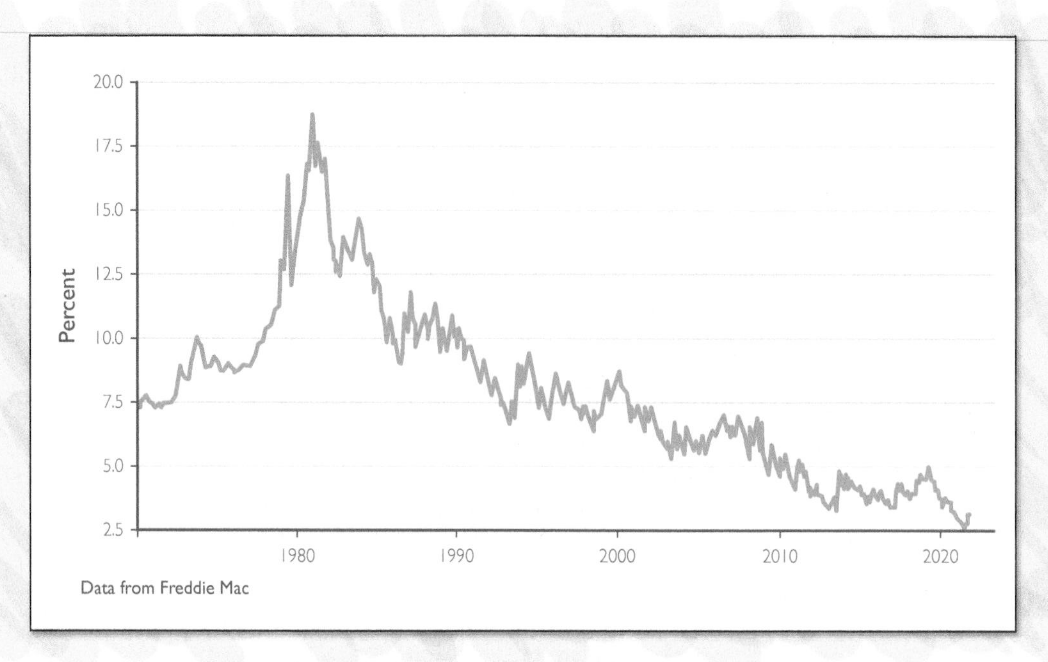

FIGURE 4.8 A graph with data from Freddie Mac of 30-year fixed rate mortgage averages in the United States, from 1970 to the present.

A home mortgage is a loan for the purchase of a home. It is very common for a mortgage to last as long as 30 years. The early mortgage payments go almost entirely toward interest, with a small part going to reduce the principal. You can see an example of this in the lower right corner of **Figure 4.9**. As a consequence, your home equity grows very slowly. For a 30-year mortgage of $150,000 at an APR of 6%, **Figure 4.10** shows the percentage of each payment that goes to interest, and **Figure 4.11** shows the equity built. Your home equity is very important because it tells you how much money you can actually keep if you sell your house, and it can also be used as collateral to borrow money.

Springside Mortgage
Customer Service: 1-800-555-1234
www.springsidemortgage.com

Jordan and Dana Smith
4700 Jones Drive
Memphis, TN 38109

Mortgage Statement
Statement Date: 3/20/2021

Account Number	1234567
Payment Due Date	4/1/2021
Amount Due	**$2,079.71**
If payment is received after 4/15/21, $160 late fee will be charged.	

Account Information	
Outstanding Principal	$264,776.43
Interest Rate (Until October 2021)	4.75%
Prepayment Penalty	$3,500.00

Explanation of Amount Due	
Principal	$386.46
Interest	$1,048.07
Escrow (for Taxes and Insurance)	$235.18
Regular Monthly Payment	**$1,669.71**
Total Fees Charged	$410.00
Total Amount Due	**$2,079.71**

FIGURE 4.9 A typical mortgage statement. Note the small amount of the payment going to reduce the principal.

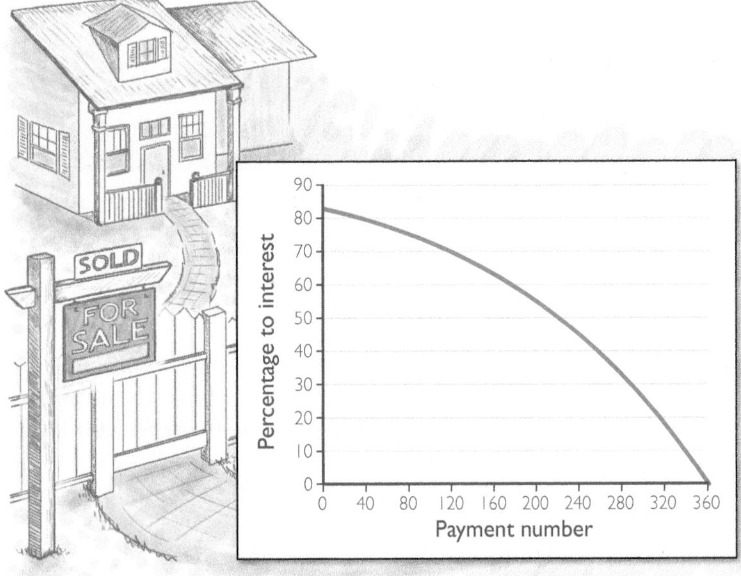

FIGURE 4.10 Percentage of payment that goes to interest: 30-year mortgage.

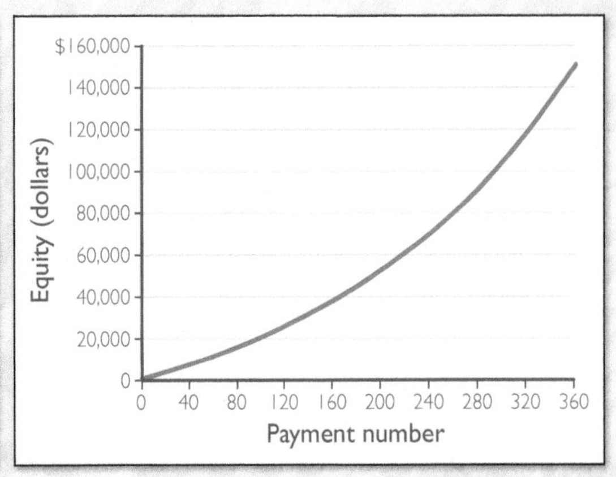

FIGURE 4.11 Equity built: 30-year mortgage.

EXAMPLE 4.16 Computing interest: 30-year mortgage

Your neighbor took out a 30-year mortgage for $300,000 at a time when the APR was 9%. She says that she will wind up paying more in interest than for the home (that is, the principal). Is that true?

SOLUTION

We first need to find the monthly payment. The monthly rate as a decimal is

$$r = \text{Monthly rate} = \frac{\text{APR}}{12} = \frac{0.09}{12} = 0.0075$$

Because the loan is for 30 years, we use $t = 30 \times 12 = 360$ months in the monthly payment formula:

$$\text{Monthly payment} = \frac{\text{Amount borrowed} \times r(1+r)^t}{((1+r)^t - 1)}$$

$$= \frac{\$300{,}000 \times 0.0075 \times 1.0075^{360}}{(1.0075^{360} - 1)}$$

$$= \$2413.87$$

She will make 360 payments of $2413.87, for a total of

$$\text{Total amount paid} = 360 \times \$2413.87 = \$868{,}993.20$$

The interest paid is the excess over $300,000, or $568,993.20. Your neighbor paid almost twice as much in interest as she did for the home.

TRY IT YOURSELF 4.16

Find the interest paid on a 25-year mortgage of $450,000 at an APR of 4.3%.

The answer is provided at the end of this section.

Now we see the effect of varying the term of the mortgage on the monthly payment.

EXAMPLE 4.17 Determining monthly payment and term: Choices for mortgages

You need to secure a loan of \$250,000 to purchase a home. Your lending institution offers you three options:

Option 1: A 30-year mortgage at 8.4% APR.

Option 2: A 20-year mortgage at 7.2% APR.

Option 3: A 30-year mortgage at 7.2% APR, including a fee of 4 *loan points*. *Note*: "Points" are a fee you pay for the loan in return for a decrease in the interest rate. In this case, a fee of 4 points means you pay 4% of the loan, or \$10,000. One way to do this is to borrow the fee from the bank by just adding the \$10,000 to the amount you borrow. The bank keeps the \$10,000 and the other \$250,000 goes to buy the home.

Determine the monthly payment for each of these options.

SOLUTION

Option 1: The monthly rate as a decimal is

$$r = \text{Monthly rate} = \frac{\text{APR}}{12} = \frac{0.084}{12} = 0.007$$

We use the monthly payment formula with $t = 360$ months:

$$\begin{aligned}\text{Monthly payment} &= \frac{\text{Amount borrowed} \times r(1+r)^t}{((1+r)^t - 1)} \\ &= \frac{\$250{,}000 \times 0.007 \times 1.007^{360}}{(1.007^{360} - 1)} \\ &= \$1904.59\end{aligned}$$

Option 2: The APR for a 20-year loan is lower. (It is common for loans with shorter terms to have lower interest rates.) An APR of 7.2% is a monthly rate of $r = 0.072/12 = 0.006$ as a decimal. We use the monthly payment formula with $t = 20 \times 12 = 240$ months:

$$\begin{aligned}\text{Monthly payment} &= \frac{\text{Amount borrowed} \times r(1+r)^t}{((1+r)^t - 1)} \\ &= \frac{\$250{,}000 \times 0.006 \times 1.006^{240}}{(1.006^{240} - 1)} \\ &= \$1968.37\end{aligned}$$

The monthly payment is about \$60 higher than for the 30-year mortgage, but you pay the loan off in 20 years rather than 30 years.

Option 3: Adding 4% to the amount we borrow gives a new loan amount of \$260,000. As with Option 2, the APR of 7.2% gives a monthly rate of $r = 0.006$. The term of 30 years means that we put $t = 360$ months into the monthly payment formula:

$$\begin{aligned}\text{Monthly payment} &= \frac{\text{Amount borrowed} \times r(1+r)^t}{((1+r)^t - 1)} \\ &= \frac{\$260{,}000 \times 0.006 \times 1.006^{360}}{(1.006^{360} - 1)} \\ &= \$1764.85\end{aligned}$$

Getting the lower interest rate makes a big difference in the monthly payment, even with the 4% added to the original loan balance. This is clearly a better choice than Option 1 if we consider only the monthly payment. Option 3 does require borrowing more money than Option 1 and could have negative consequences if you need to sell the home early. Further, comparing the amount of interest paid shows that Option 2 is the best choice from that point of view—if you can afford the monthly payment.

© Randy Glasbergen/www.glasbergen.com

"As an alternative to the traditional 30-year mortgage, we also offer an interest-only mortgage, balloon mortgage, reverse mortgage, upside down mortgage, inside out mortgage, loop-de-loop mortgage, and the spinning double axel mortgage with a triple lutz."

Adjustable-rate mortgages

In the summer of 2007, a credit crisis involving home mortgages had dramatic effects on many people and ultimately on the global economy. Several major financial institutions collapsed in the autumn of 2008, and more than 3 million people lost their homes through foreclosure.[6] There were many causes of the crisis, but one prominent cause was the increase in risky mortgage lending, a large percentage of which involved *adjustable-rate mortgages* or ARMs.

KEY CONCEPT

A fixed-rate mortgage keeps the same interest rate over the life of the loan. In the case of an adjustable-rate mortgage or ARM, the interest rate may vary over the life of the loan. The rate is often tied to the prime interest rate, which is the rate banks must pay to borrow money.

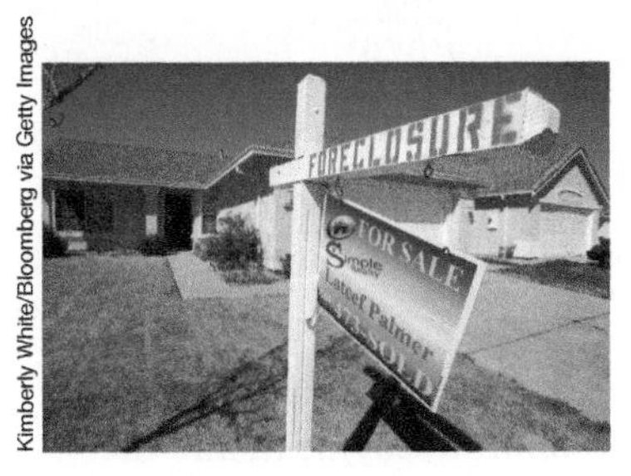

Kimberly White/Bloomberg via Getty Images

One advantage of an ARM is that the initial rate is often lower than the rate for a comparable fixed-rate mortgage. One disadvantage of an ARM is that a rising prime interest rate may cause significant increases in the monthly payment. That is what happened in 2007–2008. When housing prices began to fall and interest rates for adjustable-rate mortgages began to grow, people had higher monthly payments while the value of their homes fell. Thus, mortgage delinquencies increased. This contributed to financial markets becoming concerned about the soundness of U.S. credit, which led to slowing economic growth in the United States and Europe.

EXAMPLE 4.18 Comparing monthly payments: Fixed-rate and adjustable-rate mortgages

We want to borrow \$200,000 for a 30-year home mortgage. We have found an APR of 6.6% for a fixed-rate mortgage and an APR of 6% for an ARM. Compare the initial monthly payments for these loans.

---

[6] http://usatoday30.usatoday.com/money/economy/housing/2009-01-14-foreclosure-record-filings_N.htm

SOLUTION

Both loans have a principal of \$200,000 and a term of $t = 360$ months. For the fixed-rate mortgage, the monthly rate in decimal form is $r = 0.066/12 = 0.0055$. The monthly payment formula gives

$$\begin{aligned} \text{Monthly payment} &= \frac{\text{Amount borrowed} \times r(1+r)^t}{((1+r)^t - 1)} \\ &= \frac{\$200{,}000 \times 0.0055 \times 1.0055^{360}}{(1.0055^{360} - 1)} \\ &= \$1277.32 \end{aligned}$$

For the ARM, the initial monthly rate in decimal form is $r = 0.06/12 = 0.005$. The monthly payment formula gives

$$\begin{aligned} \text{Monthly payment} &= \frac{\text{Amount borrowed} \times r(1+r)^t}{((1+r)^t - 1)} \\ &= \frac{\$200{,}000 \times 0.005 \times 1.005^{360}}{(1.005^{360} - 1)} \\ &= \$1199.10 \end{aligned}$$

The initial monthly payment for the ARM is almost \$80 less than the payment for the fixed-rate mortgage—but the payment for the ARM could change at any time.

TRY IT YOURSELF 4.18

We want to borrow \$150,000 for a 30-year home mortgage. We have found an APR of 6% for a fixed-rate mortgage and an APR of 5.7% for an ARM. Compare the monthly payments for these loans.

The answer is provided at the end of this section.

Example 4.18 illustrates why an ARM may seem attractive. The following example illustrates one potential danger of an ARM. It is typical of what happened in 2007 to many families who took out loans in 2006 when interest rates fell to historic lows.

EXAMPLE 4.19 Using an ARM: Effect of increasing rates

Suppose a family has an annual income of \$60,000. Assume that this family secures a 30-year ARM for \$250,000 at an initial APR of 4.5%.

a. Find the family's monthly payment and the percentage of its monthly income used for the mortgage payment.

b. Now suppose that after one year, the rate adjusts to 6%. Find the family's new monthly payment, the increase in the family's monthly payment, and the percentage of its monthly income now required for the mortgage payment.

SOLUTION

a. The monthly rate as a decimal is $0.045/12 = 0.00375$, so with $t = 360$ the monthly payment on the family's home is

$$\begin{aligned} \text{Monthly payment} &= \frac{\text{Amount borrowed} \times r(1+r)^t}{((1+r)^t - 1)} \\ &= \frac{\$250{,}000 \times 0.00375 \times 1.00375^{360}}{(1.00375^{360} - 1)} \\ &= \$1266.71 \end{aligned}$$

An annual income of $60,000 is a monthly income of $5000. To find the percentage of the family's monthly income used for the mortgage payment, we calculate 1266.71/5000. The result is about 0.25, so the family is paying 25% of its monthly income for the mortgage payment.

b. The APR has increased to 6%, so $r = 0.06/12 = 0.005$. To find the new monthly payment, we use a loan period of 29 years or $29 \times 12 = 348$ months. For the principal, we use the balance owed to the bank after one year of payments. We noted earlier in this section that homeowners build negligible equity in the first year of payments, so we will get a very good estimate of the new monthly payment if we use $250,000 as the balance owed to the bank. The formula gives

$$\frac{\$250{,}000 \times 0.005 \times 1.005^{348}}{(1.005^{348} - 1)} = \$1517.51$$

Therefore, the new monthly payment is $1517.51. The difference, $1517.51–$1266.71 = $250.80, is the increase in the family's monthly payment. To find the percentage of their monthly income required for the payment, we divide $1517.51 by the monthly income of $5000. The result is approximately 0.30, or 30%.

In the example, the extra burden on family finances caused by the increase in the interest rate could lead to serious problems. Further rate adjustments could easily result in the loss of the family home.

Estimating payments on long-term loans

The payment on a long-term loan is *at least* as large as the monthly interest on the original balance. This is a pretty good estimate for mortgages in times of moderate-to-high interest rates.[7]

RULE OF THUMB 4.3 Monthly Payments for Long-Term Loans

For all loans, the monthly payment is *at least* as large as the principal times the monthly interest rate as a decimal. This is a fairly good estimate for a long-term loan with a moderate or high interest rate.

Let's see what this rule of thumb would estimate for the payment on a mortgage of $100,000 at an APR of 7.2%. The monthly rate is 7.2%/12 = 0.6%. The monthly interest on the original balance is 0.6% of $100,000:

$$\text{Monthly payment estimate} = \$100{,}000 \times 0.006 = \$600$$

If this is a 30-year mortgage, the monthly payment formula gives the value $678.79 for the actual payment. This is about 13% higher than the estimate.

We should note that most home mortgage payments also include taxes and insurance. These vary from location to location but can be significant.[8] For many, a home mortgage is the most significant investment they will ever make. It is crucial to understand clearly both the benefits and the costs of such an investment.

---

[7] *If the term is at least 30 years and the APR is at least 6%, then the actual payment is within 20% of the estimate. If the term is at least 30 years and the APR is greater than 8%, then the actual payment is within 10% of the estimate.*

[8] *See Exercise 31 (on affordability).*

ALGEBRAIC SPOTLIGHT 4.2 Derivation of the Monthly Payment Formula, Part I

First we derive a formula for the balance still owed on an installment loan after a given number of payments. Suppose we borrow B_0 dollars and make monthly payments of M dollars. Suppose further that interest accrues on the balance at a monthly rate of r as a decimal. Let B_n denote the account balance (in dollars) after n months. Our goal is to derive the formula

$$B_n = B_0(1+r)^n - M\frac{(1+r)^n - 1}{r}$$

Each month we find the new balance B_{n+1} from the old balance B_n by first adding the interest accrued rB_n and then subtracting the payment M. As a formula, this is

$$B_{n+1} = B_n + rB_n - M = B_n(1+r) - M$$

If we put $R = 1 + r$, this formula can be written more compactly as

$$B_{n+1} = B_nR - M$$

Repeated application of this formula gives

$$B_n = B_0R^n - M(1 + R + R^2 + \cdots + R^{n-1})$$

To finish, we use the *geometric sum formula*, which tells us that

$$1 + R + R^2 + \cdots + R^{n-1} = \frac{R^n - 1}{R - 1}$$

This gives

$$B_n = B_0R^n - M\frac{R^n - 1}{R - 1}$$

Finally, we recall that $R = 1 + r$ and obtain the desired formula

$$B_n = B_0(1+r)^n - M\frac{(1+r)^n - 1}{r}$$

ALGEBRAIC SPOTLIGHT 4.3 Derivation of the Monthly Payment Formula, Part II

Now we use the account balance formula to derive the monthly payment formula. Suppose we borrow B_0 dollars at a monthly interest rate of r as a decimal and we want to pay off the loan in t monthly payments. That is, we want the balance to be 0 after t payments. Using the account balance formula we derived in Algebraic Spotlight 4.2, we want to find the monthly payment M that makes $B_t = 0$. That is, we need to solve the equation

$$0 = B_t = B_0(1+r)^t - M\frac{(1+r)^t - 1}{r}$$

for M. Now

$$0 = B_0(1+r)^t - M\frac{(1+r)^t - 1}{r}$$

$$M\frac{(1+r)^t - 1}{r} = B_0(1+r)^t$$

$$M = \frac{B_0r(1+r)^t}{((1+r)^t - 1)}$$

Because B_0 is the amount borrowed and t is the term, this is the monthly payment formula we stated earlier.

WHAT DO YOU THINK?

High interest rate: You are considering taking out a loan but are concerned about the high interest rate. Should you be more concerned if you anticipate repaying the loan over 2 years or 20 years? Explain your reasoning.

Questions of this sort may invite many correct responses. We present one possibility: Assuming that interest is compounded, a high interest rate has a greater effect on a long-term loan than on a short-term loan. That is because of the exponential nature of compound interest, where interest is charged on accrued interest. The shape of an exponential graph shows that growth over the long term is very rapid. For a loan, growth is offset by payments, but the influence of high interest nonetheless causes a greater burden over long terms.

Further questions of this type may be found in the *What Do You Think?* section of exercises.

Try It Yourself answers

Try It Yourself 4.9: Using the monthly payment formula: College loan $199.08.

Try It Yourself 4.10: Computing how much I can borrow: Buying a car $13,286.65.

Try It Yourself 4.13: Estimating monthly payment: Can we afford it? No: The monthly payment is at least $500.

Try It Yourself 4.14: Making an amortization table: Buying a computer

Payment number	Payment	Applied to interest	Applied to balance owed	Outstanding balance
				$1000.00
1	$55.91	2.5% of $1000.00 = $25.00	$30.91	$969.09
2	$55.91	2.5% of $969.09 = $24.23	$31.68	$937.41
3	$55.91	2.5% of $937.41 = $23.44	$32.47	$904.94

Try It Yourself 4.16: Computing how much interest: 25-year mortgage $285,132.00

Try It Yourself 4.18: Comparing monthly payments: Fixed-rate and adjustable-rate mortgages The monthly payment of $870.60 for the ARM is almost $30 less than the payment of $899.33 for the fixed-rate mortgage.

Exercise Set 4.2

Test Your Understanding

1. For an installment loan with monthly compounding, list the factors that determine your monthly payment.

2. True or false: For a loan, your monthly payment is at least the amount borrowed divided by the number of months required to pay off the loan.

3. True or false: If you borrow to buy a car, your monthly payment is at least as large as the amount borrowed times the monthly interest rate (as a decimal).

4. If you borrow to buy a car, your equity in the car is ________.

5. True or false: For a long-term loan such as a mortgage, the total interest you pay may exceed the original principal.

6. The abbreviation ARM stands for ________.

7. True or false: An adjustable-rate mortgage may be risky because your interest rates could go up.

8. An amortization table for a loan shows you: **a.** the amount of interest you have paid each month **b.** the part of the principal you have paid each month **c.** your monthly payment **d.** all of these.

Problems

Rounding in the calculation of monthly interest rates is discouraged. Such rounding can lead to answers different from those presented here. For long-term loans, the differences may be pronounced.

9. A friend said that he borrowed $1200 for one year and his monthly payments were $90. Explain why you are certain that your friend is mistaken.

10. Bill said that he wants to borrow $6000 to buy a used car and pay off the loan in five years, but he says he can't afford more than $100 per month. Can Bill afford this car? Explain.

11. Estimating monthly payment. Use Rule of Thumb 4.2 to estimate the monthly payment on a loan of $5000 borrowed over a three-year period.

12. Estimating monthly payment. Use Rule of Thumb 4.3 to estimate the monthly payment on a loan of $200,000 at an APR of 6% over a period of 25 years.

13. Car payment. To buy a car, you borrow $20,000 with a term of five years at an APR of 6%. What is your monthly payment? How much total interest is paid?

14. Truck payment. You borrow $18,000 with a term of four years at an APR of 5% to buy a truck. What is your monthly payment? How much total interest is paid?

15. Estimating the payment. You borrow $25,000 with a term of two years at an APR of 5%. Use Rule of Thumb 4.2 to estimate your monthly payment, and compare this estimate with what the monthly payment formula gives.

16. Estimating a mortgage. Several years ago Bill got a home mortgage of $110,000 with a term of 30 years at an APR of 9%. Use Rule of Thumb 4.3 to estimate his monthly payment, and compare this estimate with what the monthly payment formula gives.

17. Affording a car. You can get a car loan with a term of three years at an APR of 5%. If you can afford a monthly payment of $450, how much can you borrow?

18. Affording a home. You find that the going rate for a home mortgage with a term of 30 years is 6.5% APR. The lending agency says that based on your income, your monthly payment can be at most $750. How much can you borrow?

19. No interest. A car dealer offers you a loan with no interest charged for a term of two years. If you need to borrow $18,000, what will your monthly payment be? Which rule of thumb is relevant here?

20. Interest paid. Assume that you take out a $2000 loan for 30 months at 8.5% APR.

a. What is the monthly payment?

b. How much of the first month's payment is interest?

c. What percentage of the first month's payment is interest? (Round your answer to two decimal places as a percentage.)

d. How much total interest will you have paid at the end of the 30 months?

21. More saving and borrowing. In this exercise, we compare saving and borrowing as in Example 4.12.

a. Suppose you borrow $10,000 for two years at an APR of 8.75%. What will your monthly payment be? How much interest will you have paid by the end of the loan?

b. Suppose it were possible to invest in a two-year $10,000 CD at an APR of 8.75% compounded monthly. How much interest would you be paid by the end of the period?

c. In part a, the financial institution loaned you $10,000 for two years, but in part b, you loaned the financial institution $10,000 for two years. What is the difference in the amount of interest paid?

22. Down payment. In this exercise, we examine the benefits of making a down payment.

a. You want to buy a car. Suppose you borrow $15,000 for two years at an APR of 6%. What will your monthly payment be? How much interest will you have paid by the end of the loan?

b. Suppose that in the situation of part a, you make a down payment of $2000. This means that you borrow only $13,000. Assume that the term is still two years at an APR of 6%. What will your monthly payment be? How much interest will you have paid by the end of the loan?

c. Compare your answers to parts a and b. What are the advantages of making a down payment? Why would a borrower not make a down payment? *Note*: One other factor to consider is the interest one would have earned on the down payment if it had been invested. Typically, the interest rate earned on investments is lower than that charged for loans.

23. Amortization table. Suppose we borrow $1500 at 4% APR and pay it off in 24 monthly payments. Make an amortization table showing payments over the first three months.

Payment number	Payment	Applied to interest	Applied to balance owed	Outstanding balance
				$1500.00
1				
2				
3				

24. Another amortization table. Suppose we borrow $100 at 5% APR and pay it off in 12 monthly payments. Make an amortization table showing payments over the first three months.

Payment number	Payment	Applied to interest	Applied to balance owed	Outstanding balance
				$100.00
1				
2				
3				

25. **Term of mortgage.** There are two common choices for the term of a home mortgage: 15 or 30 years. Suppose you need to borrow $90,000 at an APR of 6.75% to buy a home.

a. What will your monthly payment be if you opt for a 15-year mortgage?

b. What percentage of your first month's payment will be interest if you opt for a 15-year mortgage? (Round your answer to two decimal places as a percentage.)

c. How much interest will you have paid by the end of the 15-year loan?

d. What will your monthly payment be if you opt for a 30-year mortgage?

e. What percentage of your first month's payment will be interest if you opt for a 30-year mortgage? (Round your answer to two decimal places as a percentage.)

f. How much interest will you have paid by the end of the 30-year loan? Is it twice as much as for a 15-year mortgage?

26. **Formula for equity.** Here is a formula for the equity built up after k monthly payments:

$$\text{Equity} = \frac{\text{Amount borrowed} \times ((1+r)^k - 1)}{((1+r)^t - 1)}$$

where r is the monthly interest rate as a decimal and t is the term in months. In this exercise, we consider a mortgage of $100,000 at an APR of 7.2% with two different terms.

a. Assume that the term of the mortgage is 30 years. How much equity will you have halfway through the term of the loan? What percentage of the principal is this? (Round your answer to one decimal place as a percentage.)

b. Suppose now that instead of a 30-year mortgage, you have a 15-year mortgage. Find your equity halfway through the term of the loan, and find what percentage of the principal that is. (Round your answer to one decimal place as a percentage.) Compare this with the percentage you found in part a. Why should the percentage for the 15-year term be larger than the percentage for the 30-year term?

27. **Car totaled.** In order to buy a new car, you finance $20,000 with no down payment for a term of five years at an APR of 6%. After you have the car for one year, you are in an accident. No one is injured, but the car is totaled. The insurance company says that before the accident, the value of the car had decreased by 25% over the time you owned it, and the company pays you that depreciated amount after subtracting your $500 deductible.

a. What is your monthly payment for this loan?

b. How much equity have you built up after one year? *Suggestion*: Use the formula for equity stated in connection with Exercise 26.

c. How much money does the insurance company pay you? (Don't forget to subtract the deductible.)

d. Can you pay off the loan using the insurance payment, or do you still need to make payments on a car you no longer have? If you still need to make payments, how much do you still owe? (Subtract the payment from the insurance company.)

28. **Rebates.** When interest rates are low, some automobile dealers offer loans at 0% APR, as indicated in a 2021 advertisement by a prominent car dealership, offering zero percent financing or cash back deals on some models.

Zero percent financing means the obvious thing—that no interest is being charged on the loan. So if we borrow $1200 at 0% interest and pay it off over 12 months, our monthly payment will be $1200/12 = $100.

Suppose you are buying a new truck at a price of $20,000. You plan to finance your purchase with a loan that you will repay over two years. The dealer offers two options: either dealer financing with 0% interest, or a $2000 rebate on the purchase price. If you take the rebate, you will have to go to the local bank for a loan (of $18,000) at an APR of 6.5%.

a. What would your monthly payment be if you used dealer financing?

b. What would your monthly payment be if you took the rebate?

c. Should you take the dealer financing or the rebate? How much would you save over the life of the loan by taking the option you chose?

29. **Too good to be true?** A friend claims to have found a really great deal at a local loan agency without a street address: The agency claims that its rates are so low you can borrow $10,000 with a term of three years for a monthly payment of $200. Is this too good to be true? Be sure to explain your answer.

30. **Is this reasonable?** A lending agency advertises in the paper an APR of 12% on a home mortgage with a term of 30 years. The ad claims that the monthly payment on a principal of $100,000 will be $10,290. Is this claim reasonable? What should the ad have said the payment would be (to the nearest dollar)? What do you think happened here?

Exercise 31 is suitable for group work.

31. **Affordability.** Over the past 40 years, interest rates have varied widely. The rate for a 30-year mortgage reached a high of 14.75% in July 1984, and it reached 3.31%, in November, 2012. A significant impact of lower interest rates on society is that they enable more people to afford the purchase of a home. In this exercise, we consider the purchase of a home that sells for $125,000. Assume that we can make a down payment of $25,000, so we need to borrow $100,000. We assume that our annual income is $40,000 and that we have no other debt. We determine whether we can afford to buy the home at the high and low rates mentioned at the beginning of this paragraph.

a. What is our monthly income?

b. Lending agencies usually require that no more than 28% of the borrower's monthly income be spent on housing. How much does that represent in our case?

c. The amount we will spend on housing consists of our monthly mortgage payment plus property taxes and hazard insurance. Assume that property taxes plus insurance total $250 per month, and subtract this from the answer to part b to determine what monthly payment we can afford.

d. Use your answer to part c to determine how much we can borrow if the term is 30 years and the interest rate is at the historic high of 14.75%. Can we afford the home?

e. Use your answer to part c to determine how much we can borrow if the term is 30 years and the interest rate is 3.31%. Can we afford the home now?

f. What is the difference in the amount we can borrow between the rates used in parts d and e?

32–33. Payday loans. *Exercises 32 and 33 concern the In the News 4.4 article found at paydayloansonlineresource.org. The article explains what are called payday loans and points out their dangers.*

32. Suppose you borrow $100 from a payday lender for one week at a weekly rate of 10%. You'll obviously owe $110 at the end of a week. If you are unable to repay the loan, however, the lender will say that you now owe not only $110 but also 10% of that $110 at the end of the second week. How much will you owe by the end of the second week? If you *still* can't repay any of it, how much will you owe at the end of the third week?

33. Under the same scenario as in Exercise 32, it turns out that after n weeks of not repaying anything you would owe 100×1.1^n dollars. Use this formula to determine how much you would owe after 10 weeks. What total percent interest are you being charged on your 10-week loan? (Round to the nearest percent.)

34. Russian payday loans. An article in *The New York Times* on April 29, 2016, covered payday loans in Russia. According to the article, a private "microfinance industry" extends very small loans to people. What is remarkable is that the lenders sometimes hire thugs to physically assault borrowers if they do not pay. The payday loans in Russia average $125, according to United Credit Bureau statistics, and the interest rate is usually 2% per day.

a. If a person borrowed $100 at 2% per day simple interest, what would she owe after a year?

b. What would she owe if interest were compounded daily?

The following exercises are designed to be solved using technology such as calculators or computer spreadsheets.

35. Equity. You borrow $15,000 with a term of four years at an APR of 8%. Make an amortization table. How much equity have you built up halfway through the term?

36. Finding the term. You want to borrow $15,000 at an APR of 7% to buy a car, and you can afford a monthly payment of $500. To minimize the amount of interest paid, you want to take the shortest term you can. What is the shortest term you can afford? *Note*: Your answer should be a whole number of years. Rule of Thumb 4.2 should give you a rough idea of what the term will be.

Average Interest Rates for Payday Loans

Used by permission of Josh Wallach, Payday Loans Online Re Resource

If you are strapped for cash and are considering taking out a payday loan, there are several things you should first consider, such as how high the fees and interest rates associated with your loan are. Oftentimes with payday loans, the rates are much higher than other types of loans, and can end up putting you more in debt than you were to start with.

Payday loans typically range from approximately $100 to $1000, depending upon your state's legal minimum. The average loan time is two weeks, after which time you would have to repay the loan along with the fees and interest you accrued over that period. These loans usually cost 400% annual interest (APR), if not more. And the finance charge to borrow $100 ranges from $15 to $30 for two-week loans. These finance charges are sometimes accompanied by interest rates ranging from 300% to 750% APR. For loans shorter than two weeks, the APR can be even higher.

In most cases, payday loans are much more expensive than other cash loans. For example, a $500 cash advance on an average credit card that is repaid in one month would cost you $13.99 in finance charges and an annual interest rate of about 5.7%. A payday loan, on the other hand, would cost you $17.50 per $100 for borrowing the same $500, and would cost $105 if renewed once, or 400% annual interest.

Used by permission of Josh Wallach, Payday Loans Online Re Resource

Writing About Mathematics

37. History of home ownership. In the United States before the 1930s, home ownership was not standard—most people rented. In part, this was because home loans were structured differently: A large down payment was required, the term of the loan was five years or less, the regular payments went toward interest, and the principal was paid off in a lump sum at the end of the term. Home mortgages as we know them came into being through the influence of the Federal Housing Administration, established by Congress in the National Housing Act in 1934, which provided insurance to lending agencies. Find more details on the history of home mortgages, and discuss why the earlier structure of loans would discourage home ownership. Does the earlier structure have advantages?

38. Microloans and flat interest. The 2006 Nobel Peace Prize was awarded to Muhammad Yunus and the Grameen Bank he founded. The announcement of the award noted the development of *microloans* for encouraging the poor to become entrepreneurs. Microloans involve small amounts of money but fairly high interest rates. In contrast to the installment method discussed in this section, for microloans typically interest is paid at a flat rate: The amount of interest paid is not reduced as the principal is paid off. More recently such loans have come under some criticism. Investigate flat-rate loans, the reasons why microloans are often structured this way, and any downsides to them.

39. **More on payday loans.** Go on the Internet and explore some actual payday loan companies and also what regulations may exist. The Web site http://usgovinfo.about.com/od/consumerawareness/a/paydayloans.htm is a good place to learn more about payday loans.

40. **Local payday loans.** Explore some actual payday loan companies in your area. Write a report on their rates and policies. Some lenders require the borrower to allow them to withdraw payments from the borrower's bank account. See if companies in your area do this.

41. **Local auto loan rates.** Find out what some automobile loan rates are in your area. Compare them with some you find online and write a report on your findings.

42. **College loan rates.** Borrowing for college is very common. Investigate college loan rates and write a report on your findings. If you can find out what rate some of your friends are paying, you might include that.

43. **Title loans.** Look up the term *title loan*. Write a report on what title loans are and the rates and policies of companies that offer them.

What Do You Think?

44. **Is that the right payment?** You are negotiating a loan with a bank officer. The loan is at a moderate interest rate, and the term of the loan is six months. The bank officer tells you your monthly payment. Explain how you can quickly check whether or not the figure the bank officer gave you is reasonable. **Exercise 15 is relevant to this topic.**

45. **Is this true?** You are considering buying a home, but a friend tells you that by the time you finish paying for the home, the amount you pay in interest to the bank will exceed the listed price for the home. Is this an unusual situation? What factors contribute to determining the part of your total payments that is interest? **Exercise 20 is relevant to this topic.**

46. **Smart shopping.** List the factors that determine the monthly payment for a car loan. Which of these can you control? Experiment by changing each factor to see the effect. For example, if the amount you borrow is doubled, what happens to the monthly payment? Use the information you gather in this way to sketch how you would make a decision about a loan. **Exercise 22 is relevant to this topic.**

47. **Amortization table.** Explain what an amortization table is. Provide an example. **Exercise 23 is relevant to this topic.**

48. **What part of it is mine?** You are making monthly installment payments on a car. Explain what is meant by your *equity* in the car. **Exercise 26 is relevant to this topic.**

49. **Adjustable-rate mortgage.** Discuss the advantages and disadvantages of an adjustable-rate mortgage versus a fixed-rate mortgage.

50. **Is that the right payment?** You are negotiating a home mortgage that has a term of 30 years. The interest rate is moderate. You are presented with a monthly payment figure that does not include taxes and insurance. Explain how you can quickly check whether the figure the bank officer gave you is reasonable. **Exercise 30 is relevant to this topic.**

51. **Research on taxes and insurance.** When you take out a mortgage to buy a home, the lending institution will require you to have insurance and to pay your property taxes. The amount of these costs varies depending on your location. Fees to cover these costs are normally added to your monthly payment. These fees are placed in an *escrow account*, which is used to pay the bills when they come due. Figure 4.9 on page 244 shows a mortgage statement with the escrow listed. How much is it? Assume a home is assessed at $100,000. Investigate the annual cost of insurance and property taxes on this home in your area. **Exercise 31 is relevant to this topic.**

52. **Research on closing costs.** When you secure a home mortgage, the lending institution will charge you *closing costs*. These are fees in addition to the price of the home, interest on the loan, taxes, and insurance. These fees fall into several categories, and they can be substantial. Use the Internet to investigate closing costs. Report your findings.

53. **Research on Fannie and Freddie.** Fannie Mae and Freddie Mac came under scrutiny during the mortgage crisis of 2007–2008. Use the Internet to determine what these entities are and what role they play in home mortgages. Report your findings.

4.3 Saving for the long term: Build that nest egg

TAKE AWAY FROM THIS SECTION
Prepare for the future by saving now.

The following article was found on the MoneyNing Web site.

IN THE NEWS 4.5

Search

Home | TV & Video | U. S. | World | Politics | Entertainment | Tech | Travel

4 Retirement Tips Young Adults Can Implement NOW

JESSICA SOMMERFIELD

Recent studies have shown that a large percentage of young adults under 35 are declining enrollment in employer-provided retirement plans. Considering the economy, it's understandable that many young adults are finding it impossible to focus on things any further off than paying student loan debt, purchasing a home or vehicle, and getting started in their careers.

Many won't be financially able to leave their parents' home until well into their twenties. **Even if they're interested, today's young adults find it hard to sacrifice a portion of their often meager wages for retirement savings.** Especially if they're barely getting by as it is.

Considering this trend in the next generation, some companies that employ mostly younger adults aren't bothering to offer retirement plan options and are choosing instead to pay higher wages. While this may be saving companies money and allowing Generation Y or Millennial adults to channel their money into immediate debt-payoff or other savings goals, it's not encouraging the younger generations to think about and plan for the future.

Market research shows that those who are actively saving and investing for their retirement are in the 35–65 age bracket. But, according to many financial experts, this is alarmingly late in the game. **The obvious dilemma is that young adults need to start saving for retirement as soon as possible—while not floundering under other financial obligations.** It's a challenge, but it can be done.

Here are four retirement tips that young adults can start implementing now.

1. Invest early, no matter how small

Time is on your side. Because of the wonder of compound interest, as little as $200 a month invested in your twenties can build to millions by the time you retire. But realistically, how many young adults in our economy have $200 a month to put into retirement savings?

The recommended allocation for retirement savings is 15% of your income, but this is often an unrealistic number. **The key is to stop worrying about percentages and focus on saving *something*.** Even small contributions to a 401K or Roth IRA when you're young will put you much further ahead. You can always invest more heavily once you're in a better financial situation.

2. If your employer doesn't offer a 401K, open an IRA

If your employer offers a 401K, you should undoubtedly participate. Most employers match your contribution, sometimes up to 100%. **Investments are before-tax and non-taxable, unless you withdraw them early.**

If they don't offer a 401K, consider opening a Roth IRA. Although investment companies usually require a minimum amount to start an account, many are catching on to the needs of young adults and waving the minimum as long as you agree to regular contributions. Again, the amount isn't as important as the fact that you're saving something.

(*Continued*)

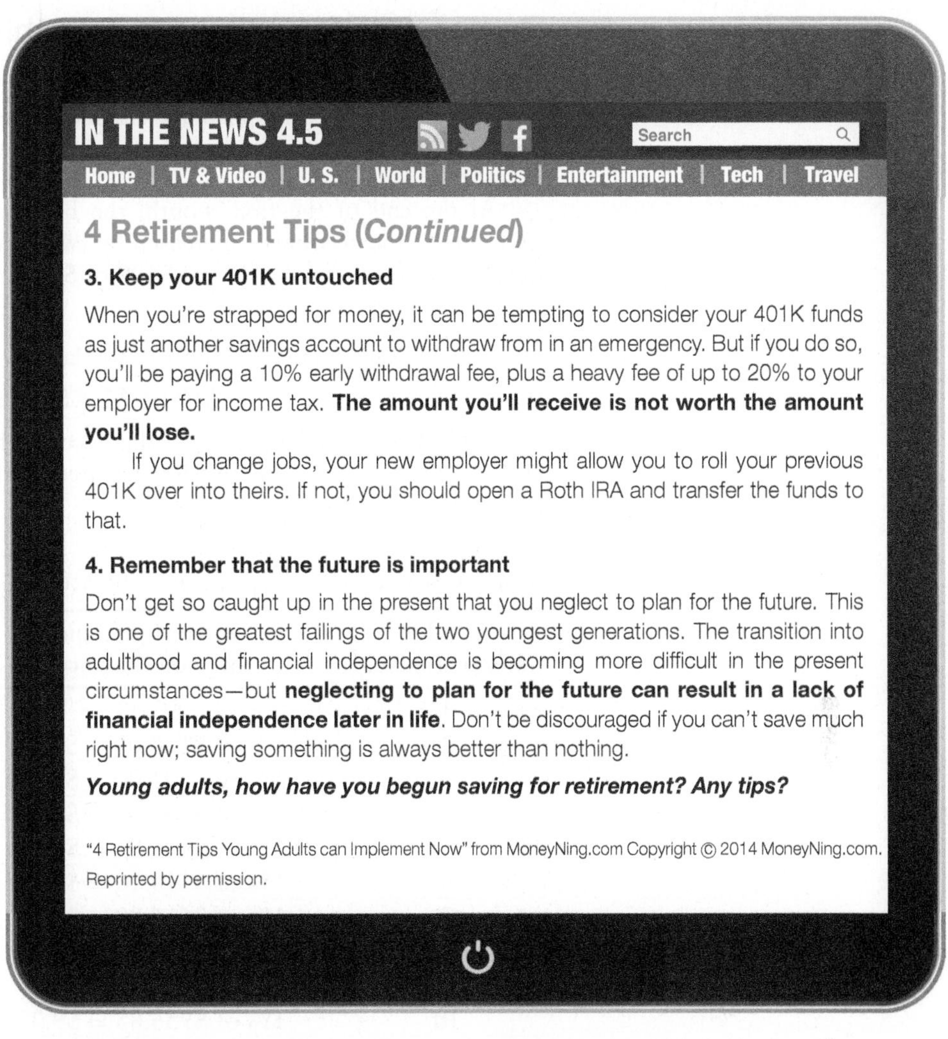
IN THE NEWS 4.5

Home | TV & Video | U. S. | World | Politics | Entertainment | Tech | Travel

4 Retirement Tips (*Continued*)

3. Keep your 401K untouched

When you're strapped for money, it can be tempting to consider your 401K funds as just another savings account to withdraw from in an emergency. But if you do so, you'll be paying a 10% early withdrawal fee, plus a heavy fee of up to 20% to your employer for income tax. **The amount you'll receive is not worth the amount you'll lose.**

If you change jobs, your new employer might allow you to roll your previous 401K over into theirs. If not, you should open a Roth IRA and transfer the funds to that.

4. Remember that the future is important

Don't get so caught up in the present that you neglect to plan for the future. This is one of the greatest failings of the two youngest generations. The transition into adulthood and financial independence is becoming more difficult in the present circumstances—but **neglecting to plan for the future can result in a lack of financial independence later in life**. Don't be discouraged if you can't save much right now; saving something is always better than nothing.

Young adults, how have you begun saving for retirement? Any tips?

In Section 4.1, we discussed how an account grows if a lump sum is invested. Many long-term savings plans such as retirement accounts combine the growth power of compound interest with that of regular contributions. Such plans can show truly remarkable growth. We look at such accounts in this section.

". . . and help my parents to pick the right investments for my college education."

Saving regular amounts monthly

Let's look at a plan where you deposit $100 to your savings account at the end of each month, and let's suppose, for simplicity, that the account pays you a monthly rate of 1% on the balance in the account. (That is an APR of 12% compounded monthly.)

At the end of the first month, the balance is $100. At the end of the second month, the account is increased by two factors—interest earned and a second deposit. Interest earned is 1% of $100 or $1.00, and the deposit is $100. This gives the new balance as

$$\begin{aligned}\text{New balance} &= \text{Previous balance} + \text{Interest} + \text{Deposit}\\ &= \$100 + \$1 + \$100 = \$201\end{aligned}$$

Table 4.2 tracks the growth of this account through 10 months. Note that at the end of each month, the interest is calculated on the previous balance and then $100 is added to the balance.

TABLE 4.2 Regular Deposits into an Account

At end of month number	Interest paid on previous balance	Deposit	Balance
1	$0.00	$100	$100.00
2	1% of $100.00 = $1.00	$100	$201.00
3	1% of $201.00 = $2.01	$100	$303.01
4	1% of $303.01 = $3.03	$100	$406.04
5	1% of $406.04 = $4.06	$100	$510.10
6	1% of $510.10 = $5.10	$100	$615.20
7	1% of $615.20 = $6.15	$100	$721.35
8	1% of $721.35 = $7.21	$100	$828.56
9	1% of $828.56 = $8.29	$100	$936.85
10	1% of $936.85 = $9.37	$100	$1046.22

EXAMPLE 4.20 Verifying a balance: Regular deposits into savings

Verify the calculation shown for month 3 of Table 4.2.

SOLUTION

We earn 1% interest on the previous balance of $201.00:

$$\text{Interest} = 0.01 \times \$201 = \$2.01$$

We add this amount plus a deposit of $100 to the previous balance:

$$\begin{aligned}\text{Balance at end of month 3} &= \text{Previous balance} + \text{Interest} + \text{Deposit}\\ &= \$201.00 + \$2.01 + \$100\\ &= \$303.01\end{aligned}$$

This agrees with the entry in Table 4.2.

TRY IT YOURSELF 4.20

Verify the calculation shown for month 4 of Table 4.2.

The answer is provided at the end of this section.

In Table 4.2, the balance after 10 months is $1046.22. There were 10 deposits made, for a total of $1000. The remaining $46.22 comes from interest.

Suppose we had deposited that entire $1000 at the beginning of the 10-month period. We use the compound interest formula (Formula 4.3 from Section 4.1) to find the balance at the end of the period. The monthly rate of 1% in decimal form is $r = 0.01$, so we find that our balance would have been

$$\begin{aligned}\text{Balance after 10 months} &= \text{Principal} \times (1 + r)^t \\ &= \$1000 \times (1 + 0.01)^{10} \\ &= \$1104.62\end{aligned}$$

That number is larger than the balance for monthly deposits because it includes interest earned on the full $1000 over the entire 10 months. Of course, for most of us it's easier to make monthly deposits of $100 than it is to come up with a lump sum of $1000 to invest.

These observations provide a helpful rule of thumb even though we don't yet have a formula for the balance.

RULE OF THUMB 4.4 Regular Deposits

Suppose we deposit money regularly into an account with a fixed interest rate.

1. The ending balance is at least as large as the total deposit.
2. The ending balance is less than the amount we would have if the entire deposit were invested initially and allowed to draw interest over the whole term.
3. These estimates are best over a short term.

Regular deposits balance formula

The formula for the account balance, assuming regular deposits at the end of each month, is as follows.[9]

FORMULA 4.8 Regular Deposits Balance Formula

$$\text{Balance after } t \text{ monthly deposits} = \frac{\text{Deposit} \times ((1 + r)^t - 1)}{r}$$

where t is the number of deposits, and r is the monthly interest rate APR/12 expressed as a decimal.

Alternatively, we can write the formula as

$$\text{Balance after deposits for } y \text{ years} = \frac{\text{Deposit} \times \left(\left(1 + \dfrac{\text{APR}}{12}\right)^{(12y)} - 1\right)}{\left(\dfrac{\text{APR}}{12}\right)}$$

where deposits are made monthly and y is the number of years. This form is equivalent since $r = \dfrac{\text{APR}}{12}$ and the number of deposits is $t = 12 \times y$.

[9] *See Exercise 23 for the adjustment necessary when deposits are made at the beginning of the month.*

The ending balance is often called the *future value* for this savings arrangement. A derivation of this formula is given in Algebraic Spotlight 4.4 at the end of this section. Throughout this section, we assume that interest is compounded monthly.

Let's check that this formula agrees with the balance at the end of the tenth month as found in Table 4.2. The deposit is \$100 each month, and the monthly rate as a decimal is $r = 0.01$. We want the balance after $t = 10$ deposits. Using Formula 4.8, we find

$$\begin{aligned}\text{Balance after 10 deposits} &= \frac{\text{Deposit} \times ((1+r)^t - 1)}{r} \\ &= \frac{\$100 \times (1.01^{10} - 1)}{0.01} \\ &= \$1046.22\end{aligned}$$

This is the same as the answer we obtained in Table 4.2.

EXAMPLE 4.21 Using the balance formula: Saving money regularly

Suppose we have a savings account earning 7% APR. We deposit \$20 to the account at the end of each month for five years. What is the future value for this savings arrangement? That is, what is the account balance after five years?

SOLUTION

We use the regular deposits balance formula (Formula 4.8). The monthly deposit is \$20, and the monthly interest rate as a decimal is

$$r = \frac{\text{APR}}{12} = \frac{0.07}{12}$$

The number of deposits is $t = 5 \times 12 = 60$. The regular deposits balance formula gives

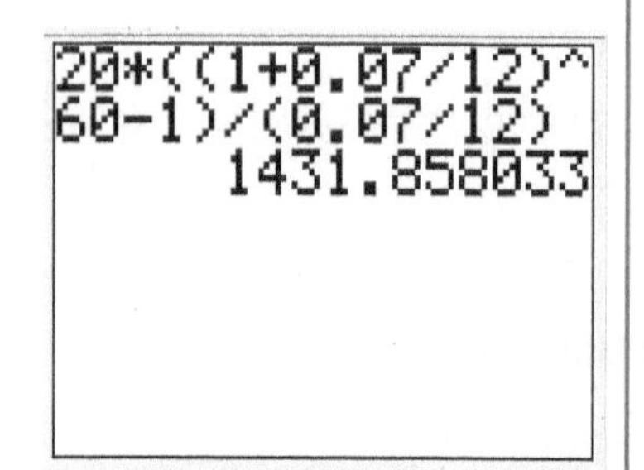

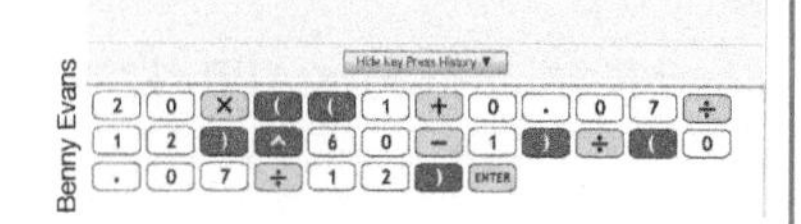

Benny Evans

$$\begin{aligned}\text{Balance after 60 deposits} &= \frac{\text{Deposit} \times ((1+r)^t - 1)}{r} \\ &= \frac{\$20 \times ((1+0.07/12)^{60} - 1)}{(0.07/12)} \\ &= \$1431.86\end{aligned}$$

The future value is \$1431.86.

TRY IT YOURSELF 4.21

Suppose we have a savings account earning 4% APR. We deposit \$40 to the account at the end of each month for 10 years. What is the future value of this savings arrangement?

The answer is provided at the end of this section.

Determining the savings needed

People approach savings in different ways: Some are committed to depositing a certain amount of money each month into a savings plan, and others save with a specific purchase in mind. Suppose you are nearing the end of your sophomore year and plan to purchase a car when you graduate in two years. If the car you have your eye on will cost \$20,000 when you graduate, you want to know how much you will have to save each month for the next two years to have \$20,000 at the end.

We can rearrange the regular deposits balance formula to tell us how much we need to deposit regularly in order to achieve a goal (that is, a future value) such as this.

FORMULA 4.9 Deposit Needed Formula

$$\text{Needed monthly deposit} = \frac{\text{Goal} \times r}{((1+r)^t - 1)}$$

where r is the monthly interest rate APR/12 as a decimal, and t is the number of deposits you will make to reach your goal.

Alternatively, we can write the formula as

$$\text{Needed monthly deposit} = \frac{\text{Goal} \times \left(\frac{\text{APR}}{12}\right)}{\left(\left(1+\frac{\text{APR}}{12}\right)^{(12y)} - 1\right)}$$

where y is the number of years to reach your goal. This form is equivalent since $r = \frac{\text{APR}}{12}$ and the number of deposits is $t = 12 \times y$.

For example, if your goal is \$20,000 to buy a car in two years, and if the APR is 6%, you can use this formula to find how much you need to deposit each month. The monthly rate as a decimal is

$$r = \frac{\text{APR}}{12} = \frac{0.06}{12} = 0.005$$

We use $t = 2 \times 12 = 24$ deposits in the formula:

$$\begin{aligned} \text{Needed deposit} &= \frac{\text{Goal} \times r}{((1+r)^t - 1)} \\ &= \frac{\$20{,}000 \times 0.005}{(1.005^{24} - 1)} \\ &= \$786.41 \end{aligned}$$

You need to deposit \$786.41 each month so you can buy that \$20,000 car when you graduate.

Student debt is today a significant issue for students and in the United States as a whole. The following excerpt of an article from the Web site Demos shows the importance of saving for college.

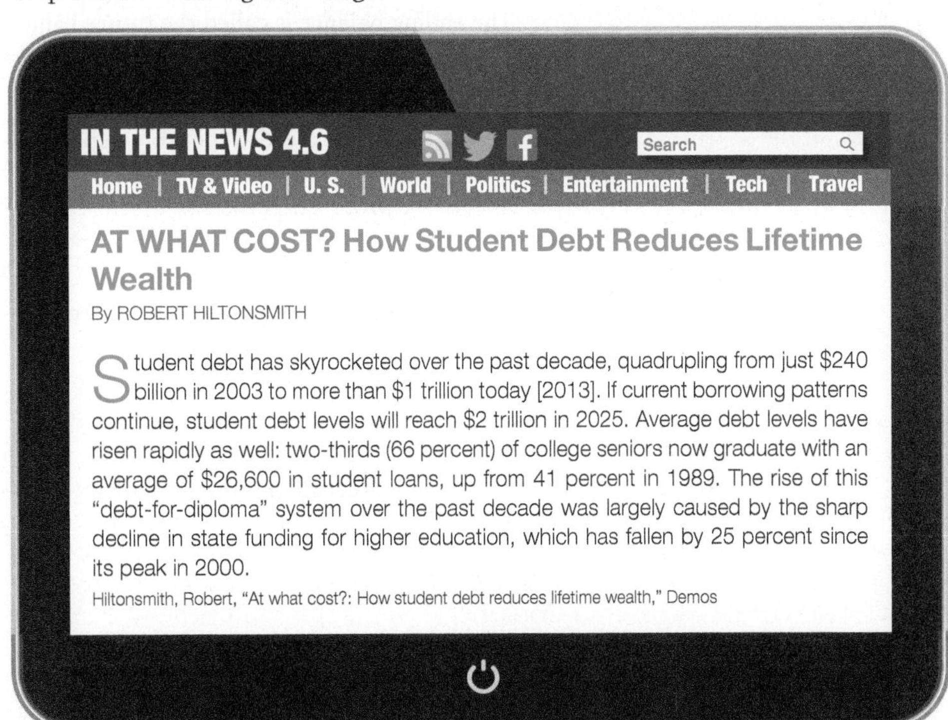

IN THE NEWS 4.6

Search

Home | TV & Video | U. S. | World | Politics | Entertainment | Tech | Travel

AT WHAT COST? How Student Debt Reduces Lifetime Wealth

By ROBERT HILTONSMITH

Student debt has skyrocketed over the past decade, quadrupling from just \$240 billion in 2003 to more than \$1 trillion today [2013]. If current borrowing patterns continue, student debt levels will reach \$2 trillion in 2025. Average debt levels have risen rapidly as well: two-thirds (66 percent) of college seniors now graduate with an average of \$26,600 in student loans, up from 41 percent in 1989. The rise of this "debt-for-diploma" system over the past decade was largely caused by the sharp decline in state funding for higher education, which has fallen by 25 percent since its peak in 2000.

Hiltonsmith, Robert, "At what cost?: How student debt reduces lifetime wealth," Demos

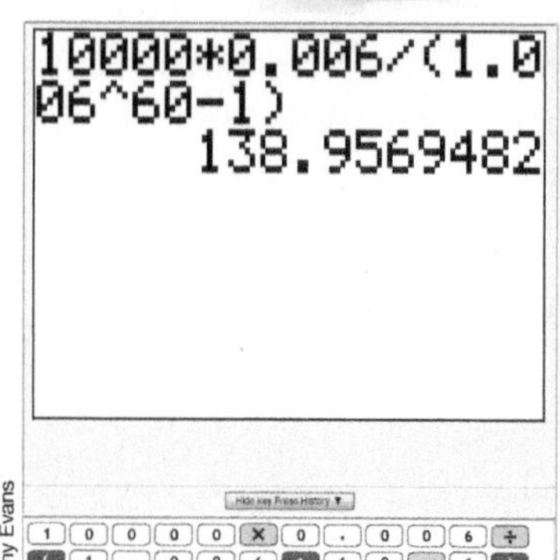

Benny Evans

EXAMPLE 4.22 Computing deposit needed: Saving for college

How much does your younger brother need to deposit each month into a savings account that pays 7.2% APR in order to have $10,000 when he starts college in five years?

SOLUTION

Your brother wants to achieve a goal of $10,000 in five years, so he uses Formula 4.9. The monthly interest rate as a decimal is

$$r = \frac{\text{APR}}{12} = \frac{0.072}{12} = 0.006$$

and the number of deposits is $t = 5 \times 12 = 60$:

$$\begin{aligned}\text{Needed deposit} &= \frac{\text{Goal} \times r}{((1+r)^t - 1)} \\ &= \frac{\$10{,}000 \times 0.006}{(1.006^{60} - 1)} \\ &= \$138.96\end{aligned}$$

He needs to deposit $138.96 each month.

TRY IT YOURSELF 4.22

How much do you need to deposit each month into a savings account that pays 9% APR in order to have $50,000 for your child to use for college in 18 years?

The answer is provided at the end of this section.

SUMMARY 4.5
Monthly Deposits

Suppose we deposit a certain amount of money at the end of each month into a savings account that pays a monthly interest rate of $r = \text{APR}/12$ as a decimal. The balance in the account after t months is given by the regular deposits balance formula (Formula 4.8):

$$\text{Balance after } t \text{ monthly deposits} = \frac{\text{Deposit} \times ((1+r)^t - 1)}{r}$$

The ending balance is called the future value for this savings arrangement.

A companion formula (Formula 4.9) gives the monthly deposit necessary to achieve a given balance:

$$\text{Needed monthly deposit} = \frac{\text{Goal} \times r}{((1+r)^t - 1)}$$

Alternatively, we can write these formulas in term of y, the number of years:

$$\text{Balance after monthly deposits for } y \text{ years} = \frac{\text{Deposit} \times \left(\left(1 + \frac{\text{APR}}{12}\right)^{(12y)} - 1\right)}{\left(\frac{\text{APR}}{12}\right)}$$

and

$$\text{Needed monthly deposit} = \frac{\text{Goal} \times \left(\frac{\text{APR}}{12}\right)}{\left(\left(1 + \frac{\text{APR}}{12}\right)^{(12y)} - 1\right)}$$

Saving for retirement

As the article at the beginning of this section points out, college students often don't think much about retirement, but early retirement planning is important.

EXAMPLE 4.23 Finding deposit needed: Retirement and varying rates

Suppose that you'd like to retire in 40 years and want to have a future value of $500,000 in a savings account. (See the article at the beginning of this section.) Also suppose that your employer makes regular monthly deposits into your retirement account.

a. If you can expect an APR of 9% for your account, how much do you need your employer to deposit each month?

b. The formulas we have been using assume that the interest rate is constant over the period in question. Over a period of 40 years, though, interest rates can vary widely. To see what difference the interest rate can make, let's assume a constant APR of 6% for your retirement account. How much do you need your employer to deposit each month under this assumption?

SOLUTION

a. We have a goal of $500,000, so we use Formula 4.9. The monthly rate as a decimal is

$$r = \text{Monthly rate} = \frac{\text{APR}}{12} = \frac{0.09}{12} = 0.0075$$

The number of deposits is $t = 40 \times 12 = 480$, so the needed deposit is

$$\text{Needed deposit} = \frac{\text{Goal} \times r}{((1+r)^t - 1)}$$

$$= \frac{\$500{,}000 \times 0.0075}{(1.0075^{480} - 1)}$$

$$= \$106.81$$

b. The computation is the same as in part a except that the new monthly rate as a decimal is

$$r = \frac{0.06}{12} = 0.005$$

We have

$$\text{Needed deposit} = \frac{\text{Goal} \times r}{((1+r)^t - 1)}$$

$$= \frac{\$500{,}000 \times 0.005}{(1.005^{480} - 1)}$$

$$= \$251.07$$

Note that the decrease in the interest rate from 9% to 6% requires that you more than double your monthly deposit if you are to reach the same goal. The effect of this possible variation in interest rates is one factor that makes financial planning for retirement complicated.

Retirement income: Annuities

How much income will you need in retirement? That's a personal matter, but we can analyze what a *nest egg* will provide.

KEY CONCEPT

Your nest egg is the balance of your retirement account at the time of retirement. The monthly yield is the amount you can withdraw from your retirement account each month.

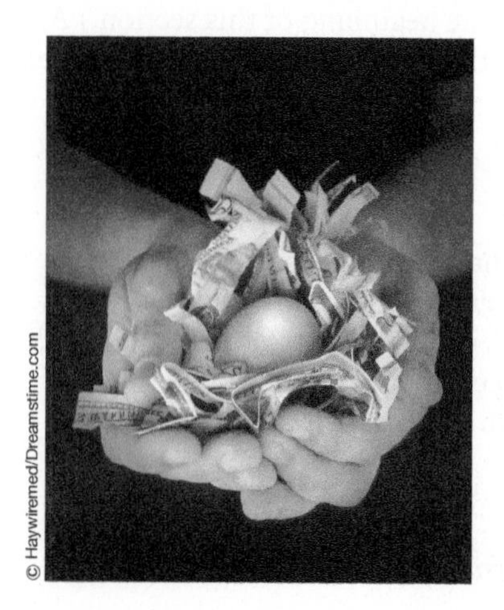
© Haywiremed/Dreamstime.com

Once you retire, there are several ways of using your retirement funds. One method is to withdraw each month only the interest accrued over that month; the principal remains the same. Under this arrangement, your nest egg will never be reduced; you'll be living off the interest alone. An arrangement like this is called a *perpetuity* because the constant income continues indefinitely. You can explore this arrangement further in Exercises 24 through 30.

With a perpetuity, the original balance at retirement remains untouched, but if we are willing to reduce the principal each month, we won't need to start with as large a nest egg for our given monthly income. In this situation, we receive a constant monthly payment, part of which represents interest and part of which represents a reduction of principal. Such an arrangement is called a *fixed-term annuity* because, unlike a perpetuity, this arrangement will necessarily end after a fixed term (when we have spent the entire principal).

KEY CONCEPT

An annuity is an arrangement that withdraws both principal and interest from your nest egg. Payments end when the principal is exhausted.

An annuity works just like the installment loans we considered in Section 4.2, only someone is paying *us* rather than the other way around.[10] In fact, the formula for the monthly payment applies in this situation, too. We can think of the institution that holds our account at retirement as having borrowed our nest egg; it will pay us back in monthly installments over the term of the annuity.

FORMULA 4.10 Annuity Yield Formula

$$\text{Monthly annuity yield} = \frac{\text{Nest egg} \times r \times (1+r)^t}{((1+r)^t - 1)}$$

where r is the monthly rate (as a decimal), and t is the term (the number of months the annuity will last).

Alternatively, we can write the formula as

$$\text{Monthly annuity yield} = \frac{\text{Nest egg} \times \frac{\text{APR}}{12} \times \left(1 + \frac{\text{APR}}{12}\right)^{(12y)}}{\left(\left(1 + \frac{\text{APR}}{12}\right)^{(12y)} - 1\right)}$$

where y is the term of the annuity in years.

[10] *We assume that the payments are made at the end of the month.*

cartoonstock.com

"We could both avoid this daily annoyance, sir, if you'd buy me an *annuity!*"

EXAMPLE 4.24 Finding annuity yield: 20-year annuity

Suppose we have a nest egg of \$800,000 with an APR of 6% compounded monthly. Find the monthly yield for a 20-year annuity.

SOLUTION

We use the annuity yield formula (Formula 4.10). With an APR of 6%, the monthly rate as a decimal is

$$r = \text{Monthly rate} = \frac{\text{APR}}{12} = \frac{0.06}{12} = 0.005$$

The term is 20 years, so we take $t = 20 \times 12 = 240$ months:

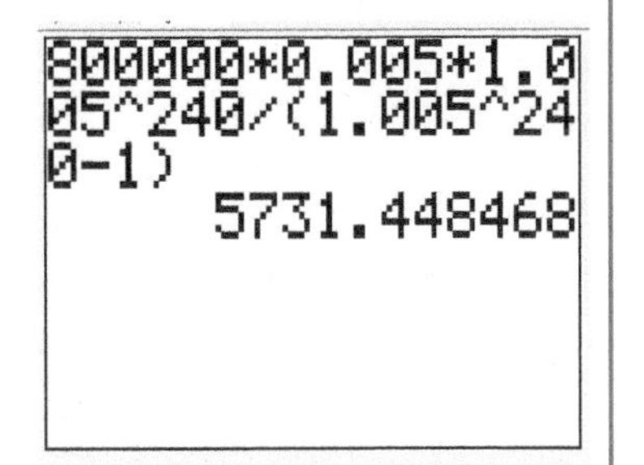

Benny Evans

$$\text{Monthly annuity yield} = \frac{\text{Nest egg} \times r \times (1+r)^t}{((1+r)^t - 1)}$$

$$= \frac{\$800{,}000 \times 0.005 \times 1.005^{240}}{(1.005^{240} - 1)}$$

$$= \$5731.45$$

TRY IT YOURSELF 4.24

Suppose we have a nest egg of \$1,000,000 with an APR of 6% compounded monthly. Find the monthly yield for a 25-year annuity.

The answer is provided at the end of this section.

How large a nest egg is needed to achieve a desired annuity yield? We can answer this question by rearranging the annuity yield formula (Formula 4.10).

FORMULA 4.11 Annuity Yield Goal

$$\text{Nest egg needed} = \frac{\text{Annuity yield goal} \times ((1+r)^t - 1)}{(r \times (1+r)^t)}$$

where r is the monthly interest rate and t is the term of the annuity in months.

Alternatively, we can write the formula as

$$\text{Nest egg needed} = \frac{\text{Annuity yield goal} \times \left(\left(1+\frac{\text{APR}}{12}\right)^{(12y)} - 1\right)}{\left(\frac{\text{APR}}{12} \times \left(1+\frac{\text{APR}}{12}\right)^{(12y)}\right)}$$

where y is the term of the annuity in years.

Here, r is the monthly rate (as a decimal), and t is the term (in months) of the annuity.

The following example shows how we use this formula.

EXAMPLE 4.25 Finding nest egg needed for annuity: Retiring on a 20-year annuity

Suppose our retirement account pays 5% APR compounded monthly. What size nest egg do we need in order to retire with a 20-year annuity that yields \$4000 per month?

SOLUTION

We want to achieve an annuity goal, so we use Formula 4.11. The monthly rate as a decimal is $r = 0.05/12$, and the term is $t = 20 \times 12 = 240$ months:

$$\begin{aligned}\text{Nest egg needed} &= \frac{\text{Annuity yield goal} \times ((1+r)^t - 1)}{(r \times (1+r)^t)} \\ &= \frac{\$4000 \times ((1+0.05/12)^{240} - 1)}{((0.05/12) \times (1+0.05/12)^{240})} \\ &= \$606{,}101.25\end{aligned}$$

```
4000*((1+0.05/12
)^240-1)/((0.05/
12)*(1+0.05/12)^
240)
          606101.2523
```

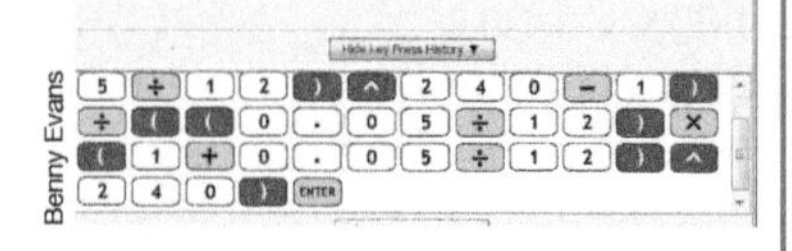

Benny Evans

TRY IT YOURSELF 4.25

Suppose our retirement account pays 9% APR compounded monthly. What size nest egg do we need in order to retire with a 25-year annuity that yields \$5000 per month?

The answer is provided at the end of this section.

The balance at retirement (the nest egg) is called the *present value* of the annuity. Future value and present value depend on perspective. When you started saving, the balance at retirement was the future value. When you actually retire, it becomes the present value.

The obvious question is, how many years should you plan for the annuity to last? If you set it up to last until you're 80 and then you live until you're 85, you're in trouble. What a retiree often wants is to have the monthly annuity payment continue for as long as he or she lives. Insurance companies offer such an arrangement, and it's called a *life annuity*. How does it work?

The insurance company makes a statistical estimate of the life expectancy of a customer, which is used in determining how much the company will probably have to pay out.[11] Some customers will live longer than the estimate (and the company

[11] *Such calculations are made by professionals known as* actuaries.

may lose money on them) but some will not (and the company will make money on them). The monthly income for a given principal is determined from this estimate using the formula for the present value of a fixed-term annuity.

SUMMARY 4.6
Retiring on an Annuity

For a fixed-term annuity, the formulas for monthly payments apply. Let r be the monthly interest rate (as a decimal) and t the term (in months) of the annuity.

1. To find the monthly yield provided by a nest egg, we use

$$\text{Monthly annuity yield} = \frac{\text{Nest egg} \times r \times (1+r)^t}{((1+r)^t - 1)}$$

2. To find the nest egg needed to provide a desired income, we use

$$\text{Nest egg needed} = \frac{\text{Annuity yield goal} \times ((1+r)^t - 1)}{(r \times (1+r)^t)}$$

The balance at retirement (the nest egg) is called the present value of the annuity.

Alternatively, we can write these formulas in terms of y, the term of the annuity in years:

$$\text{Monthly annuity yield} = \frac{\text{Nest egg} \times \frac{\text{APR}}{12} \times \left(1 + \frac{\text{APR}}{12}\right)^{(12y)}}{\left(\left(1 + \frac{\text{APR}}{12}\right)^{(12y)} - 1\right)}$$

and

$$\text{Nest egg needed} = \frac{\text{Annuity yield goal} \times \left(\left(1 + \frac{\text{APR}}{12}\right)^{(12y)} - 1\right)}{\left(\frac{\text{APR}}{12} \times \left(1 + \frac{\text{APR}}{12}\right)^{(12y)}\right)}$$

The following schematic illustrates how the formulas in this section apply to your long-term financial security. Financial advisors emphasize the importance of making an early beginning in saving for retirement.

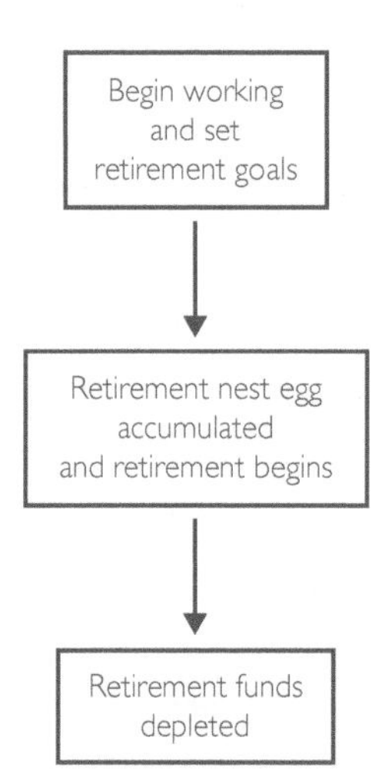

Steps to be taken at the beginning of your working life to prepare for retirement

- Set goal for monthly income at retirement.
- Use the annuity yield goal formula (Formula 4.11) to determine the nest egg needed to meet your income goal. (Take into account income from other sources, such as Social Security.)
- Use the deposit needed formula (Formula 4.9) to determine the monthly savings needed to meet your goal.

Steps to be taken at the time of retirement

- Assess the actual accumulated nest egg.
- Use the annuity yield formula (Formula 4.10) to determine your actual monthly income at retirement.

ALGEBRAIC SPOTLIGHT 4.4 Derivation of the Regular Deposits Balance Formula

Suppose that at the end of each month, we deposit money (the same amount each month) into an account that pays a monthly rate of r as a decimal. Our goal is to derive the formula

$$\text{Balance after } t \text{ deposits} = \frac{\text{Deposit} \times ((1+r)^t - 1)}{r}$$

In Algebraic Spotlight 4.2 from Section 4.2, we considered the situation where we borrow B_0 dollars and make monthly payments of M dollars at a monthly interest rate of r as a decimal. We found that the balance after t payments is

$$\text{Balance after } t \text{ payments} = B_0(1+r)^t - M\frac{(1+r)^t - 1}{r}$$

Making payments of M dollars per month subtracts money from the balance. A deposit adds money rather than subtracting it. The result is to add the term involving M instead of subtracting it:

$$\text{Balance after } t \text{ deposits} = B_0(1+r)^t + M\frac{(1+r)^t - 1}{r}$$

Now the initial balance is zero, so $B_0 = 0$. Therefore,

$$\text{Balance after } t \text{ deposits} = M\frac{(1+r)^t - 1}{r}$$

Because M is the amount we deposit, we can write the result as

$$\text{Balance after } t \text{ deposits} = \frac{\text{Deposit} \times ((1+r)^t - 1)}{r}$$

WHAT DO YOU THINK?

Your daughter's college education: You figure you will need \$50,000 to pay for your baby daughter's college education. You want to start saving now. Do you need to immediately invest \$50,000? Explain how you might be able to invest less.

Questions of this sort may invite many correct responses. We present one possibility: You can invest a good deal less than \$50,000 because your investment will be earning interest. If interest is being compounded, and if you start while your daughter is young, you may be able to invest significantly less than \$50,000. This is because the compound interest contributes to exponential growth, and as we know, exponential growth can be dramatic over the long term.

Further questions of this type may be found in the *What Do You Think?* section of exercises.

Try It Yourself answers

Try It Yourself 4.20: Verifying a balance: Regular deposits into an account $\$303.01 + 0.01 \times \$303.01 + \$100 = \$406.04$.

Try It Yourself 4.21: Using the balance formula: Saving money regularly \$5889.99.

Try It Yourself 4.22: Computing deposit needed: Saving for college \$93.22.

Try It Yourself 4.24: Finding annuity yield: 25-year annuity \$6443.01.

Try It Yourself 4.25: Finding nest egg needed for annuity: Retiring on a 25-year annuity \$595,808.11.

Exercise Set 4.3

Test Your Understanding

1. Explain how an annuity works.

2. A person's *nest egg* is: **a.** the total amount of money she has at the present **b.** the amount of money she has in savings at age 65 **c.** the amount of money she has in her retirement account at the time she retires **d.** the amount of money she has in her checking account at the time she retires.

3. True or false: In most of the formulas in this section you see $\frac{\text{APR}}{12}$. This is because the formulas are based on monthly interest rates.

4. True or false: Although compounding makes an account grow faster, it is not an important factor in saving for retirement.

5. Explain in your own words what the regular deposits balance formula (Formula 4.8) tells you.

6. Explain in your own words what the deposit needed formula (Formula 4.9) tells you.

Problems

7. **Saving for a car.** You are saving to buy a car, and you deposit \$200 at the end of each month for two years at an APR of 4.8% compounded monthly. What is the future value for this savings arrangement? That is, how much money will you have for the purchase of the car after two years?

8. **Saving for a down payment.** You want to save \$20,000 for a down payment on a home by making regular monthly deposits over five years. Take the APR to be 6%. How much money do you need to deposit each month?

9. **Planning for college.** At your child's birth, you begin contributing monthly to a college fund. The fund pays an APR of 4.8% compounded monthly. You figure that your child will need \$40,000 at age 18 to begin college. What monthly deposit is required?

10. **Savings account tabulation.** You have a savings account into which you invest \$50 at the end of every month, and the account pays you an APR of 9% compounded monthly.

a. Fill in the following table. (Don't use the regular deposits balance formula.)

At end of month number	Interest paid on prior balance	Deposit	Balance
1			
2			
3			
4			

b. Use the regular deposits balance formula (Formula 4.8) to determine the balance in the account at the end of four months. Compare this to the final balance in the table from part a.

c. Use the regular deposits balance formula to determine the balance in the account at the end of four years.

d. Use the regular deposits balance formula to determine the balance in the account at the end of 20 years.

11. **Retirement account tabulation.** You have a retirement account into which your employer invests \$75 at the end of every month, and the account pays an APR of 5.25% compounded monthly.

a. Fill in the following table. (Don't use the regular deposits balance formula.)

At end of month number	Interest paid on prior balance	Deposit	Balance
1			
2			
3			
4			

b. Use the regular deposits balance formula (Formula 4.8) to determine the balance in the account at the end of four months. Compare this to the final balance in the table from part a.

c. Use the regular deposits balance formula to determine the balance in the account at the end of four years.

d. Use the regular deposits balance formula to determine the balance in the account at the end of 20 years.

12. **Savings account estimations.** Jerry calculates that if he makes a deposit of \$5 each month at an APR of 4.8%, then at the end of two years he'll have \$100. Benny says that the correct amount is \$135. Rule of Thumb 4.4 should be helpful in this exercise.

a. What was the total amount deposited (ignoring interest earned)? Whose answer is ruled out by this calculation? Why?

b. Suppose the total amount deposited (\$5 per month for two years) is instead put as a lump sum at the beginning of the two years as principal in an account earning an APR of 4.8%. Use the compound interest formula (with monthly compounding) from Section 4.1 to determine how much would be in the account after two years. Whose answer is ruled out by this calculation? Why?

c. Find the correct balance after two years.

13–14. **Saving for a boat.** *Suppose you want to save in order to purchase a new boat. Take the APR to be 7.2% in Exercises 13 and 14.*

13. If you deposit \$250 each month, how much will you have toward the purchase of a boat after three years?

14. You want to have \$13,000 toward the purchase of a boat in three years. How much do you need to deposit each month?

15. **Fixed-term annuity.** You have a 20-year annuity with a present value (that is, nest egg) of \$425,000. If the APR is 7%, what is the monthly yield?

16. Life annuity. You have set up a life annuity with a present value of \$350,000. If your life expectancy at retirement is 21 years, what will your monthly income be? Take the APR to be 6%.

17. Nest egg. You begin working at age 25, and your employer deposits \$300 per month into a retirement account that pays an APR of 6% compounded monthly. You expect to retire at age 65.

a. What will be the size of your nest egg at age 65?

b. Suppose you are allowed to contribute \$100 each month in addition to your employer's contribution. What will be the size of your nest egg at age 65? Compare this with your answer to part a.

18. A different nest egg. You begin working at age 25, and your employer deposits \$250 each month into a retirement account that pays an APR of 6% compounded monthly. You expect to retire at age 65.

a. What will be the size of your nest egg when you retire?

b. Suppose instead that you arranged to start the regular deposits two years earlier, at age 23. What will be the size of your nest egg when you retire? Compare this with your answer from part a.

19. Planning to retire on an annuity. You plan to work for 40 years and then retire using a 25-year annuity. You want to arrange a retirement income of \$4500 per month. You have access to an account that pays an APR of 7.2% compounded monthly.

a. What size nest egg do you need to achieve the desired monthly yield?

b. What monthly deposits are required to achieve the desired monthly yield at retirement?

20. Retiring. You want to have a monthly income of \$2000 from a fixed-term annuity when you retire. Take the term of the annuity to be 20 years and assume an APR of 6% over the period of investment.

a. How large will your nest egg have to be at retirement to guarantee the income described above?

b. You plan to make regular deposits for 40 years to build up your savings to the level you determined in part a. How large must your monthly deposit be?

21. Planning for retirement. You anticipate an average income of \$10,000 per month over your working life. With this in mind, you begin planning for retirement. For this exercise, assume an APR of 5% and round all answers to the nearest whole dollar.

a. Taking into account income from Social Security, you think you will need a monthly income of about 75% of your working income. That is, you want to provide a monthly retirement income of \$7500 for 20 years. Use the annuity yield goal formula (Formula 4.11) to determine the size of the nest egg needed for retirement.

b. Use the deposit needed formula (Formula 4.9) to determine the monthly deposit required to meet the goal from part a. Assume a working life of 45 years.

22. You want to be rich. You plan to work for 50 years before retiring with ten million dollars to spend. Assuming an APR of 6%, how much money do you need to save each month? Round your answer to the nearest whole dollar.

23. Deposits at the beginning. In Section 4.3, we considered the case of regular deposits at the end of each month. If deposits are made at the beginning of each month, then the formula is a bit different. The adjusted formula is

$$\text{Balance after } t \text{ deposits} = \text{Deposit} \times (1+r) \times \frac{((1+r)^t - 1)}{r}$$

Here, r is the monthly interest rate (as a decimal), and t is the number of deposits. The extra factor of $1+r$ accounts for the interest earned on the deposit over the first month after it's made.

Suppose you deposit \$200 at the beginning of each month for five years. Take the APR to be 7.2%.

a. What is the future value? In other words, what will your account balance be at the end of the period?

b. What would be the future value if you had made the deposits at the end of each month rather than at the beginning? Explain why it is reasonable that your answer here is smaller than that from part a.

24. Retirement income: perpetuities. If a retirement fund is set up as a *perpetuity*, one withdraws each month only the interest accrued over that month; the principal remains the same. For example, suppose you have accumulated \$500,000 in an account with a monthly interest rate of 0.5%. Each month, you can withdraw \$500,000 × 0.005 = \$2500 in interest, and the nest egg will always remain at \$500,000. That is, the \$500,000 perpetuity has a monthly yield of \$2500. In general, the monthly yield for a perpetuity is given by the formula

$$\text{Monthly perpetuity yield} = \text{Nest egg} \times \text{Monthly interest rate}$$

In this formula, the monthly interest rate is expressed as a decimal.

Suppose we have a perpetuity paying an APR of 6% compounded monthly. If the value of our nest egg (that is, the present value) is \$800,000, find the amount we can withdraw each month. *Note*: First find the monthly interest rate.

25. Perpetuity yield. *Refer to Exercise 24 for background on perpetuities.* You have a perpetuity with a present value (that is, nest egg) of \$650,000. If the APR is 5% compounded monthly, what is your monthly income?

26. Another perpetuity. *Refer to Exercise 24 for background on perpetuities.* You have a perpetuity with a present value of \$900,000. If the APR is 4% compounded monthly, what is your monthly income?

27. Comparing annuities and perpetuities. *Refer to Exercise 24 for background on perpetuities.* For 40 years, you invest \$200 per month at an APR of 4.8% compounded monthly, then you retire and plan to live on your retirement nest egg.

a. How much is in your account on retirement?

b. Suppose you set up your account as a perpetuity on retirement. What will your monthly income be? (Assume that the APR remains at 4.8% compounded monthly.)

c. Suppose now you use the balance in your account for a life annuity instead of a perpetuity. If your life expectancy is 21 years, what will your monthly income be? (Again, assume that the APR remains at 4.8% compounded monthly.)

d. Compare the total amount you invested with your total return from part c. Assume that you live 21 years after retirement.

28. Perpetuity goal: How much do I need to retire? *Refer to Exercise 24 for background on perpetuities.* Here is the formula for the nest egg needed for a desired monthly yield on a perpetuity:

$$\text{Nest egg needed} = \frac{\text{Desired monthly yield}}{\text{Monthly interest rate}}$$

In this formula, the monthly interest rate is expressed as a decimal.

If your retirement account pays 5% APR with monthly compounding, what present value (that is, nest egg) is required for you to retire on a perpetuity that pays $4000 per month?

29. Desired perpetuity. *Refer to the formula in Exercise 28.* You want a perpetuity with a monthly income of $3000. If the APR is 7%, what does the present value have to be?

30. Planning to retire on a perpetuity. You plan to work for 40 years and then retire using a perpetuity. You want to arrange to have a retirement income of $4500 per month. You have access to an account that pays an APR of 7.2% compounded monthly.

a. *Refer to the formula in Exercise 28.* What size nest egg do you need to achieve the desired monthly yield?

b. What monthly deposits are required to achieve the desired monthly yield at retirement?

Exercise 31 is suitable for group work.

31. Starting early, starting late. In this exercise, we consider the effects of starting early or late to save for retirement. Assume that each account considered has an APR of 6% compounded monthly.

a. At age 20, you realize that even a modest start on saving for retirement is important. You begin depositing $50 each month into an account. What will be the value of your nest egg when you retire at age 65?

b. Against expert advice, you begin your retirement program at age 40. You plan to retire at age 65. What monthly contributions do you need to make to match the nest egg from part a?

c. Compare your answer to part b with the monthly deposit of $50 from part a. Also compare the total amount deposited in each case.

d. Let's return to the situation in part a: At age 20, you begin depositing $50 each month into an account. Now suppose that at age 40, you finally get a job where your employer puts $400 per month into an account. You continue your $50 deposits, so from age 40 on, you have two separate accounts working for you. What will be the total value of your nest egg when you retire at age 65?

32. Retiring without interest. Suppose we lived in a society without interest. At age 25, you begin putting $250 per month into a cookie jar until you retire at age 65. At age 65, you begin to withdraw $2500 per month from the cookie jar. How long will your retirement fund last?

Exercises 33 and 34 are designed to be solved using technology such as calculators or computer spreadsheets.

33. How much? You begin saving for retirement at age 25, and you plan to retire at age 65. You want to deposit a certain amount each month into an account that pays an APR of 6% compounded monthly. Make a table that shows the amount you must deposit each month in terms of the nest egg you desire to have when you retire. Include nest egg sizes from $100,000 to $1,000,000 in increments of $100,000.

34. How long? You begin working at age 25, and your employer deposits $350 each month into a retirement account that pays an APR of 6% compounded monthly. Make a table that shows the size of your nest egg in terms of the age at which you retire. Include retirement ages from 60 to 70.

Writing About Mathematics

35. History of annuities. The origins of annuities can be traced back to the ancient Romans. Look up the history of annuities and write a report on it. Include the use of annuities in Rome, in Europe during the seventeenth century, in colonial America, and in modern American society.

36. History of actuaries. Look up the history of the actuarial profession and write a report on it. Be sure to discuss what mathematics are required to become an actuary.

37. Retirement plans. Some people you know, such as family members or teachers, probably have retirement plans. Find out about their plans and write a report on what you learn.

38. Social Security. Most people count on Social Security to provide at least part of their retirement income. Write a report on the history of the Social Security system and how it works.

39. Making a plan. Use what you have learned in this section to create a hypothetical retirement plan for yourself.

What Do You Think?

40. Low interest rate. You are setting up a regular monthly deposit plan but are concerned about the low interest rate you will earn. Should you be more concerned if the plan is for 2 years or if the plan is for 20 years? Explain your reasoning. **Exercise 7 is relevant to this topic.**

41. The regular deposits formula. The regular deposits balance formula is a bit complicated. If you deposit $100 each month for 10 months, you have deposited a total of $1000. Use the regular deposits balance formula to calculate the balance after those 10 months using monthly interest rates of 10%, 1%, 0.5%, and 0.01%. Explain what you observe. **Exercise 10 is relevant to this topic.**

42. Reasonable balance? You are discussing a regular monthly deposit plan with an investment planner. The interest rate is fixed over the relatively short period of the investment. The planner tells you what ending balance you can expect. Explain how you can quickly check whether the figure the planner gave you is reasonable. **Exercise 12 is relevant to this topic.**

43. Your son's college education. You want to save money to finance your young son's college education. Explain the financial implications of waiting five more years to start saving for this expense. Explain your reasoning. **Exercise 18 is relevant to this topic.**

44. Starting early. Financial advisors counsel young people to start saving early for their retirement. Explain why an early start is important. **Exercise 31 is relevant to this topic.**

45. Reasonable annuity yield? You are discussing a 20-year annuity with an investment planner. The interest rate is fixed and relatively high. The planner tells you what monthly yield you can expect for the nest egg you have. Explain how you can quickly check whether the figure the planner gave you is reasonable. (Recall that annuities work just like the installment loans considered in Section 4.2.)

46. Annuities, plus and minus. Explain some of the advantages and disadvantages of a fixed-term annuity.

47. Perpetuities, plus and minus. *See Exercise 24 for background on perpetuities.* Explain some advantages and disadvantages of a perpetuity. **Exercise 27 is relevant to this topic.**

48. Fixed vs. life. Explain in your own words the difference between a fixed-term annuity and a life annuity.

4.4 Credit cards: Paying off consumer debt

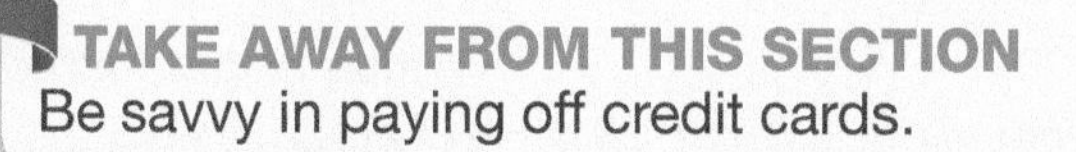

TAKE AWAY FROM THIS SECTION
Be savvy in paying off credit cards.

The accompanying excerpt is from an article in *The Chronicle of Higher Education.* The subject might be familiar to you.

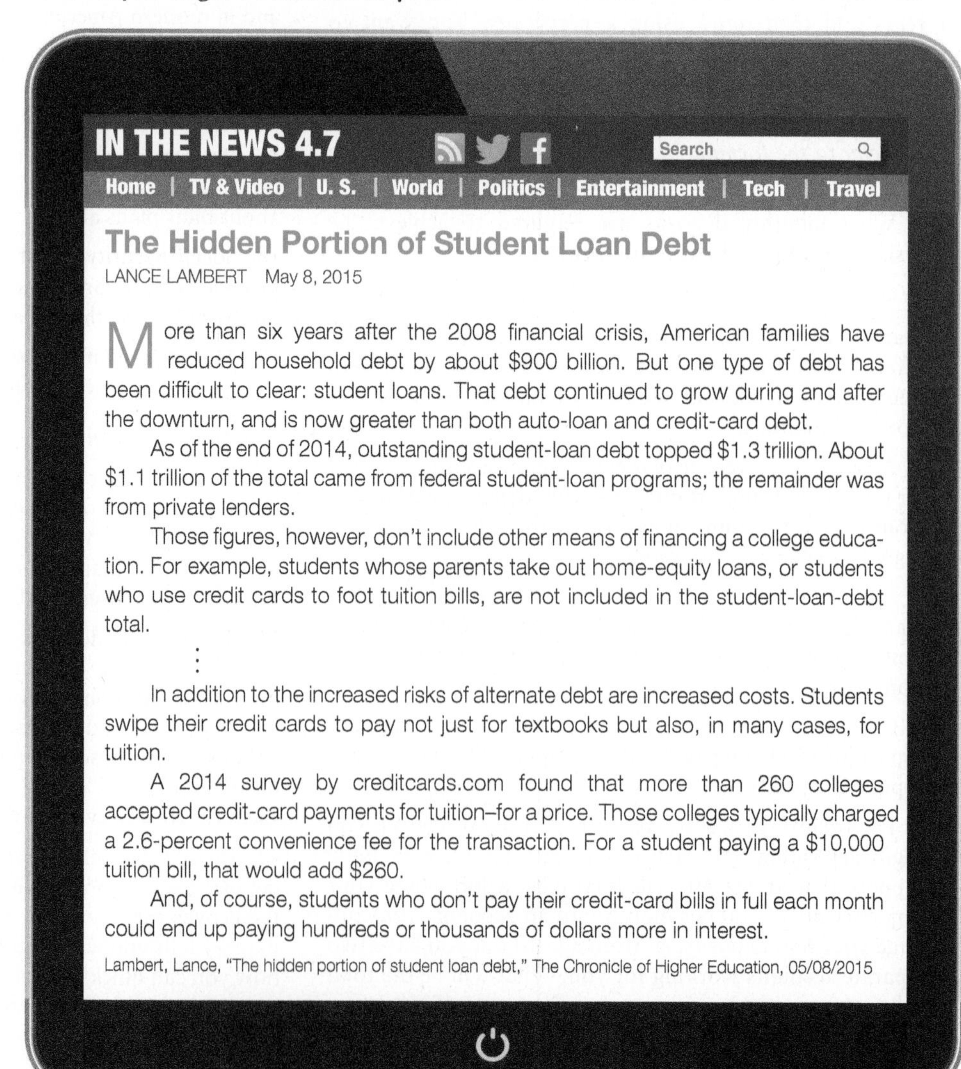

IN THE NEWS 4.7

Search

Home | TV & Video | U. S. | World | Politics | Entertainment | Tech | Travel

The Hidden Portion of Student Loan Debt

LANCE LAMBERT May 8, 2015

More than six years after the 2008 financial crisis, American families have reduced household debt by about $900 billion. But one type of debt has been difficult to clear: student loans. That debt continued to grow during and after the downturn, and is now greater than both auto-loan and credit-card debt.

As of the end of 2014, outstanding student-loan debt topped $1.3 trillion. About $1.1 trillion of the total came from federal student-loan programs; the remainder was from private lenders.

Those figures, however, don't include other means of financing a college education. For example, students whose parents take out home-equity loans, or students who use credit cards to foot tuition bills, are not included in the student-loan-debt total.

⋮

In addition to the increased risks of alternate debt are increased costs. Students swipe their credit cards to pay not just for textbooks but also, in many cases, for tuition.

A 2014 survey by creditcards.com found that more than 260 colleges accepted credit-card payments for tuition–for a price. Those colleges typically charged a 2.6-percent convenience fee for the transaction. For a student paying a $10,000 tuition bill, that would add $260.

And, of course, students who don't pay their credit-card bills in full each month could end up paying hundreds or thousands of dollars more in interest.

Lambert, Lance, "The hidden portion of student loan debt," The Chronicle of Higher Education, 05/08/2015

Credit cards are convenient and useful. They allow us to travel without carrying large sums of cash, and they sometimes allow us to defer cash payments, even interest-free, for a short time. In fact, at times, having a credit card can be almost a necessity because many hotels and rental car companies require customers to have one.

Although credit cards are convenient, they come with a potential cost. For example, many credit cards carry an APR that is much higher than other kinds of consumer loans. At the time of this writing, the site creditcards.com showed APRs on credit cards generally ranging from about 18% to almost 24% for those with poor credit. (It showed a rate of 15% for student cards.) Credit cards can differ in other ways as well. In deciding which card to use, it is very important to read the fine print and get the card that is most favorable to you.

In this section, we explore credit cards and finance charges in some detail. In particular, we look at how payments are calculated, the terminology used by credit card companies, and the implications of making only the minimum required payments.

"I found the problem. We earn money 5 days a week, but we spend money 7 days a week."

Credit card basics

All credit cards have finance charges, but most companies will waive them if you pay the full amount owed each month (see **Figure 4.12**). When finance charges are incurred, different companies calculate them in different ways. They typically explain the method they use on the monthly statement. Three methods that are used are the *previous balance method*, the *average daily balance method*, and the *adjusted balance method*.

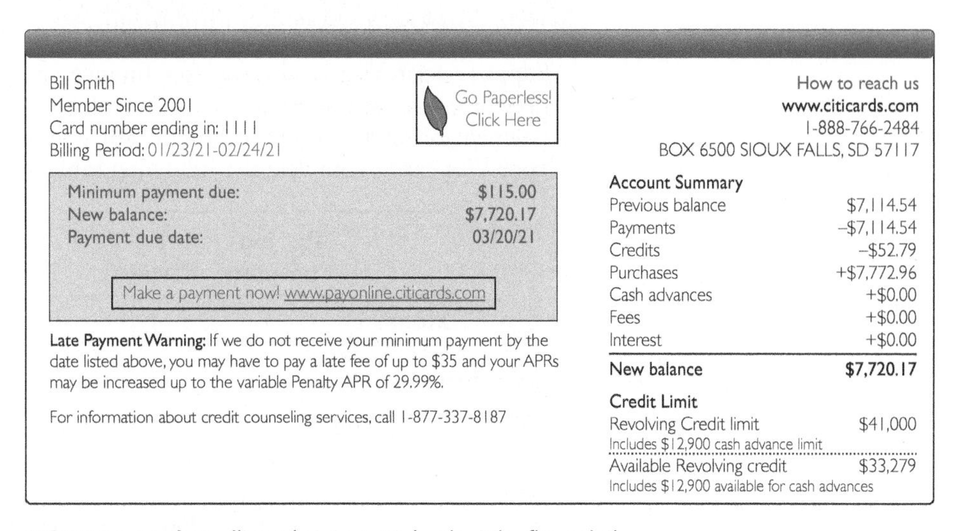

Bill Smith
Member Since 2001
Card number ending in: 1111
Billing Period: 01/23/21-02/24/21

Go Paperless! Click Here

How to reach us
www.citicards.com
1-888-766-2484
BOX 6500 SIOUX FALLS, SD 57117

Minimum payment due:	$115.00
New balance:	$7,720.17
Payment due date:	03/20/21

Make a payment now! www.payonline.citicards.com

Late Payment Warning: If we do not receive your minimum payment by the date listed above, you may have to pay a late fee of up to $35 and your APRs may be increased up to the variable Penalty APR of 29.99%.

For information about credit counseling services, call 1-877-337-8187

Account Summary	
Previous balance	$7,114.54
Payments	–$7,114.54
Credits	–$52.79
Purchases	+$7,772.96
Cash advances	+$0.00
Fees	+$0.00
Interest	+$0.00
New balance	**$7,720.17**

Credit Limit	
Revolving Credit limit	$41,000
Includes $12,900 cash advance limit	
Available Revolving credit	$33,279
Includes $12,900 available for cash advances	

FIGURE 4.12 A credit card statement. Look at the fine print!

In the previous balance method, the company calculates finance charges based on the balance at the end of the previous payment period.

In the average daily balance method, the company calculates finance charges based on the average of the daily balances over the payment period (usually one month). This means that a pair of jeans purchased early in the month will incur more finance charges than the same pair purchased late in the month. See Exercises 28 and 29 for more information on this method.

The adjusted balance method is perhaps the most common method, and it is the one we will use in this text. It starts with the balance from the previous month, subtracts payments you've made since the previous statement, and adds any new purchases. This is the amount that is subject to finance charges at the end of the month, assuming that the previous balance wasn't paid in full. After this finance charge has been added, we get the new balance. This procedure is summarized in the formula:

$$\text{New balance} = \text{Previous balance} - \text{Payments} + \text{Purchases} + \text{Finance charge}$$

The finance charge is calculated by applying the monthly interest rate (the APR divided by 12) to the amount subject to finance charges. Thus,

$$\text{Finance charge} = \frac{\text{APR}}{12} \times (\text{Previous balance} - \text{Payments} + \text{Purchases})$$

Let's look at a card that charges 21.6% APR. Suppose your previous statement showed a balance of \$300, you made a payment of \$100 in response to your previous statement, and you have new purchases of \$50.

Because the APR is 21.6%, we find the monthly interest rate using

$$\text{Monthly interest rate} = \frac{\text{APR}}{12} = \frac{21.6\%}{12} = 1.8\%$$

Now we calculate the finance charge:

$$\begin{aligned}\text{Finance charge} &= 0.018 \times (\text{Previous balance} - \text{Payments} + \text{Purchases})\\ &= 0.018 \times (\$300 - \$100 + \$50)\\ &= 0.018 \times \$250 = \$4.50\end{aligned}$$

This makes your new balance

$$\begin{aligned}\text{New balance} &= \text{Previous balance} - \text{Payments} + \text{Purchases} + \text{Finance charge}\\ &= \$300 - \$100 + \$50 + \$4.50 = \$254.50\end{aligned}$$

Your new balance is \$254.50.

EXAMPLE 4.26 Calculating finance charges: Buying clothes

Suppose your Visa card calculates finance charges using an APR of 22.8%. Your previous statement showed a balance of \$500, in response to which you made a payment of \$200. You then bought \$400 worth of clothes, which you charged to your Visa card. Complete the following table:

	Previous balance	Payments	Purchases	Finance charge	New balance
Month 1					

SOLUTION

We start by entering the information given.

	Previous balance	Payments	Purchases	Finance charge	New balance
Month 1	\$500.00	\$200.00	\$400.00		

The APR is 22.8%, and we divide by 12 to get a monthly rate of 1.9%. In decimal form, this calculation is $0.228/12 = 0.019$. Therefore,

$$\begin{aligned}\text{Finance charge} &= 0.019 \times (\text{Previous balance} - \text{Payments} + \text{Purchases}) \\ &= 0.019 \times (\$500 - \$200 + \$400) \\ &= 0.019 \times \$700 = \$13.30\end{aligned}$$

That gives a new balance of

$$\begin{aligned}\text{New balance} &= \text{Previous balance} - \text{Payments} + \text{Purchases} + \text{Finance charge} \\ &= \$500 - \$200 + \$400 + \$13.30 = \$713.30\end{aligned}$$

Your new balance is $713.30. Here is the completed table.

	Previous balance	Payments	Purchases	Finance charge	New balance
Month 1	$500.00	$200.00	$400.00	1.9% of $700 = $13.30	$713.30

TRY IT YOURSELF 4.26

This is a continuation of Example 4.26. You make a payment of $300 to reduce the $713.30 balance and then charge a TV costing $700. Complete the following table:

	Previous balance	Payments	Purchases	Finance charge	New balance
Month 1	$500.00	$200.00	$400.00	$13.30	$713.30
Month 2					

The answer is provided at the end of this section.

SUMMARY 4.7
Credit Card Basics

The formula for finding the finance charge is

$$\text{Finance charge} = \frac{\text{APR}}{12} \times (\text{Previous balance} - \text{Payments} + \text{Purchases})$$

The new balance is found by using the formula

$$\text{New balance} = \text{Previous balance} - \text{Payments} + \text{Purchases} + \text{Finance charge}$$

Making only minimum payments

Most credit cards require a minimum monthly payment, which is usually calculated as a fixed percentage of your balance or some fixed amount, whichever is more. For example, the minimum monthly payment might be 5% of the balance or $25, whichever is more. To keep things simple, we will assume in this section that the minimum monthly payment is a fixed percentage of your balance. We will see that if you make only this minimum payment, your balance will decrease very slowly and will follow an exponential pattern.

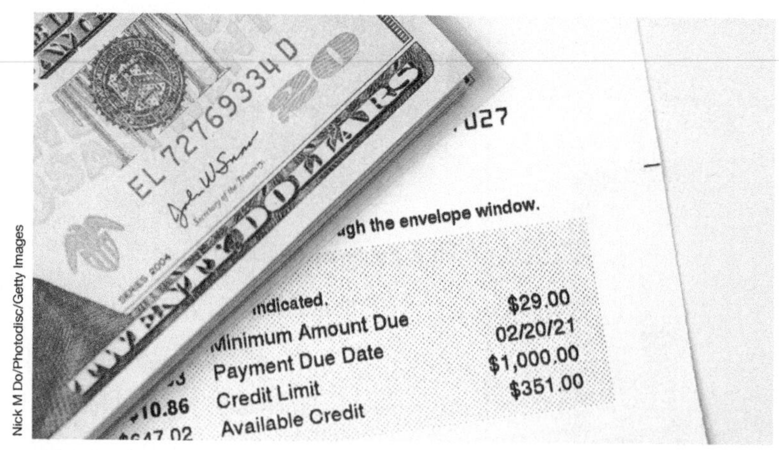

Nick M Do/Photodisc/Getty Images

Minimal payments reduce balances minimally.

EXAMPLE 4.27 Finding next month's minimum payment: One payment

We have a card with an APR of 24%. The minimum payment is 5% of the balance. Suppose we have a balance of \$400 on the card. We decide to stop charging and to pay it off by making the minimum payment each month. Calculate the new balance after we have made our first minimum payment and then calculate the minimum payment due for the next month.

SOLUTION

The first minimum payment is

$$\text{Minimum payment} = 5\% \text{ of balance} = 0.05 \times \$400 = \$20$$

The monthly interest rate is the APR divided by 12. In decimal form, this is $0.24/12 = 0.02$. Therefore, the finance charge is

$$\begin{aligned}\text{Finance charge} &= 0.02 \times (\text{Previous balance} - \text{Payments} + \text{Purchases})\\ &= 0.02 \times (\$400 - \$20 + \$0)\\ &= 0.02 \times \$380 = \$7.60\end{aligned}$$

That makes a new balance of

$$\begin{aligned}\text{New balance} &= \text{Previous balance} - \text{Payments} + \text{Purchases} + \text{Finance charge}\\ &= \$400 - \$20 + \$0 + \$7.60 = \$387.60\end{aligned}$$

The next minimum payment will be 5% of this:

$$\text{Minimum payment} = 5\% \text{ of balance} = 0.05 \times \$387.60 = \$19.38$$

TRY IT YOURSELF 4.27

This is a continuation of Example 4.27. Calculate the new balance after we have made our second minimum payment and then calculate the minimum payment due for the next month.

The answer is provided at the end of this section.

Let's pursue the situation described in the previous example. The following table covers the first four payments if we continue making the minimum 5% payment each month:

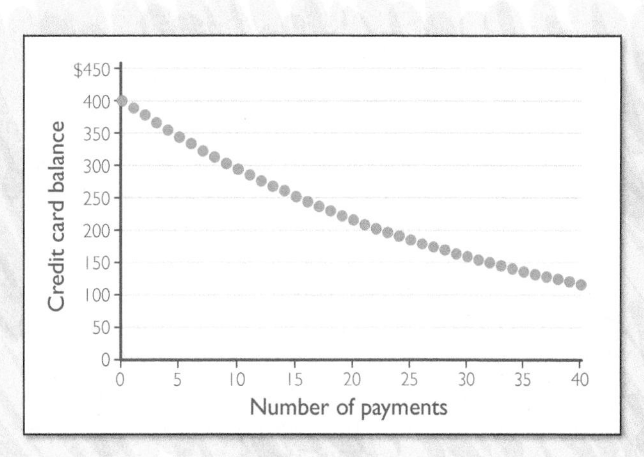

FIGURE 4.13 The balance from minimum payments is a decreasing exponential function.

Month	Previous balance	Minimum payment	Purchases	Finance charge	New balance
1	\$400.00	5% of \$400.00 = \$20.00	\$0.00	2% of \$380.00 = \$7.60	\$387.60
2	\$387.60	5% of \$387.60 = \$19.38	\$0.00	2% of \$368.22 = \$7.36	\$375.58
3	\$375.58	5% of \$375.58 = \$18.78	\$0.00	2% of \$356.80 = \$7.14	\$363.94
4	\$363.94	5% of \$363.94 = \$18.20	\$0.00	2% of \$345.74 = \$6.91	\$352.65

The table shows that the \$400 is not being paid off very quickly. In fact, after four payments, the decrease in the balance is only about \$47 (from \$400 to \$352.65). If we look closely at the table, we will see that the balances follow a pattern. To find the pattern, let's look at how the balance changes in terms of percentages:

Month 1: The new balance of \$387.60 is 96.9% of the initial balance of \$400.
Month 2: The new balance of \$375.58 is 96.9% of the Month 1 new balance of \$387.60.
Month 3: The new balance of \$363.94 is 96.9% of the Month 2 new balance of \$375.58.
Month 4: The new balance of \$352.65 is 96.9% of the Month 3 new balance of \$363.94.

The balance exhibits a constant percentage change. This makes sense: Each month, the balance is decreased by a constant percentage due to the minimum payment and increased by a constant percentage due to the finance charge. This pattern indicates that the balance is a decreasing exponential function. That conclusion is supported by the graph of the balance shown in **Figure 4.13**, which has the classic shape of exponential decay.

Because of this exponential pattern, we can find a formula for the balance on our credit card in the situation where we stop charging and pay off the balance by making the minimum payment each month.

FORMULA 4.12 Minimum Payment Balance Formula

$$\text{Balance after } t \text{ minimum payments} = \text{Initial balance} \times ((1 + r)(1 - m))^t$$

where r is the monthly interest rate and m is the minimum monthly payment as a percent of the balance. Both r and m are in decimal form. You should not round the product $(1 + r)(1 - m)$ when performing the calculation.

We provide a derivation of this formula in Algebraic Spotlight 4.5 at the end of this section.

EXAMPLE 4.28 Using the minimum payment balance formula: Balance after two years

We have a card with an APR of 20% and a minimum payment that is 4% of the balance. We have a balance of $250 on the card, and we stop charging and pay off the balance by making the minimum payment each month. Find the balance after two years of payments.

SOLUTION

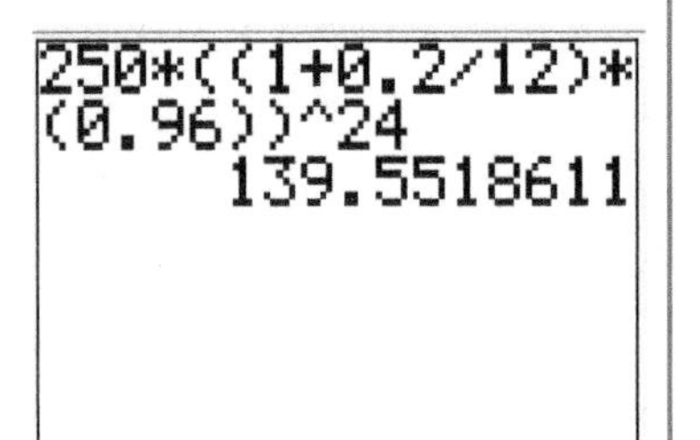

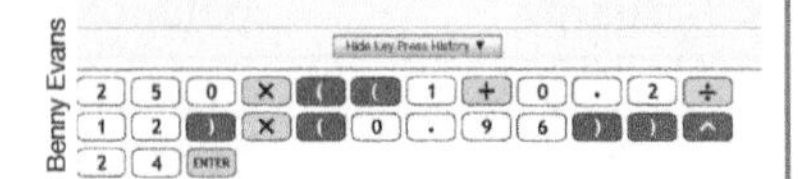

Benny Evans

The APR in decimal form is 0.2, so the monthly interest rate in decimal form is $r = 0.2/12$. To avoid rounding, we leave r in this form. The minimum payment is 4% of the new balance, so we use $m = 0.04$. The initial balance is $250. The number of payments for two years is $t = 24$. Using the minimum payment balance formula (Formula 4.12), we find

$$\begin{aligned}\text{Balance after 24 minimum payments} &= \text{Initial balance} \times ((1+r)(1-m))^t \\ &= \$250 \times ((1+0.2/12)(1-0.04))^{24} \\ &= \$250 \times ((1+0.2/12)(0.96))^{24} \\ &= \$139.55\end{aligned}$$

TRY IT YOURSELF 4.28

We have a card with an APR of 22% and a minimum payment that is 5% of the balance. We have a balance of $750 on the card, and we stop charging and pay off the balance by making the minimum payment each month. Find the balance after three years of payments.

The answer is provided at the end of this section.

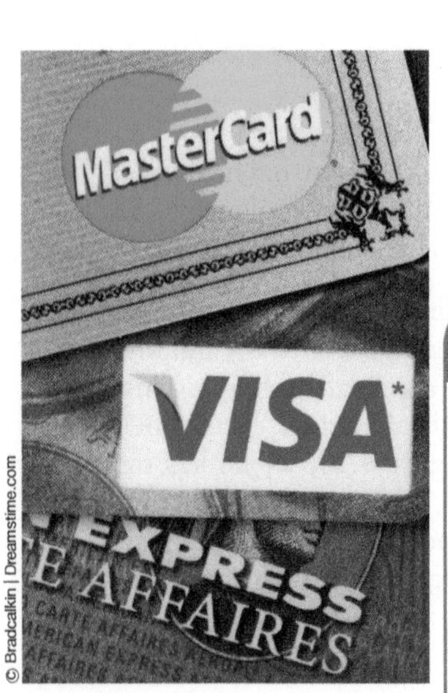

© Bradcalkin | Dreamstime.com

Wise credit card use avoids debt.

We already noted that the credit card balance is not paid off quickly when we make only the minimum payment each month. The reason for this is now clear: The balance is a decreasing exponential function, and such functions typically decrease very slowly in the long run. The next example illustrates the dangers of making only the minimum monthly payment.

EXAMPLE 4.29 Paying off your credit card balance: Long repayment

Suppose you have a balance of $10,000 on your Visa card, which has an APR of 24%. The card requires a minimum payment of 5% of the balance. You stop charging and begin making only the minimum payment until your balance is below $100.

a. Find a formula that gives your balance after t monthly payments.

b. Find your balance after five years of payments.

c. Determine how long it will take to get your balance under $100.[12]

d. Suppose that instead of the minimum payment, you want to make a fixed monthly payment so that your debt is clear in two years. How much do you pay each month?

---

[12] *It is worth noting that paying in this fashion will never actually get your balance to exactly zero. As mentioned earlier, however, the minimum monthly payment is usually stated in the form "5% of the balance or $25, whichever is larger." This method will ensure that the balance does get to zero.*

SOLUTION

a. The minimum payment as a decimal is $m = 0.05$, and the monthly rate as a decimal is $r = 0.24/12 = 0.02$. The initial balance is \$10,000. Using the minimum payment balance formula (Formula 4.12), we find

$$\begin{aligned}\text{Balance after } t \text{ minimum payments} &= \text{Initial balance} \times ((1+r)(1-m))^t \\ &= \$10{,}000 \times ((1+0.02)(1-0.05))^t \\ &= \$10{,}000 \times 0.969^t\end{aligned}$$

b. Now five years is 60 months, so we put $t = 60$ into the formula from part a:

$$\text{Balance after 60 months} = 10{,}000 \times 0.969^{60} = \$1511.56$$

After five years, we still owe more than \$1500.

c. We will show two ways to determine how long it takes to get the balance down to \$100.

Method 1: Using a logarithm: We need to solve for t the equation

$$\$100 = \$10{,}000 \times 0.969^t$$

The first step is to divide each side of the equation by \$10,000:

$$\frac{100}{10{,}000} = \frac{10{,}000}{10{,}000} \times 0.969^t$$

$$0.01 = 0.969^t$$

In Section 3.3, we learned how to solve exponential equations using logarithms. Summary 3.9 tells us that the solution for t of the equation $A = B^t$ is $t = \dfrac{\log A}{\log B}$. Using this formula with $A = 0.01$ and $B = 0.969$ gives

$$t = \frac{\log 0.01}{\log 0.969} \text{ months}$$

This is about 146.2 months. Hence, the balance will be under \$100 after 147 monthly payments, or more than 12 years of payments.

Method 2: Trial and error: If you want to avoid logarithms, you can solve this problem using trial and error with a calculator. The information in part b indicates that it will take some time for the balance to drop below \$100. So we might try 10 years, or 120 months. Computation using the formula from part a shows that after 10 years, the balance is still over \$200, so we should try a larger number of months. If we continue in this way, we find the same answer as that obtained from Method 1: The balance drops below \$100 at payment 147, which represents over 12 years of payments. (Spreadsheets and many calculators will create tables of values that make problems of this sort easy to solve.)

d. Making fixed monthly payments to clear your debt is like considering your debt as an installment loan: Just find the monthly payment if you borrow \$10,000 to buy (say) a car at an APR of 24% and pay the loan off over 24 months. We use the monthly payment formula from Section 4.2:

$$\text{Monthly payment} = \frac{\text{Amount borrowed} \times r(1+r)^t}{((1+r)^t - 1)}$$

Recall that t is the number of months taken to pay off the loan, in this case 24, and that r is the monthly rate as a decimal, which in this case is 0.02.

Hence,

$$\text{Monthly payment} = \frac{\$10{,}000 \times 0.02 \times 1.02^{24}}{(1.02^{24} - 1)} = \$528.71$$

So a payment of \$528.71 each month will clear the debt in two years.

SUMMARY 4.8
Making Minimum Payments

Suppose we have a balance on our credit card and decide to stop charging and pay off the balance by making the minimum payment each month.

1. The balance is given by the exponential formula

$$\text{Balance after } t \text{ minimum payments} = \text{Initial balance} \times ((1 + r)(1 - m))^t.$$

In this formula, r is the monthly interest rate and m is the minimum monthly payment as a percent of the balance. Both r and m are in decimal form.

2. The product $(1 + r)(1 - m)$ should not be rounded when the calculation is performed.

3. Because the balance is a decreasing exponential function, the balance decreases very slowly in the long run.

Further complications of credit card usage

Situations involving credit cards can often be even more complicated than those discussed in previous examples in this section. For instance, in each of those examples, there was a single purchase and no further usage of the card. Of course, it's more common to make purchases every month. We also assumed that your payments were made on time, but there are substantial penalties for late or missed payments.

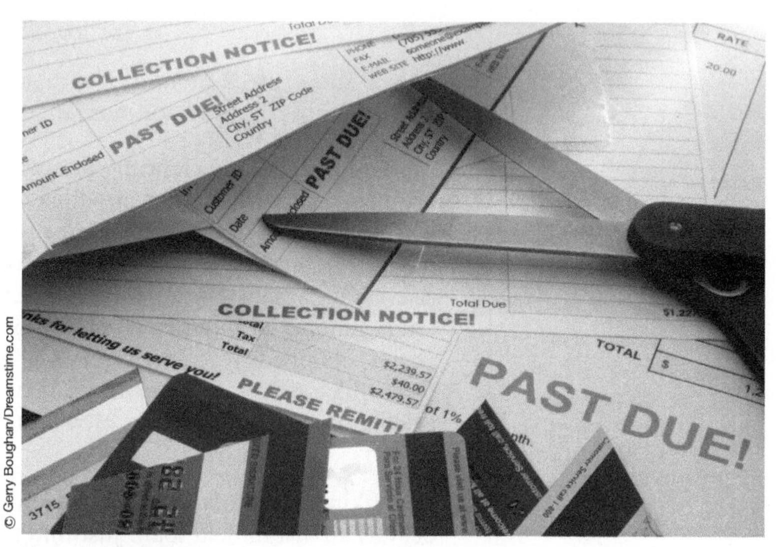

Use credit carefully.

Another complication is that credit card companies sometimes have "specials" or promotions in which you are allowed to skip a payment. And then there are *cash advances*, which are treated differently from purchases. Typically, cash advances incur finance charges immediately rather than after a month; that is, a cash advance is treated as carrying a balance immediately. Also, cash advances incur higher finance charges than purchases do.

Another complication occurs when your credit limit is reached (popularly known as "maxing out your card"). The credit limit is the maximum balance the credit card company allows you to carry. Usually, the limit is based on your credit history and your ability to pay. When you max out a credit card, the company will sometimes raise your credit limit, if you have always made required payments by its due dates.

In addition to all these complications, there are often devious hidden fees and charges. President Barack Obama signed into law the Credit Card Accountability Responsibility and Disclosure (CARD) Act of 2009. Among other things, the law requires that credit card companies inform cardholders how long it will take to pay off the balance if they make only the minimum payment. The following excerpt is from an article at the Web site of the federal Consumer Financial Protection Bureau.

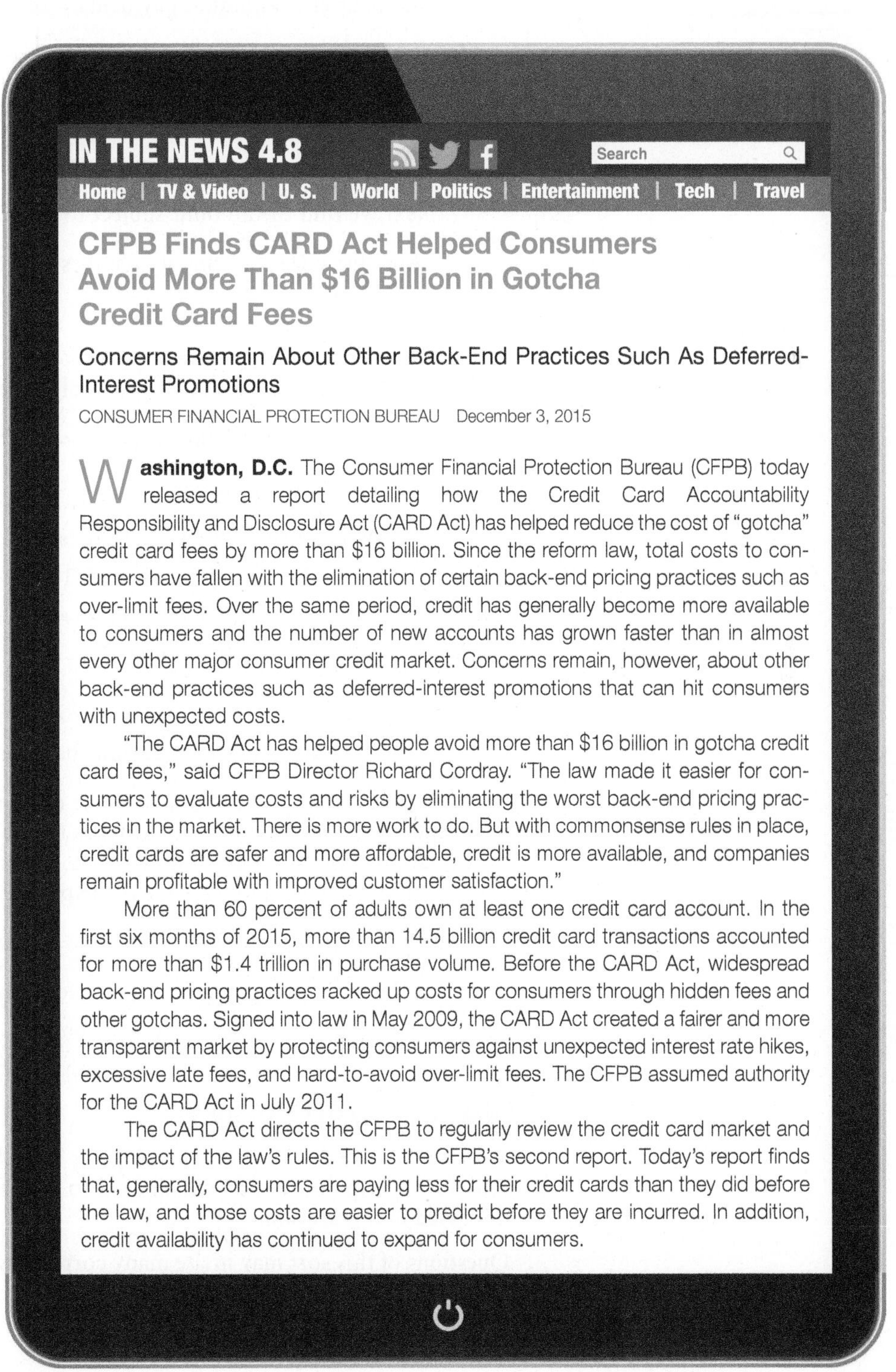

CFPB Finds CARD Act Helped Consumers Avoid More Than $16 Billion in Gotcha Credit Card Fees

Concerns Remain About Other Back-End Practices Such As Deferred-Interest Promotions

CONSUMER FINANCIAL PROTECTION BUREAU December 3, 2015

Washington, D.C. The Consumer Financial Protection Bureau (CFPB) today released a report detailing how the Credit Card Accountability Responsibility and Disclosure Act (CARD Act) has helped reduce the cost of "gotcha" credit card fees by more than $16 billion. Since the reform law, total costs to consumers have fallen with the elimination of certain back-end pricing practices such as over-limit fees. Over the same period, credit has generally become more available to consumers and the number of new accounts has grown faster than in almost every other major consumer credit market. Concerns remain, however, about other back-end practices such as deferred-interest promotions that can hit consumers with unexpected costs.

"The CARD Act has helped people avoid more than $16 billion in gotcha credit card fees," said CFPB Director Richard Cordray. "The law made it easier for consumers to evaluate costs and risks by eliminating the worst back-end pricing practices in the market. There is more work to do. But with commonsense rules in place, credit cards are safer and more affordable, credit is more available, and companies remain profitable with improved customer satisfaction."

More than 60 percent of adults own at least one credit card account. In the first six months of 2015, more than 14.5 billion credit card transactions accounted for more than $1.4 trillion in purchase volume. Before the CARD Act, widespread back-end pricing practices racked up costs for consumers through hidden fees and other gotchas. Signed into law in May 2009, the CARD Act created a fairer and more transparent market by protecting consumers against unexpected interest rate hikes, excessive late fees, and hard-to-avoid over-limit fees. The CFPB assumed authority for the CARD Act in July 2011.

The CARD Act directs the CFPB to regularly review the credit card market and the impact of the law's rules. This is the CFPB's second report. Today's report finds that, generally, consumers are paying less for their credit cards than they did before the law, and those costs are easier to predict before they are incurred. In addition, credit availability has continued to expand for consumers.

ALGEBRAIC SPOTLIGHT 4.5 Derivation of the Minimum Payment Balance Formula

Suppose a credit card has an initial balance. Assume that we incur no further charges and make only the minimum payment each month. Suppose r is the monthly interest rate and m is the minimum monthly payment as a percentage of the new balance. Both r and m are in decimal form. Our goal is to derive the formula

$$\text{Balance after } t \text{ minimum payments} = \text{Initial balance} \times ((1+r)(1-m))^t$$

Here is the derivation. Assume that we have made a series of payments, and let B denote the balance remaining. We need to calculate the new balance. First we find the minimum payment on the balance B. To do so, we multiply B by m:

$$\text{Minimum payment} = mB$$

Next we find the amount subject to finance charges:

$$\begin{aligned}\text{Amount subject to finance charges} &= \text{Previous balance} - \text{Payments}\\ &= B - mB\\ &= B(1-m)\end{aligned}$$

To calculate the finance charge, we apply the monthly rate r to this amount:

$$\text{Finance charge} = r \times B(1-m)$$

Therefore, the new balance is

$$\begin{aligned}\text{New balance} &= \text{Previous balance} - \text{Payments} + \text{Finance charge}\\ &= B(1-m) + rB(1-m)\\ &= B \times ((1-m) + r(1-m))\\ &= B \times ((1+r)(1-m))\end{aligned}$$

We can write this as

$$\text{New balance} = \text{Previous balance} \times ((1+r)(1-m))$$

Therefore, to find the new balance each month, we multiply the previous balance by $(1+r)(1-m)$. That makes the balance after t payments an exponential function of t with base $(1+r)(1-m)$. The initial value of this function is the initial balance, so we have the exponential formula

$$\text{Balance after } t \text{ minimum payments} = \text{Initial balance} \times ((1+r)(1-m))^t$$

This is the minimum payment balance formula.

WHAT DO YOU THINK?

Which credit card? You are shopping for a credit card. What features would you look for to help you decide which card to choose? You might look first at features discussed in this section.

Questions of this sort may invite many correct responses. We present one possibility: You certainly want to know the interest rate, but you also need to know how finance charges are calculated. The method of "average daily balance" is thought by some to be the fairest method of arriving at the balance on which charges are calculated. You also need to understand penalties charged by the card. They can be substantial and result in long-lasting problems. Credit cards are best used if the balance is paid in full each month, but in case that isn't possible, you need to know how the card deals with minimum payments. There are many more credit card issues that are important, including rewards programs. Recent laws are designed to limit predatory practices of credit card companies, but in the end it is up to you to "read the fine print." If you don't understand how a credit card works, don't get that card.

Further questions of this type may be found in the *What Do You Think?* section of exercises.

Try It Yourself answers

Try It Yourself 4.26: Calculating finance charges: Buying clothes

	Previous balance	Payments	Purchases	Finance charge	New balance
Month 1	$500.00	$200.00	$400.00	$13.30	$713.30
Month 2	$713.30	$300.00	$700.00	$21.15	$1134.45

Try It Yourself 4.27: Finding next month's minimum payment: One payment New balance: $375.58; minimum payment: $18.78.

Try It Yourself 4.28: Using the minimum payment balance formula: Balance after three years $227.59.

Exercise Set 4.4

Test Your Understanding

1. True or false: Most credit card companies will waive finance charges if you pay the monthly balance in full.

2. Which of the following affect finance charges on a credit card? **a.** your previous month's balance **b.** your purchases this month **c.** the amount of your payment **d.** all of the above.

3. True or false: Putting rounded numbers into financial formulas can cause significant errors.

4. True or false: The APR on most credit cards is well over 10%.

5. Which of the following should you expect to see on a credit card statement? **a.** the amount of purchases for the month **b.** the amount of finance charges for the month **c.** the balance of your account for the month **d.** the balance of your account for last month **e.** all of these.

6. If you make the minimum monthly payment of 5% on your credit card with no further purchases, your balance will: **a.** decrease exponentially **b.** decrease linearly **c.** remain about the same.

7. True or false: If the minimum monthly payment on your credit card is 5% or $25, whichever is larger, and if you make that payment each month, then with no further purchases your balance will eventually get to zero.

8. True or false: If you owe $5000 on your credit card and make only the minimum monthly payment, then with no further purchases your balance will decrease rapidly.

Problems

9. **Amount subject to finance charge.** The previous statement for your credit card had a balance of $540. You make purchases of $150 and make a payment of $60. The credit card has an APR of 22%. What is the finance charge for this month?

10. **Calculating finance charge.** The previous statement for your credit card had a balance of $650. You make a payment of $300 but you made additional purchases of $200. The card carries an APR of 24%. What is the finance charge for this month?

11. **Calculating minimum payment.** You have a credit card with an APR of 36%. The minimum payment is 10% of the balance. Suppose you have a balance of $800. You decide to stop charging and make only the minimum payment. The initial payment will be 10% of $800, or $80. Calculate the minimum payment for the next two months. (Be sure to take into account the finance charges.)

12. **Using the minimum payment balance formula.** You have a credit card with an APR of 24%. The card requires a minimum monthly payment of 5% of the balance. You have

a balance of $7500. You stop charging and make only the minimum monthly payment. What is the balance on the card after five years?

13. Calculating balances. You have a credit card with an APR of 16%. You begin with a balance of $800. In the first month you make a payment of $400 and you make charges amounting to $300. In the second month you make a payment of $300 and you make new charges of $600. Complete the following table:

	Previous balance	Payments	Purchases	Finance charge	New balance
Month 1					
Month 2					

14. Calculating balances. You have a credit card with an APR of 20%. You begin with a balance of $600. In the first month you make a payment of $400 and make new charges of $200. In the second month you make a payment of $300 and make new charges of $100. Complete the following table:

	Previous balance	Payments	Purchases	Finance charge	New balance
Month 1					
Month 2					

15. Calculating balances. You have a credit card with an APR of 22.8%. You begin with a balance of $1000, in response to which you make a payment of $200. The first month you make charges amounting to $500. You make a payment of $200 to reduce the new balance, and the second month you charge $600. Complete the following table:

	Previous balance	Payments	Purchases	Finance charge	New balance
Month 1					
Month 2					

16. Calculating balances. You have a credit card with an APR of 12%. You begin with a balance of $200, in response to which you make a payment of $75. The first month you make charges amounting to $50. You make a payment of $75 to reduce the new balance, and the second month you charge $60. Complete the following table:

	Previous balance	Payments	Purchases	Finance charge	New balance
Month 1					
Month 2					

17. A balance statement. Assume that you start with a balance of $4500 on your Visa credit card. During the first month, you charge $500, and during the second month, you charge $300. Assume that Visa has finance charges of 24% APR and that each month you make only the minimum payment of 2.5% of the balance. Complete the following table:

Month	Previous balance	Minimum payment	Purchases	Finance charge	New balance
1	$4500.00				
2					

18. A balance statement. Assume that you start with a balance of $4500 on your MasterCard. Assume that MasterCard has finance charges of 12% APR and that each month you make only the minimum payment of 3% of the balance. During the first month, you charge $300, and during the second month, you charge $600. Complete the following table:

Month	Previous balance	Minimum payment	Purchases	Finance charge	New balance
1	$4500.00				
2					

19. New balances. Assume that you have a balance of $4500 on your Discover credit card and that you make no more charges. Assume that Discover charges 21% APR and that each month you make only the minimum payment of 2.5% of the balance.

a. Find a formula for the balance after t monthly payments.

b. What will the balance be after 30 months?

c. What will the balance be after 10 years?

20. Paying tuition on your American Express card at the maximum interest rate. You have a balance of $10,000 for your tuition on your American Express credit card. Assume that you make no more charges on the card. Also assume that American Express charges 24% APR and that each month you make only the minimum payment of 2% of the balance.

a. Find a formula for the balance after t monthly payments.

b. How much will you owe after 10 years of payments?

c. How much would you owe if you made 100 years of payments?

d. Find when the balance would be less than $50.

21. Paying tuition on your American Express card. You have a balance of $10,000 for your tuition on your American Express credit card. Assume that you make no more charges on the card. Also assume that American Express charges 12% APR and that each month you make only the minimum payment of 4% of the balance.

a. Find a formula for the balance after t monthly payments.

b. How long will it take to get the balance below $50?

22. Paying off a Visa card. You have a balance of $1000 on your Visa credit card. Assume that you make no more charges on the card and that the card charges 9.9% APR and requires a minimum payment of 3% of the balance. Assume also that you make only the minimum payments.

a. Find a formula for the balance after t monthly payments.

b. Find how many months it takes to bring the balance below $50.

23. Balance below $200. You have a balance of $4000 on your credit card, and you make no more charges. Assume that the card requires a minimum payment of 5% and carries an APR of 22.8%. Assume also that you make only the minimum payments. Determine when the balance drops below $200.

24. New balances. Assume that you have a balance of $3000 on your MasterCard and that you make no more charges. Assume that MasterCard charges 12% APR and that each month you make only the minimum payment of 5% of the balance.

a. Find a formula for the balance after t monthly payments.

b. What will the balance be after 30 months?

c. What will the balance be after 10 years?

d. At what balance do you begin making payments of $20 or less?

e. Find how many months it will take to bring the remaining balance down to the value from part d.

25. Paying off a Discover card. Assume that you have a balance of $3000 on your Discover credit card and that you make no more charges. Assume that Discover charges 21% APR and that each month you make only the minimum payment of 2% of the balance.

a. Find a formula for the remaining balance after t monthly payments.

b. On what balance do you begin making payments of $50 or less?

c. Find how many months it will take to bring the remaining balance down to the value from part b.

26. Monthly payment. You have a balance of $400 on your credit card and make no more charges. Assume that the card carries an APR of 18%. Suppose you wish to pay off the card in six months by making equal payments each month. What is your monthly payment?

27. What can you afford to charge? Suppose you have a new credit card with 0% APR for a limited period. The card requires a minimum payment of 5% of the balance. You feel you can afford to pay no more than $250 each month. How much can you afford to charge? How much could you afford to charge if the minimum payment were 2% instead of 5%?

28. Average daily balance. The most common way of calculating finance charges is not the simplified one we used in this section but rather the *average daily balance*. With this method, we calculate the account balance at the end of each day of the month and take the average. That average is the amount subject to finance charges. To simplify things, we assume that the billing period is one week rather than one month. Assume that the weekly rate is 1%. You begin with a balance of $500. On day 1, you charge $75. On day 3, you make a payment of $200 and charge $100. On day 6, you charge $200.

a. Assume that finance charges are calculated using the simplified method shown in this section. Find the account balance at the end of the week.

b. Assume that finance charges are calculated using the average daily balance. Find the account balance at the end of the week.

29. More on average daily balance. The method for calculating finance charges based on the average daily balance is explained in the setting for Exercise 28. As in that exercise, to simplify things, we assume that the billing period is one week rather than one month and that the weekly rate is 1%. You begin with a balance of $1000. On day 1, you charge $200. On day 3, you charge $500. On day 6, you make a payment of $400 and you charge $100.

a. Assume that finance charges are calculated using the simplified method shown in this section. Find the account balance at the end of the week.

b. Assume that finance charges are calculated using the average daily balance. Find the account balance at the end of the week.

30. Finance charges versus minimum payments. In all of the examples in this section, the monthly finance charge is always less than the minimum payment. In fact, this is always the case. Explain what would happen if the minimum payment were less than the monthly finance charge.

Exercises 31 and 32 are designed to be solved using technology such as calculators or computer spreadsheets.

31. Paying off a Visa card—in detail. Assume that you have a balance of $1000 on your Visa credit card and that you make no more charges on the card. Assume that Visa charges 12% APR and that the minimum payment is 5% of the balance each month. Assume also that you make only the minimum payments. Make a spreadsheet listing the items below for each month until the payment falls below $20.

Month	Previous balance	Minimum payment	Purchases	Finance charge	New balance
1	$1000.00				
2					
and so on					

32. Paying off an American Express card. Assume that you have a balance of $3000 on your American Express credit card and that you make no more charges on the card. Assume that American Express charges 20.5% APR and that the minimum payment is 2% of the balance each month. Assume also that you make only the minimum payments. Make a spreadsheet listing the items below for each month until the payment drops below $50.

Month	Previous balance	Minimum payment	Purchases	Finance charge	New balance
1	$3000.00				
2					
and so on					

Writing About Mathematics

33. Average daily balance. We noted that most credit cards use the average daily balance to assess finance charges, and this method was investigated in Exercises 28 and 29. Many regard this method as fair to all, but some do not. Do some research to form your own opinion on this issue and report your conclusions.

34. Location. Many credit card companies have operations in Sioux Falls, South Dakota. (In fact, if you look at Figure 4.12 you will see a reference to that city.) Do some research to find out why and write a report on your findings.

35. Rules. We mentioned at the outset that different credit card companies may have different rules. Investigate some companies to see whether you can detect any differences other than interest rates. Write a report on your findings.

36. Chips. Credit cards have had magnetic stripes for a long time, but recently they have also come with implanted computer chips. Investigate why chips are being used and write a report on your findings.

37. A bit of history. Credit cards are relatively new. Write a brief report on the introduction of credit cards into the U.S. economy.

What Do You Think?

38. Your own credit card. Examine the monthly statement from one of your own credit cards. Note the ways it is similar to examples in the text and also note differences. Write down any parts of the statement you do not understand. **Exercise 15 is relevant to this topic.**

39. What is the APR? Credit cards often advertise the monthly interest rate rather than the APR. Use this example to explain why they do so: Suppose a credit card company advertises an interest rate of 3% per month. What is the APR for this card? Which sounds more appealing, the monthly rate or the APR?

40. Understanding repayment. A friend owes $10,000 on his credit card. He plans to stop making charges and pay off the balance. The card has a minimum payment of 5%. He figures correctly that 5% of $10,000 is $500. From that he concludes that he needs to make the 5% minimum payment for only 20 months, because $20 \times \$500 = \$10,000$. Is your friend right? If not, explain the error in his reasoning. **Exercise 20 is relevant to this topic.**

41. Purchasing a home. Why would it be unwise (even if it were allowed) to charge the purchase of a home to a credit card? Go beyond the credit limit to consider factors such as interest rates, budgeting, etc.

42. Slow repayment. In this section we looked at repaying the balance on your credit card by making the minimum payment each month. We saw that this method results in a decreasing exponential function for the amount owed. How is this situation analogous to the long-term problem of handling radioactive waste discussed in Section 3.2? How are the underlying mathematical formulas similar?

43. Company profit. Most credit card companies charge you nothing if you pay off your balance each month. Do some research on the Internet to determine how the company makes a profit from customers who pay off their balance each month.

44. Balance decreases. In the examples presented in this section, making the minimum monthly payment decreased the balance owed. Why would credit card companies want to be sure this decrease happens? Would this decrease happen if the APR were 24% and the minimum monthly payment were 1%? **Exercise 30 is relevant to this topic.**

45. Credit card rates. Credit cards sometimes come with very high interest rates. Use the Internet to investigate interest rates commonly associated with credit cards. (Try to find the interest rate associated with your own credit card.)

46. How much to pay. You have acquired an extra bit of money and you are trying to decide the best way to use it. You can pay down a credit card that has a high interest rate, or you can put the money into a savings account with a low interest rate and save it for a rainy day. Which would you choose and why? Are there additional options you might consider?

4.5 Inflation, taxes, and stocks: Managing your money

TAKE AWAY FROM THIS SECTION
Understand the impact of inflation and the significance of marginal tax rates.

The following is from the Web site of Bloomberg News.

Janet Yellen leads the faithful.

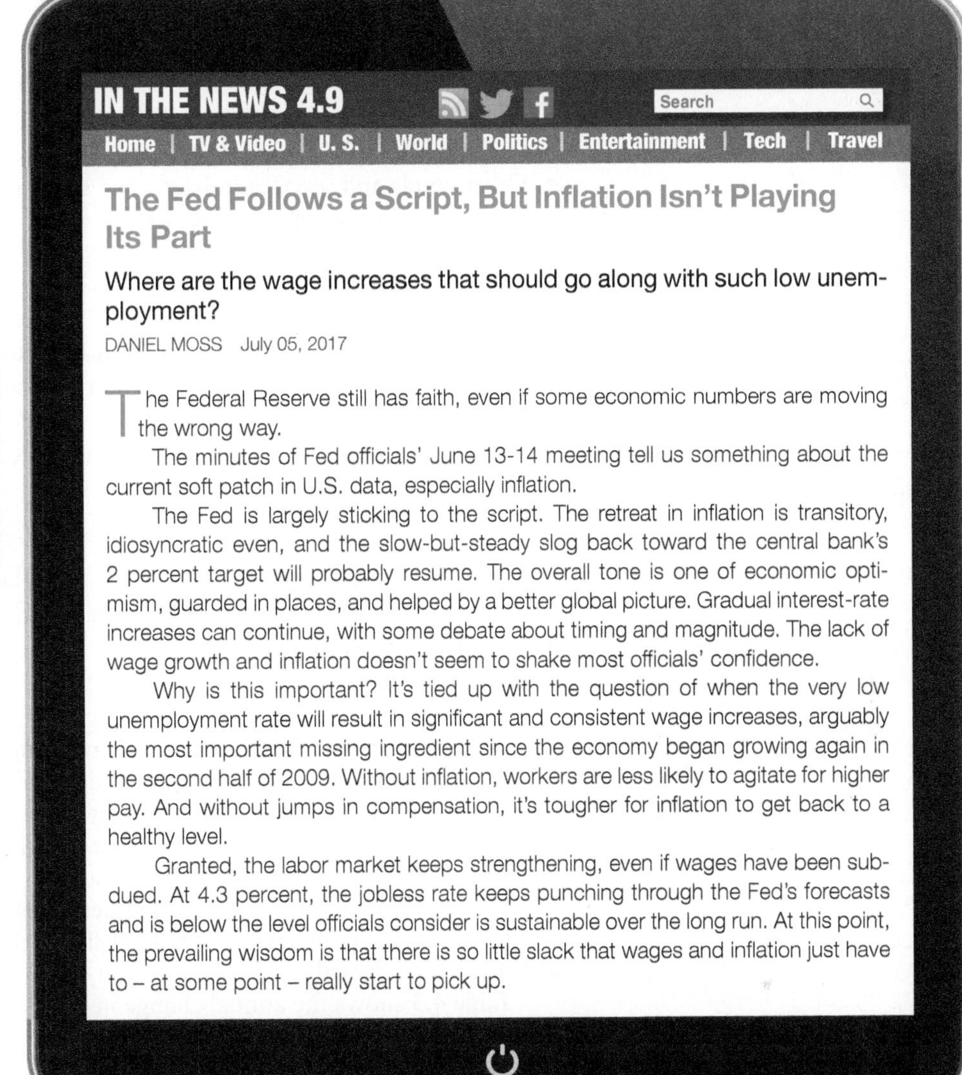

The Fed Follows a Script, But Inflation Isn't Playing Its Part

Where are the wage increases that should go along with such low unemployment?

DANIEL MOSS July 05, 2017

The Federal Reserve still has faith, even if some economic numbers are moving the wrong way.

The minutes of Fed officials' June 13-14 meeting tell us something about the current soft patch in U.S. data, especially inflation.

The Fed is largely sticking to the script. The retreat in inflation is transitory, idiosyncratic even, and the slow-but-steady slog back toward the central bank's 2 percent target will probably resume. The overall tone is one of economic optimism, guarded in places, and helped by a better global picture. Gradual interest-rate increases can continue, with some debate about timing and magnitude. The lack of wage growth and inflation doesn't seem to shake most officials' confidence.

Why is this important? It's tied up with the question of when the very low unemployment rate will result in significant and consistent wage increases, arguably the most important missing ingredient since the economy began growing again in the second half of 2009. Without inflation, workers are less likely to agitate for higher pay. And without jumps in compensation, it's tougher for inflation to get back to a healthy level.

Granted, the labor market keeps strengthening, even if wages have been subdued. At 4.3 percent, the jobless rate keeps punching through the Fed's forecasts and is below the level officials consider is sustainable over the long run. At this point, the prevailing wisdom is that there is so little slack that wages and inflation just have to – at some point – really start to pick up.

This article highlights the connection between inflation, wage growth, and interest rates in the growth of the economy. We will explore several related financial issues in this section.

CPI and the inflation rate

The preceding article states the concerns of the Federal Reserve Board about inflation. But what is inflation, and how is it measured?

Inflation is calculated using the *Consumer Price Index* (CPI), which is a measure of the price of a certain "market basket" of consumer goods and services relative to a predetermined benchmark. When the CPI goes up, we have inflation. When it goes down, we have *deflation.*

According to the U.S. Department of Labor, this "market basket" consists of commodities in the following categories:

- **Food and beverages** (breakfast cereal, milk, coffee, chicken, wine, service meals and snacks)
- **Housing** (rent of primary residence, owners' equivalent rent, fuel oil, bedroom furniture)
- **Apparel** (men's shirts and sweaters, women's dresses, jewelry)
- **Transportation** (new vehicles, airline fares, gasoline, motor vehicle insurance)

- **Medical care** (prescription drugs and medical supplies, physicians' services, eyeglasses and eye care, hospital services)
- **Recreation** (televisions, pets and pet products, sports equipment, admissions)
- **Education and communication** (college tuition, postage, telephone services, computer software and accessories)
- **Other goods and services** (tobacco and smoking products, haircuts and other personal services, funeral expenses)

KEY CONCEPT

The Consumer Price Index (CPI) is a measure of the average price paid by urban consumers for a "market basket" of consumer goods and services.

The *rate of inflation* is measured by the percentage change in the CPI.

KEY CONCEPT

An increase in prices is referred to as inflation. The rate of inflation is measured by the percentage change in the Consumer Price Index over time. When prices decrease, the percentage change is negative; this is referred to as deflation.

Inflation reflects a decline of the purchasing power of the consumer's dollar. Besides affecting how much we can afford to buy, the inflation rate has a big influence on certain government programs that affect the lives of many people.

For example, as of 2020, there were about 64.9 million Social Security beneficiaries and 39.9 million SNAP (food stamp) recipients who are affected by the CPI because their benefits are adjusted periodically to compensate for inflation. In addition, millions of military and federal civil service retirees and subsidized lunches at schools are affected.

Table 4.3 shows the annual change in prices in the United States over a 71-year period. For example, from December 1949 to December 1950, the CPI changed from 23.6 to 25.0, an increase of $25.0 - 23.6 = 1.4$. Now we find the percentage change:

$$\text{Percentage change} = \frac{\text{Change in CPI}}{\text{Previous CPI}} \times 100\% = \frac{1.4}{23.6} \times 100\%$$

This is about 5.9%, and that is the inflation rate for this period shown in the table. Usually, we round the inflation rate as a percentage to one decimal place.

EXAMPLE 4.30 Calculating inflation: CPI increase to 205

Suppose the CPI increases this year from 200 to 205. What is the rate of inflation for this year?

SOLUTION

The change in the CPI is $205 - 200 = 5$. To find the percentage change, we divide the increase of 5 by the original value of 200 and convert to a percent:

$$\text{Percentage change} = \frac{\text{Change in CPI}}{\text{Previous CPI}} \times 100\% = \frac{5}{200} \times 100\% = 2.5\%$$

Therefore, the rate of inflation is 2.5%.

TRY IT YOURSELF 4.30

Suppose the CPI increases this year from 215 to 225. What is the rate of inflation for this year?

The answer is provided at the end of this section.

TABLE 4.3 Historical Inflation

CPI in December and inflation rate					
December	CPI	Inflation rate	December	CPI	Inflation rate
1949	23.6	—	1985	109.3	3.8%
1950	25.0	5.9%	1986	110.5	1.1%
1951	26.5	6.0%	1987	115.4	4.4%
1952	26.7	0.8%	1988	120.5	4.4%
1953	26.9	0.7%	1989	126.1	4.6%
1954	26.7	−0.7%	1990	133.8	6.1%
1955	26.8	0.4%	1991	137.9	3.1%
1956	27.6	3.0%	1992	141.9	2.9%
1957	28.4	2.9%	1993	145.8	2.7%
1958	28.9	1.8%	1994	149.7	2.7%
1959	29.4	1.7%	1995	153.5	2.5%
1960	29.8	1.4%	1996	158.6	3.3%
1961	30.0	0.7%	1997	161.3	1.7%
1962	30.4	1.3%	1998	163.9	1.6%
1963	30.9	1.6%	1999	168.3	2.7%
1964	31.2	1.0%	2000	174.0	3.4%
1965	31.8	1.9%	2001	176.7	1.6%
1966	32.9	3.5%	2002	180.9	2.4%
1967	33.9	3.0%	2003	184.3	1.9%
1968	35.5	4.7%	2004	190.3	3.3%
1969	37.7	6.2%	2005	196.8	3.4%
1970	39.8	5.6%	2006	201.8	2.5%
1971	41.1	3.3%	2007	210.0	4.1%
1972	42.5	3.4%	2008	210.2	0.1%
1973	46.2	8.7%	2009	215.9	2.7%
1974	51.9	12.3%	2010	219.2	1.5%
1975	55.5	6.9%	2011	225.7	3.0%
1976	58.2	4.9%	2012	229.6	1.7%
1977	62.1	6.7%	2013	233.0	1.5%
1978	67.7	9.0%	2014	234.8	0.8%
1979	76.7	13.3%	2015	236.5	0.7%
1980	86.3	12.5%	2016	241.4	2.1%
1981	94.0	8.9%	2017	246.5	2.1%
1982	97.6	3.8%	2018	251.2	1.9%
1983	101.3	3.8%	2019	257.0	2.3%
1984	105.3	3.9%	2020	260.5	1.4%

"When you take out food, energy, taxes, insurance, housing, transportation, healthcare, and entertainment, inflation remained at a 20 year low."

Table 4.4 lists four countries and their estimated annual rates of inflation for 2013. A very high rate of inflation, as in countries like Venezuela and Zimbabwe, is referred to as *hyperinflation*.

If the rate of inflation is 10%, one may think that the buying power of a dollar has decreased by 10%, but that is not the case. Inflation is the percentage change in prices, and that is not the same as the percentage change in the value of a dollar. To see the difference, let's imagine a frightening inflation rate of 100% this year. With such a rate, an item that costs \$200 this year will cost \$400 next year. This means that my money can buy only half as much next year as it can this year. So the buying power of a dollar would decrease by 50%, not by 100%.

TABLE 4.4 Examples of Hyperinflation in 2020

Country	Estimated inflation rate
Venezuela	6500%
Zimbabwe	623%
Sudan	142%
Lebanon	85%

The following formula tells us how much the buying power of currency decreases for a given inflation rate.

FORMULA 4.13 Buying Power Formula

$$\text{Percent decrease in buying power} = \frac{100\,i}{100 + i}$$

Here i is the inflation rate expressed as a percent, not a decimal. Usually we round the decrease in buying power as a percentage to one decimal place.

The buying power formula is derived in Algebraic Spotlight 4.6 at the end of this section.

EXAMPLE 4.31 Calculating decrease in buying power: 5% inflation

Suppose the rate of inflation this year is 5%. What is the percentage decrease in the buying power of a dollar?

SOLUTION

We use the buying power formula (Formula 4.13) with $i = 5\%$:

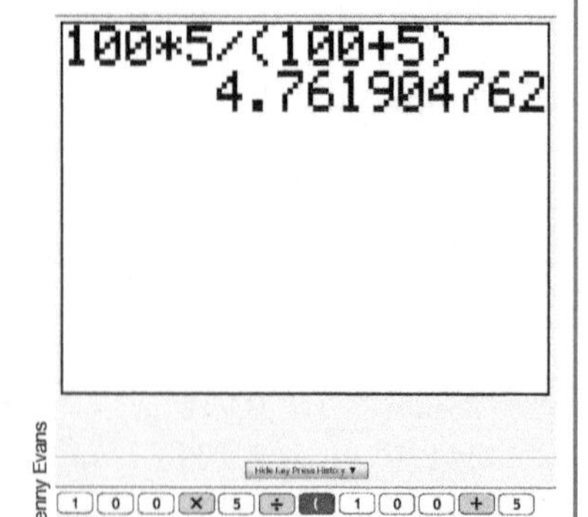

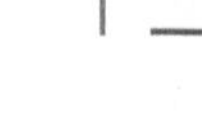

Benny Evans

$$\begin{aligned}\text{Percent decrease in buying power} &= \frac{100\,i}{100 + i}\\ &= \frac{100 \times 5}{100 + 5}\end{aligned}$$

This is about 4.8%.

TRY IT YOURSELF 4.31

According to Table 4.4, in 2020 the rate of inflation for Venezuela was 6500%. What was the percentage decrease that year in the buying power of the *bolívar* (the currency of Venezuela)?

The answer is provided at the end of this section.

A companion formula to the buying power formula (Formula 4.13) gives the inflation rate in terms of the percent decrease in buying power of currency.

FORMULA 4.14 Inflation Formula

$$\text{Percent rate of inflation} = \frac{100B}{100 - B}$$

In this formula, B is the decrease in buying power expressed as a percent, not as a decimal.

EXAMPLE 4.32 Calculating inflation: 2.5% decrease in buying power

Suppose the buying power of a dollar decreases by 2.5% this year. What is the rate of inflation this year?

SOLUTION

We use the inflation formula (Formula 4.14) with $B = 2.5\%$:

$$\text{Percent rate of inflation} = \frac{100B}{100 - B} = \frac{100 \times 2.5}{100 - 2.5}$$

This is about 2.6%.

TRY IT YOURSELF 4.32

Suppose the buying power of a dollar decreases by 5.2% this year. What is the rate of inflation this year?

The answer is provided at the end of this section.

The next example covers all of the concepts we have considered so far in this section.

EXAMPLE 4.33 Understanding inflation and buying power: Effects on goods

Parts a through c refer to Table 4.3.

a. Find the 10-year inflation rate in the United States from December 2010 to December 2020.

b. If a sofa cost \$100 in December 2008 and the price changed in accordance with the inflation rate in the table, how much did the sofa cost in December 2009?

c. If a chair cost \$50 in December 1953 and the price changed in accordance with the inflation rate in the table, how much did the chair cost in December 1954?

d. According to Table 4.4, in 2020 the rate of inflation for Lebanon was 85%. How much did the buying power of the currency, the *lira*, decrease during the year?

e. In Ethiopia, the buying power of the currency, the *birr*, decreased by 16.8% in 2020. What was Ethiopia's inflation rate for 2020?

SOLUTION

a. In 2010 the CPI was 219.2, and in 2020 it was 260.5. The increase was $260.5 - 219.2 = 41.3$, so

$$\text{Percentage change} = \frac{\text{Change in CPI}}{\text{Previous CPI}} \times 100\% = \frac{41.3}{219.2} \times 100\%$$

or about 18.8%. The 10-year inflation rate was about 18.8%.

b. According to the table, the inflation rate was 2.7% in 2009, which tells us that the price of the sofa increased by 2.7% during that year. Therefore, it cost $\$100 + 0.027 \times \$100 = \$102.70$ in December 2009.

c. The inflation rate is -0.7%, which tells us that the price of the chair decreased by 0.7% during that year. Because 0.7% of \$50.00 is \$0.35, in December 1954 the chair cost $\$50.00 - \$0.35 = \$49.65$.

d. We use the buying power formula (Formula 4.13) with $i = 85\%$:

$$\begin{aligned}\text{Percent decrease in buying power} &= \frac{100\,i}{100 + i}\\ &= \frac{100 \times 85}{100 + 85}\end{aligned}$$

or about 45.9%. The buying power of the Lebanese *lira* decreased by 45.9%.

e. We want to find the inflation rate for Ethiopia. We know that the reduction in buying power was 16.8%, so we use the inflation formula (Formula 4.14) with $B = 16.8\%$:

$$\begin{aligned}\text{Percent rate of inflation} &= \frac{100B}{100 - B}\\ &= \frac{100 \times 16.8}{100 - 16.8}\end{aligned}$$

or about 20.2%. Therefore, the rate of inflation in Ethiopia in 2020 was about 20.2%.

SUMMARY 4.9
Inflation and Reduction of Currency Buying Power

If the inflation rate is i (expressed as a percent), the change in the buying power of currency can be calculated using

$$\text{Percent decrease in buying power} = \frac{100\,i}{100 + i}$$

A companion formula gives the inflation rate in terms of the decrease B (expressed as a percent) in buying power:

$$\text{Percent rate of inflation} = \frac{100B}{100 - B}$$

Income taxes

Consider the following tax tables for the year 2020 from the Internal Revenue Service. **Table 4.5** shows tax rates for single people, and **Table 4.6** shows tax rates for married couples filing jointly. Note that the tax rates are applied to *taxable income*. The percentages in the tables are called *marginal rates*, and they apply only to earnings in excess of a certain amount. With a marginal tax rate of 12%, for example, the tax owed increases by \$0.12 for every \$1 increase in taxable income. This makes the tax owed a linear function of the taxable income within a given range of incomes. The slope is the marginal tax rate (as a decimal).

TABLE 4.5 2020 Tax Table for Singles

If Taxable Income		The Tax is		
Is over	But not over	This amount	Plus this %	Of the excess over
Schedule X—Use if your filing status is Single				
\$0	\$9875	—	10 %	\$0
9876	40,125	\$987.50	12 %	9875
40,126	85,525	4617.50	22 %	40,125
85,526	163,300	14,605.50	24 %	85,525
163,301	207,350	33,271.50	32 %	163,300
207,351	518,400	47,367.50	35 %	207,350
518,401	—	156,235.00	37 %	518,400

TABLE 4.6 2020 Tax Table for Married Couples Filing Jointly

If Taxable Income		The Tax is		
Is over	But not over	This amount	Plus this %	Of the excess over
Schedule Y-1—Use if your filing status is Married filing jointly or Qualifying widow(er)				
$0	$19,750	—	10 %	$0
19,751	80,250	$1975.00	12 %	19,750
80,251	171,050	9235.00	22 %	80,250
171,051	326,600	29,211.00	24 %	171,050
326,601	414,700	66,543.00	32 %	326,600
414,701	622,050	94,735.00	35 %	414,700
622,051	—	167,307.50	37 %	622,050

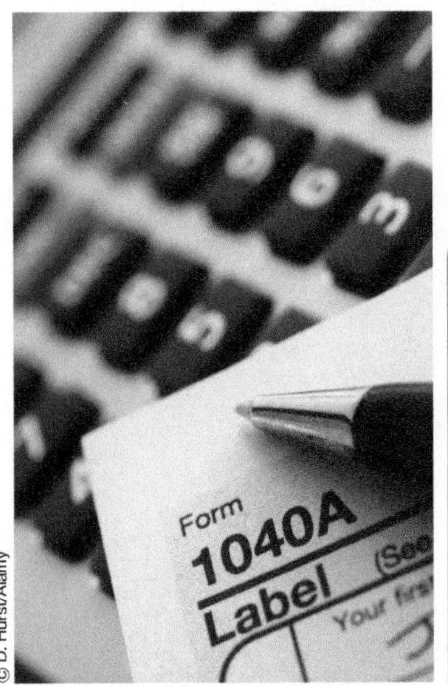

© D. Hurst/Alamy

Income taxes are part of financial planning.

Note that these marginal rates increase as you earn more and so move from one *tax bracket* to another. A system of taxation in which the marginal tax rates increase for higher incomes is referred to as a *progressive tax*.[13]

EXAMPLE 4.34 Calculating the tax: A single person

In the year 2020, Alex was single and had a taxable income of $70,000. How much tax did she owe?

SOLUTION

According to Table 4.5, Alex owed $4617.50 plus 22% of the excess taxable income over $40,125. The total tax is

$$\$4617.50 + 0.22 \times (\$70{,}000 - \$40{,}125) = \$11{,}190.00$$

TRY IT YOURSELF 4.34

In the year 2020, Bob was single and had a taxable income of $20,000. How much tax did he owe?

The answer is provided at the end of this section.

A person's taxable income is obtained by subtracting certain *deductions* from total income. Deductions may include things like state and local taxes, home mortgage interest, and charitable contributions. Some people do not have very many of these kinds of deductions. In this case, they may choose not to itemize them but rather to take what is called a "standard deduction."

EXAMPLE 4.35 Comparing taxes: The "marriage penalty"

a. In the year 2020, Ann was single. After all deductions her taxable income was $320,000. How much income tax did she owe?

b. In the year 2020, Alice and Bob were married. Together they had a taxable income of $640,000 per year and filed jointly. How much income tax did they owe?

[13] *In 1996 and 2000, magazine publisher Steve Forbes ran for president on a platform that included a 17% "flat tax." (The flat-tax concept is explored a bit more in the exercises at the end of this section.) In that system everyone would pay 17% of their taxable income no matter what that income is. In a speech Forbes said, "When we're through with Washington, the initials of the IRS will be RIP." Forbes did not win the Republican nomination.*

c. Explain what you notice about the amount of tax paid in the two situations in parts a and b.

SOLUTION

a. According to the tax table for singles in Table 4.5, Ann owed \$47,367.50 plus 35% of everything over \$207,350, which is \$86,795.00 in income tax.

b. According to the tax table for married couples shown in Table 4.6, Alice and Bob owe \$167,307.50 plus 37% of everything over \$622,050, which comes to \$173,949.00.

c. If Alice and Bob each had a taxable income of \$320,000 and could file as singles, then, using the solution to part a, they would owe $2 \times \$86,795.00 = \$173,590.00$. But because they were married and filing jointly, they had to pay \$173,949.00, which is an extra \$359 in taxes.

Marty Bucella/cartoonstock.com

"We were going over some of your returns from a past life and..."

The disparity in part c of Example 4.35 is referred to as the "marriage penalty." Many people believe the marriage penalty is unfair, but there are those who argue that it makes sense to tax a married couple more than two single people. Can you think of some arguments to support the two sides of this question? The marriage penalty has been eliminated for most middle-income earners.

Claiming deductions lowers your tax by reducing your taxable income. Another way to lower your tax is to take a *tax credit*. Here is how to apply a tax credit: Calculate the tax owed using the tax tables (making sure first to subtract from the total income any deductions) and then subtract the tax credit from the tax determined by the tables. Because a tax credit is subtracted directly from the tax you owe, a tax credit of \$1000 has a much bigger impact on lowering your taxes than a deduction of \$1000. That is the point of the following example.

EXAMPLE 4.36 Comparing deductions and credits: Differing effects

In the year 2020, Betty and Carol were single, and each had a total income of \$80,000. Betty took a deduction of \$15,000 but had no tax credits. Carol took a deduction of \$14,000 and had an education tax credit of \$1000. Compare the tax owed by Betty and Carol.

SOLUTION

The taxable income of Betty is \$80,000 – \$15,000 = \$65,000. According to the tax table in Table 4.5, Betty owes \$4617.50 plus 22% of the excess taxable income over \$40,125. That tax is

$$\$4617.50 + 0.22 \times (\$65{,}000 - \$40{,}125) = \$10{,}090$$

Betty has no tax credits, so the tax she owes is \$10,090.

The taxable income of Carol is \$80,000 – \$14,000 = \$66,000. According to the tax table in Table 4.5, before applying tax credits, Carol owes \$4617.50 plus 22% of the excess taxable income over \$40,125. That tax is

$$\$4617.50 + 0.22 \times (\$66{,}000 - \$40{,}125) = \$10{,}310$$

Carol has a tax credit of \$1000, so the tax she owes is

$$\$10{,}310 - \$1000 = \$9310$$

Betty owes

$$\$10{,}090 - \$9310 = \$780$$

more tax than Carol.

TRY IT YOURSELF 4.36

In the year 2020, Dave was single and had a total income of \$70,000. He took a deduction of \$13,000 and had a tax credit of \$1800. Calculate the tax owed by Dave.

The answer is provided at the end of this section.

In Example 4.36, the effect of replacing a \$1000 deduction by a \$1000 credit was to reduce the tax owed by \$780. This is a significant reduction in taxes and highlights the benefits of tax credits.

The Dow

In the late nineteenth century, tips and gossip caused stock prices to move because solid information was hard to come by. This prompted Charles H. Dow to introduce the Dow Jones Industrial Average (DJIA) in May 1896 as a benchmark to gauge the state of the market. The original DJIA was simply the average price of 12 stocks

xPACIFICA/Corbis Documentary/Getty Images

The New York Stock Exchange.

that Mr. Dow picked himself. Today the Dow, as it is often called, consists of 30 "blue-chip" U.S. stocks picked by the editors of the *Wall Street Journal*. For example, in June 2009 General Motors was removed from the list as it entered bankruptcy protection. Here is the list as of March 2021.

The 30 Dow Companies

- 3M Company
- American Express Company
- Amgen
- Apple Incorporated
- Boeing Company
- Caterpillar Incorporated
- Chevron Corporation
- Cisco Systems Incorporated
- Coca-Cola Company
- Dow, Inc.
- Goldman Sachs Group Incorporated
- Home Depot Incorporated
- Honeywell
- Intel Corporation
- International Business Machines Corporation
- Johnson & Johnson
- JPMorgan Chase & Company
- McDonald's Corporation
- Merck & Company Incorporated
- Microsoft Corporation
- Nike Incorporated
- Procter & Gamble Company
- Salesforce
- Travelers Companies Incorporated
- United Health Group Incorporated
- Verizon Communications Incorporated
- Visa Incorporated
- Walgreens Boots Alliance
- Wal-Mart Stores Incorporated
- Walt Disney Company

As we said earlier, the original DJIA was a true average—that is, you simply added up the stock prices of the 12 companies and divided by 12. In 1928 a divisor of 16.67 was used to adjust for mergers, takeovers, bankruptcies, stock splits, and company substitutions. Today, they add up the 30 stock prices and divide by 0.15198707565833, or equivalently, multiply by 1/0.15198707565833 or about 6.58. This means that for every $1 move in any Dow company's stock price, the average changes by about 6.58 points. (The DJIA is usually reported using two decimal places.)

EXAMPLE 4.37 Finding changes in the Dow: Disney stock goes up

Suppose the stock of Walt Disney increases in value by $3 per share. If all other Dow stock prices remain unchanged, how does this affect the DJIA?

SOLUTION

Each $1 increase causes the average to increase by about 6.58 points. So a $3 increase would cause an increase of about $3 \times 6.58 = 19.74$ points in the Dow.

TRY IT YOURSELF 4.37

Suppose the stock of Microsoft decreases in value by $4 per share. If all other Dow stock prices remain unchanged, how does this affect the DJIA?

The answer is provided at the end of this section.

The graph in **Figure 4.14** shows how the Dow has moved over the last several decades. In Exercises 30 through 33, we explore a few of the more common types of stock transactions.

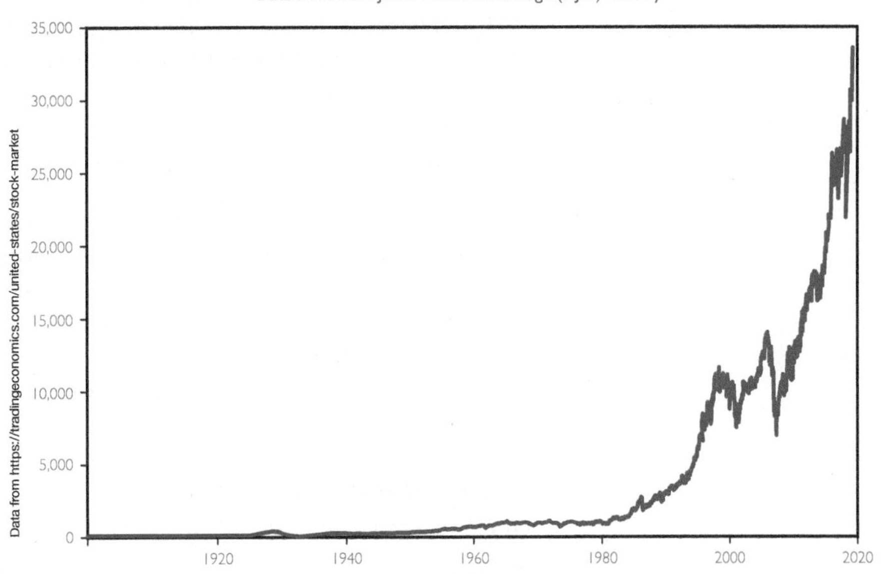

FIGURE 4.14 This graph shows the movement of the Dow.

ALGEBRAIC SPOTLIGHT 4.6 Derivation of the Buying Power Formula

Suppose the inflation rate is $i\%$ per year. We want to derive the buying power formula:

$$\text{Percent decrease in buying power} = \frac{100\,i}{100+i}$$

Suppose we could buy a commodity, say, one pound of flour, for one dollar a year ago. An inflation rate of $i\%$ tells us that today that same pound of flour would cost $1 + i/100$ dollars. To find the new buying power of the dollar, we need to know how much flour we could buy today for one dollar. Because one pound of flour costs $1 + i/100$ dollars, one dollar will buy

$$\frac{1}{1+i/100} = \frac{100}{100+i} \text{ pounds of flour}$$

This quantity represents a decrease of

$$1 - \frac{100}{100+i} = \frac{i}{100+i} \text{ pounds}$$

from the one pound we could buy with one dollar a year ago. The percentage decrease in the amount of flour we can buy for one dollar is

$$\text{Percentage decrease} = \frac{\text{Decrease}}{\text{Previous value}} \times 100\% = \frac{\frac{i}{100+i}}{1} \times 100\% = \frac{100\,i}{100+i}$$

This is the desired formula for the percentage decrease in buying power.

WHAT DO YOU THINK?

CPI and benefits: In the United States, some government entitlements, such as Social Security payments, are tied to the inflation rate. Explain why connecting retirement benefits to inflation makes sense. Find other financial arrangements that are tied to inflation.

Questions of this sort may invite many correct responses. We present one possibility: Inflation causes the cost of typical consumer items to rise. If income remains constant while inflation drives up prices, many who depend on Social Security will suffer. Increasing benefits to match inflation means that Social Security maintains a constant value to recipients. Other federal programs with benefits tied to inflation include military retirement, Veterans Disability Compensation, food stamps, Medicaid eligibility, and many more.

Further questions of this type may be found in the *What Do You Think?* section of exercises.

Try It Yourself answers

Try It Yourself 4.30: Calculating inflation: CPI increase to 225 4.7%.

Try It Yourself 4.31: Calculating decrease in buying power: 6500% inflation 98.5%.

Try It Yourself 4.32: Calculating inflation: 5.2% decrease in buying power 5.5%.

Try It Yourself 4.34: Calculating the tax: A single person $2202.50.

Try It Yourself 4.36: Comparing deductions and credits: Differing effects $6530.

Try It Yourself 4.37: Finding changes in the Dow: Microsoft goes down The DJIA decreases by 26.32 points.

Exercise Set 4.5

Test Your Understanding

1. The Consumer Price Index measures ________.

2. As inflation increases, the buying power of a dollar ________.

3. Explain in a sentence what the "marriage penalty" is.

4. The Dow Jones Industrial Average measures **a.** the value of selected stocks **b.** the growth of industry in America **c.** the gross national product.

5. True or false: To say you are in the 25% tax bracket means that your marginal tax rate is 25%.

6. The marginal tax rate measures **a.** the amount of taxes you owe **b.** the tax rate to which additional income is subject **c.** the percentage of your income that is paid in taxes **d.** none of the above.

7. The term "hyperinflation" refers to ________.

8. Explain how a tax deduction affects your income tax.

Problems

9. **Calculating inflation.** Suppose the CPI increases this year from 205 to 215. What is the rate of inflation for this year? Round your answer to the nearest tenth of a percent.

10. **Calculating decrease in buying power.** Suppose the rate of inflation this year is 3%. What is the percentage decrease in the buying power of a dollar? Round your answer to the nearest tenth of a percent.

11. **Calculating inflation from decrease in buying power.** Suppose the buying power of a dollar decreases by 4% this year. What is the rate of inflation this year? Round your answer to the nearest tenth of a percent.

12. **Calculating tax.** Use Table 4.6 to calculate the tax due from a married couple filing jointly with a taxable income of $107,000.

13. **Deductions.** If you earn $75,000 this year and have deductions of $17,000, what is your taxable income?

14. **Changes in the DOW.** Suppose the stock of the 3M Company increases by $5 per share while all other Dow stock prices remain the same. How does this affect the Dow Jones Industrial Average?

15. **Large inflation rate.** The largest annual rate of inflation in the CPI table, Table 4.3 on page 289, was 13.3% for the year 1979. What year saw the next largest rate? Why do you think the inflation rate in the table for 1949 is blank?

16. **Food inflation.** This exercise refers to *In the News 4.10*, which is from the Web site of the United States Department of Agriculture.

    a. If your total food bill was $300 in October 2015, what was it in October 2016?

    b. If your at-home food bill was $300 in October 2015, what was it in October 2016?

    c. If your away-from-home food bill was $300 in October 2015, what was it in October 2016?

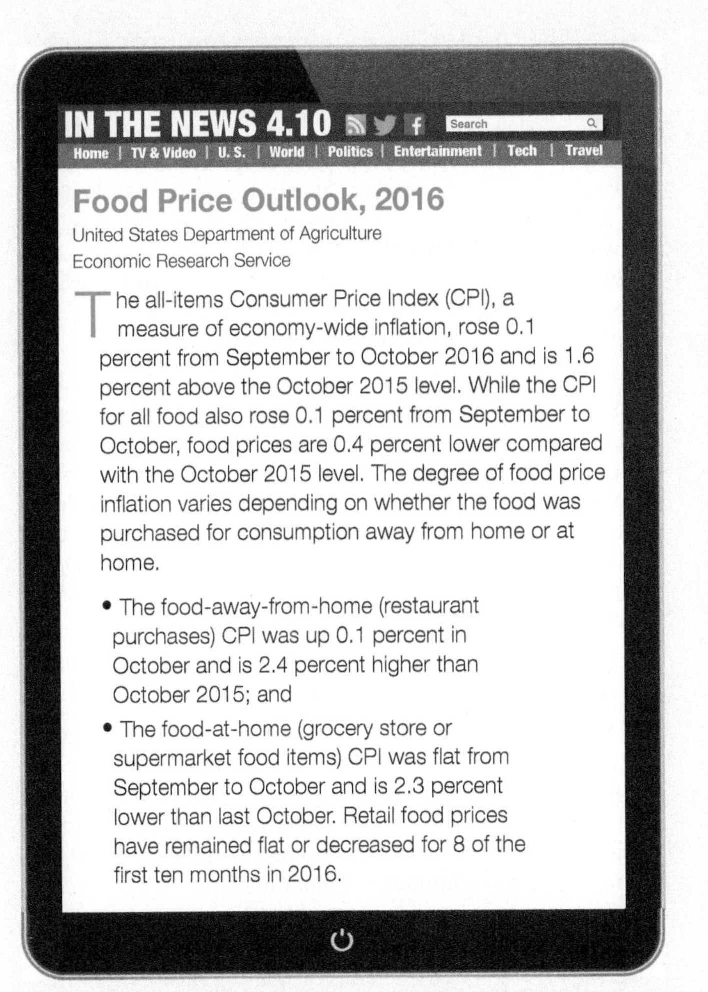
IN THE NEWS 4.10

Home | TV & Video | U. S. | World | Politics | Entertainment | Tech | Travel

Food Price Outlook, 2016

United States Department of Agriculture
Economic Research Service

The all-items Consumer Price Index (CPI), a measure of economy-wide inflation, rose 0.1 percent from September to October 2016 and is 1.6 percent above the October 2015 level. While the CPI for all food also rose 0.1 percent from September to October, food prices are 0.4 percent lower compared with the October 2015 level. The degree of food price inflation varies depending on whether the food was purchased for consumption away from home or at home.

- The food-away-from-home (restaurant purchases) CPI was up 0.1 percent in October and is 2.4 percent higher than October 2015; and
- The food-at-home (grocery store or supermarket food items) CPI was flat from September to October and is 2.3 percent lower than last October. Retail food prices have remained flat or decreased for 8 of the first ten months in 2016.

17. Inflation compounded. In this exercise, we see the cumulative effects of inflation. We refer to Table 4.3 on page 289.

a. Find the three-year inflation rate from December 1977 to December 1980.

b. Consider the one-year inflation rate for each of the three years from December 1977 to December 1980. Find the sum of these three numbers. Is the sum the same as your answer to part a?

18. More on compounding inflation. Here is a hypothetical CPI table.

Year	Hypothetical CPI	Inflation rate
1935	10	—
1936	20	%
1937	40	%
1938	80	%

a. Fill in the missing inflation rates.

b. Find the three-year inflation rate from 1935 to 1938.

c. Find the sum of the three inflation rates during the three years from 1935 to 1938. Is your answer here the same as your answer to part b?

d. Use the idea of compounding from Section 4.1 to explain the observation in part c.

19. Sudanese *pound*. Table 4.4 on page 290 shows that the inflation rate for Sudan in the year 2020 was 142%. By how much did the buying power of the Sudanese *pound* (the currency in Sudan) decrease during 2020?

20. Find the inflation rate. Suppose the buying power of a dollar went down by 60% over a period of time. What was the inflation rate during that period?

21. Continuing inflation. Suppose that prices increase 3% each year for 10 years. How much will a jacket that costs $80 today cost in 10 years? *Suggestion*: The price of the jacket increases by the same percentage each year, so the price is an exponential function of the time in years. You can think of the price as the balance in a savings account with an APY of 3% and an initial investment of $80.

22. More on continuing inflation. *This is a continuation of Exercise 21.* If prices increase 3% each year for 10 years, what is the percentage decrease in the buying power of currency over the 10-year period?

23. Flat tax. Steve Forbes ran for U.S. president in 1996 and 2000 on a platform proposing a 17% flat tax, that is, an income tax that would simply be 17% of each tax payer's taxable income. Suppose that Alice was single in the year 2020 with a taxable income of $30,000 and that Joe was single in the year 2020 with a taxable income of $300,000.

a. What was Alice's tax? Use the tax table on page 292.

b. What was Joe's tax? Use the tax table on page 292.

c. If the 17% flat tax proposed by Mr. Forbes had been in effect in 2020, what would Alice's tax have been?

d. What would Joe's tax have been under the 17% flat tax?

e. When you compare Alice and Joe, what do you think about the fairness of the flat tax versus a progressive tax?

24. More on the flat tax. Let's return to Alice and Joe from Exercise 23. We learn that Joe actually made $600,000, but his taxable income was only $300,000 because of various deductions allowed by the system in 2020. Proponents of the flat tax say that many of these deductions should be eliminated, so the 17% flat tax should be applied to Joe's entire $600,000.

a. What would Joe's tax be under the 17% flat tax?

b. How much more tax would Joe pay than under the 2020 system?

c. How much income would Joe have to make for the 17% flat tax to equal the amount he pays in the year 2020 with a taxable income of $300,000?

d. When you compare Alice and Joe now, what do you think about the fairness of the flat tax versus a progressive tax?

25. Bracket creep. At the start of 2020, your taxable income was $40,000, and you received a cost-of-living raise because of inflation. In 2020, inflation was 1.4% and your raise resulted in a 1.4% increase in your taxable income. By how much, and by what percent, did your taxes go up over what they would have been without a raise? (Assume that you were single in 2020.) *Remark*: Note that your buying power remains the same, but you're paying higher taxes. Not only that, but you're paying at a higher marginal rate! This phenomenon is known as "bracket creep," and federal tax tables are adjusted each year to account for this.

26. Deduction and credit. In the year 2020, Ethan was single and had a total income of $55,000. He took a deduction of $12,500 and had a tax credit of $1500. Calculate the tax owed by Ethan.

27. Moving DJIA. Suppose the stock of McDonald's increases in value by $2 per share. If all other Dow stock prices remain unchanged, how does this affect the DJIA?

28. Average price. What is the average price of a share of stock in the Dow list when the DJIA is 10,000? Remember that there are 30 companies.

29. Dow highlights. Use Figure 4.14 to determine in approximately what year the DJIA first reached 5000. About when did it first reach 10,000?

Exercises 30 through 33 are suitable for group work.

30–33. Stock market transactions. *There are any number of ways to make (or lose) money with stock market transactions. In Exercises 30 through 33, we will explore a few of the more common types of transactions. (Fees for such transactions will be ignored.)*

30. Market order. The simplest way to buy stock is the *market order*. Through your broker or online, you ask to buy 100 shares of stock X at market price. As soon as a seller is located, the transaction is completed at the prevailing price. That price will normally be very close to the latest quote, but the prevailing price may be different if the price fluctuates between the time you place the order and the time the transaction is completed. Suppose you place a market order for 100 shares of stock X and the transaction is completed at $44 per share. Two weeks later the stock value is $58 per share and you sell. What is your net profit?

31. Limit orders. If you want to insist on a fixed price for a transaction, you place a *limit order*. That is, you offer to buy (or sell) stock X at a certain price. If the stock can be purchased for that price, the transaction is completed. If not, no transaction occurs. Often limit orders have a certain expiration date. Suppose you place a limit order to buy 100 shares of stock X for $40 per share. When the stock purchase is completed, you immediately place a limit order to sell stock X at $52 per share. The following table shows the value of stock X. On which days are these two transactions completed, and what is your profit?

Date	Day 1	Day 2	Day 3	Day 4	Day 5	Day 6
Market price	$44	$45	$40	$48	$52	$50

32. Stop loss and trailing stops. If you own a stock, a *stop-loss* order protects you from large losses. For example, if you own 100 shares of stock X, bought at $45 per share, you might place a stop-loss order for $40 per share. This order automatically sells your stock if the price drops to $40. No matter what happens, you can't lose more than $5 per share. A similar type of order that protects profits is the *trailing stop*. The trailing-stop order sells your stock if the value goes below a certain percentage of the market price. If the market price remains the same or drops, the trailing stop doesn't change and acts like a stop-loss order. If, however, the market price goes up, the trailing stop follows it so that it protects profits. Suppose, for example, that you own 100 shares of stock X, that the market price is $40 per share, and that you place a trailing-loss order of 5%. If the price drops by $2 (5% of $40), the stock is sold. If the market value increases to $44, then you will sell the stock when it declines by 5% of $44. That is, you sell when the market price drops to $41.80. Consider the table in Exercise 31, and suppose that you purchase 100 shares of the stock and place a 5% trailing-stop order on day 1. On which day (if any) will your stock be sold?

33. Selling short. *Selling short* is the selling of stock you do not actually own but promise to deliver. Suppose you place an order to sell short 100 shares of stock X at $35. Eventually, your order must be *covered*. That is, you must sell 100 shares of stock X at a value of $35 per share. If when the order is covered, the value of the stock is less than $35 per share, then you make money; otherwise, you lose money. Suppose that on the day you must cover the sell short order, the price of stock X is $50 per share. How much money did you lose?

The following exercise is designed to be solved using technology such as calculators or computer spreadsheets.

34. Mortgage interest deduction. Interest paid on a home mortgage is normally tax deductible. That is, you can subtract the total mortgage interest paid over the year in determining your taxable income. This is one advantage of buying a home. Suppose you take out a 30-year home mortgage for $250,000 at an APR of 8% compounded monthly.

a. Determine your monthly payment using the monthly payment formula in Section 4.2.

b. Make a spreadsheet showing, for each month of the first year of your payment, the amount that represents interest, the amount toward the principal, and the balance owed.

c. Use the results of part b to find the total interest paid over the first year. (Round your answer to the nearest dollar.) That is what you get to deduct from your taxable income.

d. Suppose that your marginal tax rate is 27%. What is your actual tax savings due to mortgage payments? Does this make the $250,000 home seem a bit less expensive?

Writing About Mathematics

35. Flat tax. Do research to determine the pros and cons of a flat tax.

36. History: More on selling short. In 1992 George Soros "broke the Bank of England" by selling short the British pound. Write a brief report on his profit and exactly how he managed to make it.

37. History: The Knights Templar. The Knights Templar was a monastic order of knights founded in 1112 to protect pilgrims traveling to the Holy Land. Recent popular novels have revived interest in them. Some have characterized the Knights Templar as the first true international bankers. Report on the international aspects of their early banking activities.

38. History: The stock market. The Dow Jones Industrial Average normally fluctuates, but over the last half-century it has generally increased. Dramatic drops in the Dow (stock market crashes) can have serious effects on the economy. Report on some of the most famous of these. Be sure to include the crash of 1929.

39. History: The Federal Reserve Bank. The Federal Reserve Bank is an independent agency that regulates various aspects of American currency. Write a report on the Federal Reserve Bank. Your report should include the circumstances of its creation.

40. History: The SEC. The Securities and Exchange Commission regulates stock market trading in the United States. Write a report on the creation and function of the SEC.

41. Low and high inflation. Find current inflation rates for some countries not mentioned in this book and write a report about your findings. In particular, look for countries with very low and very high inflation.

42. Other countries' stock markets. Other countries have their own stock markets. Examine some of them and write a report about your findings. Can you see significant differences in how they perform?

43. Stagflation. Look up the meaning of *stagflation*. There was a period in history when the United States suffered from it. Write a report about this period.

44. Laffer curve. Look up the meaning of the *Laffer curve* and write a report about it.

What Do You Think?

45. Hyperinflation. Explain what is meant by the term *hyperinflation*. Use the Internet to find recent examples. **Exercise 19 is relevant to this topic.**

46. Inflation and buying power. If the inflation rate doubles, does the percentage decrease in buying power also double? If not, is the new percentage decrease in buying power more than twice the old one or less than twice the old one? Interpret your answer.

47. The Dow. The Dow is based on the prices of 30 "blue-chip" U.S. stocks. Why is it not based on the prices of all U.S. stocks?

48. The logarithmic Dow. Read Exercises 43 and 44 in Section 3.3. Compare the graphs on logarithmic and linear scales for the stock price of Google. Do the same thing for Apple and another company you find interesting. Explain what you observe.

49. Good inflation. Inflation is usually considered to be a bad thing, but some economists have recently suggested that more of it might be good because it would help with the national debt. Can you explain how inflation could make it easier for the government to pay down the national debt?

50. Deflation. Deflation is sometimes called "negative inflation" because prices go down rather than up. Deflation is considered a potentially bad thing by many economists because consumers will put off buying things, hoping for lower prices, and this behavior causes the economy to slow down. Use the Internet to find some historical examples of deflation.

51. Marriage penalty. Explain what is meant by the *marriage penalty*. Is it a fair way to tax married couples? Explain your answer.

52. Tax deduction and tax credit. Explain the difference between a tax deduction and a tax credit. Use the Internet to help you find examples of each. **Exercise 26 is relevant to this topic.**

53. Stock market. For background information, read Exercises 30 through 33. Suppose that you expect the price of a certain stock to decline in the near future. If you decide to deal in that stock, which of the following types of transactions might be to your benefit: market order, limit order, stop loss, or selling short? Explain why you made the choice you did.

CHAPTER SUMMARY

This chapter is concerned with financial transactions of two basic kinds: saving and borrowing. We also consider important financial issues related to inflation, taxes, and the stock market.

Saving money: The power of compounding

The principal in a savings account typically grows by interest earned. Interest can be credited to a savings account in two ways: as *simple interest* or as *compound interest*. For simple interest, the formula for the interest earned is

$$\text{Simple interest earned} = \text{Principal} \times \text{Yearly interest rate (as a decimal)} \times \text{Time in years}$$

Financial institutions normally compound interest and advertise the *annual percentage rate* or APR. The interest rate for a given compounding period is calculated using

$$\text{Period interest rate} = \frac{\text{APR}}{\text{Number of periods in a year}}$$

We can calculate the account balance after t periods using the compound interest formula:

$$\text{Balance after } t \text{ periods} = \text{Principal} \times (1 + r)^t$$

Here, r is the period interest rate expressed as a decimal, and it should not be rounded. In fact, it is best to do all the calculations and then round.

The *annual percentage yield* or APY is the actual percentage return in a year. It takes into account compounding of interest and is always at least as large as the APR. If n is the number of compounding periods per year,

$$\text{APY} = \left(1 + \frac{\text{APR}}{n}\right)^n - 1$$

Here, both the APR and the APY are in decimal form. The APY can be used to calculate the account balance after t years:

$$\text{Balance after } t \text{ years} = \text{Principal} \times (1 + \text{APY})^t$$

Here, the APY is in decimal form.

The *present value* of an investment is the amount we initially invest. The *future value* is the value of that investment at some specified time in the future. If the investment grows by compounding of interest, these two quantities are related by the compound interest formula. We can rearrange that formula to give the present value we need for a desired future value:

$$\text{Present value} = \frac{\text{Future value}}{(1 + r)^t}$$

In this formula, t is the total number of compounding periods, and r is the period interest rate expressed as a decimal.

The *Rule of 72* can be used to estimate how long it will take for an account growing by compounding of interest to double in size. It says that the doubling time in years can be approximated by dividing 72 by the APR, where the APR is expressed as a percentage, not as a decimal. The exact doubling time can be found using the formula:

$$\text{Number of periods to double} = \frac{\log 2}{\log (1 + r)}$$

Here, r is the period interest rate as a decimal.

Borrowing: How much car can you afford?

With an *installment loan*, you borrow money for a fixed period of time, called the *term* of the loan, and you make regular payments (usually monthly) to pay off the loan plus interest in that time. Loans for the purchase of a car or home are usually installment loans.

If you borrow an amount at a monthly interest rate r (as a decimal) with a term of t months, the monthly payment is

$$\text{Monthly payment} = \frac{\text{Amount borrowed} \times r(1 + r)^t}{((1 + r)^t - 1)}$$

It is best to do all the calculations and then round. A companion formula tells how much you can borrow for a given monthly payment:

$$\text{Amount borrowed} = \frac{\text{Monthly payment} \times ((1 + r)^t - 1)}{(r \times (1 + r)^t)}$$

For all loans, the monthly payment is *at least* the amount we would pay each month if no interest were charged, which is the amount of the loan divided by the term (in months) of the loan. This number can be used to estimate the monthly payment for a short-term loan if the APR is not large. For all loans, the monthly payment is *at least* as large as the principal times the monthly interest rate as a decimal. This number can be used to estimate the monthly payment for a long-term loan with a moderate or high interest rate.

A record of the repayment of a loan is kept in an *amortization table*. In the case of buying a home, an important thing for the borrower to know is how much *equity* he or she has in the home. The equity is the total amount that has been paid toward the principal, and an amortization table keeps track of this amount.

Some home loans are in the form of an *adjustable-rate mortgage* or ARM, where the interest rate may vary over the life of the loan. For an ARM, the initial rate is often lower than the rate for a comparable fixed-rate mortgage, but rising rates may cause significant increases in the monthly payment.

Saving for the long term: Build that nest egg

Another way to save is to deposit a certain amount into your savings account at the end of each month. If the monthly interest rate is r as a decimal, your balance is given by

$$\text{Balance after } t \text{ deposits} = \frac{\text{Deposit} \times ((1+r)^t - 1)}{r}$$

The ending balance is often called the *future value* of this savings arrangement. A companion formula gives the amount we need to deposit regularly in order to achieve a goal:

$$\text{Needed deposit} = \frac{\text{Goal} \times r}{((1+r)^t - 1)}$$

Retirees typically draw money from their nest eggs in one of two ways: either as a *perpetuity* or as an *annuity*. An annuity reduces the principal over time, but a perpetuity does not. The principal (your nest egg) is often called the *present value*.

For an annuity with a term of t months, we have the formula

$$\text{Monthly annuity yield} = \frac{\text{Nest egg} \times r(1+r)^t}{((1+r)^t - 1)}$$

In this formula, r is the monthly interest rate as a decimal. A companion formula gives the nest egg needed to achieve a desired annuity yield:

$$\text{Nest egg needed} = \frac{\text{Annuity yield goal} \times ((1+r)^t - 1)}{(r(1+r)^t)}$$

Credit cards: Paying off consumer debt

Buying a car or a home usually involves a regular monthly payment that is computed as described earlier. But another way of borrowing is by credit card. If the balance is not paid off by the due date, the account is subject to finance charges. A simplified formula for the finance charge is

$$\text{Finance charge} = \frac{\text{APR}}{12} \times (\text{Previous balance} - \text{Payments} + \text{Purchases})$$

The new balance on the credit card statement is found as follows:

$$\text{New balance} = \text{Previous balance} - \text{Payments} + \text{Purchases} + \text{Finance charge}$$

Suppose we have a balance on our credit card and decide to stop charging. If we make only the minimum payment, the balance is given by the exponential formula

$$\text{Balance after } t \text{ minimum payments} = \text{Initial balance} \times ((1+r)(1-m))^t$$

In this formula, r is the monthly interest rate and m is the minimum monthly payment as a percent of the balance. Both r and m are in decimal form. The product $(1+r)(1-m)$ should not be rounded when the calculation is performed. Because the balance is a decreasing exponential function, the balance decreases very slowly in the long run.

Inflation, taxes, and stocks: Managing your money

The *Consumer Price Index* or CPI is a measure of the average price paid by urban consumers in the United States for a "market basket" of goods and services. The *rate of inflation* is measured by the percent change in the CPI over time.

Inflation reflects a decline of the buying power of the consumer's dollar. Here is a formula that tells how much the buying power of currency decreases for a given inflation rate i (expressed as a percent):

$$\text{Percent decrease in buying power} = \frac{100i}{100 + i}$$

A key concept for understanding income taxes is the *marginal tax rate*. With a marginal tax rate of 30%, for example, the tax owed increases by \$0.30 for every \$1 increase in taxable income. Typically, those with a substantially higher taxable income have a higher marginal tax rate. To calculate our taxable income, we subtract any *deductions* from our total income. Then we can use the tax tables. To calculate the actual tax we owe, we subtract any *tax credits* from the tax determined by the tables.

The *Dow Jones Industrial Average* or DJIA is a measure of the value of leading stocks. It is found by adding the prices of 30 "blue-chip" stocks and dividing by a certain number, the *divisor*, to account for mergers, stock splits, and other factors. With the current divisor, for every \$1 move in any Dow company's stock price, the average changes by about 6.58 points.

KEY TERMS

principal, p. 216
simple interest, p. 216
compound interest, p. 217
period interest rate, p. 219
annual percentage rate (APR), p. 219
annual percentage yield (APY), p. 223
present value, p. 227
future value, p. 227
Rule of 72, p. 228
installment loan, p. 234
term, p. 234
amortization table (or amortization schedule), p. 241
equity, p. 242
fixed-rate mortgage, p. 247
adjustable-rate mortgage (ARM), p. 247
prime interest rate, p. 247
nest egg, p. 264
monthly yield, p. 264
annuity, p. 264
Consumer Price Index (CPI), p. 288
inflation, p. 288
rate of inflation, p. 288
deflation, p. 288

CHAPTER QUIZ

1. We invest \$2400 in an account that pays simple interest of 8% each year. Find the interest earned after five years.

Answer \$960

If you had difficulty with this problem, see Example 4.1.

2. Suppose it were possible to invest \$8000 in a four-year CD that pays an APR of 5.5%.

a. What is the value of the mature CD if interest is compounded annually?

b. What is the value of the mature CD if interest is compounded monthly?

Answer **a.** \$9910.60, **b.** \$9963.60

If you had difficulty with this problem, see Example 4.3.

3. We have an account that pays an APR of 9.75%. If interest is compounded quarterly, find the APY. Round your answer as a percentage to two decimal places.

Answer 10.11%

If you had difficulty with this problem, see Example 4.4.

4. How much would you need to invest now in a savings account that pays an APR of 8% compounded monthly in order to have a future value of \$6000 in a year and a half?

Answer \$5323.64

If you had difficulty with this problem, see Example 4.7.

5. Suppose an account earns an APR of 5.5% compounded monthly. Estimate the doubling time using the Rule of 72, and then calculate the exact doubling time. Round your answers to one decimal place.

Answer Rule of 72: 13.1 years; exact method: 151.6 months (about 12 years and eight months)

If you had difficulty with this problem, see Example 4.8.

6. You need to borrow $6000 to buy a car. The dealer offers an APR of 9.25% to be paid off in monthly installments over $2\frac{1}{2}$ years.

a. What is your monthly payment?

b. How much total interest did you pay?

Answer **a.** $224.78, **b.** $743.40

If you had difficulty with this problem, see Example 4.11.

7. We can afford to make payments of $125 per month for two years for a used motorcycle. We're offered a loan at an APR of 11%. What price bike should we be shopping for?

Answer $2681.95

If you had difficulty with this problem, see Example 4.10.

8. Suppose we have a savings account earning 6.25% APR. We deposit $15 into the account at the end of each month. What is the account balance after eight years?

Answer $1862.16

If you had difficulty with this problem, see Example 4.21.

9. Suppose we have a savings account earning 5.5% APR. We need to have $2000 at the end of seven years. How much should we deposit each month to attain this goal?

Answer $19.57

If you had difficulty with this problem, see Example 4.22.

10. Suppose we have a nest egg of $400,000 with an APR of 5% compounded monthly. Find the monthly yield for a 10-year annuity.

Answer $4242.62

If you had difficulty with this problem, see Example 4.24.

11. Suppose your MasterCard calculates finance charges using an APR of 16.5%. Your previous statement showed a balance of $400, toward which you made a payment of $100. You then bought $200 worth of clothes, which you charged to your card. Complete the following table:

	Previous balance	Payments	Purchases	Finance charge	New balance
Month 1					

Answer

	Previous balance	Payments	Purchases	Finance charge	New balance
Month 1	$400.00	$100.00	$200.00	$6.88	$506.88

If you had difficulty with this problem, see Example 4.26.

12. Suppose your MasterCard calculates finance charges using an APR of 16.5%. Your statement shows a balance of $900, and your minimum monthly payment is 6% of that month's balance.

a. What is your balance after a year and a half if you make no more charges and make only the minimum payment?

b. How long will it take to get your balance under $100?

Answer **a.** $377.83, **b.** 46 monthly payments

If you had difficulty with this problem, see Example 4.29.

13. Suppose the CPI increases this year from 210 to 218. What is the rate of inflation for this year?

Answer 3.8%

If you had difficulty with this problem, see Example 4.30.

14. Suppose the rate of inflation last year was 20%. What was the percentage decrease in the buying power of currency over that year?

Answer 16.7%

If you had difficulty with this problem, see Example 4.31.

Harvepino/Shutterstock

5 Introduction to Probability

The following excerpt is from an article that appeared at the Web site of the American Association for the Advancement of Science (AAAS). It refers to the Chelyabinsk meteor, an asteroid that exploded near Chelyabinsk, Russia, on February 15, 2013, and released energy about 20 times that released by the first atomic bomb.

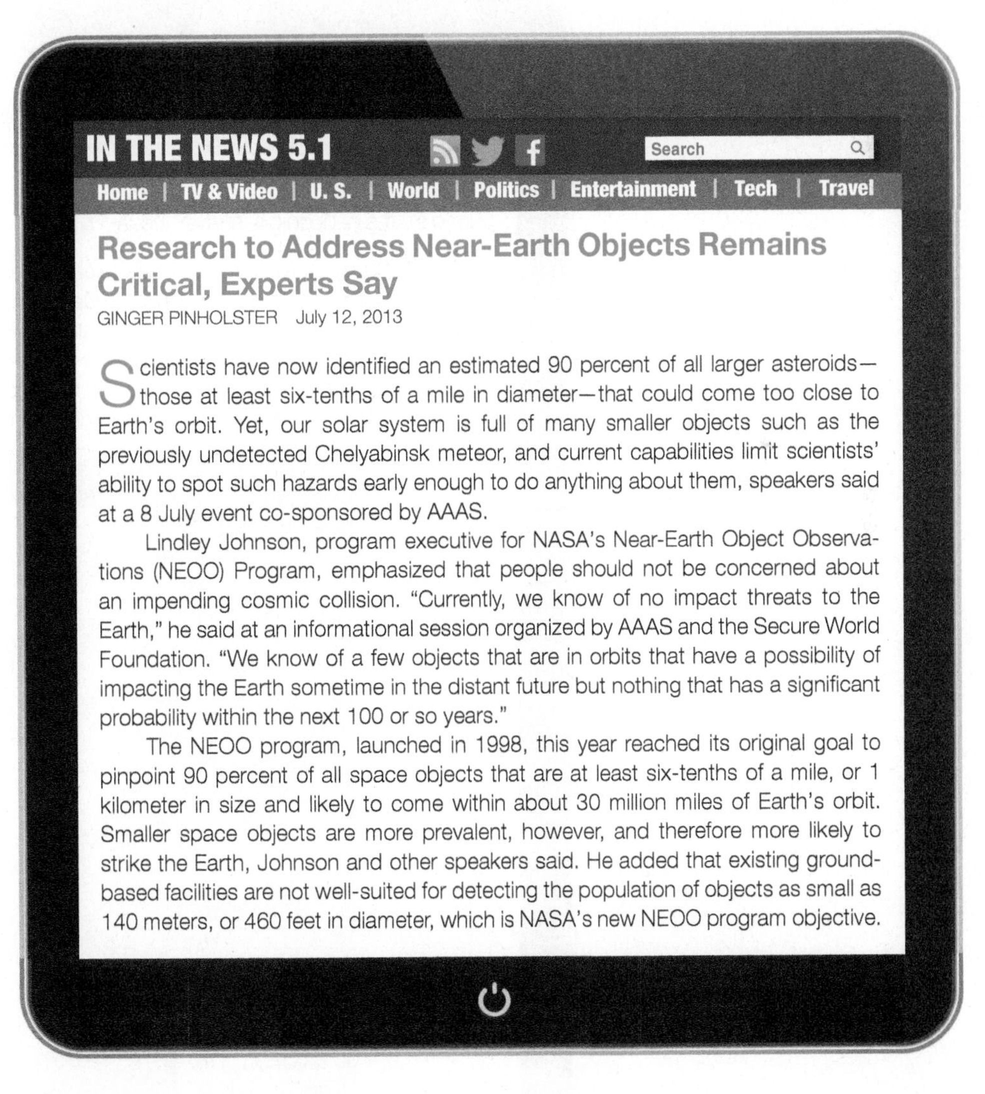

IN THE NEWS 5.1

Search

Home | TV & Video | U. S. | World | Politics | Entertainment | Tech | Travel

Research to Address Near-Earth Objects Remains Critical, Experts Say

GINGER PINHOLSTER July 12, 2013

Scientists have now identified an estimated 90 percent of all larger asteroids—those at least six-tenths of a mile in diameter—that could come too close to Earth's orbit. Yet, our solar system is full of many smaller objects such as the previously undetected Chelyabinsk meteor, and current capabilities limit scientists' ability to spot such hazards early enough to do anything about them, speakers said at a 8 July event co-sponsored by AAAS.

Lindley Johnson, program executive for NASA's Near-Earth Object Observations (NEOO) Program, emphasized that people should not be concerned about an impending cosmic collision. "Currently, we know of no impact threats to the Earth," he said at an informational session organized by AAAS and the Secure World Foundation. "We know of a few objects that are in orbits that have a possibility of impacting the Earth sometime in the distant future but nothing that has a significant probability within the next 100 or so years."

The NEOO program, launched in 1998, this year reached its original goal to pinpoint 90 percent of all space objects that are at least six-tenths of a mile, or 1 kilometer in size and likely to come within about 30 million miles of Earth's orbit. Smaller space objects are more prevalent, however, and therefore more likely to strike the Earth, Johnson and other speakers said. He added that existing ground-based facilities are not well-suited for detecting the population of objects as small as 140 meters, or 460 feet in diameter, which is NASA's new NEOO program objective.

This article both warns and reassures us. The article states that there is a possibility that an object will strike Earth sometime in the distant future but that no such event has a significant *probability* within the next century. All of NASA's highly trained scientists, mathematicians, and engineers with access to NASA's state-of-the-art equipment cannot tell us for sure whether such an event will occur. What are we to make of such information? Probabilities of this sort apply not only to asteroids but also to hurricanes, rainstorms, traffic accidents, sports contests, and winning the lottery. In this chapter, we examine the idea of probability, what it means, and how probabilities are calculated.

5.1 Calculating probabilities: How likely is it?

TAKE AWAY FROM THIS SECTION
Distinguish between the different types of probability and calculate mathematical probabilities.

Probabilities arise in many contexts. The following article illustrates one of these contexts.[1]

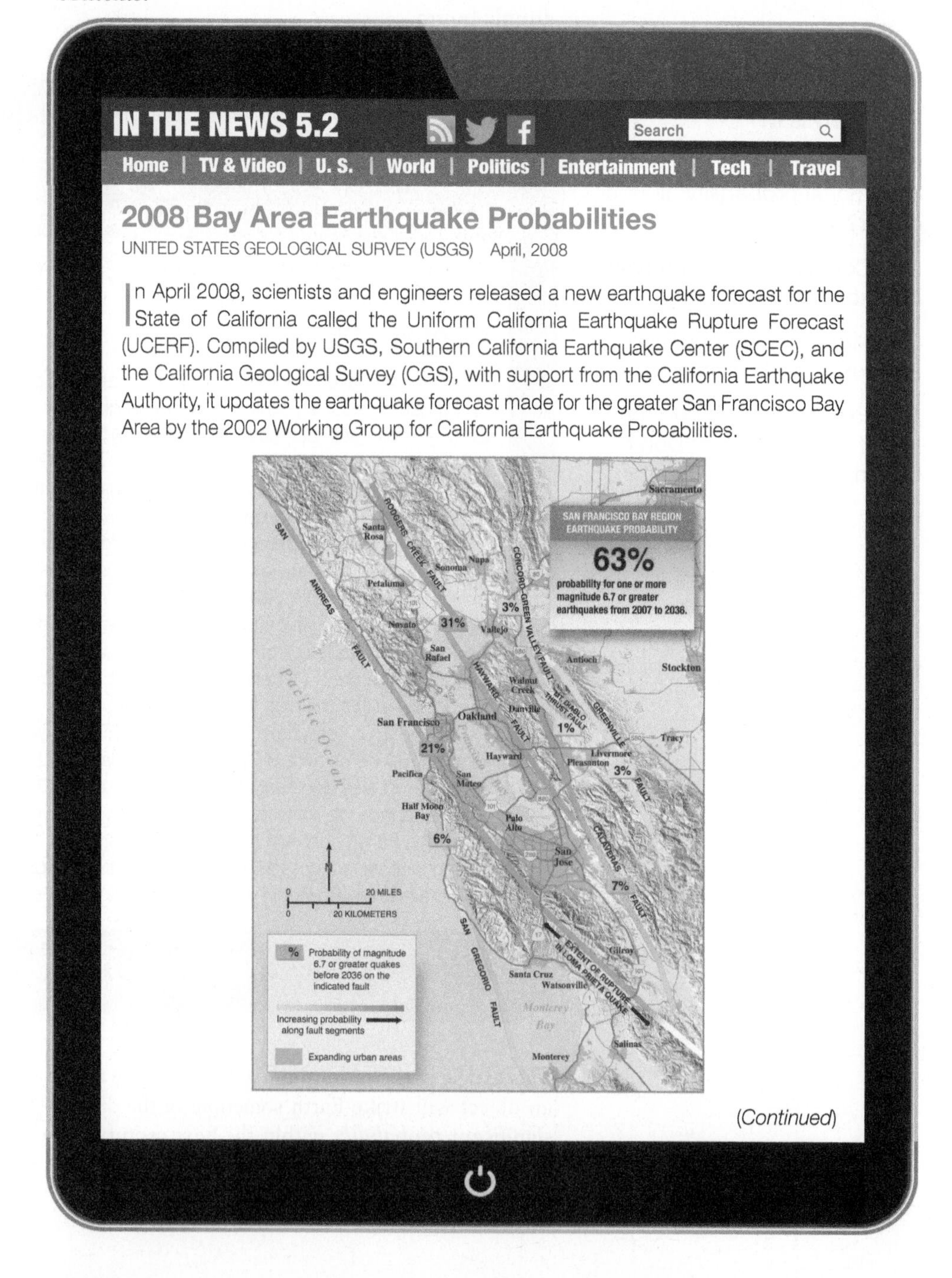

IN THE NEWS 5.2

Search

Home | TV & Video | U. S. | World | Politics | Entertainment | Tech | Travel

2008 Bay Area Earthquake Probabilities

UNITED STATES GEOLOGICAL SURVEY (USGS) April, 2008

In April 2008, scientists and engineers released a new earthquake forecast for the State of California called the Uniform California Earthquake Rupture Forecast (UCERF). Compiled by USGS, Southern California Earthquake Center (SCEC), and the California Geological Survey (CGS), with support from the California Earthquake Authority, it updates the earthquake forecast made for the greater San Francisco Bay Area by the 2002 Working Group for California Earthquake Probabilities.

(Continued)

[1] A 6.0 magnitude earthquake did in fact occur in Napa, California, on August 24, 2014.

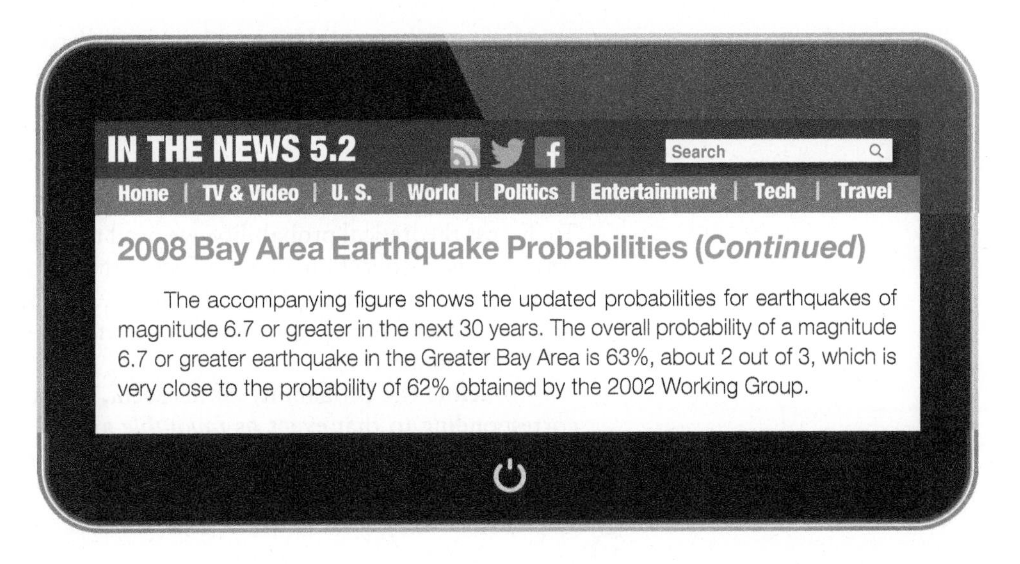
IN THE NEWS 5.2

Home | TV & Video | U. S. | World | Politics | Entertainment | Tech | Travel

2008 Bay Area Earthquake Probabilities (*Continued*)

The accompanying figure shows the updated probabilities for earthquakes of magnitude 6.7 or greater in the next 30 years. The overall probability of a magnitude 6.7 or greater earthquake in the Greater Bay Area is 63%, about 2 out of 3, which is very close to the probability of 62% obtained by the 2002 Working Group.

To appreciate the significance of the preceding article, we need to understand the meaning of "probability." We use numbers to measure all kinds of things: our 5-foot-long dining table, our 6-pound cat, the 3% annual rate of inflation in prices, and so on. Probability is just another kind of numerical measure. It expresses the likelihood that a specific event will occur.

People use the term *probability* in various ways—often casually or informally. We will begin by explaining some of these uses. Consider the following statements:

1. Your chances of being a TV star are zero.
2. If I toss a coin, the probability of getting a "head" is 1/2.
3. A recent poll tells us that if you bump into an American at random, there is a 31% chance you will be meeting a Democrat.

All of these statements have one thing in common: They assign a numerical value to the likelihood of a certain event. But they are quite different in nature. Let's take them one at a time.

Your chances of being a TV star are zero: This statement is expressing the idea that there's no way you're going to be a TV star. That's what we usually mean when we say your chances are zero. Similarly, if we say your chances of being a TV star are 100%, we are saying that you are absolutely certain to become a TV star. These statements don't give probabilities in the mathematical sense because they can't be measured with any degree of accuracy. Rather, they merely express opinions.

If I toss a coin, the probability of getting a "head" is 1/2: The probability of one-half is obtained from the knowledge that if we toss a typical coin, there are only two possible outcomes (heads or tails), and they are equally likely to occur. To say the probability is 1/2 means that if we toss a coin a number of times, we expect to obtain a head about half the time. This fact is also expressed by saying that the probability is 50%. This number is a *theoretical probability*—the type we will be most concerned with in this chapter.

A recent Gallup Poll tells us that if you bump into an American at random, there is a 31% chance you will be meeting a Democrat: This numerical value is known as an *empirical probability* because it comes from an actual measurement, as opposed to a theoretical calculation. It is obtained by polling a relatively small sample of the U.S.

population and extrapolating to estimate a percentage of the whole population. (As we will see in Chapter 6, if the sample is properly chosen, this method can be quite accurate.)

Theoretical probability

To discuss theoretical probability, we need to make clear what we mean by an event. An *event* is a collection of specified outcomes of an experiment.[2] For example, suppose we roll a standard die (with faces numbered 1 through 6). One event is that the number showing is even. This event consists of three outcomes: a 2 shows, a 4 shows, or a 6 shows.

To define the probability of an event, it is common to describe the outcomes corresponding to that event as *favorable* and all other outcomes as *unfavorable*. If the event is that an even number shows when we toss a die, there are three favorable outcomes (2, 4, or 6) and three unfavorable outcomes (1, 3, or 5).

KEY CONCEPT

If each outcome of an experiment is *equally likely*, the probability of an event is the fraction of favorable outcomes. We can express this definition using a formula:

$$\text{Probability of an event} = \frac{\text{Number of favorable outcomes}}{\text{Total number of possible equally likely outcomes}}$$

We will often shorten this to

$$\text{Probability} = \frac{\text{Favorable outcomes}}{\text{Total outcomes}}$$

It is important to recall that in this definition, each outcome is assumed to be equally likely.

Suppose again that we roll a standard six-sided die. Let's calculate the probability that we will get a 1 or a 6. There are six possible outcomes—namely, the numbers 1 through 6—and each number is equally likely to occur. Two of these outcomes are favorable: 1 and 6. So the probability is

$$\frac{\text{Favorable outcomes}}{\text{Total outcomes}} = \frac{2}{6} = \frac{1}{3}$$

What this means is that if we roll the die many times, we expect about one-third of the rolls to come up as a 1 or a 6. Now 1/3 is about 0.33 or 33%, so we can say that the probability of a 1 or a 6 is 1/3 or 0.33 or 33%. All of these numbers are equivalent ways of expressing the same ratio (although 0.33 is only approximately equal to 1/3). It is common in this setting to indicate the probability of a 1 or 6 as $P(1 \text{ or } 6)$. With this notation, we write

$$P(1 \text{ or } 6) = \frac{1}{3}$$

---

[2] *The situations we will consider all have a finite number of outcomes, so we are considering only* discrete *probabilities.*

EXAMPLE 5.1 Calculating simple probabilities: An ace

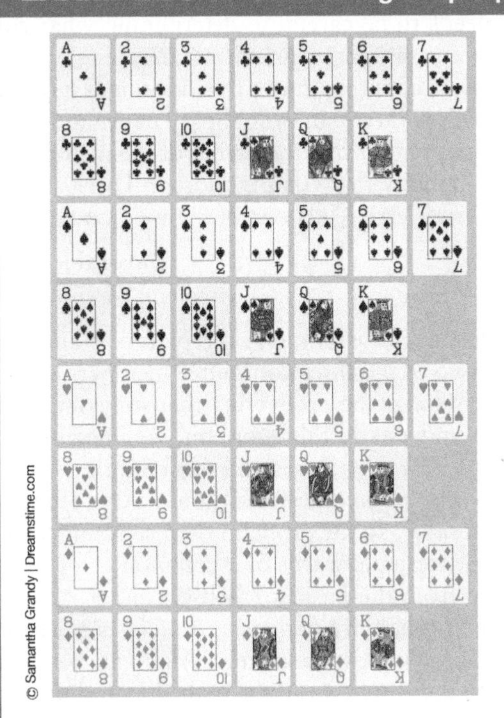
© Samantha Grandy | Dreamstime.com

If I draw a card at random from a standard deck of 52 cards, what is the probability I draw an ace?

SOLUTION

There are 52 different cards that I might draw, and these outcomes are equally likely. There are four aces, so four outcomes are favorable. Thus,

$$P(\text{Ace}) = \frac{\text{Favorable outcomes}}{\text{Total outcomes}} = \frac{4}{52} = \frac{1}{13}$$

Thus, the chance of drawing an ace is 1 in 13, or about 8%.

TRY IT YOURSELF 5.1

In a standard deck, the cards are divided into four suits, called spades, hearts, diamonds, and clubs, with 13 in each suit. If I draw a card at random from a standard deck, what is the probability I draw a diamond?

The answer is provided at the end of this section.

The probability of an event is always between 0 and 1 inclusive. (Exercise 24 asks you to provide an explanation of this.) An event has probability 0 if and only if it can never occur.[3] For example, if I draw five cards from a standard deck of cards, the probability that I get five aces is 0. At the other extreme, an event has a probability of 1 if and only if it will always occur. For example, if I flip a coin, the probability that I get either heads or tails is 1. Exercises 25 and 26 ask you to provide explanations of these statements.

SUMMARY 5.1
Probability

An event is a collection of specified outcomes of an experiment, and we describe each outcome corresponding to that event as favorable. Assume that there are a finite number of outcomes and that each outcome is equally likely.

1. The probability of an event equals the number of favorable outcomes divided by the total number of possible outcomes:

$$\text{Probability} = \frac{\text{Favorable outcomes}}{\text{Total outcomes}}$$

2. A probability may be expressed as a fraction, a decimal, or a percent.

3. A probability must be between 0 and 1 inclusive.

4. The probability of an event is 0 if and only if the event can never occur.

5. The probability of an event is 1 (or 100%) if and only if the event will always occur.

[3] *Remember that we are speaking only of* discrete *probabilities.*

Distinguishing outcomes

One needs to be careful in determining what constitutes an outcome. To illustrate this point, consider a committee consisting of three members. They need to send one member as a representative to a meeting. They decide to choose the representative by writing their three names on slips of paper and drawing a slip at random from a basket.

Suppose first that the names of the members are Alan, Benny, and Jerry. What is the probability that Alan is chosen? There are three equally likely outcomes: Alan, Benny, and Jerry. Only one of these outcomes is favorable (Alan). Thus,

$$P(\text{Alan}) = \frac{\text{Favorable outcomes}}{\text{Total outcomes}} = \frac{1}{3}$$

Now suppose that Jerry is replaced by another member whose name is also Alan. Thus, two of the three committee members are now named Alan, and one is named Benny. What is the probability that someone named Alan is chosen in this case?

To understand this situation, we imagine that members put their surnames on the slips, say, Alan Jones and Alan Smith. Now it is easy to see that there are three equally likely outcomes: Alan Jones, Alan Smith, and Benny. Two of these outcomes are favorable, so the probability of selecting a member named Alan is 2/3.

This example warns of a common error in calculating probabilities: If I choose a name from the basket, I will get either Alan or Benny. The temptation is to say that the probability of getting Alan is 1/2. The error here is that although one can think of having two outcomes, Alan and Benny, *they are not equally likely*. In fact, because there are two Alans in the hat and only one Benny, we are twice as likely to draw Alan as Benny. Often this sort of error is easy to spot, but we must be careful to avoid it. We can often avoid difficulties by distinguishing outcomes just as we used the surnames Jones and Smith to distinguish the two Alans.

EXAMPLE 5.2 Distinguishing outcomes: Coins

Suppose I flip two identical coins. What is the probability that I get two heads?

SOLUTION

We stated that the two coins are identical, but as with the "two Alans" discussed earlier, it helps to imagine that the coins are labeled to make them distinguishable. So let's imagine that one of the coins is a nickel and the other is a dime.

First we must figure out how many equally likely outcomes there are. Either coin can come up as a head (H) or a tail (T). All possibilities are listed in the table below.

Nickel	Dime
H	H
H	T
T	H
T	T

There are four equally likely outcomes: HH, HT, TH, TT. Only one of them gives two heads, so the probability of getting two heads is

$$P(\text{HH}) = \frac{\text{Favorable outcomes}}{\text{Total outcomes}} = \frac{1}{4}$$

TRY IT YOURSELF 5.2

Suppose I flip two identical coins. What is the probability that I get two tails?

The answer is provided at the end of this section.

These examples show how distinguishing things (Alan Jones and Alan Smith, or nickel and dime) that appear to be the same can ensure that all outcomes are equally likely.

EXAMPLE 5.3 Distinguishing outcomes: A traffic light

Suppose I have a 50–50 chance of getting through a certain traffic light without having to stop. I go through this light on my way to work and again on my way home.

a. What is the probability of having to stop at this light at least once on a workday?

b. What is the probability of not having to stop at all?

SOLUTION

a. A 50–50 chance means that the probability of stopping when I come to the light is 1/2 and the probability of not stopping is 1/2. For a trip from home to work and back again, the possibilities are listed in the following table.

To work	To home
Stop	Stop
Stop	Don't stop
Don't stop	Stop
Don't stop	Don't stop

This means that there are four equally likely outcomes: Stop–Stop, Stop–Don't stop, Don't stop–Stop, Don't stop–Don't stop. Three of these yield the result of stopping at least once, so the desired probability is

$$\frac{\text{Favorable outcomes}}{\text{Total outcomes}} = \frac{3}{4}$$

b. One of the four possible outcomes (Don't stop–Don't stop) corresponds to not having to stop at all, so the desired probability is

$$\frac{\text{Favorable outcomes}}{\text{Total outcomes}} = \frac{1}{4}$$

CALCULATION TIP 5.1 Equally Likely Outcomes

When calculating probabilities, take care that each of the outcomes considered is equally likely to occur. In cases involving identical items, such as coins or dice, it may help to label the items to distinguish them.

"How do you want it—the crystal mumbo-jumbo or statistical probability?"

Probability of non-occurrence

The calculations in Example 5.3 lead us to something important. There we found

$$P(\text{Stopping at least once}) = \frac{3}{4}$$

$$P(\text{Not stopping at all}) = \frac{1}{4}$$

That is, the probability of not stopping at all is 1 minus the probability of stopping at least once. This is no accident, as the following observations show: Recall that

$$\text{Probability of event occurring} = \frac{\text{Favorable outcomes}}{\text{Total outcomes}}$$

Then

$$\text{Probability of event not occurring} = \frac{\text{Unfavorable outcomes}}{\text{Total outcomes}}$$

If we add the favorable and unfavorable outcomes, we get all outcomes. Hence,

$$\text{Probability of event occurring} + \text{Probability of event not occurring} = \frac{\text{Total outcomes}}{\text{Total outcomes}} = 1$$

Rearranging this equation yields the formula in Summary 5.2.

SUMMARY 5.2
Probability of an Event Not Occurring

The probability that an event will not occur is 1 minus the probability that it will occur; if expressed as a percent, it is 100% minus the probability that it will occur. As an equation, this is

$$\text{Probability of event not occurring} = 1 - \text{Probability of event occurring}$$

EXAMPLE 5.4 Calculating probability of non-occurrence: English teachers

Several sections of English are offered. There are some English teachers I like and some I don't. I enroll in a section of English without knowing the teacher. A friend of mine has calculated that the probability that I get a teacher I like is

$$P(\text{Teacher I like}) = \frac{7}{17}$$

What is the probability that I will get a teacher whom I don't like?

SOLUTION

Our formula for non-occurrence applies:

$$\text{Probability of event not occurring} = 1 - \text{Probability of event occurring}$$

Hence,

$$\begin{aligned} P(\text{Teacher I don't like}) &= 1 - P(\text{Teacher I like}) \\ &= 1 - \frac{7}{17} \\ &= \frac{10}{17} \end{aligned}$$

TRY IT YOURSELF 5.4

I am expecting some emails. Some of these are bills and the rest are emails from friends. There is an email in my inbox. A friend has calculated that the probability that it is a bill is

$$P(\text{The email is a bill}) = \frac{15}{34}$$

What is the probability that it is an email from a friend?

The answer is provided at the end of this section.

Now we use the idea of distinguishing items in pairs with the formula for non-occurrence.

EXAMPLE 5.5 Distinguishing outcomes and non-occurence: Rolling dice

Suppose we toss a pair of standard six-sided dice.

a. What is the probability that a we get a 7 (i.e., the sum of the two faces is 7)?

b. What is the probability that we get any sum but 7?

SOLUTION

a. As we discussed previously, it is helpful to distinguish the dice. Let's imagine that one of them is red and the other is green. There are six possible outcomes for the red die (numbers 1 through 6). For each number on the red die, there are six possible numbers for the green die. That is a total of $6 \times 6 = 36$ possible outcomes: a 1 on the first die and a 1 on the second, a 1 on the first die and a 2 on the second, and so on.[4]

Now we need to know how many outcomes yield a 7 (what we would call a favorable outcome). We list the possibilities in a table.

Red die	Green die
1	6
2	5
3	4
4	3
5	2
6	1

Thus, there are six possible ways to get a 7. Therefore,

$$\text{Probability of a } 7 = \frac{\text{Favorable outcomes}}{\text{Total outcomes}} = \frac{6}{36} = \frac{1}{6}$$

This is about 0.17, or 17%.

b. We use Summary 5.2:

Probability of event not occurring = 1 − Probability of event occurring

We find

$$\text{Probability of not getting a } 7 = 1 - \frac{1}{6} = \frac{5}{6}$$

This is about 0.83, or 83%.

Note that if you tried to compute this probability directly, you would have to count all the ways of getting a sum different from 7, which would be very tedious.

Probability of disjunction

Our first calculation in this section concerned the probability that when we roll a six-sided die, we will get a 1 or a 6. This is an example of finding the probability that either one event or another occurs. In accordance with the terminology of formal logic (see Section 1.3), we call this the probability of the *disjunction* of the two events. It turns out that there is a convenient formula for finding this probability.

[4] *In Section 5.3, we explain in more detail how to find the total number of outcomes in such situations.*

KEY CONCEPT

Suppose A and B are two events. Their disjunction is the event that either A or B occurs. The probability of this disjunction is given by the formula

$$P(A \text{ or } B) = P(A) + P(B) - P(A \text{ and } B)$$

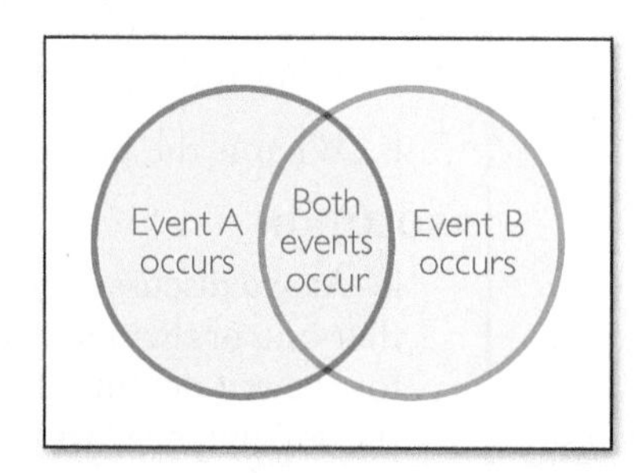

FIGURE 5.1 Adding the probabilities of events A and B counts twice the probability that both A and B occur.

The rationale for this formula comes from the fact that if you add the probability of event A and the probability of event B, you are counting the event "A and B" twice as illustrated in **Figure 5.1**. (Recall that in mathematics the disjunction allows for the possibility that both events occur.) Therefore, subtracting $P(A \text{ and } B)$ from the sum gives the correct answer.

EXAMPLE 5.6 Calculating the probability of a disjunction: A biology class

You have enrolled in a biology class and don't know yet which professor or lab instructor you will get. You like Professor Smith, but your professor will be randomly selected from a group of 10 professors. You also like Ms. Jones, who is one of five graduate assistants who will be randomly selected to guide your lab. The probability of getting both Professor Smith as your professor and Ms. Jones as your lab instructor is 1/50. What is the probability that you get at least one teacher (either professor or lab instructor) whom you like?

SOLUTION

Let A denote the event of getting Professor Smith as your professor and B the event of getting Ms. Jones as your lab instructor. We want to find $P(A \text{ or } B)$. To use the formula, we need $P(A)$, $P(B)$, and $P(A \text{ and } B)$. Because Professor Smith is one of 10 biology professors, we have

$$P(A) = \frac{1}{10}$$

Likewise, Ms. Jones is one of five potential lab instructors, so

$$P(B) = \frac{1}{5}$$

We are told that

$$P(A \text{ and } B) = \frac{1}{50}$$

Now the formula gives

$$
\begin{aligned}
P(A \text{ or } B) &= P(A) + P(B) - P(A \text{ and } B) \\
&= \frac{1}{10} + \frac{1}{5} - \frac{1}{50} \\
&= \frac{14}{50} \\
&= \frac{7}{25}
\end{aligned}
$$

Thus, the probability that you will get at least one teacher whom you like is 7/25, which is 0.28, or 28%.

TRY IT YOURSELF 5.6

You learn that two new biology professors have been hired. A friend, who is a math major, tells you that the probability of getting both Professor Smith and Ms. Jones is now 1/60. What is the new probability that you will get at least one teacher whom you like?

The answer is provided at the end of this section.

Probability with area

Let's look at a different situation in which a theoretical probability can be determined. We opened this section with an article on estimating the probability of asteroids striking Earth. Such estimates relate probability to area. To explain the connection, we examine the probability related to a man-made object striking Earth.

In 1979, Skylab, the first space station of the United States, reentered Earth's atmosphere. At the time there was widespread concern that the debris would fall in a heavily populated area, but in the end the area most affected was a sparsely populated region in western Australia. The area was roughly rectangular in shape. If pieces of Skylab fell at random over this rectangle, what is the probability that a given region would be affected? To see how to answer this question, we first consider the simpler situation in which the rectangle is divided into three regions, designated by color as follows: 1/4 of the area is red, 1/4 is blue, and 1/2 is white. See **Figure 5.2**. If a given small piece of Skylab fell at random onto this rectangle, what is the probability that it would land on the color blue?

NASA

The Skylab space laboratory.

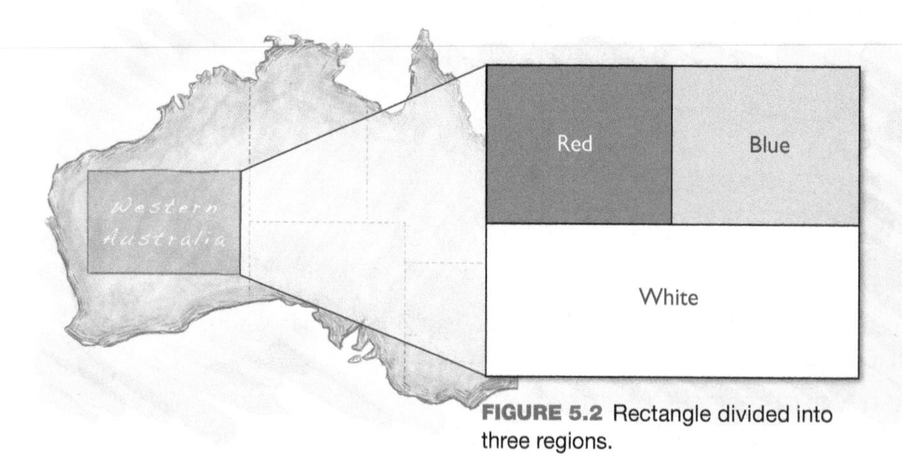

FIGURE 5.2 Rectangle divided into three regions.

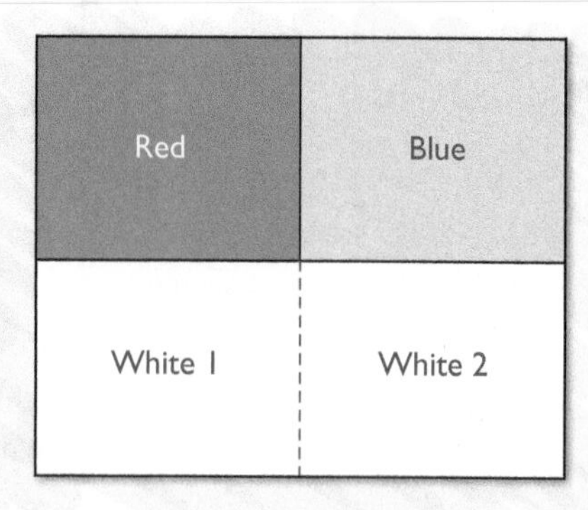

FIGURE 5.3 Cutting white into two sections.

There are three possible outcomes: The piece lands on red, blue, or white. These outcomes are *not* equally likely, however. Because a lot more of the target is white, one would expect that there is a greater likelihood of the piece landing on white than on either of the other two colors. One way of looking at this problem is to cut the white part into two sections, White 1 and White 2, as shown in **Figure 5.3**. Now because each of the four sections is the same size, it should be equally likely that the piece lands in any one of them. Because there are now four equally likely outcomes and only one of them is blue, we see that the probability of landing on blue is $1/4 = 0.25 = 25\%$. Observe that this number equals the percentage of the area that was labeled blue.

This observation holds in general. The probability of landing on a given color is the fraction of the area covered by that color. In this case, that means the probability of landing on white is $1/2$, the probability of landing on red is $1/4$, and the probability of landing on blue is $1/4$. As a practical matter, this means that an object falling to Earth at random is likely to strike an unpopulated region: Water covers about 70% of Earth's surface.

EXAMPLE 5.7 Finding probability using area: Meteor

The surface area of Earth is approximately 197 million square miles. North America covers approximately 9.37 million square miles, and South America covers approximately 6.88 million square miles. A meteor falls from the sky and strikes Earth. What is the probability that it strikes North or South America? Write your answer in decimal form rounded to three decimal places and also as a percentage.

SOLUTION

The total area covered by North and South America is $9.37 + 6.88 = 16.25$ million square miles. As a fraction of the surface area of Earth, that is $16.25/197$ or about 0.082. Hence, the probability that the meteor strikes North or South America is 0.082 or 8.2%.

TRY IT YOURSELF 5.7

Asia covers approximately 17.2 million square miles. In the situation above, what is the probability that the meteor does not strike Asia? Write your answer in decimal form rounded to three decimal places and also as a percentage.

The answer is provided at the end of this section.

Jason Persoff/Cultura/Stormdoctor/Getty Images

EXAMPLE 5.8 Applying probability: Weather in Oklahoma

When the weatherman says there is a 30% chance of rain today in Oklahoma, she is forecasting that 30% of the land mass of Oklahoma will receive rain today. Springtime in Oklahoma is a time of dangerous weather. Suppose a weatherman reports a 90% chance of severe thunderstorms in Oklahoma tomorrow. About 15% of Oklahoma's land mass is devoted to wheat production, and in the spring emerging plants can be damaged by severe weather. Assuming the weatherman's prediction is correct, is it possible that Oklahoma's burgeoning wheat crop will be totally spared?

SOLUTION

If the thunderstorms, covering 90% of the state, were to completely miss the 15% of the state growing wheat, then 105% of the land would either get rain or be planted in wheat. Because 100% of the land is all there is, this situation is impossible. We conclude that at least some of the wheat crop could be damaged.

TRY IT YOURSELF 5.8

The area of Oklahoma is 69,960 square miles. In the situation of the example, what area is expected to be subject to severe thunderstorms?

The answer is provided at the end of this section.

Defining probability empirically

Suppose I have a box containing 20 red jelly beans and 10 green ones. I select a jelly bean from the box without looking. There is a total of 30 jelly beans in the box, and 20 of them are red. Therefore, the probability of getting a red jelly bean is

$$\frac{20}{30} = \frac{2}{3}$$

or about 67%. This is a theoretical probability based on the known percentage of red jelly beans.

Now suppose we pick a jelly bean at random from the box, record its color, and toss it back into the box. If we repeat this experiment 3000 times, it is reasonable to expect that approximately 2/3 of the jelly beans chosen would be red. Thus, we would expect that about

$$3000 \times \frac{2}{3} = 2000$$

jelly beans would be red. Similarly, we would expect that approximately 1/3 of the jelly beans chosen, or 1000, would be green.

It would not be at all surprising if the numbers were something like 2005 red and 995 green, but we would definitely be surprised if the numbers were 130 red and 2870 green. In other words, we expect the actual *empirical* probabilities obtained by experiment to be fairly close to the theoretical probabilities.

This suggests that we can go backward. That is, we can do experiments to estimate the theoretical probability. Let's explore this idea.

An important feature of any manufacturing process is *quality control.* This term means just what the name implies: ensuring that high-quality products are produced. One aspect of quality control involves testing of finished products. Suppose, for example, that we make many thousands of lightbulbs each day. It is not practical to test each bulb we produce, but we can test a certain percentage of them.

Suppose that on a given day, we randomly choose 1000 bulbs for testing. We find that 64 of them are defective and 936 of them work perfectly.[5] On this basis, it is reasonable to estimate that about

$$\frac{64}{1000} = 0.064$$

[5] *These numbers are far from acceptable for any reputable producer, but we use them for illustration.*

or 6.4% of the bulbs we produce are, in fact, defective. That is, we estimate that a customer has about a 6.4% chance of purchasing a defective bulb with our name brand on it.

This is an example of calculation of an *empirical probability*. We don't know the actual probability but estimate it from a properly chosen sample.

KEY CONCEPT

The empirical probability of an event is a probability obtained by experimental evidence. It is the ratio of the number of favorable outcomes to the total number of outcomes in the experiment.

The idea seems simple enough, but the important question about an empirical probability is, how meaningful is it? For example, if I select three jelly beans from a jar and two of them are red while one is not, I surely can't deduce from this single experiment that exactly 2/3 of the jelly beans in the jar are red. However, suppose we perform this experiment numerous times and in 95% of them about 2/3 of the jelly beans we selected were red. Then, we might start believing that about 2/3 of jelly beans in the jar really are red.

So one question is, how many samples should we take before we have some confidence in our empirical probability? And is there some way to measure a level of confidence? As it turns out, these questions are at the heart of what professional polling organizations do. We will return to these questions in Section 3 of Chapter 6.

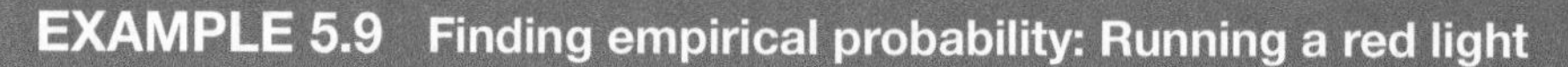

EXAMPLE 5.9 Finding empirical probability: Running a red light

Michael Krinke/iStock/Getty Images

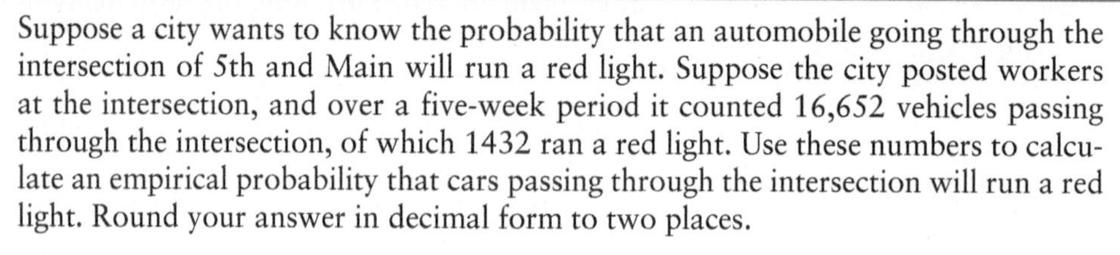

Suppose a city wants to know the probability that an automobile going through the intersection of 5th and Main will run a red light. Suppose the city posted workers at the intersection, and over a five-week period it counted 16,652 vehicles passing through the intersection, of which 1432 ran a red light. Use these numbers to calculate an empirical probability that cars passing through the intersection will run a red light. Round your answer in decimal form to two places.

SOLUTION

Now 1432 out of 16,652 ran the light. So an empirical probability is

$$\frac{1432}{16{,}652}$$

which is about 0.09, or 9%.

TRY IT YOURSELF 5.9

At another intersection over the same period, the city found that 19,221 cars passed through and that 2144 of them ran a red light. Using these numbers, calculate an empirical probability that cars passing through the intersection will run a red light. Round your answer in decimal form to two places.

The answer is provided at the end of this section.

WHAT DO YOU THINK?

Gambling: If you plan to visit a casino and participate in gambling activities, would understanding the probabilities associated with various games be helpful? If so, explain how.

Questions of this sort may invite many correct responses. We present one possibility: The first thing anyone who visits a casino needs to be aware of is that the casino knows how to calculate probabilities, and the games offered are designed so that the advantage always goes to the casino. Anyone intending to gamble in a casino should expect to lose money. Still, although the odds are always stacked in favor of the house, some games are better for the player than others. Furthermore, in many games the player has choices. Understanding probability can help you make the best choice and, at the very least, help limit your losses.

Further questions of this type may be found in the *What Do You Think?* section of exercises.

Try It Yourself answers

Try It Yourself 5.1: Calculating simple probabilities: A diamond $13/52 = 1/4$.

Try It Yourself 5.2: Distinguishing outcomes: Coins $1/4$.

Try It Yourself 5.4: Calculating probability of non-occurrence: Mailbox $19/34$.

Try It Yourself 5.6: Calculating probability of a disjunction: A biology class $16/60 = 4/15$ or about 27%.

Try It Yourself 5.7: Finding probability using area: Meteor 0.913 or 91.3%.

Try It Yourself 5.8: Applying probability: Weather in Oklahoma 62,964 square miles.

Try It Yourself 5.9: Finding empirical probability: Running a red light 0.11 or 11%.

Exercise Set 5.1

Test Your Understanding

1. We calculate the theoretical probability of an event in terms of total outcomes and favorable outcomes. The appropriate formula is ________.

2. A probability of 0 indicates an impossibility, and a probability of 1 indicates ________.

3. True or false: The disjunction of two events is the event that either occurs.

4. True or false: An event may have a negative probability if it is unlikely enough.

5. True or false: When speaking of probabilities, a "favorable event" does not necessarily mean one that is good.

6. True or false: When calculating probabilities, we should take care that each of the outcomes considered is equally likely to occur.

7. If the probability of an event occurring is p, then the probability of that event not occurring is: **a.** $1/p$ **b.** $-p$ **c.** $p-1$ **d.** $1-p$.

8. A probability may be expressed as: **a.** a decimal **b.** a percent **c.** a fraction **d.** any of these.

Problems

In your answers to these exercises, leave each probability as a fraction unless you are instructed to do otherwise.

9. Automobile accidents. According to a Fox News report,[6] the average driver can expect to have about four traffic accidents in a driving life of 50 years. According to these figures, what is the probability that an average driver will have a traffic accident this year? Report your answer as a fraction and as a percentage.

10. Boating accidents. According to the publication *Recreational Boating Statistics* from the United States Coast Guard Boating Safety Division, in 2014 there were 937 accidents reported that related to a collision with a recreational vessel. Of these, 652 resulted in injury. According to these figures, what is the probability that a collision with a recreational vessel in 2014 would result in injury? Report your answer as a fraction and as a percentage rounded to the nearest whole number.

11. Calculating simple probabilities. You draw a single card from a standard deck of 52 cards. What is the probability that it is a face card (jack, queen, or king)?

12. Distinguishing outcomes. Suppose I flip two identical coins. What is the probability that I get one head and one tail?

13. Probability of non-occurrence. You draw a single card from a standard deck of 52 cards. What is the probability that it is *not* an ace?

14. Probability of a disjunction. You have a plate of 40 cookies. Ten have chocolate chips and 15 have pecans. Of the cookies mentioned in the preceding sentence, five have both chocolate chips and pecans. You select a cookie at random. What is the probability that your cookie has chocolate chips or pecans?

15. Three ways. Express each of the following probabilities as a fraction, a decimal, and a percent:

a. 50%
b. 1/4
c. 0.05
d. 65%
e. 7/8
f. 0.37

[6] *See* www.foxbusiness.com/features/2011/06/17/heres-how-many-car-accidents-youll-have.html

16–21. Probabilities and opinions. *For Exercises 16 through 21, determine whether the statement gives a theoretical mathematical probability, an empirical mathematical probability, or just represents someone's opinion. Explain your answers.*

16. The probability of two heads on a toss of two coins is 0.25.

17. The probability of humans going to Mars is 10% at most.

18. Your chances of dating him are zero.

19. My chances of getting a call from a telemarketer during dinner time are 95%.

20. The probability of a smoker getting heart disease is 80%.

21. There's a 50–50 chance of a terrorist attack occurring in the United States sometime this year.

22. More probabilities and opinions. During the Cold War, someone from the government asked a famous mathematics professor if he could help them figure out the probability of a nuclear war between the United States and the Soviet Union. The professor said, "That's an absurd question." Can you explain why he said that? (By the way, this is a true story.)

23. Rain. Someone says that the probability of rain is always 50% because there are two outcomes—either it rains or it doesn't. Decide whether you agree or disagree with this, and explain your reasons.

24. Probability between 0 and 1. Explain why the probability of an event is always a number between 0 and 1 inclusive.

25. Zero probability. Explain why the probability of an event is 0 if and only if the event can never occur.

26. 100% probability. Explain why the probability of an event is 1 (or 100%) if and only if the event is certain to occur.

27. Red lights. A city finds empirically that an automobile going through the intersection of 5th and Main will run a red light sometime during a given day with probability 9.2%. What is the probability that an automobile will *not* run a red light at that intersection on a given day? Write your answer as a percentage.

28. Boys and girls. Alice and Bill are planning to have three children. (Assume that it is equally likely for a boy or a girl to be born.)

a. What is the probability that all three of their children will be girls?

b. What is the probability that at least one will be a girl?

c. What is the probability that all three will be of the same gender?

d. What is the probability that not all three will be of the same gender?

29. Jelly beans. Suppose there are five jelly beans in a box—two red and three green.

a. If a jelly bean is selected at random, what is the probability that it is red?

b. A friend claims that the answer to part a is 1/2 because there are two outcomes, red and green. Explain what's wrong with that argument.

30. More jelly beans. Suppose there are 15 jelly beans in a box—2 red, 3 blue, 4 white, and 6 green. A jelly bean is selected at random.

a. What is the probability that the jelly bean is red?

b. What is the probability that the jelly bean is not red?

c. What is the probability that the jelly bean is blue?

d. What is the probability that the jelly bean is red or blue?

e. What is the probability that the jelly bean is neither red nor blue?

31. Still more jelly beans. Suppose I pick a jelly bean at random from a box containing five red and seven blue ones. I record the color and put the jelly bean back in the box.

a. What is the probability of getting a red jelly bean both times if I do this twice?

b. What is the probability of getting a blue jelly bean both times if I do this twice?

c. If I do this three times, what is the probability of getting a red jelly bean each time?

d. If I do this three times, what is the probability of getting a blue jelly bean each time?

32. Tossing three coins. This exercise refers to the experiment of tossing three coins.

a. What is the total possible number of outcomes?

b. What is the probability of getting three heads?

c. What is the probability of getting one head and two tails?

d. What is the probability of getting anything but one head and two tails?

e. What is the probability of getting at least one tail?

33. Tossing three dice. You toss three dice and add up the faces that show. What is the probability that the sum will be 3, 4, or 5?

34. Passing. The probability of passing the math class of Professor Jones is 62%, the probability of passing Professor Smith's physics class 34%, and the probability of passing both is 30%. What is the probability of passing one or the other?

35. The team. The probability of making the basketball team is 32%, the probability of making the softball team is 34%, and the probability of making one or the other is 50%. What is the probability of making both teams?

36. A box. Suppose the inside bottom of a box is painted with three colors: 1/3 of the bottom area is red, 1/6 is blue, and 1/2 is white. You toss a tiny pebble into the box without aiming.

a. What is the probability that the pebble lands on the color white?

b. What is the probability that the pebble does not land on the color blue?

37. Heart disease. In one study, researchers at the Cleveland Clinic looked at more than 120,000 patients enrolled in 14 international studies in the past 10 years. Among the patients with heart problems, researchers found at least one risk factor in 84.6% of the women and 80.6% of the men.

If a woman is selected at random from the women with heart problems, what is the probability she has none of the risk factors? If a man is selected at random from the men with heart problems, what is the probability he has none of the risk factors? Write both of your answers as percentages.

38–40. Health. *Exercises 38 through 40 refer to the accompanying table. The percentages are probabilities of 15-year survival without coronary heart disease, stroke, or diabetes in men aged 50 years with selected combinations of risk factors. Here "BMI" stands for* body mass index, *an index of weight adjusted for height. The World Health Organization gives the following classifications: "normal" 18.5 to 24.9; "grade 1 overweight" 25.0 to 29.9; "grade 2 overweight" 30.0 to 39.9; "grade 3 overweight" 40.0 or higher.*

	BMI 20–24 normal weight		BMI 25–30 somewhat overweight		BMI 30+ very overweight	
	Active	Inactive	Active	Inactive	Active	Inactive
Never smoked	89%	83%	85%	77%	78%	67%
Former smokers	86%	78%	81%	71%	72%	59%
Current smokers	77%	67%	70%	56%	58%	42%

38. What is the probability that a 50-year-old man will not survive to age 65 without coronary heart disease, stroke, or diabetes if he smokes, is somewhat overweight, but is active? Write your answer as a percentage.

39. A 50-year-old man who has never smoked, has normal weight, and is active has a much higher probability of being healthy than a man who smokes, is very overweight, and is inactive. Is the probability twice as high, less than twice as high, or more than twice as high?

40. A man who used to smoke and who is very overweight and inactive is considering either starting an exercise program or dieting to lose enough weight to be somewhat overweight—but not both. According to the table, which choice would be better?

41. Skiing. Records kept by the National Ski Patrol and the National Safety Council show that, over a period of 14 years, skiers made 52.25 million visits to the slopes annually, with an average of 34 deaths each year. Assuming that these figures are current, what is the probability that a skier will die in a skiing accident in a given year? Is your answer empirical or theoretical?

42. Weather. If the forecast area is 5000 square miles and the forecast is for a 30% chance of rain, how much area should receive rain?

43. Socks. You have four socks in your drawer, two blue and two brown. You get up early in the morning while it's dark, reach into your drawer, and grab two socks without looking. What is the probability that the socks are the same color? (*Hint*: If you take one sock first, what's the probability the second sock matches it?) If instead there are 10 blue and 10 brown socks, what is the probability that the socks you choose are the same color?

44–48. Odds. *Use the following information for Exercises 44 through 48. Another way of expressing probabilities is by the use of* odds. *Suppose I roll a single die and call getting a 4 a favorable outcome. The probability of this happening is of course 1/6, which means that we expect to see a 4 once out of every six rolls. But this can also be expressed in terms of odds.*

Here's how it works: There are six possible outcomes; one of these is favorable (getting a 4) and five are unfavorable. One then says that the odds against *getting a 4 are 5 to 1. This could also be expressed by saying the* odds in favor *of getting a 4 are 1 to 5.*

In general, if we list all the possible equally likely outcomes, the odds in favor *of a favorable outcome are expressed as the number of favorable outcomes to the number of unfavorable outcomes. To find the* odds against *it, we reverse the order of these two numbers.*

44. If we draw a card at random from a standard deck, the probability of getting an ace is 1/13. What are the odds in favor of drawing an ace? What are the odds against drawing an ace?

45. If we flip a pair of fair coins, what is the probability of getting one head and one tail? What are the odds in favor of getting one head and one tail?

46. Alaska comprises about 1/5 of the total surface area of the United States. If a meteor lands at some random point in the United States, what is the probability that it will land in Alaska? What are the odds in favor of it landing in Alaska?

47. The weatherman says that there is a 30% chance of rain in your area. What are the odds against your getting rain?

48. If an event has a probability of 10%, what are the odds in favor of it, and what are the odds against it?

49–51. More odds. *Use the following information for Exercises 49 through 51. We can calculate odds if we know the probability. But we can also go the other way. Suppose the odds in favor of an event are 3 to 2. We can interpret this as saying that there are three (equally likely) favorable outcomes and two unfavorable ones. That is a total of five outcomes. So the probability of the event occurring is 3/5. In general, if the odds in favor are p to q, then the probability is $p/(p + q)$.*

49. Suppose the odds in favor of an event are 5 to 2. What is the probability that the event will occur? What is the probability that the event will *not* occur?

50. Suppose the odds against an event are 2 to 1. What is the probability that the event will occur? What is the probability that the event will *not* occur?

51. Odds in a horse race are, in fact, a reflection of money bet on the various horses rather than a statement of probability. But suppose for the moment that the money bet does indeed indicate the true mathematical odds. If Whirlaway is a 6 to 5 favorite to win, what is the probability that Whirlaway actually wins the race?

Exercises 52 through 56 involve hands-on experiments and are suitable for group work.

52. An experiment. Try an experiment like this one: Fill a box with 100 objects indistinguishable by feel, such as slips of paper. Mark 30 of them with an "X" and leave the other 70 unmarked. Now, draw a piece of paper from the box (without looking) and record whether it had an "X" or not. Return it to the box and shake up the box. Repeat this experiment 50 times.

Compare the theoretical probability of drawing a slip with an "X" to the empirical probability from your 50 trials. Discuss what you expected to happen versus what you observed. Try it again with 100 trials.

53. Another experiment. Do a survey of a number of students to find out what they are majoring in, and determine an empirical probability that a given student is a psychology major. Find out from your university administration the percentage of students who are psychology majors. Compare your empirical probability with the actual percentage of psychology majors. Discuss factors that may affect the results.

54. The Monty Hall problem. Monty Hall was the host of the popular game show *Let's Make a Deal*. At the end of each show, some lucky contestant was given the opportunity to choose one of three doors. Behind one door was a nice prize, and behind the other two were lesser prizes or even worthless items. The contestant had no information about which door might hide the prize, so his or her probability of getting the nice prize was one-third.

An interesting variant of this scenario gained notoriety when columnist Marilyn vos Savant posed a related question in her *Parade Magazine* article of September 9, 1990. Here is the question as stated by Mueser and Granberg in 1999.

> A thoroughly honest game-show host has placed a car behind one of three doors. There is a goat behind each of the other two doors. You have no prior knowledge that allows you to distinguish among the doors. "First you point toward a door," the host says, "then I'll open one of the other doors to reveal a goat. After I've shown you the goat, you may decide whether to stick with your initial choice of doors, or switch to the other remaining door. You win whatever is behind the door [you choose]."[7]

The question is, should you switch doors or stay with your original choice? Or does it matter? The fun thing about this problem is that a number of PhD mathematicians got the wrong answer!

We will present two analyses below. One is right and the other is wrong. Which is right? Explain why you chose the answer you did. If you can't decide which is right and which is wrong, you should try the experiment suggested in the next exercise.

Analysis number 1: It doesn't matter whether you switch. The probability that I guessed right to begin with is 1/3. Either of the two remaining doors is equally likely to be a winning door or a losing door. One of them is certain to be a loser, and by telling me that one of the doors is a loser, the host has given me no new information on which to base my decision. Therefore, it makes no difference whether I change or not. In either case, my probability of winning is 1/3.

Analysis number 2: It does matter. You should switch doors. If I have initially chosen the right door, then changing will cause me to lose. This happens one time out of three.

Suppose I have chosen one of the losing doors and the host shows me the other losing door. Then if I change, I win. This happens two times out of three.

So the strategy of changing allows me to win two times out of three. My probability of winning is increased from 1/3 to 2/3.

55. Experimenting with the Monty Hall problem. *This exercise is a continuation of Exercise 54.* With the help of one other person, it is not difficult to simulate the Monty Hall problem described earlier. Let one person act as host and hide pictures of one car and two goats under sheets of paper numbered 1 through 3. You be the contestant. Make your guesses and record your wins and losses if you never change and if you always change. If you repeat the game 100 times, you should have fairly good empirical estimates of the probability of winning using either strategy. Do the results make you change your mind about your answer to Exercise 54?

56. An experiment–Buffon's needle problem. The French naturalist Buffon posed the following problem—which he subsequently solved, and which bears his name today. Suppose a floor is marked by parallel lines, each a distance D apart. (Think of a hardwood floor.) Suppose we toss a needle of length L to the floor. (We assume that the length L of the needle is smaller than the distance D between the parallel lines on the floor.) What is the probability that the needle will land touching one of the parallel lines? It turns out that the correct answer is

$$\text{Probability} = \frac{2L}{\pi D}$$

One may be surprised that the number π turns up in the answer, but this fact leads us to an interesting idea that can (at least theoretically) give a novel method of approximating the decimal value of π. Get on a nice, level hardwood floor and toss a needle onto the floor 100 times. Record the number n of times the needle touches one of the parallel lines. Then $n/100$ is an empirical estimate of the probability that the needle will touch a line. This gives

$$\frac{n}{100} \approx \frac{2L}{\pi D}$$

(Here, we use the symbol $\approx$ to indicate approximate equality.) Rearranging gives

$$\pi \approx \frac{200L}{nD}$$

Perform this experiment to get an estimate for the value of π. (Do not be disappointed if your answer is not very close to the correct value. The method is correct, but in practice, it takes a lot of tosses to get anywhere close to π.)

Exercises 57 and 58 are designed to be solved using technology such as calculators or computer spreadsheets.

57. Random numbers. Simulate flipping a coin in the following way: Use a random-number generator to obtain a random sequence of 0's and 1's, with, say, 100 numbers altogether. Count how many 0's there are altogether. Is the result close to 50%? Is it exactly 50%? Does the result give a theoretical or an empirical probability? *Note*: Computer programs do not actually produce random numbers, in spite of the name, but they are close enough for our purposes.

58. More random numbers. Think of the real numbers between 0 and 10 as divided into two colors: red for those at least 0 and at most 6, blue for those greater than 6 and at most 10. Use a random-number generator to obtain a random sequence of real numbers between 0 and 10, with, say, 50 numbers altogether. (Be sure to get a random sequence of real numbers, not just a sequence of whole numbers.) Count how many "red" numbers there are altogether. What percentage of red numbers would you expect to get? Is the result of the experiment close to this?

Writing About Mathematics

59. More on the weather. In a news article, meteorologist Jeff Haby suggests that percentages in weather forecasts may provide a general idea of how likely precipitation is, but percentages don't provide enough information. He proposes alternative descriptions such as *scattered* to replace a 40% to 60% chance and *widely scattered* to replace a 20% to 30% chance. He also suggests descriptions of the nature of the precipitation (for example, severe or non-severe).

One of the weathermen on Channel 2 news in Reno, Nevada, says that he does not use percentages when forecasting precipitation. In view of the article summarized above, explain why you think the weatherman on Channel 2 would say this.

60. Terrorism. The accompanying excerpt is from an article at the Web site of the *Daily Mail*. Based on it, would you say that real probabilities are being used or just educated opinions? Explain your reasoning.

IN THE NEWS 5.3

Home | TV & Video | U. S. | World | Politics | Entertainment | Tech | Travel

'FIFTY PERCENT Chance of Another 9/11-Style Attack' in the Next Ten Years. . . but Likelihood Is MUCH Higher if the World Gets Less Stable

DAILY MAIL REPORTER September 5, 2012

There's a 50/50 chance of another catastrophic 9/11-style attack in the next ten years, and an even greater chance if the world becomes less stable.

The startling figure was floated by a pair of researchers who examined more than 13,000 lethal terrorist attacks between 1968 and 2007.

They calculated the likelihood based on the assumption that the frequency of major attacks, like earthquakes and other natural disasters, [can be modeled] using [a] mathematical power law.

61. History. Pierre de Fermat and Blaise Pascal were French mathematicians and philosophers who lived in the seventeenth century. They are credited with some of the early development of probability theory. Pascal was an interesting figure who was also very much involved with religion. Write a paper about Pascal's life and beliefs. (Exercise 51 in Section 5.3 describes a specific question about probability that attracted the interest of Pascal, and Exercise 58 in Section 5.5 is concerned with his use of a concept from probability in a religious context.)

62. History. The formulation of the modern abstract theory of probability is usually attributed to the Russian mathematician A. N. Kolmogorov (1903–1987). He lived during the reign of the brutal dictator Joseph Stalin and once wrote a paper contradicting the academician Lysenko, who was popular with Stalin. This was a very brave act indeed. Write a paper about the life and work of Kolmogorov.

63. Comte de Buffon. The Buffon needle experiment was mentioned in Exercise 56. Write a report about the man for whom it is named: Georges-Louis Leclerc, Comte de Buffon.

What Do You Think?

64. Traffic accidents. The city council has stated that the probability of a car having an accident at a particularly dangerous intersection is 0.04%. What additional information might you need to determine what kind of probability this is: empirical, theoretical, or someone's opinion? **Exercise 16 is relevant to this topic.**

65. UFO? People sometimes attach a probability to an event that has never occurred. An example might be, "There's a 20% chance that a UFO will land in the next year." Can you say what kind of probability this is: empirical, theoretical, or someone's opinion? Explain your answer. Find additional examples of probabilities of this same type. **Exercise 17 is relevant to this topic.**

66. The euro. A news article from June 2012 begins, "There is a one-in-three chance Greece will leave the euro currency union in the months ahead, according to Standard & Poor's."[8] What kind of probability is this: empirical, theoretical, or someone's opinion? Explain your answer. Find additional examples of probabilities of this same type. **Exercise 19 is relevant to this topic.**

67. Large percentage. A gloomy friend says that there is a 110% chance it will rain on his wedding day. Does this statement make sense? What do you think he means? **Exercise 26 is relevant to this topic.**

68. Making the team. At a certain school the probability of making the soccer team is 50%, and the probability of making the basketball team is 50%. A student concludes that it's a sure bet to make one of those two teams because 50% + 50% = 100%. Do you agree with this reasoning? Explain your answer. **Exercise 35 is relevant to this topic.**

69. **Horses.** On my farm there are 10 brown horses and 8 mares. Can you conclude that there are 18 horses on my farm? Explain. How does this example relate to calculating the probability of a disjunction? **Exercise 34 is relevant to this topic.**

70. **Medicine.** Is probability useful in medicine? Provide examples to support your answer. **Exercise 38 is relevant to this topic.**

71. **The weather.** The morning weather forecast is for a 70% chance of rain today. You pack your umbrella, but it does not rain. Was the forecast wrong? Explain what was meant by the rain prediction. **Exercise 42 is relevant to this topic.**

72. **Where to live.** If you choose to live in California, there is a certain probability that you will experience an earthquake. In Hawaii, you may see a tidal wave. In Oklahoma, there is a significant probability that you will witness a tornado. Have such probabilities affected your chosen living area in the past? Do you anticipate that they will in the future?

73. **Numbers.** The study of probability depends on the use of numbers to express the likelihood of an event. What is the value of assigning numbers in this way? Why can't we just say that a certain event is either likely or not likely?

5.2 Medical testing and conditional probability: Ill or not?

TAKE AWAY FROM THIS SECTION
Understand the probabilistic nature of medical test results.

The following excerpt is from an article that appeared at Tampa Bay Online.

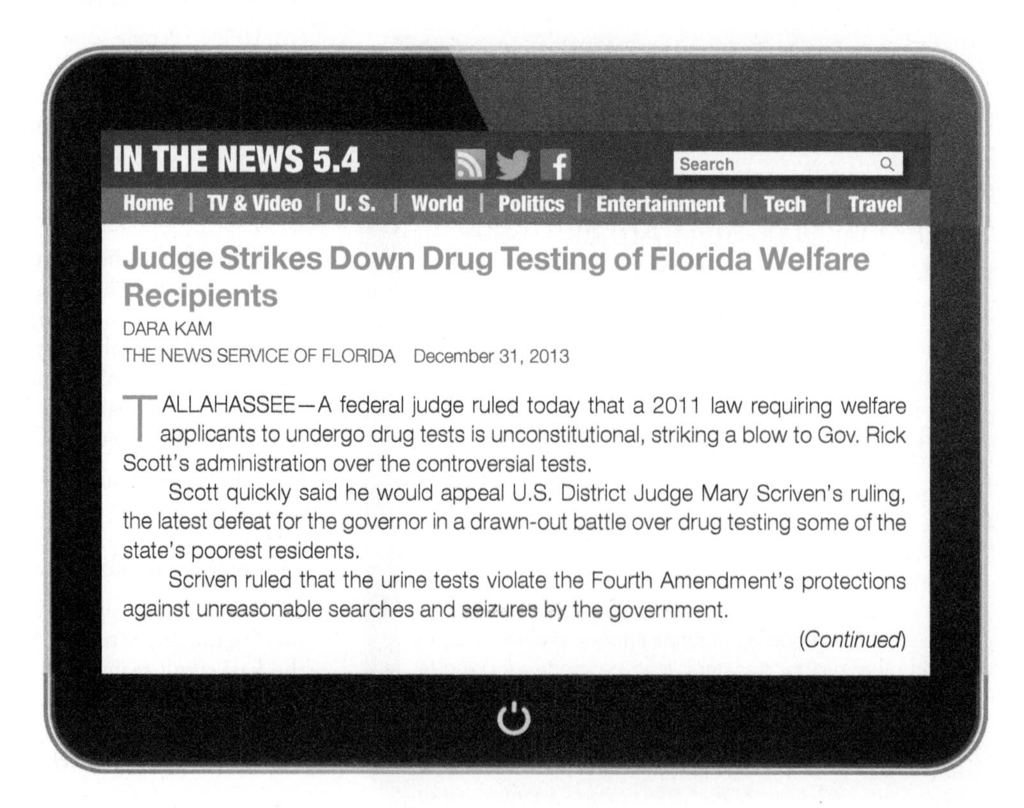

Judge Strikes Down Drug Testing of Florida Welfare Recipients

DARA KAM
THE NEWS SERVICE OF FLORIDA December 31, 2013

TALLAHASSEE—A federal judge ruled today that a 2011 law requiring welfare applicants to undergo drug tests is unconstitutional, striking a blow to Gov. Rick Scott's administration over the controversial tests.

Scott quickly said he would appeal U.S. District Judge Mary Scriven's ruling, the latest defeat for the governor in a drawn-out battle over drug testing some of the state's poorest residents.

Scriven ruled that the urine tests violate the Fourth Amendment's protections against unreasonable searches and seizures by the government.

(Continued)

[8] *Ben Rooney,* S&P: 1-in-3 chance Greece leaves euro. *CNNMoneyInvest, June 4, 2012.*

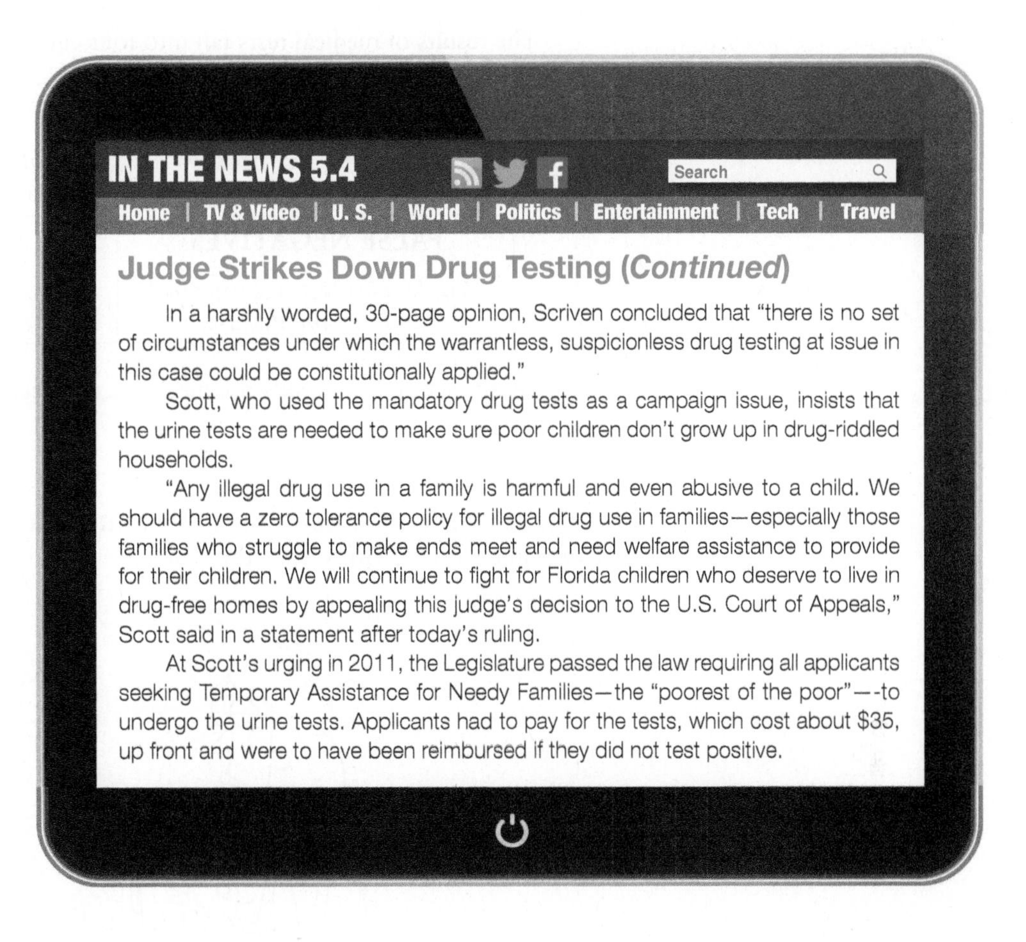

IN THE NEWS 5.4

Home | TV & Video | U.S. | World | Politics | Entertainment | Tech | Travel

Judge Strikes Down Drug Testing (*Continued*)

In a harshly worded, 30-page opinion, Scriven concluded that "there is no set of circumstances under which the warrantless, suspicionless drug testing at issue in this case could be constitutionally applied."

Scott, who used the mandatory drug tests as a campaign issue, insists that the urine tests are needed to make sure poor children don't grow up in drug-riddled households.

"Any illegal drug use in a family is harmful and even abusive to a child. We should have a zero tolerance policy for illegal drug use in families—especially those families who struggle to make ends meet and need welfare assistance to provide for their children. We will continue to fight for Florida children who deserve to live in drug-free homes by appealing this judge's decision to the U.S. Court of Appeals," Scott said in a statement after today's ruling.

At Scott's urging in 2011, the Legislature passed the law requiring all applicants seeking Temporary Assistance for Needy Families—the "poorest of the poor"--to undergo the urine tests. Applicants had to pay for the tests, which cost about $35, up front and were to have been reimbursed if they did not test positive.

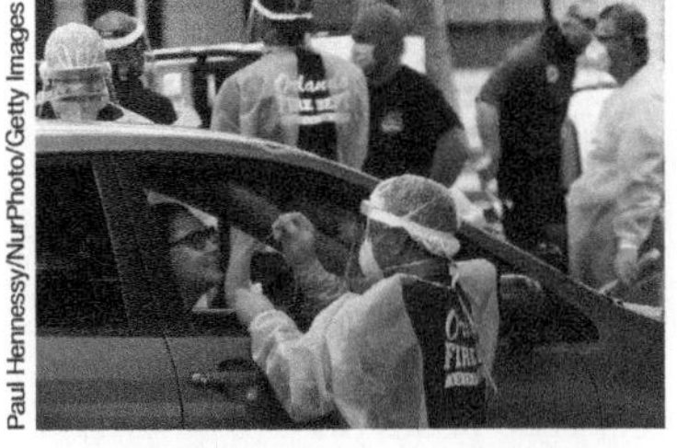

Paul Hennessy/NurPhoto/Getty Images

A nurse practitioner administers a COVID-19 swab test at a drive-thru testing site.

Whether in the context of welfare applicants being tested for drug use, Olympic athletes being tested for use of banned substances, or people being tested for COVID-19, medical tests are pervasive in modern society. Molecular tests for the coronavirus were developed early in 2020, but antibody tests and antigen tests came later. Even though public health experts called for widespread testing to slow the spread of COVID-19, production and distribution of tests was difficult. Some may be surprised to learn that medical tests are not as definitive as we might hope. In fact, they typically yield only probabilistic results. A positive result of a medical test for a disease does not tell us for certain that the patient actually has the disease—rather, it gives a probability that the disease is present. Similarly, a negative result does not guarantee that the disease is not present.

What does it really mean to say that medical tests provide probabilistic results? A patient who has tested positive for a disease would certainly like to know what the test results really mean. That is the issue that we examine in this section.

Sensitivity and specificity

Accuracy is a crucial issue for medical tests. A medical test that gives unreliable results is useless in terms of medical practice and may unduly alarm patients. There are many ways of measuring accuracy, but certainly, an accurate test should detect a disease when it is present and give negative results when the disease is not present.[9]

[9] *Although we refer to diseases, our discussion also applies to tests for banned substances, in which case having the disease means use of the substance.*

The results of medical tests fall into four categories:

TRUE POSITIVES *People who test positive and have the disease*	FALSE POSITIVES *People who test positive but do not have the disease*
FALSE NEGATIVES *People who test negative but do have the disease*	**TRUE NEGATIVES** *People who test negative and do not have the disease*

Sensitivity and *specificity* are measures of a test's accuracy in these terms.

KEY CONCEPT

The sensitivity of a test is the probability that the test will detect the disease in a person who does have the disease.

"I have a positive attitude... but it might be a false positive."

We calculate the sensitivity using the formula

$$\text{Sensitivity} = \frac{\text{True positives}}{\text{All who have the disease}} = \frac{\text{True positives}}{\text{True positives} + \text{False negatives}}$$

The sensitivity measures a test's ability to detect correctly the presence of a disease. A test with high sensitivity has relatively few false negatives. If you actually have a disease, a high-sensitivity test is very likely to detect it.

KEY CONCEPT

The specificity of a test is the probability that the test will give a negative result for a person who does not have the disease.

We calculate the specificity using the formula

$$\text{Specificity} = \frac{\text{True negatives}}{\text{All who do not have the disease}} = \frac{\text{True negatives}}{\text{True negatives} + \text{False positives}}$$

The specificity measures a test's ability to identify correctly the absence of a disease. A test with high specificity has relatively few false positives. If you are healthy, a high-specificity test is very likely to confirm that fact.

We normally write the sensitivity and specificity in percentage form and round the percentages to one decimal place when we calculate.

EXAMPLE 5.10 Estimating sensitivity and specificity: Test for COVID-19

The accompanying table of data is adapted from a study of a test for the coronavirus (COVID-19).

	Has COVID-19	Does not have COVID-19
Test positive	5510	26
Test negative	786	6505

Use these data to estimate the sensitivity and specificity of this test.

SOLUTION

A total of $5510 + 786 = 6296$ people in this group have COVID-19. Of these, 5510 test positive, so there are 5510 true positives. This indicates a sensitivity of

$$\text{Sensitivity} = \frac{\text{True positives}}{\text{All who have the disease}} = \frac{5510}{6296}$$

This is about 0.875 or 87.5%. Hence, about 87.5% of the people who have COVID-19 will test positive, so about 12.5% of the people who have COVID-19 will test negative.

A total of $26 + 6505 = 6531$ do not have COVID-19. Among these 6505 test negative, so there are 6505 true negatives. This indicates a specificity of

$$\text{Specificity} = \frac{\text{True negatives}}{\text{All who do not have the disease}} = \frac{6505}{6531}$$

This is about 0.996 or 99.6%. Thus, about 99.6% of the people who do not have COVID-19 will test negative, so about 0.4% of the people who do not have COVID-19 will test positive.

TRY IT YOURSELF 5.10

A new medical test is given to a population whose disease status is known. The results appear in the table below.

	Has disease	Does not have disease
Test positive	165	55
Test negative	29	355

Use these data to estimate the sensitivity and specificity of this test.

The answer is provided at the end of this section.

We remark that the type of table used in Example 5.10 is called a *contingency table*.

It is important to note that sensitivity and specificity depend only on the chemical and physical properties of a test and not on the population being tested. Hence, the preceding calculations give empirical estimates for theoretical probabilities. To explain this point, we consider lightbulbs of a certain brand. There is a theoretical probability that lightbulbs of this brand will burn out in the first month of use. To estimate this probability, we might observe the lights in a large building for a month to see how many burn out. That is, we can estimate the theoretical probability by calculating an empirical probability.

The next example shows how we can recover information by using the sensitivity and specificity.

EXAMPLE 5.11 Using sensitivity and specificity: Sensitivity of 93%

A certain test has a sensitivity of 93%. If 450 people who have the disease take this test, how many test positive?

SOLUTION

A sensitivity of 93% means that 93% of those who have the disease will test positive. Now 93% of 450 is about 419 people. Hence, we expect that about 419 of the 450 who have the disease will test positive.

TRY IT YOURSELF 5.11

The specificity of a test is 87%. If 300 people who do not have the disease take the test, how many will test negative?

The answer is provided at the end of this section.

The calculation of sensitivity and specificity for a medical test requires that we know whether a positive result (for example) represents a true positive or a false positive. This means that we need to be able to determine independently who in the test group is healthy and who is not. For this determination, we need an independent test that can be presumed to be accurate. Such a test is called a *gold standard*. A gold standard is very often expensive or difficult to administer. For example, in testing for HIV (a virus that causes AIDS), the gold standard would include a careful clinical examination. Researchers have developed less expensive tests for HIV, and the sensitivity and specificity provide a way to measure their accuracy in comparison with the gold standard.

Positive predictive value

A common misconception is that if one tests positive, and the test has a high sensitivity and specificity, then one should conclude that the disease is almost certainly present. But a more careful analysis shows that the situation is not so simple, and that *prevalence* plays a key role here.

KEY CONCEPT

The prevalence of a disease in a given population is the percentage of the population that has the disease.

For example, 5.1% of U.S. females aged 14 to 19 years have HPV, the most common sexually transmitted infection. That means the prevalence of HPV among women in the United States aged 14 to 19 is 5.1%. (The prevalence was much higher before the HPV vaccination was added to the routine immunization schedule.) For comparison, the prevalence of epilepsy in the United States is less than 1%. The prevalence of a disease depends on the population under consideration. For example, the prevalence of HIV in Botswana in 2008 was almost 24%, but in the United States it was less than 1%.

A patient who has tested positive for a disease wants to know whether the disease is really present or not. That is, he or she wants to know the probability that the disease is actually present, given that test results were positive. But this probability depends on the prevalence of the disease in the population being tested. A positive result for an HIV test administered among the residents of Botswana, where prevalence is high, is more likely to be accurate than is a positive result for the same test administered to residents of the United States, where prevalence is low.

We capture this notion of accuracy using the following terms: positive predictive value and negative predictive value.

KEY CONCEPT

For a population with a known prevalence, the positive predictive value (PPV) of a test is the probability that a person in the population who tests positive actually has the disease.

We calculate the positive predictive value for a given population using the formula

$$\text{PPV} = \frac{\text{True positives}}{\text{All positives}} = \frac{\text{True positives}}{\text{True positives} + \text{False positives}}$$

The *negative predictive value* is defined in a similar fashion.

KEY CONCEPT

For a population with a known prevalence, the negative predictive value (NPV) of a test is the probability that a person in the population who tests negative, in fact, does not have the disease.

We calculate the negative predictive value for a given population using the formula

$$\text{NPV} = \frac{\text{True negatives}}{\text{All negatives}} = \frac{\text{True negatives}}{\text{True negatives} + \text{False negatives}}$$

In our calculations, we normally round the PPV and NPV to one decimal place in percentage form.

EXAMPLE 5.12 Calculating PPV and NPV: Hepatitis C

GARO/Phanie/Alamy

Hepatitis C rapid antibody test

Injection drug use is among the most important risk factors for hepatitis C. Two studies were done using a hepatitis C test with 90% sensitivity and specificity. One test involved a group that was representative of the overall U.S. population, which has a chronic hepatitis C prevalence of about 1%. The other test was on a population of older and former injection drug users where the prevalence is 80%. The results are shown in the table below.

	General population		Older and former injection drug users	
	Has hep C	Does not have hep C	Has hep C	Does not have hep C
Test positive	9	99	720	20
Test negative	1	891	80	180

Calculate the PPV for each of the populations in the study.

SOLUTION

For the general population, there are 9 true positives and 99 false positives, for a total of $9 + 99 = 108$ positive results. Therefore,

$$\begin{aligned}\text{PPV} &= \frac{\text{True positives}}{\text{All positives}}\\ &= \frac{9}{108}\end{aligned}$$

This is about 0.083, so only 8.3% of those who test positive from the general population actually have hepatitis C. Applying this test to the general population would be of little value.

For the population of older and former injection drug users, there are 720 true positives and 20 false positives, for a total of $720 + 20 = 740$ positive results. Therefore,

$$\text{PPV} = \frac{\text{True positives}}{\text{All positives}} = \frac{720}{740}$$

This is about 0.973, so 97.3% of those who test positive from the population of older and former injection drug users do, in fact, have hepatitis C.

TRY IT YOURSELF 5.12

Calculate the NPV for each of the populations in this study.

The answer is provided at the end of this section.

The next example is concerned with Crohn's disease, which is an incurable inflammatory disease of the intestines.

EXAMPLE 5.13 Using prevalence and calculating PPV: Crohn's disease

Crohn's is relatively rare in the United States, where the prevalence is about 0.18%. The population of the United States is about 330 million.

a. Approximately how many people in the United States suffer from Crohn's?

b. There is a screening test for Crohn's disease that has a sensitivity of 80% and a specificity of 90.2%. Use this information to complete the following table, which shows the results we would expect if everyone in the United States were tested for Crohn's. (Report your answers in millions and round to two decimal places.)

	Has disease	Does not have disease
Test positive (millions)		
Test negative (millions)		

c. Use the information from part b to calculate the PPV of the Crohn's test for the general U.S. population.

d. There are about 3.4 million people in the United States who show symptoms of Crohn's, in the sense that their medical profile suggests the possibility of the disease. In Exercise 19, you will show that if the screening test is applied to this group, it has a PPV of about 62.7%. What conclusions can be drawn from this information along with part c?

SOLUTION

a. Because the prevalence is 0.18%, to find the number of Americans suffering from Crohn's, we multiply the total population of 330 million by 0.0018:

$$\text{Number with Crohn's} = 0.0018 \times 330$$

Thus, about 0.59 million people in the United States suffer from Crohn's.

b. Now 0.59 million people in the United States have Crohn's disease. Because the sensitivity of the test is 80%, to calculate the number of true positives, we multiply the number who have Crohn's by 0.8:

$$\text{True positives} = 0.8 \times 0.59$$

or about 0.47 million. This is the number of people who have Crohn's and test positive, so $0.59 - 0.47 = 0.12$ million Crohn's sufferers will test negative. This is the number of false negatives.

Because 0.59 million have Crohn's disease, $330 - 0.59 = 329.41$ million do not. Because the specificity of the test is 90.2%, to find the number of true negatives, we multiply 329.41 million by 0.902:

$$\text{True negatives} = 329.41 \times 0.902$$

or about 297.13 million. This is the number of people who do not have Crohn's and test negative, so $329.41 - 297.13 = 32.28$ million disease-free individuals will test positive. This is the number of false positives. The completed table is shown below.

	Has disease	Does not have disease
Test positive (millions)	0.47	32.28
Test negative (millions)	0.12	297.13

c. The total number of positive results is $0.47 + 32.28 = 32.75$ million. Because 0.47 million of these are true positives, we calculate the PPV as follows:

$$\begin{aligned} \text{PPV} &= \frac{\text{True positives}}{\text{All positives}} \\ &= \frac{0.47}{32.75} \end{aligned}$$

This is about 0.014, so the PPV for this population is 1.4%. That is, if everyone in the United States were tested for Crohn's, only 1.4% of people who test positive would actually have the disease.

d. We found in part c that if everyone in the United States were tested for Crohn's, only 1.4% of people who test positive would actually have the disease. If, on the other hand, the test is applied to the population of individuals who exhibit symptoms of the disease, then it has a PPV of 62.7%. The test is of little or no value when it is applied to the general population, but it can be an effective diagnostic tool for people who show symptoms.

TRY IT YOURSELF 5.13

Calculate the NPV for the screening test applied to the general population. Round your answer as a percentage to two decimal places.

The answer is provided at the end of this section.

Examples 5.12 and 5.13 illustrate a very important feature of medical testing. The hepatitis C test and the Crohn's test are typical of many tests that are virtually useless when applied to a population with a very low prevalence. Such tests can actually be harmful when false positives lead to inappropriate treatment of healthy individuals. For example, administering drug tests to everyone in a large high school is likely to single out many innocent people for discipline or even legal action.

This fact by no means indicates that medical testing has no value. In Examples 5.12 and 5.13, we saw that the tests could serve as a useful diagnostic tool when appropriately applied. Physicians are aware of this fact and will not administer a Crohn's test, for example, to an apparently healthy individual. On the other hand, the Crohn's test can be important for dealing with a patient complaining of Crohn's symptoms. Others, including those who may be in a position to administer certain types of tests, may not be aware of their true value.

SUMMARY 5.3
Prevalence, PPV, and NPV

1. The prevalence of a disease in a given population is the percentage of the population having the disease.

2. Among a population of individuals, the positive predictive value (PPV) of a test is the probability that a person in the population who tests positive actually has the disease. The formula is

$$\text{PPV} = \frac{\text{True positives}}{\text{All positives}} = \frac{\text{True positives}}{\text{True positives} + \text{False positives}}$$

3. Among a population of individuals, the negative predictive value (NPV) of a test is the probability that a person who tests negative actually does not have the disease. It is calculated using

$$\text{NPV} = \frac{\text{True negatives}}{\text{All negatives}} = \frac{\text{True negatives}}{\text{True negatives} + \text{False negatives}}$$

4. The NPV and PPV depend on the prevalence of the disease in the population.

Conditional probability

Positive and negative predictive values (as well as the sensitivity and specificity) are examples of *conditional probabilities.*

KEY CONCEPT

A conditional probability is the probability that one event occurs given that another has occurred.

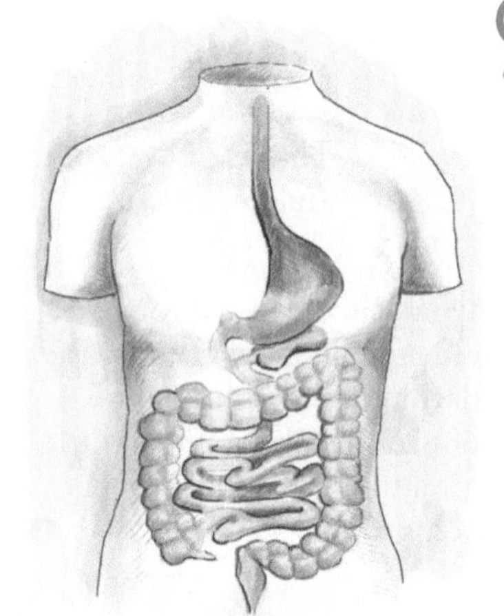

For example, suppose a card is drawn from a standard deck and laid face down on a table. What is the probability that it is a heart? There are four equally likely suits: clubs, diamonds, hearts, and spades. So the probability that the card is a heart is 1/4. Now let's change things just a little. Suppose that when the card was chosen we took a peek and noticed that it was a red card. Now, under these conditions, what is the probability that the card is a heart? What we're asking is the probability the card is a heart *given that* it is a red card. There are two equally likely red suits, hearts and diamonds. So with the additional information that the card is red, the probability that it is a heart is 1/2.

In the next example, a conditional probability is calculated in the context of medical testing.

EXAMPLE 5.14 Relating conditional probability and medical testing: TB

The accompanying table of data is adapted from a study of a test for TB among patients diagnosed with extrapulmonary TB (i.e., TB infection outside of the lungs).

	Has TB	Does not have TB
Test positive	446	15
Test negative	216	323

Calculate the conditional probability that a person tests positive given that the person has TB.

SOLUTION

There are 446 + 216 = 662 people who have TB, and 446 of this group test positive. Hence, the probability that a person tests positive given that the person has TB is

$$P(\text{Positive test given TB is present}) = \frac{\text{True positives}}{\text{All who have TB}} = \frac{446}{662}$$

This is about 0.674 or 67.4%.

TRY IT YOURSELF 5.14

Use the preceding table to find the probability that a person tests negative given that the person does not have TB.

The answer is provided at the end of this section.

You may have noticed that the conditional probability in Example 5.14 is the same as the sensitivity: Both of them can be written as

$$\frac{\text{True positives}}{\text{All who have the disease}}$$

This connection holds in general: The sensitivity is the conditional probability that a person tests positive given that the person has the disease. Similar reasoning shows that for the PPV the order is reversed: The PPV is the conditional probability that a person has the disease given that the person tests positive.

WHAT DO YOU THINK?

Should you take the test? A friend advises you that for safety's sake you should be tested for a rare disease such as lupus even though you show no symptoms of the disease and are subject to none of the common risk factors. Is this advice wise? Explain your reasoning.

Questions of this sort may invite many correct responses. We present one possibility: Medical tests are not flawless. All have false positives and false negatives. For a rare disease such as lupus, the number of true positives in the general population is very small. Even though the test has high sensitivity and specificity, the number of true positives is likely much smaller than the expected total positives from the general population. That means the positive predictive value for the general population is near zero. So if you take the test and get a positive result, it is almost certainly a false positive. As a test for the general population, the test is worthless. Nevertheless, the test is useful and very important for the tiny segment of the population for which it is indicated.

Further questions of this type may be found in the *What Do You Think?* section of exercises.

Try It Yourself answers

Try It Yourself 5.10: Estimating sensitivity and specificity: Test for disease
Sensitivity: 85.1%; Specificity: 86.6%.

Try It Yourself 5.11: Using sensitivity and specificity: Specificity of 87% 261.

Try It Yourself 5.12: Calculating PPV and NPV: Hepatitis C General population: 99.9%; older and former injection drug users: 69.2%.

Try It Yourself 5.13: Using prevalence and calculating PPV: Crohn's disease 99.96%.

Try It Yourself 5.14: Relating conditional probability and medical testing: TB
0.956 or 95.6%.

Exercise Set 5.2

Test Your Understanding

1. State in your own words the meaning of the specificity of a test.

2. If a person who has the targeted disease took a test with a high sensitivity, what result would be expected?

3. What does the positive predictive value of a test mean?

4. A person who takes a test with a high negative predictive value and tests negative should conclude ________.

5. True or false: A test with both a high sensitivity and a high specificity is virtually certain to produce accurate results.

6. Suppose you are tested for a condition. If the test says you do not have the condition, but you really do have it, this result is a: **a.** true positive **b.** true negative **c.** false positive **d.** false negative.

7. Suppose you are tested for a condition. If the test says you have the condition, and you really do have it, this result is a: **a.** true positive **b.** true negative **c.** false positive **d.** false negative.

8. Suppose you are tested for a condition. If the test says you have the condition, but you really do not have it, this result is a: **a.** true positive **b.** true negative **c.** false positive **d.** false negative.

9. If a person is tested for a condition, and the test says he or she has the condition, the person wants to know how likely it is that he or she really does have it. Which of these measures that probability? **a.** specificity **b.** sensitivity **c.** positive predictive value **d.** negative predictive value.

Problems

In these exercises, round all answers in percentage form to one decimal place unless you are instructed otherwise.

10. **A table for a medical test:** The following table shows the results of a medical test.

	Has disease	Does not have disease
Test positive	180	30
Test negative	25	400

a. Estimate the sensitivity of the test.

b. Estimate the specificity for the test.

c. Calculate the positive predictive value of the test.

d. Calculate the negative predictive value of the test.

11. **100% accurate?** Your friend suggests the following as a way of doing medical testing: Just tell each person "tested" that he or she has the disease—the test result is always positive. What is the sensitivity of this test? What is the specificity?

12. **Crohn's screening test.** The table below shows the results of a new screening test for Crohn's disease.

	Has Crohn's	Does not have Crohn's
Test positive	36	25
Test negative	9	230

a. What percentage of the individuals in the study were false negatives?

b. What percentage of the individuals in the study who had Crohn's tested negative?

c. For what percentage of the individuals in the study did the test return an incorrect result?

13. **Lupus.** You have seen in this chapter that there are limitations to using diagnostic tests. For example, the ANA (anti-nuclear antibody) test has been useful in diagnosing systemic lupus erythematosus (SLE) in patients who show multiple signs of the auto-immune disease, including chronic inflammation of various tissues in the body. But many physicians use the ANA test on patients who do not show multiple signs of the disease (sometimes called a "shotgun" approach to diagnosis). In these cases, the test results in many more false positives than true positives, meaning an overwhelming number of patients who test positive for SLE don't actually have the disease.[10]

Explain the above in terms of sensitivity, specificity, and positive predictive value.

14. **Sensitivity and specificity.** The accompanying table gives the results of a screening test for a disease. Estimate the sensitivity and specificity of the test.

	Has disease	Does not have disease
Test positive	285	280
Test negative	15	420

15. **Sensitivity and specificity.** The accompanying table gives the results of a screening test for a disease. Estimate the sensitivity and specificity of the test.

	Has disease	Does not have disease
Test positive	15	12
Test negative	5	68

16. **Extreme populations.** Suppose *everyone* in Population 1 has a certain disease and *no one* in Population 2 has the disease. What is the NPV for each of these populations?

17. **Negative predictive value.** Would you expect the NPV of a given test to be higher in a population where prevalence is high or in a population where prevalence is low?

18. **A disease.** Suppose a test for a disease has a sensitivity of 95% and a specificity of 60%. Further suppose that in a certain country with a population of 1,000,000, 30% of the population has the disease.

a. Fill in the accompanying table.

	Has disease	Does not have disease	Totals
Test positive			
Test negative			
Totals			

b. What are the PPV and NPV of this test for this country?

[10] http://application.fnu.ac.fj/classshare/Medical_Science_Resources/MBBS/MBBS1-3/PBL/Sexually%20Transmitted%20Infections/HIV-AIDS/Sensitivity%20and%20Specificity%20of%20the%20HIV%20test.pdf.

19. Crohn's disease. A screening test for Crohn's disease has a sensitivity of 80% and a specificity of 90.2%. Our goal is to calculate the PPV of this test within a population of individuals who show symptoms of the disease. Assume that this test is applied to the group of approximately 3.4 million Americans who suffer from symptoms that are very similar to those of Crohn's. Suppose also that all of the 0.59 million Crohn's sufferers belong to this group.

a. Fill in the following table based on this group of 3.4 million. (Write your answers in millions and round to two decimal places.)

	Has disease	Does not have disease	Totals
Test positive (millions)			
Test negative (millions)			
Totals			

b. Use the table in part a to show that the PPV of the test within this population of individuals showing symptoms of the disease is about 62.7%.

20. Another disease. Suppose a test for a disease has a sensitivity of 75% and a specificity of 85%. Further suppose that in a certain country with a population of 1,000,000, 20% of the population has the disease.

a. Fill in the accompanying table.

	Has disease	Does not have disease	Totals
Test positive			
Test negative			
Totals			

b. What are the PPV and NPV of this test for this country?

21. Testing for drug abuse. Here is a scenario that came up in the nationally syndicated column called "Ask Marilyn" in *Parade Magazine*.[11] Suppose a certain drug test is 95% accurate, meaning that if a person is a user, the result is positive 95% of the time, and if she or he isn't a user, it's negative 95% of the time. Assume that 5% of all people are drug users.

a. What is the sensitivity of this test?

b. What is the specificity of this test?

c. Assume that the population is 10,000, and fill in the accompanying table.

	Drug user	Not a drug user	Totals
Test positive			
Test negative			
Totals			

d. Suppose a randomly chosen person from the population in question tests positive. How likely is the individual to be a drug user? That is, what is the PPV of this test for this population?

e. If you found out a student was expelled because she tested positive for drug use with this test, would you feel that this action was justified? Explain.

f. Suppose a randomly chosen person from the population in question tests negative. How likely is the individual not to be a drug user? That is, what is the NPV of this test for this population?

g. If you found out that a student who had been accused of using drugs was exonerated because she tested negative for drug use with this test, would you feel that this action was justified? Explain.

22. Testing for Drug Abuse II. Repeat Exercise 21, this time assuming that the individual being tested is chosen from a population of which 20% are drug users. (This might be the case if the test is being applied to a prison population that has a known history of drug use.)

23. Conclusions. *This exercise uses results from Exercises 21 and 22.* In Exercise 21, it was assumed that a drug test was applied to a population with 5% users. In Exercise 22, it was assumed that this drug test was applied to a population with 20% users. What do the results suggest about how the drug test could be used most effectively?

24. Genetic testing. A certain genetic condition affects 5% of the population in a city of 10,000. Suppose there is a test for the condition that has an error rate of 1% (i.e., 1% false negatives and 1% false positives).

a. Fill in the table below.

	Has condition	Does not have condition	Totals
Test positive			
Test negative			
Totals			

b. What is the probability (as a percentage) that a person has the condition if he or she tests positive?

c. What is the probability (as a percentage) that a person does not have the condition if he or she tests negative?

25. PSA test. The following is paraphrased from a Web site that discusses the sensitivity and specificity of the PSA (prostate-specific antigen) test for prostate cancer.[12] In a larger study (243 subjects), Ward and colleagues used a cutoff ratio of 0.15 to demonstrate 78% sensitivity and 69% specificity in the diagnosis of prostate cancer. In contrast, tPSA measurements alone yielded equivalent sensitivity but only 33% specificity. These reports are encouraging, but larger clinical studies are required for determination of the optimal fPSA:tPSA cutoff ratio.

a. What does the figure 78% sensitivity in the above study mean?

b. What does the figure 69% specificity in the above study mean?

[11] *This column is summarized at the Chance Web site:* www.dartmouth.edu/~chance/course/Syllabi/mpls/handouts/section3_8.html

[12] www.scrippslabs.com/content/1995fall.pdf.

c. Assume that 10% of the male population over the age of 50 has prostate cancer. What are the PPV and NPV of this test for this population? *Suggestion*: First make a table as in Exercise 18. Because the PPV and NPV don't depend on the size of the population, you may assume any size you wish. A choice of 10,000 for the population size works well.

d. Assume that 40% of the male population over the age of 70 has prostate cancer. What are the PPV and NPV of this test for this population?

26–29. HIV. *Suppose that a certain HIV test has both a sensitivity and specificity of 99.9%. This test is applied to a population of 1,000,000 people. This information is used in Exercises 26 through 29.*

26. Suppose that 1% of the population is actually infected with HIV.

a. Calculate the PPV. *Suggestion*: First make a table as in Exercise 18.

b. Calculate the NPV.

c. How many people will test positive who are, in fact, disease-free?

27. Suppose that the population comes from the blood donor pool that has already been screened. In this population, 0.1% of the population is actually infected with HIV.

a. Calculate the PPV.

b. Calculate the NPV.

c. How many people will test positive who are, in fact, disease-free?

28. Consider a population of drug rehabilitation patients for which 10% of the population is actually infected with HIV.

a. Calculate the PPV.

b. Calculate the NPV.

c. How many people will test positive who are, in fact, disease-free?

29. Conclusions on HIV. *This exercise uses results from Exercises 26 through 28.* Consider the values of the PPV found in Exercises 26 through 28. Explain what the results tell us about testing for HIV.

30. Conditional probability and medical testing. This exercise refers to the table in Example 5.14 on page 334.

a. Find the probability that a person tests positive given that the person does not have the disease.

b. Find the probability that a person does not have the disease given that the person tests negative. Which of the four basic quantities (sensitivity, specificity, PPV, NPV) does this number represent?

31. Conditional probability. You roll a fair six-sided die and don't look at it. What is the probability that it is a 5 given that your friend looks and tells you that it is greater than 2? Leave your answer as a fraction.

32. More conditional probability. In a standard deck of cards, the jack, queen, and king are "face cards." You draw a card from a standard deck. Your friend peeks and lets you know that your card is a face card. What is the probability that it is a queen given that it is a face card? Leave your answer as a fraction.

33. Shampoo. Your wife has asked you to pick up a bottle of her favorite shampoo. When you get to the store, you can't remember which brand she uses. There are nine shampoo brands on the shelf. If you select one, what is the probability that you get the right brand of shampoo? Of the nine bottles, four are blue, three are white, and two are red. You remember that your wife's shampoo comes in a blue bottle. What is the probability that you get the right bottle with this additional information? Leave your answers as fractions.

34. Conditional probability. The total area of the United States is 3.79 million square miles, and Texas covers 0.27 million square miles. A meteor falls from the sky and strikes Earth. What is the probability that it strikes Texas given that it strikes the United States?

35. This one is tricky. A family that you haven't yet met moves in next door. You know the couple has two children, and you discover that one of them is a girl. What is the probability that the other child is a boy? That is, what is the probability that one child is male given that the other child is female? Leave your answer as a fraction.

36. Formula for conditional probability. We express the probability that event A occurs given that event B occurs as

$$P(A \text{ given } B)$$

[In many books, the notation used is $P(A|B)$.] It turns out that the following basic formula for conditional probabilities holds:

$$P(A \text{ and } B) = P(A \text{ given } B) \times P(B)$$

In this exercise, we will verify this formula in the case that A is the event that a card drawn from a standard deck is a heart and B is the event that a card drawn from a standard deck is an ace. Leave your answers as fractions.

a. Calculate $P(A \text{ and } B)$. (This is the probability that the card is the ace of hearts.)

b. Calculate $P(A \text{ given } B)$. (This is the probability that the card is a heart given that it is an ace.)

c. Calculate $P(B)$. (This is the probability that the card is an ace.)

d. Use parts a, b, and c to verify the formula above in this case.

37. Bayes's theorem. Bayes's theorem on conditional probabilities states that, if $P(B)$ is not 0, then

$$P(A \text{ given } B) = \frac{P(B \text{ given } A) \times P(A)}{P(B)}$$

Also note that Exercises 39 and 40 can be solved using this formula. Leave your answers as fractions.

Mary bakes 10 chocolate chip cookies and 10 peanut butter cookies. Bill bakes 5 chocolate chip cookies and 10 peanut butter cookies. The 35 cookies are put together and offered on a single plate. I pick a cookie at random from the plate.

a. What is the probability that the cookie is chocolate chip?

b. What is the probability that Mary baked the cookie?

c. What is the probability that the cookie is chocolate chip given that Mary baked it?

d. Use Bayes's theorem to calculate the probability that the cookie was baked by Mary given that it is chocolate chip. In this situation, Bayes's theorem tells us that

$$P(\text{Mary given Chocolate}) = \frac{P(\text{Chocolate given Mary}) \times P(\text{Mary})}{P(\text{Chocolate})}$$

e. Calculate without using Bayes's theorem the probability that the cookie was baked by Mary given that it is chocolate chip. *Suggestion*: How many of the chocolate chip cookies were baked by Mary?

38. Cookies again. As in Exercise 37, Mary bakes 10 chocolate chip cookies and 10 peanut butter cookies. Bill bakes 5 chocolate chip cookies and 10 peanut butter cookies. But now, Mary's cookies are put on one plate and Bill's on another. I pick a plate at random and then pick a cookie from that plate. Do you think the probability that cookie is chocolate chip is the same as in the scenario from Exercise 37? To help answer that question, consider the extreme case in which Mary bakes 99 chocolate chip cookies and no peanut butter cookies, while Bill bakes no chocolate chip cookies and 1 peanut butter cookie. Explain why you give the answer you do.

39. Car lots. Yesterday, car lot Alpha had two Toyotas and one Chevrolet for sale. Car lot Beta had three Toyotas and five Chevrolets for sale. This morning Alan bought a car, choosing one of the two lots at random and then choosing a car at random from that lot. Leave your answers as fractions.

a. What is the probability that Alan bought from Alpha?

b. If Alan bought from Alpha, what is the probability he bought a Toyota?

c. If Alan bought from Beta, what is the probability he bought a Toyota?

d. It can be shown that the probability that Alan bought a Toyota is 25/48. What is the probability Alan bought from Alpha given that he bought a Toyota? Give your answer both as a fraction and in decimal form rounded to two places. *Note*: In this situation, Bayes's theorem from Exercise 37 says

$$P(\text{Alpha given Toyota}) = \frac{P(\text{Toyota given Alpha}) \times P(\text{Alpha})}{P(\text{Toyota})}$$

40. Bowls of cookies. There are two bowls of cookies. Bowl Alpha has 5 chocolate chip cookies and 10 peanut butter cookies. Bowl Beta has 10 chocolate chip cookies and 10 peanut butter cookies. Jerry selects a bowl at random and chooses a cookie at random from that bowl. Leave your answers as fractions.

a. What is the probability that he chooses bowl Alpha?

b. What is the probability that he gets a chocolate chip cookie given that he chooses from bowl Alpha?

c. It can be shown that the probability that he gets a chocolate chip cookie is 5/12. What is the probability that he chose from bowl Alpha given that he got a chocolate chip cookie?
Note: In this situation, Bayes's theorem from Exercise 37 says

$$P(\text{Alpha given Chocolate}) = \frac{P(\text{Chocolate given Alpha}) \times P(\text{Alpha})}{P(\text{Chocolate})}$$

Exercise 41 is designed to be solved using technology such as a calculator or computer spreadsheet.

41. Spreadsheet. Devise a spreadsheet that will allow the user to input the prevalence of a disease in a population along with the sensitivity and specificity of a test for the disease. The spreadsheet should then complete a table similar to that in Example 5.10 on page 329 and report the PPV and NPV.

Writing About Mathematics

42. Contingency tables. Most of the tables in this section are examples of *contingency tables*. These tables came into use in the twentieth century, but the concepts in probability behind them are often traced to Thomas Bayes, who lived in Britain in the eighteenth century. Write a report about his life and work.

43. Bayesian statistics. The field of *Bayesian statistics* is named for Thomas Bayes. Write a report about Bayesian statistics.

44. Medical test. Look up a disease or condition and write a report on a medical test associated with it.

What Do You Think?

45. A worthless test. Devise a simple, but ultimately worthless, test that has 100% specificity. **Exercise 11 is relevant to this topic.**

46. Lost needle. Sometimes we use the phrase "looking for a needle in a haystack" to refer to the difficulty of locating one object in the midst of a lot of others. Explain how this phrase is relevant to the connection between the prevalence of a disease and the usefulness of testing the general population. **Exercise 20 is relevant to this topic.**

47. Test my child? A community activist who supports *zero tolerance* advocates testing every child in high school for drugs even though there is no evidence of widespread drug abuse in that school. Is this a good idea? Explain your reasoning. **Exercise 21 is relevant to this topic.**

48. Calculating sensitivity. In some exercises in this section, you are asked to use given data to estimate the sensitivity of a test. Why is the instruction to estimate rather than to calculate exactly? What information is needed in order to calculate exactly the sensitivity? **Exercise 14 is relevant to this topic.**

49. Specificity. Is it necessarily true that a large specificity (nearly equal to 100%) indicates a relatively small number of false positives? Explain your answer. **Exercise 15 is relevant to this topic.**

50. Which causes concern? If you test positive for a disease, would you be more concerned if the test is known to have a high sensitivity or a high specificity? Explain your reasoning.

51. What do they mean? Use your own words to explain the meaning of the terms *positive predictive value* and *negative predictive value*, and give some good examples. **Exercise 26 is relevant to this topic.**

52. Conditional probability. Use your own words to explain the meaning of *conditional probability*, and give some examples. **Exercise 30 is relevant to this topic.**

53. Connection. Explain the connection between conditional probability and the terms such as *sensitivity* used in medical testing.

5.3 Counting and theoretical probabilities: How many?

TAKE AWAY FROM THIS SECTION
Learn how to count outcomes without listing all of them.

The following excerpt is an article that appeared at the Web site of the *Quad City Times*.

IN THE NEWS 5.5

Search

Home | TV & Video | U. S. | World | Politics | Entertainment | Tech | Travel

How a Davenport Murder Case Turned on a Fingerprint

SUSAN DU, STEPHANIE HAINES, GIDEON RESNICK AND TORI SIMKOVIC THE MEDILL JUSTICE PROJECT December 22, 2013

On March 10, 2003, police arrived on the scene of a grisly murder in a dead-end neighborhood. . . .

Later, police found a fingerprint thought to be made in blood on [a] cigarette box's cellophane wrapping. The police used the print to place a then 29-year-old named Chad Enderle, who said he sometimes bought drugs from the victim, at the scene of the crime.

At trial, authorities highlighted seven identifying characteristics—or points of minutiae—to link the cigarette print to Enderle's right ring finger. But seven points isn't usually enough to make a match, according to several fingerprint experts who say the industry typically calls for more. The Medill Justice Project interviewed several fingerprint experts who offered conflicting conclusions.

Since Enderle's sentencing [for murder], studies, federal agencies on forensic sciences, and some judges across the country have urged greater scrutiny of fingerprint analysis. . . .

Industry challenges

Cedric Neumann, an assistant statistics professor at South Dakota State University with a Ph.D. in forensic science, said differences in the understanding of the logic behind fingerprint examination resulting from examiners' varied field and apprenticeship experiences are like "night and day."

Neumann, in the February 2012 issue of Significance, a statistical magazine, described a statistical model he developed for quantifying fingerprint characteristics.

Neumann said that although it is difficult to summarize fingerprints into sets of data that can be processed using statistics, he compared his model to that used to back DNA identification. Humans have 46 chromosomes and each chromosome has thousands of genes, yet DNA identification relies on a minimum of 16 of those genes. In Neumann's model, still being tested, the most prominent minutiae in a suspect's print would be connected to form a polygonal figure. Qualities of that polygon would be assigned numerical values that would render the probability that the suspect is the source of the crime-scene mark.

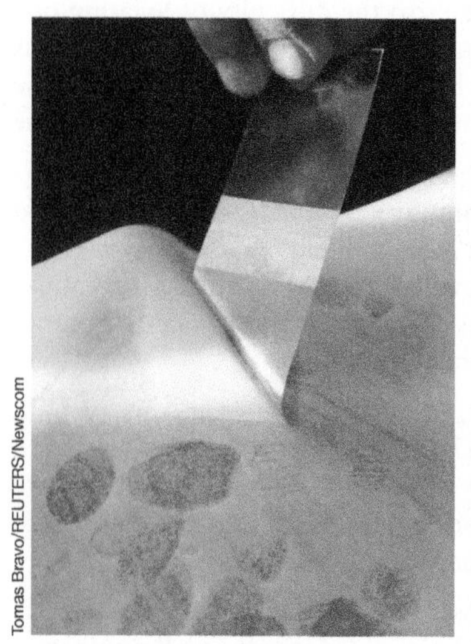

Tomas Bravo/REUTERS/Newscom

Testing for fingerprints

This article describes the use of probability calculations to aid in examining fingerprints. To calculate the necessary probabilities, we need to be able to count the number of ways in which an event can happen. In this section, we will examine some fundamental counting methods.

Simple counting

Counting the number of outcomes in the case of flipping two coins, as we did in Section 5.1, is not difficult. We distinguish the two coins (by thinking of a nickel and a dime, for example) and then simply list the four possible outcomes: HH, HT, TH, TT.

If we are flipping three coins, the number of possible outcomes is much larger. Distinguishing the coins as a penny, nickel, and dime, we find the possibilities listed below.

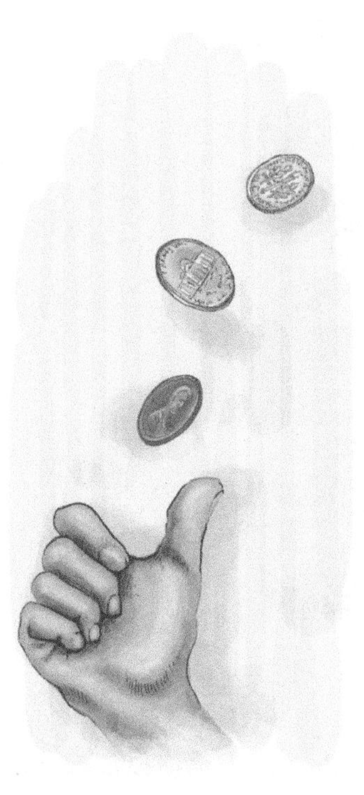

Penny	Nickel	Dime
H	H	H
H	H	T
H	T	H
H	T	T
T	H	H
T	H	T
T	T	H
T	T	T

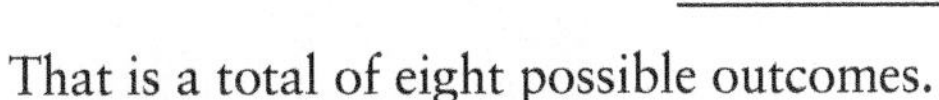

That is a total of eight possible outcomes.

If there are many coins, it is impractical to make a list of all possible outcomes. For example, if you try to make such a list for 10 coins, you are unlikely to finish it. How can we compute the number of outcomes without writing down all such sequences? We do it by a systematic counting scheme. There are two possibilities for the first coin: a head or a tail. For each of these possibilities, there are also two possibilities for the second coin. To obtain the number of possibilities for two coins, we multiply these together to get $2 \times 2 = 2^2 = 4$. (As we noted earlier, these are HH, HT, TH, TT.) If we were flipping three coins, we would continue this counting process: For each of the four possibilities for the first two coins, there are two possibilities for the third coin. Thus, the total number of outcomes for flipping three coins is $4 \times 2 = 2^3 = 8$. (This agrees with the list HHH, HHT, HTH, HTT, THH, THT, TTH, TTT we made earlier.) If we continue in this way, we see that the number of possible outcomes in tossing 10 coins is $2^{10} = 1024$.

Armed with this information, we can calculate the probability of getting all heads in the toss of 10 coins. There is only one favorable outcome—namely, all heads. Therefore,

$$P(10 \text{ heads}) = \frac{\text{Favorable outcomes}}{\text{Total outcomes}} = \frac{1}{1024}$$

This is 0.00098 or 0.098%.

EXAMPLE 5.15 Counting outcomes: Tossing coins

How many possible outcomes are there if we toss 12 coins?

SOLUTION

Reasoning as before, we see that each extra coin doubles the number of outcomes. That gives a total of $2^{12} = 4096$ possible outcomes.

TRY IT YOURSELF 5.15

How many possible outcomes are there if we toss 15 coins?

The answer is provided at the end of this section.

The Counting Principle

The method we used to count outcomes of coin tosses works in a more general setting. Its formal statement is known as the Counting Principle. Suppose we perform two experiments in succession and that for each outcome of the first experiment there is the same number of outcomes for the second. For example, think of a menu at a diner where there are five entrées, and for each choice of entrée there are three side dishes available for that choice. The Counting Principle asserts that the number of possible outcomes for the two experiments is the number of possible outcomes for the first experiment times the number of possible outcomes for the second experiment. This extends to any number of such experiments. In the case of the diner, there are $5 \times 3 = 15$ possible combinations of an entrée with a side dish.

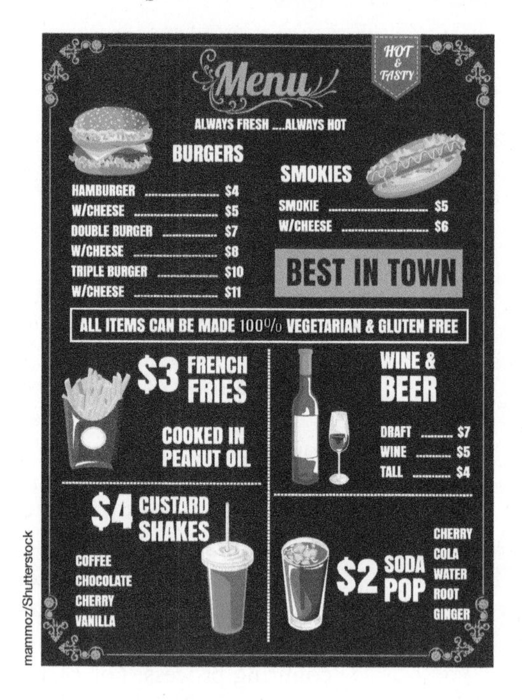

mammoz/Shutterstock

KEY CONCEPT

The Counting Principle tells us how to calculate the results of two experiments performed in succession. Suppose there are N outcomes for the first experiment. If for each outcome of the first experiment, there are M outcomes of the second experiment, then the number of possible outcomes for the two experiments is $N \times M$.

This principle extends to any number of such experiments.

Another way to think about this principle is as follows: Suppose there are two blanks

$$\_\ \_$$

that we wish to fill, and we have N choices for the first blank and M choices for the second blank:

$$\underline{N}\ \underline{M}$$

Then the total number of ways we can fill the two blanks is $N \times M$. This extends to any number of blanks.

As a simple application of the Counting Principle, let's look at product labeling. Suppose a company labels its products with a letter of the alphabet followed by a numeral (0 through 9), for example, model A4 or model D9. How many different labels can the company make? Let's think of this as filling in two blanks:

— —

A letter goes in the first blank and a numeral goes in the second blank. There are 26 letters that can go in the first blank and 10 numerals that can go in the second blank:

26 10

According to the Counting Principle, the total number of labels is the number of outcomes for the first blank times the number of outcomes for the second blank. This is $26 \times 10 = 260$.

CALCULATION TIP 5.2 Filling in Blanks

When we want to know how many ways a set of items can be arranged, it is often helpful to think of filling a list of blank spaces with the items, one at a time.

EXAMPLE 5.16 Counting by filling in blanks: Codes

How many three-letter codes can we make if the first letter is a vowel (A, E, I, O, or U)? One such code is OXJ.

SOLUTION

We think of filling in three blanks:

— — —

There are 5 letters that can go in the first blank and 26 in each of the other two:

5 26 26

Applying the Counting Principle, we find a total of $5 \times 26 \times 26 = 3380$ codes.

TRY IT YOURSELF 5.16

How many three-letter codes can we make if the first letter is a consonant rather than a vowel? One such code is KXJ.

The answer is provided at the end of this section.

The Counting Principle allows us to calculate the total number of outcomes for fairly complex events. Suppose, for example, that we toss five dice and want to know the total number of outcomes. We think of tossing the dice one at a time. Applying the Counting Principle, we see that the total number of outcomes is

Outcomes for die #1 × Outcomes for die #2 × · · · × Outcomes for die #5

Because each die has six possible outcomes, we are multiplying five 6's together, which gives the total number of outcomes as

$$6 \times 6 \times 6 \times 6 \times 6 = 6^5 = 7776$$

We can also think of this in terms of five blanks, one for each die. Each blank has six possible outcomes:

6 6 6 6 6

We find $6 \times 6 \times 6 \times 6 \times 6 = 6^5$ outcomes, just as before.

FORRAY Didier/SAGAPHOTO.COM/Alamy

EXAMPLE 5.17 Counting for complex events: License plates

Automobile license plates in the state of Nevada typically consist of three numerals followed by three letters of the alphabet, such as 072 ZXE.

a. How many such license plates are possible?

b. How many such plates are possible if we insist that on each plate no numeral can appear more than once and no letter can appear more than once?

SOLUTION

a. Let's think of a blank license tag with three slots for numbers and three slots for letters:

— — —	— — —
numbers	letters

We need to figure out how many ways we can fill in the slots. For each number slot, there are 10 possible numerals (0 through 9) to use: 10 for the first blank, 10 for the second, and 10 for the third:

10 10 10	— — —
numbers	letters

For each letter slot, there are 26 possible letters of the alphabet: 26 for the first blank, 26 for the second, and 26 for the third:

10 10 10	26 26 26
numbers	letters

Applying the Counting Principle gives

$$10 \times 10 \times 10 \times 26 \times 26 \times 26 = 17{,}576{,}000$$

ways to fill in the six slots.

By the way, the population of Nevada is approximately 2.9 million. There should be enough license plate numbers for a while!

b. As in part a, we think of a blank plate with three slots for numbers and three slots for letters. We need to count the number of ways we can fill in these slots. But this time we are not allowed to repeat letters or numbers.

For the first number slot, we have 10 choices. For the second slot, we can't repeat the number in the first blank, so we have only nine choices. For the third slot, we can't repeat the numbers used in the first two slots, so we have a choice of only eight numbers:

10 9 8	— — —
numbers	letters

Now we fill in the blanks for the letters. There are 26 ways to fill in the first blank. We can't use that letter again, so there are only 25 ways to fill in the second blank. We can't use the first 2 letters, so we have a choice of only 24 letters for the third blank:

10 9 8	26 25 24
numbers	letters

Now the Counting Principle gives

$$\text{Number of plates} = 10 \times 9 \times 8 \times 26 \times 25 \times 24 = 11{,}232{,}000$$

Applying counting to probabilities

Let's see how to apply the Counting Principle to calculate probabilities when drawing cards. Recall that there are 52 playing cards in a standard deck. They are divided into four *suits*, called spades, hearts, diamonds, and clubs. Twelve of the cards (four kings, four queens, and four jacks) are called *face cards*.

Let's use what we have learned about counting to determine in how many ways we can draw two cards from a deck in order without replacement. That is, we draw a first card from a deck and, without putting it back, draw a second card. In this situation, drawing the 2 of spades for the first card and the 3 of spades for the second is different from drawing the 3 of spades for the first card and the 2 of spades for the second. We can think of filling in two blanks. There are 52 possibilities for the first, but only 51 for the second:

$$\underline{52}\ \underline{51}$$

The Counting Principle gives $52 \times 51 = 2652$ possible outcomes. These kinds of calculations will be helpful in calculating probabilities for cards.

EXAMPLE 5.18 Counting and calculating probabilities: Draw two cards

Suppose you draw a card, put it back in the deck, and draw another. What is the probability that the first card is an ace and the second one is a jack? (Assume that the deck is shuffled after the first card is returned to the deck.)

SOLUTION

To find the probability, we have two numbers to calculate: The number of ways we can draw two cards (the total number of possible outcomes), and the number of ways we can draw an ace followed by a jack (the number of favorable outcomes).

First we count the total number of ways to draw two cards in order from a deck. Think of filling two blanks:

$$\underline{\quad}\ \underline{\quad}$$

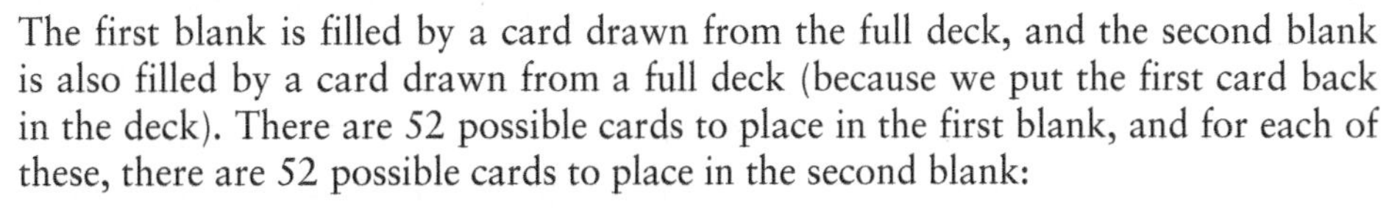

The first blank is filled by a card drawn from the full deck, and the second blank is also filled by a card drawn from a full deck (because we put the first card back in the deck). There are 52 possible cards to place in the first blank, and for each of these, there are 52 possible cards to place in the second blank:

$$\underline{52}\ \underline{52}$$

Using the Counting Principle, we find $52 \times 52 = 2704$ ways to fill the two blanks with two cards from the deck. This is the total number of possible outcomes.

Now we ask how many ways there are to fill these two blanks with an ace followed by a jack. There are four aces in the deck, so there are four choices to fill the first blank:

$$\underline{4}\ \underline{\quad}$$

After this has been done, there are four jacks with which to fill the second blank:

$$\underline{4}\ \underline{4}$$

Again, using the Counting Principle, we find $4 \times 4 = 16$ ways to fill the two blanks with an ace followed by a jack. This is the number of favorable outcomes. So the probability of drawing an ace followed by a jack is

$$\begin{aligned} P(\text{Ace followed by jack}) &= \frac{\text{Favorable outcomes}}{\text{Total outcomes}} \\ &= \frac{16}{2704} \end{aligned}$$

This is about 0.006 or 0.6%.

TRY IT YOURSELF 5.18

Suppose you draw a card and, without putting it back, draw another. What is the probability that the first card is an ace and the second one is a jack?

The answer is provided at the end of this section.

Rex May/Baloooscartoon.com

The Counting Principle applies to elections just as it does to cards.

EXAMPLE 5.19 Counting: Election outcomes

Suppose there are 10 people willing to serve as an officer for a club. It's decided just to put the 10 names in a hat and draw 3 of them out in succession. The first name drawn is declared president, the second name vice president, and the third name treasurer.

a. How many possible election outcomes are there?

b. John is a candidate. What is the probability that he will be vice president?

c. Mary and Jim are also candidates along with John. What is the probability that all three will be selected to office?

d. What is the probability that none of the three (Mary, Jim, and John) will be selected to office?

e. What is the probability that at least one of the three (Mary, Jim, and John) will not be selected? *Suggestion*: Recall from Section 5.1 that if p is the probability of an event occurring, then the probability that the event will not occur is $1 - p$.

SOLUTION

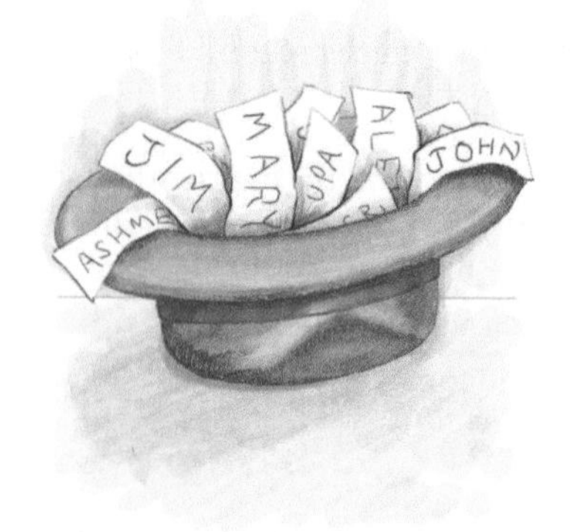

a. Imagine that we have three blanks corresponding to president, vice president, and treasurer:

$$\underset{\text{P}}{\underline{\quad}}\quad \underset{\text{VP}}{\underline{\quad}}\quad \underset{\text{T}}{\underline{\quad}}$$

There are 10 possibilities for the president blank. When that is filled, there are nine names left, so there are nine possibilities for the vice president blank. When that is filled, there are eight possibilities for the third blank:

$$\underset{\text{P}}{\underline{10}}\quad \underset{\text{VP}}{\underline{9}}\quad \underset{\text{T}}{\underline{8}}$$

The Counting Principle gives

$$\text{Number of outcomes} = 10 \times 9 \times 8 = 720$$

b. We must count the number of ways John can be vice president. In this case, John is not president, so there are nine possible names for the president blank. There is just one possibility, John, for the vice president blank, and that leaves eight possibilities for the treasurer blank:

$$\frac{9}{P} \quad \frac{1}{VP} \quad \frac{8}{T}$$

Hence,

$$\text{Number of ways John is vice president} = 9 \times 1 \times 8 = 72$$

This is the number of favorable outcomes. Using these numbers, we find the probability we want:

$$\begin{aligned} \text{Probability John is vice president} &= \frac{\text{Favorable outcomes}}{\text{Total outcomes}} \\ &= \frac{72}{720} \\ &= 0.1 \end{aligned}$$

Thus, there is a 10% chance of John being selected as vice president.

Alternatively, we can think along the following lines. All we are interested in knowing is whose name goes in the vice president blank. There are 10 equally likely possibilities. Exactly one of them is John. Thus, the probability that he is vice president is $1/10 = 0.1$.

c. We want to arrange the three names John, Jim, and Mary in the three officeholder blanks. There are three possible names for the first, two for the second, and one for the third:

$$\frac{3}{P} \quad \frac{2}{VP} \quad \frac{1}{T}$$

Thus, the total number of ways these three candidates can all win office is $3 \times 2 \times 1 = 6$. These are the favorable outcomes. We calculated in part a that there are 720 possible outcomes. Using these values, we calculate

$$\begin{aligned} \text{Probability all three selected} &= \frac{\text{Favorable outcomes}}{\text{Total outcomes}} \\ &= \frac{6}{720} \end{aligned}$$

This is about 0.008 and represents a 0.8% chance that all three are selected.

d. If these three names don't appear in the officer blanks, then there are seven possibilities for the first, six for the second, and five for the third:

$$\frac{7}{P} \quad \frac{6}{VP} \quad \frac{5}{T}$$

That is a total of $7 \times 6 \times 5 = 210$ favorable outcomes. Thus, the probability that none will be selected is

$$\begin{aligned} \text{Probability none selected} &= \frac{\text{Favorable outcomes}}{\text{Total outcomes}} \\ &= \frac{210}{720} \end{aligned}$$

This is about 0.292, and it represents a 29.2% chance that none will be selected.

e. We found in part c that the probability that all three are selected is $6/720$. The probability that at least one will not be selected is the probability that the event of all three being selected does not occur. This is 1 minus the probability that all three are selected, so the probability that at least one will not be selected is

$$1 - \frac{6}{720}$$

or about 0.992. Hence, there is a 99.2% chance that at least one will not be selected.

Independent events

Suppose we pick a person at random and consider the probability that the person is female and the probability that the person's last name is Smith. Knowing that the person is female does not affect the probability that the person's last name is Smith. These two events (picking a female and picking someone whose last name is Smith) are independent.

In contrast, suppose we pick a person at random and look at the probability that the person is female and the probability that the person's first name is Mary. If we know that the person is female, the probability that the name will be Mary may be quite small. But if we don't know whether the person is female, the probability that the name will be Mary is much smaller. These two events (picking a female and picking someone named Mary) are not independent.

KEY CONCEPT

Suppose we perform an experiment. Two events are independent if knowing that one event occurs has no effect on the probability of the occurrence of the other.

In many practical situations, it is important to determine whether or not events are independent. For example, the events of being a smoker and having heart disease are not independent. The establishment of the link between smoking and heart disease was a milestone in public health policy.

woaiss/Shutterstock

There is an important formula involving probabilities of independent events. We look at an example to illustrate it. One of my favorite stores is having a sale. When I buy something, I get a ticket at the checkout stand that gives me either 10% off, 20% off, 30% off, 40% off, or 50% off. (We suppose that each of these tickets is equally likely.) I buy a pair of jeans in the morning and a shirt in the afternoon. What is the probability that I get at least 30% off in the morning and at least 40% off in the afternoon? These are independent events because winning one discount does not give any information about the other.

First we count the total number of possible outcomes. Because there are five possibilities for the morning and five for the afternoon, the Counting Principle tells us that there are $5 \times 5 = 25$ possible outcomes. Now let's count favorable outcomes. For the morning, 30%, 40%, and 50% are favorable, and for the afternoon, only 40% and 50% are favorable. With three favorable outcomes for the morning and two for the afternoon, the Counting Principle gives $3 \times 2 = 6$ favorable outcomes. So the probability we seek is

$$\begin{aligned} P(\text{At least 30\% and At least 40\%}) &= \frac{\text{Favorable outcomes}}{\text{Total outcomes}} \\ &= \frac{6}{25} \end{aligned}$$

Now we calculate the two probabilities individually. For the morning, there are three favorable outcomes out of a total of five. So the probability of getting 30% or greater in the morning is $3/5$. For the afternoon, there are two favorable outcomes out of a total of five. Thus, the probability of getting 40% or more in the afternoon is $2/5$.

If we multiply the individual probabilities, we find the probability of 30% or more in the morning and 40% or more in the afternoon:

$$\frac{3}{5} \times \frac{2}{5} = \frac{6}{25}$$

so

$$P(\text{At least 30\%}) \times P(\text{At least 40\%}) = P(\text{At least 30\% and At least 40\%})$$

This result is no accident. Given two independent events, the probability that both occur is the product of the separate probabilities for the two events.[13] Here, each probability must be written as a decimal or a fraction, not as a percentage.

SUMMARY 5.4
Product Formula for Independent Events

Events A and B are independent exactly when the probability of Event A *and* Event B occurring is the probability of Event A times the probability of Event B. So A and B are independent exactly when

$$P(A \text{ and } B) = P(A) \times P(B)$$

Here, each probability must be written as a decimal or a fraction, not as a percentage.

The formula characterizes the situation when knowing that one event occurs has no effect on the probability of the occurrence of the other.

EXAMPLE 5.20 Using the product formula: Defective products

Suppose that 1 in 500 digital cameras is defective and 3 in 1000 printers are defective. On a shopping trip, I purchase a digital camera and a printer. What is the probability that both the camera and the printer are defective?

SOLUTION

The probability of getting a defective camera is 1/500, and the probability of getting a defective printer is 3/1000. The fact that one item is defective gives no information about the other, so the two events are independent. We use the product formula:

$$P(\text{Defective camera and Defective printer}) = \frac{1}{500} \times \frac{3}{1000} = \frac{3}{500{,}000}$$

This represents six times in a million.

TRY IT YOURSELF 5.20

Suppose that 20% of T-shirts in a stack are red and 10% of shorts in another stack are red. You pick a T-shirt and pair of shorts at random. What is the probability that both are red?

The answer is provided at the end of this section.

Let's look further at the previous example. Suppose I got home and found that my digital camera was defective. I might reason as follows: "The probability of both the camera and the printer being defective is so small that perhaps I can be assured that the printer is fine." This is a bogus argument. The probability of both being defective is indeed quite small. Nonetheless, the two events are independent, so knowing that the camera is defective does not change the probability that the printer is defective.

WHAT DO YOU THINK?

Tossing a coin: A friend tossed a fair coin 10 times, and each time it came up heads. On the next toss he is betting on tails because getting 11 heads in a row is so unlikely. Is his reasoning valid?

[13] *In fact, our initial description of independent events is useful to build intuition, but the mathematical definition of independent events is that the product formula applies.*

Questions of this sort may invite many correct responses. We present one possibility: Assuming that the coin being tossed is a fair coin, then the probability of a head on the eleventh toss is 50%. That is because coin tosses are independent events. The probability associated with the next toss has nothing to do with previous tosses. But in practice, after seeing 10 straight heads, we might be inclined to question the fairness of the coin. Perhaps our friend is thinking that 10 heads in a row from a fair coin is so unlikely that he suspects the coin is not fair. It is probably time to examine the coin closely.

Further questions of this type may be found in the *What Do You Think?* section of exercises.

Try It Yourself answers

Try It Yourself 5.15: Counting outcomes: Tossing coins $2^{15} = 32{,}768$.

Try It Yourself 5.16: Counting by filling in blanks: Codes 14,196.

Try It Yourself 5.18: Counting and calculating probabilities: Draw two cards $\frac{16}{52 \times 51}$ or about 0.60%.

Try It Yourself 5.20: Using the product formula: Clothing color 0.02 or 2%.

Exercise Set 5.3

Test Your Understanding

1. State the Counting Principle.

2. True or false: Two events are independent if knowing that one event occurs has no effect on the probability of the occurrence of the other.

3. If A and B are independent events, then the probability of A and B occurring equals ______.

Problems

In your answers to these exercises, leave each probability as a fraction unless you are instructed to do otherwise.

4. Counting outcomes. There are five possible outcomes for experiment A, and for each outcome of experiment A there are three possible outcomes for experiment B. How many possible outcomes are there if experiments A and B are performed in succession?

5. Counting outcomes. In hat A there are three different numbers. In hat B there are four distinct lower-case letters. In hat C there are five distinct upper-case letters. How many possible collections of numbers, lower-case letters, and upper-case letters are possible if I choose one item from each of the three hats?

6. Counting outcomes. There are four entrants in the 100-meter dash, five entrants in the high jump, and six entrants in the javelin throw. No one enters more than one event. The winner of each event receives a medal at the end of the contest. How many possible groups of three winners are there?

7. Counting outcomes. How many possible outcomes are there if we toss 10 dice, each of a different color?

8. Counting codes. How many three-letter codes can we make using only vowels, a, e, i, o, or u?

9. Drawing two cards. Suppose you draw a card, put it back in the deck, and draw another one. What is the probability that the first card is a diamond and the second is a spade?

10. Playing two games. You play two unrelated games, one after the other. There is a 30% chance of winning the first game, and there is a 40% chance of winning the second game. Use the product formula for independent events to find the probability that you win both games. Express your answer as a percentage.

11. Why use letters? Automobile license plates often use a mix of numerals and letters of the alphabet. Why do you suppose this is done? Why not use only numerals?

12. Bicycle speeds. A customer in a bicycle shop is looking at a bicycle with three gears on the front (technically, three chainrings) and seven gears on the rear (technically, seven sprockets). He decides to look at a different bike because he wants more than 10 speeds. Has he counted correctly?

13. Shampoo. A company wants to market different types of shampoo using color-coded bottles, lids, and label print. It has four different colors of lids, five different colors of bottles, and two different colors of print for the label available. How many different types of shampoo can the company package?

14–16. Phone numbers. *Local phone numbers consist of seven numerals, the first three of which are common to many users. Exercises 14 through 16 refer to phone numbers in various settings.*

14. A small town's phone numbers all start with 337, 352, or 363. How many phone numbers are available?

15. Until 1996 all toll-free numbers started with the area code 800. Assume that all seven digits for the rest of a number were possible, and determine how many toll-free 800 numbers there were.

16. Today there are a total of seven toll-free area codes: 800, 833, 844, 855, 866, 877, and 888. Assume that all seven digits for the rest of a number are possible, and determine how many toll-free numbers there are now.

17. I won the lottery. I just won the Powerball lottery and wanted to keep it a secret. But I told two close friends, who each told five other different people, and those in turn each told four other different people. How many people (other than me) heard the story of my lottery win? Be careful how you count.

18. Voting. I am voting in a state election. There are three candidates for governor, four candidates for lieutenant governor, three candidates for the state house of representatives, and four candidates for the senate. In how many different ways can I fill out a ballot if I vote once for each office?

19. My bicycle lock. My combination bicycle lock has four wheels, each of which has the numbers 0 through 20. Exactly one combination of numbers (in order) on the wheels (e.g., 2-17-0-12) will open the lock.

©Michael Spring/Dreamstime.com

a. How many possible combinations of numbers are there on my bicycle lock?

b. Suppose that I have forgotten my combination. At random I try 1-2-3-4. What is the probability that the lock will open with this combination?

c. Out of frustration I try at random 50 different combinations. What is the probability that 1 of these 50 efforts will open the lock?

20. Counting faces. A classic children's toy consists of 11 different blocks that combine to form a rectangle showing a man's face.[14] Each block has four sides with alternatives for that part of the face; for example, the block for the right eye might have a closed eye on one side, a partially open eye on another, and so on. A child can turn any block and so change that part of the face, and in this way many different faces can be formed.

Courtesy of www.continentalhobby.com

Changeable Charlie: A face formed by blocks.

a. How many different faces can be formed by the children's toy?

b. If a child makes one change each minute and spends 48 hours a week changing faces, how many years will it take to see every possible face?

21–30. Independent events. *In each of Exercises 21 through 30, determine whether you think the given pair of events is independent. Explain your answer.*

21. You collect Social Security benefits. You are over age 65.

22. You are in good physical condition. You work out regularly.

23. You suffer from diabetes. You are a registered Democrat.

24. You are over 6 feet tall. You prefer light-colored attire.

25. You frequent casinos. You have lost money gambling.

26. It rains today. It is cloudy today.

27. It rains today. There is a pop quiz in math today.

28. It rains today. The streets are wet today.

29. It rains today. There are flight delays at the airport today.

30. It rains today. I take my umbrella to class with me.

31. Coins. We toss a coin twice. Use the product formula for independent events to calculate the probability that we get a tail followed by a head.

32. A coin and a die. We toss a coin and then roll a die. Use the product formula for independent events to calculate the probability that we get a tail followed by a 6.

[14] *One such toy was called Changeable Charlie.*

33. Jelly beans. Assume that a box contains four red jelly beans and two green ones. We consider the event that a red bean is drawn.

a. Suppose I pick a jelly bean from the box without looking. I record the color and put the bean back in the box. Then I choose a bean again. Are the two events independent? What is the probability of getting a red bean both times?

b. Now suppose I pick a jelly bean from the same box without looking, but this time I do not put the bean back in the box. Then I choose a bean again. Are the two events independent? What is the probability of getting two red beans?

34. Two dice. This exercise refers to the toss of a pair of dice.

a. What is the probability that the total number of dots appearing on top is 5?

b. What is the probability that the total number of dots appearing on top is 7?

c. What is the probability that the total number of dots appearing on top is *not* 5?

d. What is the probability that the total number of dots appearing on top is *not* 7?

35. Marbles. Suppose an opaque jar contains 5 red marbles and 10 green marbles. This exercise refers to the experiment of picking two marbles from the jar without replacing the first one.

a. What is the probability of getting a green marble first and a red marble second?

b. What is the probability of getting a green marble and a red marble? (How is this part different from part a?)

36. More jelly beans. Suppose I pick (without looking) a jelly bean from a box containing five red beans and seven blue beans. I record the color and put the bean back in the box. Then I choose a bean again.

a. What is the probability of getting a red bean both times?

b. What is the probability of getting a blue bean both times?

37. Two cards. This exercise refers to choosing two cards from a thoroughly shuffled deck. Assume that the deck is shuffled after a card is returned to the deck.

a. If you put the first card back in the deck before you draw the next, what is the probability that the first card is a 10 and the second card is a jack?

b. If you do not put the first card back in the deck before you draw the next, what is the probability that the first card is a 10 and the second card is a jack?

c. If you put the first card back in the deck before you draw the next, what is the probability that the first card is a club and the second card is a diamond?

d. If you do not put the first card back in the deck before you draw the next, what is the probability that the first card is a club and the second card is a diamond?

e. If you do not put the first card back in the deck before you draw the next, what is the probability that both cards are diamonds?

f. If you do not put the first card back in the deck before you draw the next, what is the probability that the first card is a club and the second one is black?

38. Two cards and a die. You choose two cards from a thoroughly shuffled deck, and then you roll a die. If you do not put the first card back in the deck before you draw the next, what is the probability that the first card is a queen, the second card is a king, and the die shows a 6?

39. Traffic lights. Suppose a certain traffic light shows a red light for 45 seconds, a green light for 45 seconds, and a yellow light for 5 seconds.

a. If you look at this light at a randomly chosen instant, what is the probability that you see a red or yellow light?

b. Each day you look at this light at a randomly chosen instant. What is the probability that you see a red light five days in a row?

40. Red light. The city finds empirically that the probability that an automobile going through the intersection of 5th and Main will run a red light sometime during a given day is 9.2%, and the probability that an automobile going through the intersection of 7th and Polk will run a red light sometime during a given day is 11.7%. Suppose a car is picked at random at 5th and Main on a given day and another at random at 7th and Polk. Write each probability as a percentage rounded to one decimal place.

a. What is the probability that the car will not run a red light at 5th and Main?

b. What is the probability that the car will not run a red light at 7th and Polk?

c. What is the probability that neither car runs a red light?

41. A box. Suppose that the inside bottom of a box is painted with three colors: 1/3 of the bottom area is red, 1/6 is blue, and 1/2 is white. You toss a tiny pebble into the box without aiming and note the color on which the pebble lands. Then you toss another tiny pebble into the box without aiming and note the color on which that pebble lands.

a. Are the two trials independent?

b. What is the probability that both pebbles land on the color blue?

c. What is the probability that the first pebble lands on the color blue and the second pebble lands on red?

d. What is the probability that one of the pebbles lands on the color blue and the other pebble lands on red? (How is this part different from part c?)

42. Lightbulbs I. A room has three lightbulbs. Each one has a 10% probability of burning out within the month. Write each probability as a percentage rounded to one decimal place.

a. What is the probability that all three will burn out within the month?

b. What is the probability that all three will be lit at the end of the month?

c. What is the probability that at the end of the month at least one of the bulbs will be lit?

43. Lightbulbs II. *This exercise is a bit more challenging.* It is important that a room have light during the next month. You want to ensure that the chances of the room going dark are less than 1 in a million. How many lightbulbs should you install in the room to guarantee this if each one has a 95% probability of staying lit during the month?

44. Murphy's law. "If something can go wrong, it will go wrong." This funny saying is called Murphy's law. Let's interpret this to mean "If something can go wrong, there is a very high probability that it will *eventually* go wrong."

Suppose we look at the event of having an automobile accident at some time during a day's commute. Let's assume that the probability of having an accident on a given day is 1 in a thousand or 0.001. That is, in your town, one of every thousand cars on a given day is involved in an accident (including little fender-benders). We also assume that having (or not having) an accident on a given day is independent of having (or not having) an accident on any other given day. Suppose you commute 44 weeks per year, 5 days a week, for a total of 220 days each year. In the following parts, write each probability in decimal form rounded to three places.

a. What is the probability that you have no accident over a year's time?

b. What is the probability that you will have at least one accident over a one-year period?

c. Repeat parts a and b for a 10-year period and for a 20-year period.

d. Does your work support the idea that there is a mathematical basis for Murphy's law as we interpreted it?

45–48. Political party preference. *The following excerpt is from an article that appeared at the Web site NewsOK.*

IN THE NEWS 5.6

Search

Home | TV & Video | U. S. | World | Politics | Entertainment | Tech | Travel

In Oklahoma, Republicans Continue to Show Net Increase Over Democrats in Registered Voters

Oklahoma Republicans had a 3-1 net increase over Democrats in registered voters, registration records show.

MICHAEL MCNUTT November 2, 2012

Oklahoma, where the Republican presidential candidate four years ago won all 77 counties, continues to move toward a deeper shade of red.

Republicans showed a 3-1 net increase over Democrats in registered voters since Jan. 15, figures released Thursday by the state Election Board show. Even independents outpaced Democrats in the net increase of voters...

Democrats still have the most registered voters in the state, but have slipped to 45.6 percent of all voters. Republicans have grown to 42.4 percent of registered voters and independents have grown to 12 percent...

This information is used in Exercises 45 through 48. Write each probability as a percentage rounded to the nearest whole number.

45. If you pick a registered voter at random, what is the probability that you have chosen a Democrat? What is the probability of a Republican? What is the probability of an Independent?

46. Suppose that on 11/1/2012, you chose a registered voter in Oklahoma at random. What is the probability that you have chosen either a Democrat or a Republican?

47. Suppose you choose two people at random from among the registered voters. What is the probability that you get two independents?

48. This problem is more challenging. Suppose you choose three people at random from among the registered voters. What is the probability that you get one Republican, one Democrat, and one Independent? *Suggestion*: First assume that you pick three people one at a time, and find the probability that you get a Republican, a Democrat, and an Independent, in that order. How should you modify this number to get the requested probability?

Exercise 49 is suitable for group work.

49. Billion Dollar Bracket Challenge. In January 2014, the company Quicken Loans made news by announcing that it would pay \$1 billion to the person who submitted the perfect NCAA bracket for the men's Division I tournament that year. The company bought an insurance policy from Warren Buffett's holding company Berkshire Hathaway to cover the cost of any prize money.

a. Winning the \$1 billion prize required a contestant to complete a bracket that predicted correctly the outcome of 63 games. What is the probability that a person who predicted the winners of the 63 games by random choice would have a perfect bracket? Express your answer in terms of a power of 2.

b. The answer to part a is very small. To see how small it is, assume that everyone in the United States filled out many brackets and that all of the brackets produced in this way were different. (In fact, the contest was limited to 15 million entries altogether.) How many brackets would each person have to fill out to make sure that altogether all of the possible outcomes of the 63 games were represented? Use the fact that the population of the United States at this time was 317 million, and round your answer to the nearest billion. Note: You will need a calculator that can handle large numbers.

50. History. The Counting Principle has surely been known for a long time. In fact, it may be involved in a curious problem found in one of the oldest surviving mathematical texts, the Rhind papyrus from ancient Egypt. The papyrus itself dates from about 1650 b.c., but it may be a copy of a source 150 years older than that. The curious problem from the Rhind papyrus has the following data (where a *hekat* is a measure of volume):

Houses	7
Cats	49
Mice	343
Heads of wheat	2401
Hekat measures	16,807

One historian has suggested that the context of the table is as follows: In an estate there are seven houses, each house had seven cats, each cat could eat seven mice, each mouse could eat seven heads of wheat, and each head of wheat could produce seven hekat measures of grain.[15] Use the Counting Principle to explain how this interpretation is consistent with the numbers in the table.

51. Double sixes. Chevalier de Méré was a gambler who lived in the seventeenth century. He asked whether or not one should bet even money on the occurrence of at least one "double six" during 24 throws of a pair of dice. This question was a factor that motivated Fermat and Pascal to study probability (see Exercise 61 in Section 5.1). Answer de Méré's question: What is the probability of at least one "double six" in 24 consecutive rolls of a pair of dice? Write your answer in decimal form rounded to two places. *Hint*: First find the probability of 24 consecutive rolls of a pair of dice *without* a "double six."

Exercises 52 through 54 are designed to be solved using technology such as calculators or computer spreadsheets.

52. Factorials. Suppose you want to arrange five people in order in a waiting line. There are five choices for the person who goes first, four for the person who goes second, and so on. Thus, the number of ways to arrange five people is $5 \times 4 \times 3 \times 2 \times 1 = 120$. This sort of product arises often enough that it has a special notation:[16] We write $n!$ (read "n factorial") for $n \times (n-1) \times \cdots \times 1$. For example, $5! = 120$ according to the previous expansion. Use technology to determine how many ways there are to arrange 12 people in order in a waiting line.

53. More on factorials. *This is a continuation of Exercise 52.* Use technology to determine how many ways there are to arrange 50 people in order in a waiting line. Can you find the answer for 200 people?

54. Listing cases. Devise a computer program that will list all possible outcomes of tossing n coins. *Warning*: If you actually run your program for even moderate values of n, you will get a very long output.

Writing About Mathematics

55. The Rhind papyrus. The Rhind papyrus cited in Exercise 50 is an important document in the history of mathematics. Look up information about the Rhind papyrus and write a report.

56. Chevalier de Méré. Chevalier de Méré was mentioned in Exercise 51. Look up information about him and report your findings. Be sure to include connections with Fermat and Pascal.

What Do You Think?

57. Favorable? In calculating the probability of losing a game, you told a friend that your textbook referred to a loss as "favorable event." Your friend challenged you and said that a loss is *not* favorable. What would you say in response?

58. Passwords. When you choose passwords for Web sites, often you are prompted to use letters and numbers as well as some of the symbols across the top row of your keyboard. Why does this procedure result in a more secure password than just using the name of your pet? **Exercise 11 is relevant to this topic.**

59. The Counting Principle. Refer to the statement of the Counting Principle on page 342. Provide an example to show that $M+N$ does not give the correct number of possible outcomes for the two experiments.

60. Independent events. What does it mean for events to be *independent*? Give examples of events that are independent as well as some that are not. **Exercise 21 is relevant to this topic.**

61. Dangerous intersection. A city commission has observed that there have been accidents for each of the past 30 weeks at a certain intersection. One commissioner proposes that a traffic light be installed at that intersection. Another commissioner says that having 31 accidents in a row is very unlikely, so a traffic light is not needed. Which is right? What does this have to do with the concepts of "independent events" and "empirical probability"? **Exercise 40 is relevant to this topic.**

62. Going to St. Ives. Here is a simplified version of an old riddle. *I saw seven men going to St. Ives. Each man had seven sacks, and each sack had seven cats. Each cat had seven kits. How many people were going to St. Ives?* Of course the answer is seven, but what counting idea would you use if you were asked how many *items* were traveling to St. Ives? Can you think of other traditional riddles or fairy tales that involve this same counting idea? **Exercise 50 is relevant to this topic.**

63. Justice. Discuss how probability plays a role in our judicial system. Hint: Consider DNA testing, among other things.

64. Outcomes. In the statement of the Counting Principle, we assume that for each outcome of the first experiment there are the same number of outcomes of the second experiment. Would the Counting Principle apply without this assumption? Hint: Consider, for example, a very simple menu with two entrées. There is one side dish available for one entrée, and there are two side dishes available for the other. How many possible combinations of an entrée with a side dish are there?

65. Self-independence. Use the product formula to determine under what conditions an event A is independent of itself. Does your result make sense in terms of the initial informal description of independence in the text? *Reminder*: The solutions of the equation $x = x^2$ are $x = 0$ and $x = 1$.

[15] *See Howard Eves,* An Introduction to the History of Mathematics, *6th ed. (Philadelphia: Saunders, 1990).*

[16] *The idea of factorials is discussed more fully in Section 5.4.*

5.4 More ways of counting: Permuting and combining

TAKE AWAY FROM THIS SECTION
Efficient counting techniques apply in many practical settings.

This following article provides a basic introduction to the way genetic information is stored in DNA.

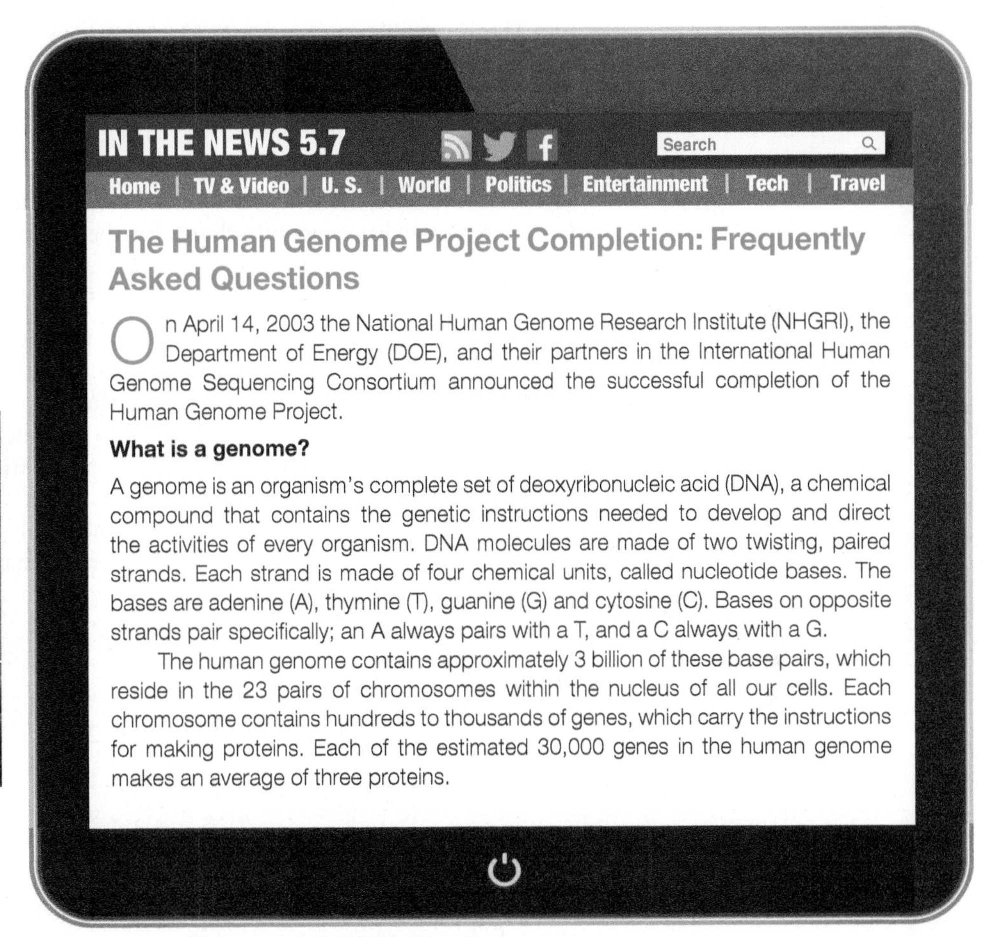

IN THE NEWS 5.7

Search

Home | TV & Video | U. S. | World | Politics | Entertainment | Tech | Travel

The Human Genome Project Completion: Frequently Asked Questions

On April 14, 2003 the National Human Genome Research Institute (NHGRI), the Department of Energy (DOE), and their partners in the International Human Genome Sequencing Consortium announced the successful completion of the Human Genome Project.

What is a genome?

A genome is an organism's complete set of deoxyribonucleic acid (DNA), a chemical compound that contains the genetic instructions needed to develop and direct the activities of every organism. DNA molecules are made of two twisting, paired strands. Each strand is made of four chemical units, called nucleotide bases. The bases are adenine (A), thymine (T), guanine (G) and cytosine (C). Bases on opposite strands pair specifically; an A always pairs with a T, and a C always with a G.

The human genome contains approximately 3 billion of these base pairs, which reside in the 23 pairs of chromosomes within the nucleus of all our cells. Each chromosome contains hundreds to thousands of genes, which carry the instructions for making proteins. Each of the estimated 30,000 genes in the human genome makes an average of three proteins.

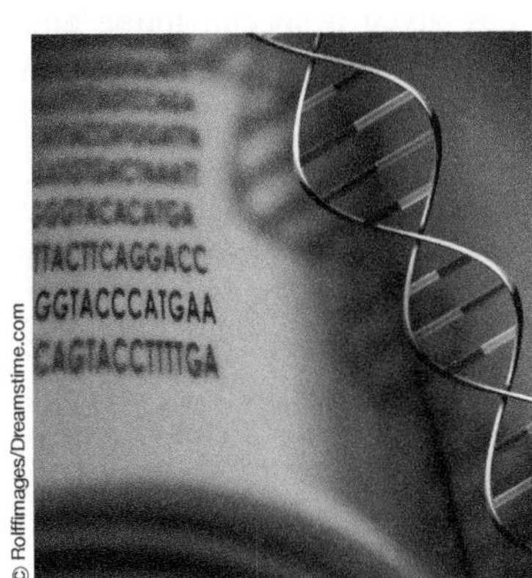

One major goal of understanding the sequence in a DNA molecule is the identification of genes that give rise to cancer. Understanding the genetics of cancer comes down to determining how four letters, A, C, G, T, go together in long strings. The number of possibilities is staggering, but there are mathematical tools to help in the enumeration. These tools, which ultimately rely on the Counting Principle, are considered in this section.

Permutations

Suppose we arrange a group of people into a line. The first in line gets first prize, the second in line gets second prize, and so on. Your place in line determines the prize you get. So the order in which people are arranged in the line is important. An arrangement where order is important is called a *permutation*.

KEY CONCEPT

A permutation of items is an arrangement of the items in a certain order. Each item can be used only once in the sequence.

Suppose, for example, that we use the letters A, B, and C to make product codes where each letter is used exactly once. How many such codes can we make? That is, how many permutations are there of the letters A, B, and C? We can list them all to get the answer. The permutations of the letters A, B, C, are ABC, ACB, BAC, BCA, CAB, CBA—six in all. When the number of items is large, it is not efficient to list all of the permutations. Instead, we need a faster way of calculating how many there are.

Think of making a permutation (or arrangement) by filling three blanks with the three letters without repetition, that is, using each letter once:

$$\_\ \_\ \_$$

There are three choices for the first blank. When that blank is filled, we have only two letters left, so there are two choices for the next blank. Now there is only one letter left, so we have just one choice for the last blank:

$$\underline{3}\ \underline{2}\ \underline{1}$$

The Counting Principle says that the total number of ways to fill all the blanks is $3 \times 2 \times 1 = 6$. This agrees with the number we found by making a list.

This kind of calculation occurs so often that it is given a special name and symbol. For a positive whole number n, the product of all the whole numbers from n down through 1 is denoted by n with an exclamation point: $n!$. This is pronounced "n factorial."

Although it may seem strange, zero factorial (0!) is always defined to be the number 1. One way to justify this is by asserting that there's just one way to arrange no items. We'll see another reason why later. Here are some factorials:

$$0! = 1$$
$$1! = 1$$
$$2! = 2 \times 1 = 2$$
$$3! = 3 \times 2 \times 1 = 6$$
$$4! = 4 \times 3 \times 2 \times 1 = 24$$

The method we used to find the number of arrangements of the letters A, B, C works in general. For example, suppose there are 10 people waiting to buy tickets to some event. In how many ways can we arrange the 10 people in a line at the ticket booth? Thinking again of filling in slots, we see that there are 10 people who could be at the head of the line. That leaves nine who might be second, eight in third place, etc. So the total number of ways to arrange the 10 people is 10!, which equals 3,628,800. That's a pretty big number.

SUMMARY 5.5
Permutations of n Items

The number of permutations of n items is the number of ways to arrange n distinct items without repetition. The Counting Principle shows that this is n factorial, which is defined as

$$n! = n \times (n-1) \times \cdots \times 2 \times 1$$

EXAMPLE 5.21 Using factorials: Arranging four letters

In how many ways can I arrange the four letters A, B, C, D using each letter only once?

SOLUTION

The number of permutations of four letters is $4! = 4 \times 3 \times 2 \times 1 = 24$.

TRY IT YOURSELF 5.21

In how many ways can I arrange the five letters A, B, C, D, E using each letter only once?

The answer is provided at the end of this section.

Factorials get big fast—as we just saw, $10! = 3,628,800$. To find large factorials on a calculator using the definition, you would have to do a lot of multiplications, which could be quite tedious. The accompanying tip should help.

CALCULATION TIP 5.3 Factorials with a Calculator or Computer

- On the TI-83 and TI-84 graphing calculators, there's a factorial command [!] located in the math probability menu. To find 10!, enter 10 MATH PRB [4].
- In the spreadsheet program Excel©, the command for n! is FACT(n). For example, to find 10!, enter FACT(10).
- You can also find factorials on the calculator that comes with Microsoft Windows©. Choose the View menu in the calculator and click on Scientific to get the full size. You can see the [n!] button in the fourth column, just above the [1/x] button.

More on arrangements

Let's revisit the idea of putting people in order and awarding prizes. But this time, let's do it under the more realistic assumption that not everyone gets a prize. We think of selecting 3 lucky winners from a group of 10 people. The first person selected gets first prize, the second gets second prize, and the third gets third prize. Such an arrangement is called *a permutation of 10 people taken 3 at a time.*

KEY CONCEPT

The number of ways to select k items from n distinct items and arrange them in order is called the number of permutations of n items taken k at a time.

The Counting Principle allows us to calculate the number of permutations of this type.

EXAMPLE 5.22 Counting permutations: Three prizes

In how many ways can we select 3 people from a group of 10 and award them first, second, and third prizes? This is the number of permutations of 10 items taken 3 at a time.

SOLUTION

We think of filling 3 slots with names selected from the 10 people:

$$\_\ \_\ \_$$

There are 10 names available for the first slot. But because nobody gets more than one prize, there are only 9 available for the second and then 8 available for the third:

$$\underline{10}\ \underline{9}\ \underline{8}$$

The Counting Principle shows that the number of ways to select the three winners is $10 \times 9 \times 8 = 720$.

TRY IT YOURSELF 5.22

In how many ways can we select four letters from the alphabet without repetition and then arrange them in a sequence?

The answer is provided at the end of this section.

Now we use our ideas about counting permutations to assign poll watchers.

EXAMPLE 5.23 Permutations: Poll watchers

A political organization wants to observe the voting procedures at five polling places. The organization has five poll watchers available.

a. In how many different ways can the poll watchers be assigned?

b. For each of the polling places, one of the available watchers lives in the precinct for that polling place. If the watchers are assigned at random, what is the probability that each one will be assigned to the polling place for the precinct where he or she lives?

c. Suppose now that the organization wants to observe only two of the polling places. In how many ways can the organization assign these places to two of the available watchers? This is the number of permutations of five items taken two at a time.

d. Assume now that there are 20 polling places and 20 poll watchers available. How many permutations are there of the 20 watchers?

Ethan Miller/Getty Images

SOLUTION

a. As we noted earlier, the number of ways to arrange five items is

$$5! = 5 \times 4 \times 3 \times 2 \times 1 = 120$$

In terms of assigning poll watchers, there are five choices for the first polling place, followed by four choices for the next place, and so on.

b. From part a there are 120 outcomes, and only one of them has each of the watchers assigned to his or her home precinct. So the probability that each one will be assigned to the polling place for the precinct where he or she lives is 1/120 or about 0.008.

c. There are five choices to watch the first polling place and four choices to watch the second polling place. That makes $5 \times 4 = 20$ different arrangements.

d. As we have noted, the number of arrangements of 20 items is 20!, which is (get ready!) 2,432,902,008,176,640,000. Yikes! This number is too big even for most calculators, which will report at best an estimate. Generally speaking, specialized mathematical software is needed to handle numbers of this size, and hand calculation is out of the question.

Calculations using factorials

We want to find a simpler way to write the number of permutations when the number of items is large. Let's look at what happens when we have lots of books to arrange on a shelf. Suppose, for example, that we have 20 different books and we want to arrange 9 of them (in order) on a shelf. That is, we want to calculate the number of permutations of 20 items taken 9 at a time. Thinking of filling in blanks, we see that there are 20 books for the first position, 19 for the second, down to 12 for the last position:

$$\underline{20}\ \underline{19}\ \underline{18} \ldots \underline{12}$$

Thus, we find

$$20 \times 19 \times 18 \times 17 \times 16 \times 15 \times 14 \times 13 \times 12 \text{ arrangements}$$

It would be very tedious to multiply all nine of these numbers, even with a calculator. But there's an easy way to find the answer on a calculator using the factorial command. Here's the trick: We write the expression

$$20 \times 19 \times 18 \times 17 \times 16 \times 15 \times 14 \times 13 \times 12$$

as

$$\frac{20 \times 19 \times 18 \times 17 \times 16 \times 15 \times 14 \times 13 \times 12 \times (11!)}{11!}$$

Now we expand the factorial in the numerator to find

$$\frac{20 \times 19 \times 18 \times 17 \times 16 \times 15 \times 14 \times 13 \times 12 \times (11 \times 10 \times 9 \times 8 \times 7 \times 6 \times 5 \times 4 \times 3 \times 2 \times 1)}{11!} = \frac{20!}{11!}$$

The result is that the number of permutations of 20 items taken 9 at a time is

$$\frac{20!}{11!} = \frac{20!}{(20-9)!}$$

We use the notation $P(20, 9)$ to denote the number of permutations of 20 items taken 9 at a time. We have found that

$$P(20, 9) = \frac{20!}{(20-9)!}$$

With the aid of a calculator, we get

$$\frac{20!}{(20-9)!} = \frac{20!}{11!} = 60,949,324,800$$

This method is often useful, and we call attention to it in the accompanying summary.

SUMMARY 5.6
Permutations of n Items Taken k at a Time

The number of ways to select k items from n items and arrange them in order is the number of **permutations of n items taken k at a time.** We use the notation $P(n, k)$ to denote this number, and we can calculate it using the *permutations formula*

$$P(n, k) = \frac{n!}{(n-k)!}$$

EXAMPLE 5.24 Using the permutations formula: Student newspaper

The editor of the student newspaper has seven slots to fill on the editorial board. Twenty-five students have applied for a slot. Use the permutations formula to find the number of ways to select seven applicants and arrange them in a sequence.

SOLUTION

We are selecting 7 of 25 applicants, so we use the permutations formula $P(n, k) = \frac{n!}{(n-k)!}$ with $n = 25$ and $k = 7$:

$$P(25, 7) = \frac{25!}{(25-7)!} = \frac{25!}{18!}$$

Using a calculator, or by a hand-calculation method that we will shortly introduce, we find that there are 2,422,728,000 ways to select seven applicants and arrange them in a sequence. We remark that many calculations involving permutations produce exceptionally large numbers.

TRY IT YOURSELF 5.24

Use the permutations formula to find the number of ways to select 10 applicants from 25 and arrange them in a sequence. Leave your answer in terms of factorials.

The answer is provided at the end of this section.

CALCULATION TIP 5.4 Permutations

The TI-83 and TI-84 graphing calculators have a permutation command $\boxed{nPr}$ located in the math probability menu along with the factorial command. To find the number of permutations of n items taken r at a time, enter the number of items n, first, then the permutation key, then the number of items to select at a time, r. For example, to find the number of permutations of 10 items taken 4 at a time ($10P4$), enter 10 MATH PRB $\boxed{2}$ 4.

In the spreadsheet Excel©, the command for the number of permutations of n items taken r at a time is PERMUT(n,r). For example, to find the number of permutations of 10 items taken 4 at a time, enter PERMUT(10,4).

Combinations

A permutation of a set of items is an arrangement of those items in a certain order. By contrast, a *combination* of a set of items is a collection that does not take into account the order in which they happen to appear.

KEY CONCEPT

A combination of a group of items is a selection from that group in which order is not taken into account. No item can be used more than once in a combination.

For example, if we consider the number of ways we can pick 5 people from a group of 10 and give the first a prize of $10, the second a prize of $9, and so on, it matters who comes first. Thus, we are interested in counting permutations. In contrast, if we select 5 people from a group of 10 and give each of them a $10 prize, order makes no difference. Everybody gets the same prize. Here, we want to find the number of combinations, not the number of permutations.

To illustrate the point further, note that CAB and BAC are two different permutations of the letters A, B, and C. However, they would be considered the same *combination*.

How many combinations of the letters A, B, C are there taken two at a time? In other words, how many ways can we select two of the three letters if it makes no difference in which order the letters are considered? Simply by listing them, we can see that there are three possible *combinations*: AB, AC, and BC.

For comparison, we note that there are six *permutations* of these three letters taken two at a time: AB, BA, AC, CA, BC, and CB. That is, there are six ways of selecting two letters from a list of three if the order in which they are chosen matters. Note that in the list of six permutations above each combination appears twice—for example, AB and BA. The number of permutations, 6, is twice the number of combinations, 3. Thus, to find the number of combinations from the number of permutations, we divide 6 by 2. This observation is the key to finding an easy way to count combinations. We find the number of permutations, then look to see how many times each different combination is listed, and divide. Expressed as a formula, this is

$$\text{Combinations} = \frac{\text{Permutations}}{\text{Repeated listings}}$$

In the case of permutations of k items, the number of times each combination is listed is $k!$. (Why?) This observation yields the formula in the following summary.

SUMMARY 5.7
Combinations

The number of ways to select k items from n items when order makes no difference is the number of **combinations of n items taken k at a time.** We use the notation $C(n, k)$ to denote this number, and we can calculate this number using the *combinations formula*:

$$\text{Combinations of } n \text{ items taken } k \text{ at a time} = \frac{n!}{k!(n-k)!}$$

EXAMPLE 5.25 Using the combinations formula: Three-person committee

Use the combinations formula to express the number of three-person committees I can select from a group of six people.

SOLUTION

This is the number of combinations of six people taken three at a time. We use the combinations formula $C(n, k) = \frac{n!}{k!(n-k)!}$ with $n = 6$ and $k = 3$:

$$\begin{aligned} C(6, 3) &= \frac{6!}{3!(6-3)!} \\ &= \frac{6!}{3!3!} \\ &= 20 \end{aligned}$$

Thus, there are 20 possible ways of selecting a three-person committee from a group of six people.

TRY IT YOURSELF 5.25

Use the combinations formula to express the number of four-person committees I can select from a group of nine people. Leave your answer in terms of factorials.

The answer is provided at the end of this section.

CALCULATION TIP 5.5 Combinations

If you enter the formula for combinations on a calculator, be sure to put parentheses around the denominator. For example, enter $\frac{8!}{5!3!}$ as 8!/(5!3!).

The TI-83 and TI-84 graphing calculators have a combination key $\boxed{nCr}$ located in the math probability menu. To find the number of combinations of n items taken r at a time, enter the number of items n first, then the combination key, then the number of items to select at one time, r. For example, to find the number of combinations of 10 items taken 4 at a time (10C4), enter 10 MATH PRB $\boxed{3}$ 4.

In the spreadsheet Excel©, the command to find the number of combinations of n items taken r at a time is COMBIN(n,r). For example, to find the number of combinations of 10 items taken 4 at a time, enter COMBIN(10,4).

Hand calculation of combinations

Sometimes we can simplify fractions involving factorials to avoid long calculations. Let's look, for example, at 55!/53!. We can save a lot of time in calculating this by hand if we notice that lots of terms cancel:

$$\begin{aligned} \frac{55!}{53!} &= \frac{55 \times 54 \times 53 \times 52 \times \cdots \times 3 \times 2 \times 1}{53 \times 52 \times \cdots \times 3 \times 2 \times 1} \\ &= \frac{55 \times 54 \times \cancel{53} \times \cancel{52} \times \cdots \times \cancel{3} \times \cancel{2} \times \cancel{1}}{\cancel{53} \times \cancel{52} \times \cdots \times \cancel{3} \times \cancel{2} \times \cancel{1}} \\ &= 55 \times 54 \\ &= 2970 \end{aligned}$$

Cancellations may be even more helpful when dealing with combinations. For example, consider the number of combinations of 10 items taken 6 at a time:

$$\begin{aligned} C(10, 6) &= \frac{10!}{6!(10-6)!} \\ &= \frac{10!}{6!4!} \end{aligned}$$

If we write this out, we find

$$\frac{10!}{6!4!} = \frac{10 \times 9 \times 8 \times 7 \times 6 \times 5 \times 4 \times 3 \times 2 \times 1}{(6 \times 5 \times 4 \times 3 \times 2 \times 1) \times (4 \times 3 \times 2 \times 1)}$$

Look at the larger factorial on the bottom (6! in this case). It will always cancel with part of the top:

$$\frac{10 \times 9 \times 8 \times 7 \times \cancel{6} \times \cancel{5} \times \cancel{4} \times \cancel{3} \times \cancel{2} \times \cancel{1}}{(\cancel{6} \times \cancel{5} \times \cancel{4} \times \cancel{3} \times \cancel{2} \times \cancel{1}) \times (4 \times 3 \times 2 \times 1)} = \frac{10 \times 9 \times 8 \times 7}{4 \times 3 \times 2 \times 1}$$

We will always get a whole number for the final answer. (Why?) Thus, the rest of the denominator will cancel as well. Note that the 4 × 2 in the denominator cancels with the 8 in the numerator, and the 3 divides into the 9:

$$\frac{10 \times \overset{3}{\cancel{9}} \times \cancel{8} \times 7}{\cancel{4} \times \cancel{3} \times \cancel{2} \times 1} = 10 \times 3 \times 7 = 210$$

Following this procedure can significantly shorten manual computations of this sort.

Remember that whenever you calculate any permutation $P(n, k)$ or combination $C(n, k)$, you should always end up with a whole number. If you wind up with a fraction, go back and check your work.

CALCULATION TIP 5.6 Combinations

If you calculate a permutation or combination with pencil and paper, you can always cancel all the terms in the denominator before doing any multiplication. Calculating combinations involving large numbers is best done with a computer or calculator.

EXAMPLE 5.26 Calculating combinations by hand: Nine items taken five at a time

Calculate by hand the number of combinations of nine items taken five at a time.

SOLUTION

The number of combinations of nine items taken five at a time is

$$C(9, 5) = \frac{9!}{5!(9-5)!} = \frac{9!}{5!4!}$$

We compute

$$\begin{aligned}
\frac{9!}{5!4!} &= \frac{9 \times 8 \times 7 \times 6 \times 5 \times 4 \times 3 \times 2 \times 1}{5 \times 4 \times 3 \times 2 \times 1 \times 4 \times 3 \times 2 \times 1} \\
&= \frac{9 \times 8 \times 7 \times 6 \times \cancel{5} \times \cancel{4} \times \cancel{3} \times \cancel{2} \times \cancel{1}}{\cancel{5} \times \cancel{4} \times \cancel{3} \times \cancel{2} \times \cancel{1} \times 4 \times 3 \times 2 \times 1} \\
&= \frac{9 \times 8 \times 7 \times 6}{4 \times 3 \times 2 \times 1} \\
&= \frac{9 \times \cancel{8} \times 7 \times 6}{\cancel{4} \times 3 \times \cancel{2} \times 1} \\
&= \frac{\overset{3}{\cancel{9}} \times 7 \times 6}{\cancel{3} \times 1} \\
&= 3 \times 7 \times 6 \\
&= 126
\end{aligned}$$

TRY IT YOURSELF 5.26

Calculate by hand the number of combinations of 10 items taken 7 at a time.

The answer is provided at the end of this section.

Here is an example of using combinations to count the number of three-judge panels used in the U.S. Ninth Circuit Court of Appeals, which covers several western states, including California.

EXAMPLE 5.27 Computing combinations: Three-judge panels

Portland Press Herald/Getty Images

The *New York Times* ran an article on June 30, 2002, with the title, "Court That Ruled on Pledge Often Runs Afoul of Justices." The court in question is the Ninth Circuit Court, which ruled in 2002 that the Pledge of Allegiance to the flag is unconstitutional because it includes the phrase "under God." The article discusses the effect of having a large number of judges, and it states, "The judges have chambers throughout the circuit and meet only rarely. Assuming there are 28 judges, there are more than 3000 possible combinations of three-judge panels."

a. Is the article correct in stating that there are more than 3000 possible combinations consisting of three-judge panels? Exactly how many three-judge panels can be formed from the 28-judge court?

b. Assume that there are 28 judges. How many 28-judge panels are there?

SOLUTION

a. We want to find the number of ways to choose 3 items from 28 items. We

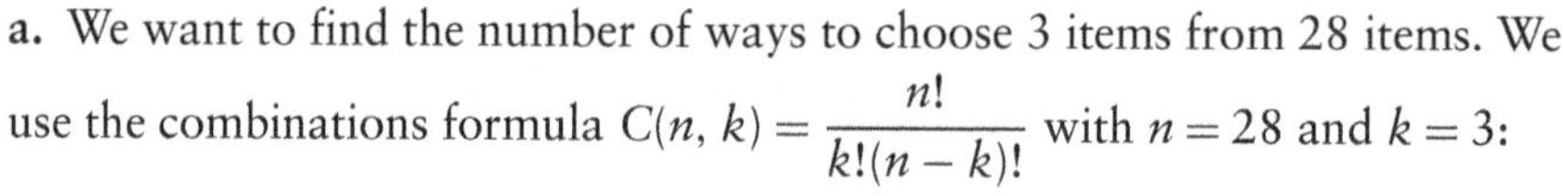

use the combinations formula $C(n, k) = \frac{n!}{k!(n-k)!}$ with $n = 28$ and $k = 3$:

$$C(28, 3) = \frac{28!}{3!(28-3)!} = \frac{28!}{3!25!}$$

This fraction can be simplified by cancellation:

$$\frac{28!}{3!25!} = \frac{28 \times 27 \times 26 \times \cancel{25} \times \cancel{24} \times \cancel{23} \times \cdots \times \cancel{1}}{(3 \times 2 \times 1) \times (\cancel{25} \times \cancel{24} \times \cancel{23} \times \cdots \times \cancel{1})}$$

$$= \frac{28 \times 27 \times 26}{3 \times 2 \times 1}$$

$$= \frac{28 \times \overset{9}{\cancel{27}} \times \overset{13}{\cancel{26}}}{\cancel{3} \times \cancel{2} \times 1}$$

$$= 3276$$

The article is correct in stating that there are more than 3000 possible three-judge panels. There are, in fact, 3276 such combinations.

b. We are really being asked, "In how many ways can we choose 28 judges from 28 judges to form a panel?" There is obviously only one way: All 28 judges go on the panel.

Let's see what happens if we apply the combinations formula to this simple case. We use the combinations formula $C(n, k) = \frac{n!}{k!(n-k)!}$ with $n = 28$ and $k = 28$:

$$C(28, 28) = \frac{28!}{28!(28-28)!} = \frac{28!}{28!0!} = 1$$

This is the same as the answer we found before. Remember that 0! is taken to be 1, and this formula is one reason why.

Probabilities with permutations or combinations

Now that we have more sophisticated ways of counting, we can calculate certain probabilities more easily.

EXAMPLE 5.28 Finding the probability: Selecting a committee

In a group of six men and four women, I select a committee of three at random. What is the probability that all three committee members are women?

SOLUTION

Recall that

$$\text{Probability} = \frac{\text{Favorable outcomes}}{\text{Total outcomes}}$$

We first find the number of favorable outcomes, which is the number of three-person committees consisting of women. There are four women, so the number of ways to select a three-woman committee is the number of combinations of four items taken three at a time. We use the combinations formula $C(n, k) = \frac{n!}{k!(n-k)!}$ with $n = 4$ and $k = 3$:

$$\begin{aligned} C(4, 3) &= \frac{4!}{3!(4-3)!} \\ &= \frac{4!}{3!1!} \\ &= 4 \end{aligned}$$

Next we find the number of ways to select a 3-person committee from the 10 people (the total number of outcomes). This is the number of combinations of 10 people taken 3 at a time. We use the combinations formula $C(n, k) = \frac{n!}{k!(n-k)!}$ with $n = 10$ and $k = 3$:

$$\begin{aligned} C(10, 3) &= \frac{10!}{3!(10-3)!} \\ &= \frac{10!}{3!7!} \\ &= 120 \end{aligned}$$

So the probability of an all-female committee is

$$\frac{\text{Favorable outcomes}}{\text{Total outcomes}} = \frac{4}{120} = \frac{1}{30}$$

This is about 0.03.

TRY IT YOURSELF 5.28

In a group of five men and five women, I select a committee of three at random. What is the probability that all three committee members are women?

The answer is provided at the end of this section.

Hand	Payout (5)	Probability
Royal Flush	4000	0.00%[17]
Straight Flush	250	0.01%
4 of a Kind	125	0.24%
Full House	45	1.15%
Flush	30	1.10%
Straight	20	1.12%
3 of a Kind	15	7.45%
Two Pair	10	12.93%
Jacks or Better	5	21.46%
Nothing	0	54.54%

This table displays some results for Deuce's Wild poker played using a popular video poker game: the hand that is dealt or held, the payout or amount of coins for that winning hand, and the probability that a hand will occur.

EXAMPLE 5.29 Finding the probability: Poker hands

There are many different kinds of poker games, but in most of them the winner is ultimately decided by the best five-card hand. Suppose you draw five cards from a full deck.

a. How many five-card hands are possible?

b. What is the probability that exactly two of the cards are kings?

c. What is the probability that all of them are hearts?

SOLUTION

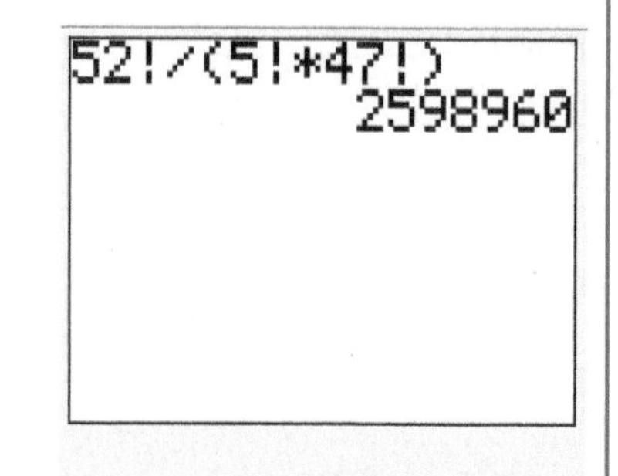

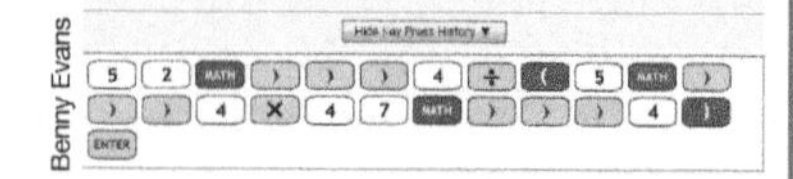

Benny Evans

a. The total number of hands is the number of combinations of 52 items taken 5 at a time. We use the combinations formula $C(n, k) = \frac{n!}{k!(n-k)!}$ with $n = 52$ and $k = 5$:

$$\begin{aligned} C(52, 5) &= \frac{52!}{5!(52-5)!} \\ &= \frac{52!}{5!47!} \\ &= 2{,}598{,}960 \end{aligned}$$

(Thus, there are more than 2.5 million different five-card poker hands.) This is the number we will use in the denominator of the remaining calculations.

b. We must count how many ways we can choose five cards from the deck so that exactly two are kings. To do this, we must choose two kings and three others. There are four kings, from which we choose two. We use the combinations formula $C(n, k) = \frac{n!}{k!(n-k)!}$ with $n = 4$ and $k = 2$:

$$\begin{aligned} C(4, 2) &= \frac{4!}{2!(4-2)!} \\ &= \frac{4!}{2!2!} \\ &= 6 \end{aligned}$$

[17] *Although the chart gives a probability of 0.00% for a Royal Flush, the probability is actually a very small positive number.*

There are 48 cards that are not kings, and we must choose 3 from these. That is, we need the number of combinations of 48 items taken 3 at a time. We use the combinations formula $C(n, k) = \dfrac{n!}{k!(n-k)!}$ with $n = 48$ and $k = 3$:

$$C(48, 3) = \frac{48!}{3!(48-3)!} = \frac{48!}{3!45!} = 17{,}296$$

The Counting Principle says that there are $6 \times 17{,}296 = 103{,}776$ ways to do both. Therefore, the probability we seek is 103,776/2,598,960. This is about 0.040 or 4.0%.

c. We need to count how many ways we can choose five cards from the deck so that all of them are hearts. There are 13 hearts, and the number of ways to choose 5 of them is the number of combinations of 13 items taken 5 at a time. We use the combinations formula

$$C(n, k) = \frac{n!}{k!(n-k)!}$$

with $n = 13$ and $k = 5$:

$$C(13, 5) = \frac{13!}{5!(13-5)!} = \frac{13!}{5!8!} = 1287$$

So the probability is 1287/2,598,960. This is about 0.0005 or 0.05%.

WHAT DO YOU THINK?

Just one more: If you use the first 10 letters of the alphabet to make a code where repetition is not allowed, there are 10! possible arrangements of those 10 letters. That means the probability of someone cracking your code by choosing a random arrangement of the 10 letters is 1 divided by 10!. How is this probability affected if you add the eleventh letter of the alphabet to your list of numbers? In general, how is a factorial affected if the number is increased by 1? What if it is increased by 2?

Questions of this sort may invite many correct responses. We present one possibility: If I add one more letter, there are 11! arrangements without repetition of the letters. Since 10! is the product of the first 10 integers and 11! is the product of the first 11 integers, 11! is 11 times 10!. Thus, adding one letter multiplies the number of possible codes by 11. The probability of someone guessing the code is 1 divided by 11! and that is one-eleventh of 1 divided by 10!. This works for any number. For example, if you go from 99 letters to 100 letters then the new probability is one-hundredth of the old probability. If you use two new letters, you reduce the probability even more. Going from 99 to 101 multiplies the number of possibilities by $100 \times 101 = 10{,}100$. Thus, the new probability is less than one ten-thousandth of the old probability.

Further questions of this type may be found in the *What Do You Think?* section of exercises.

Try It Yourself answers

Try It Yourself 5.21: Using factorials: Arranging five letters $5! = 120$.

Try It Yourself 5.22: Counting permutations: Four letters $26 \times 25 \times 24 \times 23 = 358{,}800$.

Try It Yourself 5.24: Using the permutations formula: Student newspaper $25!/15!$.

Try It Yourself 5.25: Using the combinations formula: Four-person committee $9!/(4!5!)$.

Try It Yourself 5.26: Calculating combinations by hand: Ten items taken seven at a time 120.

Try It Yourself 5.28: Finding the probability: Selecting a committee 1/12 or about 0.08.

Exercise Set 5.4

Test Your Understanding

1. True or false: Permutations take into account the order of items, whereas combinations do not.

2. True or false: If k and n are integers with k between 0 and n, then the expression $\dfrac{n!}{k!(n-k)!}$ always reduces to a whole number.

3. In mathematics, a whole number followed by an exclamation point indicates ________.

4. True or false: It is impossible for there to be more combinations of n items taken k at a time than there are permutations of n items taken k at a time.

Problems

5–10. **Calculating with factorials.** *In Exercises 5 through 10, calculate the given factorial, permutation, or combination.*

5. $4!$

6. $\dfrac{8!}{6!}$

7. $P(5, 2)$

8. $P(7, 3)$

9. $C(6, 4)$

10. $C(10, 5)$

11. **Arranging letters.** In how many ways can I arrange the six letters A, B, C, D, E, F?.

12. **Awarding prizes.** In how many ways can we select 4 people from a group of 10 people and award them first through fourth prizes?

13. **More prizes.** In how many ways can we select 6 people from a group of 20 and award first through sixth prizes? Leave your answer in terms of factorials.

14. **Selecting a committee.** In how many ways can we select a committee of 6 people from a group of 30 candidates? Leave your answer in terms of factorials.

15. **Calculating permutations.** Calculate the number of permutations of 11 items taken 5 at a time.

16. **Calculating combinations.** Calculate the number of combinations of 10 items taken 7 at a time.

17. **Candidates.** There are eight candidates running for the presidential and vice-presidential nomination. How many ways are there of selecting presidential and vice-presidential candidates from the field?

18–25. **Order or not.** *In Exercises 18 through 25, determine whether order makes a difference in the given situations.*

18. Three people are selected at random from the class and are permitted to skip the final exam.

19. We put letters and numbers on a license plate.

20. I need some change, so I pick 5 coins from a dish containing 10 coins.

21. People are lined up at the polls to vote. Some are Democrats, and some are Republicans. We are interested in the outcome of the election.

22. We select three people for a committee. The first chosen will hold the office of president, the second vice president, and the third secretary.

23. We go to the deli and buy three kinds of cheeses for next week's lunches.

24. We select four men and six women to serve on a committee.

25. We put five people on a basketball team. They get the positions of point guard, shooting guard, small forward, power forward, and center.

26. **Signals.** The king has 10 (different) colored flags that he uses to send coded messages to his general in the field. For example, red-blue-green might mean "attack at dawn" and blue-green-red may mean "retreat." He sends the message by arranging three of the flags atop the castle wall. How many coded messages can the king send?

27. **The Enigma machine.** The *Enigma machine* was used by Germany in World War II to send coded messages. It has gained fame because it was an excellent coding device for its day and because of the ultimately successful efforts of the British (with considerable aid from the Poles) to crack the Enigma code. The breaking of the code involved, among other things, some very good mathematics developed by Alan Turing and others. One part of the machine consisted of three rotors, each containing the letters A through Z. To read an encrypted message, it was necessary to determine the initial settings of the three rotors (e.g., PDX or JJN). How many different initial settings of the three rotors are there? The naval version of the Enigma machine had four rotors rather than three. How many initial settings were made possible by the rotors on the naval Enigma machine?

Hulton Archive/Getty Images

Enigma machine.

This is only the beginning of the problem of deciphering the Enigma code. Other parts of the machine allowed for many more initial settings.

28. A new Enigma machine. *This is a continuation of Exercise 27.* Suppose we invent a new Enigma machine with 10 rotors. Four of the rotors must be initially set to A, but the remaining rotors can be set to any letter from B to Z. How many initial settings are there for this new Enigma machine? *Suggestion*: First select four rotors that will be set to A. (In how many ways can you do that?) Now determine how many settings there are for the remaining six rotors. Finally, use the Counting Principle to put the two together.

29. Permutations and passcodes. In a post at the Zscaler Security Cloud blog, Satyam Tyagi discusses the security of the iPhone. He states that a four-digit passcode has 10,000 permutations and that a six-digit passcode has a million permutations.

How many permutations (in the sense we have used in this chapter) are there of four digits? Do you think this author is using the word "permutation" in the same sense as we have used it in this chapter?

30. Military draft. On December 1, 1969 (in the midst of the Vietnam War), the U.S. Selective Service System conducted a lottery to determine the order in which young men (ages 18 to 26) would be drafted into military service. The days of the year, including February 29, were assigned numbers from 1 to 366. Each such number was written on a slip of paper and placed in a plastic capsule. The 366 capsules were placed in a large bowl. Then capsules were drawn, one at a time, from the bowl. The first number drawn corresponded to September 14, and young men born on that day were the first group to be called up.

AP Images

Military draft lottery in 1969.

a. Assume that capsules were drawn at random. (This point has been the subject of much discussion.) What is the probability that a given young man would be in the first group to be called up?

b. How many different outcomes are possible for such a drawing? Leave your answer in terms of factorials.

c. Young men born on the same day of the year were distinguished according to the first letter of their last, first, and middle names. This was done by randomly selecting the 26 letters of the alphabet. How many permutations of those 26 letters are there? Leave your answer in terms of factorials.

31. Candy. On Halloween, a man presents a child with a bowl containing eight different pieces of candy. He tells her that she may have three pieces. How many choices does she have?

32. More candy. On Halloween, a man presents a child with a bowl containing eight different pieces of candy and six different pieces of gum. He tells her that she may have four of each. How many choices does she have?

33. The Ninth Circuit Court. This exercise refers to the article on the Ninth Circuit Court summarized in Example 5.27. Suppose a three-judge panel has a designated chief justice and that we count two panels with the same members as being different if they have different chief justices. That is, we count the panel Adams, Jones, Smith, with Smith as chief, as a different panel than Adams, Jones, Smith, with Jones as chief. How many such panels can be formed from the 28-judge court?

34. Committees. Suppose there is a group of seven people from which we will make a committee.

a. In how many ways can we pick a three-person committee?

b. In how many ways can we pick a four-person committee? Compare your answer with what you found in part a.

c. In how many ways can we pick a five-person committee?

d. In how many ways can we pick a two-person committee? Compare your answer with what you found in part c.

35–38. Committees of men and women. *We want to select a committee of five members from a group of eight women and six men. The order of selection is irrelevant. Exercises 35 through 38 refer to this information.*

35. What is the total number of committees we can make?

36. How many committees can we make consisting of two women and three men?

37. How many committees can we make with Jack as committee chair? (Assume that only one member has that name!)

38. How many committees can we make with fewer men than women? *Suggestion*: There could be three women and two men, four women and one man, or five women.

39. Choosing 200 from 500. Is it true that the number of ways of choosing 200 items (not considering order) from 500 items is the same as the number of ways of choosing 300 items from 500 items? Explain how you arrived at your answer.

40. Mom's Country Kitchen. A dinner order at Mom's Country Kitchen consists of a salad with dressing, an entrée, and three sides. The menu offers a salad with your choice of ranch, bleu cheese, French, or Italian dressing; a choice of entrée from chicken-fried steak, meat loaf, fried chicken, catfish, roast beef, or pork chops; and, as sides, mashed potatoes, French fries, fried okra, black-eyed peas, green beans, corn, rice, pinto beans, or squash. How many possible dinner orders are available at Mom's?

41–44. Marbles. *In Exercises 41 through 44, we have a box with eight green marbles, five red marbles, and seven blue marbles. We choose three marbles from the box at random without looking. Write each probability in decimal form rounded to three places.*

41. What is the probability that they will all be green?

42. What is the probability that they will all be red?

43. What is the probability that they will all be blue?

44. What is the probability that they will all be the same color?

45. Money. You have the following bills: a one, a two, a five, a ten, a twenty, and a hundred. In how many ways can you distribute three bills to one person and two bills to a second person?

46. Poll watchers. A certain political organization has 50 poll watchers. Of these, 15 are to be sent to Ohio, 15 to Pennsylvania, 15 to Florida, and the remaining 5 to Iowa. In how many ways can we divide the poll watchers into groups to be sent to the different states? First write your answer in terms of factorials, and then give an estimate such as "about 10^{20}." *Suggestion*: Choose 15 to go to Ohio, then choose 15 from the remaining 35 to go to Pennsylvania, and so on.

47–49. Cards. *In Exercises 47 through 49, we draw three cards from a full deck. Write each probability in decimal form rounded to four places.*

47. What is the probability that they are all red?

48. What is the probability that they are all hearts?

49. What is the probability that they are all aces?

50–52. More cards. *In Exercises 50 through 52, we draw five cards from a full deck. Write each probability in decimal form rounded to four places.*

50. What is the probability that you have at least one ace? *Suggestion*: First find the probability that you get no aces.

51. What is the probability that they are all face cards (jacks, queens, or kings)?

52. What is the probability that exactly three of them are jacks? *Suggestion*: Think of first picking three jacks. Then pick the remaining two cards at random.

Exercises 53 and 54 are more challenging and are suitable for group work. See also Exercise 55.

53. Birthdays. Consider the following problem: *Assume that Jack and Jane were both born in August and that they were born on different days. Find the probability that both were born in the first week of August.*

Here is a purported solution: The number of ways of choosing 2 days from the first 7 is $\frac{7!}{5!2!} = 21$. Next we get the number of ways of picking 2 birthdays out of 31 (because August has 31 days). There are 31 days for the first birthday and 30 for the second. That is $30 \times 31 = 930$ ways to choose 2 birthdays from 31. So the probability of both birthdays occurring in the first week is $\frac{21}{930}$.

What is wrong with this solution? What is the correct answer? *Suggestion*: Counting using permutations shouldn't be mixed with counting using combinations.

54. Changing locks. The following article is concerned with the continuing plight of evacuees from Hurricanes Katrina and Rita in 2005.

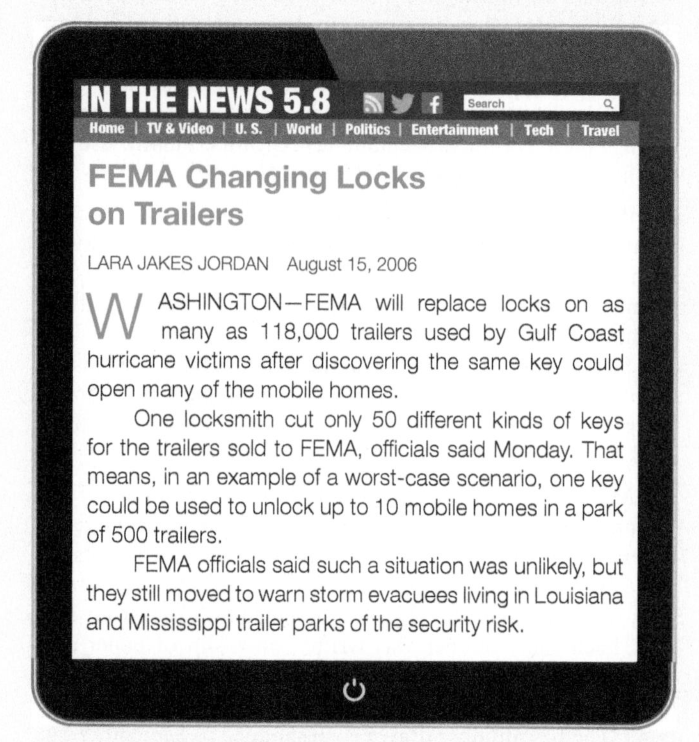

IN THE NEWS 5.8

Home | TV & Video | U. S. | World | Politics | Entertainment | Tech | Travel

FEMA Changing Locks on Trailers

LARA JAKES JORDAN August 15, 2006

WASHINGTON—FEMA will replace locks on as many as 118,000 trailers used by Gulf Coast hurricane victims after discovering the same key could open many of the mobile homes.

One locksmith cut only 50 different kinds of keys for the trailers sold to FEMA, officials said Monday. That means, in an example of a worst-case scenario, one key could be used to unlock up to 10 mobile homes in a park of 500 trailers.

FEMA officials said such a situation was unlikely, but they still moved to warn storm evacuees living in Louisiana and Mississippi trailer parks of the security risk.

This exercise refers to this article. You do not need to simplify your answers to parts a and c; just leave your answers in terms of factorials. In Exercise 55, you are asked to use technology to simplify those answers.

a. Suppose that a trailer park has 50 sets of 10 trailers, each set keyed the same. If you have one of the 50 keys and you try unlocking 34 trailers at random from the 500, what is the probability that you can open at least one of them? *Suggestion*: First calculate the probability that you can open none of them.

b. Now let's assume that each of the 50 keys fits the same number of trailers and that there are 118,000 trailers. Then each key would fit $118{,}000/50 = 2360$ trailers. If we select two trailers at random from the 118,000 trailers, what is the probability that the same key will fit them both? That is, what is the probability that the two trailers are keyed the same? Write the answer as a simple fraction and in decimal form rounded to two places.

c. Let's make the same assumptions as in part b. If we select 500 trailers at random from the 118,000, what is the probability that we get 50 sets (one for each key) of 10 trailers each that are keyed the same? (This is what the news article seems to suggest would be a "worst case.")

The following exercise is designed to be solved using technology such as a calculator or computer spreadsheet.

55. More on changing locks. Simplify the answers to parts a and c of Exercise 54. Write the probability in decimal form rounded to two places. *Warning*: For part c, this may require more sophisticated technology.

Writing About Mathematics

56. Abraham ben Meir ibn Ezra. The earliest detailed work on combinations was apparently done by Rabbi Abraham ben Meir ibn Ezra in the twelfth century in a text on astrology. Research the problems on combinations of interest to ibn Ezra, and determine some of his other scholarly pursuits.

57. The Enigma machine: history. Exercise 27 provides a very brief introduction to the German Enigma machine used in World War II. Investigate the Enigma code, including its eventual cracking by the code breakers at Bletchley Park, England, and report on your finding.

58. Alan Turing. Alan Turing was a mathematician who played a key role in cracking the Enigma code in World War II. His life story is intriguing for many reasons. Read about Alan Turing, and report your findings.

59. More on Alan Turing. One of Alan Turing's most important contributions to mathematics and computer science involved his *Turing machine*. Find out what a Turing machine is, and report your finding.

60. Factorial notation. Factorial notation has been attributed to a Frenchman, Christian Kramp. Find out what you can about him, and write a report.

What Do You Think?

61. A bicycle lock. Suppose you have a four-number bicycle lock. (Each rotor is numbered 0 to 9.) If a thief tries a random combination, how many times does he have to try in order to be sure the lock will open? Suppose the thief finds out that the four numbers in the combination are all different. Devise a systematic method that the thief could use to get the lock open. How many tries does your method require before all combinations are attempted?

62. Too big. Your friend said he was trying to use his calculator to solve a problem, and he kept getting an error message indicating that the numbers were too big. He said he was trying to find $\frac{1{,}000{,}000!}{999{,}999!}$. Can you help him?

63. Error. Describe what happens when you ask a scientific calculator or spreadsheet program for 1,000,000!. Try some other numbers such as 9999! and describe what you see. Search the Internet to see if you can find tools that can make accurate calculations of this sort, and report on your findings.

64. A simple proof. Recall that $C(n, k) = \frac{n!}{k!(n-k)!}$. It is a mathematical fact that $C(100, 20) = C(100, 80)$. Provide a simple justification for this fact in two ways: (a) Calculate by plugging in the numbers to show that the two expressions are equal, and (b) Consider how many ways you can select 20 people who will receive a prize from a group of 100 people, and compare that with how many ways you can select 80 people who will not receive a prize from a group of 100 people. Can you use either argument to formulate a general principle? Explain.

65. Permutations and combinations. How do you decide whether to approach counting problems using permutations or combinations? Give examples of two situations, one relevant to each approach.

66. Returning a card. Consider drawing one card from a standard deck of 52 cards and then drawing another card. What is the probability that the second card is the ace of spades? Does the answer depend on whether the first card was returned to the deck before the second card was drawn?

67. Listing permutations. List in a systematic way all the arrangements of the three letters A, B, C that use each letter exactly once. You should have $3! = 6$ arrangements, the number of permutations of three letters. Now consider adding the letter D. Give two explanations of the fact that there are now 4 times as many arrangements, one explanation based on how you would modify your earlier list and one explanation based on counting permutations.

68. The Enigma machine. It has been asserted that the Enigma machine code was broken by Polish cryptographers before the British at Bletchley Park broke it (see Exercise 57). Explore this story. **Exercise 27 is relevant to this topic.**

5.5 Expected value and the law of large numbers: Don't bet on it

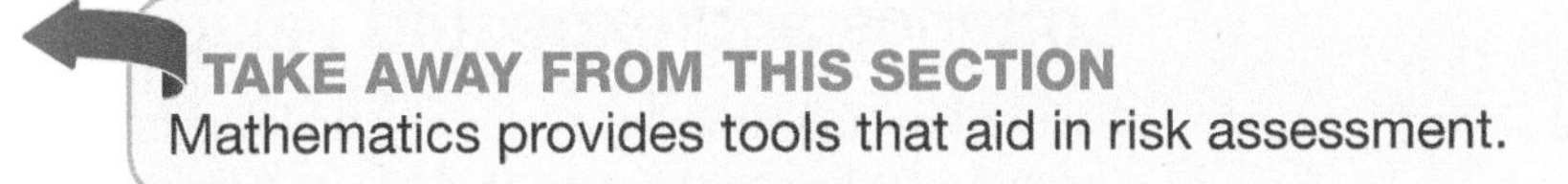

The following excerpt is from a report published by the National Gambling Impact Study Commission.

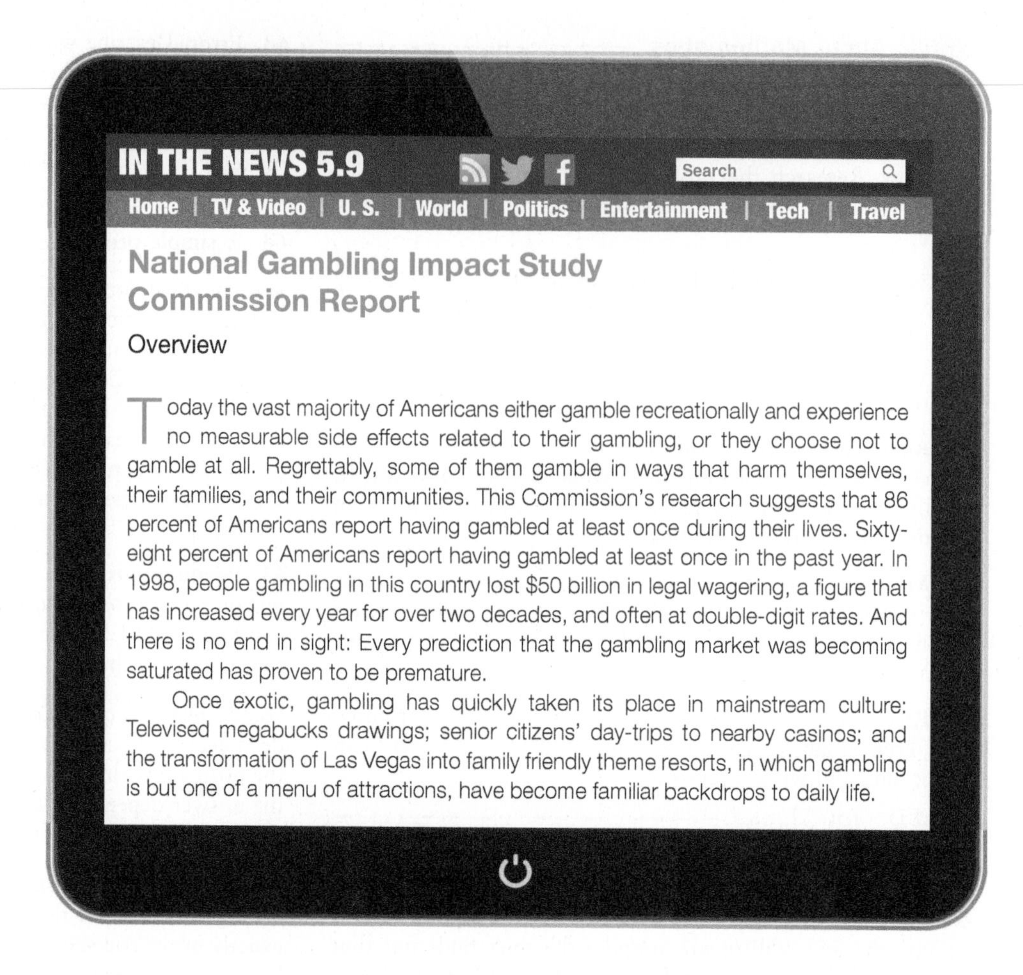

IN THE NEWS 5.9

National Gambling Impact Study Commission Report

Overview

Today the vast majority of Americans either gamble recreationally and experience no measurable side effects related to their gambling, or they choose not to gamble at all. Regrettably, some of them gamble in ways that harm themselves, their families, and their communities. This Commission's research suggests that 86 percent of Americans report having gambled at least once during their lives. Sixty-eight percent of Americans report having gambled at least once in the past year. In 1998, people gambling in this country lost $50 billion in legal wagering, a figure that has increased every year for over two decades, and often at double-digit rates. And there is no end in sight: Every prediction that the gambling market was becoming saturated has proven to be premature.

Once exotic, gambling has quickly taken its place in mainstream culture: Televised megabucks drawings; senior citizens' day-trips to nearby casinos; and the transformation of Las Vegas into family friendly theme resorts, in which gambling is but one of a menu of attractions, have become familiar backdrops to daily life.

Some form of legal gambling has spread to all but two states.[18] Examples include casinos, state lotteries, and parimutuel wagering on horse or dog racing. With gambling now so prevalent, an educated person should understand how it works. Gambling establishments have one thing in common: The odds are stacked against the player. In fact, a good title for our discussion could be, "Why you shouldn't gamble."

The mathematical concept of *expected value* is key to understanding games of chance. Theoretical expected values are connected to empirical observations through the *law of large numbers*. In this section, we will discuss these concepts. We begin with applications to gambling because this is the simplest setting for explaining the ideas we wish to explore. But of more importance to most people is the application to risk assessment for business and industry. To introduce the concept of expected value, we take a quick look at lotteries.

Lotteries and expected value

Forty-five states now have a lottery, which can be a significant source of revenue for them. According to the state of Iowa, its lottery had sales of $3.05 billion over a nine-year period. The lottery awarded $1.66 billion in prizes, spent $577.7 million on operations, and transferred $827.8 million to the state as profits (including $136.0 million in sales tax collected on the purchase of tickets). This means that only about

[18] *Hawaii and Utah are the exceptions.*

1.66 billion out of 3.05 billion, or about 54%, of the money collected was paid back to players in the form of prizes. The end result is that in the Iowa State Lottery players lose, on average, 46 cents of every dollar they spend on lottery tickets and, on average, the state takes in 46 cents of every dollar spent on lottery tickets.

"And just how are we going to win if every time I buy a ticket, you eat it?"

The Iowa State Lottery illustrates the notion of *expected value*. It's not whether you win or how many times you win, it's the overall net payback you can expect that's important. Expected value has a precise mathematical description that we will introduce shortly. But for the moment, an informal notion of expected value will suffice.

KEY CONCEPT

Informally, the expected value for you of a game is a measure of the average amount you can expect to win (or lose) per play *in the long run*.

In the Iowa State Lottery, the expected value for the state would appear to be about 46 cents for each dollar played, on the basis of the empirical information above: For each dollar played, the state expects to take in 46 cents. Often the expected value is expressed as a percentage. For the Iowa State Lottery, the state apparently has an expected value of 46%. We can also look at the expected value from the point of view of the player. The expected value for the player is *negative* 46% or −46%, which means that, *on average over time*, the players expect to lose 46 cents for each dollar wagered.

In many lotteries, only a few people ever win anything. Some win millions and most win nothing. Therefore, when we speak of the expected value for a player in the Iowa State Lottery being −46%, we don't mean that all players really lose 46% of their wager. We mean that is the *average* loss per player. Another thing to keep in mind is that expected values should be applied over the long run.

Ted S. Warren/AP Images

The third-largest jackpot in Mega Millions history was $656 million annuity value for the March 30, 2012, drawing.

EXAMPLE 5.30 Meaning of expected value: Lottery

Suppose you buy $20 worth of Iowa State Lottery tickets each week. How much should you expect to win or lose, on average, over the period of one year based on the lottery's expected value?

SOLUTION

Over the long run you would expect to lose about 46 cents of each dollar you spend on lottery tickets. Because there are 52 weeks in a year, you are betting $\$20 \times 52 = \$1040$. Therefore, you should expect to lose about $\$1040 \times 0.46 = \$478.40$.

TRY IT YOURSELF 5.30

Suppose the lottery in your state has an expected value for the player of -36%. If you spend $100 per month on lottery tickets, how much money do you expect to win or lose in an average year?

The answer is provided at the end of this section.

Expected value and the law of large numbers

Our informal definition of the expected value is that it tells the average amount you can expect to win or lose per play in the long run. But there is a *theoretical* expected value that has meaning for a single play of a game. Later we will give a formula for calculating the theoretical expected value, but first we explain the idea using slot machines.

Ralph Hagan/Cartoonstock.com

I'M NOT SURE, BUT I THINK ILLEGAL GAMBLING IS WHEN YOU WIN.

Slot machines are designed so that each outcome has a certain probability and a certain payout. According to the manufacturers, the machines are designed so that every time you pull the handle, the result is independent of whatever has happened before. The overall result is that each machine has a theoretical expected value that can be set by the manufacturer.

According to the Nevada Gaming Control Board, the payback on all slot machines in a recent year was 94.93%, which means that casinos kept, on average, about 5% of all money put into them. This suggests that the theoretical expected value for the casino on slot machines is about 5% (and therefore the expected value for the player is -5%). A return of 5% might not seem enough to support a lavish casino, but it can because of the huge amount of money involved.

The connection between the theoretical expected value and the effect of that expected value in the long run is expressed by the *law of large numbers*, which applies to games for which repeated trials are independent. (Independent events were discussed in Section 5.3.)

KEY CONCEPT

If you play a game many, many times, then according to the law of large numbers, you will almost certainly win or lose approximately what the theoretical expected value of the game says you will win or lose.

The law of large numbers is actually a mathematical theorem and not just based on empirical observation.

The law of large numbers is the mathematical principle that assures lotteries and casinos of making money even though the players are going to win sometimes. Consider again the slot machines in Nevada. Even though the expected value for the player says you would expect to lose 5 cents for each dollar you put in the machine, the occasional win is no surprise. But now suppose you play a slot machine 1000 times. The expected value tells us that you would expect to be down about 5% of \$1000, or \$50. It would not be very surprising if you were down a few dollars more than \$50 or a few dollars less, but it's very likely that you will have lost money. In the long term, the casino will come out ahead.

The fact that individual plays on a slot machine are independent means that you cannot predict what a slot machine will do based on what has happened before. Some gamblers do not behave as if they accept this. They seem to believe that if they play a slot machine many times without a payoff, the machine must be "due." This means that *surely* the machine must be ready to pay off after so many losses. This belief is the *gambler's fallacy*.

KEY CONCEPT

The gambler's fallacy is the belief that a string of losses in the past will be compensated by wins in the future.

To see why this belief is a fallacy, consider tossing a fair coin. If we toss 50 heads in a row, the probability of a tail on the 51st toss is still 1/2. In the same way, the next play of a slot machine has the same probability of a payout as any other play.

It may seem that the law of large numbers supports the gambler's fallacy. But the law of large numbers refers to the average payoff in the long run—it does not rule out streaks of losses or wins. The law of large numbers refers to future payoff, not payoff in the past. It does not mean that past losses will necessarily be made up by future wins or that past wins will be offset by future losses. It says that if you play many times from this point onward, your wins or losses will be very close to the prediction given by the expected value.

Calculating expected value

Here is how we calculate the theoretical expected value. For each winning outcome, we multiply the probability of that outcome by the profit for that outcome. Adding these numbers gives the anticipated profit. Similarly, for each losing outcome, we multiply the probability of that outcome by the loss for that outcome. Adding these numbers gives the expected loss. Then we calculate the expected value using

$$\text{Expected value} = \text{Anticipated profit} - \text{Anticipated loss}$$

For a game like the Iowa State Lottery that has only two outcomes, the expected value can be written simply as

$$\text{Expected value} = P(\text{Winning}) \times \text{Profit} - P(\text{Losing}) \times \text{Loss}$$

To illustrate the calculation, we look at a simple game. A fair die is rolled. If the result is a 1 or 2, I win \$9. If any other number comes up, I lose \$3. To calculate the expected value for me, we must first find the probability for a win and for a loss. For a win, there are two favorable outcomes (1 and 2) out of six total outcomes. Hence, the probability of a win is

$$\begin{aligned} P(\text{Winning}) &= \frac{\text{Favorable outcomes}}{\text{Total outcomes}} \\ &= \frac{2}{6} = \frac{1}{3} \end{aligned}$$

We find the probability of a loss in a similar fashion: There are four losing numbers out of six, so the probability of a loss is $4/6 = 2/3$. We use these probabilities to calculate the expected value for me:

$$\begin{aligned} \text{Expected value} &= P(\text{Winning}) \times \text{Profit} - P(\text{Losing}) \times \text{Loss} \\ &= \frac{1}{3} \times \$9 - \frac{2}{3} \times \$3 \\ &= \$1 \end{aligned}$$

giovanni mereghetti/MARKA/Alamy

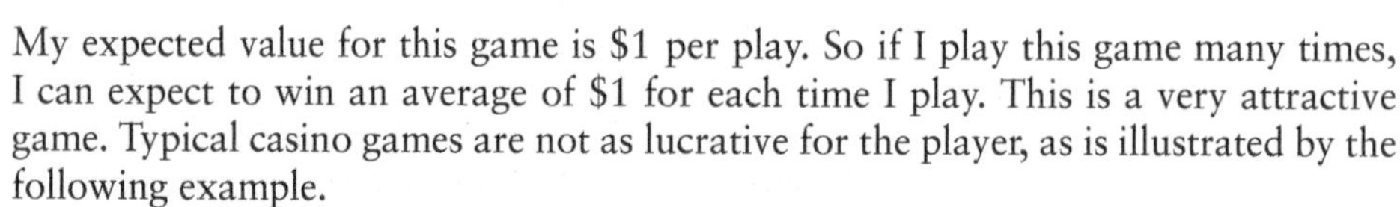

My expected value for this game is \$1 per play. So if I play this game many times, I can expect to win an average of \$1 for each time I play. This is a very attractive game. Typical casino games are not as lucrative for the player, as is illustrated by the following example.

A standard roulette wheel has numbers 1 through 36 alternately colored red and black. It also has a green 0 and a green 00 called "double zero."[19] The wheel is spun with a small white ball inside. While the wheel is spinning, the ball falls into a numbered slot, which determines the winners. Successive spins of a wheel yield independent results.

There are many wagers you can place at the roulette table. For example, you may bet on the color (red or black), or on whether the number is odd or even, or on a single number. Note that, even though 0 is in fact an even number, in roulette both 0 and 00 count as neither odd nor even. These green numbers give the edge to the house.

EXAMPLE 5.31 Calculating expected value: A roulette bet

You are playing roulette, and you bet a dollar on the number 10. If 10 comes up, you win \$35 (profit). If anything else comes up, you lose your dollar. What is your expected value for this bet?

SOLUTION

Because there are a total of 38 numbers on the roulette wheel (remember the 0 and 00), the probability of a 10 is 1/38, and this outcome wins \$35. The probability of not getting a 10 is 37/38, and this outcome loses \$1. We use these probabilities to calculate your expected value:

$$\begin{aligned} \text{Expected value} &= P(\text{Winning}) \times \text{Profit} - P(\text{Losing}) \times \text{Loss} \\ &= \frac{1}{38} \times \$35 - \frac{37}{38} \times \$1 \\ &= -\frac{2}{38} \text{ dollar} \\ &= -\frac{1}{19} \text{ dollar} \end{aligned}$$

[19] *European roulette wheels have no double zero.*

This is about −5.3 cents, so your expected value for this bet is −5.3%. This means that if you make this roulette bet many times, you can expect to lose about 5.3% of all the money you wager.

TRY IT YOURSELF 5.31

If you bet $1 on an even number, you win $1 (profit) if any of the even numbers 2 through 36 comes up, and you lose your dollar if 0, 00, or any odd number between 1 and 35 comes up. What is your expected value for this wager?

The answer is provided at the end of this section.

EXAMPLE 5.32 Calculating expected value: Roll a die

You play a gambling game with a friend where a single die is rolled, and your profit or loss depends on the number that comes up.

- If a 6 comes up, your profit is $6.00.
- If a 5 comes up, your profit is $3.00.
- If a 4 comes up, there is neither a profit nor a loss. You win 0 dollars.
- If a 1, 2, or 3 comes up, you lose $4.00

Find your expected value for this game.

SOLUTION

The method of calculating the expected value when there are multiple outcomes is the same as when there are only two, except that wins and losses may come in several pieces.

You win if there is a 5 or 6. Also, if there is a 4, you win 0 dollars. The probability for each of these outcomes is 1/6. Your expected winnings are

6 comes up		*5 comes up*		*4 comes up*		*Total anticipated profit*
$\frac{1}{6} \times 6$	$+$	$\frac{1}{6} \times 3$	$+$	$\frac{1}{6} \times 0$	$=$	$1.50

You lose $4.00 when 1, 2, or 3 comes up. The probability of a 1, 2, or 3 is 1/2. This gives

$$\text{Anticipated loss} = \frac{1}{2} \times 4 = \$2.00$$

Subtracting these gives the expected value:

$$\begin{aligned}\text{Expected value} &= \text{Anticipated profit} - \text{Anticipated loss}\\ &= 1.50 - 2.00\\ &= -0.50 \text{ dollar}\end{aligned}$$

We conclude that your expected value is −$0.50, which means that in the long run you can expect to lose about 50 cents per play.

TRY IT YOURSELF 5.32

Suppose your profit is $18.00 if a 6 comes up and 0 dollars if a 5 comes up. You lose $6.00 if a 4 or 3 comes up, and you lose $3.00 if a 1 or 2 comes up. Find your expected value for this game.

The answer is provided at the end of this section.

In addition to slot machines and roulette, casinos offer a variety of games, among which are blackjack, keno, craps, and various kinds of poker. But they all have one thing in common: The theoretical expected value for the player is negative. And no system of placing wagers can change this fact.[20]

Many people enjoy buying lottery tickets or going to casinos even though they assume they will probably lose some money. Some view it as buying entertainment, just as they might pay to go to a movie. A few will get lucky once in a while and win money, but the law of large numbers ensures that the vast majority will eventually lose and that casinos and state lotteries will be profitable for the proprietors.

Fair games

In probability theory, a game is called *fair* if its expected value is zero. Intuitively then, a fair game is one in which no participant has an advantage over others. Clearly, we would not call the game of roulette fair, because it gives the edge to the casino.

KEY CONCEPT

A game is said to be fair if the expected value is zero.

For example, suppose Sue and Alice toss a coin. Sue wins $2 from Alice if the coin comes up heads and loses $2 to Alice if it comes up tails. Let's determine whether this is a fair game.

There are two outcomes: heads (Sue wins) and tails (Sue loses). The probability that Sue wins is 1/2, in which case she wins $2. The probability that Sue loses is also 1/2, and then she loses $2. Sue's expected value is

$$\begin{aligned}\text{Expected value} &= P(\text{Winning}) \times \text{Profit} - P(\text{Losing}) \times \text{Loss}\\ &= \frac{1}{2} \times \$2 - \frac{1}{2} \times \$2\\ &= \$0\end{aligned}$$

Because the expected value is 0, this is a fair game. If Sue plays this game many times, she can expect to lose about as much as she wins. This does not mean that she is *certain* to come out even in the end, but it would be unusual for her either to win or to lose a large amount of money over a long stretch.

EXAMPLE 5.33 Making a game fair: Roll a die

In a gambling game, I roll a single die. If it comes up a 6, I win. I lose $1 otherwise. How much money should I collect on a win to make this a fair game?

SOLUTION

The probability of a win is 1/6, and the probability of a loss is 5/6. Thus, we find that my expected value is

$$\begin{aligned}\text{Expected value} &= P(\text{Winning}) \times \text{Profit} - P(\text{Losing}) \times \text{Loss}\\ &= \frac{1}{6} \times \text{Profit} - \frac{5}{6} \times \$1\end{aligned}$$

To make the game fair, the expected value should be 0. So the profit for a win should be $5.

---

[20] *"Card counting" at blackjack can produce a very small positive expected value for the player. But most casinos do not allow the practice.*

TRY IT YOURSELF 5.33

In Example 5.31, if I place a bet on the number 10 on the roulette wheel, the casino has the advantage. How much profit should I get from a $1 bet on number 10 to make the game fair?

The answer is provided at the end of this section.

Now we give an application to guessing strategies for taking tests.

EXAMPLE 5.34 Understanding fair game: Guessing on the SAT

Your parents may have had a tougher time with the SAT exam than you: When they took the exam, scoring of the exam included a penalty for guessing incorrectly. Until recently, multiple-choice questions had five possible answers. One point was added to the raw score for the correct answer, and 1/4 point was subtracted from the raw score for each incorrect answer.[21] (Supplying no answer to a problem didn't change the raw score.) In this setting, if students randomly guessed answers, what average raw score for a problem would we expect to see in the long run? In other words, what is your expected value of the raw score for a problem if the answers are filled in at random?

SOLUTION

We calculate the expected value using the formula

$$\text{Expected value} = P(\text{Winning}) \times \text{Profit} - P(\text{Losing}) \times \text{Loss}$$

When you take the SAT, winning corresponds to giving a correct answer, with a profit of 1 point, and losing corresponds to giving an incorrect answer, with a loss of 1/4 point.

Now we find the probabilities. There are five possible answers one can fill in, and one of them is correct. The probability of getting a correct answer by random guessing is thus 1/5, so the probability of an incorrect answer is 4/5.

Thus, the expected value of the score is

$$\begin{aligned}\text{Expected value} &= P(\text{Winning}) \times \text{Profit} - P(\text{Losing}) \times \text{Loss} \\ &= \frac{1}{5} \times 1 \text{ point} - \frac{4}{5} \times \frac{1}{4} \text{ point} \\ &= 0 \text{ points}\end{aligned}$$

This means that random guessing gives no advantage in the long run—which was the purpose of subtracting 1/4 point for incorrect answers. Because the expected value is 0, the awarding of points on the SAT exam was fair in the sense that we have been discussing.

TRY IT YOURSELF 5.34

Suppose that on SAT questions, you can always recognize one possible answer as incorrect. What is your expected value of the raw score for a problem if you guess at random among the remaining four choices? (Leave your answer as a fraction.) Is this "game" fair?

The answer is provided at the end of this section.

[21] *The penalty for wrong answers was discontinued in 2016.*

SUMMARY 5.8
Expected Value

The theoretical expected value for you, the player in a game, is calculated as follows: For each winning outcome, we multiply the probability of the outcome by the profit for that outcome. Adding these numbers gives the anticipated profit. Similarly, for each losing outcome, we multiply the probability of the outcome by the loss for that outcome. Adding these numbers gives the expected loss. Then

$$\text{Expected value} = \text{Anticipated profit} - \text{Anticipated loss}$$

For a game which has only two outcomes, the expected value can be written simply as

$$\text{Expected value} = P(\text{Winning}) \times \text{Profit} - P(\text{Losing}) \times \text{Loss}$$

The expected value is a measure of the average amount you can expect to win (or lose) per play *in the long run*.

If the expected value is zero, the game is *fair*.

Risk assessment

So far we have discussed the concept of expected value primarily in the context of games of chance. But this concept is relevant whenever we weigh risks and benefits by considering the likelihood of various outcomes. Applications of expected values are found in various contexts, including manufacturing, finance, economics, and the insurance industry. Blaise Pascal, a famous mathematician and religious philosopher of the seventeenth century, applied the concept of expected value to belief in the existence of God. See Exercise 58.

The concept of expected value plays a key role in the insurance industry. Companies that insure property, for example, have to pay for catastrophic damages occasionally, such as those caused by a hurricane. The companies must set the costs of premiums to compensate for such losses but also provide a profit (i.e., a positive expected value) in the long run. The experts whom they employ to figure this out are called *actuaries*.

Bill Koplitz/FEMA News Photo

Storm damage.

EXAMPLE 5.35 Calculating expected profit: Automobile insurance

The car insurance industry estimates[22] that the probability of your having an automobile collision this year is about 0.06. Repair costs vary dramatically, but for simplicity let's assume that it costs an average of \$4000 (after the deductible has been applied) to repair a car that has been in a collision. You pay \$300 this year for insurance. Can your insurance company expect business to be profitable?

[22] *See* https://www.forbes.com/sites/moneybuilder/2011/07/27/how-many-times-will-you-crash-your-car/#3515abaa4e62.

Image Source/Getty Images

SOLUTION

In order to determine profitability, we think of this scenario as a game where the insurance company wins \$300 if you have no accident and loses $4000 - 300 = 3700$ dollars if you do have an accident. The expected value for the company for this game is the profit the insurance company can reasonably expect. Note that the probability of no accident is $1 - 0.06 = 0.94$. Thus,

$$\begin{aligned}\text{Expected value} &= \text{Anticipated profit} - \text{Anticipated loss}\\ &= P(\text{No accident}) \times \$300 - P(\text{Accident}) \times \$3700\\ &= 0.94 \times \$300 - 0.06 \times \$3700\\ &= \$60\end{aligned}$$

Thus, the insurance company expects to earn a profit of about \$60 from you this year.

This example (as well as those that follow) illustrates facts that are important for the insurance industry as well as virtually every other business. In order to set prices that will ensure profitability, it is necessary to have a good idea of the probability of losses. Risk assessment involves getting a handle on expected values for your business, and it is essential to success.

EXAMPLE 5.36 Calculating expected profit: Insurance

Suppose a company charges a premium of \$150 per year for an insurance policy for storm damage to roofs. Actuarial studies show that in case of a storm, the insurance company will pay out an average of \$8000 for damage to a composition shingle roof and an average of \$12,000 for damage to a shake roof. They also determine that out of every 10,000 policies, there are 7 claims per year made on composition shingle roofs and 11 claims per year made on shake roofs. What is the company's expected value (i.e., expected profit) per year of a storm insurance policy? What annual profit can the company expect if it issues 1000 such policies?

SOLUTION

This is just like a game where the insurance company accepts a bet of \$150 from the customer and makes payouts to meet claims. To find the expected value, we multiply the probability of each outcome by the amount the company takes in (or by minus the amount if the company pays out) and add these numbers.

The probability of a composition shingle roof claim is $7/10{,}000 = 0.0007$, and the probability of a shake roof claim is $11/10{,}000 = 0.0011$. For every 10,000 policies, there are $7 + 11 = 18$ total claims. Hence, 9982 policies make no claim. So the probability of no claim is $9982/10{,}000 = 0.9982$.

Each composition shingle roof claim costs the company \$7850 (which is \$8000 minus the \$150 premium), and each shake roof claim costs \$11,850 (which is \$12,000 minus the \$150 premium). But each policy that makes no claim results in \$150 profit for the company. Thus, the expected value for the company is

$$\begin{aligned}\text{Expected value} &= P(\text{No claim}) \times \text{Profit} - P(\text{Composition claim})\\ &\quad \times \text{Cost} - P(\text{Shake claim}) \times \text{Cost}\\ &= 0.9982 \times \$150 - 0.0007 \times \$7850 - 0.0011 \times \$11{,}850\\ &= \$131.20\end{aligned}$$

This means that, on average, the insurance company earns a profit of \$131.20 on each policy it writes. If the insurance company issues 1000 such policies, it can expect an annual profit of $\$131.20 \times 1000 = \$131{,}200$.

TRY IT YOURSELF 5.36

Suppose a company charges an annual premium of $300 for a flood insurance policy. In case of a flood claim, the company will pay out an average of $200,000 per policy. Based on actuarial studies, it determines that in an average year there is one flood claim for every 5000 policies. What is the company's expected annual profit for a flood insurance policy? What annual profit can the company expect if it issues 1000 policies?

The answer is provided at the end of this section.

Here is an application of the concept of expected value to manufacturing.

EXAMPLE 5.37 Calculating expected cost: An assembly line

A company manufactures small electric motors at a cost of $50 each. Based on experience, quality control experts have come up with the following probabilities of defective motors in a given production run:

Probability of no defective motors: 30%.

Probability that 1% of motors are defective: 40%.

Probability that 2% of motors are defective: 20%.

Probability that 3% of motors are defective: 10%.

Note that the sum of the probabilities is 100%. When a defective motor is detected, it must be removed from the assembly line and replaced. This process adds an extra $15 to the cost of the replaced motor. What is the expected cost to the company of a batch of 1000 motors?

SOLUTION

You can think of a production run as a game where the company "wins" $50 for each good motor and "wins" $65 for each defective motor. The expected cost to the company is the expected value for the company of 1000 motors with the given probabilities. We proceed as follows.

If there were no defective motors in the production run, the cost to the company would be $\$50 \times 1000 = \$50{,}000$. The probability of this happening is 30%, so we multiply $50,000 by 0.3 to get $15,000.

Now suppose that 1% of the motors in the production run are defective. Then there are 10 defective motors, which costs the company $\$15 \times 10 = \$150$ extra. Therefore, the cost to the company is $50,150. The probability of this happening is 40%, so we multiply $50,150 by 0.4 to get $20,060.

Now suppose that 2% of the motors in the production run are defective. Then there are 20 defective motors, which costs the company $\$15 \times 20 = \$300$ extra. Therefore, the cost to the company is $50,300. The probability of this happening is 20%, so we multiply $50,300 by 0.2 to get $10,060.

Finally, suppose that 3% of the motors in the production run are defective. Then there are 30 defective motors, which costs the company $\$15 \times 30 = \$450$ extra. Therefore, the cost to the company is $50,450. The probability of this happening is 10%, so we multiply $50,450 by 0.1 to get $5045.

To get the expected value, we add up the results of the previous four calculations. Thus, we find

$$\text{Expected value} = \$15{,}000 + \$20{,}060 + \$10{,}060 + \$5045 = \$50{,}165$$

The expected cost to the company of a batch of 1000 motors is $50,165.

TRY IT YOURSELF 5.37

Suppose the probability of no defective motors is 40%, the probability of 1% defective motors is 30%, the probability of 2% defective motors is 15%, and the probability of 3% defective motors is 15%. Assume that the motors cost $50 each and replacement adds $15 to the cost. Find the expected cost to the company of a run of 2000 motors.

The answer is provided at the end of this section.

WHAT DO YOU THINK?

Lottery: Your friend says that the more lottery tickets he buys, the more likely he is to win a prize. Is your friend right? Is your friend more likely to make a profit by buying more tickets? Is there any strategy you can use to improve your chances of winning money from the lottery?

Questions of this sort may invite many correct responses. We present one possibility: It is certainly true that buying more tickets improves your chances of winning, but it also increases the amount of money you spend on tickets. There is no way to increase your expected winnings other than to avoid playing altogether. The lottery has a fixed expected value which is negative for the player. Each dollar you spend increases your expected losses. If you play the lottery many times, the law of large numbers virtually assures that you will lose the amount of money predicted by the expected value. One may play the lottery for fun or as a way of contributing to your state. If one does it expecting to win, disappointment is highly likely.

Further questions of this type may be found in the *What Do You Think?* section of exercises.

Try It Yourself answers

Try It Yourself 5.30: Meaning of expected value: Lottery Lose $432.

Try It Yourself 5.31: Calculating expected value: A roulette bet −5.3%.

Try It Yourself 5.32: Calculating expected value: Rolling a die $0.

Try It Yourself 5.33: Making a game fair: Another roulette bet $37.

Try It Yourself 5.34: Understanding fair game: Guessing on the SAT 1/16; no.

Try It Yourself 5.36: Calculating expected profit: Insurance $260; $260,000.

Try It Yourself 5.37: Calculating expected cost: An assembly line $100,315.

Exercise Set 5.5

Test Your Understanding

1. A *fair game* is one in which the expected value is ________.

2. True or false: Gambling games in casinos and state lotteries all have negative expected value for you, the player.

3. True or false: The idea of *expected value* is important to businesses other than gambling.

4. True or false: The law of large numbers says that the more times you play a game, the more likely you are to win.

5. State the *gambler's fallacy*.

6. Bob says, "The more lottery tickets I buy the more likely I am to win a prize." Ann says, "Buying more lottery tickets will not increase your expected value." Who is right? **a.** Bob **b.** Ann **c.** both **d.** neither.

7. If an insurance company is to be profitable, it must charge rates that will ensure that the expected value of their net income is ________.

8. True or false: The expected value is an important part of risk assessment.

Problems

In these exercises, express the expected value as a dollar amount unless otherwise instructed. If rounding is necessary, round to the nearest cent.

9–14. Calculating expected value. *In Exercises 9 through 14, a simple game is described. Find your expected value for the game, and determine whether the game is fair.*

9. A fair coin is tossed. If heads come up, your profit is \$2. If tails come up, your loss is \$2.

10. A fair coin is tossed. If heads come up, your profit is \$3. If tails come up, your loss is \$2.

11. A die is tossed. If a 5 or 6 comes up, your profit is \$6. If any other number comes up, your loss is \$3.

12. A die is tossed. If a 5 or 6 comes up, your profit is \$12. If a 3 or 4 comes up, your loss \$3. If a 1 or 2 comes up, your loss is \$6.

13. A die is tossed. If a 6 comes up, your profit is \$24. If a 5 comes up, your profit is \$6. If a 4 comes up, your profit is 0. If a 2 or 3 comes up, your loss is \$9. If a 1 comes up, your loss is \$18.

14. **A coin game.** In a game, you bet \$1 and toss two coins. If two heads come up, you win a profit of 75 cents. If two tails come up, no one wins anything. Otherwise, you lose your dollar. Don't round your answer.

15–18. Making games fair. *In Exercises 15 through 18, games are described. You are to asked to find a profit or loss that makes the game fair.*

15. A fair coin is tossed. If heads come up, your profit is \$7. In order to make the game fair, what should be your loss if tails comes up?

16. A die is tossed. If a 5 or 6 comes up, your profit is \$18. In order to make the game fair, what should be your loss if a 1, 2, 3, or 4 comes up?

17. A die is tossed. If a 6 comes up, your profit is \$24. If a 5 comes up, your profit is \$6. If a 2, 3, or 4 comes up, your loss is \$4. In order to make the game fair, what should be your loss if a 1 comes up?

18. **Make it fair.** You play a gambling game with your friend in which you win 30% of the time and lose 70% of the time. When you lose, you lose \$1. What profit should you earn when you win in order for the game to be fair?

19. **Expected value and the lottery.** Suppose your state lottery has an expected value of −34%. If you spend \$20 per month on lottery tickets, how much money would you be expected to lose in an average year?

20. **Calculating expected value.** You play a gambling game with a friend in which you roll a die. If a 1 or 2 comes up, you win \$6. Otherwise you lose \$2. What is your expected value for this game?

21. **Making a game fair.** You play a gambling game with a friend in which you roll a die. If a 1 or 2 comes up, you win \$6. How much should you lose on any other outcome in order to make this a fair game?

22. **Expected cost.** A company produces products at a cost of \$35 each. Defective products cost an additional \$15 each to repair. The probability of no defective products is 80%. The probability of 5% defective products is 20%. What is the expected cost of producing 500 products?

23. **Fair games I.** Explain why casinos do not offer fair games.

24. **Fair games II.** I told my friend Bob that I learned in math class that no casino games are fair. Bob said, "That's not true. My mom works in a casino and I know they don't cheat." What should I say to explain myself to Bob?

25. **Expected value.** Bill and Larry toss two coins. If both coins come up heads, Bill pays Larry \$4. Larry pays Bill \$1 otherwise.

a. What is the expected value of this game for Bill?

b. What is the expected value of this game for Larry?

c. Bill and Larry play the game 100 times. Who should come out ahead and by how much?

d. Bill and Larry play the game 100 times, and Larry comes out \$23 ahead. Use only this empirical information to estimate the expected value of this game for Bill and the expected value for Larry.

e. Explain the difference, in general, between the theoretical expected value and an estimate of the expected value obtained using empirical information. (You may use the game above as an example.)

26. **Double or nothing.** You are playing a game with a friend in which you toss a fair coin. Your friend pays you a nickel if it's heads, and you pay your friend a nickel if it's tails. You win 10 times in a row and want to stop, but your friend says, "Let's play one more time, double or nothing. I'm due to win." What's the expected value of this game? What is your response to your friend?

27. **Shooting free-throws.** A basketball player who typically makes 80% of his free-throws has missed his first two attempts in a game. You hear a sports commentator say that the player is therefore "due" to make his next attempt. Analyze this statement. Consider whether the gambler's fallacy applies here. Are successive free-throws independent events?

28. **Gambler's fallacy.** One sports fan, in describing how his mediocre team went from a five-game winning streak to a five-game losing streak, had this summary: "The law of averages was bound to catch up with a team like this." Is this statement an application of the law of large numbers, an example of the gambler's fallacy, or neither? Explain your answer.

29. **Martingale method.** Suppose a game pays even money. That is, if you win, you get whatever you bet, and if you lose, you lose that same amount. Here's a strategy called the *Martingale method*.

If you win, bet the same amount again. But every time you lose, you play again and double your previous bet. For example, suppose you bet a dollar in such a game. If you win, you bet \$1 again. If you lose, then you play again but this time you bet \$2. If you win, you're back to even and you bet \$1 next time. If you lose, you play again and bet \$4. If you win, you're back to even, and so forth. This strategy seems appealing because no matter how many times you lose, eventually you're bound to win one time, and that puts you back to even. If you win twice in a row, you have a profit. So in theory you can't lose in the end. But this strategy is flawed. Can you find the flaw?

30. How much will we win? We decide to open a casino where we offer only one game. We win a dollar 60% of the time, and our customer wins a dollar 40% of the time. How much profit would we expect to have if 5000 games are played?

31. Voting. A poll taken in Massachusetts in October 2012 showed Obama leading Romney by 60% to 40% in the U.S. presidential race. Let's assume that these numbers represent probabilities. I wished to make a bet with my friend on who would win Massachusetts, and we agreed to base our bet on that poll. I predicted a victory for Obama and would win a profit of $1 if I were right. Assuming we wished this to be a fair game, how much should I have agreed to pay my friend if he was right and Massachusetts went to Romney?

32. Card game. A card game goes like this: You draw a card from a 52-card deck. If it is a face card (jack, queen, or king), you win $5; otherwise, you lose $2. What is your expected value for this game?

33. Modified card game. Exercise 32 described a card game in which you draw a card from a 52-card deck. Suppose that this game is modified so that if you draw an ace, then no money changes hands. What is your expected value for this game?

34. Expected value I. What is your expected value for a game where your probability of winning is 1/6 and your profit is four times your wager? Report your answer as a percentage rounded to one decimal place.

35. Expected value II. Consider the following game: The probability of winning a profit of twice your wager is 1/7, the probability of winning a profit equal to your wager is 2/7, and all other probabilities of winning are 0. What is your expected value for this game? Report your answer as a percentage.

36. Expected value III. What is your expected value for a game whose probabilities and outcomes are given in the following table? Report your answer as a percentage.

Probability	Profit
0.2	You win 1.5 times your wager.
0.4	You win your wager.
0.2	You lose 2 times your wager.
0.2	You win nothing.

37. Expected value IV. What is your expected value for a game whose probabilities and outcomes are given in the following table? Report your answer as a percentage.

Probability	Profit
0.1	You win your wager.
0.4	You win half your wager.
0.2	You lose 3 times your wager.
0.3	You win nothing.

38. Exam score I. Each question on a multiple-choice exam has four choices. One of the choices is the correct answer, worth 5 points, another choice is wrong but still carries partial credit of 1 point, and the other two choices are worth 0 points. If a student picks answers at random, what is the expected value of his or her score for a problem? If the exam has 30 questions, what is his or her expected score?

39. Exam score II. Each question on a multiple-choice exam has five choices. One of the choices is the correct answer, worth 5 points. There are two other choices that are wrong but still carry partial credit of 2 points each. The other two choices are worth 0 points. If a student picks answers at random, what is the expected value of his or her score for a problem? If the exam has 40 questions, what is his or her expected score?

40. Insurance I. Suppose a company charges an annual premium of $450 for a fire insurance policy. In case of a fire claim, the company will pay out an average of $100,000. Based on actuarial studies, it determines that the probability of a fire claim in a year is 0.004. What is the expected annual profit of a fire insurance policy for the company? What annual profit can the company expect if it issues 1000 policies?

41. Insurance II. Suppose a company charges an annual premium of $100 for an insurance policy for minor injuries. Actuarial studies show that in case of an injury claim, the company will pay out an average of $900 for outpatient care and an average of $3000 for an overnight stay in the hospital. They also determine that, on average, each year there are five claims made that result in outpatient care for every 1000 policies and three claims made that result in an overnight stay out of every 1000 policies. What is the expected annual profit of an insurance policy for the company?

42. Tires. A company manufactures tires at a cost of $60 each. The following are probabilities of defective tires in a given production run: probability of no defective tires: 10%; probability of 1% defective tires: 30%; probability of 2% defective tires: 40%; probability of 3% defective tires: 20%. When a defective tire is detected, it must be removed from the assembly line and replaced. This process adds an extra $5 to the cost of the replaced tire. What is the expected cost to the company of a batch of 3000 tires?

43. Chairs. A company manufactures chairs at a cost of $30 each. The following are probabilities of defective chairs in a given production run: probability of 1% defective chairs: 40%; probability of 2% defective chairs: 35%; probability of 3% defective chairs: 25%. When a defective chair is detected, it must be removed from the assembly line and replaced. This process adds an extra $10 to the cost of the replaced chair. What is the expected cost to the company of a batch of 500 chairs?

44–45. Expected rate of return. *Exercises 44 and 45 refer to the following concept. In finance, the notion of expected value is used to analyze investments for which the investor has an estimate of the chances associated with various returns (and losses). For example, suppose you have the following information about one of your investments: With a probability of 0.9, the investment will return 20 cents for every dollar you invest, and with a probability of 0.1, the investment will lose 50 cents for every dollar you invest. The expected rate of return for this investment is calculated the way we calculate the expected value of a game: Multiply the probability of each outcome by the amount you earn (or by minus the amount if you lose) and add up these numbers.*

44. Calculate the expected rate of return for the investment described above.

45. Here is information about another investment: With a probability of 0.5, it will return 20 cents for every dollar you invest; with a probability of 0.3, it will return 10 cents for every dollar you invest; and otherwise, it will lose 90 cents for every dollar you invest. Calculate the expected rate of return for this investment. Is this a good investment?

46. Betting on 11. In a dice game, you win if the two dice come up 11. Otherwise, you lose $1. What should be the profit for winning to make this game fair?

47. A fair game. You roll a pair of dice. You bet $1 that you will get a 2, 3, or 12. If you win, you are paid $9 (a profit of $8). Show that this is a fair game.

48. The real game. In Exercise 47, you rolled a pair of dice, and you bet $1 that you will get a 2, 3, or 12. In the casino game called *craps*, the casino pays you $8 (a profit of $7) if these numbers come up on your "opening roll." What is your expected value for this bet expressed as a percent?

49. Betting on red. Suppose you bet $1 on red at the roulette wheel. There are 18 red numbers and 20 nonred numbers (including the green 0 and 00). You win $1 if a red number comes up, and you lose $1 if any other number comes up. What is your expected value for this wager?

50. Even or odd. In roulette, you may bet whether the number that comes up is odd (or even). (Remember that 0 does not count as either odd or even.) The wager pays even money, that is, if you win, you get whatever you bet, and if you lose, you lose that same amount. What is your expected value for this bet?

51. Roulette and gambler's fallacy. Roulette wheels in Monte Carlo have no double zero, so there are 37 numbers on a wheel.

a. What is the probability of getting an even number on one spin of a roulette wheel in Monte Carlo? Express your answer as a fraction and as a percentage rounded to one decimal place.

b. You spin a roulette wheel in Monte Carlo 26 times. What is the probability of getting an even number on all 26 spins? Express your answer as a fraction, then express the chances using a phrase such as "1 in 5 billion."

c. In 1913, at one fair roulette table in Monte Carlo, an even number came up on 26 spins in a row. What was the probability of getting an even number on the 27th spin?

d. Explain how the historical event described in part c is consistent with the law of large numbers.

52. Blackjack in Nevada. Nevada casinos offer the following wager at some blackjack tables: You are dealt two cards. If they are of the same suit, you win $3 for each $1 you bet.

a. What is the probability your two cards are the same suit?

b. What is your expected value for this wager?

c. Is this a good bet?

Exercises 53 through 56 are suitable for group work.

53–56. Powerball. *Exercises 53 through 56 refer to the following description of Powerball, a lottery in which most U.S. states participate. To play, you pay $2 and choose five white numbers between 1 and 69. Then you choose a single red ball, the Powerball, from a second set of numbers that range from 1 to 26. To win the Powerball jackpot, you must match all five white numbers, plus the Powerball number. Like other lottery games, varying amounts of money can also be won by matching fewer numbers, allowing for nine different ways to win a Powerball prize. Here they are:*

- *5 white + Powerball = Powerball jackpot prize (Probability of winning: 1 in 292 million)*
- *5 white = $1 million prize (Probability of winning: 1 in 11.7 million)*
- *4 white + Powerball = $50,000 prize (Probability of winning: 1 in 913 thousand)*
- *4 white = $100 prize (Probability of winning: 1 in 37 thousand)*
- *3 white + Powerball = $100 prize (Probability of winning: 1 in 14 thousand)*
- *3 white = $7 prize (Probability of winning: 1 in 580)*
- *2 white + Powerball = $7 prize (Probability of winning: 1 in 701)*
- *1 white + Powerball = $4 prize (Probability of winning: 1 in 92)*
- *Powerball only = $4 prize (Probability of winning: 1 in 38)*

The largest Powerball jackpot on record (as of 2021) is $1.586 billion, won in January 2016 and split by three tickets.

53. Powerball players may choose to have the numbers generated randomly by the computer—called a Quick Pick—rather than choosing the numbers themselves. Do you think it matters? Why?

54. How many possible combinations of five numbers chosen from 1 through 69 are there?

55. How many possible combinations of six Powerball numbers are there, where five white numbers are chosen from 1 through 69 and one red Powerball number is chosen from 1 through 26?

56. Suppose several people chose the same winning numbers for the jackpot. Wouldn't that cause the state to lose money? (This may require some investigation.)

Writing About Mathematics

57. History. Historically, many of the ideas from probability were developed to understand games of chance. See Exercise 51 in Section 5.3 about "double sixes." Write a paper about this historical connection.

58. History. Blaise Pascal was an eminent scientist, mathematician, and religious philosopher. His work led to the development of probability theory and the notion of expected value. In a section of his writings, he discusses belief in the existence of God in terms of expected value. Write a paper about *Pascal's wager*.

59. History. Choose one of the games discussed in this section and write a paper about its history and evolution.

60. History. If your state or a neighboring state has a lottery, write a paper about its history, the revenue it generates, and how it operates.

What Do You Think?

61. Making wagers at a casino. Mathematics does not tell you how to behave. But it can tell you the consequences of certain types of behavior. Given that the expected value of all casino games is positive for the casino, what does the law of large numbers say will happen in the long term to a person who wagers money at casinos? **Exercise 23 is relevant to this topic.**

62. Lie? A friend said he was told that the expected value of a game was \$1.00, but he played twice and lost \$1.50. He said he must have been told a lie. How would you respond? **Exercise 25 is relevant to this topic.**

63. Fair game. What does it mean to say that a game is *fair*? Does it mean that you can never lose money? **Exercise 18 is relevant to this topic.**

64. Insurance. Investigate the role of expected value in the insurance industry. Report on your findings. **Exercise 40 is relevant to this topic.**

65. Expected value and manufacturing. We have seen applications of expected value to manufacturing. After a manufacturer calculated the expected cost of a production batch, what could the manufacturer do with this information? **Exercise 42 is relevant to this topic.**

66. Roulette. Explain how the standard roulette wheel gives the edge to the casino. **Exercise 49 is relevant to this topic.**

67. Gambler's fallacy—not. If we flip a coin 10 times and each time it comes up heads, then the gambler's fallacy would lead us to bet tails for the next toss because "tails are due." For independent events, the gambler's fallacy is indeed a fallacy. But suppose you flip a coin 20 times, and each time it comes up heads. Would you conclude that heads and tails are equally likely on the next toss, or would the first 20 tosses make you suspect that the coin is in fact not a fair coin? Discuss the conditions under which you might question a game that purports to be fair. **Exercise 51 is relevant to this topic.**

68. Law of large numbers. Explain in your own words why the law of large numbers does not support the gambler's fallacy.

69. Adjusting expected value. The text states that the expected value for the Iowa State Lottery is about −46% for the player. What could the state do to increase (or decrease) this expected value?

70. Law of large numbers? On the Web, question is posed, "How can I avoid a catastrophic stock loss?" The answer provided says, "Due to the law of large numbers, when a stock suffers a big decline, it's all that more difficult to recover. Mathematically, it takes a 25% rally to break even from a 20% decline." What do you think? Is this really an example of the law of large numbers? Explain. Also, is this assertion correct: "it takes a 25% rally to break even from a 20% decline"?

CHAPTER SUMMARY

The mathematical study of probability is an attempt to model and understand uncertainty. We examine how a probability is obtained, what it means, and where it is useful in the real world.

Calculating probabilities: How likely is it?

A probability is a numerical measure of the likelihood of an event. It can be theoretical, empirical, or subjective (based merely on opinion). An example of a statement using a theoretical probability is, "The probability of a fair coin coming up heads is 50%." This represents a theoretical probability because there are two equally likely events that can occur. Therefore, in theory one of them should happen half of the time. An example of a statement using an empirical probability is, "Based on a sample of some of the lightbulbs on a production line, I estimate that the probability of a lightbulb being defective is 1%." This represents an empirical probability because it is based on experimental evidence. An example of a statement using a subjective probability is, "My probability of becoming a rock star is one in a million." This probability is based on opinion: The number is plucked out of thin air.

To calculate a theoretical probability, we designate the outcomes in which we are interested as *favorable* and all other outcomes as *unfavorable*. In the experiments we consider, we make sure that each outcome is equally likely. We have the formula

$$\text{Probability of an event} = \frac{\text{Number of favorable outcomes}}{\text{Total number of possible outcomes}}$$

For example, suppose we want to find the theoretical probability that I roll a 3 with two dice. We need to count the number of possible outcomes for the two dice, which is $6 \times 6 = 36$. The number of favorable outcomes, that is, those that give a 3, is 2 (either one and two or two and one). Then the probability is $2/36 = 1/18$. In cases involving identical items, such as coins or dice, it may help to label the items to distinguish them.

Here are two useful formulas:

$$\text{Probability of event not occurring} = 1 - \text{Probability of event occurring}$$

and

$$P(A \text{ or } B) = P(A) + P(B) - P(A \text{ and } B)$$

Medical testing and conditional probability: Ill or not?

One important application of mathematical probability involves medical testing. The terminology used in such tests includes: true positives, false positives, true negatives, false negatives, sensitivity, specificity, prevalence, positive predictive value (PPV), and negative predictive value (NPV). Two key formulas are

$$\text{PPV} = \frac{\text{True positives}}{\text{All positives}} = \frac{\text{True positives}}{\text{True positives} + \text{False positives}}$$

and

$$\text{NPV} = \frac{\text{True negatives}}{\text{All negatives}} = \frac{\text{True negatives}}{\text{True negatives} + \text{False negatives}}$$

A *conditional probability* is the probability that one event occurs given that another has occurred. Positive and negative predictive values, as well as sensitivity and specificity, are examples of conditional probabilities.

Counting and theoretical probabilities: How many?

To calculate a theoretical probability, one often needs to calculate the number of ways something can happen. The most basic way of counting for this purpose is the *Counting Principle*. It says that if we perform two experiments in succession, with N possible outcomes for the first and M possible outcomes for the second, the number of possible outcomes for the two experiments is $N \times M$.

Suppose we perform an experiment. Two events are *independent* if knowing that one event occurs has no effect on the probability of the occurrence of the other. Events A and B are independent exactly when the probability of Event A *and* Event B occurring is the probability of Event A times the probability of Event B, that is,

$$P(A \text{ and } B) = P(A) \times P(B)$$

More ways of counting: Permuting and combining

More sophisticated ways of counting that can be deduced from the Counting Principle involve permutations and combinations. A *permutation* of items is an arrangement of them in a certain order. The number of permutations of n distinct objects is $n \times (n-1) \times \cdots \times 2 \times 1$. This expression is denoted by $n!$. The number of ways to select k items from n items and arrange them in order is the number of permutations of n objects taken k at a time, denoted $P(n, k)$. It is given by the formula

$$P(n, k) = \frac{n!}{(n-k)!}$$

The number of ways to select k items from n items when order makes no difference is the number of combinations of n objects taken k at a time, denoted $C(n, k)$. The formula for calculating this is

$$C(n, k) = \frac{n!}{k!(n-k)!}$$

Expected value and the law of large numbers: Don't bet on it

The *expected value* for you of a game is a measure of the average amount you can expect to win (or lose) per play in the long run. The theoretical expected value is calculated by multiplying the probability of each outcome by the amount you win (or by minus the amount if you lose) and adding up these numbers. The expected value can be estimated from empirical information (such as the percentage of money paid out as prizes in a lottery). The concept of expected value is applied not only to games of chance but also in such areas as manufacturing, finance, economics, and insurance.

The *law of large numbers* says that if you play a game many times, then you will almost certainly win or lose approximately what the theoretical expected value of the game says you will win or lose. Because the expected value for the player for games of chance in a casino is always negative, the law of large numbers ensures that casinos will make a profit and that in the long term gamblers will lose money.

KEY TERMS

CHAPTER QUIZ

1. If I draw a card at random from a standard deck of 52 cards, what is the probability that I draw a red queen? Write your answer as a fraction.

Answer 2/52 = 1/26

If you had difficulty with this problem, see Example 5.1.

2. Suppose I flip two identical coins. What is the probability that I get a head and a tail? Write your answer as a fraction.

Answer 1/2

If you had difficulty with this problem, see Example 5.2.

3. Suppose a softball team has 25 players. Assume that 13 bat over .300, 15 are infielders, and 5 are both. What is the probability that a player selected at random is either batting over .300 or an infielder? Write your answer as a fraction.

Answer 23/25

If you had difficulty with this problem, see Example 5.6.

4. A new medical test is given to a population whose disease status is known. The results are in the table below.

	Has disease	Does not have disease
Test positive	300	100
Test negative	10	200

Use these data to estimate the sensitivity and specificity of this test. Write your answers in percentage form rounded to one decimal place.

Answer Sensitivity 96.8%; specificity 66.7%

If you had difficulty with this problem, see Example 5.10.

5. The table below shows test results for a certain population.

	Has disease	Does not have disease
Test positive	10	30
Test negative	5	40

Find the positive predictive value and negative predictive value for this test for this population. Write your answers in percentage form rounded to one decimal place.

Answer PPV 25.0%; NPV 88.9%

If you had difficulty with this problem, see Example 5.12.

6. How many four-letter codes can we make if the first letter is A, B, or C and the remaining letters are D or E? Letters may be repeated. One such code is ADDE.

Answer 24

If you had difficulty with this problem, see Example 5.16.

7. It has been found that only 1 in 100 bags of M&Ms have five more than the average number of M&Ms. Only 1 in 20 boxes of Cracker Jacks has a really good prize. I buy a box of Cracker Jacks and a bag of M&Ms. What is the probability that I get a really good prize and five more than the average number of M&Ms? Write your answer as a fraction.

Answer 1/2000

If you had difficulty with this problem, see Example 5.20.

8. In how many ways can I choose 4 people from a group of 10 and make them president, vice president, secretary, and treasurer? Write your answer in terms of factorials.

Answer 10!/6!

If you had difficulty with this problem, see Example 5.22.

9. In how many ways can I select a 4-person committee from a group of 10 people? Leave your answer in terms of factorials.

Answer 10!/(4!6!)

If you had difficulty with this problem, see Example 5.25.

10. The expected value for the player of a certain lottery is −20%. If you spend $5 each week on lottery tickets, how much money do you expect to win or lose over a 10-week period?

Answer Lose $10

If you had difficulty with this problem, see Example 5.30.

11. You roll a single die. If you roll 1 or 2, you win $10. Otherwise, you lose $4. What is your expected value for this game?

Answer 67 cents

If you had difficulty with this problem, see Example 5.31.

12. A company manufactures coats at a cost of $20 each. The following are probabilities of defective coats in a given production run: probability of no defective coats: 5%; probability of 1% defective coats: 75%; probability of 2% defective coats: 20%. When a defective coat is detected, it must be removed from the assembly line and replaced. This process adds an extra $10 to the cost of the replaced coat. What is the expected cost to the company of a batch of 4000 coats?

Answer $80,460

If you had difficulty with this problem, see Example 5.37.

POLICE
CROSS
POLICE LINE DO NOT CROSS

6 Statistics

The following excerpt is from an article that appeared at the Web site of *POLICE* magazine.

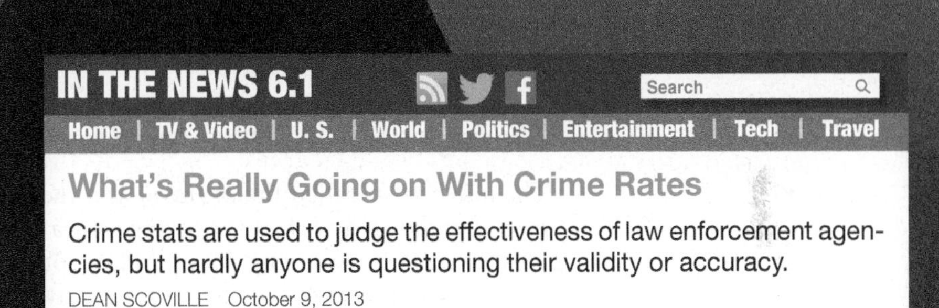

IN THE NEWS 6.1

Search

Home | TV & Video | U. S. | World | Politics | Entertainment | Tech | Travel

What's Really Going on With Crime Rates

Crime stats are used to judge the effectiveness of law enforcement agencies, but hardly anyone is questioning their validity or accuracy.

DEAN SCOVILLE October 9, 2013

"Ethics." The emphasis placed on the subject and its inclusion as part of many a law enforcement academy's curriculum is understandable; our profession seeks to bolster a character trait that, in theory, already inhabits its men and women. . .

Modeling this concept in reality, however, can be a more difficult task for law enforcement than one might expect. For all its censuring of personnel for on-duty trysts and off-duty DUIs, the profession pays curiously little attention to an ongoing transgression perpetrated within it: stat fudging.

History of Lies

Misreporting statistics about crime is not a new phenomenon. The practice may well predate the inception of the Uniform Crime Reports (UCR) in 1930.

Developed in the 1920s by the International Association of Chiefs of Police, the Uniform Crime Reports program was designed to collect, classify, and store a variety of crime data in a uniform manner to allow comparisons across the nation. The FBI took over management of the UCR program the following year, and currently works alongside over 18,000 law enforcement agencies that voluntarily provide crime data to the program. . .

The UCR is extremely valuable and often the foundation of media stories on crime rates rising or falling. It is also flawed. There has long been a notion that some of the data provided by individual agencies may be inaccurate, either through human error or human intervention.

When purposely committed, the impetus for such statistical "errors" may be a strategic feint, an attempt to create an illusion of vulnerability, or strength, depending on one's agenda. Within law enforcement agencies, that feint can be the illusion of success, particularly in metropolitan areas where a desire to attract the dollars of citizens, visitors, tourists, and businesses trumps concerns over the welfare of that patronage.

Elsewhere, creating the impression of a crime problem can have its benefits, as well.

This article refers to "statistics" and claims that law enforcement agencies misreport statistics about crime. *Statistics* is the science of gathering and analyzing data with the objective of either describing the data or drawing conclusions from them. *Descriptive statistics* refers to organizing data in a form that can be more easily understood. *Inferential statistics* refers to the process of drawing conclusions from the data, along with some estimate of how accurate those conclusions are. In this chapter, we will examine both descriptive and inferential statistics in various contexts, including polling and medical research.

6.1 Data summary and presentation: Boiling down the numbers

TAKE AWAY FROM THIS SECTION
Know the statistical terms used to summarize data.

It may be difficult to make sense of large amounts of raw data unless they are organized and summarized in some coherent way. Such a summary is often given in terms of a few key numbers.

Mean, median, and mode

The *mean* (also called the *average*) and the *median*, the middle data point, are the most common ways of summarizing a data set using a single number. The *mode*, the most frequently occurring data point, is sometimes used as well.

These numbers give us a "representative" sense of the data. They are defined as follows.

KEY CONCEPT

The mean (also called the average) of a list of numbers is the sum of the numbers divided by the number of entries in the list.

The median of a list of numbers is the middle number. It is obtained by first ordering the data from smallest to largest. If there is an odd number of data points, pick the middle value. If there is an even number of data points, take the average of the middle two numbers.

If a number in a data set occurs more frequently than any other number, it is called the mode of the data. If there are two such numbers, the data set is called bimodal. If there are more than two such numbers, the data set is multimodal.

The mean is the usual average you expect your teacher to calculate from test scores: Add the scores and divide by the number of exams. For example, if your test scores are

$$60, 90, 70, 60, 80$$

the mean is

$$\frac{60+90+70+60+80}{5} = \frac{360}{5} = 72$$

To find the median, we order the test scores from smallest to largest and pick out the middle score:

$$60, 60, \underline{70}, 80, 90$$

Thus, the median score is 70. Suppose the list included one more score, an 86. Then there would have been two middle scores:

$$60, 60, \underline{70, 80}, 86, 90$$

In this case, we find the median by averaging the middle two scores:

$$\text{Median} = \frac{70 + 80}{2} = 75$$

The mode is 60 because that score occurs more than any of the others.

EXAMPLE 6.1 Calculating the mean: Starting salaries for college graduates

The table below shows projected average starting salaries for 2020 college graduates with top-earning bachelor's degrees, listed by their major. The information is taken from the NACE Center for Career Development and Talent Acquisition.[1]

Bachelor's degree major	2020 average salary
Engineering	$69,961
Computer Science	$67,411
Math and Sciences	$62,488
Business	$57,939
Social Sciences	$57,425
Communications	$56,484
Humanities	$53,617
Agriculture & Natural Resources	$53,504

a. Find the mean of the salaries in the table.

b. Do you think the number you found in part a represents the average starting salary of 2020 college graduates?

SOLUTION

a. Using a calculator, we find the sum of the salaries in the table to be

$$69{,}961 + 67{,}411 + 62{,}488 + 57{,}939 + 57{,}425 + 56{,}484 + 53{,}617 + 53{,}504 = 478{,}829$$

There are 8 entries in the table, so we divide the sum by 8:

$$\frac{478{,}829}{8} = 59{,}853.63$$

Thus, the mean is $59,853.63.

b. In order to find the average starting salaries of all college graduates, we would need to use the starting salary for each individual graduate. This table does not include lower-earning majors such as education. Further, the number of lower-earning majors is larger than the number of majors with the largest salaries, so such an average would be affected by a relatively large number of salaries at the lower end of the scale. We conclude that the number from part a does not represent the average starting salary of 2020 graduates. In fact, the average starting salary is less than the figure of $59,853.63 from part a.

TRY IT YOURSELF 6.1

Find the median of the salaries in the table.

[1] *Taken from* www.naceweb.org/job-market/compensation/starting-salary-projections-for-top-earning-degrees-level/

One type of representative number may be preferred over another depending on the context. For example, the median is often used for home prices, and the mean is used for batting averages. The following example helps to explain this practice.

EXAMPLE 6.2 Choosing between mean and median: Home prices

The following list gives home prices (in thousands of dollars) in a small town:

80, 120, 125, 140, 180, 190, 820

The list includes the price of one luxury home. Calculate the mean and median of this data set. (Round the mean to one decimal place.) Which of the two is more appropriate for describing the housing market?

SOLUTION

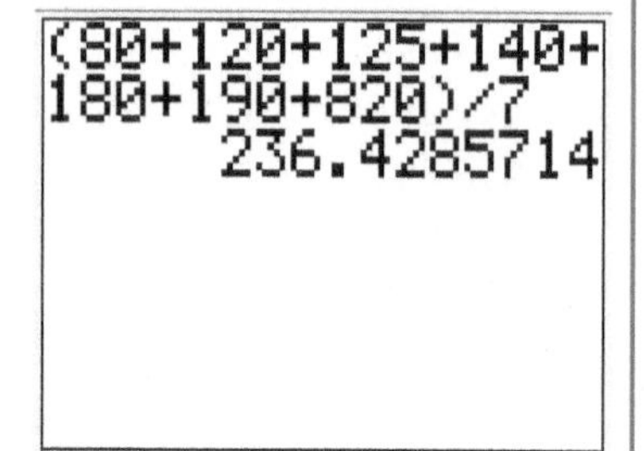

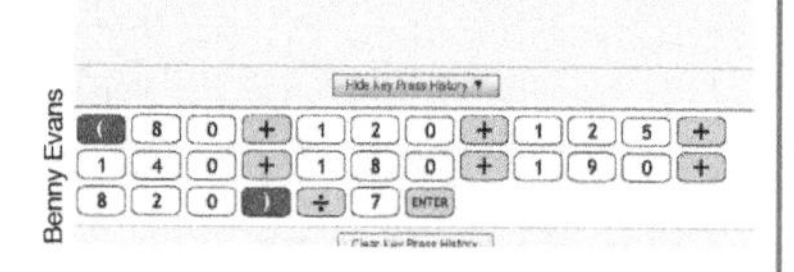

Benny Evans

To calculate the mean we add the data values and divide by the number of data points:

$$\text{Mean} = \frac{80 + 120 + 125 + 140 + 180 + 190 + 820}{7} = \frac{1655}{7}$$

or about 236.4 thousand dollars. This is the average price of a home.

The list of seven prices is arranged in order, so the median is the fourth value, 140 thousand dollars.

Note that the mean is higher than the cost of every home on the market except for one—the luxury home. The median of 140 thousand dollars is more representative of the market.

TRY IT YOURSELF 6.2

To calculate the batting average for a baseball player, we record a 1 when a batter gets a hit and a 0 when he does not. The batting average is the average of this list of 1's and 0's. (This is the number of hits divided by the number of at-bats.) The "batting median" for a player would be the median of the list of 1's and 0's recorded for a batter. All major league baseball players get hits less than half the time they go to bat. Show that this median for a major league baseball player is 0, and explain why the batting median would not be a good way to represent the batting record.

The answer is provided at the end of this section.

An *outlier* is a data point that is significantly different from most of the others. For example, the luxury home in Example 6.2 represents an outlier. In order to see how this outlier affects the mean, let's calculate the mean of home prices after omitting the home priced at 820 thousand dollars:

$$\text{New mean} = \frac{80 + 120 + 125 + 140 + 180 + 190}{6} = \frac{835}{6}$$

or about 139.2 thousand dollars. This number is far from the mean of 236.4 thousand dollars we found when the outlier of 820 thousand dollars is included.

On the other hand, if we calculate the median after omitting the home priced at 820 thousand dollars, we find the new median to be 132.5 thousand dollars. This number is not far from the median of 140 thousand dollars we found when the outlier is included.

This discussion is typical of how outliers affect averages. They often have a greater effect on the mean but a lesser effect on the median.

KEY CONCEPT

An outlier is a data point that is significantly different from most of the data. Typically, outliers have a greater effect on the mean, but a lesser effect on the median.

The skewing of average home prices by outliers is illustrated in the following article from the Web site Realtor.com.

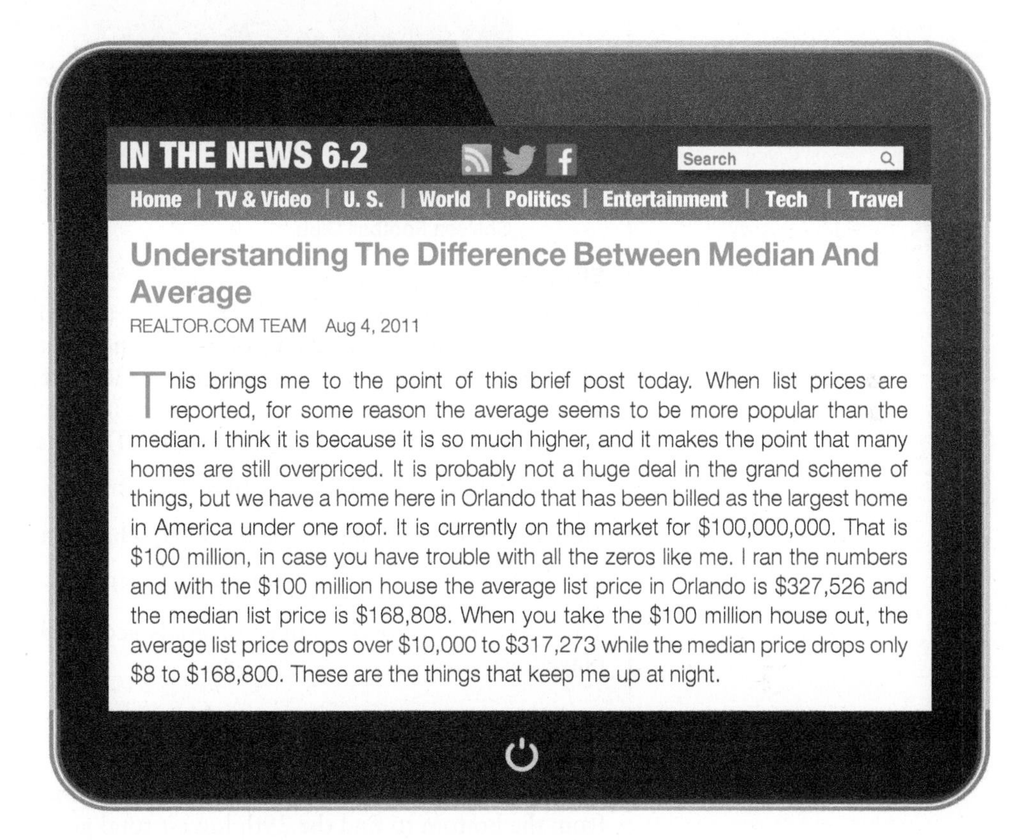
IN THE NEWS 6.2

Home | TV & Video | U. S. | World | Politics | Entertainment | Tech | Travel

Understanding The Difference Between Median And Average

REALTOR.COM TEAM Aug 4, 2011

This brings me to the point of this brief post today. When list prices are reported, for some reason the average seems to be more popular than the median. I think it is because it is so much higher, and it makes the point that many homes are still overpriced. It is probably not a huge deal in the grand scheme of things, but we have a home here in Orlando that has been billed as the largest home in America under one roof. It is currently on the market for $100,000,000. That is $100 million, in case you have trouble with all the zeros like me. I ran the numbers and with the $100 million house the average list price in Orlando is $327,526 and the median list price is $168,808. When you take the $100 million house out, the average list price drops over $10,000 to $317,273 while the median price drops only $8 to $168,800. These are the things that keep me up at night.

Frequency tables and histograms

When there are many data points, they may be presented using a *frequency table*, which shows how often each data point occurs. Frequency tables can be used to calculate the mean, median, and mode, as the next example illustrates.

EXAMPLE 6.3 Calculating mean, median, and mode: Chelsea FC

The Chelsea Football Club (FC) is a British soccer team. The accompanying table shows the goals scored (by either team) in the games played by Chelsea FC between August and May of a recent regular season. The data are arranged according to the total number of goals scored in each game.

Goals scored by either team	0	1	2	3	4	5	6	7
Number of games	5	10	14	14	10	2	1	1

Find the mean, median, and mode for the number of goals scored per game. Round the mean to one decimal place.

Paul Thomas/Getty Images

Chelsea Football Club.

SOLUTION

To find the mean, we add the data values (the total number of goals scored) and divide by the number of data points. To find the total number of goals scored, for each entry we multiply the goals scored by the corresponding number of games. Then we add. For example, there were 14 games in which exactly 2 goals were scored, so these games contribute $14 \times 2 = 28$ goals. The total number of goals scored is

$$5 \times 0 + 10 \times 1 + 14 \times 2 + 14 \times 3 + 10 \times 4 + 2 \times 5 + 1 \times 6 + 1 \times 7 = 143$$

Now we find the number of data points, which is the total number of games:

$$5 + 10 + 14 + 14 + 10 + 2 + 1 + 1 = 57$$

The mean is the total number of goals scored divided by the number of games played, so

$$\text{Mean} = \frac{143}{57}$$

or about 2.5.

Now we find the median. The number of games is 57, which is odd, so we count from the bottom to find the 29th lowest total goal score. This is 2, so

$$\text{Median} = 2$$

Because 2 and 3 occur equally often (14 times each) and more frequently than any other, this data set is bimodal. We have

$$\text{Modes} = 2 \text{ and } 3$$

Thus, on average, the teams combined to score 2.5 goals per game. Half of the games had goals totaling 2 or more (almost half had goals more than 2), and the most common numbers of goals scored in a Chelsea FC game were 2 and 3.

TRY IT YOURSELF 6.3

Find the mean, median, and mode of the data for the Chelsea FC, counting only games with no more than 2 goals scored. Round the mean to one decimal place.

The answer is provided at the end of this section.

A bar graph of a frequency table is a *histogram*. (Recall that bar graphs were discussed in Section 2.1.) For example, the graphs in **Figures 6.1 and 6.2** are histograms showing the number of games in which given total goals were scored in FC matches. The first histogram uses the frequency table in Example 6.3. For the second histogram, we group the highest score totals to make the following frequency table.

Goals scored by either team	0	1	2	3	4	5 to 7
Number of games	5	10	14	14	10	4

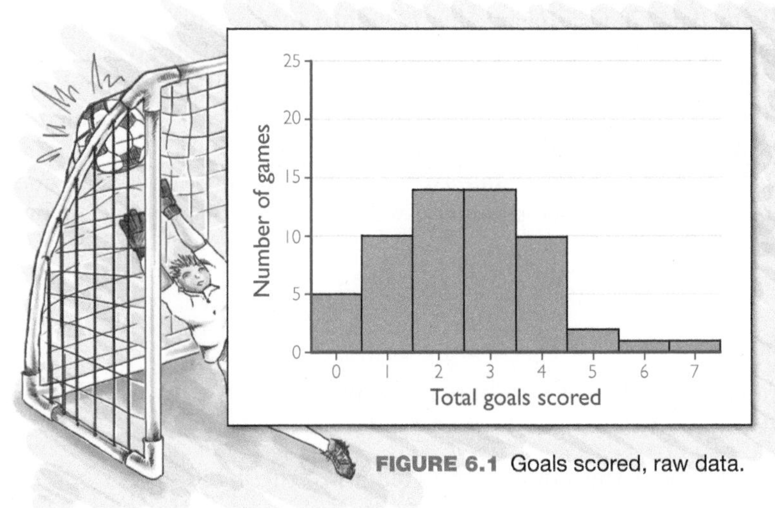

FIGURE 6.1 Goals scored, raw data.

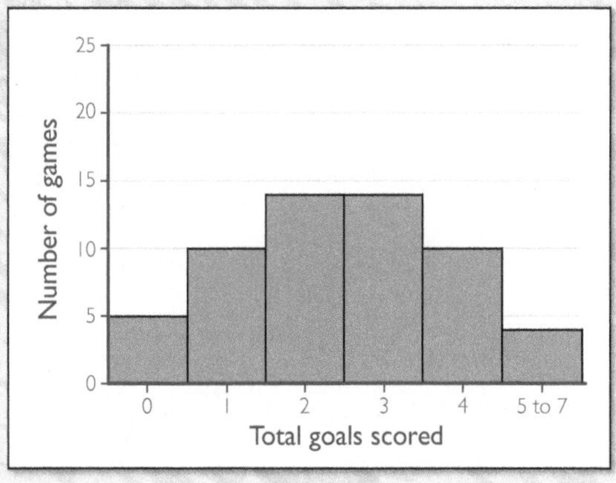

FIGURE 6.2 Goals scored, grouped data.

The grouping of categories illustrated in Figure 6.2 is common in frequency tables and histograms. To illustrate the point further, suppose we toss 1000 coins and record the percentage of heads that occur. Then we repeat this experiment 999 times for a grand total of 1000 of these 1000-coin-toss experiments. A record showing the results of each of the 1000-coin tosses would be awkward to produce and difficult to read.

In a computer simulation of these 1000 coin tosses, 23 of the tosses produced 47% heads or less. Also, 75 tosses produced 47% to 48% heads, and so on. Grouping the data in this way yields the following frequency table.

Percent heads	Less than 47%	47% to 48%	48% to 49%	49% to 50%
Number of 1000-coin tosses	23	75	140	234
Percent heads	50% to 51%	51% to 52%	52% to 53%	At least 53%
Number of 1000-coin tosses	250	157	94	27

Figure 6.3 shows a histogram for this grouping of the data. This visual presentation makes the situation even clearer. We can clearly see that the vast majority of

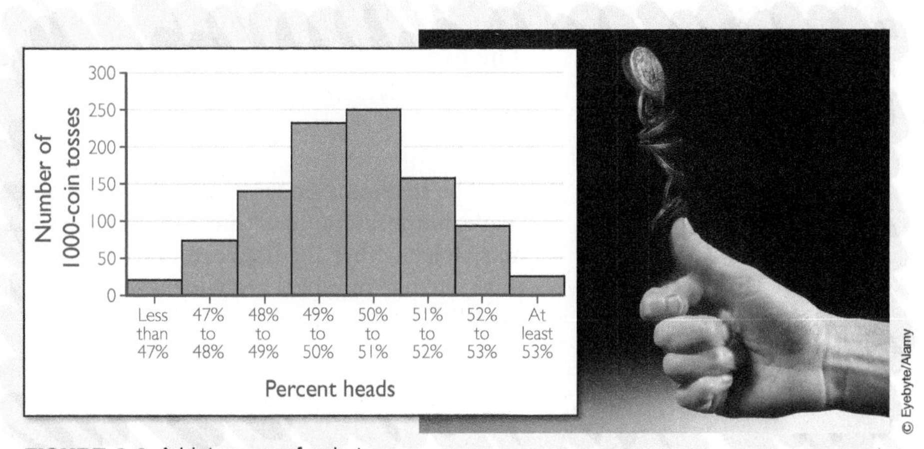

FIGURE 6.3 A histogram of coin tosses.

the 1000-coin tosses were between 47% and 53% heads. Note also how the number of heads increases near the 50% mark. This is what we would expect from tosses of fair coins.

The five-number summary

We have seen that the mean, median, and mode can often provide helpful summary information about data. But a single value is limited in what it can convey and can even be misleading. For instance, in Example 6.2 the median gives us some information about housing prices. But that number does not tell us anything about the spread of the data. Are most of the house prices clustered close to the median, or are they spread out over a wide range of prices?

A device called the *five-number summary* is helpful in answering such questions. To explain the five-number summary, we first define what is meant by a *quartile*.

KEY CONCEPT

- **The first quartile of a list of numbers is the median of the lower half of the numbers in the list.**
- **The second quartile is the same as the median of the list.**
- **The third quartile is the median of the upper half of the numbers in the list.**

If the list has an even number of entries, it is clear what we mean by the "lower half" and "upper half" of the list. If the list has an odd number of entries, eliminate the median from the list. This new list now has an even number of entries. The "lower half" and "upper half" refer to this new list.

We note that in everyday usage we may speak of a number that is *in the first quartile*. We mean that it is less than the first quartile value, which means that it is in the lowest one-quarter of the data. We say that a number is *in the second quartile* if it is between the first quartile value and the median—in the next-to-lowest one-quarter of the data. Similarly, being *in the third quartile* means being in the next-to-highest one-quarter of the data, and being *in the fourth quartile* means being in the highest one-quarter of the data.

We use quartiles to make the *five-number summary* for data.

KEY CONCEPT

The five-number summary of a list of numbers consists of the minimum, the first quartile, the median, the third quartile, and the maximum.

The five-number summary does a much better job of describing data than any single number could. If the five numbers are relatively close together, the summary tells us that the data are fairly closely bunched about the median. If, in contrast, the five numbers are widely separated, we know that the data are spread out as well.

EXAMPLE 6.4 Calculating the five-number summary: Income of celebrities

John Shearer/Getty Images

Kylie Jenner

Each year *Forbes* magazine publishes a list it calls the Celebrity 100, which shows the world's highest-paid celebrities. The accompanying table shows the top nine names on the list for 2020 and their annual incomes.

Celebrity	Income (millions of dollars)
Kylie Jenner	590
Kanye West	170
Roger Federer	106.3
Cristiano Ronaldo	105
Lionel Messi	104
Tyler Perry	97
Neymar	95.5
Howard Stern	90
Lebron James	88.2

Calculate the five-number summary for this list of incomes.

SOLUTION

The incomes are already arranged in order:

88.2, 90, 95.5, 97, <u>104</u>, 105, 106.3, 170, 590

The minimum is \$88.2 million (Lebron James), the median (the fifth in order of the nine incomes) is \$104 million, and the maximum is \$590 million (Kylie Jenner). The "lower half" of the list consists of the four numbers less than the median, which are

88.2, 90, 95.5, 97

The median of this lower half is 92.75, so the first quartile of incomes is \$92.75 million. The "upper half" of the list consists of the four numbers greater than the median, which are

105, 106.3, 170, 590

The median of this upper half is 138.15, so the third quartile of incomes is \$138.15 million. Thus, the five-number summary is

Minimum = \$88.2 million
First quartile = \$92.75 million
Median = \$104 million
Third quartile = \$138.15 million
Maximum = \$590 million

TRY IT YOURSELF 6.4

Calculate the five-number summary for the incomes in Example 6.4 if the celebrity ranked number 10, Dwayne Johnson, is added to the list. His income was \$87.5 million.

The answer is provided at the end of this section.

What does the five-number summary in Example 6.4 show? All of the incomes are between \$88.2 million and \$590 million. The median income is \$104 million, and as many of the incomes are at or above that level as are below. The lowest one-quarter of all incomes are between \$88.2 million and \$92.75 million, the next one-quarter are between \$92.75 million and the median of \$104 million, the next one-quarter are between the median of \$104 million and \$138.15 million, and the highest one-quarter are between \$138.15 million and \$590 million. The summary provided by this breakdown gives a general sense of how the incomes are spread out.

In our example, the number of data points was small, but if we were working with 100,000 data points, for example, the five-number summary would be a big help.

Boxplots

There is a commonly used pictorial display of the five-number summary known as a *boxplot* (also called a *box and whisker diagram*). **Figure 6.4** shows the basic geometric figure used in a boxplot. Note that the bottom of the figure is the minimum and the top is the maximum. The box area represents the region from the first to the third quartile. (The width of the box has no significance.) The median is marked with a horizontal line inside the box. Note that, as the picture indicates, the regions between the minimum, the first quartile, the median, the third quartile, and the maximum each represent 25% of the data. A greater length between the top and bottom "whiskers" indicates a greater spread from the maximum to the minimum, and a longer box indicates a greater spread from the first to third quartiles. The location of the median within the box gives information about how the data between the first quartile and the third quartile are distributed about the median.

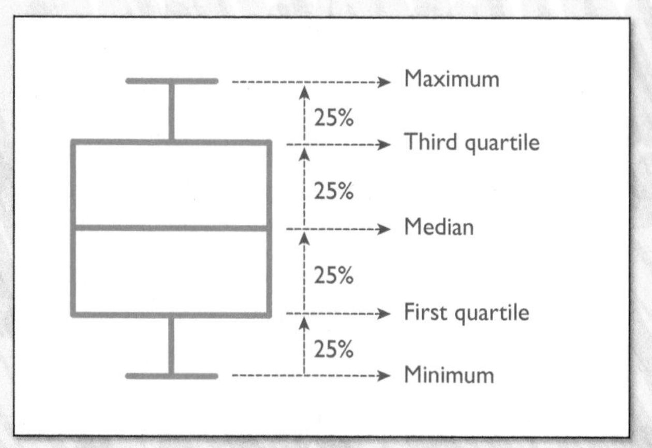

FIGURE 6.4 The basic boxplot diagram.

EXAMPLE 6.5 Making and interpreting a boxplot: Gas mileage

A report on the greenercars.org Web site shows 2016-model cars that score well in terms of environmental impact. Here are the data, with mileage measured in miles per gallon (mpg) listed in order of the Web site's "green score." (We have excluded electric vehicles.)

Model	City mileage (mpg)	Highway mileage (mpg)
Chevrolet Spark	31	53
Mazda 2	33	46
Mercedes-Benz Smart ForTwo Coupe	34	50
Volkswagen Jetta Hybrid	42	48
Toyota Camry Hybrid LE	43	39
Lexus CT 200H	43	43
Toyota Prius V	44	41
Toyota Prius C	53	40
Toyota Prius	54	39
Toyota Prius Eco	58	40

a. Find the five-number summary for city mileage.

b. Present a boxplot of city mileage.

c. Comment on how the data are distributed with respect to the maximum and minimum. *Note*: The corresponding calculations for highway mileage are left as an exercise. See Exercise 25.

SOLUTION

a. The list for city mileage, in order from lowest to highest, is

31, 33, 34, 42, 43, 43, 44, 53, 54, 58

The maximum is 58 mpg, and the minimum is 31 mpg. The two numbers in the middle are both 43, so the median is 43.

The lower half of the list is 31, 33, 34, 42, 43, and the median of this half is 34. Thus, the first quartile is 34 mpg. The upper half of the list is 43, 44, 53, 54, 58, and the median of this half is 53. Thus, the third quartile is 53 mpg.

b. The corresponding boxplot appears in **Figure 6.5**. The vertical axis is the mileage measured in miles per gallon.

c. Referring to the boxplot, we note that the minimum is not far below the first quartile. This is because the bottom three cars (in terms of city mileage) are not far apart. A similar comment applies to the maximum and the third quartile.

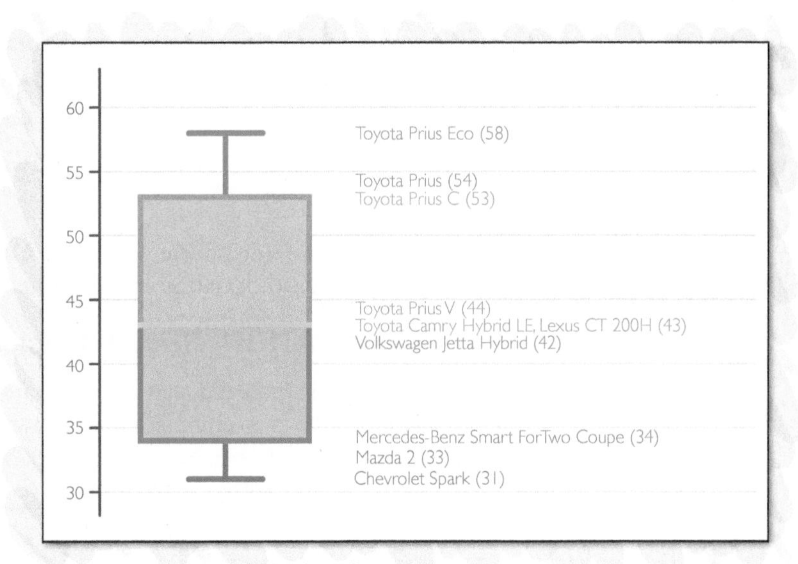

FIGURE 6.5 Boxplot for city mileage.

Standard deviation

As we have noted, data sets can be clustered together (such as 10, 11, 12, 12, 13) or spread out (such as 1, 10, 20, 100, 300). One measure of how much data are spread out is the five-number summary. Another way to measure the spread of data is the *standard deviation*. This number is commonly denoted by the lowercase Greek letter σ (pronounced "sigma").

KEY CONCEPT

The standard deviation is a measure of how much the data are spread out from the mean. The smaller the standard deviation, the more closely the data are clustered about the mean.

For example, a data set of the yearly rainfall amounts in the Gobi Desert would show values that deviate very little from the mean because there is not a significant variation in rainfall from one year to the next. In contrast, a data set of prices for new automobiles would have a wide range because high-end sports cars cost a great deal more than the average family automobile. For this data set, we would have a larger standard deviation. The standard deviation is used in many settings, and, like the mean, it is sensitive to outliers.

The unit of measurement for the standard deviation is the same as the unit for the original data. For example, if the data show car prices in dollars, then the standard deviation is measured in dollars.

Deviation from the mean refers to the difference between data and the mean of the data. For example, let's say the mean price for a new automobile is $34,000. If the price of a Mercedes is $52,000, we say that the price deviates by $18,000 from the mean because $52,000 – $34,000 = $18,000. If the price for a Hyundai Elantra is $22,000, we say that the price deviates by –$12,000 from the mean because $22,000 – $34,000 = –$12,000.

The standard deviation is used in finance to measure the risk of an investment. An investment with a high standard deviation for its returns is a volatile investment.

The formula for standard deviation looks intimidating. But, as we shall see, the actual calculation is not very difficult. Suppose the data points are

$$x_1, x_2, x_3, \ldots, x_n$$

It is common practice to use the lowercase Greek letter μ (pronounced "mew") to denote the mean. The formula for the standard deviation is[2]

$$\sigma = \sqrt{\frac{(x_1 - \mu)^2 + (x_2 - \mu)^2 + \cdots + (x_n - \mu)^2}{n}}$$

The template provided in the following Calculation Tip makes the process of calculating the standard deviation manageable.

CALCULATION TIP 6.1 Calculating Standard Deviation

To find the standard deviation of n data points, we first calculate the mean μ. The next step is to complete the following calculation template:

Data	Deviation	Square of deviation
$\vdots$	$\vdots$	$\vdots$
x_i	$x_i - \mu$	Square of second column
$\vdots$	$\vdots$	$\vdots$
		Sum of third column
		Divide the above sum by n and take the square root.

[2] *There are actually two common definitions of the standard deviation of a data set. One is the* **population** standard deviation, *denoted by σ and defined as we have it here. The other is the* **sample** standard deviation, *denoted by s and defined in the same way except that the denominator n is replaced by n – 1. The sample standard deviation s is used when the data consist of a sample from a larger population. The population standard deviation σ is used when the data are not a sample but consist of the entire "population" of data. For large data sets, their ratio is very close to 1 (so they are nearly the same). The reason statisticians have different ways of defining these is due to a technical point about using a sample standard deviation as an "unbiased estimator" for the population standard deviation.*

"It's the new keyboard for the statistics lab. Once you learn how to use it, it will make computation of the standard deviation easier."

The procedure for calculating standard deviation is illustrated in the next example.

EXAMPLE 6.6 Calculating standard deviation: Baseball pitchers

The ERA (earned run average—the lower the number, the better) histories for the pitching staffs of the New York Yankees and the Chicago Cubs are shown in the following table.

Team	ERA 2016	ERA 2017	ERA 2018	ERA 2019	ERA 2020
New York Yankees	4.16	3.72	3.78	4.31	4.35
Chicago Cubs	3.15	3.95	3.65	4.10	3.99

Calculate the mean and the standard deviation for the Yankees' ERA history. It turns out that the mean and standard deviation for the Cubs' ERA history are $\mu = 3.77$ and $\sigma = 0.34$. What comparisons between the Yankees and the Cubs can you make based on these numbers?

SOLUTION

The mean for the Yankees is

$$\mu = \frac{4.16 + 3.72 + 3.78 + 4.31 + 4.35}{5}$$

This is about 4.06.

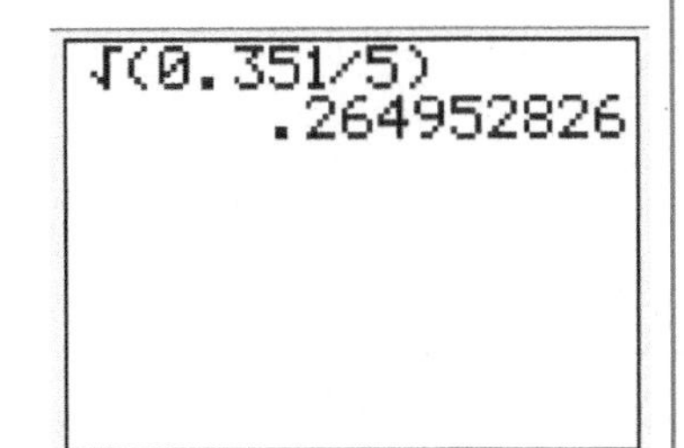

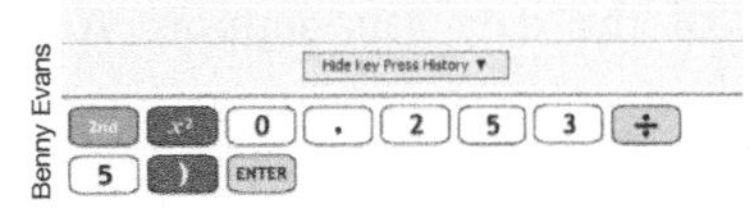

ERA x_i	Deviation $x_i - 4.06$	Square of deviation $(x_i - 4.06)^2$
4.16	$4.16 - 4.06 = 0.10$	0.010
3.72	$3.72 - 4.06 = -0.34$	0.116
3.78	$3.78 - 4.06 = -0.28$	0.078
4.31	$4.31 - 4.06 = 0.25$	0.063
4.35	$4.35 - 4.06 = 0.29$	0.084
Sum of third column		0.351
Sum divided by $n = 5$, square root		$\sigma = \sqrt{0.351/5} = 0.26$

We make the calculation of the standard deviation using Calculation Tip 1. The first column shows the data points. We find the second column by subtracting the mean $\mu = 4.06$ to find the deviations. We calculate the third column by squaring the second column to find the squares of the deviations, which we round to three decimal places.

We conclude that the mean and the standard deviation for the Yankees' ERA history are $\mu = 4.06$ and $\sigma = 0.26$.

Because the Cubs' mean of $\mu = 3.77$ is smaller than the Yankees' mean, over this period the Cubs had a better pitching record. But the Yankees' ERA had a smaller standard deviation than that of the Cubs (who had $\sigma = 0.34$). This means that the Yankees were more consistent—their numbers are not spread as far from the mean.

TRY IT YOURSELF 6.6

Verify that the mean and the standard deviation for the Cubs' ERA history are $\mu = 3.77$ and $\sigma = 0.34$.

The answer is provided at the end of this section.

In the next example, we use the concepts from this section to compare two data sets.

EXAMPLE 6.7 Using standard deviation: Free-throw percentages

The following table shows the Eastern Conference NBA team free-throw percentages at home and away for the 2015–2016 season. At the bottom of the table, we have displayed the mean and standard deviation for each data set. You will be asked to verify these values in Exercise 30.

Team	Free-throw percentage at home	Free-throw percentage away
New York	82.6	78.5
Charlotte	79.6	79.2
Atlanta	78.6	78.2
Boston	78.4	79.2
Chicago	78.3	79.1
Toronto	76.8	77.2
Indiana	76.2	77.5
Orlando	75.2	76.3
Brooklyn	75.1	76.3
Miami	74.9	74.5
Milwaukee	74.8	74.7
Cleveland	74.5	74.0
Washington	73.8	72.1
Detroit	69.8	63.5
Philadelphia	67.5	71.5
Mean	75.74	75.45
Standard deviation	3.61	4.00

What do these values for the mean and standard deviation tell us about free-throws shot at home compared with free-throws shot away from home? Does comparison of the minimum and maximum of each of the data sets support your conclusions?

SOLUTION

The means for free-throw percentages are 75.74 at home and 75.45 away, so on average the teams do slightly better at home than on the road. (This is not the case for every team.) The standard deviation of 3.61 percentage points at home, which is smaller than the standard deviation of 4.00 percentage points away from home, means that the free-throw percentages at home vary from the mean less than the free-throw percentages away.

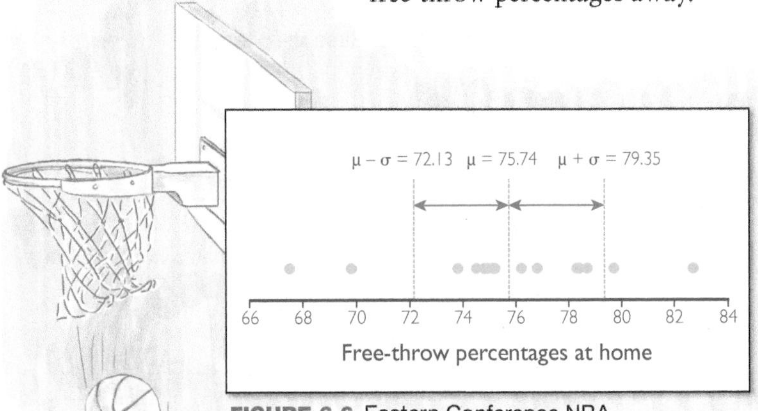

FIGURE 6.6 Eastern Conference NBA free-throw percentages at home.

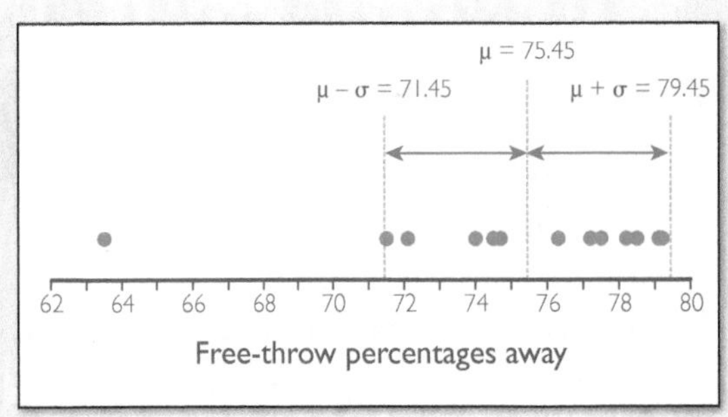

FIGURE 6.7 Eastern Conference NBA free-throw percentages away.

The difference between the maximum and minimum percentages shows the same thing: The free-throw percentages at home range from 67.5% to 82.6%, and the free-throw percentages away range from 63.5% to 79.2%. The plots of the data in **Figures 6.6 and 6.7** provide a visual verification that the data for away free-throws are more broadly dispersed than the data for home free-throws.

WHAT DO YOU THINK?

Standard deviation: You are a conservative investor. Comparing records of investment advisors, you find that both have a record of earning money for investors. The mean amounts earned are the same, but the standard deviation in amounts earned for Advisor #1 is much larger than that for Advisor #2. Which advisor would you choose? Explain your reasoning.

Questions of this sort may invite many correct responses. We present one possibility: The standard deviation is a measure of the spread of data. A large standard deviation in earnings for Advisor #1 means that many customers experience earnings that are much different from the mean. That makes it difficult to judge how your own investment will prosper. In contrast, the small standard deviation for Advisor #2 tells us that most customers earnings are near the mean. As a new customer, I have good reason to expect that my earnings will be near the mean. A conservative investor would surely choose Advisor #2.

Further questions of this type may be found in the *What Do You Think?* section of exercises.

Try It Yourself answers

Try It Yourself 6.1: Calculating the mean: Starting salaries for college graduates $57,682.

Try It Yourself 6.2: Choosing between mean and median: Batting average The median is 0 because there are more 0's than 1's in the list. This median doesn't distinguish between players.

Try It Yourself 6.3: Calculating mean, median, and mode: Chelsea FC Mean = 1.3; median = 1; mode = 2.

Try It Yourself 6.4: Calculating the five-number summary: Income of celebrities

- Minimum = $87.5 million
- First quartile = $90 million
- Median = $100.5 million
- Third quartile = $106.3 million
- Maximum = $590 million

Try It Yourself 6.6: Calculating standard deviation: Baseball pitchers Answers will vary.

Exercise Set 6.1

Test Your Understanding

1. True or false: The terms "average" and "median" mean basically the same thing.

2. A histogram is a bar graph of ______.

3. If the number 13 appears five times in a data set, and no other number appears that often, then the mode of the data set is ______.

4. Does an outlier in a data set typically have a greater effect on the median or on the mean?

5. Would you expect a data set with a small standard deviation to be closely grouped or widely dispersed?

6. Which of the following are part of a five-number summary? Indicate all correct answers. **a.** the median **b.** the mode **c.** the mean **d.** the standard deviation **e.** the first quartile **f.** the minimum.

7. How much of a data set lies between the first quartile and the median?

Problems

8–11. Analyzing data sets. *In Exercises 8 through 11, a data set is given (in some cases by a frequency table). In each case, find the mean and the five-number summary, and make a boxplot of the data. Round the mean to two decimal places.*

8. 2, 5, 6, 7, 7, 9, 11, 12, 13, 15, 16

9. 3, 6, 6, 6, 9, 12, 12, 12, 18, 20, 20

10.

Number	0	2	5	6	7	10	15
Frequency	1	3	3	2	5	1	1

11.

Number	2	5	6	10	11	13	16
Frequency	4	2	1	2	3	1	4

12–17. Choose a representative. *In Exercises 12 through 17, choose whether the mean, median, or mode is the best representative of the data. Explain your choice. Various answers may be acceptable.*

12. The square footage of homes in the area

13. The batting average of Major League Baseball players

14. The age of first-time brides in your town

15. The car prices in a used-car lot

16. The high temperature in your town on June 1 of each year

17. The wind speed at 7:00 a.m. in Galveston, Texas

18. **Homework scores.** Your first eight homework scores are 92, 86, 78, 85, 95, 81, 88, and 90. Find the mean, median, and mode of these scores. Round the mean to one decimal place.

19. **More on homework scores.** *This is a continuation of Exercise 18.* Your first eight homework scores are listed in Exercise 18. Your ninth score is 15. (It was a bad week.) Find the mean and median of these nine scores. (Again, round the mean to one decimal place.) How would dropping the lowest score affect the mean and the median?

20. **Exam scores.** Suppose that one student's four exam scores for the semester are 75, 75, 75, and 75. Suppose that another student recorded scores of 100, 0, 100, 100. What are the mean and median scores for these two students? What grade does each one earn if the cutoffs are A-90%, B-80%, C-70%, and D-60%, as determined by the mean?

It is debatable whether these students deserve the same grade. Give a good argument for why the second student deserves a "C" and a good argument for why the second student deserves a higher grade.

21. **Home prices.** Suppose we gathered the following list of home prices:

$80,000 $120,000 $120,000 $120,000 $122,000
$135,000 $135,000 $150,000 $160,000

Find the mean, median, and mode of these home prices (to the nearest thousand).

22. **Sales.** An auto dealer's sales numbers are shown in the table below. Find for each month the mean, median, and mode prices of the cars she sold. Round your answers to the nearest dollar.

	Number sold		
Price	**May**	**June**	**July**
$20,000	22	25	24
$15,000	49	24	24
$12,500	25	49	49

23. **Land areas**

a. Calculate the five-number summary of the land areas of the states in the U.S. Midwest.

State	Area (sq. miles)	State	Area (sq. miles)
Illinois	55,584	Missouri	68,886
Indiana	35,867	Nebraska	76,872
Iowa	55,869	North Dakota	68,976
Kansas	81,815	Ohio	40,948
Michigan	56,804	South Dakota	75,885
Minnesota	79,610	Wisconsin	54,310

b. Explain what the five-number summary in part a tells us about the land areas of the states in the Midwest.

c. Calculate the five-number summary of the land areas of the states in the U.S. Northeast.

State	Area (sq. miles)	State	Area (sq. miles)
Connecticut	4845	New York	47,214
Maine	30,862	Pennsylvania	44,817
Massachusetts	7840	Rhode Island	1045
New Hampshire	8968	Vermont	9250
New Jersey	7417		

d. Explain what the five-number summary in part c tells us about the land areas of the states in the Northeast.

e. Contrast the results from parts b and d.

24. **Boxplot.** A boxplot for a data set is shown in **Figure 6.8**. Estimate the minimum, first quartile, median, third quartile, and maximum of the data set.

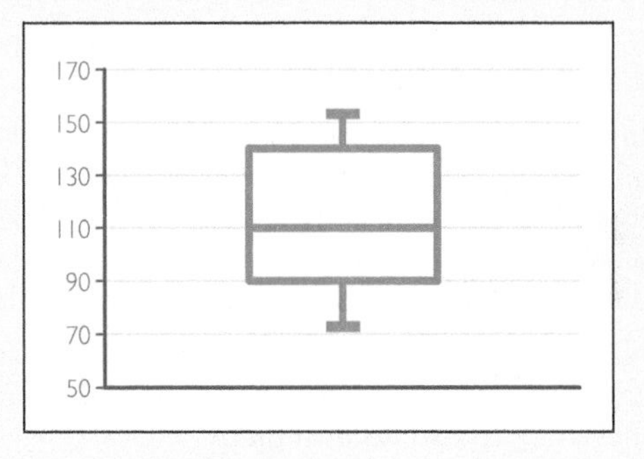

FIGURE 6.8 Boxplot.

25. **Highway mileage.** Find the five-number summary for the data on highway mileage in Example 6.5, and draw the corresponding boxplot.

26. **Math scores.** At one point, the San Diego Achievement Forum reported that during the four prior years, the percentage of students scoring in the lowest quartile in mathematics had declined by 11 percentile points, while the percentage of students in the top quartile had risen from 24 percent to 31 percent. (Presumably, this referred to quartiles in a larger group of test-takers and not just the San Diego schools.) This report seemed to suggest that things were going well. In this exercise, we write T1 for the percentage of students in the lowest quartile (i.e., in the first quartile), T2 for the percentage of students in the second quartile, T3 for the percentage of students in the third quartile, and T4 for the percentage of students in the top quartile (i.e., in the fourth quartile).

a. Suppose that the original numbers were $T1 = 25\%$, $T2 = 25\%$, $T3 = 26\%$, and $T4 = 24\%$, and that four years later the numbers were $T1 = 14\%$, $T2 = 5\%$, $T3 = 50\%$, and $T4 = 31\%$. Is this an improvement? Is there a possible downside to the new numbers in comparison to the original ones?

b. Suppose that the original numbers were $T1 = 25\%$, $T2 = 25\%$, $T3 = 26\%$, and $T4 = 24\%$, and that four years later the numbers were $T1 = 14\%$, $T2 = 50\%$, $T3 = 5\%$, and $T4 = 31\%$. Is this an improvement? Is there a possible downside to the new numbers in comparison to the original ones?

27. Mathematics faculty salaries. The first three columns of the following table are salary figures taken from an American Mathematical Society (AMS) survey of mathematics faculty from Ph.D.-granting public institutions with large PhD programs. The last two columns in the table (marked with an asterisk) were not reported by the AMS, so for the purposes of this exercise, we have just made up the (plausible) numbers in these columns.

Academic rank	First quartile	Median	Third quartile	Minimum*	Maximum*
Assistant professor	\$80,400	\$83,200	\$86,800	\$55,000	\$115,000
Associate professor	\$82,200	\$88,200	\$94,400	\$55,000	\$155,000
Professor	\$107,100	\$126,900	\$153,600	\$75,000	\$225,000

a. Make a boxplot for each of the three categories.

b. Use the boxplots you made in part a to compare the dispersal of salaries among the three categories.

28. Employee absence. A certain company recorded the number of employee absences each week over a period of 10 weeks. The result is the data list 3, 5, 1, 2, 2, 4, 7, 4, 5, 5. Find the mean and standard deviation of the number of absences per week. Round the standard deviation to two decimal places.

29. Calculation. Find the mean and standard deviation of the following list of quiz scores: 75, 88, 65, 90. Round the standard deviation to two decimal places.

30. Free-throws. In the situation of Example 6.7, verify that the means for free-throws at home and away are 75.74% and 75.45%, respectively. Verify that the standard deviations for free-throws at home and away are 3.61 percentage points and 4.00 percentage points, respectively.

31. No calculation. Suppose every student in your class gets a score of 85% on an exam. Find the mean and standard deviation of those exam scores by inspection without doing any calculations.

32. Tossing a die

a. Perform the following experiment: Roll a fair die six times. Record a 1 when the die comes up 6 and a 0 otherwise. Fill in the blanks in the following table with your results:

Roll	1	2	3	4	5	6
Data						

b. Calculate the mean and standard deviation of the data list you made in part a. Leave the mean as a fraction, and round the standard deviation in decimal form to two decimal places.

33. Spread. We have seen that the standard deviation σ measures the spread of a data set about the mean μ. *Chebyshev's inequality* gives an estimate of how well the standard deviation measures that spread. One consequence of this inequality is that for every data set at least 75% of the data points lie within two standard deviations of the mean, that is, between $\mu - 2\sigma$ and $\mu + 2\sigma$ (inclusive)[3]. For example, if $\mu = 20$ and $\sigma = 5$, then at least 75% of the data are at least $20 - 2 \times 5 = 10$ and at most $20 + 2 \times 5 = 30$.

a. Consider the following data: 5, 10, 10, 10, 10, 10, 10, 15. Find the mean and the standard deviation. How many data points does the Chebyshev inequality promise will lie within two standard deviations of the mean?

b. Our statement of this result says "at least 75%," not "exactly 75%." What percent of the data points in part a actually lies within two standard deviations of the mean?

34. An application. We have 1200 lightbulbs in our building. Over a 10-month period, we record the number of bulbs that burn out each month. The result is the data list 23, 25, 21, 33, 17, 39, 26, 24, 31, 22.

a. What is the average number of bulbs that burn out each month?

b. What is the standard deviation of these data? Round the standard deviation to one decimal place.

c. Use Chebyshev's inequality (see Exercise 33) and your answer to parts a and b to estimate how many replacement bulbs you should keep on hand so that for at least 75% of the months you don't have to acquire additional replacement bulbs.

35. Histogram. Consider an experiment in which we toss 20 coins. We perform this experiment 100 times and record the number of experiments in which we got a given number of heads. The results are given in the following table:

Number of heads	5	6	7	8	9	10	11	12	13	14	15
Number of experiments	3	5	8	11	16	18	16	13	6	2	2

Arrange the data in the following groups: less than 35% heads, at least 35% but less than 45% heads, at least 45% but less than 55% heads, at least 55% but less than 65% heads, at least 65% heads. (Here, of course, 35% refers to 35% of the 20 coins, etc.) Make a histogram showing the grouped data.

Exercises 36 through 39 are suitable for group work.

36. Your class. Collect the heights of each of your classmates. Find the mean and standard deviation of these data and make a five-number summary and box plot of it.

37. Experiment. Do the following experiment: Toss 20 coins. Perform this experiment 25 times. Record in the table below the number of experiments in which you got a given number of heads.

Number of heads	0	1	2	3	4	5	6	7	8	9	10
Number of experiments											
Number of heads	11	12	13	14	15	16	17	18	19	20	
Number of experiments											

Arrange the data in the following groups: less than 35% heads, at least 35% but less than 45% heads, at least 45% but less than 55% heads, at least 55% but less than 65% heads, at least 65% heads. (Here, of course, 35% refers to 35% of the 20 coins, etc.) Make a histogram showing the grouped data.

[3] *In the next section, we will see similar, but more precise, statements for special data sets, namely those that are* normally distributed.

38. A challenge problem. A class has an equal number of boys and girls. The boys all got 78% on a test and the girls all got 84%. What are the mean and standard deviation of the test scores for the entire class?

39. Another challenge problem. A class has 9 boys and 21 girls. The class as a whole has a GPA (grade point average) of 2.96, and the boys have a GPA of 2.40. What is the GPA of the girls?

Exercises 40 through 43 are designed to be solved using technology such as calculators or computer spreadsheets.

40. Standard deviations in Excel©. In this section, we discussed the fact that there are two versions of the standard deviation, the "sample standard deviation" and the "population standard deviation." In the Excel© spreadsheet, locate the formulas for these and use them to find both versions of the standard deviations for this list of home prices: \$80,000 \$120,000 \$120,000 \$120,000 \$122,000 \$135,000 \$135,000 \$150,000 \$160,000.

41. Computer simulation. Consider an experiment in which we toss 1000 coins. We perform the experiment 1000 times and record the percentage of heads for each experiment. Use a computer simulation to carry out this procedure. (This was done to get the data on coin tosses given on page 399 in connection with the topic of histograms.) Report your results as a histogram. How does your histogram compare with that shown in Figure 6.3?

42. Five-number summary. *This is a continuation of Exercise 41.* Consider the data on coin tosses you found in Exercise 41. Use technology to find the five-number summary for the percentage of heads obtained for each experiment.

43. Another computer simulation. Consider an experiment in which we roll 1000 standard dice. We perform the experiment 1000 times and record for each experiment the percentage of dice coming up 6. Use a computer simulation to carry out this procedure. Report your results as a histogram. How does the shape of your histogram compare with that shown in Figure 6.3?

Writing About Mathematics

44. History. The term *standard deviation* was coined by Karl Pearson (1857–1936), the founder of the world's first university statistics department. Write a brief biography of this colorful and controversial figure.

45. History. Pafnuty Lvovich Chebyshev (1821–1894) was a Russian mathematician. Investigate *Chebyshev's inequality*, sometimes called *Chebyshev's theorem*, which gives a more precise statement of how well the standard deviation measures the spread of a data set. (See Exercises 33 and 34.)

46. Chebyshev: Life and work. In Exercise 33, an inequality due to the mathematician Pafnuty Lvovich Chebyshev was introduced. Report on the life of Chebyshev and his contributions to mathematics.

47. Bacon. According to a piece from National Public Radio titled "Does Bacon Really Make Everything Better? Here's the Math," a survey found that "the average rating of a recipe with bacon is 4.26, which is only 15 percent of a standard deviation above the mean rating of 4.13." Write a report discussing what you think this finding means.

48. Coaches' salaries. A 2015 article in USA Today compared football coaches' salaries in Big Ten schools. It said that using the median salary instead of the average salary is more meaningful because Michigan coach Jim Harbaugh's salary was \$7 million and Ohio State coach Urban Meyer's salary was \$5.86 million, which skewed the average high, while the Illinois interim coach got only \$916,000, which skewed the average low. Look up the current salaries of coaches in the Big Ten or another conference. Calculate the average and median salaries and write a report on your findings.

What Do You Think?

49. Determining grades. There are lots of methods your teacher might use to come up with a final grade for you in your math course. The method might involve the mean or median of scores, and it may give different weights to quizzes, homework, and exams. It might also involve dropping low grades or rewards for attendance. Pretend you are the teacher and propose a fair method for grading your students. Explain why you chose the method you did. **Exercise 20 is relevant to this topic.**

50. Mean and median. You are looking for a home in three neighborhoods. In one neighborhood, the mean price is much smaller than the median price. In a second neighborhood, the mean and median prices are the same. In a third neighborhood, the mean price is much larger than the median price. What does this information tell you about homes in the three neighborhoods? Would you be more inclined to buy a home in one neighborhood than in the two others? Explain your reasoning. **Exercise 21 is relevant to this topic.**

51. Dress size. A dress manufacturer is trying to make sure it doesn't run out of the most popular dress size. The people in sales assemble a list of dresses sold according to dress size—for example, the list 8, 8, 10 means that two dresses of size 8 and one of size 10 were sold. Which is the most useful single number for summarizing the data in this setting: the mean, the median, or the mode?

52. Outlier. Explain what is meant by an *outlier* and find some examples in data sets from the popular media.

53. More on outliers. What effect might an outlier have on the five-number summary? **Exercise 23 is relevant to this topic.**

54. Boxplot. Explain using your own words what a boxplot is. Does it have advantages over the five-number summary? **Exercise 24 is relevant to this topic.**

55. Gas mileage. Consider data for gas mileage of vehicles. If you wanted to reduce fuel consumption, which one of the following three numbers would be the most important to increase: the minimum, the median, or the maximum? Explain your reasoning. **Exercise 25 is relevant to this topic.**

56. Chebyshev's inequality. Explain in practical terms what Chebyshev's inequality tells us about data sets. Does it make the standard deviation a more useful tool? **Exercise 33 is relevant to this topic.**

6.2 The normal distribution: Why the bell curve?

TAKE AWAY FROM THIS SECTION
Understand why the normal distribution is so important.

The following article appears at the Minitab Web site.

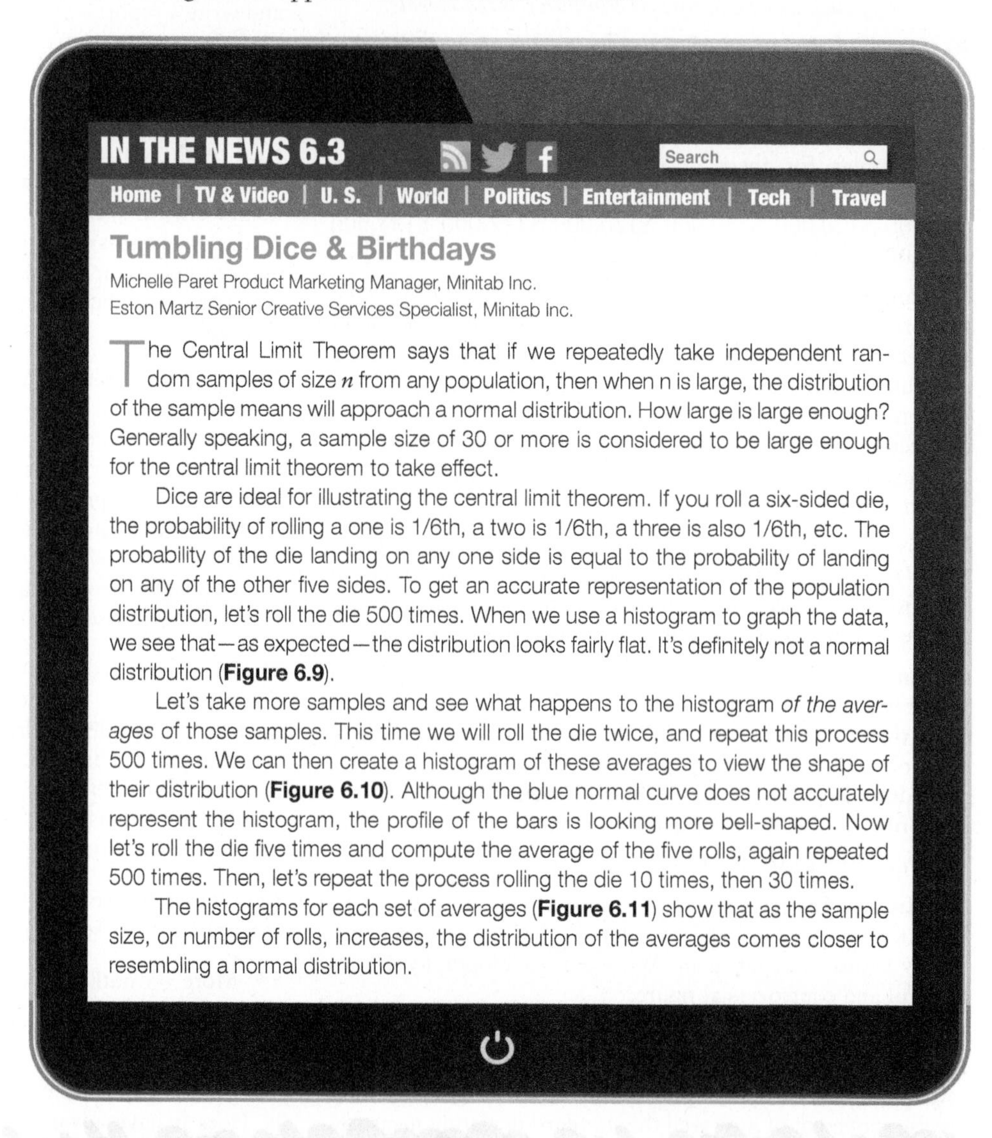

IN THE NEWS 6.3

Home | TV & Video | U. S. | World | Politics | Entertainment | Tech | Travel

Tumbling Dice & Birthdays

Michelle Paret Product Marketing Manager, Minitab Inc.
Eston Martz Senior Creative Services Specialist, Minitab Inc.

The Central Limit Theorem says that if we repeatedly take independent random samples of size n from any population, then when n is large, the distribution of the sample means will approach a normal distribution. How large is large enough? Generally speaking, a sample size of 30 or more is considered to be large enough for the central limit theorem to take effect.

Dice are ideal for illustrating the central limit theorem. If you roll a six-sided die, the probability of rolling a one is 1/6th, a two is 1/6th, a three is also 1/6th, etc. The probability of the die landing on any one side is equal to the probability of landing on any of the other five sides. To get an accurate representation of the population distribution, let's roll the die 500 times. When we use a histogram to graph the data, we see that—as expected—the distribution looks fairly flat. It's definitely not a normal distribution (**Figure 6.9**).

Let's take more samples and see what happens to the histogram *of the averages* of those samples. This time we will roll the die twice, and repeat this process 500 times. We can then create a histogram of these averages to view the shape of their distribution (**Figure 6.10**). Although the blue normal curve does not accurately represent the histogram, the profile of the bars is looking more bell-shaped. Now let's roll the die five times and compute the average of the five rolls, again repeated 500 times. Then, let's repeat the process rolling the die 10 times, then 30 times.

The histograms for each set of averages (**Figure 6.11**) show that as the sample size, or number of rolls, increases, the distribution of the averages comes closer to resembling a normal distribution.

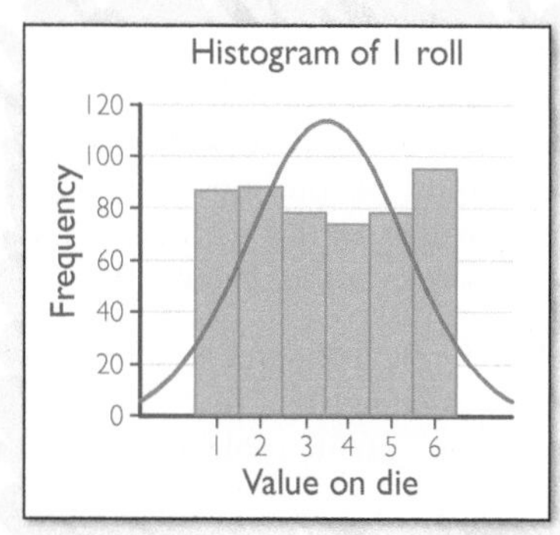

FIGURE 6.9 Because the odds of landing on all sides of a six-sided die are equal, the distribution of 500 die rolls is flat.

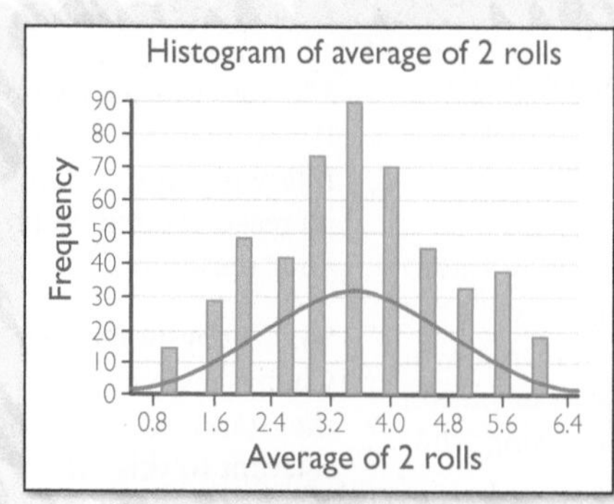

FIGURE 6.10 The distribution of 500 averages for two rolls of a die begins to resemble the familiar bell shape of a normal distribution.

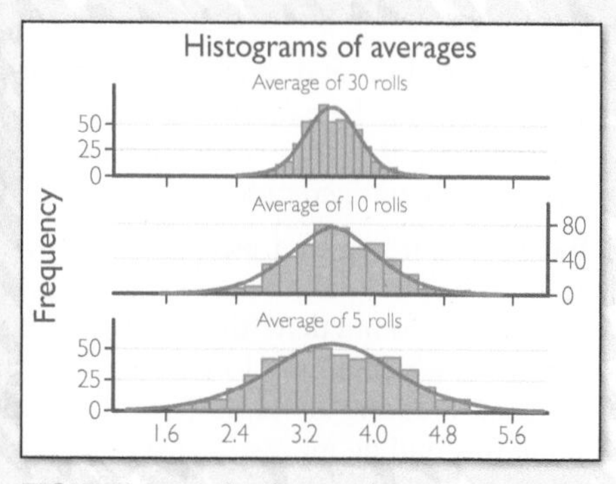

FIGURE 6.11 As the number of rolls of the die increases, the distribution of averages approaches a normal distribution.

The *distribution* of a set of data measures the frequency with which each data point occurs. In the preceding section, we introduced the histogram as a way of visualizing how data are distributed in terms of frequency. Note that each of the graphs in Figure 6.9 through Figure 6.11 is a histogram.

In the News 6.3 refers to a very important distribution, the *normal distribution*. Results of many games of chance, as illustrated in Figure 6.11, provide examples of (approximately) normally distributed data, but many other common data sets are normally distributed as well. In this section, we see the importance of the normal distribution for both descriptive and inferential statistics, as defined at the beginning of this chapter: The normal distribution provides a very efficient way of organizing many types of data, and the Central Limit Theorem concerning the normal distribution is a powerful tool for drawing conclusions from data based on sampling.

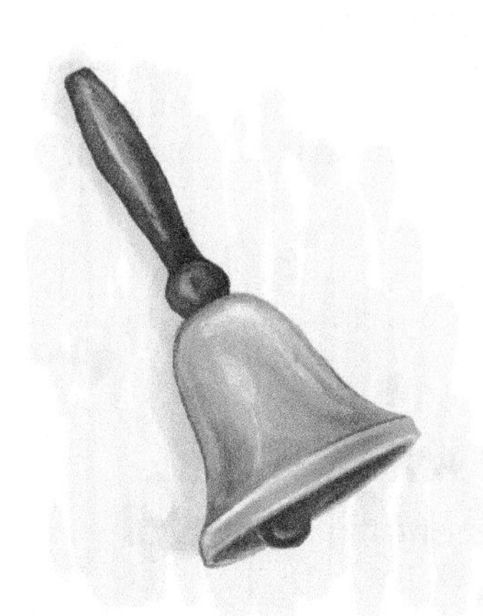

The bell-shaped curve

Figure 6.12 shows the distribution of heights of adult males in the United States. A graph shaped like this one resembles a bell—thus its common name, the *bell curve*. This bell-shaped graph is typical of normally distributed data.

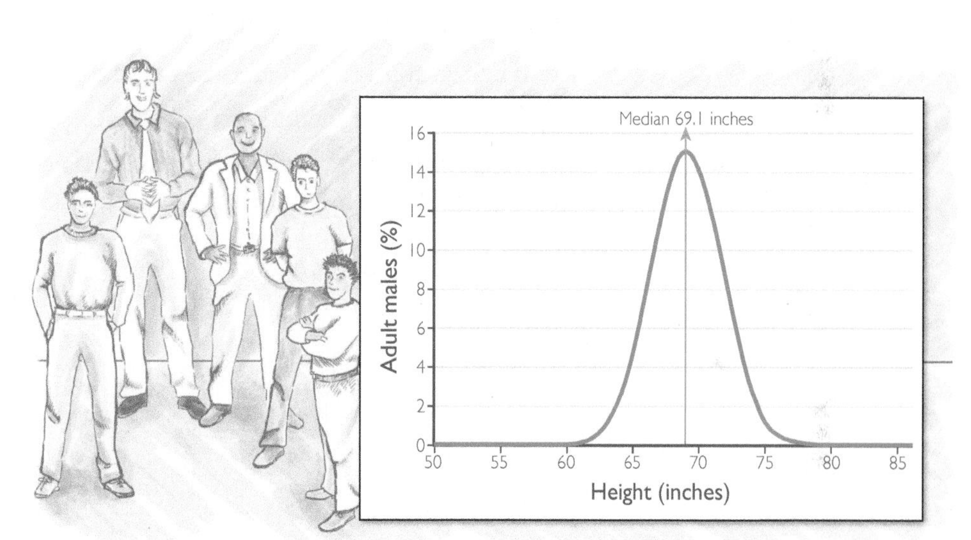

FIGURE 6.12 Heights of adult males are normally distributed.

Following are some important features of normally distributed data. These features are expressed in terms of the *mean* and *median*, which we studied in the preceding section.

The mean and median are the same: For normally distributed data, the mean and median are the same. Figure 6.12 indicates that the median height of adult males is 69.1 inches, or about 5 feet 9 inches. That means about half of males are taller than 69.1 inches and about half are shorter. Because the mean and median are the same, the average height of adult males is also 69.1 inches.

Most data are clustered about the mean: This is reflected in the fact that the vast majority of adult males are within a few inches of the mean. In fact, about 95% of adult males are within 5 inches of the mean—between 5 feet 4 inches and 6 feet 2 inches. This is shown in **Figure 6.13**. Only about 5% are taller or shorter than this. This is illustrated in **Figure 6.14**.

The bell curve is symmetric about the mean: This is reflected in the fact that the curve to the left of the mean is a mirror image of the curve to the right of the mean. In terms

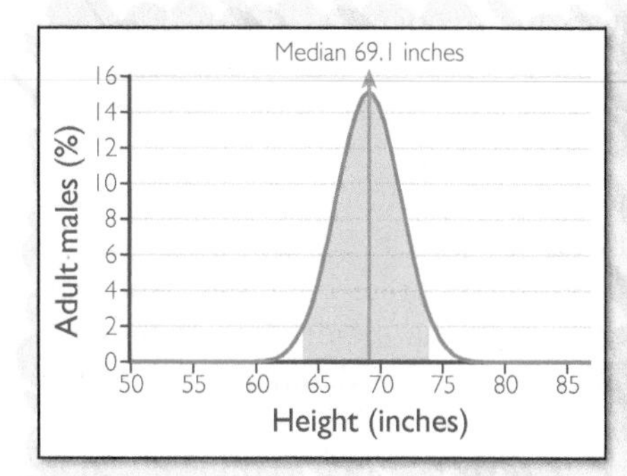

FIGURE 6.13 95% of adult males are within 5 inches of the median.

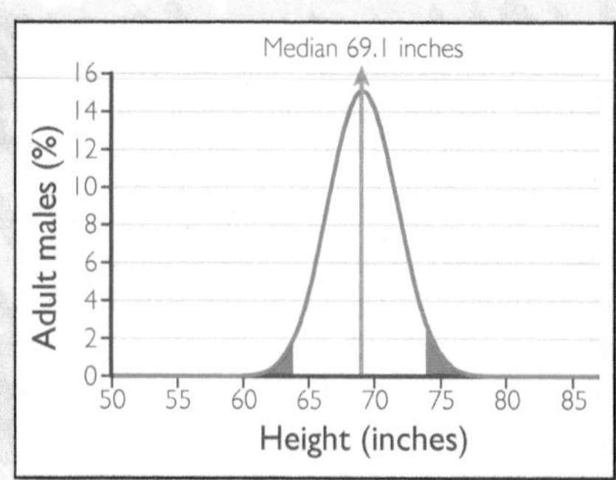

FIGURE 6.14 Relatively few men are very tall or very short.

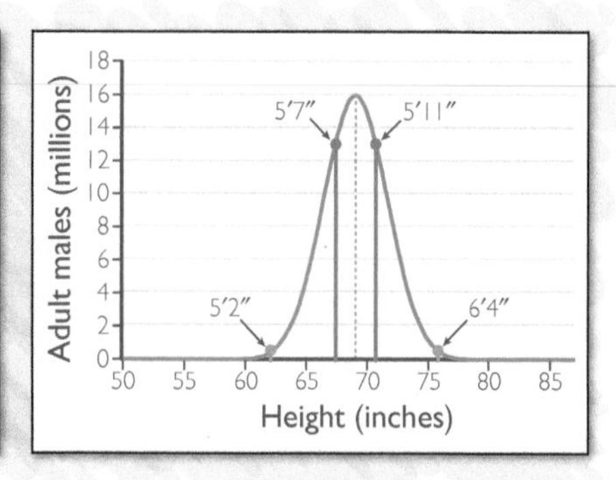

FIGURE 6.15 The bell curve is symmetric about the mean.

of heights, it means that there are about the same number of men 2 inches taller than the mean as there are men 2 inches shorter than the mean, and there are about the same number of men 6 feet 4 inches in height (7 inches above the mean) as men 5 feet 2 inches in height (7 inches below the mean). This is illustrated in **Figure 6.15**.

KEY CONCEPT

If data are normally distributed:

1. **Their graph is a bell-shaped curve.**
2. **The mean and median are the same.**
3. **Most of the data tend to be clustered relatively near the mean.**
4. **The data are symmetrically distributed above and below the mean.**

"There are lies, damn lies, and statistics. We're looking for someone who can make all three of these work for us."

Now we consider examples illustrating these properties.

EXAMPLE 6.8 Determining normality from a graph: IQ and income

Figure 6.16 shows the distribution of IQ scores, and **Figure 6.17** shows the percentage of American families and level of income. Which of these data sets appears to be normally distributed, and why?

SOLUTION

The IQ scores appear to be normally distributed because they are symmetric about the median score of 100, and most of the data are relatively close to this value. Family incomes do not appear to be normally distributed because they are not symmetric. They are skewed toward the lower end of the scale, meaning there are many more families with low incomes than with high incomes.

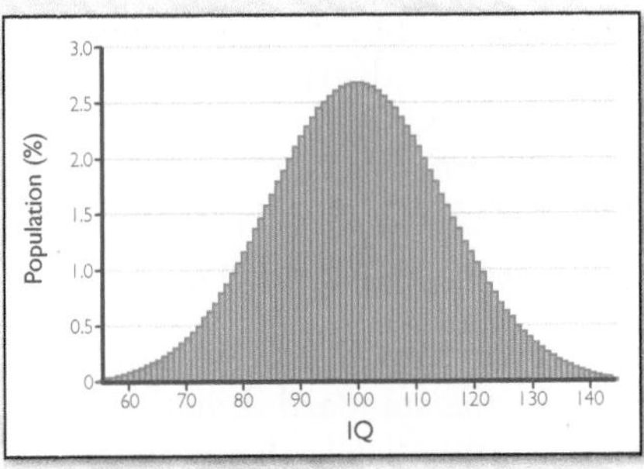

FIGURE 6.16 Percentage of population with given IQ score.

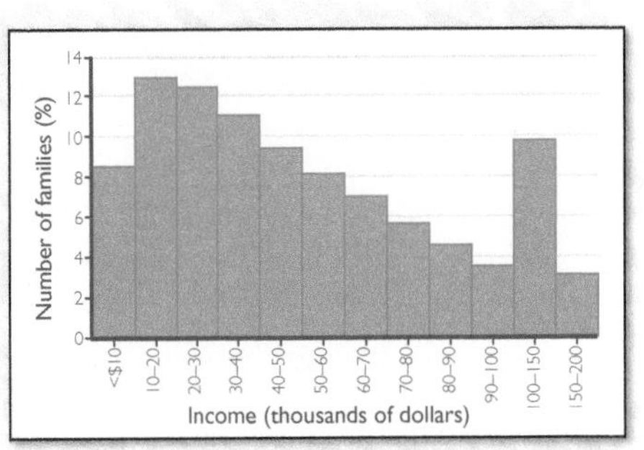

FIGURE 6.17 Percentage of families with given income.

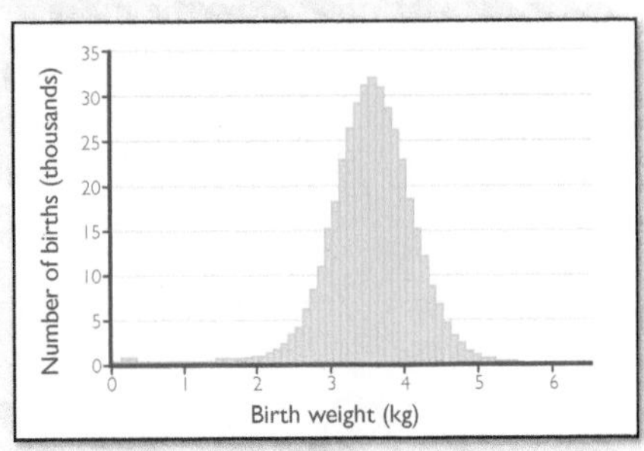

FIGURE 6.18 Number of births of a given weight in Norway.

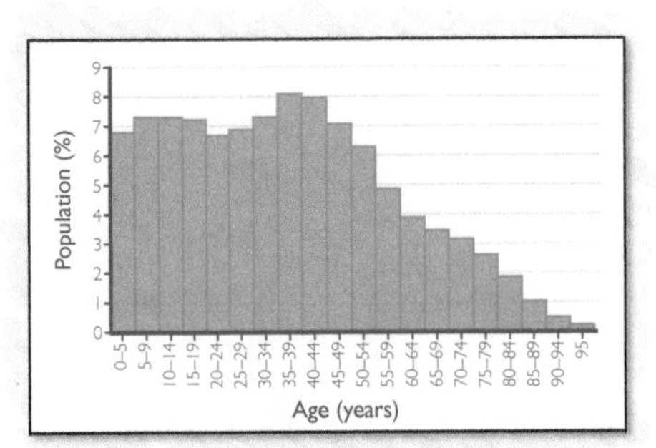

FIGURE 6.19 Percentage of Americans of a given age.

TRY IT YOURSELF 6.8

Figure 6.18 shows birthweights of Norwegian children, and **Figure 6.19** shows the distribution of ages in America in 2000. Which of these data sets appears to be normally distributed, and why?

The answer is provided at the end of this section.

Mean and standard deviation for the normal distribution

Recall that in Section 6.1 we introduced the *standard deviation* for a set of data. We noted there that the standard deviation is a measure of the spread of the data about the mean. It turns out that for a normal distribution the mean and standard deviation completely determine the bell shape for the graph of the data. Let's see why.

The mean determines the middle of the bell curve—where it peaks—and the standard deviation determines how steep the curve is. A large standard deviation results in a very wide bell, and a small standard deviation results in a thin, steep bell. **Figures 6.20 and 6.21** both show a bell curve with mean 500. In Figure 6.20, the standard deviation is 100, and the data are spread over a relatively wide range. In Figure 6.21, the standard deviation is 50. This smaller standard deviation reflects the fact that the data are bunched more tightly about the mean.

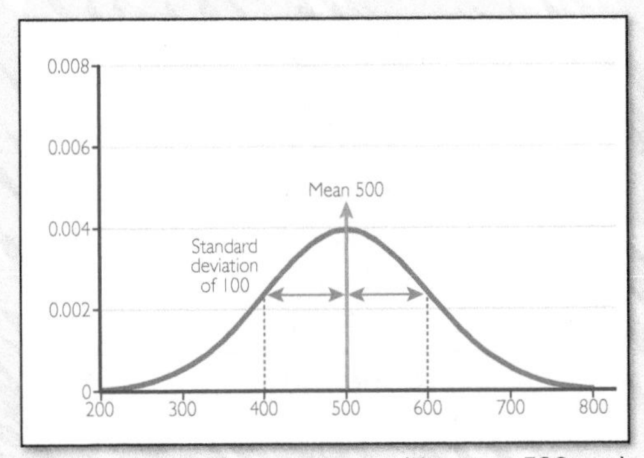

FIGURE 6.20 Normal curve with mean 500 and standard deviation 100.

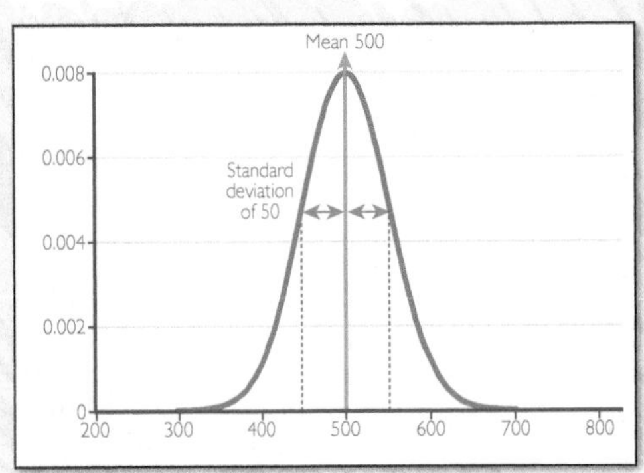

FIGURE 6.21 Normal curve with mean 500 and standard deviation 50.

The following rule of thumb gives more information on the role of the standard deviation in the normal distribution.

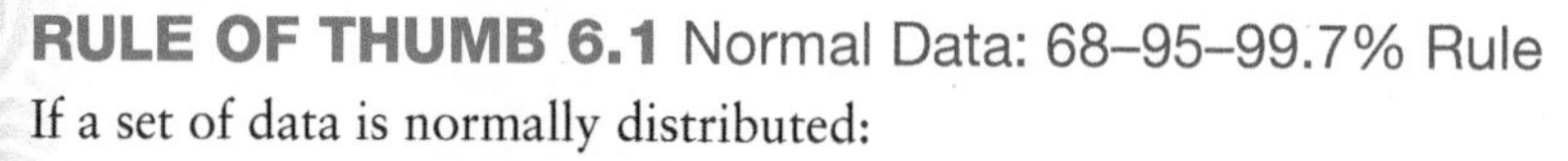

RULE OF THUMB 6.1 Normal Data: 68–95–99.7% Rule

If a set of data is normally distributed:

- About 68% of the data lie within one standard deviation of the mean (34% within one standard deviation above the mean and 34% within one standard deviation below the mean). See **Figure 6.22**.
- About 95% of the data lie within two standard deviations of the mean (47.5% within two standard deviations above the mean and 47.5% within two standard deviations below the mean). See **Figure 6.23**.
- About 99.7% of the data lie within three standard deviations of the mean (49.85% within three standard deviations above the mean and 49.85% within three standard deviations below the mean). See **Figure 6.24**.

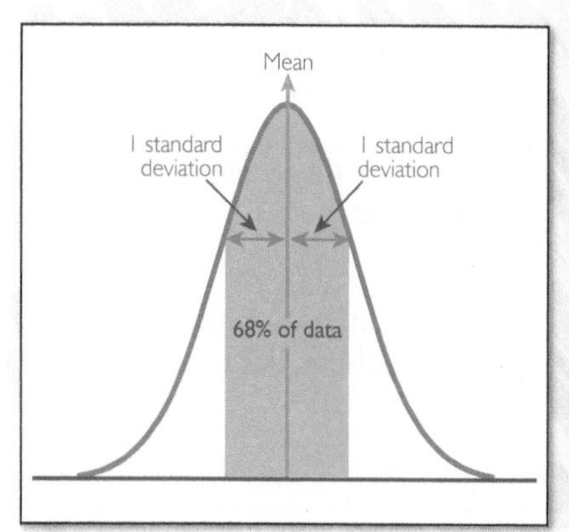

FIGURE 6.22 68% of data lie within one standard deviation of the mean.

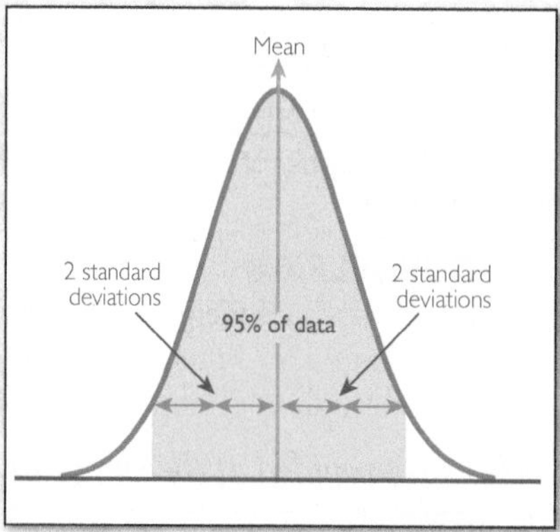

FIGURE 6.23 95% of data lie within two standard deviations of the mean.

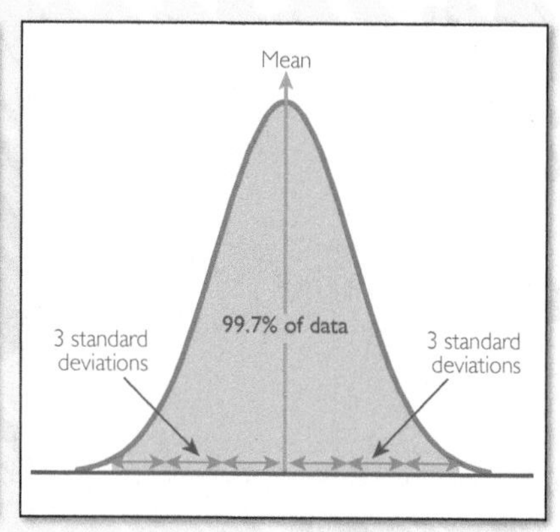

FIGURE 6.24 99.7% of the data lie within three standard deviations of the mean.

To illustrate the use of this rule of thumb, we consider scores on standardized tests. Most students are quite familiar with the ACT and SAT tests, which are widely used in college admissions. The data set in **Table 6.1** on the verbal section of the SAT Reasoning Test was obtained from the College Board Web site.

TABLE 6.1 SAT Verbal Scores

SAT Verbal	Number of students	Percent of students
750–800	25,114	1.8%
700–749	41,283	2.9%
650–699	88,799	6.3%
600–649	152,518	10.8%
550–599	204,601	14.5%
500–549	243,120	17.3%
450–499	245,615	17.5%
400–449	190,406	13.5%
350–399	116,808	8.3%
300–349	58,372	4.2%
250–299	25,332	1.8%
200–249	14,356	1 %
Total	1,406,324	100%

Mean 507
Standard deviation 111

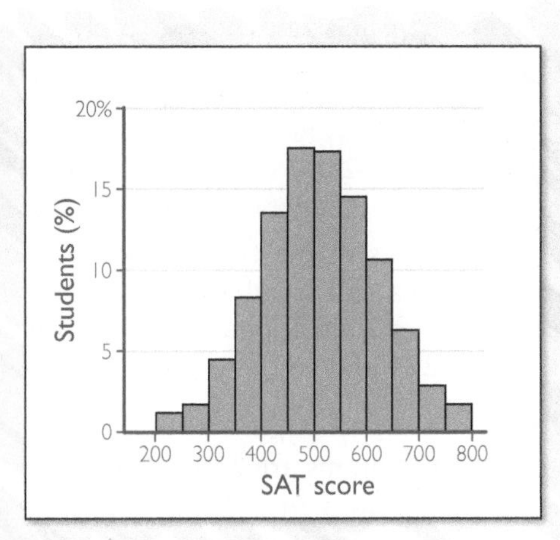

FIGURE 6.25 SAT histogram.

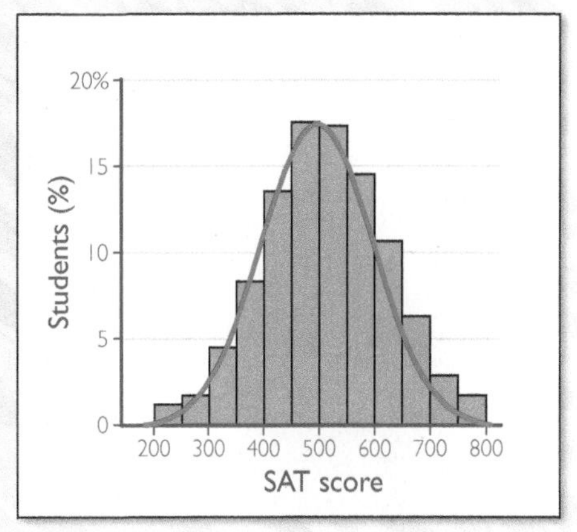

FIGURE 6.26 SAT bell curve plot.

In **Figure 6.25**, we see a histogram of these data. The bell curve associated with the data is shown in **Figure 6.26**. These figures suggest that SAT verbal scores are normally distributed. This is indeed the case, and according to Table 6.1, the mean verbal score on the SAT is 507, and the standard deviation is 111. Applying the 68–95–99.7% rule to SAT verbal scores gives the following information:

- 68% of SAT scores are within one standard deviation of the mean. This means that 68% of scores are between $507 - 111 = 396$ and $507 + 111 = 618$. Because of the symmetry of the bell curve, about half of these, 34% of the total, are above the mean (between 507 and 618), and about 34% of the scores are below the mean (between 396 and 507).
- 95% of SAT scores are within two standard deviations of the mean. This means that 95% of scores are between $507 - 2 \times 111 = 285$ and $507 + 2 \times 111 = 729$.
- 99.7% of SAT scores are within three standard deviations of the mean. This means that 99.7% of scores are between $507 - 3 \times 111 = 174$ and $507 + 3 \times 111 = 840$.[4]

EXAMPLE 6.9 Interpreting the 68–95–99.7% rule: Heights of males

We noted earlier that adult male heights in the United States are normally distributed, with a mean of 69.1 inches. The standard deviation is 2.65 inches. What does the 68–95–99.7% rule tell us about the heights of adult males?

SOLUTION

Because 68% of adult males are within one standard deviation of the mean, 68% of adult males are between $69.1 - 2.65 = 66.45$ inches (5 feet 6.45 inches) and $69.1 + 2.65 = 71.75$ inches (5 feet 11.75 inches) tall. Also, 95% of adult males are within two standard deviations of the mean. This means that 95% are between $69.1 - 2 \times 2.65 = 63.8$ inches and $69.1 + 2 \times 2.65 = 74.4$ inches tall. Furthermore, 99.7% of males are within three standard deviations of the mean. This means that 99.7% are between $69.1 - 3 \times 2.65 = 61.15$ inches and $69.1 + 3 \times 2.65 = 77.05$ inches tall.

TRY IT YOURSELF 6.9

The heights of adult American women are normally distributed, with a mean 64.1 inches and standard deviation of 2.5 inches. What does the 68–95–99.7% rule tell us about the heights of adult females?

The answer is provided at the end of this section.

[4] *Because the number of scores above 800 is so small, the College Board lumps together all scores 800 or higher and reports them as 800. Similarly, the lowest reported score is 200.*

The next example illustrates how to apply the 68–95–99.7% rule to the problem of counting the number of data points within a given range.

EXAMPLE 6.10 Counting data using the 68–95–99.7% rule: Weights of apples

The weights of apples in the fall harvest are normally distributed, with a mean weight of 200 grams and standard deviation of 12 grams. **Figure 6.27** shows the weight distribution of 2000 apples. In a supply of 2000 apples, how many will weigh between 176 and 224 grams?

SOLUTION

Apples weighing 176 grams are $200 - 176 = 24$ grams below the mean, and apples weighing 224 grams are $224 - 200 = 24$ grams above the mean. Now 24 grams represents $24/12 = 2$ standard deviations. So the weight range of 176 grams to 224 grams is within two standard deviations of the mean. Therefore, about 95% of data points will lie in this range. This means that about 95% of 2000, or 1900 apples, weigh between 176 and 224 grams. This conclusion is illustrated in **Figure 6.28.**

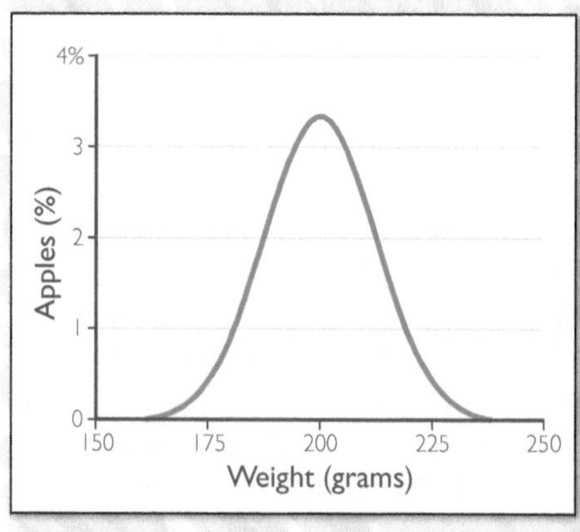

FIGURE 6.27 Apples with mean 200 grams and standard deviation 12 grams.

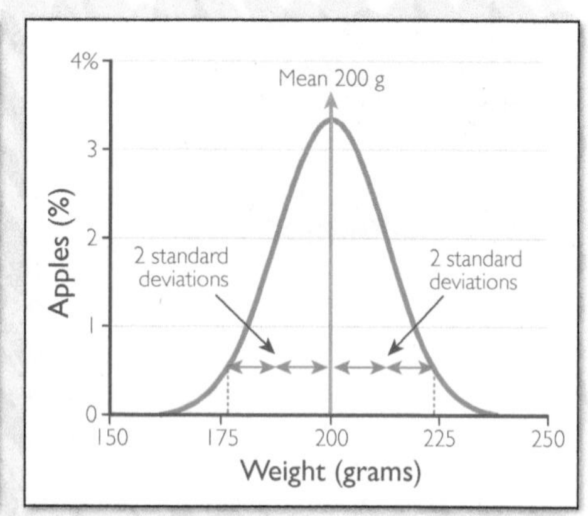

FIGURE 6.28 Apples between 176 and 224 grams.

TRY IT YOURSELF 6.10

The weights of oranges in one year's harvest are normally distributed, with a mean weight of 220 grams and standard deviation of 35 grams. In a supply of 4000 oranges, how many will we expect to weigh between 185 and 255 grams?

The answer is provided at the end of this section.

z-scores

The standard deviation is such a useful tool when dealing with normal distributions that it is often used as a unit of measurement. Instead of saying that a SAT score of 618 is 111 points above the mean of 507, often one says that the score is one standard deviation above the mean. A score of 452 is 55 points or approximately one-half of a standard deviation below the mean. When used as a unit of measurement in this way, the standard deviation is often referred to as a *z-score*.

KEY CONCEPT

In a normal distribution, the *z*-score or standard score for a data point is calculated using the formula

$$z\text{-score} = \frac{\text{Data point} - \text{Mean}}{\text{Standard deviation}}$$

"You're kidding! You count S.A.T.s?"

The z-score is normally rounded to one decimal place. Here is the interpretation of the z-score of a data point:

- A positive z-score gives the number of standard deviations that point lies above the mean.
- A negative z-score gives the number of standard deviations that point lies below the mean.

An equivalent, and often useful, version of the preceding formula is

$$\text{Data point} = \text{Mean} + z\text{-score} \times \text{Standard deviation}$$

To illustrate the calculation of z-scores, let's calculate the z-score for SAT verbal scores of 650 and 320. Recall that the mean score is 507 and the standard deviation is 111. Then

$$z\text{-score for } 650 = \frac{\text{Data point} - \text{Mean}}{\text{Standard deviation}}$$

$$= \frac{650 - 507}{111}$$

or about 1.3. Thus, the z-score for a verbal score of 650 is 1.3, which means that this score is 1.3 standard deviations above the mean. Also,

$$z\text{-score for } 320 = \frac{\text{Data point} - \text{Mean}}{\text{Standard deviation}}$$

$$= \frac{320 - 507}{111}$$

or about -1.7. The z-score for a verbal score of 320 is -1.7, which means that this score is 1.7 standard deviations below the mean. This information is illustrated in **Figure 6.29**.

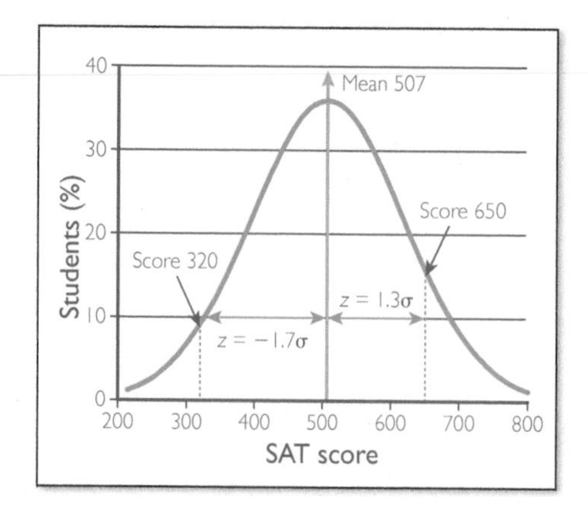

FIGURE 6.29 Illustrating z-scores.

EXAMPLE 6.11 Calculating z-scores: Weights of newborns

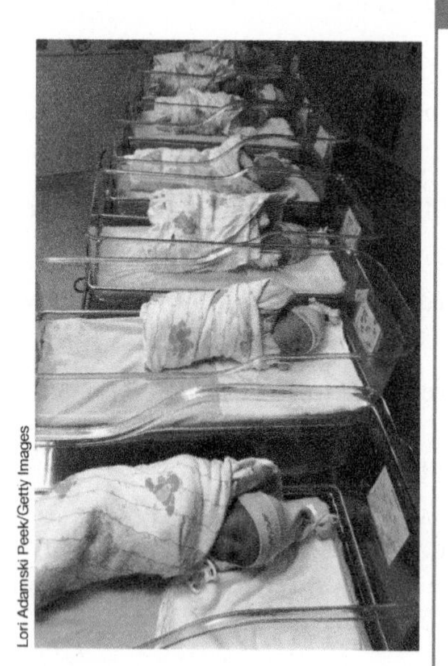

The weights of newborns in the United States are approximately normally distributed. The mean birthweight (for single births) is about 3332 grams (7 pounds 5 ounces). The standard deviation is about 530 grams. Calculate the z-score for a newborn weighing 3700 grams. (That is just over 8 pounds 2 ounces.)

SOLUTION

We use the z-score formula with a mean of 3332 grams and a standard deviation of 530 grams:

$$z\text{-score for 3700 grams} = \frac{\text{Data point} - \text{Mean}}{\text{Standard deviation}}$$

$$= \frac{3700 - 3332}{530}$$

or about 0.7. This result means that a 3700-gram newborn is a healthy 0.7 standard deviation above the mean. This fact is shown in **Figure 6.30**.

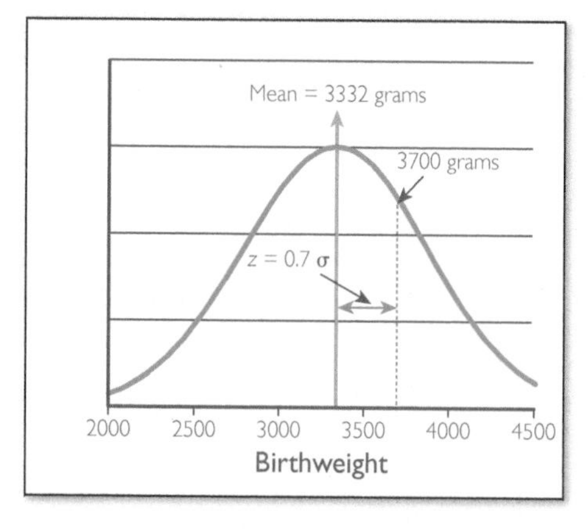

FIGURE 6.30 Weights of newborns and z-scores.

TRY IT YOURSELF 6.11

Looking at the example, calculate the z-score for a newborn weighing 3000 grams.

The answer is provided at the end of this section.

Percentile scores

Recall that in Section 6.1 we discussed *quartiles*. There is a more general term, *percentile*, that tells how a data point is positioned in relation to other data points in a normal distribution.

KEY CONCEPT

The percentile for a number relative to a list of data is the percentage of data points that are less than or equal to that number.

For example, the median is always the 50th percentile because half the data are below the median. Similarly, the first quartile is always the 25th percentile because 25% of the data are below it, and the third quartile is always the 75th percentile because 75% of the data are below it. These facts are illustrated in **Figures 6.31 through 6.33**.

It turns out that a SAT verbal score of 660 has a percentile score of 91.6%, which means that 91.6% of SAT verbal scores were 660 or lower. This percentile is shown in **Figure 6.34**.

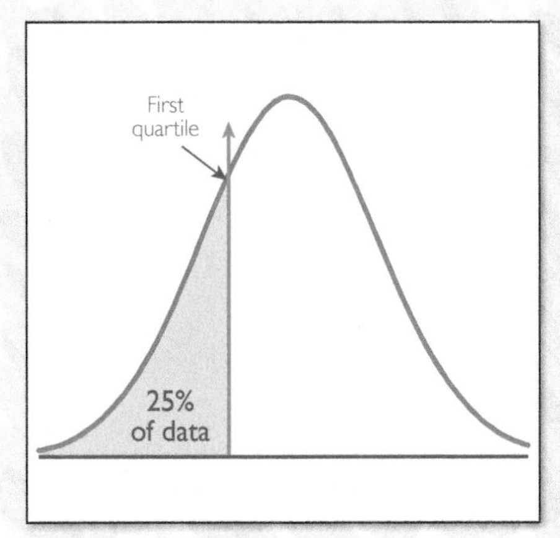

FIGURE 6.31 The percentile for the first quartile is 25%.

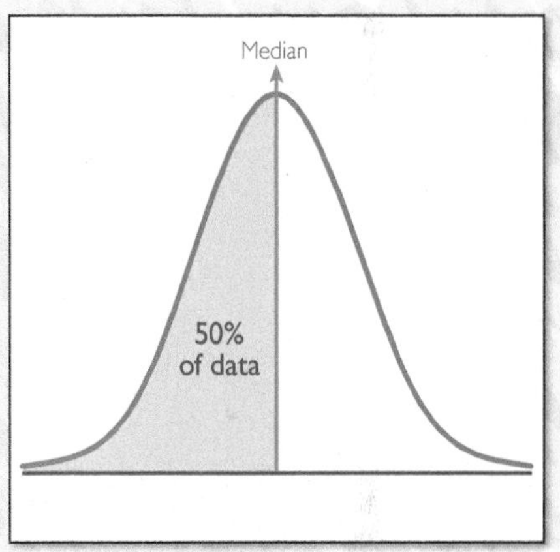

FIGURE 6.32 The percentile for the median is 50%.

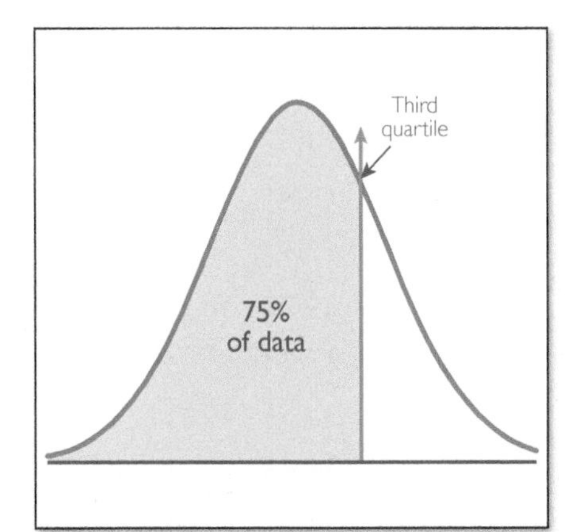

FIGURE 6.33 The percentile for the third quartile is 75%.

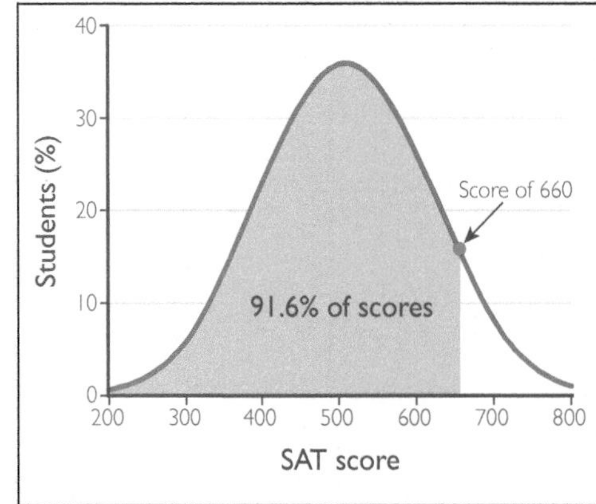

FIGURE 6.34 A score of 660 corresponds to a percentile score of 91.6%. This means that 91.6% of SAT verbal scores were 660 or lower.

Table 6.2 gives the information needed to calculate the percentile from the z-score.

For example, a z-score of 0.5 corresponds to a percentile score of 69.15. A percentile score of 93.32 corresponds to a z-score of 1.5.

Let's see how this table helps in understanding SAT verbal scores. Suppose you scored 650 on the verbal section of the SAT. How does your test score compare with scores of other students? We calculated earlier that the z-score for a SAT score of 650 is 1.3. Table 6.2 shows a percentile of 90.32 for a z-score of 1.3. We conclude that about 90.3% of all SAT verbal scores were 650 or less.

We also found that the z-score for a SAT score of 320 is −1.7. The table shows a percentile of 4.46 for a z-score of −1.7. Thus, about 4.5% of all SAT verbal scores were 320 or less. Expressed differently, this says that about 95.5% of all SAT verbal scores were higher than 320.

TABLE 6.2 Percentile from z-Score

z-score	Percentile	z-score	Percentile
−2.7	0.35	0.0	50.00
−2.6	0.47	0.1	53.98
−2.5	0.62	0.2	57.93
−2.4	0.82	0.3	61.79
−2.3	1.07	0.4	65.54
−2.2	1.39	0.5	69.15
−2.1	1.79	0.6	72.57
−2.0	2.28	0.7	75.80
−1.9	2.87	0.8	78.81
−1.8	3.59	0.9	81.59
−1.7	4.46	1.0	84.13
−1.6	5.48	1.1	86.43
−1.5	6.68	1.2	88.49
−1.4	8.08	1.3	90.32
−1.3	9.68	1.4	91.92
−1.2	11.51	1.5	93.32
−1.1	13.57	1.6	94.52
−1.0	15.87	1.7	95.54
−0.9	18.41	1.8	96.41
−0.8	21.19	1.9	97.13
−0.7	24.20	2.0	97.73
−0.6	27.43	2.1	98.21
−0.5	30.85	2.2	98.61
−0.4	34.46	2.3	98.93
−0.3	38.21	2.4	99.18
−0.2	42.07	2.5	99.38
−0.1	46.02	2.6	99.53
0.0	50.00	2.7	99.65

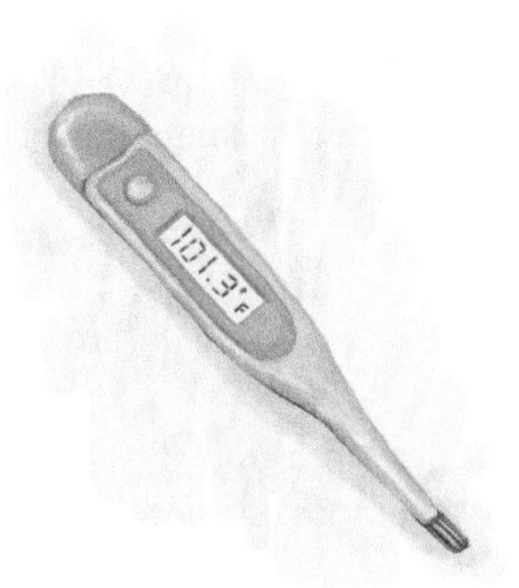

EXAMPLE 6.12 Calculating percentiles: Length of illness

The average length of illness for flu patients in a season is normally distributed, with a mean of 8 days and a standard deviation of 0.9 day. What percentage of flu patients will be ill for more than 10 days?

SOLUTION

We first calculate the z-score for 10 days:

$$z\text{-score} = \frac{\text{Data point} - \text{Mean}}{\text{Standard deviation}}$$

$$= \frac{10 - 8}{0.9}$$

or about 2.2. Table 6.2 gives a percentile of about 98.6% for this z-score. This means that about 98.6% of patients will recover in 10 days or less. Thus, only about 100% − 98.6% = 1.4% will be ill for more than 10 days.

TRY IT YOURSELF 6.12

In the preceding situation, what percentage of flu patients will recover in seven days or less?

The answer is provided at the end of this section.

Next we revisit the data on weights of newborns.

EXAMPLE 6.13 Using the normal distribution: Birthweight

Recall from Example 6.11 that the weights of newborns in the United States are approximately normally distributed. The mean birthweight (for single births) is about 3332 grams (7 pounds 5 ounces). The standard deviation is about 530 grams.

a. What percentage of newborns weigh more than 8 pounds (3636.4 grams)?

b. Low birthweight is a medical concern. The American Medical Association defines low birthweight to be 2500 grams (5 pounds 8 ounces) or less. What percentage of newborns are classified as low-birthweight babies?

SOLUTION

a. We first calculate the z-score:

$$z\text{-score} = \frac{\text{Data Point} - \text{Mean}}{\text{Standard deviation}}$$

$$= \frac{3636.4 - 3332}{530}$$

or about 0.6. Consulting Table 6.2, we find that this represents a percentile of about 72.6%. This means that about 72.6% of newborns weigh 8 pounds or less. So, about 100% − 72.6% = 27.4% of newborns weigh more than 8 pounds.

b. Again, we calculate the z-score:

$$z\text{-score} = \frac{\text{Data Point} - \text{Mean}}{\text{Standard deviation}}$$

$$= \frac{2500 - 3332}{530}$$

or about −1.6. Table 6.2 shows a percentile of about 5.5% for a z-score of −1.6. Hence, about 5.5% of newborns are classified as low-birthweight babies.

The Central Limit Theorem

In the News 6.3 reported that averages of games of chance are normally distributed. Data sets that are the result of many chance events are likely to be normally distributed. That is not just a happenstance. It is rather a mathematically proven law known as the *Central Limit Theorem*. The theorem tells why many common measurements are normally distributed.

KEY CONCEPT

The Central Limit Theorem

According to the Central Limit Theorem, percentages obtained by taking many samples of the same size from a population are approximately normally distributed.

- **The mean of the normal distribution is the mean of the whole population.**
- **If the sample size is n, the standard deviation of the normal distribution is**

$$\text{Standard deviation} = \sigma = \sqrt{\frac{p\,(100-p)}{n}} \text{ percentage points}$$

Here, p is the mean, expressed as a percentage, not a decimal.

To see an application of the Central Limit Theorem, suppose that 60% of the registered voters in the state of Oklahoma in October 2020 intended to vote for Donald Trump for president. Suppose we took a poll of 100 (randomly selected) registered Oklahoma voters and calculated the percentage of those polled who intended to vote for Trump. The Central Limit Theorem tells us that the results of such a poll are (approximately) normally distributed with mean $p = 60\%$ because that is the mean for this population.[5] The sample size is $n = 100$, so the standard deviation is

$$\begin{aligned}\sigma &= \sqrt{\frac{p(100-p)}{n}} \\ &= \sqrt{\frac{60(100-60)}{100}}\end{aligned}$$

This is about 4.9 percentage points.

How often would such a poll lead us to believe that Oklahoma would not vote for Trump? That is, what percentage of such polls would show that 50% or fewer voters intend to vote for Trump? To answer this, we first find the z-score for a sample that yields 50% for Trump:

$$\begin{aligned}z\text{-score} &= \frac{\text{Data Point} - \text{Mean}}{\text{Standard deviation}} \\ &= \frac{50-60}{4.9}\end{aligned}$$

or about -2.0. Table 6.2 gives a percentile of about 2.3% for this z-score. So only 2.3% of such polls would report 50% or less support for Donald Trump.

We will look further into polling in the next section.

[5] *Strictly speaking, we should say that if many such polls, all of the same size, were taken from the same population, then the results would be approximately normally distributed.*

EXAMPLE 6.14 Using the Central Limit Theorem: Untreated patients

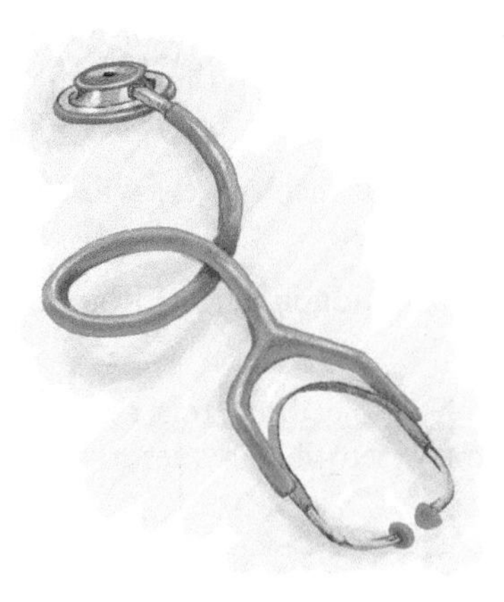

For a certain disease, 30% of untreated patients can be expected to improve within a week. We observe a population of 50 patients and record the percentage who improve within a week. According to the Central Limit Theorem, the results of such a study will be approximately normally distributed.

a. Find the mean and standard deviation for this normal distribution. Round the standard deviation to one decimal place.

b. Find the percentage of test groups of 50 patients in which more than 40% improve within a week.

SOLUTION

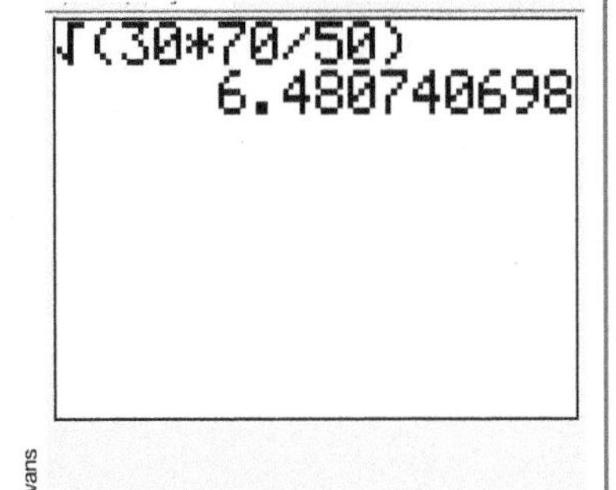

Benny Evans

a. The mean is the percentage of untreated patients who will improve within a week, which is $p = 30\%$. The sample size is $n = 50$. This gives a standard deviation of

$$\sigma = \sqrt{\frac{p(100 - p)}{n}}$$

$$= \sqrt{\frac{30 \times 70}{50}}$$

or about 6.5 percentage points.

b. We first calculate the z-score for 40%:

$$z\text{-score} = \frac{\text{Data Point} - \text{Mean}}{\text{Standard deviation}}$$

$$= \frac{40 - 30}{6.5}$$

or about 1.5. Table 6.2 gives a percentile of about 93.3%. This means that in 93.3% of test groups, we expect that 40% or fewer will improve within a week. Only 100% − 93.3% = 6.7% of test groups will show more than 40% improving within a week.

TRY IT YOURSELF 6.14

Assume that on your campus, 20% of the students are education majors. Suppose you choose a random sample of 75 students and record the percentage of education majors. The results of such a survey are approximately normally distributed.

a. Find the mean and standard deviation for this normal distribution. Round the standard deviation to one decimal place.

b. What percentage of randomly selected groups of 75 students on your campus will include 10% or fewer education majors?

The answer is provided at the end of this section.

Significance of apparently small deviations

One consequence of the Central Limit Theorem is the observation that apparently small deviations in surveys can be significant. The following example illustrates this point.

EXAMPLE 6.15 Central Limit Theorem and sampling: Allergies

Assume we know that 20% of Americans suffer from a certain type of allergy. Suppose we take a random sample of 100,000 Americans and record the percentage who suffer from this allergy.

a. The Central Limit Theorem says that percentages from such surveys will be normally distributed. What is the mean of this distribution?

b. What is the standard deviation of the normal distribution in part a? Round your answer to two decimal places.

c. Suppose we find that in a town of 100,000 people, 21% suffer from this allergy. Is this an unusual sample? What does the answer to such a question tell us about this town?

SOLUTION

a. The mean is the percentage of people from the general population who suffer from the allergy. Therefore, the mean is $p = 20\%$.

b. For a sample size of 100,000, we find

$$\sigma = \sqrt{\frac{p(100 - p)}{n}}$$

$$= \sqrt{\frac{20 \times 80}{100{,}000}}$$

or about 0.13 percentage point.

c. Our sample of 21% is one percentage point larger than the mean of 20%. That gives

$$z\text{-score} = \frac{1}{0.13}$$

or about 7.7. This score is far larger than any z-score in Table 6.2. The chance of getting a random sample 7.7 standard deviations above the mean is virtually zero. There is almost no chance that in a randomly chosen sample of this size, at least 21% will suffer from this allergy. Thus, this is a truly anomalous sample: This town is not representative of the total population of Americans. Its allergy rate is highly unusual.

To summarize: It may seem counterintuitive, but even though 21% seems to be only a little larger than 20%, a percentage that high is, in fact, a very improbable result for a truly random sample of 100,000 people. Note that it is the large size of the sample that makes the standard deviation small and thus makes such a small percentage variation unlikely.

Health issues and the Fallon leukemia cluster

In Example 6.15, we found that when the sample size is large, even small deviations from the mean are highly unlikely. In such a situation, the sample almost certainly did not happen by pure chance.

A good real-world example of this kind of phenomenon is the leukemia cluster discovered around the town of Fallon in Churchill County, Nevada.[6] (Fallon happens to be the home of the Navy's Top Gun school at the Fallon Naval Air Station.) In the summer of 2000, it was learned that five cases of leukemia in children had

[6] www.cdc.gov/nceh/clusters/fallon/

been diagnosed in Churchill County within a few months of each other. In a four-year period, 16 children had been diagnosed, all of whom had lived in Churchill County for varying lengths of time prior to diagnosis. The office of the Nevada State Epidemiologist noted that the average rate is about three childhood cases per 100,000 children. This fact suggests, based on the size of the population of Churchill County, that one case would be expected about every five years in that county.

This statistical fact alerted the authorities that it was extremely unlikely that the cases were happening by pure chance—there was probably something extraordinary causing them.

In subsequent studies, the metal tungsten has emerged as a prime suspect. University of Arizona scientists said recent tests show that Fallon has up to 13 times as much tungsten in its dust as other Nevada cities. They said tests also have found elevated levels of tungsten in tree rings in Fallon and three other towns with leukemia clusters.

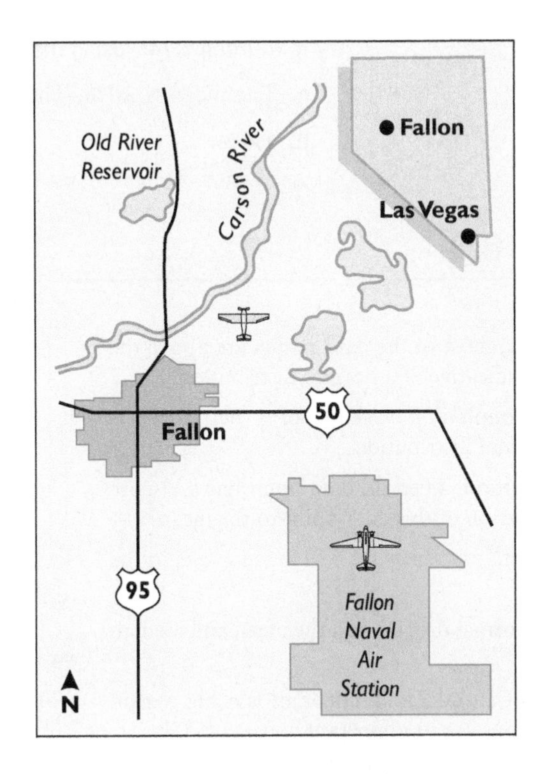

Examples like this illustrate how vital mathematics and statistics are to our well-being.

WHAT DO YOU THINK?

Shape of the bell curve: A bell curve may be tall and narrow or it may be flatter with less of a peak. What does the shape of the bell curve tell you about features of the normally distributed data they represent, such as mean, median, and standard deviation?

Questions of this sort may invite many correct responses. We present one possibility: The peak of the bell curve occurs at the median, which is the same as the mean for normal data. A flatter and wider bell indicates more data far away from the mean, and that corresponds to a large standard deviation. In contrast, a tall, narrow bell means the data are packed near the mean. That corresponds to a small standard deviation.

Further questions of this type may be found in the *What Do You Think?* section of exercises.

Try It Yourself answers

Try It Yourself 6.8: Determining normality from a graph: Birthweights and ages Birthweights of Norwegian children are normally distributed, but ages of Americans are not.

Try It Yourself 6.9: Interpreting the 68–95–99.7% rule: Heights of females 68% of adult American women are between 61.6 and 66.6 inches tall, 95% are between 59.1 and 69.1 inches, and 99.7% are between 56.6 and 71.6 inches.

Try It Yourself 6.10: Counting data using the 68–95–99.7% rule: Weights of oranges 2720 oranges.

Try It Yourself 6.11: Calculating *z*-scores: Weights of newborns -0.6.

Try It Yourself 6.12: Calculating percentiles: Length of illness About 13.6%.

Try It Yourself 6.14: Using the Central Limit Theorem: Education majors

a. The mean is 20%. The standard deviation is 4.6 percentage points.

b. 1.4%.

Exercise Set 6.2

Test Your Understanding

1. In a certain data set, most of the data points are above the mean. Is this feature indicative of a normal distribution?

2. True or false: A graph in the shape of a bell curve is characteristic of a normal distribution.

3. In a normal distribution, a certain data point has a *z*-score of -1. What is the relation of this data point to the mean?

4. What does it mean to say that your SAT test score is in the 60th percentile?

5. True or false: In a normal distribution the mean and median are always the same.

6. True or false: The Central Limit Theorem is a big reason why normal distributions are so important.

7. In the rule referring to the three numbers 68, 95, and 99.7, these numbers are: **a.** means **b.** percents **c.** standard deviations **d.** *z*-scores.

Problems

8. Which is normal? Consider the data distributions in **Figures 6.35, 6.36, 6.37, and 6.38.** Which of the distributions appear to be normal? If a distribution does not appear to be normal, explain why.

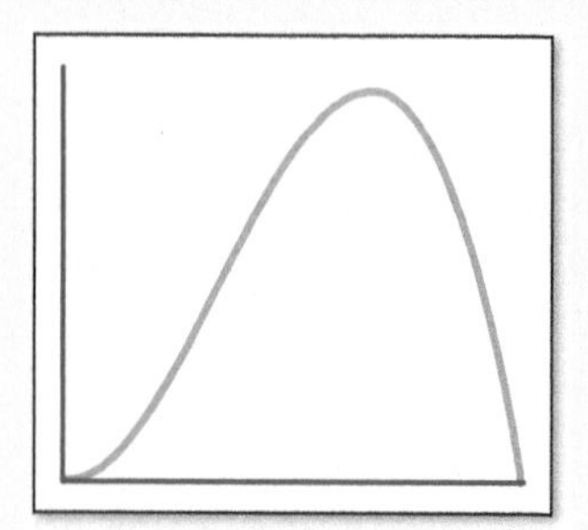

FIGURE 6.35 Distribution 1.

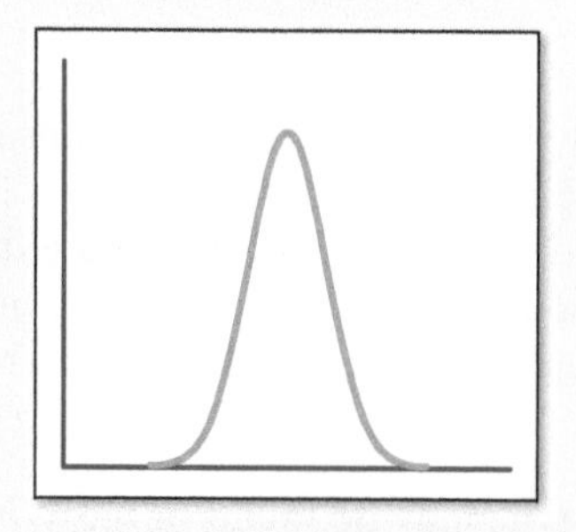

FIGURE 6.36 Distribution 2.

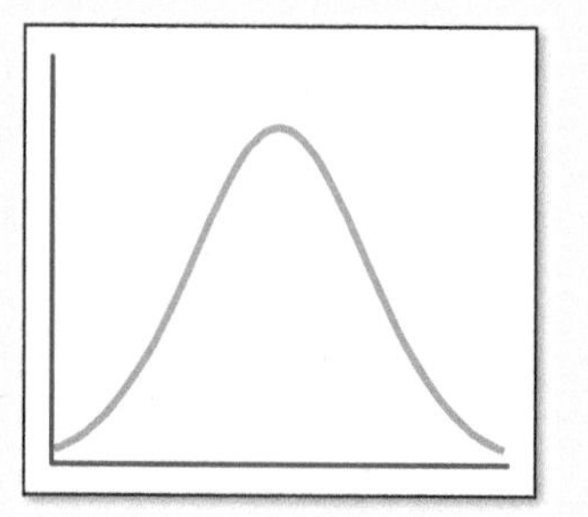

FIGURE 6.37 Distribution 3.

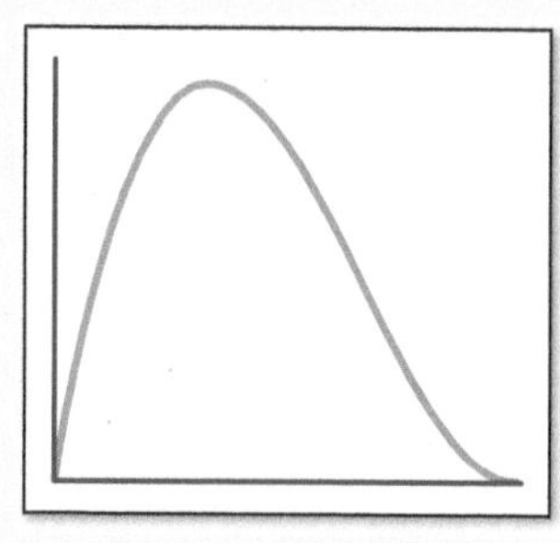

FIGURE 6.38 Distribution 4.

9. Getting heavier. There is plenty of anecdotal information, and some controversy, about whether Americans are getting fatter. But a researcher at Rockefeller University in New York claims that Americans on the whole are *not* getting fatter. He cites statistical data gathered over a 13-year period by the National Center for Health Statistics. During that time, heavier Americans, those at the top of the weight distribution, did get heavier, but only by a few pounds. The data points at the lower end of the weight distribution, representing thin Americans, did not change. There is a significant increase in weight of the extremely obese Americans, but those data points are at the very top of the distribution.

It may be reasonable to assume that body weights of Americans are normally distributed. In view of the research cited above, is it reasonable to think that the distribution of weights was normal both before and after the increase in weight of already overweight people? Explain.

10–11. It's cold in Madison, Wisconsin. *Use the following information for Exercises 10 and 11.* ***Figure 6.39*** *is a histogram of average yearly temperature in Madison, Wisconsin, over a 47-year period. It shows a mean of 40.2 degrees Fahrenheit and a standard deviation of 6.8 degrees.*

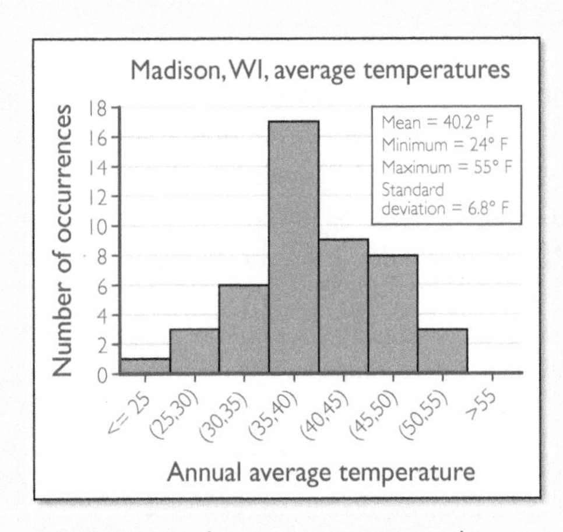

FIGURE 6.39 Average temperatures in Madison.

10. Judging simply from the appearance of the histogram, do you think average temperatures in Madison are approximately normally distributed? Explain your reasoning.

11. Challenge problem. If the data were normally distributed, then about 68% of the data would be within one standard deviation of the mean. Use the histogram to estimate the percentage of the data that are within one standard deviation of the mean. (Give your answer as a percentage to the nearest whole number.) Does your answer support the proposition that the data are approximately normally distributed?

12. Fingerprints. Ridge counts on fingerprints are approximately normally distributed, with a mean of about 140 and standard deviation of 50. What does the 68–95–99.7% rule tell us about ridge counts on fingerprints?

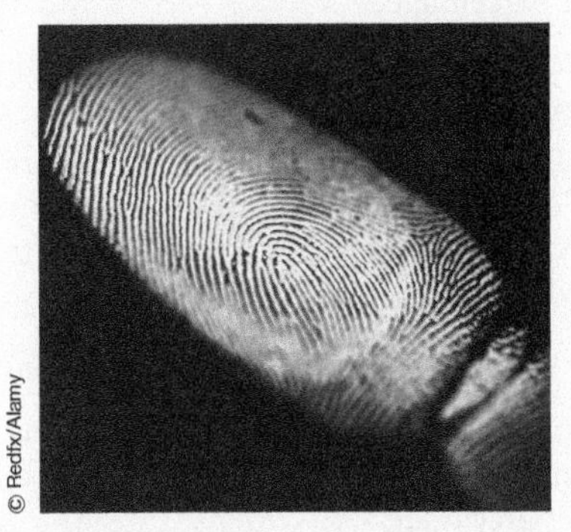

13. Diastolic blood pressure. In a study of 500 men, diastolic blood pressure was found to be approximately normally distributed, with a mean of 82 millimeters (mm) of mercury and standard deviation of 10 millimeters.

a. Use the 68–95–99.7% rule to determine what percentage of the test group had a diastolic pressure between 72 millimeters and 92 millimeters of mercury.

b. Use your answer to part a to determine how many men in the test group had a diastolic pressure between 72 millimeters and 92 millimeters of mercury.

14. Calculating the *z*-score. It is found that the number of raisins in a box of a popular cereal is normally distributed, with a mean of 133 raisins per box and standard deviation of 10 raisins. My cereal box has 157 raisins. What is the *z*-score for this box of cereal? Round your answer to one decimal place.

15. More on *z*-scores. The number of nuts in a can of mixed nuts is found to be normally distributed, with a mean of 500 nuts and standard deviation of 20 nuts. My can of mixed nuts has only 475 nuts. What is the *z*-score for this can of nuts? Round your answer to one decimal place.

16. Standardized exam scores. This exercise refers to scores on standardized exams with results that are normally distributed. Round your answers to one decimal place.

a. Suppose you have a score that puts you 1.5 standard deviations below the mean. What is your percentile score?

b. Suppose you have a score that puts you in the 60th percentile. What is your *z*-score?

c. Suppose you have a score of 360 on the verbal section of the SAT Reasoning Test. What is your *z*-score? (See Table 6.1 on page 416.) What is your percentile score?

d. Suppose you have a score of 720 on the verbal section of the SAT Reasoning Test. What is your *z*-score? (See Table 6.1 on page 416.) What is your percentile score?

17. IQ. A person's IQ (Intelligence Quotient) is supposed to be a measure of his or her "intelligence." The IQ test scores are normally distributed and are scaled to a mean of 100 and a standard deviation of 15.

a. If your IQ is 115, what is your percentile score? Round your answer to one decimal place.

b. It is often said that a genius is someone with an IQ of 140 or more. What is the *z*-score for a 140 IQ? (Round your answer to one decimal place.) What percentage of the population scores as a genius? Round your answer as a percentage to one decimal place. *Note*: The actual number is 0.4%, but you may get a slightly different value due to rounding.

18. SAT scores. Suppose that you had a score of 670 on the verbal section of the SAT Reasoning Test (see Table 6.1 on page 416). Round your answers to one decimal place.

a. What is the *z*-score for this SAT score?

b. What is your percentile score?

c. Explain in practical terms the meaning of your percentile score.

19. Birthweights in Norway. The data from the National Institute of Environmental Health Sciences reports that over a six-year period, birthweights (in grams) of newborns in Norway were distributed normally (see **Figure 6.40**), with a mean of 3668 grams and standard deviation of 511 grams.

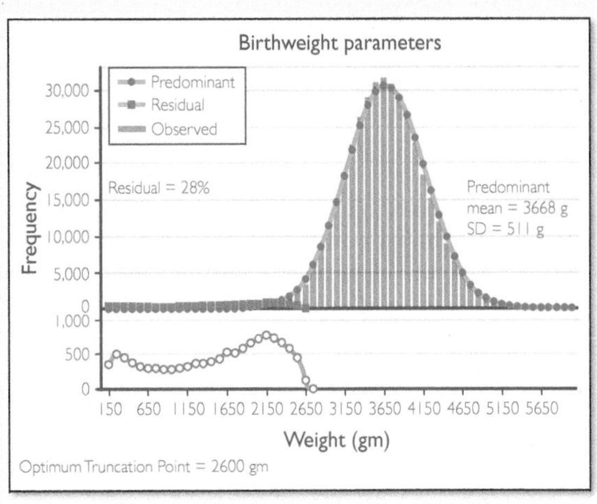

FIGURE 6.40 Birthweights in Norway.

a. What is the z-score for a newborn weighing 4000 grams? Round your answer to one decimal place.

b. What is the z-score for a newborn weighing 3000 grams? Round your answer to one decimal place.

c. Newborns weighing less than 2500 grams are classified as having a low birthweight. What percentage of newborns suffered from low birthweight? Round your answer as a percentage to the nearest whole number.

d. Would you expect the distribution to remain normal if birthweights were reported in pounds?

20. Campinas, Brazil. The average maximum monthly temperature in Campinas, Brazil, is 29.9 degrees Celsius. The standard deviation in maximum monthly temperature is 2.31 degrees. Assume that maximum monthly temperatures in Campinas are normally distributed.

a. Use Rule of Thumb 6.1 to fill in the following two blanks: "95% of the time the maximum monthly temperature is between ______ and ______." Round your answers to one decimal place.

b. What percentage of months would have a maximum temperature of 35 degrees or higher? Round your answer as a percentage to one decimal place.

21. Male height in America. The heights of adult men in America are normally distributed, with a mean of 69.1 inches and standard deviation of 2.65 inches. Round your answers as percentages to the nearest whole number.

a. **Tall.** What percentage of adult males in America is over 6 feet tall?

b. **Short.** What percentage of adult males in America is under 5 feet 7 inches tall?

c. **Dancing.** It is sometimes said that women don't like to dance with shorter men. To get a dancing partner, a 6-foot-tall woman draws from a hat the name of an adult American male. What is the probability that he will be as tall as she? Give your answer as a percentage rounded to the nearest whole number.

22. Heights of women. The heights of American women between the ages of 18 and 24 are approximately normally distributed. The mean is 64.1 inches, and the standard deviation is 2.5 inches. What percentage of such women is over 5 feet 9 inches tall?

23. Curving grades. When exam scores are low, students often ask the teacher whether he or she is going to "curve" the grades. The hope is that by curving a low score on the exam, the students will wind up getting a higher letter grade than might otherwise be expected. The term *curving* grades, or *grading on a curve*, comes from the bell curve of the normal distribution. If we assume that scores for a large number of students are distributed normally (as with SAT scores) and we also assume that the class average should be a "C," then a teacher might award grades as listed in **Table 6.3**.

Suppose a teacher curved grades using the bell curve as in Table 6.3 and that the grades were indeed normally distributed. What percent of students would get a grade of "A"? What percent of students would get a grade of "B"? Round each answer as a percentage to one decimal place. *Suggestion*: To find the percentage of students getting a grade of "B," subtract the percentage of students 0.5 standard deviation or less above the mean from the percentage of students 1.5 standard deviations or less above the mean.

TABLE 6.3 Curving Grades

A	1.5 standard deviations above the mean or higher
B	0.5 to 1.5 standard deviations above the mean
C	within 0.5 standard deviation of the mean
D	0.5 to 1.5 standard deviations below the mean
F	1.5 standard deviations below the mean or lower

24. More on curving grades. Suppose an exam had an average (mean) score of 55% and a standard deviation of 15%. If the teacher curved grades using the bell curve as in Table 6.3 from Exercise 23, what score would be necessary to receive an "A"? How about a "B"? A "C"? Round each answer as a percentage to one decimal place.

25. A medical test. It is known that in the absence of treatment, 70% of the patients with a certain illness will improve. The Central Limit Theorem tells us that the percentages of patients in groups of 400 that improve in the absence of treatment are approximately normally distributed.

a. Find the mean and standard deviation of the normal distribution given by the Central Limit Theorem. Round the standard deviation to one decimal place.

b. In what percentage of such groups (of 400) would 275 or fewer improve? Round your answer to one decimal place as a percentage.

26. Lots of coins. Suppose we toss 100 fair coins and record the percentage of heads.

a. According to the Central Limit Theorem, the percentages of heads resulting from such experiments are approximately normally distributed. Find the mean and standard deviation of this normal distribution.

b. What is the z-score for 55 heads? How about 60 heads?

c. What percentage of tosses of 100 fair coins would show more than 55 heads? Round your answer to one decimal place as a percentage.

d. What percentage of tosses of 100 fair coins would show more than 60 heads? Round your answer to one decimal place as a percentage.

27. Contracting diseases. It is known that under ordinary conditions, 15% of people will not contract a certain disease. Consider the situation in which test groups of 500 were selected and the percentages of those who did not contract the disease were recorded.

a. According to the Central Limit Theorem, percentages of groups of 500 who don't contract the disease are approximately normally distributed. Find the mean and standard deviation of this normal distribution. Round the standard deviation to one decimal place.

b. For what percentage of such groups (of 500) will 80 or fewer not contract the disease? Round your answer to one decimal place as a percentage.

28. Marbles in a jar. A large jar is filled with marbles. Of these marbles, 25% are red and the rest are blue. We randomly draw 100 marbles.

a. According to the Central Limit theorem, the percentages of red marbles in such samples are approximately normally distributed. Find the mean and standard deviation of this normal distribution. Round the standard deviation to one decimal place.

b. Our friend pulls 100 marbles from the jar, supposedly at random, and we find that 20 of them are red. Do we have reason to suspect our friend of cheating? Explain why you answered as you did.

29. Voting for president. In a certain presidential election, 38% of voting-age Americans actually voted. In our town of 200,000, 40% of voting-age citizens voted. What should we conclude regarding our town?

30. Obese Americans. About 35% of Americans over age 20 are obese.

a. If we take random samples of 500 Americans over age 20 and record the percent of people in the sample who are obese, we expect to get data that are approximately normally distributed. Find the mean and standard deviation of this normal distribution. Round the standard deviation to one decimal place.

b. Would it be unusual to find a random sample of 500 Americans over 20 in which at least 38% were obese?

31. Allergies. About 20% of a certain population suffer from some form of allergies.

a. The Central Limit Theorem tells us that percentages of allergy sufferers in random samples of 1000 people are approximately normally distributed. Find the mean and standard deviation of this normal distribution. Round the standard deviation to one decimal place.

b. How unusual would it be to find a random sample of 1000 people in which at least 25% suffered from allergies?

32. Diabetes. About 6.9% of Americans suffer from diabetes. This exercise examines the percentage of people in samples of 2000 who have diabetes.

a. The Central Limit Theorem tells us that these percentages are approximately normally distributed. Find the mean and standard deviation. Round the standard deviation to two decimal places.

b. How unusual would it be for such a sample to show at least 7.6% suffering from diabetes?

33. Tall people. Take the average height for adult males in America to be 5 feet 9 inches. Suppose for the moment that half are taller and half are shorter. Let's take samples of 1000 men and record the percentage of tall (over 5 feet 9 inches) men. Would it be unusual to find such a sample with 45% short people? Would it be unusual to find that the percentage of tall people in a sample of 1000 college basketball players is over 80%?

Exercise 34 is suitable for group work.

34. Heights in your class—an experiment. Record the heights of students in your class and make a histogram of the data. Does it appear that these data are normally distributed? Explain why you gave the answer you did.

Exercises 35 and 36 are designed to be solved using technology such as calculators or computer spreadsheets.

35. A computer simulation. Consider an experiment in which we toss 500 coins. We repeat the experiment 500 times and record the percentage of heads for each experiment. Use a computer simulation to carry out this procedure, and record your results as a histogram. Does it appear that your data are approximately normally distributed? Explain why you gave the answer you did.

36. Formula for a bell. Most people know about the number π, which is approximately 3.14159. But there is another very important number in mathematics, denoted by the letter e, which is approximately 2.71828. Plot the graph of $\frac{1}{\sqrt{2\pi}}e^{-x^2/2}$ for x between -2 and 2. Do you see a connection with the bell curve? If so, what is the mean?

Writing About Mathematics

37. History. Write a paper on the history of the bell curve and the normal distribution. Some names to consider are de Moivre, Gauss, Jouffret, and Laplace.

38. IQ and the bell curve. There are controversies associated with the connection between the bell curve and the measurement of intelligence. The book *The Bell Curve: Intelligence and Class Structure in American Life* caused quite a stir when it was published.[7] Summarize and discuss these controversies.

39. History. Proofs of the Central Limit Theorem were given early in the twentieth century. The ideas involved are much older because certain other distributions have been studied for a long time. Summarize the results of de Moivre and Laplace about the binomial distribution and its relationship to the normal distribution.

40. History. Write a paper about the men who are credited with proving the Central Limit Theorem: Lyapunov and Lindeberg.

41. Disease clusters. The Fallon leukemia cluster was discussed in this section. There have been a number of instances where statistics played an important role in establishing links between environmental contamination and disease in nearby populations. Report on at least one such case. *Suggestion*: The popular movies *Erin Brockovich* and *A Civil Action* featured such cases.

[7] *By Richard Herrnstein and Charles Murray (New York: The Free Press, 1994).*

What Do You Think?

42. Not normally distributed. Give some practical examples of data sets that do not appear to be normally distributed. In each case explain why you do not think the data are normal. **Exercise 8 is relevant to this topic.**

43. Symmetry. The bell curve is *symmetric* about the mean. Explain what this fact tells you about normally distributed data.

44. More on the shape of the bell curve. Find practical examples of bell curves that differ in their height and narrowness. In each case explain in practical terms what the shape indicates about the data.

45. Finding examples. What does the 68–95–99.7% rule tell you about normally distributed data? Find several examples of normally distributed data, and for each explain in practical terms the information this rule provided. **Exercise 12 is relevant to this topic.**

46. Percentile score. Explain what the *percentile score* on a standardized exam such as the SAT tells you. **Exercise 18 is relevant to this topic.**

47. Central Limit Theorem. A friend says the Central Limit Theorem seems to imply that all data are normally distributed. Is he correct? Explain. **Exercise 25 is relevant to this topic.**

48. z-score. Would the use of z-scores make it possible to compare students' scores on the SAT with scores on another standardized test, such as the ACT? Explain why or why not.

49. Mean. If the mean for a set of normally distributed data is 100, is it the case that half of the data have the value 100? If so, explain why. If not, explain what can be said.

50. Mode. What is the mode for a set of normally distributed data?

6.3 The statistics of polling: Can we believe the polls?

TAKE AWAY FROM THIS SECTION
Understand margins of error and confidence levels in polls.

We see polls everywhere, especially in politics. Polling is one of the most visible applications of statistics. The article on the following page appeared at the Web site of *The Wall Street Journal*. It concerns the 2016 U.S. presidential election.

Articles like this may seem straightforward, but in fact they involve some very subtle and deep mathematical ideas. Nevertheless, it is possible to gain enough understanding of the principles to be informed users of such information. That is our goal in this section.

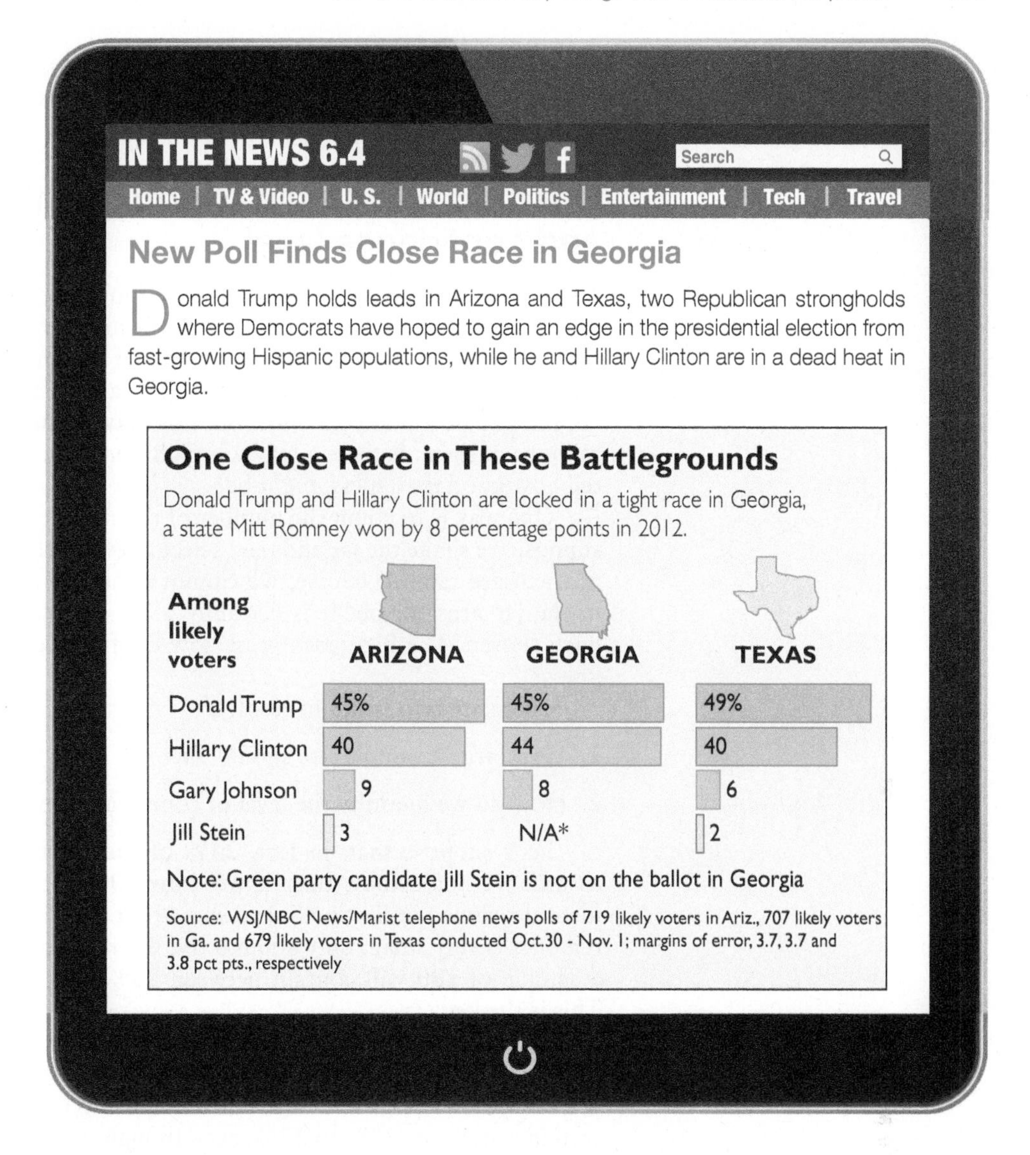

Election polling

In years past, professional polling organizations have had a good record of predicting the outcomes of elections. However, even these professionals sometimes get things wrong, as happened in the 2016 U.S. presidential race, when nearly every major poll predicted a Clinton victory. In the 2012 presidential election, many pre-election polls underestimated the popular vote for Barack Obama. In the 2000 presidential election, early projections by some news organizations led them to claim that Al Gore had won Florida, which would have secured his election. (There is also the example of "Brexit." See Exercise 39.) Nevertheless, polls are often pretty good at getting it right. How are they able to do this?

There are two important aspects of polling, one of which involves mathematical and statistical considerations. The preceding article told us how many people were interviewed in three states and mentioned something they called a "margin of error" for each. How were these numbers determined, what do they really mean, and how much faith should we put in them?

To answer these questions, we need to understand the basics of statistical inference. In Section 6.2, we saw how the Central Limit Theorem concerning the normal distribution allows us to draw inferences about a large population from samples of the population. That discussion is the background for our study of polls in this section. Here, we highlight the problem of estimating the accuracy of those inferences.

A second aspect of polling involves "methodology," that is, the way the polls are conducted, which includes how the sample is chosen. This is at least as important as the mathematical aspects.

Basic terms: Margin of error, confidence interval, and confidence level

To understand how real polls work, we look first at an idealized poll. Suppose we have an opaque jar containing a very large number of red marbles and blue marbles, and we want to know what percentage of the marbles are red. The only way to know this for certain is to actually count all the marbles to determine how many are red and how many are blue. But this may be impractical if there are hundreds or millions of them. Is there a way we might make an educated guess and, if so, can we tell how good that guess might be?

One way is to count the number of red marbles in a small sample. For example, suppose we shake the jar and then select a few marbles at random. We find that 20% of them are red. Of course, we cannot conclude that exactly 20% of the marbles in the jar are red, but it is certainly a reasonable guess based on the sample we have chosen. The big question is, how confident can we be in the accuracy of this guess?

There are two issues to resolve:

1. How large a sample size should we use?
2. How do we quantify the level of confidence we have in our answer?

Let's suppose that, in fact, 25% of the marbles in the jar are red. Let's also assume we don't know that. If we draw 100 marbles from the jar, we can use the ideas developed in Section 5.4 to show that the probability there will be exactly 25 red ones is very small, only about 0.09. But the same ideas show that the probability a sample of 100 will contain between 23% and 27% (inclusive) is 0.44 or 44%. This is the key.

The probability that a sample *exactly* reflects the population may be quite small, but the probability that it *approximately* reflects the population is much greater. We can predict with some confidence that the number of red marbles will be in a certain range, say, 23% to 27%, even though we can't be very confident of the exact number.

Now suppose we chose a marble sample of a different size than 100. What effect would that have on the probability? The following table gives the approximate probabilities that between 23% and 27% of the marbles in our sample are red for various sample sizes, denoted by n. Remember that we're assuming the jar has a vast number of marbles and that 25% of the marbles in the jar are red.

Number n selected	100	500	1000	1500	2000	2500
Probability 23% to 27%	0.44	0.72	0.87	0.93	0.96	0.98

Note that the larger the sample size, the greater the probability that our results fall within the 23% to 27% range. It makes sense that a larger sample is more representative of the whole. We saw the same phenomenon in our discussion of the Central Limit Theorem in the preceding section, where we noted that large sample sizes make even small deviations from the mean unlikely.

Let's consider a sample size of 2000. In that case, the probability is 0.96 or 96% that we will get between 23% and 27% red marbles. This means that if we were to do many such drawings, 96% of the time we would expect them to be between 23% and 27% red. That is, for a sample size of 2000, we can expect that 96% of the time we would get a result within two percentage points of the correct percentage of red marbles, 25%.

The number 2% in this example is the *margin of error*. This number is usually reported as "plus or minus 2%" or as ±2%. We will simply say 2%, with the "plus or minus" understood.

What does a margin of error of 2% mean in terms of polls? Suppose that a polling organization samples 2000 marbles and finds that 24% are red. Their report might be: "One can say with 96% confidence that the percentage of red marbles is 24%, with a margin of error of two percentage points." In other words, with 96% confidence the true percentage of red marbles is between 22% and 26%. The interval 22% to 26% is the *confidence interval*, and the number 96% is the *confidence level*. The meaning of the confidence level is that about 96 out of 100 such samples will produce a confidence interval that includes the true percentage (25%) of red marbles.

KEY CONCEPT

The margin of error of a poll expresses how close to the true result (the result for the whole population) the result of the poll can be expected to lie. To find the confidence interval, adjust the result of the poll by adding and subtracting the margin of error. The confidence level of a poll tells the percentage of such polls in which the confidence interval includes the true result.

It is very important to note that it is *possible* for the polling organization to get an "unlucky" or unrepresentative sample, say, a sample with 8% red marbles or a sample with 80% red marbles. But this is quite unlikely: For a sample size of 2000, only about 4% of the times we take our sample do we expect it to be outside the 23% to 27% range (because 100% − 96% = 4%).

"No, Kevin — there isn't any margin of error on spelling tests."

Back to the polls: Margin of error, confidence interval, and confidence level

Now let's apply what we have learned to a hypothetical idealized opinion poll. In this case, let's assume that we are dealing with people instead of marbles and, instead of differentiating on the basis of marble color, we ask people what opinion they hold on a certain question.

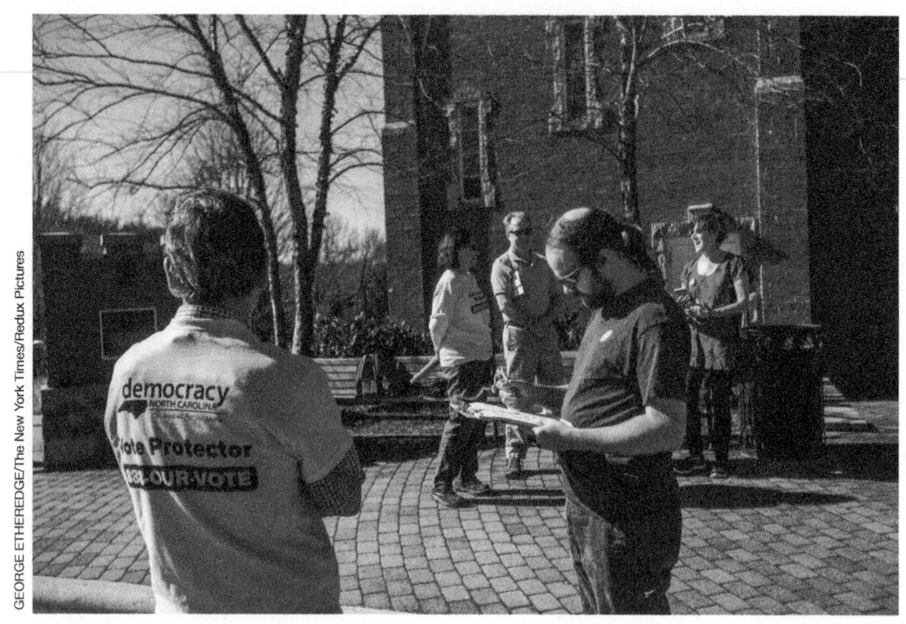

GEORGE ETHEREDGE/The New York Times/Redux Pictures

A voter fills out an exit poll survey from Democracy North Carolina outside a polling station in Asheville, N.C.

As an example, suppose a poll reports with 95% confidence that 45% of the American people support foreign aid and that the margin of error is two percentage points. (Note that 95% is a common confidence level for polls conducted by Gallup, Harris, Zogby, and other professional polling organizations.) Let's recall what this means: If we conducted this poll thousands of times, and each time we wrote down the percentage of the people in the sample who support foreign aid, we would expect about 95% of those results to be within two percentage points of the true value. In brief, we can be fairly confident that the support of the entire population for foreign aid lies in the confidence interval between 43% and 47%.

SUMMARY 6.1
Polls and Margin of Error

Suppose that, based on random sampling, a poll reports the percentage of the population having a certain property (e.g., planning to vote for a certain candidate) with a margin of error m. Assuming that this margin is based on a 95% confidence level, we can say that if we conducted this poll 100 times, then we expect about 95 of those sample results to be within m percentage points of the true percentage having that property.

We reiterate that the result from a single poll does not determine for certain the truth about the population, not even within the range given by the margin of error. In the poll about foreign aid, we cannot be certain that between 43% and 47% of the entire population really do support aid. In fact, the theory says that we should not be surprised if about 5% of the time our sample will not be within two percentage points of the true value. Indeed, it is entirely possible that 70% support aid, although the chances are very remote.

Also, opinion polls are often taken periodically over time and usually by different organizations, so you don't have to trust just one poll. Your confidence should increase if several polls report similar results.

EXAMPLE 6.16 Interpreting polls: Approval of Congress

Explain the meaning of a poll that says 33% of Americans approve of what Congress is doing, with a margin of error of 4% and confidence level of 90%.

SOLUTION

In 90% of such polls, the reported approval of Congress will be within four percentage points of the true approval level. Thus, we can be 90% confident that the true level lies in the confidence interval between 29% and 37%.

TRY IT YOURSELF 6.16

Explain the meaning of a poll that says 35% of Americans approve of what the president is doing, with a margin of error of 3% and confidence level of 93%.

The answer is provided at the end of this section.

How big should the sample be?

The next question is, how big should the sample be to ensure a certain level of confidence? Polling organizations such as the Harris Poll use rather sophisticated methods to decide how many people should be questioned (the sample size) and how they should be chosen. But there is a simple rule of thumb that gives a reasonably good estimate for a confidence level of 95%: For a sample size of n, the margin of error is approximately $100/\sqrt{n}\%$.

RULE OF THUMB 6.2 Margin of Error

For a 95% level of confidence, we can estimate the margin of error when we poll n people using

$$\text{Margin of error} \approx \frac{100}{\sqrt{n}}\%$$

Here, the symbol $\approx$ means "is approximately equal to."

Normally, when we use this estimate, we round the margin of error to one decimal place. This estimate for the margin of error is a consequence of the discussion in the preceding section. The key points are that about 95% of normal data lie within two standard deviations of the mean, and the Central Limit Theorem gives a formula for the standard deviation that depends on the sample size n.

To show how to use the formula, suppose we choose 1100 Americans at random and obtain their opinion on a certain issue. For a confidence level of 95% we get a margin of error of approximately $100/\sqrt{1100}\%$, or about 3.0%. It is perhaps surprising that we can, with 95% confidence, know the opinions of more than 280 million people within three percentage points by sampling only 1100 people.

Michael Steinebach/AFLO photos/Newscom

Kyary Pamyu Pamyu

EXAMPLE 6.17 Estimating margin of error: Fashion survey

In an Oricon fashion survey[8] of young females in Japan, Kyary Pamyu Pamyu was selected as the fashion leader among female celebrities. If 500 people were surveyed, what is the approximate margin of error for a 95% confidence level?

[8] https://onehallyu.com/topic/2028-women-choose-female-celebs-that-they-consider-fashion-leaders/

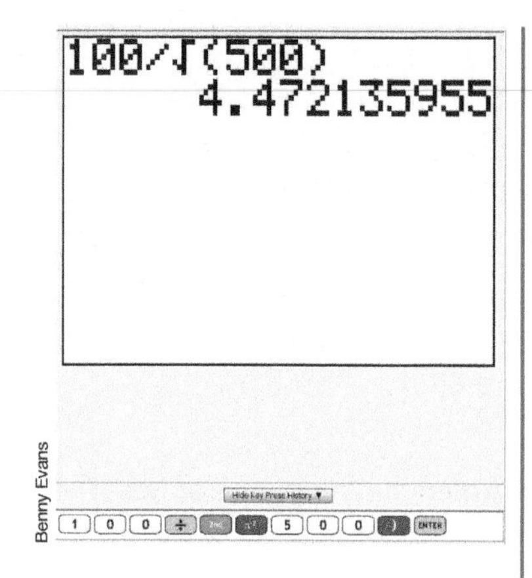

Benny Evans

SOLUTION

We use the formula

$$\text{Margin of error} \approx \frac{100}{\sqrt{n}}\%$$

with $n = 500$:

$$\text{Margin of error} \approx \frac{100}{\sqrt{500}}\%$$

Our margin of error is about 4.5%. We can be 95% confident that our poll result is within 4.5 percentage points of the true value.

TRY IT YOURSELF 6.17

If we conduct a poll of 1600 people, what is the approximate margin of error for a 95% confidence level?

The answer is provided at the end of this section.

Now we interpret the margin of error in terms of a range of estimates.

EXAMPLE 6.18 Finding confidence interval: Presidential race

In the News 6.4 at the beginning of this section gives the sample size for the poll in Arizona as 719 and the margin of error as 3.7%. Assume that the confidence level is 95%. (Because $100/\sqrt{719}$ is about 3.72, this assumption is reasonable.) The poll reported that 45% of likely voters in Arizona preferred Donald Trump. What is the confidence interval for this poll? Interpret your answer.

SOLUTION

The percentage of likely voters preferring Donald Trump was found to be 45%. Because the margin of error is 3.7%, we expect the actual percentage to be between $45-3.7 = 41.3\%$ and $45 + 3.7 = 48.7\%$. Therefore, the confidence interval is 41.3% to 48.7%.

Because the level of confidence is 95%, if we conducted this poll 100 times, then for about 95 of the results, the true value would lie in the confidence interval corresponding to the result. Therefore, we are 95% confident that the actual percentage lies between 41.3% and 48.7%.

TRY IT YOURSELF 6.18

In the same poll, 40% of those polled in Texas preferred Hillary Clinton. What is the confidence interval for this poll? Interpret your answer. (Assume that the confidence level is 95%.)

The answer is provided at the end of this section.

The next example looks at a poll concerning recovery efforts after a natural disaster.

EXAMPLE 6.19 Sample size: Hurricane Sandy aftermath

In October 2013, the Monmouth University Polling Institute surveyed 683 New Jersey residents who were displaced from their homes due to Hurricane Sandy a year earlier. The poll found that 61% of respondents were dissatisfied with New Jersey's recovery efforts so far.[9]

[9] www.nj.com/news/index.ssf/2013/10/sandy_monmouth_county_poll.html

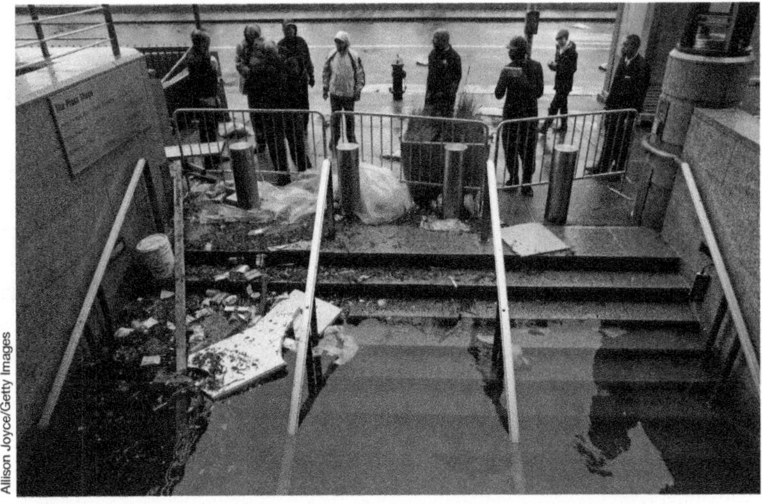

Allison Joyce/Getty Images

Damage in lower Manhattan, New York, caused by Hurricane Sandy.

a. The poll surveyed 683 people. What is the approximate margin of error for a 95% confidence interval?

b. The poll of 683 people found that 61% of respondents were dissatisfied with New Jersey's recovery efforts so far. Can we conclude with certainty that no more than 66% of displaced residents were dissatisfied?

c. Suppose instead that the poll of 683 people had found that 53% were dissatisfied. Explain what we could conclude from this result. Could we assert with confidence that a majority of displaced residents were dissatisfied with recovery efforts?

d. Suppose we wish to have a margin of error of two percentage points. Approximately how many people should we interview?

SOLUTION

a. To find the margin of error, we use the formula

$$\text{Margin of error} \approx \frac{100}{\sqrt{n}}\%$$

with $n = 683$:

$$\text{Margin of error} \approx \frac{100}{\sqrt{683}}\%$$

or about 3.8%.

b. Our answer to part a tells us that we can be 95% confident that the poll number of 61% is within 3.8 percentage points of the true percentage of all displaced residents who were dissatisfied. This means that if we conducted this poll 100 times, then about 95 of the results would be within 3.8 percentage points of the true value. Thus, it is very likely that the true value is between $61 - 3.8 = 57.2\%$ and $61 + 3.8 = 64.8\%$. Because the whole interval is below 66%, we can be quite confident (at a 95% level) that no more than 66% of displaced residents were dissatisfied. However, we cannot make this conclusion with absolute certainty.

c. We can be 95% confident that the poll number of 53% is within 3.8 percentage points of the true percentage of all displaced residents who were dissatisfied. This means that if we conducted this poll 100 times, about 95 of the results would be within 3.8 percentage points of the true value. Thus, it is very likely that the true value is between $53 - 3.8 = 49.2\%$ and

$53 + 3.8 = 56.8\%$. Most of this interval falls above 50%, so we continue to have good reason to think that a majority of displaced residents were dissatisfied. But, because a portion of the interval falls below 50%, we should be more cautious in drawing conclusions.

d. To calculate how large our sample should be, we use the formula

$$\text{Margin of error} \approx \frac{100}{\sqrt{n}}\%$$

and substitute 2% for the margin of error. Thus, we want to find n so that

$$2 = \frac{100}{\sqrt{n}}$$

Rearranging the equation to solve for n gives

$$\sqrt{n} = \frac{100}{2}$$

so

$$\sqrt{n} = 50$$

Hence,

$$n = 50^2 = 2500$$

We should interview about 2500 people. (Note that one Harris Poll with a 95% confidence level and a margin of error of 2% surveyed 2415 people—very close to the 2500 given by the formula.)

The procedure in part d of Example 6.19 can be used to estimate the sample size needed to get a given margin of error: To get a margin of error of m (as a percentage), the sample size must be about $(100/m)^2$. This assumes a 95% level of confidence. In part d of the example, we had $m = 2\%$, and this formula gives us

$$\left(\frac{100}{2}\right)^2 = 50^2 = 2500$$

RULE OF THUMB 6.3 Sample Size

For a 95% level of confidence, the sample size needed to get a margin of error of m percentage points can be approximated using

$$\text{Sample size} \approx \left(\frac{100}{m}\right)^2$$

EXAMPLE 6.20 Estimating sample size: 4% margin of error

What sample size is needed to give a margin of error of 4% with a 95% confidence level?

SOLUTION

We use the approximate formula

$$\text{Sample size} \approx \left(\frac{100}{m}\right)^2$$

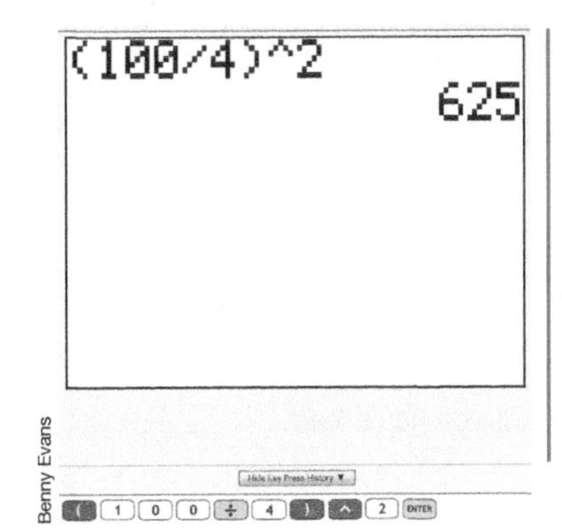

Benny Evans

and put in 4 for m:

$$\text{Sample size} \approx \left(\frac{100}{4}\right)^2 = 625$$

TRY IT YOURSELF 6.20

What sample size is needed to give a margin of error of 5% with a 95% confidence level?

The answer is provided at the end of this section.

The "statistical dead heat"

When polls show that a political race is tight, sometimes the media call the result a "statistical dead heat." The accompanying article from *Business Insider* is an example.

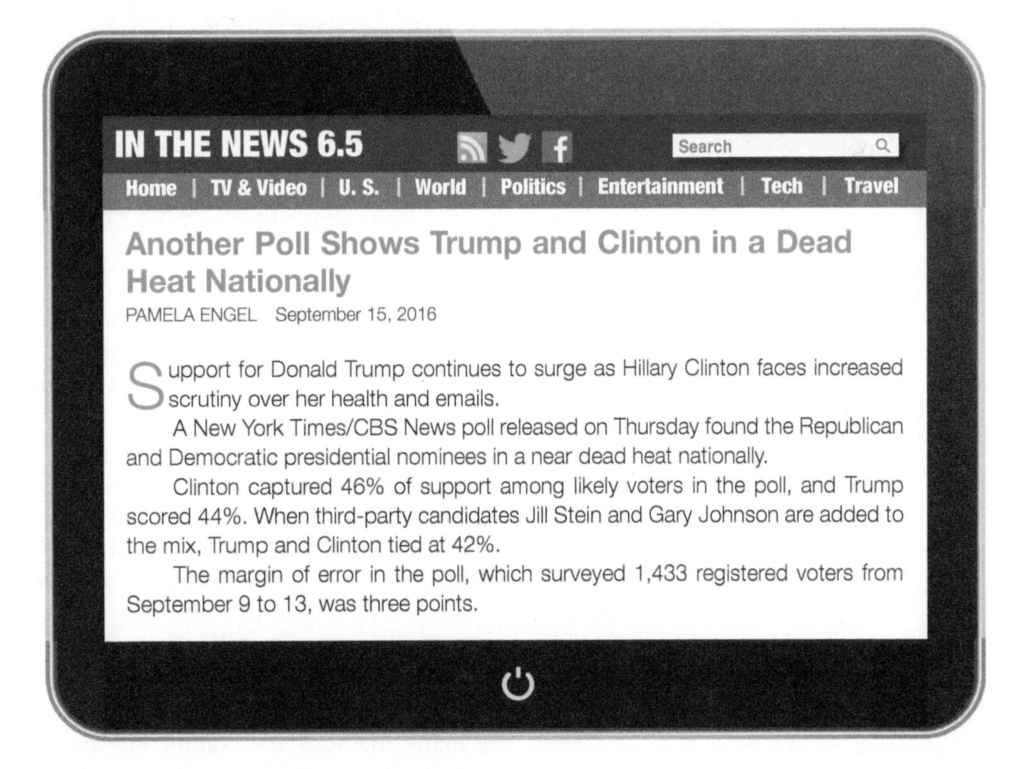

Another Poll Shows Trump and Clinton in a Dead Heat Nationally

PAMELA ENGEL September 15, 2016

Support for Donald Trump continues to surge as Hillary Clinton faces increased scrutiny over her health and emails.

A New York Times/CBS News poll released on Thursday found the Republican and Democratic presidential nominees in a near dead heat nationally.

Clinton captured 46% of support among likely voters in the poll, and Trump scored 44%. When third-party candidates Jill Stein and Gary Johnson are added to the mix, Trump and Clinton tied at 42%.

The margin of error in the poll, which surveyed 1,433 registered voters from September 9 to 13, was three points.

The article uses the term "dead heat" to characterize the poll because the difference between 44% and 46% is less than the margin of error of 3%. The reader might conclude that the poll essentially indicates a tie, but it does not. Here is why. Assume a 95% confidence level. The 3% margin of error in the poll means that we can say with 95% confidence that Clinton's poll number is within 3 percentage points of her actual percentage—so it could be as low as $46 - 3 = 43\%$ or as high as $46 + 3 = 49\%$. Similarly, we can say with 95% confidence that Trump's poll number is within 3 percentage points of his actual percentage, so it could be as high as $44 + 3 = 47\%$ or as low as $44 - 3 = 41\%$. Because the confidence intervals overlap, Trump could actually be ahead of Clinton. The fact that the difference of two percentage points between them is within the margin of error does not, however, mean that they are statistically tied. Probably Clinton does have a lead, but the level of confidence in that statement is smaller than the 95% level we have been considering. This fact is illustrated in **Figure 6.41**.

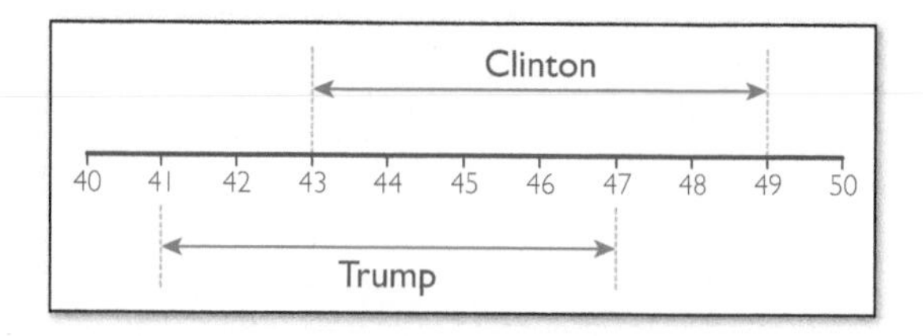

FIGURE 6.41 With 95% confidence, we can say that the actual poll numbers for Clinton and Trump lie in the respective regions. It is very likely that Clinton has a lead.

A good explanation may be found in the following excerpt from an earlier online article written by Humphrey Taylor, chairman of the Harris Poll at the time. In discussing the 2000 presidential election, it clarifies very well the idea of a statistical dead heat:

> However, even a two-point lead should never be described as a "**statistical dead heat**" as the probability is better than 50% that the lead is real. For this reason, **the words "statistical dead heat" should only be used when the candidates are actually tied.** Better still, avoid them altogether.[10]

Methodology of polling

There are many methods for conducting a poll. In most polling, the idea is to take a random representative sample of the voting population and draw conclusions from the sample; however, in practice, obtaining such a sample is a difficult thing to do. For example, if numbers are selected at random from a phone book, then the voting population that uses only cell phones is ignored. If the sample is taken from Bronx County in New York, the sample is likely dominated by Democrats. In practice, polling organizations take polls and then weight the results to try to discern a representative sample. They often get it right, but in the 2016

[10] *"Contrary to what much of the media have been saying, Gore has a statistically significant, if not huge, lead over Bush," September 18, 2000,* www.harrisinteractive.com/vault/Harris-Interactive-Poll-Research-CONTRARY-TO-WHAT-MUCH-OF-THE-MEDIA-HAVE-BEEN-SAYIN-2000-09.pdf

presidential election, almost all public polling organizations wound up underrepresenting Trump voters. As a result, Hillary Clinton was widely considered to be the likely winner right up until the voters proved otherwise. At the time of this writing, the causes of the underrepresentation by pollsters of Trump voters is still being debated.

The following excerpt from the Web site Venturebeat.com includes the methodology used in the survey.

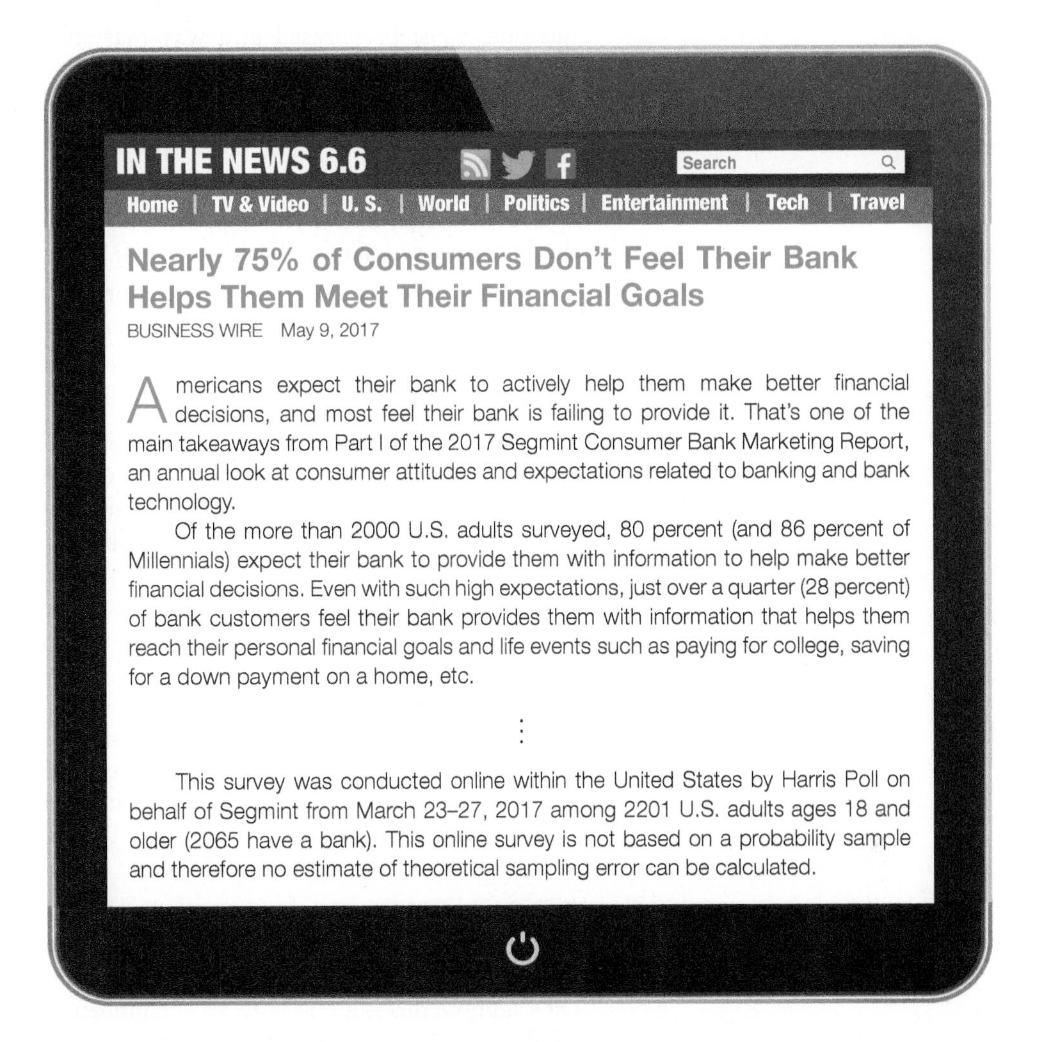

Nearly 75% of Consumers Don't Feel Their Bank Helps Them Meet Their Financial Goals

BUSINESS WIRE May 9, 2017

Americans expect their bank to actively help them make better financial decisions, and most feel their bank is failing to provide it. That's one of the main takeaways from Part I of the 2017 Segmint Consumer Bank Marketing Report, an annual look at consumer attitudes and expectations related to banking and bank technology.

Of the more than 2000 U.S. adults surveyed, 80 percent (and 86 percent of Millennials) expect their bank to provide them with information to help make better financial decisions. Even with such high expectations, just over a quarter (28 percent) of bank customers feel their bank provides them with information that helps them reach their personal financial goals and life events such as paying for college, saving for a down payment on a home, etc.

⋮

This survey was conducted online within the United States by Harris Poll on behalf of Segmint from March 23–27, 2017 among 2201 U.S. adults ages 18 and older (2065 have a bank). This online survey is not based on a probability sample and therefore no estimate of theoretical sampling error can be calculated.

The preceding excerpt notes, "This online survey is not based on a probability sample and therefore no estimate of theoretical sampling error can be calculated." A *probability sample* is one in which the participants are selected randomly, which is crucial to the mathematical basis of any poll. This particular Harris poll was conducted online, so the participants were not randomly selected. They were "self-selected," which means that they are people who happened to find the Web site and chose to respond. By contrast, one Zogby America poll states that the poll "involved 1013 likely voters selected randomly from throughout the 48 contiguous states using listed residential telephone numbers." This is arguably a probability sample, although it misses people without landline phones, for example.

Newspapers and TV shows often invite their audiences to respond to questions. It is usually pointed out that this is not a "scientific poll," which means, among other things, this is not a probability sample—the respondents are not randomly chosen and may not represent a true cross section of the whole country. For example, the audience of such a show could be predominantly conservative or predominantly liberal, which would skew the poll. Because such a poll includes only those people who watch this particular show and want to be heard, it should be viewed skeptically.

Randomness of the sample is not the only methodology problem. There may be several sources of error. These include issues such as question structure and order, which may have more effect on the results than the sample size.

Psychologists often play a role in composing the questions used in polls. The questions must be phrased in a way that elicits the desired information without causing an unwanted bias in the response. Here's an example to illustrate how the phrasing might easily affect the response. Consider how your own response might be affected by the different ways of wording the questions.

Question version 1: Do you support efforts to prevent fires in national forests?

Question version 2: Do you support efforts to prevent fires in national forests by thinning trees to maintain healthy forests?

Question version 3: Do you support efforts to prevent fires in national forests by increased road construction to allow commercial loggers to thin trees?

Question version 4: Do you support efforts to prevent fires in national forests even though fire is an essential part of maintaining a healthy ecosystem?

There are valid reasons to trust (perhaps with a grain of salt) reports from professional polling organizations such as Gallup, Harris, or Zogby that use random sampling and employ specialists from a number of areas to ensure that questions are properly composed and answers properly reported. But there is every reason to be skeptical of polls that do not use representative samples or are conducted by someone who has a point of view or vested interest to support.

WHAT DO YOU THINK?

Statistical tie: A polling service reports that one candidate leads another by 51% to 49% with a margin of error of 3 percentage points. The report goes on to say that because the difference is within the margin of error, this is a "statistical tie." What does "statistical tie" mean? Do you agree that the poll results indicate a tie? Explain your answer.

Questions of this sort may invite many correct responses. We present one possibility: We hear the term "statistical tie" often in news reports, and this is commonly taken to mean that there is no indication of a lead by either candidate. That is not correct. Let's suppose this is a poll with 90% confidence. That means 90% of polls will give the first candidate between 48% and 54%. Values larger than 51% are just as likely as values smaller than 51%, and in the range from 48% to 54% the true number is likely to be more than 50%. Thus, the first candidate likely has an actual lead. The confidence level that this occurs is, however, somewhat less than 90%.

Further questions of this type may be found in the *What Do You Think?* section of exercises.

Try It Yourself answers

Try It Yourself 6.16: Interpreting polls: Presidential approval In 93% of such polls, the reported approval rating of the president will be within three percentage points of the true approval rating. Thus, we can be 93% confident that the true approval rating is between 32% and 38%.

Try It Yourself 6.17: Estimating margin of error: Poll of 1600 2.5%.

Try It Yourself 6.18: Finding confidence interval: Presidential race 36.2% to 43.8%. If we conducted this poll 100 times, for about 95 of the results, the true value would lie in the confidence interval corresponding to the result. Therefore, we are 95% confident that the actual percentage lies between 36.2% and 43.8%.

Try It Yourself 6.20: Estimating sample size: 5% margin of error 400.

Exercise Set 6.3

Test your Understanding

1. Explain in your own words the meaning of a 90% confidence level.

2. Explain in your own words the meaning of the margin of error in polling.

3. True or false: The estimate $\frac{100}{\sqrt{n}}\%$ for the margin of error assumes a confidence level of 95%.

4. True or false: In a news story that says two candidates are in a "statistical dead heat," the writer usually means that their percentages are within the margin of error for that poll.

5. True or false: In a poll, increasing the sample size leads to a smaller confidence interval.

6. A *probability sample* is one in which the participants are: **a.** selected from a phone book **b.** selected on the Internet **c.** selected at random **d.** self-selected.

7. This section was mostly focused on the mathematics behind polling. However, as we pointed out, there are other equally important factors. Which of these is important? **a.** the geographical distribution of the sample **b.** whether the polling was conducted by phone or on the Internet **c.** the racial or ethnic makeup of the sample **d.** the way in which questions are asked **e.** all of these.

8. True or false: If, on the day before an election with two candidates, a reliable polling organization reports that, with 98% confidence, Candidate A leads Candidate B by 10 percentage points, then Candidate A is certain to win the election.

Problems

9. Interpreting polls. Explain the meaning of a poll that says 42% of Americans approve of the president's policies, with a margin of error of 3% and confidence level of 95%.

10. Finding confidence interval. A poll on a certain policy reports 62% approval with a margin of error of 4%. What is the confidence interval for this poll?

11. Estimating the margin of error. A sample size of 900 people is used for a poll. What is the approximate margin of error for a 95% confidence interval?

12. Estimating the sample size. What sample size is needed to give a margin of error of 5% with a 95% confidence interval?

13. Estimating margin of error. A polling organization conducts a poll by making a random survey of 1500 people. Estimate the margin of error at a confidence level of 95%. Round your answer as a percentage to one decimal place.

14. Estimating sample size. A polling organization conducts a poll by making a random survey and is willing to accept a margin of error of 5% at a confidence level of 95%. What should the sample size be?

15. A poll. Suppose that a news article refers to a poll that surveyed 906 registered voters and carries a margin of error of plus or minus 3.5 percentage points. Assume a 95% confidence level, and calculate the margin of error. Is your answer 3.5%? If not, how might you explain the difference between your answer and the one given in this news article?

16–18. Government surveillance. *A poll of 803 adults in the United States found that a majority—59%—said that changes should be made in government surveillance programs. The poll reported a margin of error of 3.5%. Exercises 16 through 18 refer to this poll.*

16. Use the rule of thumb for the margin of error to estimate the margin of error for this poll, assuming a 95% confidence level. (Round your answer as a percentage to one decimal place.) How does your answer compare to the margin of error reported by the poll?

17. Explain in plain English what we can conclude from this poll. Can you assert with confidence that a majority of Americans support making changes in government surveillance programs?

18. Suppose that the poll had found that 51% said that changes should be made in government surveillance programs. If the margin of error remained at 3.5%, could you assert with confidence that a majority of Americans support making changes in government surveillance programs?

19. Wisconsin poll. The 2014 race for governor of Wisconsin was a tight one. A poll of 802 registered Wisconsin voters conducted in January 2014 had Governor Scott Walker at 47% and his Democratic challenger Mary Burke at 41%. The poll also revealed that voters have strong views of Walker: 49% viewed him favorably and 44% had unfavorable views. Burke was not familiar to the majority of voters polled, but 12% had a favorable view of her and 18% had an unfavorable view. The poll had a margin of error of plus or minus 3.5 percentage points. About 40% of the poll was conducted on cellphones.

Compare the margin of error to what you obtain from the rule of thumb in this section. (Assume that the level of confidence is 95%, and round your estimate for the margin of error as a percentage to one decimal place.)

20–21. An imaginary poll. *Suppose we conduct a poll to determine American public opinion concerning our current relations with Europe. We will give the respondents three options: approve, disapprove, or no opinion. Here are some hypothetical data.*

Approve	36%
Disapprove	31%
No opinion	33%

Use these data for Exercises 20 and 21.

20. Suppose only 100 people were sampled. Estimate the margin of error for a 95% confidence level.

21. Suppose this poll has a 95% confidence level with a margin of error of 3.5%. Explain in practical terms what this means, and estimate how many people were sampled.

22. War in Afghanistan. According to a December 2013 poll conducted by CNN and ORC (Opinion Research Corporation) International, only 17% of Americans support the war in Afghanistan, which began over 12 years ago. Eighty-two percent of Americans disapprove of the war and 57% report that they don't think the war is going well. Approximately one-third think that we are winning the war. Five years prior to this poll, December 2008, 52% of Americans supported the war. United States troops were largely withdrawn from Afghanistan by the end of 2014, although a relatively small force remained in 2020.

a. Can you glean any information regarding confidence level or margin of error?

b. In light of your answer to part a, what can you really conclude from this information?

c. Suppose we learn that this poll had sampled 1300 people and had a confidence level of 95%. Estimate the margin of error. Round your answer as a percentage to one decimal place.

23. Increasing sample size. If you double the sample size in a poll, what effect does that have on the margin of error? By what factor should the sample size in a poll be increased in order to cut the margin of error in half?

24. Proportionality. One quantity is *proportional* to another if it is a nonzero, constant multiple of the other. Thus, the statement "y is proportional to x" can be expressed using the formula $y = cx$, where c is a nonzero constant.

Determine which of the following is correct, and explain your answer. For a sample size of n, the margin of error is approximately proportional to:

a. n c. $\sqrt{n}$
b. $1/n$ d. $1/\sqrt{n}$

25–26. 90% confidence levels. *There are estimates of the margin of error for confidence levels other than 95%. For a 90% confidence level and a sample size of n, the margin of error is approximately $82/\sqrt{n}\%$. Use this fact in Exercises 25 and 26.*

25. Surveying 1000 people. In this exercise, we consider surveys of 1000 people.

a. If we survey 1000 people, what is the approximate margin of error for a 90% confidence level? Round your answer as a percentage to one decimal place.

b. Suppose we conduct a poll using a representative sample of 1000 people to find whether Americans like salad. We find that 53% say they do. For a 90% confidence level, explain in plain English what we can conclude from this survey. Can you assert with confidence that a majority of Americans like salad?

c. Suppose our salad survey of 1000 people found that only 48% like salad. For a 90% confidence level, explain in plain English what we can conclude from this survey. Can you assert with confidence that less than a majority of Americans like salad?

26. Estimating sample size. Suppose we wish to have a margin of error of 2% with a 90% confidence level. Approximately how many people should we interview?

27. Sample size and margin of error. The margin of error for a given sample size presented in Rule of Thumb 6.2 is a good estimate but it is not exactly right. To get the exact number, we need to make a complicated calculation using the methods developed in Section 5.4. Suppose 25% of Americans actually believe Congress is doing a good job, but we don't know this. So we select n Americans at random and ask their opinion about Congress. The probability that our poll will show between 23% and 27% (for a margin of error of 2%) depends on the sample size n and is given in the table below. (This table is the same as the one on page 434.)

Number n selected	100	500	1000	1500	2000	2500
Probability 23% to 27%	0.44	0.72	0.87	0.93	0.96	0.98

a. For a 2% margin of error, what confidence level will a sample size of 1000 give?

b. We wish to have a margin of error of 2%. What sample size should we choose for a 72% confidence level?

28. Biased questions? Suppose you conduct a poll asking about foreign policy. You conceive of two possible ways to word the question you wish to ask:

First question: Do you approve of current U.S. policies toward Iran?

Second question: Do you approve of the president's policies toward Iran?

Do you think you would get the same result using either question? Explain why or why not.

29. Legalizing marijuana. Suppose you supported legalizing marijuana. Make up a question asking whether people supported legalizing marijuana that will bias the answer your way. Now do this supposing you did not support legalizing marijuana.

30. Tax cuts. Suppose you are in favor of a tax cut. Make up a question asking whether people support a tax cut that will bias the answer your way. Now do this supposing you are against a tax cut.

31. Polling your campus. If you poll students on your campus regarding how they intend to vote in an upcoming state election, do you believe the results would give a valid picture of what might happen in the statewide election? Explain why or why not.

32. Simulating a poll. A few million coins are lying on the floor. Half of the coins show heads, and half show tails. We think of these coins as the voting population. Coins that show heads intend to vote Democratic. Coins that show tails intend to vote Republican. We want to predict the upcoming election, so we do what the Harris Poll would do. We draw a random sample of 1000 coins from the floor and tabulate how they intend to vote.

But let's do more. Let's repeat this poll simulation 1000 times and tabulate the results. With the aid of computer simulation, we do exactly that. The number of polls showing the given percentage range of people intending to vote Democratic is shown in the accompanying table. (Here, "47% to 48%" means 470 through 479 people intending to vote Democratic, and so on.)

Percent voting Democratic	Less than 47%	47% to 48%	48% to 49%	49% to 50%
Number of polls	23	75	140	234

Percent voting Democratic	50% to 51%	51% to 52%	52% to 53%	At least 53%
Number of polls	250	157	94	27

a. What is an accurate description of the voting population (in terms of percentage voting Democratic)? (For this question, the table with the polling data is irrelevant.)

b. Use Rule of Thumb 6.2 to estimate the margin of error for each of these polls for a confidence level of 95%. Round your answer as a percentage to the nearest whole number.

c. What percentage of polls gave predictions that within three percentage points accurately describe the voting population, as given in the answer to part a? Round your answer as a percentage to the nearest whole number.

d. Explain how your answers in parts a, b, and c are consistent with each other.

e. What percentage of polls gave predictions that within two percentage points accurately describe the voting population, as given in the answer to part a? Round your answer as a percentage to the nearest whole number.

f. Make a histogram summarizing the data in the table. (See Section 6.1.)

Exercises 33 and 34 are designed to be solved using technology such as calculators or computer spreadsheets.

33. Margin of error. Make a table showing the approximate margin of error for sample sizes ranging from 1000 to 2000 in steps of 100. Assume that the level of confidence is 95%, and round each percentage to two decimal places.

34. Sample size. Make a table showing the sample size needed to get a given margin of error ranging from 2% to 3% in steps of 0.1 percentage point. Assume that the level of confidence is 95%, and round to the nearest whole number.

Writing About Mathematics

35. History of polling. Do some research on the history of polling, and report on your findings.

36. Gallup and Zogby. Gallup and Zogby are among several reputable polling organizations. Do some research on their founders, George Gallup and John Zogby, and write a paper about them.

37. Polling in troubled countries. In the subsection *Methodology of polling*, we discussed some of the many thorny issues involved in conducting a valid poll. If polling is difficult in the United States, imagine what it must be like in troubled countries like Afghanistan, Iraq, or Syria. Do some research on polling in one of these countries, and write a report on your findings.

38. History. Two famous examples of polls predicting the wrong outcome of an election are given by the American presidential elections in 1936 and 1948. Do research to discover who the candidates were, what happened in the elections, and what the polling organizations did wrong.

39. Brexit. A famous example of polls getting it wrong occurred in the United Kingdom in 2016, when citizens were asked to vote on whether they wanted to remain in the European Union. According to CNBC, "Leading up to voting day, the vast majority of polls predicted the remain side would prevail, however the final results gave the leave side a victory margin of more than one million votes."[11] Do some research to investigate how this happened, and write a paper about your findings.

What Do You Think?

40. Interpreting polls. A poll says one candidate leads another by 52% to 48%. What additional information would you need to know about this poll before you feel you can understand what it really means? **Exercise 16 is relevant to this topic.**

41. Confidence. Explain the difference between the terms *confidence interval* and *confidence level* as they are used in this section.

42. Sample size formula. Consider the approximate formula for the sample size in Rule of Thumb 6.3. Assume that we want to sample a population and have a margin of error of 3% for a 95% level of confidence. Does the sample size given by the rule of thumb depend on the size of the overall population from which the sample is taken? In other words, does it matter whether the overall population is 1 million or 100 million? Does your answer agree with what intuition would suggest? Explain. **Exercise 21 is relevant to this topic.**

43. TV poll. A TV talk show asked its listeners to call in their vote in a race between candidates X and Y. The result was that 63% of the callers voted for X and 37% for Y. A friend said that she was quite confident X would win because of this poll. Would you agree? Explain.

44. Margin formula. Consider the formula in Rule of Thumb 6.2 for approximating the margin of error. What assumption is made about the sampling in order for the formula to hold? For example, would this formula apply to a poll conducted online?

[11] www.cnbc.com/2016/07/04/why-the-majority-of-brexit-polls-were-wrong.html

45. How should the question be asked? When a pollster samples opinions, great care must be taken to minimize bias introduced by the wording of questions. Give examples of some subtle ways that wording may bias questions. Do you think it is possible to ask questions in such a way that the wording introduces no bias at all? **Exercise 28 is relevant to this topic.**

46. Push polls. Reputable polling organizations work very hard to produce unbiased, accurate results. But sometimes polls are conducted by organizations with a different goal. As an extreme example, an automobile company might ask, "Do you prefer our high-quality cars or the inferior cars the competition sells?" Then they might report the results as, "70% of people polled prefer our cars." Polls that are designed to push a certain point of view rather than gain information are called *push polls*. Can you find actual examples of push polls?

47. Try it yourself. Conduct a poll of students on your campus regarding an issue relevant to current campus life. Report the results, including margin of error and confidence level. What issues did you encounter, and how did you deal with them? You might consider such things as question structure and how to make your sample as nearly random as possible.

48. Polling works. Often people question how polls can be accurate when they sample relatively few people. Use the concepts of this section and the preceding one to explain how polling a relatively small sample can give fairly accurate results.

6.4 Statistical inference and clinical trials: Effective drugs?

TAKE AWAY FROM THIS SECTION
Understand statistical significance and p-values.

We have already examined the role of statistical inference in polling. Another important use of statistics is in medical research, where statistical inference is used to answer questions about the efficacy of medical treatments and drugs.

The following article is an announcement of a clinical trial to determine the effects of psoriasis treatment on cardiometabolic diseases.

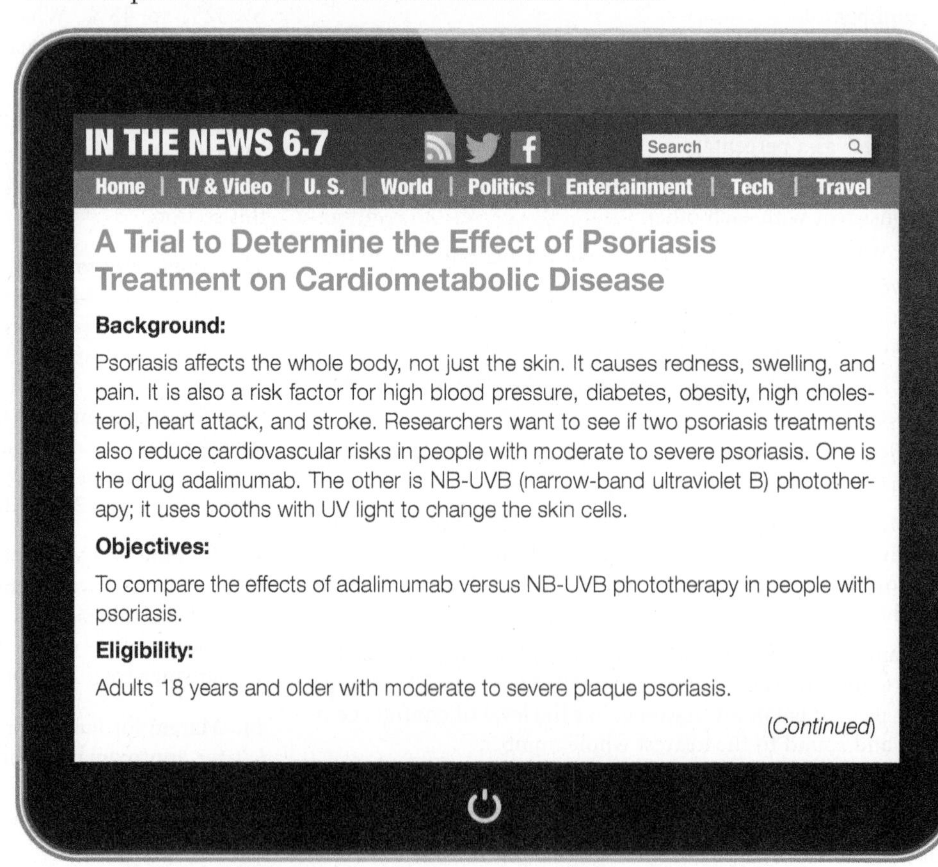

IN THE NEWS 6.7

Search

Home | TV & Video | U. S. | World | Politics | Entertainment | Tech | Travel

A Trial to Determine the Effect of Psoriasis Treatment on Cardiometabolic Disease

Background:

Psoriasis affects the whole body, not just the skin. It causes redness, swelling, and pain. It is also a risk factor for high blood pressure, diabetes, obesity, high cholesterol, heart attack, and stroke. Researchers want to see if two psoriasis treatments also reduce cardiovascular risks in people with moderate to severe psoriasis. One is the drug adalimumab. The other is NB-UVB (narrow-band ultraviolet B) phototherapy; it uses booths with UV light to change the skin cells.

Objectives:

To compare the effects of adalimumab versus NB-UVB phototherapy in people with psoriasis.

Eligibility:

Adults 18 years and older with moderate to severe plaque psoriasis.

(Continued)

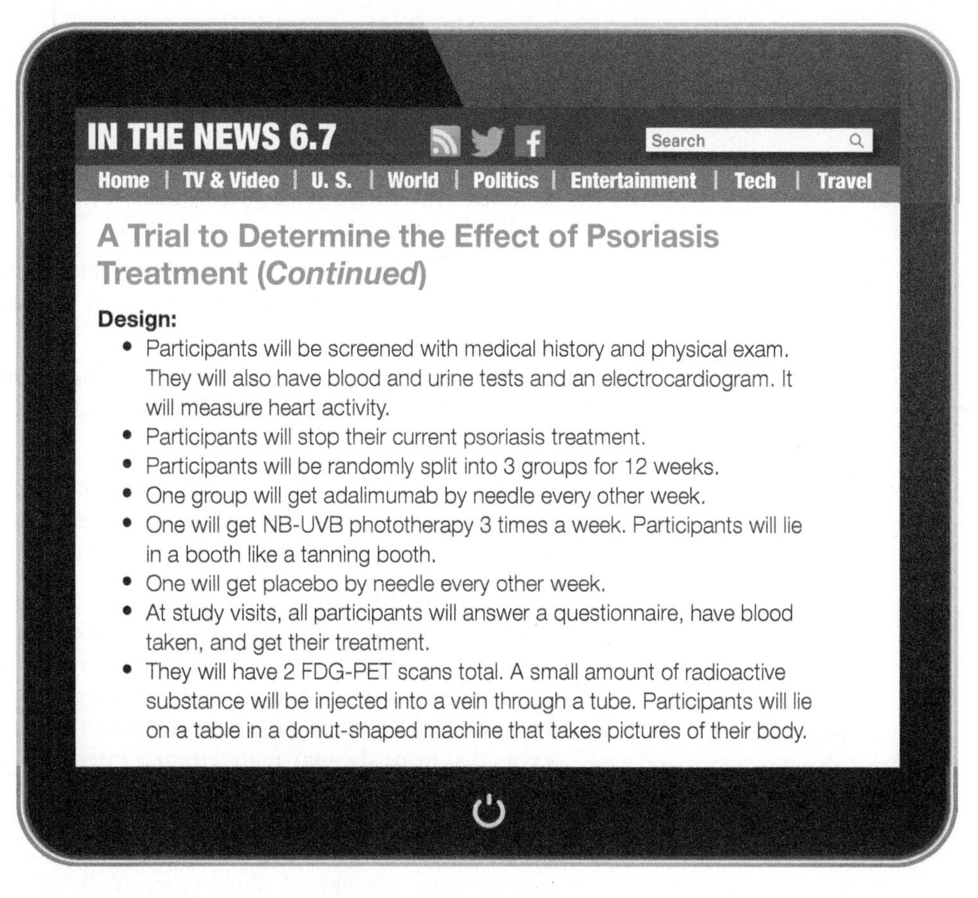

IN THE NEWS 6.7

Search

Home | TV & Video | U. S. | World | Politics | Entertainment | Tech | Travel

A Trial to Determine the Effect of Psoriasis Treatment (*Continued*)

Design:

- Participants will be screened with medical history and physical exam. They will also have blood and urine tests and an electrocardiogram. It will measure heart activity.
- Participants will stop their current psoriasis treatment.
- Participants will be randomly split into 3 groups for 12 weeks.
- One group will get adalimumab by needle every other week.
- One will get NB-UVB phototherapy 3 times a week. Participants will lie in a booth like a tanning booth.
- One will get placebo by needle every other week.
- At study visits, all participants will answer a questionnaire, have blood taken, and get their treatment.
- They will have 2 FDG-PET scans total. A small amount of radioactive substance will be injected into a vein through a tube. Participants will lie on a table in a donut-shaped machine that takes pictures of their body.

Experimental design

There are many factors involved in the setup of an experiment or clinical trial. In this case, the purpose of the trial is to determine the efficacy (or lack of efficacy) of a drug.

In a typical trial, one group of subjects is given the actual drug or treatment and another group is given a *placebo*. A placebo is a benign substance that contains no medication and is outwardly indistinguishable from the real drug. One kind of placebo might be a simple sugar pill. Of course, the subjects do not know which they are getting.[12]

"The test results are in for our new drug. 9 out of 10 doctors recommended the placebo."

[12] *It's interesting that taking a placebo can create an expectation of improved health and thus result in positive medical benefits. This is called the "placebo effect."*

In some studies, the second group will be given a standard treatment rather than a placebo to compare the effect of a known drug with a proposed new drug. In these cases, the known drug will be disguised to look like the proposed drug.

Here are the typical key ingredients when planning a clinical trial:

1. Researchers first decide how much of a difference between treatment groups is medically important.

2. Then they decide how many subjects should be enrolled in the trial. The sample should include enough participants for the researchers to have confidence in the results.

3. Next, part of the group is given the drug or treatment to be tested, and the other subjects are given the placebo. The group given the placebo is usually called the *control group*.

4. Finally, the researchers compare the subjects in the two groups to see whether there is a significant difference between their conditions after the trial. The key word here is "significant." That's where statistics comes in.

So far so good, but just as in polling, there are many possible pitfalls in such a study. One such pitfall is that the observer or the subject can inadvertently affect the results. For example, suppose a researcher has worked so hard to develop a drug that he or she has come to believe in it, and just *really* wants it to work. If the researcher knows which subjects have been given the drug and which the placebo, the chance exists that he or she may inadvertently interpret the condition of the subjects in favor of the drug.

That is why clinical trials are usually *double blind*, an experiment in which neither the subjects nor the researchers know who is getting the drug and who is getting the placebo. A third party is the only one who knows which patient is receiving what.

Statistical significance and p-values

It happens by chance that some sick people just get better with no treatment at all and some get worse. Therefore, it would be extremely unusual for every subject in the group who received the drug being tested to get better and everyone in the control group to get worse. Usually, there will be some of each in both groups.

The trick is to figure out whether the difference (if any) between the results for the treated group and the control group is "significant" or could be happening simply by chance. Statistics provides a way to compare the results of the study with what would happen due to chance and thus to determine whether the results are *statistically significant*.

KEY CONCEPT

The results of a clinical trial are considered to be statistically significant if they are unlikely to have occurred by chance alone.

In this Key Concept, we have not said what we mean by "unlikely." We will explain this point shortly. First we present examples related to statistical significance.

EXAMPLE 6.21 Determining statistical significance: Coins and trucks

For each of these situations, determine whether the results described appear to be statistically significant.

a. You flip a coin 40 times and get heads each time.

b. You are driving and are passed by two pickup trucks in a row.

SOLUTION

a. By the Counting Principle from Section 5.3, the probability of getting 40 heads in a row by flipping a fair coin is $1/2^{40}$. This number is so unbelievably small that this result is very unlikely to have occurred by chance alone. Therefore, this result is statistically significant.

b. Pickup trucks are quite common in many areas, so being passed by two in a row does not seem terribly unlikely. This result does not seem statistically significant.

TRY IT YOURSELF 6.21

For each of these situations, determine whether the results described appear to be statistically significant.

a. You flip a coin twice and get heads both times.

b. You are driving and you pass three 1951 Studebaker cars in a row.

The answer is provided at the end of this section.

As the preceding examples indicate, statistical significance corresponds to our common-sense notion of being highly unusual. Although these examples are fairly straightforward, the statistical significance of results of clinical trials can be more subtle. For example, consider the following table showing the results of a hypothetical trial:

	Treated group	Control group
Improved	53	50
Got worse	47	50

Certainly, the treated group improved a little more than the control group, but is this difference statistically significant? The answer is not obvious. Statisticians involved with clinical trials determine the answer by calculating something called the *p-value*. The *p*-value measures how likely it is that the difference between the two groups in a clinical trial would occur by chance alone if the treatment had no effect. The *p*-value is a way to quantify the word "unlikely" that appeared in our explanation of the term *statistically significant.*

KEY CONCEPT

The *p*-value measures the probability that the outcome of a clinical trial would occur by chance alone if the treatment had no effect. A small *p*-value is usually interpreted as evidence that the event in question is unlikely to be due to chance alone. The results are usually accepted as statistically significant if the *p*-value is $0.05 = 5\%$ or smaller.

Here is a very important real example involving clinical trials.

EXAMPLE 6.22 Determining statistical significance: Vioxx and *p*-values

The drug Vioxx was once used to treat arthritis. One published study compared the incidence of heart disease in patients treated with Vioxx versus patients treated with Aleve.[13] (Here, the group treated with Aleve is considered to be the control group.)

[13] *Mukherjee, D., Nissen, S.E., and Topol, E.J., "Risk of Cardiovascular Events Associated With Selective COX-2 Inhibitors,"* Journal of American Medical Association *(2001), 954–959.*

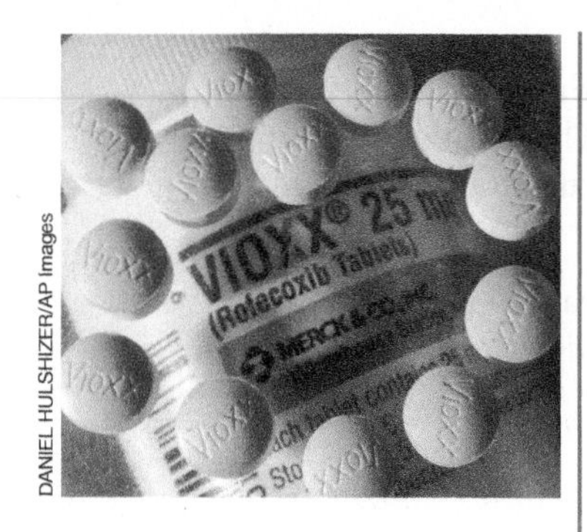

DANIEL HULSHIZER/AP Images

The study showed a much greater incidence of heart disease for the Vioxx patients, with a p-value of 0.002. Explain the meaning of the p-value in this test. Would the result of this study normally be accepted as statistically significant?

SOLUTION

If Vioxx and Aleve had the same effects, the probability that the difference in results between the Vioxx and the Aleve patients would occur by chance alone is 0.002 or 2/10th of 1%. Because this probability is much less than the accepted threshold of 0.05, this result is certainly statistically significant.

TRY IT YOURSELF 6.22

The same study compared the effect of Vioxx with the effect of a placebo and also found a greater incidence of heart disease for the Vioxx patients, with a p-value of 0.04. Explain the meaning of the p-value in this case. Would the result of this study normally be accepted as statistically significant?

The answer is provided at the end of this section.

As the example indicates, in practice, one computes a p-value that measures the probability that the difference between the subjects in the control and test groups would occur by pure chance if the treatment had no effect. To explain in detail how this is done would take us too far afield, so we will use a simplified example to give the basic idea.

Idealized trials

Suppose we test a drug that we believe may prevent colds. There is no control group in this case, but suppose we know that, on average, 50% of people will get colds this winter. Suppose we treat 100 patients with this drug and find that 55 did not get a cold. Would we conclude that the drug is effective in preventing colds? Or did the fact that 55 patients did not get a cold just happen by chance?

Here's the idea: Without treatment, 50% of people will get colds, and from this fact it follows that the probability of 55 or more out of 100 subjects not getting a cold is 0.184 or 18.4%.[14] The number 0.184 is the p-value for this experiment. It says that in 18.4% of observations of 100 untreated subjects, we expect that 55 or more of the subjects will not get a cold. Therefore, having 55 or more stay healthy is not a highly unlikely event—we expect it to occur almost 20% of the time even without treatment. The benefits of the test drug would be in doubt if only 55 out of 100 patients we actually treated did not get a cold.

Suppose, however, that 65 of the patients taking the test drug did not get a cold. The p-value for this outcome turns out to be $0.002 = 0.2\%$. This is the probability that 65 or more patients will not get a cold if the drug had no effect. We'd be very surprised to see that many patients not get a cold merely by chance. If 65 of the patients taking the test drug did not get a cold, it is reasonable to believe that something other than chance is at work.

Having a small p-value associated with a study suggests that the outcome is due to something other than chance alone, but it doesn't necessarily say what that something is. For this reason, it's important for the researchers to take care to avoid factors, other than the drug or procedure being tested, that might cause the two groups to differ. If this is done, then having a small p-value suggests that the drug or procedure was probably responsible.

---

[14] *This number can be found using the ideas developed in Section 5.4.*

In the end, no matter how good the study, there is always a probability, small though it may be, that the conclusion suggested by the results is incorrect. That is one reason why independent researchers might repeat a clinical trial. If a small p-value is obtained in several different trials, then we have much greater confidence in the result. This is important—after all, these results decide which drugs get on the market and which do not.

Correlation and causation

An important issue in medicine is deciding the cause of a certain disease or condition. How do we tell whether one condition actually causes another, or whether the two conditions are simply associated? Mathematically speaking, we would like to describe whether a change in one variable is related to changes in other variables. This is the context for the term *correlation*.

KEY CONCEPT

We say that two numerical variables are positively correlated if an increase in one of them accompanies an increase in the other. Similarly, two numerical variables are said to be negatively correlated if an increase in one of them accompanies a decrease in the other. If neither of these is true, the variables are called uncorrelated.

For example, the time required for a trip and the speed you drive are negatively correlated, because an increase in travel time accompanies a decrease in speed. By contrast, the price of a pizza and its diameter are positively correlated because an increase in price usually accompanies an increase in diameter.

In these examples, the correlations are the result of real relationships between the variables, but some correlations are the result of chance or coincidence, and we have to be careful not to draw unwarranted conclusions. Furthermore, even when two variables are connected in some way, we should not necessarily assume that one causes the other. For example, there is a positive correlation between ice cream sales and the number of shark attacks on swimmers. Both of these tend to rise during warm weather when more people swim and buy ice cream, but certainly neither causes the other.

EXAMPLE 6.23 Understanding correlation as an idea: A balloon

Determine whether the diameter of a balloon and its volume are positively correlated, negatively correlated, or uncorrelated.

SOLUTION

As diameter increases, so does the volume. Hence, these variables are positively correlated.

TRY IT YOURSELF 6.23

Determine whether the outdoor temperature and your home heating bill are positively correlated, negatively correlated, or uncorrelated.

The answer is provided at the end of this section.

There are statistical methods that quantify the extent to which two variables are, in fact, correlated. In particular, the method called *linear regression* quantifies the extent to which two variables are *linearly correlated*. We have already introduced linear regression in the context of trend lines in Section 3.1.

The following data show weekly expenditures in British pounds of households in the given region:[15]

Region	Alcohol	Tobacco
North	6.47	4.03
Yorkshire	6.13	3.76
Northeast	6.19	3.77
East Midlands	4.89	3.34
West Midlands	5.63	3.47
East Anglia	4.52	2.92
Southeast	5.89	3.20
Southwest	4.79	2.71
Wales	5.27	3.53
Scotland	6.08	4.51

We want to determine whether there is a linear correlation between the two variables, spending on alcohol and spending on tobacco. First we make a scatterplot of the data, as shown in **Figure 6.42**. The plot suggests a positive correlation between the two variables: For the most part, an increase in one accompanies an increase in the other. To determine whether this is a *linear* correlation, we add the trend (or regression) line, as seen in **Figure 6.43**. Note that the points seem roughly to fall along this line. This fact leads us to suspect that there is a linear relation between the two variables. In summary, we have evidence that a positive linear correlation exists between spending on alcohol in a region and spending on tobacco in that region.

Figure 6.43 also shows the equation of the trend line. We round the coefficients to get $y = 0.612x + 0.108$. Here, x represents spending on alcohol and y represents spending on tobacco, both in British pounds. This equation lets us do more than just establish a link between the variables. It quantifies, approximately, how the two variables are related. For example, if in a given region, households spend an average of 4 pounds per week on alcohol, we would expect $0.612 \times 4 + 0.108 = 2.556$, or about 2.56 pounds per week to be spent on tobacco. Nevertheless, the data does not tell us that the consumption of alcohol causes tobacco usage, or vice versa.

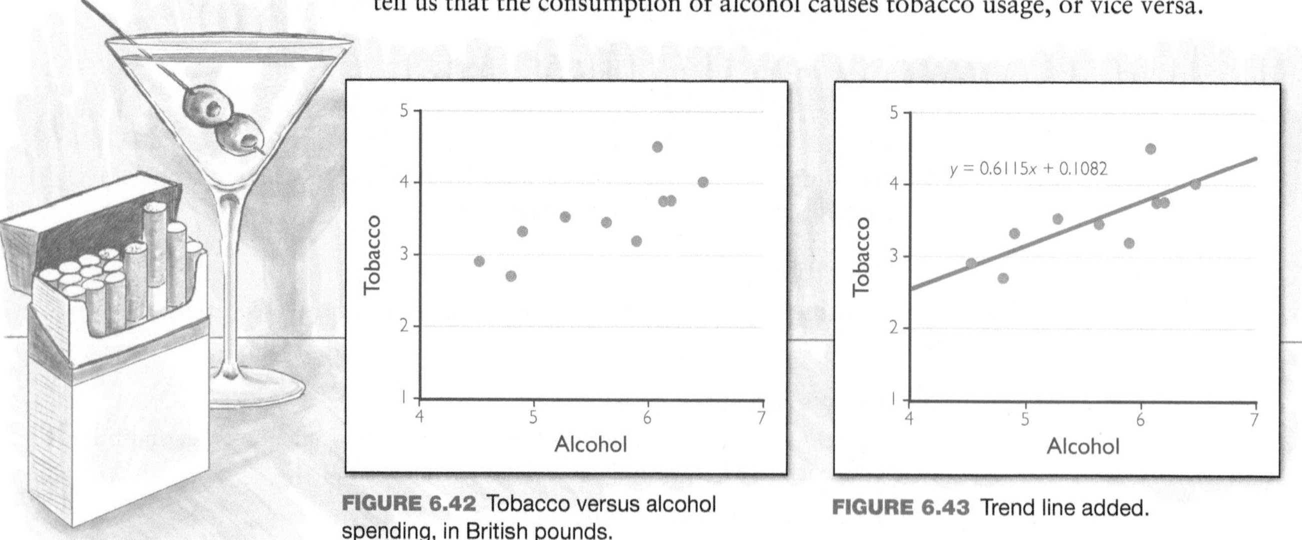

FIGURE 6.42 Tobacco versus alcohol spending, in British pounds.

FIGURE 6.43 Trend line added.

[15]*Taken from* Family Expenditure Survey, *Department of Employment (British Official Statistics). We have omitted the data for Northern Ireland.*

EXAMPLE 6.24 Determining correlation from data: Dental care

Consider the following data table comparing by state the percentages of people who visited the dentist this past year and the percentages of people over age 65 who have had all of their teeth extracted.

State	Dental visit	All teeth extracted
Oklahoma	58.0%	28.3%
Nevada	66.2%	18.4%
Alaska	66.9%	23.6%
Florida	68.7%	17.4%
New York	71.8%	17.5%
Connecticut	80.5%	12.8%

Figure 6.44 shows a plot of the data along with the trend line. Do these data show a correlation between people visiting the dentist within the past year and the elderly having all their teeth extracted? If so, what type of correlation exists?

SOLUTION

In general, as the percent visiting the dentist increases, the percent having had all their teeth extracted decreases. Therefore, the two variables are correlated, and the correlation is negative. The graph suggests that the correlation is linear.

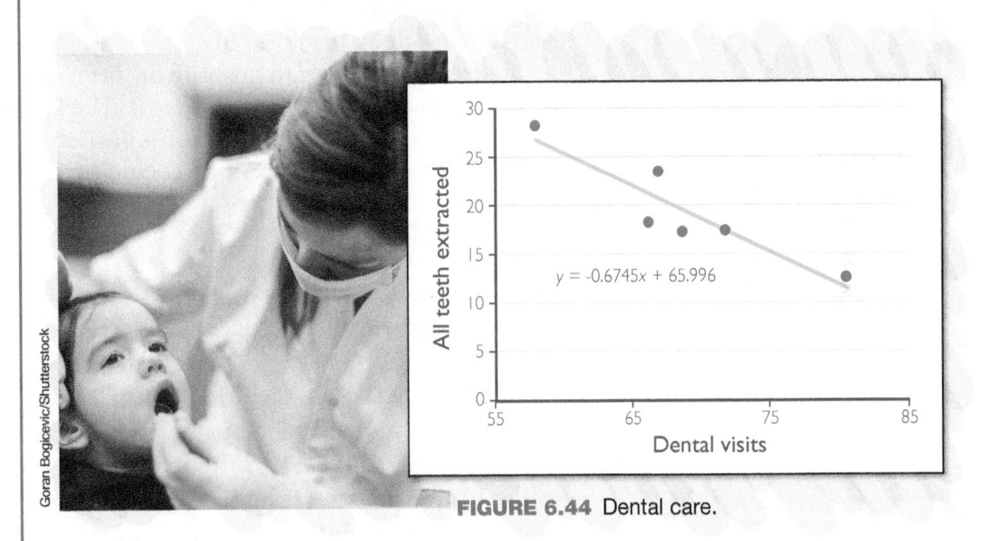

FIGURE 6.44 Dental care.

TRY IT YOURSELF 6.24

The following table shows the running speed of various animals versus their length. We discussed these data in Section 2.1 and in Section 3.1.

Animal	Length (inches)	Speed (feet per second)
Deermouse	3.5	8.2
Chipmunk	6.3	15.7
Desert crested lizard	9.4	24.0
Grey squirrel	9.8	24.9
Red fox	24.0	65.6
Cheetah	47.0	95.1

Figure 6.45 shows a plot of the data along with the trend line. Are length and speed correlated? If so, what type of correlation is there?

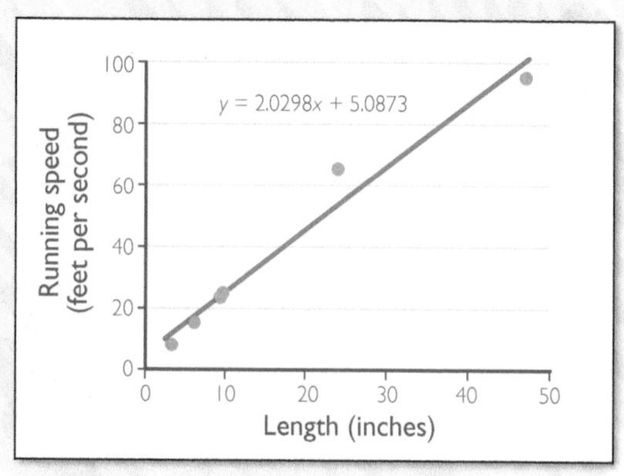

FIGURE 6.45 Length and speed.

The answer is provided at the end of this section.

Exactly how strong is the evidence of a linear correlation between alcohol and tobacco expenditures or between seeing the dentist regularly and having had all teeth extracted? For alcohol and tobacco purchases, consider again Figure 6.43. If the points were to lie exactly on the line, anyone would be convinced of a strong linear relationship. In fact, the points roughly follow the trend line, but most of them do not fall on the line. How do we decide how strong the correlation is?

Statisticians attempt to quantify the degree of linear correlation by using a number known as the *correlation coefficient*. The formula for calculating the correlation coefficient is complicated, and the job is best left to your calculator or computer. In the case of Figure 6.43, the correlation coefficient is 0.78. What does this number mean?

KEY CONCEPT

The correlation coefficient always lies between -1 and 1 inclusive. The closer a correlation coefficient is to 1, the greater the degree of positive linear correlation, with 1 indicating perfect positive linear correlation (the points lie exactly on the increasing trend line). The closer a correlation coefficient is to -1, the greater the degree of negative linear correlation, with -1 indicating perfect negative linear correlation (the points lie exactly on the decreasing trend line). A correlation coefficient near 0 indicates little if any *linear* correlation.

Note that when the coefficient is near 0, we can't conclude that there is no relationship between the variables. We can conclude only that no *linear* relationship exists.

The sign of the correlation coefficient matches the sign of the slope of the trend line. A positive correlation is indicated by a positive slope and an increasing line, and a negative correlation is indicated by a negative slope and a decreasing line. The graphs in **Figures 6.46 through 6.49** may help.

Interpreting correlation coefficients partly requires judgment and experience, just as it does with p-values and (as we saw in Section 6.3) confidence levels. After all, the 95% confidence level is rather arbitrary. One rule of thumb is that a correlation coefficient whose magnitude is 0.8 or more indicates strong linear correlation, a coefficient whose magnitude is between 0.5 and 0.8 indicates moderate linear correlation, and a coefficient whose magnitude is less than 0.5 indicates weak

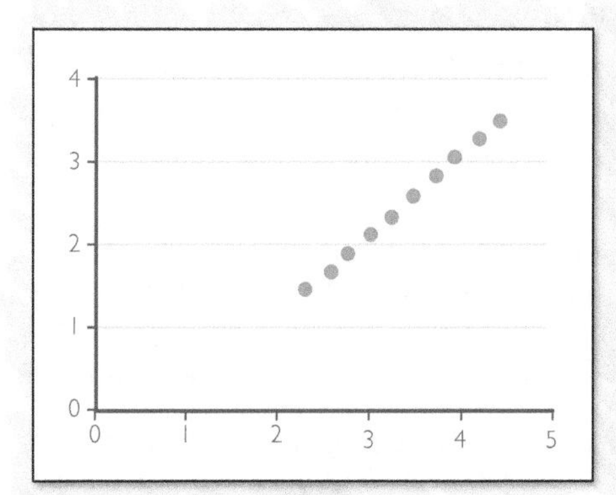

FIGURE 6.46 Correlation coefficient of 1 shows a strong linear relation with positive slope.

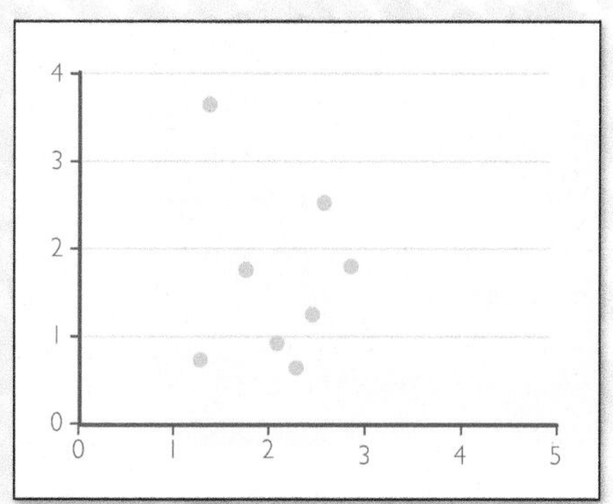

FIGURE 6.47 Correlation coefficient of 0.02 shows little if any linear relation.

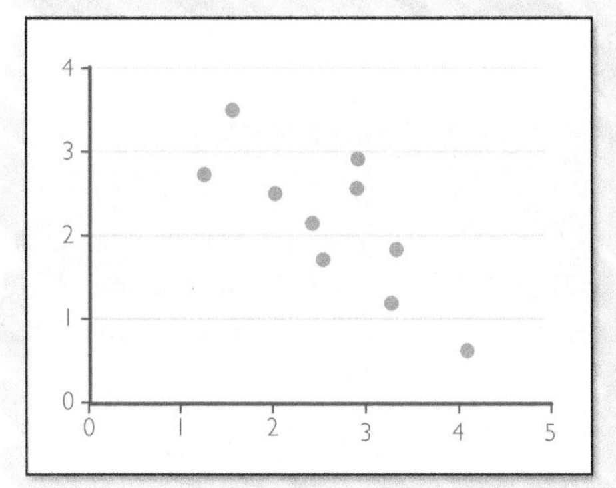

FIGURE 6.48 Correlation coefficient of -0.77 shows a moderately strong linear relation with negative slope.

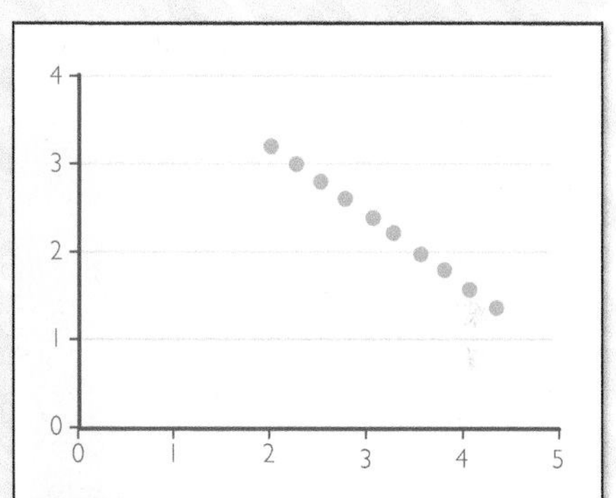

FIGURE 6.49 Correlation coefficient of -1 shows a strong linear relation with negative slope.

linear correlation at best. By this measure, the correlation coefficient of 0.78 in the example concerning alcohol and tobacco spending is fairly good evidence of linear correlation.

RULE OF THUMB 6.4 Correlation Coefficient

A correlation coefficient whose magnitude is 0.8 or more indicates strong linear correlation, a coefficient whose magnitude is between 0.5 and 0.8 indicates moderate linear correlation, and a coefficient whose magnitude is less than 0.5 indicates weak linear correlation at best.

EXAMPLE 6.25 Interpreting correlation: Diversifying index funds

An *index fund* is a type of mutual fund that invests in a large number of diversified stocks. An index fund that invests in U.S. stocks is commonly considered to be a fairly safe investment. But if U.S. stock prices fall overall, it's possible that you would have been better off if you had also invested in a fund that is not strongly correlated with U.S. stocks.

One calculation shows that the correlation coefficient between the prices of U.S. stocks in the S&P 500 Index and stocks in the London stock exchange called the FTSE 100 averaged 0.848. Does this correlation coefficient indicate that there is a great advantage in adding a London stock index fund to your portfolio?

Chris Radcliffe/Bloomberg/Getty Images

SOLUTION

A correlation coefficient of 0.848 indicates a strong positive linear correlation. This says that a London index fund's prices will rise and fall, more or less, in a similar way to a U.S. index fund, so there is no great advantage in adding this investment.

TRY IT YOURSELF 6.25

Another calculation shows that the correlation coefficient between the prices of U.S. stocks and Japanese stocks (the S&P 500 Index and the Nikkei 225) averaged 0.609. Does this correlation coefficient indicate that there is an advantage in adding a Japanese stock index fund to your portfolio?

The answer is provided at the end of this section.

Causation

Now we return to the subject of causation and its relation to correlation. Consider the excerpt from a news article.

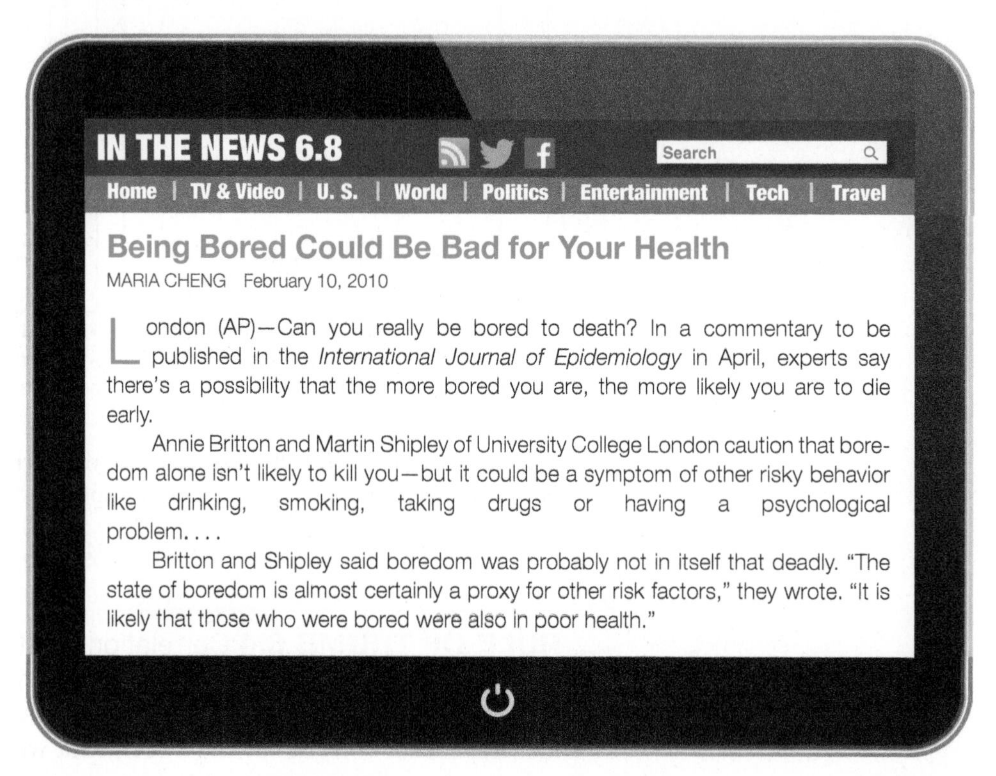

IN THE NEWS 6.8

Search

Home | TV & Video | U. S. | World | Politics | Entertainment | Tech | Travel

Being Bored Could Be Bad for Your Health

MARIA CHENG February 10, 2010

London (AP)—Can you really be bored to death? In a commentary to be published in the *International Journal of Epidemiology* in April, experts say there's a possibility that the more bored you are, the more likely you are to die early.

Annie Britton and Martin Shipley of University College London caution that boredom alone isn't likely to kill you—but it could be a symptom of other risky behavior like drinking, smoking, taking drugs or having a psychological problem. . . .

Britton and Shipley said boredom was probably not in itself that deadly. "The state of boredom is almost certainly a proxy for other risk factors," they wrote. "It is likely that those who were bored were also in poor health."

This article suggests that there is a link (a positive correlation) between boredom and early death. By noting that boredom is probably only a symptom associated with other risk factors, the article warns us against concluding that boredom is a *cause* of early death.

We have already noted that determining whether one event causes another event is much more difficult than determining mere correlation. Two phenomena can be correlated without one of them causing the other. When two quantities are correlated, there are three possibilities: One causes the other, both are caused by something else, or the correlation may simply be coincidence—just pure chance.

Here is an example: Suppose a mother notices that from time to time her baby gets a runny nose and a cough. The cough seems to come first most of the time, so she concludes that the cough causes the runny nose. In fact, the most likely explanation is that both are caused by a third factor: The baby has a cold. That they occur together, and thus are correlated, does not mean that one must cause the other.

For another example, consider again the data on tobacco and alcohol consumption in British households for various regions of Britain (page 454). Consumption of tobacco and consumption of alcohol in a region are certainly positively correlated because, in general, an increase in one corresponds to an increase in the other. (In fact, the correlation coefficient of 0.78 confirms the positive correlation.) It's possible that one of these somehow causes the other. But it's also possible that people who have a tendency to use one of these products also have a tendency to use the other.

A classic example of an argument about correlation versus causation is the connection between smoking and lung cancer. Many studies have shown that smoking and lung cancer are positively correlated. Although it is indisputable that smokers have a much higher rate of lung cancer than non-smokers, the tobacco industry has argued in the past that this does not necessarily tell us that smoking *causes* lung cancer. They have suggested that it may be like the baby with a cough and runny nose. The conditions may occur together, but how can we say one causes the other? Maybe, they argue, there is a gene or some unknown behavior that increases the propensity for both smoking and lung cancer.

Showing causation requires much more evidence than showing correlation, but enough research has been done to convince the medical community that smoking does, in fact, cause lung cancer.

WHAT DO YOU THINK?

Placebo: Why must patients in a clinical trial not be told if they are getting a placebo? Won't their condition just persist with a placebo, whether they are told or not?

Questions of this sort may invite many correct responses. We present one possibility: First of all, patients are unlikely to participate in an experiment where they are told they are getting a placebo. Although some may consent to take a harmless pill, few would submit to an injection they know is useless. But there is a second reason. There is currently a lively debate among medical professionals regarding the effectiveness of placebo treatment. Many reputable scientists believe that a sugar pill can relieve pain if the patient is told that the pill is indeed an effective remedy, but it is not clear whether a patient's belief in the effectiveness of a treatment has an actual physical effect. A number of physicians report that they routinely administer placebo treatment to their patients. In a clinical trial it is crucial to remove as much uncertainty as possible.

Further questions of this type may be found in the *What Do You Think?* section of exercises.

Try It Yourself answers

Try It Yourself 6.21: Determining statistical significance: Coins and cars

a. The probability of getting heads twice in a row is $1/4$, so this result is not statistically significant.

b. Certainly, 1951 Studebaker cars are few and far between, so it would be unusual enough to pass even one. Passing three such cars in a row is surely statistically significant.

Try It Yourself 6.22: Determining statistical significance: Vioxx and p-values If the Vioxx had no effect, the probability that the difference in results between the Vioxx and the placebo patients occurred by chance alone is 0.04 or 4%. Because this is less than the accepted threshold of 0.05, this result is statistically significant.

Try It Yourself 6.23: Understanding correlation as an idea: Heating bill They are negatively correlated.

Try It Yourself 6.24: Determining correlation from data: Running speed The two variables have a positive linear correlation.

Try It Yourself 6.25: Interpreting correlation: Diversifying index funds. Because there is a moderate positive linear correlation, there is no great advantage.

Exercise Set 6.4

Test Your Understanding

1. Explain in your own words the meaning of the p-value.

2. True or false: p-values are used to determine statistical significance.

3. True or false: If two events are positively correlated, then one of them must have caused the other.

4. True or false: If two variables are negatively correlated, then one of them tends to decrease as the other increases.

5. Explain in your own words what a double-blind trial is.

6. In a drug study, some people get the drug, and some do not. Those who do not get the drug are called the ______________ group.

7. When two variables have a correlation coefficient that is nearly zero, the two variables have: **a.** a positive correlation **b.** a negative correlation **c.** little if any linear correlation.

8. A *placebo* is: **a.** a drug that is being tested **b.** a benign substance **c.** a double-blind study.

Problems

9. **Determining statistical significance.** In a drug trial, 95% of patients taking the drug showed improvement while only 10% of patients who did not take the drug showed improvement. Would you consider this result to be statistically significant? Explain your reasoning.

10. **Statistical significance and p-values.** In a drug trial, patients showed improvement with a p-value of 0.02. Explain the meaning of the p-value in this trial. Would such a result normally be considered to be statistically significant?

11. **Ethics.** Suppose we want to test the idea that unregulated diets for children lead to serious health problems. Comment on the ethics of the following experimental design. We select two groups of small children. We let one group eat whatever they want whenever they want it. The second group eats what nutritionists and physicians tell us is a healthy diet. At the end of a few years, we analyze the health of each group.

Can you think of a more ethical design?[16]

12. **p-value.** A study to test the efficacy of a drug for the treatment of allergic rhinitis (inflammation of the nasal mucous membranes) reported a p-value as 0.012. Interpret carefully what this means.

13. **An interesting drug.** Does anything appear to be wrong in the following news release? Explain. *A drug company announced the results of a clinical trial for a certain drug for hangnails. Patients showed statistically significant improvement (p-value > 0.05) in the treatment group as compared to the control group.*

14. **What caused it?** In a clinical trial for patients with hangnails, the test group soaked their hands in warm water and were given an experimental drug. They showed statistically significant improvement (p-value < 0.03) as compared to the control group, which did nothing to treat their hangnails. The president of the drug company sponsoring this trial asserted that this p-value shows that the new drug had a 97% probability of helping hangnails. Discuss the validity of the president's comment.

15. **Sunburn.** The CEO of a certain drug company announced that the results of a clinical trial for an experimental drug in patients with sunburn showed statistically significant improvement. His justification for this claim was the following fact: 55% of the subjects treated with the drug got better in two days, but it is well known that, on average, only 50% of sunburn victims get better in two days. Discuss the validity of the CEO's comment. *Suggestion*: To find the p-value, see the discussion on page 452 of a test drug for preventing colds.

16. **What's the cause?** A certain drug company announced that the results of a clinical trial for a new drug in patients with sunburn showed statistically significant improvement (p-value $= 0.04$). The president of the company asserted that this shows that 96% of the subjects in the trial were helped by the drug. Discuss the validity of the president's comment.

17. **Reducing ovarian cancer risk.** This exercise refers to In the News 6.9, which is an excerpt from the Web site of the National Cancer Institute.

According to the report, women who take aspirin daily may reduce their risk of ovarian cancer. But the report states that the finding for NSAIDs "did not fall in a range that was significant statistically." Explain in your own words what this statement means and what implications it has.

IN THE NEWS 6.9

Search

Home | TV & Video | U. S. | World | Politics | Entertainment | Tech | Travel

NIH Study Finds Regular Aspirin Use May Reduce Ovarian Cancer Risk

Women who take aspirin daily may reduce their risk of ovarian cancer by 20 percent, according to a study by scientists at the National Cancer Institute (NCI), part of the National Institutes of Health. However, further research is needed before clinical recommendations can be made. The study was published February 6, 2014, in the Journal of the National Cancer Institute.

. . .

Among study participants who reported whether or not they used aspirin regularly: 18 percent used aspirin, 24 percent used non-aspirin NSAIDs [nonsteroidal anti-inflammatory drugs], and 16 percent used acetaminophen. The researchers determined that participants who reported daily aspirin use had a 20 percent lower risk of ovarian cancer than those who used aspirin less than once per week. For non-aspirin NSAIDs, which include a wide variety of drugs, the picture was less clear: the scientists observed a 10 percent lower ovarian cancer risk among women who used NSAIDs at least once per week compared with those who used NSAIDs less frequently. However, this finding did not fall in a range that was significant statistically. In contrast to the findings for aspirin and NSAIDs, use of acetaminophen, which is not an anti-inflammatory agent, was not associated with reduced ovarian cancer risk.

[16] *Many such totally ethical studies have been done, and the results surprise no one.*

18–19. Smarter people. *Exercises 18 and 19 refer to the following information. A 2011 article available from the National Institutes of Health discusses the benefits of breastfeeding during the first year of an infant's life. The article cites controversial studies of the effects of breastfeeding on cognitive development and IQ. Part of the controversy is due to the methodologies used in the studies, and the real possibility that women who do breastfeed are inherently different, in terms of their own habits and lifestyles, from those who do not breastfeed. For instance, in one key study of almost 500 infants of non-smoking women who gave their children only breast milk, tests showed between 2.1 and 3.8 higher IQ points in these children than those who were not breastfed exclusively (they were fed breast milk plus infant formula). In this and similar studies, the overall IQ advantage from breastfeeding appears to be small, but the effect is highly significant from a public health perspective. The controversy surrounding the benefits of breastfeeding requires one to critically consider the cause-and-effect relationships between breastfeeding and cognitive outcomes.*

18. One paragraph in the article states, "the above arguments require one to critically consider the cause-and-effect relationships between breastfeeding and cognitive outcomes." Use the concepts developed in this section to explain what this paragraph means.

19. Using your own words, explain what the next-to-last sentence from the summary above means.

20. Calculating p-values. Suppose that for a certain illness, the probability is 45% that a given patient will improve without treatment. Then the probability that at least n out of 20 patients will improve without treatment is given in the following table:

n	10	11	12	13	14	15	16
Probability at least n improve	0.41	0.25	0.13	0.058	0.021	0.0064	0.0015

We give an experimental drug to 20 patients who have this illness.

a. Suppose 11 patients show improvement. What is the p-value?

b. Suppose we count the test significant if the p-value is 0.05 or less. How many patients must show improvement in order to make the test statistically significant?

c. Suppose 16 of the 20 subjects showed improvement. In light of the design of the test, would it be appropriate to recommend this drug as a "cure" for this disease?

21. Correlation as an idea. For each pair of variables, determine whether they are positively correlated, negatively correlated, or uncorrelated.

a. Brightness of an oncoming car's headlights and your distance from the oncoming car

b. The number of home runs hit by Barry Bonds in a year and my water bill for that year

c. Damage in a city at the epicenter of an earthquake and the Richter scale reading of the earthquake

d. A golfer's tournament ranking and her score

22. Intensity of light. If one moves away from a light source, the intensity of light striking one's eye decreases. **Figure 6.50** shows a graph of the relationship between the intensity of light and distance from the source. (We have not included the units because they are not relevant here.)

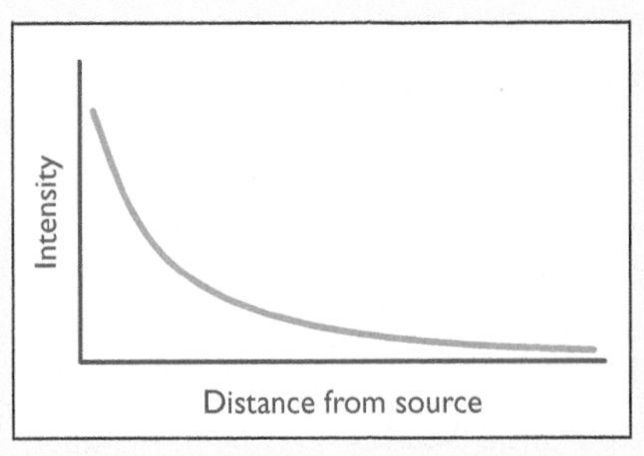

FIGURE 6.50 Intensity versus distance.

Is there a correlation between the intensity of light and distance from the source? If so, is the correlation positive or negative?

23. Correlation coefficient. Assume that **Figures 6.51 through 6.54** show data sets with correlation coefficients 0.99, 0.58, −0.10, and −0.63. Which figure shows data with which correlation coefficient?

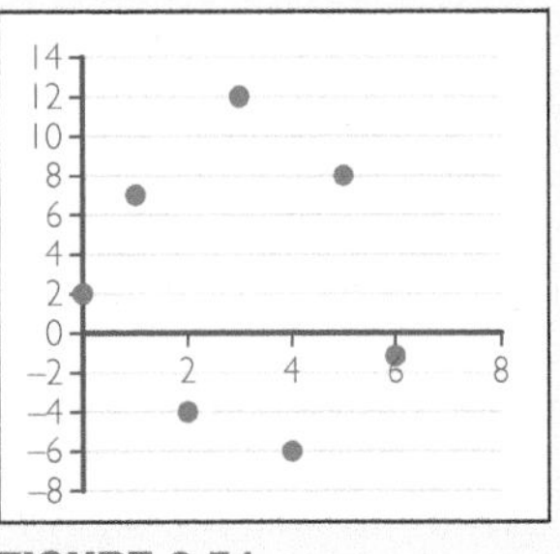

FIGURE 6.51

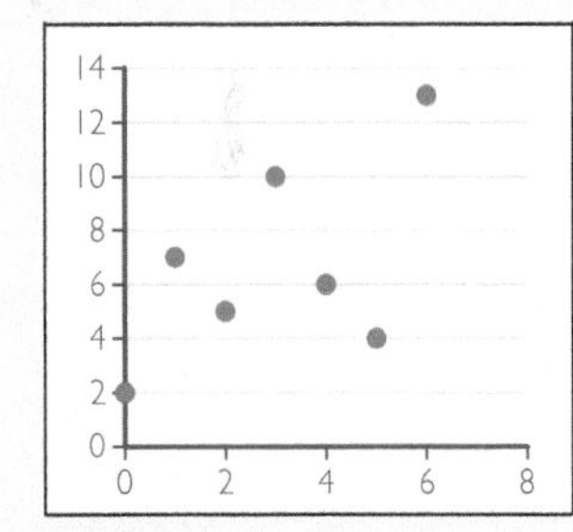

FIGURE 6.52

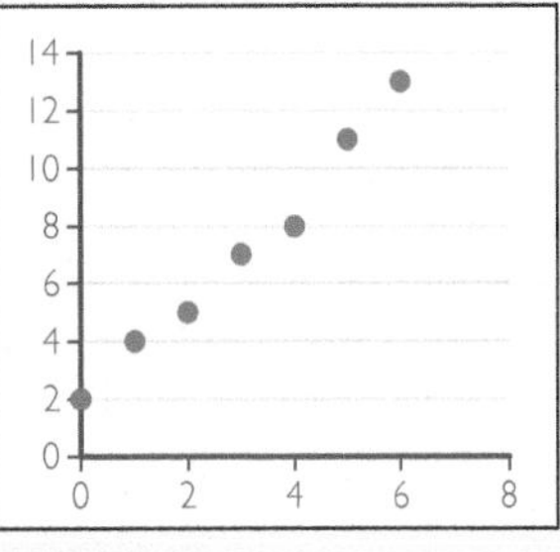

FIGURE 6.53

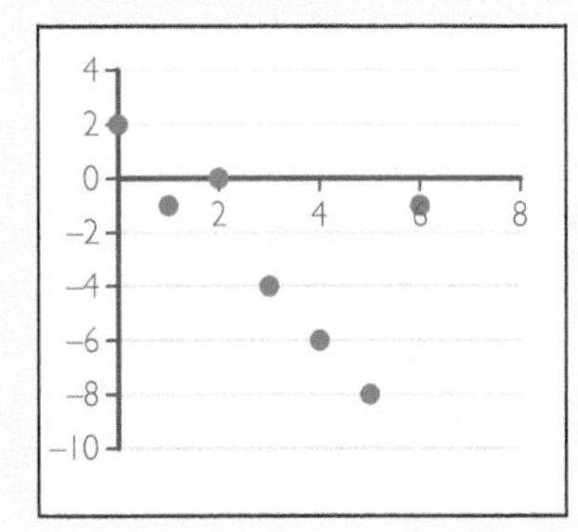

FIGURE 6.54

24. Advertising expense. Figure 6.55 shows the effectiveness of spending on advertising by major American retail companies.

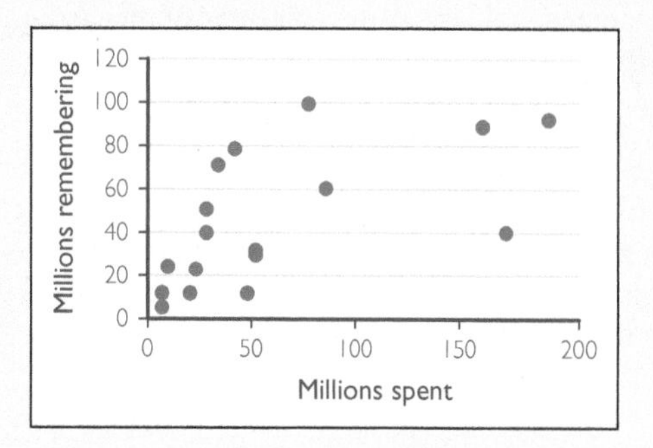

FIGURE 6.55 Spending and effectiveness.

a. Judging solely on the basis of the figure, does there appear to be a positive correlation between spending and effective advertising?

b. In Figure 6.56 we have added the trend line. As the figure indicates, the equation for the trend line is $y = 0.3632x + 22.163$, where y is the number (in millions) remembering and x is the amount (in millions of dollars) spent on advertising. The next-to-last point on the graph (well below the line) corresponds to Ford Motor Company. Does it appear that Ford was getting good value from its advertising dollars?

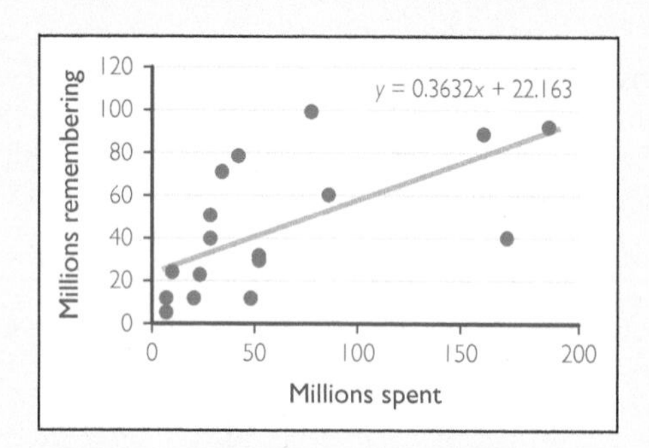

FIGURE 6.56 Regression line added.

c. Use the regression line in Figure 6.56 to estimate how many millions of people Ford should have expected to remember one of its ads. (See part b.) Round your answer to the nearest 10 million of those remembering.

d. The correlation coefficient was reported as 0.65. Comment on the hypothesis that there is a straight-line relation between advertising spending and effectiveness.

25. A study? Many times you hear testimonials regarding the health benefits of certain activities, foods, or herbal supplements. Suppose your trusted friend Jane tells you that in spite of a good diet and exercise and treatment from her physician, her blood pressure remained high. Then one year ago, she started a daily intake of 2 ounces of dried lemon peel. Her blood pressure returned to normal within 3 months and has remained so.

Discuss the pros and cons of Jane's testimonial regarding the curative properties of dried lemon peel in light of the concepts presented in this section.

26. Vegetarians. The following excerpt comes from an article published by the Vegetarian Society of the United Kingdom.

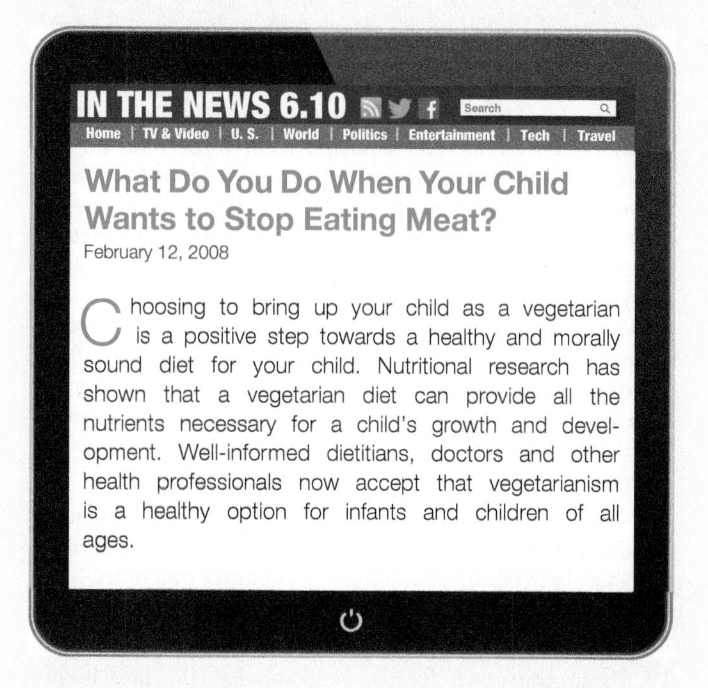

What Do You Do When Your Child Wants to Stop Eating Meat?

February 12, 2008

Choosing to bring up your child as a vegetarian is a positive step towards a healthy and morally sound diet for your child. Nutritional research has shown that a vegetarian diet can provide all the nutrients necessary for a child's growth and development. Well-informed dietitians, doctors and other health professionals now accept that vegetarianism is a healthy option for infants and children of all ages.

Your good friend Larry tells you that this article has convinced him to raise his child as a vegetarian. Discuss what you would say to Larry in light of the concepts presented in this section.

27. Potatoes contribute to athletics. Suppose a study showed that more than 90% of gold medal winners from the 2020 Summer Olympics in Tokyo eat potatoes as a regular part of their diet. In light of the concepts presented in this section, discuss whether it's proper to conclude that eating lots of potatoes will improve one's athletic abilities.

28. Flu. Suppose a certain medical study shows that more than 90% of patients with mild to moderate flu recover within 10 to 12 days if they are exposed to fresh, clean air. What can we conclude from this information about the relationship between breathing clean air and curing the flu? Explain the reasons for your answer.

Exercises 29 through 31 are designed to be solved using technology such as calculators or computer spreadsheets.

29. Sales and price. The following table shows the number of units of a certain item sold by a store over a month in terms of the price in dollars charged by the store.

Price in dollars	1.00	1.10	1.20	1.30	1.40
Number sold	320	305	270	265	220

Use technology to find the trend line and correlation coefficient for the data. (Round the correlation coefficient to two decimal places.) Plot the data along with the trend line. What type of correlation is there between the price and the number sold? Interpret your answer.

30. Calculating p-values. If a fair coin is tossed 100 times, the probability that one gets at least 60 heads is given by

$$\frac{100!}{60!40!}0.5^{60}0.5^{40} + \frac{100!}{61!39!}0.5^{61}0.5^{39}$$
$$+ \frac{100!}{62!38!}0.5^{62}0.5^{38} + \cdots + \frac{100!}{100!0!}0.5^{100}0.5^{0}$$
$$= 0.5^{100}\left(\frac{100!}{60!40!} + \frac{100!}{61!39!} + \frac{100!}{62!38!} + \cdots + \frac{100!}{100!0!}\right).$$

This is the p-value if you toss 100 coins and get 60 heads. Use technology to calculate this number. Round your answer to three decimal places.

31. Confirming example. Consider the table of data on page 454 showing weekly expenditures on alcohol and tobacco. Use technology to find the trend line and correlation coefficient for the data. Round the coefficients for the trend line and the correlation coefficient to two decimal places.

Writing About Mathematics

32. A research project. DPT (diphtheria, pertussis, tetanus) injections for infants are strongly recommended by the medical community. Sadly, many parents report cases of severe illness or even death following the injection. As a potential parent, you may well be in a position to make this decision for your child. Use the Web and other available sources to research this issue carefully and write a brief report on your conclusions. Pay particular attention to the design of any of these studies.

33. A research project. Statisticians are quick to point out that p-values are sometimes misunderstood. For example, in a drug trial, the p-value is *not* the probability that the drug is ineffective. You will find many references to the incorrect use of p-values on the Web. Investigate these and write a report on your findings.

34. A research project. There has been a great deal of controversy concerning a possible link between autism and certain vaccinations. Use the Internet and other available sources to research this issue carefully and write a brief report on your conclusions.

35. History. Sir Ronald A. Fisher was a towering figure in the history of statistics. He helped develop statistics as a separate discipline, thought through the concepts involved in randomized experiments, and contributed in a fundamental way to the notions of inference we discuss in this chapter. Write a brief biography of Fisher highlighting his contributions to statistical inference.

36. Tyson Foods v. Bouaphakeo. In March 2016, the United States Supreme Court ruled on a case in which thousands of workers at an Iowa pork processing plant banded together in a single lawsuit to recover overtime pay from Tyson Foods. Justice Anthony Kennedy, writing for the majority in the 6–2 decision, said the plaintiffs were entitled to rely on statistics to prove their case. He wrote, "A representative or statistical sample, like all evidence, is a means to establish or defend against liability." Do some research to learn more details about this case, which is called *Tyson Foods v. Bouaphakeo,* No. 14-1146, and write a report on it.

37. Correlation and causation. Make a list of several pairs of events that, in your observation, often occur together, but for which no causal relationship exists. In each case, explain the reasons why you believe there is no causation.

What Do You Think?

38. Double blind. Why must doctors supervising a clinical trial not be told which patients are getting a placebo?

39. Diet. An ad for a weight-loss drug claims that clinical trials showed that their drug was highly effective, when combined with a program of exercise and healthy diet. Does this convince you that the drug works? Explain.

40. Statistically significant. Give an example of a result that is not statistically significant. Explain the connection between the notion of statistical significance and p-values.

41. p-values. Explain in practical terms the meaning of p-values. **Exercises 10 and 12 are relevant to this topic.**

42. Correlation and causation. Discuss the relationship between causation and correlation. Provide examples of variables that are positively correlated but for which causation is in doubt. **Exercise 21 is relevant to this topic.**

43. Correlation and mortgage. You borrow money to buy a home, and you repay the loan over 30 years. Consider two variables: the amount you still owe on your home and the number of monthly payments you have made. Are these variables positively correlated or negatively correlated? Do you expect them to have a strong linear correlation? *Reminder*: In the early years, almost all of your monthly payment goes toward interest, not toward loan repayment. That proportion changes dramatically in later years.

44. Choosing a treatment. You are considering treating a medical condition using a certain drug. It has been advertised extensively on TV, and the advertisements are accompanied by reliable testimonials. But the drug has been the subject of extensive double-blind studies by medical professionals, and the *New England Journal of Medicine* reports that the drug is ineffective and possibly harmful. Would you use the drug? Explain your answer.

45. Linear regression. You assert that there is a linear correlation between two variables. A classmate says that you couldn't really know because linear regression can be performed on any pair of variables. How would you respond?

46. Colds and coins. In what ways is medical testing similar to, or different from, flipping coins? **Exercise 30 is relevant to this topic.**

CHAPTER SUMMARY

Descriptive statistics refers to organizing data in a form that can be clearly understood and that describes what the data show. *Inferential statistics* refers to the attempt to draw valid conclusions from the data, along with some estimate of how accurate those conclusions are. We study both in this chapter.

Data summary and presentation: Boiling down the numbers

Four important measures in descriptive statistics are the *mean, median, mode,* and *standard deviation.* The mean (also called the average) of a list of numbers is their sum divided by the number of entries in the list. The median of a list of numbers is the "middle" of the list. There are as many numbers above it as there are below it. The mode is the most frequently occurring number in the list (if there is one). The standard deviation σ of a list of numbers measures how much the data are spread out from the mean. It is calculated as follows: Subtract the mean from each number in the list, square these differences, add all these squares, divide by the number of entries in the list, and then take the square root. As a formula, this is

$$\sigma = \sqrt{\frac{(x_1 - \mu)^2 + (x_2 - \mu)^2 + \cdots + (x_n - \mu)^2}{n}}$$

where the data points are $x_1, x_2, x_3, \ldots, x_n$ and μ is the mean.

A *histogram* is a bar graph that shows frequencies with which certain data appear. Histograms can be especially useful when we are dealing with large data sets.

A common method of describing data is the *five-number summary.* It consists of the minimum, the first quartile, the median, the third quartile, and the maximum. The first quartile is the median of the lower half of the list of numbers, and the third quartile is the median of the upper half. A *boxplot* is a pictorial display of a five-number summary.

The normal distribution: Why the bell curve?

The *normal distribution* arises naturally in analyzing data from standardized exam scores, heights of individuals, and many other situations. A plot of normally distributed data shows the classic *bell-shaped curve.* For normally distributed data, the mean and median are the same, and in fact the data are symmetrically distributed about the mean. Also, most of the data tend to be clustered relatively near the mean.

A normal distribution is completely determined by its mean and standard deviation. The size of the standard deviation tells us how closely the data are "bunched" about the mean. A larger standard deviation means a "fatter" curve, as shown in **Figure 6.57**, and a smaller standard deviation means a "thinner" curve, as shown in **Figure 6.58**, where the data are more closely bunched about the mean.

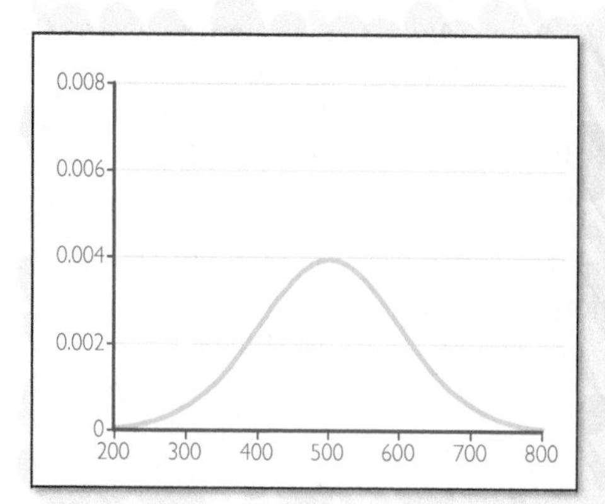

FIGURE 6.57 Normal distribution with mean 500 and standard deviation 100.

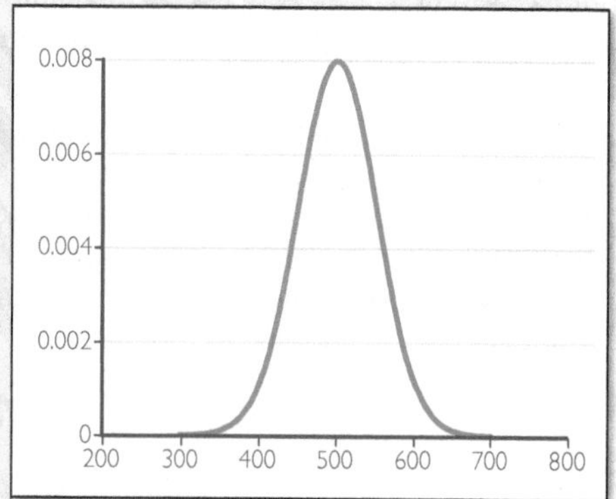

FIGURE 6.58 Normal distribution with mean 500 and standard deviation 50.

More precisely, for a normal distribution:

- About 68% of the data lie within one standard deviation of the mean.
- About 95% of the data lie within two standard deviations of the mean.
- About 99.7% of the data lie within three standard deviations of the mean.

For example, for the verbal section of the SAT Reasoning Test, scores were normally distributed with mean 507 and standard deviation 111. Thus, about 68% of scores fell between $507 - 111 = 396$ and $507 + 111 = 618$.

For a more exact tabulation of how data are distributed about the mean, we use *z-scores*. The z-score for a data point is the number of standard deviations that the data point lies above or below the mean. The z-score, together with a standard table of z-scores and percentiles, allows easy and complete analysis of normally distributed data.

One of the most striking facts about normal distributions is that, according to the *Central Limit Theorem*, data that consist of percentages of population samples (all of the same size) are approximately normally distributed. This allows us to use the normal distribution to analyze virtually any type of data from repeated samples.

The statistics of polling: Can we believe the polls?

Inferential statistics is used in attempting to draw conclusions about an entire population by collecting data from a small sample of the population. One of the most common applications in the media is polling.

Polling involves a *margin of error*, a *confidence level*, and a *confidence interval*. To say that a poll has a margin of error of 3% with a confidence level of 95% means that if we conducted this poll 100 times, then we expect about 95 of those sample results to be within three percentage points of the true percentage having the given property. To find the confidence interval, adjust the result of the poll by adding and subtracting the margin of error.

For a 95% level of confidence, the margin of error when we poll n people is about $100/\sqrt{n}\%$, and the sample size needed to get a margin of error of m percentage points is about $(100/m)^2$.

There are significant and difficult questions of methodology involved in obtaining accurate results from polling.

Statistical inference and clinical trials: Effective drugs?

Another important use of inferential statistics is in clinical trials. Medical testing uses the notions of *statistical significance* and *p-values*. If the results of a clinical trial are unlikely to have occurred by chance alone, the results are considered to be statistically significant. Roughly speaking, the p-value measures the probability that the outcome of a clinical trial would occur by pure chance if the treatment had no effect. A small p-value is usually interpreted as evidence that it is unlikely the results are due to chance alone. The results are usually considered statistically significant if the p-value is 0.05 or smaller.

Two numerical variables are *positively correlated* if an increase in one of them accompanies an increase in the other. Two numerical variables are *negatively correlated* if an increase in one of them accompanies a decrease in the other. If neither of these is true, the variables are *uncorrelated*. Linear regression can quantify the extent to which two variables are *linearly* correlated.

Correlation can be the result of real connections between two variables, but some correlations are the result of chance or coincidence. Even when two variables are connected in some way, we should not assume causation.

KEY TERMS

CHAPTER QUIZ

1. The goals scored (by either team) in the games played by a soccer team are listed below according to the total number of goals scored in each game. Find the mean, median, and mode for these data. Round the mean to one decimal place.

Goals scored by either team	0	1	2	3	4	5	6	7	8
Number of games	7	14	11	12	3	2	1	2	2

Answer Mean: 2.4; median: 2; mode: 1.

If you had difficulty with this problem, see Example 6.3.

2. Calculate the five-number summary for this list of automobile prices:

$27,000 $19,500 $24,500 $25,600 $17,000
$32,700 $18,000 $27,800 $29,000 $30,200

Answer The minimum is $17,000; the first quartile is $19,500; the median is $26,300; the third quartile is $29,000; the maximum is $32,700.

If you had difficulty with this problem, see Example 6.4.

3. Calculate the mean and standard deviation for this list of automobile prices:

$27,000 $19,500 $24,500 $25,600 $17,000
$32,700 $18,000 $27,800 $29,000 $30,200

Answer Mean: $25,130; standard deviation: $5078.

If you had difficulty with this problem, see Example 6.6.

4. The weights of oranges in a harvest are normally distributed, with a mean weight of 150 grams and standard deviation of 10 grams. In a supply of 1000 oranges, how many will weigh between 140 and 160 grams? How many will weigh between 130 and 170 grams?

Answer 680 oranges between 140 and 160 grams; 950 oranges between 130 and 170 grams.

If you had difficulty with this problem, see Example 6.10.

5. The weights of oranges in a harvest are normally distributed, with a mean weight of 150 grams and standard deviation of 10 grams. Calculate the z-score for an orange weighing

a. 160 grams

b. 135 grams

Round your answers to one decimal place.

Answer

a. 1

b. −1.5

If you had difficulty with this problem, see Example 6.11.

6. For a certain disease, 40% of untreated patients can be expected to improve within a week. We observe a population of 80 untreated patients and record the percentage who improve within one week. According to the Central Limit Theorem, such percentages are approximately normally distributed. Find the mean and standard deviation of this normal distribution. Round the standard deviation to two decimal places.

Answer The mean is 40%. The standard deviation is $\sqrt{(40 \times 60)/80}$ or about 5.48 percentage points.

If you had difficulty with this problem, see Example 6.14.

7. Use the mean and standard deviation calculated in the preceding exercise to find the percentage of test groups of 80 untreated patients in which 50% or less improve within a week. Round your answer as a percentage to one decimal place.

Answer z-score $= 10/5.48$ or about 1.8 standard deviations above the mean. Table 6.2 (page 422) gives a corresponding percentage of about 96.4%.

If you had difficulty with this problem, see Example 6.14.

8. Explain the meaning of a poll that says 55% of Americans approve of the job the president is doing, with a margin of error of 3% and a confidence level of 95%.

Answer In 95% of such polls, the reported approval of the president will be within three percentage points of the true approval level. Thus, we can be 95% confident that the true level is between 52% and 58%.

If you had difficulty with this problem, see Example 6.16.

9. A poll asked 800 people their choice for mayor. What is the approximate margin of error for a 95% confidence level? Round your answer as a percentage to one decimal place.

Answer 3.5%.

If you had difficulty with this problem, see Example 6.17.

10. What sample size is needed to give a margin of error of 2.5% with a 95% confidence level?

Answer 1600.

If you had difficulty with this problem, see Example 6.20.

11. For each of these situations, determine whether the results described appear to be statistically significant.
a. You are walking down the street and encounter three people in a row with blond hair.

b. You cut a deck of playing cards 30 times and get an ace every time.

Answer
a. Not statistically significant

b. Statistically significant

If you had difficulty with this problem, see Example 6.21.

12. A clinical trial for an experimental drug finds that the drug is effective with a p-value of 0.08. Explain the meaning of the p-value in this case. Would this result normally be accepted as statistically significant?

Answer The probability that the outcome of the clinical trial would occur by chance alone if the drug had no effect is 0.08 (8%). This result would not normally be accepted as statistically significant because the p-value is larger than 0.05.

If you had difficulty with this problem, see Example 6.22.

13. Decide whether the following variables are positively correlated, negatively correlated, or uncorrelated.
a. Your grades and the number of hours per week you study

b. Your grades and the number of hours per week you party

c. Your grades and the number of cookies per week you eat

d. The daily average temperature and sales of cell phones

e. The daily average temperature and sales of overcoats

f. The daily average temperature and sales of shorts

Answer Positively correlated: a and f; negatively correlated: b and e; uncorrelated: c and d.

If you had difficulty with this problem, see Example 6.23.

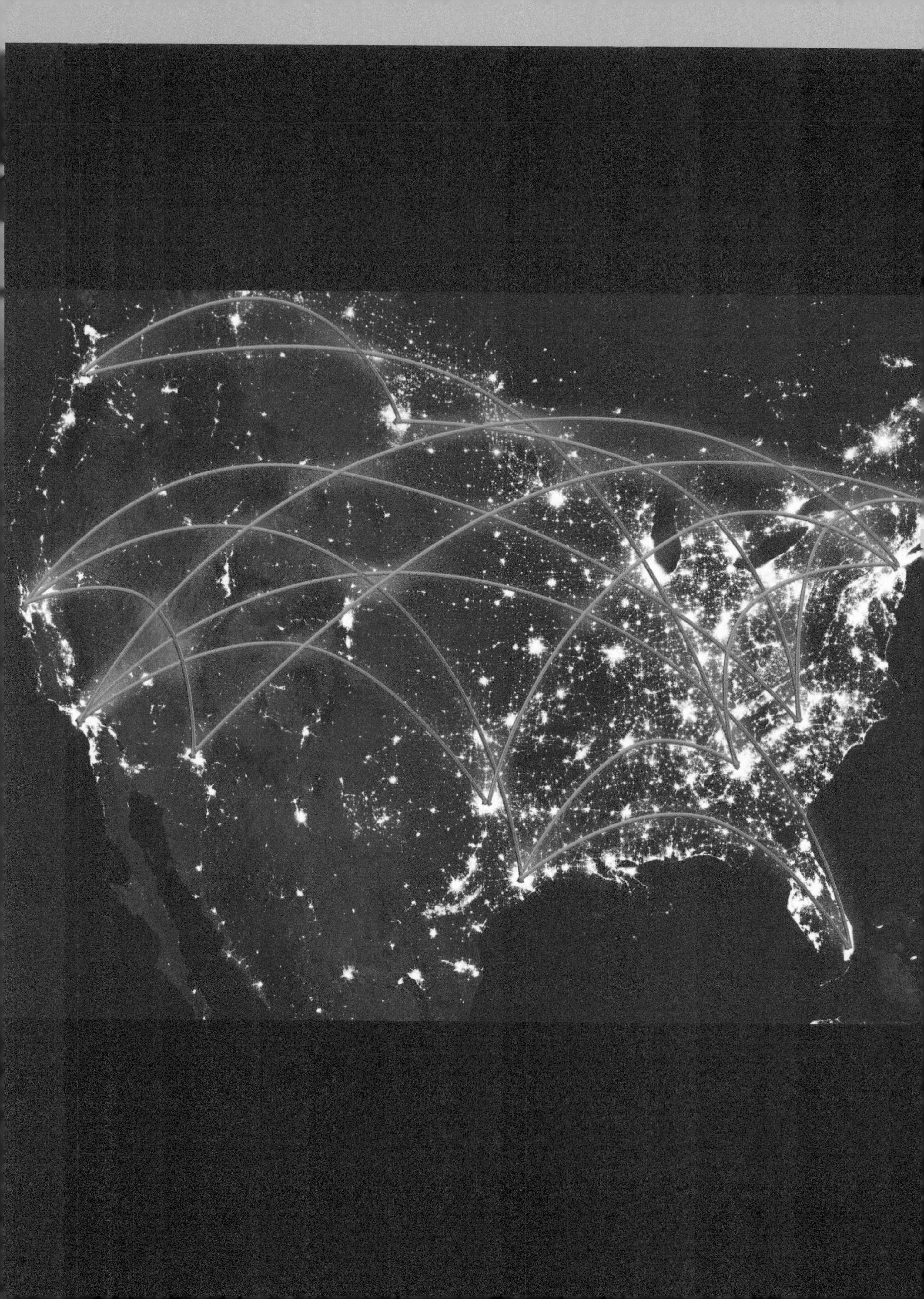

7 Graph Theory

7.1 Modeling with graphs and finding Euler circuits

7.2 Hamilton circuits and traveling salesmen: Efficient routes

7.3 Trees: Viral e-mails and spell checkers

David Malan/Photographer's Choice RF/Getty Images

The following appeared in a report published in 2010 by *The National Security Agency's Review of Emerging Technologies*.

IN THE NEWS 7.1

Search

Home | TV & Video | U. S. | World | Politics | Entertainment | Tech | Travel

Revealing Social Networks of Spammers

Spam doesn't really need an introduction—anyone who owns an email address likely receives spam emails every day. However, spam is much more than just an annoyance. Spam's hidden economic cost for companies in wasted storage, bandwidth, technical support, and most important, the loss of employee productivity, is astronomical.

Understanding the behavior of spammers is one of the benefits of a global approach to fighting spam. But what do the social networks of spammers look like? In particular, how well organized are spammers? Do they operate alone, or in groups?

The social network of spammers can be represented as a graph consisting of nodes and edges, as shown in the figure.

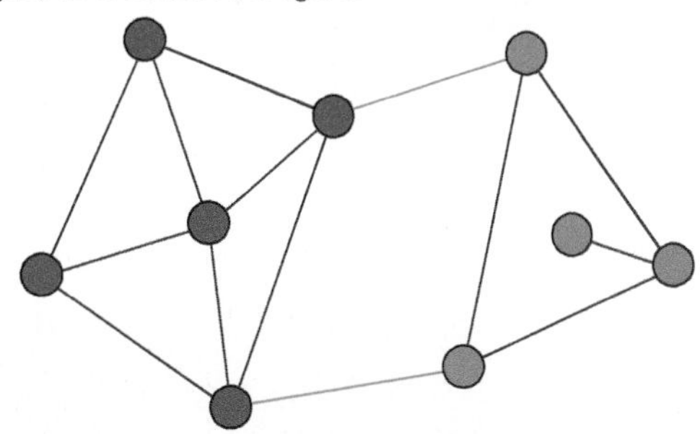

The nodes correspond to spammers, and an edge between two nodes corresponds to a social relationship between the corresponding spammers. A social relationship can be inferred by the use of common resources or by similar behavior patterns over time.

Communities in a social network emerge by partitioning the graph into groups of nodes. Sets of nodes in the same group are highly similar and sets of nodes in different groups are not similar. The main challenge in constructing the graph is choosing the edges because we cannot observe relationships among spammers. This problem does not arise in most other community detection studies. For example, in a friendship or collaboration networks, users willingly participate in the study, and information on relationships among members of the network is readily available.

This article shows one application of the mathematical concept of a *graph*, which in the simplest terms tells whether certain objects are somehow connected to each other.[1] For example, we can model Facebook using a graph if we say that two users are connected when they are friends. Graphs can be represented pictorially using dots and segments, as in the background of the image in **Figure 7.1**. The dots are called *vertices*, and the segments connecting them are called *edges*.

Graph theory has important real-world applications, and in this chapter we will examine some of the basic ideas behind them. Serious applications of graph theory involve highly complex situations, such as fast routing of dynamic content on the World Wide Web, with huge numbers of vertices and edges that can be dealt with only by very powerful computers. We will focus on everyday applications of graphs but also will explain their importance to business, government, and industry.

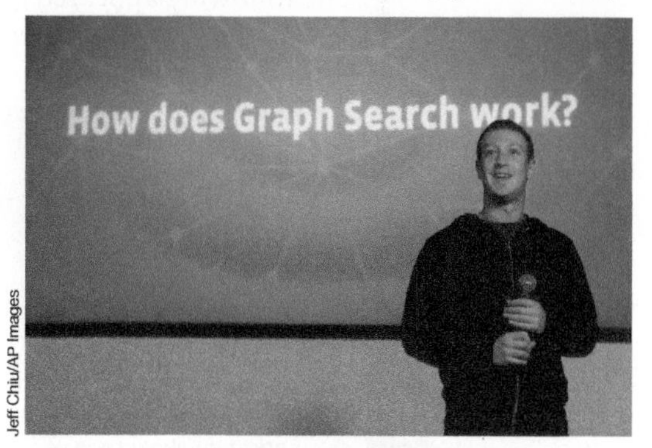

Jeff Chiu/AP Images

FIGURE 7.1 Facebook CEO Mark Zuckerberg speaks at Facebook headquarters in Menlo Park, California, on Tuesday, January 15, 2013.

7.1 Modeling with graphs and finding Euler circuits

TAKE AWAY FROM THIS SECTION
Know how to use graphs as models and how to determine efficient paths.

The following excerpt is from the Web site of the Aerospace Industries Association.

IN THE NEWS 7.2

Search

Home | TV & Video | U. S. | World | Politics | Entertainment | Tech | Travel

NextGen: The Future of Flying

The Next Generation Air Transportation System (NextGen) transforms the National Airspace System to meet future safety, security, capacity and environmental needs. The full implementation of NextGen will fundamentally change air traffic management by combining new technologies for surveillance, navigation and communications with procedural changes and airfield development.

(Continued)

[1] *In earlier chapters, we talked about a graph as a pictorial representation of data or a formula. In this chapter, we are going to use the term in a completely different way.*

IN THE NEWS 7.2

Home | TV & Video | U. S. | World | Politics | Entertainment | Tech | Travel

NextGen: The Future of Flying (*Continued*)

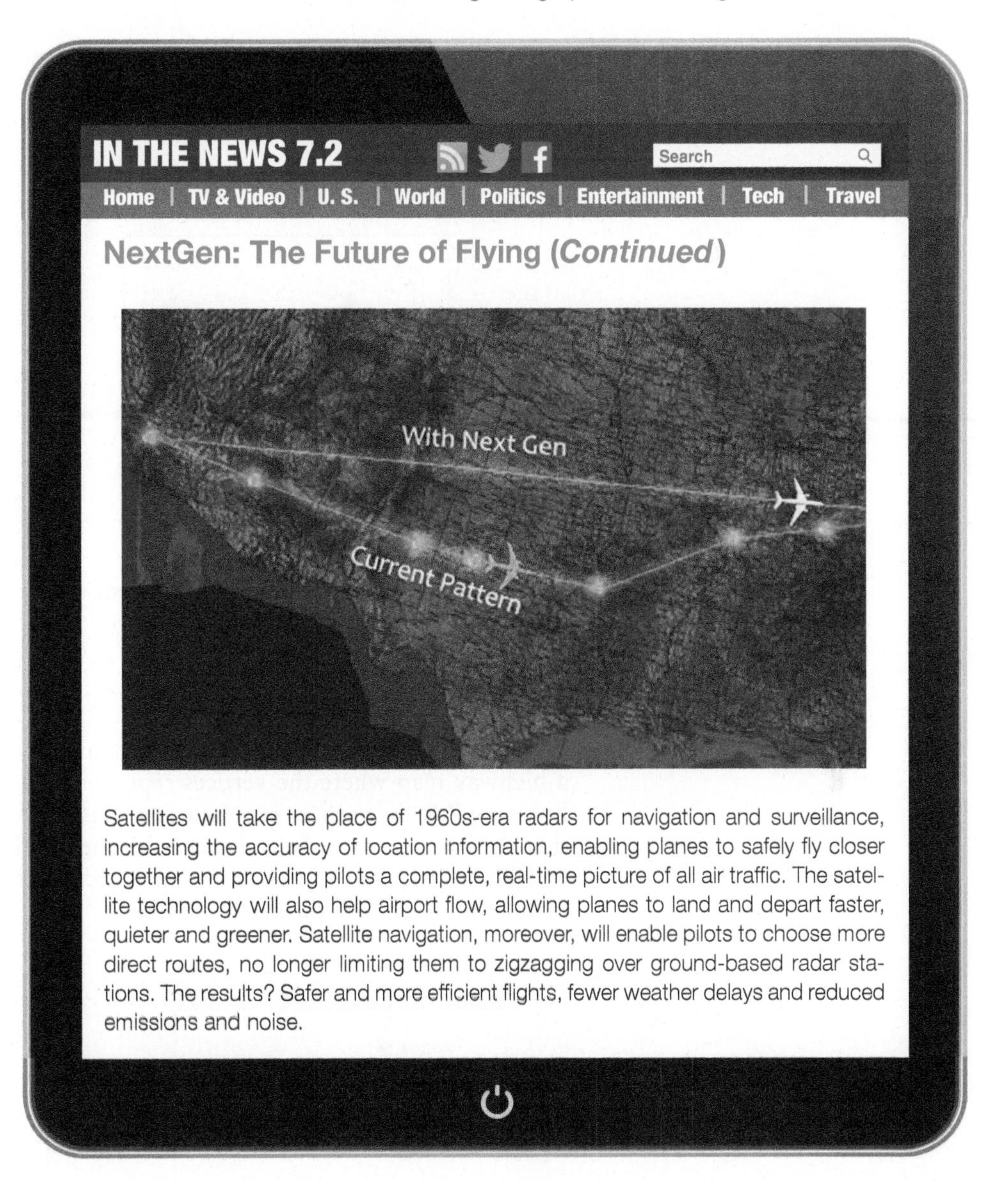

Satellites will take the place of 1960s-era radars for navigation and surveillance, increasing the accuracy of location information, enabling planes to safely fly closer together and providing pilots a complete, real-time picture of all air traffic. The satellite technology will also help airport flow, allowing planes to land and depart faster, quieter and greener. Satellite navigation, moreover, will enable pilots to choose more direct routes, no longer limiting them to zigzagging over ground-based radar stations. The results? Safer and more efficient flights, fewer weather delays and reduced emissions and noise.

Efficient routes for airplanes save money and get passengers to their destinations in a timely manner. Efficient routing is important for many business applications, from trucking to mail delivery. Typical options may consist of cities to pass through or highway routes to utilize. In this section, we study how to use graphs to find efficient routes.

Before we begin this study, let's agree that the graphs we will discuss have two properties: First, an edge cannot start and end at the same vertex. (See **Figure 7.2.**)

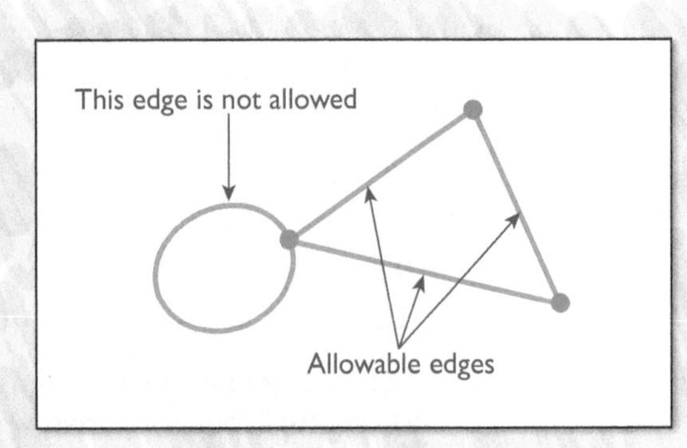

FIGURE 7.2 An edge is not allowed to start and end at the same vertex.

Second, the graphs are *connected*, that is, each pair of vertices can be joined by a sequential collection of edges, called a *path*. The graph in **Figure 7.3** is connected, but the one in **Figure 7.4** is not connected—it comes in two pieces.

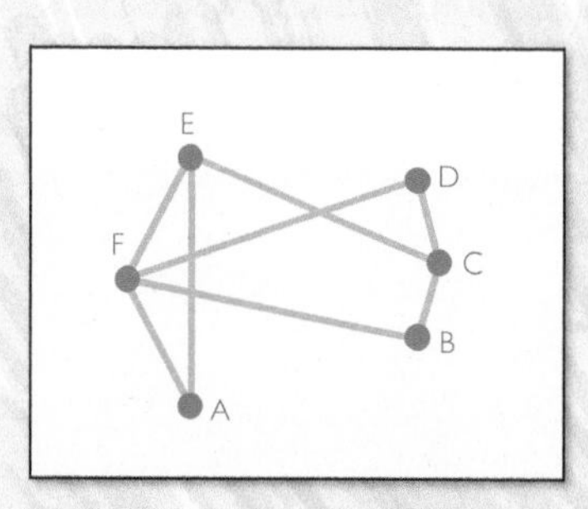

FIGURE 7.3 A connected graph.

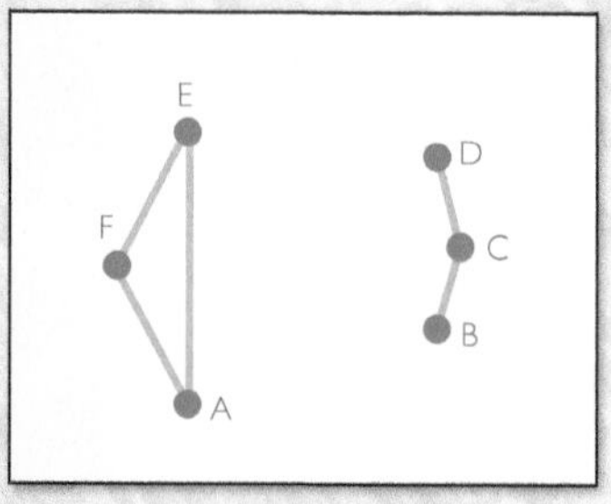

FIGURE 7.4 A disconnected graph.

The first step in our study of efficient routes is learning how to use graphs to make models of real situations.

Modeling with graphs

Graphs can be used to represent many situations. For example, **Figure 7.5** shows a highway map where the vertices represent towns and the edges represent roads. In **Figure 7.6**, the vertices represent some users of Facebook, and the edges indicate that the users are friends on Facebook. The figure tells us, for example, that Mary and Jill are friends because an edge connects them. Mary and Tom are not friends because no edge joins those two vertices.

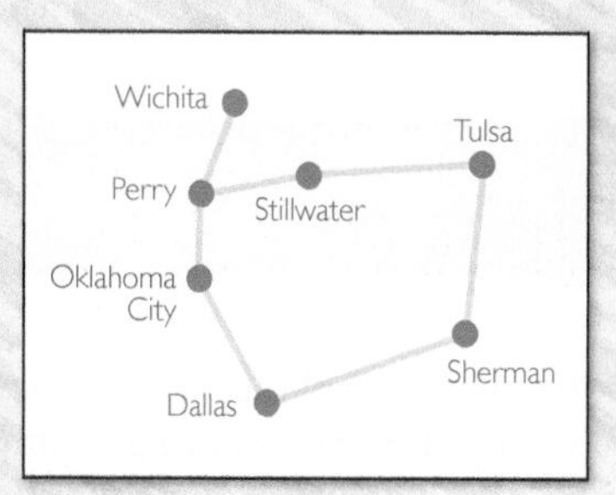

FIGURE 7.5 A graph representing towns and roads.

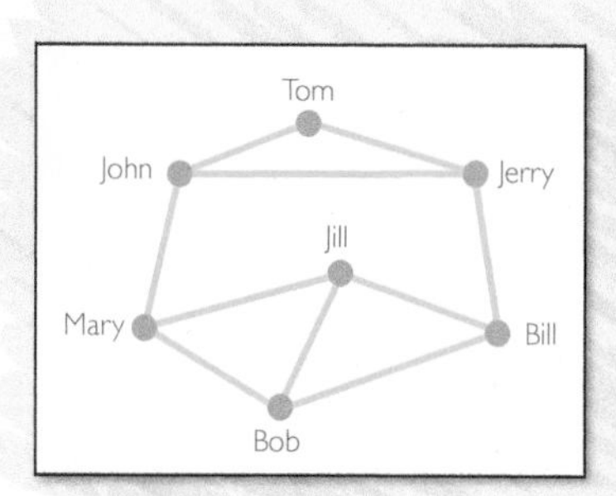

FIGURE 7.6 Facebook friends.

EXAMPLE 7.1 Making models: Little League

The Little League in our town has two divisions, the Red Division and the Blue Division. The Red Division consists of the Cubs, Lions, Bears, and Tigers. The Blue Division consists of the Red Sox, Yankees, and Giants. The inter-division schedule this season is:

Cubs play Red Sox, Yankees, and Giants.

Lions play Red Sox and Yankees.

Bears play Red Sox and Yankees.

Tigers play Red Sox and Giants.

Make a graphical representation of this schedule by letting the vertices represent the teams and connecting two vertices with an edge if they are scheduled to play this season.

SOLUTION

The appropriate graph is shown in **Figure 7.7**. Note, for example, that the edge joining the Cubs and Giants indicates they have a game scheduled. There is no edge joining the Tigers with the Yankees because they do not have a game scheduled.

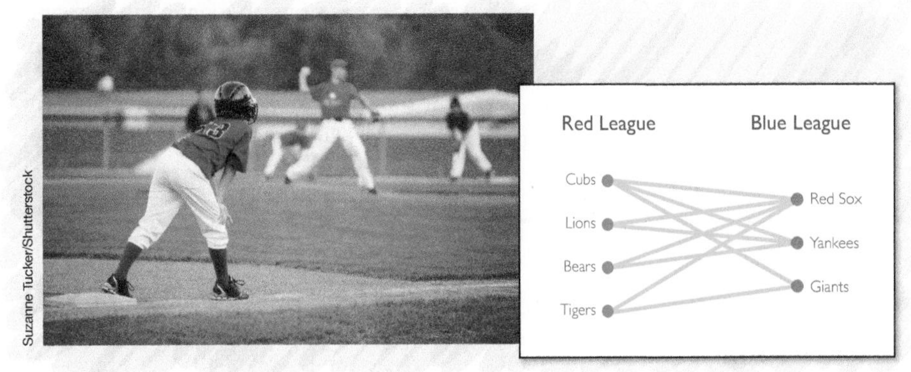

FIGURE 7.7 A Little League schedule.

TRY IT YOURSELF 7.1

Ferry routes connect Portsmouth and Poole, England, with Le Havre, Caen, Cherbourg, and St. Malo, France, and Bilbao, Spain (see the map below).

Portsmouth has ferries to and from Le Havre, Caen, Cherbourg, St. Malo, and Bilbao.

Poole has ferries to and from Cherbourg and St. Malo.

Make a graph that shows these ferry routes.

The answer is provided at the end of this section.

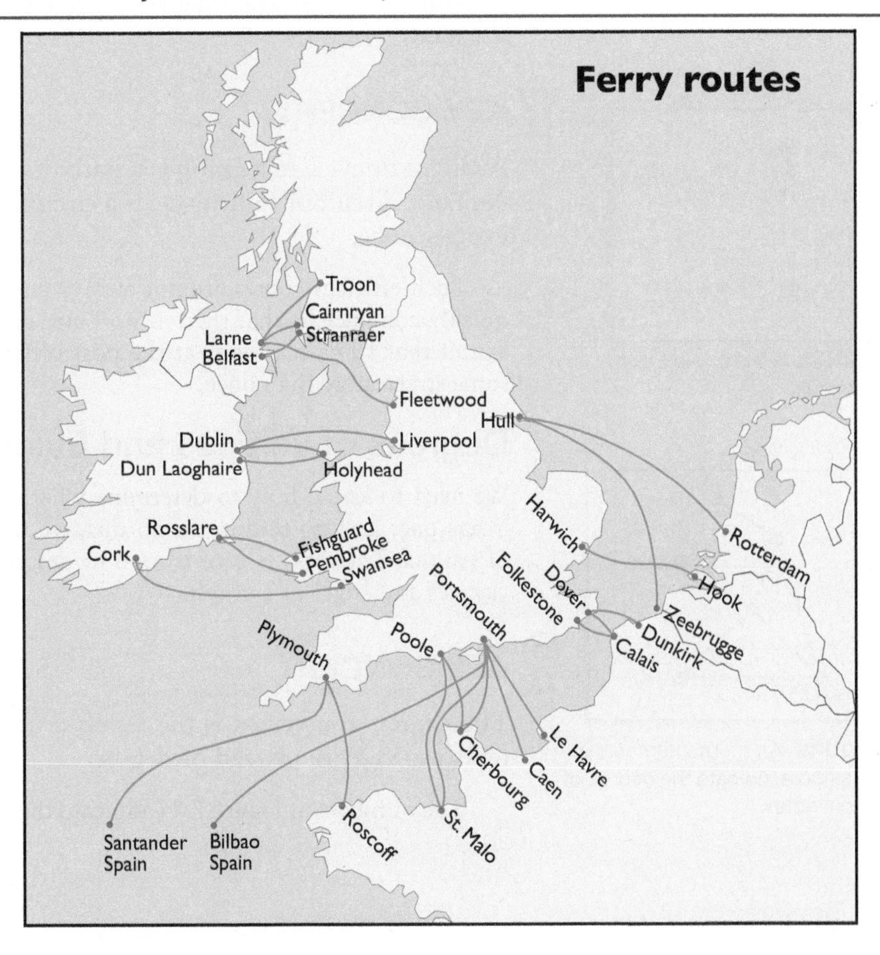

Euler circuits

We noted at the beginning of this section that graphs can be used to find efficient routes. Let's consider the example of mail delivery. The edges in **Figure 7.8** represent streets along which mail must be delivered, and the vertices represent the post office and street intersections. We have labeled one vertex as "post office" and the others as A through F. An efficient route for the mail carrier starts at the post office, travels along each street exactly once, and ends up back at the post office. Such a route is shown in **Figure 7.9**. We indicate the route by listing the vertices as we encounter them along the route:

Post office-A-B-C-D-B-E-F-Post office

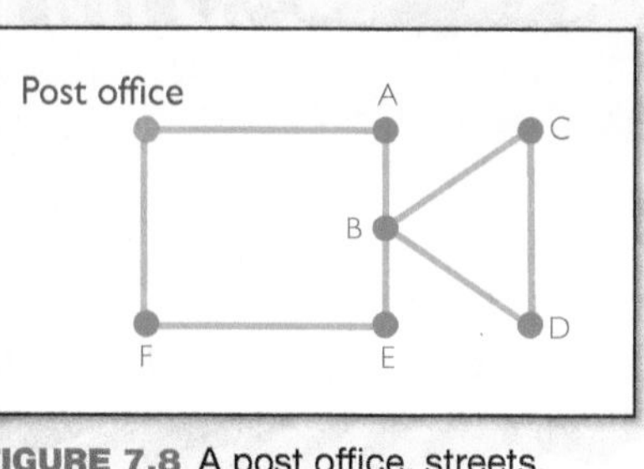

FIGURE 7.8 A post office, streets, and intersections.

FIGURE 7.9 An efficient mail route.

The path for this postal route starts and ends at the same vertex and traverses each edge exactly once. Such a path is known as an *Euler circuit*, or *Euler cycle*, after the eighteenth-century Swiss mathematician Leonhard Euler (pronounced "oiler"). Euler is considered to be the father of modern graph theory. Euler circuits represent efficient paths for postal delivery, street sweepers, canvassers, and lots of other situations.

KEY CONCEPT

A circuit or cycle in a graph is a path that begins and ends at the same vertex. An Euler circuit or Euler cycle is a circuit that traverses each edge of the graph exactly once.

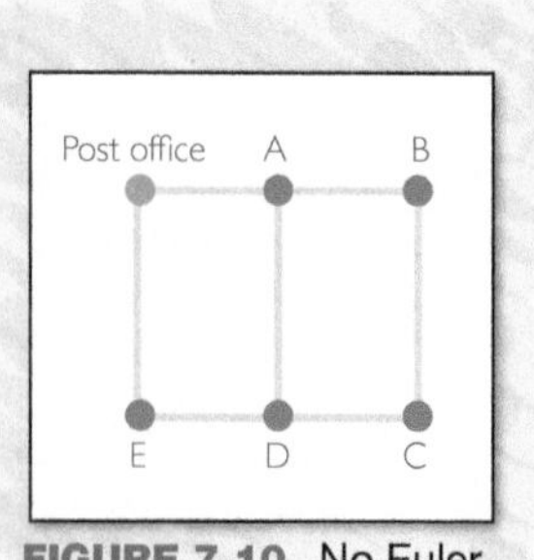

FIGURE 7.10 No Euler circuit.

Such efficient routes are not always possible. For example, trial and error can quickly convince us that there is no Euler circuit for the graph in **Figure 7.10**. Any circuit that starts and ends at the post office must either omit a street or traverse some street more than once.

Degrees of vertices and Euler's theorem

We need to know how to determine whether a graph has an Euler circuit and, if it has one, how to find it. To do this, we need to look more closely at properties of graphs. The term *degree* is used to describe an important relationship between vertices and edges of a graph.

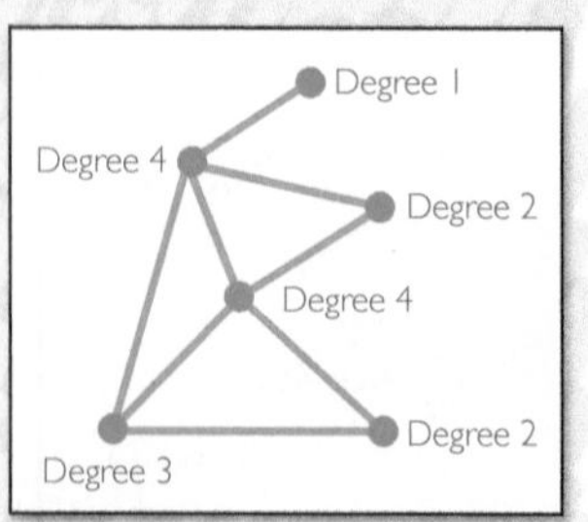

FIGURE 7.11 The numbers in this figure indicate the degree of each vertex.

KEY CONCEPT

The degree of a vertex is the number of edges that touch that vertex. Some texts use valence instead of degree.

The numbers in **Figure 7.11** indicate the degree of each vertex.

EXAMPLE 7.2 Finding the degree of a vertex: An air route map

The graph in **Figure 7.12** shows a simplified air route map. It shows connections that are available between various cities. In the context of this graph, what is the meaning of the degree of a vertex? Find the degree of each of the vertices.

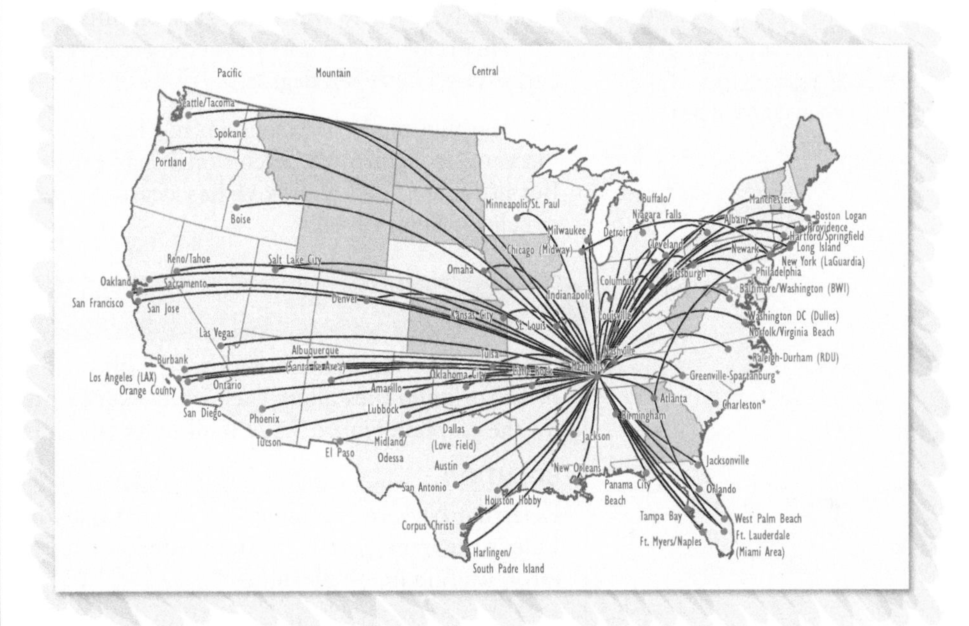

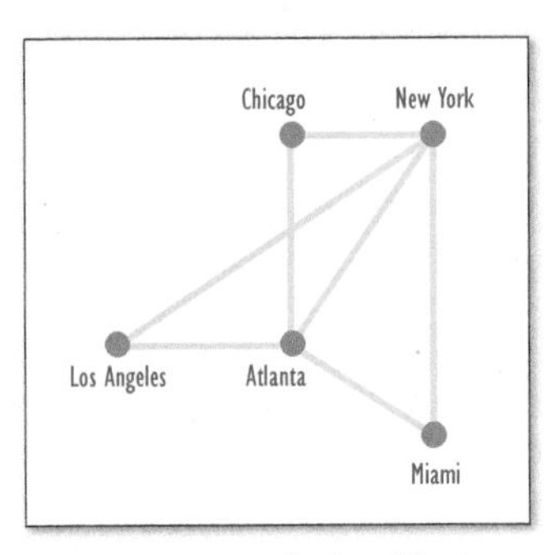

FIGURE 7.12 A simplified air route map.

SOLUTION

Because each edge indicates a direct connection between cities, the degree of each vertex is the number of direct flights available from the given city. The degrees of the vertices are: Atlanta: degree 4; Chicago: degree 2; Los Angeles: degree 2; Miami: degree 2; New York: degree 4.

TRY IT YOURSELF 7.2

Figure 7.13 shows a simplified map of Amtrak rail service. Find the degree of each vertex.

The answer is provided at the end of this section.

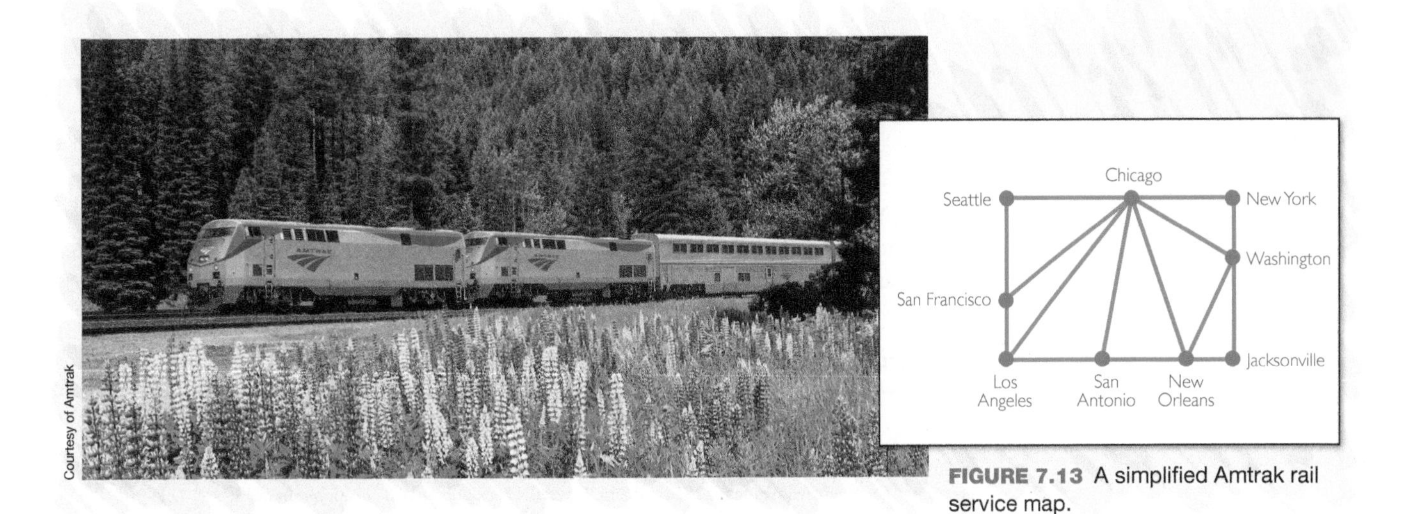

Courtesy of Amtrak

FIGURE 7.13 A simplified Amtrak rail service map.

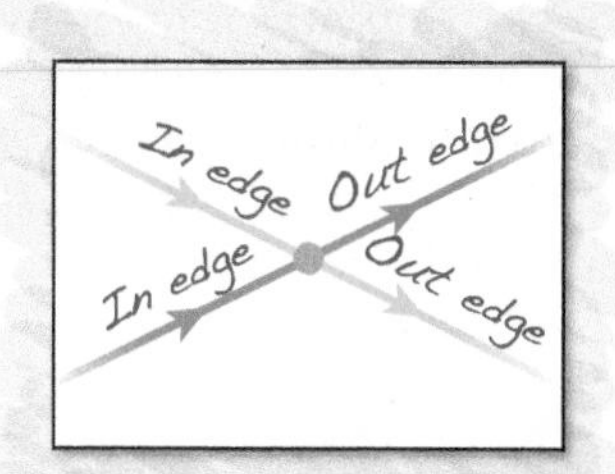

FIGURE 7.14 Edges of an Euler circuit come in pairs.

There is a crucial relationship between Euler circuits and the degrees of vertices. An Euler circuit goes into and then out of each vertex and uses each edge exactly once, as is illustrated in **Figure 7.14**. Because the edges meeting a vertex come in pairs (in-edges paired with out-edges), the number of edges at each vertex must be a multiple of 2. That is, the degree of each vertex is even.

In the eighteenth century, Leonhard Euler discovered that this degree condition completely characterizes when such circuits exist.

Theorem[2] (Euler, 1736): A connected graph has an Euler circuit precisely when every vertex has even degree.

The two mail routes shown in Figures 7.8 and 7.10 provide evidence for this theorem: In Figure 7.8, each vertex has even degree, and there is an Euler circuit. But in Figure 7.10, vertex A has degree 3, which is odd. Thus, there is no Euler circuit for Figure 7.10.

EXAMPLE 7.3 Applying Euler's theorem: Scenic trips

Does the air route map in Figure 7.12 allow for a scenic trip that starts and ends at New York and flies along each route exactly once? That is, is there an Euler circuit for the graph in Figure 7.12? If an Euler circuit exists, use trial and error to find one.

SOLUTION

Referring back to Example 7.2, we see that the degree of each vertex is even. So Euler's theorem guarantees the existence of an Euler circuit. Proceeding by trial and error, we find one such route:

New York – Miami – Atlanta – New York – Chicago – Atlanta –
Los Angeles – New York

This route is shown in **Figure 7.15**. We note that this is not the only correct answer: There are many other Euler circuits for this graph, and hence many scenic routes starting and ending at New York that use each airway exactly once.

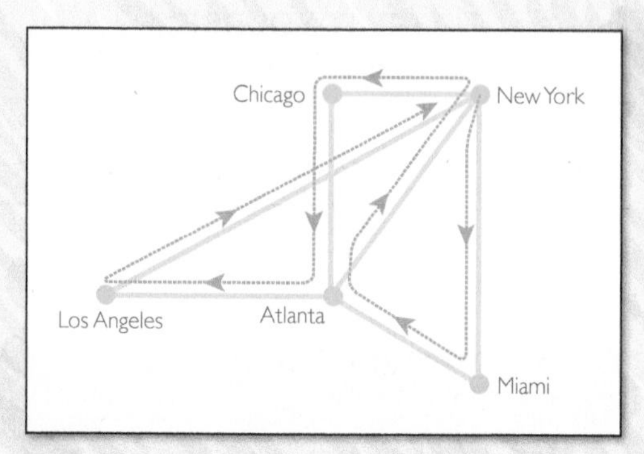

FIGURE 7.15 A scenic route that is an Euler circuit.

TRY IT YOURSELF 7.3

Does the Amtrak rail service map in Figure 7.13 allow for a scenic rail trip that begins and ends in New York and travels along each track route exactly once? In other words, is there an Euler circuit for this graph?

The answer is provided at the end of this section.

---

[2] *"Solutio Problematis ad Geometriam Situs Pertinentis (The solution of a problem relating to the geometry of position),"* Commentarii Academiae Scientiarum Imperialis Petropolitanae *8 (1736; published 1741), 128–140.*

EXAMPLE 7.4 Applying Euler's theorem: The seven bridges of Königsberg

Königsberg is an old city that is now part of Russia and has been renamed Kaliningrad. It includes both banks of the Pregel River as well as two islands in the river. The banks and islands were connected by seven bridges, as shown in **Figure 7.16**. Tradition has it that the people of Königsberg enjoyed walking and wanted to figure out how to start at home, go for a walk that crossed each bridge exactly once, and arrive back home. Nobody could figure out how to do it. It was Leonhard Euler's solution of this problem that began the modern theory of graphs.

a. Make a graph that models Königsberg and its seven bridges.

b. Determine whether the desired walking path is possible.

FIGURE 7.16 Left: Königsberg (modern-day Kaliningrad). Right: The bridges of Königsberg.

SOLUTION

a. To make the model, we place a vertex on each bank of the river and on both islands. We connect these vertices by paths across the seven bridges. This is shown in **Figure 7.17**. We then obtain the graph shown in **Figure 7.18** by showing only the vertices and edges we have drawn. The degree of each vertex is noted in that figure.

b. The desired walking path would be an Euler circuit for the graph in Figure 7.18. But because this graph has a vertex of odd degree, it has no Euler circuit. The desired walking path in Königsberg does not exist.

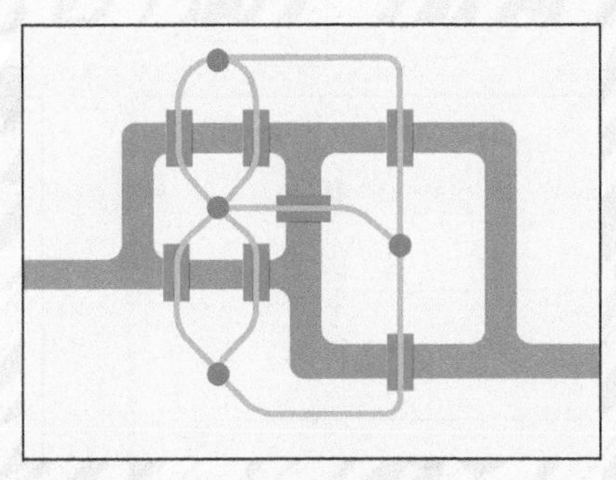

FIGURE 7.17 Preparing a graph model.

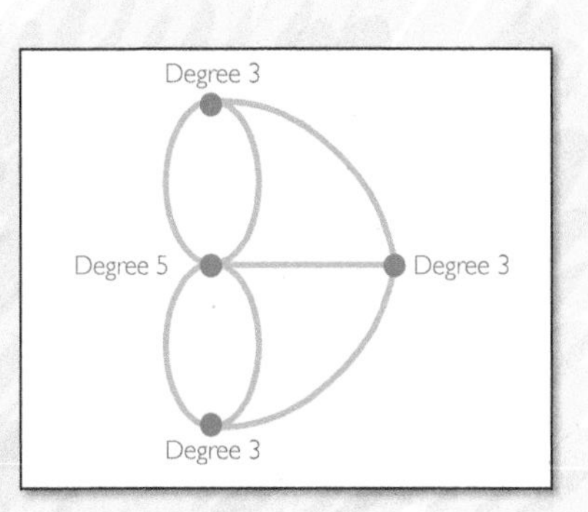

FIGURE 7.18 A graphical representation of the seven bridges.

SUMMARY 7.1
Graphs and Euler Circuits

1. A graph is a collection of vertices, some (or all) of which are joined by edges. Generally, we consider only connected graphs, and we do not allow edges to start and end at the same vertex.

2. A circuit in a graph is a path (a sequential collection of edges) that begins and ends at the same vertex. An Euler circuit is a circuit that uses each edge exactly once.

3. The degree of a vertex is the number of edges touching it.

4. A connected graph has an Euler circuit precisely when each vertex has even degree.

Methods for making Euler circuits

We know that a graph has an Euler circuit if and only if each of its vertices has even degree. This degree condition is easy to check, but how can we actually find an Euler circuit? For "small" graphs this can be done by trial and error, but for a graph with hundreds or even thousands of vertices, trial and error is a hopeless method for finding Euler circuits.

Fortunately, there are several systematic methods for constructing Euler circuits. We will present two such methods here. The first works by combining smaller circuits to produce larger ones.

To illustrate the method, consider the graph in **Figure 7.19**. It has two obvious circuits, which we have marked as *cycle 1* and *cycle 2*. These two circuits meet at vertex A, but they share no common edge. We break the two circuits at this vertex and rejoin them, as shown in **Figure 7.20**. The result is that both circuits combine to make the longer circuit illustrated in **Figure 7.21**. This gives an Euler circuit. For a larger graph, we may have to put together a number of smaller circuits to obtain an Euler circuit.

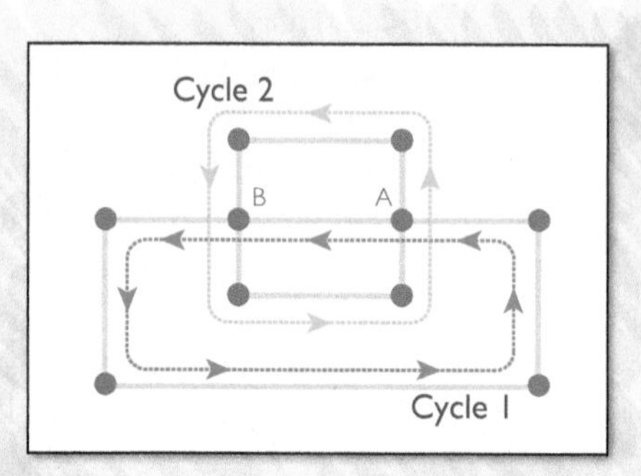

FIGURE 7.19 A graph with two circuits.

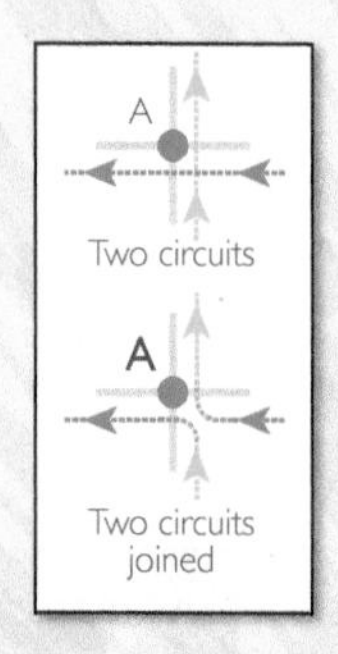

FIGURE 7.20 Breaking and rerouting at A.

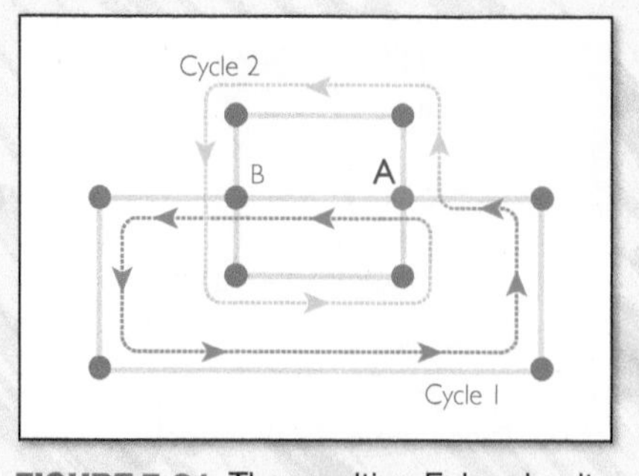

FIGURE 7.21 The resulting Euler circuit.

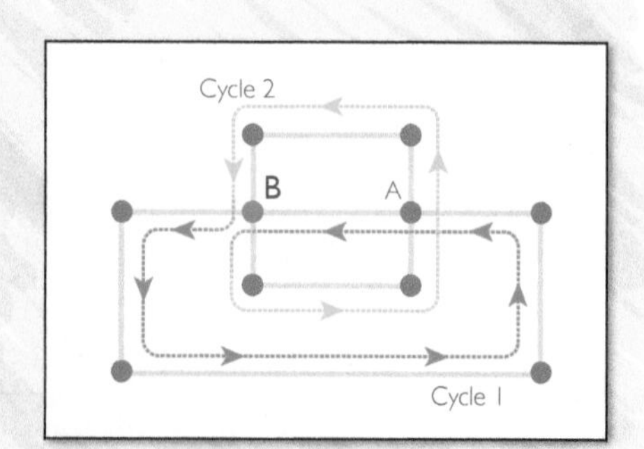

FIGURE 7.22 Breaking and rerouting at B.

We note that the two circuits in Figure 7.19 also meet at vertex B. We can find a different Euler circuit by breaking the circuits and rejoining at B rather than A. This is shown in **Figure 7.22**. When circuits meet at several vertices, we can break and reroute the circuits at any of these vertices.

A second method for making Euler circuits is known as *Fleury's algorithm*. For this method, we remove edges from the graph one at a time and add them to a path that will grow to become an Euler circuit. To lengthen this path, we add edges by repeatedly applying the following two-step procedure:

Step 1: Remove an edge from the graph and add it to the path so that the path is lengthened. But do not choose an edge whose removal will cut the graph into two pieces unless there is no other choice.

Step 2: Delete from the graph any isolated vertices resulting from the first step.

Repeat these two steps until an Euler circuit is constructed. Let's apply Fleury's algorithm to the graph in **Figure 7.23**. Starting at the vertex A, we have selected the three red edges to begin Fleury's algorithm. In **Figure 7.24**, we have used these edges to start an Euler circuit.

We have arrived at vertex D, and from here we might select any of the yellow edges in **Figure 7.25**. But removal of the edge DC would result in the disconnected graph shown in **Figure 7.26**.

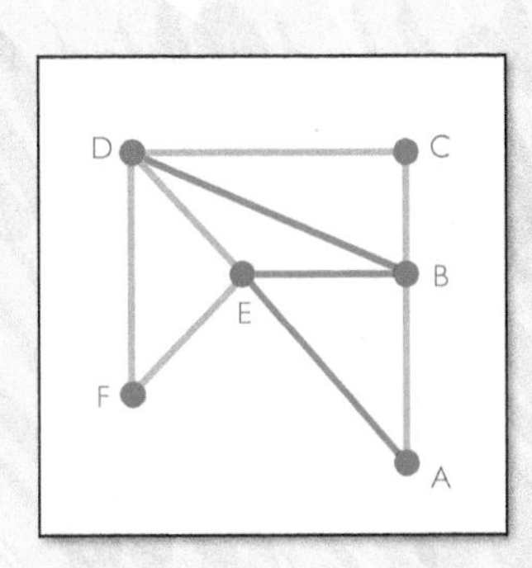

FIGURE 7.23 Selecting edges to start Fleury's algorithm.

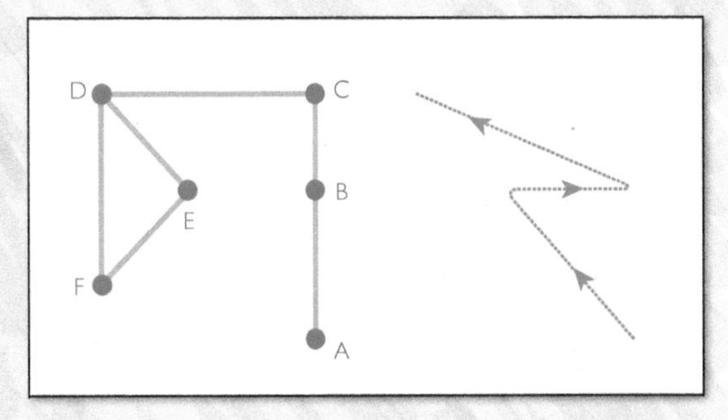

FIGURE 7.24 The beginning of an Euler circuit.

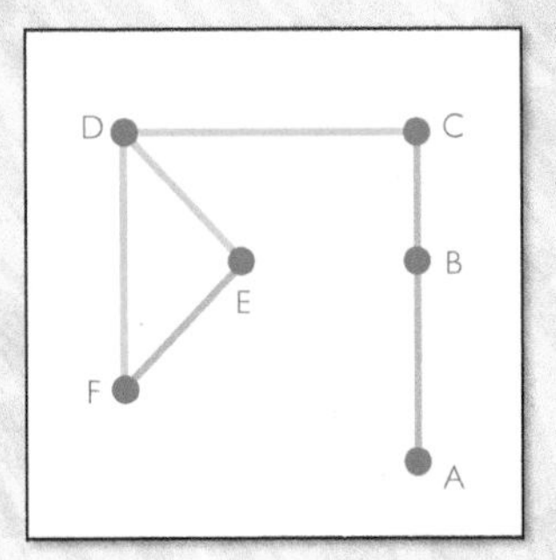

FIGURE 7.25 The three yellow edges are possibilities for continuing past vertex D.

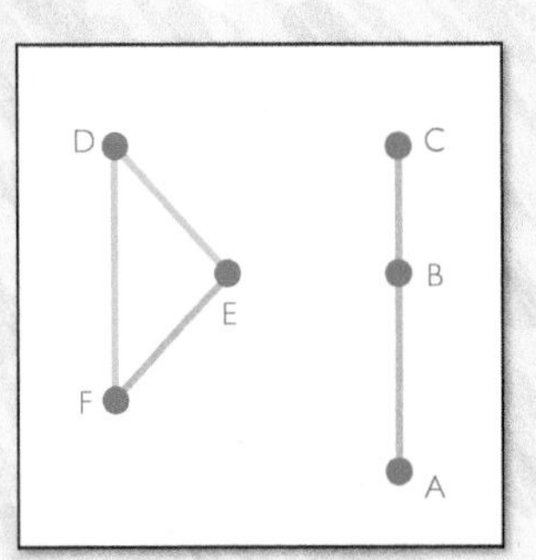

FIGURE 7.26 Removal of DC would result in a disconnected graph.

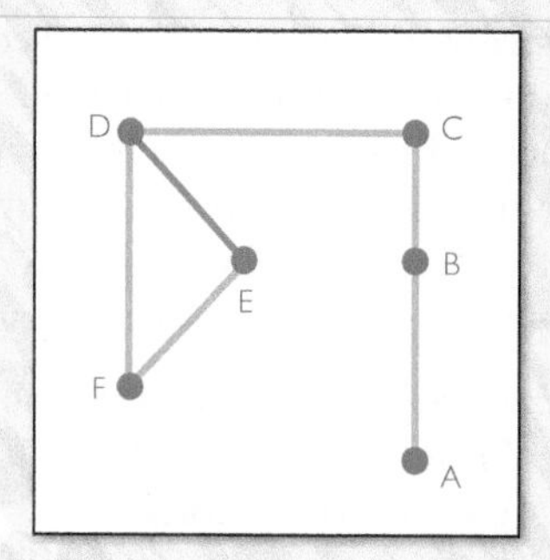

FIGURE 7.27 The edge *DE* is selected.

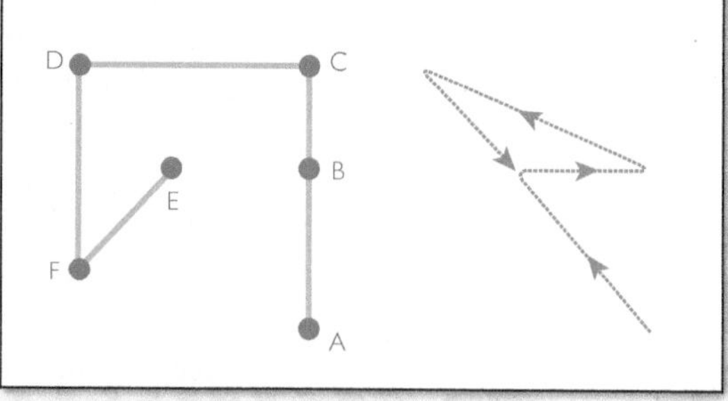

FIGURE 7.28 The edge *DE* is added to the path.

So we can choose either of the edges *DF* or *DE*, but not *DC*. In **Figure 7.27**, we have selected the edge *DE* and added that edge to the path in **Figure 7.28**.

We have arrived at the vertex *E*. Removal of the edge *EF*, as shown in **Figure 7.29**, will result in a disconnected graph, but there is no other choice, so we do it. We add this edge to the path, as shown in **Figure 7.30**. Note that this leaves the vertex *E* with no edges attached. We delete this isolated vertex and continue.

From this point on, we have no decisions to make, and the completed Euler circuit is shown in **Figure 7.31**.

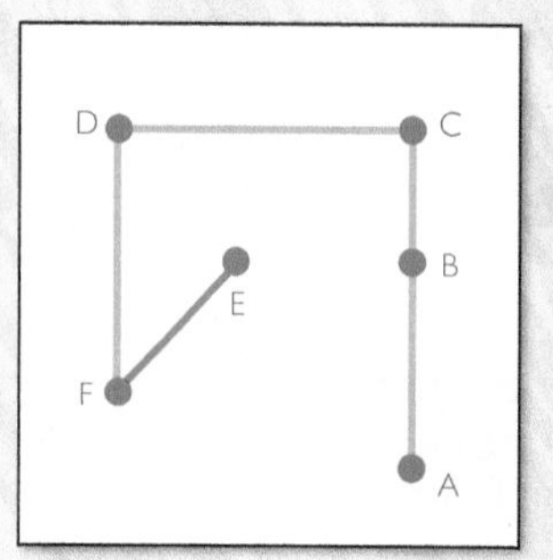

FIGURE 7.29 *EF* is the only choice even though its removal leaves an isolated vertex.

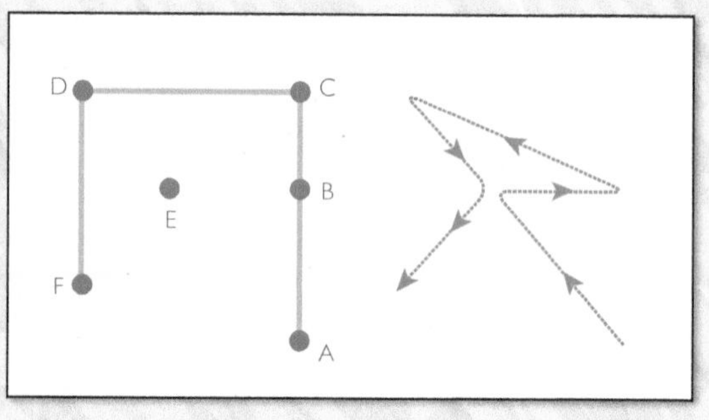

FIGURE 7.30 *EF* added to the path.

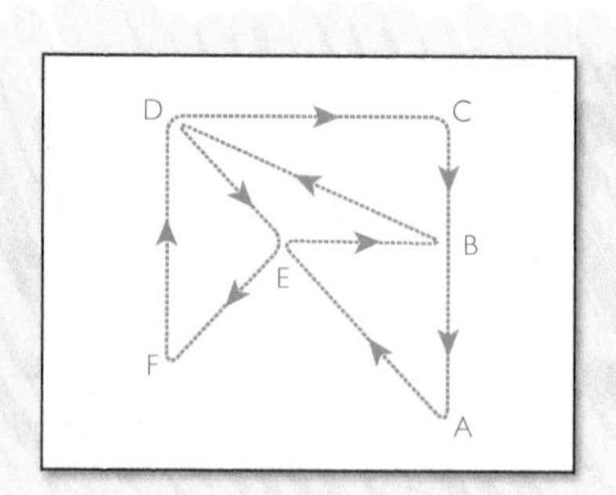

FIGURE 7.31 The completed Euler circuit.

EXAMPLE 7.5 Finding Euler circuits: A subway tour

Figure 7.32 shows a part of the New York City subway. Is it possible to take a tour of this part of the subway line that starts and ends at the 57 St & 7 Ave station and traverses each rail section exactly once? In other words, is there an Euler circuit for this graph? If an Euler circuit exists, find one.

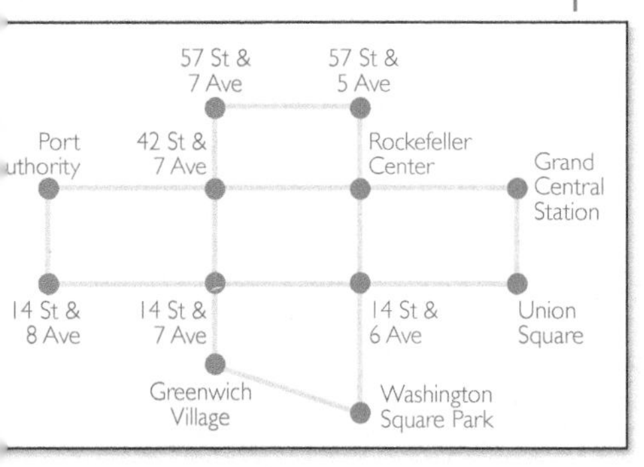

URE 7.32 A part of the New York City way.

SOLUTION

Each of the vertices has degree 2 or 4. Because each vertex has even degree, we are guaranteed an Euler circuit. One could find an Euler circuit by trial and error, but we apply the method of breaking and rerouting. In **Figure 7.33**, we have marked two circuits that share no common edges.

Circuit 1: 57 St & 7 Ave - 57 St & 5 Ave - Rockefeller Center - 14 St & 6 Ave - Washington Square Park - Greenwich Village - 14 St & 7 Ave - 42 St & 7 Ave - 57 St & 7 Ave

Circuit 2: Port Authority - 42 St & 7 Ave - Rockefeller Center - Grand Central Station - Union Square - 14 St & 6 Ave - 14 St & 7 Ave - 14 St & 8 Ave - Port Authority

In **Figure 7.34**, we have broken the circuits at Rockefeller Center and rerouted to combine the two circuits into one. The result is the required subway tour:
Euler circuit: 57 St & 7 Ave - 57 St & 5 Ave - Rockefeller Center - Grand Central Station - Union Square - 14 St & 6 Ave - 14 St & 7 Ave - 14 St & 8 Ave - Port Authority - 42 St & 7 Ave - Rockefeller Center - 14 St & 6 Ave - Washington Square Park - Greenwich Village - 14 St & 7 Ave - 42 St & 7 Ave - 57 St & 7 Ave

You should verify that Fleury's algorithm can also be used to produce an Euler circuit.

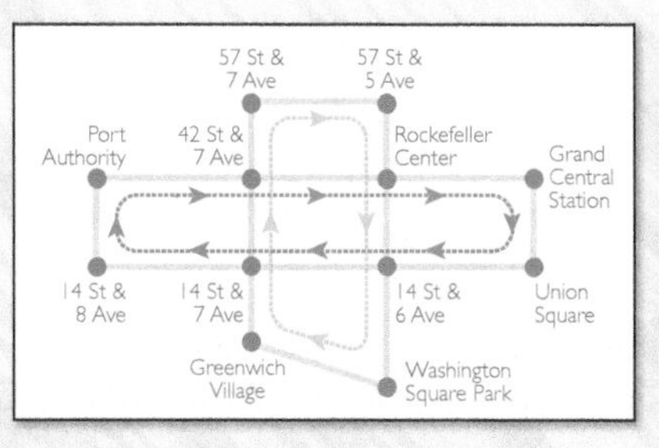

FIGURE 7.33 Two circuits.

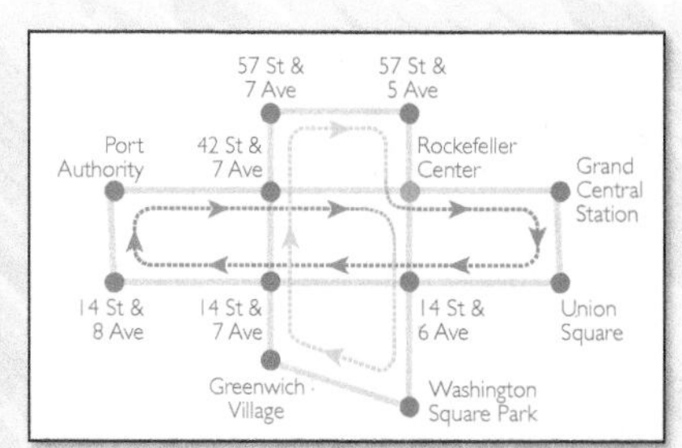

FIGURE 7.34 Patching together to make an Euler circuit.

TRY IT YOURSELF 7.5

Find the Euler circuit obtained by breaking and rerouting circuit 1 and circuit 2 at 14 St & 7 Ave rather than at Rockefeller Center.

The answer is provided at the end of this section.

SUMMARY 7.2
Making Euler Circuits

Two circuits with a vertex in common (but no edge in common) can be put together to make a single longer circuit. Using this method repeatedly allows us to put smaller circuits together to form an Euler circuit when one exists.

Alternatively, we can make Euler circuits using Fleury's algorithm.

Eulerizing graphs

We have seen that in many practical settings, a most efficient route is an Euler circuit—if there is one. If there is no Euler circuit, any circuit traversing each edge must include some backtracking of edges, that is, going back along some paths we have already taken. We still want a most efficient route possible in the sense that we have to backtrack as little as possible. We do this by *Eulerizing* graphs, temporarily adding duplicate edges.

We know that if all vertices have even degree, then we can find an Euler circuit. We will show how to find an efficient path when there are exactly two odd-degree vertices. The graph in **Figure 7.35** shows a part of the shuttle route in Yosemite National Park. Note that the vertices at Yosemite Creek and the Ansel Adams Gallery have degree 3 and all other vertices have even degree. Because there are odd-degree vertices, there is no Euler circuit: It is not possible to start and end at Yosemite Creek and travel each shuttle route without backtracking. We want to find a route that involves the least possible backtracking. The first step in our search for an efficient route is to pick a shortest[3] (in terms of the number of edges) path joining the odd-degree vertices, in this case Yosemite Creek and Ansel Adams Gallery. There is more than one choice for this, and any of them will serve. We have selected one shortest path from Yosemite Creek to the Ansel Adams Gallery and highlighted it in orange in Figure 7.35. The second step in our search is to add duplicate edges, as shown in **Figure 7.36**. Observe that after the addition of the duplicate edges, all vertices have even degree. Hence, there is an Euler circuit, as shown in Figure 7.36. We complete the process by collapsing the duplicate edges back to their originals. The result is a circuit starting and ending at Yosemite Creek that backtracks as little as possible (twice in this case), as shown in **Figure 7.37**.

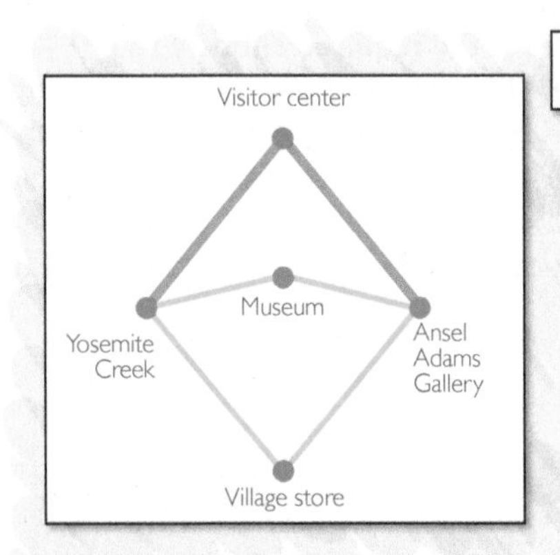

FIGURE 7.35 Step 1: Find shortest path between odd-degree vertices.

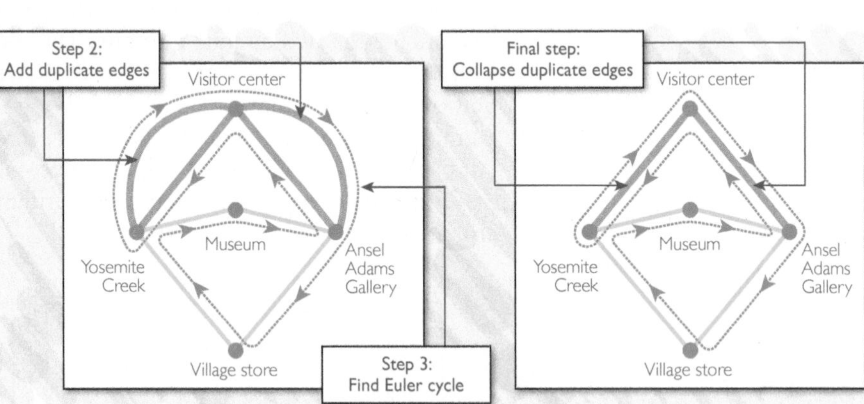

FIGURE 7.36 Step 2: Add duplicate edges. Step 3: Find an Euler circuit.

FIGURE 7.37 Final step: Collapse the duplicate edges.

Best route: Yosemite Creek - Visitor center - Ansel Adams Gallery - Village store - Yosemite Creek - Museum - Ansel Adams Gallery - Visitor center - Yosemite Creek

[3] *In large graphs, even the seemingly simple problem of finding a shortest path can be problematic.*

EXAMPLE 7.6 Eulerizing graphs: Streets in our town

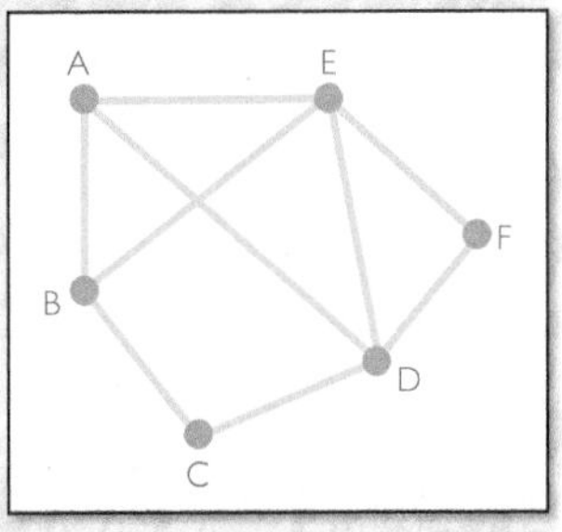

FIGURE 7.38 Street map for Example 7.6.

The graph in **Figure 7.38** represents the streets and intersections in our town. We wish to start at vertex *A* and paint a centerline on each street. Find a route (starting and ending at *A*) that backtracks as little as possible.

SOLUTION

We note that every vertex has even degree except vertices *A* and *B*, which have degree 3. In **Figure 7.39**, we have added a duplicate edge between *A* and *B* and indicated an Euler circuit. Collapsing the duplicate edge, as shown in **Figure 7.40**, we find the efficient path *A-B-C-D-F-E-A-D-E-B-A* for the painters. It backtracks only once: from *A* to *B* and back.

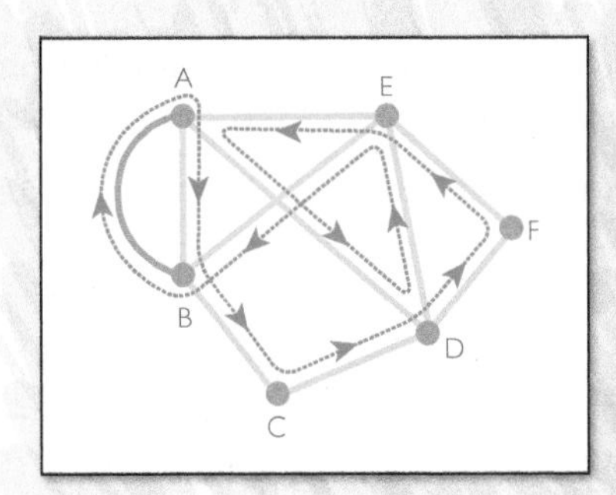

FIGURE 7.39 Adding a duplicate edge and indicating an Euler circuit.

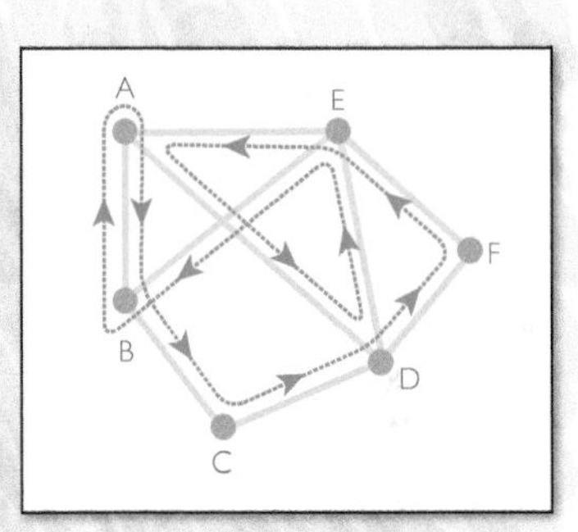

FIGURE 7.40 Collapsing the duplicate edge to get an efficient path.

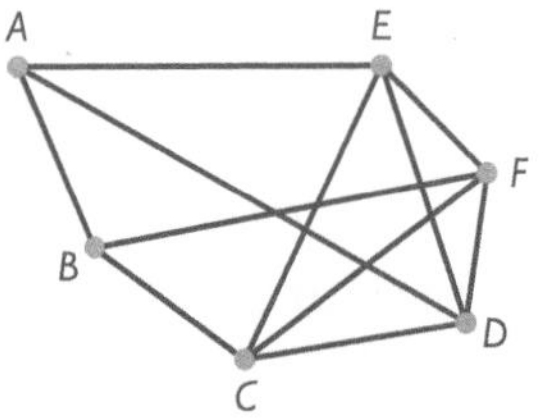

TRY IT YOURSELF 7.6

The graph in **Figure 7.41** shows the streets and intersections in our town. We wish to start at vertex *A* and paint a centerline on each street. Find a route (starting and ending at *A*) that backtracks as little as possible.

FIGURE 7.41 Street map for Try It Yourself 7.6.

The answer is provided at the end of this section.

WHAT DO YOU THINK?

Trial and error: A fellow student says that there's no need to learn one of the two methods for making an Euler circuit—trial and error will always work. How would you respond?

Questions of this sort may invite many correct responses. We present one possibility: Trial and error works just fine for relatively small graphs. But applications where Euler circuits are really needed, such as mail routes, may involve a map of a city represented by a graph with hundreds or thousands of edges. For such large graphs, trial and error is a hopeless approach. Computers handle graphs of this type, and computers need effective plans (called *algorithms*) such as the procedures presented in this section.

Further questions of this type may be found in the *What Do You Think?* section of exercises.

Try It Yourself answers

Try It Yourself 7.1: Making models: Ferry routes We use vertices for each of the towns. Edges indicate ferry routes.

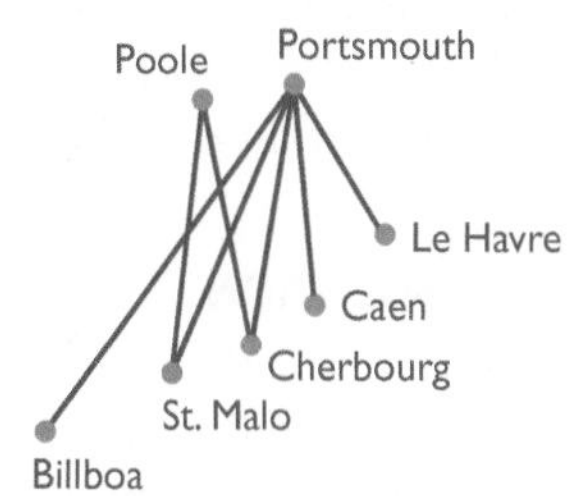

Try It Yourself 7.2: Finding the degree of a vertex: Rail service Chicago: degree 7; Jacksonville: degree 2; Los Angeles: degree 3; New Orleans: degree 4; New York: degree 2; San Antonio: degree 3; San Francisco: degree 3; Seattle: degree 2; Washington: degree 4.

Try It Yourself 7.3: Applying Euler's theorem: A rail service map Some of the vertices have an odd degree, so it is not possible to take the desired trip.

Try It Yourself 7.5: Finding Euler circuits: A subway tour The required Euler circuit is

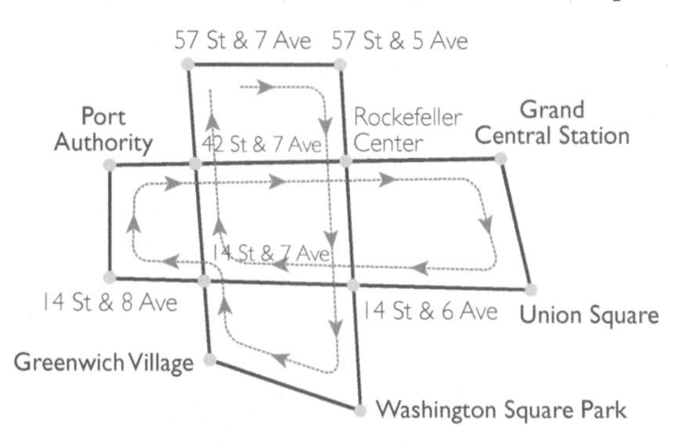

Try It Yourself 7.6: Eulerizing graphs: Streets in our town An Eulerized graph is

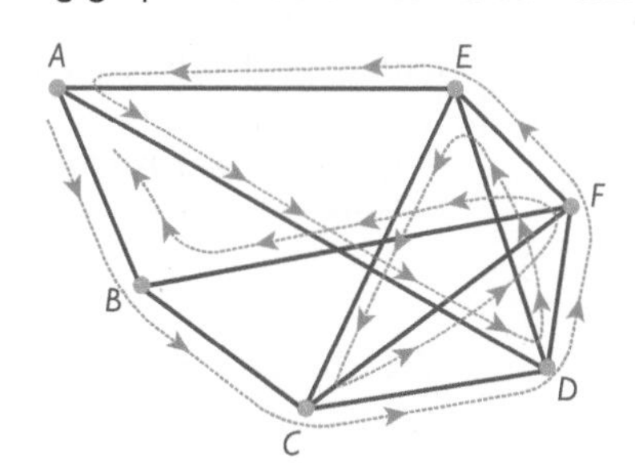

Exercise Set 7.1

Test Your Understanding

1. A graph consists of two things: ______________ and ______________.

2. True or false: A circuit is a path that begins and ends at the same vertex.

3. Explain in your own words what an Euler circuit is.

4. The degree of a vertex is ______________.

5. A graph has an Euler circuit provided that the degree of each vertex is ______________.

6. True or false: We *Eulerize* graphs by adding duplicates to certain edges.

Problems

7. **Connectedness.** All of the graphs we have considered are connected, in the sense that each pair of vertices can be joined by a path. One application of graphs that we gave at the beginning of this section involves efficient routes for airplanes. Explain what the property of connectedness means in that context. Also explain why it is important for graphs representing routes for airplanes to be connected.

8. Staying connected. For some applications, it is important that the associated graph have no edge whose removal disconnects the graph. In other words, each pair of vertices can be joined by a path even after any single edge is removed. Explain why this property is important in the case when the graph represents electric power transmission lines forming an electrical grid. (The edges represent power lines, and the vertices represent power plants, substations, consumers, etc.) How is this property of a graph related to the existence of redundant power lines between points?

9. Little League. The Little League in our town consists of the Red Division—the Cubs, Lions, Bears, and Tigers—and the Blue Division—the Red Sox, Yankees, and Giants. The inter-division schedule this season is:

Cubs play Red Sox, Yankees, and Giants.
Lions play Red Sox and Giants.
Bears play Red Sox and Yankees.
Tigers play Red Sox, Yankees, and Giants.

Make a graph to model this season's Little League schedule.

10. More Little League. The Little League in our town consists of the Red Division—the Cubs, Lions, Bears, and Tigers—and the Blue Division—the Red Sox, Yankees, and Giants. This season each team is to play exactly three teams from the other division. Can you make a graph to model this schedule? Is such a schedule possible?

11. A computer network. Each of the offices 501, 502, 503, 504, 505, 506, and 507 has a computer. Some of these are connected to others:

The computer in 501 connects to computers in 502, 503, 505, and 507.

The computer in 502 connects to the computer in 501, 503, and 504.

The computer in 503 connects to all other computers.

Make a graph that shows the office computer network.

12. Bridges. Figure 7.42 shows islands in a river and bridges connecting them. Make a graph model of walking paths using vertices to indicate land masses and edges to indicate paths across bridges.

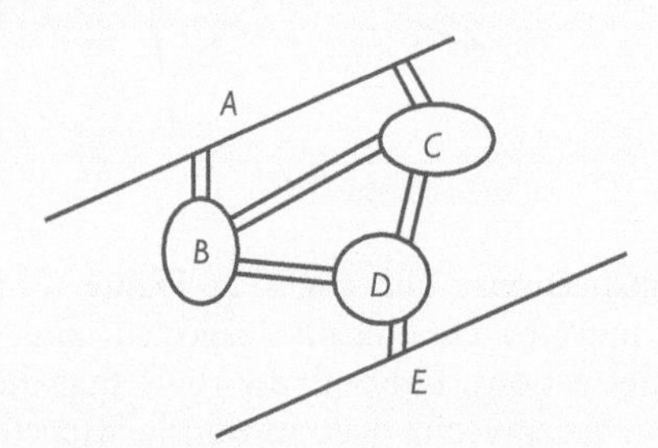

FIGURE 7.42 Islands and bridges for Exercise 12.

13. Birds. Here are characteristics of some birds:

The ruby-throated hummingbird is a hummingbird that lives in the eastern and central United States.

The calliope hummingbird is a hummingbird that lives in the northwestern United States.

The red-shafted flicker is a woodpecker that lives in the western United States.

The red-bellied woodpecker is a woodpecker that lives in the eastern United States.

The cinnamon teal is a duck that lives in the western United States.

The hooded merganser is a duck that lives in the eastern United States.

Make a graph with these birds as vertices. Use an edge to indicate that either the birds belong to the same family or their habitats overlap. For example, the ruby-throated and calliope are joined by an edge because they are both hummingbirds. The red-shafted flicker and calliope hummingbird are joined by an edge because their habitats overlap.

14. A computer network. The computers in a mall are connected.

Central security connects to all computers.

The food court computer is connected to central security and custodial services.

The custodial services computer is connected to central security, food court, and Five Below.

The Five Below computer is connected to central security and custodial services.

Make a graph modeling the mall computer network.

15. Degrees of vertices. Find the degrees of each of the vertices in **Figure 7.43.**

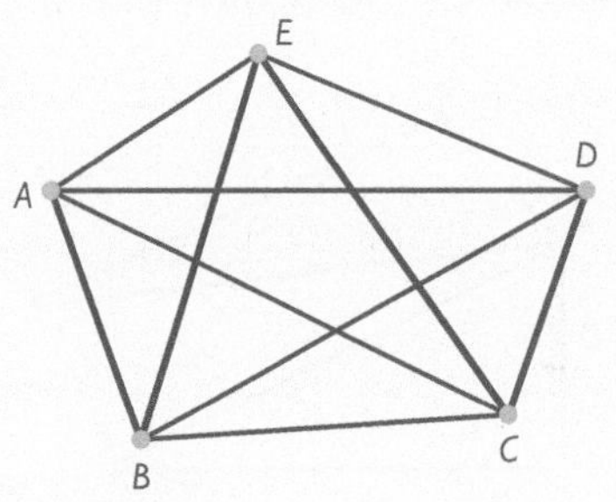

FIGURE 7.43

16. More degrees of vertices. Find the degrees of each of the vertices in **Figure 7.44.**

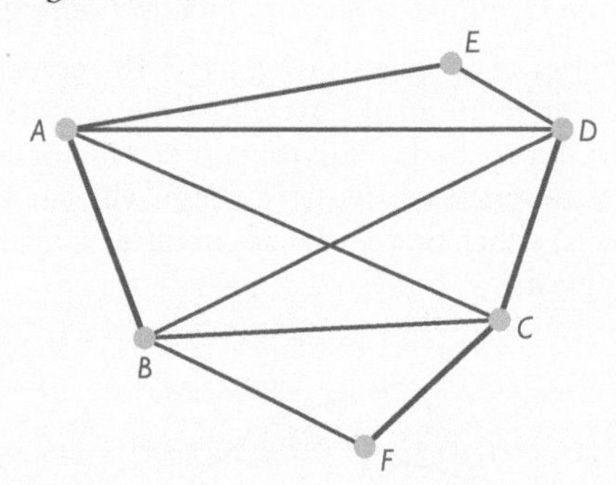

FIGURE 7.44

17. Interpreting Euler. The edges in a certain graph represent phone lines that must be maintained, and the vertices represent junctions. Explain why a worker maintaining the lines would like to find an Euler circuit for this graph.

18–22. Euler circuits. *In Exercises 18 through 22, determine whether the given graph has an Euler circuit. If an Euler circuit exists, find one.*

18. Refer to the graph in Figure 7.43.

19. Refer to the graph in Figure 7.44.

20. Refer to the graph in **Figure 7.45**.

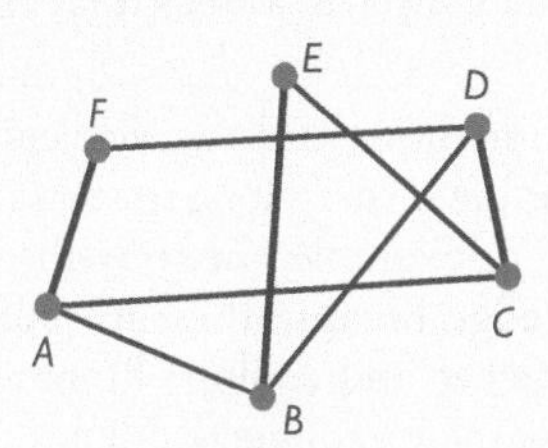

FIGURE 7.45

21. Refer to the graph in **Figure 7.46**.

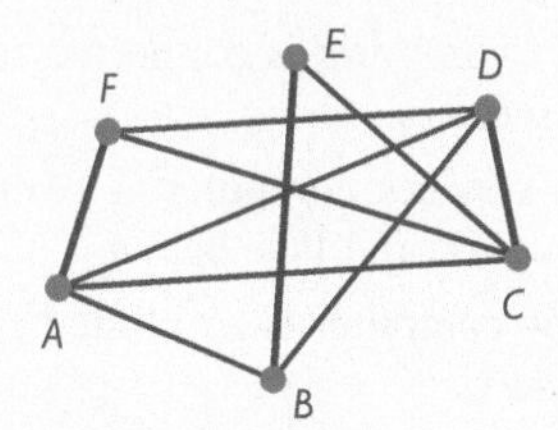

FIGURE 7.46

22. Refer to the graph in **Figure 7.47**.

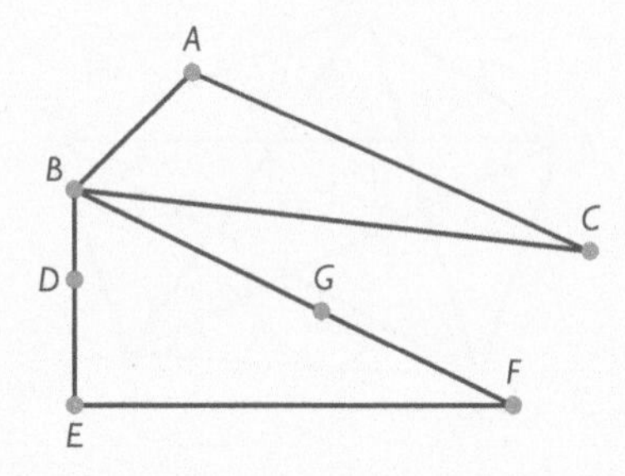

FIGURE 7.47

23–33. Euler circuits in practical settings. *Exercises 23 through 33 involve Euler circuits in practical settings.*

23. Mail routes. The edges in **Figure 7.48** represent streets along which there are mail boxes, and the vertices represent intersections. Either find a mail route that will not require any retracing by the mail carrier or explain why no such route exists. That is, either find an Euler circuit or explain why no such circuit exists.

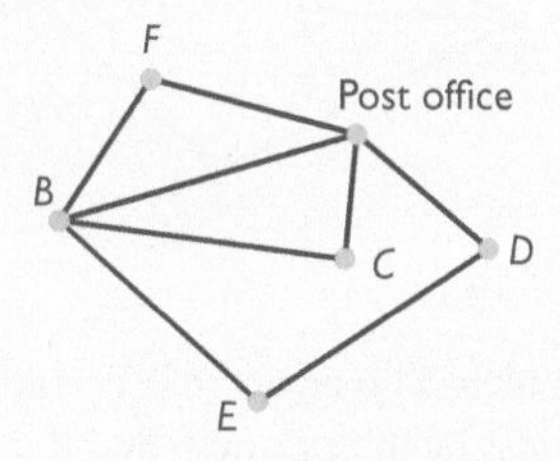

FIGURE 7.48

24. Snow plows. The edges in **Figure 7.49** represent streets that have to be cleared of snow, and the vertices represent intersections. Either find a route for the snow plow that will not require it to travel over any streets that have already been cleared of snow or explain why no such route exists. That is, either find an Euler circuit or explain why no such circuit exists.

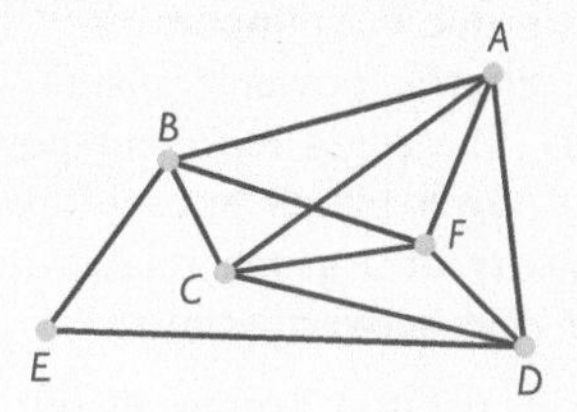

FIGURE 7.49

25. Paving streets. The edges in **Figure 7.50** represent unpaved roads, and the vertices represent intersections. Either find a route that will not require the paving machine to move along already paved streets or show that no such route exists. That is, either find an Euler circuit or explain why no such circuit exists.

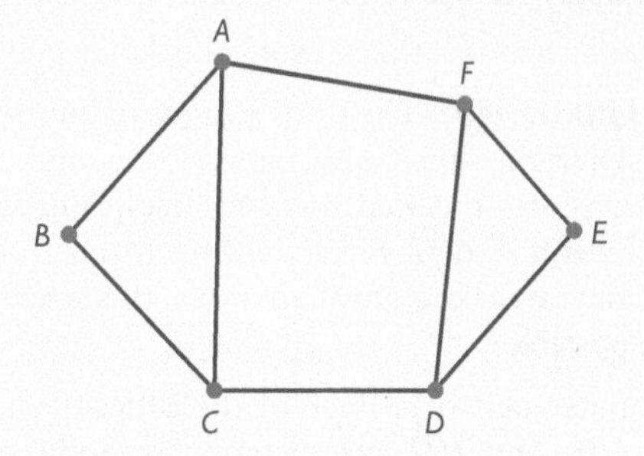

FIGURE 7.50

26. Utility ditches. The edges in **Figure 7.51** represent ditches that must be dug for utilities, and the vertices represent homes. Either find a route that will not require the ditch digger to travel along ditches that have already been dug or show that no such route exists. That is, either find an Euler circuit or explain why no such circuit exists.

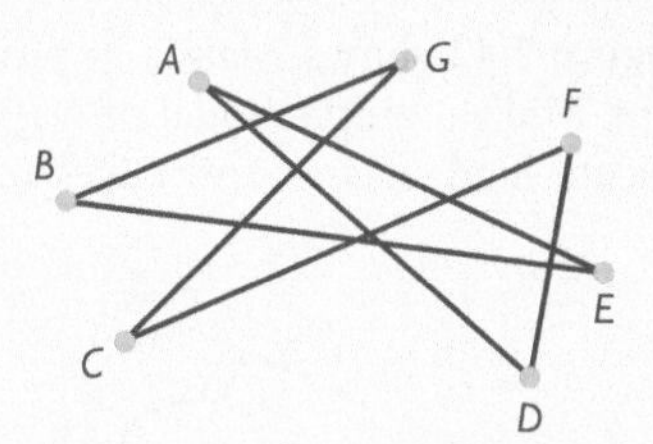

FIGURE 7.51

27. Street inspections. The edges in **Figure 7.52** represent streets that must be periodically inspected, and the vertices represent intersections. Either find a route that the inspectors can take to avoid traveling over previously inspected streets or show that no such route exists. That is, either find an Euler circuit or explain why no such circuit exists.

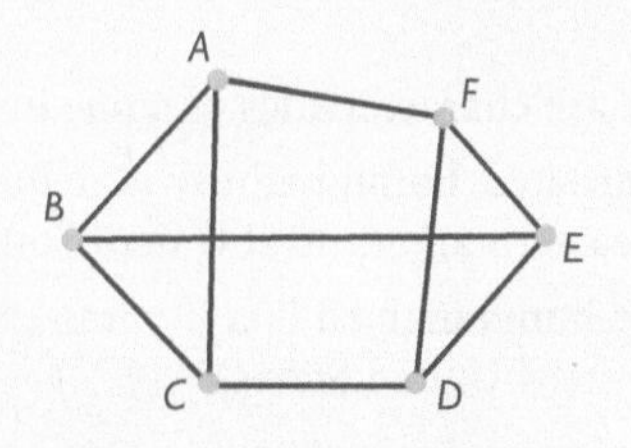

FIGURE 7.52

28. Repairing hiking paths. The edges in **Figure 7.53** represent hiking paths that must be repaired, and the vertices represent intersections. Either find a route that the crews can follow to avoid traveling over already repaired trails or show that no such route exists. That is, either find an Euler circuit or explain why no such circuit exists.

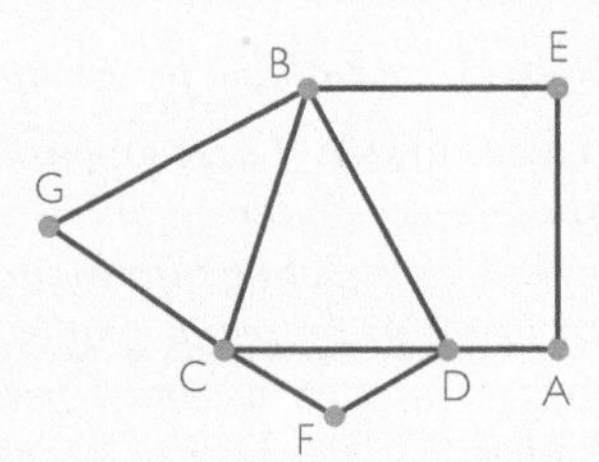

FIGURE 7.53

29. Package delivery. The edges in **Figure 7.54** represent streets along which packages must be delivered, and the vertices represent intersections. Either find a path that will allow the delivery truck to avoid streets where packages have already been delivered or show that no such path exists. That is, either find an Euler circuit or explain why no such circuit exists.

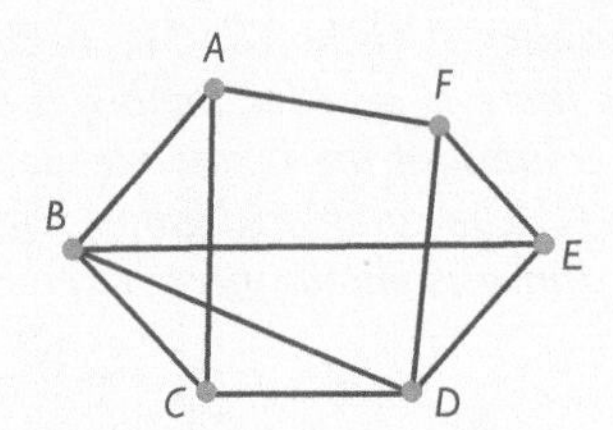

FIGURE 7.54

30. Police beats. The edges in **Figure 7.55** represent streets along which a policeman must walk his beat, and the vertices represent intersections. Either find a route that will allow the policeman to avoid retracing steps or show that no such route exists. That is, either find an Euler circuit or explain why no such circuit exists.

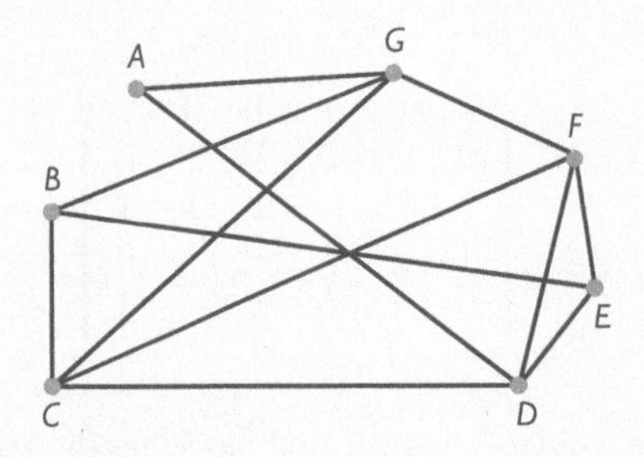

FIGURE 7.55

31. Fire inspections. The edges in **Figure 7.56** represent streets along which businesses must be inspected for fire safety, and the vertices represent intersections. Either find a route that allows the inspector to avoid retracing steps or show that no such route exists. That is, either find an Euler circuit or explain why no such circuit exists.

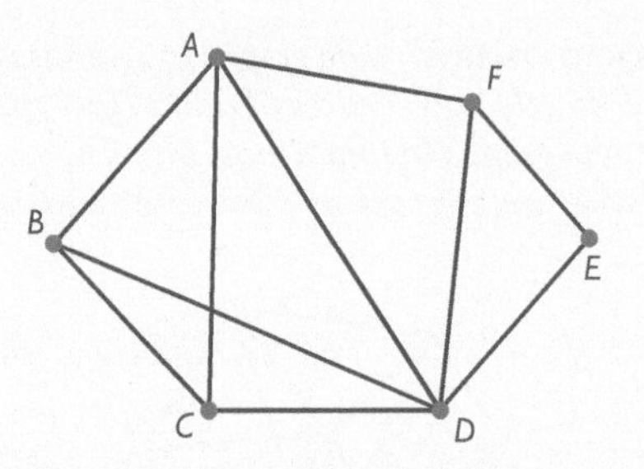

FIGURE 7.56

32. Phone line maintenance. The edges in **Figure 7.57** represent phone lines that must be periodically maintained, and the vertices represent junctions. Either find a maintenance path that does not require retracing or show that no such path exists. That is, either find an Euler circuit or explain why no such circuit exists.

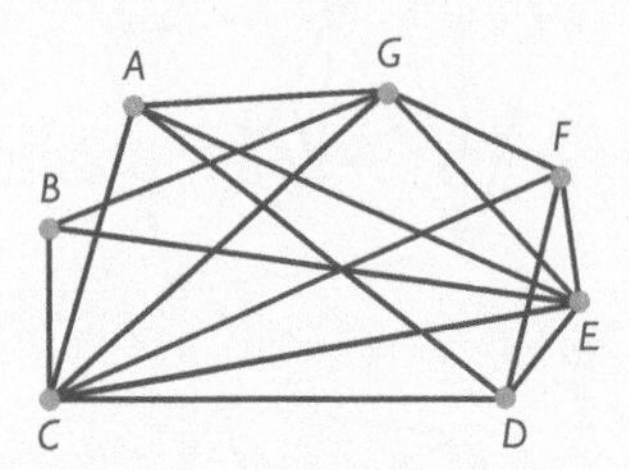

FIGURE 7.57

33. Inspecting for storm damage. The edges in **Figure 7.58** represent streets that must be inspected for storm damage, and the vertices represent intersections. Either find a route that will allow the inspector to avoid retracing or show that no such route exists. That is, either find an Euler circuit or explain why no such circuit exists.

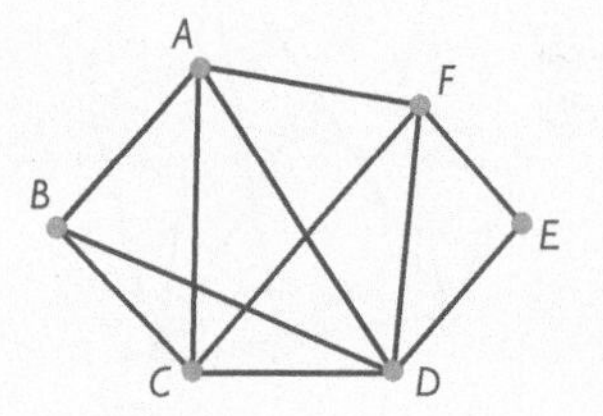

FIGURE 7.58

34. Making Euler circuits. The graph in **Figure 7.59** represents streets and intersections. We have marked three circuits, labeled cycle 1, cycle 2, and cycle 3. Patch these cycles together at vertices E and G to make an Euler circuit.

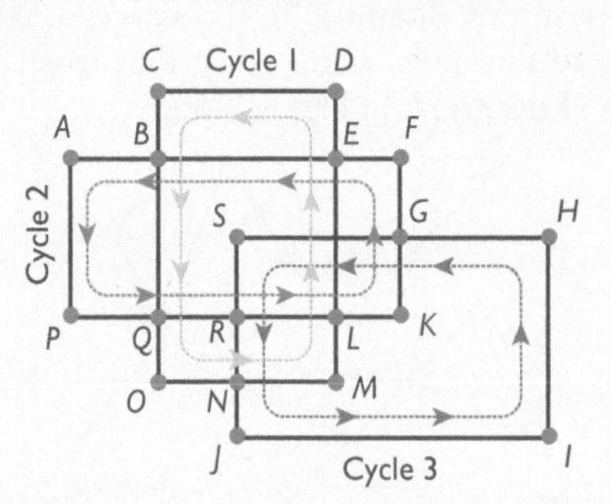

FIGURE 7.59

35–38. Efficient circuits. *The graphs in Exercises 35 through 38 each have exactly two vertices of odd degree, namely A and B. Eulerize these graphs to find a most efficient path that starts and ends at the same vertex and traverses each edge at least once.*

35.

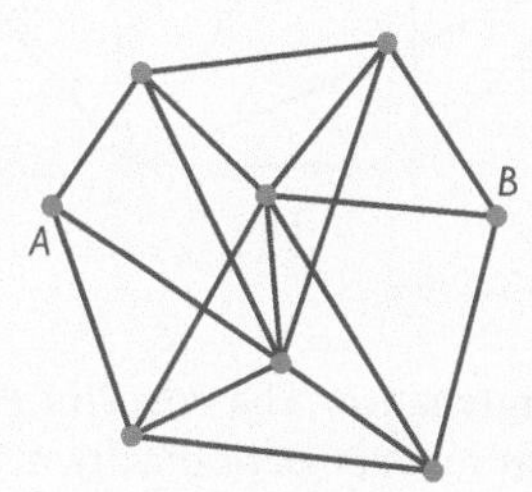

36.

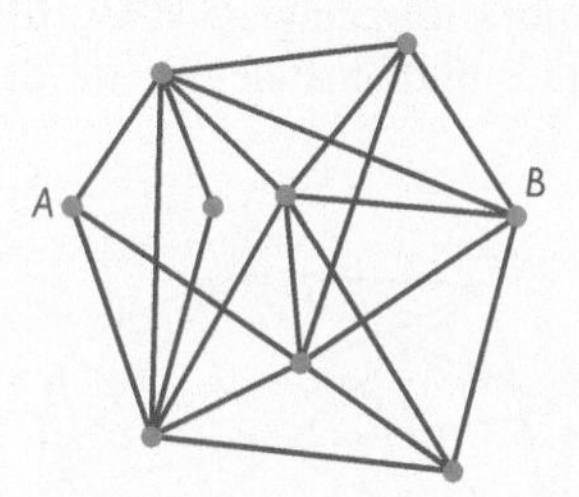

37.

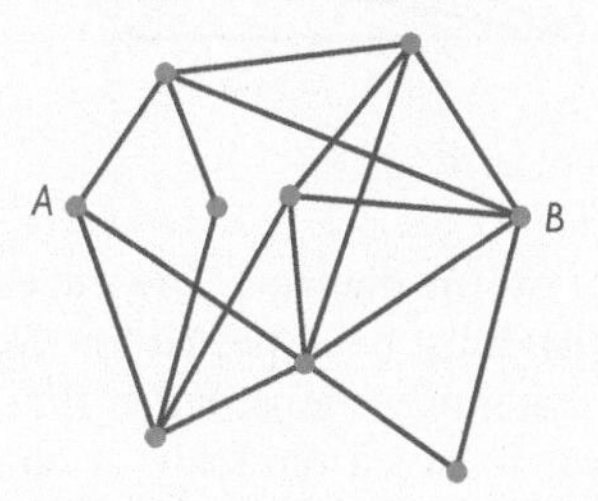

38.

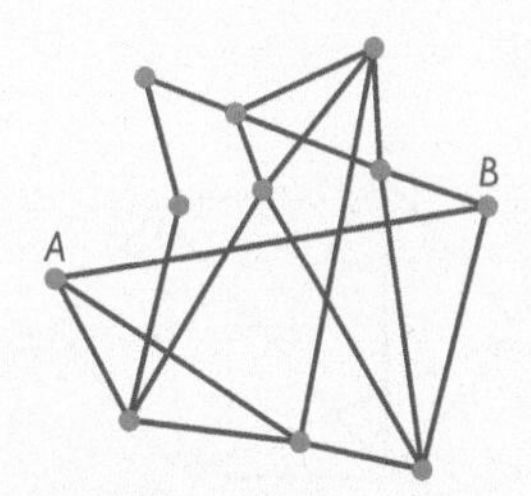

39. Euler paths. In some graphs where no Euler circuit exists, there is instead an *Euler path*. That is a path that starts at one vertex, traverses each edge exactly once, and ends at another vertex. Euler showed in 1736 that such a graph has exactly two vertices of odd degree (the beginning and ending points of the Euler path). It turns out that Fleury's algorithm produces an Euler path in this situation. In this exercise, we will show another way to find Euler paths when they exist. We will use the example of the graph in **Figure 7.60**.

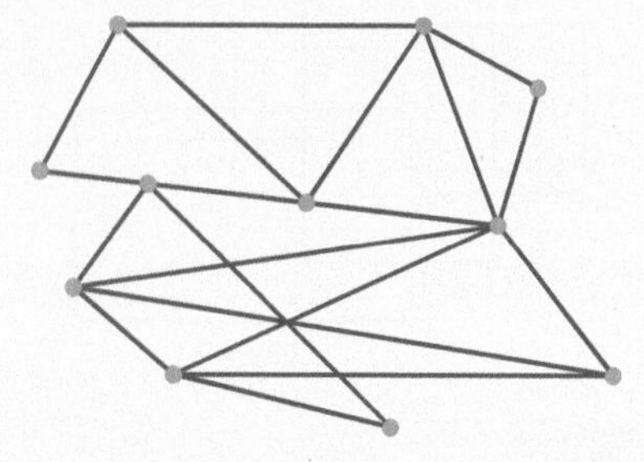

FIGURE 7.60 A graph with an Euler path.

a. In Figure 7.60, identify the two vertices of odd degree.

b. Make a new graph by adding an edge joining the two vertices of odd degree in part a.

c. Find an Euler circuit for the graph you made in part b.

d. Use the Euler circuit you found in part c to find an Euler path for the graph in Figure 7.60.

Exercises 40 through 43 are suitable for group work.

40–43. Talking to a computer. *Large graphs must be handled by computers. Although computer graphics can be stunning, computers have difficulty extracting information from the pictures we have been using to represent graphs. So if we want to feed a graph to a computer, we must present it in some way other than a picture. One way of doing this is via an incidence matrix, which is an array of dashes, 1's, and 0's. We will explain what we mean with an example:*

	A	B	C
A	–	0	1
B	0	–	1
C	1	1	–

In this matrix, the letters A, B, C represent the vertices of the graph. The dashes on the diagonal are just placeholders, and we ignore them. The "0" in the B column of the A row means there is no edge joining A to B. The "1" in the C column of the A row means there is an edge joining A to C. Similarly, the "1" in the C column of the B row means that there is an edge joining B to C. The completed graph represented by the incidence matrix above is shown here. Exercises 40 through 43 use this idea.

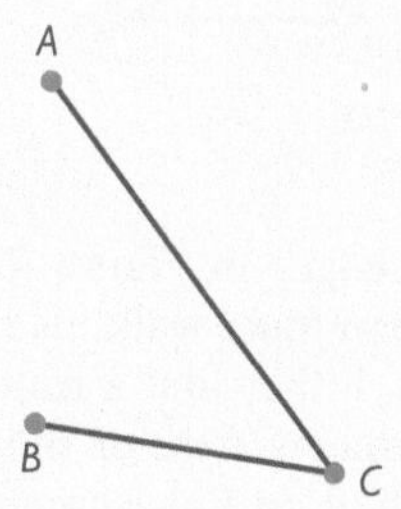

40. Make the graph represented by the following incidence matrix:

	A	B	C	D	E	F
A	–	0	1	1	0	1
B	0	–	0	1	1	0
C	1	0	–	1	1	1
D	1	1	1	–	1	0
E	0	1	1	1	–	1
F	1	0	1	0	1	–

41. Make the incidence matrix that describes the graph shown below.

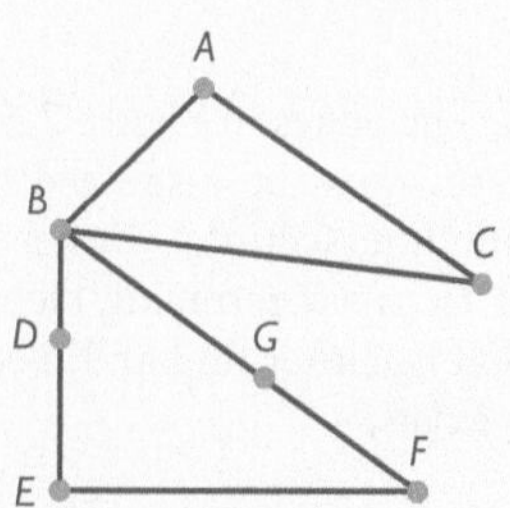

42. Can you make a graph with the following incidence matrix? Explain any difficulties you encounter.

$$\begin{array}{c} \\ A \\ B \\ C \\ D \\ E \\ F \end{array} \begin{array}{c} \begin{array}{cccccc} A & B & C & D & E & F \end{array} \\ \begin{pmatrix} - & 0 & 1 & 1 & 0 & 1 \\ 0 & - & 0 & 1 & 1 & 0 \\ 1 & 0 & - & 1 & 1 & 1 \\ 1 & 1 & 1 & - & 1 & 0 \\ 0 & 1 & 1 & 1 & - & 1 \\ 1 & 1 & 1 & 1 & 1 & - \end{pmatrix} \end{array}$$

43. Considering the results of the previous exercise, can you suggest a property that all incidence matrices of graphs must have?

Writing About Mathematics

44. Finding practical applications. Describe some practical situations where finding Euler circuits may be important. In each case, show a representative graph.

45. Life of Euler. Write a paper on the life of Leonhard Euler.

46. The four color problem. One of the most celebrated mathematics problems in history was called the *four color problem.* It was first solved in 1976, so it is now called the four color *theorem*. Write a paper about the history of this theorem. Explain what it says and how it can be stated in terms of graphs.

47. Friends on Facebook. Make a graph where the vertices are you and some of your Facebook friends. Connect friends by an edge to complete the graph.

What Do You Think?

48. Comparing complexity. Graph theory can be used to model Internet connections and routing of truck traffic. Which do you think would be more complex? Why?

49. Three dimensions. It's easy to imagine a graph in three dimensions. Such a graph would just be a collection of dots connected by segments. Can you think of a situation in which a three-dimensional graph would be a natural model?

50. Graphs and biology. Use the Internet to find an application of graph theory to biology, and write a brief report on your findings.

51. Vertices and edges. Give several examples of graphs, and for each graph make two calculations:

Calculation 1: Add up the degrees of all the vertices in the graph.

Calculation 2: Count the number of edges in the graph.

Do your calculations lead you to any ideas regarding graphs? Can you show that your idea is always correct?

52. A road map. You are looking at a local road map. In what situations might you be interested in an Euler circuit for the map? What businesses might have need of such a circuit? **Exercise 23 is relevant to this topic.**

53. The best you can do. If there is no Euler circuit, we have shown how to make efficient routes when there are exactly two odd-degree vertices. Use the Web to investigate ways of making efficient routes when there are more than two odd-degree vertices. **Exercise 35 is relevant to this topic.**

54. Social media. The text pointed out that graphs can be used to model connections between users of Facebook. Can you think of other social media that can be modeled using graphs?

55. Web links. Many Web sites have hyperlinks, also called links, to other sites. Consider a graph in which the vertices are Web sites and the edges indicate that one Web site has a link to another. Could you tell by looking at the graph which site is the source (that is, which one originates the link) for an edge? If not, how could you mark the graph to indicate which is the source? Use the Internet to investigate *directed graphs*.

56. More on Web links. Consider the graph in which the vertices are all of the sites on the Web and the edges indicate that one Web site has a hyperlink to another. Does that graph have any circuits? Is that graph connected, in the sense that each pair of vertices can be joined by a path? Does that graph have an edge that starts and ends at the same vertex?

7.2 Hamilton circuits and traveling salesmen: Efficient routes

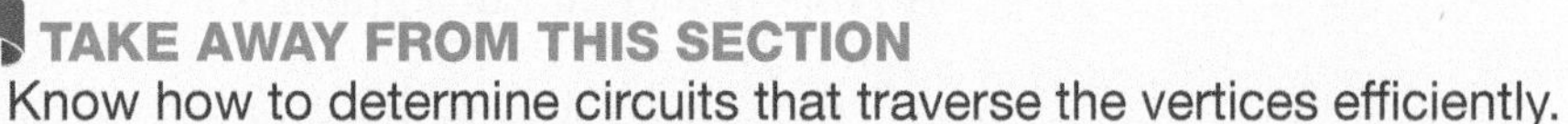

TAKE AWAY FROM THIS SECTION
Know how to determine circuits that traverse the vertices efficiently.

The following article appeared in *Scientific American*.

John Taggart/Bloomberg/Getty Images

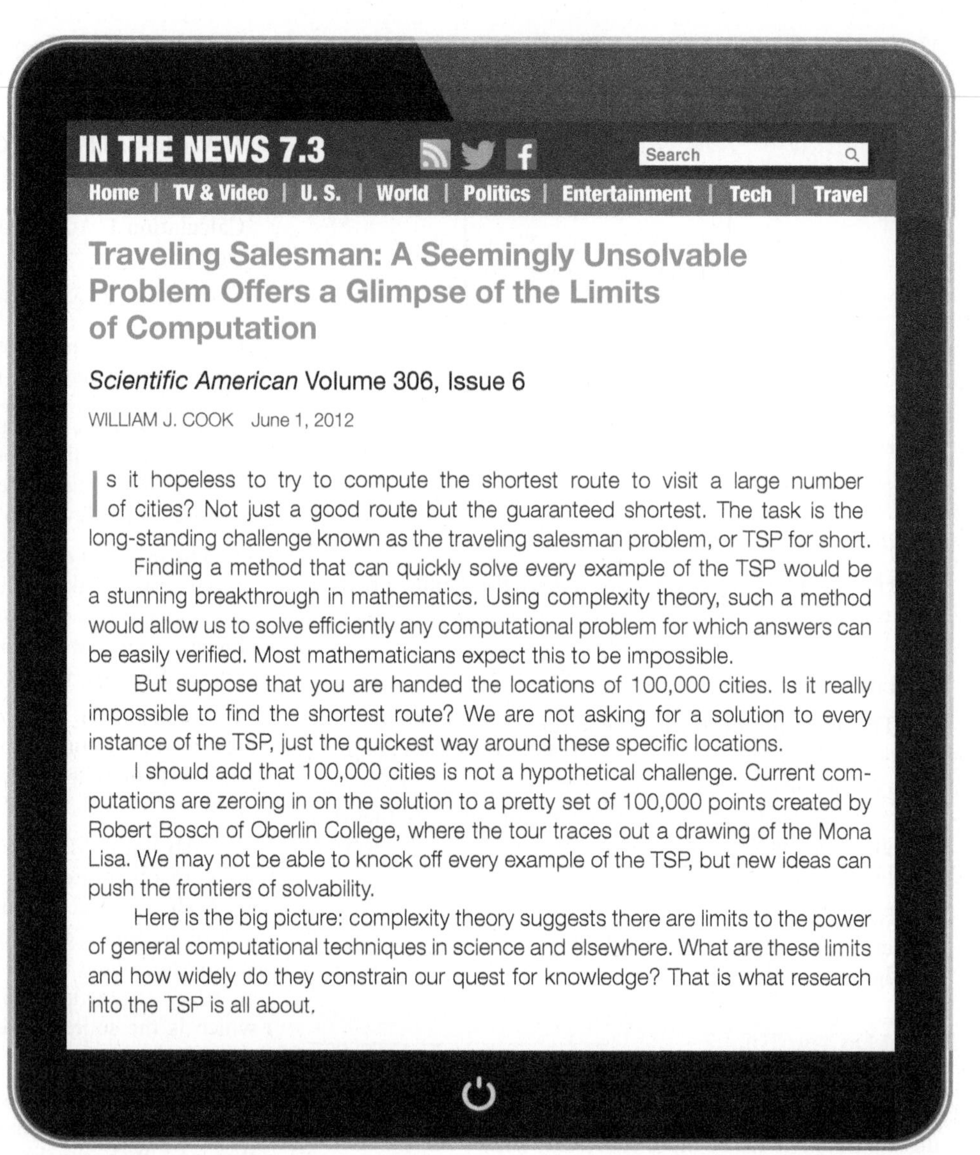
IN THE NEWS 7.3

Home | TV & Video | U. S. | World | Politics | Entertainment | Tech | Travel

Traveling Salesman: A Seemingly Unsolvable Problem Offers a Glimpse of the Limits of Computation

Scientific American Volume 306, Issue 6

WILLIAM J. COOK June 1, 2012

Is it hopeless to try to compute the shortest route to visit a large number of cities? Not just a good route but the guaranteed shortest. The task is the long-standing challenge known as the traveling salesman problem, or TSP for short.

Finding a method that can quickly solve every example of the TSP would be a stunning breakthrough in mathematics. Using complexity theory, such a method would allow us to solve efficiently any computational problem for which answers can be easily verified. Most mathematicians expect this to be impossible.

But suppose that you are handed the locations of 100,000 cities. Is it really impossible to find the shortest route? We are not asking for a solution to every instance of the TSP, just the quickest way around these specific locations.

I should add that 100,000 cities is not a hypothetical challenge. Current computations are zeroing in on the solution to a pretty set of 100,000 points created by Robert Bosch of Oberlin College, where the tour traces out a drawing of the Mona Lisa. We may not be able to knock off every example of the TSP, but new ideas can push the frontiers of solvability.

Here is the big picture: complexity theory suggests there are limits to the power of general computational techniques in science and elsewhere. What are these limits and how widely do they constrain our quest for knowledge? That is what research into the TSP is all about.

In the preceding section, we studied Euler circuits for graphs. The path for the *traveling salesman problem* described in the article above is a different type of circuit, called a *Hamilton circuit*. With an Euler circuit, our goal is to traverse efficiently the *edges* by using each edge exactly once. For a Hamilton circuit, we want to traverse efficiently the *vertices*; that is, we want a path starting and ending at the same place that visits each vertex exactly once (though it may miss some edges altogether).

KEY CONCEPT

A Hamilton circuit in a graph is a circuit that visits each vertex exactly once.

To compare the two types of circuits, we consider a highway map. An Euler circuit would be appropriate if you wish to travel along each *road* exactly once, as shown in **Figure 7.61**. A Hamilton circuit would be appropriate if you wish to visit each *city* exactly once, as shown in **Figure 7.62**. Note that in a Hamilton circuit, some edges may not be used at all.

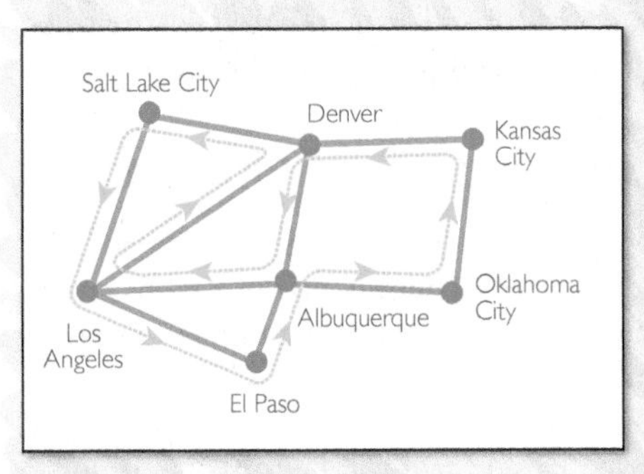

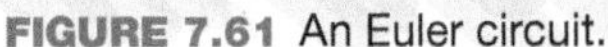
FIGURE 7.61 An Euler circuit.

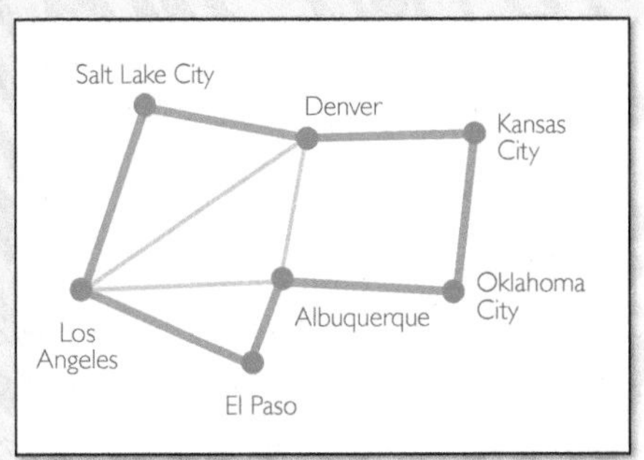

FIGURE 7.62 A Hamilton circuit shown in heavier edges.

A graph may have many Hamilton circuits, just as it may have many Euler circuits. For example, **Figures 7.63 and 7.64** show two different Hamilton circuits for the same graph. Thus, when you complete an exercise or follow one of our examples, you may well find a correct answer that is different from the circuit that we found.

Some graphs have no Hamilton circuit. A little trial and error will quickly convince you that the graph in **Figure 7.65** has no Hamilton circuit: Any circuit containing all the vertices must visit the middle vertex twice.

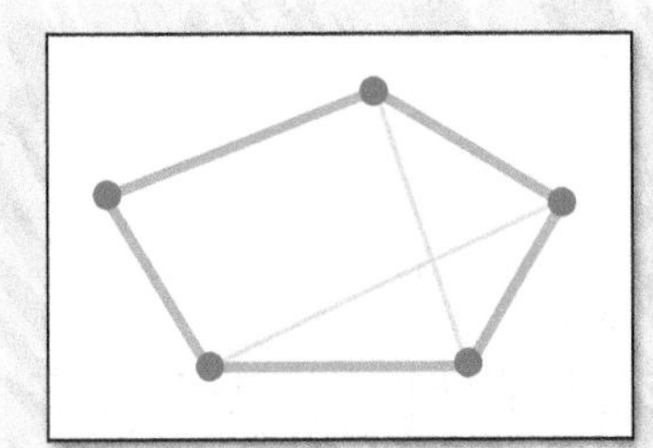

FIGURE 7.63 One Hamilton circuit.

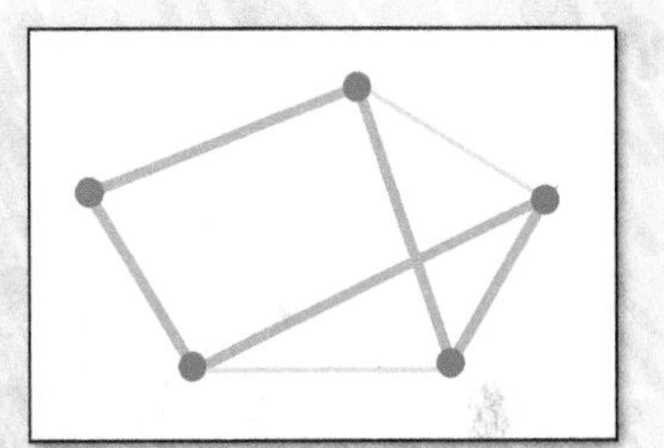

FIGURE 7.64 Another Hamilton circuit.

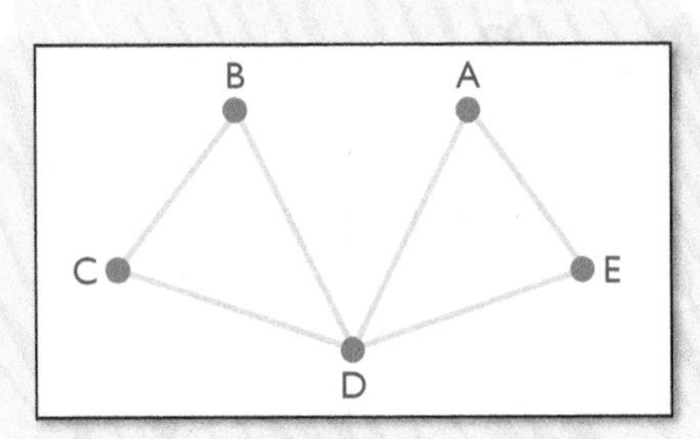

FIGURE 7.65 A graph with no Hamilton circuit.

Hamilton circuits are very important in routing problems that arise with airlines, delivery services, Internet routing, etc. For example, suppose a delivery driver has dozens of packages to be delivered to various destinations in a day. She would want to travel to each destination exactly once and not go to (or pass) one destination many times. If the vertices are the destinations, along with headquarters, and the edges are the road routes between them, the desired efficient route is a Hamilton circuit from headquarters and back.

Making Hamilton circuits

We know that connected graphs have Euler circuits precisely when each vertex is of even degree. There is no such easy indicator for when a Hamilton circuit exists. Furthermore, even when Hamilton circuits exist, they may be much more difficult to find than Euler circuits.

Although there is no nice recipe to tell us how to make Hamilton circuits, at least one observation is helpful when vertices of degree 2 are present. If we want to visit a city with only one road in and one road out, then any Hamilton circuit must use both roads. In general, if a graph has a vertex of degree 2, then each edge meeting that vertex must be part of any Hamilton circuit.

CALCULATION TIP 7.1 Vertices of Degree 2 and Hamilton Circuits

If a graph has a vertex of degree 2, then each edge meeting that vertex must be part of any Hamilton circuit.

Let's see how this observation can help us make Hamilton circuits. Suppose we need an efficient route that allows us to make deliveries from our terminal to the bank, the hardware store, Starbucks, Walmart, the Shoe Box, and Old Navy. That is, we want to find a Hamilton circuit for the graph in **Figure 7.66**. The vertices at the hardware store and at Walmart have degree 2. Hence, any Hamilton circuit must use the four roads highlighted in **Figure 7.67**.

It is now easy to see how to complete the Hamilton circuit, as shown in **Figure 7.68**.

It is helpful to keep in mind that as we try to extend a path to a Hamilton circuit, we cannot return to any vertex already used (except at the last step, when the path returns to its starting point). Therefore, no smaller circuit can be part of any Hamilton circuit.

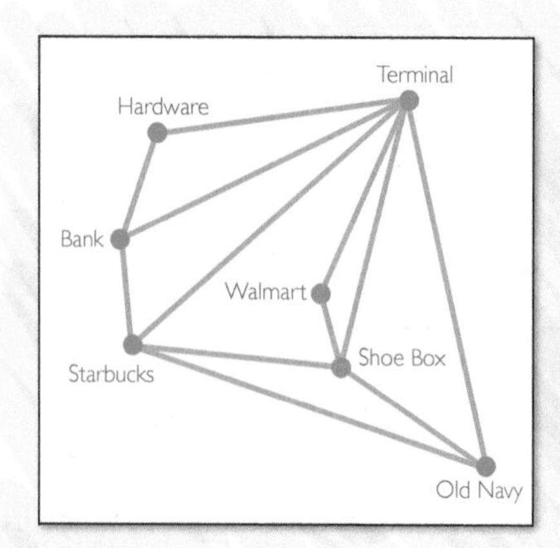

FIGURE 7.66 A delivery route problem.

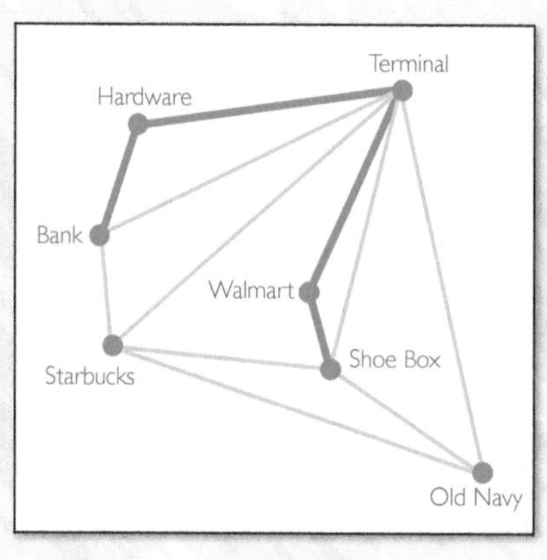

FIGURE 7.67 We must use edges that meet vertices of degree 2.

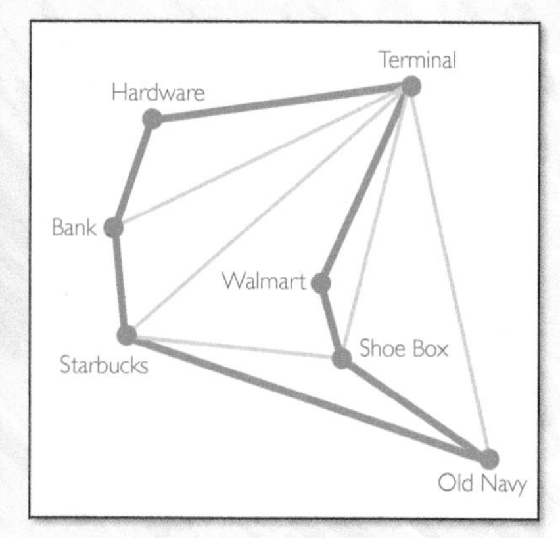

FIGURE 7.68 Completing the delivery route.

EXAMPLE 7.7 Finding a Hamilton circuit: Hiking trails

Figure 7.69 shows hiking trails and points of interest in Yellowstone National Park.

We want to find a path that starts and ends at Shoshone Lake and visits each point of interest exactly once. That is, we want to find a Hamilton circuit for the graph. Either find a Hamilton circuit or explain why no such circuit exists.

SOLUTION

Note that the vertices Grizzly Lake, Mammoth, Norris, and Lewis Lake all have degree 2. Thus, in any Hamilton circuit, we must use the hiking paths that touch these vertices. The corresponding edges are highlighted in **Figure 7.70.** The result is that we already have a circuit at the top of Figure 7.70. Such a circuit cannot be part of any larger Hamilton circuit, and we conclude that it is not possible to take the desired walking path. We will have to settle for a path that visits at least one point of interest more than once.

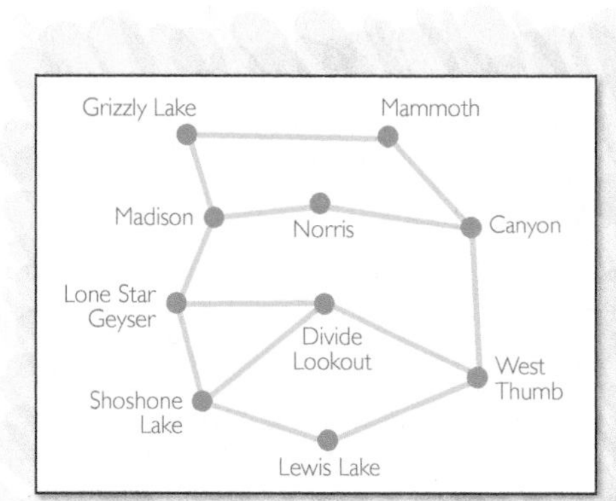

FIGURE 7.69 Hiking paths.

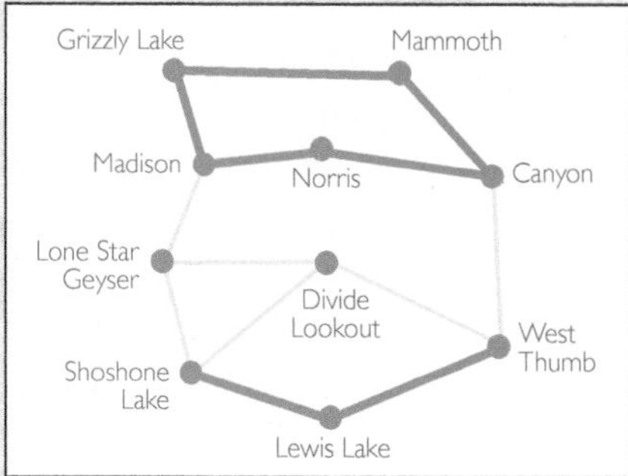

FIGURE 7.70 Edges meeting vertices of degree 2 must be used.

TRY IT YOURSELF 7.7

Figure 7.71 shows a terminal and businesses from which items must be picked up. Find a route, if one exists, that starts and ends at the terminal and visits each business exactly once. That is, either find a Hamilton circuit or explain why no such circuit exists.

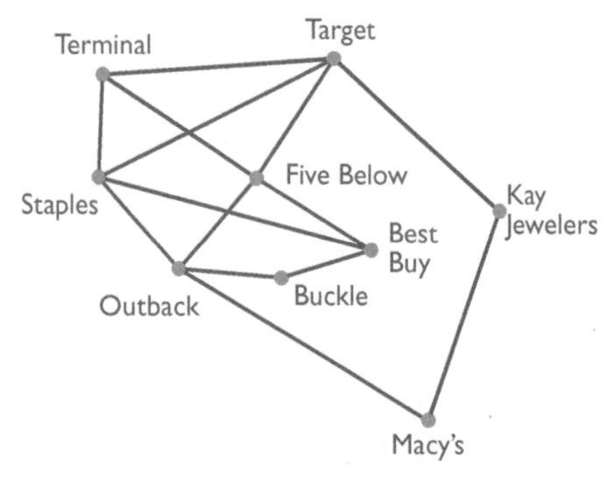

FIGURE 7.71 Terminal and businesses.

The answer is provided at the end of this section.

SUMMARY 7.3
Hamilton Circuits

1. A Hamilton circuit is a circuit that visits each vertex exactly once.

2. There is no set procedure for determining whether Hamilton circuits exist. Here is one helpful observation: If a graph has a vertex of degree 2, then each edge meeting that vertex must be part of any Hamilton circuit.

3. A Hamilton circuit cannot contain a smaller circuit.

The traveling salesman problem

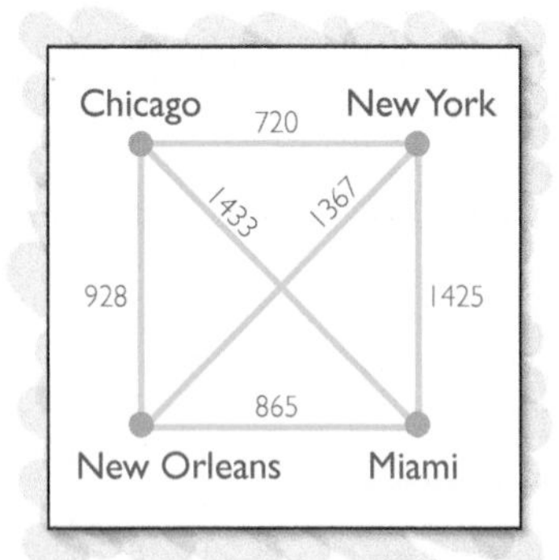

FIGURE 7.72 A simple map.

Consider a trucking company that wants to send a truck from its home base in New York to Chicago, Miami, and New Orleans, and then home again. A simplified map with mileage is shown in **Figure 7.72**. The company could save a lot of money by using a shortest route. This is an example of the famous *traveling salesman problem*, which seeks to find a shortest route that visits each city once and then returns home. In other words, the problem is to find a "shortest" Hamilton circuit. The traveling salesman problem has many important applications.

We can solve the traveling salesman problem in this case by listing the six possible routes and calculating the mileage for each:

- New York - Miami - Chicago - New Orleans - New York: Distance 5153 miles.
- New York - New Orleans - Chicago - Miami - New York: Distance 5153 miles. (*Note*: This is the reverse of the preceding trip.)
- New York - Miami - New Orleans - Chicago - New York: Distance 3938 miles.
- New York - Chicago - New Orleans - Miami - New York: Distance 3938 miles. (*Note*: This is the reverse of the preceding trip.)
- New York - Chicago - Miami - New Orleans - New York: Distance 4385 miles.
- New York - New Orleans - Miami - Chicago - New York: Distance 4385 miles. (*Note*: This is the reverse of the preceding trip.)

The shortest route is then New York - Miami - New Orleans - Chicago - New York or its reverse. It is worth noting that the mileage saved over alternative routes is significant.

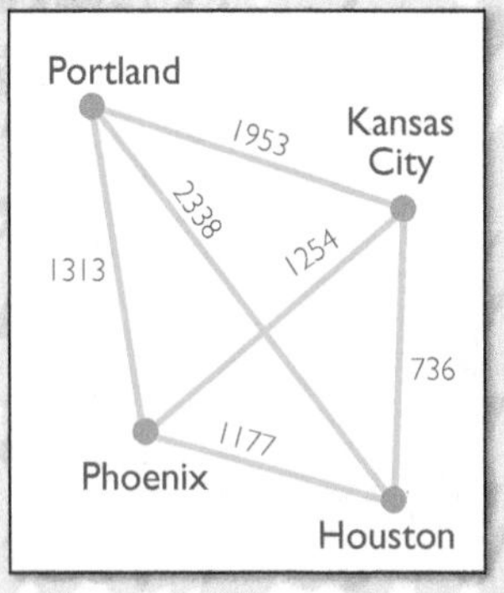

FIGURE 7.73 Delivery map.

EXAMPLE 7.8 Finding by hand a shortest route: Delivery truck

The graph in **Figure 7.73** shows a delivery map for a trucking firm based in Kansas City. The firm needs a shortest route that will start and end in Kansas City and make stops in Houston, Phoenix, and Portland. That is, the trucking firm needs a solution of the traveling salesman problem for this map. Calculate the mileage for each possible route to find the solution.

SOLUTION

There are six possible routes altogether, but we need to list only three because we gain no new information by looking at the reverse of a route:

- Kansas City - Houston - Phoenix - Portland - Kansas City: Distance 5179 miles.
- Kansas City - Phoenix - Houston - Portland - Kansas City: Distance 6722 miles.
- Kansas City - Phoenix - Portland - Houston - Kansas City: Distance 5641 miles.

The shortest route of 5179 miles is the first one listed.

TRY IT YOURSELF 7.8

For the graph in **Figure 7.74**, find a shortest delivery route that starts and ends at the bakery and makes deliveries to the coffee shop, the restaurant, and the convenience store. That is, solve the traveling salesman problem for this graph.

The answer is provided at the end of this section.

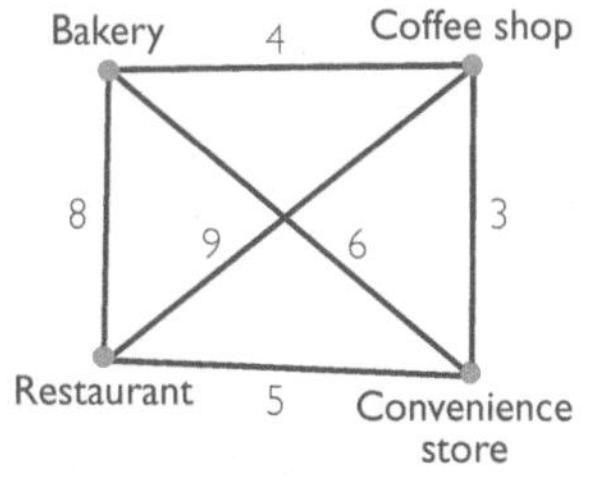

FIGURE 7.74 Delivery map.

The complexity of the traveling salesman problem

In each of the preceding traveling salesman examples, each destination is connected to every other by a direct route (an edge). In general, the traveling salesman problem applies to *complete graphs*, which are graphs where each vertex is connected to every other vertex by an edge. A complete graph on three vertices is a triangle. Complete graphs on four and five vertices are shown in **Figures 7.75 and 7.76.**

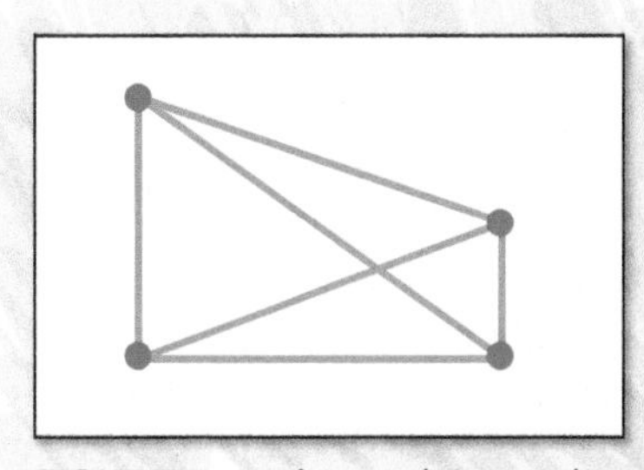

FIGURE 7.75 A complete graph on four vertices.

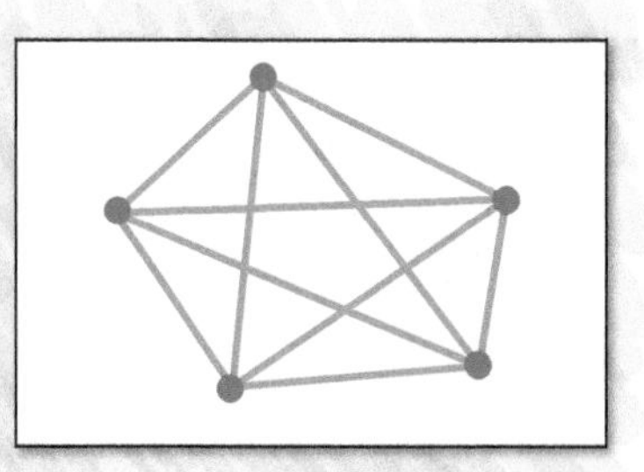

FIGURE 7.76 A complete graph on five vertices.

KEY CONCEPT

In a complete graph, each vertex is connected to every other vertex by an edge. The traveling salesman problem applies to complete graphs for which a distance (more generally, a value) is assigned to each edge. The problem is to find a shortest Hamilton circuit.

An example of the traveling salesman problem for five cities is shown in **Figure 7.77.** If we start at Seattle, it turns out that we need to check 12 separate routes (Hamilton circuits) to solve the problem for this map.

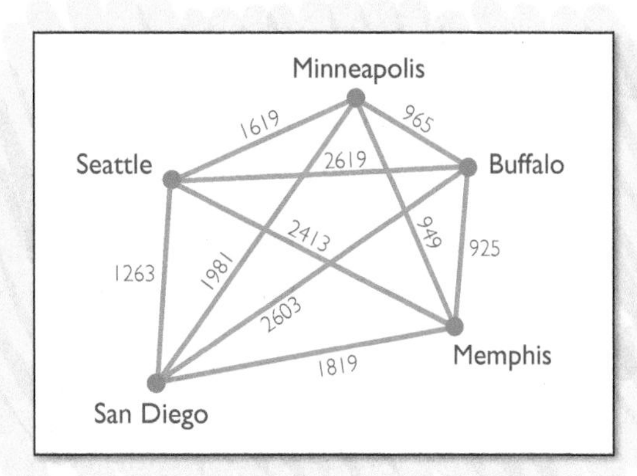

FIGURE 7.77 A traveling salesman problem for five cities.

The number of Hamilton circuits for the traveling salesman problem starting at a given city, not counting reverse routes, is shown in the table on the next page.[4]

[4] *See Exercise 50.*

Number of cities	Hamilton circuits
4	3
5	12
6	60
7	360
8	2520
9	20,160
10	181,440
11	1,814,400

Indeed, on a complete graph with n vertices, there are $(n-1)!/2$ Hamilton circuits starting at a given vertex, not counting reverse routes. (Recall that $k! = k \times (k-1) \times \cdots \times 1$. For example, $3! = 3 \times 2 \times 1 = 6$.)

This table means, for example, that if we deliver to 11 or more locations, we have to check more than a million routes to find a shortest one. For 100 cities, there are about 10^{158} possible tours. As a consequence, even a myriad of the very fastest computers could not solve the traveling salesman problem for 100 cities within a reasonable time frame. The result is that solving the traveling salesman problem for a large number of cities by listing each possible route is totally impractical.[5]

In 1962, Procter & Gamble held a contest to find a shortest route connecting 33 U.S. cities. There are 131,565,418,466,846,765,083,609,006,080,000,000 possible routes for 33 cities. Amazingly, the contest did have some winners!

Nearest-neighbor algorithm

Solving the traveling salesman problem by listing each route is not practical if the number of cities is large. To get around this difficulty, researchers have developed a number of clever algorithms that find Hamilton circuits that are "nearly optimal." One of these is the *nearest-neighbor algorithm*, which starts at a vertex and makes a Hamilton circuit. At each step in the construction, we travel to the nearest vertex that has not already been used. If there are two or more vertices equally nearby, we just pick any one of those nearest vertices.

KEY CONCEPT

The nearest-neighbor algorithm constructs a Hamilton circuit in a complete graph by starting at a vertex. At each step, it travels to the nearest vertex not already visited (except at the final step, where it returns to the starting point). If there are two or more vertices equally nearby, any one of them may be selected.

Let's see how this algorithm would work starting from Seattle for the map in **Figure 7.78**. The nearest city from Seattle is San Diego, which is 1263 miles away. From there, the nearest city is Memphis, 1819 miles away. The nearest city (not already visited) from Memphis is Buffalo. From Buffalo, we have no choice but to travel to Minneapolis and then to Seattle. The resulting route is shown in **Figure 7.79**. The total mileage for this route is 6591 miles. This turns out to be the shortest Hamilton circuit; however, as we shall see, the nearest-neighbor algorithm does not always produce a shortest route.

[5] *There is a very good history of the traveling salesman problem at* www.tsp.gatech.edu/index.html. *A poster for the contest next described is shown at that site.*

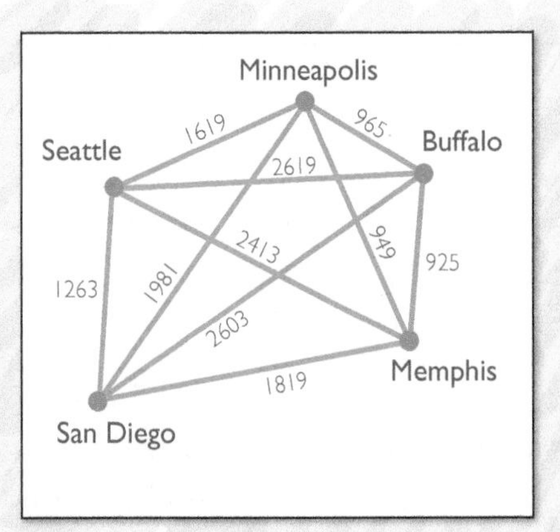

FIGURE 7.78 Traveling salesman problem for five cities.

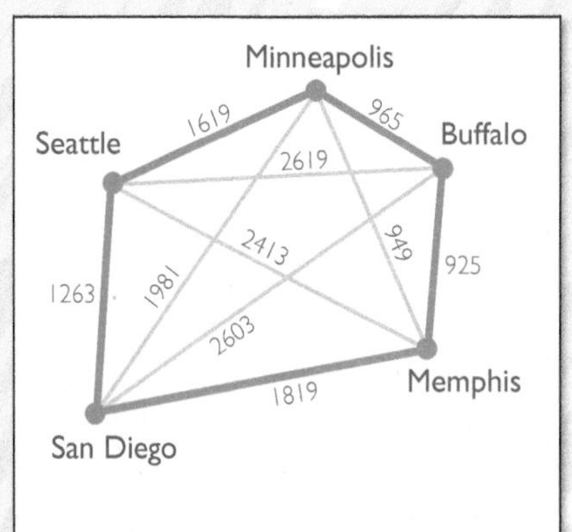

FIGURE 7.79 Result of the nearest-neighbor algorithm.

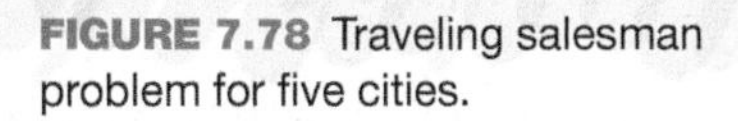

EXAMPLE 7.9 Using the nearest-neighbor algorithm: Connecting schools

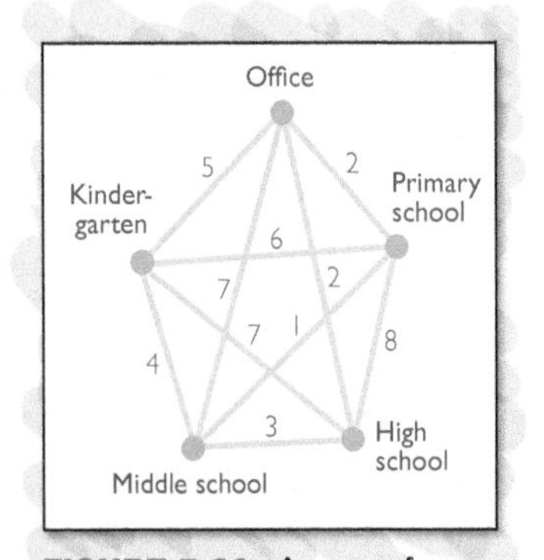

FIGURE 7.80 A map of schools.

The local school board needs to visit the high school, middle school, primary school, and kindergarten each day. A map illustrating the office and schools, along with distances in miles, is shown in **Figure 7.80**.

a. Find a shortest path by listing all possible routes starting and ending at the office.

b. Use the nearest-neighbor algorithm starting at the office to approximate a shortest Hamilton circuit.

SOLUTION

a. There are 12 possible routes if we do not include reverse routes. We provide a visual display of each of these routes in **Figures 7.81** through **7.92**. The shortest Hamilton circuit, as shown in Figure 7.90, is the 16-mile route shown: Office - High school - Kindergarten - Middle school - Primary school - Office. We emphasize once more that there is no known way to get the shortest route without listing all the possibilities, and when even only a moderate number of vertices are involved, the list becomes extraordinarily long. This is what makes the problem so difficult.

b. From the office, there are two nearest schools—both the primary school and high school are 2 miles away. We are free to choose either route. Suppose we choose to go first to the primary school. The closest school from the primary school is the middle school 1 mile away. The next nearest school (not already visited) is the high school.

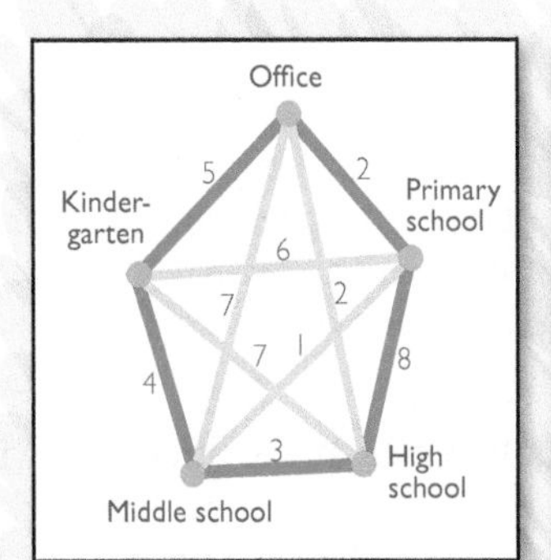

FIGURE 7.81 22 miles.

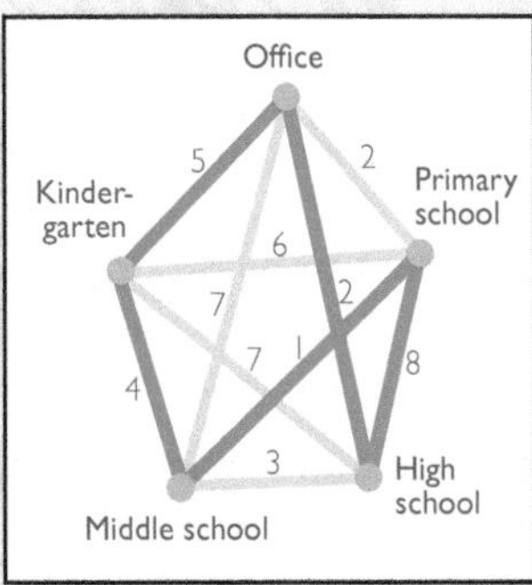

FIGURE 7.82 20 miles.

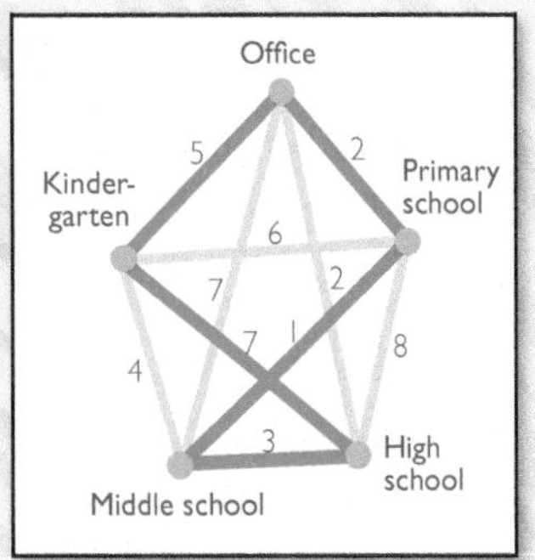

FIGURE 7.83 18 miles. (This is one of two routes that are found by the nearest-neighbor algorithm in part b.)

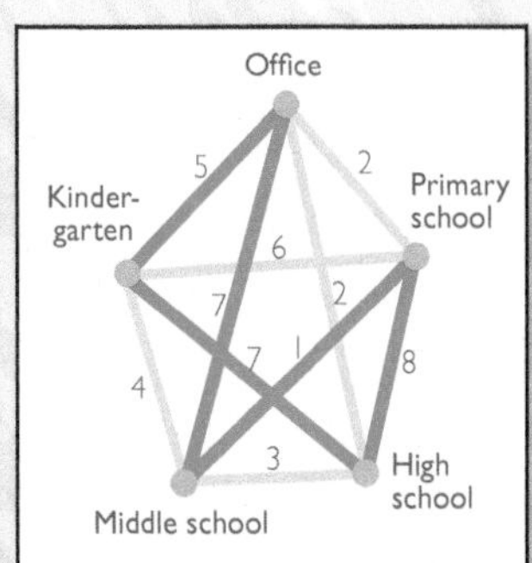

FIGURE 7.84 28 miles.

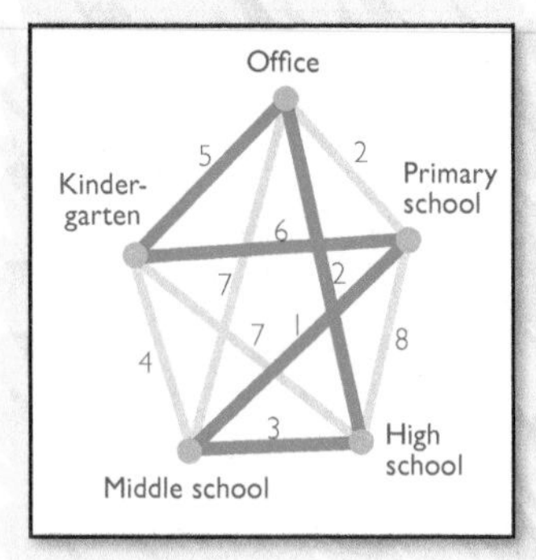

FIGURE 7.85 17 miles. (This is one of two routes that are found by the nearest-neighbor algorithm in part b.)

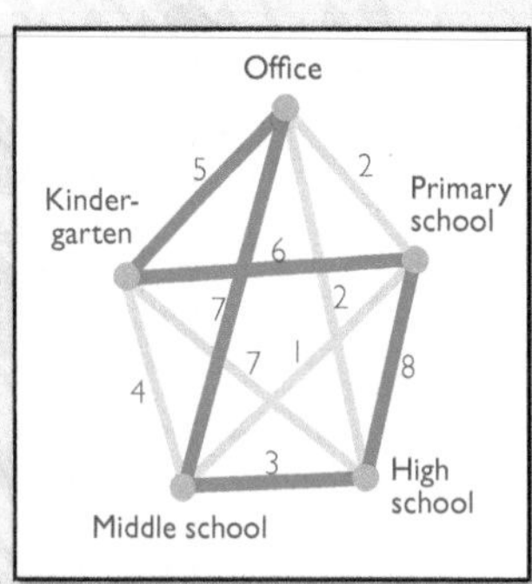

FIGURE 7.86 29 miles.

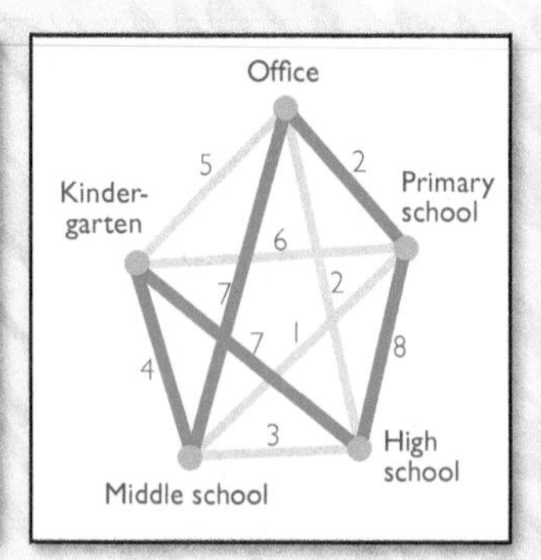

FIGURE 7.87 28 miles.

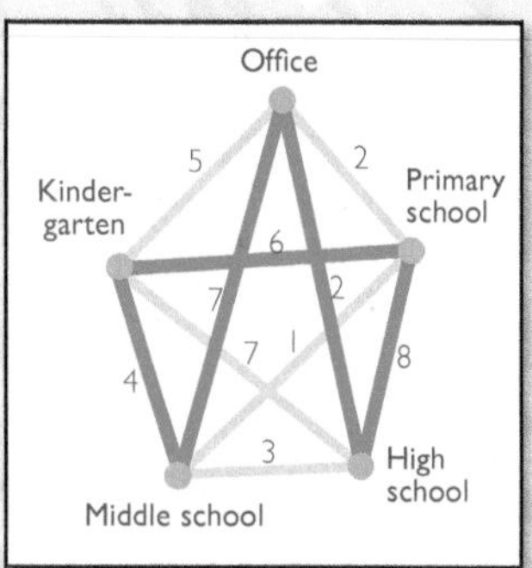

FIGURE 7.88 27 miles.

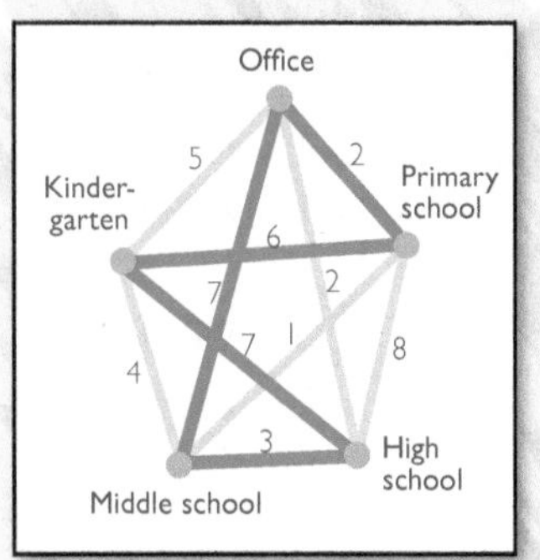

FIGURE 7.89 25 miles.

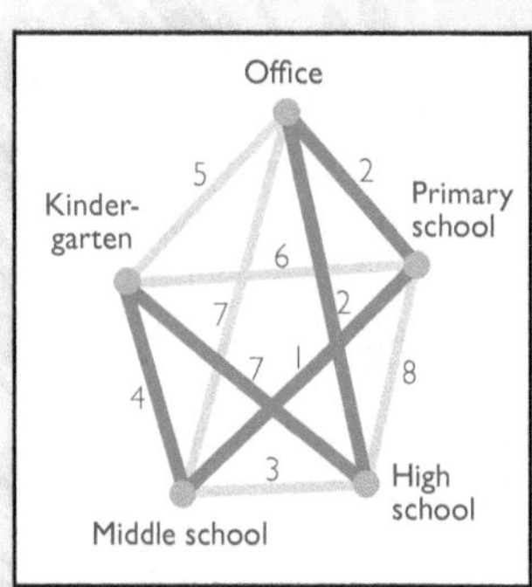

FIGURE 7.90 16 miles.

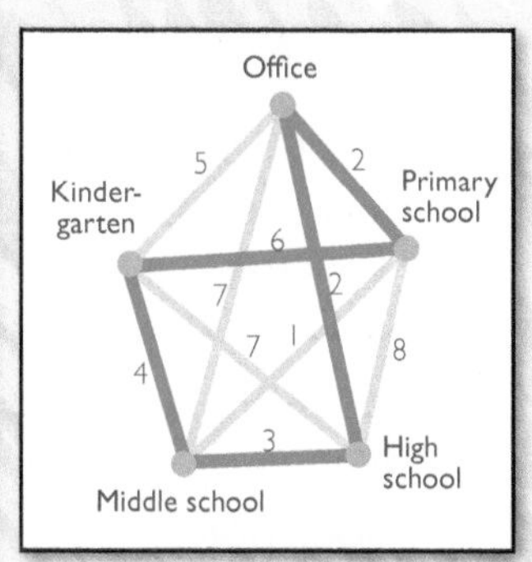

FIGURE 7.91 17 miles.

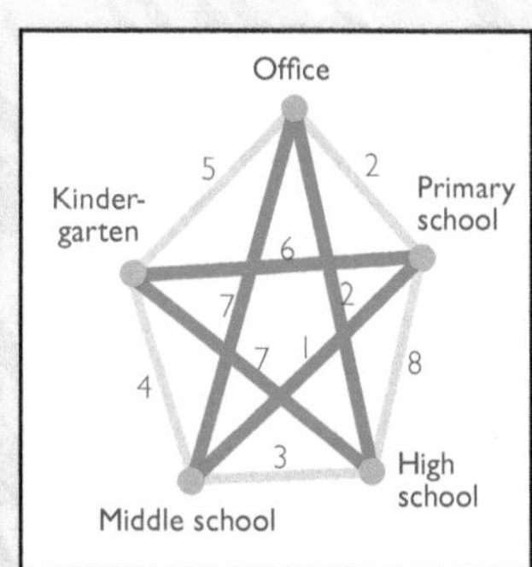

FIGURE 7.92 23 miles.

We complete the route by traveling to the kindergarten and then back to the office. The resulting route, which is shown in Figure 7.83, is Office - Primary school - Middle school - High school - Kindergarten - Office. This route is 18 miles long.

The reader can easily verify that if we had chosen to go first to the high school rather than the primary school, we would have gotten the route shown in Figure 7.85. That route is 17 miles long. We note that both of these routes are longer than the shortest route of 16 miles found in part a.

This example shows that the nearest-neighbor algorithm can fail to find a shortest route.

The cheapest link algorithm

Another method used to find approximate solutions of the traveling salesman problem is the *cheapest link algorithm*. This procedure begins by selecting the shortest edge in the graph and continues by selecting the shortest edge that has not already been chosen. (If there is more than one shortest edge, choose either.) But at each step, we must be careful not to violate the following rules:

Rule 1: Do not choose an edge that results in a circuit except at the final step, when a Hamilton circuit is constructed.

Rule 2: Do not choose an edge that would result in a vertex of degree 3.

The cheapest link algorithm is just as easy to implement as the nearest-neighbor algorithm. Let's see how it works with the school route problem from Example 7.9.

We first select the shortest edge in the graph, which is the 1-mile edge from the primary school to the middle school. This is shown in **Figure 7.93**. Next there are two edges of length 2. We can choose both of them without violating either Rule 1 or Rule 2. The pieces thus far assembled are shown in **Figure 7.94**.

The next shortest edge is the 3-mile route joining the middle school to the high school. But we can't use this edge because it would make the circuit shown in **Figure 7.95**. Instead, we choose the 4-mile route from the middle school to the kindergarten, as shown in **Figure 7.96**.

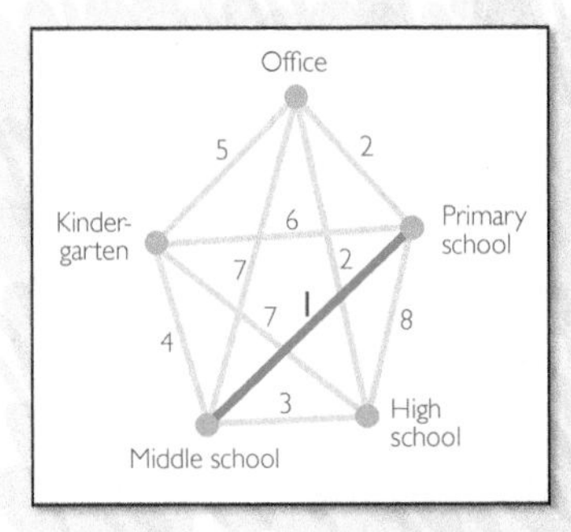

FIGURE 7.93 Selecting the shortest edge in the graph.

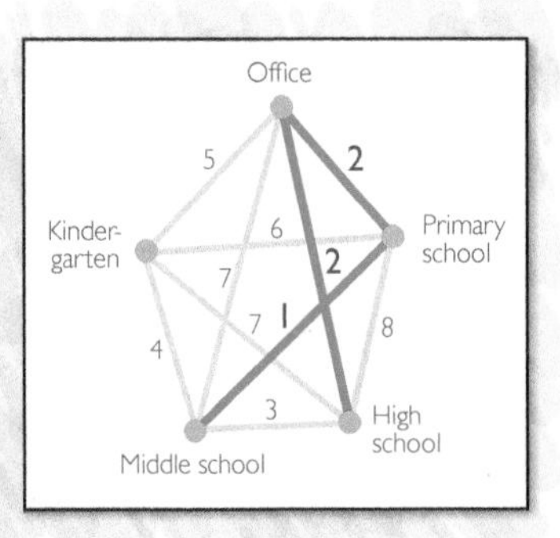

FIGURE 7.94 Adding two more edges.

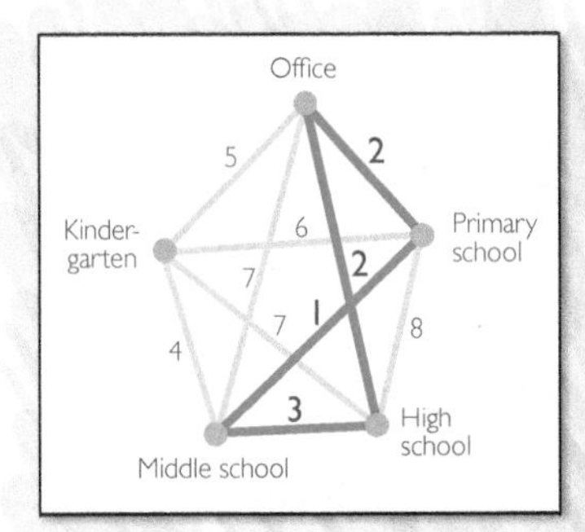

FIGURE 7.95 Choosing the 3-mile edge results in a circuit.

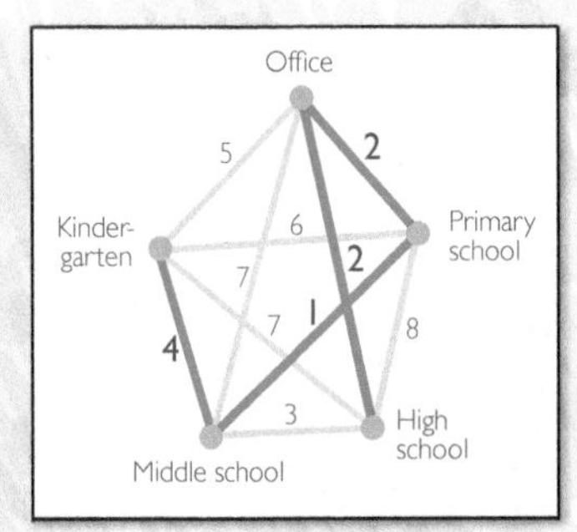

FIGURE 7.96 Use the 4-mile edge instead.

The next shortest edge is the 5-mile route from the kindergarten to the office. But **Figure 7.97** shows that this choice violates both Rule 1 and Rule 2. The only other choice is the 7-mile route from the kindergarten to the high school, which results in the Hamilton circuit shown in **Figure 7.98**. We note that this is the shortest possible

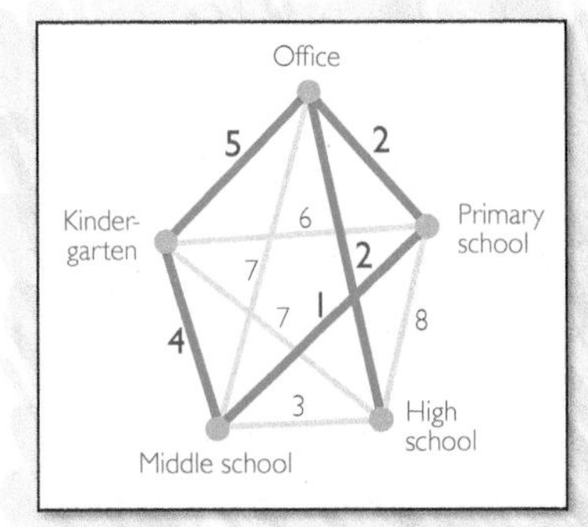

FIGURE 7.97 The edge from kindergarten to office violates both Rule 1 and Rule 2.

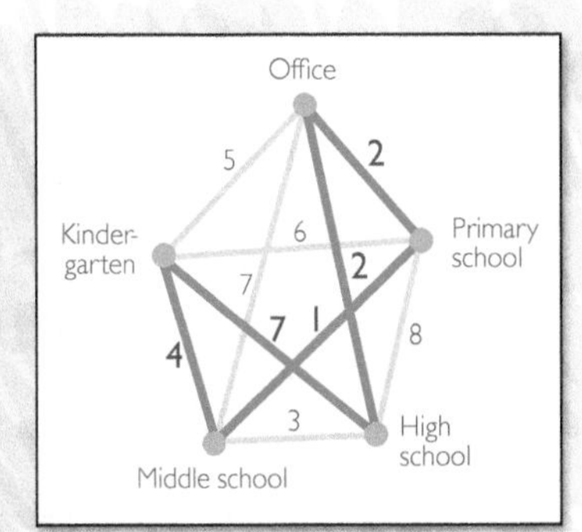

FIGURE 7.98 The completed Hamilton circuit.

route, as we found in Example 7.9. But the cheapest link algorithm does not always produce the shortest path, as the next example shows.

EXAMPLE 7.10 Applying the cheapest link algorithm: A trucking company

We run a trucking company that makes deliveries in Seattle, Minneapolis, Buffalo, Memphis, and San Diego. The map is shown in **Figure 7.99**. Use the cheapest link algorithm to find an approximate solution of the traveling salesman problem.

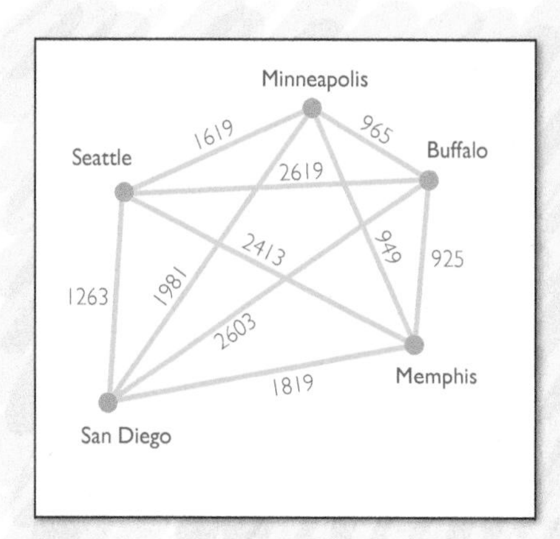

FIGURE 7.99 Five cities.

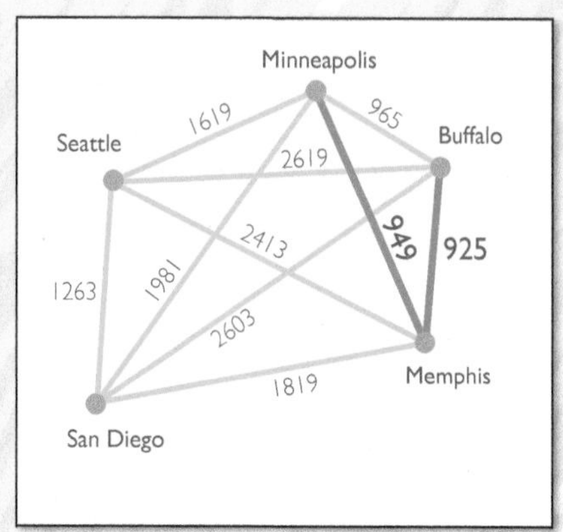

FIGURE 7.100 Choosing the first two segments of the route.

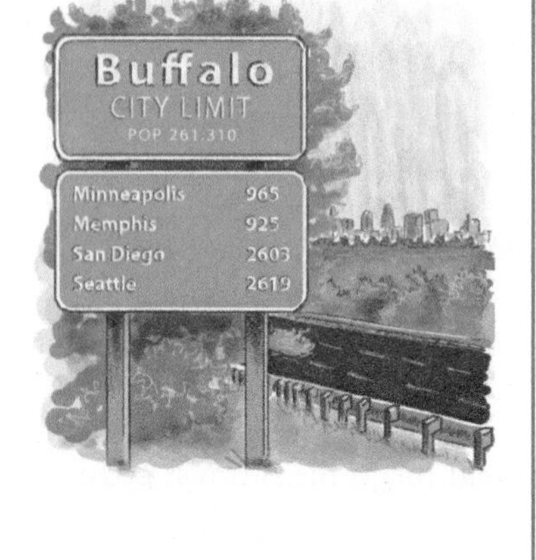

SOLUTION

We begin by choosing the two shortest edges in the graph, as shown in **Figure 7.100**.

The next shortest edge is the 965-mile route from Minneapolis to Buffalo. But we can't use this route because it would result in the circuit shown in **Figure 7.101**. We choose instead the next shortest edge, the 1263-mile route from San Diego to Seattle. The edges from Seattle to Minneapolis and San Diego to Buffalo complete the route, as shown in **Figure 7.102**.

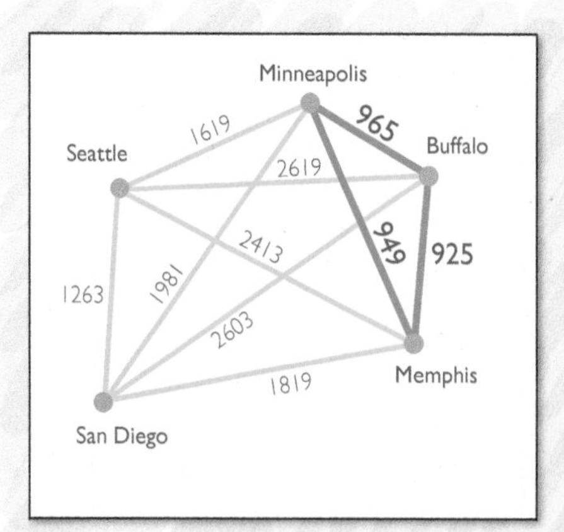

FIGURE 7.101 Adding the edge from Minneapolis to Buffalo makes a circuit.

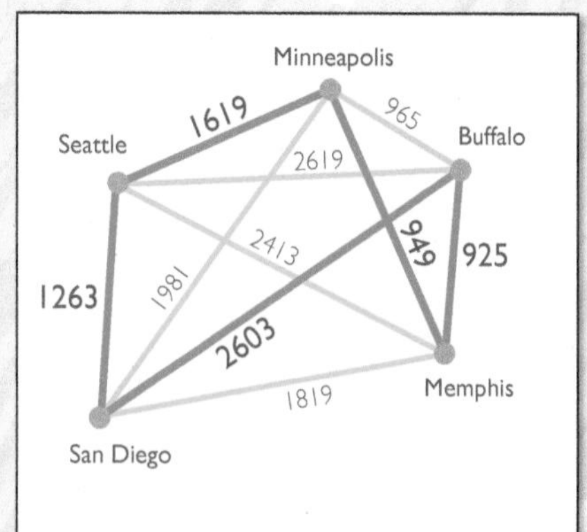

FIGURE 7.102 The completed route.

This route has a total length of $925 + 949 + 1263 + 1619 + 2603 = 7359$ miles. This is longer than the shortest route of 6591 miles that we found earlier using the nearest-neighbor algorithm (see Figure 7.79).

TRY IT YOURSELF 7.10

Figure 7.103 shows a map of the cities where our air freight company makes deliveries. Use the cheapest link algorithm to find an approximate solution of the traveling salesman problem.

The answer is provided at the end of this section.

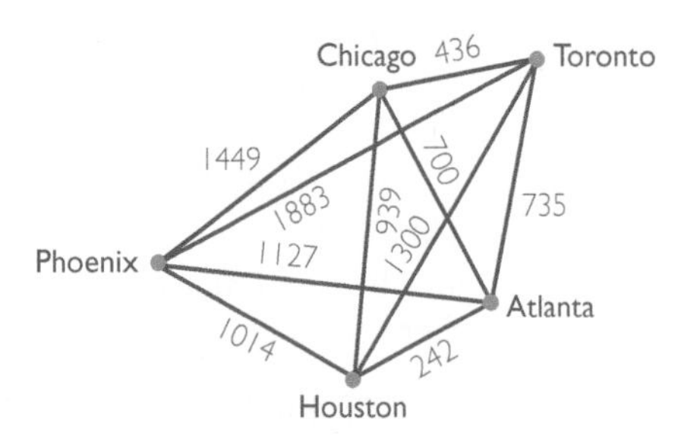

FIGURE 7.103 An air freight map.

SUMMARY 7.4
The Traveling Salesman Problem

1. A complete graph is one in which each pair of vertices is joined by an edge.
2. If a distance (more generally, a value) is assigned to each edge, the traveling salesman problem is to find a shortest Hamilton circuit.
3. Solving the traveling salesman problem is difficult when there are more than just a few vertices. However, there are algorithms that can give an approximate solution. One such is the nearest-neighbor algorithm. Another is the cheapest link algorithm.

The two algorithms that we have presented have the advantage of being simple, and they provide a flavor of how algorithms work. But, in fact, they are of questionable effectiveness in real situations. Much better (but more complex) algorithms are available that provide practical approximations to the optimal solution.

WHAT DO YOU THINK?

Hamilton circuits and Euler circuits: Explain the differences between a Hamilton circuit and an Euler circuit. Make two graphs to show that you can have one when the other does not exist.

Questions of this sort may invite many correct responses. We present one possibility: Both Hamilton and Euler circuits involve a circuit that begins and ends at the same place. An Euler circuit visits each edge exactly once but may visit some vertices more than once. A typical application is a mail route where the postman needs to travel along each road exactly once. A Hamilton circuit visits each vertex exactly once but may miss some edges altogether. This might be of interest to a trucking company that has to make one stop at each client business.

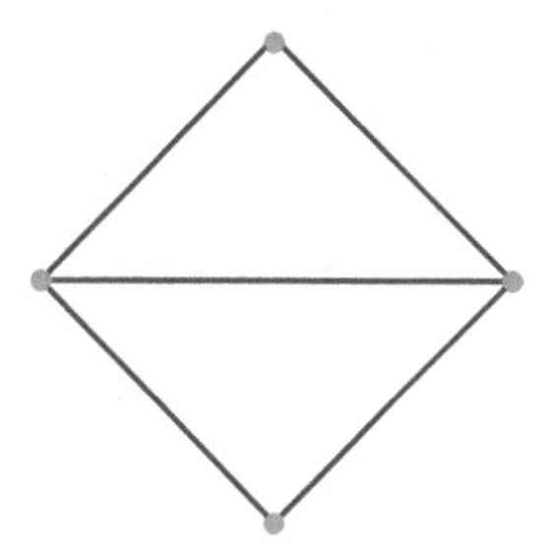

A graph with a Hamilton circuit but no Euler circuit.

A graph with an Euler circuit but no Hamilton circuit.

The figure on the left has no Euler circuit because it has vertices of odd degree. The outer circuit is a Hamilton circuit. The "bow tie" in the figure on the right has an obvious Euler circuit, but every circuit must visit the center vertex twice. Hence there is no Hamilton circuit.

Further questions of this type may be found in the *What Do You Think?* section of exercises.

Try It Yourself answers

Try It Yourself 7.7: Finding a Hamilton circuit: Visiting stores One Hamilton circuit is shown using highlighted edges.

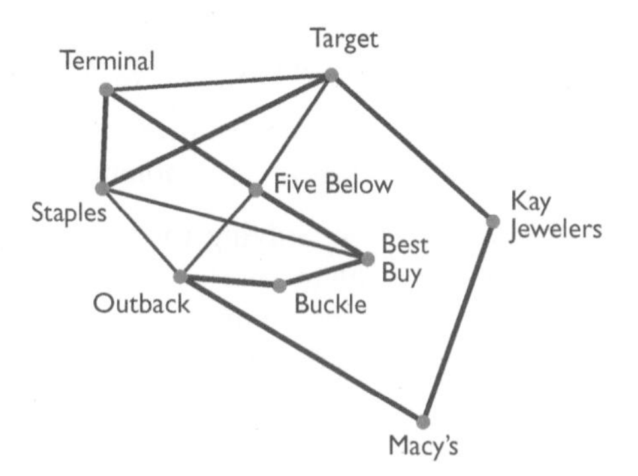

Try It Yourself 7.8: Finding by hand a shortest route: Delivery truck The route Bakery - Coffee shop - Convenience store - Restaurant - Bakery (or its reverse) has the minimal length of 20 miles.

Try It Yourself 7.10: Applying the cheapest link algorithm: Air freight deliveries The route below has a length of 4275 miles.

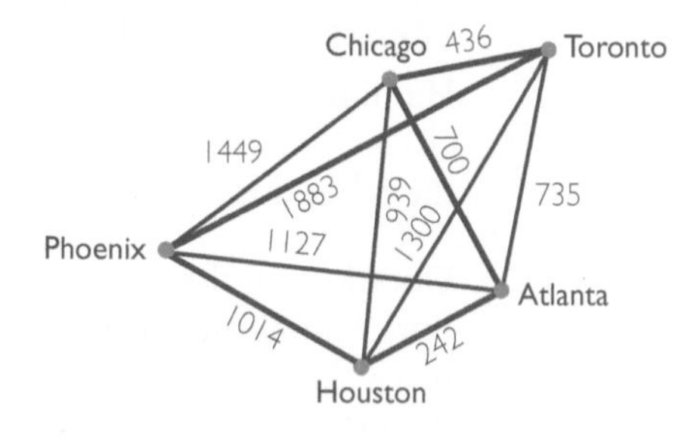

Exercise Set 7.2

Test Your Understanding

1. State in your own words what a Hamilton circuit is.

2. Which of the following are true for the traveling salesman problem? **a.** The traveling salesman problem asks how to find the most efficient Hamilton circuit in a complete graph for which a value is assigned to each edge. **b.** Trial and error is the only known procedure that always finds an optimal solution. **c.** The cheapest link algorithm is an example of a procedure designed to approximate an optimal solution. **d.** all of the above.

3. True or false: The nearest-neighbor algorithm always finds the most efficient Hamilton circuit in a complete graph where values are assigned to edges.

4. True or false: Modern computers can solve the traveling salesman problem even for very large, complete graphs.

5. True or false: A graph never contains more than one Hamilton circuit.

6. True or false: The traveling salesman problem does not have practical applications.

Problems

7. Euler or Hamilton? The edges in a certain graph represent unpaved roads, and the vertices represent intersections. We want to pave all of the streets and not require the paving machine to move along streets already paved. Should we look for an Euler circuit or a Hamilton circuit?

8. Euler or Hamilton? The edges in a certain graph represent roads, and the vertices represent intersections. We want to inspect the traffic signs at each intersection and not visit any intersection twice. Should we look for an Euler circuit or a Hamilton circuit?

9. Euler but not Hamilton? Consider the graph in **Figure 7.104**. We noted on page 491 that this graph (which is the same as the graph in Figure 7.65) has no Hamilton circuit. Does this graph have an Euler circuit?

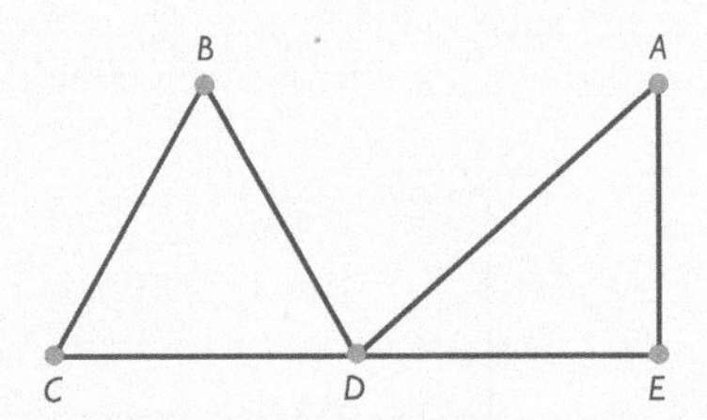

FIGURE 7.104 A graph with no Hamilton circuit.

10. Hamilton but not Euler? The graph in **Figure 7.105** has a Hamilton circuit, as is shown by the heavier edges. Does this graph have an Euler circuit?

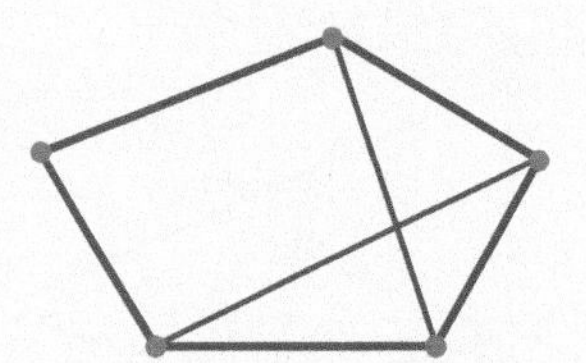

FIGURE 7.105 A graph with a Hamilton circuit.

11. A park. Figure 7.106 shows hiking trails and points of interest in a national park. Either find a route that begins at the park entrance, visits each point of interest once, and returns to the park entrance, or show that no such route exists. That is, either find a Hamilton circuit or explain why no such circuit exists.

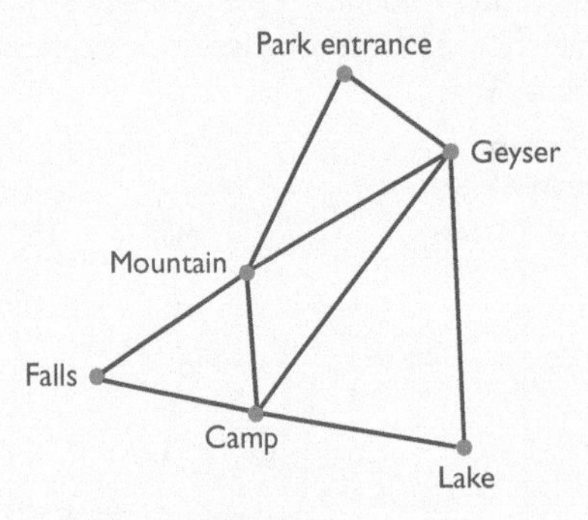

FIGURE 7.106 A park.

12. A bakery. Figure 7.107 shows a bakery and several sites where baked goods are to be delivered. Either find a route that begins and ends at the bakery and passes each store exactly once, or show that no such route exists. That is, either find a Hamilton circuit or explain why no such circuit exists.

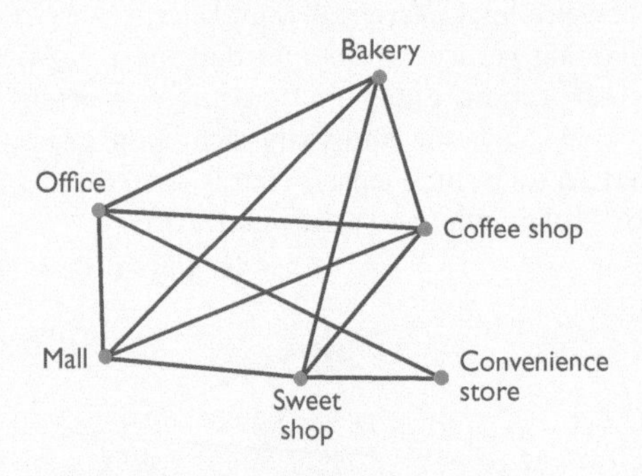

FIGURE 7.107 A bakery.

13. Computer network. **Figure 7.108** shows a computer network. We want to route a message starting from computer A, passing to each other computer on the network exactly once, and returning to A. Either find such a path or show that no such path exists. That is, either find a Hamilton circuit or explain why no such circuit exists.

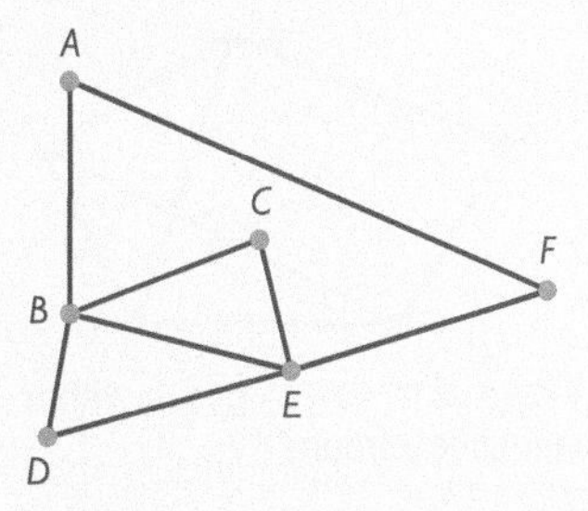

FIGURE 7.108 Computer network.

14. Visiting historical sites. **Figure 7.109** shows some historical sites in Philadelphia. We want to take a walking tour beginning and ending at Franklin Square and visiting each site exactly once. Either find such a route or show that none exists. That is, either find a Hamilton circuit or explain why no such circuit exists.

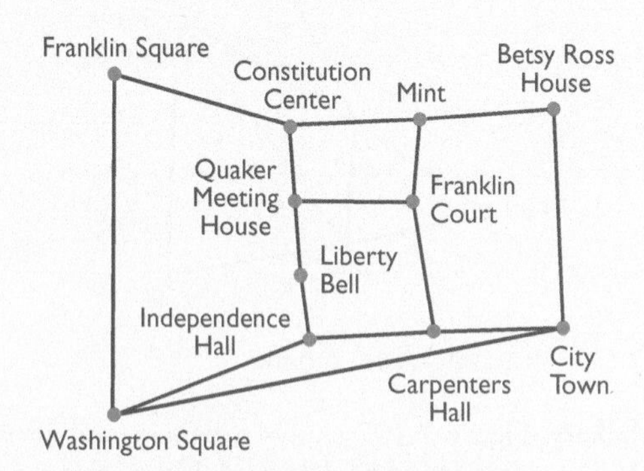

FIGURE 7.109 Visiting historical sites.

15. Internet connections. The edges in **Figure 7.110** represent fiber-optic lines. Vertex A is an Internet service provider. The remaining vertices are homes that must be wired for an Internet connection. Either find a route that begins and ends at the provider location and visits each home exactly once, or show that no such route exists. That is, either find a Hamilton circuit or explain why no such circuit exists.

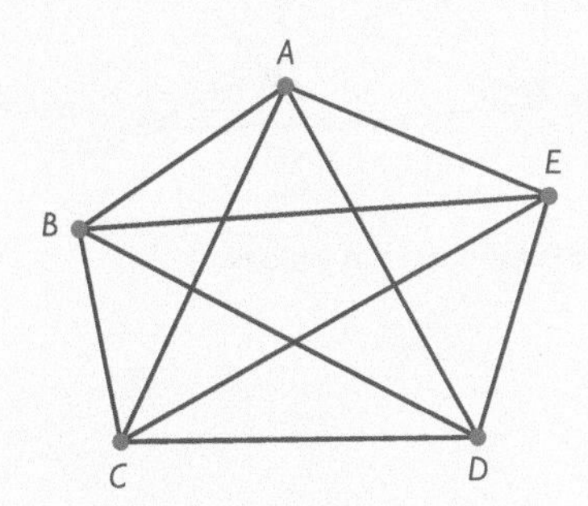

FIGURE 7.110 Internet connections.

16. Tour of western states. The vertices in **Figure 7.111** represent western states. Vertices are connected by an edge if the two states share a common border. We wish to take a scenic tour of the western United States. Either find a path that begins and ends at Washington and visits each state exactly once, or show that no such path exists. That is, either find a Hamilton circuit or explain why no such circuit exists.

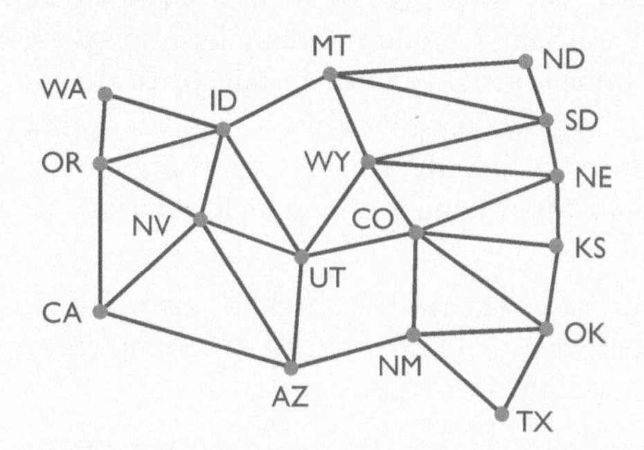

FIGURE 7.111 Tour of western states.

17. Tour of southern states. The vertices in **Figure 7.112** represent southern states. Vertices are connected by an edge if the two states share a common border. We wish to take a scenic tour of the southern United States. Either find a path that begins and ends at Missouri and visits each state exactly once, or show that no such path exists. That is, either find a Hamilton circuit or explain why no such circuit exists.

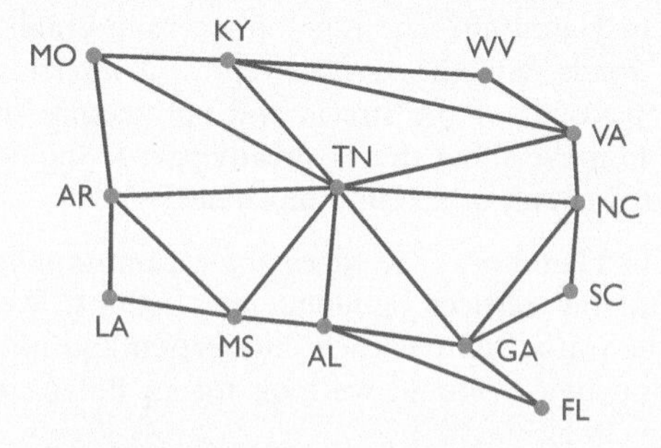

FIGURE 7.112 Tour of southern states.

18. Viewing holiday decorations. The edges in **Figure 7.113** represent streets. Vertex A is your home, and the remaining vertices are neighborhood homes that are displaying holiday decorations. Either find a path that begins and ends at your own home and passes each neighborhood home exactly once, or show that no such path exists. That is, either find a Hamilton circuit or explain why no such circuit exists.

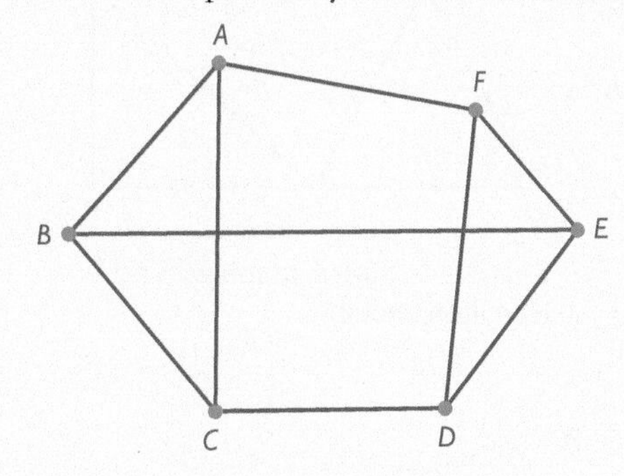

FIGURE 7.113 Viewing holiday decorations.

19. Holiday shopping. **Figure 7.114** shows a shopping center. The vertices are stores, and the edges are sidewalks. We wish to visit each store for holiday shopping. Either find a route that begins and ends at T.J. Maxx and visits each store once, or show that no such route exists. That is, either find a Hamilton circuit or explain why no such circuit exists.

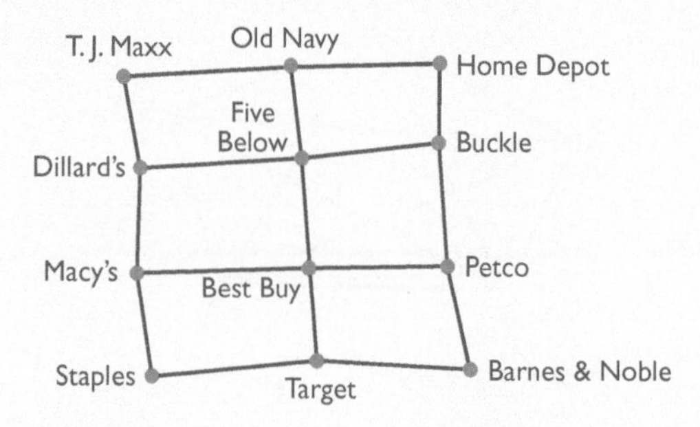

FIGURE 7.114 Holiday shopping.

20. Checking in-home care patients. The edges in **Figure 7.115** represent roads. The vertex marked "office" is the headquarters of an in-home care service. The remaining vertices are homes where patients must be visited each day. Either find a route beginning and ending at headquarters and visiting each patient home exactly once, or show that no such route exists. That is, either find a Hamilton circuit or explain why no such circuit exists.

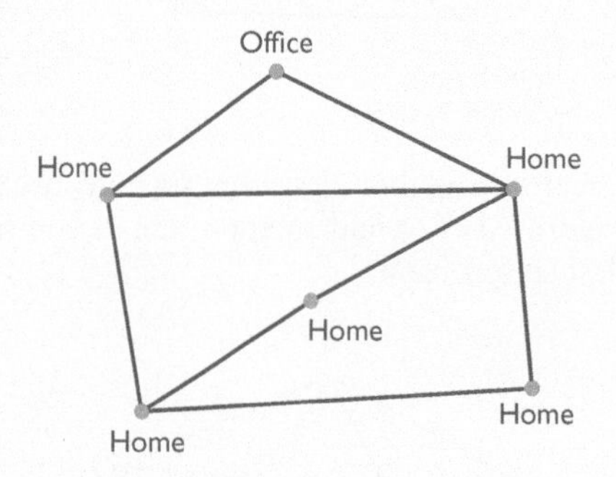

FIGURE 7.115 Checking in-home care patients.

21. Meals on Wheels. The edges in **Figure 7.116** represent roads. Meals on Wheels is an organization that provides meals for the elderly. The remaining vertices are homes of elderly people to which hot meals are to be delivered. Either find a path that begins and ends at the kitchen and passes each home exactly once, or show that no such path exists. That is, either find a Hamilton circuit or explain why no such circuit exists.

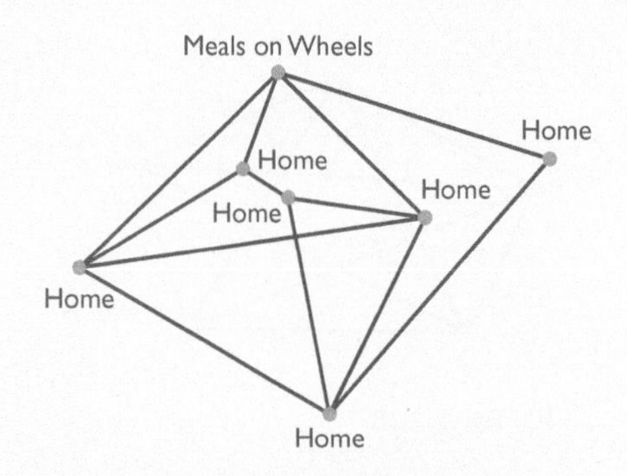

FIGURE 7.116 Meals on Wheels.

22. Routing a message. The vertices in **Figure 7.117** represent computers, and the edges represent connections. To check the network setup, we want to send a test message that starts at computer A, passes through each other computer exactly once, and successfully returns to computer A. Either find a workable path or show that no such path exists. That is, either find a Hamilton circuit or explain why no such circuit exists.

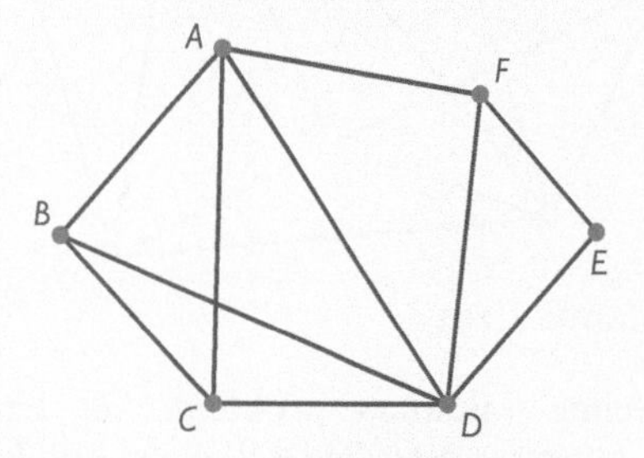

FIGURE 7.117 Routing a message.

23–26. Hamilton circuits. *In Exercises 23 through 26, either find a Hamilton circuit or show that none exists.*

23. Either find a Hamilton circuit or show that none exists for the graph in **Figure 7.118**.

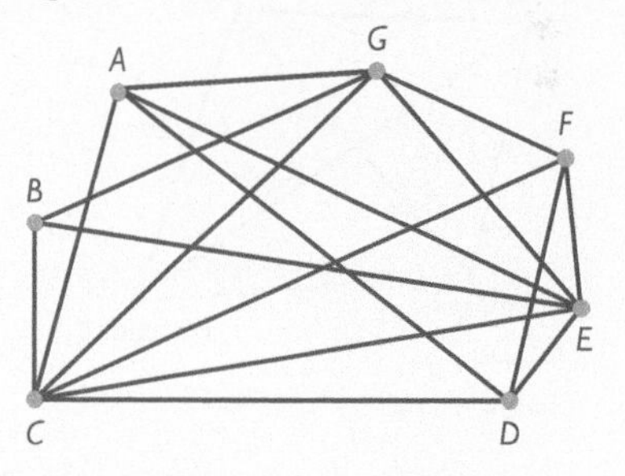

FIGURE 7.118

24. Either find a Hamilton circuit or show that none exists for the graph in **Figure 7.119**.

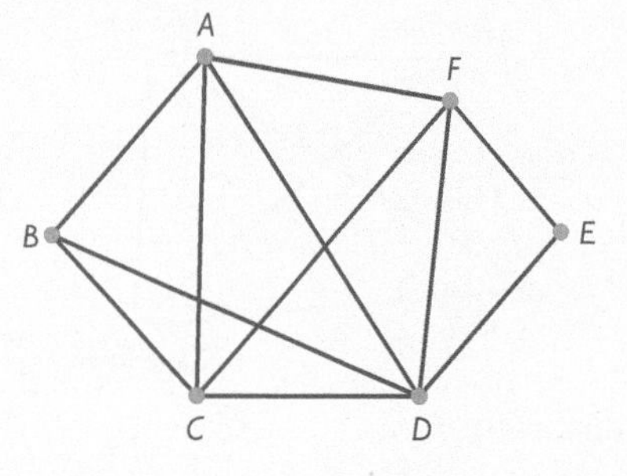

FIGURE 7.119

25. Either find a Hamilton circuit or show that none exists for the graph in **Figure 7.120**.

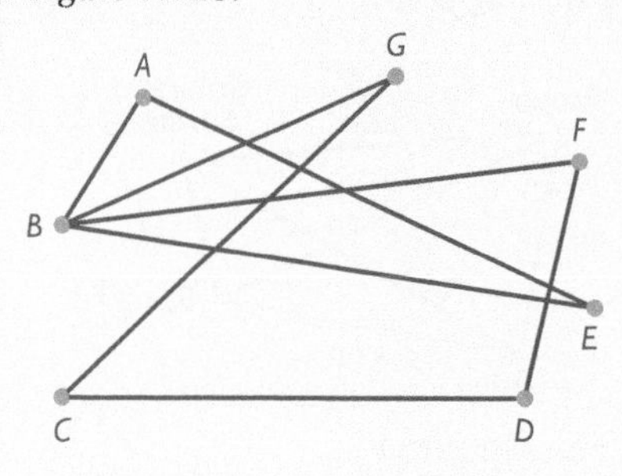

FIGURE 7.120

26. Either find a Hamilton circuit or show that none exists for the graph in **Figure 7.121**.

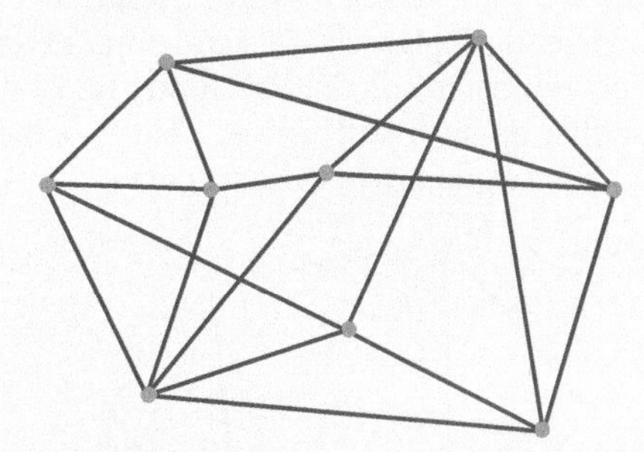

FIGURE 7.121

27–29. Traveling salesman problem. *In Exercises 27 through 29, solve the traveling salesman problem for the given map.*

27. Solve the traveling salesman problem for the map in **Figure 7.122** by calculating the mileage for each possible route. Use Ft. Wayne as a starting point.

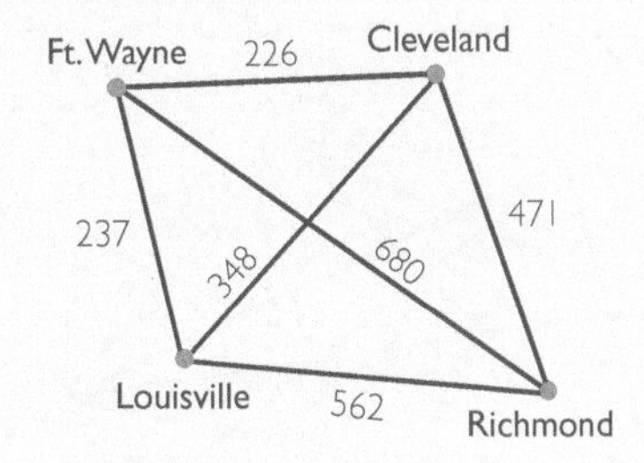

FIGURE 7.122

28. Solve the traveling salesman problem for the map in **Figure 7.123** by calculating the mileage for each possible route. Use New York as a starting point.

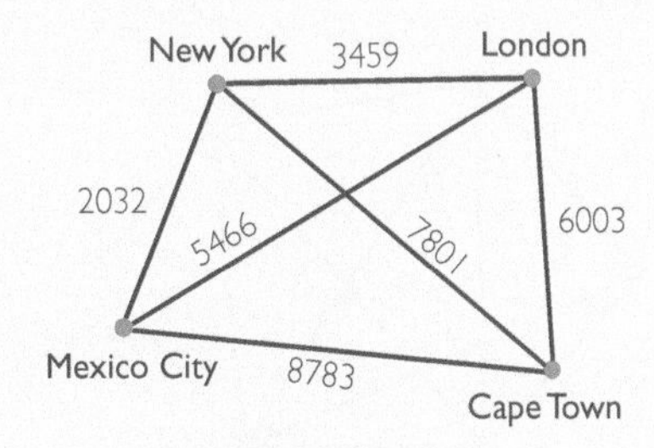

FIGURE 7.123

29. Solve the traveling salesman problem for the map in **Figure 7.124** by calculating the mileage for each possible route. Use Provo as a starting point.

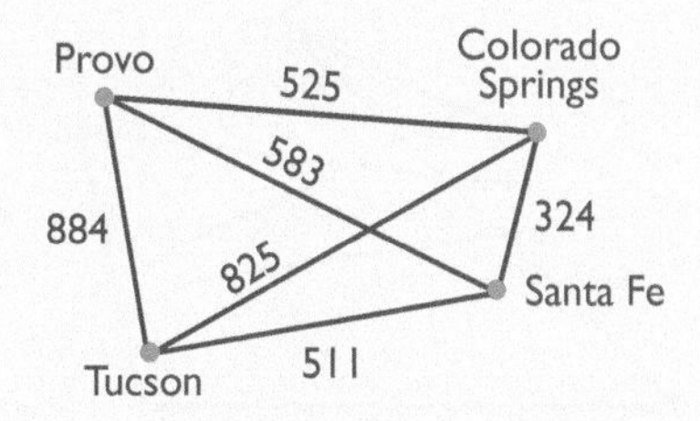

FIGURE 7.124

30–35. Nearest-neighbor algorithm. *In Exercises 30 through 35, use the nearest-neighbor algorithm.*

30. Use the nearest-neighbor algorithm starting at vertex *A* of the map in **Figure 7.125** to find an approximate solution of the traveling salesman problem.

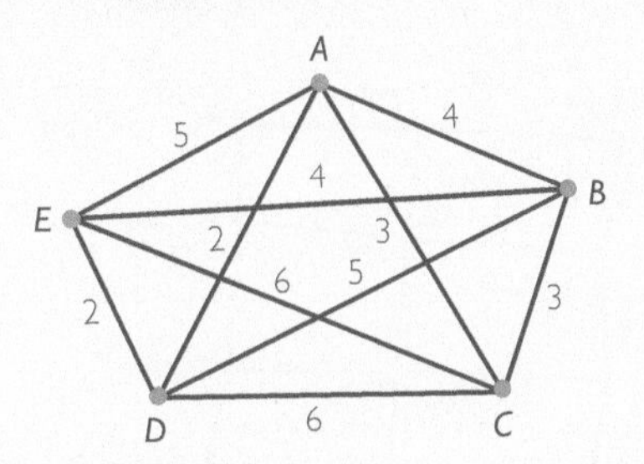

FIGURE 7.125

31. Use the nearest-neighbor algorithm starting at vertex *A* of the map in **Figure 7.126** to find an approximate solution of the traveling salesman problem.

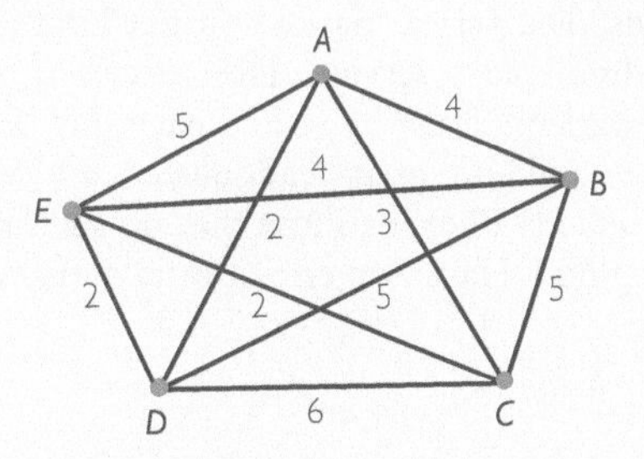

FIGURE 7.126

32. Use the nearest-neighbor algorithm starting at vertex *A* of the map in **Figure 7.127** to find an approximate solution of the traveling salesman problem.

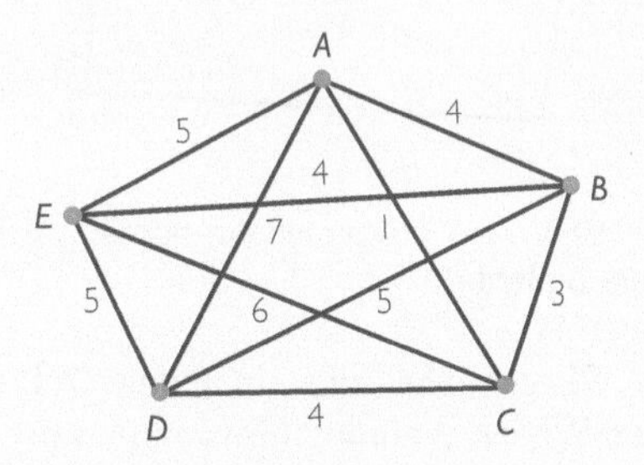

FIGURE 7.127

33. Use the nearest-neighbor algorithm starting at vertex *A* of the map in **Figure 7.128** to find an approximate solution of the traveling salesman problem.

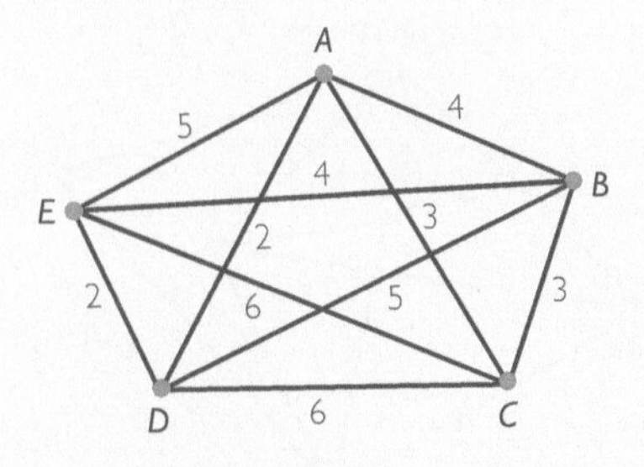

FIGURE 7.128

34. Use the nearest-neighbor algorithm starting at vertex A of the map in **Figure 7.129** to find an approximate solution of the traveling salesman problem.

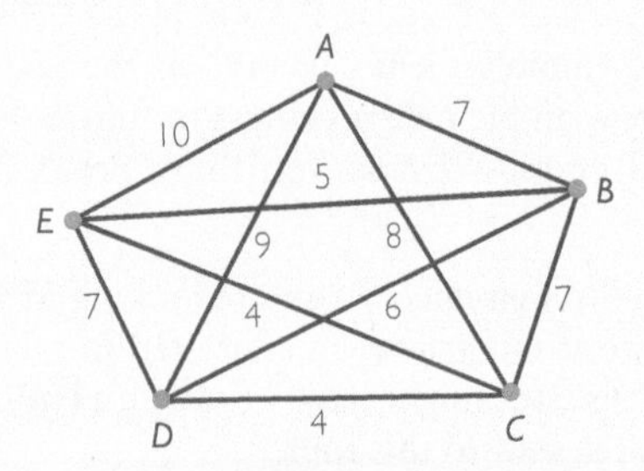

FIGURE 7.129

35. Use the nearest-neighbor algorithm starting at vertex A of the map in **Figure 7.130** to find an approximate solution of the traveling salesman problem.

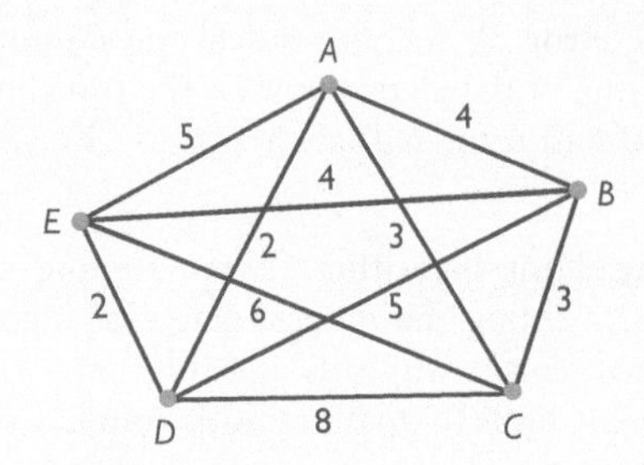

FIGURE 7.130

36–41. Cheapest link algorithm. *In Exercises 36 through 41, use the cheapest link algorithm.*

36. Use the cheapest link algorithm to find an approximate solution of the traveling salesman problem for the map in Figure 7.125.

37. Use the cheapest link algorithm to find an approximate solution of the traveling salesman problem for the map in Figure 7.126.

38. Use the cheapest link algorithm to find an approximate solution of the traveling salesman problem for the map in Figure 7.127.

39. Use the cheapest link algorithm to find an approximate solution of the traveling salesman problem for the map in Figure 7.128.

40. Use the cheapest link algorithm to find an approximate solution of the traveling salesman problem for the map in Figure 7.129.

41. Use the cheapest link algorithm to find an approximate solution of the traveling salesman problem for the map in Figure 7.130.

42. Nearest-neighbor failure. This exercise provides an example like Example 7.9 where the nearest-neighbor algorithm fails to find a shortest route. Consider the map shown in **Figure 7.131.**

a. Use the nearest-neighbor algorithm starting at A to find an approximate shortest route. How long is the route you found?

b. By trial and error, find a shortest route. How does its length compare with that in part a?

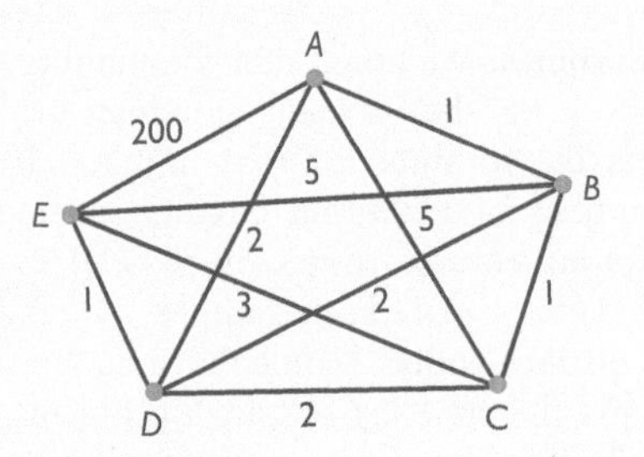

FIGURE 7.131

43. Starting point for nearest-neighbor algorithm. This exercise refers to the map in Figure 7.131. In each part, use the nearest-neighbor algorithm starting at the indicated vertex to find an approximate solution of the traveling salesman problem.

a. Start at A.
b. Start at B.
c. Start at C.
d. Start at D.
e. Start at E.
f. Compare your answers to parts a through e. What does the result say about the choice of starting point for the nearest-neighbor algorithm?

Exercises 44 through 48 are suitable for group work.

44–48. An improved nearest-neighbor algorithm. *An improvement of the nearest-neighbor algorithm goes as follows. Apply the nearest-neighbor algorithm once for each vertex by starting the algorithm at that vertex. Take the best of these routes.*

This information is needed for Exercises 44 through 48. In Exercises 44 through 47, apply the improved nearest-neighbor algorithm to the indicated graph.

44. Apply the improved nearest-neighbor algorithm to the graph in Figure 7.125 to find an approximate solution of the traveling salesman problem. Report the length of the shortest circuit you find.

45. Apply the improved nearest-neighbor algorithm to the graph in Figure 7.126 to find an approximate solution of the traveling salesman problem. Report the length of the shortest circuit you find.

46. Apply the improved nearest-neighbor algorithm to the graph in Figure 7.127 to find an approximate solution of the traveling salesman problem. Report the length of the shortest circuit you find.

47. Apply the improved nearest-neighbor algorithm to the graph in Figure 7.128 to find an approximate solution of the traveling salesman problem. Report the length of the shortest circuit you find.

48. Find an example of a complete graph with distances for which the improved nearest-neighbor algorithm does not find the shortest path.

49. Research problem. Locate four cities in the United States and find the distances between each using an atlas or the Internet. Solve the traveling salesman problem for these four cities.

The following exercise is designed to be solved using technology such as a calculator or computer spreadsheet.

50. Counting routes. We noted that the number of Hamilton circuits to check for the traveling salesman problem is very large. Here is the formula for that number: If there are n cities, the number of Hamilton circuits starting at a given city, not counting reverse routes, is $(n-1)!/2$. (Recall that $k! = k \times (k-1) \times \cdots \times 1$. For example, $3! = 3 \times 2 \times 1 = 6$.) How many distinct routes would have to be compared in solving by hand the traveling salesman problem for 50 cities? What about 70 cities? Round your answer to the nearest power of 10.

Writing About Mathematics

51. Finding practical applications. Make a list of several practical situations where finding Hamilton circuits may be important. For each item on your list, include a representative graph.

52. History. Hamilton circuits are named for Sir William Rowan Hamilton. Write a report on his scientific contributions.

53. History. Write a report on the history of the traveling salesman problem.

54. Efficient algorithms. For many years, researchers have been looking for efficient algorithms for attacking the traveling salesman problem. Seek out the most recent information you can on such algorithms, and write a report on what you find.

What Do You Think?

55. Why is it important? If a friend of yours said she had just heard that something called the traveling salesman problem is very important, how would you explain to her why it is important?

56. Algorithms. Algorithms have been mentioned several times in connection with graph theory. Explain in your own words what an algorithm is. **Exercise 36 is relevant to this topic.**

57. Procter & Gamble. On page 496 of the text, it is stated that Procter & Gamble held a contest to find a shortest route connecting 33 cities. Why do you think Procter & Gamble would care?

58. Why Hamilton circuits? You are looking at a local map. In what situations might you be interested in a Hamilton circuit? What businesses might have need of a Hamilton circuit? **Exercise 21 is relevant to this topic.**

59. Trucking companies. We have indicated that trucking companies might be interested in Hamilton circuits. But factors other than just the shortest mileage may come in to play. What are some of these factors? Contact a local trucking company and find out how they use efficient routes. **Exercise 16 is relevant to this topic.**

60. Trial and error. A fellow student says that there's no need to learn one of the algorithms for the traveling salesman problem—trial and error will always work. How would you respond?

61. Nearest-neighbor algorithm. What are the choices one makes in implementing the nearest-neighbor algorithm? Use your answer to explain why this algorithm can fail to find a shortest path—or even to give a unique result. **Exercise 48 is relevant to this topic.**

62. Cheapest link algorithm. What are the choices one makes in implementing the cheapest link algorithm? Use your answer to explain why this algorithm can fail to find a shortest path—or even to give a unique result.

63. Complete graphs. The text states that the graphs to which the traveling salesman problem applies are complete graphs. Why do you think that is the case?

7.3 Trees: Viral e-mails and spell checkers

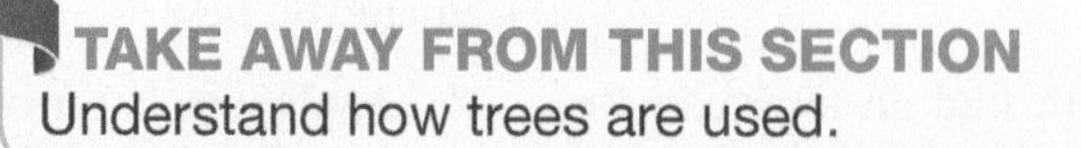

TAKE AWAY FROM THIS SECTION
Understand how trees are used.

A *tree* is a special kind of graph. In this section we will see how trees are useful in some practical situations, such as the way spell checkers work. The following article is from the Web site of the National Science Foundation.

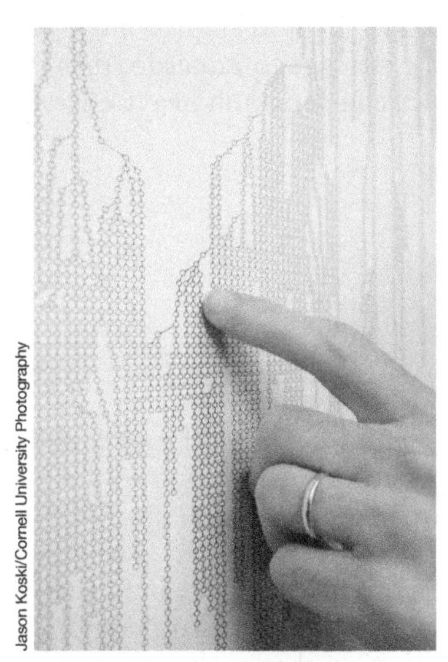

A part of the tree diagram used to analyze how e-mail petitions traveled to people's inboxes.

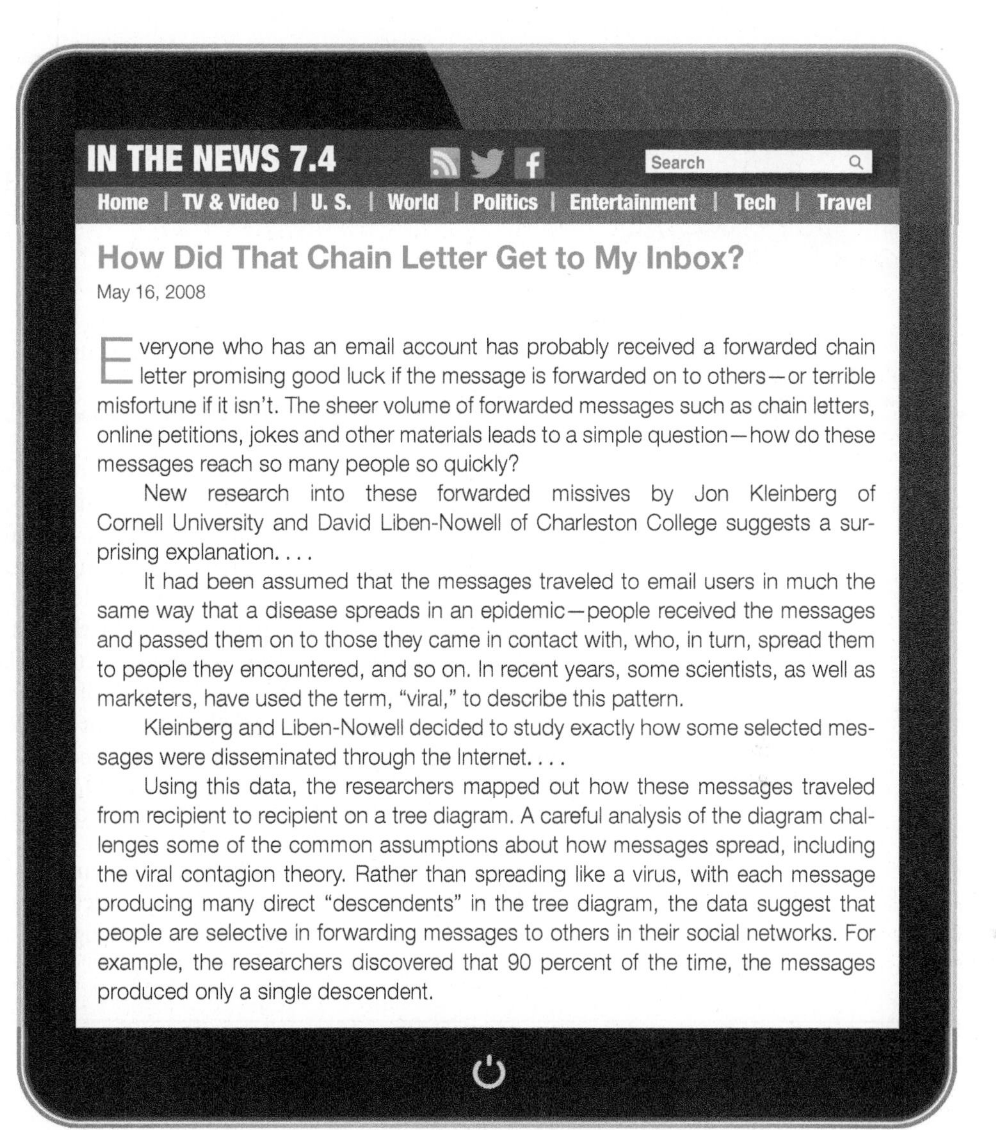

IN THE NEWS 7.4

Home | TV & Video | U. S. | World | Politics | Entertainment | Tech | Travel

How Did That Chain Letter Get to My Inbox?

May 16, 2008

Everyone who has an email account has probably received a forwarded chain letter promising good luck if the message is forwarded on to others—or terrible misfortune if it isn't. The sheer volume of forwarded messages such as chain letters, online petitions, jokes and other materials leads to a simple question—how do these messages reach so many people so quickly?

New research into these forwarded missives by Jon Kleinberg of Cornell University and David Liben-Nowell of Charleston College suggests a surprising explanation. . . .

It had been assumed that the messages traveled to email users in much the same way that a disease spreads in an epidemic—people received the messages and passed them on to those they came in contact with, who, in turn, spread them to people they encountered, and so on. In recent years, some scientists, as well as marketers, have used the term, "viral," to describe this pattern.

Kleinberg and Liben-Nowell decided to study exactly how some selected messages were disseminated through the Internet. . . .

Using this data, the researchers mapped out how these messages traveled from recipient to recipient on a tree diagram. A careful analysis of the diagram challenges some of the common assumptions about how messages spread, including the viral contagion theory. Rather than spreading like a virus, with each message producing many direct "descendents" in the tree diagram, the data suggest that people are selective in forwarding messages to others in their social networks. For example, the researchers discovered that 90 percent of the time, the messages produced only a single descendent.

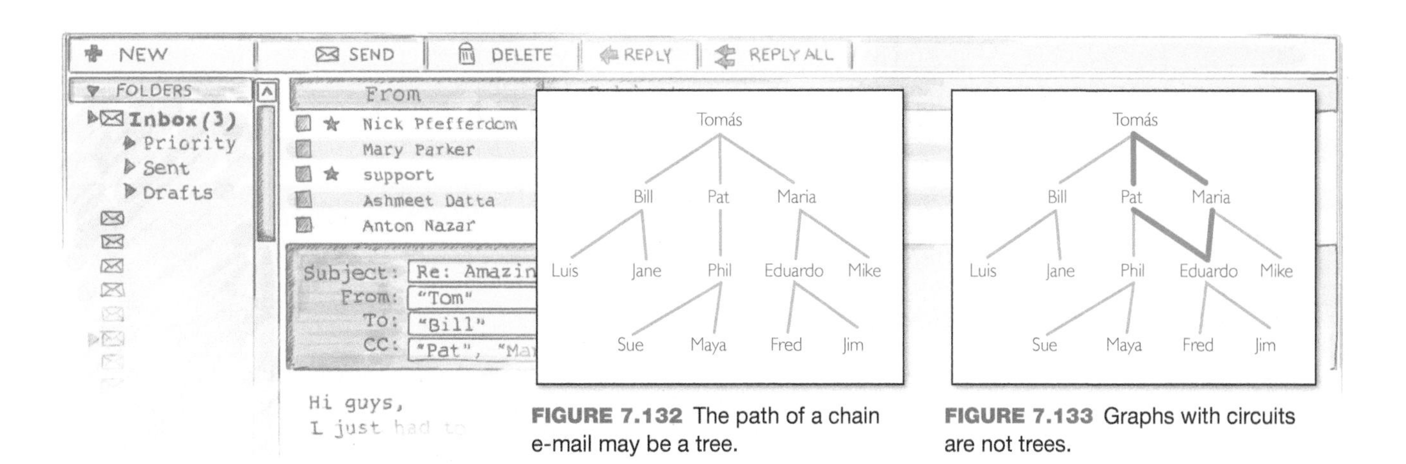

FIGURE 7.132 The path of a chain e-mail may be a tree.

FIGURE 7.133 Graphs with circuits are not trees.

The path typically taken by a chain letter or chain e-mail is a special kind of graph known as a *tree*. Suppose, for example, that you send an e-mail to some of your friends, who in turn forward it to some of their friends, and so on. The resulting path of the chain e-mail might look like the picture in **Figure 7.132**.

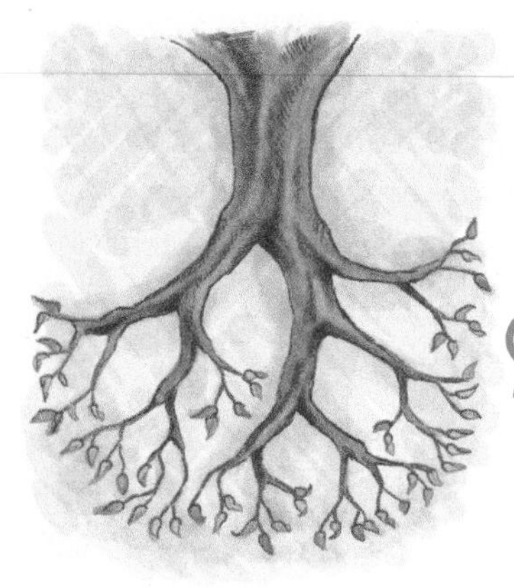

Informally, such a graph resembles an upside-down tree, in the sense that it has a starting vertex and the edges make branches that don't grow back together. To say that the branches don't grow back together is to say that the graph contains no circuits. If, for example, both Pat and Maria send the message to Eduardo, then the resulting graph is not a tree because it contains the circuit shown in heavy edges in **Figure 7.133**.

KEY CONCEPT

A tree is a graph that contains no circuits.

For the trees we consider, we always choose a starting vertex, typically drawn at the top.

The following terminology is a whimsical mixture of botanical and familial terms that are standard (we did not invent them). The vertex at the top of the tree is called the *root*. For example, Tomás is the root of the e-mail tree, as shown in **Figure 7.134**. There is a path leading from the root to any other vertex. The vertex in the path just before a given vertex is called its *parent*, and any vertex immediately after is its *child*. We see in Figure 7.134 that Maria is the parent of Eduardo and Mike, so Eduardo and Mike are children of Maria. Vertices that are not parents are *leaves*. These are typically at the bottom of the tree, as shown in Figure 7.134.

Every vertex of a tree is either a parent or a leaf, as shown in **Figure 7.135**. This gives us our first important formula for trees:

$$\text{Number of vertices} = \text{Number of parents} + \text{Number of leaves}$$

A tree also has *levels*, which are shown in **Figure 7.136**. The root is level 0. The children of the root are level 1. The grandchildren are level 2, and so on. The largest level of the tree is the *height*. Thus, the height of the tree in Figure 7.136 is 3.

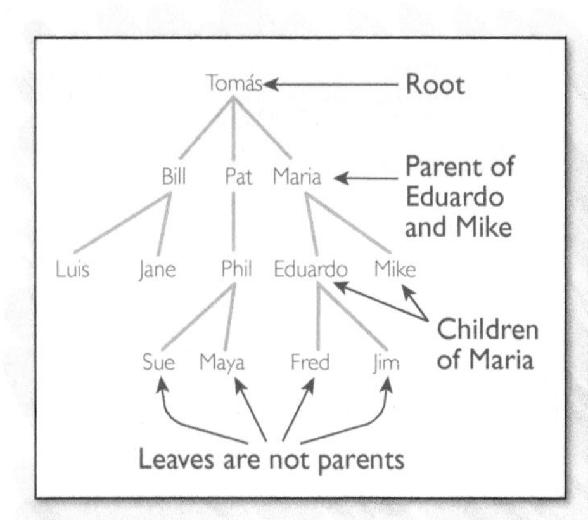

FIGURE 7.134 Root, parents, and children in trees.

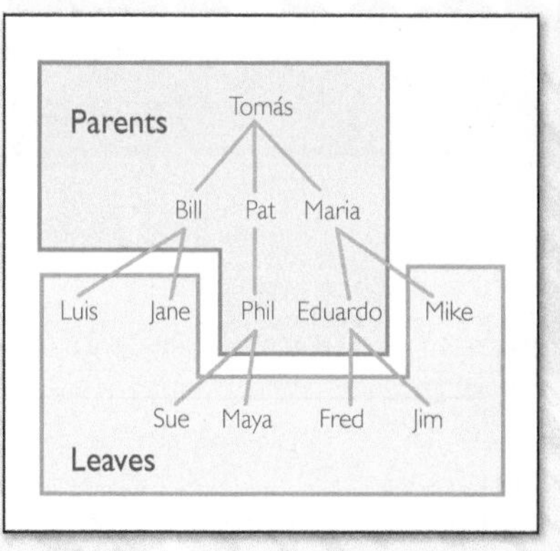

FIGURE 7.135 Vertices separated into parents and leaves.

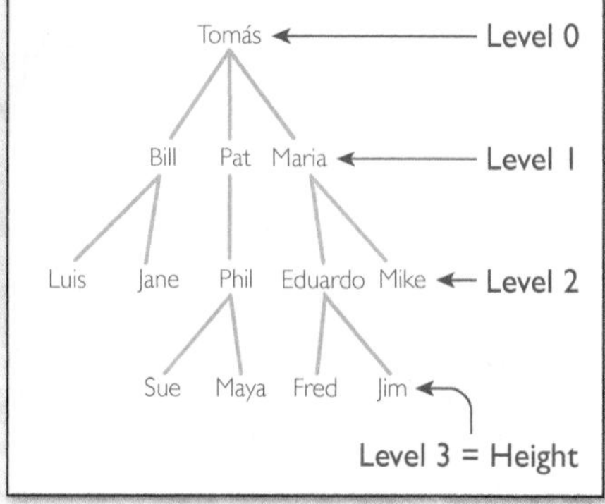

FIGURE 7.136 Levels of a tree.

Binary trees

Perhaps one of the most common trees we see in everyday life, and certainly the one with the most hype, is the tree formed by the NCAA basketball tournament brackets.[6] A piece of the 2021 tree is shown in **Figure 7.137**. This particular type of tree is known as a *binary tree* because each parent vertex has exactly two children.[7]

[6] *In our discussion of this tournament, we ignore the "First Four" play-in games.*

[7] *In some texts, a binary tree is one where parents have* at most *two children. If each parent has* exactly *two children, those texts call the tree a* full binary tree.

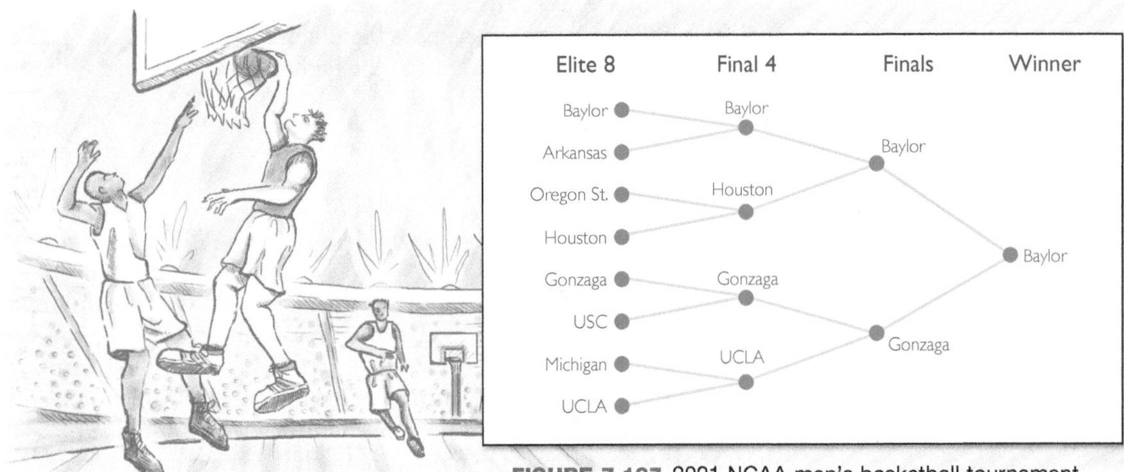

FIGURE 7.137 2021 NCAA men's basketball tournament is a binary tree.

KEY CONCEPT

In a binary tree, each parent has exactly two children.

Because each parent in a binary tree has exactly two children, the number of children is twice the number of parents. But the only vertex in a tree that is not a child is the root. So, in a binary tree

$$\text{Number of vertices} = 2 \times \text{Number of parents} + 1$$

EXAMPLE 7.11 Counting in a binary tree: PTA and phone calls

There are 127 members of a PTA group. A school-closure alert plan requires one member to phone two others with the information. Each PTA member who receives a call phones two members who have not yet gotten the news. This continues until all members have the message. How many PTA members make phone calls?

SOLUTION

We represent the phone calls with a binary tree. Each vertex represents a PTA member, and edges indicate phone calls. A part of this tree is shown in **Figure 7.138**.

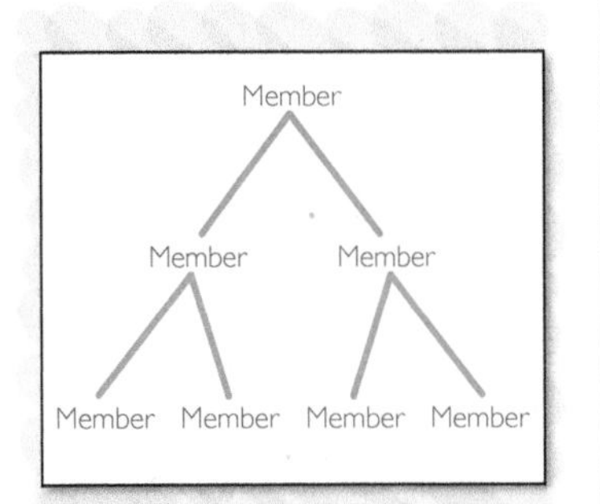

FIGURE 7.138 A phone-call binary tree.

The parent vertices in the tree are the PTA members who make phone calls. The total number of vertices is the number of PTA members, 127. We can use this information to find the number of parent vertices:

$$\begin{aligned} \text{Number of vertices} &= 2 \times \text{Number of parents} + 1 \\ 127 &= 2 \times \text{Number of parents} + 1 \\ 126 &= 2 \times \text{Number of parents} \\ 63 &= \text{Number of parents} \end{aligned}$$

Thus, 63 members make phone calls.

TRY IT YOURSELF 7.11

How many members make no phone calls?

The answer is provided at the end of this section.

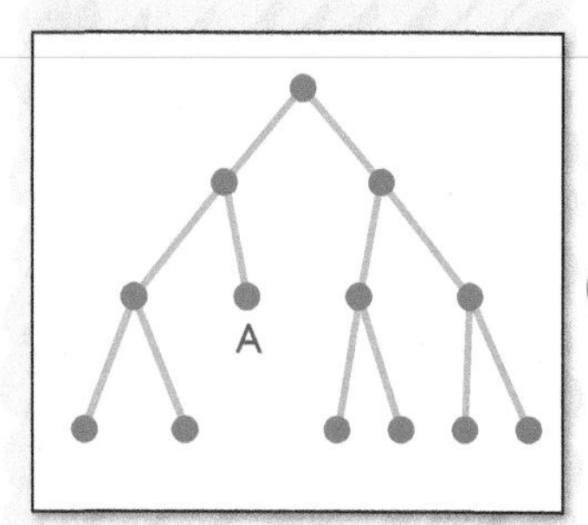

FIGURE 7.139 This binary tree is not complete.

The NCAA bracket in Figure 7.137 is an example of a *complete* binary tree, which has as many vertices as possible. **Figure 7.139**, in contrast, shows a tree that is not complete. The leaf A is not at the highest level. Intuitively, the tree has missing branches.

KEY CONCEPT

A complete binary tree has all the leaves at the highest level.

In a complete binary tree, each parent has exactly two children, and the only leaves are at the highest level. Hence, going from one level to the next doubles the number of vertices. At level 0 there is 1 vertex, the root. There are 2 vertices at level 1, 4 at level 2, 8 at level 3, and so on (see **Figure 7.140**). In general, there are 2^L vertices at level L. Note that in a tournament bracket like the one in **Figure 7.141**, the leaves are the entrants in the tournament, and the parent vertices represent games played.

Thinking of binary trees as tournaments is often useful. For example, there are 64 teams in the NCAA basketball tournament. Each game produces exactly one loser, and every team loses except for the tournament champion. Hence, there are 63 games played altogether. Because each game played represents a parent in the tournament tree, there are 63 parents in the tree.

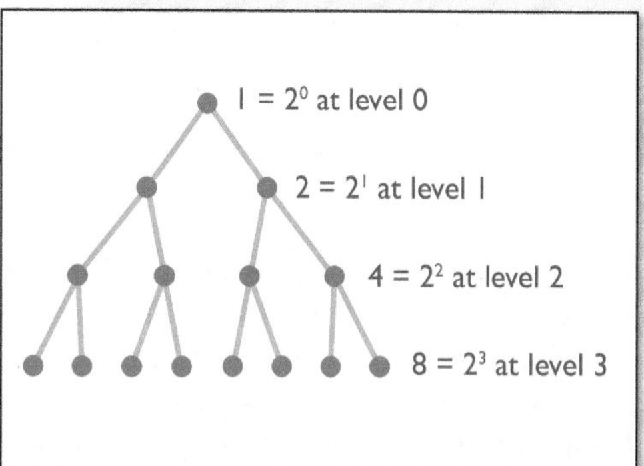

FIGURE 7.140 In a complete binary tree are 2^L vertices at level L.

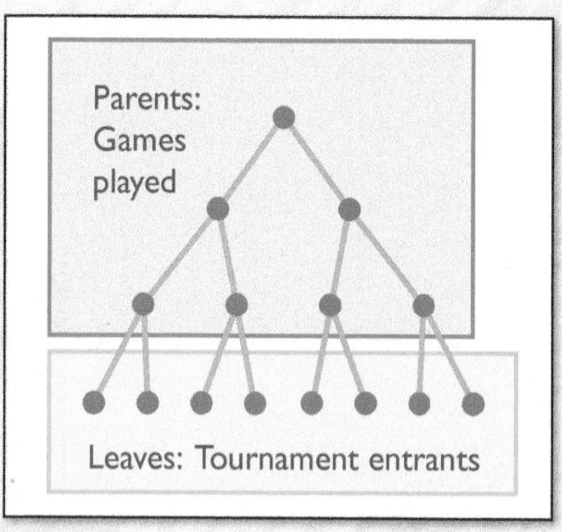

FIGURE 7.141 Entrants and games played in a tournament.

This reasoning allows us to count vertices in any complete binary tree. In a complete binary tree of height H, there are 2^H vertices at the highest level. These are the leaves. We think of the leaves as tournament entrants and the parents as games played. Each game produces one loser, and every entrant except one loses a game. Hence,

$$\begin{aligned}\text{Number of parents} &= \text{Number of games played}\\ &= \text{Number of losers}\\ &= \text{Number of entrants} - 1\end{aligned}$$

Thus,

$$\text{Number of parents} = 2^H - 1$$

Now we can find the total number of vertices in a complete binary tree of height H:

$$\begin{aligned}\text{Number of vertices} &= 2 \times \text{Number of parents} + 1 \\ &= 2(2^H - 1) + 1 \\ &= 2 \times 2^H - 2 + 1\end{aligned}$$

Hence,

$$\text{Number of vertices} = 2^{H+1} - 1$$

SUMMARY 7.5
Formulas for Trees

- In any tree,

$$\text{Number of vertices} = \text{Number of parents} + \text{Number of leaves}$$

- In a binary tree,

$$\text{Number of vertices} = 2 \times \text{Number of parents} + 1$$

- In a complete binary tree, there are 2^L vertices at level L.
- In a complete binary tree of height H:

$$\begin{aligned}\text{Number of leaves} &= 2^H \\ \text{Number of parents} &= 2^H - 1 \\ \text{Number of vertices} &= 2^{H+1} - 1\end{aligned}$$

EXAMPLE 7.12 Applying formulas for trees: A Ponzi scheme

A shady broker offers a dubious investment opportunity. In week 0, he sells a \$10,000 note promising to repay \$11,000 in one week. In week 1, he sells two such notes and uses the proceeds to pay off the week 0 investor. In week 2, he sells four notes and pays off the two week 1 investors. Each week, he sells twice as many notes as the week before and uses the proceeds to pay off last week's investors. An investment scam of this sort is known as a *Ponzi scheme*.

a. How many notes did the broker sell in week 10?

b. What was the total number of notes sold by week 10?

c. How much money was collected by week 10?

d. How much money was paid out by week 10?

e. In week 10, the broker took his profits and left the country. How much money did investors lose?

Kristin Callahan © Everett Collection Inc/Alamy

In 2009, Bernard Madoff pleaded guilty to operating a multi-billion-dollar Ponzi scheme. Many fortunes were lost when the scheme collapsed.

SOLUTION

a. We can model this scheme with a complete binary tree of height $H = 10$. Each vertex represents a sale. The number of sales in week 10 is the number of leaves, which is

$$\begin{aligned}\text{Number of leaves} &= 2^H \\ &= 2^{10} \\ &= 1024\end{aligned}$$

Thus, 1024 notes were sold in week 10.

b. The number of notes sold is the total number of vertices, which is

$$\begin{aligned}\text{Number of vertices} &= 2^{H+1} - 1\\ &= 2^{11} - 1\\ &= 2047\end{aligned}$$

Thus, 2047 notes were sold by week 10.

c. The broker sold 2047 notes at \$10,000 each. Therefore, he collected a total of 2047 × \$10,000 = \$20,470,000.

d. The broker paid off the notes sold through week 9. The number of notes sold through week 9 is the number of vertices in a complete binary tree of height 9. Thus,

$$\begin{aligned}\text{Number sold through week 9} &= 2^{9+1} - 1\\ &= 1023 \text{ notes}\end{aligned}$$

He paid \$11,000 to each of these. That is a total of 1023 × \$11,000 = \$11,253,000.

e. He cheated investors out of \$20,470,000 − \$11,253,000 = \$9,217,000.

Spell checkers

Modern word processors come equipped with spell checkers that include a dictionary of as many as 100,000 or so words, and most cell phones have a spell checker built into "autocorrect." When we type a word, the spell checker instantly compares that word with the words in its dictionary and highlights the word if there is no match. Did the spell checker instantly make 100,000 comparisons? No, even very fast computers aren't that fast. At the heart of the process is a binary tree. To illustrate the process, let's adopt a very simple dictionary containing only the following list of 13 one-letter "words":

A, C, E, G, I, K, M, O, Q, S, U, W, Y

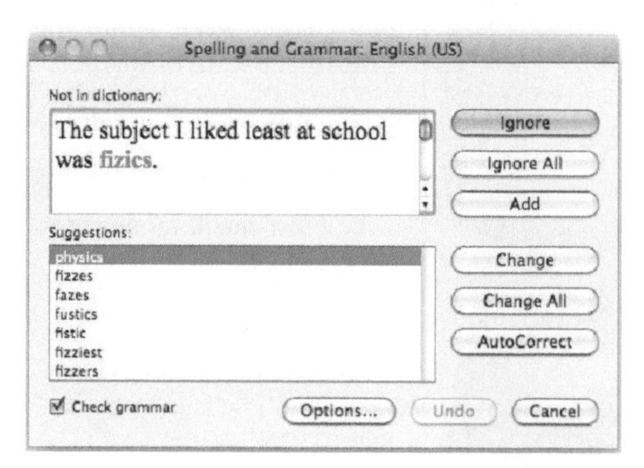

A typical spell checker in action.

We have put a box around M because it is the middle "word" in the list, and it will be the root of our tree—the place where we start our search. If we type in a letter, we will first compare it with M. If the letter we type happens to be M, we have a valid word. If it precedes M in the alphabet, we move to the left in our list; otherwise, we move to the right. These two possibilities correspond to two branches from the root.

Now we consider the two branches. Again, we go to the middle to make our comparison:

A, C, E, G, I, K, M, O, Q, S, U, W, Y

In the left branch A, C, E, G, I, K, there are two letters, E and G, that have equal claim to the middle position. We adopt the convention that if there are two middle

"Must have been an old can of alphabet soup. No spell checker.

words, we will take the one on the left. Thus, we chose E to make our spell-checking tree. In the right branch, there are also two letters that can claim the middle, and we chose the one on the left. Thus, we use S. We go to the middle of each remaining piece to make the next comparison.

The completed spell-checking tree is shown in **Figure 7.142.** At each stage, we move to the right if the letter we typed is "bigger" and to the left if it is "smaller," and we report a valid word if it is the same. Note that the tree is binary except at A and O. The path we would take if we typed in J, which is not in our list and so not a valid word, is highlighted in **Figure 7.143.** Note that we need to make a total of only four comparisons (not 13) to find out that J is not in our dictionary.

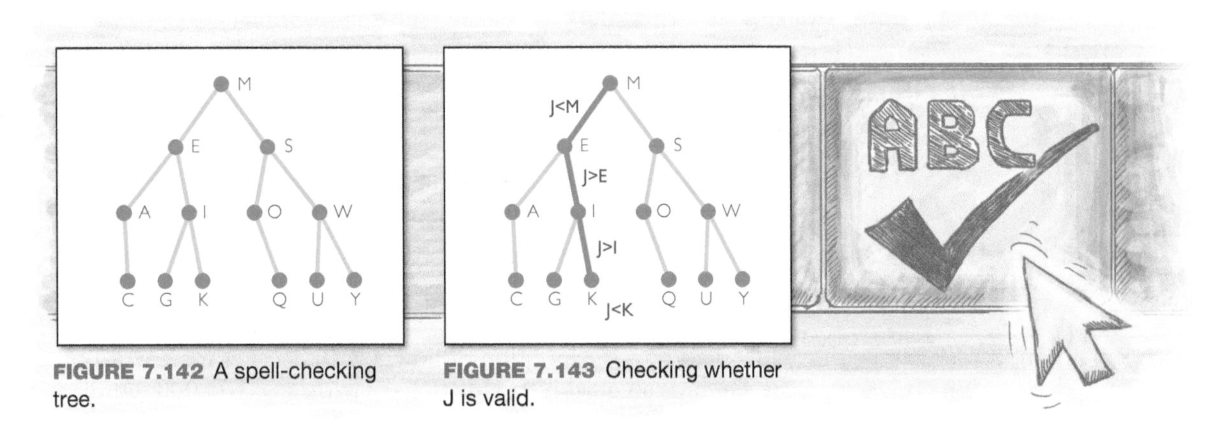

FIGURE 7.142 A spell-checking tree.

FIGURE 7.143 Checking whether J is valid.

EXAMPLE 7.13 Making a spell checker

Make a spell-checking tree for the following dictionary:

apple, boy, cat, dog, ear, fox, girl, hat, jet, kit, love, mom, net, out, pet

How many checks (or comparisons) are needed to test whether "cow" is a valid word? Show the path you take on the spell-checking tree to check whether "cow" is a valid word.

SOLUTION

We go to the middle of the list, which is "hat":

apple, boy, cat, dog, ear, fox, girl, hat, jet, kit, love, mom, net, out, pet

Thus, "hat" goes at the top of our tree.

Next we go to the middle of the left and right sections:

apple, boy, cat, dog, ear, fox, girl, hat, jet, kit, love, mom, net, out, pet

We get "dog" and "mom." These go on the first level of our tree. We continue to go to the middle of each piece to get the next comparison word. The completed tree is shown in **Figure 7.144**.

Four checks are needed to spell-check "cow," and because it matches no word in the tree, we conclude that "cow" is not a valid word in our limited dictionary. The path of the check is shown in **Figure 7.145**.

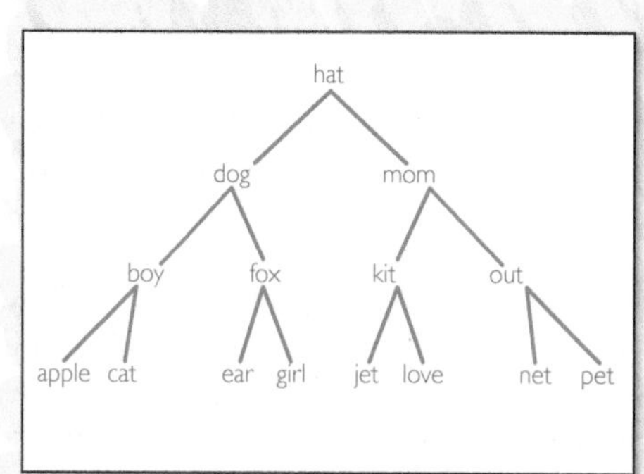

FIGURE 7.144 The spell-checking tree.

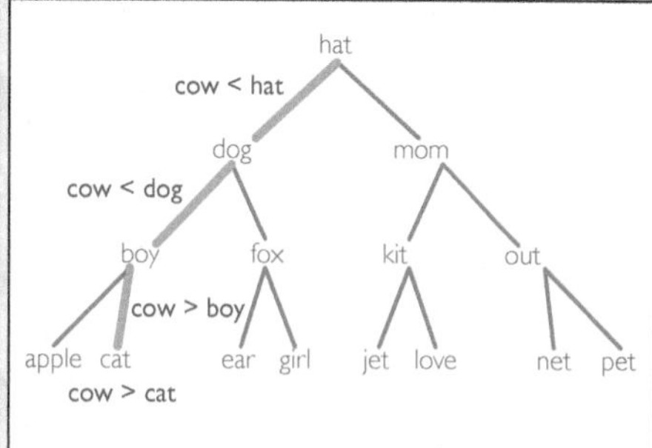

FIGURE 7.145 Checking "cow."

TRY IT YOURSELF 7.13

Make a spell-checking tree for the following dictionary:

box, cap, car, dip, eat, fur, him, hit, ink, joy, lid, man, now, pan, rat

Show the path you take on the spell-checking tree to check whether "it" is a valid word.

The answer is provided at the end of this section.

Efficiency of spell checkers

Now we return to the question of how a spell checker for a word processor is able to check words so quickly in a dictionary with, say, 100,000 words. In terms of the associated tree, the answer lies in the fact that the height is much smaller than the number of vertices.

We have seen that a complete binary tree with height H has $2^{H+1} - 1$ vertices. In a spell-checking tree, we make checks at level 0, level 1, . . . , level H. That makes a total of $H + 1$ checks. We conclude that a spell-checking tree requiring a maximum of $c = H + 1$ checks can hold up to $2^c - 1$ words. Now we calculate that

$$2^{17} - 1 = 131{,}071$$

This means that we can check a word against a 100,000-word dictionary easily—we need at most 17 checks. With only three more checks, we can accommodate a 1-million-word dictionary. That is why spell checkers work so fast. We note that this kind of binary branching is a formalized version of what we all do when we look up a word in the dictionary.

SUMMARY 7.6
Spell Checking

A dictionary of $2^c - 1$ words can be accommodated using no more than c checks.

EXAMPLE 7.14 Analyzing a spell checker

a. You want to use a dictionary so that you have to check any word at most five times. How big a dictionary can you accommodate?

b. Estimate the number of checks needed if you have a 10,000-word dictionary.

SOLUTION

a. We know that with at most c checks allowed, a dictionary of $2^c - 1$ words can be accommodated. We want five checks, so we use $c = 5$ in this formula. We calculate that $2^5 - 1$ equals 31. This means that we can accommodate a dictionary with 31 words.

b. We want to choose the smallest c so that $2^c - 1$ is at least 10,000. We proceed by trial and error (with the help of a calculator) and find that $2^{14} - 1$ is greater than 10,000 but $2^{13} - 1$ is not. Thus at most, 14 checks are needed.

TRY IT YOURSELF 7.14

Estimate the number of checks needed for a spell-checking tree if we have a 1000-word dictionary.

The answer is provided at the end of this section.

WHAT DO YOU THINK?

Spell checkers: A friend is amazed that spell checkers are so fast—after all, there are lots of words in the dictionary. In your own words, explain how spell checkers are so efficient.

Questions of this sort may invite many correct responses. We present one possibility: Spell checkers work much in the same way humans look up words in a dictionary. We don't check the word *heaume* by comparing every word in the dictionary. Rather, we open the dictionary to a likely spot. If we find ourselves in the "f" section, we move forward. If instead we land in the "k" section, we move back. We keep up this "move-up/move-back" routine until we find that a heaume is a thirteenth-century helmet. This procedure gets us to the word we want quickly, and it is exactly how spell checkers work.

Further questions of this type may be found in the *What Do You Think?* section of exercises.

Try It Yourself answers

Try It Yourself 7.11: Counting in a binary tree: PTA and phone calls 64 members do not make phone calls.

Try It Yourself 7.13: Making a spell checker Here is the spell-checking tree.

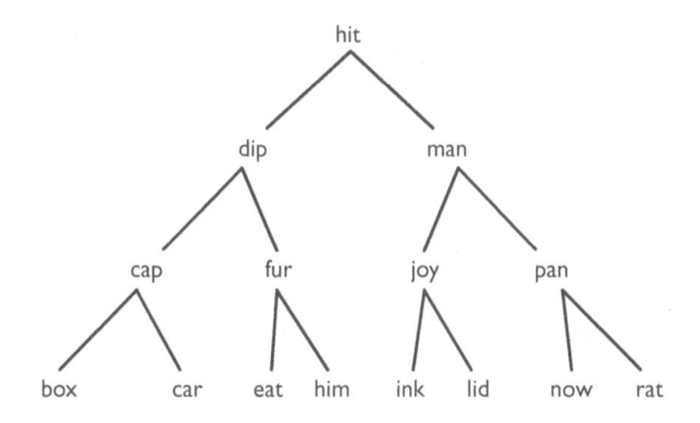

This is the path for checking "it."

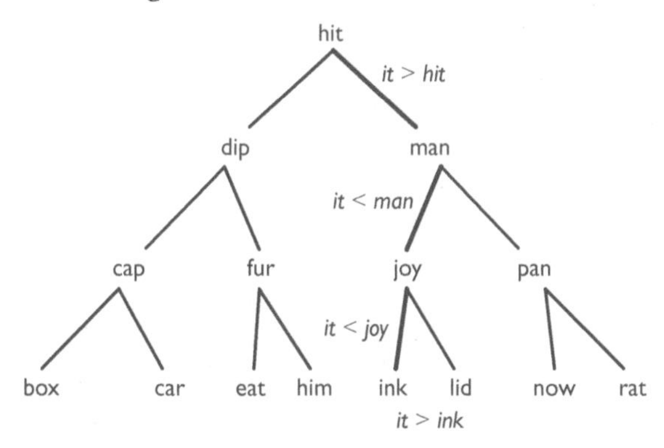

Try It Yourself 7.14: Analyzing a spell checker 10 checks.

Exercise Set 7.3

Test Your Understanding

1. A tree is a graph that: **a.** branches out from one vertex **b.** has edges that go from one vertex to another below it **c.** has no circuits.

2. There are four names that apply to the vertices of a tree. Which of these are *not* among them? **a.** bud **b.** parent **c.** child **d.** leaf **e.** limb **f.** root

3. A binary tree is one in which ______________.

4. A leaf of a tree is a vertex that is: **a.** at the bottom of the tree **b.** not a parent **c.** part of a circuit

5. True or false: Modern spell checkers compare a candidate with each word in the dictionary until a match is found or the candidate is determined to be misspelled.

6. State in your own words the meaning of a Ponzi scheme.

Problems

7. **Making a tree.** Make a tree with seven vertices and exactly four parents.

8. **Few leaves.** Make a tree with 10 vertices and the fewest possible leaves.

9. **Many leaves.** Make a tree with 10 vertices and the most possible leaves.

10. **A binary tree.** Is it possible to make a binary tree with 18 vertices?

11. **Another binary tree.** Is it possible to make a binary tree using 256 vertices?

12–14. **Nomenclature.** *The tree in **Figure 7.146** is used in Exercises 12 through 14.*

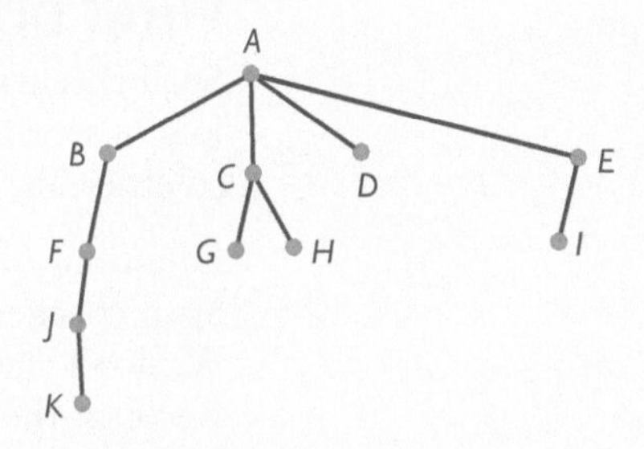

FIGURE 7.146

12. List all parents in Figure 7.146.

13. List all leaves in Figure 7.146.

14. List all children in Figure 7.146.

15. **Making a spell-checking tree.** Make a spell-checking tree for the dictionary:

and, bake, can, dig, even, fax, get, hive, in, is, it, jump, kite, loud, mate, never, open, quiet, rat, saw, top, under, vow, xray, yes, zebra

16. A spell checker. Make a spell-checking tree using the following dictionary:

air, bake, cow, din, eat, fire, game, home, it, jab, kick, let, moon, now, over, pal, quit, rain, sand

How many checks are needed to test whether "hope" is a valid word? How many checks are needed to test the word "cow"?

17. A bigger spell checker. We have a dictionary consisting of 100 words. If we use a binary tree as a spell checker, estimate how many checks might be required to test whether "name" is a valid word.

18. How big is the dictionary? List the sizes of dictionaries that can be accommodated by binary spell checkers if the number of checks required is between 3 and 10.

19. How many checks? Estimate how many checks are required if our dictionary contains 50,000 words.

20. A story. I tell a story to two of my friends. Each of these is allowed to repeat the story to exactly two others (or not pass the story on). This process continues: Each time the story is told, it is told to two new people, and no one hears the story twice. Sometime later, 253 people know the story. How many people have related the story to others? *Suggestion*: Use the formula

$$\text{Number of vertices} = 2 \times \text{Number of parents} + 1$$

21. A chain e-mail. For a chain e-mail, each mailer sends messages to two new people. Some recipients continue the chain, and some don't. No one gets the same message twice. Sometime later, there are a total of 1000 message recipients. (That means your graph model will have 1001 vertices because the initiator of the chain e-mail is not one of the 1000 recipients.) How many people received messages but did not mail anything? *Suggestion*: Use the formulas

$$\text{Number of vertices} = 2 \times \text{Number of parents} + 1$$

and

$$\text{Number of vertices} = \text{Number of parents} + \text{Number of leaves}$$

22. Lucky nights. One night I put a silver dollar in a slot machine and got back two silver dollars. On the second night, I put these two dollars back in the machine one at a time, and each time two silver dollars came out. This pattern continued through the fifth night, after which I took my winnings and went home.

a. How much money did I take home?

b. How many silver dollars were fed into the machine?

23. Quality testing. *The second part of this exercise uses counting techniques from Section 5.4.* A certain computer chip is subjected to 10 separate tests before it can be sold. We represent the testing with a binary tree. The first vertex is test number 1. It meets two edges. The right-hand branch represents a "pass," and the left-hand branch represents a "fail." We then administer the next test and branch right or left depending on the performance of the chip. This continues until all 10 tests are administered. How many leaves are there in the completed tree? Suppose a chip must pass at least seven of the tests before it can be marketed. How many leaves of the tree represent market-ready chips? *Suggestion*: Think of a chip with 10 blanks on it. We will stamp the blanks P or F. For a chip that passes exactly seven tests, we want to choose seven of the blanks on which to stamp a P.

24. An e-mail chain. An e-mail chain is initiated. On day 1, e-mails are sent to two people, and each is asked to forward the message to two new recipients the next day. Assume that each recipient complies with the request and that no one receives e-mails from two separate sources. How many people have been notified by day 6? Include the initiator of the e-mail chain.

25. The CFP. Since the 2014 season, the national champion for the NCAA Division I Football Bowl Subdivision has been determined by a system known as the College Football Playoff (CFP). Under the CFP, there is a playoff involving four teams. Many observers, including U.S. President Barack Obama, have advocated that eight teams be chosen for a championship tournament. Under that system, how many games would be played before a national champion is determined?

26. Division III. In NCAA Division III football, there is a year-end tournament involving 32 teams. How many teams play in the fourth round of the tournament?

27. Expanding the NCAA basketball tournament. Suppose we decide to expand the NCAA basketball tournament to include 256 teams.

a. How many games are played?

b. How many games are played in the third round?

c. How many teams lose in the first four rounds?

d. What round corresponds to the Elite Eight? (The Elite Eight are the final eight teams left in the tournament.)

28. Finishing your word. Some computer applications try to anticipate words you are typing and finish them automatically. For example, in your iPhone application, when you type `elec`, your phone may immediately expand this to `electric company`. This trick is accomplished using a tree (which may be constructed by the software based on your previous typing). Suppose our computer knows that there are six words you are likely to type in the current field:

elephant, element, elfin, elvis, escalator, elevator

We make a *typing tree* associated with this list as shown in the figure below.

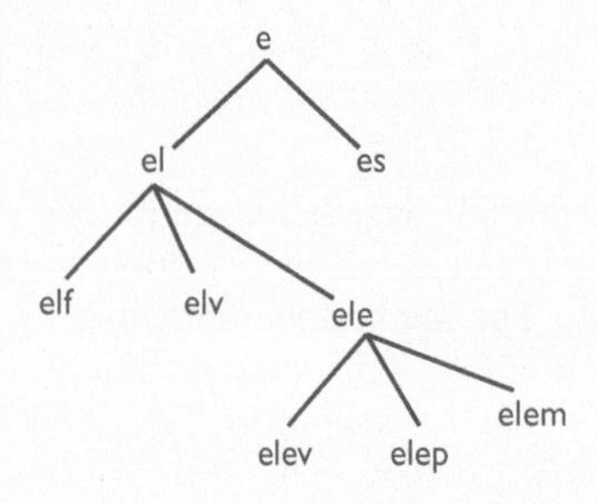

The vertices of the tree are made up of letter sequences that begin the test words. Note that the root has one letter, the level 1 vertices have two letters, and so on. When a leaf such as `elf` is reached, the computer knows that, if you intended to type one of the six likely words, then it must be *elfin*. Once the computer has constructed such trees, then the trick of finishing your words is easy.

Suppose your computer knows that you are likely to type one of the following words in the current field:

computer, container, comet, call, cask, carpet,
chess, cheese, checker, chase, chin, chili

Make a typing tree for these words.

Exercise 29 is suitable for group work.

29. Testing for connectivity. A human being can quickly ascertain whether or not a graph is connected (i.e., comes in one piece) if she or he sees a picture of it. But computers deal with graphs that are not in picture form and that contain many thousands of vertices and edges. In such a case, it is not so easy to determine connectedness. Here we illustrate a method that computers use to solve this problem.

In the context of Exercises 40 through 43 in Section 7.1, we introduced the idea of an *incidence matrix* as a way to represent a graph that a computer can understand without a picture. Here is a review in terms of the following incidence matrix:

	A	*B*	*C*	*D*	*E*	*F*
A	–	1	1	0	0	0
B	1	–	1	1	1	0
C	1	1	–	0	1	1
D	0	1	0	–	0	0
E	0	1	1	0	–	0
F	0	0	1	0	0	–

The letters *A*, *B*, *C*, *D*, *E*, *F* represent the vertices of the graph. The dashes on the diagonal are just placeholders, and we ignore them. The 1 in the *B* column of the *A* row means there is an edge joining *A* to *B*. The 1 in the *C* column of the *A* row means there is an edge joining *A* to *C*. The 0 in the *D* row of the *A* column means there is no edge joining *A* to *D*. Similarly, the 1 in the *B* column of the *C* row means there is an edge joining *B* to *C*. The completed graph represented by the incidence matrix above is shown next.

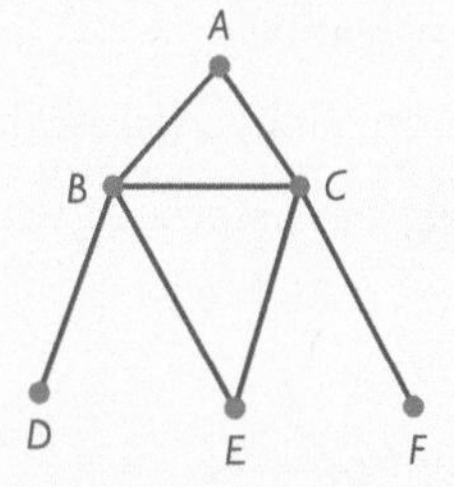

We want to determine whether a graph represented by an incidence matrix is connected, but we don't want to reproduce the entire graph. The method we present is known as a *tree search* because it employs the construction of a tree. We will show how it works using the preceding example. Begin with any vertex you like, say, *A*, and start the construction of a tree by listing the vertices adjacent to *A*. We get the list *B*, *C*. The partial tree associated with this list is shown below.

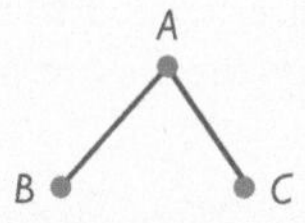

Note that we do not include the edge joining *B* to *C*. We are interested only in connecting to new vertices, not what connections may or may not exist among vertices already listed. Next we go to the first vertex of this list. We look only for new (other than *A*, *B*, or *C*) vertices adjacent to *B*. We find the new vertices *D* and *E* and thus obtain the following partial tree:

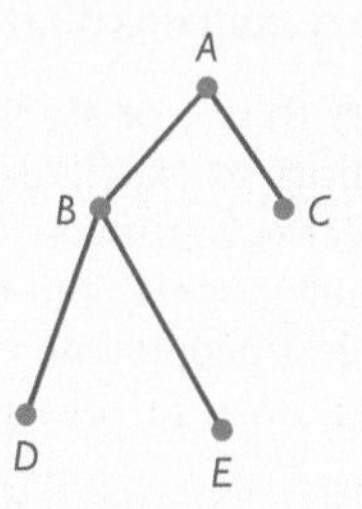

Finally, we look for new vertices adjacent to *C*, and we find only *F*. Our final tree is shown below.

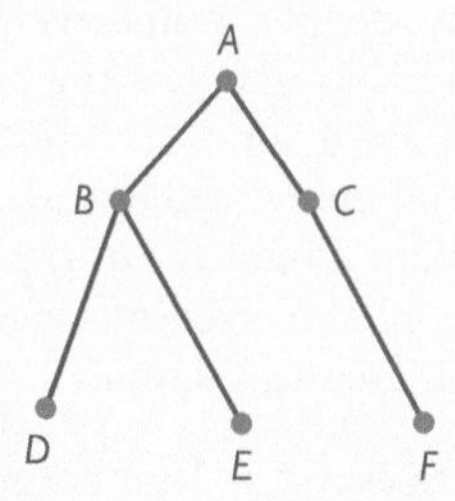

This search tree will either produce a tree including all the vertices (as it did here) or terminate without including all the vertices. In the first case, the graph is connected, and in the second case, it is not.

For the following incidence matrix, make a search tree starting at the vertex *A* to determine whether the graph represented is connected.

	A	*B*	*C*	*D*	*E*	*F*	*G*	*H*	*I*	*J*
A	–	1	1	0	0	0	0	0	0	0
B	1	–	1	1	1	1	0	0	0	0
C	1	1	–	1	0	0	1	0	0	0
D	0	1	1	–	0	1	0	0	0	0
E	0	1	0	0	–	0	0	0	0	0
F	0	1	0	1	0	–	1	0	0	0
G	0	0	1	0	0	1	–	0	0	0
H	0	0	0	0	0	0	0	–	1	1
I	0	0	0	0	0	0	0	1	–	1
J	0	0	0	0	0	0	0	1	1	–

The following exercise is designed to be solved using technology such as a calculator or computer spreadsheet.

30. A very large dictionary. What size dictionary can be accommodated by a spell checker that uses 33 checks? What size dictionary can be accommodated by a spell checker that uses 100 checks? We know names for numbers like millions, billions, or trillions. Do you have a name for the number of words you found for 100 checks?

Writing About Mathematics

31. History. Write a report on the history of spell checkers.

32. Autocorrect. One kind of spell checking is called *autocorrect*. Write a report on the history of autocorrect.

33. History. Minimal-cost spanning trees are very important in many applications. A number of efficient algorithms have been developed to find them. Report on the development of methods for solving this problem.

34. Charles Ponzi. The term *Ponzi scheme* was coined to describe the investment scam of Charles Ponzi. Report on the investment scheme operated by Ponzi.

What Do You Think?

35. Exponential growth. Section 3.2 discussed the concept of *exponential growth*. Does the number of vertices in a binary tree grow exponentially? Explain.

36. Computer graphics. Binary trees are used in three-dimensional computer graphics in the form of data structures known as *BSP trees*. Use the Internet to investigate BSP trees and write a brief report. What does *BSP* stand for?

37. Number checkers. Suppose that instead of a dictionary of words, you had a long list of numbers—for example, 7, 11, 15, 20, 21, . . . Explain how the concept of a binary-tree spell checker applies to searching such a list for a given number.

38. Other trees. In this chapter we have mentioned trees in the context of tournaments, spell checkers, chain mailings, Ponzi schemes, and more. Find other applications for trees. *Suggestion*: You might start with genealogies.

39. Social Security. In a recent U.S. presidential primary, some candidates referred to Social Security as a Ponzi scheme. Is this an accurate criticism of Social Security? You may need to investigate carefully how Social Security works.

40. NCAA basketball tournament. Some complain that too many teams are left out of the NCAA basketball tournament. How would the number of teams included change if there were one additional round of games? What if there were two additional rounds? Three? **Exercise 27 is relevant to this topic.**

41. Number of paths. How many different paths are there from the root of a tree to a given vertex? *Suggestion*: Look again at the definition of a tree.

42. Complete binary trees. Is it easier to count the number of leaves in a complete binary tree than in a binary tree that is not complete? Why or why not?

43. Root. Consider a graph that contains no circuits. Could any vertex be used as the root for the corresponding tree? Explain why in some situations one choice of root might make more sense than another.

CHAPTER SUMMARY

In this chapter, a *graph* is understood to be a collection of points joined by a number of line segments. The points are called *vertices*, and the segments connecting them are called *edges*. A highway map is a typical example of a graph.

Modeling with graphs and finding Euler circuits

Graphs can be helpful in determining efficient routes. One such route, an *Euler circuit*, begins and ends at home and traverses each edge exactly once. It is the type of route a mail carrier might use. A connected graph has an Euler circuit precisely when each vertex has even order. There are fairly straightforward procedures for constructing Euler circuits. One such procedure is simply to combine shorter circuits into longer ones. Another such procedure is *Fleury's algorithm*. If a graph does not have an Euler circuit, a route that backtracks as little as possible is desired. Such routes can often be found if the graph is first *Eulerized* by temporarily adding duplicate edges.

Hamilton circuits and traveling salesmen: Efficient routes

A driver of a delivery truck desires a route that starts and ends at home and visits each destination exactly once. Such a route is an example of a *Hamilton circuit*. In general, it is not easy to determine whether a graph has a Hamilton circuit. Even if a Hamilton circuit is known to exist, finding it can be problematic. There are guides that can help. For example, if a graph has a vertex of order 2, then each edge meeting that vertex must be part of any Hamilton circuit. But in practice, the search for a Hamilton circuit may involve a good deal of trial and error.

Some of the most important routing problems involve minimum-distance circuits that visit each city exactly once. Many such problems can be phrased in terms of *complete graphs*, graphs in which each pair of vertices is joined by an edge. If we assign distances (more generally, values) to each edge, then the *traveling salesman problem* is the problem of finding a shortest Hamilton circuit. This simple-sounding problem turns out to be surprisingly difficult. Mathematicians believe that there is no "quick" way to find a best route. In practice, no quick solution is known, and computers are used in some cases to find the solution. Generally speaking, we must settle for an approximately shortest route. One procedure for finding such a route is the *nearest-neighbor algorithm*: At each step, it travels to the nearest vertex not already visited (except at the final step, where it returns to the starting point). Another method used to find approximate solutions of the traveling salesman problem is the *cheapest link algorithm*. Algorithms that are much better than these two are available, but the better algorithms are more complex.

Trees: Viral e-mails and spell checkers

Special counting rules apply to *trees*, which are graphs with no circuits. The most obvious occurrences of trees in popular culture are tournament brackets, such as those for the NCAA basketball tournament. In this context, they serve as useful guides. But more important trees lie just below the surface of much of our daily activity. A spell checker, for example, is a tree functioning inside a computer that saves the authors of this text much embarrassment. A typical spell checker may know 100,000 or more words. A tree structure allows the computer to determine whether a typed word is in its dictionary using no more than 17 comparisons, not 100,000 comparisons. Trees also play important roles in many business and industrial settings.

KEY TERMS

circuit (cycle), p. 474
Euler circuit (Euler cycle), p. 474
degree (valence), p. 474
Hamilton circuit, p. 490
complete graph, p. 495
traveling salesman problem, p. 495
nearest-neighbor algorithm, p. 496
tree, p. 510
binary tree, p. 511
complete binary tree, p. 512

CHAPTER QUIZ

1. For the graph in **Figure 7.147**, find the degree of each vertex.

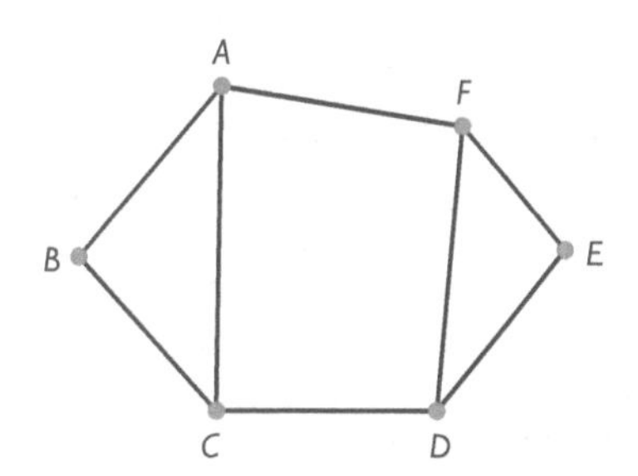

FIGURE 7.147 Finding degrees.

Answer *A*: 3; *B*: 2; *C*: 3; *D*: 3; *E*: 2; *F*: 3.

If you had difficulty with this problem, see Example 7.2.

2. The edges in **Figure 7.148** represent electrical lines that must be inspected, and the vertices represent junctions. Is there a route that will not require the inspector to travel along already inspected lines? That is, is there an Euler circuit for the graph in Figure 7.148? Be sure to explain your answer.

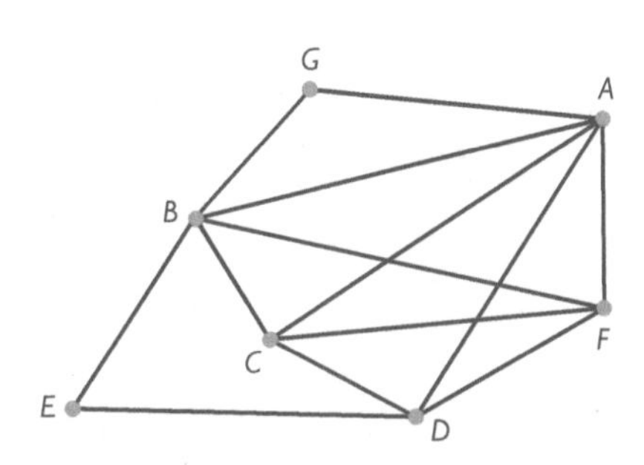

FIGURE 7.148 Inspecting electrical lines.

Answer No Euler circuit exists because some of the vertices have odd order.

If you had difficulty with this problem, see Example 7.3.

3. The edges in **Figure 7.149** represent streets that must be swept, and the vertices represent intersections. Is there a route for the street sweeper that will not require traversing already clean streets? In other words, is there an Euler circuit for the graph in Figure 7.149? If an Euler circuit exists, find one.

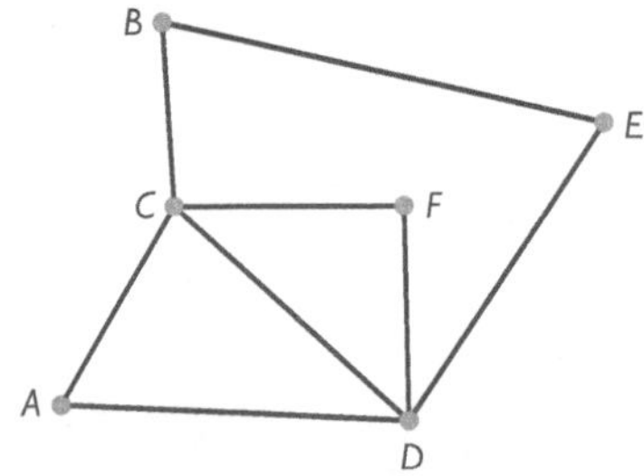

FIGURE 7.149 Street sweepers.

Answer One Euler circuit is *A-C-D-F-C-B-E-D-A*.

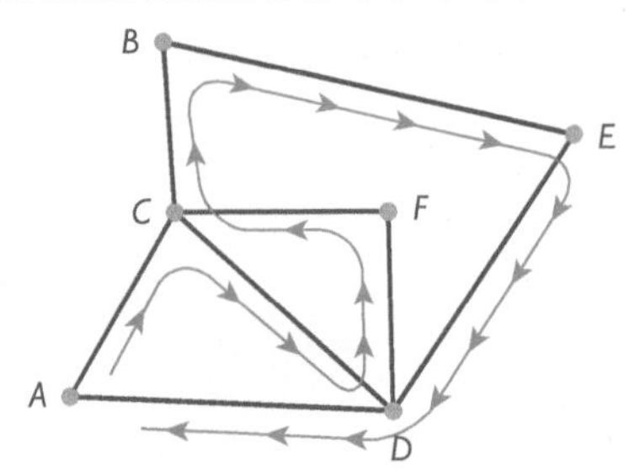

If you had difficulty with this problem, see Example 7.5.

4. For the graph in **Figure 7.150**, either find a Hamilton circuit or explain why no such circuit exists.

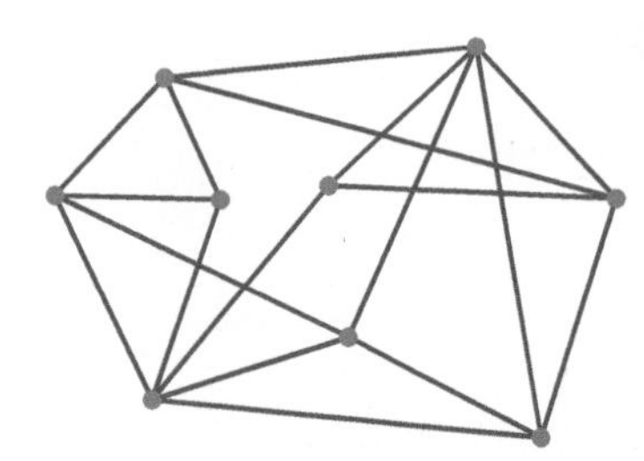

FIGURE 7.150 Hamilton circuit?

Answer One Hamilton circuit is

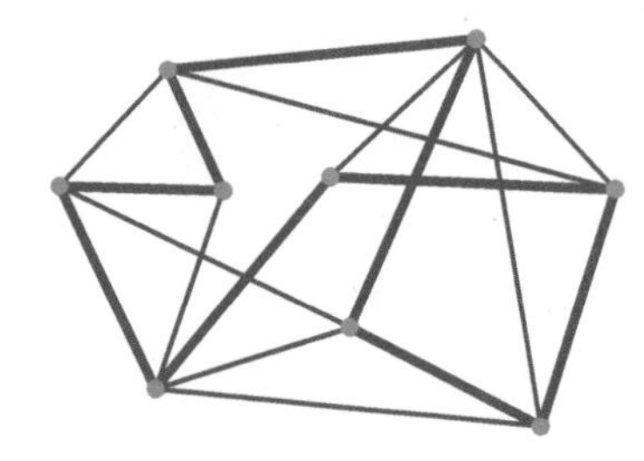

If you had difficulty with this problem, see Example 7.7.

5. Solve the traveling salesman problem for the map in **Figure 7.151** by calculating the mileage for each possible route. Use Oklahoma City as a starting point.

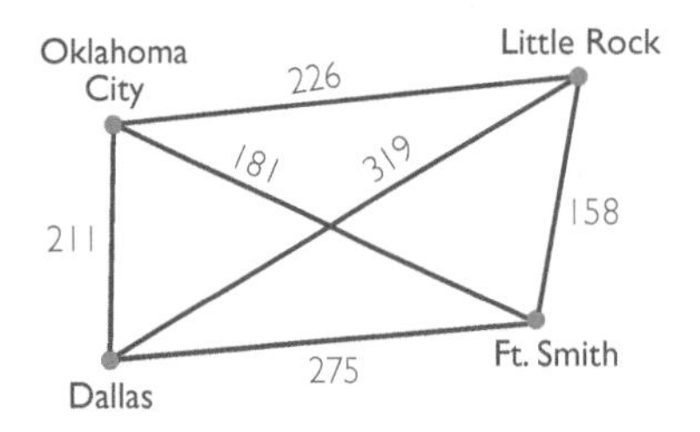

FIGURE 7.151 Traveling salesman.

Answer Oklahoma City - Ft. Smith - Little Rock - Dallas - Oklahoma City; distance: 869 miles.

If you had difficulty with this problem, see Example 7.8.

6. There are 31 members of an organization. The plan for disseminating news requires one member to contact two others with the information. Each member who receives the news contacts two members who have not yet gotten the news. This continues until all members have the information. Use the formula

$$\text{Number of vertices} = 2 \times \text{Number of parents} + 1$$

to determine how many members contact others.

Answer 15.

If you had difficulty with this problem, see Example 7.11.

7. Make a spell-checking tree for the following dictionary:

aardvark, bear, cat, deer, emu, fox, giraffe, hound, impala, javelina, krill

How many checks are needed to test whether "dove" is a valid word?

Answer For "dove": four checks.

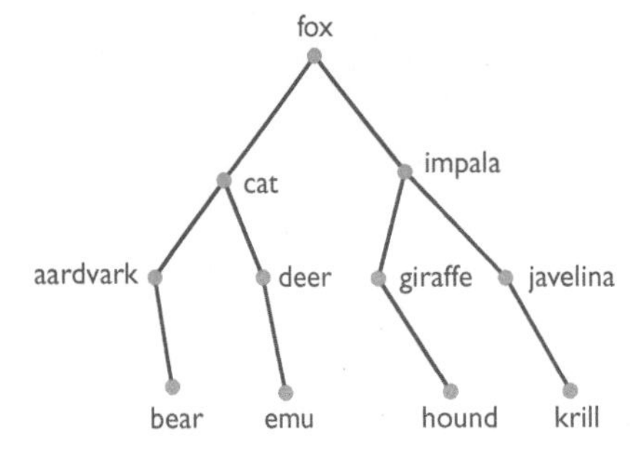

If you had difficulty with this problem, see Example 7.13.

8. We have a dictionary consisting of 300 words. If we use a binary tree as a spell checker, how many checks might be required to test whether "gnu" is a valid word?

Answer Nine checks.

If you had difficulty with this problem, see Example 7.14.

Democrats **How USA voted 2020** Republicans

The winning Electoral Votes

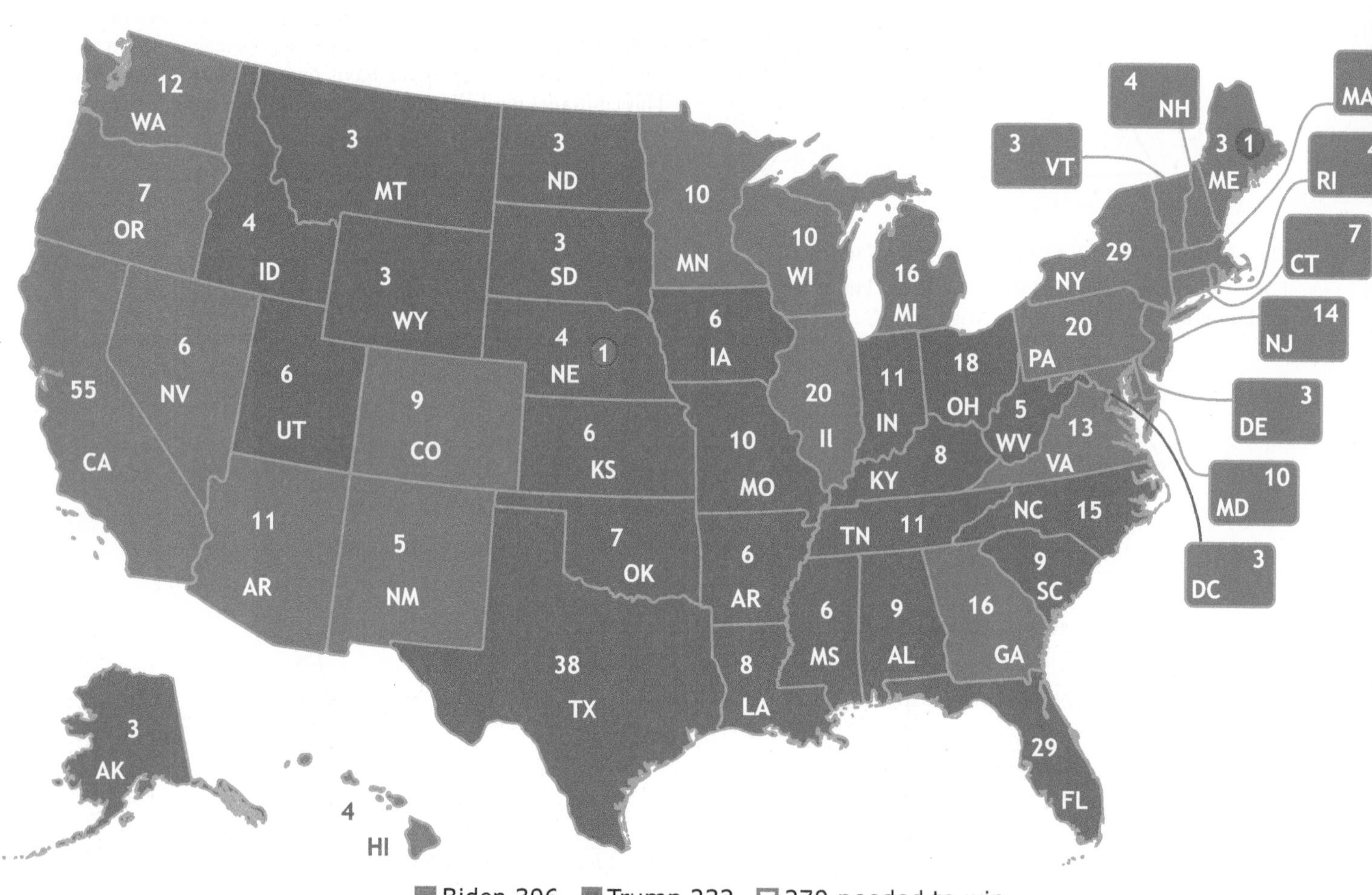

8 Voting and Social Choice

Voting rules and their consequences are not always as straightforward as one might first imagine, especially when some participants may have more than one vote. A prime example is the U.S. Electoral College. In electing a president of the United States, each state receives a number of electoral votes equal to the number of representatives and senators from that state. If a candidate wins the majority of the popular vote in a state, the candidate normally receives all the electoral votes from that state (the current exceptions are Maine and Nebraska). The following article from the Pew Research Center concerns the controversy arising from this system.

IN THE NEWS 8.1

Search

Home | TV & Video | U. S. | World | Politics | Entertainment | Tech | Travel

Majority of Americans continue to favor moving away from Electoral College

BRADLEY JONES January 27, 2021

Prior to the 2020 election, many observers noted that – if Donald Trump were to win – his most likely path toward victory would involve him winning the Electoral College while losing the popular vote (as was the case in 2016). This did not happen, but the current political geography of the United States continues to allow for the possibility that the winner of the popular vote may not be able to secure enough Electoral College votes to win the office.

Joe Biden won the popular vote by a margin of about 7 million votes and 4.5 percentage points overall (51.4% of all votes cast across the country were for Biden, 46.9% for Trump). That ultimately translated into an even greater share of the votes in the Electoral College, but – for the second straight election – the outcome in the Electoral College was determined by a relatively small number of voters in a handful of swing states.

Democrats and Republicans differ over scrapping Electoral College for popular vote to elect president.

A modest majority of Americans continue to favor changing the way presidents are elected, as they did in a January 2020 survey: 55% in the new poll say the system should be changed so that the winner of the popular vote nationwide wins the presidency, while 43% favor keeping the Electoral College system. The current balance of opinion is little changed over the last few years.

Attitudes about the Electoral College remain deeply divided along partisan lines. Democrats and Democratic-leaning independents – especially liberal Democrats – say they would prefer changing the system to be based on the popular vote (71% of Democrats overall, including 82% of liberal Democrats, say this). Republicans and Republican leaners – especially conservative Republicans – prefer keeping the current system where the winner of the Electoral College takes office (61% of Republicans overall, including 71% of conservative Republicans, say this).

Source: Pew Research Center (www.pewresearch.org/fact-tank/2021/01/27/majority-of-americans-continue-to-favor-moving-away-from-electoral-college/)

Questions of how a group should make decisions arise not only in weighty situations such as the election of a president but also in everyday situations, for example, choosing which movie a group of friends will rent for the night. Other situations in which making a decision is difficult come up in the context of equitable division. Typical examples include inheritance and property settlements associated with divorces. Determining how to divide assets fairly is often difficult and sometimes contentious.

A particularly sticky division problem comes up when delegates to the House of Representatives are assigned. The issue of how many representatives each state should get was a problem when our country was founded, and it remains an area of controversy today. We consider all of these topics in this chapter.

8.1 Measuring voting power: Does my vote count?

TAKE AWAY FROM THIS SECTION
A voter's power is not merely proportional to the number of votes he or she has. Two ways of measuring true voting power are the Banzhaf index and the Shapley–Shubik index.

The following is an excerpt from a post on the website Governing.com:

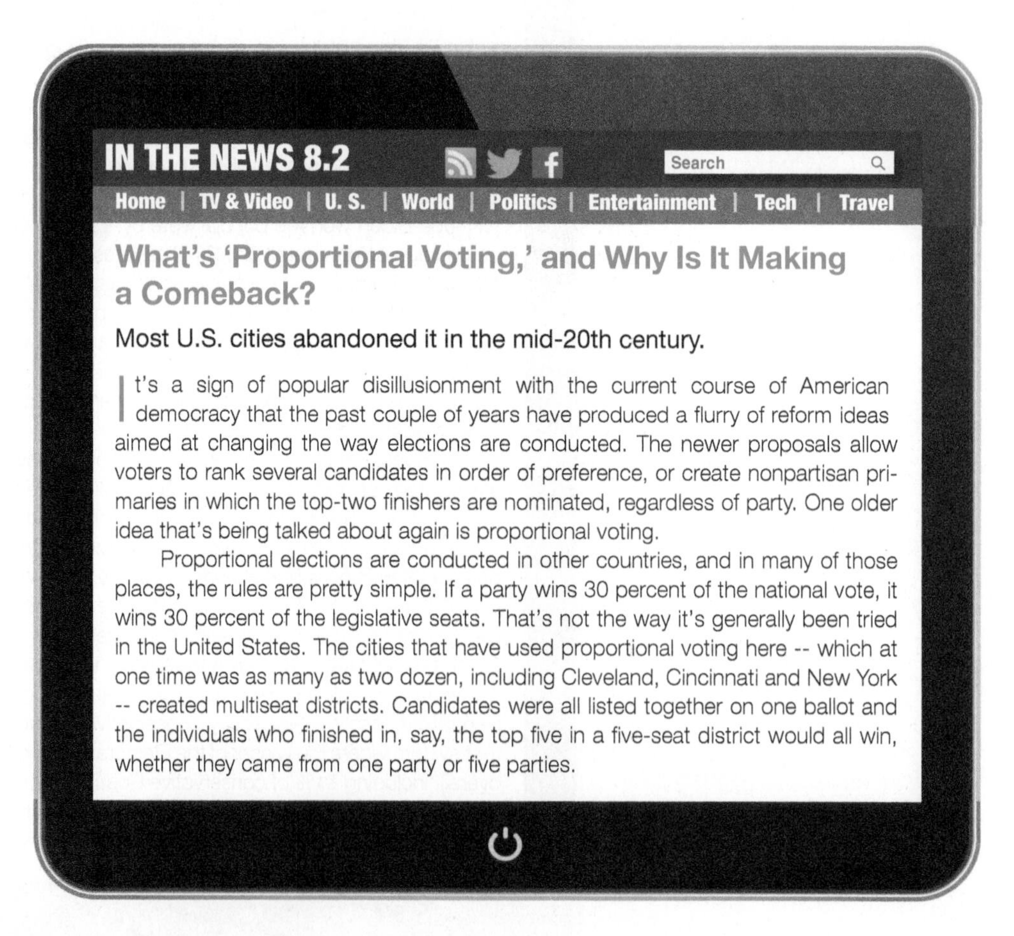

IN THE NEWS 8.2

Search

Home | TV & Video | U. S. | World | Politics | Entertainment | Tech | Travel

What's 'Proportional Voting,' and Why Is It Making a Comeback?

Most U.S. cities abandoned it in the mid-20th century.

It's a sign of popular disillusionment with the current course of American democracy that the past couple of years have produced a flurry of reform ideas aimed at changing the way elections are conducted. The newer proposals allow voters to rank several candidates in order of preference, or create nonpartisan primaries in which the top-two finishers are nominated, regardless of party. One older idea that's being talked about again is proportional voting.

Proportional elections are conducted in other countries, and in many of those places, the rules are pretty simple. If a party wins 30 percent of the national vote, it wins 30 percent of the legislative seats. That's not the way it's generally been tried in the United States. The cities that have used proportional voting here -- which at one time was as many as two dozen, including Cleveland, Cincinnati and New York -- created multiseat districts. Candidates were all listed together on one ballot and the individuals who finished in, say, the top five in a five-seat district would all win, whether they came from one party or five parties.

In the proportional voting scheme discussed in the article above, coalitions, or voting blocs, emerge as important electoral tools. One question that arises is how to measure the relative power of voters or voting blocs. Voting alliances or *coalitions* play an important role here. It may seem that a voter's power should be proportional to the number of his or her votes, but that is not the case. Just because one entity (a person, state, or coalition) has more votes than another does not mean that entity has more power. In this section, one way we will measure voting power is in terms of coalitions of voters.

Here's an example to illustrate the importance of voters joining together to form an alliance or coalition. Suppose the board of directors of ACE Computer Corporation has three members. Ann controls 19 shares, Beth controls 18 shares, and Cate controls only 3 shares. They want to elect a chairperson, and the election requires a simple majority of the voting shares. Because there are 40 shares altogether, 21 shares are required for a majority.

How powerful are these voters? It might seem on the surface that Ann is the most powerful but that Cate doesn't have much power at all because she has only 3 shares out of a total of 40. A good case can be made, however, that all three members have equal power. Let's see how to make that case.

None of the members has enough shares (21) to win. The only way for any member to be elected is to convince one of the other members to throw support to her. If Cate can convince Ann to support her, she will have $3 + 19 = 22$ votes, so she will be elected. If Cate can convince Beth to support her, she will have $3 + 18 = 21$ votes, so she will be elected. Thus, if Cate can get the support of either Ann or Beth, then she will be elected, even though she controls only 3 shares. Similarly, if Beth can convince either Cate or Ann to support her, she will win. This observation suggests that all three members have equal power.

This example illustrates that voting power is not simply proportional to the number of votes a person or bloc has. Coalitions play an important role in voting power, and we now focus on the question of measuring voting power in terms of coalitions.

Voting power and coalitions

One way to measure voting power is the *Banzhaf index*. This index is named after John F. Banzhaf III, an attorney who first published his method in 1965.[1]

A group of voters who vote the same way is called a *voting coalition*. (Each coalition includes at least one voter.) In the example involving the board of ACE Corporation, no one voter could win alone, but coalitions consisting of two or three voters could. Banzhaf devised a way to measure the relative power of voters in terms of *winning coalitions*.[2] We will describe his scheme shortly, but first we have to explain the term *quota*, the number of votes necessary to win.

---

[1] *"Weighted Voting Doesn't Work: A Mathematical Analysis,"* Rutgers Law Review *19 (1965), 317–343.*

[2] *In the real-life situation addressed by Banzhaf in his original paper, votes were allocated among six district supervisors; however, three of the supervisors had no power because their votes were never necessary to determine the outcome. See Exercise 32 for the details.*

KEY CONCEPT

In a voting system, the number of votes necessary to win the election is called the quota.

In many cases, the quota is just a simple majority, which requires more than half of the total number of votes cast. For instance, in the previous example there were 40 votes altogether, so the quota for a simple majority is 21. In the case of a cloture vote in the U.S. Senate, however, the quota is 60 votes out of 100. To ratify a constitutional amendment in the United States requires (among other things) approval by three-quarters of the states, which makes a quota of 38 states.

In this section, you should assume (unless you are instructed otherwise) that the quota is a simple majority. The *methods* we use are unaffected by the choice of quota. Changing the quota can, however, affect the power of voters. See Exercise 20.

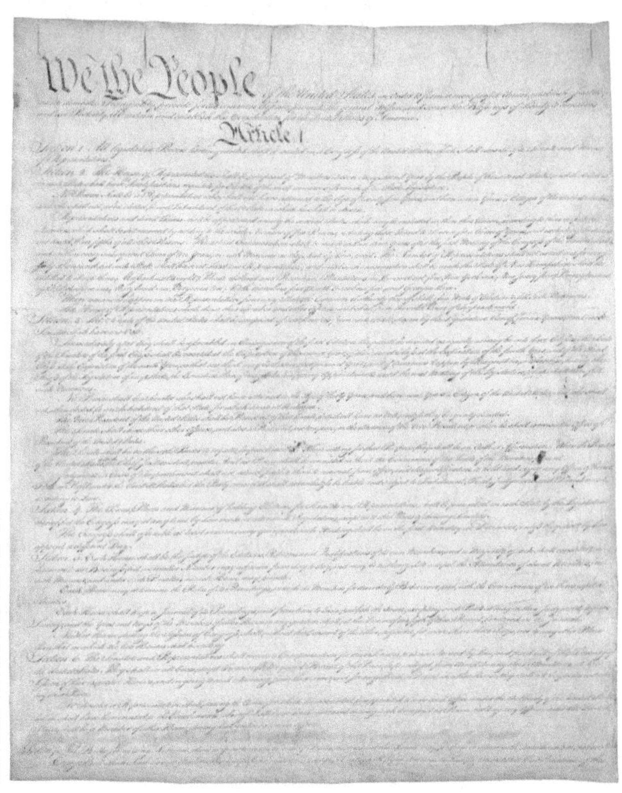
We the People

Article. I

National Archives and Records Administration

The U.S. Constitution.

Article V

The Congress, whenever two thirds of both Houses shall deem it necessary, shall propose Amendments to this Constitution, or, on the Application of the Legislatures of two thirds of the several States, shall call a Convention for proposing Amendments, which, in either Case, shall be valid to all Intents and Purposes, as Part of this Constitution, when ratified by the Legislatures of three fourths of the several States, or by Conventions in three fourths thereof, as the one or the other Mode of Ratification may be proposed by the Congress; Provided that no Amendment which may be made prior to the Year One thousand eight hundred and eight shall in any Manner affect the first and fourth Clauses in the Ninth Section of the first Article; and that no State, without its Consent, shall be deprived of its equal Suffrage in the Senate.

Winning coalitions and critical voters

We have noted that some voters have more sway in determining outcomes than others and that voters may increase their influence by banding together into coalitions. Banzhaf devised a way to measure the relative power of voters in terms of coalitions. The first step in describing his method is to look more closely at coalitions.

KEY CONCEPT

A set of voters with enough votes to determine the outcome of an election is a winning coalition; otherwise, it is a losing coalition.

In the example involving the ACE Corporation, the winning coalitions consist of any two voters or all three. Here is another example.

EXAMPLE 8.1 Finding the winning coalitions: County convention

Let's suppose there are three delegates to a county convention: Abe has 4 votes from his precinct, Ben has 3 votes, and Condi has 1 vote. A simple majority of the votes wins.

a. What is the quota?

b. Make a table listing all of the coalitions of voters. Designate which of them are winning coalitions.

SOLUTION

a. There are 8 votes, so the quota for a simple majority is 5 votes.

b. The accompanying table shows all possible coalitions together with the votes that each coalition controls. The last column indicates whether or not the coalition controls a majority (at least 5) of the votes and therefore is a winning coalition.

Number of votes			Total votes	Winning coalition?
4	3	1		
Abe	Ben	Condi	8	Yes
Abe	Ben		7	Yes
Abe		Condi	5	Yes
Abe			4	No
	Ben	Condi	4	No
	Ben		3	No
		Condi	1	No

TRY IT YOURSELF 8.1

Suppose that Abe has 5 votes and that Ben and Condi have 3 votes each. A simple majority is required to win. Find the quota and make a table listing all of the coalitions of voters. Designate which of them are winning coalitions.

The answer is provided at the end of this section.

Now we focus attention on the voters who are crucial components of winning coalitions.

KEY CONCEPT

A member of a winning coalition is said to be a critical voter for that coalition if the coalition is no longer a winning one when that voter is removed.

Let's determine the critical voters for each of the three winning coalitions in Example 8.1. These winning coalitions appear in the top three rows of the table.

Number of votes			Total	Winning
4	3	1	votes	coalition?
Abe	Ben	Condi	8	Yes
Abe	Ben		7	Yes
Abe		Condi	5	Yes

Abe, Ben, Condi (8 total votes): If Abe is removed, the remaining coalition controls only 4 votes—not a majority. So, removing Abe changes the coalition from a winning one to a losing one. Therefore, Abe is a critical voter for this coalition.

Removing either Ben or Condi from the coalition leaves at least 5 votes controlled by the remaining coalition, so the remaining coalition is still a winning coalition. Therefore, neither Ben nor Condi is a critical voter in this coalition.

Abe, Ben (7 total votes): Removing either voter makes it a losing coalition, so both Abe and Ben are critical for this coalition.

Abe, Condi (5 total votes): As with the preceding coalition, removing either voter makes it a losing coalition. So, both Abe and Condi are critical for this coalition.

We summarize all of this information in a *coalition table* (see **Table 8.1**). It lists the critical voters.

TABLE 8.1 Coalition Table

Number of votes			Total	Winning	Critical
4	3	1	votes	coalition?	voters
Abe	Ben	Condi	8	Yes	Abe
Abe	Ben		7	Yes	Abe, Ben
Abe		Condi	5	Yes	Abe, Condi
Abe			4	No	Not applicable
	Ben	Condi	4	No	Not applicable
	Ben		3	No	Not applicable
		Condi	1	No	Not applicable

SUMMARY 8.1
Winning Coalitions and Critical Voters

A set of voters with enough votes to determine the outcome of an election is a *winning coalition.* A voter in a winning coalition is *critical* for that coalition if the coalition is no longer a winning one when that voter is removed. We can summarize the essential information about coalitions in a *coalition table.*

Counting coalitions

To make a coalition table, we had to list all the different coalitions of voters. Suppose there are 7 voters. How many different coalitions are possible? Each of the 7 voters is either in the coalition or not in it, so there are 2^7 possibilities for voters to be either in or not in a coalition.[3] This includes all the voters not being in, which isn't a coalition, so there are $2^7 - 1$ coalitions possible. In general, if there are n voters, then there are $2^n - 1$ possible coalitions. In the case of voters Abe, Ben, and Condi that we considered in Example 8.1, the number of voters is $n = 3$, so there should be $2^3 - 1 = 7$ coalitions—just as indicated in the coalition table, Table 8.1. Of course, for large collections of voters, it is impractical to list all the possible coalitions.

SUMMARY 8.2
Number of Coalitions

For n voters, there are $2^n - 1$ possible coalitions (each of which includes at least 1 voter).

The Banzhaf power index

Now we define the Banzhaf index for a voter in terms of the number of instances in which that voter is critical.

KEY CONCEPT

A voter's Banzhaf power index is the number of times that voter is critical in some winning coalition divided by the total number of instances of critical voters. The index is expressed as a fraction or as a percentage.

We illustrate this definition by computing the Banzhaf index for each voter in the situation we have been considering.

EXAMPLE 8.2 Computing the Banzhaf index: County convention

Use the coalition table, Table 8.1, to compute the Banzhaf index for each delegate to the county convention described in Example 8.1.

SOLUTION

From the coalition table, we see that overall there were 5 instances of critical voters. This means that to find the Banzhaf index of a voter, we first find the number of times that voter is critical and then divide by 5.

Abe was critical in 3 of these 5 instances, so he has a Banzhaf index of 3/5 or 60%. Ben was critical in only 1 of these 5 instances, so he has a Banzhaf index of 1/5 or 20%. Condi also was critical in just 1 of these 5 instances, so she also has a Banzhaf index of 1/5 or 20%.

TRY IT YOURSELF 8.2

Make a coalition table for the board of ACE Computer Corporation, where Ann controls 19 shares, Beth controls 18 shares, and Cate controls only 3 shares. Find the Banzhaf index of each voter.

The answer is provided at the end of this section.

[3] *Here, we are using the Counting Principle from Section 3 of Chapter 5.*

Let's make a couple of observations based on the preceding example. First, note that even though Abe has only one more vote than Ben, he has three times the voting power of Ben (as measured by the Banzhaf index). Also note that Ben has 3 votes to Condi's 1 vote, but they have the same voting power. Once again, this indicates that simply knowing how many votes a voter has does not really tell us much about his or her voting power. The Banzhaf index gives us a way to measure the real power of voters.

EXAMPLE 8.3 Calculating the Banzhaf index: The 1932 Democratic National Convention

At the 1932 Democratic National Convention, there were several candidates for the presidential nomination, the most prominent of whom was Franklin D. Roosevelt. In those days it was not uncommon for the nomination to be decided by negotiations among the various blocs of delegates, which effectively formed coalitions. In this convention, there were four rounds of voting before the final nominee, Roosevelt, was elected. For the purposes of this example, we will examine only the first round of voting. In this round, Roosevelt received 666.25 votes, Al Smith (a former governor of New York) received 201.75 votes, John Nance Garner (the Speaker of the U.S. House of Representatives and later vice president) received 90.25 votes, and the other seven candidates combined received 195.75 votes. To win the nomination required 770 votes.

Using the four blocs of votes listed above, answer the following questions.

a. What is the quota?

b. Determine the winning coalitions.

c. Determine the critical voting blocs in each winning coalition.

d. Determine the Banzhaf index of each bloc.

SOLUTION

a. Because 770 votes were required to win the nomination, that is the quota. Note that this number is considerably more than a simple majority of the 1154 total votes cast.

Bettmann/Getty Images

Franklin D. Roosevelt speaking at the 1932 Democratic National Convention in Chicago, Illinois.

b. There are $2^4 - 1 = 15$ possible coalitions. The accompanying table lists them along with the total number of votes for each. (We use R for Roosevelt, S for Smith, G for Garner, and O for others.) The winning coalitions are those with a total of 770 or more votes.

666.25	201.75	195.75	90.25	Total votes	Winning coalition?
R	S	O	G	1154	Yes
R	S	O		1063.75	Yes
R	S		G	958.25	Yes
R	S			868	Yes
R		O	G	952.25	Yes
R		O		862	Yes
R			G	756.5	No
R				666.25	No
	S	O	G	487.75	No
	S	O		397.5	No
	S		G	292	No
	S			201.75	No
		O	G	286	No
		O		195.75	No
			G	90.25	No

c. The accompanying table lists only the winning coalitions along with the critical voting blocs in each case.

Winning Coalitions Only

Votes					
666.25	201.75	195.75	90.25	Total votes	Critical voters
R	S	O	G	1154	R
R	S	O		1063.75	R
R	S		G	958.25	R, S
R	S			868	R, S
R		O	G	952.25	R, O
R		O		862	R, O

d. To calculate the Banzhaf index of a voter, we divide the number of times that voter is critical in some winning coalition by the total number of instances of critical voters. In this case, there are 10 instances in which a voting bloc is critical. This means that to find the Banzhaf index of a voting bloc, we first find the number of times that bloc is critical and then divide by 10.

Let's do the computation for each bloc:

Roosevelt is critical in 6 of these 10 instances. Therefore, this Banzhaf index is $6/10 = 60\%$.

Smith is critical in 2 of these 10 instances. Therefore, this Banzhaf index is $2/10 = 20\%$.

"**Other**" is critical in 2 of these 10 instances. Therefore, this Banzhaf index is also $2/10 = 20\%$.

Garner is not critical in any of these 10 instances. Therefore, this Banzhaf index is 0%.

The calculations in Example 8.3 show that, as measured by the Banzhaf index, the most powerful candidate by far was Roosevelt. They also show that although Smith and "Other" had different numbers of votes, they had the same power.

SUMMARY 8.3
Banzhaf Power Index

A voter's *Banzhaf index* is the number of times that voter is critical in some winning coalition divided by the total number of instances of critical voters. The index is expressed as a fraction or as a percentage.

Swing voters and the Shapley–Shubik power index

In a seminal paper, Lloyd Shapley and Martin Shubik[4] devised a method of measuring voting power that is somewhat different from the Banzhaf index. The Banzhaf index is based on the idea of a voter who is *critical* for a coalition, and the Shapley–Shubik index is based on the concept of a *swing voter*. To describe the Shapley–Shubik index, we assume that the voters vote in order and that their votes are added as they are cast.

We start with the first voter. If that voter has enough votes to decide the election (because her votes meet the quota), she is the swing voter. If not, add the second voter's votes. If the total of the two meets the quota, the second voter is the swing. If not, add a third voter's votes, and so on until the total of the votes cast meets the quota. The deciding voter for this order of voting is the *swing voter*.

KEY CONCEPT

Suppose that voters vote in order and that their votes are added as they are cast. The swing voter is the one whose votes make the total meet the quota and thus decides the outcome. Which is the swing voter depends on the order in which votes are cast.

EXAMPLE 8.4 Finding the swing voter: Council of the European Union

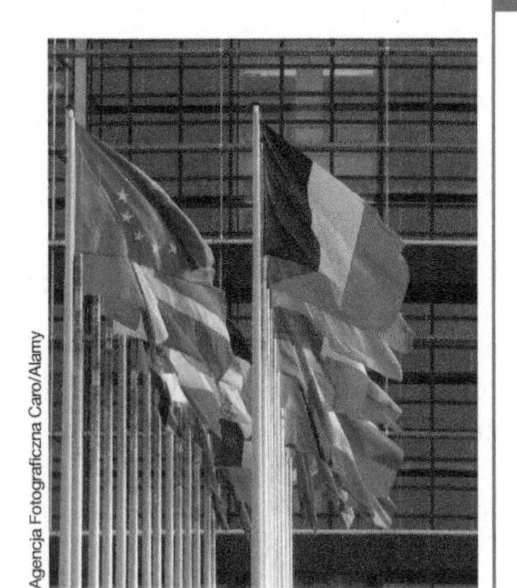

Agencja Fotograficzna Caro/Alamy

Flags of the European Union.

Between 2001 and 2014, member states of the European Union had votes on the Council assigned by the 2001 Treaty of Nice. The number of votes was roughly determined by a country's population but progressively weighted in favor of smaller countries. Ireland had 7 votes, Cyprus had 4 votes, and Malta had 3 votes. Suppose that these three countries serve as a committee where a simple majority is required to win, so the quota is 8 votes. Make a table listing all the permutations[5] of the voters and the swing voter in each case.

SOLUTION

There are $n!$ different permutations[6] of n objects, where $n! = n \times (n-1) \times \cdots \times 1$. Our three objects are Ireland, Cyprus, and Malta and $3! = 3 \times 2 \times 1 = 6$, so there will be six rows in our table. We let the first row represent the voting order Ireland, Cyprus, Malta. Then Ireland, with 7 votes, doesn't have a majority, but the 4 votes of Cyprus added to Ireland's do give a majority. Therefore, Cyprus is the swing voter for this order, and we indicate this in the last column of the first row.

---

[4] *Shapley and Shubik, prominent game theorists, were portrayed in the book and movie* A Beautiful Mind. *See their paper "A Method for Evaluating the Distribution of Power in a Committee System,"* American Political Science Review *48 (1954), 787–792. Shapley was a winner of the 2012 Nobel Prize in Economics and died March 12, 2016.*

[5] *Recall from Section 4 of Chapter 5 that a permutation of objects is just an arrangement of them in a certain order.*

[6] *For more information on permutations, see Section 4 of Chapter 5.*

If we continue in this way for each row, we obtain the following table:

Order of voters			Swing
Ireland (7)	Cyprus (4)	Malta (3)	Cyprus
Ireland (7)	Malta (3)	Cyprus (4)	Malta
Cyprus (4)	Ireland (7)	Malta (3)	Ireland
Cyprus (4)	Malta (3)	Ireland (7)	Ireland
Malta (3)	Ireland (7)	Cyprus (4)	Ireland
Malta (3)	Cyprus (4)	Ireland (7)	Ireland

TRY IT YOURSELF 8.4

Germany had 29 votes on the European Union Council, Spain had 27, and Sweden had 10. Suppose these three countries serve as a committee where a simple majority of 34 votes is required to win. Make a table listing all the permutations of the voters and the swing voter in each case.

The answer is provided at the end of this section.

Note that in Example 8.4, the second voter is not always the swing voter. The third voter is the swing in the fourth and sixth rows of the table. Note, too, that Ireland is the swing voter four out of six times. This proportion is the key to the measurement we will now introduce.

Definition of the Shapley–Shubik power index

The Shapley–Shubik index measures the distribution of power by observing the proportion of times a given voter is the swing.

KEY CONCEPT

The Shapley–Shubik power index of a given voter is calculated as the fraction (or percentage) of all permutations of the voters in which that voter is the swing.

EXAMPLE 8.5 Computing the Shapley–Shubik index: Committee of the Council

Compute the Shapley–Shubik power index for the committee of Ireland, Cyprus, and Malta from Example 8.4.

SOLUTION

For each voter, we calculate the fraction of all permutations of the voters in which that voter is the swing. There are six permutations. Ireland is the swing in 4 of the 6 cases, so the index for Ireland is $4/6 = 2/3$, or about 66.67%. Cyprus is the swing in only 1 of the 6 cases, so the index for Cyprus is $1/6$, or about 16.67%. Malta also is the swing in only 1 of the 6 cases, so the index for Malta is also $1/6$, or about 16.67%.

TRY IT YOURSELF 8.5

Compute the Shapley–Shubik power index for the committee of Germany, Spain, and Sweden from Try It Yourself 8.4 on p. 535.

The answer is provided at the end of this section.

It is instructive to compare the two voting power measurements we have studied. The reader should verify that the *Banzhaf* indices for the committee of Ireland, Cyprus, and Malta are Ireland 60%, Cyprus 20%, and Malta 20%. The two voting power indices give similar but not identical results. It is interesting to see that both of these measures tell us that Cyprus and Malta have equal power on the committee even though Cyprus has 4 votes and Malta has only 3. This fact might not be obvious at first glance, especially in a more complex situation, but it is made transparent by the assignment of these power indices.

Now we consider a situation in which there are four voters.

EXAMPLE 8.6 Calculating the Shapley–Shubik index: Four states

The number of electoral votes allotted to the states varies from 55 (California) to 3 (Alaska and several others). A few medium-sized states can have a large influence on the presidential election. Assume that a simple majority from the following four states would determine the outcome: Virginia with 13 electoral votes, Wisconsin with 10, Colorado with 9, and Iowa with 6 (these numbers are before the 2020 census).

Stefani Reynolds/Bloomberg/Getty Images

Joe Biden speaks during a campaign event on August 12, 2020.

SAUL LOEB/AFP/Getty Images

Donald Trump waves at a campaign rally on October 21, 2020.

a. How many permutations of these four states are there?

b. What is the quota?

c. Make a table listing all of the permutations of the states and the swing voter in each case.

d. Find the Shapley–Shubik index for each state.

SOLUTION

a. The number of permutations of four items is $4! = 4 \times 3 \times 2 \times 1 = 24$.

b. There are 38 votes, and it requires a majority to win, so the quota is 20.

c. The table is shown on the next page. We have abbreviated the state names and written the number of votes in parentheses. The swing state is in the last column. For example, looking at the first row, we see that Virginia's 13 electoral votes are not enough to make 20, but Wisconsin's 10 added to it exceeds 20. Therefore, Wisconsin is the swing voter in this order. Looking at the fifth row, we see that Virginia's 13 electoral votes plus Iowa's 6 are not enough to make 20, but Wisconsin's 10 added to these exceeds 20. Therefore, Wisconsin is the swing voter in this order too.

States				Swing
VA (13)	WI (10)	CO (9)	IA (6)	WI
VA (13)	WI (10)	IA (6)	CO (9)	WI
VA (13)	CO (9)	WI (10)	IA (6)	CO
VA (13)	CO (9)	IA (6)	WI (10)	CO
VA (13)	IA (6)	WI (10)	CO (9)	WI
VA (13)	IA (6)	CO (9)	WI (10)	CO
WI (10)	VA (13)	CO (9)	IA (6)	VA
WI (10)	VA (13)	IA (6)	CO (9)	VA
WI (10)	CO (9)	VA (13)	IA (6)	VA
WI (10)	CO (9)	IA (6)	VA (13)	IA
WI (10)	IA (6)	VA (13)	CO (9)	VA
WI (10)	IA (6)	CO (9)	VA (13)	CO
CO (9)	VA (13)	WI (10)	IA (6)	VA
CO (9)	VA (13)	IA (6)	WI (10)	VA
CO (9)	WI (10)	VA (13)	IA (6)	VA
CO (9)	WI (10)	IA (6)	VA (13)	IA
CO (9)	IA (6)	VA (13)	WI (10)	VA
CO (9)	IA (6)	WI (10)	VA (13)	WI
IA (6)	VA (13)	WI (10)	CO (9)	WI
IA (6)	VA (13)	CO (9)	WI (10)	CO
IA (6)	WI (10)	VA (13)	CO (9)	VA
IA (6)	WI (10)	CO (9)	VA (13)	CO
IA (6)	CO (9)	VA (13)	WI (10)	VA
IA (6)	CO (9)	WI (10)	VA (13)	WI

d. Virginia is the swing in 10 cases, Wisconsin and Colorado are the swing in 6 cases each, and Iowa is the swing in 2 cases. Therefore, the power indices are Virginia: 10/24, or about 41.67%; Wisconsin: 6/24, or 25%; Colorado: 6/24, or 25%; and Iowa: 2/24, or about 8.33%.

It is interesting to note that in this example, Virginia has about 2.2 times the number of votes that Iowa has, but it has 5 times the voting power as measured by the Shapley–Shubik index.

SUMMARY 8.4
The Shapley–Shubik Power Index

The Shapley–Shubik power index of a given voter is calculated as the fraction (or percentage) of all permutations of the voters in which that voter is the swing.

The practical value of measures of voting power is addressed in the following quote from the previously mentioned paper of Shapley and Shubik:

> It can easily happen that the mathematical structure of a voting system conceals a bias in power distribution unsuspected and unintended by the authors.... Can a consistent criterion for "fair representation" be found? ... The method of measuring "power" which we present in this paper is intended as a first step in the attack on these problems.

Thus, the Shapley–Shubik index gives at the very least a method of revealing possibly hidden bias or complexity in a system.

CALCULATION TIP 8.1 Computer Software

Computing either the Banzhaf index or the Shapley–Shubik index by hand is impractical even for moderately large collections of voters. Computer software that performs the calculations (as long as the number of voters is not terribly large) is freely available. One such program for the Banzhaf index is called ipdirect, and one for the Shapley–Shubik index is called ssdirect. Both programs can be found at http://homepages.warwick.ac.uk/~ecaae/index.html.

WHAT DO YOU THINK?

Power: A friend doesn't understand why you need to study voting power in your math class. He thinks that the voting bloc with the most votes has the most power. How would you explain, using your own words, that the subject is more complicated?

Questions of this sort may invite many correct responses. We present one possibility: In a simple setting, a majority vote among three people wins. Suppose Jim has 4 votes, Mary has 3, and Bob has 2. No single voter can manage a majority of the 9 votes. But any pair does hold a majority. Thus, Jim with 4 votes holds exactly the same sway as does Bob with 2 votes or Mary with 3. Suppose, however, that Mary loses 1 vote to Jim so that he has 5 votes while both Mary and Bob have 2. Then Jim has the majority vote all by himself while Bob's and Mary's votes count for nothing at all. Thus, the transfer of a single vote can dramatically change the influence of voters. These examples make clear that small changes in the distribution of votes can have enormous effects on voting power.

Further questions of this type may be found in the *What Do You Think?* section of exercises.

Try It Yourself answers

Try It Yourself 8.1: Finding the winning coalitions: County convention The quota is 6 votes.

Number of votes			Total	Winning
5	3	3	votes	coalition?
Abe	Ben	Condi	11	Yes
Abe	Ben		8	Yes
Abe		Condi	8	Yes
Abe			5	No
	Ben	Condi	6	Yes
	Ben		3	No
		Condi	3	No

Try It Yourself 8.2: Computing the Banzhaf index: ACE Computer Corporation

Number of votes			Total	Winning	Critical
19	18	3	votes	coalition?	voters
Ann	Beth	Cate	40	Yes	None
Ann	Beth		37	Yes	Ann, Beth
Ann		Cate	22	Yes	Ann, Cate
Ann			19	No	Not applicable
	Beth	Cate	21	Yes	Beth, Cate
	Beth		18	No	Not applicable
		Cate	3	No	Not applicable

Each voter has an index of $2/6 = 1/3$, or about 33.33%.

Try It Yourself 8.4: Finding the swing voter: Council of the European Union

Order of voters			Swing
Germany (29)	Spain (27)	Sweden (10)	Spain
Germany (29)	Sweden (10)	Spain (27)	Sweden
Spain (27)	Germany (29)	Sweden (10)	Germany
Spain (27)	Sweden (10)	Germany (29)	Sweden
Sweden (10)	Germany (29)	Spain (27)	Germany
Sweden (10)	Spain (27)	Germany (29)	Spain

Try It Yourself 8.5: Computing the Shapley–Shubik index: Council of the European Union
Germany: 1/3; Spain: 1/3; Sweden: 1/3.

Exercise Set 8.1

Test Your Understanding

1. The Banzhaf index is designed to measure ________.

2. For the Banzhaf index, a voting coalition is ________.

3. Which of the following are important in the calculation of the Banzhaf index? **a.** a coalition table. **b.** critical voters. **c.** the number of winning coalitions.

4. True or false: The Shapley–Shubik power index assumes that votes are taken in order.

5. For the Shapley–Shubik index, the swing voter is ________.

6. True or false: Neither the Banzhaf index nor the Shapley–Shubik index can be practically calculated by hand, even for only moderately large groups of voters.

Problems

Leave each power index as a fraction unless you are directed to do otherwise.

7. **How many coalitions?** In a collection of 5 voters, how many different coalitions are possible?

8. **How many coalitions?** In a collection of 10 voters, how many different coalitions are possible?

9. **How many permutations?** In a collection of 5 voters, how many different permutations are possible?

10. **How many permutations?** In a collection of 10 voters, how many different permutations are possible?

11. **State coalitions.** Each of the 50 states is assigned a certain number of electoral votes. How many possible coalitions (each coalition including at least 1 state) are there?

a. 2^{50}
b. $2^{50} - 1$
c. $50! - 1$
d. $50!$
e. 50^2

12. **State orders.** Each of the 50 states is assigned a certain number of electoral votes. How many vote orders would we need to consider in order to calculate the Shapley–Shubik index for each state?

a. 2^{50}
b. $2^{50} - 1$
c. $50! - 1$
d. $50!$
e. 50^2

13. **A strategy.** Your political opponent has put together a coalition of voters who control sufficient votes to defeat you in the upcoming election. You wish to improve your chances by persuading some of the coalition members to switch their support to you. Should you talk with critical voters in the coalition or with noncritical voters in the coalition? Explain why you gave the answer you did.

14. **Two voters.** Suppose there are only two voters. Anne controls 152 votes, and the quota is 141 votes. What is Anne's Banzhaf index? What is the Banzhaf index of the other voter?

15. **Two voters again.** Suppose there are only two voters. Anne controls 152 votes, and the quota is 141 votes. What is Anne's Shapley–Shubik index? What is the Shapley–Shubik index of the other voter?

16. **One person, one vote.** In a group of 100 voters, each person has 1 vote, and the quota is a simple majority. What is the Banzhaf index for each voter? *Hint:* Each voter should have the same index.

17. **One person, one vote again.** In a group of 100 voters, each person has 1 vote, and the quota is a simple majority. What is the Shapley–Shubik index for each voter? *Hint:* Each voter should have the same index.

18. **Three delegates.** Suppose there are three delegates to a county convention. Abe has 100 votes, Ben has 91 votes, and Condi has 10 votes. A simple majority wins.

a. What is the quota?

b. Make a table listing all of the coalitions and the critical voter in each case.

c. Find the Banzhaf index for each voter.

d. Make a table listing all of the permutations and the swing voter in each case.

e. Find the Shapley–Shubik index for each voter.

19. **Four voters.** Suppose there are four voters: A with 13 votes, B with 6 votes, C with 5 votes, and D with 2 votes. Suppose that a simple majority is required to win.

a. What is the quota?

b. How many coalitions are there?

c. Make a coalition table.

d. Find the Banzhaf index for each voter.

e. How many permutations of four voters are there?

f. Make a table listing all of the permutations of the voters and the swing voter in each case.

g. Find the Shapley–Shubik index for each voter.

20. Ace Solar. Suppose there are four stockholders in a meeting of the Ace Solar Corporation and that each person gets as many votes as the number of his or her shares. Assume that Abe has 49 shares, Ben has 48 shares, Condi has 4 shares, and Doris has 3 shares.

a. Assume that a simple majority is required to prevail in a vote. Make a table listing all of the permutations of the voters and the swing voter in each case, and calculate the Shapley–Shubik index for each voter.

b. Assume now that a two-thirds majority is required to prevail in a vote, so the quota is 70. Calculate the Shapley–Shubik index for each voter.

c. Does Condi's voting power increase or decrease with the change from a simple majority to a two-thirds majority?

21. Five voters. Suppose there are five voters, A, B, C, D, and E, with 6, 5, 4, 3, and 2 votes, respectively. Suppose that a simple majority is required to win.

a. What is the quota?

b. How many permutations of five voters are there?

c. How many coalitions are there?

d. Fill out the accompanying coalition table.

Votes					Total votes	Winning coalition?	Critical voters
6	5	4	3	2			
A	B	C	D	E			
A	B	C	D				
A	B	C		E			
A	B	C					
A	B		D	E			
A	B		D				
A	B			E			
A	B						
A		C	D	E			
A		C	D				
A		C		E			
A		C					
A			D	E			
A			D				
A				E			
A							
	B	C	D	E			
	B	C	D				
	B	C		E			
	B	C					
	B		D	E			
	B		D				
	B			E			
	B						
		C	D	E			
		C	D				
		C		E			
		C					
			D	E			
			D				
				E			

e. Find the Banzhaf index for each voter.

22. Four states. This exercise refers to Example 8.6, which discusses four states influential for a presidential election. Recall that Virginia had 13 electoral votes, Wisconsin had 10, Colorado had 9, and Iowa had 6. Assume again that a simple majority vote from only these four states would determine the election, so the quota is 20. Find the Banzhaf index for each of these four states.

23. Party switch. In 2001, the U.S. Senate had 50 Republicans and 50 Democrats. On May 24 of that year, James Jeffords, a Republican from Vermont, announced his intention to leave the Republican Party and declare himself an Independent. That made the Senate consist of 49 Republicans, 50 Democrats, and 1 Independent. For a simple majority vote, make a table listing all of the permutations of the three voting blocs and the swing voter in each case. What is the Shapley–Shubik index for each voting bloc?

24. Party switch 2. Repeat Exercise 23, this time making a coalition table. What is the Bahzhaf index for each voting bloc?

25. Electoral College. Suppose that there are only three hypothetical states with a distribution of popular and electoral votes as shown in the accompanying table. (The number of electoral votes for a state is the size of its congressional representation.)

	State A	State B	State C	Total
Population	100	200	400	700
Senators	2	2	2	6
Representatives	1	2	4	7
Electoral votes	3	4	6	13

a. Find the Shapley–Shubik index for each state using the popular vote and then using the electoral vote. Assume that a simple majority is required in each case.

b. Use your answers to part a to explain why states with small populations resist changing the Electoral College system.

26. The Security Council. The United Nations Security Council has five permanent members (China, France, Great Britain, Russia, and the United States) and 10 rotating members. Each member has 1 vote, and 9 votes are required to pass a measure. But each permanent member has a veto. That is, if a permanent member votes "no," the measure fails. Although this is not a true weighted voting system, it is equivalent to one.

a. Suppose that each rotating member has 1 vote and that each permanent member has 7 votes, for a total of 45 votes. If the quota is set at 39 votes, explain why a measure cannot pass if a permanent member votes "no."

b. Consider the weighted system in part a. Explain why a measure will pass if nine members of the council, including all of the permanent members, support it.

c. Explain why the weighted voting system described in part a is equivalent to the actual United Nations Security Council voting system. *Note*: The Shapley–Shubik index for this weighted voting system can be calculated using a computer program. The result is 19.63% for each permanent member and 0.19% for each rotating member.

d. In parts a through c, we failed to mention something: In reality, a permanent member of the Security Council may abstain, which does not count as a veto. Suppose France abstains and no one else does. Then 9 votes, including those of all four of the remaining permanent members, are required to pass a measure. Devise an equivalent weighted voting system for these 14 members, as was done previously. That is, determine how many votes should be allotted to each of the remaining four permanent members and what the quota should be. (Assign 1 vote to each rotating member.) *Note*: The Shapley–Shubik index for this weighted voting system can be calculated using a computer program. The result is 23.25% for each permanent member and 0.70% for each rotating member.

Exercises 27 and 28 may be suitable for group work.

27. A research project. Voting coalitions are a vital part of parliamentary systems. A nice example is the Israeli Knesset, where there are several parties and coalitions often are necessary for forming a government. Write a report on the Knesset or other parliaments. Focus on coalitions.

28. A research project. At national party conventions, state delegations usually vote for the presidential nominee in alphabetical order. Find other situations in which votes are taken in a particular order.

Exercises 29 through 31 are designed to be solved using technology such as calculators or computer spreadsheets.

29. Three voters.

a. Produce a spreadsheet that computes the Banzhaf index in the case when there are three voters and the quota is a simple majority. The spreadsheet should allow the user to enter the number of votes controlled by each voter. Do the same for the Shapley–Shubik index.

b. Use your spreadsheet to check the computations of the Shapley–Shubik index for some of the previous examples and exercises.

c. Compare the results given by your spreadsheet with those given by the software mentioned in Calculation Tip 8.1.

30. Four voters.

a. Produce a spreadsheet that computes the Banzhaf index in the case when there are four voters. The spreadsheet should allow the user to enter the number of votes controlled by each voter, as well as the quota.

b. Solve Example 8.3 (on the 1932 Democratic National Convention) using your spreadsheet.

c. Experiment with different vote totals for the four blocs (Roosevelt, Smith, Garner, and "Other") in Example 8.3 and report how the power index varies.

d. Compare the results given by your spreadsheet with those given by the software mentioned in Calculation Tip 8.1.

31. Paradox. Attempting to measure voting power can give rise to paradoxes.[7] Two voting schemes, each with a total of 11 votes, are given. Assume that a two-thirds majority is required to win. Express each value of the Banzhaf index as a percentage rounded to two decimal places.

a. How many votes are required to win?

b. Suppose that A has 5 votes; B has 3 votes; and C, D, and E have 1 vote each. Use a computer program to find the Banzhaf index for each voter.

c. Suppose that, instead of the votes in part b, A gives away 1 vote to B. So A has 4 votes; B has 4 votes; and C, D, and E still have 1 vote each. Use a computer program to find the Banzhaf index for each voter.

d. Did the Banzhaf index for A increase or decrease after giving away 1 vote? Does this make sense?

e. Explain why C, D, and E have no voting power in the voting scheme of part c. Use this observation to compute by hand your answers to part c.

Writing About Mathematics

32. History. When John Banzhaf introduced his index of voting power in 1965, he analyzed the voting structure of the Nassau County Board of Supervisors. (Nassau County is located in the western part of Long Island, New York.) That board had divided 115 votes among the districts based on population. The following table shows the vote allocation based on the population in 1964. In the table, Hempstead #1 represents the presiding supervisor, always from Hempstead, and Hempstead #2 represents the second supervisor from Hempstead.

District	Number of votes
Hempstead #1	31
Hempstead #2	31
Oyster Bay	28
North Hempstead	21
Glen Cove	2
Long Beach	2

a. Assume that a simple majority is required to win a vote. Explain why North Hempstead, Glen Cove, and Long Beach have zero power (in the sense that their votes are never necessary to determine the outcome of an election). Do not answer this question by calculating the Banzhaf index.

b. Even though the number of votes was allocated according to the size of a district, clearly voting power was not allocated that way. This voting system was modified several times over the years because of legal challenges. Investigate this history and write a brief summary, including the final resolution.

[7] *See Dan S. Felsenthal and Moshe Machover,* The Measurement of Voting Power: Theory and Practice, Problems and Paradoxes *(Cheltenham, UK: Edward Elgar, 1998).*

33. History. This exercise refers to the article by Shapley and Shubik in which they first defined their index.

a. The following quotation is from the introductory portion of the Shapley–Shubik article. "It is possible to buy votes in most corporations by purchasing common stock. If their policies are entirely controlled by simple majority votes, then there is no more power to be gained after one share more than 50% has been acquired." How does this observation support the importance of the concept of a swing voter?

b. Here is another quotation from the introductory portion of the Shapley–Shubik article. "Put in crude economic terms, . . . if votes of senators were for sale, it might be worthwhile buying forty-nine of them, but the market value of the fiftieth (to the same customer) would be zero." How do the numbers 49 and 50 in the statement reflect the fact that the paper was published in 1954? How would you update the statement to be true now?

34. History 2. This exercise refers to the following quotation from the article by Shapley and Shubik in which they first defined their index.

> To illustrate, take Congress without the provision for overriding the President's veto by means of two-thirds majorities. This is now a pure tricameral system with chamber sizes of 1, 97, and 435. The values come out to be slightly under 50% for the President, and approximately 25% each for the Senate and House, with the House slightly less than the Senate.

a. Note that the figure of 97 refers to the Senate. Explain why it was valid in 1954 (when the paper was published). *Hint*: Consider the role of the vice president.

b. Does it make sense that the president's power index is nearly 50%? Explain.

35. Your own example. Find a situation, preferably a local one, where calculation of either the Banzhaf or Shapley–Shubik index is appropriate. Perform the appropriate calculation, and explain what the results of your calculation mean.

36. Lloyd Shapley. Write a brief paper about the career of Lloyd Shapley.

What Do You Think?

37. Banzhaf and coalitions. A fellow student asserts, "The Banzhaf index for a voter depends on how good the voter is at forming coalitions." Is that an accurate statement? Explain. **Exercise 18 is relevant to this topic.**

38. Shapley–Shubik and order. A fellow student asserts, "The Shapley–Shubik index depends on the order in which the voters vote." Is that an accurate statement? Explain. **Exercise 19 is relevant to this topic.**

39. Secret ballot voting. The Shapley–Shubik index is defined under the assumption that the voters vote in order. But if the vote is by secret ballot, then votes are neither cast nor counted in any specific order. Does this fact affect the validity of the Shapley–Shubik index? **Exercise 20 is relevant to this topic.**

40. Parliamentary government. Under a parliamentary system (such as that in the United Kingdom or Israel), we often hear of a *coalition government*. Does this usage of the term *coalition* agree with the usage in this section of the text? **Exercise 21 is relevant to this topic.**

41. Weighted voting. Give some real-world examples of weighted voting, where voters may have several votes each. **Exercise 26 is relevant to this topic.**

42. Dummy. A voter whose votes have no effect on the outcome of an election is sometimes called a *dummy voter*. Assume that a simple majority is required to win. Ann has 4 votes, Beth has 1, and Cate has 1. Which of the three, if any, are dummy voters?

43. More on the dummy. A voter whose votes have no effect on the outcome of an election is sometimes called a *dummy voter*. What is the Banzhaf index for a dummy voter? Is every voter with that index a dummy voter? Explain your answers. **Exercise 32 is relevant to this topic.**

44. Bias. Explain how the use of the power indices described in this section can reveal bias that may not be apparent in a voting system. **Exercise 32 is relevant to this topic.**

45. Different answers. Can you identify practical examples where the Banzhaf index and the Shapley–Shubik index are different?

8.2 Voting systems: How do we choose a winner?

TAKE AWAY FROM THIS SECTION
There is no perfect voting system if there are three or more candidates. There are many different voting systems in use.

The following excerpt is from www.governing.com/topics/elections/gov-maine-ranked-choice-voting-2016-ballot-measure.html.

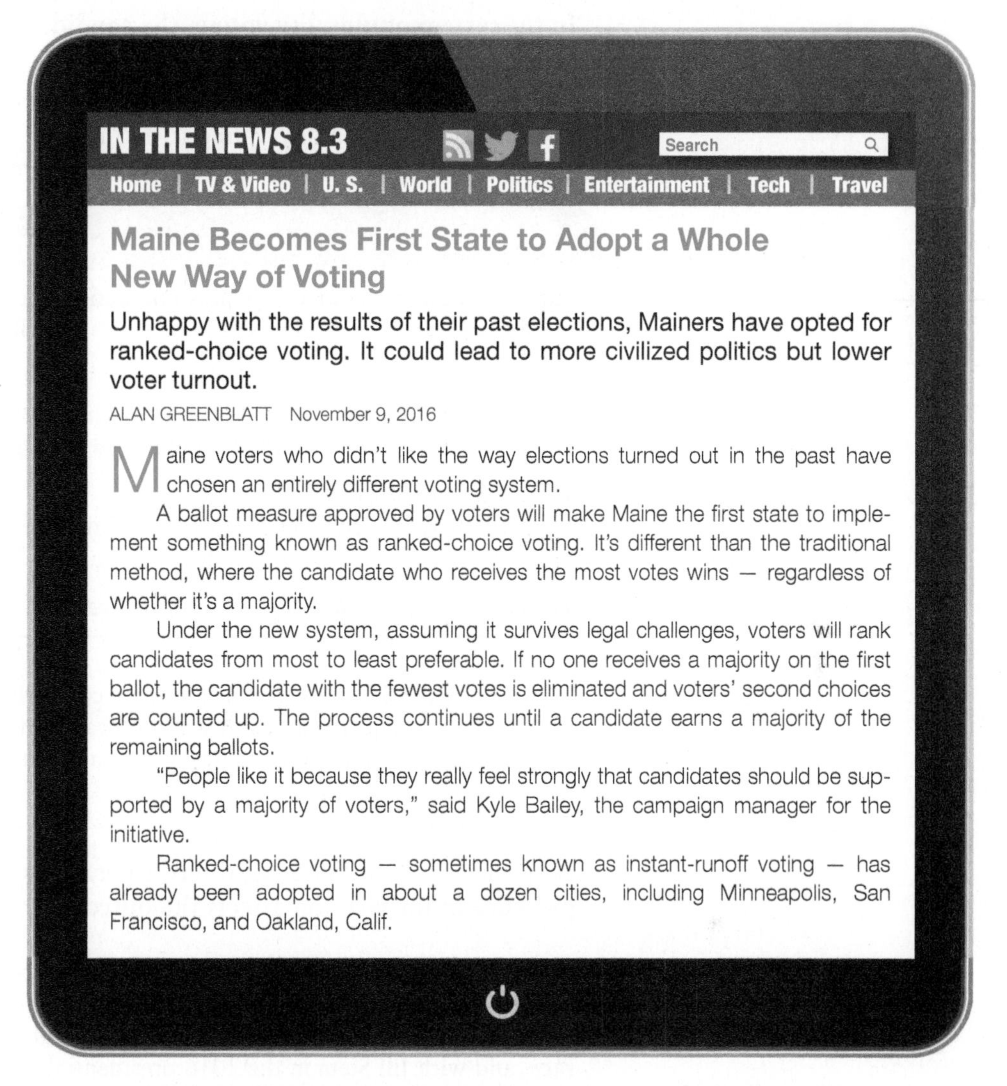

IN THE NEWS 8.3

Home | TV & Video | U. S. | World | Politics | Entertainment | Tech | Travel

Maine Becomes First State to Adopt a Whole New Way of Voting

Unhappy with the results of their past elections, Mainers have opted for ranked-choice voting. It could lead to more civilized politics but lower voter turnout.

ALAN GREENBLATT November 9, 2016

Maine voters who didn't like the way elections turned out in the past have chosen an entirely different voting system.

A ballot measure approved by voters will make Maine the first state to implement something known as ranked-choice voting. It's different than the traditional method, where the candidate who receives the most votes wins — regardless of whether it's a majority.

Under the new system, assuming it survives legal challenges, voters will rank candidates from most to least preferable. If no one receives a majority on the first ballot, the candidate with the fewest votes is eliminated and voters' second choices are counted up. The process continues until a candidate earns a majority of the remaining ballots.

"People like it because they really feel strongly that candidates should be supported by a majority of voters," said Kyle Bailey, the campaign manager for the initiative.

Ranked-choice voting — sometimes known as instant-runoff voting — has already been adopted in about a dozen cities, including Minneapolis, San Francisco, and Oakland, Calif.

One of the hallmarks of a democracy is that its leaders are chosen by the "people," but there are many voting schemes by which this may be accomplished. This excerpt shows how controversial such schemes can be. It refers to ranked-choice voting and majority voting.

If a candidate receives a majority of the votes cast, many would agree that he or she should be the winner. Even this is not as simple as it sounds, as the Electoral College system for the U.S. president shows. And even if we agree that a candidate who receives a majority of the votes should be the winner, it may be that no one achieves that goal. When there are more than two candidates, the problem of determining the "people's choice" is difficult, to say the least. Indeed, a case can be made that for three or more candidates, there is no completely satisfactory system for determining a winner, as we will see.

There are many voting systems currently in use. In this section, we discuss several popular ones, including the method described in the preceding excerpt. We also examine the troubling paradox of Condorcet and Arrow's impossibility theorem.

Plurality voting and spoilers

A *voting system* is a set of rules under which a winner is determined. The simplest voting system for three or more candidates is *plurality* voting.

KEY CONCEPT

In the system of plurality voting, the candidate receiving more votes than any other candidate wins.

On the surface, this system seems quite reasonable, but there are difficulties. One objection to plurality voting is that if there are a number of candidates, the winner could be determined by only a small percentage of the electorate.

EXAMPLE 8.7 Examining plurality voting: Four candidates

Suppose there are four candidates competing for 100 votes. What is the smallest number of votes a candidate could receive and still win a plurality?

SOLUTION

If the four candidates received an equal number of votes, they would have $100/4 = 25$ each. A candidate could have a plurality with as few as 26 votes.

TRY IT YOURSELF 8.7

Suppose there are five candidates competing for 100 votes. What is the smallest number of votes a candidate could receive and still win a plurality?

The answer is provided at the end of this section.

In Example 8.7, note that a candidate could receive only 26% of the votes cast, far less than a majority, and still win a plurality of the votes.

With three or more candidates, there can be a *spoiler*.

KEY CONCEPT

A spoiler is a candidate who has no realistic chance of winning but whose presence in the election affects the outcome.

Typically, a spoiler splits the vote of a more popular candidate who may otherwise have won a majority. Many argue that this was the case with Ross Perot in the 1992 and 1996 U.S. presidential races, with Ralph Nader in the 2000 presidential race, and with Jill Stein in the 2016 presidential race. (Although the president is not determined by popular vote, a state's electoral votes are, and thus a spoiler can have a real effect.)

Shelly Katz/The LIFE Images Collection/Getty Images

H. Ross Perot was considered by many to be a spoiler candidate in the 1996 presidential election.

EXAMPLE 8.8 Understanding plurality voting and spoilers: The 1996 U.S. presidential election

There were three major candidates in the 1996 U.S. presidential race: Bill Clinton (Democrat), Robert Dole (Republican), and H. Ross Perot (Reform Party). Here is the outcome for the state of Florida, which is always important in presidential races.

Candidate	Votes
Clinton	2,546,870
Dole	2,244,536
Perot	483,870
Others	28,518
Total votes cast	5,303,794

a. What percentage of all votes cast in Florida were for Clinton? Did any candidate achieve a majority of votes cast?

b. Florida determines the winner of a presidential election by plurality, so Clinton was declared the winner of all 25 of Florida's electoral votes. Let's suppose that the vote was conducted with only the top three candidates and that all their supporters continued to vote for them. Would it be possible for any of the three to achieve a majority if all the votes cast for "Others" were awarded to him?

c. It was speculated that most Perot voters would have voted for Dole if Perot were not in the race. Let's suppose that 83% of Perot voters would have voted for Dole and 17% for Clinton. How many votes would Clinton have had and how many would Dole have had? Who would have won the election? Would you call Perot a spoiler in this case?

SOLUTION

a. Of the 5,303,794 votes cast, Clinton received 2,546,870 votes, which is 2,546,870/5,303,794, or about 48.0% of the votes. This is not a majority. Clinton won more votes than any other candidate, so no candidate received a majority of votes cast.

b. Suppose that all of the 28,518 "Other" votes went to Clinton. Then he would have received 2,575,388 votes. But half of 5,303,794 votes is 2,651,897, so he still would not have had a majority. Certainly the same is true of the other two candidates as well.

c. If 83% of Perot's votes went to Dole, then Dole would have received 83% of 483,870, or an extra 401,612 votes, for a total of 2,646,148. Clinton would have received 17% of 483,870, or an extra 82,258 votes, for a total of 2,629,128. Therefore, Dole would have won with 17,020 more votes than Clinton. If this were indeed true, then Perot would have been a spoiler.

Because of the problems with plurality voting we have mentioned, it is desirable to have an alternative way of choosing a winner.

Runoffs, preferential voting, and the Borda count: Alternatives to plurality-rule voting

There are many alternatives to plurality-rule voting. Some of them use runoff elections in case no candidate garners a majority. In one such method, the *top-two runoff system*, there is a second election with the two highest vote-getters. This method is used in many state elections. Another runoff method is the *elimination runoff system*, in which the lowest vote-getter is eliminated and a new vote is tallied among the remaining candidates. This process continues until someone achieves a majority.

In practice, to avoid several rounds of voting, a voter lists candidate preferences on a single *ranked* ballot. Such a ballot is used in the voting for the Heisman Trophy, awarded annually to the most outstanding collegiate football player in the United States: Each ballot allows a first-choice, second-choice, and third-choice vote.

A ranked ballot contains much more information than a ballot that allows only one choice. In particular, a ranked ballot shows all the preferences of the voter, that is, which candidate is preferable to which other candidates. *Preferential voting systems* usually use ranked ballots to avoid the need for repeated votes. Different preferential voting systems use the preferences indicated on the ranked ballots in different ways, sometimes with quite different outcomes.

The elimination runoff system is also called the *Hare system*.

Variations on this voting system are used today in the city of San Francisco and in Australia (as well as other countries). The excerpt opening this section refers to such a system as *ranked-choice voting* or *instant runoff voting*. If there are only three candidates, the elimination runoff system is equivalent to the top-two runoff system. If there are four or more candidates, these two systems can produce surprisingly different outcomes.

KEY CONCEPT

In the top-two runoff system, if no candidate receives a majority then there is a new election with only the two highest vote-getters.

In the elimination runoff system, if no candidate receives a majority, then the lowest vote-getter is eliminated and a vote is taken again among those who are left. This process continues until someone achieves a majority.

These two systems are examples of preferential voting systems, in which voters express their ranked preferences between various candidates, usually using a ranked ballot.

EXAMPLE 8.9 Finding the winner in the elimination runoff system: 10 voters, 3 candidates

Consider the following ranked ballot outcome for 10 voters choosing among 3 candidates: Alfred, Betty, and Gabby.

Rank	4 voters	4 voters	2 voters
First choice	Alfred	Gabby	Betty
Second choice	Betty	Alfred	Gabby
Third choice	Gabby	Betty	Alfred

As an example of how to interpret this table, look at the row labeled "First choice." The first-choice votes were 4 for Alfred, 4 for Gabby, and 2 for Betty. The fourth column, labeled "2 voters," indicates that two voters chose Betty first, Gabby second, and Alfred third.

Determine the winner under the elimination runoff system. (Because there are only three candidates, this is equivalent to the top-two runoff system.)

SOLUTION

No candidate has a majority of the first-choice votes. Betty has the fewest first-choice votes (2), so she is eliminated in the first round. For the second round, we first remove Betty from the original table:

Rank	4 voters	4 voters	2 voters
First choice	Alfred	Gabby	~~Betty~~
Second choice	~~Betty~~	Alfred	Gabby
Third choice	Gabby	~~Betty~~	Alfred

Now with Betty eliminated, for the four voters represented in the second column, Alfred is the first choice and Gabby is the second choice. Similarly, for the four voters represented in the third column, Gabby is the first choice and Alfred is the second choice. For the two voters represented in the fourth column, Gabby is the first choice and Alfred is the second choice among the remaining candidates. The ballot outcomes for the second round are as follows:

Rank	4 voters	4 voters	2 voters
First choice	Alfred	Gabby	Gabby
Second choice	Gabby	Alfred	Alfred

Now the first-choice votes are 4 votes for Alfred and $4 + 2 = 6$ votes for Gabby. In this runoff, Gabby has a majority of the votes and is the winner.

TRY IT YOURSELF 8.9

Consider the following ranked ballot outcome for 10 voters choosing among 3 candidates: Alfred, Betty, and Gabby.

Rank	5 voters	3 voters	2 voters
First choice	Alfred	Betty	Gabby
Second choice	Betty	Gabby	Alfred
Third choice	Gabby	Alfred	Betty

Determine the winner under the elimination runoff system.

The answer is provided at the end of this section.

There are other methods that use ranked ballots. In 1770, Jean-Charles de Borda proposed a system of this type, and it bears his name today.

KEY CONCEPT

The *Borda count* assigns for each ballot zero points to the choice ranked last, 1 point to the next higher choice, and so on. The Borda winner is the candidate with the highest total Borda count.

Although the Borda count system takes into account strength of feelings about the voting choices, one drawback of this system is that it is possible for a candidate to win a majority of first-place votes and yet lose the election. The following example illustrates this fact.

EXAMPLE 8.10 Using the Borda count: Our first choice loses

A group of five people needs to make a decision on which food to order. To make the choice, everybody marks a ranked ballot for pizza, burgers, and tacos. The number 2 indicates a first-place vote, the number 1 a second-place vote, and so on. The table below shows the five ballots cast.

	Pizza	Tacos	Burgers
Ballot 1	2	1	0
Ballot 2	2	1	0
Ballot 3	2	1	0
Ballot 4	0	2	1
Ballot 5	0	2	1

a. Did one of the food items receive a majority of first-place votes?

b. Use the Borda count to determine what the group should order.

SOLUTION

a. First-place votes are indicated by the number 2. There were three first-place votes for pizza, which is a majority of the five first-place votes.

b. For pizza, the Borda count is $2 + 2 + 2 + 0 + 0 = 6$. The Borda count for tacos is $1 + 1 + 1 + 2 + 2 = 7$, and for burgers, the count is $1 + 1 = 2$. According to the Borda count, the group should order tacos. This is true in spite of the fact that pizza got a majority of first-choice votes.

We mentioned earlier that the voting for the Heisman Trophy uses a ranked ballot. A first, second, and third choice is indicated on each ballot. In the next example, we consider the race for the 2020 Heisman.

EXAMPLE 8.11 Finding the Borda count and Borda winner: Heisman Trophy

Here are the vote counts of the top three finalists for the 2020 Heisman.

Player	First-place votes	Second-place votes	Third-place votes
DeVonta Smith (Alabama)	447	221	73
Trevor Lawrence (Clemson)	222	176	169
Mac Jones (Alabama)	138	248	220

DeVonta Smith, of the University of Alabama, won the Heisman trophy in 2020.

Determine the Borda counts and the Borda winner.[8]

[8] *The actual system used to award the Heisman is a slight modification of the Borda count. See Exercise 46. The outcome here is not affected.*

SOLUTION

The Borda count assigns numbers 2, 1, and 0 to the choices. We incorporate this information into the original table:

Player	First-place votes	Second-place votes	Third-place votes
DeVonta Smith (Alabama)	447	221	73
Trevor Lawrence (Clemson)	222	176	169
Mac Jones (Alabama)	138	248	220
Borda count value	2	1	0

To find the Borda count for DeVonta Smith, we assign 2 points to each of his 447 first-place votes, 1 point to each of his 221 second-place votes, and 0 points to each of his 73 third-place votes. Then we add to find the total number of points. Here are the results:

$$\text{Borda count for DeVonta Smith} = 447 \times 2 + 221 \times 1 + 73 \times 0 = 1115$$

$$\text{Borda count for Trevor Lawrence} = 222 \times 2 + 176 \times 1 + 169 \times 0 = 620$$

$$\text{Borda count for Mac Jones} = 138 \times 2 + 248 \times 1 + 220 \times 0 = 524$$

DeVonta Smith has the highest Borda count, so he is the Borda winner—and the winner of the Heisman Trophy in 2020.

TRY IT YOURSELF 8.11

Here are the vote counts of the top three finalists for the 2019 Heisman:

Player	First-place votes	Second-place votes	Third-place votes
Joe Burrow (LSU)	841	41	3
Jalen Hurts (Oklahoma)	12	231	264
Chase Young (Ohio State)	6	271	187

Determine the Borda counts and the Borda winner.

The answer is provided at the end of this section.

The next example shows that the four voting systems we have considered can give four different outcomes for an election.

EXAMPLE 8.12 Differing voting systems give differing outcomes

Consider the following ranked ballot outcome for 100 voters choosing among four options: A, B, C, and D.

Rank	28 voters	25 voters	24 voters	23 voters
First choice	A	B	C	D
Second choice	D	C	D	C
Third choice	B	D	B	B
Fourth choice	C	A	A	A

a. Who wins under plurality voting?

b. Who wins under the top-two runoff system?

c. Who wins under the elimination runoff system?

d. Who wins under the Borda count system?

SOLUTION

a. To determine the winner under plurality voting, we must assume that if voters could choose only one name, they would select their first choice. The result of such a vote would be

Candidate	A	B	C	D
Votes	28	25	24	23

Candidate A has the most first-choice votes, so A wins under plurality voting.

b. As we saw in part a, the top two vote-getters on an unranked ballot are A with 28 votes and B with 25 votes. To determine the outcome of a runoff between A and B, we use the preferences shown on the ballots, but we consider only A and B. That is, we eliminate C and D from the original table showing the voter preferences and look at the resulting preferences as expressed by the ranking on the ballots.

Rank	28 voters	25 voters	24 voters	23 voters
Adjusted first choice	A	B	B	B
Adjusted second choice	B	A	A	A

In this runoff, B has 72 first-choice votes. This is a clear majority, so B is the winner under the top-two runoff system.

c. In the first round, D has the fewest votes (23) and is eliminated. If we remove D from the original table, the combined ballot outcomes are

Rank	28 voters	25 voters	24 voters	23 voters
First choice	A	B	C	C
Second choice	B	C	B	B
Third choice	C	A	A	A

Now the first-choice votes are 28 votes for A, 25 votes for B, and 47 votes for C. Therefore, B (having the fewest votes) is eliminated. If we remove B from the table, the combined ballot outcomes are

Ranks	28 voters	25 voters	24 voters	23 voters
First choice	A	C	C	C
Second choice	C	A	A	A

In this runoff, C has 72 votes and so is the winner under the elimination runoff system.

d. The Borda count assigns numbers 3, 2, 1, and 0 to the choices. We incorporate this information into the original table:

Rank	Borda count value	28 voters	25 voters	24 voters	23 voters
First choice	3	A	B	C	D
Second choice	2	D	C	D	C
Third choice	1	B	D	B	B
Fourth choice	0	C	A	A	A

Here are the Borda counts:

$$\text{Borda count for A} = 28 \times 3 + 25 \times 0 + 24 \times 0 + 23 \times 0 = 84$$
$$\text{Borda count for B} = 28 \times 1 + 25 \times 3 + 24 \times 1 + 23 \times 1 = 150$$
$$\text{Borda count for C} = 28 \times 0 + 25 \times 2 + 24 \times 3 + 23 \times 2 = 168$$
$$\text{Borda count for D} = 28 \times 2 + 25 \times 1 + 24 \times 2 + 23 \times 3 = 198$$

Hence, D is the winner based on the Borda count.

Example 8.12 is not just a curiosity. It shows some of the difficulties in determining exactly who is the "people's choice." All four of the candidates in this example have a reasonable argument that they deserve to be named the winner.

Mike Keefe/Cagle Cartoons

SUMMARY 8.5
Common Preferential Voting Systems

A *voting system* is a scheme for determining the winner of an election when there are three or more candidates. Common systems that are in use today include variants of the following.

Plurality: The candidate with the most votes wins.

Top-two runoff: If no one garners a majority of the votes, a second election is held with the top two vote-getters as the only candidates.

Elimination runoff: Successive elections are held where the candidate with the smallest number of votes is eliminated. This continues until there is a majority winner.

Borda count: Voters rank the candidates first to last. The last place candidate gets 0 points, the next 1 point, and so on. The candidate with the most points wins.

Condorcet's paradox

Normally, an individual's preferences are *transitive*. That is, if I prefer Pepsi over Coke and I prefer Coke over Dr. Pepper, then it seems obvious that, if I am given a choice between Pepsi and Dr. Pepper, I'll choose Pepsi.

It may seem paradoxical, but this transitivity is not necessarily true if the preferences are expressed by groups instead of individuals. We will illustrate this by the following example.

Suppose there are three voters and three candidates, say, A, B, and C. The voters' preferences are shown in the accompanying table.

Preferences	Voter 1	Voter 2	Voter 3
First choice	A	B	C
Second choice	B	C	A
Third choice	C	A	B

Now a majority of voters (voters 1 and 3) prefer candidate A over candidate B, and a majority of voters (voters 1 and 2) prefer candidate B over candidate C. And yet, paradoxically, a majority of voters do not prefer candidate A over candidate C; in fact, a majority of voters (voters 2 and 3) prefer candidate C over candidate A.

In 1785, the Marquis de Condorcet published an essay that described this paradox, which is now named after him. He referred to this phenomenon as the *intransitivity of majority preferences*.

The existence of this paradox suggests that we focus on a candidate who would win in any head-to-head contest with every other candidate. The name given to such candidates honors Condorcet.

KEY CONCEPT

In an election, a Condorcet winner is a candidate who beats each of the other candidates in a one-on-one election.

Note that in the preceding simple example, there is no Condorcet winner.

EXAMPLE 8.13 Finding the Condorcet winner: The Gentoo Foundation

The Gentoo Foundation, which distributes the software Gentoo Linux, uses a Condorcet voting process for elections to its board of trustees.[9] Suppose in such an election there are seven voters and three candidates, say, A, B, and C. The voters' preferences are shown in the accompanying table.

Preferences	3 voters	2 voters	2 voters
First choice	A	C	C
Second choice	B	B	A
Third choice	C	A	B

Is there a Condorcet winner? If so, which candidate is it?

SOLUTION

We find the results of each head-to-head contest. Consider a contest between A and B. There are $3 + 2 = 5$ voters who rank A over B and only two voters who rank B over A. Therefore, A wins the one-on-one contest with B. Here are results of the other two head-to-head contests:

A and C: C wins by 1 (4 to 3).
B and C: C wins by 1 (4 to 3).

There is a Condorcet winner, and it is C.

[9] *See* `wiki.gentoo.org/wiki/Project:Elections`

TRY IT YOURSELF 8.13

In an election, there are seven voters and three candidates, say, A, B, and C. The voters' preferences are shown in the accompanying table.

Preferences	3 voters	2 voters	2 voters
First choice	B	B	C
Second choice	A	C	A
Third choice	C	A	B

Is there a Condorcet winner? If so, which candidate is it?

The answer is provided at the end of this section.

Even if there is a Condorcet winner, that candidate may lose the election in all the voting systems we have discussed. This is a further troubling aspect of common voting schemes, and we illustrate it in the following example.

EXAMPLE 8.14 Different voting systems and the Condorcet winner

In a certain election, there are seven voters and four candidates, say, A, B, C, and D. The tally of ranked ballots is as follows:

Rank	Voter 1	Voter 2	Voter 3	Voter 4	Voter 5	Voter 6	Voter 7
First choice	A	A	B	C	D	A	C
Second choice	B	D	A	B	B	D	B
Third choice	C	B	C	A	A	B	A
Fourth choice	D	C	D	D	C	C	D

a. Who wins in the plurality system?

b. Who wins in the top-two runoff system?

c. Who wins in the elimination runoff system?

d. Who wins the Borda count?

e. Is there a Condorcet winner? If so, which candidate is it?

SOLUTION

a. To determine the winner in the plurality system, we must assume that if voters could choose only one name, they would select their first choice. The result of such a vote would be

Rank	Voter 1	Voter 2	Voter 3	Voter 4	Voter 5	Voter 6	Voter 7
First choice	A	A	B	C	D	A	C

The 7 votes are distributed as follows: Candidate A has 3 votes, B has 1 vote, C has 2 votes, and D has 1 vote. Candidate A has the most votes and wins in the plurality system.

b. The first-place winner is A. Candidate C is second with 2 votes. In a runoff between A and C, A wins 5 votes to 2. So, A wins the top-two runoff.

c. Because B and D get only 1 vote each, they are eliminated and A wins, just as in part b.

d. We calculate the Borda count as follows:

$$\text{Borda count for A} = 3 \times 3 + 1 \times 2 + 3 \times 1 = 14$$
$$\text{Borda count for B} = 1 \times 3 + 4 \times 2 + 2 \times 1 = 13$$
$$\text{Borda count for C} = 2 \times 3 + 2 \times 1 = 8$$
$$\text{Borda count for D} = 1 \times 3 + 2 \times 2 = 7$$

So, A wins the Borda count.

e. There is a Condorcet winner, and it is B, as the following list of head-to-head outcomes shows:

B beats A (4 to 3)
B beats C (5 to 2)
B beats D (4 to 3)

In conclusion, B wins in any head-to-head contest but loses to A in the top-two runoff, elimination runoff, Borda count, and plurality systems.

SUMMARY 8.6
Condorcet Winners

In an election, a *Condorcet winner* is a candidate who would beat all others in a one-on-one election. Some elections do not produce a Condorcet winner.

Arrow's impossibility theorem

Given the problems with the various voting systems described in this section, it's reasonable to ask whether all voting systems for three or more candidates are fraught with such problems. Is there a voting system we haven't considered that in some sense is "the best" or maybe the most "fair"? Of course, this depends on what you mean by *best* or *fair*. For example, if it means that the candidate with the most votes wins, then you should be happy with plurality voting.

The notion of "best" in reference to voting procedures has been a subject of debate and research.[10] In general terms, the answer to the question of what is best or most fair lies in the expectation that voting systems have certain properties that can be viewed as reasonable.

One such expectation for a fair voting system is the *Condorcet winner criterion*.

KEY CONCEPT

The Condorcet winner criterion says that if there is a Condorcet winner–that is, a candidate preferred over all others in one-on-one comparisons–then he or she should be the winner of the election.

Another apparently reasonable property of a voting system involves the effect of alternatives that seem irrelevant. Suppose I am asked to choose between hamburgers and hot dogs as my favorite food, and I choose hamburgers. Later pizza is added to the list as a third choice for me to rank. Even if I like pizza far more than either hamburgers or hot dogs, I will still prefer hamburgers over hot dogs. No matter where I rank pizza on my list, it will not change the fact that hamburgers will be ranked over hot dogs. Pizza is an irrelevant alternative—that is, liking or disliking pizza is irrelevant to the question of whether I prefer hamburgers to hot dogs.

---

[10] *Alan Taylor's book* Mathematics and Politics: Strategy, Voting, Power and Proof *(New York: Springer-Verlag, 1995) gives a good overview of the many ideas in this fruitful area of debate.*

KEY CONCEPT

The condition of independence of irrelevant alternatives says the following: Suppose candidate A wins an election and candidate B loses. Then suppose there is another election in which no voter changes his or her relative preference of A to B. Then B should still lose to A no matter what else may have happened concerning other candidates.

Of course, as we saw with Condorcet's paradox, what is reasonable for an individual may be surprisingly unreasonable for a group. Here is an example where the introduction of an "irrelevant alternative" changes the outcome of an election: Suppose that there are 10 voters and that the winner is the one with a plurality of votes. Assume that with candidates A and B in the race, the outcome is as follows:

Rank	6 voters	4 voters
First choice	A	B
Second choice	B	A

Candidate A wins not only the plurality vote but also the majority vote by 6 to 4. Now suppose that candidate C enters the race and that the new 10-vote tally is

Rank	3 voters	3 voters	4 voters
First choice	C	A	B
Second choice	A	B	A
Third choice	B	C	C

This time, candidate B wins with a plurality of 4 votes. Note that none of the voters in the elections changed their preference between A and B. Nonetheless, the introduction of the irrelevant alternative of candidate C changed the election results from A to B.

In this case, we see that the irrelevant alternative was nothing more than a spoiler. So the criterion of independence of irrelevant alternatives perhaps seems a bit less realistic than on first consideration. If the irrelevant alternative is, for example, everyone's first choice, then the irrelevant alternative would win under any system, but that would not violate the independence of irrelevant alternatives. The presence of a spoiler—as in the preceding example, changing the outcome from A to B—can violate the independence of irrelevant alternatives.

The economist Kenneth Arrow studied the properties of voting systems. He was seeking "fair" systems—in particular, systems that would have the reasonable Condorcet winner criterion and also independence of irrelevant alternatives. These two properties are quite different, but surely both seem fair. In point of fact, in 1950, Arrow proved the following startling theorem.[11]

SUMMARY 8.7
Arrow's Impossibility Theorem

If there are three or more candidates, there is no voting system (other than a dictatorship) for which both the Condorcet winner criterion and independence of irrelevant alternatives hold.

This is quite surprising for at least two reasons: First, the condition of independence of irrelevant alternatives seems reasonable. Second, any voting system can be adjusted to satisfy the Condorcet winner criterion by saying that, if there is a

[11] *Kenneth Arrow, "A Difficulty in the Concept of Social Welfare,"* Journal of Political Economy *58 (1950), 328–346.*

Condorcet winner, then that candidate wins, and if there isn't, then we carry on with the voting system as originally designed.

Arrow's theorem can be thought of as saying that there is no "best" voting system for three or more candidates. Therefore, we must consider the strengths and weaknesses of any voting system and decide what criteria *seem* the most fair to the voters.

WHAT DO YOU THINK?

Arrow's theorem: A classmate says, "Arrow's impossibility theorem shows that no election is fair. What's the point in having an election at all?" How would you respond?

Questions of this sort may invite many correct responses. We present one possibility: A famous Winston Churchill quote: "It has been said that democracy is the worst form of government except all the others that have been tried from time to time." Nothing associated with government is likely to work out perfectly, and Arrow's theorem assures that in the case of voting. But society must have some way of choosing its laws, policies, and leaders, and some form of voting, though imperfect, is probably better than any other known method. The debate about which voting scheme works best is likely to continue. Improving the system is a better idea than giving up.

Further questions of this type may be found in the *What Do You Think?* section of exercises.

Try It Yourself answers

Try It Yourself 8.7: Examining plurality voting: Five candidates 21.

Try It Yourself 8.9: Finding the winner in the elimination runoff system: 10 voters, three candidates Alfred.

Try It Yourself 8.11: Finding the Borda count and Borda winner: Heisman Trophy Burrow: 1723; Hurts: 255; Young: 283. Burrow is the Borda winner.

Try It Yourself 8.13: Finding the Condorcet winner: Seven voters, 3 candidates Yes: B.

Exercise Set 8.2

Test Your Understanding

1. Explain how the winner is determined in plurality voting.

2. Explain the meaning of a spoiler in an election.

3. True or false: In the top-two runoff system, if one candidate receives a majority of votes, then that candidate is elected, and there is no runoff.

4. Explain in your own words how the elimination runoff system works.

5. True or false: In an election using the elimination runoff system, the candidate receiving the most votes on the first ballot always wins.

6. True or false: When the Borda count is used, the person receiving the most first-place votes wins.

7. In an election, a Condorcet winner is ________.

8. True or false: If there is a Condorcet winner, then that candidate wins the election in every voting scheme.

9. True or false: In an election using ranked ballots, different voting systems may determine different winners.

10. True or false: Arrow's impossibility theorem tells us that we should not use voting to choose office holders.

Problems

11. Plurality voting. Suppose there are 3 candidates competing for 900 votes. What is the smallest number of votes a candidate could receive and still win a plurality?

12. Spoiler? In a certain election Candidate A received 3000 votes, Candidate B received 2000 votes, and Candidate C received 400 votes. Could Candidate C reasonably be thought of as a spoiler?

13. The majority loses? In Example 8.10, we saw an outcome involving a ranked ballot in which one candidate received a majority of first-place votes but lost the Borda count. Is it possible to have a two-candidate race in which one candidate receives a majority of first-place votes but loses the Borda count?

14. Condorcet winners. In a ranked ballot tally, is it possible that there is more than one Condorcet winner? Justify your answer.

15. Bush versus Clinton. The popular vote in the 1992 presidential election was

Candidate	Votes
Clinton	44,909,806
Bush	39,104,550
Perot	19,734,821
Other	665,746

a. Conventional wisdom is that Perot took more votes from Bush than from Clinton, but there is no real consensus on the effect of Perot's candidacy. Suppose that 60% of Perot voters had voted for Bush instead and the rest had voted for Clinton. Would Bush have had a plurality of the popular vote? If 65% of Perot voters had voted for Bush instead, would Bush have had a plurality of the popular vote?

b. In your opinion, was Perot a spoiler in this election?

16–20. Who wins? *Use the following information for Exercises 16 through 20. The vote tally of ranked ballots for a slate of four candidates is as follows:*

Rank	3 voters	3 voters	2 voters	4 voters	4 voters
First choice	A	C	B	C	D
Second choice	B	D	A	B	B
Third choice	C	B	C	A	A
Fourth choice	D	A	D	D	C

16. Who wins a plurality vote?

17. Who wins a top-two runoff?

18. Who wins an elimination runoff?

19. Who wins the Borda count?

20. Is there a Condorcet winner? If so, who?

21–25. Who wins? *Use the following information for Exercises 21 through 25. The vote tally of ranked ballots for a slate of four candidates is as follows:*

Rank	2 voters	1 voter	3 voters	4 voters	4 voters	1 voter
First choice	C	A	B	D	D	B
Second choice	B	C	C	A	B	A
Third choice	D	B	A	B	A	D
Fourth choice	A	D	D	C	C	C

21. Who is the plurality winner?

22. Who is the top-two runoff winner?

23. Who is the elimination runoff winner?

24. Who is the Borda winner?

25. Is there a Condorcet winner? If so, who?

26–30. Who wins? *Use the following information for Exercises 26 through 30. The vote tally of ranked ballots for a slate of four candidates is as follows:*

Rank	4 voters	5 voters	1 voter	4 voters	4 voters	2 voters
First choice	C	A	B	D	D	B
Second choice	B	C	C	A	B	A
Third choice	D	B	A	B	A	D
Fourth choice	A	D	D	C	C	C

26. Who is the plurality winner?

27. Who is the top-two runoff winner?

28. Who is the elimination runoff winner?

29. Who is the Borda winner?

30. Is there a Condorcet winner? If so, who?

31–35. Who wins? *Use the following information for Exercises 31 through 35. The vote tally of ranked ballots for a slate of four candidates is as follows:*

Rank	4 voters	10 voters	2 voters	4 voters	3 voters
First choice	A	C	B	A	B
Second choice	B	A	D	C	A
Third choice	C	D	C	B	C
Fourth choice	D	B	A	D	D

31. Who is the plurality winner?

32. Who is the top-two runoff winner?

33. Who is the elimination runoff winner?

34. Who is the Borda winner?

35. Is there a Condorcet winner? If so who?

36–40. Who wins? *Use the following information for Exercises 36 through 40. The tally of ranked ballots is shown below.*

Rank	2 voters	2 voters	3 voters	2 voters	2 voters
First choice	A	C	B	A	B
Second choice	B	A	D	C	C
Third choice	C	B	C	B	A
Fourth choice	D	D	A	D	D

36. Who is the plurality winner?

37. Who is the top-two runoff winner?

38. Who is the elimination runoff winner?

39. Who is the Borda winner?

40. Is there a Condorcet winner? If so who?

41–45. Who wins? *Use the following information for Exercises 41 through 45. The vote tally of ranked ballots for a slate of four candidates is as follows:*

Rank	4 voters	5 voters	3 voters	4 voters	6 voters
First choice	A	C	B	A	B
Second choice	B	A	D	C	C
Third choice	C	D	C	B	A
Fourth choice	D	B	A	D	D

41. Who is the plurality winner?

42. Who is the top-two runoff winner?

43. Who is the elimination runoff winner?

44. Who is the Borda winner?

45. Is there a Condorcet winner? If so who?

46. Heisman Trophy. In Example 8.11, we considered the following vote counts of the top three finalists for the 2020 Heisman Trophy:

Player	First-place votes	Second-place votes	Third-place votes
DeVonta Smith (Alabama)	447	221	73
Trevor Lawrence (Clemson)	222	176	169
Mac Jones (Alabama)	138	248	220

The actual system used to determine the winner of the Heisman assigns 3 points to each first-place vote, 2 points to each second-place vote, and 1 point to each third-place vote. Use this method to determine the point total for each of the finalists and find the winner of the Heisman.

47. Florida again. Let's suppose the voting tally in the state of Florida for the 2000 presidential election represents voting blocs in the state. Let's classify them as follows:

Voting bloc	Votes
Republican	2,912,790
Democrat	2,912,253
Independent	97,488
Other	40,579

Assume that the quota is determined by a majority vote. Leave each index as a fraction.

a. Find the Banzhaf index for each voting bloc. *Suggestion:* This will be easier if you first observe that the smallest bloc has no voting power.

b. Find the Shapley–Shubik index for each bloc.

c. Interpret what your answers to parts a and b say about the relative power of the four blocs.

48. Bush versus Gore. In the 2000 presidential election, the official popular vote tallies were as follows:

Candidate	Popular vote
Al Gore	51,003,926
George W. Bush	50,460,110
Others	3,898,559

Gore won a plurality of the popular vote, but Bush won the electoral vote and so won the presidency. Let's treat these results as voting blocs. Assume that the quota is determined by a majority vote. Leave each index as a fraction.

a. What is the Shapley–Shubik index for each bloc?

b. What is the Banzhaf index for each bloc?

c. Interpret what your answers to parts a and b say about the relative power of the three blocs.

Writing About Mathematics

49. The tyranny of the majority. There are critiques of majority rule from Plato,[12] de Tocqueville,[13] religious communities,[14] and contemporary authors,[15] each for quite different reasons. Winston Churchill is quoted as saying, "it has been said that democracy is the worst form of Government except for all those other forms that have been tried from time to time." Report on the concerns of scholars regarding the concept of majority rule.

50. Constitutional limits. Majority rule does not mean that the majority totally runs the show. For example, in the United States, it is not lawful for a majority to disenfranchise a minority. There are many constitutional limits to what a majority may impose on a minority. Report on a few of these.

51. History. It has happened five times in U.S. history that a presidential candidate lost the popular vote but won the presidency. Report on these five incidents.

52. History. The 1960 presidential race between Richard Nixon and John F. Kennedy was another very close race. In this election, Illinois was crucial. There were allegations of voter fraud in Illinois that, if proven valid, could have changed the results of the election. Report on the 1960 presidential race in Illinois.

53. History. The year 1960 was the first time presidential candidates faced each other in televised debates. The debates stirred controversy for any number of reasons. Report on the 1960 presidential debates.

54. History. In the United States, the winner of a presidential race is the candidate who receives a majority of the electoral vote. If no candidate achieves a majority, the House of Representatives decides the winner of the election. This happened only once, in 1824. Report on the 1824 presidential election.

55. Arrow. Write a brief paper on the career of Kenneth Arrow.

[12] The Republic, *for example.*

[13] Democracy in America, *1835.*

[14] *M. Sheeran,* Beyond Majority Rule, *Annual Meeting of the Religious Society of Friends, Philadelphia, 1983.*

[15] *For example, Lani Guinier,* The Tyranny of the Majority: Fundamental Fairness in Representative Democracy *(New York: Free Press, 1994).*

56. Borda and Condorcet. The names of Borda and Condorcet arose in this section. Choose one of these men and write a brief paper on him.

What Do You Think?

57. Spoiler. A political commentator claims that a certain candidate was a spoiler in a recent election. Would it be easy to verify that claim? Why or why not? **Exercise 15 is relevant to this topic.**

58. Plurality. How many votes do you have to get to win an election in a plurality voting system? **Exercise 16 is relevant to this topic.**

59. Ranked ballot. What are the advantages of having a ranked ballot? What are the disadvantages?

60. Runoff. Compare the top-two runoff system with the elimination runoff system. Be sure to discuss what advantage one system has over the other. **Exercise 17 is relevant to this topic.**

61. Paradox. What is the paradox in Condorcet's paradox? **Exercise 20 is relevant to this topic.**

62. Plurality problems. What are some problems with plurality voting? Are these problems addressed by the other systems we considered? **Exercise 21 is relevant to this topic.**

63. Tie vote. Is it possible to have a vote for three candidates using a ranked ballot that produces a tie in each of the following systems, all at the same time: plurality, top-two runoff, elimination runoff, and Borda count?

64. Irrelevant alternative. Explain what is meant by an *irrelevant alternative* in an election. Can you provide examples in recent presidential elections?

65. Interpreting Arrow's theorem. One could interpret Arrow's impossibility theorem as an argument for choosing dictatorship as a form of government. Is this interpretation reasonable? Explain your answer. Do you think dictatorship is a good form of government?

8.3 Fair division: What is a fair share?

TAKE AWAY FROM THIS SECTION
Fair division of assets can be difficult, but there are mathematically sound ways to divide fairly.

The following article is from the *Daily News*.

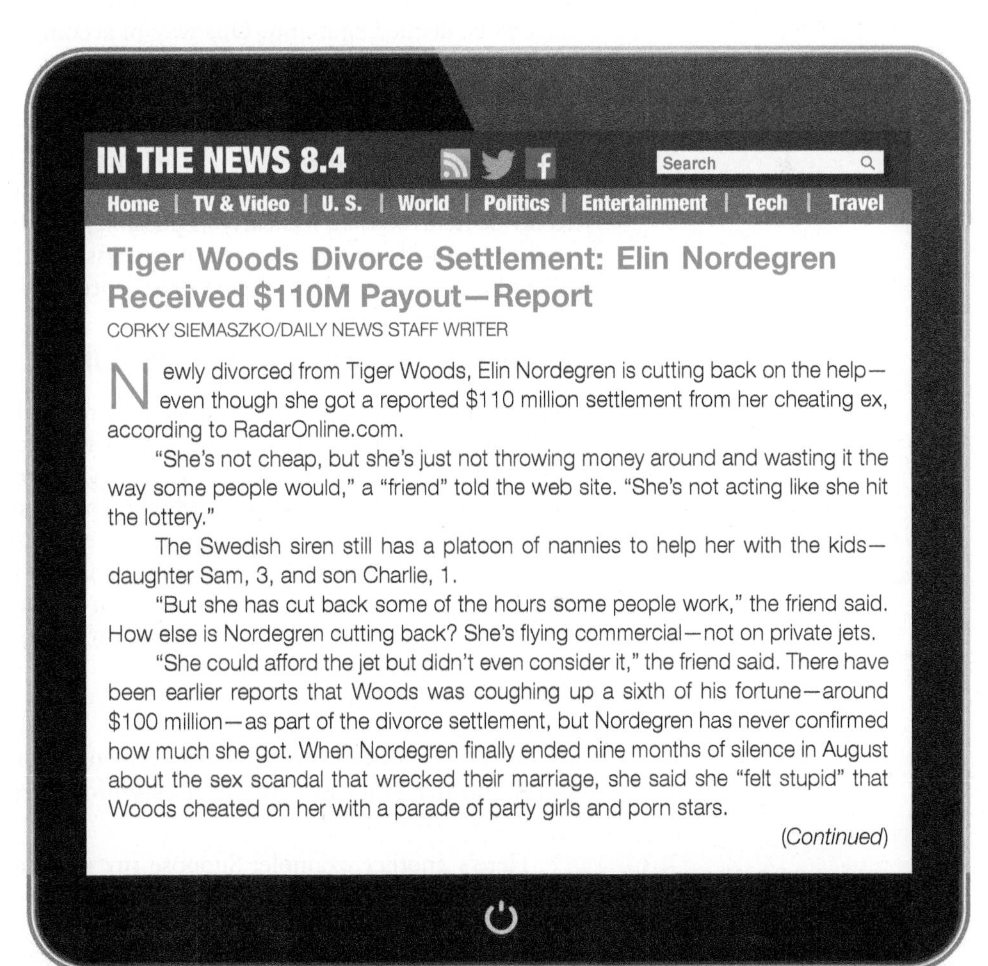

IN THE NEWS 8.4

Search

Home | TV & Video | U. S. | World | Politics | Entertainment | Tech | Travel

Tiger Woods Divorce Settlement: Elin Nordegren Received $110M Payout—Report

CORKY SIEMASZKO/DAILY NEWS STAFF WRITER

Newly divorced from Tiger Woods, Elin Nordegren is cutting back on the help—even though she got a reported $110 million settlement from her cheating ex, according to RadarOnline.com.

"She's not cheap, but she's just not throwing money around and wasting it the way some people would," a "friend" told the web site. "She's not acting like she hit the lottery."

The Swedish siren still has a platoon of nannies to help her with the kids—daughter Sam, 3, and son Charlie, 1.

"But she has cut back some of the hours some people work," the friend said. How else is Nordegren cutting back? She's flying commercial—not on private jets.

"She could afford the jet but didn't even consider it," the friend said. There have been earlier reports that Woods was coughing up a sixth of his fortune—around $100 million—as part of the divorce settlement, but Nordegren has never confirmed how much she got. When Nordegren finally ended nine months of silence in August about the sex scandal that wrecked their marriage, she said she "felt stupid" that Woods cheated on her with a parade of party girls and porn stars.

(Continued)

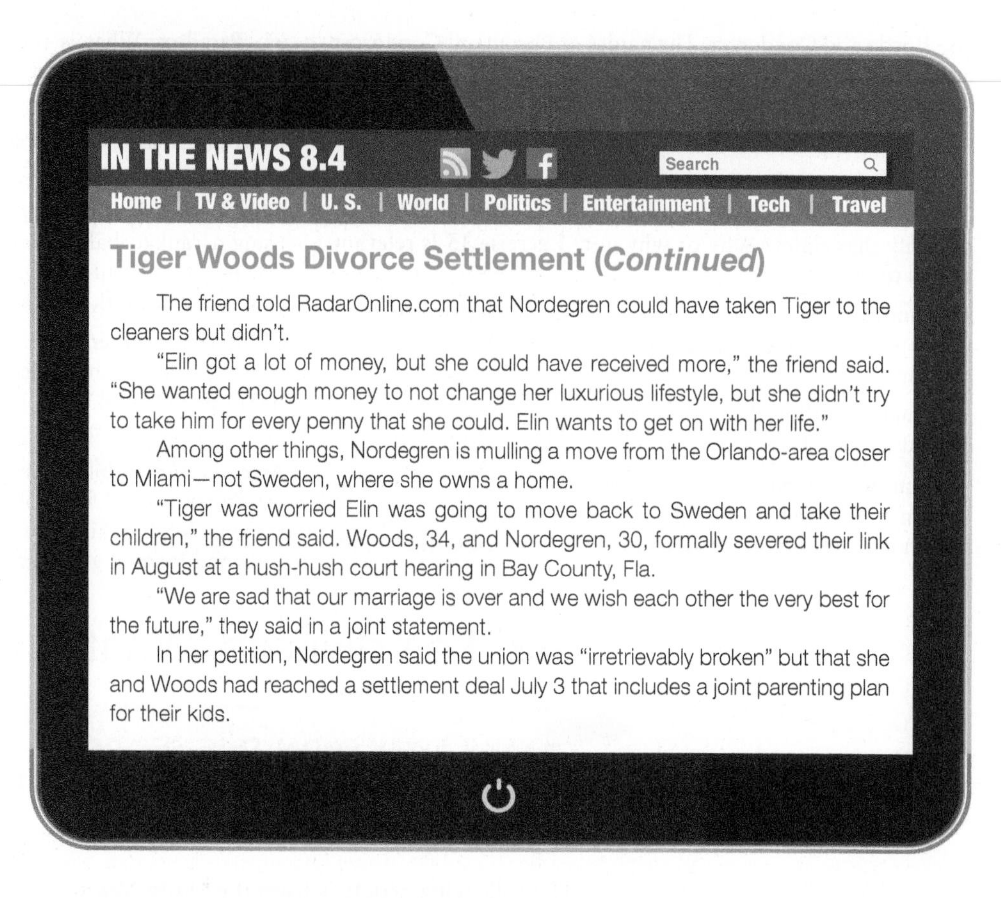
IN THE NEWS 8.4

Search

Home | TV & Video | U. S. | World | Politics | Entertainment | Tech | Travel

Tiger Woods Divorce Settlement (*Continued*)

The friend told RadarOnline.com that Nordegren could have taken Tiger to the cleaners but didn't.

"Elin got a lot of money, but she could have received more," the friend said. "She wanted enough money to not change her luxurious lifestyle, but she didn't try to take him for every penny that she could. Elin wants to get on with her life."

Among other things, Nordegren is mulling a move from the Orlando-area closer to Miami—not Sweden, where she owns a home.

"Tiger was worried Elin was going to move back to Sweden and take their children," the friend said. Woods, 34, and Nordegren, 30, formally severed their link in August at a hush-hush court hearing in Bay County, Fla.

"We are sad that our marriage is over and we wish each other the very best for the future," they said in a joint statement.

In her petition, Nordegren said the union was "irretrievably broken" but that she and Woods had reached a settlement deal July 3 that includes a joint parenting plan for their kids.

There are many situations, such as divorces and inheritances, in which goods have to be divided equitably. One way of accomplishing this is to liquidate all assets and divide the resulting cash. But in many cases, liquidation is neither practical nor desirable. For example, if a small business is part of an inheritance, one heir may have an interest in taking over the business rather than selling it. There is also the issue of value other than cash. For example, my father's old guitar has little cash value but holds great sentimental value to me. In this section, we will look at ways to divide items so that, as nearly as possible, all parties see the division as fair. In the process, we will see that the idea of fairness is difficult to define. See also Exercises 27 and 28, where the idea of envy-free procedures is explored.

Divide-and-choose procedure

Perhaps the simplest, and sometimes most practical, fair-division scheme is the *divide-and-choose* procedure. Suppose you and I wish to share a chocolate chip cookie. I could break the cookie into two pieces and give you one of the pieces, but you might not be happy with the outcome. Maybe you're not happy because I gave you the smaller piece, or maybe the pieces are of equal size but I gave you the half with fewer chocolate chips. One simple solution is that I break the cookie but you choose which piece you want (or *vice versa*—you divide and I decide). If I intentionally break the cookie "unfairly," I am setting myself up to get the less desirable piece.

KEY CONCEPT

In the divide-and-choose procedure, one person divides the items into two parts and the other person chooses which part he or she wants.

Here's another example: Suppose my sister and I inherit a jewelry collection that we want to divide. She could separate the jewelry into two piles and let me take the pile I like best. While this may not result in an equal division, it is likely that both of us will at least consider it fair. Perhaps my sister is particularly fond of a necklace that Mom wore. When she divides the jewelry, she may intentionally put the necklace in a small pile, knowing that I will choose the larger pile (which has the greater cash value). Sister is happy because she gets to wear Mom's necklace, and I am happy because I got a larger cash value than I expected.

This division procedure can be extended easily to three people. The following extension is the *lone-divider method*. Suppose now that my brother, my sister, and I want to divide fairly a jewelry collection. I divide the jewelry into three piles that I consider to be of equal value. There are piles 1, 2, and 3. My brother and my sister each must surely consider that at least one pile has at least one-third of the total value. Suppose my brother likes pile number 1 and my sister likes pile number 2. Then brother and sister each take the pile they like, and I get pile number 3. Suppose, however, that each of them likes the same pile, say, pile 1. Then I choose whichever of piles 2 and 3 that I like. The remaining two piles are put back together, and brother and sister use the divide-and-choose method for two persons to divide what's left. Just as in the case of Mom's necklace, various strategies may be employed to get the desired outcome.

In Exercises 25 and 26, you will be asked to extend this procedure to more than three parties. Related procedures are the *lone-chooser method* (Exercise 23) and the *last-diminisher method* (Exercise 24).

Adjusted winner procedure

In practice, most fair-division problems are much more complex than dividing a cookie. There are a number of ways of resolving them. One that works well when only two people are involved is the *adjusted winner procedure*.

KEY CONCEPT

In the adjusted winner procedure, two people assign points to bid on each item, assigning a total of 100 points. Initial division of the items gives each item to the person offering the highest bid. The division is then adjusted based on the ratios of the bids for each item so that ultimately each person receives a group of items whose bid totals are the same for each person.

To see how it works, let's suppose my sister and I want to divide an inheritance. The assets consist of a guitar, a jewelry collection, a car, a small library, and a certain amount of cash. To start the procedure, each of us takes 100 points and divides those points among the assets according to how we individually value the items. Suppose our point values are as follows:

My points	Item	Sister's points
35	Guitar	10
10	Jewelry	10
20	Car	40
15	Library	10
20	Cash	30

Initially, each item goes to the person who bid the most for it. So, in the initial round, I get the guitar and the library for a point total of 50. My sister gets the car and the cash for a point total of 70. We give the tied item, the jewelry, to the point leader—my sister. She now has a total of 80 points. This is not a fair division. In a sense, I am only 50% satisfied, but my sister is 80% satisfied. We need to adjust to even the totals.

To make the score come out even, my sister must give me some of her property. Here is how we do it. For each item currently belonging to my sister (jewelry, car, and cash), we calculate the following fraction:

$$\frac{\text{Sister's bid}}{\text{My bid}}$$

Note that my sister's bid goes on top because she is the leader (in terms of point totals), and we are considering the items she owns. This fraction serves as a measure of the relative value placed by each of us on the items. We calculate these fractions:

$$\text{Ratio for jewelry} = \frac{10}{10} = 1$$

$$\text{Ratio for car} = \frac{40}{20} = 2.0$$

$$\text{Ratio for cash} = \frac{30}{20} = 1.5$$

We arrange these ratios in increasing order:

Jewelry	Cash	Car
1	1.5	2

Now we begin transferring items from my sister to me in this order until we reach an item that changes the point leader. We first transfer the jewelry. This gives me a point total of 60 and my sister a point total of 70. Transferring the next item, the cash, from my sister to me would make me the point leader. The cash is the *critical item*, and we will transfer just enough of it from my sister to me to make the score come out even. We want to take p percent of the cash from my sister and give it to me. That will reduce her point score by p percent of her bid for the cash and increase my score by p percent of my bid for the cash. We want the resulting point totals to be the same. Expressed as an equation, this is

$$\text{My score} + p \text{ percent of my cash bid} = \text{Sister's score} - p \text{ percent of her cash bid}$$

Writing p in decimal form gives the equation

$$60 + 20p = 70 - 30p$$

We solve this equation as follows:

$$60 + 20p = 70 - 30p \quad \text{Add } 30p \text{ to each side.}$$

$$60 + 50p = 70 \quad \text{Subtract 60 from each side.}$$

$$50p = 10 \quad \text{Divide both sides by 50.}$$

$$p = 0.20$$

So, in the end, I get the guitar, the jewelry, the library, and 20% of the cash. My sister gets the car and the remaining 80% of the cash. Now my sister's total points are $40 + 30 \times 0.80 = 64$, and my total points are $35 + 10 + 15 + 20 \times 0.20 = 64$, so each one of us is 64% satisfied with the outcome. This is a fair division, according to each person's valuation of the items.

SUMMARY 8.8
Adjusted Winner Procedure

Each of two people makes a bid totaling 100 points on a list of items to be divided.

Step 1: Initial division of items: Items go to the person with the highest bid. Tied items are held for the moment. The person with the highest score is the *leader*, and the person with the lowest score is the *trailer*.

Step 2: Tied items: Tied items go to the leader.

Step 3: Calculate leader/trailer ratios: For each item belonging to the leader, calculate the ratio

$$\frac{\text{Leader's bid}}{\text{Trailer's bid}}$$

Step 4: Transference of some items from leader to trailer: Transfer items from the leader to the trailer in order of increasing ratios as long as doing so does not change the lead. The first item whose transference would change the lead is the *critical item*.

Step 5: Division of the critical item: Take p percent of the critical item from the leader and give it to the trailer. Here, p in decimal form is the solution of the equation.

$$\text{Trailer's score} + p \times \text{Trailer's bid} = \text{Leader's score} - p \times \text{Leader's bid}$$

EXAMPLE 8.15 Using the adjusted winner procedure: An inheritance

Keith Bell/Shutterstock

The adjusted winner procedure can be used to divide an inheritance when items have sentimental value that would not be reflected by simply using dollar values. Use the adjusted winner procedure to divide the listed items fairly when the following bids are made:

John	Item	Faye
50	Vacation condominium	65
20	Red 1962 GT Hawk	15
20	Family silver set	15
10	Dad's Yale cap and gown	5

SOLUTION

Initially, John gets the Hawk, the silver, and the cap and gown, and Faye gets the condominium. John has a score of 50, and Faye has a score of 65. Faye is the leader. We must use Faye's only item, the condo, to even the score. (The condo is the critical item.) We are going to take p percent of the condo from Faye and give it to John. We find the decimal form of p by solving the equation from Step 5 of Summary 8.8:

$$\text{John's score} + p \times \text{John's bid} = \text{Faye's score} - p \times \text{Faye's bid}$$

$$50 + 50p = 65 - 65p \quad \text{Add } 65p \text{ to each side.}$$

$$50 + 115p = 65 \quad \text{Subtract 50 from each side.}$$

$$115p = 15 \quad \text{Divide by 115.}$$

$$p = \frac{15}{115} \quad \text{or about 0.13.}$$

So we take 13% of the condo from Faye and give it to John. In the end, Faye gets 87% ownership in the condo, while John gets the Hawk, the silver, the cap and gown, and 13% ownership in the condo. The shared ownership could be accomplished, for example, by giving John use of the condo for 7 weeks (13% of 52 weeks) and Faye use of it for the remaining 45 weeks each year.

TRY IT YOURSELF 8.15

Redo the preceding example using the following bids:

John	Item	Faye
60	Vacation condominium	45
15	Red 1962 GT Hawk	20
15	Family silver set	20
10	Dad's Yale cap and gown	15

The answer is provided at the end of this section.

There are situations in which we transfer more than one item. We look at one of these scenarios in the following example.

EXAMPLE 8.16 Using the adjusted winner procedure: Multiple transfers

Use the adjusted winner procedure to divide the listed items fairly when the following bids are made:

Anne	Item	Becky
50	Laptop	40
5	DVDs	5
6	MP3 player	5
31	Tablet	10
8	TV	40

SOLUTION

Initially, Anne gets the laptop, the MP3 player, and the tablet for a point total of 87. Becky gets the TV for a point total of 40. Anne is the leader, so she gets the tied item, the DVDs. Anne's new point total is 92. Because Anne has the lead, we look at ratios of bids for items she owns, with Anne's bid in the numerator:

$$\text{Ratio for laptop} = \frac{50}{40} = 1.25$$

$$\text{Ratio for DVDs} = \frac{5}{5} = 1$$

$$\text{Ratio for MP3 player} = \frac{6}{5} = 1.2$$

$$\text{Ratio for tablet} = \frac{31}{10} = 3.1$$

We arrange these ratios in increasing order:

DVDs	MP3 player	Laptop	Tablet
1	1.2	1.25	3.1

Starting from the left, we transfer as many items as we can from Anne to Becky without changing the point leader. We transfer the DVDs and the MP3 player, giving Anne a new score of 81 and Becky a new score of 50. The next item on the list is the laptop. Its transference would change the leader, so this is the critical item. We want to transfer p percent of the laptop from Anne to Becky. We use the equation from Step 5 of Summary 8.8:

$$\text{Becky's score} + p \times \text{Becky's bid} = \text{Anne's score} - p \times \text{Anne's bid}$$
$$50 + 40p = 81 - 50p$$
$$90p = 31$$
$$p = \frac{31}{90} \text{ or about } 0.34$$

In the final division, Becky gets the DVDs, the MP3 player, the TV, and 34% of the laptop, and Anne gets the tablet and the remaining 66% of the laptop.

TRY IT YOURSELF 8.16

Use the adjusted winner procedure to divide the listed items fairly when the following bids are made:

Alfred	Item	Betty
90	Computer	90
9	Cables	8
1	Printer paper	2

The answer is provided at the end of this section.

The adjusted winner procedure has several advantages. Under the assumption that neither person has a change of heart about the original allocation of points, the following can be shown to be true:

1. Both people receive the same number of points.
2. Neither person would be happier if the division were reversed.
3. No other division would make one person better off without making the other worse off.

This means that the allocation is as fair as anyone knows how to make it. It may or may not result in two happy campers, but from an objective point of view, it is a fair division.

The adjusted winner procedure does have several drawbacks. One drawback is that it is only for division between two people or parties. Another drawback is that the procedure is useful only if there are two or more items to be divided. If there is only one item to be divided, each person must bid all 100 points for the item and, ultimately, each person will be given 50% ownership of that one item. A third drawback is that the procedure may require that ownership of an item be divided in some way between the people involved, or that some similar arrangement be made for one of the items. Such an arrangement may be problematic if, for example, the item has sentimental value and neither side wants to relinquish it.

The Knaster procedure

The *Knaster procedure* is a method for dividing items among several parties, and it doesn't require dividing ownership of items. The procedure is based on having the parties assign monetary value to each item, their *bid*. The bidding is without knowledge of others' bids. In the end, all receive at least their fair share of their own bids. For example, if there are five parties, then each party receives at least one-fifth of its bid on each item. This procedure is also known as the *method of sealed bids*.

KEY CONCEPT

In the Knaster procedure, or method of sealed bids, each of three or more people assigns a dollar value to each item through secret bidding. Each item ultimately goes to the highest bidder while cash is used to equalize the distribution of the items according to each bidder's value of the item.

Sealed bid real estate sales in Dubai.

To illustrate the procedure, let's first assume that there is only one item to be divided, an automobile. There are four people involved, Abe, Ben, Caleb, and Dan, and each is entitled to one-fourth of the value (to them) of the car. Each bids the value he places on the car, and each is going to wind up with a value of at least one-fourth of his bid. Suppose the bids are as follows:

Person	Abe	Ben	Caleb	Dan
Bid	\$30,000	\$23,000	\$28,000	\$25,000

The highest bidder, Abe, gets the car. But that is the entire estate, and Abe is only entitled to one-fourth of it. He corrects it with cash. The value of the car to him is \$30,000, so, to begin, he puts the \$30,000 into a *kitty*.

Now each of the four is entitled to one-fourth of the car, so each withdraws from the kitty one-fourth of his assigned value of the car. So, Abe takes one-fourth of \$30,000, which is \$7500. Similarly, Ben takes one-fourth of \$23,000, which is \$5750, Caleb takes \$7000, and Dan takes \$6250. Note that the kitty is not being divided evenly between the four. The reason is that they did not each assign the same value to the car. Each person is being given one-fourth of the value he assigned to the car. The total withdrawals from the \$30,000 kitty are \$7500 + 5750 + 7000 + 6250 = \$26,500. That leaves a kitty of \$3500 to be divided equally among the four people, \$875 for each. Each person is receiving one-fourth of the value he assigned to the car, plus \$875.

The following table summarizes the complete transaction:

Person Bid	Abe \$30,000	Ben \$23,000	Caleb \$28,000	Dan \$25,000	Kitty
Car award	Car				0
To kitty	−30,000				30,000
From kitty	7500	5750	7000	6250	3500
From kitty	875	875	875	875	0
Final share	Car less \$21,625	\$6625	\$7875	\$7125	\$0

One important result of this procedure is that each person gets at least a fourth of his own evaluation of the worth of the car.

If there is more than one item to be divided, the Knaster procedure can be applied to each item sequentially, as illustrated in Example 8.17.

EXAMPLE 8.17 Employing the Knaster procedure with three heirs: Keeping the family farm

Julie, Anne, and Steve have inherited the family farm. They want the farm to remain in the family, and they want to divvy up the parts of the farm rather than try to share ownership in each part. Using the Knaster procedure, they make sealed bids for the three parts of the farm: the farmhouse, the wheat fields, and the cattle operation. Here are the bids:

Off the Mark by Mark Parisi/Atlantic Feature Syndicate

Bid	Julie	Anne	Steve
Farmhouse	$120,000	$60,000	$75,000
Wheat fields	210,000	450,000	390,000
Cattle operation	300,000	750,000	600,000

Determine the fair division of the farm using the Knaster procedure.

SOLUTION

Each part of the farm is given to the highest bidder, so Julie will get the farmhouse, Anne will get both the wheat fields and the cattle operation, and Steve will not get any of the farm (although he will get cash). We begin with the farmhouse, which goes to Julie. Because she valued the farmhouse at $120,000, she puts $120,000 in the kitty. Julie, Anne, and Steve are each entitled to one-third of their respective bids for the farmhouse, so Julie takes from the kitty one-third of $120,000, or $40,000, Anne takes from the kitty one-third of $60,000, or $20,000, and Steve takes one-third of $75,000, or $25,000. That leaves $120,000 – 40,000 – 20,000 – 25,000 = $35,000 in the kitty, as shown below.

ND700/Shutterstock

Bid	Julie	Anne	Steve	Kitty
Farmhouse	$120,000	$60,000	$75,000	
Wheat fields	210,000	450,000	390,000	
Cattle operation	300,000	750,000	600,000	
Farmhouse award	Farmhouse			0
To kitty	−120,000			120,000
From kitty	40,000	20,000	25,000	35,000

Similarly, Anne placed the highest bid for the wheat fields, which she gets. Because she valued the fields at $450,000 she puts $450,000 in the kitty. Julie, Anne, and Steve are each entitled to one-third of their respective bids for the wheat fields, so Julie takes from the kitty one-third of $210,000, or $70,000, Anne takes one-third of $450,000, or $150,000, and Steve takes one-third of $390,000, or $130,000. The net effect is to add $450,000 – 150,000 – 70,000 – 130,000 = $100,000 to the $35,000 already in the kitty, as shown on the next page.

Bid	Julie	Anne	Steve	Kitty
Farmhouse	$120,000	$60,000	$75,000	
Wheat fields	210,000	450,000	390,000	
Cattle operation	300,000	750,000	600,000	
Farmhouse award	Farmhouse			0
To kitty	−120,000			120,000
From kitty	40,000	20,000	25,000	35,000
Wheat fields award		Wheat fields		35,000
To kitty		−450,000		485,000
From kitty	70,000	150,000	130,000	135,000

Anne also had the highest bid for the cattle operation, which she gets. Because she valued the cattle operation at $750,000, she puts $750,000 in the kitty. Julie, Anne, and Steve are each entitled to one-third of their respective bids for the cattle operation, so Julie takes from the kitty one-third of $300,000, or $100,000, Anne takes one-third of $750,000, or $250,000, and Steve takes one-third of $600,000, or $200,000. The net effect is to add $750,000 – 250,000 – 100,000 – 200,000 = $200,000 to the $135,000 already in the kitty, as shown below. Finally, the balance of $335,000 left in the kitty is divided equally among the three heirs, so each gets approximately $111,667, as shown below.

Bid	Julie	Anne	Steve	Kitty
Farmhouse	$120,000	$60,000	$75,000	
Wheat fields	210,000	450,000	390,000	
Cattle operation	300,000	750,000	600,000	
Farmhouse award	Farmhouse			0
To kitty	−120,000			120,000
From kitty	40,000	20,000	25,000	35,000
Wheat fields award		Wheat fields		35,000
To kitty		−450,000		485,000
From kitty	70,000	150,000	130,000	135,000
Cattle operation award		Cattle operation		135,000
To kitty		−750,000		885,000
From kitty	100,000	250,000	200,000	335,000
Kitty distribution	111,667	111,667	111,667	

Summing the columns, we find that Julie gets the farmhouse and cash totaling

$$-\$120{,}000 + 40{,}000 + 70{,}000 + 100{,}000 + 111{,}667 = \$201{,}667$$

Anne gets the wheat fields, the cattle operation, and cash totaling

$$\$20{,}000 - \$450{,}000 + 150{,}000 - 750{,}000 + 250{,}000 + 111{,}667 = -\$668{,}333$$

That is, Anne needs to pay $668,333. Steve gets no property, but he does get cash totaling

$$\$25{,}000 + 130{,}000 + 200{,}000 + 111{,}667 = \$466{,}667$$

TRY IT YOURSELF 8.17

Determine the fair division of the farm using the Knaster procedure if the bids are as follows:

Bid	Julie	Anne	Steve
Farmhouse	$120,000	$60,000	$75,000
Wheat fields	210,000	450,000	480,000
Cattle operation	330,000	780,000	690,000

The answer is provided at the end of this section.

SUMMARY 8.9
Knaster Procedure or Method of Sealed Bids

Step 1: Each person bids a dollar amount for each item to be distributed. The bidding is without knowledge of others' bids—hence the phrase "sealed bids."

Step 2: For each item separately:

a. The item is given to the person with the highest bid, who is called the winner. If two or more people make the same high bid, a winner is chosen at random from the high bidders.

b. The winner contributes to the kitty his or her bid.

c. Each person, including the winner, receives from the kitty their share of their bid for the item (which is their bid divided by the number of bidders).

Step 3: The money left in the kitty is distributed equally to each bidder.

One disadvantage of the Knaster procedure is that it may require participants to have a large amount of cash available.

WHAT DO YOU THINK?

No division: You need to divide some items fairly, but in this case it doesn't make sense to divide ownership of individual items. Which of the fair-division methods discussed in the text would be best? Give an example.

Questions of this sort may invite many correct responses. We present one possibility: The adjusted winner procedure often requires ownership in some items to be split, but the Knaster procedure does not. Thus, the Knaster procedure would be better for items such as an automobile, which would be difficult to share, or collections whose value might by diminished by division. The downside of the Knaster procedure is that it requires participants to have a certain amount of cash available.

Further questions of this type may be found in the *What Do You Think?* section of exercises.

Try It Yourself answers

Try It Yourself 8.15: Using the adjusted winner procedure: An inheritance John gets 95% ownership in the vacation condominium and Faye gets the red 1962 GT Hawk, family silver set, Dad's Yale cap and gown, and 5% ownership in the vacation condominium.

Try It Yourself 8.16: Using the adjusted winner procedure: Multiple transfers Alfred gets the cables and 46% ownership of the computer, and Betty gets the paper and 54% ownership of the computer.

Try It Yourself 8.17: Employing the Knaster procedure with three heirs: Keeping the family farm Julie gets the farmhouse plus $205,000, Anne gets the cattle operation and pays $245,000, and Steve gets the wheat fields plus $40,000.

Exercise Set 8.3

Test Your Understanding

1. Explain how the divide-and-choose procedure works.

2. True or false: In the adjusted winner procedure, tied items go initially to the leader.

3. If items are to be divided among several parties, which is more appropriate, the adjusted winner procedure or the Knaster procedure?

4. True or false: In the Knaster procedure, each person receives at a minimum their own fair share of their bid for an item.

Problems

In each case, round any answer in percentage form to one decimal place unless you are directed to do otherwise.

5. **Strategy.** Suppose Abe, Caleb, and Mary wish to divide an inheritance of jewelry fairly using the lone-divider method. Mary is particularly fond of a pair of earrings. If she performs the first division, what strategy should she use to give her the best chance of getting the earrings?

6. **Strategy.** Suppose Abe, Caleb, and Mary wish to divide an inheritance of jewelry fairly using the lone-divider method. Mary is particularly fond of a pair of earrings. If she does not perform the first division, what strategy should she use to give her the best chance of getting the earrings?

7–15. **Adjusted winner.** *In Exercises 7 through 15, use the adjusted winner procedure to divide items if the initial bids are as shown.*

7. Carla and Daniel divide an art collection.

Daniel	Item	Carla
20	Drawings	25
40	Paintings	30
40	Art supplies	45

8. Mike and Simon divide automobiles.

Mike	Item	Simon
50	BMW	30
25	Mercedes	30
25	Lexus	40

9. Anna and Ashley divide a small library.

Anna	Item	Ashley
10	Math books	10
10	Craft books	30
80	Travel books	60

10. Monica and Angel divide items associated with travel.

Monica	Item	Angel
35	Plane tickets	30
40	Luggage	30
25	Clothes	40

11. Acom divides into two companies, Bcom and Ccom. Bcom and Ccom divide some of the assets of Acom.

Bcom	Item	Ccom
15	Appliance dealership	20
5	Coffeeshop	10
35	Computer outlet	20
20	Bakery	20
25	Cleaning service	30

12. David and Mike divide household items.

David	Item	Mike
40	Dishes	40
10	Cooking utensils	30
20	Crockery	10
25	Furniture	10
5	Linens	10

13. Ben and Ruth divide a Southwest art collection.

Ben	Item	Ruth
30	Hopi kachinas	20
20	Navajo jewelry	30
30	Zuni pottery	20
15	Apache blankets	20
5	Pima sand paintings	10

14. **What to do?** How would you fairly divide things if the initial bids are as follows?

Person 1	Item	Person 2
50	Item 1	50
30	Item 2	30
20	Item 3	20

15. **What to do?** How would you fairly divide things if the initial bids are as follows?

Person 1	Item	Person 2
25	Item 1	25
25	Item 2	25
25	Item 3	25
25	Item 4	25

16–22. The Knaster procedure. *In Exercises 16 through 22, use the Knaster procedure to fairly divide property with the given bids.*

16. Oscar, Ralph, Mary, and Wanda wish to divide a business property.

Person	Oscar	Ralph	Mary	Wanda
Bid	$33,000	$40,000	$38,000	$32,000

17. J. C., Trevenor, Alta, and Loraine inherit a home and wish to divide it.

Person	J. C.	Trevenor	Alta	Loraine
Bid	$330,000	$340,000	$380,000	$320,000

18. Shirley, Brooke, Tommy, and Jessie inherit a plot of land that they wish to divide.

Person	Shirley	Brooke	Tommy	Jessie
Bid	$45,000	$48,000	$43,000	$48,000

19. Lee, Max, Anne, and Becky wish to divide a painting.

Person	Lee	Max	Anne	Becky
Bid	$100	$150	$90	$120

20. Ann, Liz, Jeanie, and Lorie wish to divide an art collection.

Person	Ann	Liz	Jeanie	Lorie
Bid	$33,000	$38,000	$38,000	$32,000

21. James and Gary want to divide three small businesses.

Bid	James	Gary
Business 1	$12,000	$22,000
Business 2	22,000	2000
Business 3	8000	12,000

22. A conglomerate splits into three companies and has to divide the distribution rights for a clothing line and a food line.

Bid	Company 1	Company 2	Company 3
Clothing line	$12,000,000	$21,000,000	$6,000,000
Food line	21,000,000	6,000,000	6,000,000

23. Lone-chooser method. The *lone-chooser method* is an adaptation of the divide-and-choose procedure for three people. For the purposes of illustration, we will use a pie. The first and second persons use the divide-and-choose procedure to divide the pie into two pieces. Now each person divides his or her share of the pie into three pieces. The third person chooses one slice from each half. The first two people each keep what is left of their half.

a. Explain why each person believes that he or she has a fair portion of the pie at the end of this procedure.

b. Propose a division scheme of this sort for four people.

24. Last-diminisher method. The *last-diminisher method* works for items that can be reasonably divided in virtually any way we wish—land or a cake, for example. The method would make no sense for a home. We will describe the method for three people, Anne, Becky, and Alice, who are placed in this order randomly. Let's suppose they want to divide up cookie dough from a mixing bowl. They proceed as follows:

Step 1: Anne scoops what she believes is a fair portion of the cookie dough into her own bowl.

Step 2: Becky has two options: She can pass if she thinks that Anne's portion is fair or less than fair. Becky's other option is to take Anne's bowl of dough from her and then return some of the dough to the mixing bowl, thereby reducing the amount to what she considers a fair portion. In this case, we say she *diminishes* the dough.

Step 3: Alice has the same kind of options as Becky. She may pass or take Becky's bowl, and in the latter case she returns some of the dough to the mixing bowl.

The last person of the three to take and diminish keeps the bowl she is holding. (If no one diminished, then Anne, the one who started, would keep the bowl.) Let's say that person is Becky. The remaining two, Anne and Alice, use the divide-and-choose procedure for two to allocate the rest of the dough.

Explain why each of Anne, Becky, and Alice believes she has a fair share of the cookie dough at the end of this procedure.

25. Divide and choose for four. You want to divide a seashell collection among four people. Suggest a divide-and-choose procedure that you think is fair.

26. Five people. Propose a divide-and-choose procedure for five people.

Exercises 27 and 28 are suitable for group work.

27–28. Envy. *A division procedure is envy-free if, when all is said and done, no one has reason to wish for someone else's division. In Exercises 27 and 28, we investigate this idea.*

27. Explain why the divide-and-choose procedure for two players is envy-free.

28. The point of this exercise is that the lone-divider method need not be envy-free. Abe, Ben, and Dan are dividing three chocolate chip cookies: cookie 1 with 10 chocolate chips, cookie 2 with 9 chocolate chips, and cookie 3 with no chocolate chips. Abe doesn't care one way or the other about chocolate chips. As far as he is concerned, the three cookies are of equal value. Ben and Dan care only about chocolate chips—they are going to pick the chips out of the dough and eat only the chocolate.

In the rest of the exercise, we assume the following: Abe, Ben, and Dan use the lone-divider method to divide the three cookies. Abe starts. The three cookies have equal value as far as he can see, so he just lays out the three cookies for Ben and Dan. There are various possible outcomes after this.

a. Let's suppose that Ben and Dan both prefer cookie 1. (This seems reasonable.) According to the procedure, Abe can take either cookie 2 or cookie 3. Suppose he takes cookie 2. Then Ben and Dan divide cookie 1 and cookie 3 between themselves using the divide-and-choose procedure for two people. (The cookies can be broken!) Explain why either Ben or Dan will envy Abe.

b. In part a, we sketched one possible outcome. There may be reasons for Ben and Dan not to state a preference for cookie 1. List all of the possible preferences the two of them can state. Show that if they state different preferences from each other, then one of them will envy the other. Also show that if they state the same preference as each other, then Abe can settle on a choice that will make Ben or Dan envy him.

The following exercise is designed to be solved using technology such as a calculator or computer spreadsheet.

29. Knaster procedure. Make a spreadsheet that executes the Knaster procedure.

Writing About Mathematics

30. History. In 1945, the Polish mathematician Bronislaw Knaster introduced the procedure that bears his name. In fact, the study of fair division and the related mathematical area called *game theory* developed rapidly in a period starting before World War II and extending to the early stages of the Cold War. Research the relevance of game theory to military strategy in general and in particular to the doctrine of mutually assured destruction.

31. Knaster. The Knaster procedure is named for the Polish mathematician Bronislaw Knaster. Write a brief paper on him.

32. Experimenting with divide and choose. Start with a collection of items that you will pretend to divide with a friend. Use the divide-and-choose procedure and report on both your own and your friend's satisfaction with the results.

33. Experimenting with the adjusted winner procedure. Start with a collection of items that you will pretend to divide with a friend. Use the adjusted winner procedure and report on both your own and your friend's reaction to the results.

34. Experimenting with the Knaster procedure. Start with a collection of items that you will pretend to divide with a group of your friends. Use the Knaster procedure and report on both your own and your friends' reactions to the results.

What Do You Think?

35. Values. You need to divide some valuable items fairly. It makes sense to assign numerical values to the items, but the participants don't have much cash available. Which of the fair-division methods discussed in the text would be best?

36. No values. You need to divide some items fairly, but in this case it doesn't make sense to assign numerical values to the items. Which of the fair-division methods discussed in the text would be best?

37. Sealed bids. In the Knaster procedure, why is it important that the bids be sealed? Do the bids have to be sealed in the adjusted winner procedure? Explain your answer.

38. Knaster. In the Knaster procedure, why is it a disadvantage that the method may require participants to have a large amount of cash available?

39. Divide and choose. In the description of the divide-and-choose procedure for two people, the text says that if one person divides the items "unfairly" that person could get the less desirable piece. What if the two people have different ideas about what items are valuable? Would this mean that the word "unfairly" in the text doesn't make sense? **Exercise 27 is relevant to this topic.**

40. Fair? List some factors you would consider in determining whether a procedure for dividing items is fair. **Exercise 28 is relevant to this topic.**

41. Divide and choose for more than two. Can you devise what you consider to be a fair divide-and-choose procedure for three people? **Exercise 23 is relevant to this topic.**

42. Try it yourself. To get a real feeling for how the adjusted winner procedure works, it is a good idea to try it yourself. Make up a list of at least 10 items that you and a friend will divide using the adjusted winner procedure. Can you devise a strategy that will give you an advantage?

43. Try it yourself. To get a real feeling for how the Knaster procedure works, it is a good idea to try it yourself. Make up a list of at least 10 items that you and several friends will divide using the Knaster procedure. Can you devise a strategy that will give you an advantage?

8.4 Apportionment: Am I represented?

TAKE AWAY FROM THIS SECTION
Understand the variety of ways representatives can be apportioned.

The following excerpt from RealClearPolitics.com discusses the apportionment of seats in the U.S. House of Representatives. The "nasty formula" mentioned in the second paragraph is the Huntington-Hill method, which is one of the apportionment methods discussed in this section.[16]

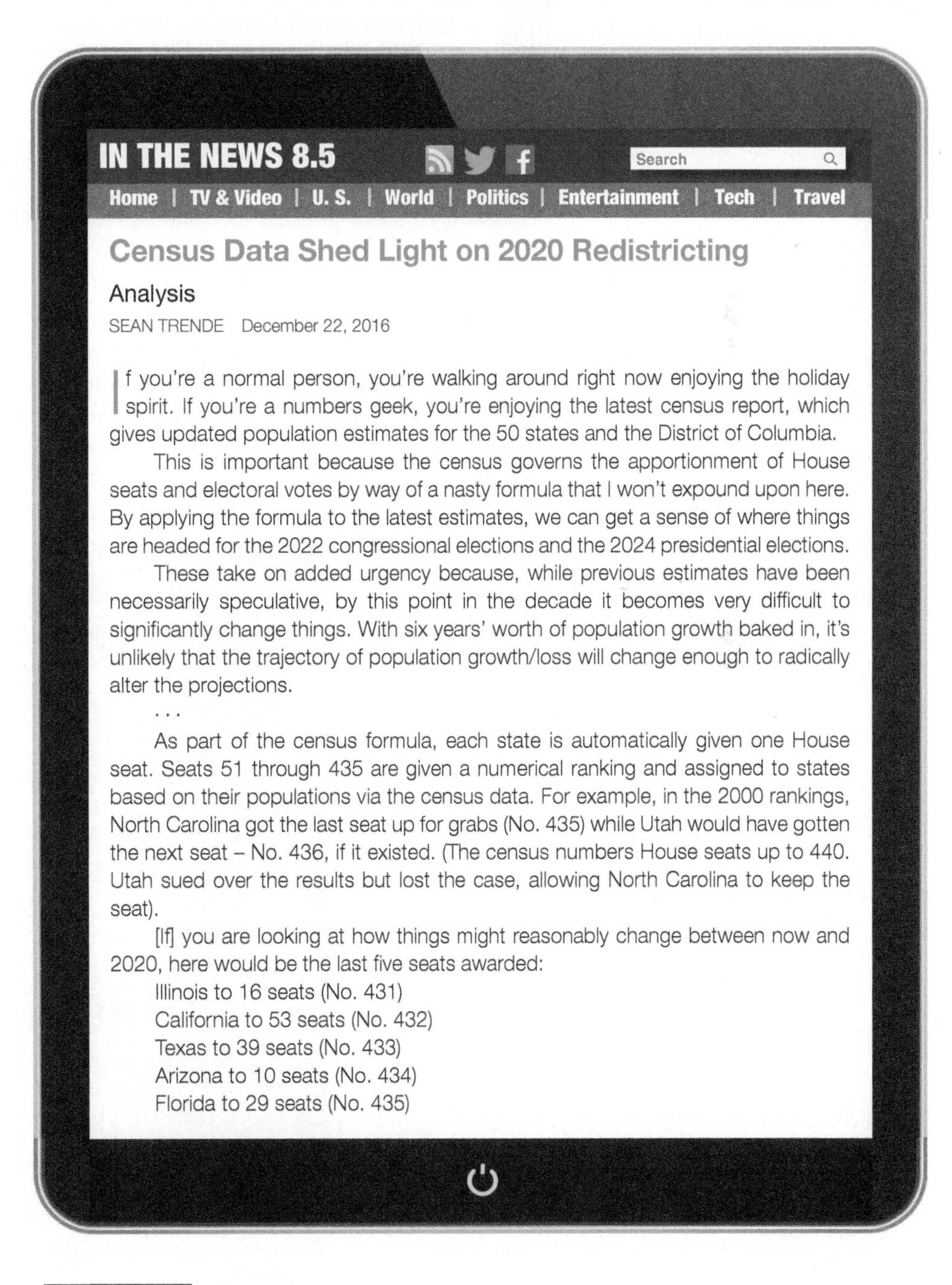

IN THE NEWS 8.5

Home | TV & Video | U. S. | World | Politics | Entertainment | Tech | Travel

Census Data Shed Light on 2020 Redistricting

Analysis

SEAN TRENDE December 22, 2016

If you're a normal person, you're walking around right now enjoying the holiday spirit. If you're a numbers geek, you're enjoying the latest census report, which gives updated population estimates for the 50 states and the District of Columbia.

This is important because the census governs the apportionment of House seats and electoral votes by way of a nasty formula that I won't expound upon here. By applying the formula to the latest estimates, we can get a sense of where things are headed for the 2022 congressional elections and the 2024 presidential elections.

These take on added urgency because, while previous estimates have been necessarily speculative, by this point in the decade it becomes very difficult to significantly change things. With six years' worth of population growth baked in, it's unlikely that the trajectory of population growth/loss will change enough to radically alter the projections.

. . .

As part of the census formula, each state is automatically given one House seat. Seats 51 through 435 are given a numerical ranking and assigned to states based on their populations via the census data. For example, in the 2000 rankings, North Carolina got the last seat up for grabs (No. 435) while Utah would have gotten the next seat – No. 436, if it existed. (The census numbers House seats up to 440. Utah sued over the results but lost the case, allowing North Carolina to keep the seat).

[If] you are looking at how things might reasonably change between now and 2020, here would be the last five seats awarded:

Illinois to 16 seats (No. 431)
California to 53 seats (No. 432)
Texas to 39 seats (No. 433)
Arizona to 10 seats (No. 434)
Florida to 29 seats (No. 435)

[16] *The original article supplies the following link to an explanation of the formula:* www.census.gov/population/apportionment/about/computing.html

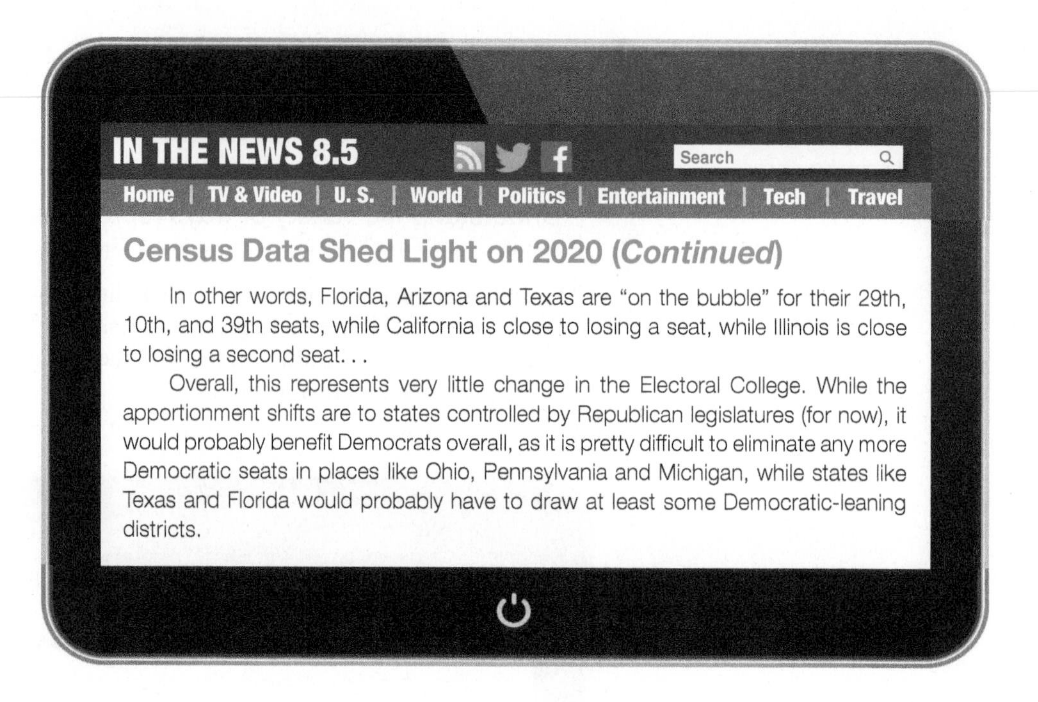
IN THE NEWS 8.5

Census Data Shed Light on 2020 (*Continued*)

In other words, Florida, Arizona and Texas are "on the bubble" for their 29th, 10th, and 39th seats, while California is close to losing a seat, while Illinois is close to losing a second seat. . .

Overall, this represents very little change in the Electoral College. While the apportionment shifts are to states controlled by Republican legislatures (for now), it would probably benefit Democrats overall, as it is pretty difficult to eliminate any more Democratic seats in places like Ohio, Pennsylvania and Michigan, while states like Texas and Florida would probably have to draw at least some Democratic-leaning districts.

You may be surprised to learn that the U.S. Constitution lays out some basic rules for the makeup of the House of Representatives, but leaves both the size of the House and the exact number of representatives for each state to be determined by Congress. This is the problem of *apportionment* and, as this article shows, the issue remains important to this day. The Constitution gives these rules in Article 1, Section 2: "Representatives and direct taxes shall be apportioned among the several

Albert Knapp/Alamy

The first Congress met in Congress Hall in Philadelphia.

states which may be included within this union, according to their respective numbers. . . ." The numbers—that is, the population of each state—are established by the census: "The actual enumeration shall be made within three years after the first meeting of the Congress of the United States, and within every subsequent term of ten years, in such manner as they shall by law direct." The history of the census is fascinating in itself. The nation has addressed questions such as how to count slaves, people overseas, those in the military, native Americans, and noncitizens. In particular, before the Civil War, only three-fifths of the population of slaves was counted for apportionment purposes.

The Constitution specifies a maximum size for the House of Representatives: "The number of Representatives shall not exceed one for every thirty thousand, but each state shall have at least one Representative." It specifies the initial size (65 members) and distribution of the House. Congress regularly increased the size of the House to account for increasing population until 1911, when the size was fixed at 435. The size has remained at this level since then.[17]

There has been considerable political controversy about these issues since the beginning of the republic. We highly recommend the comprehensive treatment of apportionment in *Fair Representation* by Michel Balinski and H. Peyton Young[18] and the 2001 Congressional Research Service Report for Congress on proposals for change in the House of Representatives apportionment formula.[19]

Apportionment: The problem

To understand some of the difficulties with apportionment, we look at the 1810 House of Representatives. The census of 1810 gave the population of the United States for apportionment purposes as 6,584,255. The size of the House was set at 181 members. To find the ideal size of a congressional district, we divide the U.S. population by the size of the House. Then we divide the population of each state by the size of a congressional district to determine that state's proper share of the House, called its *quota*.

KEY CONCEPT

To find the size of the ideal district, we use the formula

$$\text{Ideal district size} = \frac{\text{U.S. population}}{\text{House size}}$$

The ideal district size is also known as the standard divisor. We use this number to determine each state's quota:

$$\text{State's quota} = \frac{\text{State's population}}{\text{Ideal district size}}$$

[17] *It was temporarily increased to 437 in 1959 when Alaska and Hawaii were admitted to the Union.*

[18] *Michel Balinski and H. Peyton Young,* Fair Representation: Meeting the Ideal of One Man, One Vote *(New Haven, CT, and London: Yale University Press, 1982).*

[19] *David C. Huckabee,* The House of Representatives Apportionment Formula: An Analysis of Proposals for Change and Their Impact on States, *Congressional Research Service, 2001;* `http://usinfo.org/enus/government/branches/docs/housefrm.pdf`

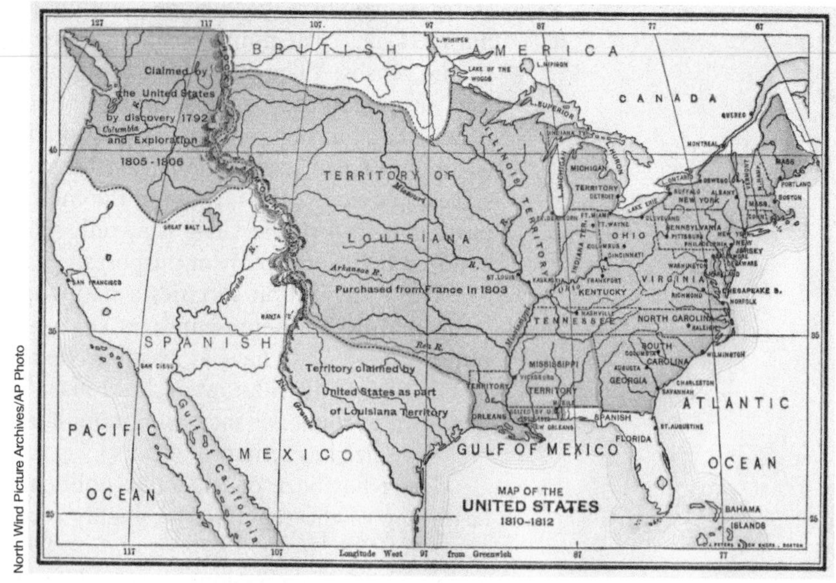

North Wind Picture Archives/AP Photo

The United States in 1810.

For the population in 1810 with the House at 181, we find an ideal district size of

$$\text{Ideal district size} = \frac{6{,}584{,}255}{181}$$

or about 36,377.099. For each state, we divide the population by the ideal district size to calculate the quota. For example, Connecticut had a population of 261,818, so its quota is

$$\text{Quota for Connecticut} = \frac{261{,}818}{36{,}377.099}$$

or about 7.197. Here are the results of the 1810 census and the quota (to three decimal places) for each state with the ideal district size of 36,377.099. We have added a column giving the quota rounded to the nearest whole-number.

State	Population	Quota	Rounded quota
Connecticut	261,818	7.197	7
Delaware	71,004	1.952	2
Georgia	210,346	5.782	6
Kentucky	374,287	10.289	10
Maryland	335,946	9.235	9
Massachusetts	700,745	19.263	19
New Hampshire	214,460	5.895	6
New Jersey	241,222	6.631	7
New York	953,043	26.199	26
North Carolina	487,971	13.414	13
Ohio	230,760	6.344	6
Pennsylvania	809,773	22.261	22
Rhode Island	76,888	2.114	2
South Carolina	336,569	9.252	9
Tennessee	243,913	6.705	7
Vermont	217,895	5.990	6
Virginia	817,615	22.476	22
Total	**6,584,255**	**181**	**179**

The question is what to do with the fractional parts. For example, North Carolina's quota is 13.414. Should North Carolina get 13 representatives, or should it get 14? This is no small issue—the number of representatives allocated to a state makes a big difference in its congressional voting power.

The obvious solution is simply to round each quota to the nearest whole number, as we have done in the last column of the table. One difficulty with this solution is shown in the total. Rounding each state's quota to the nearest whole number gives a total of only 179 representatives. There are two representatives not assigned to any state. This shows why the easy solution of rounding to the nearest whole number by itself would not be a consistently reliable method for apportionment. In general terms, the problem of apportionment is how to deal with the fractions.

Since 1790, several apportionment methods for the U.S. House of Representatives have been developed and either used to allocate seats after various censuses or cited in court cases, even as recently as 2001 (after the 2000 census). These methods fall into two categories: those that rank the fractions, and those that round the fractions.

Hamilton's solution: Ranking the fractions

The first apportionment bill was passed by Congress in 1792. It was based on a method developed by Alexander Hamilton. This method calculates the quotas for each state and then makes an initial allocation to each state by rounding down to the whole-number part of the quota.[20] The sum of these whole numbers is less than the size of the House, so the leftover members are allocated, one per state, in order from the highest fractional part down. For this reason, this method is also called the *method of largest fractional remainders*.[21]

Let's see the results given by this method for the 1810 Congress.

State	Quota	Quota rounded down: Initial seats	Fractional part of quota	Added seats	Final seats
Connecticut	7.197	7	0.197		7
Delaware	1.952	1	**0.952**	1	2
Georgia	5.782	5	**0.782**	1	6
Kentucky	10.289	10	0.289		10
Maryland	9.235	9	0.235		9
Massachusetts	19.263	19	0.263		19
New Hampshire	5.895	5	**0.895**	1	6
New Jersey	6.631	6	**0.631**	1	7
New York	26.199	26	0.199		26
North Carolina	13.414	13	**0.414**	1	14
Ohio	6.344	6	0.344		6
Pennsylvania	22.261	22	0.261		22
Rhode Island	2.114	2	0.114		2
South Carolina	9.252	9	0.252		9
Tennessee	6.705	6	**0.705**	1	7
Vermont	5.990	5	**0.990**	1	6
Virginia	22.476	22	**0.476**	1	23
Total	**181**	**173**		**8**	**181**

[20] *In practice, this number is never allowed to be 0.*

[21] *Hamilton's method is also known as the Vinton method or Hamilton–Vinton method because Representative Samuel Vinton was a strong advocate for using this method for the 1850 census.*

Using the whole-number part of the quota for initial allocation, we see that 173 seats are allocated. This leaves $181 - 173 = 8$ leftover seats. These seats are given to the 8 states for which the fractional parts of the quotas are the greatest. These states correspond to the boldface numbers in the column giving the fractional parts. The final allocations are shown in the last column of the table.

SUMMARY 8.10
Hamilton's Method for Apportionment

Step 1: We calculate the quota for each state by dividing its population by the ideal district size.

Step 2: We give to each state the number of representatives corresponding to the whole-number part of its quota.

Step 3: We allocate the leftover seats—that is, the difference between the House size and the total given in Step 2—as follows: Rank the states by the size of the fractional part of the quota, from greatest to least, and give one leftover member to each state in that order until the leftovers are exhausted.

EXAMPLE 8.18 Apportioning by Hamilton's method: City councilors

Our city has four districts: North, East, West, and South. There are required to be 10 city councilors allocated according to population, with at least one from each district. Given the populations below, calculate the quota for each district. (You need keep only one place beyond the decimal point.) Then use Hamilton's method to determine the number of councilors to represent each district.

District	Population
North	3900
East	4800
West	18,000
South	3300
Total	30,000

SOLUTION

We have 10 seats to fill and a total population of 30,000, so the ideal district size is

$$\begin{aligned}\text{Ideal district size} &= \frac{\text{Total population}}{\text{Number of seats}}\\ &= \frac{30{,}000}{10}\\ &= 3000\end{aligned}$$

We divide the population in each district by this district size, and the resulting quota is shown in the second column of the accompanying table.

District	Quota	Quota rounded down: Initial seats	Fractional part of quota	Added seats	Final seats
North	1.3	1	0.3		1
East	1.6	1	**0.6**	1	2
West	6.0	6	0.0		6
South	1.1	1	0.1		1
Total	10	9		1	10

The third column of the table shows that the whole-number parts sum to 9, so there is one leftover councilor to allocate. Because the quota of 1.6 has the largest fractional part (0.6), the East district gets that councilor. The final result: North and South should each get 1 councilor, East should get 2 councilors, and West should get 6 councilors.

TRY IT YOURSELF 8.18

For the same city, districts, and populations, use Hamilton's method to apportion a total of 20 councilors. You need keep only one place beyond the decimal point.

The answer is provided at the end of this section.

The congressional bill of March 26, 1792, that allocated the U.S. House of Representatives according to Hamilton's method was vetoed by President George Washington. Indeed, this was the very first presidential veto. Some of the reasons for Washington's veto are investigated in Exercise 16. Some additional problems with Hamilton's method are investigated in Exercises 17 and 18. Hamilton's method was not used until it was revived for the 1850 census. Instead, a method devised by Thomas Jefferson was used beginning with the 1790 census.

Jefferson's solution: Adjusting the divisor

Thomas Jefferson, disturbed by the fractions, proposed a different solution. The first step is to round down (but not less than 1, because each state is guaranteed at least

On April 5, 1792, George Washington cast the first presidential veto.

one representative) using the whole-number part of the quota just as with the Hamilton method. But rather than rank the fractional parts, Jefferson proposed that the divisor be adjusted until the sum of the rounded-down quotas is the correct size—the legislated size of the House of Representatives. There are complicated formulas that can tell us how to adjust the divisor but, with the aid of a spreadsheet, trial-and-error is a suitable method. Let's consider again the allocation of seats for 1810. We start with the divisor as the ideal district size of 36,377.099 and round each quota down as our first allocation effort. As the fourth column of the accompanying table shows, this allocates only 173 seats. In the fifth column, we decrease the divisor to 34,000 to increase the number of seats allocated. For example, with a divisor of 34,000, Connecticut's adjusted quota is $261,818/34,000 = 7.701$ to three decimal places, which rounds down to 7. The overall result is a total of 187 seats allocated—too many. So, we increase the divisor to 35,000 and try again. The result is a sum of 181, as is shown in the last column of the table. Because this is the correct number of seats, we have found the final allocation.

State	Population	Quota using divisor 36,377.10	First try: Quota rounded down	Second try: Decrease divisor to 34,000, round down	Third try: Increase divisor to 35,000, round down
Connecticut	261,818	7.197	7	7	7
Delaware	71,004	1.952	1	2	2
Georgia	210,346	5.782	5	6	6
Kentucky	374,287	10.289	10	11	10
Maryland	335,946	9.235	9	9	9
Massachusetts	700,745	19.263	19	20	20
New Hampshire	214,460	5.895	5	6	6
New Jersey	241,222	6.631	6	7	6
New York	953,043	26.199	26	28	27
North Carolina	487,971	13.414	13	14	13
Ohio	230,760	6.344	6	6	6
Pennsylvania	809,773	22.261	22	23	23
Rhode Island	76,888	2.114	2	2	2
South Carolina	336,569	9.252	9	9	9
Tennessee	243,913	6.705	6	7	6
Vermont	217,895	5.990	5	6	6
Virginia	817,615	22.476	22	24	23
Total	**6,584,255**	**181**	**173**	**187**	**181**

Comparing this apportionment with Hamilton's, we see that they are very similar. But Jefferson's method gives one less delegate each to New Jersey, North Carolina, and Tennessee and gives one additional delegate each to Massachusetts, New York, and Pennsylvania.

As we noted, trial-and-error (with the help of a spreadsheet) is an effective way to adjust divisors to find an allocation that yields the correct House size. Finding the proper divisor is usually not difficult because there is often a wide range of workable answers. In the preceding table, any divisor between 34,856 and 35,037 will produce the correct size.

SUMMARY 8.11
Jefferson's Method for Apportionment

Step 1: We start using the ideal district size as the divisor.

Step 2: We calculate the quota for each state by dividing its population by the divisor.

Step 3: We round down each quota to the nearest whole number (but not less than 1) and sum the rounded quotas.

Step 4: If the sum from Step 3 is larger than the size of the House, we increase the divisor and repeat Steps 2 and 3. If the sum is too small, we decrease the divisor and repeat Steps 2 and 3. We continue this process until a divisor is found for which the sum of the rounded-down quotas (but not less than 1) is the House size. The result is the apportionment given by Jefferson's method.

EXAMPLE 8.19 Apportioning by Jefferson's method: City councilors

Recall the situation studied in Example 8.18: Our city has the four districts North, East, West, and South. There are required to be 10 city councilors allocated according to population, with at least one from each district. Given the populations on the next page, apportion the number of councilors to represent each district using Jefferson's method. Quotas may be reported using one place beyond the decimal point.

District	Population
North	3900
East	4800
West	18,000
South	3300
Total	30,000

SOLUTION

We have 10 councilors and a total population of 30,000, so our ideal district size is 30,000/10 = 3000. We use this as our first divisor. Dividing gives quotas of 1.3, 1.6, 6, and 1.1, which round down to 1, 1, 6, and 1 (see the accompanying table). These sum to 9, so we need to adjust the divisor down.

District	Population	Quota using divisor 3000	First try: Quota rounded down	Second try: Decrease divisor to 2500, round down
North	3900	1.3	1	1
East	4800	1.6	1	1
West	18,000	6.0	6	7
South	3300	1.1	1	1
Total	30,000	10	9	10

Trying a divisor of 2500 gives quotas of 1.6, 1.9, 7.2, and 1.3, which round down to 1, 1, 7, and 1. These sum to 10, so North, East, and South should each get 1 councilor, and West should get 7 councilors. Note that this distribution differs from that of Hamilton's method in Example 8.18.

TRY IT YOURSELF 8.19

For the same city, districts, and populations, use Jefferson's method to apportion a total of 8 councilors. Quotas may be reported using one place beyond the decimal point.

The answer is provided at the end of this section.

Jefferson's method was adopted by Congress for the Third Congress (1793) and used with each census through 1830. Even from the beginning, it was clear that small states may well be disproportionately underrepresented by this method. For 1800, Jefferson's method awards Delaware only a single representative based on its apportionment population[22] of 61,812—far more than the ideal district size of 34,680 at that time. As Jefferson's method continued to be applied for the censuses through 1830, it became more and more evident that the method favored larger states over smaller states. Note for the 1810 results that, compared with Hamilton's method, Jefferson's method gives extra representatives to the larger states of Massachusetts, New York, and Pennsylvania and takes them away from the smaller New Jersey, North Carolina, and Tennessee.

In addition to the problem of bias in favor of larger states, there was the question of *staying within the quota*. In Exercise 40, you will be asked to verify that in the 1820 census New York had a quota of 32.503. We would thus expect New York to get either 32 or 33 representatives. In fact, Jefferson's method gives New York 34 representatives, thus *violating quota*. Exercise 31 gives a simple situation where quota is violated by Jefferson's method. By contrast, Hamilton's method always stays within quota.

[22] *The actual population of Delaware was enumerated in the census as 64,273; for apportionment purposes, however, each of Delaware's 6153 slaves was counted as three-fifths of a person, as noted at the beginning of this section.*

KEY CONCEPT

A desirable trait of any apportionment method is to stay within quota, meaning that the final apportionment for each state would be within one of the quota. That is, the final apportionment should be the quota either rounded down or rounded up. Apportionment methods that don't stay within quota are said to violate quota.

Return to Hamilton's method: Paradoxes

From 1850 through the end of the nineteenth century, Congress nominally required that Hamilton's method be used to apportion representatives. In practice, Congress would begin with apportionment according to Hamilton's method and then add members not following the method. Because presidential elections are determined by the Electoral College, whose membership by state depends on apportionment, Samuel Tilden would have won the presidency over Rutherford B. Hayes in 1876 had Hamilton's method been consistently used. In three situations, however, Hamilton's method presents troubling and paradoxical results.

The first problem is an apparent inconsistency in allocating members: It is possible for a state to lose a member if the House size is increased by one member! A similar observation was made based on the census of 1870, but it was the 1880 census that brought the issue to the political forefront. While doing some calculations to determine the House size, the chief clerk of the Census Office noticed the *Alabama paradox*: For a House size of 299, Alabama would be allocated 8 representatives. For a House size of 300, Alabama would be allocated only 7 representatives. The census data for 1880 and the calculations made by the clerk appear in **Table 8.2**. They show that if the House size increased from 299 to 300, then Illinois and Texas would each gain a seat and Alabama would lose a seat.

EXAMPLE 8.20 Exploring the Alabama paradox: 1880 census

a. For a House size of 299, determine the ideal district size and verify the quotas for Alabama and Illinois.

b. How do the fractional parts of the quotas for Alabama and Illinois compare for 299 members?

c. How do the fractional parts of the quotas for Alabama and Illinois compare for 300 members?

d. Use parts b and c to give a plausible explanation for the paradox.

SOLUTION

a. The table on the next page shows that the total population is 49,371,340. For a House size of 299, the ideal district size is the total population divided by 299, which is 49,371,340/299 or about 165,121.539. To verify the quotas, we divide the state population by the ideal district size. For Alabama, the quota is therefore $1,262,505/165,121.539 = 7.646$ to three decimal places. For Illinois, we have a quota of 3,077,871/165,121.539 or about 18.640.

b. With 299 members, Alabama's fractional part is 0.646, and the fractional part for Illinois is 0.640. Alabama has the larger fractional part.

c. The table shows that with 300 members, the fractional part for Alabama is 0.671 and the fractional part for Illinois is 0.702. Illinois has the larger fractional part.

continued on page 584

TABLE 8.2 Alabama Paradox: 1880 Census, Hamilton's Method

		House size of 299				House size of 300			
State	Population	Quota	Initial seats	Added seats	Total seats	Quota	Initial seats	Added seats	Total seats
Alabama	1,262,505	7.646	7	1	8	7.671	7		7
Arizona	802,525	4.860	4	1	5	4.876	4	1	5
California	864,694	5.237	5		5	5.254	5		5
Colorado	194,327	1.177	1		1	1.181	1		1
Connecticut	622,700	3.771	3	1	4	3.784	3	1	4
Delaware	146,608	0.888	1		1	0.891	1		1
Florida	269,493	1.632	1		1	1.638	1		1
Georgia	1,542,180	9.340	9		9	9.371	9		9
Illinois	3,077,871	18.640	18		18	18.702	18	1	19
Indiana	1,978,301	11.981	11	1	12	12.021	12		12
Iowa	1,624,615	9.839	9	1	10	9.872	9	1	10
Kansas	996,096	6.033	6		6	6.053	6		6
Kentucky	1,648,690	9.985	9	1	10	10.018	10		10
Louisiana	939,946	5.692	5	1	6	5.711	5	1	6
Maine	648,936	3.930	3	1	4	3.943	3	1	4
Maryland	934,943	5.662	5	1	6	5.681	5	1	6
Massachusetts	1,783,085	10.799	10	1	11	10.835	10	1	11
Michigan	1,636,937	9.914	9	1	10	9.947	9	1	10
Minnesota	780,773	4.728	4	1	5	4.744	4	1	5
Mississippi	1,131,597	6.853	6	1	7	6.876	6	1	7
Missouri	2,168,380	13.132	13		13	13.176	13		13
Nebraska	452,402	2.740	2	1	3	2.749	2	1	3
Nevada	62,266	0.377	1		1	0.378	1		1
New Hampshire	346,991	2.101	2		2	2.108	2		2
New Jersey	1,131,116	6.850	6	1	7	6.873	6	1	7
New York	5,082,871	30.783	30	1	31	30.886	30	1	31
North Carolina	1,399,750	8.477	8		8	8.505	8		8
Ohio	3,198,062	19.368	19		19	19.433	19		19
Oregon	174,768	1.058	1		1	1.062	1		1
Pennsylvania	4,282,891	25.938	25	1	26	26.025	26		26
Rhode Island	276,531	1.675	1	1	2	1.680	1	1	2
South Carolina	995,577	6.029	6		6	6.050	6		6
Tennessee	1,542,359	9.341	9		9	9.372	9		9
Texas	1,591,749	9.640	9		9	9.672	9	1	10
Vermont	332,286	2.012	2		2	2.019	2		2
Virginia	1,512,565	9.160	9		9	9.191	9		9
West Virginia	618,457	3.745	3	1	4	3.758	3	1	4
Wisconsin	1,315,497	7.967	7	1	8	7.993	7	1	8
Total	49,371,340	299	279	20	299	300	282	18	300

d. From parts b and c, we see that increasing the size of the House from 299 to 300 causes the fractional part of Alabama's quota to become smaller than the fractional part for Illinois. Because Hamilton's method allocates leftover seats according to the size of that fractional part, it seems reasonable that Illinois could gain a seat and Alabama could lose a seat in the process.

TRY IT YOURSELF 8.20

For a House size of 300, determine the ideal district size and verify the quotas for Alabama and Illinois.

The answer is provided at the end of this section.

Another troubling consequence of Hamilton's method is the *population paradox*. In the 1900 census, Virginia's population was 1,854,184 and growing at 1.07% per year, whereas Maine's population was 694,466 and growing at 0.67% per year. The quota for Virginia was 9.599, which Hamilton's method rounded up to 10, and Maine's quota was 3.595, which was rounded down to 3. Note how close the fractional parts are: 0.599 and 0.595. Because the nation as a whole was growing faster than either state and Virginia had a larger population, after one year Virginia's fractional part of 0.599 would slip below that of Maine. Then Maine's quota could be rounded up and Virginia's rounded down. This could happen despite the fact that Virginia was growing faster than Maine.

A third problematic consequence of Hamilton's method is the *new states paradox*. When Oklahoma became a state in 1907, 5 seats were added to the House. This was the number of seats to which Oklahoma would be entitled, based on its population then. Oklahoma was indeed awarded 5 seats, and one would assume that the number of representatives for the other states would be unchanged. If Hamilton's method had been used, however, it would have changed the number of representatives for Maine and New York.

These paradoxes were sufficiently troubling to Congress that in time new methods were sought that would avoid real and perceived problems with Hamilton's method.

More adjusted divisor methods: Adams and Webster

Problems with Hamilton's method led Congress to revive Jefferson's method and modify it in the hopes of making it better. Any system for apportionment that rounds the quotas by some method and then adjusts the divisor until the House is of the correct size is an *adjusted divisor method*. Jefferson's method is one of several adjusted divisor methods. All of the methods follow the same steps as in Summary 8.11, except that in Step 3 they vary in the method by which they round. Jefferson's method rounds all the quotas down (but not less than 1).

In the 1820s and 1830s, the U.S. population was moving westward, adding states west of the Appalachians, and the proportion of the population living in New England decreased. These and other factors inspired John Quincy Adams to reexamine Jefferson's method. Adams believed that Jefferson's rounding down of quotas left some people unrepresented and failed to meet the constitutional intent that all the people be represented. He suggested an adjusted divisor method in which all quotas are rounded *up*. We noted earlier that Jefferson's method favored larger states, and it comes as no surprise that Adams's method favors smaller states. Like Jefferson's method, Adams's method could violate quota. Adams's method was never formally used by Congress.

Jefferson's method always rounds down and Adams's method always rounds up. A third adjusted divisor method was proposed by Daniel Webster. Webster's method rounds to the nearest whole number (rather than always up or always down) before adjusting the divisor. This method was used by Congress with the 1840 census and revived for use in the censuses of 1900, 1910, and 1930.

"The latest polls show that you need to fool more of the people more of the time."

SUMMARY 8.12 Three Adjusted Divisor Methods for Apportionment

1. We calculate ideal district size by dividing the total population by the size of the House. This is used as the first divisor.

2. We calculate the quota for each state by dividing its population by the divisor.

3. We round each quota to a whole number as follows:

Jefferson's method: We round down (but not less than 1).

Adams' method: We round up.

Webster's method: We round to the nearest whole number (up if the fractional part is 0.5 or greater and down otherwise, but not less than 1).

Then we sum the resulting rounded quotas.

4. If the sum from Step 3 is larger than the size of the House, we increase the divisor and repeat Steps 2 and 3. If the sum is too small, we decrease the size of the divisor and repeat Steps 2 and 3. We continue this process until a divisor is found for which the sum of the rounded quotas is the House size. This is the final apportionment.

EXAMPLE 8.21 Comparing methods: Montana's loss is Florida's gain

According to the 2010 census, this was the population of four states:

State	Population
Montana	994,416
Alaska	721,523
Florida	18,900,773
West Virginia	1,859,815
Total	**22,476,527**

Congress allocated 32 seats in total to these states.

a. Calculate the ideal district size and each state's quota. Keep three places beyond the decimal point.

b. Calculate the apportionment of the 32 seats according to Hamilton's method.

c. Calculate the apportionment of the 32 seats according to Jefferson's method.

d. Calculate the apportionment of the 32 seats according to Adams's method.

e. Calculate the apportionment of the 32 seats according to Webster's method.

SOLUTION

a. The ideal district size is the total population divided by the number of seats, or $22{,}476{,}527/32 = 702{,}391.469$. The quota for each state is calculated by dividing the state's population by this divisor:

State	Population	Quota
Montana	994,416	1.416
Alaska	721,523	1.027
Florida	18,900,773	26.909
West Virginia	1,859,815	2.648
Total	22,476,527	32

b. For Hamilton's method, we make an initial allocation by rounding the quotas down. That gives an initial allocation of 30 seats. The remaining 2 seats are given to the states having the largest fractional parts of their quotas—Florida and West Virginia. The final allocation is shown in the last column of the accompanying table.

State	Population	Quota	Quota rounded down: Initial seats	Fractional part of quota	Added seats	Final seats
Montana	994,416	1.416	1	0.416		1
Alaska	721,523	1.027	1	0.027		1
Florida	18,900,773	26.909	26	0.909	1	27
West Virginia	1,859,815	2.648	2	0.648	1	3
Total	22,476,527	32	30		2	32

According to Hamilton's method, Montana should get 1 seat, Alaska 1 seat, Florida 27 seats, and West Virginia 3 seats.

c. For Jefferson's method, we begin as in part b by rounding the quota down as an initial allocation. Because the total from the initial allocation is less than 32, the total number of seats, we adjust the divisor to a smaller number. Using an adjusted divisor of 700,000 and calculating the corresponding quotas and then rounding down, we obtain 31 seats, so we try an even smaller divisor of 675,000.

State	Population	Quota using divisor 702,391.469	First try: Quota rounded down	Second try: Decrease divisor to 700,000, round down	Another try: Decrease divisor to 675,000, round down
Montana	994,416	1.416	1	1	1
Alaska	721,523	1.027	1	1	1
Florida	18,900,773	26.909	26	27	28
West Virginia	1,859,815	2.648	2	2	2
Total	22,476,527	32	30	31	32

According to Jefferson's method, Montana should get 1 seat, Alaska 1 seat, Florida 28 seats, and West Virginia 2 seats.

d. For Adams's method, we make an initial allocation by rounding the quotas up. Because the total from the initial allocation is more than 32 (see the accompanying table), the total number of seats, we adjust the divisor to a larger number. Using an adjusted divisor of 750,000, calculating the corresponding quotas, and then rounding up, we obtain 32 seats, the desired number. The last column of the table below shows this result.

State	Population	Quota using divisor 702,391.469	First try: Quota rounded up	Another try: Increase divisor to 750,000, round up
Montana	994,416	1.416	2	2
Alaska	721,523	1.027	2	1
Florida	18,900,773	26.909	27	26
West Virginia	1,859,815	2.648	3	3
Total	22,476,527	32	34	32

According to Adams's method, Montana should get 2 seats, Alaska 1 seat, Florida 26 seats, and West Virginia 3 seats.

e. For Webster's method, we make an initial allocation by rounding the quotas to the nearest whole number; that is, we round down if the fractional part of the quota is less than 0.5 and round up otherwise. Because the total from the initial allocation is 32 (see the accompanying table), no further adjustments are needed: According to Webster's method, Montana should get 1 seat, Alaska 1 seat, Florida 27 seats, and West Virginia 3 seats.

State	Population	Quota using divisor 702,391.469	First try: Quota rounded to nearest whole number
Montana	994,416	1.416	1
Alaska	721,523	1.027	1
Florida	18,900,773	26.909	27
West Virginia	1,859,815	2.648	3
Total	22,476,527	32	32

Note that the four methods used in Example 8.21 give three different answers.

The Huntington–Hill method

The Huntington–Hill method is an adjusted divisor method that deserves special attention because it is the one in use today. Huntington–Hill follows the steps of the other adjusted divisor methods (Jefferson, Adams, Webster), but rounding is done using the *geometric mean*: If n is the whole-number part of the quotient (initially the quota), the geometric mean we use is $\sqrt{n(n+1)}$. We round up if the quotient is at least this large; otherwise, we round down. **Table 8.3** on the next page shows some values of the geometric mean.

For example, for $n = 1$, the geometric mean is about 1.414. This means that a quotient of 1.433 (whose whole-number part is 1) is rounded up to 2 by Huntington–Hill. In contrast, the geometric mean for 5 is 5.477. So a quotient of 5.433 is rounded down to 5 by Huntington–Hill. Both quotients have the same fractional part, 0.433, but they are rounded differently. Note that Huntington–Hill gives a benefit to smaller states because it rounds the small quotient 1.433 up, but it rounds the larger quotient 5.433 down.

TABLE 8.3 Geometric Means

Whole number	Geometric mean	Whole number	Geometric mean	Whole number	Geometric mean
1	1.414	11	11.489	21	21.494
2	2.449	12	12.490	22	22.494
3	3.464	13	13.491	23	23.495
4	4.472	14	14.491	24	24.495
5	5.477	15	15.492	25	25.495
6	6.481	16	16.492	26	26.495
7	7.483	17	17.493	27	27.495
8	8.485	18	18.493	28	28.496
9	9.487	19	19.494	29	29.496
10	10.488	20	20.494	30	30.496

EXAMPLE 8.22 Using Huntington–Hill: Four states

Use the Huntington–Hill method to allocate 32 seats to the four states in Example 8.21.

SOLUTION

We begin by showing the quota using the ideal district size of 702,391.469 and the corresponding geometric mean (from Table 8.3). We round down if the quota is less than the geometric mean; otherwise, we round up.

State	Population	Quota using divisor 702,391.469	Geometric mean	First try: Rounded quota
Montana	994,416	1.416	1.414	2
Alaska	721,523	1.027	1.414	1
Florida	18,900,773	26.909	26.495	27
West Virginia	1,859,815	2.648	2.449	3
Total	22,476,527	32		33

In the final column of the table, we see that we have allocated a total of 33 seats—too many. So we use a larger divisor of 710,000.

State	Population	Quota using divisor 702,391.469	Geometric mean	First try: Rounded quota	Second try: Quotient using divisor of 710,000	Rounded quotient
Montana	994,416	1.416	1.414	2	1.401	1
Alaska	721,523	1.027	1.414	1	1.016	1
Florida	18,900,773	26.909	26.495	27	26.621	27
West Virginia	1,859,815	2.648	2.449	3	2.619	3
Total	22,476,527	32		33		32

The last column of the table shows that this allocates a total of 32 seats, which is the required number. So Huntington–Hill awards 1 seat to Montana, 1 to Alaska, 27 to Florida, and 3 to West Virginia.

TRY IT YOURSELF 8.22

Use the Huntington–Hill method to allocate the 10 councilors from Example 8.18.

The answer is provided at the end of this section.

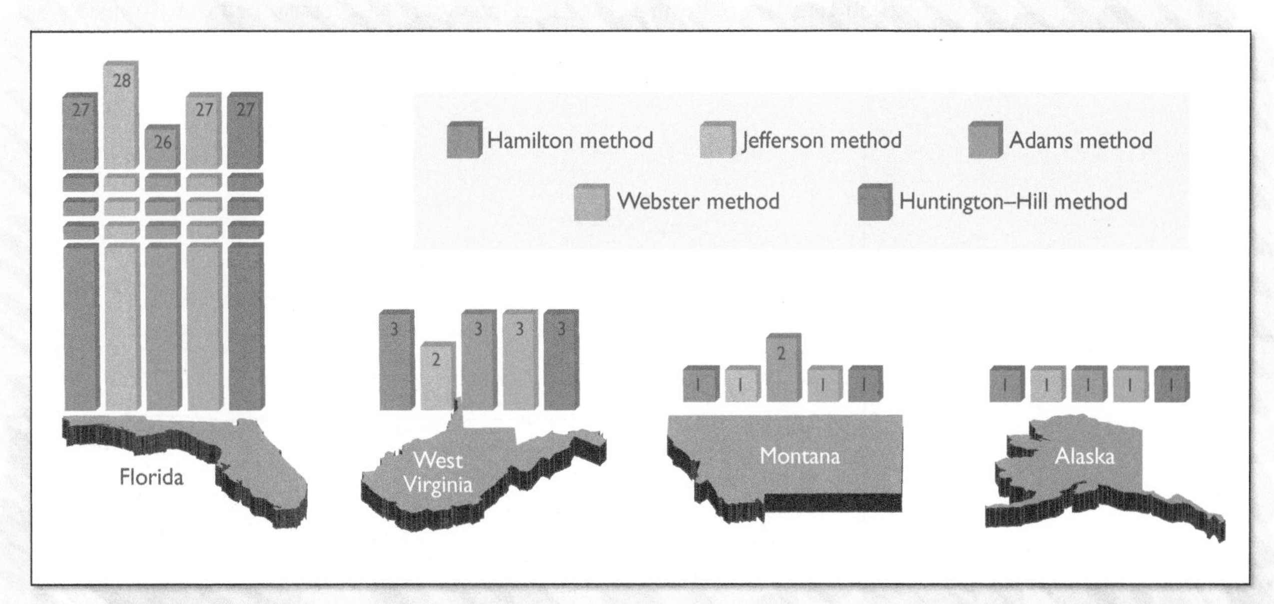

FIGURE 8.1 Results of various apportionment methods.

Figure 8.1 illustrates the differing effects of the five methods from Examples 8.21 and 8.22 on allocation of seats to Montana, Alaska, Florida, and West Virginia. Since each of these methods has historically been used at various times by Congress, these differing effects raise the question, "Which is fair?"

Is it fair? What is the perfect method?

We've seen five different methods for apportionment: Hamilton's method and four adjusted divisor methods. Currently, Congress uses the Huntington–Hill method, which rounds according to the geometric mean. Is this method fair? The answer surely depends on what criteria are used to define "fair." Of these methods, only Hamilton's method never violates quota, and yet it suffers from the Alabama, population, and new states paradoxes. Only adjusted divisor methods avoid the population paradox, but all adjusted divisor methods violate quota.[23] In fact, it can be shown that no method both avoids the population paradox and never violates quota. There is no perfect method.

All adjusted divisor methods violate quota, but by how much? Of all the methods, Webster's violates quota the least: The probability of violating quota is only 0.61 time per 1000 times, that is, about once per 1640 apportionments or once per 16,400 years on average (because each apportionment occurs once every 10 years).[24]

From the very beginning, there has been tension between smaller states and larger states because the different methods differ in their effect, depending on the size of the state. Michel Balinski and H. Peyton Young have analyzed the period from 1790 to 1970 and calculated the bias of the adjusted divisor methods toward smaller states.[25] In **Figure 8.2** on the next page, the vertical axis indicates percentage bias toward smaller states.

As we can see from Figure 8.2, the Huntington–Hill method (called the Hill method in the graph) has a slight bias of about 3% in favor of smaller states, whereas Webster's method is the closest to being unbiased. Congress continues to use the

[23] *See Balinski and Young,* op. cit., *p. 79.*

[24] *See Balinski and Young,* op. cit., *p. 81.*

[25] *See Balinski and Young,* op. cit., *p. 75.*

Huntington–Hill method, despite lawsuits filed by states after both the 1990 and 2000 censuses. Although the Supreme Court has determined that Congress has the discretion to choose an apportionment method, we can anticipate attempts to change apportionment methods in the future.

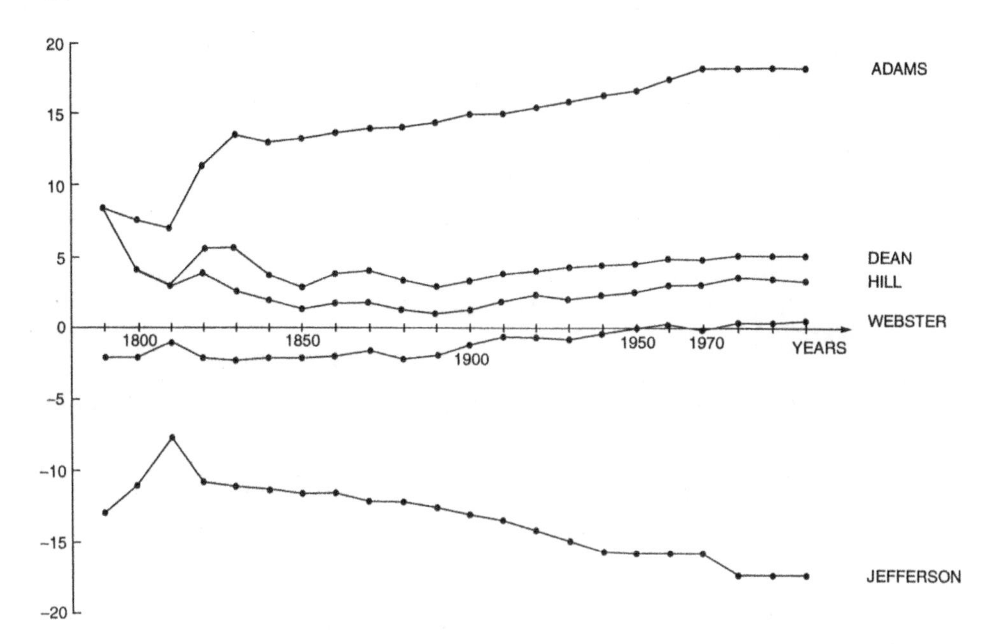

FIGURE 8.2 Biases of adjusted divisor methods.

WHAT DO YOU THINK?

Violate quota? What does it mean to *violate quota*, and why is it important for a method to stay within quota?

Questions of this sort may invite many correct responses. We present one possibility: The quota for a state is the population divided by ideal district size. In ideal conditions it represents the total number of representatives a state should get. Unfortunately, the quota is almost never a whole number, and the problem is what to do with the fractional part of the quota. Many solutions have been proposed, but most agree that the number of representatives a state gets should be no more than 1 away from its quota. If a state's number of representatives is outside this range, then quota is said to have been violated. When quota is violated, a state is either getting fewer representatives than it deserves, or it gets an unfair advantage by being awarded too many.

Further questions of this type may be found in the *What Do You Think?* section of exercises.

Try It Yourself answers

Try It Yourself 8.18: Apportioning by Hamilton's method: City councilors North and East each get 3, West gets 12, and South gets 2.

Try It Yourself 8.19: Apportioning by Jefferson's method: City councilors Using a divisor of around 3500, we see that North, East, and South each get 1 councilor and West gets 5 councilors.

Try It Yourself 8.20: Exploring the Alabama paradox: 1880 census The ideal district size is 49,371,340/300 or 164,571.133. The Alabama quota is 1,262,505/164,571.133 = 7.671 to three places. The Illinois quota is 3,077,871/164,571.133 = 18.702 to three places.

Try It Yourself 8.22: Using Huntington–Hill: City councilors 1 each to North and South, 2 to East, and 6 to West.

Exercise Set 8.4

Test Your Understanding

1. Explain how the ideal district size is calculated.

2. A state's quota is determined by **a.** the state's population divided by the ideal district size **b.** an act of Congress **c.** the state's land mass **d.** a vote of the people.

3. The fundamental issue with apportionment is **a.** the relationship between large states and small states **b.** how the fractional parts of a state's quota should affect allocation of representatives **c.** the ideal district size of some states is too large **d.** how representatives are allocated when a new state joins the Union.

4. Explain in your own words how Hamilton's method works.

5. Explain how Jefferson's method works.

6. The Alabama paradox is associated with Hamilton's method. The paradox is that **a.** increasing a state's population can decrease its number of representatives **b.** increasing the House size can decrease a state's number of representatives **c.** the adjusted divisor becomes zero **d.** several of the quota fractions may be the same.

7. In the context of apportionment, what does it mean to violate quota?

8. The apportionment method used today is called ________.

Problems

9. **District size in 1790.** The U.S. population in 1790 was 3,615,920, and the House size was set at 105. What was the ideal congressional district size in 1790? Use three decimal places in your answer.

10. **District size in 2010.** In 2010, there were 435 members of the House of Representatives. The U.S. population in 2010 was 309,183,463. What was the ideal congressional district size in 2010? Use three decimal places in your answer.

11. **New York's quota in 1790.** The district size in 1790 was 34,437. The population of New York in 1790 was 331,589. What was New York's quota in 1790? Use two decimal places in your answer.

12. **Virginia's quota in 1790.** The district size in 1790 was 34,437. The population of Virginia in 1790 was 630,560. What was Virginia's quota in 1790? Use two decimal places in your answer.

13. **California's quota in 2010.** In 2010, the U.S. population was 309,183,463. There were 435 members of the House of Representatives. The population of California in 2010 was 37,341,989. What was California's quota in 2010? Use two decimal places for your answer.

14. **Minnesota's quota in 2010.** In 2010, the U.S. population was 309,183,463. There were 435 members of the House of Representatives. The population of Minnesota in 2010 was 5,314,879. What was Minnesota's quota in 2010? Use two decimal places for your answer.

15. **An interesting House size.** The U.S. Constitution states that the minimum size of a congressional district is 30,000 people. The U.S. population in 2020 was about 330 million. If each district size were 30,000 people, how many House members would we have had in 2020? Round your answer to the nearest whole number.

16. **Washington's veto.** President Washington vetoed the Hamilton apportionment bill. Part of his reasoning involved the state of Delaware. Hamilton's plan would have awarded Delaware 2 House seats. The population of Delaware in 1790 was 55,540.

a. With 2 House seats, what would have been the size of a congressional district in Delaware?

b. Compare your answer to part a with the constitutional minimum for the number of people in a congressional district.

17. **Hamilton's method and ties.** One problem with Hamilton's method could arise if there are ties in the quotas. (This would be highly unlikely to happen in practice.) Suppose a country consisted of State 1 with population 140, State 2 also with population 140, and State 3 with population 720. That is a total population of 1000. Suppose the House size is 10.

a. What is the ideal district size?

b. Calculate the quota for each of the three states.

c. Explain the difficulty Hamilton's method has in allotting the 10 House seats.

18. **Hamilton's method and small states.** When a state's population is too small, Hamilton's method (if unaltered) may not assign it any representatives. Suppose, for example, that the country consists of two states: State 1 with a population of 9 and State 2 with a population of 91. There are 5 seats to be assigned. How does Hamilton's method assign these 5 seats? It should be noted that Hamilton's method was later modified to account for this difficulty.

19–24. **A small town.** *Use the following information for Exercises 19 through 24. A small town elects a total of 12 board members from four districts: North, South, East, and West. The populations of each district are shown below.*

District	Population
North	100
South	150
East	200
West	700

19. Find the ideal district size for this small town. Use two decimal places in your answer.

20. *This exercise uses the result of Exercise 19.* Find the distribution of the 12 board members using Hamilton's method.

21. *This exercise uses the result of Exercise 19.* Find the distribution of the 12 board members using Jefferson's method.

22. *This exercise uses the result of Exercise 19.* Find the distribution of the 12 board members using Adams's method.

23. *This exercise uses the result of Exercise 19.* Find the distribution of the 12 board members using Webster's method.

24. *This exercise uses the result of Exercise 19.* Find the distribution of the 12 board members using the Huntington–Hill method.

25–30. A larger town. *Use the following information for Exercises 25 through 30. A town elects a total of 15 board members from four districts: North, South, East, and West. The populations of each district are shown below.*

District	Population
North	145
South	150
East	326
West	735

25. What is the ideal district size for this town? Use one decimal place in your answer.

26. *This exercise uses the result of Exercise 25.* Find the distribution of the 15 board members using Hamilton's method.

27. *This exercise uses the result of Exercise 25.* Find the distribution of the 15 board members using Adams's method.

28. *This exercise uses the result of Exercise 25.* Find the distribution of the 15 board members using Jefferson's method.

29. *This exercise uses the result of Exercise 25.* Find the distribution of the 15 board members using the Huntington–Hill method.

30. *This exercise uses the result of Exercise 25.* Find the distribution of the 15 board members using Webster's method.

31. Violating quota. This exercise shows how Jefferson's method can violate quota. The following table shows the population for a nation consisting of State 1, State 2, and State 3. The quota for a House size of 25 is also given.

State	Population	Quota
State 1	200	3.571
State 2	200	3.571
State 3	1000	17.857
Total	**1400**	**25**

a. Show that 52 works as a divisor for Jefferson's method.

b. For which state is quota violated?

32–34. The trouble with a new state. *Use the following information for Exercises 32 through 34. Suppose that Oklahoma, Texas, Arkansas, and New Mexico form a Board of Representatives to attract business to the region. There are to be 125 members of the board, and the membership from each state was determined by its population in 2009. The populations of these states are given in the table below.*

State	Population
Oklahoma	3,687,050
Texas	24,782,302
Arkansas	2,889,450
New Mexico	2,009,971
Total	**33,368,773**

32. Find the ideal district size for the coalition. Use three decimal places in your answer.

33. *This exercise uses the result of Exercise 32.* Determine the distribution of board members among the four states if Hamilton's method is used.

34. Kansas, which has a population of 2,805,747, joins the coalition. Because the population of Kansas is close to that of Arkansas, it is decided that 11 new board members should be added, for a new total membership of 136.

a. Determine the distribution of board members among the five states if the 136-member board is selected using Hamilton's method.

b. Do Kansas and Arkansas have the same number of board members?

c. *This part uses the result of Exercise 33.* Which of the four original states gained a board member with the addition of Kansas to the coalition?

35. Dean's method. Dean's method is an adjusted divisor method. It follows the steps of the other adjusted divisor methods, but rounding is done using the *harmonic mean*: If n is the whole-number part of the quotient (initially the quota), the harmonic mean we use is $\frac{n(n+1)}{n+1/2}$. We round up if the quotient is at least this large; otherwise, we round down. The following table provides a few helpful values for the harmonic mean:

Whole number	Harmonic mean
1	1.333
2	2.400
26	26.491

What allocation does Dean's method give for the four states in Example 8.21?

The following exercises are designed to be solved using technology such as calculators or computer spreadsheets and are suitable for group work.

36–39. The 1790 Congress. *The table below gives state populations in 1790, when the size of the House was 105. The ideal district size in 1790 was 34,437, and the table shows the quotas to three decimal places. This information is used in Exercises 36 through 39.*

State	Population	Quota
Connecticut	236,841	6.878
Delaware	55,540	1.613
Georgia	70,835	2.057
Kentucky	68,705	1.995
Maryland	278,514	8.088
Massachusetts	475,327	13.803
New Hampshire	141,822	4.118
New Jersey	179,570	5.214
New York	331,589	9.629
North Carolina	353,523	10.266
Pennsylvania	432,879	12.570
Rhode Island	68,446	1.988
South Carolina	206,236	5.989
Vermont	85,533	2.484
Virginia	630,560	18.311
Total	**3,615,920**	**105**

36. Find each state's representation for 1790 using Hamilton's apportionment method.

37. What allocation would Jefferson's method give for 1790?

38. What allocation would Adams's method give for 1790?

39. What allocation would Webster's method give for 1790?

40. **Jefferson's method for 1820: violating quota.** The population figures for 1820 are given in the following table.

   a. Use Jefferson's method to calculate the apportionment of 213 representatives for 1820.

   b. Verify that quota is violated for New York.

   c. For what other states is quota violated?

State	Population	State	Population
New York	1,368,775	Connecticut	275,208
Pennsylvania	1,049,313	New Jersey	274,551
Virginia	895,303	New Hampshire	244,161
Ohio	581,434	Vermont	235,764
North Carolina	556,821	Indiana	147,102
Massachusetts	523,287	Louisiana	125,779
Kentucky	513,623	Alabama	111,147
South Carolina	399,351	Rhode Island	83,038
Tennessee	390,769	Delaware	70,943
Maryland	364,389	Missouri	62,496
Maine	298,335	Mississippi	62,320
Georgia	281,126	Illinois	54,843
		Total	**8,969,878**

Writing About Mathematics

41. **Your state.** Find out how many representatives your state has in the House and make a list of the representatives, along with his or her party affiliation.

42. **Calculating quota.** Determine the current ideal district size. Then find your state's current population and its quota. Is quota violated for your state?

43. **Alexander Hamilton.** Write a report on the life of Alexander Hamilton and his contributions to the early development of the United States.

44. **Thomas Jefferson.** One of the best known of our Founding Fathers is Thomas Jefferson. Report on his life and explain why he is so highly regarded by Americans.

What Do You Think?

45. **Ideal.** How is the ideal congressional district size determined? What could Congress do to adjust this number? **Exercise 9 is relevant to this topic.**

46. **Quota.** How is the congressional quota for a state determined? If the ideal district size and each state's quota are not rounded, what is the sum of the state quotas? **Exercise 11 is relevant to this topic.**

47. **Roots.** Which apportionment method uses square roots? Explain in terms of the rounding procedure why this method gives some advantage to smaller states.

48. **Fair?** List some factors you would consider in determining whether a method for apportionment is fair.

49. **Washington's veto.** The very first presidential veto was cast by George Washington when he vetoed the Hamilton apportionment bill. Do some research to find out why Washington vetoed the bill. Some of the reasons are presented in this section. **Exercise 16 is relevant to this topic.**

50. **Paradox.** Explain where the paradox lies in each of the three paradoxes mentioned in this section (Alabama, population, new states). **Exercise 34 is relevant to this topic.**

51. **Census.** Sketch the history of the census, including the way slaves were counted.

52. **Other applications.** Do the ideas of this section apply to situations other than choosing representatives? If so, give some practical examples.

53. **How it's done today.** Explain in your own words how the U.S. House of Representatives is apportioned today. Do you think the method is fair?

CHAPTER SUMMARY

This chapter revolves around questions of how a group should make decisions. How to measure voting power and how to evaluate voting systems are two topics from the political process. Other situations in which making a decision is difficult arise in the context of equitable division, such as dividing an inheritance. One important division problem is the apportionment of delegates to the U.S. House of Representatives among the 50 states. In this chapter, we consider all of these topics.

Measuring voting power: Does my vote count?

In many situations, voters or voting blocs control multiple votes. Intuitively, it would seem that in such situations a voter's power is directly proportional to the number of votes controlled. But *voting power* is more complicated than may appear at first blush. Voting power is hard to define, but we discuss two common methods of measuring it.

The number of votes required to decide the outcome of an election is the *quota*. Often the quota is a simple majority, which requires more than half of the total number of votes cast.

Consider the case in which Voter A has 20 votes, Voter B has 20 votes, and Voter C has 1 vote. It appears on the surface that Voters A and B have more power than Voter C. But if a simple majority of 21 votes is required, then any pair of the three voters can combine to win an election regardless of how the third votes. One might reasonably claim that the three voters actually have equal power.

One measure of voting power is the *Banzhaf power index*. A coalition of voters is a *winning coalition* if it controls sufficient votes to determine the outcome of an election. A voter in a winning coalition is *critical* for that coalition if his or her absence would prevent the coalition from being a winner. We summarize the information for the preceding example in a *coalition table*.

Number of votes			Total	Winning	Critical
20	20	1	votes	coalition?	voters
A	B	C	41	Yes	None
A	B		40	Yes	A, B
A		C	21	Yes	A, C
A			20	No	Not applicable
	B	C	21	Yes	B, C
	B		20	No	Not applicable
		C	1	No	Not applicable

The Banzhaf index of a voter is the number of times that voter is critical in a winning coalition divided by the total number of instances of critical voters. So the Banzhaf index of each of the three voters in the example is $2/6 = 1/3$ or about 33.33%. This indicates that they, indeed, have equal voting power.

The *Shapley-Shubik power index* is another measure of voting power. If voters cast their votes in order, then one specific voter, the *swing voter*, may cast a vote that lifts the total to the quota. The Shapley–Shubik power index for a voter is the fraction of all possible orderings in which that voter is the swing. For the preceding example, we list the possible orderings of the three voters and the swing voter.

Voting order			Swing
A	B	C	B
A	C	B	C
B	A	C	A
B	C	A	C
C	A	B	A
C	B	A	B

Each voter is the swing voter in 2 of the 6 orderings, so the Shapley–Shubik index of each is $2/6 = 1/3$, or about 33.33%. This indicates once again that the three voters have equal power.

The Banzhaf and Shapley–Shubik indices are not always equal. But each can point out bias or complexity in a system.

Voting systems: How do we choose a winner?

There are several ways in which a winner is selected in elections that involve more than two candidates, and there may be different winners depending on the voting system used. Among these systems are *plurality* voting, the *top-two runoff*, the *elimination runoff*, and the *Borda count*.

Suppose, for example, that the ranked ballot total for a certain election is

Rank	28 votes	25 votes	24 votes	23 votes
First choice	A	B	C	D
Second choice	D	C	D	C
Third choice	B	D	B	B
Fourth choice	C	A	A	A

For this ballot total, A wins the plurality, B wins the top-two runoff, C wins the elimination runoff, and D wins the Borda count. Each of the four candidates has a reasonable claim to be the "people's choice."

A *Condorcet* winner of an election is a candidate who would beat all others in a head-to-head election. Some elections have a Condorcet winner, and some don't. Even if there is a Condorcet winner, none of the systems described in this chapter can guarantee that the Condorcet winner will win the election.

If the voters prefer one candidate over another, the introduction of a third candidate should not reverse the result. That is the principle of *independence of irrelevant alternatives. Arrow's impossibility theorem* tells us that no voting system (other than a dictatorship) that always selects the Condorcet winner can have this property. One interpretation of this result is that no voting system is perfect, and we should be aware that the voting system may play as large a role as the vote total in selecting the winner.

Fair division: What is a fair share?

The problem of fair division arises in many settings, including divorces and inheritances. The simplest kind of fair division scheme is the *divide-and-choose* procedure for two people. One participant divides and the other chooses the share he or she likes best. Fair division does not necessarily mean equal division.

When several items are to be divided between two parties, the *adjusted winner procedure* may be used. In this scheme, each party assigns a value to each item so that the total value for each party is 100 points. The procedure uses the point distribution to arrive at a fair division. The adjusted winner procedure has some advantages:

1. Both parties receive the same point value.
2. Neither party would be happier if the division were reversed.
3. No other division would make one party better off without making the other worse off.

One difficulty with this scheme is that it may require sharing of items that are not easily divisible.

A division scheme that can be used for multiple parties is the *Knaster procedure*. It is normally applied to one item at a time. In this procedure, each party bids on the item. The highest bidder gets the item but must then contribute cash to be divided among all the parties. The division of the cash takes into account each party's valuation. One advantage of this procedure is that each party gets at least a fixed fraction of its valuation of each item. For example, if there are five parties, then each party gets at least one-fifth of its valuation of each item. However, this procedure may require each party to have relatively large amounts of cash on hand.

In general, fair division is a complex issue with no simple solutions that work in all cases.

Apportionment: Am I represented?

The problem of *apportionment* is concerned with the allocation of representatives for a governing body among several groups. The main focus of this section is on historical approaches to apportionment of seats in the U.S. House of Representatives among the states.

Congress sets the size of the House of Representatives. To find the ideal size of a congressional district, we divide the U.S. population by the size of the House. Then we divide the population of each state by the size of a congressional district to determine that state's proper share of the House, called its *quota*. In general, the quota will not be a whole number, and the varying approaches to apportionment differ in the way they treat the fractional parts.

Hamilton's method initially gives each state the number of representatives corresponding to the whole-number part of its quota. (In practice, this number is never allowed to be 0.) If the sum of the resulting quotas is less than the House size, the leftover seats are distributed, one per state, in order from the highest fractional part down.

For *Jefferson's method* of apportionment, we round each quota down (but not less than 1). If the sum of the initial rounded quotas is larger than the size of the House, we increase the divisor, recalculate the quotas, round, and check the sum again. If the sum of the initial rounded quotas is too small, we decrease the divisor and repeat those steps. We continue this process until a divisor is found for which the sum of the rounded-down quotas (but not less than 1) is the House size.

Other methods of apportionment are similar to Jefferson's in that they are *adjusted divisor methods*. They differ only in the way rounding is done. *Adams's method* rounds the quotas up, *Webster's method* rounds to the nearest whole number, and the *Huntington–Hill method* rounds in terms of the geometric mean. Currently, Congress uses the Huntington–Hill method for apportionment.

There is no perfect method of apportionment. Hamilton's method suffers from the *Alabama*, *population*, and *new states* paradoxes. The final apportionment for a state should be the quota either rounded down or rounded up, thus *staying within quota*. But all adjusted divisor methods fail this test: They *violate quota*.

KEY TERMS

quota, p. 528
winning coalition, p. 529
losing coalition, p. 529
critical voter, p. 530
Banzhaf power index, p. 531
swing voter, p. 534
Shapley–Shubik power index, p. 535
plurality, p. 544
spoiler, p. 544
top-two runoff system, p. 546
elimination runoff system, p. 546
preferential voting systems, p. 546
ranked ballot, p. 546
Borda winner, p. 547
Condorcet winner, p. 552
Condorcet winner criterion, p. 554
independence of irrelevant alternatives, p. 555
divide-and-choose procedure, p. 560
adjusted winner procedure, p. 561
Knaster procedure (or method of sealed bids), p. 565
ideal district size, p. 575
standard divisor, p. 575
state's quota, p. 575
stay within quota, p. 582
violate quota, p. 582

CHAPTER QUIZ

1. Suppose there are three delegates to a small convention. Alfred has 10 votes, Betty has 7 votes, and Gabby has 3 votes. A simple majority wins. Find the quota, make a table listing all of the coalitions and the critical voter in each case, and find the Banzhaf index for each voter.

Answer Quota: 11 votes. We use A for Alfred, B for Betty, and G for Gabby.

Number of votes			Total	Winning	Critical
10	7	3	votes	coalition?	voters
A	B	G	20	Yes	A
A	B		17	Yes	A, B
A		G	13	Yes	A, G
A			10	No	Not applicable
	B	G	10	No	Not applicable
	B		7	No	Not applicable
		G	3	No	Not applicable

Alfred: 3/5; Betty: 1/5; Gabby: 1/5.
If you had difficulty with this problem, see Example 8.2.

2. There are three voters: A with 10 votes, B with 8 votes, and C with 3 votes. A simple majority wins. Find the quota, make a table listing all of the permutations of the voters and the swing voter in each case, and calculate the Shapley–Shubik index for each voter.

Answer Quota: 11 votes.

Order of voters			Swing
A (10)	B (8)	C (3)	B
A (10)	C (3)	B (8)	C
B (8)	A (10)	C (3)	A
B (8)	C (3)	A (10)	C
C (3)	A (10)	B (8)	A
C (3)	B (8)	A (10)	B

A: 1/3; B: 1/3; C: 1/3.
If you had difficulty with this problem, see Example 8.5.

3. Consider the following ranked ballot outcome for 14 voters choosing among three candidates: Alfred, Betty, and Gabby.

Rank	7 voters	4 voters	3 voters
First choice	Alfred	Betty	Gabby
Second choice	Betty	Gabby	Alfred
Third choice	Gabby	Alfred	Betty

Determine the winner under the elimination runoff system and the winner using the Borda count.

Answer Elimination runoff: Alfred; Borda count: Alfred.
If you had difficulty with this problem, see Example 8.9.

4. In an election, there are 15 voters and 3 candidates, say, Abe, Ben, and Condi. The voters' preferences are shown in the accompanying table.

Preferences	7 voters	5 voters	3 voters
First choice	Abe	Ben	Condi
Second choice	Ben	Condi	Abe
Third choice	Condi	Abe	Ben

Is there a Condorcet winner? If so, which candidate is it?

Answer No.
If you had difficulty with this problem, see Example 8.13.

5. Use the adjusted winner method to divide the listed items fairly when the following bids are made:

Michael	Item	Juanita
40	Lakeside cottage	30
30	Penthouse apartment	25
30	Ferrari	45

Answer Michael gets the cottage and 55% of the apartment. Juanita gets the Ferrari and 45% of the apartment.
If you had difficulty with this problem, see Example 8.15.

6. Ann, Beth, Cate, and Dan wish to divide a computer. Use the Knaster procedure to fairly divide the computer with the following bids:

Person	Ann	Beth	Cate	Dan
Bid	\$1200	\$2000	\$1200	\$1600

Answer Ann: \$425; Beth: computer less \$1375; Cate: \$425; Dan: \$525.
If you had difficulty with this problem, see Example 8.17.

7. Our city has four districts: North, East, West, and South. There are required to be 15 city councilors allocated according to population, with at least one from each district. Given the following populations, calculate the quota for each district. (You need keep only one place beyond the decimal point.) Then use Hamilton's method to determine the number of councilors to represent each district.

District	Population
North	4200
East	13,500
West	7600
South	9200
Total	34,500

Answer

District	Quota	Quota rounded down: Initial seats	Fractional part of quota	Added seats	Final seats
North	1.8	1	0.8	1	2
East	5.9	5	0.9	1	6
West	3.3	3	0.3		3
South	4.0	4	0.0		4
Total	15	13		2	15

North should get 2 councilors, East should get 6, West should get 3, and South should get 4.
If you had difficulty with this problem, see Example 8.18.

8. Our city has four districts: North, East, West, and South. There are required to be 15 city councilors allocated according to population, with at least one from each district. Given the following populations, use Jefferson's method to determine the number of councilors to represent each district. (You need keep only one place beyond the decimal point in each quotient.)

District	Population
North	4200
East	13,500
West	7600
South	9200
Total	34,500

Answer

District	Quota using divisor 2300	First try: Quota rounded down	Second try: Decrease divisor to 2000, round down
North	1.8	1	2
East	5.9	5	6
West	3.3	3	3
South	4.0	4	4
Total	15	13	15

North should get 2 councilors, East should get 6, West should get 3, and South should get 4.
If you had difficulty with this problem, see Example 8.19.

THE MILLENIUM

9 Geometry

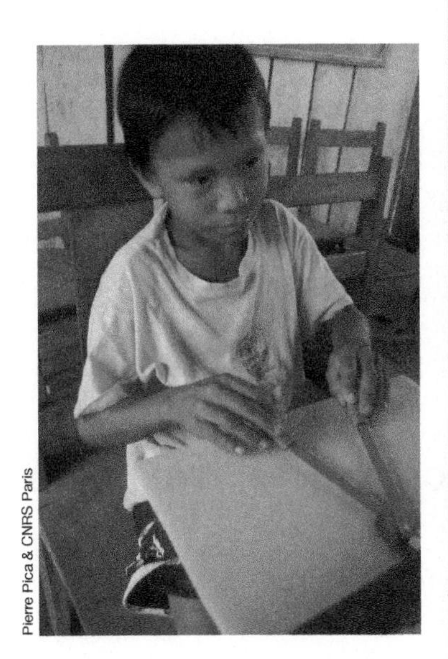

You might be surprised by the following article from the *Proceedings of the National Academy of Sciences*. It indicates that all of us may have an innate knowledge of geometry, even though we are not conscious of it.

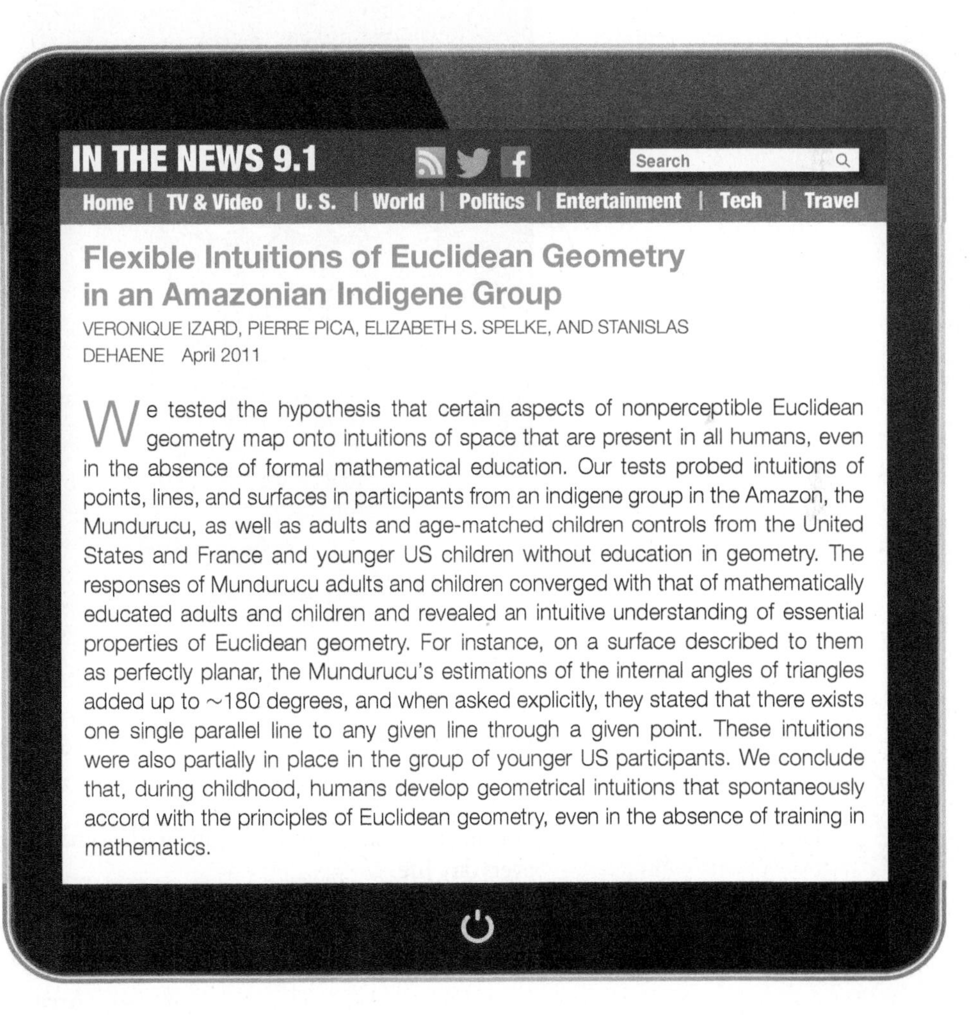

IN THE NEWS 9.1

Search

Home | TV & Video | U. S. | World | Politics | Entertainment | Tech | Travel

Flexible Intuitions of Euclidean Geometry in an Amazonian Indigene Group

VERONIQUE IZARD, PIERRE PICA, ELIZABETH S. SPELKE, AND STANISLAS DEHAENE April 2011

We tested the hypothesis that certain aspects of nonperceptible Euclidean geometry map onto intuitions of space that are present in all humans, even in the absence of formal mathematical education. Our tests probed intuitions of points, lines, and surfaces in participants from an indigene group in the Amazon, the Mundurucu, as well as adults and age-matched children controls from the United States and France and younger US children without education in geometry. The responses of Mundurucu adults and children converged with that of mathematically educated adults and children and revealed an intuitive understanding of essential properties of Euclidean geometry. For instance, on a surface described to them as perfectly planar, the Mundurucu's estimations of the internal angles of triangles added up to ~180 degrees, and when asked explicitly, they stated that there exists one single parallel line to any given line through a given point. These intuitions were also partially in place in the group of younger US participants. We conclude that, during childhood, humans develop geometrical intuitions that spontaneously accord with the principles of Euclidean geometry, even in the absence of training in mathematics.

In this chapter, we examine topics from geometry that are encountered in everyday life. In the first section, we study the measurement of geometric figures. In the next section, we examine how the notion of geometric similarity is related to the idea of proportionality. In the last section, we consider symmetries of shapes and the notion of tiling the plane.

9.1 Perimeter, area, and volume: How do I measure?

TAKE AWAY FROM THIS SECTION
Calculate perimeters, areas, volumes, and surface areas of familiar figures.

Geometric shapes do not concern only mathematicians. They are part of the language of architects, engineers, and designers. Think of the Coliseum in Rome, the spiral-shaped Hirshhorn Museum in Washington, DC, and all the baseball, soccer, basketball, and football fields you visited or played on!

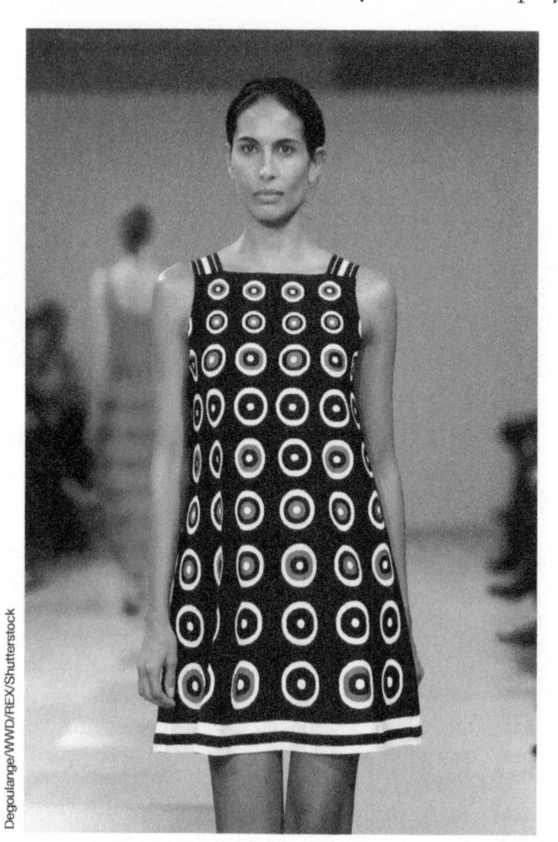
Degoulange/WWD/REX/Shutterstock

Geometric shapes are also prominent in fashion, although their popularity ebbs and flows with the times. Geometric patterns convey a strong statement, usually either a throwback to the 1960s or a postmodern look.

Geometric shapes are our concern in this section. We will see how to apply some basic formulas for measuring various figures in practical settings.

Finding the area: Reminders about circles, rectangles, and triangles

For most of us, geometry begins with some basic formulas that have applications in everyday life.

Circles: A circle consists of all points that are a fixed distance, the *radius*, from a fixed point, the *center*. For a circle of radius r (see **Figure 9.1**):

$$\text{Area} = \pi r^2$$

$$\text{Circumference} = 2\pi r$$

Recall that the number π is approximately 3.14159.

The formula for the circumference of a circle can be rewritten as

$$\frac{\text{Circumference}}{\text{Diameter}} = \pi$$

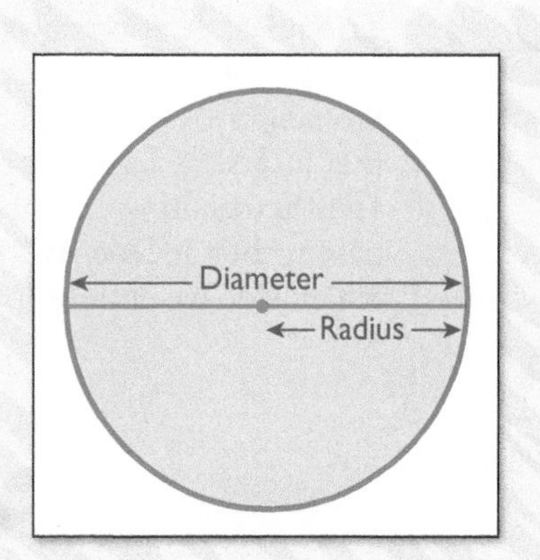

FIGURE 9.1 Radius and diameter of a circle.

This tells us that the ratio of circumference to diameter is always the same no matter which circle you study. The ratio is the same for the equator of Earth and the equator of a baseball (**Figure 9.2**).

FIGURE 9.2 Circumference/Diameter is the same for Earth and a baseball.

Rectangles: A rectangle is a figure with four right angles whose opposite sides are straight and of equal length. (See **Figure 9.3**.) Here are two basic formulas:

$$\text{Area} = \text{Length} \times \text{Width}$$

$$\text{Perimeter} = 2 \times \text{Length} + 2 \times \text{Width}$$

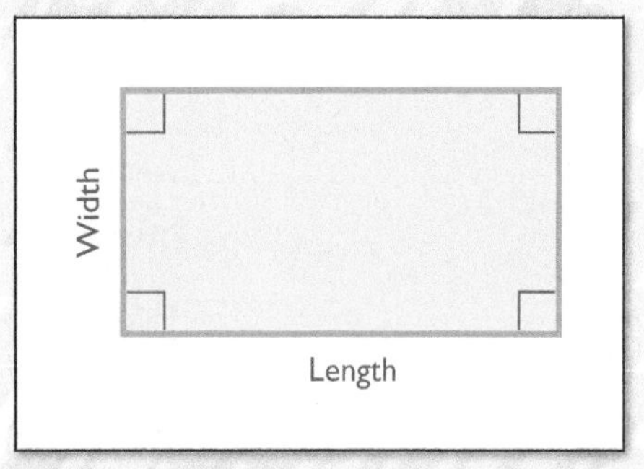

FIGURE 9.3 Opposite sides of a rectangle are of equal length.

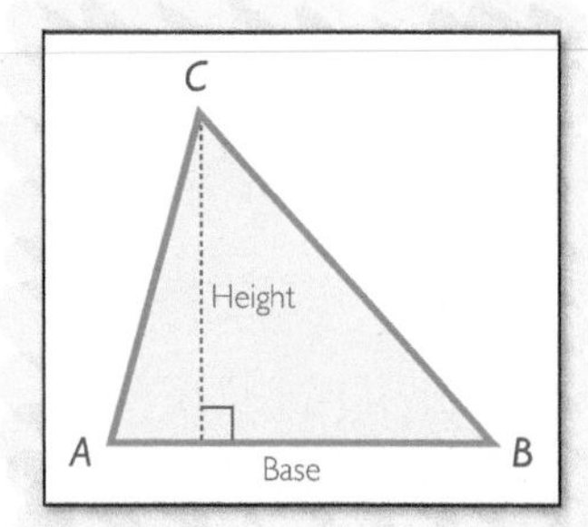

FIGURE 9.4 Base with height inside the triangle.

Triangles: A triangle is a figure with three straight sides. To find the area of a triangle, we select any one of the three sides and label it the *base*. Then we find the *height* by starting at the vertex opposite the base and drawing a line segment that meets the base in a right angle. In **Figure 9.4**, we have chosen the base to be AB, and the resulting height is "inside" the triangle. In **Figure 9.5**, we have chosen the base to be the side AC, which results in a height "outside" the triangle. No matter which side we choose for the base, the area is always given by the formula

$$\text{Area} = \frac{1}{2}\,\text{Base} \times \text{Height}$$

KEY CONCEPT

The perimeter of a geometric figure is the distance around it. In the case of a circle, the perimeter is referred to as its circumference and equals 2π times the radius. In the case of a rectangle, the perimeter is the sum of the lengths of its four sides. In the case of a triangle, the perimeter is the sum of the lengths of its three sides.

The area of a geometric figure measures the region enclosed by the figure. In the case of a circle, the area is π times the radius squared. In the case of a rectangle, the area is the product of the length and the width. In the case of a triangle, the area is one-half the product of the base times the height.

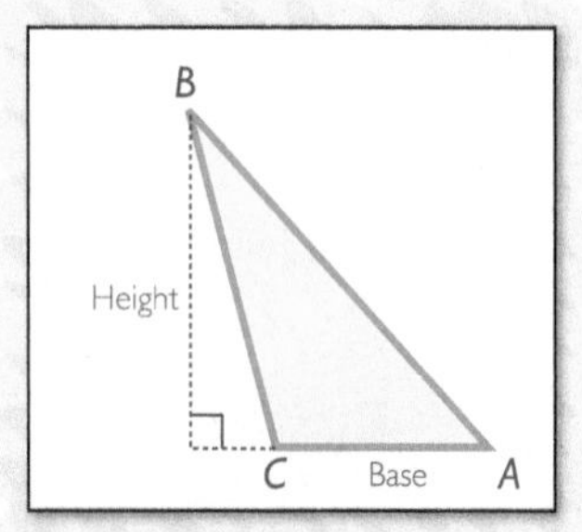

FIGURE 9.5 Different base with height outside the triangle.

Right triangles: A *right triangle* is a triangle with one 90-degree (or right) angle. The traditional names for sides of a right triangle are illustrated in **Figure 9.6.** One of the most familiar of all facts in geometry is the famous theorem of Pythagoras.

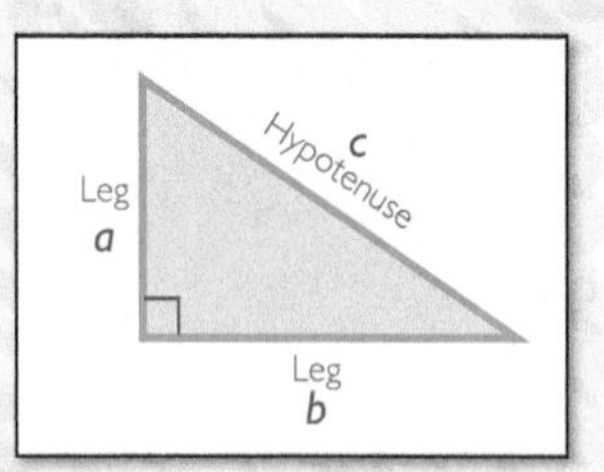

FIGURE 9.6 The sides of a right triangle.

Pythagorean theorem: For the right triangle shown in Figure 9.6,

$$a^2 + b^2 = c^2$$

In this formula, c always represents the *hypotenuse*, which is the side opposite the right angle. The other two sides of a right triangle are the *legs*. The converse of the Pythagorean theorem is true: If a triangle has three sides of lengths a, b, and c such that $a^2 + b^2 = c^2$, then that triangle is a right triangle with hypotenuse c.

KEY CONCEPT

The Pythagorean theorem states that for a right triangle, the square of the length of the hypotenuse c equals the sum of the squares of the lengths of the two legs, a and b, so $a^2 + b^2 = c^2$.

Applications of basic geometry formulas

Now we see some ways these formulas are used in practice.

EXAMPLE 9.1 Using circumference and diameter: Runners

Suppose two runners A and B are side by side, 2 feet apart, on a circular track, as seen in **Figure 9.7**. If they both run one lap, staying in their lanes, how much farther did the outside runner B go than the inside runner A? (Note that no information was given about the diameter of the track.)

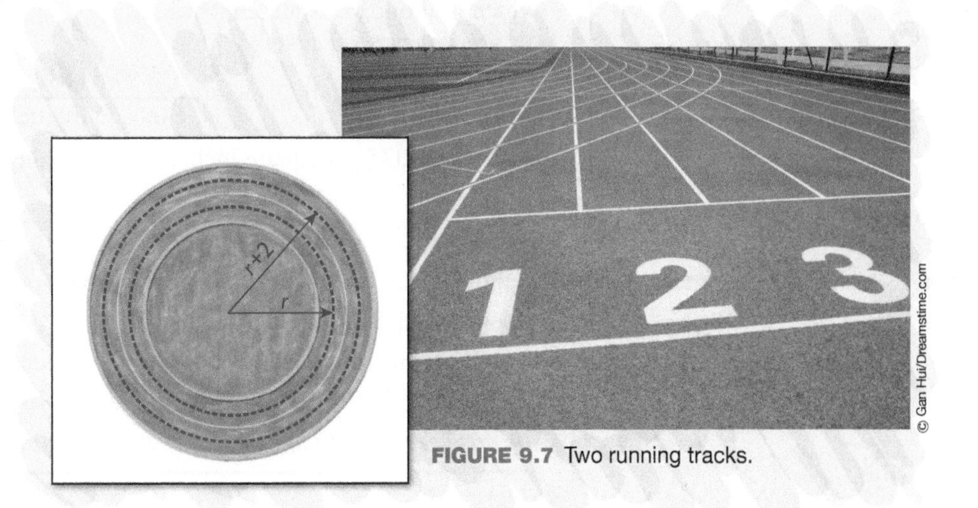

FIGURE 9.7 Two running tracks.

SOLUTION

Let r denote the radius of the inside lane, where A is running. Then $r + 2$ is the radius of the outside lane, where B is running. The distance covered by runner A is

$$\text{Length of inside track} = 2\pi r$$

The distance covered by B is

$$\begin{aligned}\text{Length of outside track} &= 2\pi(r+2)\\ &= 2\pi r + 2\pi \times 2\\ &= \text{Length of inside track} + 4\pi\end{aligned}$$

Therefore, the distances covered by the two runners differ by 4π or about 12.6 feet. At first glance, it may appear that a longer inside track would cause a greater difference in the distance the runners travel. But the difference is the same, 12.6 feet, whether the inside track is 100 yards in diameter or 100 miles in diameter.

TRY IT YOURSELF 9.1

Suppose the runners are side by side, 3 feet apart. How much farther does the outside runner go?

The answer is provided at the end of this section.

Next we use the formula for the area of a circle.

EXAMPLE 9.2 Using area and diameter: Pizza

At a local Italian restaurant, a 16-inch-diameter pizza costs \$15 and a 12-inch-diameter pizza costs \$10. Which one is the better value (i.e., costs less per square inch)?

SOLUTION

First we find the area of each pizza in terms of its radius r using the formula

$$\text{Area} = \pi r^2$$

The radius of the 16-inch-diameter pizza is $r = 8$ inches. Hence, its area is

$$\text{Area of 16-inch-diameter pizza} = \pi \times (8 \text{ inches})^2 = 64\pi \text{ square inches}$$

(which is about 201 square inches). The radius of the 12-inch-diameter pizza is $r = 6$ inches, so its area is

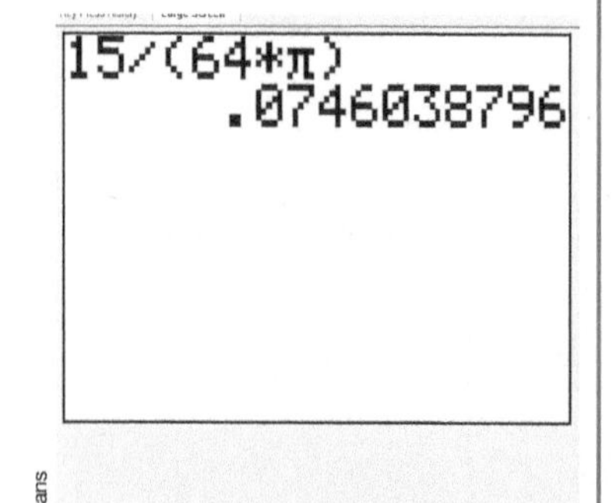

Benny Evans

$$\text{Area of 12-inch-diameter pizza} = \pi \times (6 \text{ inches})^2 = 36\pi \text{ square inches}$$

(which is about 113 square inches).

The larger pizza costs[1]

$$\frac{\$15}{64\pi \text{ square inches}}$$

or about \$0.075 per square inch, and the smaller pizza costs

$$\frac{\$10}{36\pi \text{ square inches}}$$

or about \$0.088 per square inch. Thus, the larger pizza is the better value.

TRY IT YOURSELF 9.2

Pizza from this restaurant comes with a thin cheese ring around its perimeter. How long are the cheese rings of the two pizzas in the preceding example?

The answer is provided at the end of this section.

Now we apply the Pythagorean theorem.

EXAMPLE 9.3 Applying the Pythagorean theorem: Slap shot

Hockey player A is located 5 vertical yards and 7 horizontal yards from the goal, and hockey player B is 8 vertical yards and 3 horizontal yards from the goal. (See **Figure 9.8**.) Which player would have the longer shot at the goal?

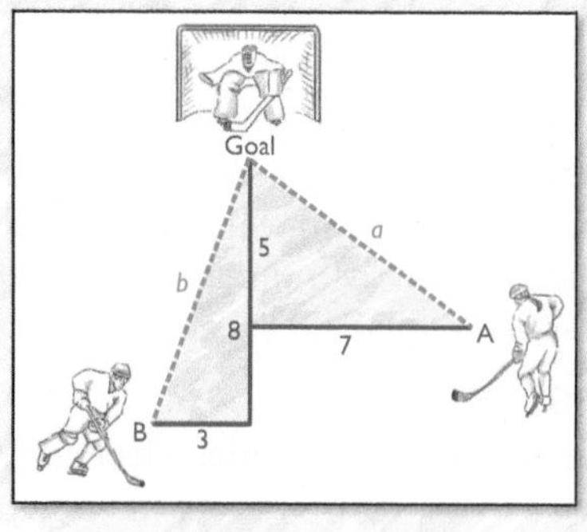

© Iurii Osadchi/Dreamstime.com

FIGURE 9.8 Two hockey players.

[1] *On your calculator, enter this as* $15/(64 * \pi)$.

SOLUTION

The distance for player A is marked a in the figure. It represents the hypotenuse of a right triangle. Similarly, b is the distance for player B, and it represents the hypotenuse of a right triangle. We use the Pythagorean theorem:

$$a^2 = 5^2 + 7^2 = 74$$
$$a = \sqrt{74} \text{ or about 8.6 yards}$$
$$b^2 = 8^2 + 3^2 = 73$$
$$b = \sqrt{73} \text{ or about 8.5 yards}$$

Player A has the longer shot.

TRY IT YOURSELF 9.3

Which player has the longer shot if they are positioned as in **Figure 9.9**?

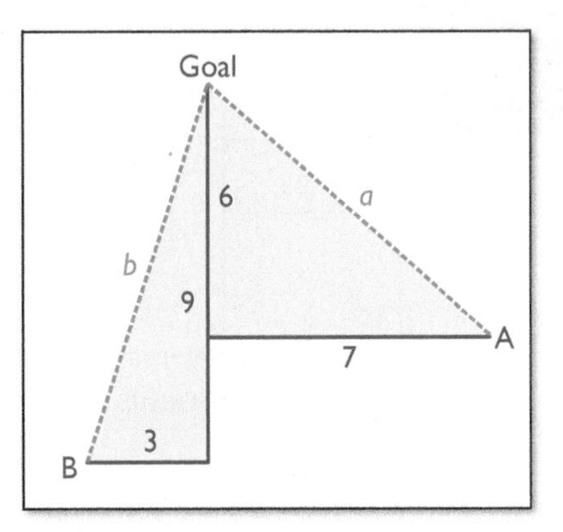

FIGURE 9.9 Positions for Try it Yourself 9.3.

The answer is provided at the end of this section.

Heron's formula

We know how to find the area of a triangle if we know the lengths of the base and height. Heron's formula gives the area of a triangle if we know the lengths of all three sides. Referring to the triangle shown in **Figure 9.10** with sides a, b, and c, we calculate one-half of the perimeter, the *semi-perimeter*, $S = \frac{1}{2}(a + b + c)$. Heron's formula says that the area is given by

$$\text{Area} = \sqrt{S(S-a)(S-b)(S-c)}$$

For the triangle in **Figure 9.11**, which has sides 5, 6, and 9, we find that $S = \frac{1}{2}(5 + 6 + 9) = 10$. Thus, the area is

$$\text{Area} = \sqrt{10(10-5)(10-6)(10-9)} = \sqrt{200}$$

or about 14.1 square units.

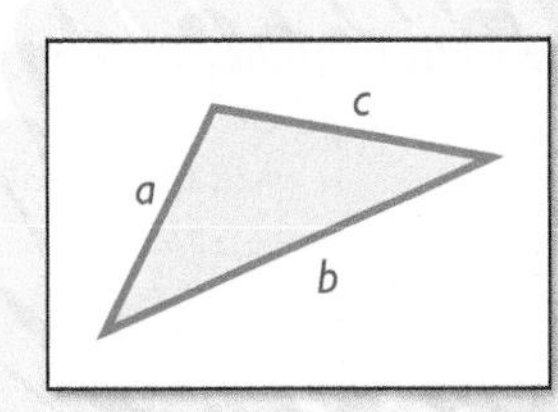

FIGURE 9.10 A triangle with given sides.

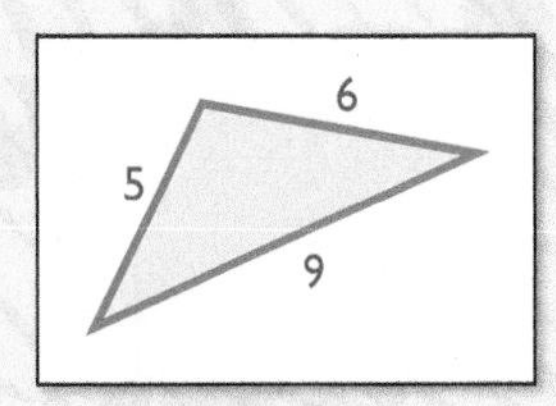

FIGURE 9.11 Applying Heron's formula.

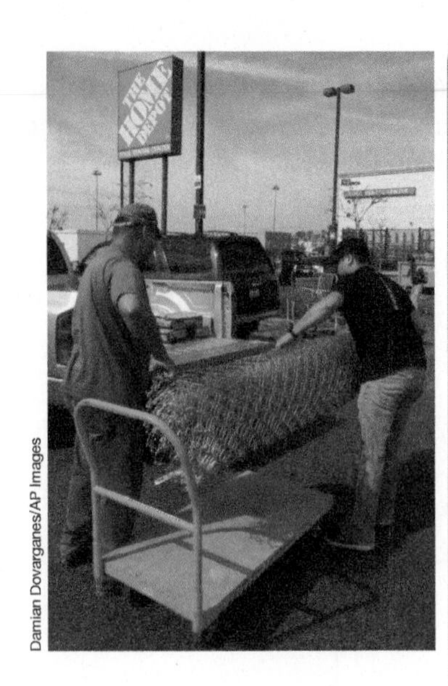

Damian Dovarganes/AP Images

EXAMPLE 9.4 Applying geometric formulas: Getting ready for spring

My lawn has the shape shown in **Figure 9.12**, where the dimensions are in feet.

a. I want to fence my lawn this spring. How many feet of fencing will be required?

b. I want to fertilize my lawn with bags of fertilizer that cover 500 square feet each. How many bags of fertilizer do I need to buy?

SOLUTION

a. The lawn has an irregular shape, but by adding the line segments shown in **Figure 9.13**, we can divide it into three pieces: a right triangle, a rectangle, and an oblique triangle (a triangle with no right angle).

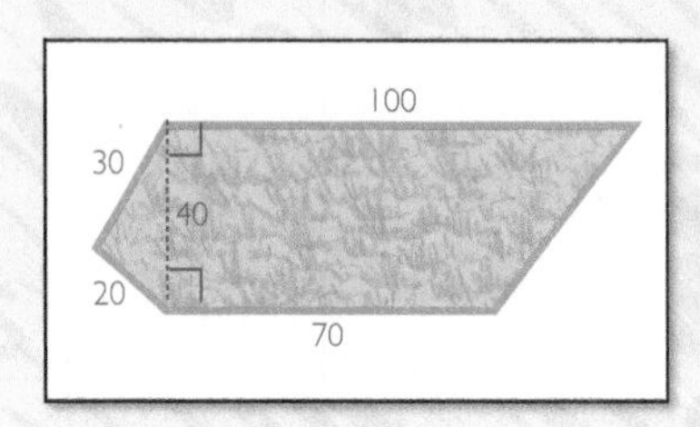

FIGURE 9.12 My lawn.

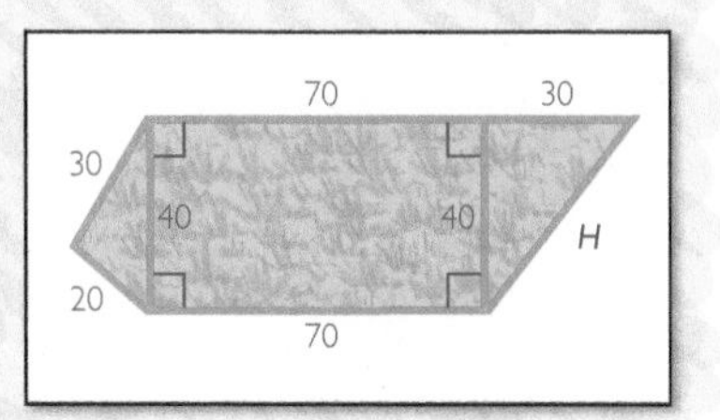

FIGURE 9.13 Dividing into a rectangle and two triangles.

The length of fence we need is the perimeter. To find the perimeter, we need to find the hypotenuse (marked H in Figure 9.13) of the right triangle. The legs have length 30 feet and 40 feet. We find H using the Pythagorean theorem:

$$c^2 = a^2 + b^2$$
$$H^2 = 30^2 + 40^2$$
$$H^2 = 2500$$
$$H = \sqrt{2500}$$
$$H = 50 \text{ feet}$$

Now we calculate the perimeter of the lawn as

$$\text{Perimeter} = 30 + 20 + 70 + 50 + 30 + 70 = 270 \text{ feet}$$

b. Look again at Figure 9.13. The area of the lawn is the area of the rectangle plus the area of the right triangle plus the area of the oblique triangle. For the rectangle, we find

$$\begin{aligned} \text{Area of rectangle} &= \text{Length} \times \text{Width} \\ &= 70 \times 40 \\ &= 2800 \text{ square feet} \end{aligned}$$

For the right triangle, we find

$$\begin{aligned} \text{Area of right triangle} &= \frac{1}{2}\text{ Base} \times \text{Height} \\ &= \frac{1}{2} \times 30 \times 40 \\ &= 600 \text{ square feet} \end{aligned}$$

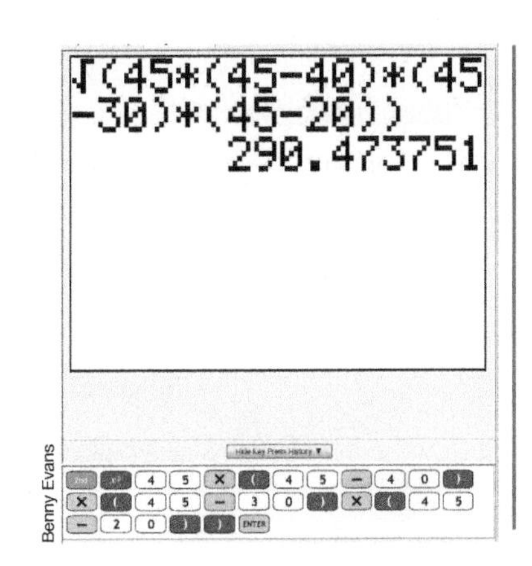

We use Heron's formula to calculate the area of the oblique triangle. The semi-perimeter is $\frac{1}{2}(40 + 30 + 20) = 45$, so the area is

$$\begin{aligned}\text{Area of oblique triangle} &= \sqrt{S(S-a)(S-b)(S-c)}\\ &= \sqrt{45(45-40)(45-30)(45-20)}\\ &= \sqrt{84{,}375}\end{aligned}$$

This is about 290 square feet.

Thus, the total area of the lawn is $2800 + 600 + 290 = 3690$ square feet. Each bag of fertilizer covers 500 square feet. Dividing 3690 by 500 gives about 7.4. Therefore, I will need to buy eight bags of fertilizer.

SUMMARY 9.1
Perimeter and Area

- For a rectangle:

$$\begin{aligned}\text{Area} &= \text{Length} \times \text{Width}\\ \text{Perimeter} &= 2 \times \text{Length} + 2 \times \text{Width}\end{aligned}$$

- For a circle of radius r:

$$\begin{aligned}\text{Area} &= \pi r^2\\ \text{Circumference} &= 2\pi r\end{aligned}$$

- For a triangle with sides $a,\ b,\ c$:

$$\begin{aligned}\text{Perimeter} &= a + b + c\\ \text{Area} &= \frac{1}{2}\ \text{Base} \times \text{Height}\end{aligned}$$

If S is half the perimeter

$$\text{Area} = \sqrt{S(S-a)(S-b)(S-c)}$$

Three-dimensional objects

So far we have discussed geometric objects lying in a plane, but many objects that we want to measure are not flat. Now we turn to the study of three-dimensional objects. First we consider volume.

The simplest volume to compute is that of a box. Labeling the dimensions of a box as *length, width*, and *height* as shown in **Figure 9.14**, we calculate the volume using

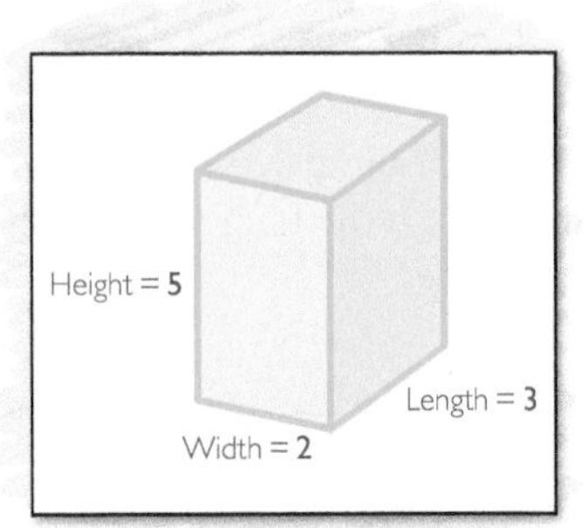

FIGURE 9.14 Dimensions of a box.

$$\text{Volume of a box} = \text{Length} \times \text{Width} \times \text{Height}$$

For example, the volume of the box in Figure 9.14 is

$$2 \text{ units} \times 3 \text{ units} \times 5 \text{ units} = 30 \text{ cubic units}$$

Note that the volume of a box can also be expressed as the area of the base times the height:

$$\text{Volume} = \text{Area of base} \times \text{Height}$$

This formula also holds for any three-dimensional object with uniform cross sections, such as a cylinder. A cylinder has uniform circular cross sections, and a box has uniform rectangular cross sections.

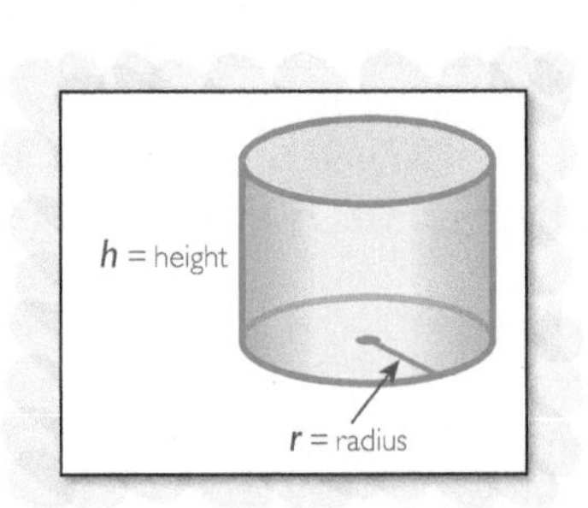

FIGURE 9.15 A cylinder.

For the cylinder shown in **Figure 9.15**, the radius is r, so the area of the base is πr^2. Using h for the height, we obtain

$$\text{Volume} = \text{Area of base} \times \text{Height} = \pi r^2 h$$

EXAMPLE 9.5 Calculating the volume of a cylinder: A swimming pool

What is the volume of a cylindrical wading pool that is 6 feet across and 15 inches high? Report your answer to the nearest cubic inch.

SOLUTION

Because the pool is 6 feet across, the diameter of the circular base is 6 feet. The diameter is given in feet and the height is given in inches, so we need to convert units for one of them. Because the final answer is supposed to be in cubic inches, we change the diameter of 6 feet to 72 inches. Then the radius is $r = 36$ inches, and the height is $h = 15$ inches. The volume of the pool is

$$\begin{aligned}\text{Volume} &= \text{Area of base} \times \text{Height} \\ &= \pi r^2 h \\ &= \pi(36 \text{ inches})^2 \times 15 \text{ inches}\end{aligned}$$

This is about 61,073 cubic inches.

TRY IT YOURSELF 9.5

Suppose you want to fill the same pool to a depth of 12 inches and your hose puts out 475 cubic inches per minute (about 2 gallons per minute). How long will it take you to fill the pool? Report your answer to the nearest minute.

The answer is provided at the end of this section.

Now we apply the formula for the volume of a box.

EXAMPLE 9.6 Calculating the volume of a box: Cake batter

Cake batter rises when baked. The best results for baking cakes occur when the batter fills the pan to no more than two-thirds of the height of the pan. Suppose we have 7 cups of batter, which is about 101 cubic inches. We have a pan that is 2 inches high and has a square bottom that is 9 inches by 9 inches. Is this pan large enough?

SOLUTION

This pan forms the shape of a box, and we want to find the volume based on two-thirds of the full height of 2 inches. Thus we use a height of

$$\text{Height} = \frac{2}{3} \times 2 = \frac{4}{3} \text{ inches}$$

We find

$$\begin{aligned}\text{Volume} &= \text{Length} \times \text{Width} \times \text{Height}\\ &= 9 \text{ inches} \times 9 \text{ inches} \times \frac{4}{3} \text{ inches}\\ &= 108 \text{ cubic inches}\end{aligned}$$

This pan will easily hold batter with a volume of 101 cubic inches.

TRY IT YOURSELF 9.6

We have a round pan that is 10 inches across and 2 inches high. To bake a cake, we need to fill it to only two-thirds of the height. Is this pan large enough for baking 101 cubic inches of batter?

The answer is provided at the end of this section.

The volume of a ball of radius r is $\frac{4}{3}\pi r^3$. For example, Earth has a radius of about 4000 miles. So its volume is

$$\begin{aligned}\text{Volume of Earth} &= \frac{4}{3}\pi r^3\\ &= \frac{4}{3}\pi(4000)^3\end{aligned}$$

or about 268,082,573,106 cubic miles. The volumes of other geometric objects are addressed in the exercises.

The analogue of perimeter in three-dimensional objects is surface area. We get the surface area of a box with a lid by summing the areas of the six sides of the box. The surface area of a ball of radius r is given by $4\pi r^2$. Looking once again at Earth, we find

$$\begin{aligned}\text{Surface area of Earth} &= 4\pi r^2\\ &= 4\pi(4000)^2\end{aligned}$$

or about 201,061,930 square miles.

To find the surface area of a cylinder of radius r and height h (excluding the top and bottom), think of the cylinder as a can that we split lengthwise and roll out flat. See **Figure 9.16**. This gives a rectangle with width h and length equal to the circumference of the circular base. The circumference of the base is $2\pi r$, so the surface area of the cylinder (excluding the top and bottom) is

$$\text{Surface area of cylinder} = 2\pi rh$$

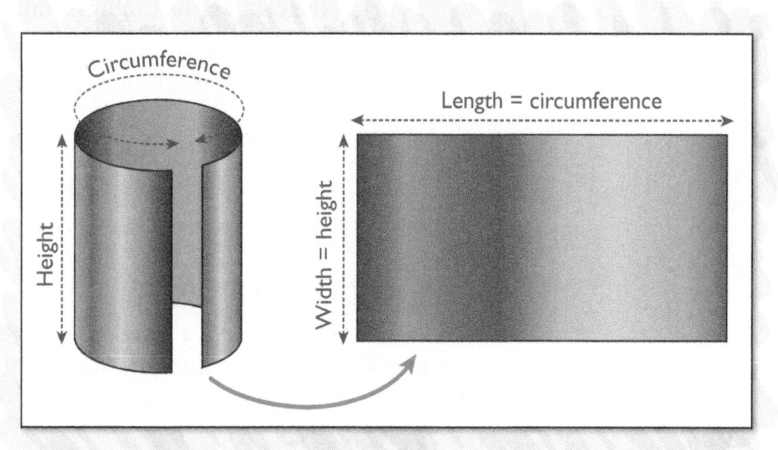

FIGURE 9.16 Splitting the cylindrical side of a can and rolling it out flat.

EXAMPLE 9.7 Calculating volume and surface area: A can

A tin can has a radius of 1 inch and a height of 6 inches.

a. How much liquid will the can hold? Round your answer in cubic inches to one decimal place.

b. How much metal is needed to make the can? Round your answer in square inches to one decimal place.

SOLUTION

a. The base is a circle of radius 1, so its area is

$$\text{Area of base} = \pi \times 1^2 = \pi \text{ square inches}$$

The height is 6 inches, so the volume is

$$\text{Volume} = \text{Area of base} \times \text{Height} = \pi \times 6 \text{ cubic inches}$$

This is about 18.8 cubic inches.

b. The metal needed to make the can consists of the top, bottom, and cylindrical side. We already found that the base has area π square inches, so

$$\text{Area of top and bottom} = 2\pi \text{ square inches}$$

To find the area of the cylindrical side, we use the formula $2\pi rh$:

$$\text{Area of side} = 2\pi \times 1 \times 6 = 12\pi \text{ square inches}$$

The total area includes the top and bottom of the can, for a total area of $2\pi + 12\pi = 14\pi$ or about 44.0 square inches.

TRY IT YOURSELF 9.7

A can of tuna fish has a diameter of $3\frac{1}{4}$ inches and a height of $1\frac{3}{8}$ inches. Find its volume. (Round your answer in cubic inches to one decimal place.) How much metal is needed to make the can? (Round your answer in square inches to one decimal place.)

The answer is provided at the end of this section.

SUMMARY 9.2
Volumes and Surface Areas

- The volume of a box is:
  Volume of a box = Length × Width × Height.
- The volume of a cylinder is: Area of base × Height.
  If the cylinder has radius r and height h, this equals $\pi r^2 h$.
- The surface area of a cylinder (excluding the top and bottom) of radius r and height h is $2\pi rh$.
- The volume of a ball of radius r is $\frac{4}{3}\pi r^3$.
- The surface area of a ball of radius r is $4\pi r^2$.

WHAT DO YOU THINK?

What shape? You have a fixed amount of material to build a totally enclosed container. What shape would you use to encompass the largest volume? Explain why

you believe the shape you chose to be the best one. If you have trouble, try first to figure out what shape in the plane with a fixed perimeter encloses the most area.
Questions of this sort may invite many correct responses. We present one possibility: Perhaps the easiest way to solve this is to think backward. Let's start with any fixed volume. We are going to use the same enclosing material to make the volume as large as possible. We assume the boundary is totally flexible but not expandable. We may change the shape of the enclosure but not its volume. We blow in air to stretch the enclosure as much as possible. This is similar to blowing up a balloon, and we get the same shape: a sphere.

Further questions of this type may be found in the *What Do You Think?* section of exercises.

Try It Yourself answers

Try It Yourself 9.1: Using circumference and diameter: Runners The difference in the distances is 6π or about 18.8 feet.

Try It Yourself 9.2: Using area and diameter: Pizza The perimeter of the 12-inch pizza is 37.7 inches. The perimeter of the 16-inch pizza is 50.3 inches.

Try It Yourself 9.3: Applying the Pythagorean theorem: Slap shot Player B has the longer shot.

Try It Yourself 9.5: Calculating the volume of a cylinder: A swimming pool 103 minutes, or 1 hour and 43 minutes.

Try It Yourself 9.6: Calculating the volume of a box: Cake batter Multiplying the height by 2/3 gives a volume of about 105 cubic inches, so the pan is large enough.

Try It Yourself 9.7: Calculating volume and surface area: A can The volume is 11.4 cubic inches; the amount of metal needed is 30.6 square inches.

Exercise Set 9.1

Test Your Understanding

1. The Pythagorean theorem shows the relationship that exists between: **a.** the three sides of a right triangle **b.** the circumference and diameter of a circle **c.** circles and squares **d.** area and volume.

2. True or false: Heron's formula shows how to calculate the area of a triangle from the lengths of the sides.

3. The ratio of the circumference of any circle to its diameter is: **a.** the same number no matter how large or small the circle **b.** the number π **c.** both of the above.

4. True or false: Areas of irregular regions can sometimes be found by dividing the region into rectangles and triangles.

Problems

5. Pool fence. A circular swimming pool is 10 feet across and enclosed by a fence. How long is the fence?

6. Pool cover. A circular swimming pool is 15 feet across and has a cover. How much material is in the pool cover?

7. Finding the perimeter. Find the perimeter of the figure in Figure 9.17.

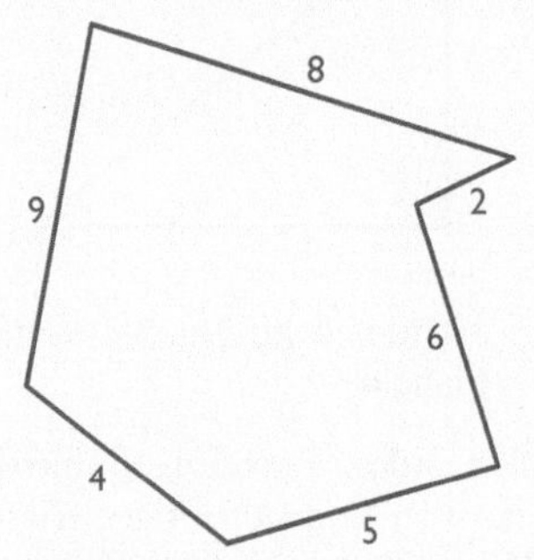

FIGURE 9.17

8. Finding the area. Find the area of a right triangle whose legs have lengths 8 and 15.

9. Finding the hypotenuse. Find the length of the hypotenuse of a right triangle whose legs have lengths 3 and 7.

10. Finding a leg. The hypotenuse of a right triangle has length 7, and a leg has length 3. Find the length of the other leg.

11. A basketball problem. In basketball, free-throws are taken from the free-throw line, which is 13 feet from a point directly below the basket. The basket is 10 feet above the floor. These dimensions are shown in **Figure 9.18**. How far is it from the free-throw line to the basket?

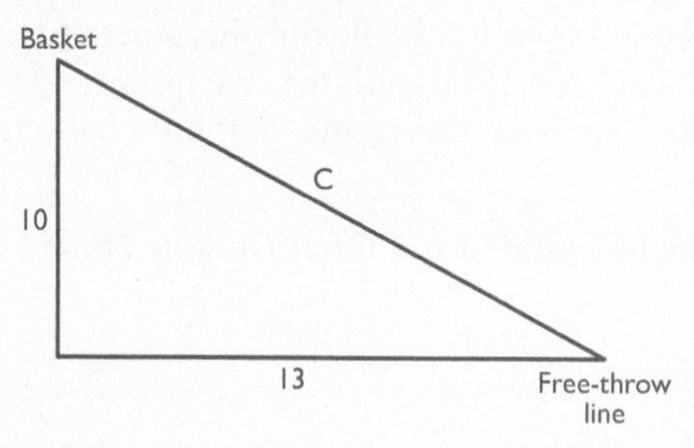

FIGURE 9.18

12. Distance between towns. Town B is 5 miles due north of town A, and town C is due east of town A. It is 9 miles from town B to town C. Find the distance from town A to town C, marked b in **Figure 9.19**.

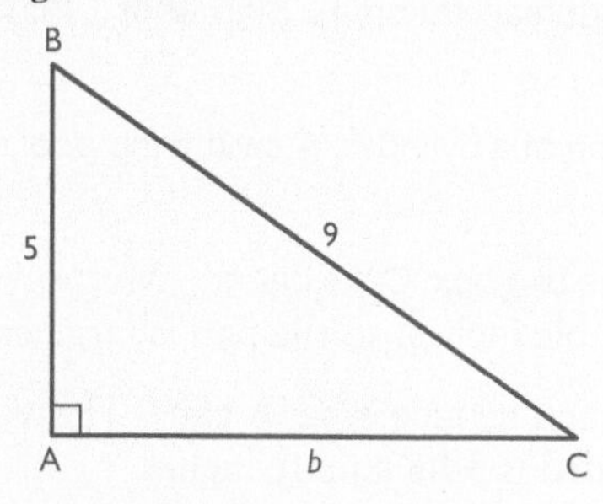

FIGURE 9.19

13. Area of a right triangle. Verify that the triangle with sides 3, 4, and 5 shown in **Figure 9.20** is a right triangle, and find its area. This is the *3-4-5 right triangle*.

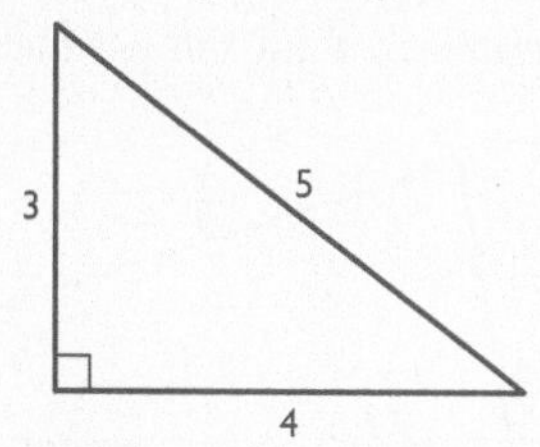

FIGURE 9.20 The 3-4-5 right triangle.

14. Whole-number sides. Exercise 13 introduced the 3-4-5 right triangle. Can you find other right triangles with whole-number sides? *Hint*: You are looking for whole numbers a, b, and c so that $a^2 + b^2 = c^2$.

15. Finding the area. The hypotenuse of a right triangle has length 11, and a leg has length 4. Find the area of the triangle.

16–19. Heron's formula. *Exercises 16 through 19 use Heron's formula.*

16. Find the area of a triangle with sides 7, 8, and 9.

17. Find the area of a garden that is a triangle with sides 4 feet, 5 feet, and 7 feet.

18. Find the area of a triangle with sides 6 yards, 6 yards, and 7 yards.

19. One triangular plot has sides 4 yards, 5 yards, and 6 yards. Another has sides 3 yards, 6 yards, and 7 yards. Which plot encloses the larger area?

20. Making drapes. A window in your home has the shape shown in **Figure 9.21**. The top is a semi-circle (half of a circle).

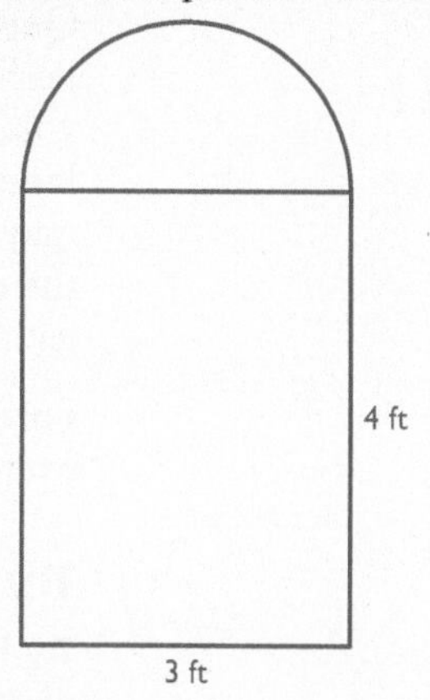

FIGURE 9.21 A window.

To buy material for drapes, you need to know the area of the window. What is it?

21. Trimming a window. How many feet of moulding do we need to make a border (excluding the diameter of the half-circle) for the window in Figure 9.21?

22. Carpet for a ramp. A ramp is 3 feet wide. It begins at the floor and rises by 1 foot over a horizontal distance of 10 feet. (See **Figure 9.22**.) How much carpet is needed to cover the top of the ramp?

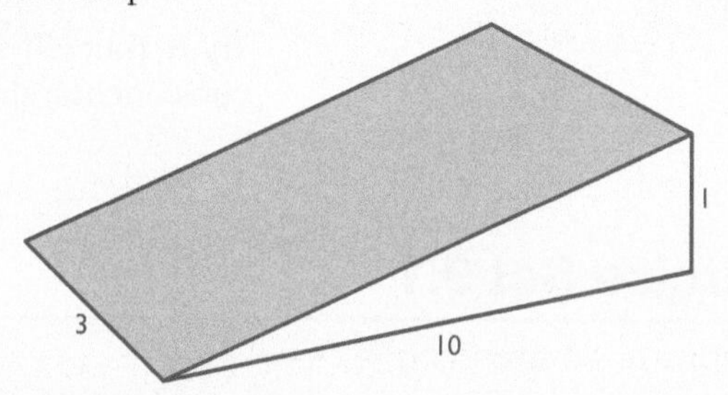

FIGURE 9.22 A ramp.

23. Isosceles triangles. An *isosceles* triangle has two equal sides. In an isosceles triangle, consider the line segment starting at the vertex that touches both of the equal sides and meeting the opposite side in a right angle. It is a fact that this segment cuts the opposite side (the base) in half. Use this fact to find the area and perimeter of the isosceles triangle in **Figure 9.23**. (The base is 10 units, and the height is 20 units.) *Suggestion*: First use the Pythagorean theorem to find the lengths of the unknown sides of the triangle.

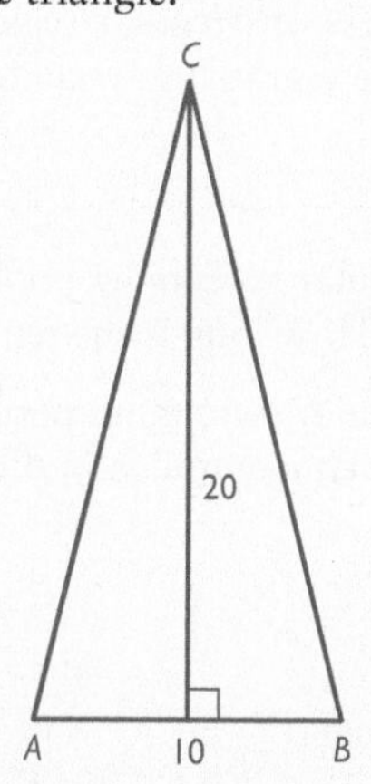

FIGURE 9.23 Isosceles triangle.

24–27. Perimeter and area of a triangle. *In Exercises 24 through 27, use the Pythagorean theorem to find the missing lengths and then find the area and perimeter. (Do not include the height in the perimeter.)*

24. Refer to the triangle in **Figure 9.24**.

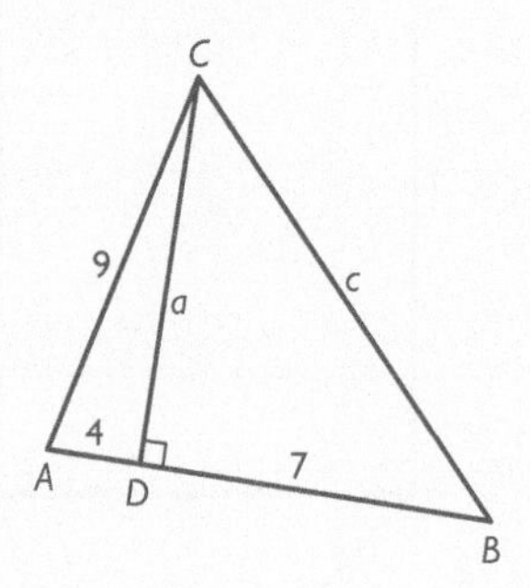

FIGURE 9.24

25. Refer to the triangle in **Figure 9.25**.

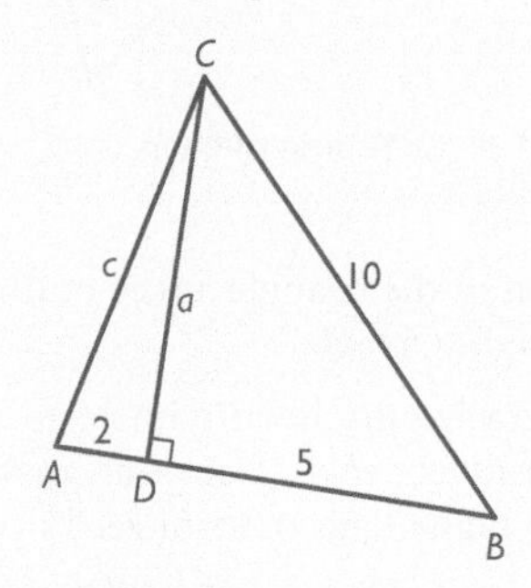

FIGURE 9.25

26. Refer to the triangle in **Figure 9.26**.

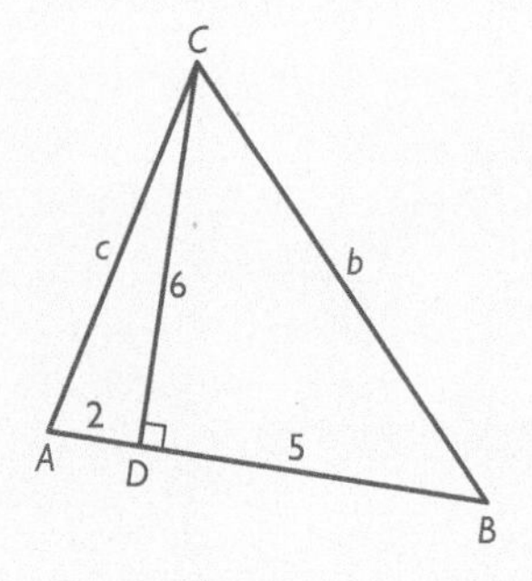

FIGURE 9.26

27. Refer to the triangle in **Figure 9.27**.

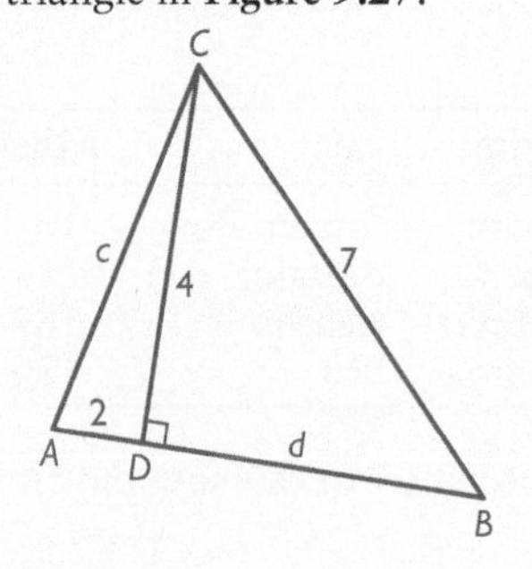

FIGURE 9.27

28. Pitcher foam. A restaurant serves soda pop in cylindrical pitchers that are 5 inches in diameter and 8 inches tall. If the pitcher has a 1-inch head of foam, how much soda is lost as a result?

29. Soda and soup. *This is a continuation of Exercise 28.* A typical soup can is about 2.5 inches in diameter and 3.5 inches tall. Compare the amount of soda lost in Exercise 28 with the volume of a soup can.

30–33. A soup can. *A soup can has a diameter of $2\frac{5}{8}$ inches and a height of $3\frac{3}{4}$ inches. Exercises 30 through 33 refer to this can.*

30. Find the volume of the soup can.

31. How many square inches of paper are required to make the label on the soup can?

32. How many square inches of metal are required to make the soup can?

33. When you open the soup can, how far does the can opener travel?

34. Distance from Dallas to Fort Smith. Dallas, Texas, is 298 miles due south of Oklahoma City, Oklahoma. Fort Smith, Arkansas, is 176 miles due east of Oklahoma City. How many miles long is a plane trip from Dallas directly to Fort Smith?

35. Houston to Denver. Houston, Texas, is 652 miles due south of Kansas City, Missouri. Denver, Colorado, is 558 miles due west of Kansas City. How many miles long is a plane trip from Houston directly to Denver? Round your answer to the nearest mile.

36. Big-screen TVs. The size of a television set is usually classified by measuring its diagonal. Suppose that one giant television measures 5 feet long by 3 feet high, and another is a 4.5-foot square.

a. Which television has the larger diagonal?

b. Which television has the larger screen in terms of area?

37. Monitor sizes. The size of a computer monitor (also called a display) is usually measured in terms of the diagonal of the rectangle.

a. A monitor measures 12.6 inches by 16.8 inches. What is the length of the diagonal?

b. A monitor measures 10.3 inches by 18.3 inches. What is the length of the diagonal?

c. Calculate the areas of the rectangles described in parts a and b. (Round your answers to the nearest whole number.) In light of your answers, does measurement of a monitor in terms of the diagonal give complete information about the size?

38. Pizza coupon. At your local pizza place, a large pizza costs \$10.99. You have a coupon saying you can buy two medium pizzas for \$12.99. A medium pizza has a diameter of 12 inches, and a large pizza has a diameter of 14 inches. Which is the better value (measured in dollars per square inch), one large pizza or two medium pizzas with the coupon?

39–40. Taxes and square footage. *Property taxes on a home are often based on its valuation, determined by the number of square feet of living space. Incorrect measurements of this area can significantly affect your property taxes. Exercises 39 and 40 concern square footage of a home.*

39. Suppose the floor plan for a home is shown in **Figure 9.28.** (The curved part is a half-circle.) How many square feet of floor space does this home have?

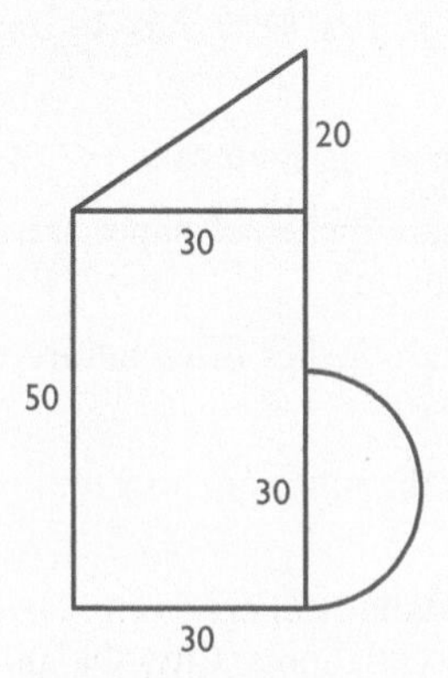

FIGURE 9.28 Floor plan.

40. The floor plan for a home is shown in **Figure 9.29**. How many square feet of floor space does this home have?

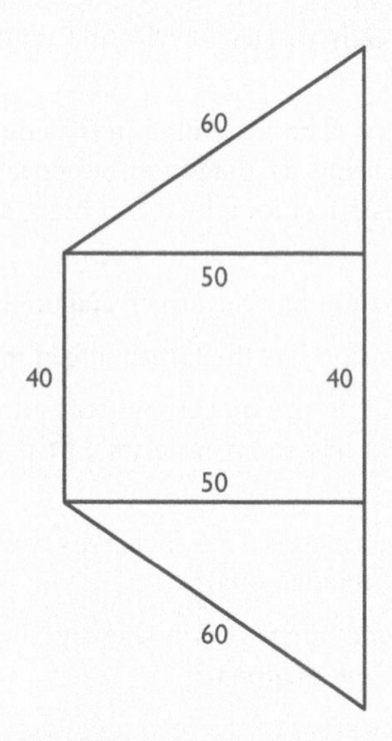

FIGURE 9.29 Floor plan.

41. Getting ready for spring. My lawn has the shape shown in **Figure 9.30**, where the dimensions are in feet.

a. Find the length marked p.

b. Recall that opposite sides of a rectangle are equal. Use this fact to find the length marked q.

c. I want to fence my lawn this spring. Use your answers to parts a and b to determine how many feet of fencing will be required.

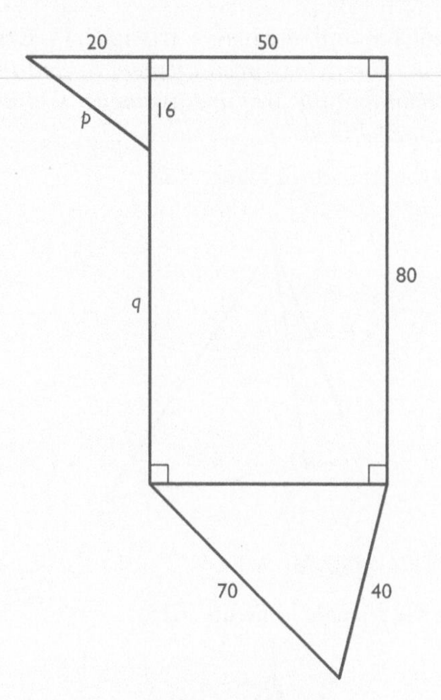

FIGURE 9.30 Shape of lawn.

d. Find the area of the triangle at the bottom of the figure. (This is not a right triangle.)

e. I want to fertilize my lawn with bags of fertilizer that cover 500 square feet each. Use your answer to part d to determine how many bags of fertilizer I need to buy.

42. Area of Oklahoma. Like many states, Oklahoma has an irregular shape, as shown in **Figure 9.31**. This makes its area difficult to calculate exactly. The following table shows distances between certain towns in Oklahoma.

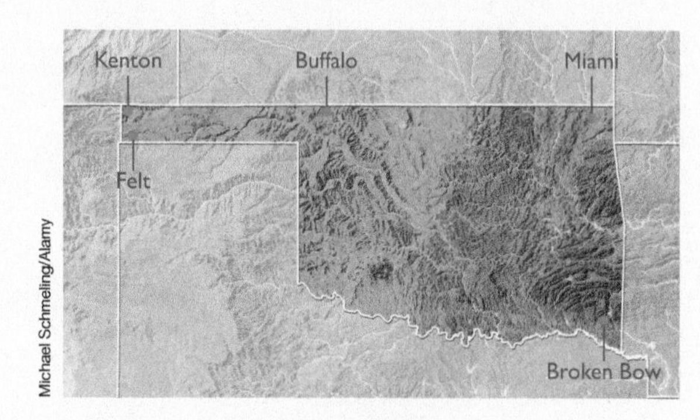

FIGURE 9.31 Map of Oklahoma.

From	To	Miles
Miami	Broken Bow	193
Miami	Buffalo	256
Buffalo	Kenton	187
Kenton	Felt	20

Use the given information to estimate the area of the state of Oklahoma.

43. Perimeter of Oklahoma. *This is a continuation of Exercise 42.* Suppose we wished to set up a string of cameras every 10 miles around the border of Oklahoma to monitor the migration of birds across state lines. Use the information given in Exercise 42 to estimate the number of cameras needed.

44. More on the area of Oklahoma. *This is a continuation of Exercise 42.* The actual area of the state of Oklahoma is 69,960 miles. Explain how you would make a more accurate calculation than was possible in Exercise 42 if you had a detailed map of the state.

45. Area of Nevada. Like many states, Nevada has an irregular shape, as shown in **Figure 9.32**. This makes its area difficult to calculate exactly. The following table shows distances between certain towns in Nevada.

From	To	Miles
Carson City	Ely	258
Carson City	Needles, CA	426
Ely	Jackpot	166
Ely	Needles, CA	339

Use the given information to estimate the area of the state of Nevada.

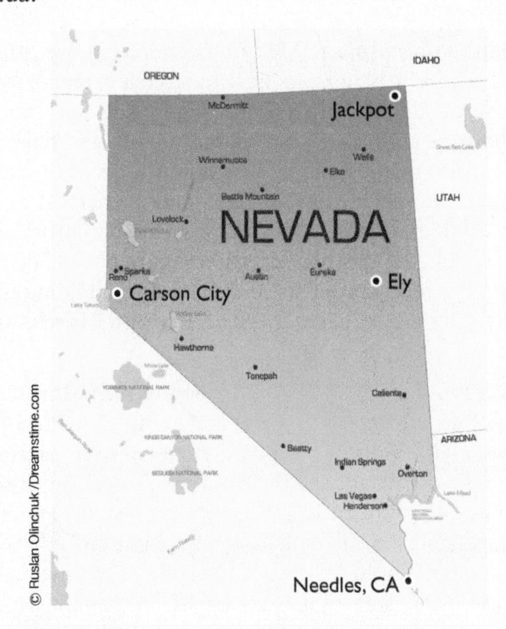

FIGURE 9.32 Map of Nevada.

46. Perimeter of Nevada. *This is a continuation of Exercise 45.* Suppose we wished to set up a string of cameras every 10 miles around the border of Nevada to monitor the migration of birds across state lines. Use the information given in Exercise 45 to estimate the number of cameras needed.

47. More on area of Nevada. *This is a continuation of Exercise 45.* The actual area of the state of Nevada is 110,567 square miles. Explain how you would make a more accurate calculation than was possible in Exercise 45 if you had a detailed map of the state.

48. Area of your state. Locate a good map of your state and use it to estimate its area.

49. Perimeter of your state. Locate a good map of your state and use it to estimate its perimeter.

50. Cement for a sidewalk. You need to order cement to make a concrete sidewalk 3 feet wide and 100 feet long. Assume that the depth of the concrete is 4 inches. How much cement is needed, in cubic feet? Cement is usually ordered in cubic yards. How many cubic yards of cement do you need?

51. Gravel for a driveway. You need to order gravel for your driveway, which is 8 feet wide and 230 feet long. Assume that you would like the gravel to be 6 inches deep. How much gravel is needed, in cubic feet? Gravel is ordered in cubic yards. How many cubic yards of gravel do you need?

52. Water for an in-ground pool. Your in-ground pool is rectangular and has a depth of 5 feet. If the width of the pool is 15 feet and the length is 25 feet, how much water does the pool hold?

53. A baseball. The diameter of a baseball is 2.9 inches. What is the volume of a baseball? A baseball is covered in cowhide. How many square inches are needed to make the cover of a baseball?

54. A bigger baseball. *This is a continuation of Exercise 53.* If you had twice the leather needed to cover a baseball, could you cover a ball with twice the radius? *Suggestion*: Calculate the surface area of a ball of radius 2.90 inches.

55. A basketball. The diameter of a standard basketball is 9.39 inches. Find the volume. An NBA basketball is made of "composite leather," a synthetic material designed to withstand slam dunks. How many square inches are required to make a basketball?

56. A bigger basketball. *This is a continuation of Exercise 55.* If you had twice the composite leather needed to cover a basketball, could you cover a ball with twice the radius? *Suggestion*: Calculate the surface area of a ball of radius 9.39 inches.

57–60. Cones. *To measure a cone as shown in* **Figure 9.33**, *we need to know its height h and the radius r of its top. For a cone of these dimensions, we have*

$$\text{Volume} = \frac{1}{3}\pi r^2 h$$

$$\text{Surface area} = \pi r\sqrt{r^2 + h^2}$$

Note that the surface area does not include the top circle of the cone. Exercises 57 through 60 use these formulas.

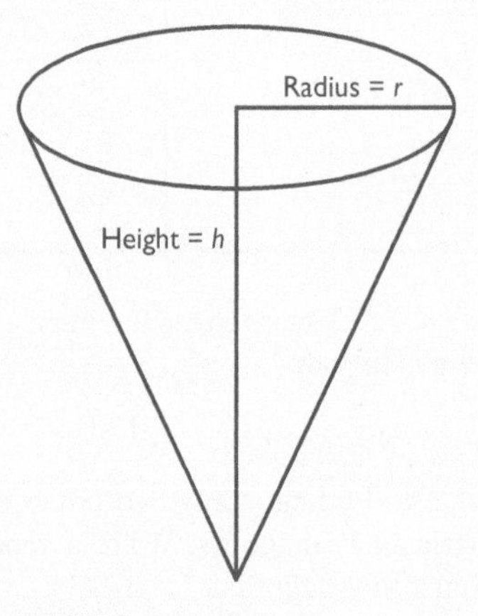

FIGURE 9.33 A cone.

57. A cone for ice cream has a top diameter of 2.25 inches and a height of 4.5 inches. How much ice cream will this cone hold?

58. A (solid) conical spire is made of concrete. The spire has a radius of 3 feet and a height of 10 feet. How much concrete is needed to make the spire?

59. How many square feet of paint are needed to paint a cone with a top diameter of 5 feet and a height of 6 feet, assuming that you will paint the entire cone including the top circle?

60. A waffle cone has a top diameter of 2.25 inches and a height of 4.5 inches. What is the area of the dough needed to make the cone?

Exercise 61 is suitable for group work.

61. A demonstration of the Pythagorean theorem. This exercise uses a puzzle to demonstrate why the Pythagorean theorem is true. Trace the square shown in **Figure 9.34** onto your own paper. Cut out the square and then cut the square along the three interior lines shown. You should have three pieces of a puzzle. Reassemble the three pieces into the figure shown in **Figure 9.35**. Explain how this demonstrates why the Pythagorean theorem is true.

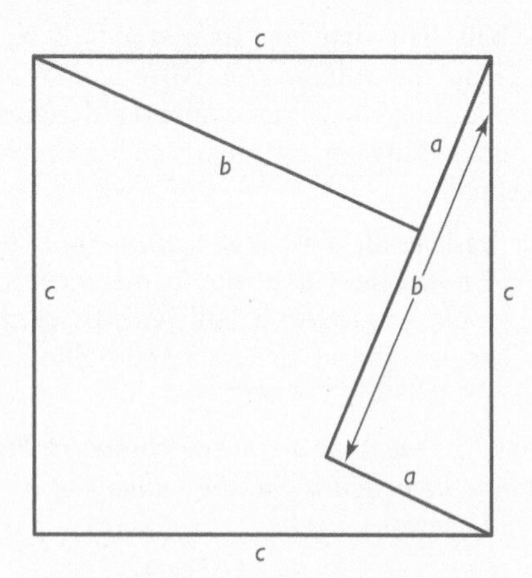

FIGURE 9.34 Trace and cut to make a Pythagorean puzzle.

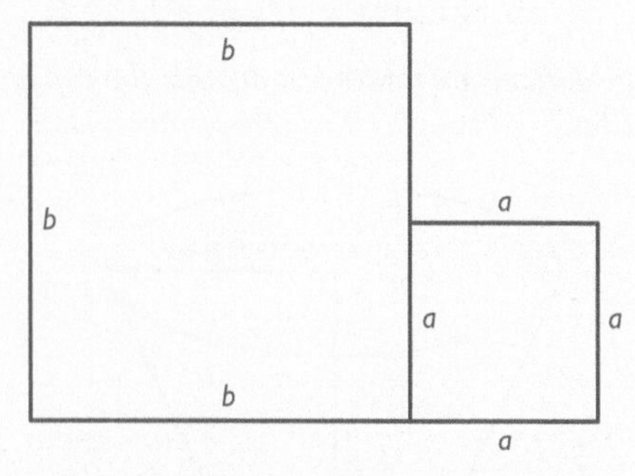

FIGURE 9.35 Reassemble the puzzle to make this picture.

Writing About Mathematics

62. Pythagoras. The Pythagorean theorem is named for the Greek mathematician Pythagoras. Write a report on the life and contributions of Pythagoras.

63. Straight lines. We are all familiar with the geometric statement that the shortest distance between two points is a straight line. Explain how to find the shortest distance that a ship would travel from New York to Lisbon. Before you begin, use the Internet to find out what a *great circle* is.

What Do You Think?

64. The number π. The number π comes up often in relation to geometric figures, and in fact throughout the field of mathematics and its applications. Mathematicians have known about this number for a very long time. But accurate estimates of its numerical value and many of its properties as a number have been clearly understood only in the past couple of centuries. Investigate historical attempts to understand this number. **Exercise 5 is relevant to this topic.**

65. Rectangle. Does the perimeter of a rectangle determine its area? This question can also be asked the other way around: Does the area of a rectangle determine its perimeter? What if it is a square? Explain your answer and include examples.

66. A cube. Does the surface area of a box determine its volume? What if it is a cube—where all three dimensions (length, width, and height) are the same? Explain your answer and include examples.

67. An interesting project. Many measurements are difficult to make because it is not practical to stretch a tape between certain points. Here is an example. Figure out how far it is from a bottom corner of your classroom to the diagonally opposite corner at the ceiling. This is not a distance along a wall or floor, but such measurements may be helpful to you. You may find the Pythagorean theorem to be useful. Once you have completed this project, try to figure out the distance from a ground-level corner of your classroom building to the diagonally opposite corner at the roof. **Exercise 9 is relevant to this topic.**

68. Scarecrow. In the 1939 film *The Wizard of Oz*, the Scarecrow states the following as a fact about geometry: "The sum of the square roots of any two sides of an isosceles triangle is equal to the square root of the remaining side." An *isosceles* triangle has two sides of equal length. To test the statement of the Scarecrow, consider the case of a right triangle whose legs have equal length. What is the length of the hypotenuse in this case? Is it true that the sum of the square roots of the two legs is equal to the square root of the hypotenuse? If not, how can you adjust the Scarecrow's statement to make it true for isosceles right triangles? Does your adjusted statement hold for *all* isosceles triangles? **Exercise 23 is relevant to this topic.**

69. A television screen. Television screens are often described in advertisements by the diagonal measurement. Is this a fair description of the size of a television screen? Explain your answer and include examples. **Exercise 36 is relevant to this topic.**

70. A brain teaser. A long rope is stretched tightly around the equator of Earth. How much additional rope would be needed if the rope ring is to be suspended a uniform distance of

1 foot above the equator? How would things change if we were discussing the equator of Mars rather than Earth? If you need help, look at Example 9.1.

71. Heron's formula. One reason that Heron's formula looks complicated is the square root, but that part of the formula is suggested by analyzing the units: If the sides a, b, c of the triangle are measured in feet, what are the units for the semi-perimeter S? Then what are the units for $S(S-a)(S-b)(S-c)$? What should the units for the area be? Use your answers to explain why the square root is reasonable. **Exercise 17 is relevant to this topic.**

72. Cylinder. Recall that the volume of a cylinder of radius r and height h is $\pi r^2 h$. Start with a cylinder having radius 1 foot and height 1 foot. If the radius is doubled from 1 foot to 2 feet, but the height remains at 1 foot, what happens to the volume? If the height is doubled from 1 foot to 2 feet, but the radius remains at 1 foot, what happens to the volume? Which has a larger effect on the volume, doubling the radius or doubling the height? Explain your answer by looking at the way r and h appear in the formula for the volume.

9.2 Proportionality and similarity: Changing the scale

TAKE AWAY FROM THIS SECTION
Understand the concept of proportionality and how a change in the dimensions of a figure affects its area or volume.

The subject of proportion is of interest not only to mathematicians but also to artists and architects. One focus of interest is the golden ratio, which appears in the following excerpt.

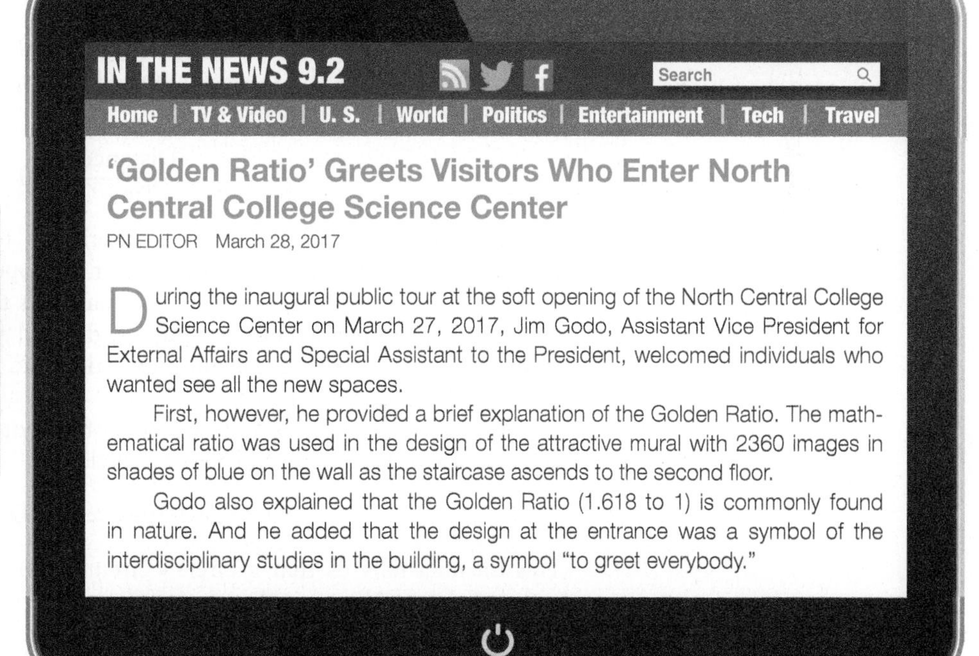

IN THE NEWS 9.2

Search

Home | TV & Video | U. S. | World | Politics | Entertainment | Tech | Travel

'Golden Ratio' Greets Visitors Who Enter North Central College Science Center

PN EDITOR March 28, 2017

During the inaugural public tour at the soft opening of the North Central College Science Center on March 27, 2017, Jim Godo, Assistant Vice President for External Affairs and Special Assistant to the President, welcomed individuals who wanted see all the new spaces.

First, however, he provided a brief explanation of the Golden Ratio. The mathematical ratio was used in the design of the attractive mural with 2360 images in shades of blue on the wall as the staircase ascends to the second floor.

Godo also explained that the Golden Ratio (1.618 to 1) is commonly found in nature. And he added that the design at the entrance was a symbol of the interdisciplinary studies in the building, a symbol "to greet everybody."

The opening article is concerned with ratio and proportion in a geometric setting. Proportionality relationships are quite common in many settings, both numerical and geometrical. For example, they arise when we convert from one unit of measurement to another. In this section, our purpose is to examine proportionality and similarity relationships that arise in geometry. In the process we will look more closely at the golden ratio, which is at the heart of much controversy regarding the interplay of art and mathematics.

Definition and examples

To say that one variable quantity is proportional to another simply means that one is always a constant multiple of the other.

KEY CONCEPT

One variable quantity is (directly) proportional to another if it is always a fixed (nonzero) constant multiple of the other. Expressed in terms of a formula, this says that the quantity A is proportional to the quantity B if there is a nonzero constant c such that $A = cB$. The number c is the constant of proportionality.

Note that if $A = cB$, we can turn the relationship around and write $B = \frac{1}{c}A$, which says that B is proportional to A with constant of proportionality $\frac{1}{c}$.

Another way to write the equation $A = cB$ is $\frac{A}{B} = c$. Expressed in this way, the proportionality formula tells us that two quantities are proportional if their ratio is constant.

For example, if you are driving at a constant speed of 60 miles per hour, then the distance you travel is proportional to the time you spend driving. If distance is measured in miles and time is in hours,

$$\text{Distance} = 60 \times \text{Time}$$

The constant of proportionality is your speed, 60 miles per hour. Note that distance traveled increases by the same amount, 60 miles, for each extra hour of travel. We can also write the relationship as

$$\frac{\text{Distance}}{\text{Time}} = 60$$

Note that distance traveled is proportional to time only when your speed is constant. If your speed varies, distance is not proportional to time.

You can probably think of lots of quantities that are not proportional to each other. One example would be your height and your age. Although your height increases as you get older, it is not proportional to your age because you don't grow by the same amount each year.

One of the most important instances of proportionality in geometry is the familiar formula for the circumference of a circle:

$$\text{Circumference} = \pi \times \text{Diameter}$$

This formula tells us that a circle's circumference is proportional to its diameter and that the number π is the constant of proportionality. Expressed as a ratio, this relationship is

$$\frac{\text{Circumference}}{\text{Diameter}} = \pi$$

It was pointed out in Section 9.1 that if we divide the circumference of a very small circle by its diameter, we get exactly the same value, π, as we would if we divided the length of the equator of Earth by its diameter. This is another way of saying that the circumference is proportional to the diameter.

In contrast, the radius r and area A of a circle are related by

$$A = \pi r^2$$

Here, A is not a constant multiple of r, so the area is not proportional to the radius. Because the diameter is twice the radius, the area of a circle is not proportional to its diameter either.

EXAMPLE 9.8 Applying proportionality: Forestry

Biologists and foresters use the number of the trees and their diameters as one measure of the condition and age of a forest. Measuring the diameter of a tree directly is not easy, especially for a large tree. However, the circumference is easy to measure simply by running a tape measure around the tree. Is the diameter of a tree proportional to its circumference? (Assume that the cross section is circular in shape, as in **Figure 9.36**.) If so, what is the constant of proportionality?

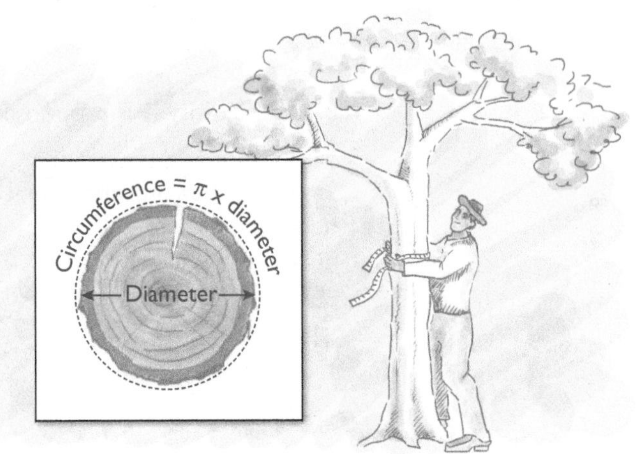

FIGURE 9.36 Foresters find the diameter of trees by measuring the circumference.

SOLUTION

We know that circumference is proportional to diameter, and the relationship is

$$\text{Circumference} = \pi \times \text{Diameter}$$

We can rearrange this formula to find

$$\text{Diameter} = \frac{1}{\pi} \times \text{Circumference}$$

Thus, the diameter is proportional to the circumference, and the constant of proportionality is $1/\pi$.

TRY IT YOURSELF 9.8

We saw down a tree, and the top of the resulting stump is circular in shape. The area of this circle is the cross-sectional area of the base of the tree. Is the cross-sectional area of the base proportional to the circumference?

The answer is provided at the end of this section.

Properties of proportionality

We can discover an important property of proportionality by further consideration of tree measurement, as in the previous example. Foresters determine the diameter of a tree by measuring its circumference. What happens to the circumference of a tree if the diameter is doubled? We can find out by multiplying both sides of the equation

$$\text{Circumference} = \pi \times \text{Diameter}$$

by 2. The result is

$$2 \times \text{Circumference} = \pi \times (2 \times \text{Diameter})$$

Thus, doubling the diameter causes the circumference to double as well. Conversely, a tree with twice the circumference of another has twice the diameter. The same is true for any multiple—tripling the diameter causes the circumference to triple, and halving the diameter cuts the circumference in half. It turns out that this fact is true for any proportionality relation, and it is another characterization of proportionality.

SUMMARY 9.3
Properties of Proportionality

1. Two quantities are proportional when one is a (nonzero) constant multiple of the other. This multiple is the constant of proportionality. Another way to say this is that two quantities are proportional when their ratio is always the same.

2. If a quantity A is proportional to another quantity B with proportionality constant c, then B is proportional to A with proportionality constant $1/c$.

3. If one of two proportional quantities is multiplied by a certain factor, the other one is also multiplied by that factor. That is, if B is proportional to A, then multiplying A by k has the result of multiplying B by k. In particular, if one quantity doubles, then the other will double as well.

EXAMPLE 9.9 Comparing volumes using proportionality: Two cans

It is a fact that the volume of a cylindrical can of diameter 3 inches is proportional to the height of the can. One such can is 4 inches high, and another is 12 inches high. How do their volumes compare?

SOLUTION

If one of two proportional quantities is multiplied by a certain factor, the other one is also multiplied by that factor. A change from 4 inches to 12 inches is a tripling of the height. Therefore, the volume of the taller can is three times that of the shorter one.

TRY IT YOURSELF 9.9

Two cans have diameter 3 inches. If one can has a height of 10 inches and the second can has a height of 2 inches, how does the volume of the second can relate to that of the first?

The answer is provided at the end of this section.

Another example involves a square. Recall that the perimeter of a square of side S is given by $P = 4S$. This formula says that the perimeter of a square is proportional to the length of one side, with constant of proportionality 4. Assuming that the cost of a square wooden picture frame is proportional to the length of wood used, we can say that the cost of wood for framing the picture is proportional to the length of a side.

The golden ratio ϕ

The article at the beginning of this section refers to a quantity called the "golden ratio," or "golden mean." This special number, denoted by the Greek letter ϕ (phi, pronounced "fie"), is approximately 1.62. Discussion of the golden ratio arises in a variety of contexts, including art and architecture.

Some claim that awareness of the golden ratio can be found in the Egyptian pyramids and in the writings of Pythagoras, but the first known direct mention of the number comes from Euclid (around 300 b.c.), who wrote

> If a straight line is cut in extreme and mean ratio, then as the whole is to the greater segment, the greater segment is to the lesser segment.

Radius Images/Alamy

The great pyramids of Egypt.

This somewhat confusing statement can be restated to give a definition of the golden ratio.

KEY CONCEPT

Cut the segment in Figure 9.37 so that the ratios $\frac{\text{Longer}}{\text{Shorter}}$ and $\frac{\text{Whole}}{\text{Longer}}$ are equal. This common ratio is the golden ratio.

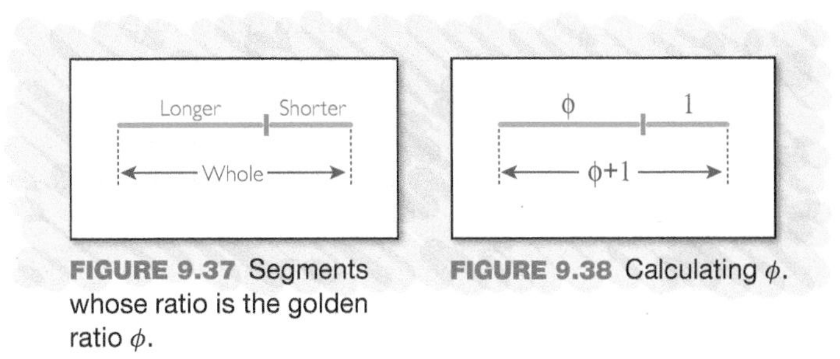

FIGURE 9.37 Segments whose ratio is the golden ratio ϕ.

FIGURE 9.38 Calculating ϕ.

This definition can be used to derive the formula

$$\phi = \frac{1 + \sqrt{5}}{2}$$

which is about 1.62. A derivation is given in Algebraic Spotlight 9.1.

ALGEBRAIC SPOTLIGHT 9.1 Formula for the Golden Ratio

A formula for the golden ratio can be found with a bit of algebra. If we let the shorter length be 1 unit and the longer be ϕ units, as shown in **Figure 9.37** and **Figure 9.38**, we find:

$$\frac{\text{Longer}}{\text{Shorter}} = \frac{\text{Whole}}{\text{Longer}}$$
$$\frac{\phi}{1} = \frac{\phi + 1}{\phi}$$
$$\phi^2 = \phi + 1$$
$$\phi^2 - \phi - 1 = 0$$

We can use the *quadratic formula* from algebra to solve this equation for ϕ. Recall that the quadratic formula says that if $ax^2 + bx + c = 0$, then $x = \frac{-b \pm \sqrt{b^2 - 4ac}}{2a}$. Our equation is $\phi^2 - \phi - 1 = 0$, so in the quadratic formula we put ϕ in place of x and use $a = 1$, $b = -1$, and $c = -1$. The quadratic formula allows us to conclude that

$$\phi = \frac{1 \pm \sqrt{(-1)^2 - 4 \times 1 \times (-1)}}{2 \times 1} = \frac{1 \pm \sqrt{5}}{2}$$

We discard the negative solution. The positive solution is

$$\phi = \frac{1 + \sqrt{5}}{2},$$

which is about 1.62.

The number ϕ arises in a number of settings, but perhaps its best-known occurrence is in the context of a *golden rectangle*.

KEY CONCEPT

A golden rectangle is a rectangle for which the ratio of the length to the width is the golden ratio ϕ. That is, among golden rectangles the length is proportional to the width with constant of proportionality ϕ.

One golden rectangle is shown in **Figure 9.39**.

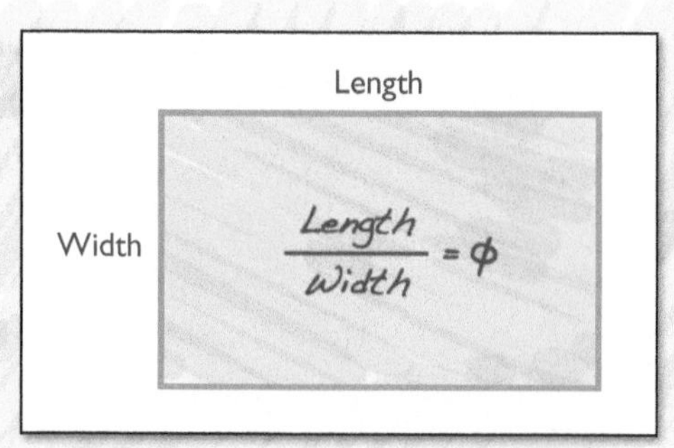

FIGURE 9.39 A golden rectangle: $\frac{\text{Length}}{\text{Width}} = \phi$.

EXAMPLE 9.10 Constructing a golden rectangle: 3 feet wide

The width (the shorter side) of a rectangle is 3 feet. What would the length have to be to make this a golden rectangle?

SOLUTION

Because

$$\frac{\text{Length}}{\text{Width}} = \phi$$

we have

$$\text{Length} = \text{Width} \times \phi$$

Because the width is 3 feet, the length is $3 \times \phi$ feet. This is about 3×1.62 or roughly 4.9 feet.

TRY IT YOURSELF 9.10

The longer side of a golden rectangle is 5 feet. How long is the shorter side?

The answer is provided at the end of this section.

Many claim to see evidence of the golden ratio in the works of da Vinci.

Golden rectangles in art and architecture

Much has been written concerning the aesthetic qualities of ϕ. The relative dimensions of golden rectangles are thought by many to make them the most visually pleasing of rectangles.

In the 1860s, the German physicist and psychologist Gustav Theodor Fechner conducted the following experiment. Ten rectangles varying in their length-to-width ratios were placed in front of a subject, who was asked to select the most pleasing one. The results showed that 76% of all choices centered on the three rectangles having ratios of 1.75, 1.62, and 1.50, with a peak at 1.62. These results, together with many other measurements of picture sizes, book sizes, and architecture, were reported by Fechner in his book *Vorschule der Aesthetik* (School for Aesthetics). He claimed that his observations supported the proposition that the golden rectangle is pleasing to the human eye.

The article in Exercise 62 makes clear the divide that exists regarding people's perceptions of the importance of ϕ. Some people claim the golden rectangle is evident in some of Leonardo da Vinci's paintings, including the *Mona Lisa*. But the article in Exercise 62 disputes that claim. We will not settle the issue here, but we can say that there is some evidence that the famous Spanish artist Salvador Dalí expressly used the golden rectangle in his painting *The Sacrament of the Last Supper*. The number ϕ also played a role in the popular novel *The Da Vinci Code* (2003).

Similar triangles

Unlike golden rectangles, *similar triangles* are not likely to become part of the popular culture. But they are examples of proportionality and form the basis of a great deal of important mathematics, including trigonometry. They are also very important in applications of mathematics.

 **KEY CONCEPT**

Two triangles are similar if all corresponding angles have the same measure.

Figure 9.40 shows similar triangles. For comparison, **Figure 9.41** shows triangles that are not similar.

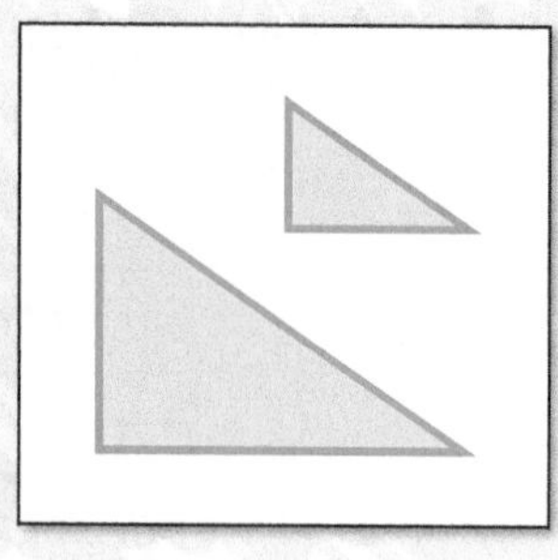

FIGURE 9.40 Similar triangles: Same angle measures.

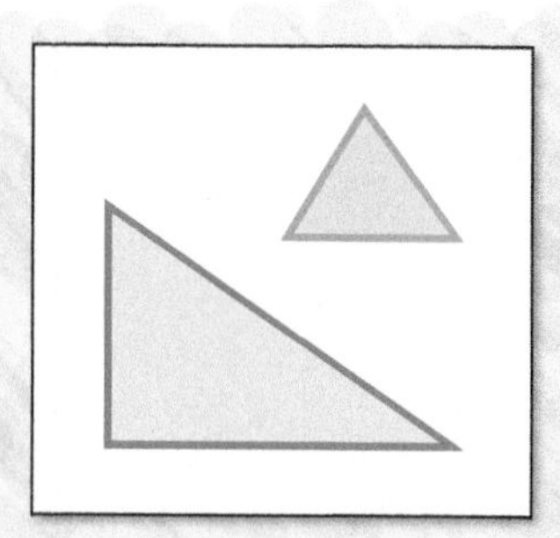

FIGURE 9.41 Triangles that are not similar: Different angle measures.

As is made clear from these figures, similar triangles have the same shape. The most important feature of similar triangles is that the ratios of corresponding sides are the same. That is, the lengths of corresponding sides are proportional. This key fact is illustrated in **Figure 9.42.** In the notation of that figure, the similarity relationship can be expressed using the equations

$$\frac{A}{a} = \frac{B}{b} = \frac{C}{c}$$

A typical use of similarity can be illustrated using **Figure 9.43.** Let's find the missing lengths B and c in the figure. First we find B:

$$\frac{A}{a} = \frac{B}{b}$$
$$\frac{12}{3} = \frac{B}{2}$$
$$4 = \frac{B}{2}$$
$$8 = B$$

In a similar fashion, we find the length c:

$$\frac{A}{a} = \frac{C}{c}$$
$$4 = \frac{16}{c}$$
$$4c = 16$$
$$c = 4$$

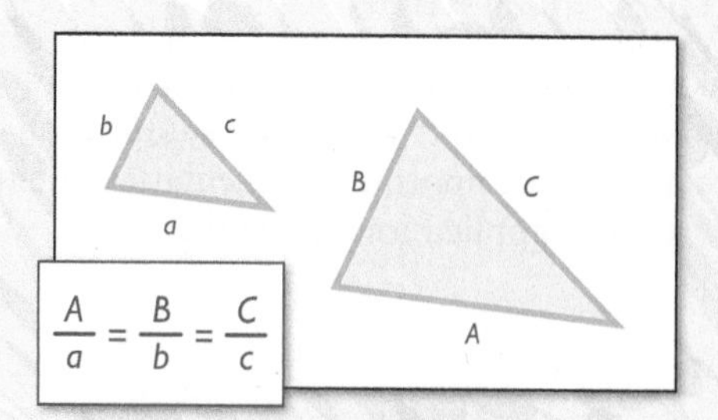

FIGURE 9.42 In similar triangles, ratios of corresponding sides are equal.

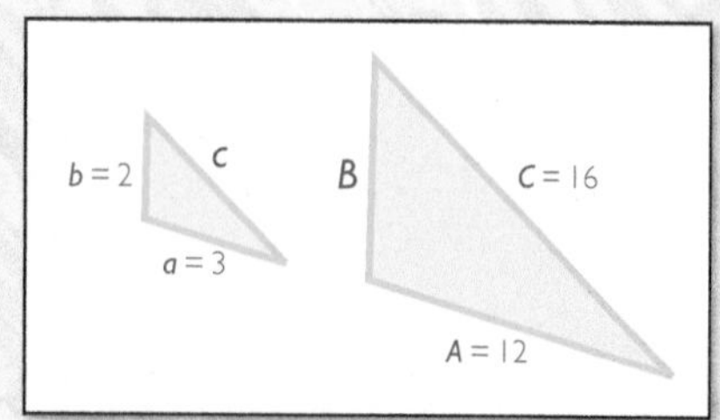

FIGURE 9.43 An example of similarity.

An alternative way of thinking about two similar triangles is to think of them as having sides marked using the same numbers but with different units of length. For example, a triangle with sides of length 3 inches, 4 inches, and 5 inches is similar to a triangle with sides of length 3 feet, 4 feet, and 5 feet. Both are similar to a triangle with sides of length 3 miles, 4 miles, and 5 miles.

It is a fact that the three angles of a triangle must add up to 180 degrees. Consequently, if two corresponding angles of a pair of triangles have the same measure, so does the third, and the triangles are similar.

SUMMARY 9.4
Similar Triangles

1. Triangles are similar if all corresponding angles have the same measure.
2. For similar triangles, ratios of corresponding sides are the same. Therefore, the lengths of corresponding sides are proportional.
3. If two of the three pairs of corresponding angles of a triangle have the same measure, the triangles are similar.

EXAMPLE 9.11 Using similar triangles: Shadows

Suppose we find that a tall building casts a shadow of length 25 feet. At the same time, a 6-foot-tall man casts a shadow 2 feet long. How tall is the building?

Shadow of the Empire State Building.

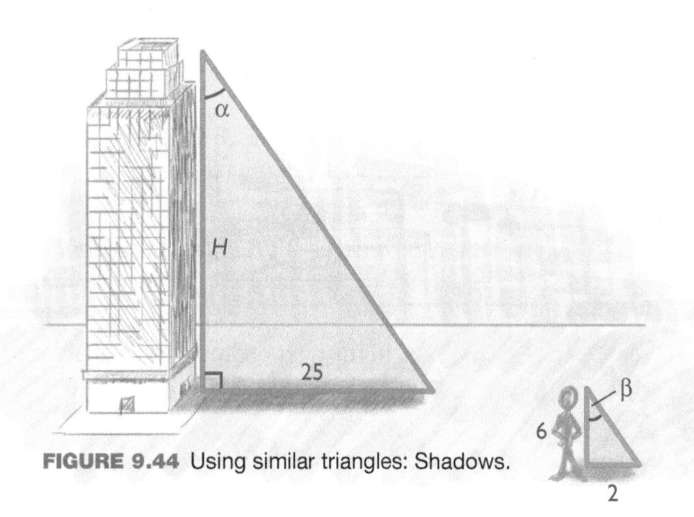

FIGURE 9.44 Using similar triangles: Shadows.

SOLUTION

We can represent the man and the building using vertical lines. Thus, the two triangles in **Figure 9.44** each have a right angle at their base. Because we are measuring the shadows at the same time of day, the angles of sunlight with the vertical (marked α and β) are the same for both the man and the building. Because the two triangles have two pairs of equal angles, the triangles are similar. Hence, the ratios of corresponding sides are the same. In the figure, we have marked the height of the building as H. Therefore,

$$\frac{H}{6} = \frac{25}{2}$$
$$H = 6 \times \frac{25}{2}$$
$$H = 75$$

The building is 75 feet tall.

TRY IT YOURSELF 9.11

A 5-foot-tall woman casts a shadow 3 feet long. What will be the length of the shadow of a 12-foot-tall tree at the same time of day?

The answer is provided at the end of this section.

Other types of proportionality

We have already noted that the area A of a circle is *not* proportional to the radius r. Because $A = \pi r^2$, the area is a constant multiple of the *square* of the radius, however. Therefore, we may say that A is proportional to r^2. We express this relationship verbally by saying that *the area of the circle is proportional to the square of the radius, with proportionality constant* π. Another way to look at this relationship is to note that the ratio A/r^2 is a constant, the number π, no matter how large or small the circle might be.

EXAMPLE 9.12 Comparing areas: Radius doubled

If the radius of a circle is doubled, what happens to the area?

SOLUTION

If the radius is r and the area is A, the relationship is given by the formula $A = \pi r^2$. If the radius is doubled, we replace r in the expression πr^2 by $2r$. This gives a new area of $\pi(2r)^2 = 4\pi r^2 = 4A$. The result is that the area is multiplied by 4.

TRY IT YOURSELF 9.12

Suppose the diameter of a pizza is reduced by half. What happens to the amount of pizza?

The answer is provided at the end of this section.

In the next example, we consider proportionality relationships involving volume and surface area.

EXAMPLE 9.13 Applying proportionality relationships: Mailing boxes

Suppose we have several boxes of different sizes, each of which is in the shape of a cube.

a. Is the volume of a box proportional to the length of one side of the box, the square of the length, or the cube of the length? Find the constant of proportionality.

b. One of our boxes has sides that are twice as long as those of another box. How much more does the larger box hold than the smaller box?

c. Is the surface area of a box proportional to the length of one side of the box, the square of the length, or the cube of the length?

d. The cost of wrapping paper for a box is proportional to the surface area. What happens to the cost of wrapping paper if the sides of the cube are doubled in length?

SOLUTION

a. Let x denote the length of one side of the cube, and let V denote the volume. Because the box is cubical in shape, the length, width, and height are all the same, x. Therefore,

$$V = \text{Length} \times \text{Width} \times \text{Height}$$
$$V = x \times x \times x$$
$$V = x^3$$

Thus, the volume is proportional to the cube of the length, and the constant of proportionality is 1.

b. As in part a, we denote the length of a side by x. If the sides are doubled, we replace x by $2x$. This gives a new volume of

$$(2x)^3 = 2^3x^3 = 8x^3$$

The other volume was x^3. Thus, the larger volume is 8 times the smaller volume.

c. The amount of paper needed is the surface area of the box. The surface of a cube has six square "faces." If x represents the length of a side, the area of one face is $x \times x = x^2$. Therefore, the area of the surface of the cube is $6x^2$. This means that the surface area is proportional to the square of the length, with constant of proportionality 6.

d. We know that the cost is proportional to the surface area. To find the surface area of the larger cube, we replace x by $2x$ in the formula we used in part c:

$$\text{New surface area} = 6 \times (2x)^2$$
$$= 4(6x^2)$$

The larger surface area is 4 times the other surface area. So 4 times as much wrapping paper will be needed for the larger cube as for the smaller cube. Therefore, the cost of wrapping paper is multiplied by 4 when the sides are doubled in length.

The idea of proportionality occurs often in biology. For example, as a simple model, biologists may say that the weight of an animal is proportional to the cube of its height. Models like this show that animals grown to an enormous size (as often represented in science fiction) would collapse under their own weight. That is the point of the next example.

Bettmann/Getty Images

Classic poster of King Kong with Fay Wray (from the original movie, released April 7, 1933).

EXAMPLE 9.14 Applying proportionality relationships: Overgrown apes

In a science fiction movie, an ape has grown to 100 times its usual height.

a. Assume the weight of an ape is proportional to the cube of its height. How does the weight of the overgrown ape compare to its original weight?

b. Assume the cross-sectional area of a limb is proportional to the square of the height. How does the cross-sectional area of a limb of the overgrown ape compare to the original area?

c. The pressure on a limb is the weight divided by the cross-sectional area. Use parts a and b to determine how the pressure on a limb of the overgrown ape compares to the original pressure.

SOLUTION

a. Let h denote the height of an ape and W the weight. Because the weight is proportional to the cube of the height, we have the formula

$$W = ch^3$$

Here, c is the constant of proportionality. If the height is multiplied by 100, we replace h by $100h$. This gives a new weight of

$$c(100h)^3 = 100^3 \times ch^3$$

The original weight was ch^3. Thus, the new weight is 100^3 times the original weight. In other words, the weight increases by a factor of $100^3 = 1{,}000{,}000$ (one million).

b. Let A denote the cross-sectional area of a limb and, as in part a, let h denote the height. Because the cross-sectional area is proportional to the square of the height, we have the formula

$$A = Ch^2$$

Here, C is the constant of proportionality. If the height is multiplied by 100, we replace h by $100h$. This gives a new area of

$$C(100h)^2 = 100^2 \times Ch^2$$

The original area was Ch^2. Thus, the new area is 100^2 times the original area. In other words, the area increases by a factor of $100^2 = 10{,}000$ (ten thousand).

c. Because the pressure on a limb is the weight divided by the cross-sectional area, the pressure will increase by a factor of

$$\frac{\text{Factor by which weight increases}}{\text{Factor by which area increases}} = \frac{100^3}{100^2} = 100$$

The pressure on a limb of the overgrown ape is 100 times the original pressure.

The tremendous increase in pressure on a limb means that the overgrown ape would collapse under its own weight. Science fiction aside, such *scaling arguments* are used by biologists to study the significance of the size and shape of organisms. Exercise 56 provides one example.

WHAT DO YOU THINK?

A bigger pizza: A pizza shop offers, for double the regular price, either to double the thickness of a pizza or to double the radius of a pizza. Which is the better deal in terms of the volume of food per dollar spent? *Suggestion*: Consider the formula for the volume of a cylinder, and think of the thickness of the pizza as the height of the cylinder.

Questions of this sort may invite many correct responses. We present one possibility: Use r for the radius of the pizza and t for the thickness. The pizza itself is a cylinder, so its volume is given by

$$V = \pi r^2 t$$

If we double the thickness,

$$\text{New volume } = \pi r^2(2t) = 2V$$

Thus, doubling the thickness doubles the volume. If we double the radius,

$$\text{New volume } = \pi(2r)^2 t = 4\pi r^2 t = 4V$$

So doubling the radius gives 4 times the volume. We should take the doubled-radius pizza.

Further questions of this type may be found in the *What Do You Think?* section of exercises.

Try It Yourself answers

Try It Yourself 9.8: Applying proportionality: Forestry No.

Try It Yourself 9.9: Comparing volumes using proportionality: Two cans The volume of the second can is one-fifth that of the first can.

Try It Yourself 9.10: Constructing a golden rectangle: 5 feet long $5/\phi$ or about 3.1 feet.

Try It Yourself 9.11: Using similar triangles: Shadows 7.2 feet.

Try It Yourself 9.12: Comparing areas: Radius halved The area is one-fourth of what it was, which means we get one-fourth the pizza.

Exercise Set 9.2

Test Your Understanding

1. Two variable quantities are proportional provided ________.

2. True or false: If quantity A is proportional to quantity B, then quantity B is proportional to quantity A.

3. If quantity A is proportional to quantity B, and if quantity A is doubled, then what happens to quantity B?

4. The golden rectangle is a rectangle where the ratio of length to width is ________.

5. Two triangles are similar provided ________.

6. If triangles are similar, then the lengths of corresponding sides are ________.

Problems

7. Comparing volumes. If the base and width of a box remain the same, then the volume is proportional to the height. If one such box is 6 inches tall and another is 24 inches tall, how do their volumes compare?

8. A golden rectangle. The width (shorter side) of a rectangle is 5 feet. What would the length have to be to make this a golden rectangle?

9. Comparing areas. If the radius of a circle is tripled, what happens to the area?

10. The surface area of a sphere. The surface area of a sphere of radius r is given by $S = 4\pi r^2$. If the radius is multiplied by 2, how is the surface area affected?

11. Luggage size. The Web site of Southwest Airlines says that the airline allows a maximum weight of 50 pounds and a maximum size of 62 inches per checked piece of luggage. By "size," the statement means the sum Length + Width + Height. (Charges apply if these are exceeded.) Is the size proportional to the length? Is the size proportional to the volume of the suitcase?

12–21. Which are proportional? *Exercises 12 through 21 ask about proportionality relationships. For those that describe proportionality relationships, give the constant of proportionality. In some cases, your answer may depend on your interpretation of the situation.*

12. If each step in a staircase rises 8 inches, is the height of the staircase above the base proportional to the number of steps?

13. Suppose that hamburger meat costs $3.90 per pound. Is the cost of the hamburger meat proportional to its weight?

14. Is the volume of a balloon proportional to the cube of its diameter?

15. Is the volume of a balloon proportional to its diameter?

16. Is the area of a pizza proportional to its diameter?

17. Is the volume of a pizza with diameter 12 inches proportional to its thickness?

18. Is the area of land in a square lot proportional to the length of a side?

19. Is the area of a rectangular TV screen proportional to the length of its diagonal?

20. Consider a box with a square base and a height of 2 feet. Is the volume proportional to the area of the base?

21. The surface area S of a sphere of radius r is given by $S = 4\pi r^2$. Is the surface area of a balloon proportional to the square of its diameter?

22. Circle. Recall that the area of a circle of radius r is given by the formula $A = \pi r^2$. Your friend says that this formula shows that the area is proportional to the radius with proportionality constant πr because the area is πr times r. Is your friend correct? Explain.

23–30. The golden ratio. *Exercises 23 through 30 concern the golden ratio ϕ.*

23. The shorter side of a golden rectangle is 5 inches. How long is the longer side? Round your answer to one decimal place.

24. The longer side of a golden rectangle is 10 inches. How long is the shorter side? Round your answer to one decimal place.

25. Measure your television screen. What is the ratio of the longer side to the shorter? Is the screen a golden rectangle?

26. Measure your computer screen. What is the ratio of the longer side to the shorter? Is the screen a golden rectangle?

27. Is this textbook a golden rectangle?

28. It has been observed that the distance from a person's shoulder to the fingertips divided by the distance from the elbow to the fingertips is close to ϕ. Take these measurements of your own arm and calculate the ratio. Is your answer close to ϕ?

29. It has been observed that a person's height divided by the distance from the person's belly button to the floor is close to ϕ. Take these measurements of your own body and calculate the ratio. Is your answer close to ϕ?

30. It has been observed that the distance from a person's hip to the floor divided by the distance from the knee to the floor is close to ϕ. Take these measurements of your own leg and calculate the ratio. Is your answer close to ϕ?

31–35. Similar triangles. *Exercises 31 through 35 deal with similarity of triangles.*

31. The triangles in **Figure 9.45** are similar. Find the sides labeled A and B.

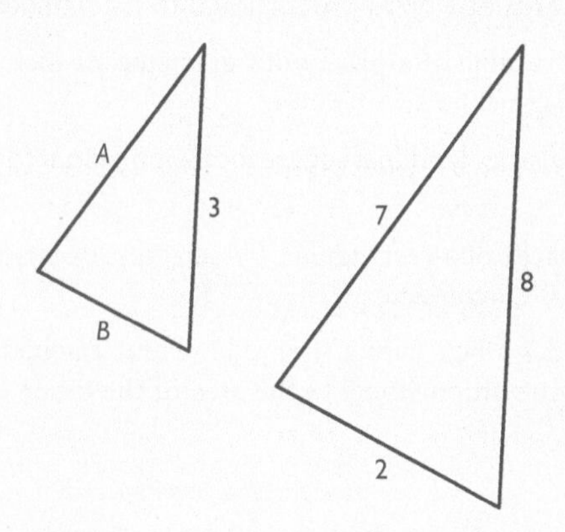

FIGURE 9.45

32. The triangles in **Figure 9.46** are similar. Find the sides labeled A and b.

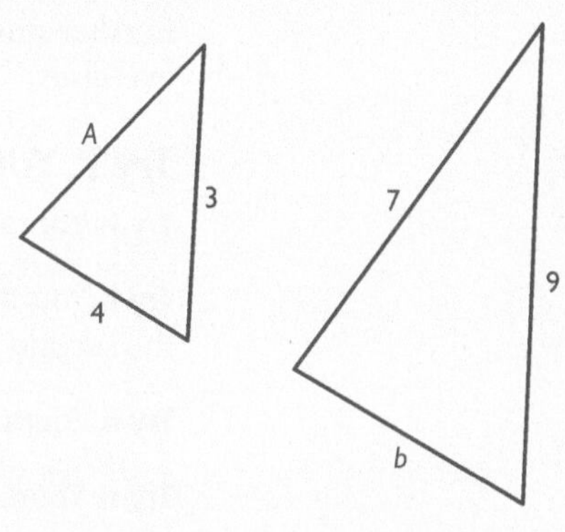

FIGURE 9.46

33. A pole casts a 4-foot shadow. At the same time of day, a 6-foot-tall man casts a 1-foot shadow. How tall is the pole?

34. A 10-foot-tall tree casts a 4-foot shadow, and at the same time of day a girl casts a 2-foot shadow. How tall is the girl?

35. A tennis player stands 39 feet behind the net, which is 3.5 feet tall. He extends his racket to a height of 9 feet and hits a serve. The ball passes 6 inches above the net. How far beyond the net does the tennis ball strike the court? *Suggestion*: If d is the distance past the net, the distance from the server to where the ball strikes the court is $39 + d$.

36. Doubling an equilateral triange. It is a fact that the area of an equilateral triangle is proportional to the square of a side. If all sides of an equilateral triangle are doubled, what happens to the area?

37. Boxes of fixed height. Suppose a box has a fixed height of 9 feet and a square base. Explain why the length of one of the sides of the base is proportional to the square root of the volume. What is the constant of proportionality?

38–41. Volume of a ball. *The volume V of a ball of radius r is $V = \frac{4}{3}\pi r^3$, and the surface area S is given by $S = 4\pi r^2$. Exercises 38 through 41 make use of these formulas.*

38. The amount of gas in a balloon of radius r is proportional to:

a. r **b.** r^2 **c.** r^3 **d.** π

What is the constant of proportionality?

39. A child wants a balloon of twice the radius of the offered balloon. How does the increase in radius affect the amount of gas in the balloon?

40. The amount of paint needed to color a styrofoam ball of radius r is proportional to:

a. r **b.** r^2 **c.** r^3 **d.** π

What is the constant of proportionality?

41. A softball has a radius 1.33 times that of a baseball. How many times as much material is required to cover a softball than a baseball?

42–50. A cylinder. *Recall that a cylinder with radius r and height h has volume* $V = \pi r^2 h$ *and surface area* $S = 2\pi rh$ *(not counting the top and bottom). Exercises 42 through 50 make use of these formulas.*

42. If the height of a cylindrical can is held constant, the volume is proportional to:

a. r b. r^2 c. h d. π

What is the constant of proportionality?

43. Two soup cans have a height of 4 inches. One can is twice the radius of the other. How does the amount of soup in the cans compare?

44. Two soup cans have a radius of 2 inches. One can is twice the height of the other. How does the amount of soup in the cans compare?

45. If the radius is held constant, the volume of a cylindrical can is proportional to:

a. r b. r^2 c. h d. π

What is the constant of proportionality?

46. If the cylinder is a can of tuna and the radius is held constant, the area of the label is proportional to:

a. r b. r^2 c. h d. π

What is the constant of proportionality?

47. Two tuna cans have a radius of 2 inches. One is twice as tall as the other. How do their labels compare in area?

48. If the height is held constant, the surface area of a can is proportional to:

a. r b. r^2 c. h d. π

What is the constant of proportionality?

49. Two soup cans have a height of 4 inches. One can has 3 times the radius of the other. How do their labels compare in area?

50. If we include the top and bottom of a cylindrical can, the total surface area is $A = 2\pi rh + 2\pi r^2$. If the height is held constant, the total surface area is proportional to:

a. r b. r^2 c. h d. π e. none of these

51–55. A cone. *Suppose a cone has height h and the radius of its circular base is r. Then the volume is given by* $V = \frac{1}{3}\pi r^2 h$. *The surface area (not counting the circular base) is given by* $A = \pi r\sqrt{r^2 + h^2}$. *Exercises 51 through 55 make use of these formulas.*

51. Varying radius. If the height of a cone is held constant, the volume is proportional to:

a. r b. r^2 c. h d. π e. none of these

What is the constant of proportionality?

52. Two ice cream cones have the same height. One's radius is twice the radius of the other. Compare the amounts of ice cream the cones can hold.

53. If the radius of a cone is held constant, the volume is proportional to:

a. r b. r^2 c. h d. π e. none of these

What is the constant of proportionality?

54. Two ice cream cones have the same radius. One has twice the height of the other. Compare the amounts of ice cream the cones can hold.

55. If the radius of the circular base is held constant, the surface area is proportional to:

a. r b. r^2 c. h d. π e. $r^2 + h^2$ f. none of these

56. Basal metabolic rate. The *basal metabolic rate* (BMR) of an animal is a measure of the amount of energy it needs for survival. An animal with a larger BMR will need more food to survive. One simple model of the BMR states that among animals of a similar shape, the rate is proportional to the square of the length. (This model assumes that the BMR is proportional to the surface area.) If one animal is 10 times as long as another, how do the BMR values compare?

Writing About Mathematics

57. An excerpt. An excerpt from the 2003 best-selling novel *The Da Vinci Code* goes as follows:

> Plants, animals and even human beings all possessed dimensional properties that adhered with eerie exactitude to the ratio of PHI [sic, ϕ] to 1.... So the ancients assumed PHI must have been preordained by the Creator....
> Langdon grinned. "Ever study the relationship between female and males in a honeybee community?"
> "Sure. The female bees always outnumber the male bees."
> "And did you know that if you divide the number of female bees by the number of male bees in any beehive in the world, you always get the same number?"
> "You do?"
> "Yup. PHI."
> The girl gasped. "No way!"

There are various resources, including the Web, available for doing research about bees. Does your research bear out the claim that bees "adhered with eerie exactitude to the ratio of ϕ to 1"?

58. Art. Do some research on the use of the golden rectangle in art and write a report.

59. Architecture. Do some research on the use of the golden rectangle in architecture and write a report.

60. History. The idea of proportionality has a long history in geometry. Explain how the idea of proportion is defined by the Greek scholar Eudoxus and used in Euclid's *Elements*.

61. History. Scaling arguments such as those used in Example 9.14 were given by Galileo in his 1638 book *Two New Sciences*. Summarize the arguments used by Galileo and explain how he applied them.

62. A different point of view. *In The News 9.3*, on the next page, is an excerpt from the full article that appears at www.maa.org/external_archive/devlin/devlin_06_04.html. It refers to the 2003 novel *The Da Vinci Code*, which was later made into a movie. Although the article was written in 2004, it presents a skeptical view of some of the common stories about the golden ratio, a view that is still relevant.

Try to find other sources that either perpetuate stories about the golden ratio in art and architecture or that back up the skepticism. Write a report on your findings.

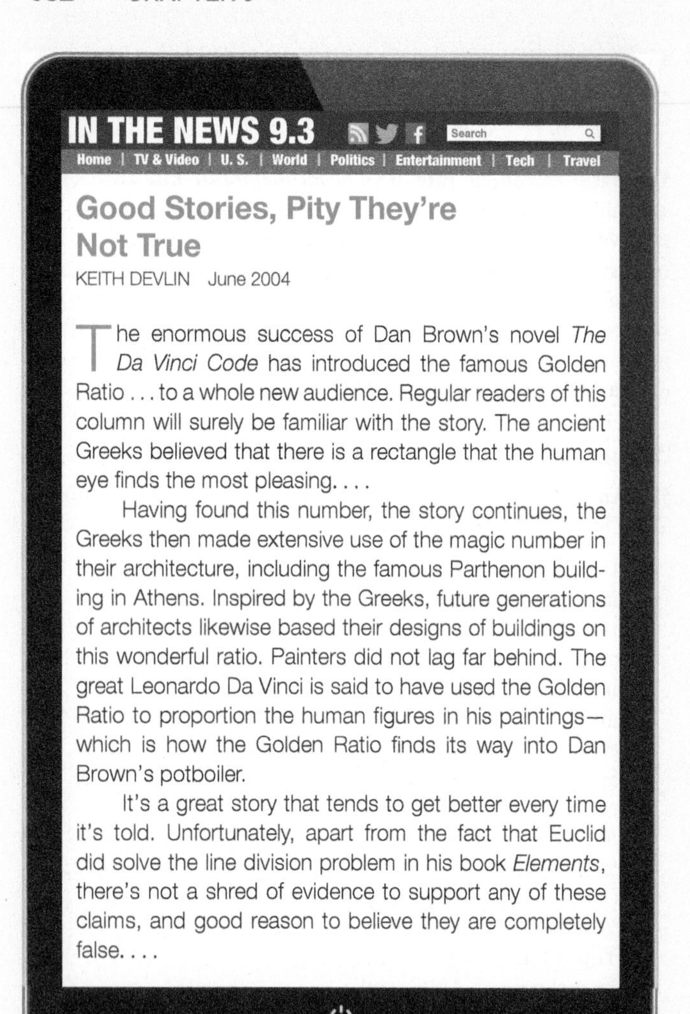

IN THE NEWS 9.3

Home | TV & Video | U. S. | World | Politics | Entertainment | Tech | Travel

Good Stories, Pity They're Not True

KEITH DEVLIN June 2004

The enormous success of Dan Brown's novel *The Da Vinci Code* has introduced the famous Golden Ratio . . . to a whole new audience. Regular readers of this column will surely be familiar with the story. The ancient Greeks believed that there is a rectangle that the human eye finds the most pleasing. . . .

Having found this number, the story continues, the Greeks then made extensive use of the magic number in their architecture, including the famous Parthenon building in Athens. Inspired by the Greeks, future generations of architects likewise based their designs of buildings on this wonderful ratio. Painters did not lag far behind. The great Leonardo Da Vinci is said to have used the Golden Ratio to proportion the human figures in his paintings—which is how the Golden Ratio finds its way into Dan Brown's potboiler.

It's a great story that tends to get better every time it's told. Unfortunately, apart from the fact that Euclid did solve the line division problem in his book *Elements*, there's not a shred of evidence to support any of these claims, and good reason to believe they are completely false. . . .

What Do You Think?

63. Proportionality. Proportionality is discussed in this section. An important related concept is *inverse proportionality*. Do research to find out what that term means, and give practical examples.

64. How areas change. How is the area of a rectangle affected as the length and width increase? You might begin your investigation by considering the following questions. What happens if both are doubled? What happens if either the length or the width is doubled but the other stays the same? **Exercise 18 is relevant to this topic.**

65. How the volume of a box changes. How is the volume of a box affected if the sides change by the same factor? First determine what happens if all the sides are doubled. Then consider what happens if all the sides are multiplied by 3, 4, . . . **Exercise 20 is relevant to this topic.**

66. The golden rectangle. Much has been written regarding the role of the golden rectangle in science, art, and philosophy. There are lots of opinions, but there is little agreement. Do your own research regarding the golden rectangle and report on your findings. **Exercise 57 is relevant to this topic.**

67. Similarity and surveying. Similarity of triangles plays an important role in surveying. Give examples of how you might use similarity to produce measurements that are difficult to make otherwise. For example, explain how to use a shadow to determine the height of a building. **Exercise 33 is relevant to this topic.**

68. Square. Show that the perimeter of an equilateral triangle is proportional to the length of a side, and find the constant of proportionality. Show that the perimeter of a square is proportional to the length of a side, and find the constant of proportionality. Can you generalize these two results to other figures? **Exercise 36 is relevant to this topic.**

69. Representation. Certain voting systems incorporate the idea of *proportional representation*. Do some research to determine what this term means. Is this usage consistent with the way the term *proportionality* is defined in this section?

70. Directly proportional. Sometimes in the popular media, a quantity A is said to be *directly proportional* to a quantity B if A increases when B increases. (In the terminology of Section 6.4, such quantities are *positively correlated*.) For example, you might read, "The amount of ice cream sold is directly proportional to the temperature outdoors." Is the popular usage of this term consistent with the usage in this section? Explain your answer by giving examples.

71. Applying proportionality. The concepts of proportionality and similarity are used by cartographers to make maps, and by architects and others to create scale drawings. Investigate the use of proportionality and similarity in these areas. To get started, answer the following question: Is a triangle whose sides have length 2, 7, and 8 miles similar to a triangle whose sides have length 2, 7, and 8 inches?

9.3 Symmetries and tilings: Form and patterns

TAKE AWAY FROM THIS SECTION
Various types of symmetries appear in art, architecture, and nature. Rotational symmetry and reflectional symmetry are two of these types.

The chief forms of beauty are order and symmetry and definiteness, which the mathematical sciences demonstrate in a special degree.
—Aristotle

Symmetry is among the most important properties of a geometric object. It may not be easy to define symmetry, but the idea is intuitive and has an aesthetic appeal. The following is an excerpt from an article published by the *Proceedings of The Royal Society* that appeared on the Pub Med Central (PMC) section of the National Institutes of Health (NIH) Web site.

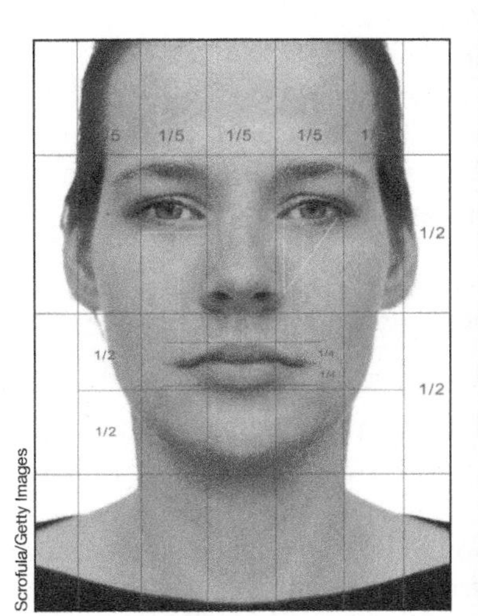

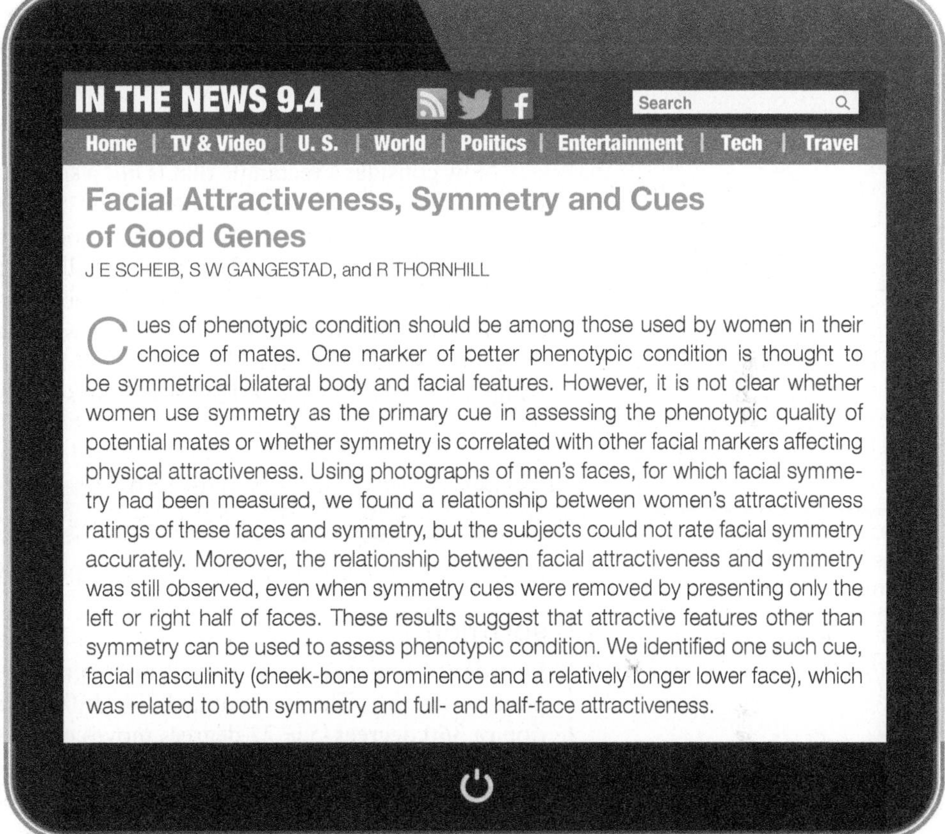
IN THE NEWS 9.4

Search

Home | TV & Video | U. S. | World | Politics | Entertainment | Tech | Travel

Facial Attractiveness, Symmetry and Cues of Good Genes

J E SCHEIB, S W GANGESTAD, and R THORNHILL

Cues of phenotypic condition should be among those used by women in their choice of mates. One marker of better phenotypic condition is thought to be symmetrical bilateral body and facial features. However, it is not clear whether women use symmetry as the primary cue in assessing the phenotypic quality of potential mates or whether symmetry is correlated with other facial markers affecting physical attractiveness. Using photographs of men's faces, for which facial symmetry had been measured, we found a relationship between women's attractiveness ratings of these faces and symmetry, but the subjects could not rate facial symmetry accurately. Moreover, the relationship between facial attractiveness and symmetry was still observed, even when symmetry cues were removed by presenting only the left or right half of faces. These results suggest that attractive features other than symmetry can be used to assess phenotypic condition. We identified one such cue, facial masculinity (cheek-bone prominence and a relatively longer lower face), which was related to both symmetry and full- and half-face attractiveness.

Rotational symmetries in the plane

We will examine two types of symmetry: rotational symmetry and reflectional symmetry.

KEY CONCEPT

A planar figure has rotational symmetry about a point if it remains exactly the same after a rotation about that point of less than 360 degrees.

Let's agree at the outset that when we speak of a "rotation," we always mean it to be in a clockwise direction. If we rotate a square about its center by 90 degrees, the result is a square that is an exact copy of the one we started with. Therefore, we say that a square has 90-degree *rotational symmetry.* In fact, a rotation of a square returns an exact copy of the square if and only if we rotate it by a multiple of 90 degrees. For example, if the square is rotated by 45 degrees, it is tilted and does not remain the same. These facts are illustrated in **Figure 9.47**.

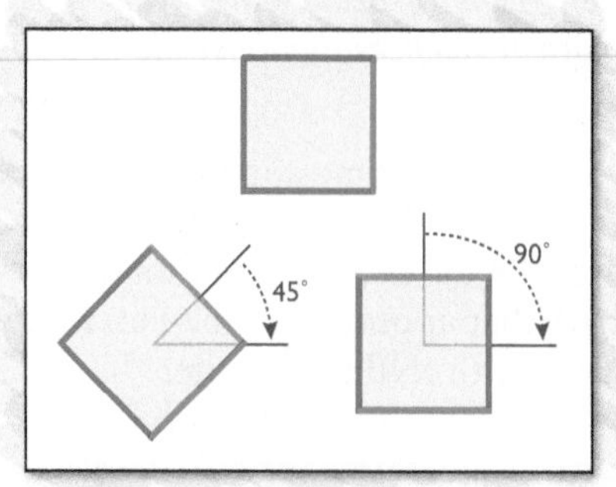

FIGURE 9.47 A square has 90-degree rotational symmetry but not 45-degree rotational symmetry.

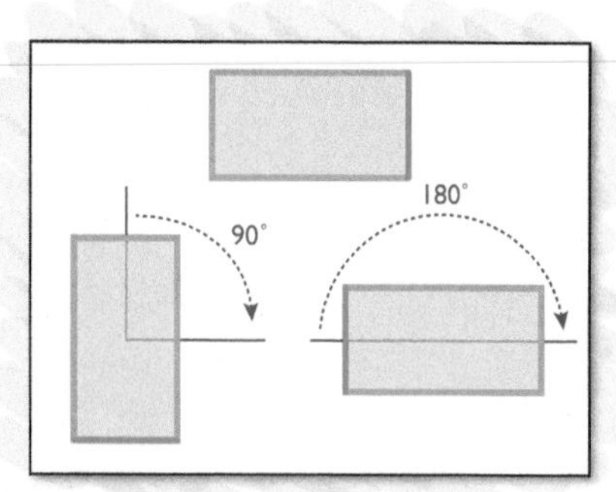

FIGURE 9.48 A rectangle has 180-degree rotational symmetry but not 90-degree rotational symmetry.

Now consider a rectangle that is *not* a square. This object does not have 90 degrees rotational symmetry because rotation by 90 degrees does not leave the rectangle exactly the same as it was originally. It does have 180-degree rotational symmetry about its center, however. These facts are illustrated in **Figure 9.48**.

Another example of rotational symmetry is provided by a circle. A circle has *complete rotational symmetry* because if we rotate it any number of degrees about the center of the circle, the result is exactly the same as the original circle. By way of comparison, the triangle in **Figure 9.49** has no rotational symmetries at all.

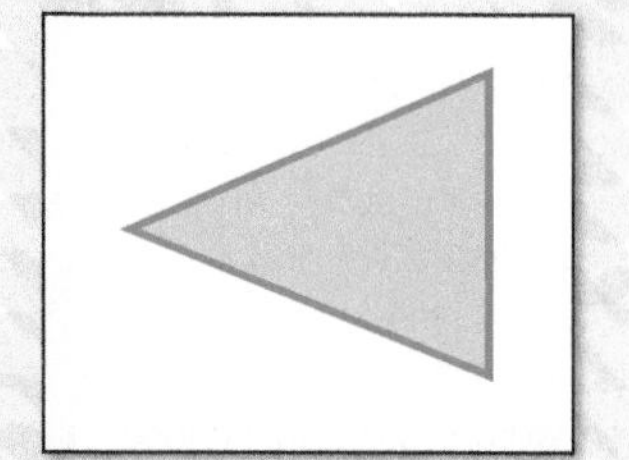

FIGURE 9.49 This triangle has no rotational symmetries.

EXAMPLE 9.15 Finding rotational symmetries: Pentagram

Find the rotational symmetries of the five-pointed star, or pentagram, in **Figure 9.50**.

SOLUTION

In order to preserve the star, we must rotate through an angle that takes one star point to another. The star points divide 360 degrees into five equal angles, so a rotation of 360 degrees/5 = 72 degrees moves one star point to the next. The pentagram has 72-degree rotational symmetry about its center. The other rotational symmetries are multiples of 72 degrees, namely 2×72 degrees = 144 degrees, 3×72 degrees = 216 degrees, and 4×72 degrees = 288 degrees.

FIGURE 9.50 A pentagram.

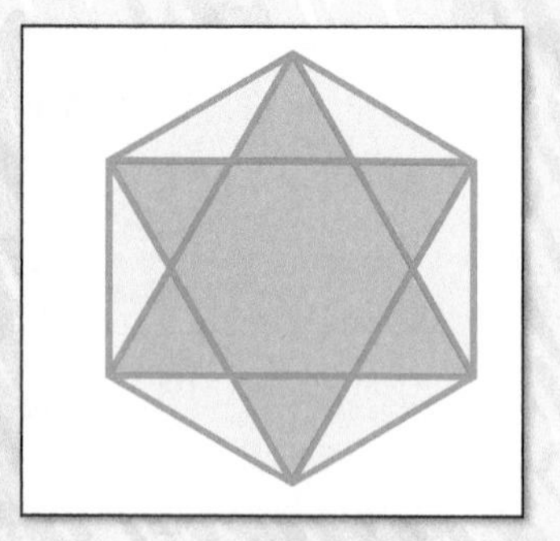

FIGURE 9.51 A six-pointed star.

TRY IT YOURSELF 9.15

Find the rotational symmetries of the six-pointed star shown in **Figure 9.51**.

The answer is provided at the end of this section.

Many familiar objects have rotational symmetries, as the following photos show.

Mableen/Getty Images

Ole Schoener/Shutterstock

© Novadomus/Dreamstime.com

The pictured starfish has 60-degree rotational symmetry, the lunaria has 90-degree rotational symmetry, and the dahlia has many more rotational symmetries.

Reflectional symmetry of planar figures

When we reflect a figure in the plane about a line in the plane, we can think of letting the line serve as an axis and flipping the plane about that line. If we imagine a mirror placed along the line, the reflection of the figure about the line is the mirror image of the figure. **Figure 9.52** illustrates the reflection of a shape about the line L.

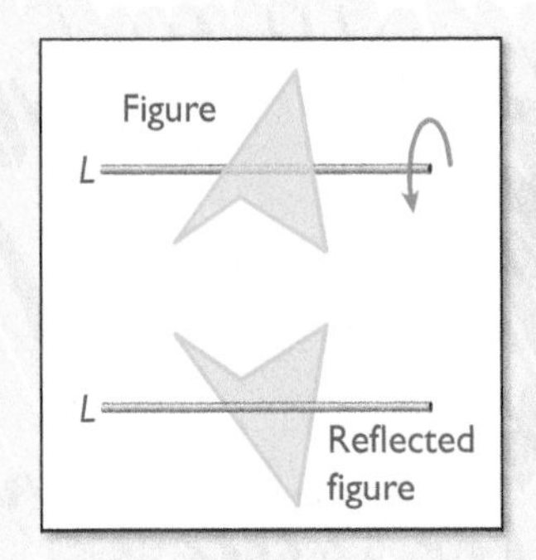

FIGURE 9.52 Reflection is the same as flipping the plane about a line.

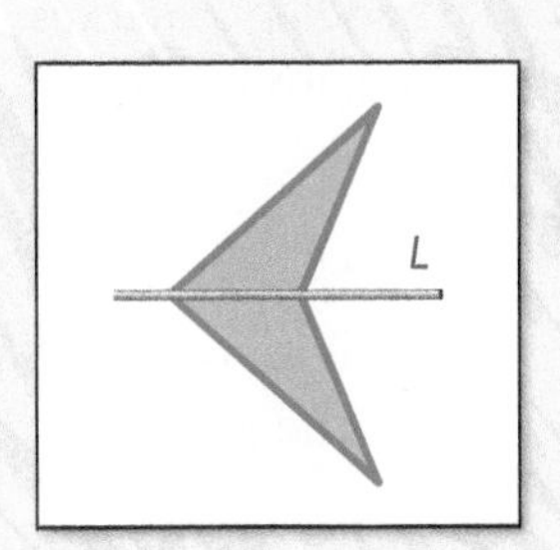

FIGURE 9.53 This shape has reflectional symmetry about the line shown.

FIGURE 9.54 This star has reflectional symmetry about five different lines.

KEY CONCEPT

A planar figure has reflectional symmetry about a line L if the figure is identical to its reflection through L.

Cartoonstock.com

"You're a nice guy and all that, but I'm looking for somebody more *symmetrical.*"

For example, the dart shape in **Figure 9.53** has reflectional symmetry about the line shown. Similarly, the triangle in Figure 9.49 has reflectional symmetry about a horizontal line that bisects the angle at the left-hand side of the picture. Note that neither of these figures shows any rotational symmetry. The five-pointed star in **Figure 9.54** has reflectional symmetry about each of the five lines shown there.

Reflectional symmetries are shown in many familiar settings, including the faces of animals.

Faces show reflectional symmetry.

Many flags such as the Macedonian flag show reflectional symmetry through both horizontal and vertical lines.

EXAMPLE 9.16 Finding symmetries: A given shape

What are the rotational and reflectional symmetries of the shape in **Figure 9.55**?

SOLUTION

There is reflectional symmetry about both horizontal and vertical lines. The shape has a rotational symmetry of 180 degrees.

TRY IT YOURSELF 9.16

What are the rotational and reflectional symmetries of the shape in **Figure 9.56**?

FIGURE 9.55

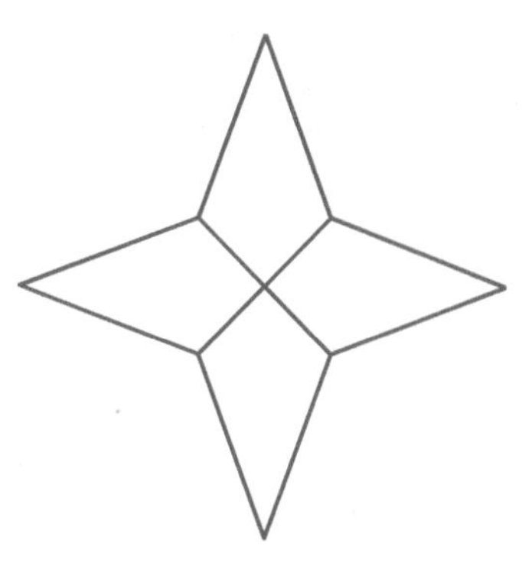

FIGURE 9.56

The answer is provided at the end of this section.

Regular tilings of the plane

We often come across repeated patterns of polygons (closed figures made of three or more line segments), especially in the form of ceramics or porcelain as decorative building materials. A polygon that is repeated and covers a plane with no overlaps or gaps in between the polygons is called a *tiling* or *tessellation* of the plane.

KEY CONCEPT

A tiling or tessellation of the plane is a pattern of repeated figures that cover up the plane.

Simple examples are the tilings shown in **Figure 9.57**, which uses one square tile, and **Figure 9.58**, which uses one square and two different rectangular tiles. Perhaps one of these matches the configuration in your bathroom floor.

FIGURE 9.57 A tiling by squares.

FIGURE 9.58 A tiling by squares and two rectangles.

A *regular polygon* is a polygon in which all sides are of equal length and all angles have equal measure. Such polygons can be used to make a special kind of tiling.

KEY CONCEPT

A regular tiling is a tiling of the plane that consists of repeated copies of a single regular polygon, meeting edge to edge, so that every vertex has the same configuration.

The tiling in Figure 9.57 is regular, but the tiling in Figure 9.58 is not regular because different vertices have different configurations. The fact that each vertex of a regular tiling must have the same configuration lets us quickly narrow the list of polygons that might regularly tile the plane. At each vertex, the polygons must fit together, and their angles must add up to 360 degrees. This is illustrated in **Figure 9.59**, where the vertices of six polygons fit together and have angles adding up to 360 degrees. For example, if the tile is a square, then it takes four 90-degree angles to add up to 360 degrees. That means that Figure 9.57 shows the only way to make a regular tiling of the plane by squares. If the angle of a regular polygon is greater than 120 degrees, it cannot make a regular tiling of the plane. This is illustrated in **Figure 9.60**, which shows that if two angles are greater than 120 degrees, then there isn't room for a third angle of that size because the sum of three or more such angles would exceed 360 degrees.

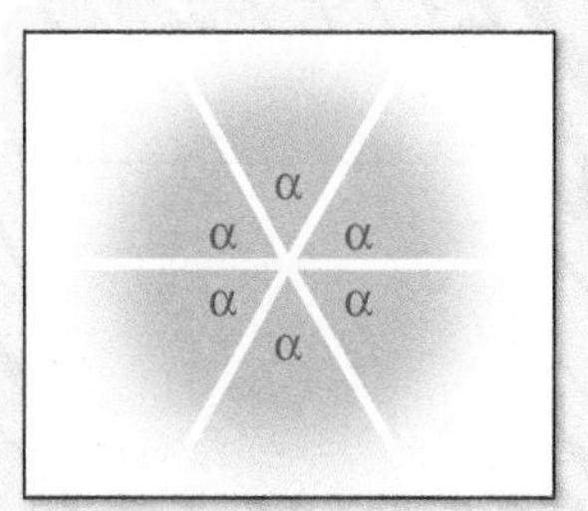

FIGURE 9.59 In a regular tiling, the angles that meet at a vertex must sum to 360 degrees. If there are six polygons, the angles must be 60 degrees.

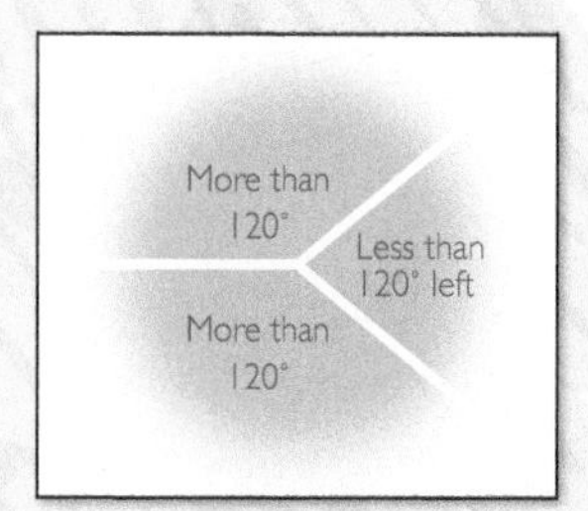

FIGURE 9.60 If a polygon has an angle of more than 120 degrees, a regular tiling cannot be created.

To see which other regular polygons might be used to make a regular tiling of the plane, let's first list the sizes of angles in some regular polygons.

Regular polygon	Angle size
Equilateral triangle	60°
Square	90°
Pentagon	108°
Hexagon	120°

Regular polygons with more than six sides have angles larger than 120 degrees. Hence, they cannot regularly tile the plane.

EXAMPLE 9.17 Finding regular tilings: Equilateral triangle

Show how to use an equilateral triangle to make a regular tiling of the plane.

SOLUTION

The tiling is shown in **Figure 9.61**.

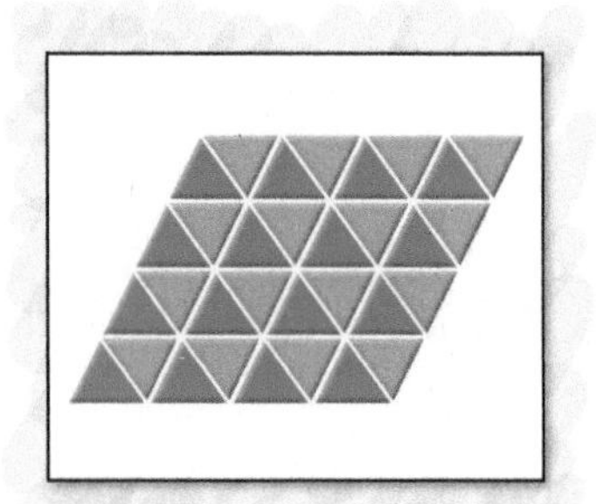

FIGURE 9.61 A regular tiling with equilateral triangles.

TRY IT YOURSELF 9.17

Find a regular tiling of the plane using regular hexagons.

The answer is provided at the end of this section.

As a contrast to the preceding example, you are asked in Exercise 20 to show that there is no regular tiling of the plane by a regular pentagon.

Irregular tilings

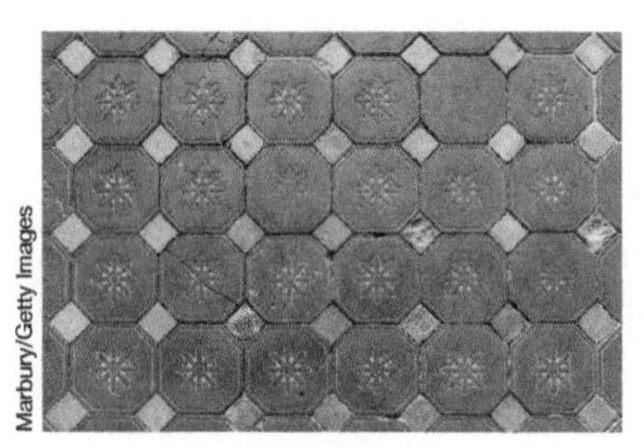

Marbury/Getty Images

A semi-regular tiling.

Regular tilings must satisfy two conditions: They use a single regular polygon, and the same configuration of edges must occur at each vertex. Tilings that are not regular fail to meet one of these two conditions. One classic example of a tiling that is not regular is shown in the accompanying photo of floor tiles. This tiling does satisfy the second condition, so it is *semi-regular*.

Figure 9.62 shows how to cut a regular hexagon into three (nonregular) pentagons. We saw in Try It Yourself 9.17 that the plane can be tiled by regular hexagons. Then such a tiling automatically gives an irregular tiling by pentagons. See **Figure 9.63**.

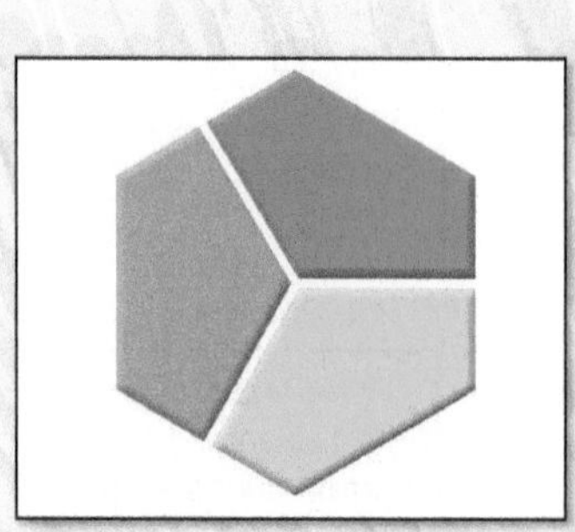

FIGURE 9.62 Cutting a regular hexagon into three (irregular) pentagons gives an irregular tiling.

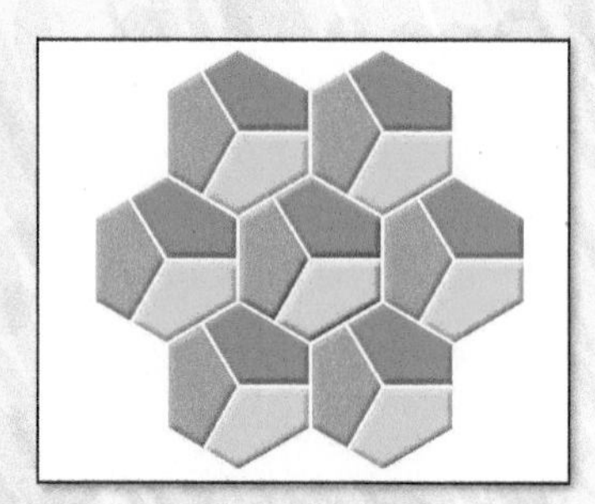

FIGURE 9.63 Using a hexagon tiling to make a pentagon tiling.

FIGURE 9.64 A triangle.

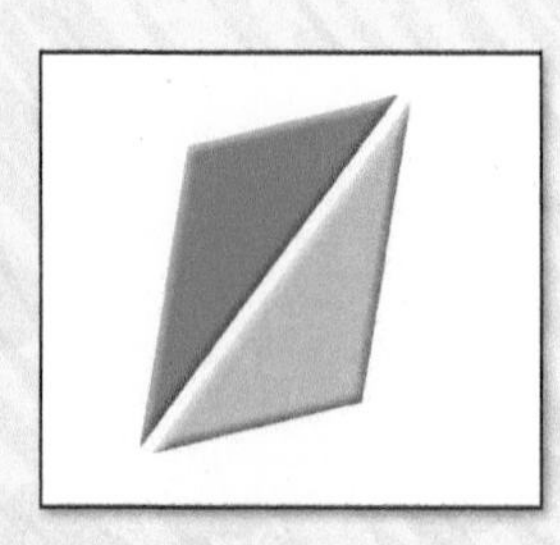

FIGURE 9.65 Two copies of a triangle make a parallelogram.

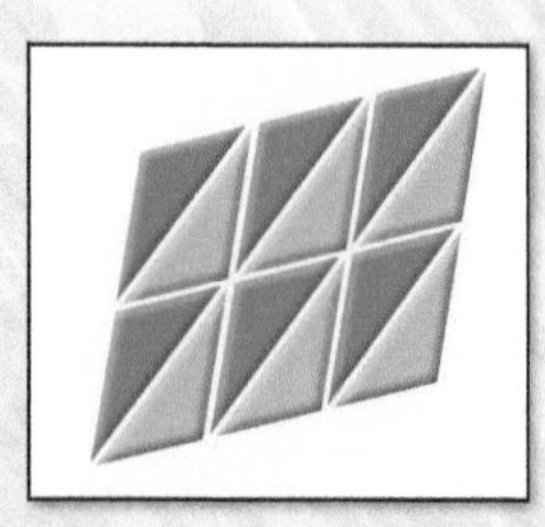

FIGURE 9.66 Using the parallelogram to make a triangle tiling.

The idea suggested by Figure 9.63—that we can make new tilings by subdividing old ones—is fruitful. We will use it to show how to use *any* triangle to tile the plane. We begin with the triangle shown in **Figure 9.64**. The first step is to note that we can put together two copies of this triangle to make a parallelogram, as shown in **Figure 9.65**. Now we use parallelograms to tile the plane as shown in **Figure 9.66**, and this gives the required tiling by triangles.

EXAMPLE 9.18 Making new tilings from old: Three given pieces

Show that the three pieces in **Figure 9.67** can be used to tile the plane.

SOLUTION

Note that the three pieces go together to make the regular hexagon shown in **Figure 9.68**. We know that regular hexagons tile the plane, and that gives a tiling using the three pieces shown.

FIGURE 9.67 Pieces for Example 9.18.

FIGURE 9.68 Pieces go together to make a regular hexagon.

TRY IT YOURSELF 9.18

Show that the three pieces in **Figure 9.69** can be used to tile the plane.

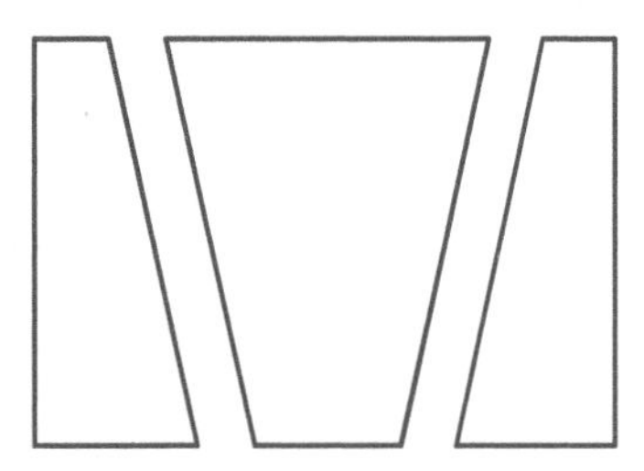

FIGURE 9.69

The answer is provided at the end of this section.

Escher tilings

Some of the best known tilings of the plane are those of M. C. Escher. Several of his famous tilings are displayed here. They are interesting from both a mathematical and an artistic point of view. They are much more elaborate than any tilings we have produced, but they are constructed using similar ideas. Let's show how we can make tilings using unusual shapes.

Some famous Escher prints.

FIGURE 9.70 A regular tiling by squares changed into an irregular tiling.

Let's start with the regular tiling by squares shown on the left in **Figure 9.70**. Now from each square, let's remove a puzzle tab from the bottom and add it to the top of the square. The resulting tiling piece is the irregular shape shown at the bottom of Figure 9.70. We get the new tiling of the plane shown on the right in Figure 9.70.

A little thought can lead one to produce truly unusual tiling pieces. Many of Escher's prints use this idea as well as more sophisticated ones. Some start with hexagons rather than squares.

WHAT DO YOU THINK?

Rotation and reflection: A classmate says that if a shape has a rotational symmetry of 180 degrees, then it has a reflectional symmetry. Is that true? Explain your answer.

Questions of this sort may invite many correct responses. We present one possibility: This is not true. Think of a capital S. If we rotate the S by 180° we get the S back again. But if we reflect through either a vertical or a horizontal line we get a backward letter S. No other reflections work either, so the letter does not have reflectional symmetry.

Further questions of this type may be found in the *What Do You Think?* section of exercises.

Try It Yourself answers

Try It Yourself 9.15: Finding rotational symmetries: Six-pointed star It has 60-degree rotational symmetry (as well as 120-degree, 180-degree, 240-degree, and 300-degree).

Try It Yourself 9.16: Finding symmetries: A given shape 90-degree rotational symmetry (as well as 180-degree and 270-degree). Reflectional symmetry through vertical and horizontal lines as well as two 45-degree lines.

Try It Yourself 9.17: Finding regular tilings: Hexagon

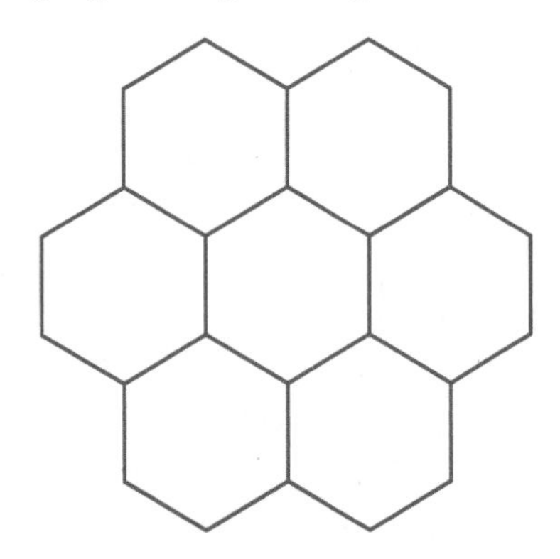

Try It Yourself 9.18: Making new tilings from old: Three given pieces The three pieces fit together to make a rectangle, which can tile the plane.

Exercise Set 9.3

Test Your Understanding

1. A planar figure has rotational symmetry of d degrees about a point, provided ______________.

2. True or false: A figure may have reflectional symmetry about more than one line.

3. What is special about a *regular* polygon?

4. True or false: There is no regular tiling of the plane by a regular polygon with more than 6 sides.

Problems

5. **Complete rotational symmetry.** Find a planar figure other than a circle that has complete rotational symmetry.

6–11. Making figures with given symmetry. *In Exercises 6 through 11, find an example of a figure with the given symmetry.*

6. 45-degree rotational symmetry

7. 30-degree rotational symmetry

8. n-degree rotational symmetry

9. 180-degree rotational symmetry but no reflectional symmetries

10. Reflectional symmetries through exactly one line

11. Reflectional symmetries through exactly two lines

12–16. Many rotations. *Exercises 12 through 16 make use of the following information. If R_θ denotes a clockwise rotation of the plane through an angle of θ degrees, we use R_θ^n to mean that we make a clockwise rotation of the plane through an angle of θ degrees n times. For example, R_{45}^2 means a clockwise rotation of 45 degrees followed by a second clockwise rotation of 45 degrees. The result is a rotation through 90 degrees. This means*

$$R_{45}^2 = R_{90}$$

12. What is the result of R_{45}^8?

13. Find θ between 0 and 360 degrees so that $R_{60}^8 = R_\theta$.

14. Find θ between 0 and 360 degrees so that $R_{30}^{15} = R_\theta$.

15. For what value of n will R_2^n result in leaving each point in the plane fixed?

16. Find θ between 0 and 360 degrees so that $R_{10}^{255} = R_\theta$.

17. **A hexagon.** What are the reflectional and rotational symmetries of a regular hexagon?

18. **An octagon.** What are the reflectional and rotational symmetries of a regular octagon?

19. **Tiling with a rhombus.** Show how to make a tiling of the plane with a rhombus, which is a square pushed over (as shown in **Figure 9.71**).

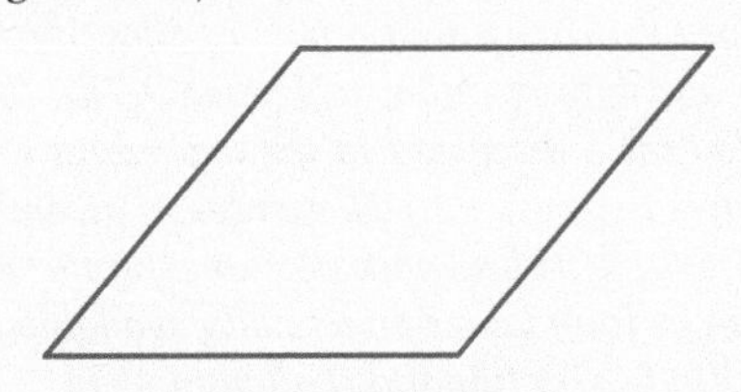

FIGURE 9.71 A rhombus.

20. Tiling with a pentagon. Explain why there is no regular tiling of the plane using regular pentagons. *Suggestion*: Each angle of a regular pentagon is 108 degrees. How many of these would need to fit together at a vertex?

21. Tiling with hexagons and triangles. Show how to tile the plane using regular hexagons and triangles. *Suggestion*: Put the hexagons together so that triangular spaces are left.

22. Tiling with unusual shapes. Show how to tile the plane using the shape in **Figure 9.72**.

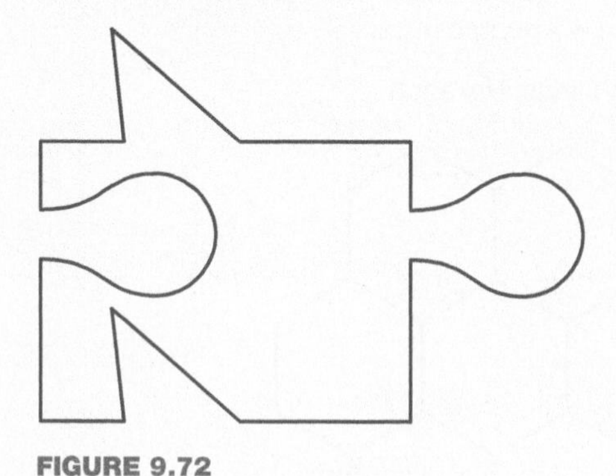

FIGURE 9.72

23. Tiling the plane. Show how to tile the plane using the shape in **Figure 9.73**.

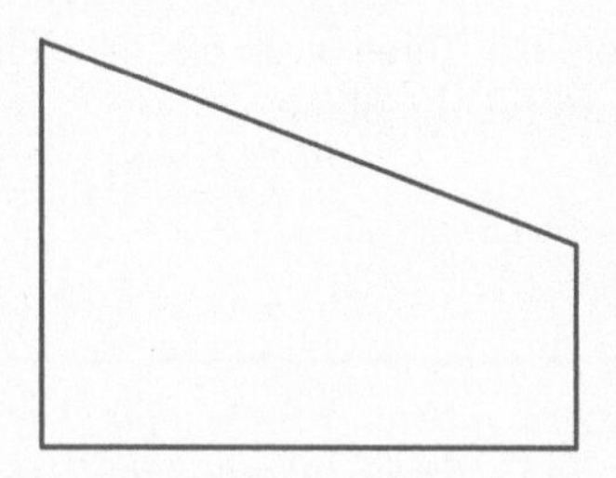

FIGURE 9.73

24. Tiling the plane. Show how to tile the plane using the shape in **Figure 9.74**.

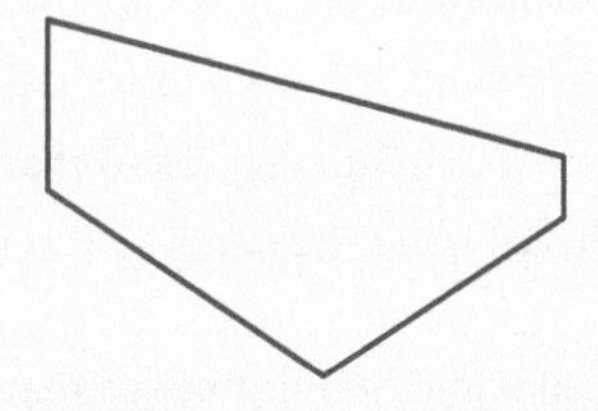

FIGURE 9.74

25–28. Penrose tiles. *Exercises 25 through 28 make use of the following information. The tilings of the plane that are most familiar to us are periodic in that they are made up of the repetition of a single pattern. For example, the tiling in* ***Figure 9.75*** *is periodic, and the repeated pattern is outlined in that figure. But the tiling in* ***Figure 9.76*** *is not periodic. In 1974, Roger Penrose found a pair of tiles, darts (shown in* ***Figure 9.77****) and kites (shown in* ***Figure 9.78****). (The distance ϕ is the golden ratio discussed in the preceding section and is approximately equal to 1.62.) These tiles, if assembled according to some simple rules, always make nonperiodic tilings. These tiles can be found as commercially available puzzles, and many people find them fun and challenging.*

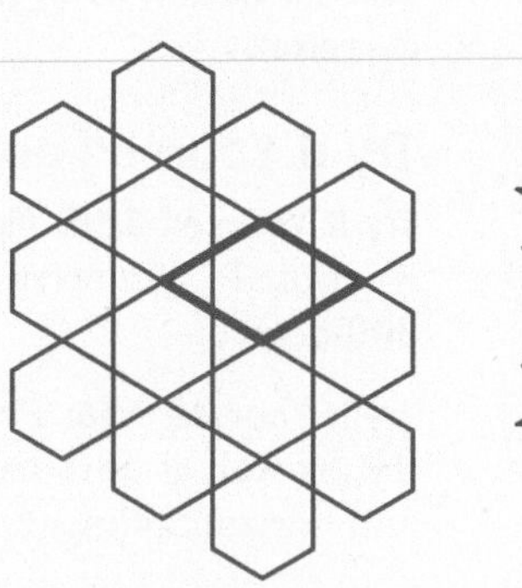

FIGURE 9.75 A periodic tiling.

FIGURE 9.76 A nonperiodic tiling.

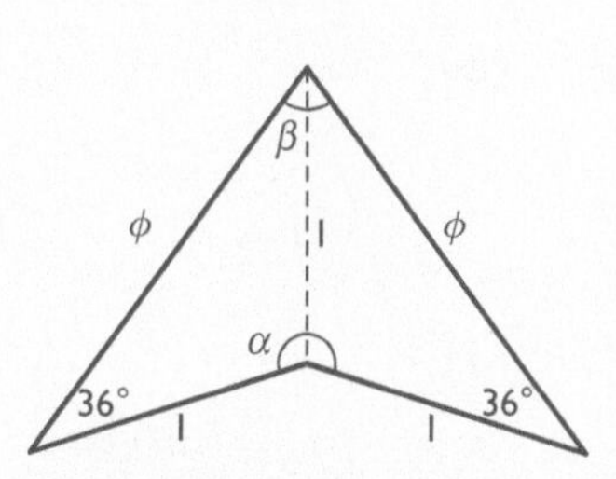

FIGURE 9.77 A dart.

FIGURE 9.78 A kite.

25. Our goal here is to explain why the tiling in Figure 9.76 is nonperiodic.

a. How many points of rotational symmetry does the tiling in Figure 9.76 have? That is, how many points can we take for the center of a rotation less than 360 degrees that leaves the figure unchanged?

b. Can a tiling with exactly one point of rotational symmetry be periodic?

c. Explain using parts a and b why the tiling in Figure 9.76 is not periodic.

26. Find the angles marked α, β, γ, and δ in Figures 9.77 and 9.78. Recall the following facts from geometry:

- The angle sum of every triangle is 180 degrees.
- Recall that an isosceles triangle has two equal sides. It is a fact that the "base angles" of an isosceles triangle have the same degree measure. (See **Figure 9.79**. The angles marked "1" and "2" are base angles because they are opposite the equal sides.)

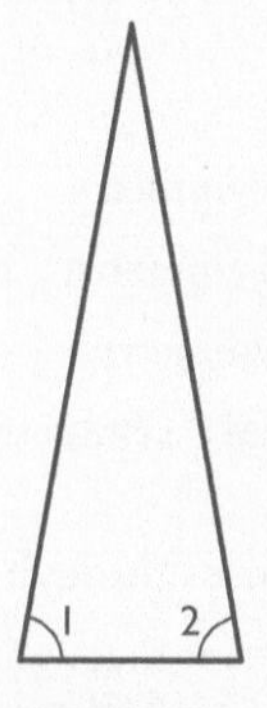

FIGURE 9.79 Base angles of an isosceles triangle.

27. Penrose tiling: A periodic tiling with kites and darts. Show that a kite and a dart can be fitted together to make a rhombus, each side of which is ϕ. Explain why this gives a periodic tiling of the plane using kites and darts.

28. Penrose tiling: Experimenting with kites and darts. Kites and darts come with assembly rules, which are as follows:

Assembly Rule 1: Pieces must be assembled along matching edges.

Assembly Rule 2: The kites and darts in **Figures 9.80** and **9.81** have certain angles marked with dots. Dotted angles must be adjacent. **Figure 9.82** shows an assembly that is allowed because the dots match. **Figure 9.83** shows an assembly that is not allowed because the dots do not match.

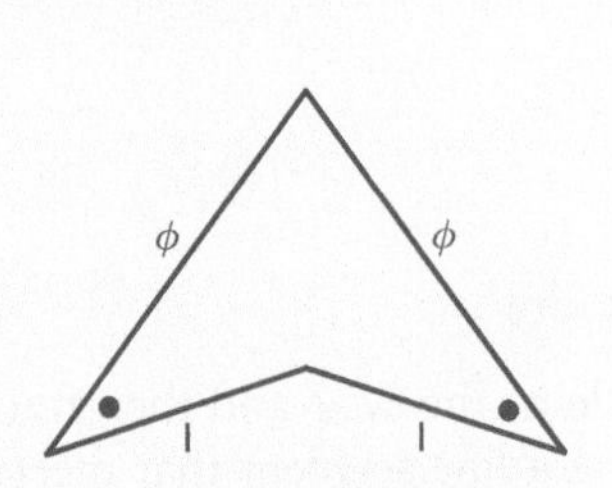

FIGURE 9.80 A dart with dots.

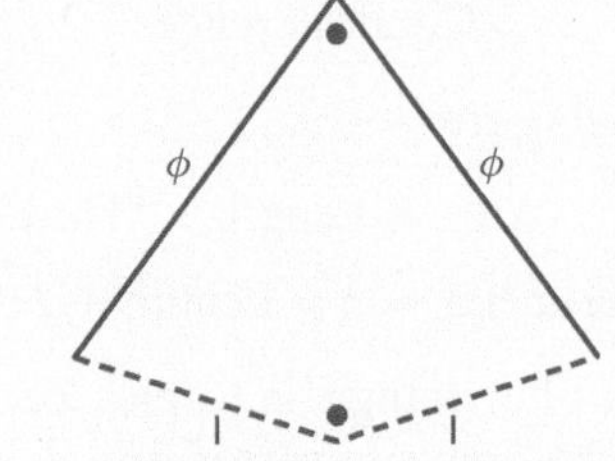

FIGURE 9.81 A kite with dots.

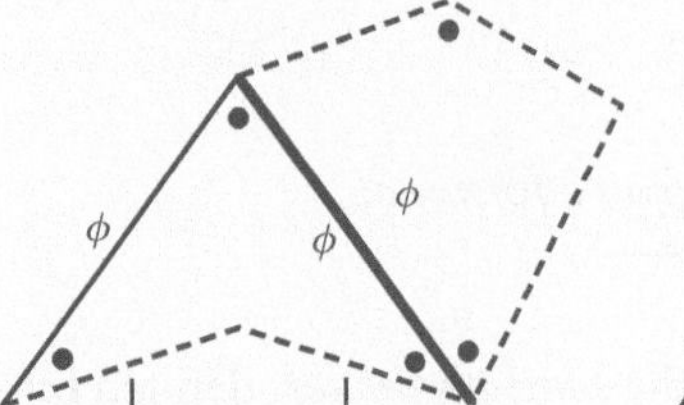

FIGURE 9.82 An allowable assembly.

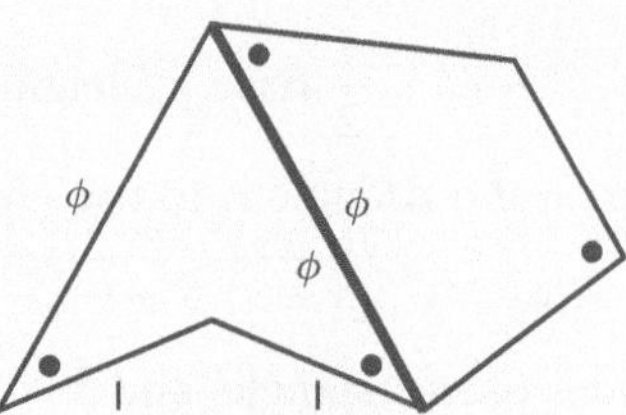

FIGURE 9.83 An assembly that is not allowed.

Fit together a number of kites and darts to begin a tiling of the plane. It is a fact that any such tiling you make will be nonperiodic. An example of such a nonperiodic tiling is shown in **Figure 9.84**.

FIGURE 9.84 A nonperiodic tiling using kites and darts by Roger Penrose.

Writing About Mathematics

29. M. C. Escher. Write a report on the contributions of M. C. Escher to both mathematics and art.

30. Roger Penrose. Sir Roger Penrose is famous not only for his nonperiodic tilings of the plane but also for his work as a mathematical physicist. Read about him on the Internet and report on his life and accomplishments.

What Do You Think?

31. Rotational symmetry. In the definition of *rotational symmetry* on page 633, the rotation is required to be less than 360 degrees. What figures have 360-degree rotational symmetry? What can you say about rotational symmetry with rotations greater than 360 degrees? **Exercise 6 is relevant to this topic.**

32. Reflect. Can a figure have reflectional symmetry about a line that does not touch the figure? Explain. **Exercise 10 is relevant to this topic.**

33. Jigsaw. Do the pieces of a jigsaw puzzle make a regular tiling? Explain. **Exercise 20 is relevant to this topic.**

34. Is symmetry beautiful? Some sources, such as the article from *In The News 9.4*, suggest that part of the human perception of beauty is tied to symmetry. Do some research on this topic and report on what you find. Is your own conception of beauty related to symmetry?

35. Escher prints. Many prints by M. C. Escher are of interest from both mathematical and artistic perspectives. Find examples on the Web, and report on your reaction to their artistic quality.

36. Do it yourself. A number of tilings of the plane are discussed in this section, and many others are available through other sources. It can be fun to come up with your own, and interesting tilings can be found by anyone. Make several examples that you like. **Exercise 24 is relevant to this topic.**

37. Tilings by pentagons. Four types of pentagons that tile the plane were found in the 1970s by Marjorie Rice, an amateur mathematician. Look on the Internet to find the methods she used, and summarize them. Why do you think advanced mathematics is not necessary to make such discoveries?

38. Reflection and rotation. A classmate says that if a shape has a reflectional symmetry, then it has a rotational symmetry of 180 degrees. Is that true? Explain your answer.

39. Regular polygons. Look at the table on page 637 showing angle sizes for regular polygons. Note that the sum of the angles for an equilateral triangle is 180 degrees because there are three angles, each of size 60 degrees. Find the sum of the angles for the square, the regular pentagon, and the regular hexagon. What pattern do you see? Use this pattern to find the angle size for a regular heptagon (which has seven sides).

CHAPTER SUMMARY

In this chapter, we consider geometric topics that are encountered in everyday life. We study how to measure geometric objects and how the notion of geometric similarity is related to the idea of proportionality. We also consider symmetries of shapes and the notion of tiling the plane.

Perimeter, area, and volume: How do I measure?

Two-dimensional geometric objects can be measured in terms of perimeter and area. The most common figures are circles, rectangles, and triangles. For each of these objects, there are standard formulas for computing area and perimeter. The area and circumference of a circle of radius r are given by the formulas

$$\text{Area} = \pi r^2$$
$$\text{Circumference} = 2\pi r$$

For a rectangle, the formulas are

$$\text{Area} = \text{Length} \times \text{Width}$$
$$\text{Perimeter} = 2 \times \text{Length} + 2 \times \text{Width}$$

One way to find the area of a triangle is to pick one side as the *base* and then find the *height* by starting at the opposite vertex and drawing a line segment that meets the base in a right angle. The area is given by the formula

$$\text{Area} = \frac{1}{2}\,\text{Base} \times \text{Height}$$

Another way to find the area of a triangle is to use *Heron's formula*:

$$\text{Area} = \sqrt{S(S-a)(S-b)(S-c)}$$

Here, a, b, and c are the sides of the triangle, and S is the *semi-perimeter*, defined by $S = \frac{1}{2}(a+b+c)$.

The *Pythagorean theorem* relates the legs a, b of a right triangle with the hypotenuse c. It states that

$$a^2 + b^2 = c^2$$

Conversely, if any triangle has three sides of lengths a, b, and c such that $a^2 + b^2 = c^2$, then that triangle is a right triangle with hypotenuse c.

Three-dimensional geometric objects can be measured in terms of volume and surface area. For a box, we have the formula

$$\text{Volume} = \text{Length} \times \text{Width} \times \text{Height}$$

For a cylinder with radius r and height h, the volume is

$$\text{Volume} = \pi r^2 h$$

and the surface area (excluding the top and bottom) is given by

$$\text{Surface area} = 2\pi r h$$

There is a general formula for the volume of any three-dimensional object with a uniform cross section:

$$\text{Volume} = \text{Area of base} \times \text{Height}$$

For a ball of radius r, the volume is

$$\text{Volume} = \frac{4}{3}\pi r^3$$

and the surface area is given by

$$\text{Surface area} = 4\pi r^2$$

Proportionality and similarity: Changing the scale

Proportionality relationships are common in both numerical and geometrical settings. One quantity is *proportional* to another if it is always a fixed (nonzero) constant multiple of the other. This multiple is the *constant of proportionality*. An alternative definition of proportionality requires the ratio of the two quantities to be constant.

If one of two proportional quantities is multiplied by a certain factor, the other one is also multiplied by that factor. For example, if one quantity doubles, then the other doubles as well. This property is useful when we compare the measurements (such as area) of geometrically similar objects.

The *golden ratio* is a special number that arises in geometry. It is denoted by the Greek letter ϕ and is approximately equal to 1.62. Here is one way to define the golden ratio: Divide a line segment into two pieces, one longer and one shorter, so that the ratios $\frac{\text{Longer}}{\text{Shorter}}$ and $\frac{\text{Whole}}{\text{Longer}}$ are equal. This common ratio is by definition the golden ratio.

The number ϕ appears in geometry in *golden rectangles*, which are rectangles for which the ratio of the length to the width is ϕ. The golden ratio appears in other ways as well. The significance of the golden ratio and golden rectangles in science and art is the subject of some debate.

In geometry, one important application of proportionality involves similar triangles. Two triangles are *similar* if all corresponding angles have the same measure. The most important feature of similar triangles is that the ratios of corresponding sides are the same.

Sometimes a quantity is proportional to the square or some other power of another quantity. For example, the area of a circle is proportional to the square of the radius, with constant of proportionality π.

Symmetries and tilings: Form and patterns

One of the most important properties a geometric object can have is symmetry. We consider two types of symmetry: rotational and reflectional. A planar figure has *rotational symmetry* if it remains exactly the same after a rotation of less than 360-degrees. For example, a square has 90-degree rotational symmetry, and a rectangle has 180-degree rotational symmetry.

A planar figure has *reflectional symmetry* about a line if the figure is identical to its reflection through that line. If we imagine a mirror placed along the line, the mirror image of the figure is the same as the original figure.

A *tiling* of the plane is a pattern of repeated figures that cover up the plane. Such a tiling is *regular* if it consists of repeated copies of a single regular polygon, meeting edge to edge, so that every vertex has the same configuration. For example, squares can be used to make a regular tiling of the plane.

Irregular tilings are often encountered as well. They can be obtained by subdividing existing tilings. The tilings of M. C. Escher are interesting from both a mathematical and an artistic point of view.

KEY TERMS

perimeter, p. 602
circumference, p. 602
area, p. 602
Pythagorean theorem, p. 603
directly proportional, p. 618
constant of proportionality, p. 618
golden ratio, p. 621
golden rectangle, p. 622
similar, p. 623
rotational symmetry, p. 633
reflectional symmetry about a line, p. 635
tiling or tessellation, p. 636
regular tiling, p. 637

CHAPTER QUIZ

1. The net on a tennis court is 3.5 feet high. The baseline is 39 feet from the bottom of the net. How far is the baseline from the top of the net?

Answer 39.2 feet.

If you had difficulty with this problem, see Example 9.3.

2. Use Heron's formula to find the area of a triangle with sides 11, 14, and 16.

Answer 75.5 square units.

If you had difficulty with this problem, see Example 9.4.

3. A trunk is 2 feet by 3 feet by 4 feet. What is the volume of the trunk?

Answer 24 cubic feet.

If you had difficulty with this problem, see Example 9.6.

4. The volume of concrete needed to make a sidewalk is proportional to the length of the sidewalk. (Assume that the sidewalk has a fixed width and depth.) Six cubic yards of concrete are required to make a sidewalk that is 100 feet long. How much concrete is needed to make a sidewalk that is 25 feet long?

Answer 1.5 cubic yards.

If you had difficulty with this problem, see Example 9.9.

5. A 5-foot-tall girl casts a 3-foot shadow at the same time that a building casts a 12-foot shadow. How tall is the building?

Answer 20 feet tall.

If you had difficulty with this problem, see Example 9.11.

6. The volume of a cylindrical water tank of height 30 feet is proportional to the square of the diameter. Two tanks have height 30 feet, and the diameter of one is 3 times the diameter of the other. How do the volumes compare?

Answer The larger tank has 9 times the volume of the smaller.

If you had difficulty with this problem, see Example 9.12.

7. Find the rotational symmetries of an equilateral triangle.

Answer It has 120-degree rotational symmetry (as well as 240-degree).

If you had difficulty with this problem, see Example 9.15.

8. Find the reflectional symmetries of an equilateral triangle.

Answer It has reflectional symmetry about three lines: the lines from a vertex to the midpoint of the opposite side.

If you had difficulty with this problem, see Example 9.16.

Appendix 1: Unit Conversion

In this appendix, we discuss conversion of physical units (e.g., feet to inches) and currency conversion.

Physical units

The most common system of weights and measures in the United States is informally called the "U.S. Customary System" or the "English system." Most of the rest of the world uses what we usually call the "metric system." The official name of this system of weights and measures is Le Système International d'Unités, or "SI" for short.

Some basic relationships for unit conversions are contained in Tables A1.1 and A1.2.

TABLE A1.1

Conversion Among Familiar English Units		Conversion Factors	
Length			
1 foot	12 inches	inches = 12 × feet	feet = $\frac{1}{12}$ × inches
1 yard	3 feet	feet = 3 × yards	yards = $\frac{1}{3}$ × feet
1 (statute) mile	5280 feet	feet = 5280 × miles	miles = $\frac{1}{5280}$ × feet
1 nautical mile	1.15 (statute) miles	miles = 1.15 × nautical miles	nautical miles = $\frac{1}{1.15}$ × miles
1 lightyear	5.88×10^{12} miles	miles = 5.88×10^{12} × lightyears	lightyears = $\frac{1}{5.88} \times 10^{-12}$ × miles
1 parsec	3.26 lightyears	lightyears = 3.26 × parsecs	parsecs = $\frac{1}{3.26}$ × lightyears
Area			
1 acre	43,560 square feet	square feet = 43,560 × acres	acres = $\frac{1}{43,560}$ × square feet
Volume			
1 tablespoon	3 teaspoons	teaspoons = 3 × tablespoons	tablespoons = $\frac{1}{3}$ × teaspoons
1 fluid ounce	2 tablespoons	tablespoons = 2 × fluid ounces	fluid ounces = $\frac{1}{2}$ × tablespoons
1 cup	8 fluid ounces	fluid ounces = 8 × cups	cups = $\frac{1}{8}$ × fluid ounces
1 pint	2 cups	cups = 2 × pints	pints = $\frac{1}{2}$ × cups
1 quart	2 pints	pints = 2 × quarts	quarts = $\frac{1}{2}$ × pints
1 gallon	4 quarts	quarts = 4 × gallons	gallons = $\frac{1}{4}$ × quarts
Weight			
1 pound	16 ounces	ounces = 16 × pounds	pounds = $\frac{1}{16}$ × ounces
1 ton	2000 pounds	pounds = 2000 × tons	tons = $\frac{1}{2000}$ × pounds

TABLE A1.2

English/Metric Conversions		Conversion Factors	
Length			
1 inch	2.54 centimeters	centimeters = 2.54 × inches	inches = $\frac{1}{2.54}$ × centimeters
1 foot	30.48 centimeters	centimeters = 30.48 × feet	feet = $\frac{1}{30.48}$ × centimeters
1 meter (100 centimeters)	3.28 feet	feet = 3.28 × meters	meters = $\frac{1}{3.28}$ × feet
1 kilometer (1000 meters)	0.62 mile	miles = 0.62 × kilometers	kilometers = $\frac{1}{0.62}$ × miles
Volume			
1 liter	1.06 quarts	quarts = 1.06 × liters	liters = $\frac{1}{1.06}$ × quarts
1 gallon	3.79 liters	liters = 3.79 × gallons	gallons = $\frac{1}{3.79}$ × liters
1 milliliter (mL)	0.001 liter	liters = 0.001 × milliliters	milliliters = 1000 × liters
1 cubic centimeter (cc)	0.001 liter or 1 milliliter	liters = 0.001 × cubic centimeters	cubic centimeters = 1000 × liters
Weight/mass			
1 ounce	28.35 grams	grams = 28.35 × ounces	ounces = $\frac{1}{28.35}$ × grams
1 pound	453.59 grams	grams = 453.59 × pounds	pounds = $\frac{1}{453.59}$ × grams
1 kilogram (1000 grams)	2.20 pounds	pounds = 2.20 × kilograms	kilograms = $\frac{1}{2.20}$ × pounds
1 milligram (mg)	0.001 gram	grams = 0.001 × milligrams	milligrams = 1000 × grams
Other conversions (assuming standard temperature and pressure)			
1 liter of water	mass of 1 kilogram		
1 cc of water	mass of 1 gram		
1 mL of water	mass of 1 gram		

TABLE A1.3

Metric Prefix	Meaning	Example
kilo	1000	1 kilogram is 1000 grams.
deci	1/10	1 deciliter is one-tenth of a liter.
centi	1/100	1 centimeter is one-hundredth of a meter.
milli	1/1000	1 milligram is one-thousandth of a gram. 1 milliliter is one-thousandth of a liter.

Conversion from one metric unit to another is just a matter of multiplying by the appropriate power of 10 indicated by the metric prefix. Common prefixes are shown in Table A1.3.

The following tips may be helpful in using these tables.

Tip 1: Conversion factors: To convert one unit to another, we multiply by a number called the *conversion factor*, which is included in the preceding tables. For example, to convert feet to inches, we multiply length in feet by 12 because there are 12 inches in a foot. Here, the conversion factor is 12 inches per foot. This is indicated in Table A1.1 by the entry

$$\text{inches} = 12 \times \text{feet}$$

Informally, the conversion factor from unit A to unit B is the number of B's in a single A. For example, from Table A1.1 we find that there are 5280 feet in a mile. Thus, 5280 feet per mile is the conversion factor for changing miles to feet. This is indicated in Table A1.1 by the entry

$$\text{feet} = 5280 \times \text{miles}$$

For example, 5 miles is $5 \times 5280 = 26{,}400$ feet.

Tip 2: Reversing conversions: If instead of converting from miles to feet, we convert from feet to miles, the conversion factor is the reciprocal, $\frac{1}{5280}$ miles per foot. For example, 2400 feet is $2400 \times \frac{1}{5280}$ or about 0.45 mile. Similarly, Table A1.2 tells us that there are 2.20 pounds in a kilogram, so to convert from kilograms to pounds, we multiply by 2.20. To convert from pounds to kilograms, we use the reciprocal $\frac{1}{2.20}$:

$$\text{Conversion factor for kilograms to pounds} = 2.20\ \frac{\text{pounds}}{\text{kilogram}}$$

so

$$\text{Conversion factor for pounds to kilograms} = \frac{1}{2.20}\ \frac{\text{kilograms}}{\text{pound}}$$

(This is about 0.45 kilogram per pound.) This is shown in Table A1.2 as the companion entries

$$\text{pounds} = 2.20 \times \text{kilograms} \qquad \text{kilograms} = \frac{1}{2.20} \times \text{pounds}$$

For example, to convert 180 pounds to kilograms, we proceed as follows:

$$\begin{aligned} 180 \text{ pounds} &= 180 \text{ pounds} \times \frac{1}{2.20}\ \frac{\text{kilograms}}{\text{pound}} \\ &= 180 \times \frac{1}{2.20} \cancel{\text{pounds}} \frac{\text{kilograms}}{\cancel{\text{pound}}} \\ &= \frac{180}{2.20} \text{ kilograms} \end{aligned}$$

This is about 82 kilograms.

In general, we find the conversion factor for *B* units to *A* units by taking the reciprocal of the conversion factor for *A* units to *B* units.

Tip 3: Dimensional analysis: Sometimes it's easy to get confused when doing unit conversions—do I multiply or divide? The units themselves can help you decide. This approach is called *dimensional analysis*. You can see a case of it in Tip 2 where "pounds" were canceled. Let's consider how we convert yards to feet.

The word "per" always means division, so we can write the conversion factor for yards to feet, 3 feet per yard, as $3\ \frac{\text{feet}}{\text{yard}}$. Let's see how this observation helps in converting yards to feet. A football field is 100 yards long. Here's how to convert that length to feet:

$$\begin{aligned} 100 \text{ yards} &= 100 \text{ yards} \times 3\ \frac{\text{feet}}{\text{yard}} \\ &= 100 \times 3 \cancel{\text{yards}} \frac{\text{feet}}{\cancel{\text{yard}}} \\ &= 300 \text{ feet} \end{aligned}$$

Note that the "yards" units cancel, leaving "feet" as desired.

This dimensional analysis is even more helpful when we do complex unit conversions. For example, let's convert 60 miles per hour to feet per second. We can perform the conversion in steps: First change miles to feet and then change hours to seconds.

To convert miles to feet, we use the conversion factor of 5280 feet per mile from Table A1.1:

$$\begin{aligned} 60 \text{ miles} &= 60 \text{ miles } \times 5280 \frac{\text{feet}}{\text{mile}} \\ &= 60 \times 5280 \,\cancel{\text{miles}} \times \frac{\text{feet}}{\cancel{\text{mile}}} \\ &= 60 \times 5280 \text{ feet} \end{aligned}$$

Note how, after we cancel, the final units come out to be "feet," which is what we intended. The fact that this happened confirms that we're on the right track.

We proceed similarly to convert 1 hour to seconds:

$$\begin{aligned} 1 \text{ hour} &= 1 \text{ hour } \times 60 \frac{\text{minutes}}{\text{hour}} \times 60 \frac{\text{seconds}}{\text{minute}} \\ &= 1 \times 60 \times 60 \,\cancel{\text{hour}} \times \frac{\cancel{\text{minutes}}}{\cancel{\text{hour}}} \times \frac{\text{seconds}}{\cancel{\text{minute}}} \\ &= 60 \times 60 \text{ seconds} \end{aligned}$$

Note again how, after we cancel "minutes" and "hours," the final units come out to be "seconds," which is what we intended.

Therefore,

$$\begin{aligned} 60 \frac{\text{miles}}{\text{hour}} &= \frac{60 \times 5280 \text{ feet}}{60 \times 60 \text{ seconds}} \\ &= 88 \frac{\text{feet}}{\text{second}} \end{aligned}$$

EXAMPLE A1.1 Dosage calculation

The correct dosage of a certain drug is 4 mg per kilogram of body weight. What is the correct dosage for a woman weighing 135 pounds?

SOLUTION

First we convert the weight from pounds to kilograms. Consulting Table A1.1, we find

$$\text{kilograms} = \frac{1}{2.20} \times \text{pounds}$$

So, to get kilograms we multiply 135 by 1/2.20:

$$\text{kilograms} = \frac{1}{2.20} \times 135$$

We get the dosage by multiplying by 4 milligrams per kilogram.

$$\begin{aligned} \text{Dosage} &= 4 \frac{\text{milligrams}}{\cancel{\text{kilogram}}} \times \frac{1}{2.20} \times 135 \,\cancel{\text{kilograms}} \\ &= 245.45 \text{ milligrams} \end{aligned}$$

Thus, the proper dosage is 245.45 milligrams.

Currency conversion

Unit conversion involving currency arises in connection with currency exchange. For example, at the time of this writing, the value of one U.S. dollar is 0.8850 euro. Hence, the *exchange rate* for converting dollars to euros is 0.8850 euro per dollar.

For example, to convert \$25 to euros, we proceed as follows:

$$25 \text{ dollars} = 25 \text{ dollars} \times 0.8850 \frac{\text{euro}}{\text{dollar}}$$
$$= 25 \times 0.8850 \cancel{\text{dollars}} \times \frac{\text{euro}}{\cancel{\text{dollar}}}$$
$$= 25 \times 0.8850 \text{ euros}$$

This is about 22.13 euros.

To find the exchange rate for conversion in the other direction, we use reciprocals. For example, under an exchange rate of 0.8850 euro per dollar, the exchange rate for converting euros to dollars is $\frac{1}{0.8850}$ or about 1.13 U.S. dollars per euro.

Practice Exercises

In these exercises, round all answers to one decimal place unless you are instructed otherwise.

1. Quarter pounders. How many grams of meat are in a "quarter pounder"?

2. Speed limit. In Canada the speed limit is measured in kilometers per hour (abbreviated km/h), and it's common to see 100 km/h speed limit signs in Canada. How fast is 100 km/h in miles per hour?

3. Cola. You can get a quart bottle of cola for \$1.50 or a liter bottle for \$1.55. Which is the better buy in terms of price per volume?

4–5. Cans and bottles. *Soft drinks are often sold in six-packs of 12-ounce cans and in 2-liter bottles. A liter is about 33.8 fluid ounces.*

4. Which is the greater volume: a six-pack or 2 liters?

5. A store offers a 2-liter bottle of soft drink for \$1.15 and a six-pack of 12-ounce cans for \$1.20. Which is the better value (based on the price per ounce)?

6. Currency conversion. As of this writing, 1 U.S. dollar is worth 19.92 Mexican pesos. What is the exchange rate for converting dollars to pesos? What is the exchange rate for converting pesos to dollars? Round your answer to three decimal places.

7. Currency conversion. At the time of this writing 1 U.S. dollar is worth 1.28 Canadian dollars. An automobile in Toronto sells for 23,000 Canadian dollars. The same car sells in Detroit for 19,000 U.S. dollars. Which is the better buy?

8. Conversion factors. What is the conversion factor for kilometers to miles? What is the conversion factor for miles to kilometers? Round your answer to two decimal places.

9. Conversion factors. What is the conversion factor for parsecs to lightyears? What is the conversion factor for lightyears to parsecs? Round your answer to two decimal places.

10. Star Wars. In the 1977 film *Star Wars Episode IV: A New Hope*, the character Han Solo says that a certain starship "made the Kessel Run in less than 12 parsecs." Do the units here make sense?*

11. Cubic centimeters. The quantity of fluid administered in an inoculation is usually measured in cubic centimeters, abbreviated cc. One liter is 1000 cc, and 1 liter is about 33.8 fluid ounces. How many cubic centimeters are in 1 fluid ounce?

12. Football field. How many meters long is a football field? (A football field is 100 yards long.)

13. Meters and miles. Which is longer, a 1500-meter race or a 1-mile race? By how much?

14. Meters and yards. Which is longer, a 100-meter dash or a 100-yard dash? By how much? Round your answer to the nearest yard.

15. Records. In 1996 Ato Boldon of UCLA ran the 100-meter dash in 9.92 seconds. In 1969 John Carlos of San Jose State ran the 100-yard dash in 9.1 seconds. Which runner had the faster average speed? What was his speed in yards per second? Round your answer to two decimal places.

16. Gasoline. My car gets 26 miles per gallon, and gasoline costs \$2.30 per gallon. If we consider only the cost of gasoline, how much does it cost to drive each mile? Round your answer in dollars to two decimal places.

17. Speed. A car is traveling at 70 miles per hour. What is its speed in feet per second?

18. Gravity. In the metric system the acceleration due to gravity near Earth's surface is about 9.81 meters per second per second. What is the acceleration due to gravity as measured in the English system, that is, in feet per second?

19. Speed of light. The speed of light is 300 million meters per second. What is the speed of light measured in miles per hour? (Round your answer to the nearest million miles per hour.) How long does it take light to travel the distance of 93 million miles from the sun to Earth? (Round your answer to the nearest minute.)

20–21. Lightning. *The speed of sound in air on a certain stormy day is about 1130 feet per second. This fact is used in Exercises 20 and 21.*

20. What is the speed of sound measured in miles per second? In miles per hour? Round the former answer to three decimal places and the latter answer to the nearest whole number.

---

* *Various explanations of this can be found on the Web.*

21. You see a flash of lightning striking some distance away, and five seconds later you hear the clap of thunder. How many miles away did the lightning strike? Round your answer to the nearest mile. *Note*: The speed of light is so much faster than the speed of sound that you can assume that the strike occurs at the instant you see it.

22–24. A drug bust. *The following information was reported in a Virginia newspaper. A sheriff in Bedford County, Virginia, reported that investigators found 6.5 kilograms of cocaine in a car belonging to a man who lives in Roanoke, Virginia. This is a significant amount of cocaine, having a street value of about $1.2 million dollars. This information is used in Exercises 22 through 24.*

22. How many pounds of cocaine were seized in this bust?

23. What is the street value of a pound of cocaine? What is the street value of an ounce of cocaine? In each case, round your answer in dollars to the nearest whole number.

24. Assume that a typical dose contains half a gram of pure cocaine. How much would a dose cost? Round your answer to the nearest dollar.

25. A project, the Gimli glider. There are several examples of errors involving unit conversion that had serious consequences. One such story with a happy ending involved the *Gimli glider*, a commercial aircraft that ran out of fuel at an altitude of about 28,000 feet. The crew glided the aircraft to a landing, and no one was seriously injured. Do a research project on the Gimli glider. Pay special attention to how unit conversion played a part in the incident.

Brief Answers to Practice Exercises

1. 113.4 grams

2. 62 miles per hour

3. The liter bottle

4. A six-pack

5. The six-pack

6. 19.92 pesos per dollar; 0.050 dollar per peso

7. The car in Toronto

8. 0.62 mile per kilometer; 1.61 kilometers per mile

9. 3.26 lightyears per parsec; 0.31 parsec per lightyear

10. No, a parsec is a measure of distance.

11. 29.6 cc

12. 91.5 meters

13. A 1-mile race is 360 feet longer.

14. A 100-meter dash is about 9 yards longer.

15. Ato Boldon; 11.02 yards per second

16. $0.09

17. 102.7 feet per second

18. 32.2 feet per second per second

19. 671 million miles per hour; 8 minutes

20. 0.214 mile per second; 770 miles per hour

21. 1 mile

22. 14.3 pounds

23. $83,916; $5245

24. $92

25. Answers for this research project will vary.

Appendix 2: Exponents and Scientific Notation

Exponents

The following is a summary of the basic rules for exponents.

Negative exponents: a^{-n} is the reciprocal of a^n.

Definition	Example
$a^{-n} = \frac{1}{a^n}$	$2^{-3} = \frac{1}{2^3} = \frac{1}{8}$

Zero exponent: If a is not 0, then a^0 is 1.

Definition	Example
$a^0 = 1$	$2^0 = 1$

Basic properties of exponents:

Property	Example
$a^p a^q = a^{p+q}$	$2^2 \times 2^3 = 2^{2+3} = 2^5 = 32$
$\frac{a^p}{a^q} = a^{p-q}$	$\frac{3^6}{3^4} = 3^{6-4} = 3^2 = 9$
$(a^p)^q = a^{pq}$	$(2^3)^2 = 2^{3\times 2} = 2^6 = 64$

Powers of 10

Powers of 10 provide an easy way to represent both large and small numbers. If n is a positive whole number, we find 10^n by starting with 1 and moving the decimal point n places to the right. For example,

For 10^3	*3 tells how to move the decimal point*	1.000 (→)	*right shift by 3 places results in*	1000

A negative power of 10 causes the decimal point to be moved to the *left*. If n is a positive whole number, we find 10^{-n} by starting with 1 and moving the decimal point n places to the left. For example,

For 10^{-3}	*−3 tells how to move the decimal point*	0001.0 (←)	*left shift by 3 places results in*	0.001

Scientific notation

When we multiply by a power of 10, the effect is to move the decimal point, just as when powers of 10 stand alone. For example,

For 3.45×10^3	*3 tells how to move the decimal point*	3.450 (→)	*right shift by 3 places results in*	3450
For 3.45×10^{-3}	*−3 tells how to move the decimal point*	003.45 (←)	*left shift by 3 places results in*	0.00345

This observation leads to the idea of expressing numbers in *scientific notation.*

KEY CONCEPT

A positive number is said to be expressed in scientific notation if it is written as a power of 10 multiplied by a number that is less than 10 and greater than or equal to 1.

For example, to write 24,500 in scientific notation, we start with 2.45 because that is between 1 and 10. To get 24,500 from 2.45, we need to move the decimal place four places to the right. That is the power of 10 that we use. The result is

24,500 expressed in scientific notation is 2.54×10^4

To write 0.000245 in scientific notation, we start once more with 2.54. This time we need to move the decimal point four places to the left. That means we use an exponent of −4:

0.000245 expressed in scientific notation is 2.45×10^{-4}

CALCULATION TIP A2.1

Calculators often resort to scientific notation when they encounter very large or very small numbers. They usually do this by expressing the power of 10 with the letter "E" or "e." For example, if I enter $2{,}000{,}000^{10}$ in my Microsoft Windows© calculator, I see in the display

$$1.024\text{e}+63$$

This means 1.024×10^{63}, which is scientific notation.

If I enter 2^{-100} in my calculator, I see in the display

$$7.8886090522101180541172856528279\text{e}-31$$

Rounding off to one place, this means 7.9×10^{-31}, which is scientific notation.

Scientific notation is useful in dealing with very large and very small numbers. For example, according to the U.S. Treasury Department, the national debt as of 2020, was 27.8 *trillion* dollars. To write 27.8 trillion or 27,800,000,000,000 in scientific notation, we start with 2.78 because that number is between 1 and 10. To get 27,800,000,000,000 from 2.78, we must move the decimal point 13 places to the right. This is the power of 10 we use:

$$27.8 \text{ trillion } = 27{,}800{,}000{,}000{,}000 = 2.78 \times 10^{13}$$

In chemistry there is a very large number called *Avogadro's number*. It is the number of atoms in 12 grams of carbon-12 and is about 602,000,000,000,000, 000,000,000 (21 zeros in a row). To express this number in scientific notation, we start with 6.02 because it is between 1 and 10. We need to move the decimal point 23 places to the right, so this is the power of 10 that we use. The result is 6.02×10^{23}.

As another example, the wavelength of blue light is 0.000000475 meters (6 zeros after the decimal place). To express this number in scientific notation, we begin with 4.75 because it is between 1 and 10. We need to move the decimal seven places to the left, so we use -7 for the power of 10. The result is 4.75×10^{-7}.

Scientific notation not only gives us an easy way to express very large and very small numbers, but also makes it easier for us to do calculations based on the rules of exponents. For example, light travels about 186 thousand miles per second. Let's figure out how far it will travel in a year (i.e., the number of miles in a lightyear). In scientific notation, 186 thousand is 1.86×10^5. There are $60 \times 60 \times 24 \times 365$ or about 31.5 million seconds in a year. That is 3.15×10^7. So in a year, light will travel about

$$\begin{aligned} 1.86 \times 10^5 \times 3.15 \times 10^7 &= (1.86 \times 3.15) \times (10^5 \times 10^7) \\ &= (1.86 \times 3.15) \times 10^{5+7} \\ &= 5.859 \times 10^{12} \text{ miles} \end{aligned}$$

That is almost 6 trillion miles.

Practice Exercises

1. Earth to moon. The distance from the center of Earth to the center of the moon is about 239,000 miles. Express this number in scientific notation.

2–9. Scientific notation. *In Exercises 2 through 9, express the given number in scientific notation.*

2. Two hundred million

3. 953,000

4. Twenty-eight billion

5. Twenty and a half trillion

6. 0.12 (twelve hundredths)

7. 0.0003 (three ten-thousandths)

8. Fifteen millionths

9. One-hundred-and-eight billionths

10–12. **World population.** *As of this writing, the world's population was about 7.8 billion, and the population of Canada was 38 million. This information is used in Exercises 10 through 12.*

10. Write in scientific notation the world population.

11. Write in scientific notation the population of Canada.

12. What percentage of world population is the population of Canada? Round your answer to one decimal place.

Brief Answers to Practice Exercises

1. 2.39×10^5
2. 2×10^8
3. 9.53×10^5
4. 2.8×10^{10}
5. 2.05×10^{13}
6. 1.2×10^{-1}
7. 3×10^{-4}
8. 1.5×10^{-5}
9. 1.08×10^{-7}
10. 7.8×10^9
11. 3.8×10^7
12. 0.5%

Appendix 3: Calculators, Parentheses, and Rounding

Parentheses

Difficulties with calculators often stem from improper use of parentheses. When they occur in a formula, their use is essential, and sometimes additional parentheses must be supplied.

For example, expressions like $(1 + 0.06/12)^{36}$ arise in Chapter 4 when we consider compound interest. In such expressions, the parentheses are crucial. If you omit them, you may get the wrong answer. Let's see what happens. The correct calculator entry is

[( ] 1 [+] .06 [÷] 12 [ )] [∧] 36

The calculator's answer is 1.196680525. If the parentheses are omitted, the expression becomes $1 + 0.06/12^{36}$, which we enter as

1 [+] .06 [÷] 12 [∧] 36

The calculator's answer is 1. As we can see, the parentheses make a big difference.

Sometimes when parentheses do not appear, we must supply them. For example, no parentheses appear in the expression $\dfrac{1.24 + 3.75}{1.62}$. But when we enter this expression, we must enclose $1.24 + 3.75$ in parentheses to tell the calculator that the entire expression goes in the numerator. The correct entry is

[( ] 1.24 [+] 3.75 [ )] [÷] 1.62

The calculator gives an answer of 3.080246914.

If we forget to use parentheses and enter 1.24 [+] 3.75 [÷] 1.62, the calculator thinks only 3.75 (not the entire expression $1.24 + 3.75$) goes in the numerator. The result is a wrong answer: 3.554814815.

EXAMPLE A3.1 Using parentheses

Supply parentheses for proper entry of the following expressions into a calculator.

a. $\dfrac{2}{3.6 + 5.7}$

b. $2^{1.7 \times 3.9}$

SOLUTION

a. We need to enclose $3.6 + 5.7$ in parentheses to indicate that the entire expression is included in the denominator of the fraction. The proper entry is

2 [÷] [(] 3.6 [+] 5.7 [)]

The result is 0.2150537634.

b. We need to enclose 1.7×3.9 in parentheses to indicate that the entire expression is the exponent. The proper entry is

2 [∧] [(] 1.7 [×] 3.9 [)]

The result is 99.04415959.

Rounding

It is often unnecessary to report all the digits given as an answer by a calculator, so it is common practice to shorten the answer by *rounding*.

First we need to decide how many digits to keep when we round. There is no set rule for determining this number. Generally, in this text we keep one digit beyond the decimal point. In certain applications, though, it is appropriate to keep fewer or more digits. For example, when we report the balance of an account in dollars, we usually keep two digits beyond the decimal point, that is, we round to the nearest cent. Another consideration in rounding is the accuracy of the data on which we base our calculation. We should not report more digits in our final result than the number of digits in the original data.

The usual method for rounding a number begins with the digit to be rounded. That digit is increased by 1 if the digit to its right is 5 or greater and is left unchanged if the digit to its right is less than 5. All the digits to the right of the rounded digit are dropped.

For example, let's round 23.684 to one decimal place. That means we are going to keep only one digit beyond the decimal point. We focus on the 6. Because the next digit to its right (8) is 5 or larger, we increase the 6 to 7 and drop all the digits to the right of it. This results in 23.7.

If we choose to round 23.684 to two decimal places, we are going to keep two digits beyond the decimal point. We focus on the 8. The next digit is a 4, which is less than 5. So we leave the 8 unchanged and delete the remaining digits. This results in 23.68.

Although it is often appropriate to round final answers, rounding in the middle of a calculation can lead to errors. We were cautioned about this in Calculation Tip 4.1 in Section 4.1. Let's look a bit more closely at this situation.

Here is a typical financial calculation. It gives the balance in dollars of an account after 5 years of daily compounding at 6% annual interest:

$$\text{Balance} = 10{,}000\left(1 + \frac{0.06}{365}\right)^{1825}$$

The right way to calculate it: Use parentheses and enter the formula as it reads. Here is the sequence of keystrokes:

10000 [×] [(] 1 [+] .06 [÷] 365 [)] [∧] 1825

What you should see in the calculator display is

```
10000*(1+.06/365)^1825
            13498.25527
```

Rounding to the nearest cent yields 13,498.26 dollars.

Warning! The wrong way to calculate it: It is not uncommon for students to calculate $0.06/365 = 0.00016438\cdots$, round this number to 0.00016, and plug that result into the formula. This is a bad idea. Here's what happens if you do this: You are calculating $10{,}000(1.00016)^{1825}$. This gives 13,390.7174, which rounds to 13,390.72 dollars. The difference between this answer and the correct answer above is \$107.54. The reason for this discrepancy is that your calculator stores a lot of decimal places that can have a big effect on the final answer. It's the rounding of $0.06/365 = 0.00016438\cdots$ to 0.00016 that makes the \$107.54 error.

The moral of the story is that it is best to avoid rounding numbers in the middle of a calculation. Avoiding rounding not only improves accuracy but also is less work. Next we see one way to avoid rounding in the middle.

Chain calculations

For complicated formulas, it is possible to make calculations in steps without intermediate rounding. To illustrate the procedure, let's recall the monthly payment formula from Section 4.2:

$$\text{Monthly payment} = \frac{\text{Amount borrowed} \times r(1+r)^t}{((1+r)^t - 1)}$$

Here, t is the term in months and $r = \text{APR}/12$ is the monthly interest rate as a decimal. Let's do the calculation when the amount borrowed is \$10,000, $t = 30$ months, and the monthly rate is $r = 0.05/12$. With these values, the formula gives

$$\text{Monthly payment} = \frac{10000 \times 0.05/12 \times (1+0.05/12)^{30}}{\left((1+0.05/12)^{30} - 1\right)}$$

If we choose to enter the entire expression all at once, we must be very careful to get all the parentheses right:

10000 [×] .05 [÷] 12 [×] [(] 1 + .05 [÷] 12 [)] [∧] 30
[÷] [(] [(] 1 [+] .05 [÷] 12 [)] [∧] 30 [−] 1 [)]

It is difficult to enter this expression without making a mistake. Here is a method that makes entering a bit easier. First calculate $(1 + .05/12)^{30}$ using

[(] 1 [+] .05 [÷] 12 [)] [∧] 30

The calculator displays 1.132854218, and it automatically stores the result. We can recall it using the keystrokes [2nd] [ans] . So we can now enter the formula as

10000 [×] .05 [÷] 12 [×] [2nd] [ans] [÷] [(] [2nd] [ans] [−] 1 [)]

The answer, rounded to the nearest cent, is \$355.29.

Practice Exercises

1–7. Using parentheses. *Exercises 1 through 7 focus on the use of parentheses. In each case, show how parentheses should be used to calculate the given expression and then find the value rounded to one decimal place.*

1. $\frac{3.1 + 5.8}{2.6}$

2. $\frac{3.1}{5.8 - 2.6}$

3. $\frac{3.1 - 1.3}{5.8 + 2.6}$

4. $3^{1.1-2.2}$

5. $3 - \frac{1.6 + 3.7}{4.3}$

6. Calculate the following three expressions. The only difference is placement of parentheses. Compare the results.
   a. $2 \times 3 + 4$
   b. $2 \times (3 + 4)$
   c. $(2 \times 3) + 4$

7. Calculate the following four expressions. The only difference is placement of parentheses. Compare the results.
   a. $2 \times 3 + 4^2$
   b. $(2 \times 3 + 4)^2$
   c. $2 \times (3 + 4)^2$
   d. $(2 \times 3) + 4^2$

8. Round the number 14.2361172 to one decimal place.

9. Round the number 14.2361172 to two decimal places.

10. Round the number 14.2361172 to three decimal places.

11. **Intermediate rounding.** This exercise shows problems that can arise using improper calculation techniques.
   a. Calculate $\$100{,}000(1+0.07/12)^{100}$ to the nearest cent.
   b. Observe that $0.07/12 = 0.0058333333\cdots$. Round this to 0.006, and use that result to calculate $\$100{,}000(1 + 0.07/12)^{100}$ to the nearest cent.
   c. What is the difference between your two answers?

Brief Answers to Practice Exercises

1. $(3.1 + 5.8) \div 2.6 = 3.4$
2. $3.1 \div (5.8 - 2.6) = 1.0$
3. $(3.1 - 1.3) \div (5.8 + 2.6) = 0.2$
4. $3 \wedge (1.1 - 2.2) = 0.3$
5. $3 - (1.6 + 3.7) \div 4.3 = 1.8$
6. a. 10
   b. 14
   c. 10
7. a. 22
   b. 100
   c. 98
   d. 22
8. 14.2
9. 14.24
10. 14.236
11. a. $178,896.73
   b. $181,885.50
   c. $2988.77

Appendix 4: Basic Math

This appendix is designed to review some basic calculations that often give students trouble.

Decimals and fractions

We can always turn a fraction into a decimal using long division, though we often shortcut this operation using a calculator. But it is also true that decimals can be represented as fractions. To see this fact, let's recall the basics of decimal notation. Each digit to the right of the decimal point is a specific place holder.

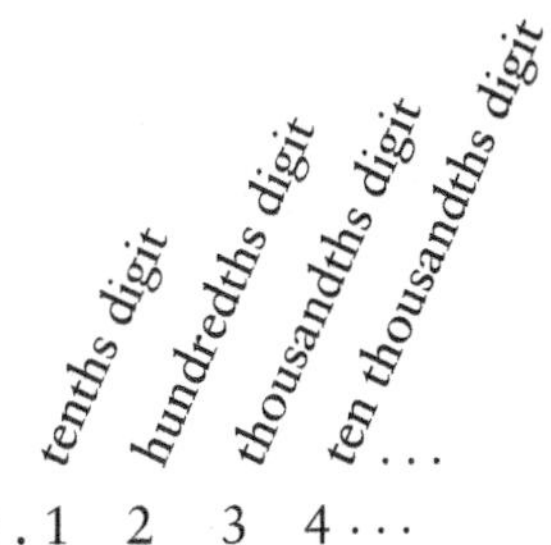

This diagram is a guide to expressing decimals as fractions.

$$\text{In } 0.1, \text{ the 1 is in the 10th position. Hence } 0.1 = \frac{1}{10}$$

$$\text{In } 0.12, \text{ the 2 is in the 100th position. Hence } 0.12 = \frac{12}{100}$$

$$\text{In } 0.123, \text{ the 3 is in the 1000th position. Hence } 0.123 = \frac{123}{1000}$$

$$\text{In } 0.1234, \text{ the 4 is in the 10000th position. Hence } 0.1234 = \frac{1234}{10000}$$

EXAMPLE A4.1 Equivalence of fractions and decimals

a. Use long division to express $\frac{3}{4}$ as a decimal.

b. Express 3.625 as a fraction.

SOLUTION

a. The long-division procedure is

$$\begin{array}{r} 0.75 \\ 4\,\overline{)3.00} \\ \underline{2.8} \\ 0.20 \\ \underline{0.20} \\ 0 \end{array}$$

Hence $\frac{3}{4} = 0.75$.

b. In the decimal 3.625, the last digit, the 5, is in the 1000th position. Hence

$$3.625 = \frac{3625}{1000}$$

We can simplify this fraction:

$$\frac{3625}{1000} = \frac{29 \times 125}{8 \times 125} = \frac{29 \times \cancel{125}}{8 \times \cancel{125}} = \frac{29}{8}$$

Using percentages

There is a close connection between percentages and decimals. If we want 25% of a quantity, we multiply it by the decimal 0.25, which is $\frac{25}{100}$. We can think of obtaining 0.25 from 25% by dropping the % symbol and moving the decimal point two places to the left. If we want to express a decimal as a percentage we reverse this procedure: To express 1.2 as a percentage we calculate 100 times 1.2 to obtain 120%.

Quick Review

If r is a number, then $r\%$ is a way of expressing the fraction $\frac{r}{100}$. This fraction is expressed in decimal form by moving the decimal point of r two places to the left. We often refer to this associated decimal as *the decimal form of the percentage.*

If we are given a number in decimal form and are asked to express it as a percentage we reverse this procedure, multiplying the number by 100 followed by a percent sign.

Here are some common ways in which percentages are used.

What is $r\%$ of a quantity? To calculate $r\%$ of a quantity, we multiply the quantity by the decimal form of the percent.

A Typical Example

In a drug trial involving 50 people, 74% showed positive effects from the drug. How many people showed positive effects?

SOLUTION

The number of people who showed positive effects is 74% of 50. To find 74% of 50, we multiply 50 by the decimal form of 74%, which is 0.74:

$$\text{Number who showed positive effects } = 50 \times 0.74 = 37 \text{ people}$$

What is the result of an $r\%$ increase or decrease? To find the effect of an $r\%$ increase or decrease of a quantity, we first calculate $r\%$ of the quantity and then add or subtract as appropriate.

A Typical Example

a. An investment of \$1500 has increased in value by 12%. What is the current value of the investment?

b. An investment of \$1500 has decreased in value by 12%. What is the current value?

SOLUTION

a. We will show two ways to find the answer. First we calculate the increase, 12% of 1500:

$$\text{Increase} = 12\% \text{ of } 1500 = 0.12 \times 1500 = 180 \text{ dollars}$$

The value of the investment has increased by \$180, so its current value is $1500 + 180 = 1680$ dollars.

Here is another way to find the answer. Because the value of the investment has increased by 12%, its current value is 112% of its old value:

$$\text{Current value} = 112\% \text{ of old value } = 1.12 \times 1500 = 1680 \text{ dollars}$$

b. Because the investment decreased by 12%, we subtract 12% of 1500 from the investment value to obtain $1500 - 180 = 1320$ dollars.

Another way to answer this question is to think of the new value as 88% of the old value and then calculate $0.88 \times 1500 = 1320$ dollars.

What percentage is value A of value B? To find the decimal form of the percentage, we divide value A by value B.

A Typical Example

In a survey of 800 people, 644 approved of the current mayor. Judging from this sample, what is the mayor's percentage approval?

SOLUTION

We are being asked to find 644 as a percentage of 800, so we first divide 644 by 800 to find the decimal form:

$$\frac{644}{800} = 0.805$$

To find the associated percentage, we multiply by 100 (or move the decimal point two places to the right). The resulting answer is an 80.5% approval rating.

What is the percentage increase or decrease? If a quantity increases from one value to another, then the change is the difference. This difference, expressed as a percentage of the original value, is the percentage increase. If a quantity decreases from one value to another, then the difference, expressed as a percentage of the original value, is the percentage decrease.

A Typical Example

a. The population of a town increased from 15,400 to 18,942. What is the percentage increase?

b. The population of a town decreased from 18,942 to 15,400. What is the percentage decrease?

SOLUTION

a. The actual increase is the difference $18{,}942 - 15{,}400 = 3542$. Thus, to calculate the percentage increase we need to know what percentage 3542 is of 15,400. As a decimal this is the quotient

$$\frac{3542}{15{,}400} = 0.23$$

This is the decimal form of 23%. Hence, the population increased by 23%.

b. The actual decrease is the difference $18{,}942 - 15{,}400 = 3542$. Thus, to calculate the percentage decrease we need to know what percentage 3542 is of 18,942. As a decimal this is the quotient

$$\frac{3542}{18{,}942}$$

or about 0.187. This is the decimal form of 18.7%. Hence, the population decreased by 18.7%.

Practice Exercises

1–4. Fractions to decimals.

1. Write $\frac{4}{5}$ as a decimal.

2. Write $\frac{5}{4}$ as a decimal.

3. Write $\frac{6}{7}$ as a decimal. Round to two decimal places.

4. Write $\frac{7}{3}$ as a decimal. Round to two decimal places.

5–6. Decimals to fractions.

5. Write 3.72 as a fraction in reduced form.

6. Write 0.04 as a fraction in reduced form.

7–8. Percentages to decimals.

7. Write the decimal form of 34.5%.

8. Write the decimal form of 1/2%.

9–11. Decimals to percentages.

9. The number 0.36 is the decimal form of what percentage?

10. The number 0.055 is the decimal form of what percentage?

11. The number 3.1 is the decimal form of what percentage?

12. Find the percentage. Express these numbers as percentages: 0.1, 0.01, and 1.0.

13. What percentage is it? Candidate 1 received 320 of 500 votes cast. What percentage of voters voted for Candidate 1?

14. What percentage is it? Assume that 32 of 60 students in a class are female. What percentage of students in the class are female? Round your answer to the nearest tenth of a percent.

15. What is the result of a percentage increase? A city of 433,000 experiences a population increase of 12%. What is the new population of the city?

16. What is the result of a percentage decrease? An investment of $500 loses 15% of its value. What is the new value of the investment?

17. An item on sale. An appliance marked at $535 is on sale for 15% off. What is the sale price of the item?

18. What is the percentage increase? A company buys items at a wholesale cost of $35. The items are sold at the retail price of $50. What is the percentage price mark-up for this item? Round your answer to the nearest tenth of a percent.

19. What is the percentage decrease? An item that normally costs $42.50 is on sale for $38.00. What is the percentage savings? Round your answer to the nearest tenth of a percent.

Brief Answers to Practice Exercises

1. 0.8	**2.** 1.25	**3.** 0.86	**4.** 2.33	**5.** $\frac{93}{25}$
6. $\frac{1}{25}$	**7.** 0.345	**8.** 0.005	**9.** 36%	**10.** 5.5%
11. 310%	**12.** 10%, 1%, 100%	**13.** 64%	**14.** 53.3%	**15.** 484,960
16. $425	**17.** $454.75	**18.** 42.9%	**19.** 10.6%	

Appendix 5: Problem Solving

The book *How to Solve It* by George Pólya is probably the best-known source of methods for solving mathematical problems. The ideas here are drawn from Pólya's work, but the book itself is highly recommended.

Students, perhaps anxious to finish a homework assignment, often want to get straight to the answer. This is rarely a good idea. Successful problem solving is a *process*, and it typically requires patience and dedication. Pólya offers four basic principles to guide the process.

First principle: Understand the problem

Many times we struggle with problems because we don't have a clear understanding of what we are being asked to do. Progress simply isn't possible without this. Recommended approaches include the following:

- Note specifically what you are asked to show.
- Get clarification regarding any terms in the problem that you don't understand.
- Restate the problem in your own words.
- Draw a picture or a diagram when that is appropriate.

Let's look back at part 2 of Example 1.5, where we are asked to classify the fallacy *My dad is a professor of physics, and he says Dobermans make better watchdogs than collies.* We can't make progress on this problem unless we know what is being asked. What is a fallacy anyway, and what does it mean to classify one? Upon reading the section we find the answers to both of these questions. The important idea here is that we must address these issues before we can proceed. They should be dealt with before we even look at the italicized sentence we are being asked to classify. This particular problem almost solves itself once we understand the question, and we are led quickly to classify this as a fallacy of false authority.

Here is another example. Suppose you are asked to find the perimeter of a rectangle. Simply drawing a labeled picture of a rectangle, as shown, can contribute greatly to a clear understanding of what is being asked and may lead to a focus on the elements that lead to a solution.

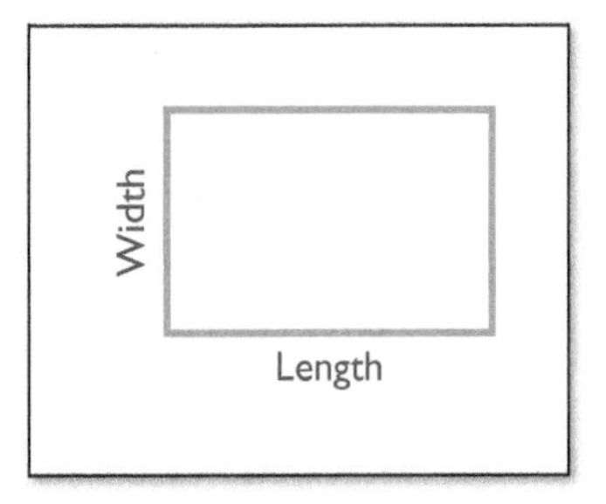

Second principle: Devise a plan

Once the problem is understood and a clear goal is in mind, we need a plan that will lead to that goal. Such a plan may include the following:

- List the pieces that must go together to make the solution.
- Eliminate possibilities.
- Guess and then check your guess.
- Consider special cases.
- Identify formulas that may apply.
- Propose an equation to solve.

Refer to Example 3.12, where we are asked to find an exponential formula that describes the amount of amoxicillin in the bloodstream t hours after injection. Also, we are asked to determine whether a second injection is necessary after five hours. If we have properly followed the first principle, here is an orderly plan for finding an exponential formula.

Step 1: Identify the initial value.

Step 2: Identify the base.

Step 3: Use the base and initial value to construct the exponential formula.

Once we have the formula, we can use it to find the level of amoxicillin in the bloodstream after five hours.

Let's look at Example 4.16: *Your neighbor took out a 30-year mortgage for $300,000 at a time when the APR was 9%. She says that she will wind up paying more in interest than for the home (that is, the principal). Is that true?*

Certainly the first principle applies. We will make no progress at all without knowing what a mortgage is and how to use the APR. It may be helpful to restate the problem: "Will she pay more than $300,000 in interest?"

Once we have a clear understanding of the problem, we need to devise a plan for solving it. Here is a possible outline.

Step 1: Find the total amount that she paid.

Step 2: Calculate the difference between the total amount paid and the principal. That difference is the interest paid.

Step 3: Compare the interest paid with the principal.

In this case it is probably helpful to enhance **Step 1**.

Enhancement of Step 1:

Step 1a: Calculate the monthly interest rate as a decimal.

Step 1b: Use the monthly payment formula to find the amount paid each month.

Step 1c: Multiply the monthly payment by the number of months to find the total amount paid.

It is a good idea to actually write down the plan of attack.

There is no single strategy that will lead you to the solution of all problems, but understanding the problem and devising a plan for solution are key ingredients in the process.

Third principle: Execute the plan you devised

This is the stage of problem solving that is often the focus of students, and sometimes students skip the first two steps and try to start here. That is a poor strategy. But if the first two principles have been followed, this may be the easiest piece of the problem-solving process.

Executing our plan may require a bit of tenacity, though. Let's continue solving the mortgage problem just discussed. Part of our plan requires us to use a rather complicated formula:

$$\text{Monthly payment} = \frac{\text{Amount borrowed} \times r(1+r)^t}{((1+r)^t - 1)} = \frac{\$300{,}000 \times 0.0075 \times 1.0075^{360}}{(1.0075^{360} - 1)}$$

We will have to evaluate this expression using a calculator or computer, and it is very easy to enter this expression improperly. If you calculate a monthly payment of 47 cents or 47 thousand dollars, you can be pretty sure that something is wrong. The point is that *you should not give up*. Try entering the formula again. If persistence doesn't result in a reasonable answer, it may be time to re-examine your plan of attack. You are not alone when this happens. It happens to the authors of this text, and it happens to the strongest mathematicians on the planet. In fact, this is a key to how all problems in mathematics are solved. If your idea doesn't work out, look for your error and try again—and again.

Fourth principle: Review and extend

When you have successfully completed the problem, pause to reflect on what happened. For example, in solving the mortgage problem, did you learn something that may be important in your own future? Was persistence helpful? Did you gain some facility with the calculator? Is there something about the methods you used that can be helpful in attacking future problems?

When all is said and done, the one thing that contributes most to successful problem solving is experience doing it. The more problems you successfully attack and solve, the better you will get at solving problems. Even failure to solve a problem can be helpful if you gain something from the experience. Finally, there are more problems than just mathematics problems, and these same principles can be helpful in dealing with many issues encountered in daily life. Most importantly: don't give up. If you don't succeed, try again.

Brief Answers

Exercise Set 1.1

Test Your Understanding

1. Answers will vary.
2. True
3. False
4. True

Problems

5. 75%
7. 1600
9. It is not possible.
11. It is possible, and it is an example of Simpson's paradox.
13. It is possible, and it is not an example of Simpson's paradox.
15. It is possible, and it is not an example of Simpson's paradox.
17. a. Batter 1: 0.300 for 2019, 0.250 for 2020, 0.299 for two-year period. Batter 2: 0.400 for 2019, 0.260 for 2020, 0.269 for two-year period.
    b. Batter 2
    c. Batter 2
    d. Batter 1
19. a.

	Percentage males hired	Percentage females hired
Hardware	75%	75%
Ladies' apparel	10%	10%

    b. Answers will vary.
    c. Answers will vary.
    d. Answers will vary.
21. a.

All participants		
	Total	% with insurance
Not traced	1174	17%
Five-year group	416	11%

    b. The percentages are not similar. Answers will vary.
    c.

White participants		
	Total	% with insurance
Not traced	126	83%
Five-year group	12	83%

    d.

Black participants		
	Total	% with insurance
Not traced	1048	9%
Five-year group	404	9%

    e. They are the same in both cases.
    f. Answers will vary.

Exercise Set 1.2

Test Your Understanding

1. Answers will vary.
2. a. a distorted or different position
3. An inaccurate or incomplete list of alternatives
4. In a fallacy of relevance, the premises are logically irrelevant to the conclusion. In a fallacy of presumption, false or misleading assumptions are made.
5. Deductive reasoning
6. Inductive reasoning

Problems

7. False authority
9. The premises are that all dogs go to heaven and my terrier is a dog. The conclusion is that my terrier will go to heaven.
11. The premise is that the mayor was caught in a lie. The conclusion is that nobody believes the mayor anymore.
13. The premise is that people have freedom of choice. The conclusion is that they will choose peace.
15. The premise is that bad times have a scientific value. The conclusion is that these are occasions a good learner would not miss.

17. The argument is invalid.

19. The argument is invalid.

21. The argument is invalid.

23. Dismissal based on personal attack

25. False cause

27. False dilemma

29. Appeal to common practice

31. Appeal to ignorance

33. Straw man

35. False dilemma

37. Circular reasoning

39. Straw man

41. Answers will vary.

43. Answers will vary.

45. The number of blocks is the cube (the third power) of the size of a side. 125,000 blocks fit in a cube 50 inches on each side.

47. These are the positive odd integers in order.

Exercise Set 1.3

Test Your Understanding

1. 1-b, 2-a, 3-d, 4-c

2. c. either p or q must be true, and possibly both

3. True

4. True

5. p is true but q is false

6. $q \rightarrow p$

Problems

7. (NOT p) $\rightarrow$ (NOT q)

9. (France OR England) AND (eighteenth OR nineteenth century)

11. If I don't buy this MP3 player, then I can't listen to my favorite songs. No.

13. Answers will vary.

15. $q \rightarrow$ (NOT p)

17. $q \rightarrow$ (NOT p)

19. p: "We pass my bill"; r: "The economy will recover"; $p \rightarrow r$

21. c: "I want cereal for breakfast"; e: "I want eggs for breakfast"; c AND e

23. c: "You clean your room"; f: "I'll tell your father"; (NOT c) $\rightarrow f$

25. d: "I'll go downtown with you"; m: "We can go to a movie"; r: "We can go to a restaurant"; (m OR r) $\rightarrow d$

27. "It is not the case that he's American or Canadian." "He's not American and he's not Canadian."

29. False 31. True 33. True

35. True 37. False 39. True

41. True 43. True

45. If you clean your room, I won't tell your father.

47. If you don't agree to go to a movie or a restaurant, then I won't go downtown with you.

49. If it is not a math course, then it is not important.

51. If you don't take your medicine, you won't get well.

53. If I told your father, then you didn't clean your room.

55. If I went downtown with you, then you agreed to go to a movie or a restaurant.

57. All important courses are math courses.

59. If you get well, then you took your medicine.

61. If you have not chosen the side of the oppressor, you are not neutral in situations of injustice.

63. If you do not get arrested, then you did not drink and drive.

65. If I didn't tell your father, then you did clean your room.

67. If I did not go downtown with you, then you did not agree to go to a movie or a restaurant.

69. If you do not drive a Porsche, then you are not my friend.

71. If radar does not see thunderstorms, then they are not approaching.

73.

p	q	(p OR q)	$\rightarrow$
T	T	T	T
T	F	T	T
F	T	T	T
F	F	F	T

75.

p	q	(p OR q)	AND
T	T	T	T
T	F	T	T
F	T	T	F
F	F	F	F

77. Their truth values are both (F, T, T, T).

79. 0

81. 1

Exercise Set 1.4

Test Your Understanding

1. they have no elements in common
2. **b.** every element of A is also an element of B
3. elements that are common to both of the sets represented by the circles
4. A is a subset of B.

Problems

5. {5, 6, 7, 8, 9}
7. {b, c, d, f, g, h, j, k, l, m, n, p, q, r, s, t, v, w, x, y, z}
9. B and C are subsets of A. B and C are disjoint.
11. A is a subset of B, A is a subset of C, and B is a subset of C.

13. 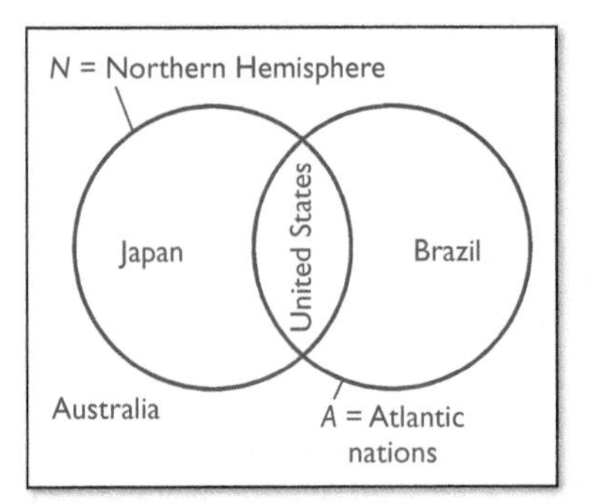

15. 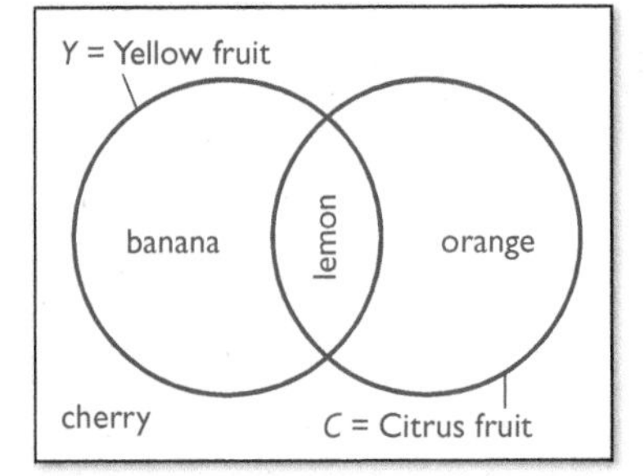

17. 68
19. 851
21. 991
23. 1989
25. 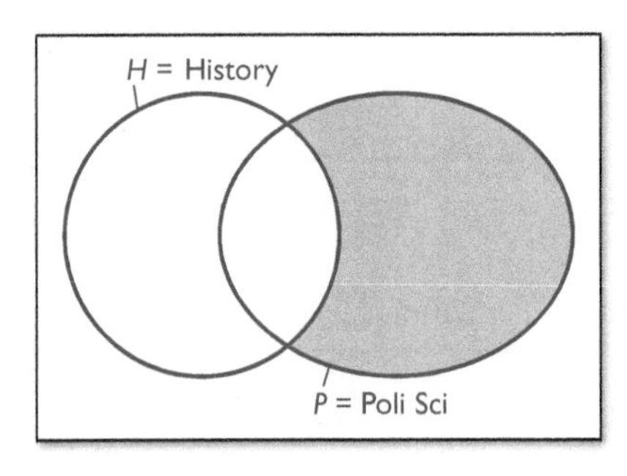

27. 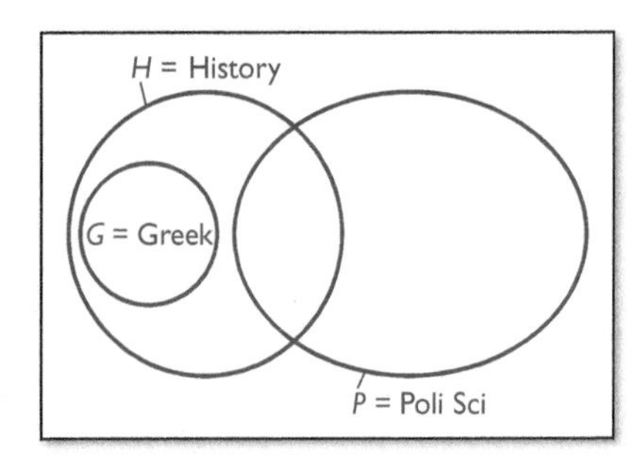

29. **a.** Answers will vary.
    b. False positives: 2790; true negatives: 6210; false negatives: 220. Missing: true positives.
    c. True positives: 780
    d. 78%

31. 70 warm days

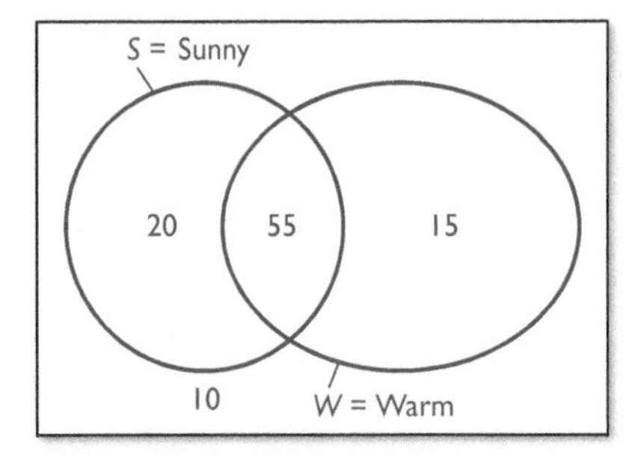

33. 1765 freshmen

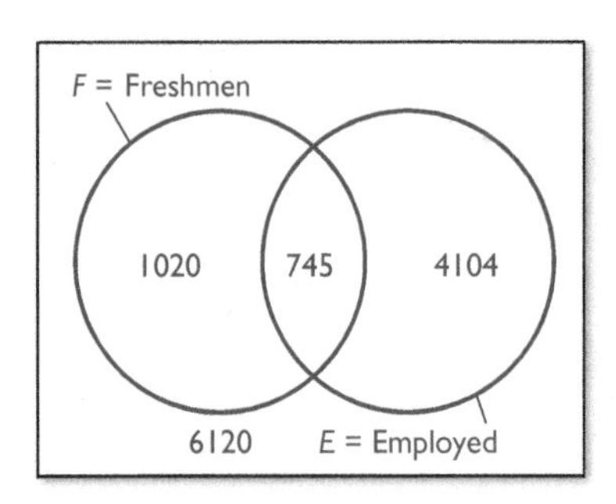

35. **a.** 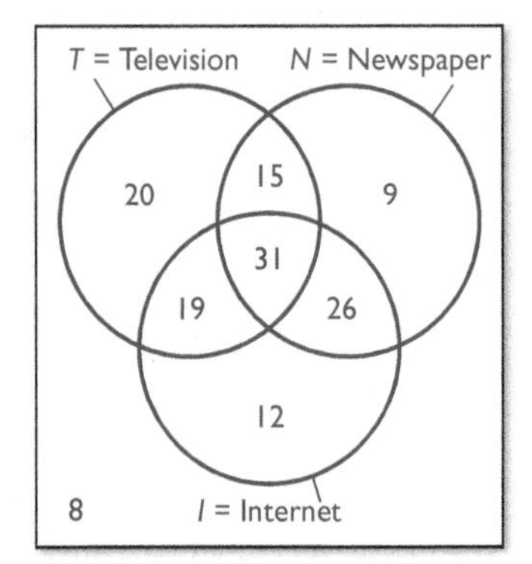

    b. 140
    c. 41
    d. 88
    e. 55

Exercise Set 1.5

Test Your Understanding

1. Answers will vary.
2. Answers will vary.
3. **b.** A liter is about the same as a quart.
4. **c.** a very small number
5. **a.** multiply the price per square foot by 9
6. Answers will vary.

Problems

7. The bacterium is 50 times as large as the virus.
9. \$10,000
11. 10 million
13. 96,907 miles. That's nearly four times around the world.
15. 39,878 cubic feet
17. 9.5 years
19. 12 users per second
21. About 25 minutes
23. Answers will vary.
25. \$100
27. Estimate 1.00%; exact figure 0.97%
29. The carpet costing \$12 per square yard
31. 29,794 people per square mile
33. 10^{21}

Exercise Set 2.1

Test Your Understanding

1. True
2. **d.** all of the above
3. **a.** the change in the function value divided by the change in the independent variable
4. Interpolation
5. Extrapolation
6. False

Problems

7. **a.**

Period	2017–18	2018–19	2019–20	2020–21
Percentage increase	10%	10%	20%	10%

b.

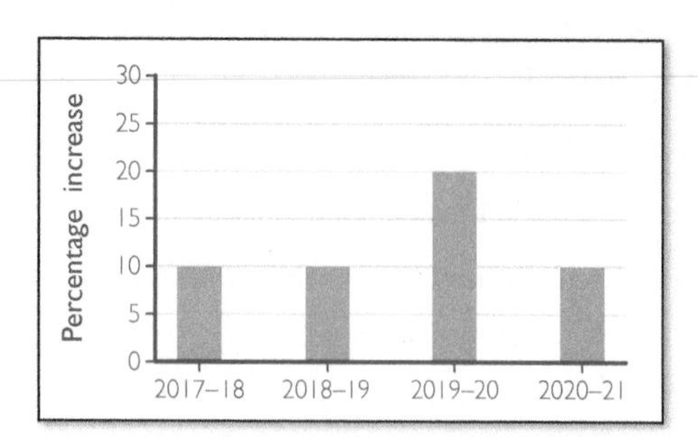

c. 2019–20

9. Cases per day; the number of new cases of the disease we expect each day.

11. One choice: Let the year be the independent variable and the number of home runs the dependent variable. Then the function gives the number of home runs in a given year.

13. **a.** 64.5%

 b. 65.1%

15. **a.** The percentage increased.

 b. 2 percentage points per year

 c. 52%

 d. The estimate in part c is greater than the actual percentage. The percentage grew faster from 2012 to 2015 than it did from 2010 to 2012.

17. **a.** 1996: 2.9%; 1997: 3.0%; 1998: 3.2%; 1999: 3.3%. (Some variation in answers is acceptable.)

 b. \$308

 c. \$98

 d. \$200

19. **a.** Enrollment in French decreased from 1995 to 1998 and increased after that.

 b. 1995–1998: −2.10 thousand per year; 1998–2002: 0.73 thousand per year; 2002–2006: 1.10 thousand per year; 2006–2009: 3.33 thousand per year

 c.

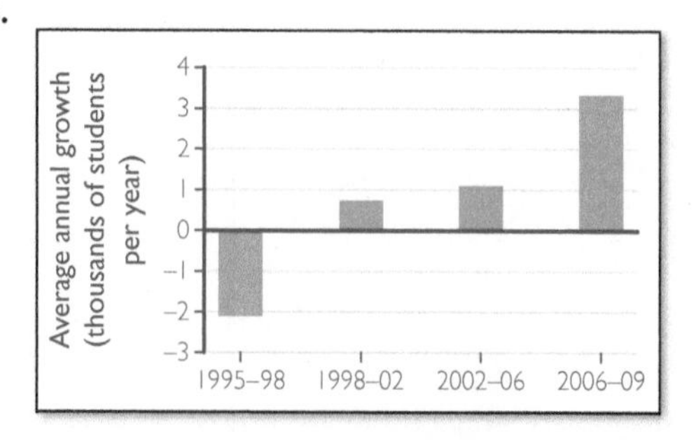

21. a.

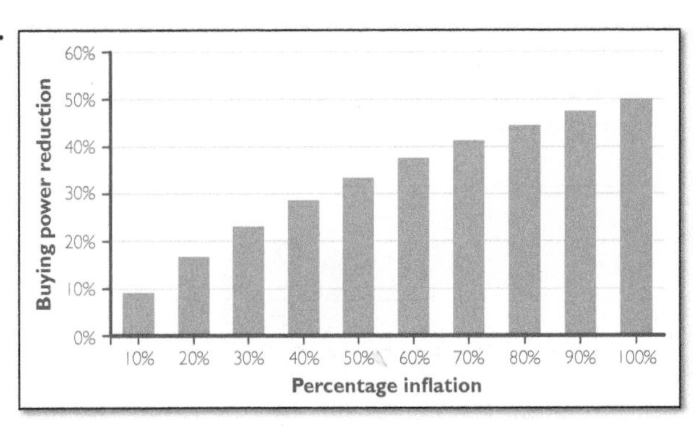

b. 33.5%

Exercise Set 2.2

Test Your Understanding

1. A-II, B-III, C-I
2. A-II, B-I
3. c. interpolating using the average growth rate
4. True

Problems

5. Expenses decrease from 2010 to 2013, then increase until 2016, then stay the same until 2018 when they increase through 2020. The growth rate is negative from 2010 to 2013, positive from 2013 to 2016, zero from 2016 to 2018, and positive from 2018 to 2020.
7. The population started at about 15,000 and then decreased, but less and less so, almost disappearing by 2020.
9. Bar graph:

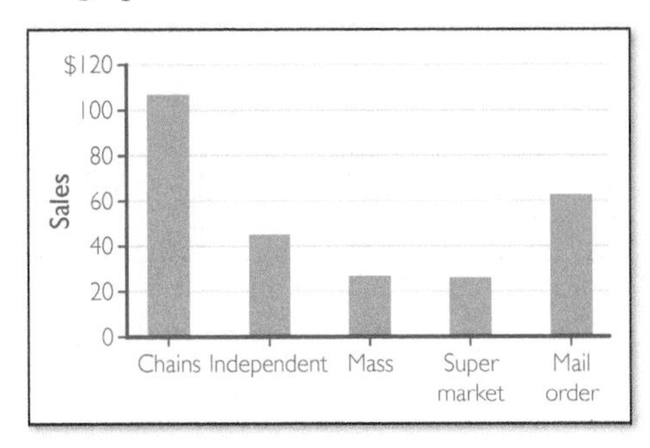

Scatterplot:

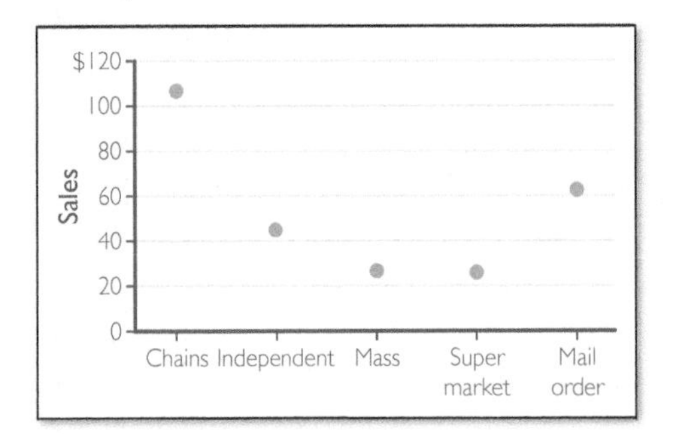

Line graph:

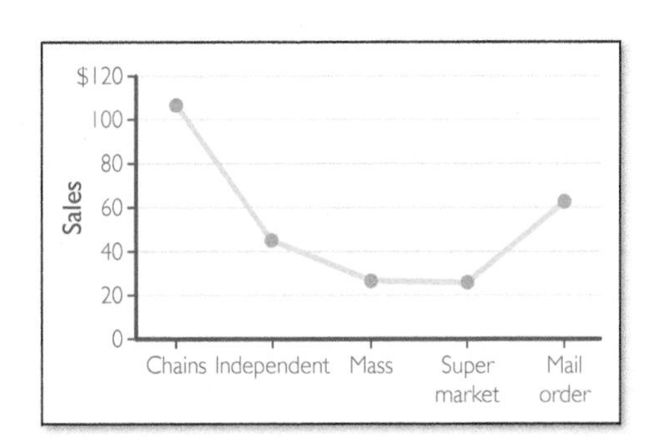

Answers will vary.

11. One possibility is

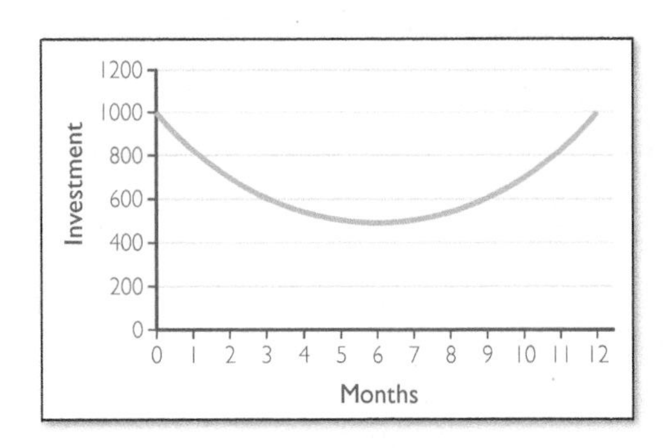

13. One possibility is

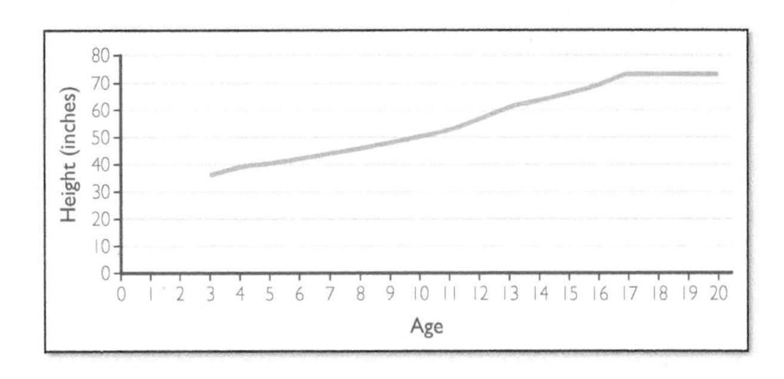

15. One possibility is

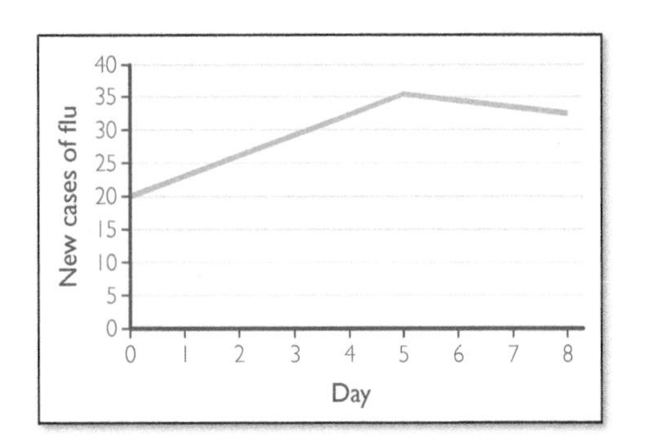

17. One possibility is

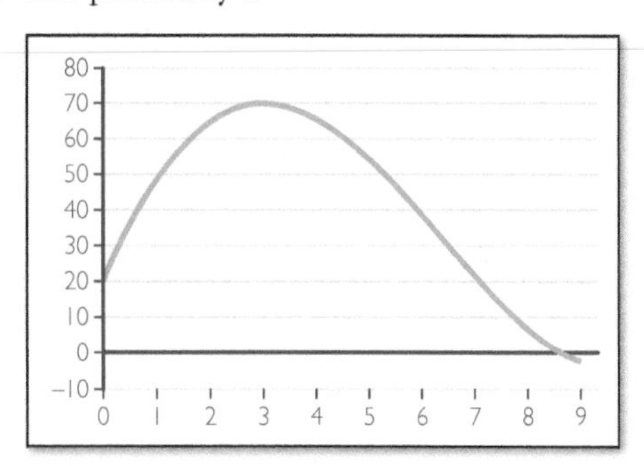

19. One possibility is

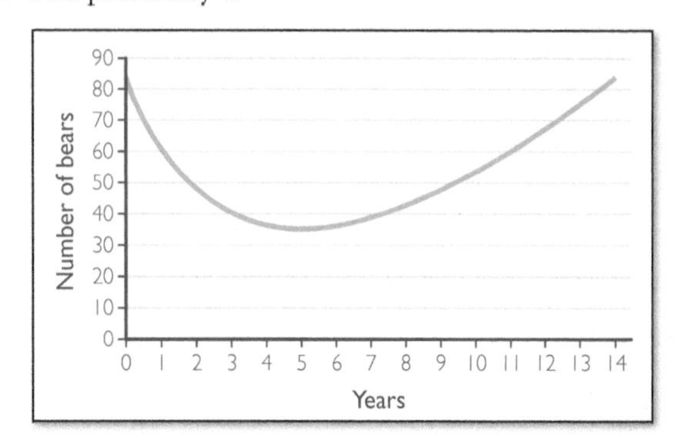

21. Enrollments decreased from 1968 to about 1980 and from about 1990 to about 1995.

23. 1960–1965; the slope then is larger.

25. Mystery curve 1: growth; Mystery curve 2: sales. Answers will vary.

Exercise Set 2.3

Test Your Understanding

1. **d.** all of the above
2. It emphasizes the change.
3. False
4. True
5. **a.** show how constituent parts make up a whole
6. True

Problems

7. A narrow range on the vertical axis should be used. One possibility is

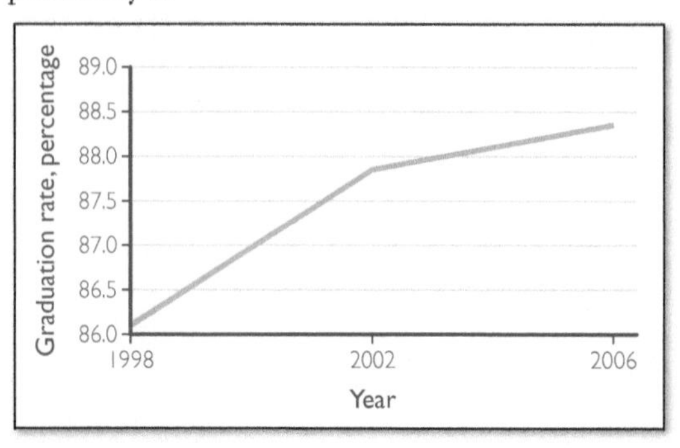

9. We can emphasize the changes by choosing to display a range of numbers on the vertical axis that is much closer to the data values. In the following figure, we have used a range on the vertical axis of 195 to 275.

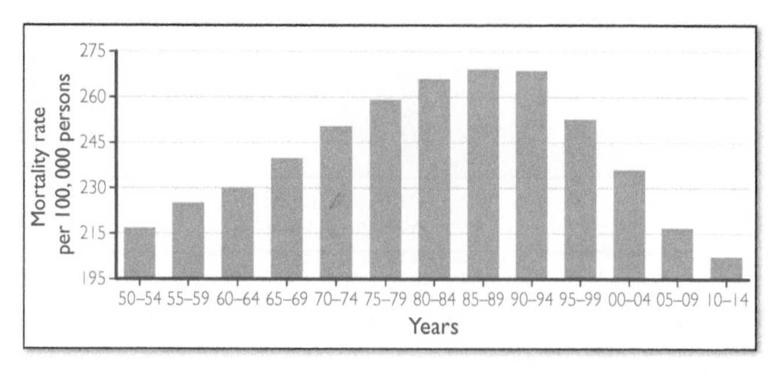

11. The fact that nearly 60% of employees are in the unknown category makes the graph of questionable value.

13. The lines connecting the dots do not represent any conceivable data. A table, or perhaps a bar graph or scatterplot, would be better.

15. Answers will vary. Knowing the number of drivers in each category would be important.

17. **a.** From 1945 to 1990
b. 0.29 thousand dollars per year, or 290 dollars per year
c. 1.14 thousand dollars per year, or 1140 dollars per year
d. Answers will vary.

19. Answers will vary.

21.

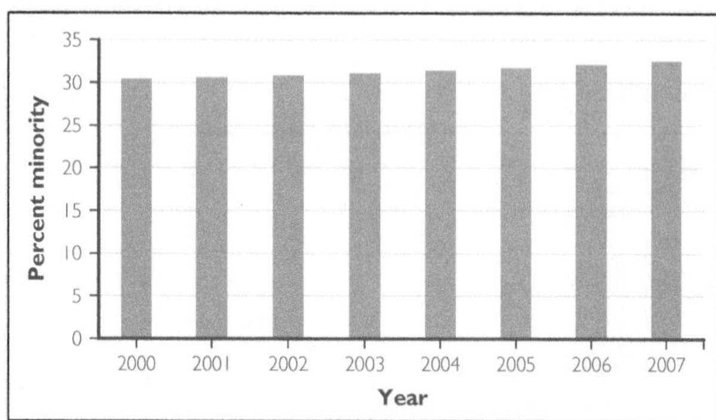

23.

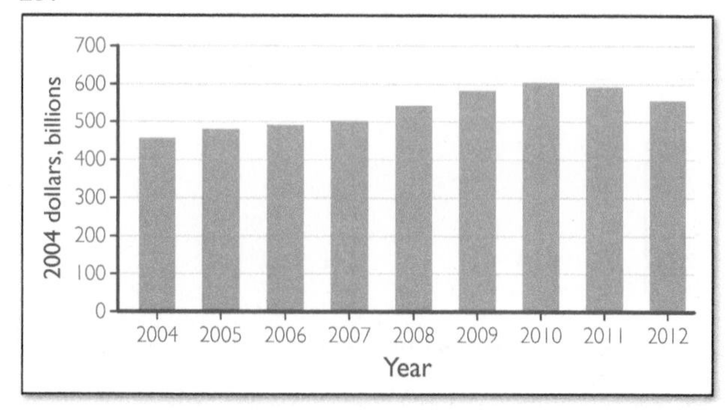

Answers will vary.

25. a.

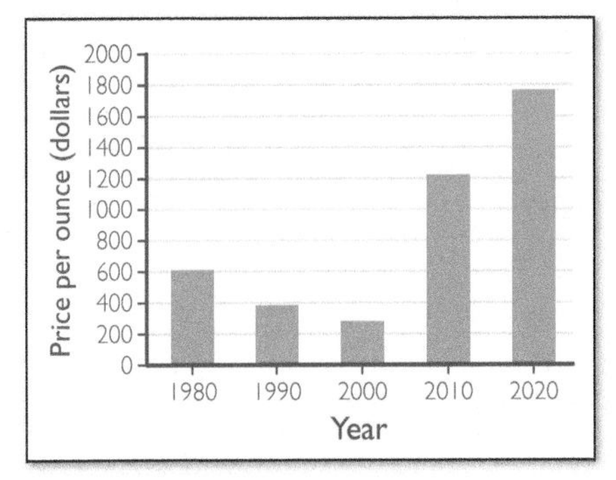

b.

Year	Price (2020 dollars)
1980	\$1923.44
1990	\$759.35
2000	\$418.67
2010	\$1457.19
2020	\$1769.64

c.

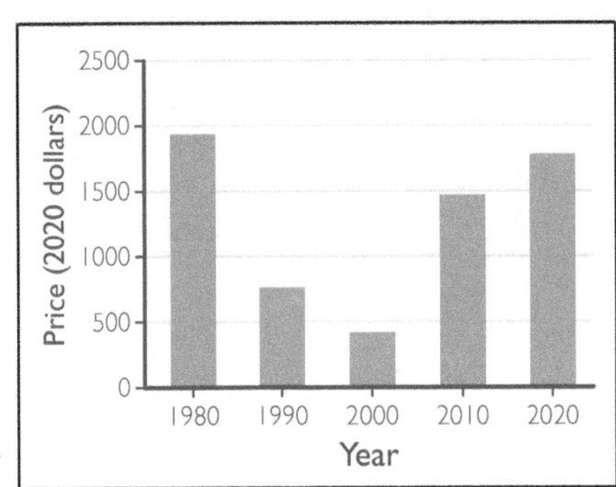

27. a.

Year	Price (2020 dollars)
1980	\$94.20
1990	\$247.50
2000	\$487.50
2010	\$1190.00
2020	\$3488.00

b.

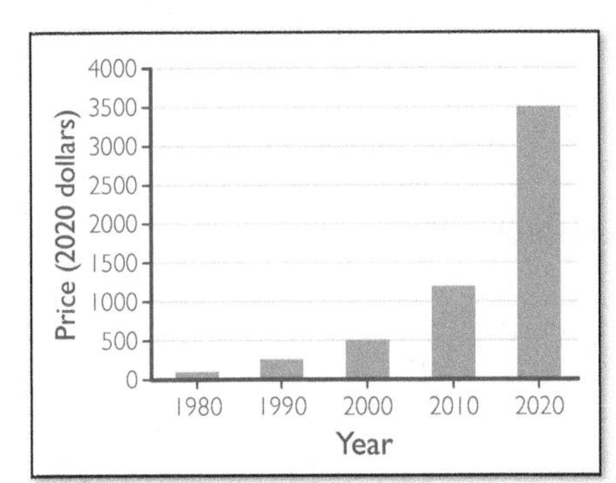

c. Tickets were about 37 times as expensive in 2020.

29. The vertical axis represents percentage of the total by weight.

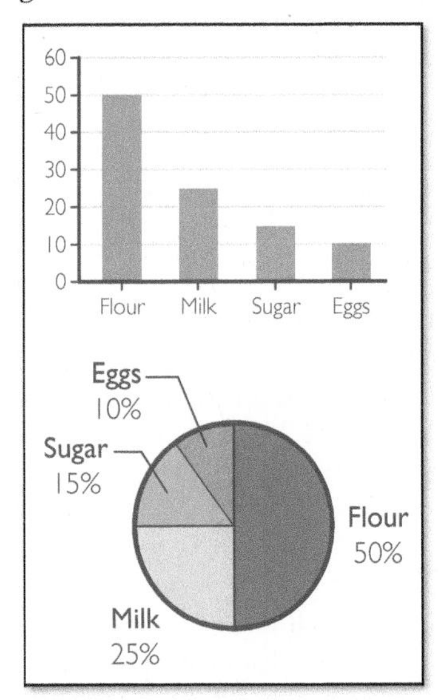

Exercise Set 3.1

Test Your Understanding

1. Add m to this year's amount.
2. **d.** all of the above
3. the change in the function value divided by the change in the independent variable
4. True
5. True
6. False

Problems

7. Initial value: 5 millimeters; growth rate: 2.1 millimeters per week. The flower was 5 millimeters tall at the beginning of the month and grows by 2.1 millimeters per week.
9. Constant growth rate, so linear; growth rate: 2 inches per year; initial value: 40 inches; $H = 2t + 40$
11. -3 pounds per square inch per hour
13. Yes
15. No
17. No
19. No
21. Yes. If M is the number of miles and g is the number of gallons, then $M = 25g$.
23. No
25. It is linear. $T = 0.25I + 5156.25$
27. The growth rate is 200 feet per mile. The formula is $E = 200h + 3500$.

29. a. The growth rate is 17 meters per second per kilometer, so the growth rate is constant.
b. The slope is 17 meters per second per kilometer. The initial value is 1534 meters per second. Answers will vary.
c. 34 meters per second
d. 1568 meters per second
e. $S = 17D + 1534$
f. 1568 meters per second

31. a. There is a constant growth rate of 0.178 trillion dollars per year. Answers will vary.
b. If C is the amount collected (in trillions of dollars) and t is the time in years since 2003, then $C = 0.178t + 1.952$.
c. 2.486 trillion dollars

33. No. Answers will vary.

35. a. There is a constant growth rate of 6 patients diagnosed per 5 days, or 1.2 patients diagnosed per day.
b. 1.2 patients diagnosed per day. Answers will vary.
c. $F = 1.2t + 35$ if F denotes the number of diagnosed flu cases and t the time in days.
d. About 55 cases

37. 4.20 units per year. Answers will vary.

39. 264.0

41. a. From $73,300 through $74,700, there is $30 additional tax due over each span. From $74,900 through $75,500, there is $50 additional tax due over each span.
b. From $73,300 through $74,700, and from $74,900 through $75,500
c. 15% and 25%

43. 2020. Answers will vary.

45. 106.8 years; no

Exercise Set 3.2

Test Your Understanding

1. Multiply this year's amount by b.
2. 2
3. True
4. b. $1 + r$
5. c. $1 - r$
6. half of the substance to decay

Problems

7. 300% increase per day

9. $B = 500 \times 1.15^t$ dollars

11. $B = 300 \times 1.02^t$ dollars

13. $D = D_0 \times 0.95^t$

15. a. $B = 1000 \times 0.97^t$
b. $858.73

17. Exponential

19. Exponential

21. Linear

23. Linear

25. Answers will vary.

27. 0.2%; exponential since constant percentage growth

29. $3.73 billion; $6.06 billion

31. a. 164 degrees b. 211 degrees

33. Exponential since constant percentage growth. If N is the population and t is time in years since the start, then $N = 500 \times 1.02^t$. After 5 years: 552.

35. Exponential since constant percentage growth; 337 million

37. 305%

39. a. Measure the thickness of a ream and divide by 500, the number of sheets.
b. If T denotes the thickness in millimeters, then $T = 0.1 \times 2^f$. If T denotes the thickness in kilometers, then $T = 0.0000001 \times 2^f$.
c. $0.0000001 \times 2^{50} = 112{,}589{,}991$ kilometers

41. $126.94

43. $23.42

45. At age 55: $531,759.45; at age 65: $3,174,243.86. At age 55: option 1; at age 65: option 2

47. a. 1.95 parts per million b. 1.23 parts per million

49. a. 2.08 b. 24%

51. a. 10 years b. 10 years

Exercise Set 3.3

Test Your Understanding

1. the power to which 10 must be raised in order to equal x
2. $10^t = x$
3. True
4. a. Magnitude = log(Relative intensity)
5. False
6. b. Decibels = 10log(Relative intensity)

Problems

7. a. 4 b. 9 c. −1

9. 3.9

11. Relative intensity is multiplied by 4.0.

13. 22.5

15. Answers will vary.

17. 100 times

19. The one in Alaska was about twice as intense.

21. $1.76 \times 10^{16} = 17,600,000,000,000,000$ joules

23. 1.2 units larger, so 63.0 times as much energy

25. 83 decibels

27. 75 decibels

29. 1.16

31. 1000 times

33. After 153 months, or 12 years and 9 months

35. 2.32 half-lives; 69.6 years

37. **a.** $C = 100 \times 1.025^t$
b. 28 years **c.** 56 years

39. **a.** \$348.68 **b.** 6.6 years

41. **a.** 0.8
b. 0.1
c. 0.009
d. No

43. Answers will vary.

Exercise Set 3.4

Test Your Understanding

1. parabola
2. the quadratic formula
3. **d.** at most two solutions
4. $x = -\dfrac{b}{2a}$
5. **b.** the minimum point

Problems

7. **a.** $x = 1$ or -2
b. $y = 6$ or -3
c. $r = 5$ or -2
d. $x = 5/2$ or -2
e. $x = -2 \pm \sqrt{5}$

9. $x = (1 + \sqrt{5})/2$

11. 1.25 seconds

13. 3 seconds

15. 1 second and 5 seconds; one going up, one going down

17. 2000 feet

19. 500 feet

21. $s = 33$ miles per hour

23. 17 inches by 17 inches

25. 34.1 feet by 29.3 feet or 5.9 feet by 170.7 feet

27. 3.4 yards by 7.4 yards

29. 1.5 feet

31. 13.25 feet

33. No

35. 16; $(x + 4)^2$

Exercise Set 4.1

Test Your Understanding

1. False
2. True
3. interest is periodically added to the balance
4. the number of periods in a year
5. False
6. False
7. **b.** how long it takes money earning compound interest to double
8. **d.** the initial amount that is deposited

Problems

9. **a.** 2%

11. \$1200.00

13. **a.** \$2382.03
b. \$2393.36

15. \$50; \$2050

17.

Quarter	Interest	Balance
1	\$60.00	\$3060.00
2	\$61.20	\$3121.20
3	\$62.42	\$3183.62
4	\$63.67	\$3247.29

19. **a.**

Quarter	Interest	Balance
1	\$50.00	\$2050.00
2	\$51.25	\$2101.25
3	\$52.53	\$2153.78
4	\$53.84	\$2207.62

b. \$207.62
c. Simple interest would yield \$200 in interest. Semi-annual compounding would yield \$205 in interest.

21. For simple interest, the balance is \$1120.00. For monthly compounding, the balance is \$1127.46.

23. Option 2 is better if you retire at age 65. Option 1 is better if you retire at age 55.

25. \$7607.64

27. First quarter: \$2040.00. Second quarter: \$2080.80. Third quarter: \$2122.42. Fourth quarter: \$2164.87. Total interest \$164.87. The APY is 8.24%.

29. a. $412.16; 8.24%
    b. $5400. We know that compounding is taking place because, according to the table, the APY is higher than the APR.
    c. The APY is 4.06%, which is higher than the value in the table.

31. 7.50%

33. $2451.36. This is how much the investment will be worth after 10 years.

35. It doubles in 8 years and doubles again in 8 more years.

37. 11 years and 7 months. This is 5 months less than the estimate of 12 years given by the Rule of 72.

39. 8.19%

Exercise Set 4.2

Test Your Understanding

1. The amount borrowed (the principal), the APR, and the term of the loan
2. True
3. True
4. the part of the principal that you have paid
5. True
6. adjustable-rate mortgage
7. True
8. **d.** all of these

Problems

9. Payments of $90 per month for 12 months will not even repay the principal.

11. $138.89

13. $386.66; $3199.60

15. The rule of thumb says that the payment is at least $1041.67. The monthly payment is $1096.78.

17. $15,014.57

19. Rule of Thumb 4.2 gives the correct answer of $750.00 per month.

21. a. $455.70; $936.80
    b. $1904.90
    c. The interest earned on the CD is $968.10 more.

23.

Payment number	Payment	Applied to interest	Applied to balance owed	Outstanding balance
				$1500.00
1	$65.14	$5.00	$60.14	$1439.86
2	$65.14	$4.80	$60.34	$1379.52
3	$65.14	$4.60	$60.54	$1318.98

25. a. $796.42
    b. 63.57%
    c. $53,355.60
    d. $583.74
    e. 86.73%
    f. You will have paid $120,146.40 in interest. It is more than twice as much as for the 15-year loan.

27. a. $386.66
    b. $3536.06
    c. $14,500
    d. You still owe $1963.94.

29. Yes, this is too good to be true: By Rule of Thumb 4.2, the payment must be at least $277.78.

31. a. $3333.33
    b. $933.33
    c. $683.33
    d. We can afford to borrow $54,909.07, or about $54,909, so we cannot afford the home.
    e. We can afford to borrow $155,831.35, or about $155,831, so we can afford the home.
    f. about $100,922

33. $259.37; 159%

35. The amortization schedule begins as follows:

Month	Payment	To interest	To balance owed	Balance
				$15,000.00
1	$366.19	$100.00	$266.19	$14,733.81
2	$366.19	$ 98.23	$267.96	$14,465.85

Equity after 24 payments is about $6903.

Exercise Set 4.3

Test Your Understanding

1. Money is invested in a fund that provides a fixed income for a fixed period.
2. **c.** the amount of money she has in her retirement account at the time she retires
3. True
4. False
5. It gives the balance of an account when you make regular monthly deposits.
6. It gives the monthly deposit needed to achieve a goal in a given number of years.

Problems

7. $5027.42

9. $116.91

11. a.

Month	Interest	Deposit	Balance
1	$0.00	$75.00	$75.00
2	$0.33	$75.00	$150.33
3	$0.66	$75.00	$225.99
4	$0.99	$75.00	$301.98

b. The formula gives $301.97; the table gives $301.98.
c. $3996.22
d. $31,733.38

13. $10,012.57

15. $3295.02

17. a. $597,447.22
b. $796,596.29. This is $199,149.07 more than the previous amount.

19. a. $625,357.24
b. $225.20

21. a. $1,136,440
b. $561

23. a. $14,479.30
b. $14,392.95. In part a, the interest on the first deposit becomes part of the balance, and we earn interest on that.

25. $2708.33

27. a. $289,744.28
b. $1158.98
c. $1827.13
d. We invested $96,000. Our return is $460,436.76. The amount that our investment yields during our retirement is nearly 5 times the amount we invested.

29. $514,285.71

31. a. $137,799.63
b. $198.85
c. Starting at age 40 requires a monthly deposit of nearly 4 times that for starting at age 20. Starting at age 20 requires $27,000. Starting at age 40 requires $59,655, over twice as much.
d. $414,997.22

33.

Nest Egg	Deposit
$100,000	$50.21
$200,000	$100.43
$300,000	$150.64
$400,000	$200.85
$500,000	$251.07
$600,000	$301.28
$700,000	$351.50
$800,000	$401.71
$900,000	$451.92
$1,000,000	$502.14

Exercise Set 4.4

Test Your Understanding

1. True
2. d. all of the above
3. True
4. True
5. e. all of these
6. a. decrease exponentially
7. True
8. False

Problems

9. $11.55

11. $74.16; $68.75

13. We list only previous balance, finance charge, and new balance.

Previous balance	Charge	New balance
$800.00	$9.33	$709.33
$709.33	$13.46	$1022.79

15. We list only previous balance, finance charge, and new balance.

Previous balance	Charge	New balance
$1000.00	$24.70	$1324.70
$1324.70	$32.77	$1757.47

17. We present an abbreviated table.

Previous balance	Payment	Charge
$4500.00	$112.50	$97.75
$4985.25	$124.63	$103.21

19. a. Balance $= 4500 \times (1.0175 \times 0.975)^t$ dollars
b. $3543.10
c. $1729.40

21. a. Balance $= 10,000 \times (1.01 \times 0.96)^t$ dollars
b. 172 months, or 14 years and 4 months

23. 93 months, or 7 years and 9 months

25. a. Balance $= 3000 \times (1.0175 \times 0.98)^t$ dollars
b. $2500
c. 64 months, or 5 years and 4 months

27. $5000; $12,500

29. a. $1414.00
b. $1414.71

31. We present only a portion of the table.

Month	Previous balance	Payment	Finance charge
21	\$437.42	\$21.87	\$4.16
22	\$419.71	\$20.99	\$3.99
23	\$402.71	\$20.14	\$3.83

The next payment is \$19.32.

Exercise Set 4.5

Test Your Understanding

1. the average price paid by urban consumers for a "market basket" of consumer goods and services
2. decreases
3. Under some conditions, taxes are higher on a married couple (filing jointly) than they would be if the two people were single.
4. **a.** the value of selected stocks
5. True
6. **b.** the tax rate to which additional income is subject
7. a very high rate of inflation
8. A tax deduction reduces your taxable income.

Problems

9. 4.9%

11. 4.2%

13. \$58,000.00

15. 1980. There was no CPI figure in 1948 from which to compute a percent change.

17. **a.** 39.0% **b.** 34.8%; no

19. 58.7%

21. \$107.51

23. a. \$3402.50
 b. \$79,795
 c. \$5100
 d. \$51,000
 e. Answers will vary.

25. \$110.70; 2.4%

27. The DJIA increases by 13.16 points.

29. 1996; 1999

31. The stock is bought on day 3 and sold on day 5. The profit is \$1200.

33. \$1500

Exercise Set 5.1

Test Your Understanding

1. $\text{Probability} = \dfrac{\text{Favorable outcomes}}{\text{Total outcomes}}$
2. a certainty (the event will always occur)
3. True
4. False
5. True
6. True
7. **d.** $1 - p$
8. **d.** all of these

Problems

9. $4/50 = 2/25$; 8%

11. $12/52 = 3/13$

13. $48/52 = 12/13$

15. a. $1/2$; 0.5
 b. 0.25; 25%
 c. $1/20$; 5%
 d. $65/100 = 13/20$; 0.65
 e. 0.875; 87.5%
 f. $37/100$; 37%

17. opinion

19. empirical

21. opinion

23. They are not always equally likely outcomes. It depends on the weather conditions.

25. A fraction is 0 exactly when the numerator is 0.

27. 90.8%

29. **a.** $2/5$ **b.** Those outcomes are not equally likely.

31. **a.** $25/144$ **b.** $49/144$
 c. $125/1728$ **d.** $343/1728$

33. $10/216$

35. 16%

37. Women: 15.4%; men: 19.4%

39. more than twice as high

41. $\dfrac{34}{52{,}250{,}000}$; empirical

43. $1/3$; $9/19$

45. $1/2$; 1 to 1

47. 7 to 3

49. $5/7$; $2/7$

51. $6/11$

53. Answers will vary.

55. Answers will vary.

57. Answers will vary.

Exercise Set 5.2

Test Your Understanding

1. The specificity is the probability that the test will give a negative result for a person who does not have the disease.
2. a positive test result
3. It is the probability that a person who tests positive actually has the disease.
4. that he or she is very unlikely to have the disease
5. False
6. d. false negative
7. a. true positive
8. c. false positive
9. c. positive predictive value

Problems

11. 100%; 0%
13. Answers will vary.
15. Sensitivity is 75.0%. Specificity is 85.0%.
17. NPV is higher if prevalence is low.
19. a.

	Ill	Healthy	Total
Positive	0.47	0.28	0.75
Negative	0.12	2.53	2.65
Total	0.59	2.81	3.40

b. PPV is 0.47/0.75, so about 0.627 or 62.7%

21. a. 95%
b. 95%
c.

	User	Not user	Total
Positive	475	475	950
Negative	25	9025	9050
Total	500	9500	10,000

d. exactly 50%
e. Answers will vary.
f. 99.7%
g. Answers will vary.

23. Apply the test to groups with known high likelihood of drug use.
25. a. 78% of those with prostate cancer test positive.
b. 69% of those without prostate cancer test negative.
c. PPV: 21.8%; NPV: 96.6%
d. PPV: 62.7%; NPV: 82.5%
27. a. exactly 50%
b. 100.0%
c. 999
29. Answers will vary.
31. 1/4
33. 1/9 if you can't remember which color, and 1/4 if you remember the bottle is blue.
35. 2/3
37. a. $15/35 = 3/7$
b. $20/35 = 4/7$
c. $10/20 = 1/2$
d. 2/3
e. $10/15 = 2/3$
39. a. 1/2
b. 2/3
c. 3/8
d. 16/25 or 0.64
41. Answers will vary.

Exercise Set 5.3

Test Your Understanding

1. If there are N possible outcomes for experiment number 1 and M possible outcomes for experiment number 2, then there are $N \times M$ possible outcomes if the two experiments are performed in succession.
2. True
3. the probability of event A occurring times the probability of event B occurring

Problems

5. 60
7. 6^{10} or 60,466,176
9. $(13 \times 13)/(52 \times 52) = 1/16$
11. You can make many more plates using numerals and letters than with only numerals.
13. 40
15. 10,000,000
17. 52
19. a. $21^4 = 194,481$
b. 1/194,481
c. 50/194,481
21. No
23. Yes
25. No
27. Yes
29. No
31. 1/4

33. a. Yes; 4/9 b. No; 2/5

35. a. $(10 \times 5)/(15 \times 14) = 5/21$ b. 10/21

37. a. $4^2/52^2 = 1/169$ b. 4/663
c. $13^2/52^2 = 1/16$ d. 13/204
e. $(13 \times 12)/(52 \times 51) = 1/17$ f. 25/204

39. a. $50/95 = 10/19$ b. $(45/95)^5$

41. a. Yes b. 1/36
c. 1/18 d. 1/9

43. 5

45. Democrat: 46%; Republican: 42%; Independent: 12%

47. 1%

49. a. $1/2^{63}$ b. 29 billion

51. Probability of at least one double six: $1 - (35/36)^{24}$ or 0.49

53. These numbers are so large that only specialized mathematical software can calculate them exactly. For 50 people, the correct answer is 30,414,093, 201,713, 378,043,612,608,166,064,768,844,377, 641,568,960, 512,000,000,000,000 or about 3.041×10^{64}. For 200, the answer is too large to reproduce here. The answer is approximately 7.887×10^{374}.

Exercise Set 5.4

Test Your Understanding

1. True
2. True
3. a factorial
4. True

Problems

5. 24

7. 20

9. 15

11. 720

13. 20!/14!

15. 55,440

17. 56

19. Yes

21. No

23. No

25. Yes

27. $26^3 = 17{,}576$; $26^4 = 456{,}976$

29. 10,000; answers will vary

31. 56

33. 9828

35. 2002

37. 715

39. Yes

41. 0.049

43. 0.031

45. 60

47. 0.1176

49. 0.0002

51. 0.0003

53. When we counted the number of ways to select 2 days from 7, we used combinations, which do not consider order. But when we counted the number of ways to choose 2 birthdays from 31, we used permutations, which do take order into account. The correct answer is 21/465.

55. Part a: 0.51; part c: 0.00 (less than 4.6×10^{-44})

Exercise Set 5.5

Test Your Understanding

1. Zero
2. True
3. True
4. False
5. The gambler's fallacy is the belief that a string of losses in the past will be compensated by wins in the future.
6. c. both
7. Positive
8. True

Problems

9. Expected value: \$0. The game is fair.

11. Expected value: \$0. The game is fair.

13. Expected value: −\$1. The game is not fair.

15. \$7

17. $18

19. $81.60

21. $3.00

23. They would not make a profit.

25. **a.** −$0.25 **b.** $0.25
    c. Larry, by $25.00
    d. −$0.23; $0.23
    e. Answers will vary.

27. Answers will vary.

29. The strategy assumes you have an essentially infinite amount of money so that you can double your wager many times. Also, casinos place a limit on how large a wager they will accept. If you lose too many times in a row, you will not be able to place a large enough wager.

31. $1.50

33. −$0.23

35. 0%

37. −30%

39. 1.8; 72

41. $86.50

43. $15,092.50

45. −$0.05; no

47. The expected value is $0.

49. −$0.05

51. **a.** 18/37; 48.6%
    b. $(18/37)^{26}$; 1 in 100 million
    c. 18/37, or 48.6%
    d. Answers will vary.

53. No. The numbers are chosen randomly.

55. $26 \times 11{,}238{,}513 = 292{,}201{,}338$

Exercise Set 6.1

Test Your Understanding

1. False
2. a frequency table
3. 13
4. on the mean
5. closely grouped
6. a, e, f
7. 25%

Problems

9. Mean: 11.27
   Five-number summary: minimum: 3; first quartile: 6; median: 12; third quartile: 18; maximum: 20

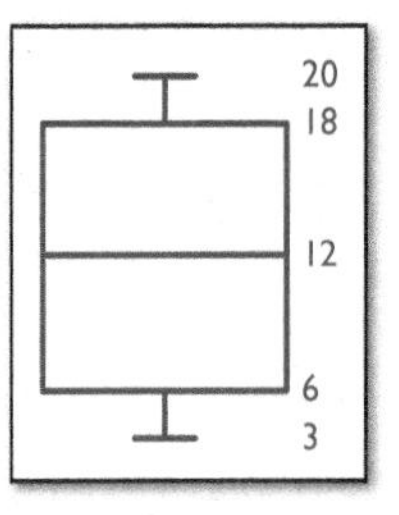

11. Mean: 9.06
    Five-number summary: minimum: 2; first quartile: 3.5; median: 10; third quartile: 14.5; maximum: 16

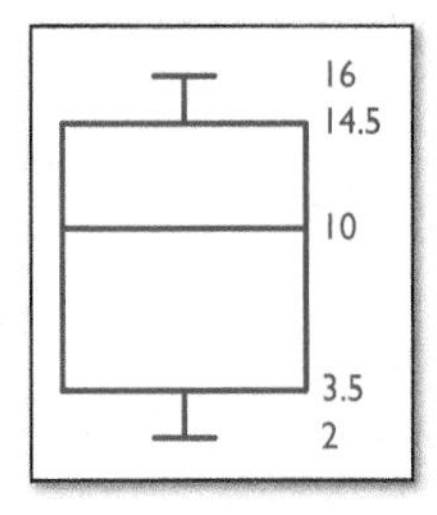

13. Answers will vary.

15. Answers will vary.

17. Answers will vary.

19. Mean: 78.9; median: 86. Dropping the lowest score gives the data in Exercise 18. This significantly raises the mean but has relatively little effect on the median.

21. Mean: $127,000; median: $122,000; mode: $120,000

23. **a.** Minimum: 35,867 sq. miles; first quartile: 54,947 sq. miles; median: 62,845 sq. miles; third quartile: 76,378.5 sq. miles; maximum: 81,815 sq. miles
    b. Answers will vary.
    c. Minimum: 1045 sq. miles; first quartile: 6131 sq. miles; median: 8968 sq. miles; third quartile: 37,839.5 sq. miles; maximum: 47,214 sq. miles
    d. Answers will vary.
    e. Answers will vary.

25. Minimum: 39 mpg; first quartile: 40 mpg; median: 42 mpg; third quartile: 48 mpg; maximum: 53 mpg.

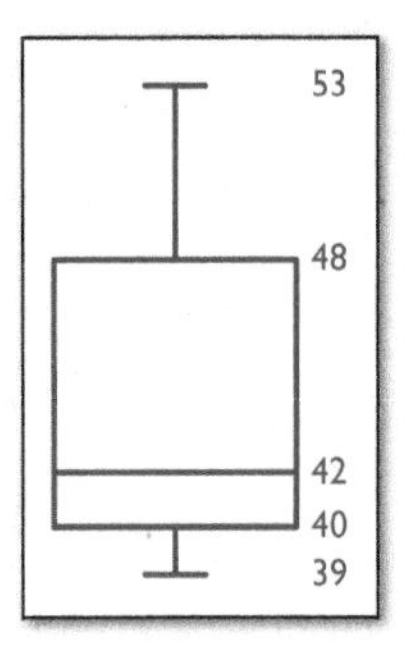

27. a. The vertical axis is the salary in dollars.

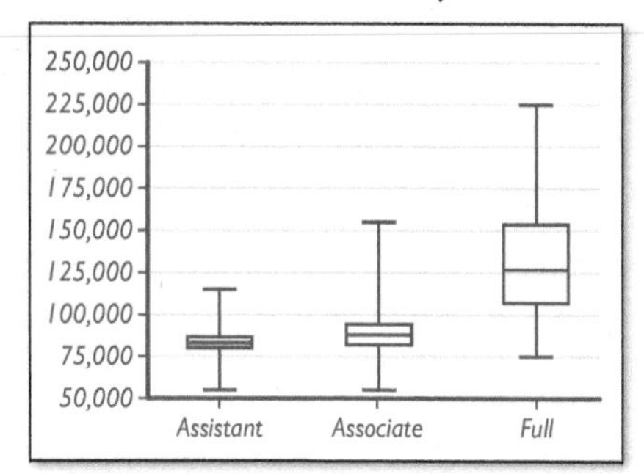

b. Answers will vary.

29. Mean: 79.5; standard deviation: 10.16

31. Mean: 85%; standard deviation: 0

33. a. Mean: 10; standard deviation: 2.5; number of points: 6
b. 100%

35.

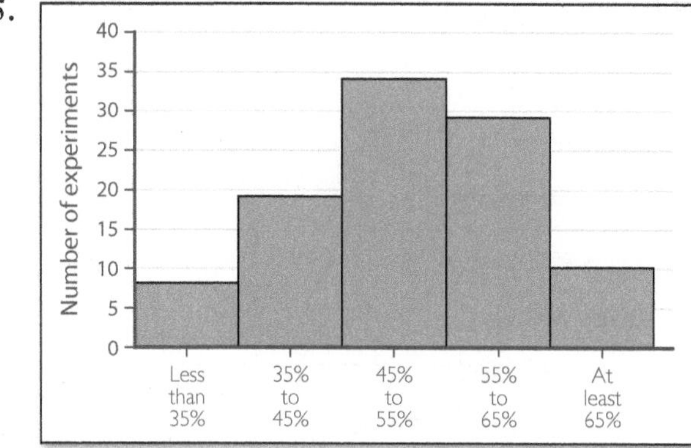

37. Answers will vary.

39. 3.20

41. Answers will vary.

43. Answers will vary.

Exercise Set 6.2

Test Your Understanding

1. No
2. True
3. It is 1 standard deviation below the mean.
4. It means that you scored higher than 60% of test takers.
5. True
6. True
7. b. percents

Problems

9. Answers will vary.

11. 70%. Yes.

13. a. 68% b. 340

15. −1.3

17. a. 84.1
b. The z-score is 2.7. For score 140 or higher, either 0.3% or 0.4% is acceptable.

19. a. 0.6 b. −1.3 c. 1%

21. a. 14% b. 21% c. 14%

23. A: 6.7%; B: 24.2%

25. a. Mean 70%, standard deviation 2.3
b. 30.9%

27. a. Mean: 15%; standard deviation: 1.6
b. 72.6%

29. Our town is extremely unusual. It's almost 20 standard deviations above the mean.

31. a. Mean: 20%; standard deviation: 1.3.
b. Extremely unusual. This is almost 4 standard deviations above the mean.

33. Extremely unusual; no

35. Answers will vary.

Exercise Set 6.3

Test Your Understanding

1. Answers will vary.
2. Answers will vary.
3. True
4. True
5. True
6. c. selected at random
7. e. all of these
8. False

Problems

9. In 95% of such polls, the reported approval of the president's policies will be between 39% and 45%.

11. 3.3%

13. 2.6%

15. Margin of error is about 3.3%. Perhaps the author of the article simply rounded 3.3 up to 3.5.

17. Answers will vary. Yes: At a 95% confidence level, we can say that between 55.5% and 62.5% support making changes in government surveillance programs.

19. The Rule of Thumb gives the same margin of error, 3.5%.

21. Answers will vary. About 816 people were sampled.

23. Doubling the sample size multiplies the margin of error by a factor of $1/\sqrt{2}$ or about 0.7. Cutting the margin of error in half requires quadrupling the sample size.

25. a. 2.6%
   b. With 90% confidence, between 50.4% and 55.6% like salad. Yes.
   c. With 90% confidence, between 45.4% and 50.6% like salad. The majority level is within the margin of error, so we cannot say with 90% confidence that a majority like salad.

27. a. 87%
   b. 500

29. Answers will vary.

31. Answers will vary.

33.

Size	% Error
1000	3.16
1100	3.02
1200	2.89
1300	2.77
1400	2.67
1500	2.58
1600	2.50
1700	2.43
1800	2.36
1900	2.29
2000	2.24

Exercise Set 6.4

Test Your Understanding

1. Answers will vary.
2. True
3. False
4. True
5. Answers will vary.
6. Control
7. c. little if any linear correlation
8. b. a benign substance

Problems

9. Yes; answers will vary.

11. Answers will vary.

13. Small p-values, not large ones, indicate statistical significance.

15. The p-value is 0.184, which is not normally considered statistically significant.

17. Answers will vary.

19. Answers will vary.

21. a. These variables are negatively correlated: As distance increases, brightness decreases.
   b. These variables are uncorrelated.
   c. These variables are positively correlated: Earthquakes with larger readings on the Richter scale generally cause more damage.
   d. These variables are negatively correlated: A golfer wins by getting the lowest score. So the lower a golfer's score, the higher her ranking.

23. Figure 6.51: −0.10; Figure 6.52: 0.58; Figure 6.53: 0.99; Figure 6.54: −0.63

25. Answers will vary

27. Answers will vary.

29. The trend line is $y = -240x + 564$, where y is the number sold and x is the price in dollars. The correlation coefficient is −0.97. This represents a strong negative linear correlation. The number sold decreases as the price increases, and the relationship is close to linear.

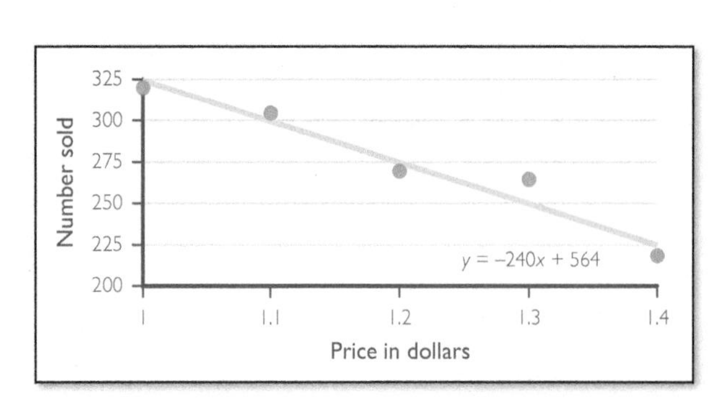

31. Answers will vary.

Exercise Set 7.1

Test Your Understanding

1. edges and vertices
2. True
3. It is a circuit that traverses each edge exactly once.
4. the number of edges that touch the vertex
5. Even
6. True

Problems

7. Answers will vary.

9.
Red Division Blue Division
Cubs
Lions
Bears
Tigers
Red Sox
Yankees
Giants

11.
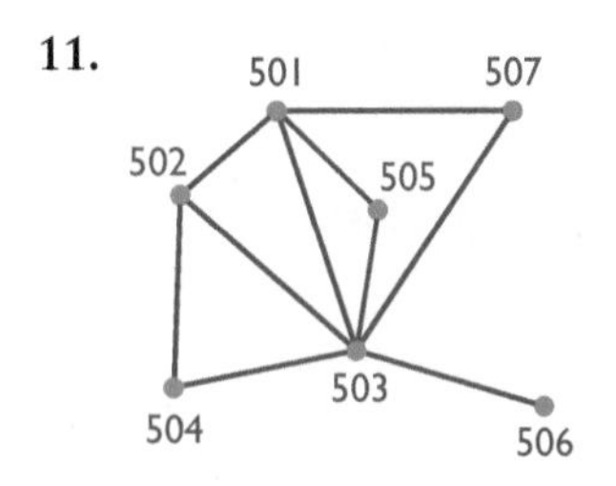

13.
Ruby-throated
Cinnamon teal
Colliope
Hooded Merganser
Red-shafted
Red-bellied

15. Each vertex has degree 4.

17. The worker would like to have a route traversing each edge (line) exactly once.

19. One Euler circuit is *A-B-F-C-D-E-A-D-B-C-A*.

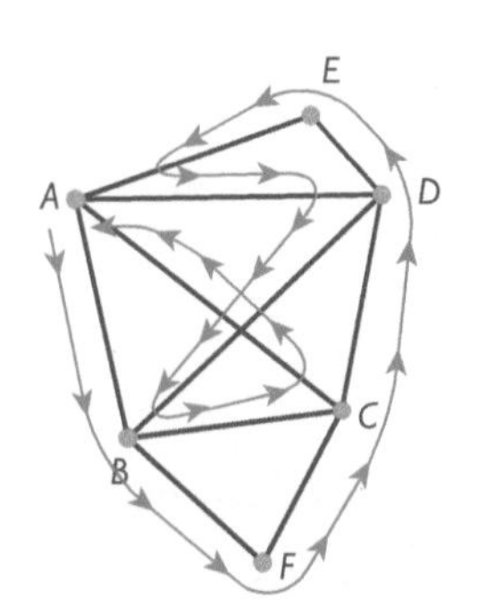

21. No Euler circuit exists.

23. One Euler circuit is Post Office-*F-B-E-D*-Post Office-*C-B*-Post Office.

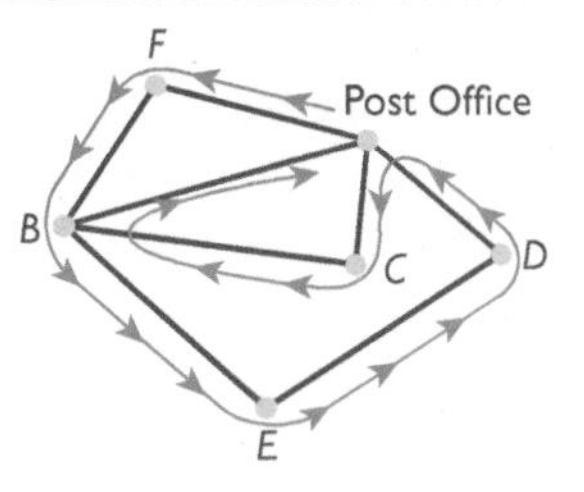

25. No Euler circuit exists.

27. No Euler circuit exists

29. No Euler circuit exists.

31. No Euler circuit exists.

33. No Euler circuit exists.

35. One correct answer is

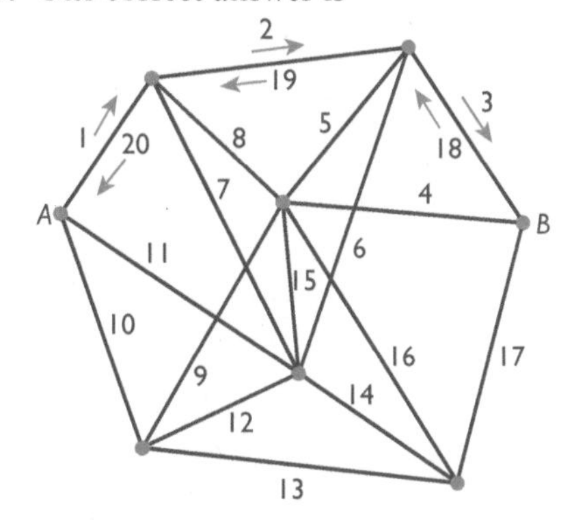

37. One correct answer is

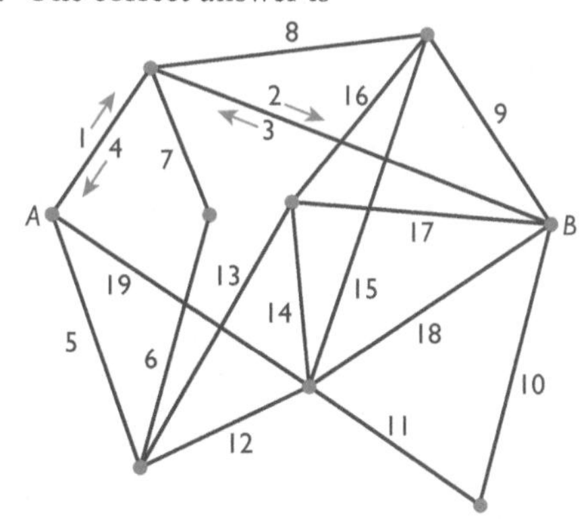

39. **a.** The vertices of odd degree are marked *A* and *B*.

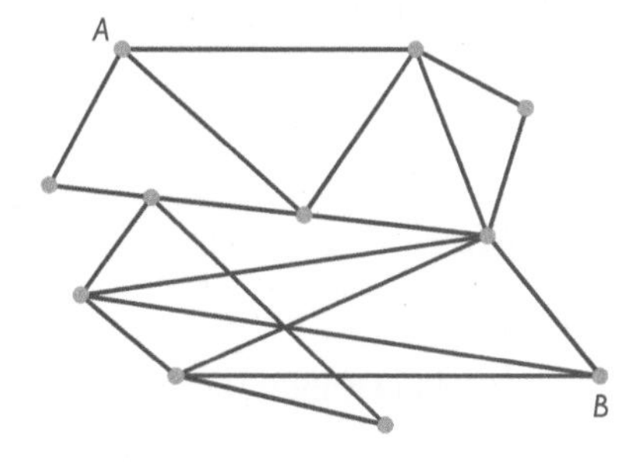

b.

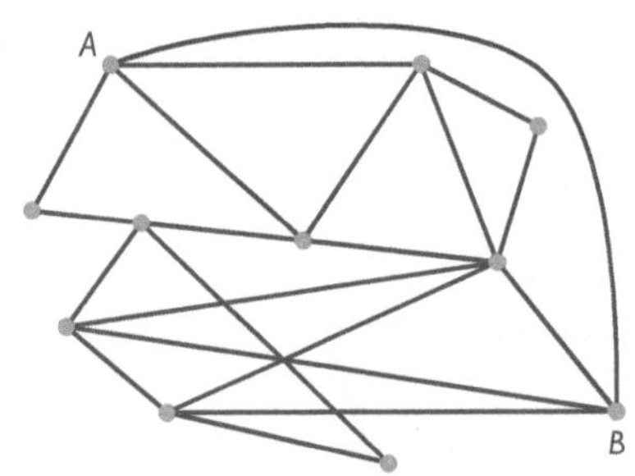

c. One Euler circuit is

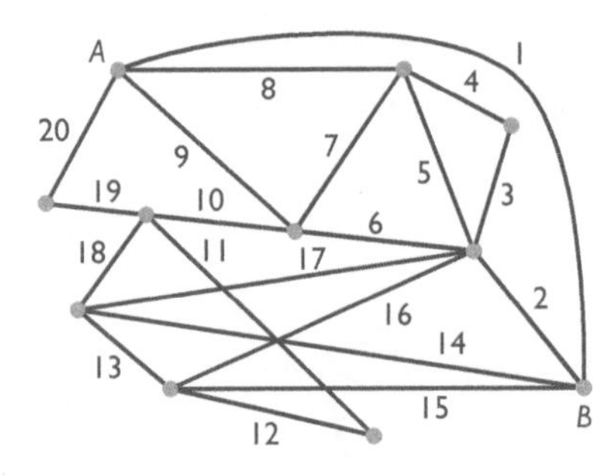

d. Delete the first edge from the Euler circuit to get the Euler path.

41.

	A	B	C	D	E	F	G
A	–	1	1	0	0	0	0
B	1	–	1	1	0	0	1
C	1	1	–	0	0	0	0
D	0	1	0	–	1	0	0
E	0	0	0	1	–	1	0
F	0	0	0	0	1	–	1
G	0	1	0	0	0	1	–

43. They are symmetric around the diagonal of dashes.

Exercise Set 7.2

Test Your Understanding

1. It is a circuit that visits each vertex exactly once.
2. **d.** all of the above
3. False
4. False
5. False
6. False

Problems

7. Euler circuit
9. Yes
11. The highlighted edges show one Hamilton circuit.

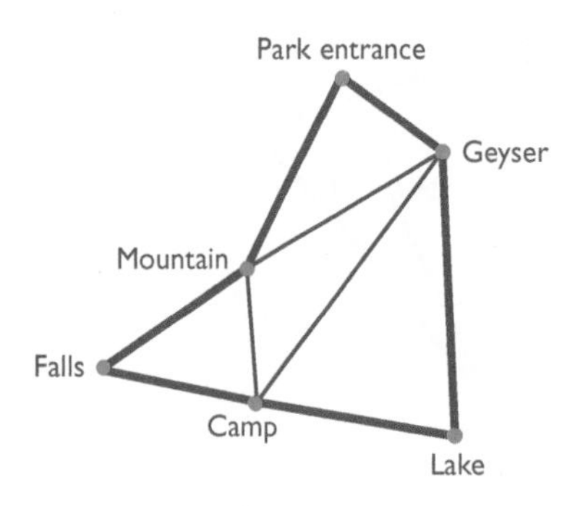

13. No Hamilton circuit exists.
15. The highlighted edges show one Hamilton circuit.

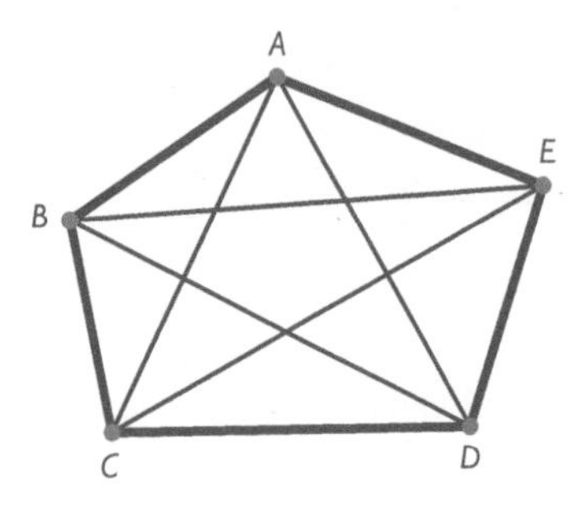

17. The highlighted edges show one Hamilton circuit.

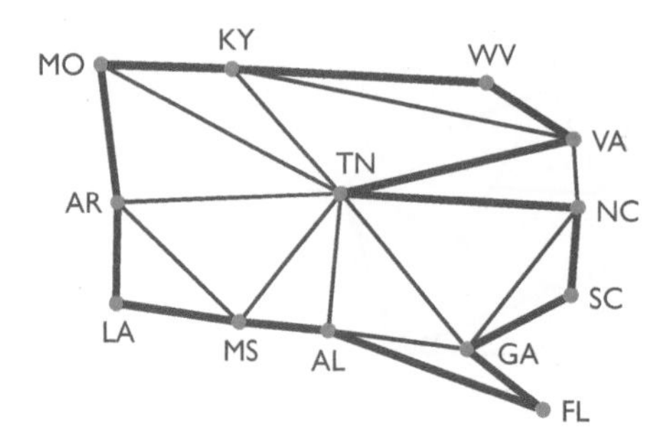

19. The highlighted edges show one Hamilton circuit.

21. The highlighted edges show one Hamilton circuit.

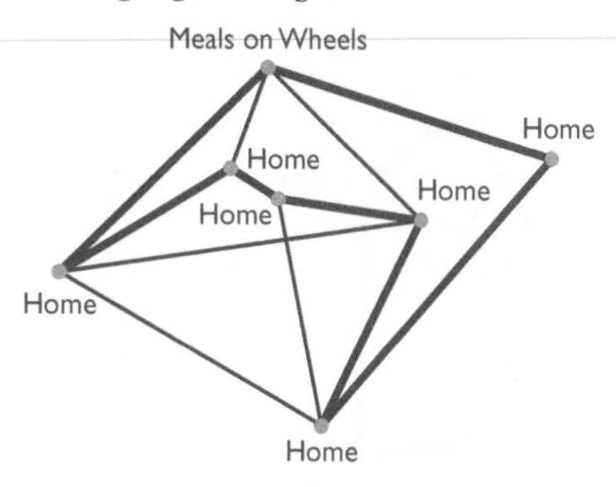

23. The highlighted edges show one Hamilton circuit.

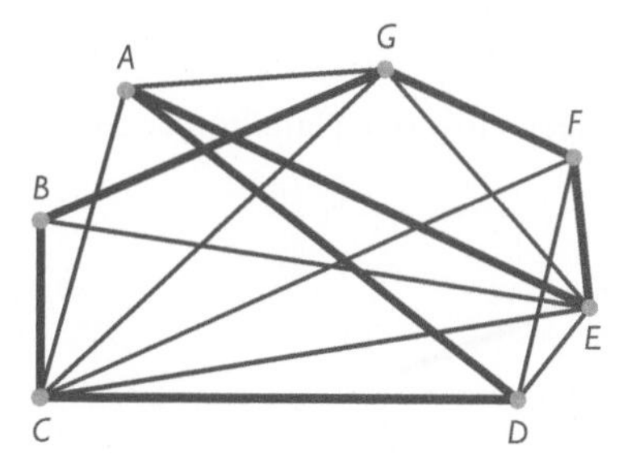

25. No Hamilton circuit exists.

27. Ft. Wayne - Cleveland - Richmond - Louisville - Ft. Wayne: Distance 1496 miles

29. Provo - Colorado Springs - Santa Fe - Tucson - Provo: Distance 2244 miles

31. *A-D-E-C-B-A* has length 15.

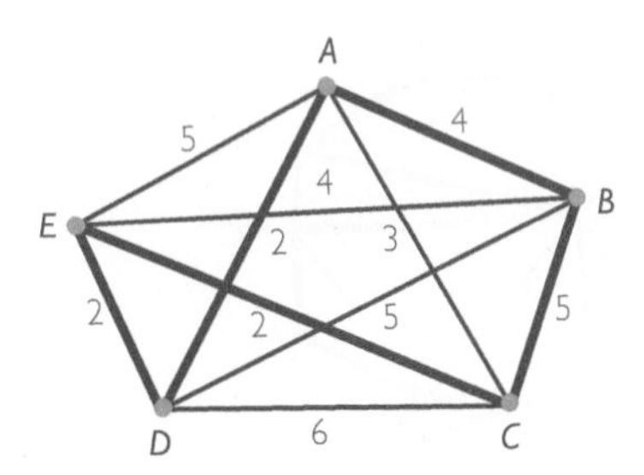

33. *A-D-E-B-C-A* has length 14.

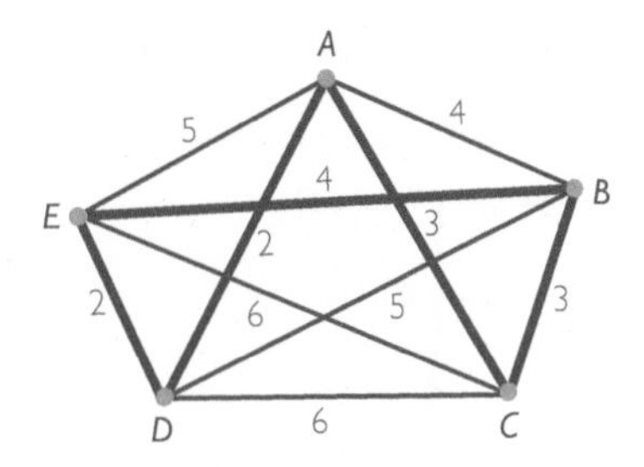

35. *A-D-E-B-C-A* has length 14.

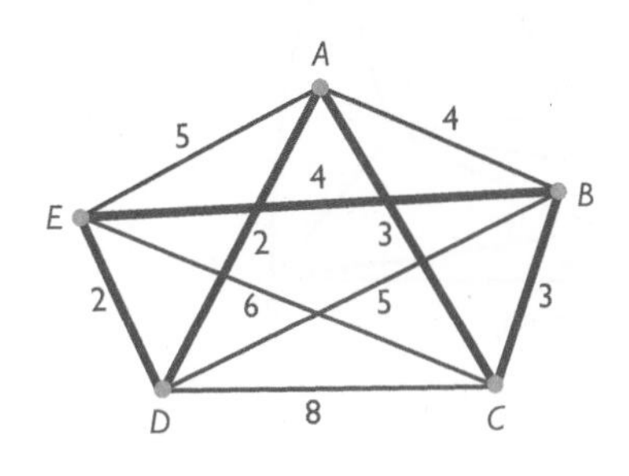

37. *A-D-E-C-B-A* has length 15.

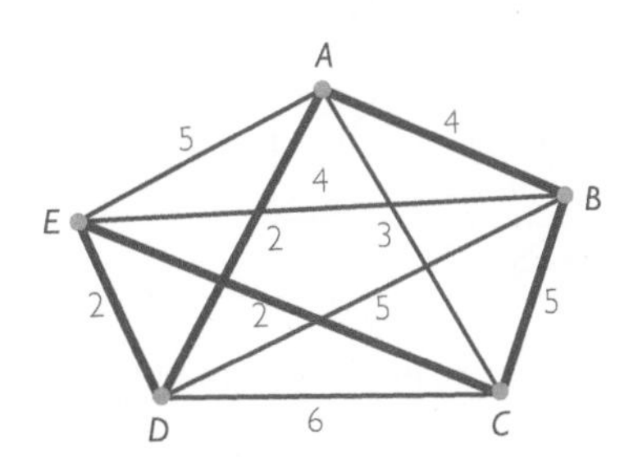

39. *A-D-E-B-C-A* has length 14.

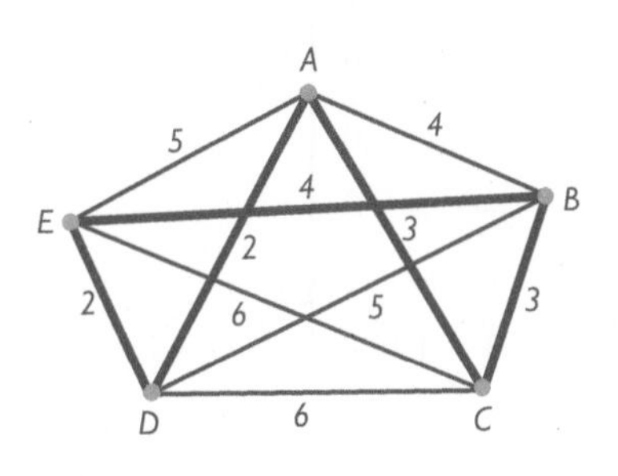

41. *A-D-E-B-C-A* has length 14.

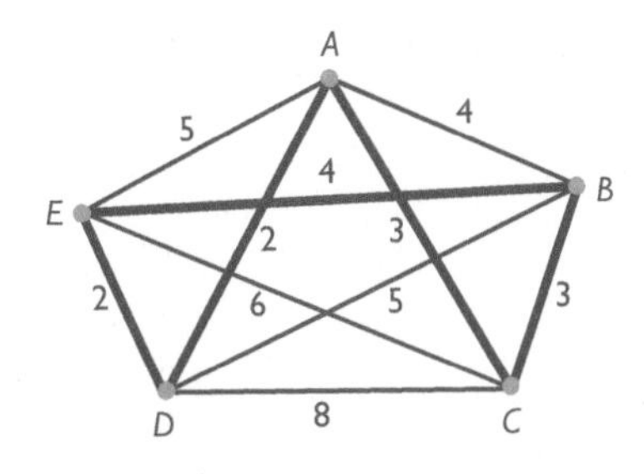

43. **a.** *A-B-C-D-E-A* has length 205.
b. *B-A-D-E-C-B* has length 8 or *B-C-D-E-A-B* has length 205.
c. *C-B-A-D-E-C* has length 8.
d. *D-E-C-B-A-D* has length 8.
e. *E-D-C-B-A-E* has length 205 or *E-D-B-C-A-E* has length 209 or *E-D-B-A-C-E* has length 12 or *E-D-A-B-C-E* has length 8.
f. Different starting points may yield different answers.

45. 15

47. 14

49. Answers will vary.

Exercise Set 7.3

Test Your Understanding

1. **c.** has no circuits

2. **a.** bud and **e.** limb

3. Each parent has exactly two children.

4. **b.** not a parent

5. False

6. It is a fraudulent scheme in which investors are paid from the proceeds of other investors.

Problems

7. One solution is

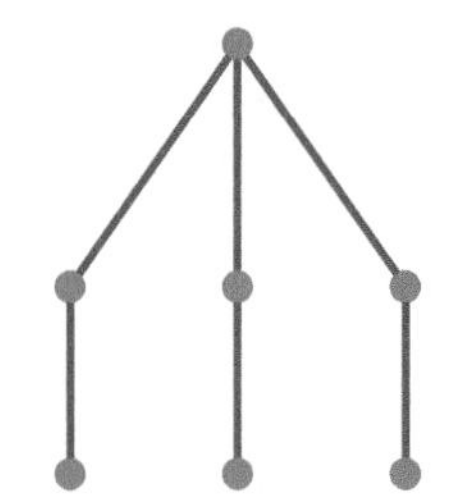

9.

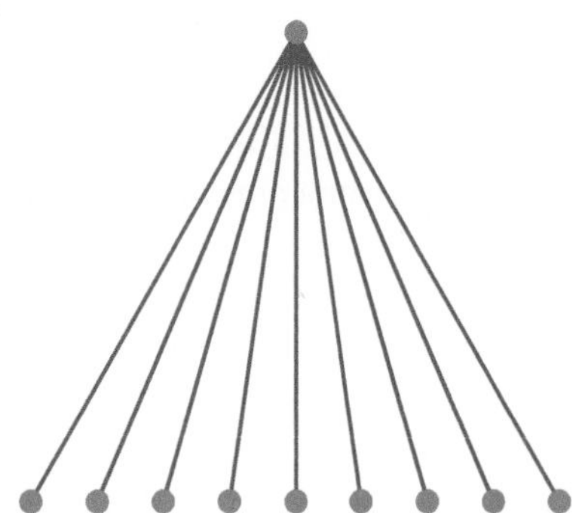

11. No

13. *D*, *G*, *H*, *I*, *K*

15.

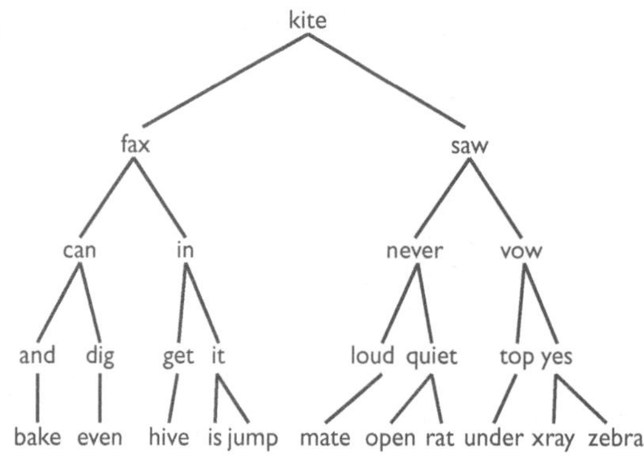

17. 7 checks

19. 16 checks

21. 501

23. 1024 leaves; 176 show at least 7 passes.

25. 7 games

27. **a.** 255 **b.** 32 **c.** 240 **d.** Round 6

29. The graph is not connected. One answer for the result of a tree search is the following graph. (We omit the smaller piece, which shows *H* connected to *I* and *J*.)

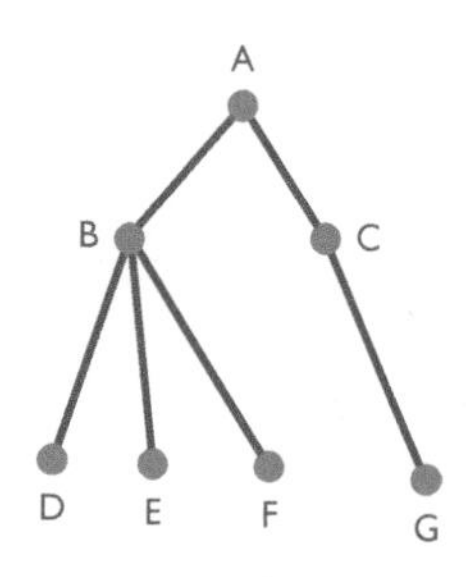

Exercise Set 8.1

Test Your Understanding

1. voting power

2. a group of voters who vote the same way

3. **a.** a coalition table and **b.** critical voters

4. True

5. the deciding voter in a particular voting order

6. True

Problems

7. $2^5 - 1 = 31$

9. $5! = 120$

11. (b) $2^{50} - 1$

13. Critical voters

15. Anne 1, other voter 0

17. 1/100

19. a. 14 **b.** 15

c. We list only winning coalitions.

Coalition	Votes	Critical voters
A B C D	26	A
A B C	24	A
A B D	21	A
A B	19	A, B
A C D	20	A
A C	18	A, C
A D	15	A, D

d. A: 7/10; B: 1/10; C: 1/10; D: 1/10

e. 24

f.

Order	Swing
A B C D	B
A C B D	C
C A B D	A
C B A D	A
B C A D	A
B A C D	A
D C A B	A
D A C B	A
A D C B	D
A C D B	C
C A D B	A
C D A B	A
B A D C	A
B D A C	A
D B A C	A
D A B C	A
A D B C	D
A B D C	B
C D B A	A
C B D A	A
B C D A	A
B D C A	A
D B C A	A
D C B A	A

g. A: $18/24 = 3/4$; B, C, D each $2/24 = 1/12$

21. a. 11

b. 120

c. 31

d. We list only the winning coalitions with critical voters.

Coalition	Votes	Critical voters
A B C	15	A, B
A B D E	16	A
A B D	14	A, B
A B E	13	A, B
A B	11	A, B
A C D E	15	A
A C D	13	A, C, D
A C E	12	A, C, E
A D E	11	A, D, E
B C D E	14	B, C
B C D	12	B, C, D
B C E	11	B, C, E

e. A: 1/3, B: 7/27, C: 5/27, D: 1/9, E: 1/9

23.

R	D	I	D
R	I	D	D
D	R	I	R
D	I	R	I
I	R	D	D
I	D	R	D

Democrats: $4/6 = 2/3$, Republicans: $1/6$, Independent: $1/6$

25. a. Using the popular vote, the indices are State A: $0/6 = 0$, State B: $0/6 = 0$, State C: $6/6 = 1$. Using the electoral vote, the index is $1/3$ for each state.

b. The power indices for small states are larger under the Electoral College system than they would be using the popular vote.

27. Answers will vary.

29. Answers will vary.

31. a. 8 votes

b. A: 47.37%, B: 36.84%, C: 5.26%, D: 5.26%, E: 5.26%

c. A: 50%, B: 50%, C: 0%, D: 0%, E: 0%

d. The index for A increased. Answers will vary.

e. The 8-vote quota will be reached exactly when both A and B are in support. The votes of the remaining voters don't matter. In calculating the Banzhaf index, we can discard voters with no power.

Exercise Set 8.2

Test Your Understanding

1. The winner is the candidate with more votes than any other candidate.

2. A spoiler is a candidate who has no realistic chance of winning but who affects the outcome.

3. True
4. If no candidate receives a majority of votes, then the candidate with the lowest vote total is eliminated, and another vote is taken. The process continues until someone achieves a majority.
5. False
6. False
7. a candidate who beats each of the other candidates in a one-on-one election
8. False
9. True
10. False

Problems

11. 301
13. No
15. **a.** No; yes
    b. Answers will vary.
17. *C*
19. *B*
21. *D*
23. *D*
25. *D*
27. *D*
29. *B*
31. *C*
33. *C*
35. *C*
37. *A*
39. *B*
41. *B*
43. *A*
45. No
47. **a.** Republican: 1/3, Democrat: 1/3, Independent: 1/3, Other: 0
    b. Republican: 1/3, Democrat: 1/3, Independent: 1/3, Other: 0
    c. The first three blocks have equal power and the fourth has no power.

Exercise Set 8.3

Test Your Understanding

1. One person divides the items into two parts, and the other person chooses a part.
2. True
3. The Knaster procedure
4. True

Problems

5. Put the earrings in a small pile.
7. Daniel gets the paintings and 35.3% of the art supplies. Carla gets the drawings and 64.7% of the art supplies.
9. Anna gets 71.4% of the travel books. Ashley gets the math books, craft books, and 28.6% of the travel books.
11. Bcom gets the computer outlet, the bakery, and 9.1% of the cleaning service. Ccom gets the appliance dealership, the coffee shop, and 90.9% of the cleaning service.
13. Ben gets the Hopi kachinas and the Zuni pottery. Ruth gets the Navajo jewelry, the Apache blankets, and the Pima sand paintings. There is no need to divide any of the items.
15. Various answers are possible. One answer is: Person 1 gets items 1 and 2 and person 2 gets items 3 and 4.
17. J.C.: $91,875; Trevenor: $94,375; Alta: home less $275,625; Loraine: $89,375
19. Lee: $33.75; Max: painting less $103.75; Anne: $31.25; Becky: $38.75
21. James: Business 2 and $7500; Gary: Business 1 and Business 3 less $7500
23. **a.** Answers will vary.
    b. Answers will vary.
25. Answers will vary.
27. Answers will vary.
29. Answers will vary.

Exercise Set 8.4

Test Your Understanding

1. Divide the U.S. population by the size of the House.
2. **a.** the state's population divided by the ideal district size
3. **b.** how the fractional parts of a state's quota should affect allocation of representatives
4. The initial allocation is based on quotas rounded down to the nearest whole number. Leftover members are allocated in order from the highest fractional part of quotas down.
5. The divisor, the ideal district size, is adjusted so that allocation based on rounded-down quotas yields the proper number of representatives.
6. **b.** increasing the House size can decrease a state's number of representatives

7. The number of representatives allocated to some state is more than 1 away from that state's quota.

8. The Huntington-Hill method

Problems

9. 34,437.333

11. 9.63

13. 52.54

15. 11,000

17. **a.** 100
    b. States 1 and 2 have a quota of 1.4. State 3 has a quota of 7.2.
    c. Hamilton's method gives 2 seats to each of States 1 and 2 and 7 seats to State 3. That is a total of 11 seats.

19. 95.83

21. North 1, South 1, East 2, West 8

23. North 1, South 2, East 2, West 7

25. 90.4

27. North 2, South 2, East 4, West 7

29. North 2, South 2, East 3, West 8

31. **a.** A divisor of 52 gives an allocation of 3 seats for each of States 1 and 2 and 19 seats for State 3.
    b. State 3

33. Oklahoma 14, Texas 93, Arkansas 11, New Mexico 7

35. Montana should get 2 seats, Alaska 1 seat, Florida 26 seats, and West Virginia 3 seats, using Dean's method.

37.

State	Seats
Connecticut	7
Delaware	1
Georgia	2
Kentucky	2
Maryland	8
Massachusetts	14
New Hampshire	4
New Jersey	5
New York	10
North Carolina	10
Pennsylvania	13
Rhode Island	2
South Carolina	6
Vermont	2
Virginia	19

39.

State	Seats
Connecticut	7
Delaware	2
Georgia	2
Kentucky	2
Maryland	8
Massachusetts	14
New Hampshire	4
New Jersey	5
New York	10
North Carolina	10
Pennsylvania	13
Rhode Island	2
South Carolina	6
Vermont	2
Virginia	18

Exercise Set 9.1

Test Your Understanding

1. **a.** the three sides of a right triangle

2. True

3. **c.** both of the above

4. True

Problems

5. 31.4 feet

7. 34 units

9. $\sqrt{58}$ or about 7.6 units

11. $\sqrt{269}$ or about 16.4 feet

13. Answers will vary; 6 square units

15. 20.5 square units

17. 9.8 square feet

19. For the 4-5-6 triangle, the area is about 9.9 square yards. For the 3-6-7 triangle, the area is about 8.9 square yards. The 4-5-6 triangle is larger.

21. 15.7 feet

23. Area is 100 square units. Perimeter is 51.2 units.

25. Area is $7\sqrt{75}/2$ or about 30.3 square units. Perimeter is $17 + \sqrt{79}$ or about 25.9 units.

27. Area is $4(2 + \sqrt{33})/2$ or about 15.5 square units. Perimeter is $9 + \sqrt{20} + \sqrt{33}$ or about 19.2 units.

29. The can holds 17.2 cubic inches. Therefore, the amount of lost soda is more than that.

31. Area: 30.9 square inches

33. Circumference: 8.2 inches

35. 858 miles

37. a. 21 inches
b. 21.0 inches
c. 212 square inches and 188 square inches; no

39. 2153.4 square feet

41. a. $\sqrt{656}$ or about 25.6 feet
b. 64 feet
c. 349.6 feet
d. 979.8 square feet
e. 11

43. Answers will vary.

45. 86,559 square miles

47. Answers will vary.

49. Answers will vary.

51. 920 cubic feet, 34.1 cubic yards

53. Volume: 12.8 cubic inches; Area: 26.4 square inches

55. Volume: 433.5 cubic inches; Area: 277.0 square inches

57. Volume: 6.0 cubic inches

59. 70.7 square feet

61. Answers will vary.

Exercise Set 9.2

Test Your Understanding

1. one is a (nonzero) multiple of the other

2. True

3. It is doubled.

4. The golden ratio ϕ

5. all corresponding angles have the same measure

6. Proportional

Problems

7. The taller has 4 times the volume.

9. Area is multiplied by 9.

11. No; no

13. Yes; 3.90 dollars per pound if the cost is measured in dollars and the weight is measured in pounds

15. No

17. Yes; the surface area of the pizza

19. No

21. Yes; π

23. 8.1 inches

25. Answers will vary.

27. Answers will vary.

29. Answers will vary.

31. $A = 21/8$; $B = 3/4$

33. 24 feet

35. 31.2 feet

37. Answers will vary. The constant of proportionality is 1/3.

39. 8 times as much gas is needed.

41. 1.33^2, or about 1.77, times as much

43. The larger can holds 4 times as much soup as the smaller can.

45. c. h; πr^2

47. The area of the label for the larger can is twice the area of the label for the smaller can.

49. The area of the label for the larger can is three times the area of the label for the smaller can.

51. b. r^2; $\pi h/3$

53. c. h; $\pi r^2/3$

55. f. none of these

Exercise Set 9.3

Test Your Understanding

1. it remains exactly the same after a rotation about that point of d degrees

2. True

3. All sides have equal length, and all angles have equal measure.

4. True

Problems

5. Answers will vary.

7. Answers will vary.

9. Answers will vary.

11. Answers will vary.

13. $\theta = 120$ degrees

15. $n = 180$

17. Reflectional symmetries through lines through opposite angles and lines through the mid-points of opposite sides (6 lines altogether). Rotational symmetry of 60 degrees (as well as 120, 180, 240, and 300 degrees).

19. One tiling is:

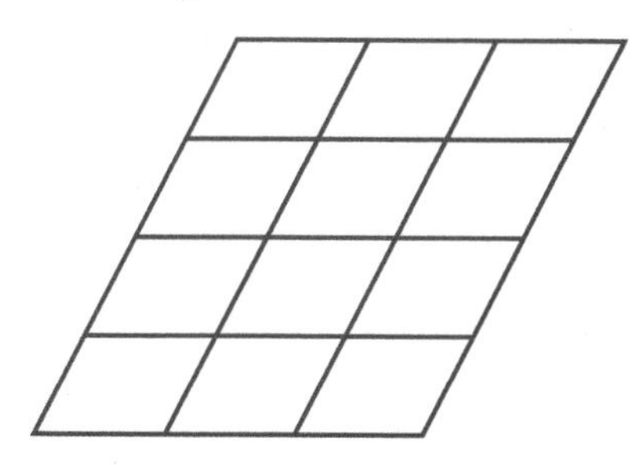

21. Answers will vary.

23. Two copies go together to make a rectangle.

25. Answers will vary.

27.

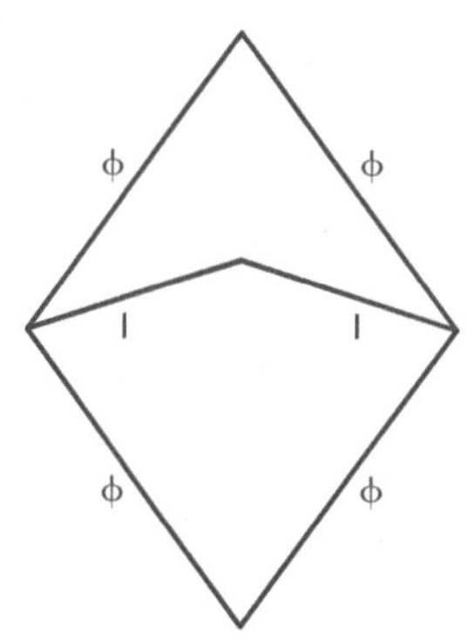

Answers will vary.

Credits

In the News, Illustrations, and Tables

Ch. 1

p. 3: (In the News 1.1) Chronicle Staff, "Yes, Colleges Do Teach Critical-Thinking Skills, Study Finds," *The Chronicle of Higher Education*, 10/20/15.

p. 4: (In the News 1.2) Smith, David, "A Great Example of Simpson's Paradox: U.S. Median Wage Decline," *Revolutionary Analytics*, 07/30/13.

p. 16: (In the News 1.3) Teresi, Brittni, "Use Debate Time to Focus on What Matters Most," *The Philadelphia Inquirer*, 10/17/16.

p. 29: (In the News 1.4) Wilson, Polly, "Biocomputer Decides When to Administer Drugs," Chemistryworld.com, 12/23/13, www.chemistryworld.com/news/biocomputer-decides-when-to-administer-drugs/6922.article.

p. 45: (In the News 1.5) *Charting My Own Course* on December 7, 2011. Reprinted with permission from the author.

p. 61: (In the News 1.6) From "The Budget and Economic Outlook: 2016 to 2026," Congressional Budget Office, 01/25/16.

Ch. 2

p. 79: (In the News 2.1) Hockenson, Lauren, and Molla, Rani, "Facebook Grows Daily Active Users by 25 Percent, Mobile Users by 45 Percent," *Gigaom*, 10/30/13.

p. 80: (In the News 2.2) "NCAA Graduation Rates Improving," *Associated Press*, 10/25/12.

p. 113: (In the News 2.4) "The Most Misleading Charts of 2015, Fixed," by Keith Collins, *Quartz*, December 23, 2015.

p. 116: (In the News 2.5) Gollob, Beth, "State ACT Scores Improve, Still Low," *NewsOK*, 08/16/06.

Ch. 3

p. 137: (In the News 3.1) Feeding America, "Every Single County in America Is Facing Hunger," *PR Newswire*, 04/28/16.

p. 138: (In the News 3.2) National Snow and Ice Data Center, "Sled Dog Days of Summer," *Arctic Sea Ice News and Analysis*, 08/06/14, http://nsidc.org/arcticseaicenews/2014/08/sled-dog-days-of-summer/

p. 157: (In the News 3.3) Butler, Desmond, and Ghirda, Vadim, "AP Investigation: Nuclear Black Market Seeks IS Extremists," *Associated Press*, 10/07/15.

p. 178: (In the News 3.4) "Record Number of Oklahoma Tremors Raises Possibility of Damaging Earthquakes," USGS, 05/02/14, http://earthquake.usgs.gov/contactus/golden/newsrelease_05022014.php

p. 195: (In the News 3.5) Crawford, Mark, "Solar Power Shines," American Society of Mechanical Engineers.

Ch. 4

p. 213: (In the News 4.1) "How U.S. Teens Compare with Their Global Peers in Financial Literacy," CBS News, May 25, 2017.

p. 214: (In the News 4.2) "Americans Still Aren't Saving Enough," by Melanie Hicken. Courtesy of CNN.

p. 233–234: (In the News 4.3) "Paying for College in Installments." Copyright © 2017 by COLLEGEdata. Reprinted by permission from www.collegedata.com/cs/content/content_payarticle_tmpl.jhtml?articleId=10070

p. 254: (In the News 4.4) "Average Interest Rates for Payday Loans." Used by permission of Josh Wallach, Payday Loans Online Resource

p. 256: (In the News 4.5) Sommerfield, Jessica, "4 Retirement Tips Young Adults Can Implement Now," MoneyNing.com. Copyright © 2014. Reprinted by permission from http://moneyning.com/retirement/4-retirement-tips-young-adults-can-implement-now/

p. 261: (In the News 4.6) Hiltonsmith, Robert, "AT WHAT COST? How Student Debt Reduces Lifetime Wealth," Demos.

p. 272: (In the News 4.7) Lambert, Lance, "The Hidden Portion of Student Loan Debt," *The Chronicle of Higher Education*, 05/08/15.

p. 281: (In the News 4.8) "CFPB Finds CARD Act Helped Consumers Avoid More Than $16 Billion in Gotcha Credit Card Fees," Consumer Financial Protection Bureau, 12/03/15, www.consumerfinance.gov/about-us/newsroom/cfpb-finds-card-act-helped-consumers-avoid-more-than-16-billion-in-gotcha-credit-card-fees/

p. 287: (In the News 4.9) Moss, Daniel, "The Fed Follows a Script, But Inflation Isn't Playing its Part," Bloomberg, 07/05/17.

p. 299: (In the News 4.10) "Food Price Outlook, 2016," USDA, www.ers.usda.gov/data-products/food-price-outlook/summary-findings.aspx

Ch. 5

p. 307: (In the News 5.1) Pinholster,Ginger, "Research to Address Near-Earth Objects Remains Critical, Experts Say," 07/12/13. Copyright © 2013. Reprinted with permission from AAAS.

p. 308: (In the News 5.2) "2008 Bay Area Earthquake Probabilities," United States Geological Survey, 04/08, http://earthquake.usgs.gov/regional/nca/ucerf/

p. 324: (Exercise 54) Copyright © 1990 Marilyn vos Savant. Initially published in *Parade Magazine*. All rights reserved. Reprinted by permission.

p. 325: (In the News 5.3) "'Fifty Percent Chance of Another 9/11-style attack' in the Next Ten Years... But Likelihood Is Much Higher If the World Gets Less Stable," 09/05/12, Copyright © *The Daily Mail*. Reprinted by permission of Associated Newspapers Limited.

p. 326: (In the News 5.4) Kam, Dara, "Judge Strikes Down Drug Testing of Florida Welfare Recipients," 12/31/13. Used by permission of The News Service of Florida.

p. 340: (In the News 5.5) Du, Susan, Haines, Stephanie, Resnick, Gideon, and Simkovic, Tori, "How a Davenport Murder Case Turned on a Fingerprint," 12/22/13. Used by permission of The Medill Justice Project.

p. 353: (In the News 5.6) McNutt, Michael, "In Oklahoma, Republicans Continue to Show Net Increase Over Democrats in Registered Voters," *NewsOk*, 11/02/12.

p. 355: (In the News 5.7) "The Human Genome Project Completion: Frequently Asked Questions," www.genome.gov/11006943/

p. 370: (In the News 5.8) Jordan, Lara Jakes, "FEMA Changing Locks on Trailers," Associated Press, 8/15/06. Used with permission of The Associated Press, Copyright © 2014.

p. 372: (In the News 5.9) National Gambling Impact Study, http://govinfo.library.unt.edu/ngisc/reports/1.pdf

Ch. 6

p. 393: (In the News 6.1) From Dean Scoville, "What's Really Going on with Crime Rates: Crime stats are used to judge the effectiveness of law enforcement agencies, but hardly anyone is questioning their validity or accuracy," (October 09, 2013). Reprinted by permission of *POLICE Magazine* and Bobit Business Media.

p. 397: (In the News 6.2) "Understanding the Difference Between Median and Average," Realtor.com, August 4, 2011.

p. 412: (In the News 6.3) "Tumbling Dice & Birthdays," Michelle Paret, Minitab, Inc. Used by permission of Minitab Inc.

p. 433: (In the News 6.4) Republished with permission of the *Wall Street Journal*, from WSJ/NBC News/Marist telephone news polls of 719 likely voters in Ariz., 707 likely voteres in Ga. And 679 likely voters in Texas conducted Oct. 30-Nov. 1; margins of error, 3.7, 3.7 and 3.8 pct. Pts., respectively; permission conveyed through Copyright Clearance Center, Inc.

p. 441: (In the News 6.5) Engel, Pamela, "Another Poll Shows Trump and Clinton in a Dead Heat Nationally," *Business Insider*, 9/15/16. Copyrighted 2017. Business Insider, Inc. 128366:0617SH.

p. 443: (In the News 6.6) "Nearly 75% of Consumers Don't Feel Their Bank Helps Them Meet Their Financial Goals," 5/9/17, Business Wire.

p. 448: (In the News 6.7) "A Trial to Determine the Effect of Psoriasis Treatment on Cardiometabolic Disease."

p. 458: (In the News 6.8) Cheng, Maria, "Being Bored Could Be Bad for Your Health," Associated Press, 02/10/10. Used with permission of The Associated Press, Copyright © 2014.

p. 460: (In the News 6.9) "NIH Study Finds Regular Aspirin Use May Reduce Ovarian Cancer Risk," NCI Press Release, www.nih.gov/news-events/news-releases/nih-study-finds-regular-aspirin-use-may-reduce-ovarian-cancer-risk

p. 461: (Exercises 18–19) "Breastfeeding Is a Dynamic Biological Process—Not Simple a Meal at the Breast," www.ncbi.nlm.nih.gov/pmc/articles/PMC3199546

p. 462: (In the News 6.10) "What Do You Do When Your Child Wants to Stop Eating Meat?," *The Guardian*, 02/12/08, Copyright © Guardian News & Media Ltd., 2008

Ch. 7

p. 469: (In the News 7.1) "Revealing Social Networks of Spammers," The National Security Agency's Review of Emerging Technologies, 2010, www.nsa.gov/research/tnw/tnw183/articles/pdfs/TNW_18_3_Web.pdf

p. 470: (In the News 7.2) "NextGen: The Future of Flying." Copyright © Aerospace Industries Association. Reprinted by permission.

p. 490: (In the News 7.3) Cook, William J., "Traveling Salesman: A Seemingly Unsolvable Problem Offers a Glimpse of the Limits of Computation," *Scientific American* 306:6. Reproduced with permission. Copyright © 2012 Scientific American, a division of Nature America, Inc. All rights reserved.

p. 509: (In the News 7.4) "How Did That Chain Letter Get to My Inbox?," 05/16/08, www.nsf.gov/news/news_summ.jsp?cntn_id=111580

Ch. 8

p. 525: (In the News 8.1) Jones, Bradley, "Majority of Americans Continue to Favor Moving away from Electoral College," Pew Research Center, January 27, 2021. Used with permission.

p. 526: (In the News 8.2) "What's 'Proportional Voting,' and Why Is It Making a Comeback?" by Alan Greenblatt, September 2017, Governing.com

p. 543: (In the News 8.3) Greenblatt, Alan, "Maine Becomes First State to Adopt a Whole New Way of Voting," Governing.com, 11/09/16.

p. 559: (In the News 8.4) From Corky Siemaszko, "Tiger Woods Divorce Settlement: Elin Nordegren Received $110M payout - report." © Daily News, L.P. (New York). Used with permission.

p. 573: (In the News 8.5) "Census Data Shed Light on 2020 Redistricting" by Sean Trende, December 22, 2016, www.realclearpolitics.com/articles/2016/12/22/census_data_shed_light_on_2020_redistricting__132623.html

Ch. 9

p. 599: (In the News 9.1) Izard, Veronique, Pica, Pierre, Spelke, Elizabeth S., and Dehaene, Stanislas, "Flexible Intuitions of Euclidean Geometry in an Amazonian Indigene Group," *Proceedings of the National Academy Sciences of the United States of America*, 108(24), 9782–9787. Reprinted by permission of PNAS.

p. 617: (In the News 9.2) "'Golden Ratio' Greets Visitors Who Enter North Central College Science Center," *Positively Naperville*, 03/28/17.

p. 631: (Exercise 57) Brown, D. (2003). *The Da Vinci Code: A Novel.* New York: Doubleday.

p. 632: (In the News 9.3) Devlin, Keith, "Good Stories, Pity They're Not True." Copyright © 2004 Mathematical Association of America. Reprinted by permission.

p. 633: (In the News 9.4) Scheib, J. E., Gangestad, S. W., and Thornill, R., "Facial Attractiveness, Symmetry, and Cues of Good Genes," *Proceedings of The Royal Society*, www.ncbi.nlm.nih.gov/pmc/articles/PMC1690211/

INDEX

D

E

L

M

N

W

Z